is

Now you will find Saunders College Publishing's distinguished innovation, leadership, and support under a different name . . . a new brand that continues our unsurpassed quality, service, and commitment to education.

We are combining the strengths of our college imprints into one worldwide brand: Harcourt

Our mission is to make learning accessible to anyone, anywhere, anytime—reinforcing our commitment to lifelong learning.

We are now Harcourt College Publishers. Ask for us by name.

One Company
"Where Learning Comes to Life."

www.harcourtcollege.com
www.harcourt.com

Asking About Life

Second Edition

Allan J. Tobin
*University of California,
Los Angeles*

Jennie Dusheck
*University of California,
Santa Cruz*

HARCOURT COLLEGE PUBLISHERS

Fort Worth Philadelphia San Diego New York Orlando Austin
San Antonio Toronto Montreal London Sydney Tokyo

Publisher: Emily Barrosse
Executive Editor: Nedah Rose
Senior Marketing Strategist: Kathleen Sharp
Senior Developmental Editor: Lee Marcott
Senior Project Editor: Bonnie Boehme
Senior Production Manager: Charlene Catlett Squibb
Senior Art Director: Caroline McGowan
Text Designers: Ruth A. Hoover and Kim Menning

Cover: African elephants near Mount Kilimanjaro, Tanzania *(Renee Lynn/Stone Images)*

Frontispiece: Copyright © 1999 Mauritius/Nawrocki Stock Photo, Inc. All Rights Reserved.

ASKING ABOUT LIFE, Second Edition
ISBN: 0-03-033636-8
Library of Congress Catolog Card Number: 99-068520

Copyright © 2001, 1998 by Harcourt, Inc.

All rights reserved. No part of this publication may be reproduced or transmitted in any form or by any means, electronic or mechanical, including photocopy, recording, or any information storage and retrieval system, without permission in writing from the publisher.

Requests for permission to make copies of any part of the work should be mailed to: Permissions Department, Harcourt, Inc., 6277 Sea Harbor Drive, Orlando, Florida 32887-6777.

Copyright © 2001 Digital:Convergence Corporation. All rights reserved.

Digital:Convergence, :C, :CRQ, :CueCat, Slant Code, the :CueCat reader, "It's your cue," "Engineering Your Digital Future" and all marks as indicated are trademarks of Digital:Convergence Corporation.

Address for domestic orders:
Harcourt College Publishers, 6277 Sea Harbor Drive, Orlando, FL 32887-6777
1-800-782-4479
e-mail collegesales@harcourt.com

Address for international orders:
International Customer Service, Harcourt, Inc.
6277 Sea Harbor Drive, Orlando, FL 32887-6777
(407) 345-3800
Fax (407) 345-4060
e-mail hbintl@harcourt.com

Address for editorial correspondence:
Harcourt College Publishers, Public Ledger Building, Suite 1250, 150 S. Independence Mall West, Philadelphia, PA 19106-3412

Web Site Address:
http://www.harcourtcollege.com

Printed in the United States of America

0123456789   032   10 987654321

Preface

Many good conversations start with a question. What do you think of the new ballpark downtown? Is the new sociology professor a good lecturer? Science is one long conversation, like a thousand-year cocktail party. And, as at any party, the topics of conversation often start with questions. Why are plants green? How do cells replicate?

The history of science reveals that the answers to such questions change over time, but the questions themselves, if they are good ones, remain the same. One good question has been, "How do new species form?" Ever since biologists realized that species do form, we have been trying to find out *how* they do it. But it is a question with many overlapping answers. The answers that seemed correct in 1880 or 1960 continue to be refined, expanded, or even overthrown.

In this new edition of *Asking About Life,* we present some new answers and some new approaches. But our starting point, that science is about curiosity, remains the same. We emphasize *how* and *why* scientists ask questions, how they test hypotheses, and how they reach conclusions. As much as possible, rather than merely presenting dry conclusions, we show our readers how science actually works. For example, we present the many pieces of evidence that convinced scientists that all life evolved from simple organisms. And we present the experimental steps that revealed how plants use sunlight to build sugar molecules, during photosynthesis.

Although our philosophy remains the same, we have improved the book in several important ways. Above all else, we have continued to refine and clarify our written explanations. As we write each sentence, we think first and foremost about how we can help our readers better understand biology.

We have added new metaphors and analogies, new lead-in stories, and new art. Some of our metaphors appear in the text alone, while others appear in figures as art or photos. As in the first edition, we compare chromosomes to socks and the functional groups of molecules to the attachments on a Swiss Army knife. But we have introduced new metaphors, such as the windmill in Chapter 25 that illustrates the idea that organisms recycle materials but not energy. Good visual metaphors help readers remember important concepts.

We regard illustration as a teaching tool in its own right, not just a reinforcement of the text. We have continued to work closely with nationally prominent illustrator Elizabeth Morales, whose skill, attention, and insight have contributed not only to the art manuscript but to the text as well. Morales, the current president of the Guild of Natural Science Illustrators, has been our art developmental editor since the first art meetings for the first edition. She has created a warm, friendly style that perfectly complements the conversational tone of our text. Her clean designs and precise illustrations greatly clarify sometimes difficult material. In addition, we retain many of the beautiful pieces done by natural science illustrator Elizabeth McClelland for the first edition.

We have updated and expanded our coverage of ecology. We think students need to know, for example, that ecology now includes global oceanic and atmospheric studies. Readers can also learn that population biologists study not only questions about how populations regulate their numbers (or fail to), but also the sad story, repeated all too often in modern times, of how species go extinct.

We have introduced a new section on statistics in Chapter 1. We hope this brief overview will be especially useful in

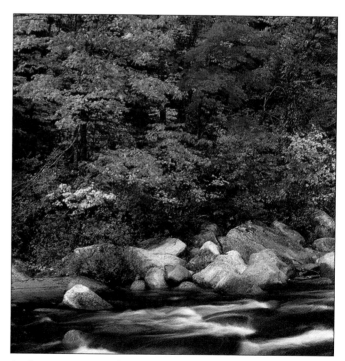

Stephen J. Krasemann/Photo Researchers, Inc.

classes where students encounter the first rudiments of statistics in a lab component of a course. Our purpose, however, is not to teach statistics *per se,* but to help students see why statistics is an essential analytical tool in biology and how it affects experimental design. We hope our brief introduction to the concepts of sample size and variation will prompt students to ask skeptical questions about science stories in the news.

Despite all these new features, the basic vision of *Asking About Life* remains the same. We continue to present biology in the context of the development of biological ideas, never as disconnected facts. We present a relatively few ideas in some depth rather than many ideas superficially.

Asking About Life consistently emphasizes the importance of questions and the process of finding answers. To remind ourselves and our readers of this emphasis, headings and subheadings are more often questions than statements. In addition, at the end of most subsections we summarize the main points covered. The question heading and the summary statement together act as a reality check for both readers and the authors. In our first edition, many readers pointed out that we didn't always answer the questions in the headings. Did we ask the wrong question? Or, did we not answer our own question? Truth to tell, sometimes the former, sometimes the latter. In this edition, we have taken special pains to improve our questions, as well as answers.

Fixing our mistakes has convinced us more than ever that asking explicit questions is the best way to talk about science. We have done our best to improve the question headings. But we welcome sharp-eyed readers to catch us up now and again. Readers of this edition who do not think we have answered the question in the heading should alert us at jennie.dusheck@pobox.com. We look forward to hearing from you.

Refining questions and fixing mistakes is the way science works. What scientist hasn't been brought up short by an insightful comment from a colleague? The way questions, or hypotheses, are phrased determines how they can be answered. Sometimes scientists ask the wrong questions. Sometimes they come up with the wrong answers to the right questions. The process is a delightfully social one, in which everyone can play a key role.

Throughout *Asking About Life,* we have emphasized the passionate engagement of individual scientists. Each chapter begins with a story illustrating how an individual biologist or group of biologists pursued a scientific question—often in the face of intense intellectual and social adversity. We tell the story, for example, of how Barry Marshall convinced first himself and then others that bacteria, not "frustrated personalities," cause ulcers; and how Rosalind Franklin struggled in deep social isolation to elucidate the structure of DNA. Our anecdotes are about real people—their triumphs, their frustrations, their genius, and their persistence. Biology is a story, and, as such, it must be presented as continuously as possible.

In every kind of learning, story and context give shape and meaning to dry facts and ideas. When we are told that someone named Charles Drew invented a way to supply clean blood plasma, a new technology that saved many lives, we may remember this fact briefly. But what if, as happens in Chapter 37, we also learn that Drew was one of only a handful of African-American physicians in the United States in the 1940s? At that time, black physicians did little or no research, but were mostly relegated to quiet country practices in the rural South. Yet Drew ran the "Plasma for Britain" program for the American Red Cross during World War II and invented a way to preserve blood plasma so that it could be shipped across the Atlantic to Great Britain. His work saved the lives of thousands of Londoners injured by Nazi bombs. Midway through the program, however, the American Red Cross decided to refuse blood from any American donor whose skin was black, and Drew resigned in protest. An eminent professor of surgery at Howard University, Drew suffered serious injuries in an auto accident a few years later. Within hours of the accident, he was dead. He bled to death in a small Southern emergency room where neither whole blood nor plasma could save him.

Life's too short not to hear the whole story. Our philosophy of telling the stories of biology sometimes runs counter to current trends. Many textbooks break up information into modules in which each bit of information is presented dictionary-fashion, as if unrelated to the information in the rest of the book. Important ideas may be marooned on illustration islands. The result is a series of disconnected facts. Publishers say that such an approach is necessary because today's students belong to the "visual-information generation" and are incapable of sustained reading or synthesis.

We have more faith in students. We know they enjoy reading, provided the reading is interesting and rewarding. Indeed, student reviewers have raved about the clarity of the writing and the engaging stories that reveal scientists as ordinary human beings.

In the evolution section, for example, we mention the intense frustration Charles Darwin experienced while struggling to distinguish among different species of barnacles. Each species seemed to blend into the next. In the next chapter, on classification, we show readers that biologists have been arguing about the definition of a species for nearly 200 years. In the context of Darwin's difficulties, the highly politicized debate over whether the red wolf is a species (that deserves legal protection) or a hybrid (that doesn't) takes on a different and deeper meaning. At the same time, the red wolf story brings to life what might otherwise be an abstract discussion of classification.

While we stress the continuity of ideas within a discipline, however, we know that every instructor takes a different approach to teaching biology. Therefore, each unit of the book is understandable on its own terms. Nothing prevents instructors from teaching the units in a different order.

The first edition of *Asking About Life* was well received. The Italian publisher Edizioni Bruno Mondadori selected *Asking About Life* for translation into Italian. And, in 1999, the

Text and Academic Authors association conferred on the book an award of excellence for best new textbook in life sciences. But we are most proud of the many instructors and students who have told us they love our book. Thanks to them for their remarkable support.

Who Made This Book?

Our developmental editor, Lee Marcott, continues to be an active participant in the work on *Asking About Life*. In addition to editing the manuscript, and choosing excellent reviewers, Lee has also tactfully managed the two authors—pushing us to meet deadlines, organizing us, and giving us pep talks.

Photo researcher Amy Ellis Dunleavy's novel photo ideas continue to give the book much of its personality. Project editor Bonnie Boehme kept the flow of paper moving smoothly during production and broke up many a logjam with quick-witted solutions and generous contributions of her own time. Art director Caroline McGowan's consistent sense of design gives the book much of its unity and elegance, and her sharp eye caught mistakes in biology as well as the occasional misplaced "leader line" or missing label.

We also thank production manager Charlene Catlett Squibb, manager of art and design Carol Bleistine, and designers Ruth Hoover and Kim Menning. And we deeply appreciate the work of many others who have contributed to this book—including Donald Jackson, Michael Brown, Edward Murphy, Elizabeth Widdicombe, Julie Alexander, and Edith Beard Brady. We thank especially publisher Emily Barrosse for her special combination of toughness and warmth and executive editor Nedah Rose for her cool head and patience. Senior marketing strategist Kathleen Sharp and field editorial specialist Dave Theisen have made creative contributions to the sales and marketing efforts of both the first and second editions.

Special thanks also to Naomi Cappuccino of Carleton University, Mary Kay Cassani of Edison Community College, Kerry S. Kilburn of Old Dominion University, and Jorge Rey of the University of Florida, all of whom helped us with the ecology section in different ways. We also thank science writer Marina Chicurel, who wrote the new story about Stanley Prusiner (Chapter 3); science writer Liese Greensfelder, who revised the discussion of phylloxera in Chapter 31; Marni Fylling, who wrote and illustrated the accompanying lab manual; Harry W. Greene of Cornell and Daniel J. Meinhardt of St. Olaf College, both of whom helped ease us through the subtleties of cladistics; Nancy Segal of Cal State Fullerton, who provided the striking pictures of twins in Chapter 9 and suggested changes to the text; Barney Schlinger and Gordon Fain, of UCLA, Warren Burggren, of the University of North Texas, and Michael Stryker of UCSF for their help in refocusing some of the physiology chapters; John E. Wilkes of the University of California, Santa Cruz, who read and skillfully copyedited many chapters of this edition; Primavera Hernandez, Annie Gallo, April Hosein, and Vanessa Flores, who helped with the book in dozens of ways; and the students in Life Sciences 2 at UCLA, whose comments, suggestions, and enthusiasm were especially valuable in preparing the second edition. Finally, we thank Richard Robinson, Mary K. Miller, Karen Scanlon, and Natalie Peretti for their work on the Web site that accompanies this book.

We thank all the reviewers who took the time to read and comment on this manuscript—correcting our errors, asking thought-provoking questions, and suggesting examples, alternative wordings, or new ways of thinking. Although we have met only a few of our reviewers in person, working with them has been a rewarding intellectual experience. Both the process of writing this book and the resulting book itself would not have been the same without the reviewers.

What Kinds of Pedagogy Does *Asking About Life* Employ?

Asking About Life has a variety of features designed to engage the reader and to aid student learning. Each chapter begins with a story about a piece of research that draws students into the subject of the chapter and also introduces the key questions and ideas that are discussed. The chapter-opening image is described at the end of each chapter in a new feature called **About the Chapter-Opening Image.**

At the start of each chapter, readers will find a list of **Key Concepts,** which are the most basic ideas covered in the chapter. Most headings are posed in the form of questions. These question headings focus the reader's attention on the most significant question to be explored in that section. Following most subsections is a **Summary Statement**—a brief summary of the take-home message. Summary Statements provide students with a reality check. If the student doesn't understand the summary statement, that is a cue to study the preceding material more closely.

Visual metaphors rendered as photos or illustrations drive home key points introduced in the text. Drawings and photographs support concepts covered in the text and help students visualize the structures of objects as diverse as molecules and ecological communities. Photographs of structures that are too small to be seen with the naked eye are accompanied by size bars to give a sense of scale. And **tables** and **graphs** summarize key facts and additional material. Most chapters feature one or more **Boxes,** which introduce science in the news or discuss special topics in greater depth. Figure legends are designed to stand alone, so that a student flipping through the chapter for the first time, glancing at the diagrams and reading the legends, will learn something.

The second edition of *Asking About Life* introduces **Web Bits,** short essays on topics that go beyond the book and are related to the chapter. They are listed at the end of each chapter along with a barcode that may be scanned with a :CueCat™ to take you to their location on the book's Web site.

Throughout the text, **boldface key terms** help students to locate key terms and their definitions. At the end of each chapter, all of the boldface terms are used again in a highly compressed summary called the **Study Outline with Key Terms.** The Study Outline provides students with another opportunity to check their understanding of the chapter. If they encounter terms they don't remember or ideas that seem unfamiliar, they can return to the main text and illustrations.

Following the Study Outline is a set of **Review and Thought Questions.** Our Thought Questions are especially engaging, frequently bringing ideas in the chapter into the everyday world. At the end of the book, a periodic table (Appendix A), a table of standard weights and measures (Appendix B), and a **Glossary** provide useful reference information for students.

Supplements

To further facilitate learning and teaching, we provide a carefully designed supplements package for students and instructors.

Marni Fylling's **Laboratory Manual to accompany *Asking About Life*,** Second Edition, covers laboratory topics drawn from six units of the text. Each of the 15 labs accompanies a specific chapter (or chapters) from the book. These interactive labs use familiar materials to give students a hands-on approach to basic biological principles. Author Marni Fylling is an accomplished artist/illustrator, and the manual is fully illustrated with nearly 150 full-color line drawings and photographs.

The **Study Guide,** by Lori Garrett of Danville Area Community College, includes the Key Concepts, an Extended Chapter Outline that gives an overview of the most important topics covered in the chapter, Vocabulary Building exercises, and Chapter Tests. Each Chapter Test has four parts: Multiple Choice; Matching; Short Answer; and Critical Thinking—Using Your Knowledge. All answers are provided, with the exception of the Critical Thinking—Using Your Knowledge questions.

A Guide to *Asking About Life* for Teachers and TAs by Donald Cronkite of Hope College is an instructor's manual for teaching biology using the inquiry-based approach of the textbook. It includes suggested syllabi for teaching using various themes, such as biodiversity, cellular and molecular approaches, and social issues, as well as one- and two-semester courses. Chapters offer intriguing demonstrations, interactive exercises, and group learning. The Preface provides a listing of where the various types of exercises can be found in the manual. Answers to the Review and Discussion Questions from the textbook also appear in the teaching guide.

The **Test Bank,** by Alma Moon Novotny of the University of St. Thomas, and Frederick Peabody of University of South Dakota, consists of 2950 questions of assorted type (multiple choice, fill-in-the-blank, and short-answer essay questions). These are organized by the main chapter headings. The **Computerized Test Bank** is available for both Windows™ and Macintosh® platforms. The computer program allows the instructor to sort the questions by chapter, section head, and question type.

The *Asking About Life* supplements package also includes a set of 200 **Overhead Transparencies** consisting of drawings from the book and a **Student Art Notebook,** which is a set of 300 unlabeled, full-color line drawings from the text. Our Custom Overhead Transparency Program allows instructors to choose additional four-color images from the text.

An **Instructor's Resource CD-ROM for Biology 2001** features selected photographs and all the line art illustrations from *Asking About Life,* Second Edition, and can be used with PowerPoint™ and Persuasion™ on both Macintosh® and Windows™ platforms. A variety of file formats include Web-ready PDFs, print-ready PDFs, and files for use in creating custom overhead transparencies and custom PowerPoint™ presentations. Eric Rabitoy and Terry Damron of Citrus College have created a **PowerPoint™ Presentation** that is available at the *Asking About Life* Web site for downloading as well as on the IRCD-ROM.

The **Process of Science: Discovering Biology™** CD-ROM reflects the spirit of inquiry that characterizes *Asking About Life,* Second Edition. It allows students to explore the discoveries of some of the most important concepts in biology. In the *Interactive Investigations,* students retrace the steps of scientists' experiments and discoveries using the scientific process as their road map. An *Investigator's Notepad* allows the students to track their progress through each investigation, to pose new questions for themselves to pursue, or to initiate a discussion with the instructor or other students via an Internet connection. *Concept Tutorials* provide students with essential background information in general biology for the course and the investigations.

The CD-ROM set **Introduction to Biology,** created by Archipelago Distributed Learning, consists of four CD-ROM disks with an **Instructor's Guide** and a **Student User Guide.** The course covers all the topics that would be covered in a two-term introduction to general biology and can be used as an ancillary to any general biology text, such as *Asking About Life.* Heavily illustrated and complete with animations and videoclips it can be used by students for self-study or as projection software for the instructor. A demonstration of the course is posted at the *Asking About Life* site.

A **WebCT** course for *Asking About Life,* Second Edition, is in development for the text, the first semester of which will be ready for use in Fall, 2000. A demonstration of Chapter 4 from the WebCT course can be accessed at the book site.

Please visit our *Asking About Life* Web site at http://www.harcourtcollege.com/lifesci/aa2/

Students will find:

- A set of Web links for each chapter, including a bibliography of books, readings, and other media for each unit of the text.

- Sets of quiz questions and answers for self-testing of each chapter.
- On-line glossary of words in *Asking About Life*.
- View-only versions of the PowerPoint™ Presentation.
- **Web Bits**—short articles that expand ideas from certain chapters. (These are indicated by an iconic spider web.)
- Samples of the **Lab Manual, Study Guide,** and **Student Art Notebook** will be posted so that students are aware that these study aids are available to them for purchase.

Instructors will find:

- The **Teacher's Guide** posted chapter-by-chapter.
- Sample syllabi for one- and two-semester courses and courses emphasizing a particular theme or approach.
- Answers to the Review and Thought Questions in the text.
- Downloadable **PowerPoint™ Presentations** for all the line drawings in the text.
- **Test Bank** for the Second Edition.
- Image bank of the line art illustrations from the text.
- WebCT demo of Chapter 4.

Harcourt College Publishers may provide complimentary instructional aids and supplements or supplemental packages to those adopters qualified under our adoption policy. Please contact your sales representative for more information. If, as an adopter or potential user, you receive supplements you do not need, please return them to your sales representative or send them to:

Attn: Returns Department
Troy Warehouse
465 South Lincoln Drive
Troy, MO 63379

Reviewers of the Second Edition

Susan Bandoni, *SUNY Geneseo*
Claudia Barreto, *University of Wisconsin, Milwaukee*
Mark S. Blackmore, *Valdosta State University*
Steve Blumenshine, *Arkansas State University*
Susan L. Bower, *Pasadena City College*
Sara Brenizer, *Shelton State Community College*
Arthur L. Buikema, Jr., *Virginia Polytechnic Institute and State University*
Mary Kay Cassani, *Edison Community College, Lee Campus*
Kerry L. Cheesman, *Capital University*
Harold Cones, *Christopher Newport University*
Walter Conley, *St. Petersburg Junior College*
Patricia B. Cox, *University of Tennessee*
Charles Creutz, *University of Toledo*
Forbes Davidson, *Mesa State College*
Jerry D. Davis, *University of Wisconsin, LaCrosse*
David DeGroote, *St. Cloud State University*
Jean DeSaix, *University of North Carolina, Chapel Hill*
Donald Deters, *Bowling Green State University*
Matthew Douglas, *Grand Rapids Community College* and *the University of Kansas*
Patrick Duffie, *Loyola University, Chicago*
Thomas C. Emmel, *University of Florida*
Kathy McCann Evans, *Reading Area Community College*
Robert Evans, *State University of New Jersey Rutgers, Camden*
Victor Fet, *Marshall University*
Richard F. Firenze, *Broome Community College*
James Fitch, *Jones Junior College*
Jennifer Fritz, *University of Texas, Austin*
Shirley Porteous-Gafford, *Fresno City College*
Wendy Jean Garrison, *University of Mississippi*
Robert George, *University of Wyoming*
Karen F. Greif, *Bryn Mawr College*
Charles J. Grossman, *Xavier University*

Lonnie J. Guralnick, *Western Oregon University*
Betsy Harris, *Appalachian State University*
Patricia Hauslein, *St. Cloud State University*
Chris Haynes, *Shelton State Community College*
James A. Hewlett, *Finger Lakes Community College*
Nan Ho, *Las Positas College*
Margaret Hollyday, *Bryn Mawr College*
Stephen Hudson, *Furman University*
Charles W. Jacobs, *Henry Ford Community College*
John D. Jenkin, *Blinn College, Bryan*
Victoria Johnson, *San Jose State University*
Diane Auer Jones, *Community College of Baltimore County, Catonsville Campus*
Marlene Kayne, *College of New Jersey*
Chris Kellner, *Arkansas Tech University*
Robert M. Kitchin, *University of Wyoming*
Dan E. Krane, *Wright State University*
Keith A. Krapf, *John A. Logan College*
Cheryl Laursen, *Eastern Illinois University*
Siu-Lam Lee, *University of Massachusetts, Lowell*
Thomas Lord, *Indiana University of Pennsylvania*
Jon H. Lowrance, *Lipscomb University*
Craig E. Martin, *University of Kansas*
Timothy D. Metz, *Campbell University*
Alison M. Mostrom, *University of the Sciences in Philadelphia*
Anne-Marie Murray, *Arkansas Tech University*
Ken Nadler, *Michigan State University*
Judy H. Niehaus, *Radford University*
Debra Pearce, *Northern Kentucky University*
Andrew J. Penniman, *Georgia Perimeter College*
Helen K. Pigage, *U.S. Air Force Academy*
Jay Pitocchelli, *Saint Anselm College*
David Polcyn, *California State University, San Bernardino*
Steven M. Pomarico, *Louisiana State University*

Carol Pou, *St. Cloud State University*
Paul A. Rab, *Sinclair Community College*
Eric Rabitoy, *Citrus College*
Lynda Randa, *College of DuPage*
David Ribble, *Trinity University*
Laurel Roberts, *University of Pittsburgh*
Frank A. Romano, III, *Jacksonville State University*
Christopher S. Sacchi, *Kutztown University*
K. Sata Sathasivan, *University of Texas, Austin*
Robert M. Schoch, *Boston University*
Brian W. Schwartz, *Columbus State University*
Angelica P. Seitz, *Auburn University*
David Seigler, *University of Illinois at Urbana-Champaign*
Gail Stratton, *University of Mississippi*
Robert J. Swanson, *North Hennepin Community College*
Sandra J. Turner, *St. Cloud State University*
Linda Tyson, *Santa Fe Community College*
Jack Waber, *West Chester University*
Ken Revis-Wagner, *Clemson University*
Timothy S. Wakefield, *Auburn University*
Tom Weeks, *University of Wisconsin, LaCrosse*
J.D. Wilhide, *Arkansas State University*
Edmund B. Wodehouse, *Skyline College*
Anne E. Zayaitz, *Kutztown University*
Henry H. Ziller, *Southeastern Louisiana University*

The following reviewers graciously shared their expertise with the authors in the effort to maintain currency and accuracy across the many sub-disciplines in biology:

Warren Burggren, *University of North Texas*
Naomi Cappuccino, *Carleton University*
Richard P. Elinson, *University of Toronto*
Carol Erickson, *University of California, Davis*
Gordon Fain, *University of California, Los Angeles*
Robert Full, *University of California, Berkeley*
Harry W. Greene, *Cornell University*
Susan Hughmanick, *University of California, Santa Cruz*
Laura J. Jenski, *Indiana University–Purdue University at Indianapolis*
Peter Kareiva, *University of Washington*
Daniel J. Meinhardt, *St. Olaf College*
Barney Schlinger, *University of California, Los Angeles*
Cynthia V. Sommer, *University of Wisconsin, Milwaukee*
Michael P. Stryker, *University of California, San Francisco*
Chris Tarp, *Contra Costa College*
Robert Thornton, *University of California, Davis*
Robert Turgeon, *Cornell University*
Philip C. Withers, *University of Western Australia*

Reviewers of the First Edition

Juan Aninao, *Dominican College of San Rafael*
Edwin A. Arnfield, *Macomb Community College*
Linda W. Barham, *Meridian Community College*
George W. Barlow, *University of California, Berkeley*
Brenda Blackwelder, *Central Piedmont Community College*
Mildred Brammer, *Ithaca College*
Richard B. Brugam, *Southern Illinois University, Edwardsville*
William Bowen, *Jacksonville State University*
Bradford Boyer, *Suffolk County Community College*
Hara Dracon Charlier, *Miami University*
H. Tak Cheung, *Illinois State University*
Karen Crombie, *Fresno City College*
Donald Cronkite, *Hope College*
Tom Daniel, *University of Washington*
Darleen A. DeMason, *University of California, Riverside*
Leah Devlin, *Pennsylvania State University, Abington College*
Ernest F. DuBrul, *University of Toledo*
Peter Ducey, *State University of New York at Cortland*
Steven H. Everhart, *Campbell University*
Lynn Fancher, *College of DuPage*
Cynthia Fitch, *Seattle Pacific University*
Dietrich Foerstel, *Champlain Regional College & Bishop's University*
Sally Frost-Mason, *University of Kansas*
Jack Gallagher, *William Rainey Harper College*
Lori K. Garrett, *Danville Area Community College*
Ben R. Golden, *Kennesaw State College*

Glenn A. Gorelick, *Citrus College*
Nels H. Granholm, *South Dakota State University*
Herbert H. Grossman, *The Pennsylvania State University*
Lonnie J. Guralnick, *Western Oregon State College*
Ross Hamilton, *Okaloosa-Walton Community College*
Richard Harrison, *Cornell University*
Wiley Henderson, *Alabama A&M University*
Bob Highley, *Bergen Community College*
Kathleen L. Hornberger, *Widener University*
Linda Hsu, *Seton Hall University*
David Inouye, *University of Maryland*
William A. Jensen, *The Ohio State University*
J. Morris Johnson, *West Oregon State University*
Peter Kareiva, *University of Washington*
James Karr, *University of Washington*
Tanseem Khaleel, *Montana State University, Billings*
Robert Kitchin, *University of Wyoming*
Ross Koning, *Eastern Connecticut State University*
Dan Krane, *Wright State College*
James W. Langdon, *University of South Alabama*
Anton Lawson, *Arizona State College*
Charles Leavell, *Fullerton Community College*
Kathleen Lively, *Marquette University*
Melanie Loo, *California State University, Sacramento*
Linda A. Malmgren, *Franklin Pierce College*
Nilo Marin, *Broward Community College*
Theresa Martin, *College of San Mateo*

Dorrie Matthews, *Sage Jr. College of Albany*
Gary F. McCracken, *University of Tennessee, Knoxville*
Robert J. McDonough, *DeKalb College*
Joseph McGrellis, *Atlantic Community College*
John Mertz, *Delaware Valley College*
Debbie Meuler, *Cardinal Stritch College*
Robert Morris, *Widener University*
Alison M. Mostrom, *Philadelphia College of Pharmacy and Science*
David M. Ogilvie, *The University of Western Ontario*
Bruce Parker, *Utah Valley State College*
Lee R. Parker, *California Polytechnic State University, San Luis Obispo*
Frederick Peabody, *University of South Dakota*
Ed Perry, *Faulkner State Community College*
Gary Pettibone, *State College of New York at Buffalo*
Jay Pitocchelli, *St. Anselm College*
David M. Polcyn, *California State University, San Bernadino*
Shirley Porteous-Gafford, *Fresno City College*
Paul F. Ramp, *University of Tennessee, Knoxville*
Franklin Robinson, *Mountain Empire Community College*
Lyndell Robinson, *Lincoln Land Community College*

Earle Rowe, *Walters State Community College*
Donna Rowell, *Holmes Community College*
Andrew Scala, *Dutchess Community College*
Shirley Seagle, *Jacksonville State University*
Phillip R. Shelp, *Brookhaven College*
John Simpson, *Gadsden State College*
Linda Simpson, *University of North Carolina, Charlotte*
Michael E. Smith, *Valdosta State University*
Fred L. Spangler, *University of Wisconsin, Oshkosh*
Steven Spilatro, *Marietta College*
Herbert Stewart, *Florida Atlantic University*
Cindy Stokes, *Kennesaw State College*
Jeffrey Thompson, *California State University, San Bernadino*
Janice Toyoshima, *Bakersfield College*
John Tramontano, *Orange County Community College*
Michael J. Ulrich, *Elon College*
Kristin Vessay, *Bowling Green University*
James A. Winsor, *The Pennsylvania State University, Altoona Campus*
Daniel Wivagg, *Baylor University*
Tom Worcester, *Mt. Hood Community College*

For George Dusheck and Eve Tobin
And in Loving Memory of Nina Dusheck and Maurice Tobin

About the Authors

Allan J. Tobin Is Director of the UCLA Brain Research Institute. He holds the Eleanor Leslie Chair in Neuroscience at UCLA, where he is both Professor of Neurology and Professor of Physiological Science. Tobin is also Scientific Director of the Hereditary Disease Foundation (HDF), where he helped organize the consortium that identified the gene responsible for Huntington's disease. Both at UCLA and at HDF, he has encouraged close interactions among cell and molecular biologists, geneticists, physiologists, and clinicians, with the goal of bringing basic science discoveries from the laboratory bench to the patient's bedside.

Tobin's undergraduate degree is in literature and biology (MIT, 1963), and his doctoral degree is in biophysics, with an emphasis on physical biochemistry (Harvard, 1969). Tobin did postdoctoral work at the Weizmann Institute of Science, in Israel, and at MIT. At UCLA, his active research laboratory studies the production and action of GABA, the major inhibitory signal in the brain. These studies may eventually lead to new therapeutic approaches to epilepsy, Huntington's disease, and juvenile diabetes. Tobin is the recipient of a Jacob Javits Neuroscience Investigator Award from the National Institute of Neurological Disorders and Stroke.

For more than 35 years, Tobin has taught introductory courses in general biology, cell biology, molecular biology, developmental biology, and neuroscience. He is the recipient of a UCLA Faculty Teaching and Service Award and is regarded as an excellent and highly interactive teacher.

Jennie Dusheck Is a writer living in Santa Cruz, California. She is a lecturer in science writing and illustration at the University of California, Santa Cruz, and a member of the National Association of Science Writers. Her undergraduate degree is in zoology (U.C. Berkeley, 1978) and her master's degree (by thesis) is in zoology (U.C. Davis, 1983). She also holds a certificate in science writing from U.C. Santa Cruz (1985).

Dusheck has written for *Science News, Science* magazine, and other publications. From 1985 to 1993, she worked as a Principal Editor at U.C. Santa Cruz. She has received several national awards, including the Gold Medal for Best-in-Category from the National Council for the Advancement and Support of Education. Besides *Asking About Life,* she recently coauthored *Life Science,* a middle school text in the Holt Science & Technology Series.

In her previous life as a biologist, Dusheck studied the effects of light intensity on bird song; social behavior in field mice; food preferences of deer, cattle, and skipper butterflies; as well as axis formation in *Xenopus laevis*. While working for the Department of Molecular Biology at U.C. Berkeley, she designed and wrote the protocol for a Space Shuttle experiment that sent live frog embryos into space in the fall of 1992. She has taught university labs in introductory zoology, embryology, and comparative anatomy.

Contents

1 The Unity and Diversity of Life — 2

I Chemistry and Cell Biology — 21
2 The Chemical Foundations of Life — 22
3 Biological Molecules Great and Small — 46
4 Why Are All Organisms Made of Cells? — 70
5 Directions and Rates of Biochemical Processes — 102
6 How Do Organisms Supply Themselves with Energy? — 118
7 Photosynthesis: How Do Organisms Get Energy from the Sun? — 140

II Genetics: The Continuity of Life — 161
8 Cell Reproduction — 162
9 From Meiosis to Mendel — 180
10 The Structure, Replication, and Repair of DNA — 208
11 How Are Genes Expressed? — 234
12 Jumping Genes and Other Unconventional Genetic Systems — 258
13 Genetic Engineering and Recombinant DNA — 276
14 Human Genetics — 296

III Evolution — 317
15 What Is the Evidence for Evolution? — 318
16 Microevolution: How Does a Population Evolve? — 344
17 Macroevolution: How Do Species Evolve? — 368
18 How Did the First Organisms Evolve? — 398

IV Diversity — 417
19 Classification: What's in a Name? — 418
20 Prokaryotes: Success Stories, Ancient and Modern — 432
21 Classifying the Protists and Multicellular Fungi — 452
22 How Did Plants Adapt to Dry Land? — 476
23 Protostome Animals: Most Animals Form Mouth First — 496
24 Deuterostome Animals: Echinoderms and Chordates — 522

V Ecology   543

25	Ecosystems	544
26	Terrestrial and Aquatic Ecosystems	562
27	Communities: How Do Species Interact?	584
28	Populations: Extinctions and Explosions	602
29	The Ecology of Animal Behavior	626

Appendix A Periodic Table of the Elements   **A-1**

Appendix B Units of Measure: The Metric System   **A-2**

Glossary   **G-1**

Index   **I-1**

Asking About Life
Second Edition

1 The Unity and Diversity of Life

Enough To Give You an Ulcer

In 1984, an obscure Australian physician named Barry Marshall secretly performed a dangerous experiment that would ultimately deprive some of the world's largest corporations of billions of dollars. Marshall's radical idea was born in 1979, when his colleague and friend J. Robin Warren noticed that samples of stomach tissue taken from ulcer patients were often infected with bacteria.

To find one bacterial infection in the stomach would have been strange; to find dozens was bizarre. The human stomach secretes acid so concentrated that few organisms survive it for more than a few minutes, let alone live and reproduce in it. Yet Warren found bacteria flourishing there.

Warren's discovery suggested an alternative to doctors' longstanding belief that ulcers are caused by excess stomach acid. Ulcers, every medical textbook reported, were caused by the oversecretion of stomach acid in people with overanxious, frustrated personalities. Such personality problems were thought to be aggravated by the stressful pace of modern life. But if a bacterium could infect the stomach, Marshall and Warren realized, maybe it could cause ulcers. Intrigued by this idea, Marshall began ordering biopsies for any patient who had stomach problems (Figure 1-1). He found that nearly every patient with ulcers was infected with the same bacterium.

The most common kind of ulcer is a peptic ulcer, an open wound located where the stomach joins the small intestine, at the bottom of the stomach. The word "peptic" comes from the Greek word *peptein*, to digest. Nearly one in ten adults has a peptic ulcer. Some people with ulcers feel no discomfort. But most feel at least mild pain, and many suffer excruciating pain for weeks at a time throughout their adult lives. In rare cases, blood may pour from the wound so freely that the person bleeds to death. The standard treatment for ulcers had always been a bland diet consisting of eggs, milk, creams, custards, and overcooked cereals. In addition, doctors often prescribed tranquilizers, psychotherapy, and, in severe cases, surgery. Mainly, however, doctors prescribed antacids—lots of antacids.

Until recently, prescription antacids were the biggest-selling prescription drugs in the world. In 1992, Americans alone bought $4.4 billion worth. These drugs are remarkably effective at controlling the secretion of stomach acid, but remarkably ineffective at controlling ulcers. Ninety-five percent of ulcer patients have a new ulcer within 2 years of treatment. That means people who had ulcers took the $100-a-month antacids almost continuously. In a lifetime, an ulcer-sufferer could spend tens of thousands of dollars on antacids.

Yet, if ulcers were caused by a bacterium, as Marshall was suggesting, then a simple 2-week course of antibiotics could cure an ulcer forever. Millions of people could be saved from a lifetime of suffering. If Marshall's hunch was right, he had very good news, although not for the companies selling antacids.

In 1983, Marshall decided to present his hypothesis at a scientific conference in Brussels, Belgium. But he had a problem. None of the other researchers had ever heard of him. He was barely out of medical school and he had no credentials at all as a researcher. His idea seemed to have about as much chance of winning recognition as bacteria had of flourishing in the acidic environment of the stomach.

Predictably, Marshall's presentation was a disaster. He was unknown; he was young, inexperienced, and overexcited. Worst of all, he had what looked like a screwball idea. "He didn't have the de-

Key Concepts

- Scientific models suggest testable hypotheses that help answer questions and suggest entirely new questions.
- A good experiment includes one or more controls and a large enough sample to compensate for variation in the data.
- All living organisms reproduce, evolve, respond to their environment, and consist of a single cell or ordered arrangements of many cells.
- All organisms on Earth have descended from a common ancestor through a process called evolution.

meanor of a scientist," recalled Martin Blaser, professor of medicine at Vanderbilt University. "He was strutting around the stage. I thought, 'this guy is nuts'." When Marshall's presentation was over, his audience of eminent medical researchers shifted uneasily in their seats, embarrassed. A few laughed. They couldn't believe he was serious. Most bacteria can barely survive a brief passage through the stomach. How could they flourish there for months or years?

In any case, Marshall had no scientific evidence to back up his claim. Maybe, his audience told him, the bacteria had contaminated the stomach samples after the stomach tissues had been removed. Or maybe the bacteria were harmless and unrelated to the ulcers. Or maybe the bacteria were only able to grow in the stomach because the ulcer was there. The audience peppered him with challenging questions, and Marshall realized he couldn't answer any of them.

The only way to settle all these questions was to study the bacterium in an experimental animal. But to do that, Marshall would need to find an animal whose stomach could be infected with the bacterium. After returning to Australia, he began feeding rats the still-unnamed organism. The bacteria all died in the rats' stomachs without having any effect. He fed the stuff to pigs, with the same result. Now he began to wonder, Could the bacteria really infect a stomach? Maybe the researchers in Brussels had been right to laugh at him.

Desperate to prove that he was no nut, Marshall planned a highly unusual "experiment." He told no one ahead of time—not the medical ethics board at the hospital, not his wife. They wouldn't have approved, he knew. He began with a stomach exam and biopsy to make sure his stomach was healthy. Then he made himself an "ulcer bug" cocktail containing at least a billion bacteria, grown from bacteria he had isolated from biopsies of ulcer patients. Then, in a few swift gulps, he drank it down. The cocktail was enough, he hoped, to infect his stomach.

Nothing happened. Days passed. Then, 8 days later, nausea woke him early and he vomited. For another week he was tired, irritable, and hoarse. He had headaches and foul breath. A second stomach exam and biopsy showed that his stomach was inflamed and swarming with bacteria.

By the third week, Marshall was lucky enough to have recovered completely. In the April 15, 1985, issue of *The Medical Journal of Australia*, Marshall reported the results of his trial. He had not proved that the bacterium could cause ulcers, or even that it could infect the stomach for years at a time. But he had done something that strongly suggested that the bacterium, still unnamed, could infect a healthy human stomach—one that didn't already have an ulcer.

His disastrous debut in Brussels had ensured that other researchers would remember him. They soon began to take an interest in Marshall's idea. Mainly, they were interested in proving him *wrong*. Yet, by the end of the 1980s, the evidence that the bacterium could infect the stomach was unassailable. The bacterium, they found, has a twisted, helical shape and usually lives in the "pylorus," near the bottom of the stomach. *Helicobacter pylori*, as it was finally named in 1989, twisted down into the mucous lining of the stomach and settled there.

By 1993, medical researchers had found that about 80 percent of all ulcers were caused by *Helicobacter pylori*. Nearly all could be swiftly cured with ordinary antibiotics.

At first, drug companies resisted Marshall's idea. It was true that people treated with $20 worth of antibiotics had a relapse rate of only 5 or 10 percent. It was true that millions of ulcer-sufferers would each spend thousands of dollars less on prescription antacids. But, gradually, they began to see a silver lining in the cloud the Australian had created.

The good news, for drug companies, was that infection by *H. pylori* was one of the most common bacterial infections in the world.

Figure 1-1 An independent thinker. Barry Marshall showed that bacteria, not cranky personalities, cause ulcers. *(Courtesy, Dr. Barry Marshall)*

As many as half of all people worldwide harbor the infection. And studies of war veterans showed that people infected by *H. pylori* had six times the rate of stomach cancer as other people. Here was a potential market for antibiotics consisting of half the population of the world, some 3 billion people.

Drug companies hastily developed diagnostic tests for *H. pylori* and new combinations of antibiotics to treat the infection. Biotechnology companies were trying to develop a vaccine, to be given in childhood, that would protect against *H. pylori* infection, ulcers, and maybe even stomach cancer—the world's second leading cause of cancer death after lung cancer.

Early in 1994, the National Institutes of Health declared antibiotics the official treatment for most ulcers. Eleven years after Marshall first presented his hypothesis, doctors could offer their ulcer patients a permanent cure. Even then, change came slowly. As late as 1996, only one-third of doctors in the United States were prescribing antibiotics for ulcer patients. The rest continued to prescribe antacids only. Today, antibiotics are standard treatment for this widespread infection, and every family doctor knows what causes ulcers.

In the end, Marshall won full acceptance as a scientist. Ultimately, he moved to the United States and joined the faculty at the University of Virginia Medical School. Marshall succeeded because he possessed many of the attributes of a good scientist. He had curiosity, intelligence, vision, and the dogged determination to pursue an idea—even when his stubbornness made him appear foolish. Perhaps most importantly, Marshall displayed an unusual independence of thought that allowed him to pursue an idea unimaginable to more dogmatic thinkers.

In this book we will meet many scientists who are as curious, independent, and stubborn as Marshall. Some are impulsive, like Marshall. Others take years to reach their conclusions. Most work long hours for years on end. A very few seem merely to play at science, reaping brilliant discoveries from a few hours' work. We'll see some of them risk their reputations to defend the ideas they believe in. A few even lose their lives. Right or wrong, all are fascinated by questions about what makes living things tick. Good biologists, like other scientists, are people who are intensely engaged with life. And, like all people, they work in intense social and political environments.

In this first chapter of *Asking About Life*, we will examine two important aspects of biology. We will see how biologists try to answer questions about life, and we will try to define life itself.

HOW DO BIOLOGISTS ASK QUESTIONS?

From early childhood, everyone asks questions about life: What makes me alive? What goes wrong when I am sick? Where did I come from? How am I like other living things?

The beginning of all scientific inquiry is curiosity. Something in the world captures our attention, and we begin to ask questions. The complex lives of bees may inspire in us pure wonderment. Or perhaps the slow unfolding of a rose arrests our eye. Sometimes, a practical problem commands our attention. Maybe we would like to see a dying friend get well or a hungry village grow enough food to feed its children. Whatever engages our minds—whether beauty, complexity, or misfortune—can be the beginning of scientific inquiry.

The physicist Albert Einstein attributed his scientific achievements to his childlike curiosity. Einstein asked the kinds of questions that most of us ask when we are children. But unlike most people, he continued asking questions when he had the intellectual power to answer them.

Many scientists pride themselves on their childlike curiosity. But scientific inquiry is more than asking questions. Science is curiosity that is controlled and channeled. For example, when Barry Marshall asked whether a bacterium could cause ulcers, he knew it was a question that could be answered. Good scientists try to channel their energy into questions that can be answered.

Do Scientists Use the Scientific Method?

In the first part of this chapter, we will discuss how scientists choose questions to ask and how scientists answer those questions. Because science has been so successful in increasing knowledge, many philosophers and historians have tried to describe how scientists work. One description of the way scientists work is the **scientific method,** a set of formal rules for expressing, testing, and eliminating ideas. In reality, scientists do not think of themselves as following the rules of this or any other single "method." Rather, they pursue knowledge in a variety of individual and creative ways. Nonetheless, we can give a general description of how most scientists go about learning about the world.

The "scientific method" is a formal set of rules for forming and testing hypotheses. Most scientists use the scientific method only loosely and often unconsciously.

How do scientists begin?

A scientist's first step in understanding something is to focus on a single question or small set of questions. Because science, especially biology, works in tiny steps, scientists rarely answer big, general questions all at once. A scientist does not ask, How do animals move? and expect to come up with a simple answer to that question.

When we ask how something works, whether it be a piano or a pancreas, we generally want to know how the parts operate. For example, consider the unbending of an arm. We can ask, What structures change as an arm unbends? What forces cause the movement? These are specific questions that can be answered.

The second step in understanding is *observation*. Looking, hearing, smelling, and touching may all contribute to our observations. For example, during a pushup, one can feel the bulging and hardening of the triceps, the long muscle at the back side of the upper arm.

Figure 1-2 **Observing the human arm.** The upper arm consists of a long bone called the humerus and the lower arm consists of two shorter bones, the ulna and the radius. The ulna pivots on the end of the humerus. The triceps muscle pulls the end of the ulna and swings the lower arm in a wide arc.

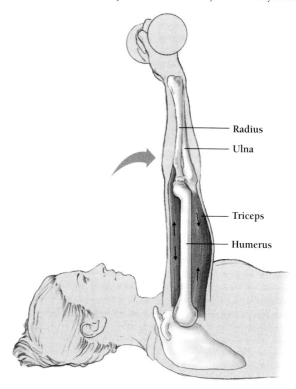

Now, let's examine (observe) the internal structure of the arm (Figure 1-2). The skeleton of the upper arm consists of a single, long bone, called the "humerus," between the shoulder and the forearm. The forearm consists of two smaller bones, the radius and the ulna. The ulna pivots, rocking across the elbow end of the humerus. The triceps muscle connects to the elbow end of the ulna and to the shoulder.

Now we are ready for the next step. We need a **model**—a highly simplified (often cartoonlike) view of how the arm works when we unbend it. Let's imagine that the arm is like a seesaw on a children's playground (Figure 1-3). When a child sits on one end of a seesaw, the child's body exerts a force that pushes the seesaw down. As a result, the other end of the seesaw moves up. The ulna pivots on the tip of the humerus in the same way that a seesaw pivots on its base. When one end of the ulna moves up, the other end moves down.

In science, the first step is to ask a sharply focused question about some process. The next step is to observe closely, looking for a familiar pattern of behavior. The next step is to devise a model that suggests how the phenomenon works.

What does a hypothesis do?

A model must be consistent with what we already know. But it must also suggest things we might not have thought of otherwise. A good model should suggest a **hypothesis**—an educated guess about the way a process works that allows us to make predictions. Scientists could not work without hypotheses. Hypotheses are as essential to the daily functioning of science as vitamins are to the normal functioning of our bodies. So where do scientists get hypotheses?

The seesaw model suggests a good hypothesis. If the ulna is like a seesaw, some force must cause it to rock back and forth during a pushup. Maybe when the triceps muscle contracts, it pulls on the end of the ulna. The elbow end would move toward the shoulder, and the wrist end would move away (Figure 1-3).

If that's true, we can predict that if we try to unbend our arm without using our triceps muscle, we won't be able to do it. We could perform an experiment to test this prediction. We could, for example, temporarily immobilize the triceps with an injection of some paralyzing drug.

In general, if the predictions of a hypothesis are right, we begin to guess that we are on the right track. On the other hand, if the predictions of a hypothesis are wrong, then both the hypothesis and the model it is based on are likely wrong.

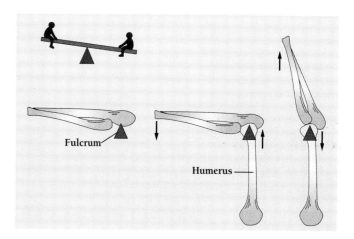

Figure 1-3 **It rocks.** Like a children's seesaw, the ulna (in the lower arm) pivots on the tip of the humerus (in the upper arm).

A good model also suggests how other similar processes work. For example, if the arm works like a seesaw when the arm is unbending (as in a pushup), then maybe it also works like a seesaw when the arm is bending (as in a chin-up). Perhaps during a chin-up some force pulls the wrist end of the ulna. Our model of the unbending of the arm accomplishes two important things: (1) it allows us to make predictions and (2) it suggests how other body movements might work.

A hypothesis is an informed guess about the way a process works that enables a scientist to predict what will happen in different situations. It is a product of logic, previous knowledge, insight, and creativity.

What is a testable hypothesis?

We can quickly generate lots of models and hypotheses to explain any observation. If we note that ants in the kitchen are crowding around the sink, we might hypothesize: (1) that they are looking for water, (2) that they are looking for food particles, (3) that they are looking for a new nest site, or (4) that they want to clean the dishes for us. Clearly, some hypotheses are better than others. Scientists need ways of deciding which hypotheses to explore.

In general, a hypothesis is valuable only when it is **testable**, meaning that someone can devise an experiment that would disprove the hypothesis if it were incorrect. In the last section we predicted that someone whose triceps were temporarily paralyzed would not be able to do a pushup. This is a test of the hypothesis that the triceps provides the force that powers a pushup.

Whether a hypothesis is testable does not depend on whether it is correct. The hypothesis that bears eat berries is testable. The hypothesis that bears eat rocks is equally testable. Testable (or *disprovable*) means that we can make a prediction that we can test by experiment.

The hypothesis that extraterrestrial beings influence the movement of ants in your kitchen is untestable—but *not* because it is wrong. We have no way to prove that it is wrong. This hypothesis is untestable—and therefore of no interest to science—because designing an experiment in which the influence of the alleged extraterrestrials is ruled out is impossible. We can't prevent their alleged influence because we have no information about what kinds of influence extraterrestrials might exert. We have no reliable observations of extraterrestrials. The same arguments apply to fairies, gnomes, elves, and deities.

A hypothesis is valuable only if it is testable.

How do we decide if a hypothesis might be correct?

We can often prove that a hypothesis is wrong, but we can never prove that it is right. This is because a prediction can turn out right—even when the hypothesis is wrong. We could hypothesize that a great runner named Jackie can run the 100-m dash faster than anyone else in the world. If we find even one person who can run the 100-m dash faster than Jackie, we have disproved our hypothesis. That part is easy.

But what if we cannot find anyone who can run faster? How can we be sure that there isn't some still-undiscovered person who can run faster? The answer is that we can never be sure. In most cases, an experiment cannot prove that a hypothesis is right. An experiment can only disprove a hypothesis.

Even disproving a hypothesis can be hard. Suppose we hypothesize that bears eat berries. We predict that we will find berry skins in bears' stools. And we test this prediction by examining stools from 100 bears. We find grass, insect parts, and even small plastic bags, but no sign of berries. Does this mean the hypothesis was wrong? Maybe not. Our results might mean only that berries were not in season. Maybe we should repeat our study in the fall. We might also wonder if berry skins are digested completely and leave no trace in the stool. We could test that hypothesis by feeding berries to captive bears. In this way, one experiment leads to another.

Any experiment can fail to give the predicted result for reasons that have nothing to do with the hypothesis being tested. The prediction might have been wrong because of faulty logic, or the experiment might have been poorly designed or executed. It is the scientist's job to figure out why the model's prediction was incorrect and then to come up with either a better model or a better experiment.

When scientists test a hypothesis, they can sometimes prove that the hypothesis is wrong, but they can never prove that it is right.

What makes a good experiment?

An **experiment** is a way of testing a hypothesis or of searching for some unknown effect. An experiment is a procedure that is carried out under "controlled" conditions.

Suppose, for example, that we observe that people who live in countries where cinnamon and other spices are used extensively are shorter than other people. Based on this observation, we hypothesize that cinnamon inhibits growth. We decide to test this hypothesis by measuring cinnamon's effect on the growth of laboratory mice. We choose mice because experimenting on humans would be unethical and because mice are easy to keep.

We inject baby mice with extract of cinnamon to see if the cinnamon slows their growth. We find that their growth is slow compared with that of mice in other studies. Can we conclude that cinnamon slowed their growth? We cannot. Maybe we were not feeding the mice enough food. Maybe the room they are in is too hot. Maybe chemicals in the extract other than the cinnamon are slowing their growth. The outcome of this experiment could depend on many causes besides the cinnamon.

To ensure that the outcome of an experiment depends only on the proposed cause, biologists compare the results of every experiment with the results of a "control" experiment. A **control** is a version of the experiment in which everything is the same except for the one thing being tested. To test our cinnamon hypothesis properly, for example, we need at least two controls. First, we need, in our own lab, a group of mice that receive no injection. These will show us how fast mice normally grow under the conditions in our laboratory, not someone else's. Second, we need a group of mice each of which receives an injection containing everything in the extract except the cinnamon. This second control group will show whether something else in the extract slows the growth of baby mice.

The design of a control is critically important and not always obvious. For example, the trauma of being handled might be enough to slow a mouse's growth. To make sure that is not the case, we might want to include a third control in which researchers merely pretend to give baby mice injections—sticking them with a needle, but not injecting anything. This might sound silly, but we know that both humans and other animals respond in both positive and negative ways to handling. It is good science to test anything that may affect our results. Because an experiment must distinguish causes from irrelevant factors, an ideal experiment allows a researcher to evaluate the effects of one factor at a time.

When Barry Marshall swallowed his mixture of stomach bacteria, he was performing an experiment—but only just barely. It was not a good, controlled experiment. His symptoms—nausea, tiredness, and bad breath—might have been caused by the flu, a sinus infection, or any number of things. Even the inflammation of his stomach shortly after he drank the ulcer bug cocktail might have been a coincidence.

Experiments on a single individual may be strongly suggestive, as Marshall's was. But such experiments do not provide results that mean anything scientifically. To do the experiment properly, Marshall would have had to persuade at least three groups of people to participate. The first group, the "experimental" group, would have gotten the ulcer bug cocktail. The second group would have gotten a different bacterium—one not believed to cause ulcers—to rule out the possibility that any concentrated mixture of bacteria can make someone sick. The third group would have gotten the same mixture except without any bacteria, to rule out the possibility that the cocktail—even without any bacteria—is itself a sickening mixture.

An experiment must include carefully designed controls.

How Do Scientists Design Experiments?

Designing experiments is the greatest challenge to a scientist's ingenuity and the greatest source of excitement. Experimental design takes skill and artistry. Like a custom paint job on a car or the cut of a special dress, great experiments are the product of both mastery and inspiration. One of the most famous stories of inspiration in science comes from the German physiologist Otto Loewi.

Loewi was trying to understand how nerves control muscles. He had been studying the heart, a large muscle that contracts with each beat, and a nerve called the vagus, which controls how fast the heart beats. When Loewi stimulated a frog's vagus nerve with electricity, the frog's heart rate slowed.

But Loewi didn't know if the nerve slowed the heart through direct contact or by means of some chemical. He suspected that nerves make a chemical that stimulates muscles. If that were true, then a muscle should respond to the right chemical even if the nerve weren't there. But Loewi had no idea what kind of chemical might do that. And he didn't see how he could possibly test his hypothesis without knowing.

In fact, 17 years passed before he thought of a way. Loewi himself later described how the experiment finally came to him:

> The night before Easter Sunday of 1920 I awoke, turned on the light, and jotted down a few notes on a tiny slip of thin paper. Then I fell asleep again. It occurred to me at six o'clock in the morning that during the night I had written down something most important, but I was unable to decipher the scrawl. The next night, at three o'clock the idea returned. It was the design of an experiment to determine whether or not the hypothesis of chemical transmission that I had uttered seventeen years ago was correct. I got up immediately, went to the laboratory, and performed a simple experiment on a frog heart according to the nocturnal design.

In hindsight, Loewi's 3 a.m. experiment seems straightforward, almost routine. Loewi isolated two frog hearts, one with the vagus nerve still attached, the other without the nerve. He stimulated the vagus nerve of the first heart, collected some of the fluid surrounding it, and transferred the fluid to the second heart, the one with no nerve. The heart slowed.

The dramatic result showed that the slowing was a response to some substance produced by the vagus nerve. Loewi called this substance "vagus stuff," which we now know as acetylcholine—an important chemical signal that nerves use to communicate. Loewi eventually received a Nobel prize for his work.

Designing experiments to test a hypothesis requires both creativity and logic.

Why Do Biologists Study Groups of Organisms?

We have talked about experiments in groups of mice or other organisms. Why can't scientists just measure growth rate in one mouse with cinnamon and one mouse without? Why do scientists study groups of organisms?

The answer is that every organism is unique. Every fruit fly is unique and every cat is unique (Figure 1-4). So even are identical twins and the individual clones of bacteria or sheep. For example, one pair of identical twin girls differed in height by 7 inches as adults. The difference apparently resulted from prenatal trauma in one of the twins. Other identical twins

BOX 1.1

Reductionism

One way to study a thing is to look at its parts. If we look at the wheels of a bicycle, for example, we can see that they turn and are attached to the bicycle's frame by means of intricately designed hubs. Much of biology involves studying the parts of organisms, whether their organs, their cells, or their molecules. For every level of organization, biologists seek to explain structure and function in terms of the next finer level of structure. We can, for example, understand something about the action of arm muscles by studying the anatomy of the bones and muscles. The effort to understand the

Figure A How does muscle contract? We can learn more about how muscles work by looking at their structure. A microscope shows that muscles consist of bundles of fibers that are made up of still thinner fibers that are made of muscle cells. These cells themselves consist of bundles of even thinner fibers called myofibrils. The myofibril is the unit of contraction of a muscle. Each one consists of sections called sarcomeres. When a sarcomere contracts, minute fibers called "thick filaments" and "thin filaments" slide past each other in the same way as the two parts of an extendible ladder. When a muscle contracts, all of the sarcomeres in all of the myofibrils in the muscle contract together.

may have deformities on opposite sides of their faces, heart defects in one but not the other, and so on. Variation among organisms is the rule. Whether we are talking about humans, mice, worms, oak trees, dandelions, mushrooms, or even one-celled organisms, every individual is unique.

Some of this individuality is due to genetic differences. Some of it is due to the impact of the environment. An *H. pylori* bacterium on the wrong end of your stomach may not get the same nutrients as a bacterium in the prime position down by the pylorus.

whole in terms of the parts is called **reductionism**.

Understanding the parts of a process can help explain larger, more complex biological processes. Many of the questions that we ask in this book concern the relationships of structures and processes. The questions are often the same: What are the involved structures? What are they made of? How do they capture energy? What controls their operation? The answers to our questions may lie at the level of ecosystems, populations, individual organisms, cells, or even individual molecules.

The reductionist approach to science has been immensely successful. The most successful and enthusiastic reductionists are molecular biologists, who seek to answer a wide variety of questions just by looking at the structures of molecules (Figure A). These questions range from the inner workings of cells to the evolution of whole families of organisms over millions of years.

But reductionism has definite limits. All of the properties of an object are not explainable in terms of the object's parts. To understand how the circulatory system works, we need to study the veins, arteries, and heart in whole animals. Looking at the individual molecules of the veins and blood would tell us nothing in the absence of a good understanding of how blood circulates in the body.

A whole is greater than the sum of its parts. The painting in Figure B illustrates this principle simply and beautifully. When we examine the painting under high magnification, we see only an abstract pattern of colored dots. It is only when we step back and look at the whole picture at once that we see a river, trees, and flowered hillsides. The landscape is an **emergent property,** a characteristic that arises only at complex levels of organization. Living organisms are no different.

To discover how bees pollinate flowers, we need to study hives of bees and fields of flowers, not their molecules. Just as a painting is much more than a collection of brush strokes, a group of organisms is more than a collection of individuals. A detailed knowledge of the behavior of solitary humans, for example, would never allow us to predict such bizarre group behaviors as war and Tupperware parties.

Figure B Art as emergent property. A landscape painting is an emergent property—a characteristic that arises only at complex levels of organization. When we look at the individual brushstrokes, we see only an abstract pattern of colored dots. When we step back and look at the whole picture at once, we see a landscape— "La Seine à Herblay," by Maximilien Luce (Musée d'Orsay, Paris). *(Erich Lessing/ Art Resource)*

Because each individual organism is unique, each one can react to an experiment differently from every other individual. For example, every mouse in the cinnamon-extract experiment could grow at a different rate. Individual mice, like individual humans, grow at different rates. We'd be amazed if every mouse in the experiment grew at exactly the same rate, cinnamon or no cinnamon. Because individuals vary, biologists always experiment with groups of organisms rather than single individuals. Then they compare the average of the experimental group with that of the control group.

Figure 1-4 Why are individuals unique? When organisms reproduce, the genetic information in the DNA allows the offspring to be both like their parents and unlike their parents. Here, the offspring of a cat show a diversity of phenotypes because each has a unique set of genes inherited from mother and father. Our uniqueness comes not only from our genes, however, but strongly from our environment. For example, even genetically identical twins can differ dramatically if the environment of the womb favors one over the other. *(Francois Gohier/Photo Researchers, Inc.)*

How many individuals should an experiment include?

How many individuals an experiment should include depends on how much individual variation exists. For example, a group of genetically identical mice raised in a laboratory under identical conditions would be expected to show only minimal variation. Imagine that all the mice are genetically identical, that they all receive the same amount of food and water each day, and that the temperature and humidity in their cages was identical. A biologist wouldn't need many mice to show an effect on growth rate—ten in each group might be enough.

Now imagine a group of wild mice living outdoors. Each mouse is genetically unique. Each one lives a slightly different life from the others. Some get more milk from their mother from the start. Some contract a disease that the others don't get. The list of possible sources of variation is endless. The result is mice whose growth rates vary by as much as 50 percent. In this case, detecting a 10 percent difference between the experimental mice and control mice would be difficult. Like wild mice, human populations vary enormously, which is why the best studies of humans often involve thousands of individuals.

The greater the variation, the more individuals we need for an experiment. But we can only find out how many individuals we need by first measuring variation. To measure variation, we have to do a simple version of the experiment. Fortunately, there is a shortcut. A biologist might need hundreds of mice. But thanks to statistics, it's possible to measure a difference between two groups, even when both groups vary enormously.

How Do Biologists Use Statistics To Plan and Evaluate Experiments?

Biologists often plan experiments and evaluate the results of experiments with **statistics**—the mathematics of collecting and analyzing numerical data. Statistics is a powerful tool with two formal uses. (1) Statistics can help us estimate the actual (or "true") value of something. (2) Statistics can help us test a hypothesis. Statistics cannot do experiments for us. But it can measure the probability that what we measure is what it seems to be and not a random effect. The arrangement of data into graphs and pictures can also suggest new questions and new hypotheses.

How do scientists sample a population?

Statistics relies on certain assumptions. But statistics always states its assumptions clearly and then examines those assumptions. If we study the effects of aspirin on heart attack rates in groups of men, we are really interested in the effect of aspirin on *all* men. The men in the experiment are supposed to represent all men. Statisticians call all men the **population** of men. The group selected for the experiment is a **sample** of the population. In statistics, we want to be able to assume that the sample represents the population.

One way of ensuring that the sample represents the population is to take a **random sample.** A sample is random when we ignore any differences among the members of a population while choosing the sample. Suppose we have a population of 100 men, one-third of whom eventually have heart attacks (Figure 1-5). One-fourth of the 100 men agree to volunteer for a study of heart attacks and aspirin.

Figure 1-5A shows the whole population. Figure 1-5B shows what happens if you choose a sample of 25 men randomly. Figure 1-5C shows what happens if you choose a sample based on some difference. In this case, only volunteers, who do not represent the population, were chosen. Apparently, the volunteers are more likely to have heart attacks than the general population. Maybe the volunteers were men who volunteered for the study because they knew they were at risk for heart attacks. Compared to the general population, they may include more smokers, more older men, and more sedentary men than the population at large.

A sample can also fail to represent the population because the sample size is too small. Figure 1-5D shows two samples of only 10 men each. In neither one does the number of heart attack victims represent the rate in the population. This difference is due to random errors.

If we flip a coin 100 times, we will get equal or almost equal numbers of heads and tails. We might get 52:48 or 47:53, but the ratio will be close to 50:50. On the other hand, if we flip the coin only 10 times, we will probably not get a ratio very close to 50:50. We would have a good chance of getting 2, 2, 3, or 4 heads instead of 5. But chances are we wouldn't get exactly 5 heads and 5 tails. Random effects are big in small samples and tiny in big samples.

In a biological experiment, a sample should represent the whole population of organisms. This can happen only if the sample is chosen randomly and the sample is large enough.

Biologists are primarily interested in two statistical measures. The first measure is called the *mean*, which is just the average value of something. The second measure is the *standard deviation*, which is an estimate of the variation in the data. If we were studying heart attack rates in men, we would want to know the mean heart attack rate both in the experimental group and in each of the control groups. We would also need to measure the standard deviation in each sample. Both the mean and the standard deviation are essential to thoughtful evaluation of the experimental results.

How does statistics help biologists estimate value?

The **mean** is the average value, or arithmetic average, that we all learned about in elementary school. Real-life studies have shown that the mean heart attack rate in the population of men over 50 years old who are taking aspirin is lower than that in the population of men over 50 not taking aspirin. Because this is known to be true, millions of older men now take an aspirin every other day to reduce their chance of getting a heart attack.

How can biologists tell if two groups differ?

It might seem as if all we need to know is whether the mean heart attack rate of aspirin-takers is lower than that of the non-aspirin-takers. If it is lower, then all men over 50 should take their aspirin. But it turns out that we need to know more.

Imagine that the sample size is too small and the sample mean is different from the population mean. In that case, the mean for the whole population of aspirin-takers could actually be the *same* as the mean for all non-aspirin-takers. In fact, things could be much worse. The aspirin-takers could have a *higher* rate of heart attack. Meanwhile, the sample means could show the opposite. Indeed, in medical studies of just a few people, this often happens. When you read a newspaper article about the latest health findings, always look to see whether the study included thousands of people or just a few dozen. If the article doesn't say, don't take the conclusions too seriously.

Fortunately, statisticians have devised ways of checking for this problem. In order to find out if two means are truly different, we have to estimate the variation in the data. We have to find out, in other words, how much individual variation

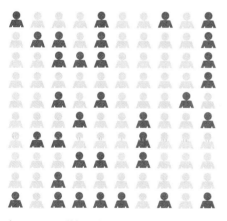

A. 33/100 will have heart attacks.   33%

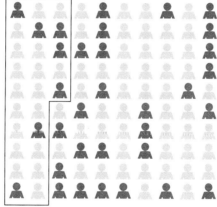

B. 7/25 will have heart attacks.   28%

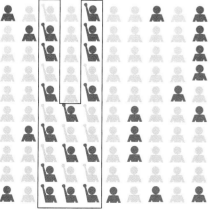

C. 15/25 will have heart attacks.   60%

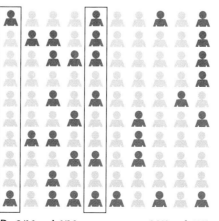

D. 2/10 and 6/10 will have heart attacks.   20% and 60%

Figure 1-5 Sampling a population. A. A population of 100 men includes 33 who have heart attacks. B. A random sample of 25 men yields 7 who later have heart attacks. This rate, 28 percent, approximately reflects the heart attack rate in the general population, but with some sampling error. C. 25 men volunteer for a study of heart attacks, but the heart attack rate in the volunteers—15 out of 25, or 60 percent—does not reflect the rate in the whole population. The volunteers are not a random sample. D. Two small samples of just 10 men each show a larger sampling error than in B. In one case, the heart attack rate is 2/10, or 20 percent. In the other case, it is 6/10, or 60 percent. Neither is a good measure of the overall heart attack rate in the general population.

there is. Statistics measures variation by measuring how far each data point is from the mean. When those distances are averaged, we get a measure called the **standard deviation,** which summarizes how much the data deviate from the mean.

This sounds more complex than it is. Here's a simple example. Suppose we want to determine the average length of earthworms we have just dug up from our back yard. We start with only three worms whose lengths are 3 cm, 5 cm, and 7 cm. We say we have three "data points." The mean worm length is simply the average of those measurements, or 5 cm. We can then measure the deviation from the mean as follows:

$$7 \text{ cm} - 5 \text{ cm} = 2 \text{ cm}$$
$$5 \text{ cm} - 5 \text{ cm} = 0 \text{ cm}$$
$$5 \text{ cm} - 3 \text{ cm} = 2 \text{ cm}$$

To calculate standard deviation, statisticians square each of these differences, then add them all together, and then divide by the number of data points minus 1. That would be $(2^2 + 0^2 + 2^2)/(3 - 1) = 8/2 = 4$. (We leave out the "cm" here for simplicity.) Then, because they did all that squaring, statisticians bring the number back to reality by taking the square root of 4, which is 2. In this case, the standard deviation is the mean difference from the average.

We express this by saying that the length of the earthworms in our sample is 5 cm "plus or minus" 2 cm.

$$5 \text{ cm} \pm 2 \text{ cm}$$

The standard deviation here is huge, a full 40 percent of the mean. We can see that our sample of earthworms is highly variable. But does the population vary that much? If we sampled 100 worms, would the mean still be 5 cm $\pm$ 2 cm? Not necessarily. We may have accidentally picked the very longest and shortest worms. Indeed, the mean length may not be 5 cm at all. It may be something quite different. If the standard deviation is large, we know that our estimates are unreliable. Only with a much larger sample size could we accurately measure the mean of the population.

Imagine, on the other hand, that the standard deviation had been very small. Suppose the three earthworms were 5.1 cm, 5.0 cm, and 4.9 cm. We would be more confident that our 5-cm sample mean was close to the population mean. And if we had 20 earthworms whose lengths were so similar, we could be extremely confident that 5 cm represented the population mean.

Calculating standard deviations by hand for lots of data is not something anybody does for fun. We can see intuitively, however, that 2 cm is probably a good estimate of how much variation there is in the group of earthworms that measure 3 cm, 5 cm, and 7 cm.

Once we know the standard deviation, we can calculate how big a sample we need to tell if the two population means are truly different. If the standard deviation is a big number, our sample size may be too small. In that case, we should repeat the experiment using the right sample size. In this case, we need to measure many more earthworms. Alternatively, if the standard deviation is small enough, then we may already have enough data. We can then calculate the probability that the two population means are different. Statisticians do this all the time. They might say, The probability that the aspirin group has a lower heart attack rate than the controls is 95 percent. By this, they mean that there is only a 5 percent chance that the aspirin-takers have either more heart attacks or the same number of heart attacks as the controls.

Can the mean and standard deviation explain all data sets?

Knowing the standard deviation is essential. But it is accurate only in groups of numbers that are distributed around the mean "normally." A **normal distribution** is one in which there are equal numbers of data points on each side of the mean. When graphed, they are shaped something like a bell (Figure 1-6). Such distributions are surprisingly common in biology. In Chapter 16, Figure 16-6 shows a graph of student heights. That's a normal distribution.

When data have a normal distribution, biologists can assume that about 68 percent of the data will fall within one standard deviation of the mean, and about 95 percent will fall within two standard deviations (Figure 1-7). Once we know that, we can test the hypothesis that the two means are different.

We will stop now. But just remember this. If the numbers in a population are normally distributed (shaped like a bell curve) and if you know the mean and standard deviation of the two samples (experimental and control), then you can calculate the probability that the mean for the experimentals is different from the mean for the controls—in the whole population. This aspect of statistics is a fantastically useful and powerful tool.

In the course of calculating these things, you may learn that the sample sizes that you took were too small because the variation is huge. In this case, you need more data. Or you may learn that your data are not normally distributed, in which case you need more complicated statistics to test your hy-

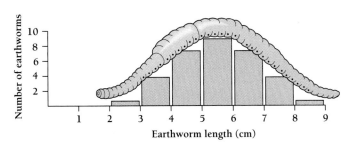

Figure 1-6 A normal distribution of worms. Worms can be long or short, but most are medium, as shown in this graph of worm lengths. As many worms are 1 cm shorter than the average (5 cm) as are 1 cm longer than average. This equal distribution of lengths on both sides of the mean is what makes this a "normal distribution." If most of the worms were either very short or very long and few were average, the distribution would not be normal.

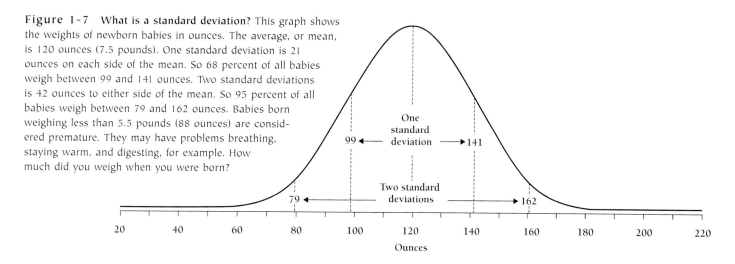

Figure 1-7 **What is a standard deviation?** This graph shows the weights of newborn babies in ounces. The average, or mean, is 120 ounces (7.5 pounds). One standard deviation is 21 ounces on each side of the mean. So 68 percent of all babies weigh between 99 and 141 ounces. Two standard deviations is 42 ounces to either side of the mean. So 95 percent of all babies weigh between 79 and 162 ounces. Babies born weighing less than 5.5 pounds (88 ounces) are considered premature. They may have problems breathing, staying warm, and digesting, for example. How much did you weigh when you were born?

pothesis. But don't worry. You don't have to calculate anything right now.

What Is a Theory?

When a series of experiments fails to disprove the predictions of a hypothesis, scientists gain confidence in the hypothesis and its underlying model. In addition, the more things a model explains, the more important the model becomes to other scientists.

Once scientists have tested a model for a group of related processes, they may decide that they can accept a more general hypothesis. If we continued to study the movements of muscles and bones, for example, we might conclude that our limbs work mostly by simple mechanical laws. Scientists often call a group of related hypotheses a **theory**—a system of statements and ideas that explains a group of related facts or phenomena. For example, "atomic theory" includes many ideas, all of which are based on the conclusion that matter is made of tiny entities called atoms. "Evolutionary theory" includes many ideas, all of which are based on the conclusion that living organisms are related to one another and have descended from a common ancestor.

Scientists use the word "theory" differently from many nonscientists. A nonscientist might use the word "theory" for any half-baked idea. Joe has a "theory" that if he reads every other paragraph in this book, he'll get an A on the final. To a scientist, that's only a hypothesis, and one that needs testing. It doesn't rate as a formal theory. A scientific theory is a set of interconnected hypotheses that have withstood rigorous experimental testing.

A set of related hypotheses that consistently resist scientists' attempts to disprove them may become recognized as a formal theory.

HOW ARE ALL ORGANISMS ALIKE?

Everywhere on Earth, there is life: from the depths of the Pacific Ocean to the top of Mt. Everest, from the frozen wastes of the Arctic Ocean to the tropical rain forests of the Amazon and the arid Sahara Desert. For each living thing, the same questions arise: How is it put together? How does it work? How did it get here? In answering these questions, biologists have accumulated millions of individual facts (only a tiny fraction of which we will study). Tying these disconnected facts together, however, are a few basic themes. Once we grasp these few generalities, the study of life becomes much easier.

Since **biology** [Greek, *bios* = life + *logos* = word, argument] is the study of life, the first general question we must answer is, What is life? Most of the time we can distinguish the living from the nonliving as easily as we can tell a live bear from a teddy bear. Although agreeing on a definition of life that is both precise and general is harder, we can agree on some common characteristics of all living things. Figure 1-8 lists and illustrates eight features of all living organisms.

In summary, all living things are organized into cells and other parts, perform chemical reactions, obtain energy from their surroundings, change with time, respond to their environments, reproduce, and share a common evolutionary history.

How Do Organisms Self-Regulate?

To survive, organisms need to keep their insides separate from the outside world. The first step to doing this is to create a barrier. In the case of our bodies, it is our skin. In the case of our cells, it is the cell membrane.

The second step is to actively resist change. The process of resisting change is called **homeostasis** [Greek, *homeo* = same + *stasis* = standing still]. Homeostasis always involves two stages: (1) detecting change and (2) counteracting such change. Counteracting change is called *negative feedback*.

A. Living things consist of organized parts, as illustrated by this beautiful chambered nautilus. The word "organism" has the same root as "organization."

B. All organisms are made of cells, as illustrated by this plant embryo. Just as the atom is a unit of matter, a cell is the unit of life.

C. Organisms perform chemical reactions. Honeybees convert flower nectar into honey.

D. Living organisms obtain energy from their surroundings. Sunlight is the ultimate source of energy for most organisms. Plants obtain energy directly from the sun. Animals obtain energy from plants, or from animals that have eaten plants.

E. Organisms respond to their environments. The plants in this pot are bending toward a sunlit window.

F. Organisms mature over time. A tadpole maintains its organization while developing into a frog.

G. Organisims reproduce. Each organism, including each of these rabbits, comes from another organism or from a pair of organisms.

H. Organisms share a common evolutionary history. All organisms on Earth today are descended from organisms that lived in earlier times. Some have changed dramatically. Others have remained the same. The horseshoe crab of Atlantic beaches shown here is nearly identical to one that lived 200 million years ago.

Figure 1-8 **Some characteristics of living organisms.** (A, J.H. Robinson/Animals Animals; B, John D. Cunningham/Visuals Unlimited; C, M.A. Chappell/Animals Animals; D, Otto Willner/OKAPIA/Photo Researchers, Inc.; E, N. Pecnik/Visuals Unlimited; F, Dan Suzio/Photo Researchers, Inc.; G, Jane Burton/Bruce Coleman, Inc.; H, [top] John Cancalosi/Peter Arnold, Inc., [bottom] Alan Desbonnet/Visuals Unlimited)

One example of the use of negative feedback in homeostasis is temperature regulation. Birds and mammals maintain a remarkably constant internal temperature. A person can be exposed to temperatures as low as 55°F or as high as 140°F and still maintain a body temperature of between 98°F and 99°F.

When our temperature rises, we perspire, drink water, or look for a cool place to rest. When our body temperature drops, we conserve heat by not perspiring and by reducing blood flow to the skin, hands, and feet. In addition, we increase heat production by moving around, shivering, or increasing our metabolic rate. And, of course, we can also put on warm clothes or turn up the heater. All of these responses, whether physiological or behavioral, are homeostatic responses that keep the temperature of the body between 98°F and 99°F.

Using negative feedback, animals regulate such things as body temperature, thirst, hunger, and even sperm production. Plants regulate their intake of water and carbon by means of negative feedback. In all organisms, cells regulate the production of proteins and other molecules using negative feedback.

All organisms resist change through homeostasis and negative-feedback mechanisms.

What Kinds of Cells Are Organisms Made Of?

Whether dandelion or dachshund, *Helicobacter pylori* or bathroom mold, we organisms are all made of cells. A cell is a tiny mass of water, protein, and other molecules, all surrounded by a thin membrane. Indeed, all cells are made from the same few molecular building blocks. But a cell is much more than a bag of protein and water. It can do all of the things listed in Figure 1-8, including reproduce itself.

Cells are wonderfully diverse and specialized. In our own skin, flat "epithelial" cells protect our body, "follicle" cells produce hairs, and nerve cells carry messages. Plant cells are similarly specialized. Tough outer cells protect the plant's body, others transform light energy into sugar, and still others form passages that carry sugar, water, and other materials throughout the plant. There is no easy way to count all the different kinds of cells. Mammals alone have at least 200 different kinds of cells (Figure 1-9).

Nonetheless, we can group cells into just two basic kinds. Our own cells are eukaryotic cells. **Eukaryotic** cells [Greek, *eu* = true + *karyon* = nucleus] contain a **nucleus,** which is a package of genetic material enclosed in a thin membrane. Eukaryotic cells contain about 20 kinds of little membrane-enclosed parts called **organelles** [little organs]. These include the energy-supplying mitochondria and chloroplasts (in plants only). Plants, animals, fungi, and the microscopic protists are all made of eukaryotic cells.

The second kind of cell is the **prokaryotic** cell [Greek, *pro* = before]. Prokaryotic cells contain no nuclei and no membrane-enclosed organelles. They have genetic material, but they are not enclosed in a membrane. All prokaryotic cells are kinds of bacteria and all bacteria are prokaryotic cells. So far, biologists have named only 4800 species of prokaryotes, compared to more than a million species of eukaryotes.

Yet if the tiny prokaryotes are few in kind, they are great in number. More bacterial cells live in our intestines, for example, than we have cells in our bodies. And it is from the prokaryotes that all eukaryotes—including humans—are descended.

A.   50 μm
B.   50 μm
C.   50 μm

Figure 1-9 Every organism is composed of many kinds of cells. A. Human skin cells. B. Nerve cells from the spinal cord. C. Cells of heart muscle. *(A, G.W. Willis, M.D./Biological Photo Service; B, Manfred Kage/Peter Arnold, Inc.; C, Ed Reschke/Peter Arnold, Inc.)*

Eukaryotic cells have a nucleus and other organelles, each surrounded by a thin membrane. Prokaryotic cells lack a nucleus or other organelles.

HOW ARE ORGANISMS DIFFERENT FROM ONE ANOTHER?

Organisms come in many sizes, shapes, and colors. They range in size from *Chlamydia*, one of the smallest species of bacteria, to California's coastal redwoods, the most massive organisms in the world. Organisms range in shape from streamlined dolphins to prickly cacti, and in color from the softest brown moth to the brightest red poppy. The more we look, the more different forms of life we see. Biologists now divide these many organisms into three large categories, called **domains** (Figure 1-10).

The first two domains, **Archaea** and **Eubacteria**, comprise all the prokaryotes, commonly called bacteria. The great majority of these simple, single-celled organisms are free living. Only a handful of bacteria infect humans or other organisms. Each domain is further divided into one or more **kingdoms** of bacteria.

The third and last domain is **Eukarya**, which consists of four kingdoms. The two biggest kingdoms are the **Plantae** and the **Animalia**, the plants and animals. All plants and animals are multicellular. The 248,000 species of plants range from tiny mosses and ferns to giant redwood and mahogany trees.

Animals are even more diverse, with well over a million known species. Yet, fewer than 1 percent of animal species are what many people consider animals—birds, mammals, fish, reptiles, and amphibians. About 200,000 animal species are kinds of sponges, jellyfish, worms, spiders, snails, or sea urchins. But the vast majority of animals, some 800,000 species, are insects. Beetles alone constitute about 30 percent of all known animals, with some 300,000 to 350,000 species.

The third kingdom in the domain Eukarya is the **Fungi**, with just under 70,000 named species. The funguses include both multicellular organisms (such as mushrooms) and single-celled organisms (such as yeast). The fourth kingdom is the **Protista**, most of which are single celled. These tiny organisms include more than 30,000 species. Protists include different kinds of amebas, such as those that live in pond water; dinoflagellates, some of which cause poisonous red tides in coastal waters; the delicate and lacy foraminifera, and numberless algae, water molds, and slime molds.

Altogether, the named species of organisms within these three domains total about 1.4 million. Biologists believe, however, that most living species have yet to be named. The world may hold as many as 10 million or even 100 million species. No one really knows how many. Whatever the current total, however, it is only a fraction of all the species that have ever lived. Life first appeared on Earth almost 4 billion years ago. Fewer than 1 percent of all species that have ever lived are still living today. The vast majority of species—in all their diversity—went extinct millions of years ago.

Life is enormously diverse. About 1.4 million living species have names, but millions more have no names or are already extinct.

How Do Organisms Become Different from One Another?

Even though organisms are enormously diverse, all share the same chemical building blocks and the same cellular organization. Such common characteristics have persuaded biologists that all organisms are genetically related and are descended from a common ancestor. But how have related organisms become so different? How can plants and animals have evolved from the same common ancestor?

Scientists have discovered that during the last 4.5 billion years, our planet has changed from a lifeless sphere of rock and water to a rich, green world alive with fabulously diverse organisms. This transformation is the result of **evolution**—the process by which species have arisen and changed as they descended from common ancestors.

Why do children look like their parents?

Evolutionary change occurs because offspring inherit traits from their parents. A visitor to a museum may be surprised to find that a cat living in ancient Egypt 6000 years ago looked just like a house cat from Houston. Species change very little from one generation to the next. Just as today's children look much like those born 30 years ago, they also closely resemble children born 3000 years ago.

Children look like their parents because every individual inherits a set of **genes** [Greek, *gen* = to produce], which carry detailed information about the molecular parts of an organism. The preservation of similarity depends on the way that parents transmit genes to their offspring.

Genes ensure continuity from generation to generation.

What are genes made of?

As we will discuss in detail later in this book, biologists have learned that organisms store genetic information in long, thin molecules called **deoxyribonucleic acid (DNA)**. DNA contains coded instructions for building different kinds of proteins. A cell can read each gene and translate it into a protein. The DNA of humans, for example, contains about 30,000 genes, most of which code for proteins. A set of genes—called a *genome*—is in many ways like the list of ingredients in a recipe. But the genes are a recipe for building an organism. Instead of "two eggs," the genome says, "42,000 molecules of the protein albumin."

Genes are the means by which organisms transmit the information needed to build and maintain new individuals.

Species Differ in Their Adaptations to Distinct Environments

The members of a species share most of the same genes (and proteins). As a result, all the members of a species share characteristic **adaptations,** special structures or behaviors that fit individuals for life in a particular environment. All adaptations are inherited. Adaptations increase the chance that an individual will live and reproduce. In Figure 1-10, the hard, carapacelike upper wings of the spotted lady beetle have evolved to protect the delicate flying wings beneath. Its habit of attacking and eating aphids provides it with a steady supply of nutrients and energy.

Size, form, color, internal structure, and special chemical properties all contribute to an organism's ability to survive in a particular environment and to reproduce more of its own kind. Despite the differences among individuals within a species, all the members of a species are similarly adapted for a particular way of life. Given the opportunity, all cows stand around eating grass, all polar bears hunt seals, all kittens pounce on small objects, and all human children laugh and chase one another.

Life is diverse because different species have different sets of adaptations. The adaptations prompt biologists to ask questions such as, How does a particular structure or behavior help this organism to survive and reproduce? Put more concretely,

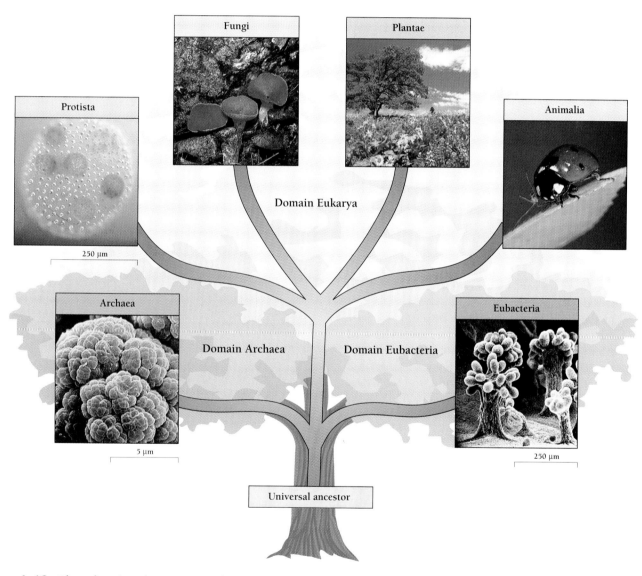

Figure 1-10 **Three domains of organisms.** At bottom are the first two domains, two very different kinds of bacteria. The first domain, the Archaea, is represented by a colony of methanobacteria, *Methanosarcina*. The second domain, the Eubacteria, is represented here by aggregating myxobacteria. The third domain, the Eukarya, includes four kingdoms. The seven-spotted lady beetle represents the Animalia. An oak tree surrounded by wildflowers (arrowleaf balsam root, paintbrush, and lupine) represents the Plantae. The stalked scarlet cup, *Sarcoscypha occidentalis*, represents the Fungi. A colony of one-celled organisms called *Volvox* represents the Protista. (Lady beetle, Runk/Schoenberger from Grant Heilman; oak tree, F. Stuart Westmorland/Photo Researchers, Inc.; stalked scarlet cup, R.M. Meadows/Peter Arnold, Inc.; Volvox, Biological Photo Service; myxobacteria, Patricia L. Grilione/Phototake; Methanosarcina, R. Robinson/Visuals Unlimited)

What good is an elephant's trunk, a mallard's mating dance, or a cactus's thick stems? Of ourselves we might ask, Of what use are a large brain and hairless skin? These are the same general questions that we might ask about the parts of a sewing machine or a car engine: What does it do? How else could it be done?

Because every species has so many adaptations, the same general questions arise again and again. After a while, biology students might begin to wonder how such a variety of adaptations came to be in the first place. As we mentioned earlier, all organisms appear to share a common origin. Biologists must therefore ask how individual species became so different from one another and why there are so many different species. The question, Why are organisms so diverse? becomes, How did so many different species derive from a single kind of organism?

Every species has a unique set of adaptations.

How Do We Know That the Diversity of Life Resulted from Evolution?

Until the middle of the 19th century, most Europeans believed that the Earth was only about 6000 years old and that modern organisms were no different from those present at the beginning of life on Earth. After all, ancient Egyptian paintings and mummies revealed creatures and plants indistinguishable from those of modern times (Figure 1-11).

However, the discovery of **fossils**—objects dug from the ground that contain remains or traces of past life—strongly suggested that organisms very different from the ones we know today once lived on Earth (Figure 1-12).

Early in the 19th century, geologists began to recognize that the Earth was far older than 6000 years. They understood for the first time that over millions of years, huge parts of the Earth's surface had moved about, new mountain ranges had arisen, and wind and water had shifted land from one place to another. Organisms that had been near the surface became covered over with new layers of mud and other materials that later turned to stone. The successive layers of fossil-rich rock produced by these processes, geologists realized, contained the history of the Earth and its inhabitants.

Fossils represent extinct forms of life that are often different from anything living today.

Charles Darwin provided evidence for evolution

By the early 19th century, many scientists realized two things: first, that some species of plants and animals had disappeared long ago and, second, that new species had appeared to take their places. Many scientists suspected that species of organisms changed over time, *evolving* from one species into another. At first, no one understood how this could possibly have occurred. Then, in 1859, the English biologist Charles Darwin (1809–1882) published a groundbreaking book titled *On the Origin of Species*, in which he suggested a way that evolution could work. Darwin called his theory *evolution through natural selection.*

Figure 1-11 **Organisms may remain unchanged for many generations.** This hippopotamus mother, carved more than 2000 years ago, closely resembles hippopotami that live today. *(Larry Tackett/Tom Stack & Associates)*

Figure 1-12 **Parasaurolophus.** The unearthing of fossils in the 18th century suggested that strange organisms had once lived on Earth. Other geologic evidence suggested that the Earth was unimaginably old and that the layers of the Earth's crust contained the history of the Earth and its inhabitants. *(Francois Gohier/Photo Researchers, Inc.)*

Darwin's model for how evolution could work came from farming. For thousands of years, plant and animal breeders have bred new varieties of corn, wheat, dogs, or horses. Their method is simple: a breeder selects certain individuals to breed—the most productive wheat plants, the most intelligent dogs, the fastest horses. Slow race horses and other animals with undesirable characteristics were not allowed to breed.

The power of this strategy is undeniable. Every kind of dog, from chihuahua to mastiff, is the result of selection from a line of animals descended from the ordinary wolf. Similarly, a surprising number of vegetables have been bred from a single plant, the wild sea cabbage of the Mediterranean. Figure 1-13 shows the results of selection in wild sea cabbage to produce broccoli, cabbage, Brussels sprouts, and more. By selecting for leaves, gardeners developed cabbage. By selecting for flower buds, gardeners developed broccoli.

Darwin saw that a similar kind of selection, which he called natural selection, could operate in nature. **Natural selection** is the superior survival and reproduction of some individuals that have certain inherited traits compared to others lacking those traits.

Figure 1-13 European gardeners developed a great variety of vegetables from a single species of plant, *Brassica oleracea*. Shown above are cabbage, cauliflower, broccoli, Brussels sprouts, kale, and kohlrabi. In the case of kale and cabbage, farmers selected only those plants with large, edible leaves. To create broccoli and cauliflower, they selected plants with the most flower buds. For Brussels sprouts, they chose those with the largest, most edible leaf buds. In kohlrabi, breeders selected plants with the biggest, most edible stems. What other plant parts could be selected for?

While artificial selection by breeders depends on the effects of human preferences, natural selection depends on the ever-changing demands of an organism's natural environment. For example, organisms whose adaptations improve their ability to escape predators or to compete for food or sunlight usually live longer and produce more offspring than organisms without these adaptations. To the extent that a trait increases an organism's ability to survive and reproduce, that trait appears more often in the next generation. In this way, Darwin suggested, one organism can gradually evolve into another.

Charles Darwin supplied extensive evidence for the idea that species have evolved from common ancestors. He also suggested that species have evolved by means of natural selection.

How are traits passed from one generation to the next?

Darwin understood how evolution might work, but he had no idea how living organisms pass traits to their offspring or why the offspring are different from their parents and different from each other. In short, Darwin did not know why individuals are unique or how they inherit traits from their parents. Darwin knew neither why we look like our parents nor why we look slightly different from our parents. The question of how heredity works puzzled and frustrated him for the rest of his life. Ironically, the answer lay close at hand, although the answer would remain hidden for another 30 years.

In 1866, just 7 years after Darwin published *On the Origin of Species,* the Austrian naturalist monk Gregor Johann Mendel published a paper describing the laws of heredity. Mendel's work laid the groundwork for the modern science of genetics. Mendel's discovery was profound, yet his paper was virtually unknown until 1900, when his work was rediscovered (Chapter 9).

Gregor Mendel discovered the genetic rules that underlie natural variation.

We have now seen how biologists define life and how they try to answer questions about living organisms. We have also seen, briefly, how living organisms evolve. In the next few chapters, we will survey the chemistry of life, view the structure of cells, and see how cells obtain energy to fuel life.

As you read, keep in mind that in the next 20 years, Earth's rapidly growing human population and dwindling resources will force the world's leaders to make many hard decisions. To participate as citizens in these decisions, we need to do more than simply learn the facts of biology. We need to know how to scrutinize those facts as they come to us from television and radio, the Internet, or a newspaper or magazine. We should know how to ask what biologists and other scientists mean when they describe their findings. To do that, we need to know how scientists think and how they themselves ask questions.

Study Outline with Key Terms

The **scientific method** is a formal set of rules intended to describe the way scientists work. Most scientists, however, do not stick to a rigid set of rules. Both thinking up questions and finding ways to answer those questions require inspiration and creativity, as well as hard work, patience, and a methodical approach. In general, a scientist begins by narrowing an area of interest to a specific question, then observing a phenomenon.

A scientist may sometimes visualize the phenomenon by constructing a **model**. A **hypothesis** is a formal guess at the answer to the question, often stimulated by a model. A good hypothesis allows a scientist to make a **testable** prediction about the way the process or structure will act under specific circumstances. A set of hypotheses that consistently resists scientists' attempts to disprove them may become accepted as a **theory**.

An **experiment** is a way of testing the predictions of a hypothesis or, sometimes, of searching for some unknown effect. An experiment is a procedure that is carried out under "controlled" conditions. An experiment can prove a hypothesis wrong but not right. Essential elements of most biology experiments are the **control** experiment and **statistical** analysis.

Meaningful statistics requires a **sample** that represents the whole **population** of organisms. A sample can represent the population only if the sample is a **random** sample and the sample is large enough to overcome random variation. Statisticians can determine if the sample represents the population using the **mean** and **standard deviation**. Statistics is most easily done on data with a **normal distribution**, or bell curve.

Biology is the study of living organisms. All organisms are composed of one or more cells, which are themselves alive, and which all come from other cells. All are organized into parts, obtain energy from their surroundings, perform chemical reactions, mature over time, reproduce, share a common evolutionary history, and respond to their environments. Mechanisms that maintain a constant internal environment are a form of **homeostasis**.

Eukaryotic cells contain a **nucleus** and other **organelles**, each of which is enclosed by a membrane. **Prokaryotic** cells have no membrane-enclosed nucleus. The millions of species of organisms are organized by biologists into three **domains: Archaea, Eubacteria,** and **Eukarya.** Eukarya comprise just four **kingdoms: Animalia, Plantae, Fungi,** and **Protista.**

Genes, which are made of long, thin molecules of **deoxyribonucleic acid (DNA),** ensure continuity from generation to generation. Traits that are inherited in the genes and help a species thrive are **adaptive.** The discovery of **fossils** suggested to 18th- and 19th-century scientists that organisms had changed, or evolved, over long periods. Charles Darwin assembled compelling evidence for the idea of **evolution** and suggested a mechanism, called **natural selection.** Gregor Mendel discovered and articulated the first rules of genetic inheritance.

Review and Thought Questions

Review Questions

1. Explain the differences between a hypothesis and a theory.
2. Explain why a hypothesis can sometimes be proved wrong but can never be proved right.
3. Give an example of an untestable hypothesis. Why is it untestable? What would you need to know or do to test it?
4. Suppose you decide to test the hypothesis that your favorite soft drink contains substances that cause cancer. You plan to feed the drink to 100 mice and then count the number of tumors they develop. Describe an appropriate control experiment.
5. Describe eight distinguishing characteristics of living things. Give an example of each one.
6. Define "homeostasis." List two examples of homeostasis.
7. Explain what is meant by "natural selection." What is "selected"?

Thought Questions

8. The ants in your kitchen are driving you nuts. Using the "scientific method," how would you find out where they were entering your house? How would you determine why they were entering your house? Think of several alternative hypotheses. Now design a series of experiments whose results will either support or eliminate the various alternative explanations.
9. A virus is a bit of DNA surrounded by a protein coat. A virus can attach to the wall of a cell and inject the DNA into the cell. Inside the cell, the viral DNA forces the cell to make new viruses. The cell makes so many new viruses that the cell eventually bursts open, releasing thousands of new viruses, which go on to infect other cells. Is the virus alive? Using the definitions of life given at the beginning of this chapter, explain why a virus is alive or why a virus is not alive.
10. From your own experience, describe a moment of inspiration—intellectual or otherwise. How did it feel? How did it change you?

About the Chapter-Opening Image

The amazing story of how we came to know that the twisted bacterium *Helicobacter pylori* causes ulcers is a classic example of how science works.

On-line materials relating to this chapter are on the World Wide Web at **http://www.harcourtcollege.com/lifesci/aal2/**

Web Bit 1-1 Dorothy Crowfoot Hodgkin

Web Links

Quizzing & Testing

part I

Chemistry and Cell Biology

Chapter 2
The Chemical Foundations of Life

Chapter 3
Biological Molecules Great and Small

Chapter 4
Why Are All Organisms Made of Cells?

Chapter 5
Directions and Rates of Biochemical Processes

Chapter 6
How Do Organisms Supply Themselves with Energy?

Chapter 7
Photosynthesis: How Do Organisms Get Energy from the Sun?

Photo: Scanning electron micrograph of the mucous membrane lining the stomach. (P. M. Motta, K. R. Porter, and P. M. Andrews/Science Photo Library/Photo Researchers, Inc.)

2 The Chemical Foundations of Life

Just Say NO

Alfred Nobel, one of the names most associated with scientific discovery, was not a scientist but an inventor who made a fortune from the invention of dynamite. Nobel devoted his wealth to establish the Nobel prizes, now given to recognize extraordinary achievements in peace, chemistry, physiology and medicine, physics, literature, and economics. Nobel wanted to use his fortune—much of it made from the military use of dynamite—to promote peace and peaceful achievement. Strangely, the 1998 Nobel Prize in Physiology and Medicine actually rewarded insights that came from the biological use of Nobel's troubling explosive.

The active chemical in dynamite is *nitroglycerin*, a colorless oily liquid. Nitroglycerin is so unstable that just dropping it causes an explosion. Nobel's invention consisted of a way of stabilizing nitroglycerin by mixing it with porous soil, so that it could be safely transported.

Despite the dangers of pure nitroglycerin, however, physicians have been prescribing it for more than 100 years. In small doses, nitroglycerin reduces blood pressure, and it is especially effective in treating the chest pain associated with heart disease (*angina*). Nobel's own doctor recommended nitroglycerin treatment for his heart condition, but (perhaps understandably) Nobel refused.

Almost 80 years later, in the early 1970s, biologist Ferid Murad, then at the University of Virginia, was trying to learn exactly how nitroglycerin actually helped the heart. In 1977, Murad discovered that nitroglycerin releases nitric oxide, a gas that is a well-known by-product of automobile engines. Nitric oxide, he argued, is the agent that caused blood vessels to dilate, thereby lowering blood pressure. But nitric oxide—abbreviated as NO by chemists—is a major component of smog, a noxious chemical that causes irritation and burning in the lungs. No one believed that it could possibly be an effective biological agent. According to Murad's former professor, "People used to say 'C'mon, Fred [as they called him]. Why don't you work on something important?'"

About the same time, Robert Furchgott, working at the State of New York Health Science Center at Brooklyn, was trying to understand the effect of other drugs on blood vessels. Furchgott's work was extremely frustrating, since the same chemical sometimes caused vessels to contract and other times caused them to dilate. Trying to get to the bottom of these contradictions, Furchgott demonstrated that dilation-causing drugs worked only when the lining of the blood vessels, the *endothelium*, was intact. He concluded that the drugs were somehow inducing the endothelium to release an unknown chemical, which he called "endothelium-derived relaxing factor," or EDRF. In 1986, Furchgott, Louis Ignarro (from the University of California, Los Angeles),

Key Concepts

- An atom is the fundamental unit of an element, and a molecule is the fundamental unit of a compound.
- In an ionic bond, electrical attraction holds atoms together. In a covalent bond, atoms share electrons.
- If the atoms in a covalent bond do not share electrons equally, the resulting molecule will be polar.
- Individual molecules may also interact with each other via hydrogen bonds, hydrophobic interactions, or van der Waals interactions.

and Salvador Moncada (from University College London) found that EDRF was, in fact, identical to nitric oxide. It was the same molecule that Murad had already implicated in the action of nitroglycerin.

The effect of these announcements was electrifying. Furchgott, Ignarro, and Moncada were claiming—against all precedent—that gas molecules could carry messages from one cell to another. Although biologists knew of some bacteria that produced NO, skeptics argued that there was no known mechanism by which eukaryotes could produce the toxic substance. Besides, they argued, NO itself was unstable and was destroyed within 10 seconds after it appeared. How could it be a biologically important regulator? The critics dismissed NO as "hot air."

But the researchers persisted. They learned how cells produced NO and how it acted. Pharmaceutical companies began to develop medicines that increase NO production for heart patients. Perhaps most famously, the anti-impotence drug Viagra was shown to act by increasing the action of NO, allowing more blood flow to the penis. The increased blood flow enables many men suffering from impotence to achieve a normal erection.

Nitric oxide is not always beneficial. In people suffering from septic or toxic shock, for example, white blood cells pump out too much NO in response to a bacterial infection. The resulting rapid drop in blood pressure can be fatal. Therefore, many researchers are trying to discover ways of reducing as well as increasing NO production. On the other hand, since white blood cells also use NO to defend against tumor cells, scientists are using new knowledge of NO to develop new anti-cancer drugs. By 1992, the influential journal *Science* had named NO the "molecule of the year," and the 1998 Nobel prize (to Furchgott, Ignarro, and Murad) again underscored its importance.

Nitric oxide is invisible but it still has weight and occupies space. It is a particular form of matter. Understanding the composition and the transformations of matter is the business of chemistry, the subject of this chapter.

> "'C'mon, Fred. Why don't you work on something important?'"

WHAT IS MATTER?

Anyone whose car has ever run out of gas knows that when the gauge says empty, the gas is gone. Where does it go? A car's engine burns (or "oxidizes") these chemicals into water and carbon dioxide (the same stuff that makes the bubbles in soft drinks). This chemical transformation happens only when the engine provides a spark (from a spark plug) to get things going. Our bodies perform similar chemical reactions, though without spark plugs. Like a car engine, we convert chemicals into water and carbon dioxide. But we also use the molecules in our food to make thousands of other chemicals, which are the stuff of our bodies.

Chemistry is the science dealing with the properties and the transformations of all forms of matter—whether gasoline or food, people or planets. **Matter** is anything that occupies space and has weight (or "mass"). Matter can exist in different states or phases—solid, liquid, or gas.

Every bit of matter consists of one or more *pure substances,* each of which has its own unique properties. **Chemical reactions** transform pure substances into other substances, without changing the total amount of matter. ("Biochemical reactions" are those chemical reactions that occur in organisms.)

Pure substances that cannot be converted into simpler substances in chemical reactions are called **elements.** Hydrogen, helium, oxygen, and carbon are elements. So are silver and gold, the carbon in a pencil, the copper in a wire, and the mercury in a thermometer. Altogether, scientists know of 92 naturally occurring elements. Elements are composed of tiny particles called **atoms**—the smallest units that still have the properties of an element. Atoms themselves are made of smaller "subatomic" particles, including protons, neutrons, and electrons.

Each element has a standard one- or two-letter abbreviation. These usually correspond to the beginning of the element's English name, though sometimes it's the Latin or Greek name. "C" stands for carbon and "O" for oxygen, while "Na"

Figure 2-1 **Humans, like other organisms, are primarily water.** The commonest two atoms in our bodies, then, are hydrogen and oxygen (H_2O). Hydrogen is a much lighter atom than oxygen, so most of our mass is made up of oxygen atoms. Next comes carbon, which, like hydrogen and oxygen, is in virtually every other molecule in our bodies. *(Roy Morsch/The Stock Market)*

stands for sodium [Latin, *natrium*]. Four elements—carbon (C), oxygen (O), nitrogen (N), and hydrogen (H)—make up more than 99 percent of the matter in organisms. Figure 2-1 shows the main elements of the human body.

Chemical reactions cannot change or destroy elements. Carbon cannot be converted to oxygen. Silver cannot be converted to gold. However, elements can be created or changed by means of **nuclear reactions,** the high-energy transformations of individual atoms that occur in nuclear reactors or in the bowels of stars such as our sun.

Although oxygen is an element, we do not breathe oxygen in its atomic form. Instead, we breathe "molecular" oxygen, which is two atoms of oxygen bound together by a "chemical bond." A **molecule** is a stable assembly of two or more atoms.

Most molecules are **compounds**—pure substances forged in chemical reactions from two or more *different* elements. Water, for example, is a compound of the elements hydrogen and oxygen (Figure 2-2A). Carbon dioxide is a compound of carbon and oxygen.

A compound is unlike the elements that compose it. The properties of water are different from those of either hydrogen alone or oxygen alone, and the properties of carbon dioxide are different from those of carbon and oxygen.

A compound is made of a single kind of molecule. Chemists designate the composition of a molecule by writing the number of atoms of each type as subscripts to the right of the symbol for each element. Water is always H_2O, carbon dioxide is CO_2, and ordinary table sugar (sucrose) is $C_{12}H_{22}O_{11}$.

In our day-to-day lives, we rarely encounter either compounds or elements in pure form. Most matter is a **mixture** of different compounds (Figure 2-2A). Sand and salt poured together, for example, is a mixture. One component of a mixture does not affect the properties of the other components. If we pour salt into boiling water to make spaghetti, the salt is still salt and the water is still water. If we boil away all of the water, the salt forms crystals at the bottom of the pot, and we can retrieve the same salt we poured into the pot.

Unlike the components of a compound, which are always present in the same fixed ratio, the components of a mixture may be present in any ratio. A molecule of water always contains twice as much hydrogen as oxygen. But you can mix a little sugar into your cup of coffee, or a lot.

Elements are pure substances that cannot be converted into simpler substances. An atom is the fundamental unit of an element, and a molecule is the fundamental unit of a compound.

What Is Special About the Chemistry of Life?

Until the 16th century, chemistry as we know it did not exist. Alchemists combined and recombined various compounds and elements, hoping to convert lead and other cheap metals into silver and gold (Figure 2-3). Alchemists did not yet recognize that the elements could not be changed one into another by ordinary means. Some alchemy was science of a sort, but much of it was simple superstition. None of the alchemists in the Middle Ages possessed a modern understanding of what matter was and why it acted as it did. In their efforts, however, they learned a lot about the properties and transformations of matter.

A still older kind of chemistry was *metallurgy*—the study of the properties and transformations of metals. Thousands of years ago, bronze tools replaced stone tools and, later, iron tools replaced bronze tools. Steel tools have, in turn, replaced iron

Figure 2-2 **Atom, molecule, compound, and mixture.** A. Two atoms of the same element, such as two oxygen atoms, may form a molecule. But most molecules are compounds forged from two or more different elements. Water is a compound of hydrogen and oxygen, and carbon dioxide is a compound of carbon and oxygen. Most substances are mixtures of different compounds. Sparkling water, for example, is a mixture of carbon dioxide and water. B. An atomic force microscope reveals the arrangement of individual atoms of copper. (B, Xueping Xu. From Science, vol. 258, p. 788, 1992)

ones. Metallurgy was, and is, so important that its mastery remains a measure of the progress of a civilization.

In the 1500s, the chemistry of medicines emerged as a serious discipline. Then, bit by bit, Renaissance scientists and philosophers began reinterpreting the chemical facts discovered by the alchemists. Not until the 19th century, however, did scientists begin to understand the chemical nature of matter. Today, at the beginning of the 21st century, chemistry extends beyond metals and medicines to all matter, including the substances that make up organisms. One of the great achievements of 20th-century biology was to explain the chemistry that underlies all cellular processes—from the workings of a gene to the contraction of a muscle.

One of the great questions in biology was whether living matter was in any way different from nonliving matter. Are a horse and a plow made of the same basic stuff? In this chapter, we will see that the molecules of living organisms indeed obey the same chemical laws as the molecules of nonliving matter.

From a chemist's point of view, just two aspects of living matter distinguish it from nonliving matter—the elaborate organization of the molecules and the limited number of kinds of molecules that comprise living matter. All organisms consist of the same few kinds of small *organic* (carbon-containing) molecules, which we may think of as "building blocks." Just as a good cook can make a dazzling array of cookies, cakes,

Figure 2-3 *The Alchemist,* painted by David Teniers the Younger in 1648. Alchemists of the medieval period and later attempted to turn lead and other cheap metals into gold and silver. (Scala/Art Resource, NY)

and pies from just a few ingredients, so organisms can produce extraordinary structures from a small number of standard molecules.

The chemistry of living matter is the same as that of nonliving matter.

WHAT DETERMINES THE PROPERTIES OF AN ATOM?

By the early 19th century, chemists were able to distinguish compounds—which could be broken down into simpler substances—from elements—which could not. They then began to ask, Does the way in which a substance breaks down reveal something about its basic structure? Are there units that cannot break down further?

To answer these questions, the English chemist John Dalton revived an idea, first developed by the Greek philosopher Democritus in the 5th century B.C., that matter is actually built from small particles. Dalton proposed that each element is composed of tiny particles called "atoms." Only in the past 20 years has it been possible to see atoms, using the *atomic force microscope,* which provides a detailed image of the surfaces of solids. Figure 2-2B shows rows of individual copper atoms aligned on a surface.

The atoms of each element have a characteristic weight. For example, an atom of carbon weighs 12 times as much as an atom of hydrogen. Chemists and physicists prefer to talk of the "mass" of an atom, rather than its weight. **Mass** is the amount of matter in something. Weight results from the action of gravity on mass. You may weigh 125 pounds on Earth and zero pounds in space, but your mass remains the same no matter where you are. Mass is closely related to weight, but the mass of an object, unlike its weight, is constant.

What Are Atoms Made Of?

In the late 18th and early 19th centuries, chemists discovered that they could measure the relative mass of each element in any compound. Such measurements revealed a fascinating relationship: The mass of every kind of atom is an almost exact multiple of the mass of a hydrogen atom. An atom of helium has four times as much mass as an atom of hydrogen. An atom of lithium has six times as much mass as hydrogen. Carbon has 12 times the mass of hydrogen; oxygen, 16 times; and nitrogen, 14 times.

This pattern suggested that the larger atoms were composed of groups of hydrogen atoms. We now know that atoms are not actually made of hydrogen atoms but of subatomic particles, each of which has about the same mass as a hydrogen atom. An atom of carbon is composed of 12 particles each about the size of a hydrogen atom, whereas an atom of oxygen is composed of 16 such particles.

The masses of atoms are extremely small. An oxygen atom, for example, weighs less than 3×10^{-23} gram: that's 0.00000000000000000000003 gram. (A gram is about as heavy as one-third of an animal cracker.) For convenience, chemists have defined a measuring unit for the mass of an atom or a molecule that is much smaller than a gram. A dalton, named after the chemist John Dalton, is approximately the mass of a hydrogen atom. An atom of hydrogen has a mass of 1 dalton, and an atom of oxygen has a mass of 16 daltons.

Figure 2-4 **Rutherford's experiment.** Rutherford did not perform his now famous experiment himself. He delegated it to a student, to whom he initially did not even speak. Rutherford later wrote, "One day Geiger came to me and said, 'don't you think that young Marsden, whom I am training in radioactive methods, ought to begin a small research?' . . . I said, 'Why not let him see if any alpha particles can be scattered through a large angle?' I may tell you in confidence that I did not believe that there would be, since we knew that the alpha particle was a very fast massive particle, with a great deal of energy, and you could show that . . . the chance of an alpha particle being scattered backwards was very small. Then I remember two or three days later Geiger coming to me in great excitement and saying, 'We have been able to get some of the alpha particles coming backwards . . .' It was quite the most incredible event that has ever happened to me in my life. It was almost as incredible as if you fired a 15-inch shell at a piece of tissue paper and it came back and hit you." (*Rutherford and the Nature of the Atom,* by E.N. da Costa Andrade, Doubleday, Garden City, NY, 1964; quoted by S. Weinberg, *The Discovery of Subatomic Particles,* Scientific American Library, New York, 1983, p. 124.)

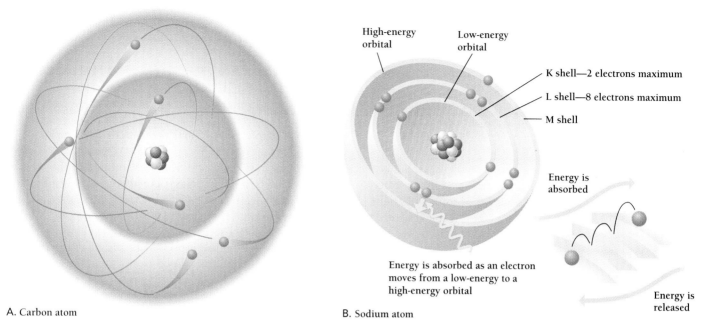

Figure 2-5 The structure of an atom. A. A carbon atom. By convention, orbitals are drawn as flat rings on the page as in B, but in reality, electron orbitals are diffuse, three-dimensional spheres (or other shapes), as shown here. B. A sodium atom. Orbitals are grouped into shells, each of which has a different energy. Closest to the nucleus is the K shell. It has the lowest energy and holds a single orbital with a capacity for two electrons. Farther out is the L shell, which contains four orbitals with a capacity for eight electrons. Each orbital has a defined shape, which describes the most probable locations of the electrons within them.

The number of daltons in 1 gram is 6×10^{23} and is called "Avogadro's number" [after Amadeo Avogadro, an 18th-century Italian lawyer and physicist]. To get an idea how large this number is, imagine that hydrogen atoms were huge—the size of unpopped popcorn kernels—and that you poured Avogadro's number of kernels out of an airplane window. You would find that Avogadro's number of kernels would cover the whole United States to a depth of 9 miles.

Since a molecule consists of a fixed number of atoms, every molecule has a characteristic molecular mass, usually called its **molecular weight.** A molecule of water (H_2O), for example, has a molecular weight of 18:

$$18 = 16 \text{ (oxygen)} + 1 \text{ (hydrogen)} + 1 \text{ (hydrogen)}$$

Chemists and biologists also say that a water molecule has a mass of 18 daltons.

Atoms are made of subatomic particles similar in size to one hydrogen atom.

What Is the Internal Structure of an Atom?

Once physicists knew that atoms were made up of hydrogen-sized particles, they began to wonder about those particles. What were the hydrogen-sized particles, and how did they fit together to make an atom? The first important finding about atomic structure was that atoms are almost entirely empty space. In 1910, a New Zealand physicist working in England devised an ingenious way to probe the structure of the atom. Ernest Rutherford's approach, diagrammed in Figure 2-4, was like searching for a stone in a haystack by shooting bullets into the haystack. If a bullet goes straight through the haystack, it means that it has failed to hit anything solid. If it comes out at an angle, then it must have bounced off something. The "bullets" Rutherford used were small, positively charged alpha (α) particles, produced from the radioactive decay of radium. The "haystack" was a thin piece of gold foil.

Rutherford found that his α particles passed right through most parts of a gold atom, only bouncing off a tiny, central part of each atom. He proposed that the atom is mostly empty space, with almost all of its mass concentrated in a small heavy core, or **nucleus,** at its center.

The diameter of the nucleus is only 1/10,000 that of the whole atom. Imagine a Ping-Pong ball (the nucleus) at the center of two vacant city blocks (the atom), and you will have an idea of the proportions of an atom. Because the tiny nucleus reflected Rutherford's positively charged particles, he reasoned that the nucleus must itself have a positive charge that repelled his "bullets." The positive charge of the nucleus is balanced by negatively charged particles, called **electrons,** which surround it (Figure 2-5). Rutherford imagined that the negative electrons orbit the positive nucleus in the same way that the planets orbit the sun.

In fact, later researchers confirmed that the nucleus contains particles called **protons.** The positive charge of a proton exactly balances the negative charge of an electron. A neutral (uncharged) atom has equal numbers of protons and electrons. The nucleus of an atom also contains **neutrons,** which have no electric charge.

The mass of an atom depends almost entirely on the total number of protons and neutrons in the nucleus. The tiny electrons have little mass. Each has just 1/2000th the mass of a proton.

The atoms of each element always have the same number of protons. Oxygen, for example, always has eight protons. However, some forms of an element, called **isotopes,** may have different numbers of neutrons. Carbon, for example, has three common isotopes: carbon 12, carbon 13, and carbon 14. The atoms of all three isotopes have six protons and six electrons and the same chemical properties. However, each has a different number of neutrons and, therefore, different masses: carbon 12 has six neutrons, carbon 13 has seven, and carbon 14 has eight. The three isotopes of carbon are also written ^{12}C, ^{13}C, and ^{14}C.

Some isotopes are unstable and tend to break down, or *decay*. The nucleus of ^{14}C, for example, which consists of six protons and eight neutrons, is unstable. It is more likely to come apart than a ^{12}C nucleus, which consists of six protons and six neutrons. Every second, the nuclei of about four ^{14}C atoms in 1 trillion (10^{12}) break up. Each decay event consists of a neutron decomposing into a proton and an electron, which then speeds away from the nucleus (Figure 2-6A).

An electron ejected from the nucleus in this manner is called a "beta (β) particle" and is easily detected and counted. It is distinct from the six electrons that surrounded the ^{14}C nucleus. The nucleus now contains a total of seven neutrons (having lost one). Because it has gained a proton, the carbon nucleus is now a nitrogen (N) nucleus, but the new nitrogen (^{14}N) nucleus does acquire a seventh electron outside its nucleus.

Isotopes that release beta particles are called **radioactive isotopes.** The decay process occurs at a constant rate that is characteristic of each radioactive isotope. The time required for half the atoms of a radioactive isotope to decay is called that isotope's **half-life** (Figure 2-6B). Half-lives for different isotopes range from as little as 0.3 microsecond to as long as

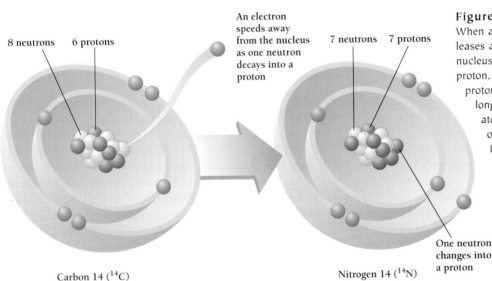

Figure 2-6 Half-life. A. Breakdown of ^{14}C. When a neutron decays into a proton, it releases an electron (a beta particle). The altered nucleus has one fewer neutron and one more proton, for a total of seven neutrons and seven protons. A nucleus with seven protons is no longer carbon, however, but nitrogen. The atom then acquires an extra electron in its outer shell and stays electrically neutral. B. Half-life of chocolates. (*B, Paraskevas Photography*)

BOX 2.1

The Uses and Dangers of Radioisotopes

Most of the isotopes found in nature are stable forms. Many of the isotopes of the heaviest elements, however, are unstable, or radioactive. Radioisotopes break down spontaneously into entirely different elements. As they break down, or decay, they release high-energy radiation.

The radiation from radioisotopes can be used to kill cancer cells. Radiation is now a standard therapy for many forms of cancer. Radioisotopes can also be used as tracers. For example, because the radioactive form of carbon (^{14}C) has the same properties as the regular form (^{12}C), organisms process both forms in the same way. Researchers can use sensitive instruments or photographic films to trace the movements of ^{14}C and answer many questions about molecular processes.

The Polish-born French chemist Marie Curie did some of the earliest and best work on radioisotopes. Marie Curie and her husband, Pierre, shared the 1903 Nobel Prize for Chemistry with H. Becquerel for the discovery of radioactivity. After Pierre's death in 1906, Marie Curie assumed his professorship at the Sorbonne and went on to win a second Nobel prize, in 1911, for the isolation of pure radium. Her subsequent work laid the groundwork for the discovery of the neutron and the synthesis of artificial radioactive elements. This latter work was carried out by Curie's daughter Irène Joliot-Curie and her husband, Frèdèric Joliot. In 1934, Marie Curie died of leukemia at age 67, a result of her lifelong exposure to radiation. The following year, Irène and Frèdèric won the Nobel Prize for Chemistry. Irène Curie also died of leukemia, in 1956, at age 59, and her husband Frèdèric assumed her professorship at the University of Paris.

Figure A Marie Curie and her two daughters, Irène and Eva. *(Culver Pictures)*

Radium became a major weapon in the war on cancer. No one, especially Marie Curie, wanted to believe that radium could also cause harm. She hid the cataracts that clouded her eyes lest they implicate her beloved radium as harmful, and she could not believe that radium had anything to do with the early deaths of four of her colleagues who also worked with radium. Even today, her desk and laboratory notebooks are still highly radioactive, testimony to the massive amounts of radium that she accumulated for study.

Perhaps the saddest result of the discovery of radium was the inadvertent poisoning of young working women during World War I. Glowing radium on watch dials helped the Allied effort by allowing soldiers to read their watches in the dark and to synchronize their attacks. Young women in U.S. factories were painting radium onto watch dials, drawing their brushes to fine points between their lips. In doing so, they were ingesting radium. Because radium chemically resembles calcium, the workers were also incorporating it into their teeth and bones. One worker later recalled that after work, "We'd get the neighborhood kids to watch Alice glow in a dark closet. They'd get a good laugh out of that." Only 10 years later did the women start to get sick, with cancers of their jaws and mouths and the failure to make blood cells. Finally, the dangers of radiation became known.

$2,000,000,000,000,000$ (2×10^{15}) years. The half-life of ^{14}C is about 5760 years, while that of uranium 235 (^{235}U) is about 704 million years.

The predictable decay of isotopes is extremely useful to researchers. Isotopes can be used, for example, to determine the age of rocks and fossils; to track individual molecules in viruses, cells, organisms, or ecosystems; and even to identify the origins of products ranging from brand-name beverages to the explosive materials used in bombs.

Atoms are mostly empty space. They consist of a tiny central nucleus orbited by even tinier electrons. The nucleus has a positive charge, balanced by the negative charge of the electrons.

What does radioactivity do to living organisms?

Beta particles released by radioactive isotopes are a form of radiation that damages and kills cells. Alpha particles, beta particles, and x rays, for example, are all different forms of *ionizing radiation*. Ionizing radiation causes the water molecules in tissues and cells to break down into *free radicals*. Free radicals are chemically active, oxygen-containing molecules that easily react with and damage other molecules, such as proteins and DNA. Free radicals can kill cells outright or alter their ability to function. The most lasting and serious effects occur, however, when DNA is damaged.

Cells whose DNA has been altered by radiation cannot divide properly. Ionizing radiation has its greatest effect in tissues whose cells need constant replacement—cells that line the mouth, the stomach, and the intestines; skin cells that produce hair; bone marrow cells that produce red and white blood cells. As a result, people and other mammals exposed to intense doses of ionizing radiation—from nuclear weapons used in World War II, from accidents, or as therapy for cancer—suffer nausea, vomiting, diarrhea, hair loss, and anemia.

Because ionizing radiation can damage the DNA in cells and prevent them from dividing, it is commonly used to treat cancers. In some cases, however, DNA damage can itself cause the failure of normal cell regulation, thereby leading to a subsequent bout of cancer. An entire medical discipline—radiation oncology—is devoted to understanding this trade-off in order to maximize the effectiveness of radiation treatment of cancer (Box 2.1).

Where Are the Electrons in an Atom?

Although Rutherford envisioned the electrons neatly orbiting the nucleus like tiny planets, physicists have found that electrons actually move around the nucleus randomly, like a cloud of gnats. Although we still speak of electron **orbitals,** the distributions of electrons more closely resemble clouds. Like clouds, orbitals have fuzzy boundaries but definite shapes (Figure 2-5).

An electron in a particular orbital has a characteristic **energy**—the capacity to perform work. Each orbital has a different amount of energy. When an electron moves from a high-energy orbital to a low-energy orbital, an atom may release energy, often in the form of light. Conversely, when an atom absorbs energy, one or more electrons may move from a low-energy orbital to a high-energy one. We can visualize the energy of electrons as steps on a staircase. A ball on a staircase tends to bounce down the stairs. Just so, electrons tend to fall into orbitals with the lowest energy.

Two rules determine the distribution of electrons in an atom: (1) electrons occupy orbitals with the lowest possible energy, and (2) each orbital may contain only two electrons.

The atoms that are most common in organisms contain fewer than 18 electrons. The electrons of each of these atoms occupy only the nine orbitals with the lowest energies. Chemists group orbitals into three electron **shells**—groups of orbitals whose electrons have nearly equal energy. The first shell has just one orbital containing two places for electrons. The second shell includes four orbitals, with room for eight electrons. The third shell includes 9 orbitals, with room for 18 electrons. Most biologically important atoms (except for some metals), however, use only the four innermost orbitals of the third shell, with a capacity of eight electrons.

The arrangement of electrons in the outermost shell of each atom is the basis for the Periodic Table (Figure 2-7). The elements in each column have the same number of electrons in their outermost shell. These outer electrons are those that interact with other atoms and determine the chemical properties of an atom.

Electrons occupy orbitals with the lowest possible energy, and each orbital may contain only two electrons.

WHAT HOLDS MOLECULES TOGETHER?

An atom is most stable when its outermost shell contains eight electrons (or an "octet" of electrons). (This generalization is called the "octet rule.") The atoms of some elements—such as neon and argon—already contain eight electrons in their outer shells and are chemically unreactive. Most atoms, however, do not have full octets. These atoms can attain an octet by sharing electrons with other atoms, by stealing electrons from other atoms, or by losing electrons to other atoms, as illustrated in Figure 2-8. Such arrangements are the basis of chemical bonds.

A few molecules, however, have atoms that do not obey the octet rule. The most significant rule-breaking molecule of this sort is nitric oxide, discussed at the beginning of this chapter.

Covalent and Ionic Bonds Hold Atoms Together Strongly

Two kinds of bonds—covalent bonds and ionic bonds—hold atoms together especially strongly and are difficult to break. Such "strong" bonds hold together all the major molecules of life—from the carbohydrates in a loaf of bread to the DNA in your cells.

Two atoms that share one or more electrons are said to be held together by a **covalent bond.** An atom that gains or loses electrons becomes an **ion**—an atom or a molecule with a net electrical charge. Ions with opposite charges may be drawn together to form an **ionic bond.**

What holds two atoms together in a covalent bond?

The molecules of life are mainly built from just six kinds of atoms: hydrogen (H), carbon (C), nitrogen (N), oxygen (O), phosphorus (P), and sulfur (S). One way to remember this list is to arrange the symbols for each element into the nonsense word "SPONCH." The outer shells of these six atoms determine their ability to form molecules. Carbon, for example, has only six electrons—two in the first shell and four in the second (Figure 2-8A). To fill its outer shell with eight electrons, it must find other atoms with which to share an additional four electrons. These four shared electrons can form four covalent bonds, as in the case of methane (CH_4), the simplest organic compound (Figure 2-8B).

Two atoms can sometimes share more than one pair of electrons. For example, carbon may share two pairs of electrons with a single other carbon atom in a **double covalent bond.** The molecule ethylene (C_2H_4), important in the ripening of apples, bananas, and other fruits, has a double bond between its two carbon atoms. Some other molecules, including the nitrogen (N_2) in our atmosphere, contain **triple covalent bonds,** in which two nitrogen atoms share three pairs of electrons.

A covalent bond holds two atoms together because they share electrons. Most of the molecules in organisms are made from just six kinds of atoms.

Figure 2-7 The first three periods in the Periodic Table of the elements. The chemical and physical properties of the elements depend on the atomic numbers *(left)* and the atomic mass *(right)* of the nucleus. The atomic number is the number of protons in the nucleus. The atomic mass is the total number of protons plus neutrons. On the far right are elements such as helium and argon, whose electron orbitals are full. These "noble gases" remain aloof from other elements and rarely form chemical compounds. On the far left are hydrogen, sodium, and other highly reactive elements. The atoms of these elements contain a single electron in their outer orbital and are "eager" to bond with other atoms. In the middle, near the top, are carbon, nitrogen, and oxygen, all fairly reactive, but not explosively so. These three elements, together with hydrogen, form the basis of all organic molecules. *(Silicon chip, Alfred Pasleka/Peter Arnold, Inc.; neon sign, Allan Kaye/DRK Photo; blimp, Gianni Tortoli, Photo Researchers, Inc.; aluminum can, Paraskevas Photography; fireworks, AlaskaStock)*

What determines the shape of a molecule?

The covalent bonds within a molecule have fixed directions and lengths, which result from the characteristics of electron orbitals. For example, the distance between a carbon atom and a hydrogen atom is the same whether in a complex molecule (such as DNA) or a simple one (such as methane) (Figure 2-8). Similarly, the angles between bonds are always the same.

Because bond lengths and angles are constant, every molecule has a definite size and shape. A water molecule, for example, always has the same boomerang shape. We can calculate the shape of a molecule by making a three-dimensional model, based on what we know about the positions of the atoms and the bond lengths and angles. Chemists and biologists represent the structures of molecules in a variety of ways—on a two-dimensional page, and in several types of three-dimensional models (Figure 2-8).

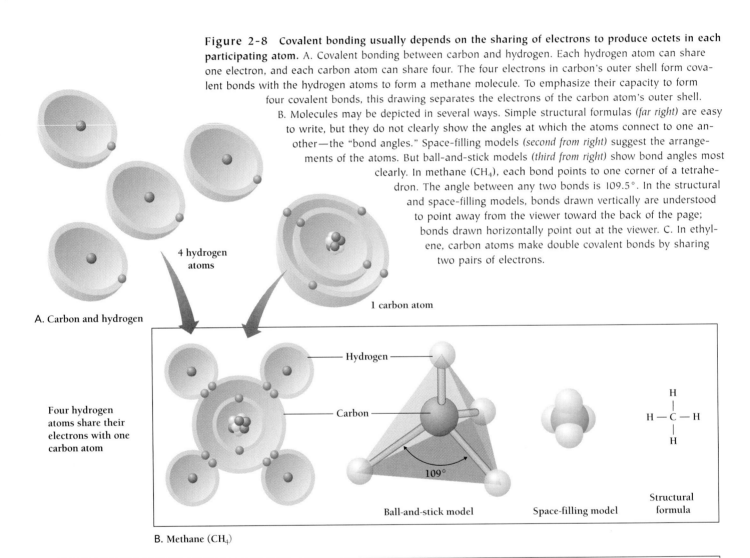

Figure 2-8 Covalent bonding usually depends on the sharing of electrons to produce octets in each participating atom. A. Covalent bonding between carbon and hydrogen. Each hydrogen atom can share one electron, and each carbon atom can share four. The four electrons in carbon's outer shell form covalent bonds with the hydrogen atoms to form a methane molecule. To emphasize their capacity to form four covalent bonds, this drawing separates the electrons of the carbon atom's outer shell. B. Molecules may be depicted in several ways. Simple structural formulas (*far right*) are easy to write, but they do not clearly show the angles at which the atoms connect to one another—the "bond angles." Space-filling models (*second from right*) suggest the arrangements of the atoms. But ball-and-stick models (*third from right*) show bond angles most clearly. In methane (CH_4), each bond points to one corner of a tetrahedron. The angle between any two bonds is 109.5°. In the structural and space-filling models, bonds drawn vertically are understood to point away from the viewer toward the back of the page; bonds drawn horizontally point out at the viewer. C. In ethylene, carbon atoms make double covalent bonds by sharing two pairs of electrons.

Because bond lengths and angles are constant, every molecule has a definite size and shape.

What holds two atoms together in an ionic bond?

The most familiar ionic bonds are those in salts, such as ordinary table salt—sodium chloride. The sodium atoms and the chlorine atoms in sodium chloride have stable electron configurations, with eight electrons in their outer shells (Figure 2-9). Instead of sharing electrons, however, the sodium atom gives away an electron and the chlorine atom takes one. The result is positively and negatively charged ions. The positively charged sodium ions and the negatively charged chloride ions (abbreviated Na^+ and Cl^-) are held together by simple electrical attraction, in ionic bonds. Electrostatic attraction is the same thing that makes your socks cling to your shirts when they come out of the dryer.

Figure 2-9 Table salt. A. Ionic bonding between sodium and chloride ions in a crystal of sodium chloride. B. Reaction of sodium and chloride to form salt. C. Salt crystals. *(C, Bruce Iverson)*

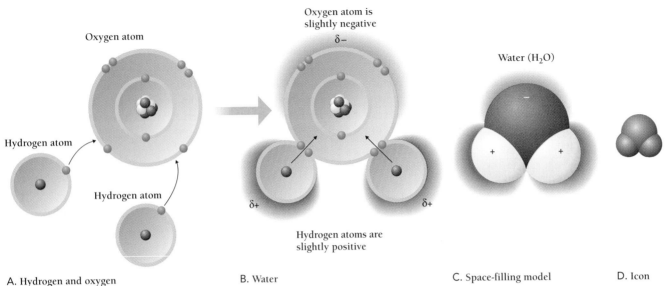

Figure 2-10 The polarity of water. A. Two atoms of hydrogen and one atom of oxygen share electrons in two covalent bonds to form a molecule of water. B. Because the oxygen atom is more electronegative than the hydrogen atoms, the electrons spend more time hovering around the oxygen end of the water molecule. As a result, the oxygen end has a slightly negative charge, while the hydrogens have a slight positive charge. Such a partial charge, less than one full electron, is symbolized by δ. C. Space-filling model of water molecule. D. Icon for water. *(C, Tom Pantages/Phototake)*

Electrical forces hold together ions of opposite charge, forming a complex of positive and negative ions rather than a true molecule. In a salt crystal, for example, each Na^+ ion is surrounded by Cl^- ions, and each Cl^- ion, by Na^+ ions (Figure 2-9). If salt is dissolved in water, however, each ion is surrounded by water molecules, as we discuss later.

An ionic bond is a simple electrical attraction between two oppositely charged ions.

Why are some molecules polar?

A chlorine atom has a high tendency to gain an electron to form a chloride ion, while a sodium atom has a high tendency to lose an electron to form a sodium ion. A carbon atom, on the other hand, tends to share its outer electrons to form covalent bonds. The American chemist Linus Pauling realized that the atoms of each element have a characteristic **electronegativity**—the tendency to gain electrons. Oxygen and chlorine, for example, have a great tendency to gain electrons and are among the most electronegative elements. In contrast, sodium, potassium, and lithium are among the least electronegative elements—they have the greatest tendency to lose electrons. In between are elements such as carbon that have no tendency to lose or gain electrons. (We can think of electronegativity as a measure of electron greediness.)

Knowing the electronegativity of two atoms allows one to predict whether a bond between them will be covalent or ionic. The larger the difference in the electronegativities of two atoms, the more likely they are to form an ionic rather than a covalent bond. Sodium and chlorine, for example, have a large difference in electronegativities and therefore form ionic bonds. Carbon and nitrogen, on the other hand, have similar, moderate electronegativities, and they usually form covalent bonds.

Even in a covalent bond, however, atoms may not share electrons equally. When atoms differ in electronegativity, they do not share electrons equally. Instead, the diffuse clouds of shared electrons tilt toward the more electronegative atoms. In a water molecule, for example, an oxygen atom shares electrons with two hydrogen atoms. But the shared electrons are more concentrated around the oxygen nucleus than around the two hydrogen nuclei (Figure 2-10). As a result, the oxygen atom has a slight negative charge, and the two hydrogen atoms have a slight positive charge.

Molecules that have uneven distributions of electrical charge are said to be **polar,** since they have positive and negative poles in the same way that a magnet has two poles (Figure 2-10). When a polar molecule, such as water, comes close to an ion or to another polar molecule, its negative pole points toward the other molecule's positive pole, and its positive pole toward a nearby negative pole. Molecules with approximately uniform charge distributions are said to be **nonpolar.**

Atoms with great differences in electronegativity tend to form ionic bonds. Atoms with only slight differences in electronegativity tend to form covalent bonds. Molecules with uneven charge distributions are said to be polar.

What Kinds of Forces Hold Separate Molecules Together?

The forces that hold atoms together in covalent bonds are strong. Breaking them requires lots of energy. Electrostatic interactions in salt crystals (ionic bonds) are also strong. Most of the chemistry that concerns biologists, however, happens in or around water. In such an **aqueous** (watery) environment, a variety of weak interactions bind molecules together. Although individually each interaction is weak, the summation of many weak interactions determines the structure of many complex molecules.

Why do nonpolar molecules cling together in an aqueous solution?

When a bartender mixes an alcoholic drink, such as brandy and soda water, the result is a smooth mixture, uniform throughout. Yet, when a cook mixes oil and vinegar to make salad dressing, the oil runs together in globs and floats to the top of the vinegar (Figure 2-11). Why is this? The answer is that polar molecules, such as alcohol, mix well with water and are said to be **hydrophilic** [Greek, *hydro* = water + *phili* = love], or water loving. But because water molecules attract and surround themselves with other polar molecules, nonpolar molecules, such as oil, seem to be repelled by water. We say that nonpolar molecules are **hydrophobic** [Greek, *hydro* = water + *phobos* = fear], or water fearing. In water, hydrophobic molecules are driven together by their mutual tendency to avoid water. The aggregation of nonpolar molecules is called a **hydrophobic interaction** (Figure 2-11).

The properties of polar and nonpolar molecules play a key role in life. The tendency of nonpolar molecules to separate from water underlies the formation of the membranes that separate the inside of a cell from the outside (Chapter 4).

Oils and other hydrophobic molecules cling together when confronted with an aqueous environment.

Van der Waals attractions reinforce hydrophobic interactions

Because electrons move around rapidly, the charges on the different parts of a molecule change constantly. At a given instant, such fluctuations may temporarily produce positive and negative poles, even in a nonpolar molecule. A temporarily polar molecule may, at the instant of polarity, induce a charge in an adjacent molecule, so that it too becomes temporarily polar. For a moment, then, the two molecules are attracted to

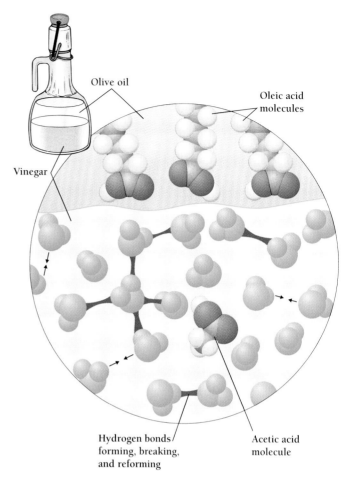

Figure 2-11 Fear of water. Separation of oil and vinegar in a salad dressing. Vinegar, a dilute solution of acetic acid, is mostly water. In hydrophobic interactions, water molecules attract and surround themselves with other polar molecules (including water) but repel nonpolar molecules such as oils. Nonpolar molecules are driven together by their mutual tendency to avoid water.

each other. Such **van der Waals interactions** operate only over very short distances and tend to reinforce the hydrophobic interactions among nonpolar molecules in water. Van der Waals interactions are important in maintaining the three-dimensional arrangements of proteins and other large molecules, such as DNA.

If two molecules are already very close, van der Waals attractions can pull them closer together.

What are hydrogen bonds?

When a hydrogen atom attaches to a highly electronegative atom, such as oxygen or nitrogen, the resulting covalent bond is polar. In this case, the hydrogen atom acquires a slight positive charge. Such a hydrogen atom can then participate in a

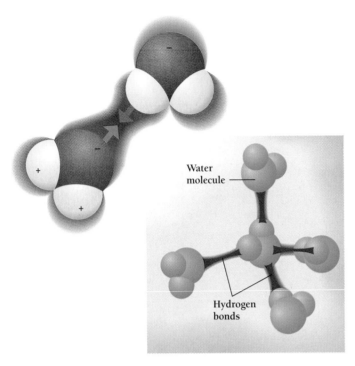

Figure 2-12 **Hydrogen bonds between water molecules.** The hydrogen atoms of one water molecule are attracted to the oxygen atom of another water molecule.

hydrogen bond—a weak attraction to a negatively charged atom in another molecule (Figure 2-12). The most common hydrogen bonds are those between water molecules, but other hydrogen bonds also play a critical role in the structure of proteins and DNA.

The polarity (or nonpolarity) of molecules underlies their noncovalent interactions.

HOW IS WATER ESPECIALLY WELL SUITED FOR ITS ROLE IN LIFE?

Life and water are intimately connected (Figure 2-13). As we discuss in Chapter 18, life probably originated in water. Wherever life is found, there is water, and wherever liquid water is found, there is life. Water makes up more than 70 percent of the material of living organisms themselves and covers more than 75 percent of the Earth's surface. It is the medium in which most cells are constantly bathed and the major component of cells themselves. Not only do most biochemical reactions occur in water, but water itself participates in many biochemical reactions. Although water is common—at least on Earth—its properties are highly unusual. Water's special attributes make it uniquely fit for its important role in life.

Why Does Ice Float?

Most solids sink into their corresponding liquids, but ice floats on water (Table 2-1A). Because ice floats, lakes and ponds freeze from the top down, instead of from the bottom up. The surface ice of a wintry pond, such as that in Figure 2-14, insulates the liquid water below the surface from the freezing air above. As a result, ponds, lakes, and rivers remain liquid at the bottom, allowing fish and other organisms to survive through long winters.

Ice has a lower density than liquid water because the water molecules in ice form a rigid lattice or crystal that holds them somewhat apart from one another: Each water molecule is hydrogen bonded to four other water molecules. In liquid water, the network of water molecules constantly changes, and there is less open space between the molecules.

Ice floats on water because ice has a lower density than water.

How Does Water Resist Changes in Temperature?

In our daily lives, we know that temperature is a measure of how warm or cold something is. But to a scientist, temperature is a measure of the motion of molecules.

Molecules move constantly. In solids, molecules merely jiggle. In gases, molecules whiz around at mind-boggling speeds, bumping into other molecules. At room temperature, for example, the average speed of an oxygen molecule in the air is about 500 meters per second (more than 1000 mph). The higher the temperature, the more rapidly molecules move about. Heat is energy that increases the random movement of molecules.

For most substances, any heat is converted into the more rapid motion of whole molecules, that is, into increased tem-

Figure 2-13 **Water is essential to every organism.** Killer whales (*Orcinus orca*) spend their entire lives in the ocean. (*Francois Gohier/Photo Researchers, Inc.*)

perature. In water, however, additional heat is converted not only into the increased motion of individual molecules but also into a more rapid exchange of hydrogen ions between adjacent molecules. As a result, for a given amount of additional heat, water temperature increases much less than in other substances, that is, water has a high **heat capacity.**

One consequence of water's heat capacity is that the oceans heat up and cool down very slowly, moderating the climates of nearby land. In winter, the interiors of the world's continents disappear under a layer of snow, while temperatures in coastal areas remain relatively mild. Conversely, in summer, the continental interiors reach blazing temperatures, while coastal areas remain cool. Every large body of water has a similar influence on the land nearby.

Water not only cools the coasts of continents, it also cools the human body. To keep our body temperature within the narrow range that many biochemical reactions demand, we must rid ourselves of excess heat. Humans do so largely by perspiring. As the water in sweat evaporates, it carries the heat away. Dogs accomplish the same thing by panting (Figure 2-15).

When a substance changes from a liquid to a gas, it absorbs heat. As water evaporates, it absorbs a lot of heat—more than twice as much as the same amount of ethanol, for example. (Ethanol is also called "ethyl alcohol," or just plain "alcohol" when in wine, beer, or spirits.) Water absorbs a lot of heat when it turns into water vapor because each of the hydrogen bonds that hold the water molecules together must be broken.

Figure 2-14 **Pond life under winter ice.** Because water is denser than ice, the ice floats and covers lakes and ponds in the winter. This layer of ice insulates the water below from freezing air temperatures. Fish, frogs, insects, copepods, protists, and other life forms survive the winter, all because water is denser than ice.

TABLE 2-1 HOW DOES WATER COMPARE WITH ETHANOL AND OLEIC ACID?

	Water	Ethanol	Oleic acid
A. **Density.** In water, the liquid phase is more dense than the solid phase (ice), which is why ice floats. In most substances, including ethanol and oleic acid, the solid phase is denser than the liquid phase.			
B. **Cohesion.** In water, the attraction of the water molecules for one another creates a web of molecules that resists the sinking of the spoon. Neither ethanol nor oleic acid is as cohesive as water, and the spoon sinks.			
C. **Adhesion.** Water molecules' polarity enables them to form stronger interactions with the molecules in paper, so water rises faster than does ethanol or oleic acid.			
D. **Ability to dissolve other substances.** Water dissolves table sugar (sucrose) more quickly than does ethanol or oleic acid.			
E. If you were trapped on a desert island which of these would you choose?			

(A–D, Charles D. Winters; E, © 1994 Zefa Germany/The Stock Market; © 1997 Paraskevas Photography; © Camerique/The Picture Cube.)

Water's enormous heat capacity stabilizes temperatures in oceans, lakes, coastal areas, and even within individuals.

Why Do Water Molecules Cling to One Another?

An attraction between molecules of the same substance is called **cohesion.** Water and chewing gum are both highly cohesive. Under tension, they stretch but do not break. The cohesion of water molecules also creates high **surface tension**—the tendency of a substance to form a smooth round surface. Because water molecules cohere, a water surface resists pressure. For example, as illustrated in Table 2-1B, the bowl of a teaspoon balanced on the edge of a teacup will rest on the surface of the tea as though floating. But a cup of alcohol or olive oil will not support the same spoon.

Water's cohesiveness is caused by hydrogen bonds among its molecules. The attraction among water molecules is so great that water has the highest surface tension of any liquid except for liquid mercury. The surface of water acts like a thin skin. Some insects, such as the water strider, take advantage of water's surface tension and walk easily over the surface of a pond.

Hydrogen bonding makes water molecules cohere, creating surface tension.

Why Do Water Molecules Cling to Other Substances?

An attraction of one substance to another is called **adhesion.** Water clings to and "wets" surfaces, such as glass, that are composed of polar or charged molecules. Water's adhesiveness allows a close interaction between it and many other biological surfaces.

Because water both clings to itself (cohesion) and clings to other surfaces (adhesion), it readily enters and climbs tiny tubes, in a process called **capillary action.** If we place fine glass tubes (capillaries) into beakers of water, ethanol, and olive oil, the water moves upward a fair distance, the ethanol moves less, and the olive oil not at all (Table 2-1C). Such capillary action is important in the movement of water through soil, and up the narrow tubes inside of trees and other plants. Water's cohesiveness and adhesiveness help trees raise water from roots deep in the ground to leaves hundreds of feet in the air (Figure 2-16).

Water's adhesiveness results from the tendency of water molecules to form hydrogen bonds with other polar or charged molecules. Water molecules stick to glass capillaries because glass contains charges on its surface. Similarly, paper towels, which are made of long polar molecules called cellulose, absorb water very well. In contrast, olive oil, which is mostly nonpolar, is relatively poorly absorbed by the cellulose in paper towels. That's why cleaning up water is so much easier than cleaning up olive oil.

Hydrogen bonding allows water to adhere to other compounds.

Why Is Water Such a Powerful Solvent?

Most people think of a solvent as the fluid that the dry cleaner uses to clean a wool coat, or maybe the gasoline you use to get grease off the wheels of your car. In fact, a **solvent** is any fluid in which other substances dissolve. Anyone who has poured salt into boiling water or stirred sugar into a cup of hot coffee will have noticed how rapidly salt and sugar dissolve in water. Water is a powerful solvent. It is an especially good solvent for ions, such as those of table salt, and for polar molecules, such as sugar. Salt and sugar, which dissolve so well in water, dissolve poorly in ethanol and hardly at all in olive oil, which is nonpolar (Table 2-1D).

The substances that dissolve within a solvent such as water are called **solutes.** All salts are composed of ions, and all are extremely soluble in water. When sodium chloride dissolves in water, the highly polar water molecules surround the Na^+

Figure 2-15 **Why are dogs' noses wet?** Water's enormous heat capacity makes it a useful coolant. As the water in a dog's nose evaporates, it carries heat away, cooling the inside of the nose. On a hot day, overheated blood from the body flows through the cold nose before reaching the brain, so that the dog's brain tissues remain cool. *(John Cancalosi/DRK Photo)*

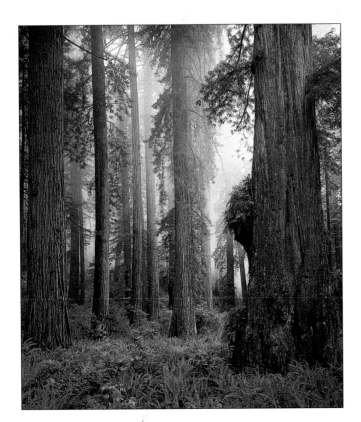

Figure 2-16 **Only because water is so cohesive can trees raise it hundreds of feet off the ground.** Individual California coast redwoods (*Sequoia sempervirens*) may grow to heights of 360 feet, taller than a 25-story building. *(Larry Ulrich/DRK Photo)*

and Cl^- ions like devoted fans crowding around a sports star (Figure 2-17). This shell of water molecules shields the positive and negative ions from each other, allowing the ions to remain farther away from each other than they can in their solid form (crystals), where they have a regular arrangement.

Sugars and other polar molecules also dissolve well in water, but for a slightly different reason. When water molecules surround a polar molecule such as sugar, hydrogen bonds form between the water molecules and the electronegative oxygen atoms of the sugar. These water molecules isolate the molecules of sugar from one another, breaking the attachments that keep the sugar molecules in a crystal.

The polar properties of water explain why some molecules are hydrophilic and others are hydrophobic. Hydrophilic molecules are polar or charged molecules that interact strongly with water molecules. Nonpolar molecules are hydrophobic and do not interact with water molecules. Oil, a mixture of nonpolar molecules, does not readily mix with water but it mixes easily with nonpolar solvents. We can summarize the tendency of polar molecules to dissolve in water and nonpolar molecules to dissolve in oil with the phrase "like dissolves like."

Many biologically important molecules contain both hydrophilic and hydrophobic regions and are said to be **amphipathic** [Greek, *amphi* = both + *pathos* = feeling]. When an amphipathic molecule is placed in water, its hydrophobic regions avoid the water, and its hydrophilic regions bond with water (Figure 2-18). Mayonnaise, for example, consists of tiny oil droplets suspended in a watery solution.

All **detergents** are made of amphipathic molecules that interact both with water and with hydrophobic molecules such as oil. Detergents remove oil from dirty dishes and clothes by making the oil soluble in water.

Water is the most important polar molecule both within cells and in their external environments. Water molecules interact strongly with each other and with other charged and polar molecules. These interactions, and the lack of interactions with nonpolar molecules, are the most important factors in establishing biological structures.

Water's polar properties make it an excellent solvent for ions and polar molecules.

Water Participates in Many Biochemical Reactions

Besides serving as the medium in which most biochemical reactions occur, water itself participates in many reactions. For example, when we break down the large carbohydrate molecules in spaghetti into the small sugar molecules that we use for fuel, we do so by adding water molecules. The negatively charged oxygen atom of a water molecule can easily approach positively charged atoms in other molecules. The oxygen may then form a covalent bond with the other molecule by simultaneously breaking its covalent bond with one of its hydrogen atoms.

How Do Water Molecules Change in Solution?

Recall that the electrons that form the chemical bonds in a water molecule are closer to the oxygen atom than to the hydrogen atoms. Sometimes the electrons get so close to the oxygen atom that one of the hydrogen atoms has no electron at all and becomes a naked nucleus, or **hydrogen ion** (H^+). A hydrogen ion without an electron is just a proton, and it can jump back and forth between the oxygen atoms of two different water molecules (Figure 2-19). When that happens, one water molecule can end up with an extra proton and the other water molecule ends up missing a proton. The deprived water molecule is called a **hydroxide ion** (OH^-) and has a charge of -1. The water molecule with the extra hydrogen ion (H^+) is called a **hydronium ion** (H_3O^+) and has a charge of $+1$. We can represent this exchange of hydrogen ions by the following chemical equation:

$$H_2O + H_2O \longrightarrow OH^- + H_3O^+$$

water + water ⟶ hydroxide ion + hydronium ion

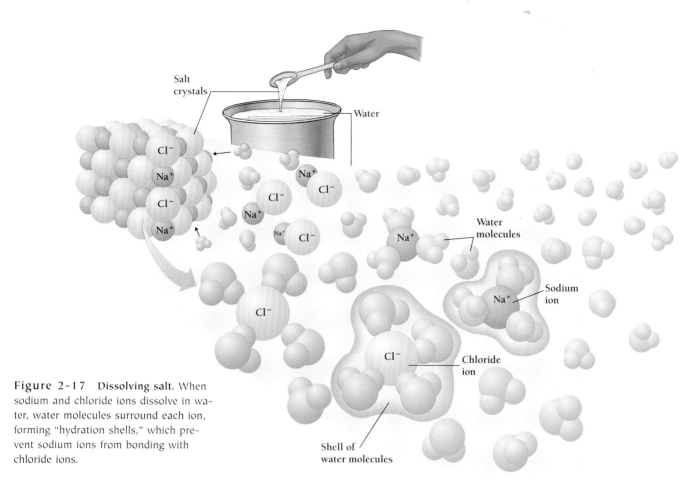

Figure 2-17 **Dissolving salt.** When sodium and chloride ions dissolve in water, water molecules surround each ion, forming "hydration shells," which prevent sodium ions from bonding with chloride ions.

Notice that each side of the arrow has equal numbers of hydrogen atoms and oxygen atoms and that the charges are balanced. For the sake of brevity, we can leave one of the water molecules out of the equation. We can represent the shuttling of hydrogen ions as the splitting of water into a hydroxide ion and a naked hydrogen ion:

$$H_2O \longleftrightarrow H^+ + OH^-$$

water ⟷ hydrogen ion + hydroxide ion

Figure 2-18 **Amphipathic molecules contain a hydrophobic tail and hydrophilic head.** A ball of these molecules, called a "micelle," forms when water molecules cluster around the hydrophilic heads and the hydrophobic tails hide together in the middle.

The opposite charges of hydrogen ions and hydroxide ions ensure that they do not go far. They quickly recombine with other hydroxide ions and hydrogen ions to form uncharged water molecules again. Water, then, constantly splits into hydrogen ions and hydroxide ions, and the ions constantly recombine to form water.

When the splitting of water into hydrogen ions and hydroxide ions exactly balances the recombining of hydrogen ions and hydroxide ions back into water, the number of ions at any instant is always the same, and we say that the solution is "at equilibrium." Such a balance between forward and reverse reactions is called **chemical equilibrium** [Latin, *aequus* = equal + *libra* = balance].

In pure water at equilibrium, the number of hydrogen ions exactly equals the number of hydroxide ions. However, in most solutions, the number of hydrogen ions differs from the number of hydroxide ions. Vinegar has a million times more hydrogen ions than hydroxide ions. Stomach acid is even stronger. It contains about one trillion times more hydrogen ions than hydroxide ions.

The **concentration** of hydrogen ions—the number of hydrogen ions in a given amount of liquid—varies so widely in different solutions that chemists have defined a scale, called **pH,** to measure the concentration of hydrogen ions in a solu-

Figure 2-19 Juggling protons. A. Hydrogen nuclei, or protons, commonly jump back and forth between water molecules. B. When a hydrogen nucleus, or proton, jumps from one water molecule to another and stays, a hydronium ion (H_3O^+) and a hydroxide ion (OH^-) result.

tion (Figure 2-20). The pH scale is based on the logarithm (to the base 10) of the concentration of hydrogen ions. We can think of pH as standing for "power of hydrogen." But the logarithm is actually a negative number. So a pH of 7, for example, represents a concentration of hydrogen ions equal to 10^{-7} moles per liter (a *mole* is Avogadro's number of molecules, so moles per liter is a unit of concentration). The middle of the pH scale is 7, which represents a neutral solution, such as water, in which H^+ and OH^- concentrations are the same. Acid solutions have a lower pH. For example, the pH of vinegar is 4 (10^{-4} moles per liter). The pH of the hydrochloric acid secreted by the stomach is less than 1 (10^{-1} moles per liter), or 1/10 mole per liter—very concentrated. In short, the greater the concentration of hydrogen ions, the lower the pH value. A tenfold increase in concentration means a decrease of one pH unit.

A molecule that can give up a hydrogen ion is called an **acid**. A molecule that can accept a hydrogen ion from an acid is called a **base**. An acid dissolved in water contributes hydrogen ions to water molecules and raises the concentration of hydrogen ions, which decreases the pH. Conversely, bases dissolved in water result in a decreased concentration of hydrogen ions and increased pH. Bases, then, have pH values higher than 7.

Water molecules split into hydrogen ions and hydroxide ions. The exact balance between the two kinds of ions determines the pH (the relative acidity) of the solution.

Why Is pH Important to Organisms?

Changing the pH of a solution affects the properties of other molecules. For example, the characteristic sour taste of acids results from the interaction of hydrogen ions with molecules on the tongue that report taste to the brain. Similarly, basic

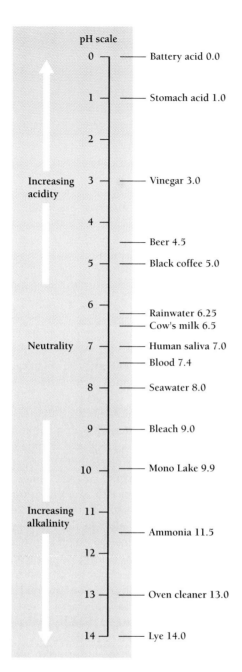

Figure 2-20 The pH scale. The pH scale measures acidity and alkalinity. Pure water, with a pH of 7.0, is considered neutral. Anything above 7.0 is alkaline and anything below 7.0 is acid.

solutions feel slippery because they change the characteristics of molecules on the skin.

When a molecule donates or accepts a hydrogen ion, its net charge changes. Such changes alter a molecule's interactions with other molecules and ions. As a result, pH dramatically affects both the structure and the chemical reactivity of most biologically important molecules.

In humans, the pH of the blood is normally 7.4. If the pH of the blood decreases until it is even slightly acidic—to pH 6.95—the nervous system becomes unresponsive, and coma and death soon follow. Alternatively, if the blood pH increases (becomes "alkaline") to as little as pH 7.7, the nervous system becomes overreactive, and muscle spasms and convulsions begin. Normally, such disasters are prevented by a number of physiological adaptations, which include the regulation of the heart rate, changes in the rate and depth of breathing, and the secretion of hydrogen ions by the kidneys.

However, a wide variety of illnesses, injuries, and drugs can cause acid or alkaline blood. One of the most common causes of acid blood, for example, is severe diarrhea. Basic blood can result when a physician gives a patient excessive amounts of diuretics (drugs that cause the body to eliminate large amounts of water and hydrogen ions) or if a person takes too much bicarbonate of soda for an upset stomach.

Changes in pH do not have to be dangerous. Many organisms use pH to control normal processes. For example, a sperm in the testes remains motionless, saving energy for the time when it is near an egg. Yet something must activate the sperm when the moment is right. In sea urchins, that something is a decrease in pH. The pH of the sea urchin's testes and sperm is about 7.2, very slightly basic. When the sea urchin releases its sperm into seawater, which has a pH of about 8.0, the sperm become more basic (about 7.6), which activates the sperm. Off they swim.

Both inside and outside of cells, pH powerfully influences the chemistry of life.

Buffers: How Do Organisms Resist Changes in pH?

Because pH dramatically affects both the structure and the chemical reactivity of most biologically important molecules, cells have mechanisms that maintain nearly constant pH. The interior contents of cells almost always have a pH of about 7.

Organisms use both chemical and physiological devices to maintain constant pH. The chemical strategy is to employ **buffers**—molecules that easily interconvert between acidic and basic forms by donating or accepting hydrogen ions. In human blood, the main buffer is bicarbonate (which is in chemical equilibrium with carbonic acid).

Many chemical reactions in organisms involve exchanges of hydrogen ions. For example, when a person exercises, the muscles produce carbon dioxide (CO_2), which the blood carries to the lungs. Most of the carbon dioxide in the blood, however, combines with water to form carbonic acid (H_2CO_3), which tends to break down into a hydrogen ion and a bicarbonate ion (HCO_3^-):

$$CO_2 + H_2O \longrightarrow H_2CO_3 \longrightarrow$$
carbon dioxide + water ⟶ carbonic acid

$$H^+ + HCO_3^-$$
hydrogen ion + bicarbonate ion

When the bicarbonate ions reach the lungs, they reacquire a hydrogen ion and reform carbon dioxide and water:

$$H^+ + HCO_3^- \longrightarrow H_2CO_3 \longrightarrow CO_2 + H_2O$$

hydrogen ion + bicarbonate ion ⟶ carbonic acid

carbon dioxide + water

The equilibrium between carbonic acid and bicarbonate helps maintain a constant pH. When the digestion of food produces extra hydrogen ions, bicarbonate ions can absorb them by becoming carbonic acid, and the pH of the blood changes little. Similarly, hydrogen ions provided by carbonic acid can "soak up" hydroxide ions added to the blood. In both cases, the presence of the buffer minimizes changes in the total concentration of hydrogen ions. Buffers such as carbonic acid/bicarbonate reduce, but do not eliminate, the pH changes of organisms and cells.

Organisms resist changes in pH with chemical buffers and other physiological mechanisms.

The biological functions of molecules are determined by their structures. Since water is the universal medium of life, the attraction or avoidance of water by molecules or parts of molecules is the basis of the organization of cells and organisms. The interactions of molecules with water and with each other, in turn, depend on the properties and the arrangements of the atoms that compose them.

Study Outline with Key Terms

Chemistry is the study of the properties and transformations of matter. All **matter** is composed of **atoms.** An atom is the smallest particle of an element that has the properties of that element. **Elements** are forms of matter that cannot be broken down to simpler substances. **Compounds** can be broken down into elements. Most matter consists of **mixtures** of compounds.

Compounds consist of **molecules,** which are fixed arrangements of two or more atoms, with a characteristic **molecular weight.** The atoms of each element have characteristic properties that determine what kinds of molecules can be formed. **Chemical reactions** change the arrangements of atoms in molecules.

Atoms are made of still smaller particles. Almost all of an atom's **mass,** or weight, is contained in its **nucleus.** An atom's nucleus contains both protons and **neutrons. Nuclear reactions** change the number of protons and neutrons in a nucleus.

Typical atoms are electrically neutral; the positive charge of the **protons** in the nucleus is balanced by an equal number of negatively charged **electrons.** Electrons are outside the nucleus and occupy most of the space of the atom.

The atoms of an element all have the same atomic number, but they may have different mass numbers, that is, different numbers of neutrons. The different kinds of atoms of a single element are called **isotopes. Radioactive isotopes** spontaneously disintegrate, with a characteristic **half-life,** giving off radioactive particles that can be detected photographically or electronically.

Electrons are arranged in **orbitals,** each of which can hold a different number of electrons. Groups of orbitals, in turn, form **shells,** which differ in **energy.** The chemical properties of an atom are largely determined by the number of electrons in the outermost shell.

In forming chemical bonds, electrons may be rearranged in two distinct ways: (1) by sharing with other atoms to form **covalent bonds,** or (2) by donating or accepting electrons to form **ionic bonds.** Carbon atoms, unlike most other atoms, can form **double,** or even **triple,** covalent bonds. Individual atoms have different tendencies to gain or to lose electrons. Atoms that are most apt to gain electrons are said to be the most **electronegative.** Differences in the electronegativity of atoms in a molecule determine the extent to which the electrons in a chemical bond are equally distributed. Molecules with uneven distributions of electrons are said to be **polar;** those with uniform charge distributions are said to be **nonpolar.** Individual molecules may interact via several types of relatively weak forces—**hydrogen bonds, hydrophobic interactions,** and **van der Waals interactions.**

Water is the most abundant molecule in all organisms, and its unusual properties are especially suited for its role in life. Among these properties are its **cohesiveness** (which leads to water's strong **surface tension**), its higher density as a liquid than as a solid, its high **heat capacity,** its **adhesiveness** (which leads to water's **capillary action**), its ability to serve as a powerful **solvent,** its ability to absorb heat, and its participation in many biochemical reactions.

The properties of water result from its polarity. The polarity of water permits interaction with **ions** and polar molecules in **aqueous** solutions. **Hydrophilic solutes** readily dissolve in water. Nonpolar molecules do not dissolve readily in water, and they are said to be **hydrophobic.** Many biologically important molecules are **amphipathic,** containing both hydrophilic and hydrophobic regions. **Detergents** are amphipathic molecules that allow hydrophobic molecules to dissolve in water.

The hydrogen nuclei that participate in hydrogen bonds can also jump between water molecules. This ability leads to water's high heat capacity.

A molecule that can donate a **hydrogen ion** to another molecule is called an **acid.** A molecule that can accept a hydrogen ion from an acid is called a **base.** Some of the time, the jumping hydrogen ions create **hydroxide ions** and **hydronium ions,** which lie in **chemical equilibrium** with water. Water is thus both an acid and a base. The **pH** scale measures the **concentration** of H^+ and OH^- ions in a solution. Changes in pH alter the charge on molecules and affect their chemical and physical properties. The fluids of organisms contain **buffers,** which resist changes in pH. Cells and organisms also have physiological adaptations that maintain nearly constant pH.

Review and Thought Questions

Review Questions
1. Define the words "element," "atom," "compound," and "molecule."
2. What is an isotope? What is meant by the term "radioactive isotope"? Why are only some isotopes radioactive?
3. What is an electron orbital?
4. Explain the difference between a covalent bond and an ionic bond. How is each formed? Name a common compound that exhibits each type of bond.
5. Explain how a covalent bond can be polar or nonpolar. Give examples of each type of bond. Name a polar molecule that is of great importance to organisms and their environments.
6. Describe each of the following: hydrogen bonds, hydrophobic interactions, and van der Waals interactions. Which of these reactions are likely to occur between (1) polar compounds or (2) nonpolar compounds?
7. Define "cohesion" and "surface tension." What is the relationship between these two properties of water?
8. How does adhesion differ from cohesion? What is a capillary?
9. Give examples of polar and ionic compounds that dissolve easily in water as well as nonpolar ones that do not. Which type of molecule would you use to make waterproof bags?
10. What type of bond is formed between adjacent water molecules? Why do they cohere? Is this more similar to a covalent or ionic bond?
11. What types of bonds tend to produce (1) hydrophilic substances or (2) hydrophobic ones?
12. What is a hydrogen ion? If a hydrogen ion is formed from a water molecule, what other ion is also formed? If a hydrogen ion is formed idiot from HCl, what other ion is also formed? How is the pH of water defined?

Thought Questions
13. Explain how the results of Rutherford's experiment caused him to propose that the nucleus of an atom was very small and dense and that most of the atom was nearly empty space.
14. The tendency of living things to remain in a narrow range of temperatures is part of homeostasis—one of the characteristics of living things. How do the properties of water help organisms regulate temperature? Consider organisms that live in water, then organisms, such as humans, that live outside water.

About the Chapter-Opening Image
Alfred Nobel invented a way of packaging highly explosive nitroglycerin into dynamite, but he would not take nitroglycerin as a medicine for his heart condition. We now know that nitroglycerin, as a medicine, releases nitric oxide (NO), which dilates blood vessels.

On-line materials relating to this chapter are on the World Wide Web at **http://www.harcourtcollege.com/lifesci/aal2/**

3 Biological Molecules Great and Small

Prions: A Slow-Growing Idea

In July of 1972, a young medical resident at a San Francisco hospital admitted an elderly woman whose memory was so poor that she was having difficulty doing everyday chores. The resident was 30-year-old Stanley Prusiner, who would someday win a Nobel prize for his work on the kind of disease she suffered from. But that was far into the future. Prusiner soon learned there was nothing he could do to help the woman.

She had a rare illness of the brain called Creutzfeldt-Jakob disease. No one knew what caused it or how to cure it. Her body's defenses mounted no attack against it. After only 7 weeks, she died. An autopsy showed that her brain was riddled with holes, like a sponge. Intrigued, Prusiner resolved to read everything he could about the mysterious killer. He soon learned that it was one of a group of fatal brain diseases known as transmissible spongiform encephalopathies (TSEs).

One TSE, called kuru, was killing people in Papua New Guinea who had honored the dead by eating their brains. Another, called scrapie, damaged the brains of goats and sheep. Scrapie, Prusiner learned, caused sheep to lose coordination and scratch themselves so viciously that they scraped off their wool.

What caused these different forms of TSEs? One clue was that tissues from an animal with a TSE could be used to infect another animal. From this, biologists concluded that some infectious agent—perhaps a bacterium or a virus—was causing the disease. Most researchers concluded that TSEs were caused by "slow viruses," which, like the AIDS virus, take a long time to produce symptoms. But no one had been able to isolate the presumed virus. Prusiner soon encountered a radically different hypothesis about TSEs—an idea that slowly took over his life (Figure 3-1).

In 1967, radiobiologist Tikvah Alper had published a scientific paper on scrapie so controversial that scientists around the world buried her in mail. Indeed, the number of letters reached the point that mail room employees at Hammersmith Hospital, in London, refused to deliver any more. They told Alper to come get the bags of mail herself.

Alper's results implied that scrapie couldn't be caused by a virus or any known organism. Her experiments had shown that tissues infected with scrapie could infect animals even after apparently all DNA and RNA had been destroyed by radiation. Viruses, like bacteria and other organisms, rely on the nucleic acids DNA and RNA to carry genetic information. To destroy their DNA or RNA was to destroy their ability to infect cells.

Alper proposed that whatever was causing scrapie had no genes, or didn't need them. But it was heresy to suggest that a virus or any infectious microbe could exist and reproduce without genes. And it was heresy to suggest that anything else besides a microbe or a virus could infect the body and spread.

Alper knew what her experiments had shown, however, and she was not one to shy away from controversy. When at first her paper was rejected for publication, she sent an outraged letter to the editor of the magazine and convinced him to revoke his decision. "Tikvah is remembered by most people for her forceful discussion style, which could be quite alarming," two of Alper's colleagues later wrote in the *International Journal of Radiation Biology*. Indeed, at 79, Alper sent a burglar fleeing from her home after tackling him and grappling for his gun.

Key Concepts

- Organisms build most structures from just four molecular building blocks: lipids, sugars, amino acids, and nucleotides.

- Functional groups of atoms determine the chemical properties of each building block.

- Dehydration reactions, which remove the equivalent of one molecule of water, join small building blocks into macromolecules.

- The three-dimensional structure of a protein depends on interactions among the functional groups of its amino acid building blocks.

But as much as Alper would have liked to pursue her work on scrapie and convince her colleagues of her results, she was unable to do so. The Agricultural Research Council of Great Britain refused to fund additional research. The council thought scrapie was of little financial importance. Disappointed but not defeated, Alper turned her attention to her many other interesting research projects.

Later events showed that the council's decision to cut funding for scrapie research was a mistake. In the 1980s, an outbreak in England of the scrapie-like TSE called "mad cow disease" shut down the English beef industry. Nearly 2 million beef cattle and other farm animals had to be slaughtered at a cost of more than $6 billion. Despite these precautions, more than 40 people have so far died of mad cow disease, apparently from having eaten tainted beef years earlier.

Like scrapie itself, however, Alper's discovery would not go away. There was, after all, a possible explanation for her idea that neither virus nor bacterium caused the disease. Shortly after her article was published, J.S. Griffith, of Bedford College in London, proposed that proteins could sometimes fold into unusual shapes and then coax normal proteins into similar misfoldings. Could scrapie proteins cause disease by forcing the proteins in the brains of sheep to misfold? Almost no one thought so.

In 1974, Stanley Prusiner had just begun working as an assistant professor at the University of California at San Francisco (UCSF). Inspired by his early brush with Creutzfeldt-Jakob disease, he had chosen scrapie as his research subject. Despite Alper's findings, Prusiner assumed that the cause was some sort of virus.

But as Prusiner began purifying and testing extracts from scrapie tissues, the evidence gradually forced him to change his mind. The more he repeated Alper's experiments and refined them, the more he came to believe that she must be right. When he damaged the nucleic acids in scrapie extracts, they continued to infect animals. But when he denatured—or unfolded—scrapie's proteins, the extracts lost their ability to infect. These results could only mean that the infectious agent was a protein. But for something to cause an infection, it must make more of itself. And everyone knew that proteins were incapable of reproducing themselves.

Even as Prusiner became more fascinated by his findings, the agency that had supported his work cut his research grant. Almost simultaneously, the university informed him that he would not be given tenure. Prusiner was devastated. He was losing his job, his lab, and the grant he needed to pay for his research. As he later wrote, "Everything seemed to be going wrong, including the conclusions of my research studies."

Yet the moral support of a few of his closest colleagues carried him through. Skilled at selling his ideas, Prusiner soon

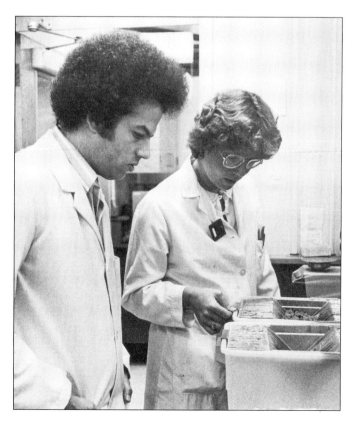

Figure 3-1 **Stanley Prusiner.** Prusiner's research on the brain disease scrapie has persuaded most scientists that the way proteins fold can carry biological information and even cause disease. Yet despite Prusiner's Nobel prize for this remarkable work, some biologists still doubt that renegade proteins cause scrapie. (*Courtesy of Stanley Prusiner, UCSF*)

Figure 3-2 **An α helix and a β sheet.** In the prion protein PrP, the difference between an α helix and a β sheet can spell the difference between life and death. When regions of the protein that are normally α helices (*left*) fold into β sheets (*right*), the result may be a fatal disease of the brain. (*Courtesy of Stanley Prusiner, UCSF*)

convinced two private foundations to fund his work. And, amazingly, even his tenure decision was reversed. He would be allowed to stay at UCSF.

By 1982, Prusiner still lacked a convincing explanation for his data. Nonetheless, he boldly proposed that scrapie was transmitted by infectious proteins he called *prions* (pronounced "pree-onz"). Many scientists said his claim was premature, even brazen.

Former allies turned against him. Frank Masiarz, a researcher who had worked closely with Prusiner, quit. As Masiarz told *Discover* magazine in 1986, "There's no point in creating a name for something that we don't know even exists yet.... [Prusiner] tends to jump to conclusions, to give less credence to facts that would discount his preferred interpretation of the results." Feeling betrayed, Prusiner refused to talk to the press for a decade after the publication of the *Discover* article.

But like Tikvah Alper, Prusiner was determined to change the minds of those around him. His research team succeeded in isolating the prion protein, which they named PrP. The next step was to describe its structure. All proteins are made of long chains of 20 small molecules called amino acids that are strung together like beads on a string. These long chains fold into the complex shapes of functioning proteins.

Working with Leroy Hood of the California Institute of Technology and Charles Weissman of the University of Zurich, the group determined the order of amino acids in the PrP protein. Prusiner and other colleagues then showed that PrP could fold up in at least two ways. One form was *scrapie PrP*, found in the brains of sick sheep, and the other was *cellular PrP*, a protein made in the cells of healthy animals. When Prusiner and Fred Cohen, also of UCSF, compared the shapes of cellular and scrapie PrP, they discovered an important difference. Where the normal, cellular PrP was twisted into spirals called "α helices" (alpha helices), scrapie PrP was folded into parallel strands called "β pleated sheets" (beta pleated sheets), as shown in Figure 3-2.

Was it possible that scrapie PrP was compelling the body's own PrP to misfold and become scrapie PrP? Prusiner and Cohen hypothesized that scrapie PrP could turn the α helices of cellular PrP into β sheets. Working with short pieces of PrP, they found that β sheets could indeed nudge α helices into becoming β sheets. Amazingly, other biologists were even able to transform full-length cellular PrP into scrapie PrP by mixing cellular PrP and scrapie PrP together.

These astonishing experiments brought to light a completely new phenomenon, the ability of a protein (scrapie PrP) to alter the three-dimensional structure of another protein. The experiments shattered the established view that genes alone determine the three-dimensional structures of proteins and carry biological information. Proteins too could convey information.

On October 6, 1997, a young technician walked into Prusiner's usually quiet lab to start his day's work and stumbled on a boisterous party. "People were shouting and drinking champagne. They practically tackled me," he recalled. After more than two decades of struggling against the scientific establishment, Prusiner had won one of the highest honors in science, a Nobel prize.

For more than 20 years, Prusiner had probed the enigmatic nature of the scrapie agent. Like Alper, Prusiner faced disbelief from his colleagues and difficulties getting funded. Ultimately, however, Prusiner uncovered what most agree is the explanation underlying scrapie's odd nature. In the process, he changed our understanding of how biological information can be transmitted.

Following a winding path of experiments, Prusiner realized that scrapie's secret lay in the folding of a protein. By coercing proteins into adopting altered shapes, scrapie proteins can carry and transmit biological information free of nucleic acids. But even today, many questions remain. Like most scientific adventures, the story keeps evolving and reshaping itself as scientists add new findings to their growing collection of clues.

Some nagging questions still lingered when Prusiner was awarded the Nobel prize for discovering a new class of disease-causing agents and for showing how they could infect an organism. Most importantly, no one had yet infected an animal with lab-made PrP. Some researchers continue to insist that the real cause of TSEs could be viruses that are exceptionally good at protecting and repairing their damaged nucleic acids. Scrapie comes in different forms, and some biologists doubt that different shapes of a single protein can explain the different strains of scrapie. Such biologists claim that Prusiner's Nobel prize was premature at best and an outright error at worst.

Even Tikvah Alper has disagreed with Prusiner. In 1993, Alper began to revisit her early work on scrapie. Only 2 weeks before her death, in 1995, she was still writing to scientific journals. She argued that the scrapie agent was neither a protein nor a virus, but a rogue piece of membrane from a cell. But few people have taken her suggestion seriously.

The study of protein folding is now one of the most exciting fields in biology. Understanding how and why proteins fold the way they do is critical to understanding the physiology and genetics of cells and whole organisms. Yet, to understand what causes a protein to fold, we need to understand the chemistry of the smaller molecules that make up proteins. In this chapter, we look at the way cells build small building blocks into larger molecules.

HOW DO ORGANISMS BUILD BIOLOGICAL MOLECULES?

If we measure the sizes of all the molecules in a cell, we discover a surprising thing. Cells have many small molecules, with molecular weights less than 300, and many large molecules, with molecular weights greater than 10,000. But cells have very few molecules with intermediate sizes.

How Big Are Biological Molecules?

Biologists call large molecules **macromolecules** [Greek, *macro* = large]. Macromolecules consist of small molecules linked together in long chains of small building blocks.

Nearly all molecules found in living organisms share one important trait. They are made of chains of carbon atoms. In fact, living organisms are associated so closely with carbon that all carbon-and-hydrogen-containing compounds are called **organic** molecules (even when they have nothing to do with organisms). **Organic chemistry** is the study of the structures and reactions of carbon-and-hydrogen compounds. **Biochemistry** is the study of the chemistry of living organisms.

The carbon-based building blocks of life are simple and universal. Even though there are a nearly infinite number of carbon-containing molecules, organisms use only a few. We are all of us made of these few building blocks. In this chapter, we will discuss 35 small molecules that together form almost all biological structures.

The 35 building blocks fall into four classes: sugars, amino acids, nucleotides, and lipids. Sugars join together to form long-chain macromolecules called polysaccharides, like the starch in bread and potatoes. Amino acids form polypeptide chains, which make up proteins.

Nucleotides form long chains called nucleic acids, such as the DNA that our genes are made of. Lipids, which include fats and oils, do not form long chains. However, they do form sheets of thin membrane. Along with other molecules, lipid membranes form the membranes that surround all cells and regulate their interactions with their environment.

All organisms are made of the same four kinds of molecules: sugars, amino acids, lipids, and nucleotides.

Why Are Biological Structures Made from So Few Building Blocks?

Given the enormous diversity of organic molecules, the relatively small number that are used by organisms comes as a surprise. But why should there be so few building blocks? For example, Why do almost all organisms use the sugar glucose as their principal energy source? And why do all of the millions of species on this planet use the same 20 amino acids to make proteins?

The most likely explanation for this sameness is that we are all related. We all use the same building blocks because all life on Earth has a common origin. The same molecules are used again and again because organisms inherit from their ancestors the tools to use these few building blocks.

All living organisms are made of the same few small molecules.

What Kinds of Structures Do Carbon Atoms Form?

Because all the building blocks of life include carbon, we begin by discussing carbon. Carbon atoms often form long chains of atoms. Each carbon bonds to the next by single or double covalent bonds. Each carbon atom can bond with up to four other atoms.

Along the sides of a simple carbon chain, hydrogen atoms fill the remaining slots. Such long chains of carbon and hydrogen are called **hydrocarbon chains.** One simple hydrocarbon chain is octane, a component of gasoline, which consists of 8 carbon and 18 hydrogen atoms (Fig. 3-3A). Most of the 35 building blocks are based on hydrocarbon chains.

Hydrocarbon chains can also form rings. Because carbon can form covalent bonds only at certain angles, most carbon rings have either five or six atoms. A ring sometimes includes other types of atoms, often oxygen or nitrogen.

When the atoms of a ring are connected by alternating single and double bonds, the rings lie flat, like the benzene ring shown in Figure 3-3B. Such flat rings belong to a special group of organic molecules called **aromatic** compounds.

Aromatics include, for example, acetylsalicylic acid (aspirin), benzene (an industrial solvent), and dimethylpyrazine (a component of chocolate). Some aromatic compounds consist of carbon atoms only, and some include one or two nitrogen atoms in their rings. Many are quite literally aromatic,

A. Octane (C$_8$H$_{18}$)

B. Benzene (C$_6$H$_6$)

Figure 3-3 A chain of carbon atoms forms the skeleton of many building blocks. A. Octane, a component of gasoline, has 8 carbons, with 18 hydrogen atoms hanging off the sides. Inside a working automobile engine, carbon-carbon bonds in octane are first broken, using the energy supplied by the spark plug. Following this, new bonds form with oxygen, releasing stored potential energy. B. Many building blocks consist of aromatic rings, like that of benzene. Aromatic rings have alternating double bonds that make them lie flat.

including such strong-smelling compounds as the oil that makes almonds bitter, the quinine in tonic water, and the benzene, toluene, and xylene that together give high-performance gasoline its zing. Because many of the building blocks of life are aromatics, they are also planar. Such aromatics include the nucleotides that make up our DNA.

Not all carbon rings lie flat. When the atoms of a ring are connected by single covalent bonds (instead of double bonds), the ring can bend into a structure that looks a little like a chair or a boat (Figure 3-4).

Carbon atoms form long hydrocarbon chains and five- or six-carbon rings.

Functional Groups: How Do the Small Molecules Differ from One Another?

The structure of a molecule determines its biological function. Just as the parts of a car or a bicycle must fit together for the machine to work, the molecules in a cell must also fit together or interact for the cell to work. Every molecule has a distinctive shape and charge distribution. And every molecule interacts with water and other molecules in unique ways.

The special properties of a molecule depend on the arrangement of its atoms. In analyzing the structures of many organic molecules, chemists have repeatedly found standard small groupings of atoms called **functional groups,** which help predict the properties of a molecule.

Functional groups are comparable to the standardized parts of a Swiss Army knife, which may include, for example, a knife, a corkscrew, and a can opener (Figure 3-5). A Swiss Army knife may have only one or two such attachments or it may have many, in a variety of combinations. In the same way, a biological molecule may have a few functional groups or many, in a variety of combinations. The exact combination of functional groups on a biological molecule profoundly affects the way the molecule behaves.

Figure 3-5 shows the structures of four simple molecules: ethane, ethanol, acetic acid, and aminoethane. Each molecule contains two carbon atoms and different numbers of hydrogen atoms. Of these molecules, ethane has no functional groups, while each of the other three molecules has a different functional group. These are a hydroxyl group, a carboxyl group, and an amino group.

Ethane has the simplest structure: two carbon atoms, each with three hydrogens attached. It is an odorless gas that makes

Glucose molecule (C$_6$H$_{12}$O$_6$)

Figure 3-4 Most rings bend. Like benzene, the simple sugar glucose consists of a ring of atoms. But while benzene's double bonds keep it flat, glucose's single bonds easily twist around, allowing the molecule to fold into a "chair" structure *(left)* or a "boat" structure *(right)*. The green hexagon is our icon for glucose.

up about 9 percent of natural gas, the kind we use to heat homes (Figure 3-5A). In high concentrations, ethane has a narcotic effect, but it is not particularly toxic.

Ethanol—the kind of alcohol found in beer, wine, and liquor—looks like ethane except that in place of one hydrogen atom is a **hydroxyl,** or OH group (Figure 3-5B). That single oxygen atom makes ethanol very different from ethane. Ethanol (or ethyl alcohol) is a sweet-smelling liquid that is comparatively toxic. It is used as an antiseptic to kill both prokaryotic and eukaryotic cells. Ethanol is also used as an industrial solvent, as an additive in gasoline, and, in veterinary medicine, to kill nerve tissue.

Ethanol's most familiar use is in beverages such as beer and liquor. In small amounts, it gives a pleasant high for a short period. In large amounts, ethanol causes nausea, vomiting, mental excitement or depression, loss of coordination, stupor, coma, and even death.

The hydroxyl group in alcohol is an extremely common functional group, which easily forms hydrogen bonds. Because oxygen is more electronegative than hydrogen, electrons in a hydroxyl group spend more time near oxygen than hydrogen. This leaves the hydrogen with a slight positive charge, so it readily forms a hydrogen bond with the slightly negative oxygen atom in a water molecule. As a result, ethanol is extremely hydrophilic and dissolves readily in water. In contrast, ethane, which has no hydroxyl group, does not interact much with water. Ethane bonds are all nonpolar, and ethane molecules do not interact much, even with one another. Most molecules that contain an OH group attached to a carbon atom—including other alcohols and also sugars—have properties similar to ethanol's.

We can take ethane and lop off one carbon with its three hydrogens and substitute a **carboxyl** group (—COOH). The result is acetic acid, the acid that gives vinegar its familiar sour smell and sharp bite (Figure 3-5C). Just as the OH group characterizes alcohols and sugars, the carboxyl group characterizes organic acids.

The carboxyl group is acidic because it gives up its hydrogen easily. Both of the two oxygens in a carboxyl group pull electrons toward them so strongly that the hydrogen atom actually loses the electron that binds it to the carboxyl group. The hydrogen ion breaks away, leaving behind a COO$^-$ ion.

Suppose now that, instead of a hydrogen atom or a hydroxyl or carboxyl group, we substitute an **amino** group, a single atom of nitrogen bonded to two atoms of hydrogen (NH_2). An amino group tends to make molecules basic because it attracts hydrogen ions, forming NH_3^+ ions. Substituting an amino group for one of the hydrogen atoms in ethane creates aminoethane (Figure 3-5D). Like ethanol, aminoethane is a liquid, but it is otherwise very different. Highly alkaline, aminoethane's powerful vapors smell intensely of ammonia and can severely irritate the skin, eyes, and mucous membranes. Aminoethane is 20 times more toxic than ethanol.

Aminoethane belongs to a class of compounds called amines. Amines all have amino functional groups and nearly all smell terrible or at least unpleasant. Two of the most offensive amines are putrescene and cadaverine, which together give memorable odors to rotting fish, aging corpses, urine, semen, and bad breath. Table 3-1 shows the properties and structures of half a dozen of the most important functional groups. Many more functional groups exist that are not shown.

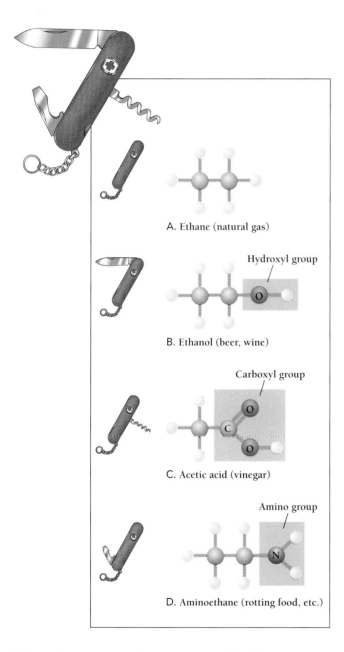

A. Ethane (natural gas)

B. Ethanol (beer, wine) — Hydroxyl group

C. Acetic acid (vinegar) — Carboxyl group

D. Aminoethane (rotting food, etc.) — Amino group

Figure 3-5 Functional groups are like the different attachments on a Swiss Army knife. Functional groups determine the structure and function of a molecule. A. In the nontoxic gas ethane, only hydrogen atoms attach to the two-carbon chain. B. In ethanol, a hydroxyl (OH) group transforms the carbon chain into an alcohol, a toxic liquid. C. In acetic acid, a carboxyl (COOH) group transforms the two-carbon chain into an organic acid. D. In aminoethane, an amino group (NH_2) makes the molecule alkaline, or basic. Like many other amines, aminoethane is strong smelling and toxic.

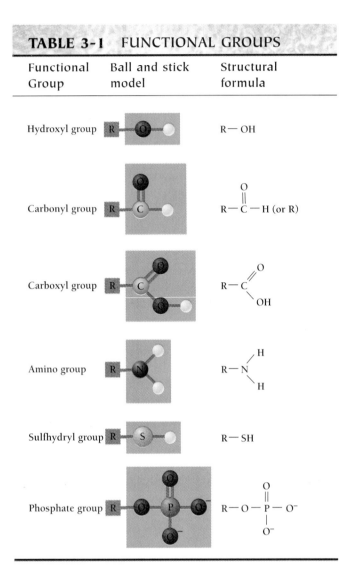

A small number of functional groups determine the properties of an array of biological molecules.

How Do Cells Build Complex Molecules?

We are now in a position to understand the properties of many organic molecules. Each building block has a carbon skeleton to which are attached hydrogen atoms or functional groups. These groups determine the properties of the molecule.

Some of a molecule's most important properties depend on how it interacts with water. Recall from Chapter 2 that whether a molecule is hydrophobic, hydrophilic, or amphipathic depends on the polarity of the molecule's bonds. Polar bonds have a positive charge at one end that can form hydrogen bonds with the negative oxygen in a water molecule. This is why polar molecules dissolve easily in water.

As we mentioned earlier, almost all the structures of a cell are built from combinations of the 35 small-molecule building blocks listed in Table 3-2. These building blocks fall into four classes: (1) **lipids,** fatty or oily nonpolar compounds that tend to dissolve in oils, but not in water; (2) **sugars,** molecules that have two hydrogen and one oxygen atom (water) for every atom of carbon; (3) **amino acids,** molecules that contain both amino and carboxyl groups; and (4) **nucleotides,** molecules that each consist of a nitrogen-containing aromatic ring, a sugar, and a molecule of phosphoric acid.

Of the four kinds of building blocks, three kinds link end to end in long chains, just like a child's plastic "pop beads." Sugars link together to form polysaccharides, nucleotides form nucleic acids, and amino acids form proteins. Only lipids do not form long chains.

In proteins and nucleic acids, the chains are simple, unbranched strings of small molecules (Figure 3-6A). In contrast, polysaccharide chains formed from sugars are sometimes highly branched (Figure 3-6B). Even though proteins and nucleic acids are unbranched, they form much more complex molecules than polysaccharides. Protein chains can coil and fold into complex three-dimensional shapes. As Stanley Prusiner showed with scrapie PrP, how a protein folds is crucial to the way it behaves in a living organism.

Cells build nearly all complex molecules from four types of building blocks.

What Links the Building Blocks of Life?

When we digest a slice of bread, we break its carbohydrates into sugars and its proteins into amino acids. At the same time, we also break down and rebuild the proteins of our own skin, muscles, and bones. All organisms continually break down macromolecules and reuse the building blocks.

Organisms have to assemble and disassemble macromolecules easily. The bonds that hold macromolecules together must be strong enough so that the macromolecules will not fall apart. But the bonds must be weak enough so that organisms can easily take them apart. Like children's Lego bricks, the building blocks of life are easily put together and easily taken apart.

Amazingly, biological building blocks snap together just as easily as Lego bricks. The building blocks of all the major macromolecules join by the same simple chemical reaction. In every case, enzymes remove two hydrogen atoms and one oxygen atom from between pairs of building blocks, forming a bond. Removing two hydrogens and an oxygen—the equivalent of a molecule of water—is called a **dehydration reaction** (Figure 3-7A).

The linking of two building blocks by a dehydration reaction is also sometimes called a "condensation" reaction. Amino acids are "condensed" into polypeptides in the same sense that milk is condensed by the removal of water. In each case, something is made more compact by removing water.

Figure 3-6 Chains of small molecules form proteins, nucleic acids, and polysaccharides. A. Proteins and nucleic acids are simple, unbranched chains. B. Polysaccharides are more complex branching chains of simple sugars.

To take apart macromolecules, cells reverse the dehydration process. Enzymes detach each small molecule from a macromolecule by adding one molecule of water, a process called **hydrolysis** [Greek, *hydro* = water + *lysis* = breaking] (Figure 3-7B).

Although all the building blocks are joined by similar dehydration reactions, the exact bonds that form are different in each case. For example, sugars form *glycosidic bonds,* while amino acids form *peptide bonds.*

Enzymes link two building blocks by taking away the equivalent of one water molecule. Adding a water molecule breaks a building block from a chain.

LIPIDS INCLUDE A VARIETY OF SMALL NONPOLAR COMPOUNDS

Look around your kitchen and you'll find lipids. Oils, fats, and candle wax, to name a few, are all lipids. Because lipids contain more chemical energy per gram than other biological molecules, lipids often serve as energy stores.

One characteristic of lipids is that they dissolve poorly in water but dissolve well in nonpolar solvents such as gasoline or olive oil. The reason that lipids do not dissolve in water is that they have few functional groups that are polar. The limited solubility of lipids in water explains why removing a butter or gravy stain often requires dry cleaning, that is, treatment with nonpolar organic solvents such as benzene.

But not all lipids are hydrophobic. Some are amphipathic: they contain a polar functional group as well as nonpolar chains of carbon and hydrogen atoms. An important class of amphipathic lipids is the **fatty acids**—a major component of the membranes that enclose cells.

Fatty acids differ from one another both in the number of carbon atoms they contain and in the numbers of single and double bonds. The hydrocarbon chain in a fatty acid consists of carbon atoms linked either to one or two other carbon atoms and to hydrogen atoms. When all the bonds between carbon atoms are single bonds, the hydrocarbon chain contains the maximum number of hydrogen atoms and is said to be **saturated**

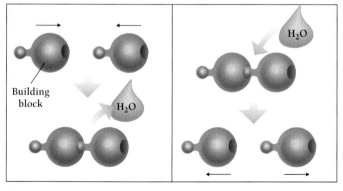

A. Dehydration   B. Hydrolysis

Figure 3-7 Dehydration and hydrolysis. Two simple reactions allow molecular building blocks to snap together or apart. A. In a dehydration reaction, building blocks join together when a hydrogen is lost from one molecule and a hydroxyl is lost from the other. Each time two building blocks join, the equivalent of one molecule of water is lost. B. In a hydrolysis reaction, the addition of one molecule of water breaks the bond between each pair of small molecules.

(Text continued on page 56)

TABLE 3-2 THE BUILDING BLOCKS OF LIFE

Like children's Lego™ bricks, building block molecules easily snap together to form larger molecules. The Lego™ icon represents the 35 building blocks of life—20 amino acids, 6 lipids, 5 nucleotide bases, 3 simple sugars, and a phosphate group.

Amino acid

Amino acids with basic side chains

Lysine (Lys, K)   Arginine (Arg, R)   Histidine (His, H)

Amino acids with acidic side chains

Aspartic acid (Asp, D)   Glutamic acid (Glu, E)

Amino acids with uncharged polar side chains

Asparagine (Asn, N)   Glutamine (Gln, Q)   Serine (Ser, S)   Threonine (Thr, T)   Tyrosine (Tyr, Y)

Amino acids with nonpolar side chains

Glycine (Gly, G)   Alanine (Ala, A)   Valine (Val, V)   Leucine (Leu, L)   Isoleucine (Ile, I)

Phenylalanine (Phe, F)   Methionine (Met, M)   Cysteine (Cys, C)   Proline (Pro, P)   Tryptophan (Trp, W)

Glucose

Ribose

Deoxyribose

Bases

Pyrimidines (single ring)

Cytosine   Uracil   Thymine

Purines (two rings)

Adenine   Guanine

Phosphate group

Cholesterol

Stearic acid (saturated)

Oleic acid (unsaturated)

Phosphatidylcholine

Phosphatidylinositol

Glycerol

LIPIDS

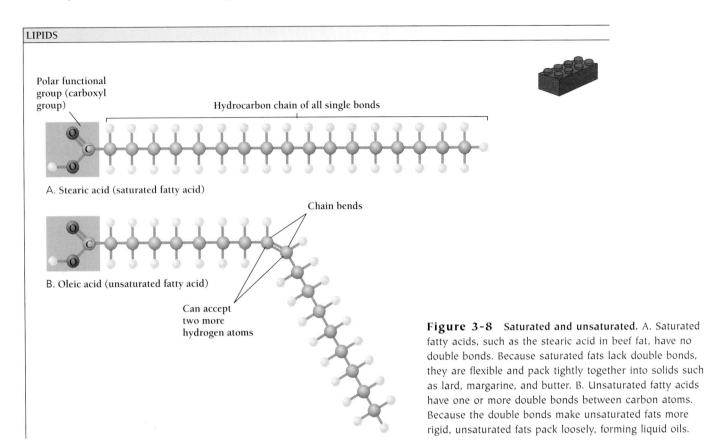

Figure 3-8 Saturated and unsaturated. A. Saturated fatty acids, such as the stearic acid in beef fat, have no double bonds. Because saturated fats lack double bonds, they are flexible and pack tightly together into solids such as lard, margarine, and butter. B. Unsaturated fatty acids have one or more double bonds between carbon atoms. Because the double bonds make unsaturated fats more rigid, unsaturated fats pack loosely, forming liquid oils.

with hydrogen atoms. Stearic acid, an important component in beef fat, is a highly saturated fatty acid (Figure 3-8A).

When a fatty acid contains one or more double bonds between adjacent carbons, and is therefore missing hydrogen atoms, the fatty acid is said to be **unsaturated.** Oleic acid, an important component of vegetable oils, is unsaturated because it has double bonds (Figure 3-8B). A fatty acid with just one double bond is said to be monounsaturated. A fatty acid with many double bonds is said to be **polyunsaturated.**

Fats, waxes, and other lipids that have saturated hydrocarbon chains are solid at room temperature. In contrast, polyunsaturated lipids tend to be liquid at room temperature, and we call them oils. The reason is in the bonds. The single C — C bonds in saturated fats rotate freely, allowing the chains to pack tightly together to form a solid. In contrast, the double C = C bonds in saturated lipids are more rigid and tend to form kinky polycarbon chains that do not pack well. The result is a fluid, or oil. In general, animals contain more fats and plants contain more oils.

Food manufacturers sometimes add hydrogen atoms to polyunsaturated oils in order to make them solid at room temperature. Cookies, chips, and other products made with such "hydrogenated" fats tend to be crisper than if made with oil. Butter would work, but hydrogenated fats are less expensive.

Both naturally and artificially saturated fats seem to increase cholesterol levels and overall risk of heart disease. Nutritionists generally recommend that we avoid saturated fats and stick to diets rich in polyunsaturated and monounsaturated oils, such as those found in nuts, grains, and olive oil. However, this area of research is surprisingly complex. For example, even though stearic acid, found in beef and chocolate, is a saturated fat, it apparently does not increase blood cholesterol levels.

Lipids, which include both fats and oils, are usually oily to the touch, insoluble in water, and high in energy. One class of lipids, the fatty acids, may be more or less "saturated."

How Are Fatty Acids Linked Together?

Glycerol is a simple three-carbon molecule with three hydroxyl groups. Each hydroxyl can attach to the carboxyl carbon of a fatty acid by means of a dehydration reaction (Figure 3-9A). A glycerol with all three hydroxyl groups attached to fatty acids, so that the glycerol has three, long hydrocarbon "streamers," is a **triacylglycerol** (Figure 3-9B). When a triacylglycerol (also called a triglyceride) forms, the hydrophilic character of the carboxyl group is lost. As a result, triacylglycerols are even less soluble in water than fatty acids. Most of the fats that humans and other animals store, as well as the oils of plants, consist of triacylglycerols.

LIPIDS

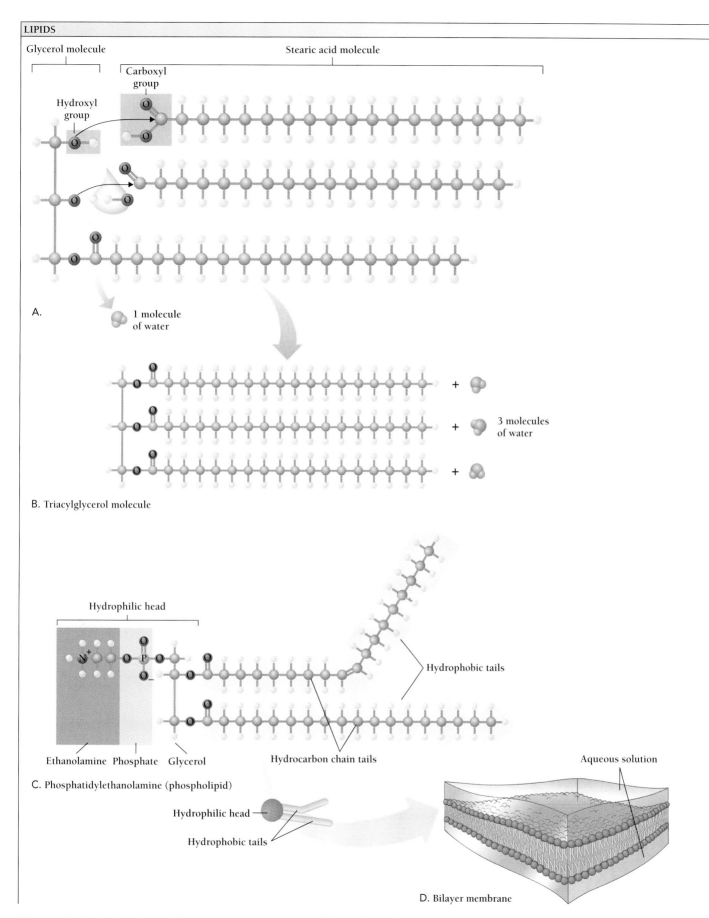

Figure 3-9 Putting fatty acids to work. A and B. A triacylglycerol (or triglyceride) forms when three fatty acid chains attach to glycerol. One molecule of water is lost at each attachment. C. A phospholipid forms when just two fatty acids attach to glycerol. D. Phospholipids form membranes in which the hydrophilic heads face out into the aqueous environment and the hydrophobic tails hide together in the middle.

BOX 3.1

The Biochemistry of Cholesterol

A quick scan of any health magazine would leave most people with the impression that cholesterol is always bad for you. In fact, you can't live without it. The outer membranes of your cells are rich in this lipid, having as much as one cholesterol molecule for every phospholipid molecule. Cholesterol's bulky shape contributes to membrane fluidity, in the same way the double bonds in phospholipids' fatty acid chains contribute to the fluidity of vegetable oils. Unlike phospholipids, however, cholesterol is hydrophobic, which means that, by itself, it doesn't dissolve well in blood.

Cholesterol also serves as a chemical precursor to the sex hormones progesterone, testosterone, and estradiol (a form of estrogen). All steroids contain the large, nonpolar ring structure of cholesterol and are fat soluble (hydrophobic).

Cholesterol comes from two sources: the liver and the diet. The liver makes cholesterol as a component of bile, a detergentlike substance secreted into the upper intestine to aid digestion. In addition, most Americans consume about 550 mg of cholesterol a day, all of which comes from butter, eggs, meat, and other animal products. Some of this cholesterol passes through the digestive tract into the feces, and some gets absorbed into the blood. The blood plasma of a typical college-aged American contains about 180 mg/deciliter (1 deciliter = 100 ml) of cholesterol. Total plasma cholesterol rises slowly with age to peak at about 230 mg/dl in men and 250 mg/dl in women by age 55.

Cholesterol doesn't travel alone in the blood. If it did, its hydrophobic nature would cause it to form big clumps, like oil in a bottle of salad dressing. Instead it travels in lipoprotein complexes. The most common are called low-density lipoproteins, or LDLs. These are responsible for delivering cholesterol to cells and tissues where it is needed. High-density lipoprotein, HDLs, remove cholesterol from cells and transport it to the liver, where it is secreted into the digestive juices, or "bile."

Sometimes people call LDL cholesterol "bad" and HDL cholesterol "good." This is because if not enough HDLs are present to remove cholesterol, the excess builds up along the walls of arteries, contributing to deposits called "plaques." Plaques cause hardening of the arteries, or atherosclerosis (Figure A). Atherosclerosis reduces blood flow to the organs, including the heart, and damages the arteries. Tiny bits of plaque may break off and lodge in an artery, blocking blood flow to the heart and causing a heart attack.

While limiting cholesterol intake is a good idea, one of the best indicators of the chance of developing heart disease from cholesterol deposits in the arteries is not total cholesterol, but the ratio of bad to good cholesterol. The average American has an LDL-to-HDL ratio of 5:1. Reduce that ratio to 3.5:1, by increasing plasma HDL, and you'll cut your risk of heart disease in

---

A glycerol molecule with a phosphate group attached in place of the third hydroxyl group and with just two fatty acid chains instead of three is a **phospholipid** (Figure 3-9C). Phospholipids are highly amphipathic. Each one has a hydrophobic "tail," which consists of the hydrocarbon chains of two fatty acids, and a hydrophilic "head," which contains the charged phosphate group.

In water, the hydrophobic tails of phospholipids clump together to form an oily interior, while the hydrophilic heads point outward into the surrounding water. Both the outer

Figure 3-10 Cholesterol, a building block of life. A. The basic four-ring structure of cholesterol. B. With the substitution of a few functional groups, cells transform cholesterol into progesterone, a female sex hormone. C. Progesterone becomes testosterone, a sex hormone that gives males many of their most striking characteristics. D. A few more changes, and testosterone becomes estradiol, a powerful form of the female hormone estrogen.

Figure A Atherosclerosis results when plaque deposits narrow and block our arteries. (a) One artery is clear. (b) The other is blocked by plaque. *(a, Cabisco/Visuals Unlimited; b, Ober/Visuals Unlimited)*

half. An even better predictor of heart disease are high levels of IDLs, intermediate-density lipoproteins, which, for now, few doctors measure.

How can people increase HDLs? The following items all tend to raise HDLs: female sex hormones; regular aerobic exercise; monounsaturated fats, such as olive oil (instead of, not in addition to, saturated fats); weight loss; and certain forms of fiber. On the other hand, tobacco smoke lowers plasma HDL levels and also contributes to heart disease in other ways.

But don't assume science has found all the answers. Research in the area of diet and heart disease raises baffling new questions almost daily. For example, while high-fat diets are clearly bad for us, people on extremely low-fat diets have increased numbers of both LDLs and IDLs. And low-fat meals can have other effects. In another study, 12 overweight teenagers were given either a high-fat meal or a low-fat meal, of equal calories. Over the next five hours, the adolescents were allowed to eat whatever they liked. Those who had eaten the low-fat meal ate twice as many calories as those who had eaten the high-fat meal. Apparently, the super-low-fat meal made them hungrier.

---

membranes of cells and their interior membranes are composed primarily of phospholipids. All of these membranes have a hydrophobic interior and a hydrophilic exterior (Figure 3-9D).

Another class of lipids is the **steroids**—hydrocarbon chains with four interconnected rings (Figure 3-10). Many of the hormones responsible for sexual development and reproductive functions in animals are steroids. The starting point for the synthesis of all steroids is **cholesterol**—a hydrocarbon with the same pattern of four interconnected rings (Box 3.1). Small differences in the attached functional groups can make enormous differences in a steroid's biological properties. For example, cells make the female hormone progesterone from cholesterol by replacing an OH group with an oxygen atom with a double bond, along with some other changes (Figure 3-10A and B). Cells then make testosterone from the progesterone and then estrogen from the testosterone, as needed (Figure 3-10C and D). All four lipids have profoundly different effects in the body, yet differ by only a few functional groups.

Most of the fatty acids in organisms are linked to glycerol to form lipids as diverse as triacylglycerols, phospholipids, and steroids.

HOW DO SUGARS FORM POLYSACCHARIDES?

We have all been told to avoid eating too much sugar. But while this may be sound advice, avoiding all sugar is neither possible nor desirable. Sugar is the fundamental energy storage molecule of all living organisms.

The smallest sugars are "simple sugars," or **monosaccharides** [Greek, *mono* = one + *saccharine* = sugar), which can have from three to nine carbon atoms. Two of the most important simple sugars are **glucose** and fructose, which have six carbon atoms each (Figure 3-11). Cells link sugar molecules using a dehydration reaction. The result is a **glycosidic bond,** in which one oxygen atom forms a bridge between carbon atoms on two adjacent sugar molecules. Two sugars can be linked to form a **disaccharide** [Greek, *di* = two + *saccharine* = sugar), such as **sucrose,** or table sugar (Figure 3-11). Several sugars can form short chains, called **oligosaccharides** [Greek, *oligos* = few + *saccharine* = sugar]. The longest chains of sugars, called **polysaccharides** [Greek, *poly* = many + *saccharine* = sugar], may include thousands of simple sugars. Polysaccharides include the starch in bread and the cellulose in wood and paper.

Besides glucose, two other important sugars are **ribose** and **deoxyribose,** simple five-carbon sugars that combine with

SUGARS

Figure 3-11 Life is sweet. A simple dehydration reaction joins the two simple sugars glucose and fructose, forming the disaccharide sucrose—regular table sugar.

other small molecules to form the building blocks for DNA (deoxyribonucleic acid) and RNA (ribonucleic acid), as shown in Table 3-2.

All sugars and polysaccharides are **carbohydrates,** molecules that contain the equivalent of one water molecule for every carbon atom. Most of the carbons in a carbohydrate have a hydrogen atom on one side and a hydroxyl group on the other, giving big carbohydrates hundreds or even thousands of hydroxyl groups. Because these hydroxyl groups easily form hydrogen bonds with water molecules, most carbohydrates dissolve well in water. Stir some sugar into a cup of hot tea and you'll see.

How Do Organisms Use Sugars To Store Energy?

Organisms use monosaccharides and disaccharides to store energy for a few hours, and they use longer polysaccharides, such as starch, to store energy for weeks, months, or years. In the same way, we keep our money in cash if we want to buy something right away, but put it in the bank if we want to save our money for later.

The single most important simple sugar is glucose, what we sometimes call blood sugar. Almost all biochemical reactions are in some way connected to glucose. Synthesizing glucose is a way of storing energy. Breaking down glucose is a way of releasing energy. The energy from glucose drives everything that happens in an organism from cell division to long division. So, linking long chains of glucose molecules is just another way for organisms to store energy.

Animals store energy in a long chain of glucose molecules called **glycogen.** Because glycogen has many short branches, it is easy for enzymes to quickly pick off many glucose molecules at a time. Instead of breaking down the whole molecule to get energy, enzymes harvest glucose from the ends of the branches as needed.

Plants store energy as **starch,** another large polysaccharide made of glucose molecules. Potatoes, pasta, bread, rice, and many other familiar foods contain lots of energy-rich starch, which may be either branched or unbranched. Two of the most common kinds of starch are amylopectin and amylose. Smaller amylose consists of hundreds to thousands of glucose molecules in simple, unbranched chains. Larger amylopectin consists of up to 50,000 glucose molecules in highly branched chains (Figure 3-12).

How Else Do Organisms Use Polysaccharides?

Plants and animals use polysaccharides not only as a way to store energy, but also as a building material. Arthropods such as butterflies and crabs construct skeletons from the polysaccharide **chitin.** Plants build their bodies using **cellulose**—the major component of wood, paper, cotton, and linen.

The structure and properties of polysaccharides depend both on the kinds of sugars from which they are made and on the way in which the sugars are joined. Although cells can use different sugars to make polysaccharides, many of the most important consist entirely of glucose. Starch, cellulose, and glycogen are all made of glucose alone. The differences in structure come from the way the glucose molecules are linked.

For us humans, the most important difference between cellulose and starch is that—due to the arrangement of the glycosidic bonds between the glucose molecules—one polysaccharide is digestible and the other is not. In cellulose, the links between glucose are oriented differently from those in starch. The result is an extremely straight—and strong—chain of glucose molecules that makes an excellent building material for plants. But the digestive enzymes that work well on starches cannot break the bonds of cellulose. Different enzymes are needed to digest cellulose.

Because most animal cells do not have enzymes that break the links in cellulose, they cannot digest it. The few animals that can digest cellulose—cows and termites, for example—do so with the help of microorganisms that have the needed enzymes. For the rest of us, the cellulose in bran flakes and prunes is just "roughage" or fiber—a water-absorbing material

Figure 3-12 *An apple contains three important polysaccharides.* The starch in most plants consists mainly of two polysaccharides, amylopectin and amylose. Amylopectin molecules are huge, branching chains consisting of up to 50,000 glucose building blocks. Amylose molecules are smaller simple chains of a few hundred or a few thousand glucose molecules. An apple also contains cellulose, an indigestible polysaccharide. The bonds in cellulose are different from those in starch, and we do not have the enzymes necessary to break them. (*Photo, Harcourt Photo Library*)

that keeps things moving through the digestive tract. Indeed, cellulose is one of those no-calorie foods that the food industry uses to thicken milkshakes or to make low-calorie breads. Because we cannot digest it, these foods are said to be "low in calories." But if we were cows, these foods would have plenty of calories.

One of the most important uses for sugars is in short polysaccharides called oligosaccharides. Oligosaccharides are often tacked on to proteins to form **glycoproteins,** which just means "sugar protein." Glycoproteins are commonly found on the outsides of cell membranes. Like big signs, they help cells to recognize one another and communicate. Inside cells, glycoproteins tacked onto newly made proteins act as address labels that tell a cell where to ship the new proteins.

HOW DO NUCLEOTIDES LINK TOGETHER TO FORM NUCLEIC ACIDS?

Nucleotides are the building blocks of RNA (ribonucleic acid) and DNA (deoxyribonucleic acid), the two kinds of molecules that carry genetic information. Each nucleotide building block consists of three parts: a sugar (ribose or deoxyribose), one or more phosphate groups, and a nitrogenous base.

Most cells use just five nitrogenous bases (Figure 3-13A). Each **nitrogenous base** consists of either one or two aromatic rings, each of which includes two nitrogens. Attached to these rings are hydrogen atoms and various polar functional groups. Like all bases, nitrogenous bases accept hydrogen ions.

The five bases are divided into two types: the two-ring **purines**—adenine and guanine—and the one-ring **pyrimidines**—cytosine, uracil, and thymine. Uracil and thymine are almost identical, differing by the presence of a methyl group ($—CH_3$) in thymine.

Nucleotides serve many important biological roles. First and foremost, they are the building blocks of RNA and DNA (Figure 3-13B). In addition, nucleotides with one, two, or three phosphate groups play a central role in the production of energy within cells. **ATP (adenosine triphosphate),** which has three phosphates, is the immediate source of energy for all biological processes. Energy from the breakdown of glucose is stored in ATP. Cells then release energy from ATP by hydrolyzing its phosphate-phosphate bonds. The resulting energy drives all the chemical reactions of a cell (Figure 3-14).

A nucleotide consists of a sugar, a phosphate, and a base. Nucleotides are essential to life both as the units of DNA and RNA and for energy storage.

How Are the Nucleotides of a Nucleic Acid Linked Together?

Like sugars, nucleotides can connect one to another by a simple dehydration reaction (Figure 3-15). With the removal of one molecule of water, nucleotides form **phosphodiester bonds,** which connect the nucleotides into long, unbranched chains called **nucleic acids.** Two nucleic acids—DNA and RNA—are so important in the storage and transmission of genetic information that this book devotes eight chapters to them.

DNA serves as a library of information for every cell. Along the thin strands of DNA are genes that carry instructions for building every protein a cell will ever need. The genes of all organisms are made of DNA. The only genes that are not DNA are those of a few viruses—including those responsible for influenza and AIDS—whose genes are made of RNA. But viruses are not organisms, for they are not made of cells. In cells, RNA acts only as an interpreter, carrying information from the DNA to the rest of the cell.

In a chain of nucleotides, the sugar and phosphate parts of each nucleotide link together alternately to form a sugar-phosphate "backbone." RNA consists of a single long backbone. DNA usually consists of two parallel backbones connected to each other by thousands of hydrogen bonds

NUCLEOTIDES

A. Five nitrogenous bases

B. Nucleotide (AMP)

Figure 3-13 The five nitrogenous bases. A. The purines have double rings of four nitrogen atoms and five carbon atoms. The pyrimidines have one ring of two nitrogen atoms and four carbon atoms. B. Each nitrogenous base can join a sugar and a phosphate group to form a nucleotide, such as adenosine monophosphate (AMP). AMP is one of just four nucleotide building blocks that form long chains of DNA.

NUCLEOTIDES

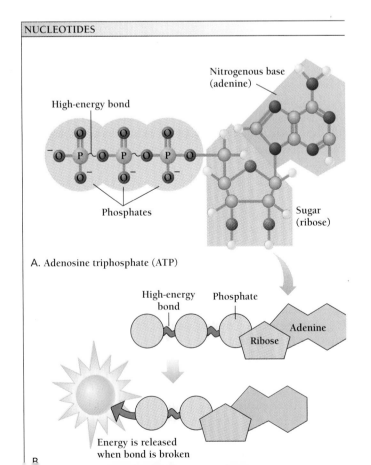

between adjacent bases. The result is a molecule that resembles a long ladder, with each rung consisting of a pair of nucleotide bases (Figure 3-16). The whole ladder is twisted along its long axis to form a "double helix."

DNA and RNA each consist of just four kinds of nucleotide bases. Three bases—adenine, guanine, and cytosine—occur in both DNA and RNA. But DNA's fourth base is thymine, and RNA's fourth base is uracil. The structures of RNA and DNA are important to the way that they function as information molecules. In Chapters 10 and 11, we discuss the structure and function of DNA and RNA in great detail.

The long chains of nucleotides in the information molecules DNA and RNA are linked—sugar-phosphate-sugar—by phosphodiester bonds.

Figure 3-14 For all organisms, ATP is the common currency of energy. A. The nucleotide ATP consists of the nitrogenous base adenine, the sugar ribose, and three phosphate groups. B. ATP releases energy for use in chemical reactions when one of the phosphate groups breaks off.

Nucleotide Collection

NUCLEOTIDES

A. Formation of a phosphodiester bond

Figure 3-15 Building a library of information.
A. All organisms store genetic information in long chains of nucleotides—either DNA or RNA. The nucleotide building blocks join by a dehydration reaction to form a phosphodiester bond, losing one water molecule in the process. B. RNA consists of a single chain of nucleotides.

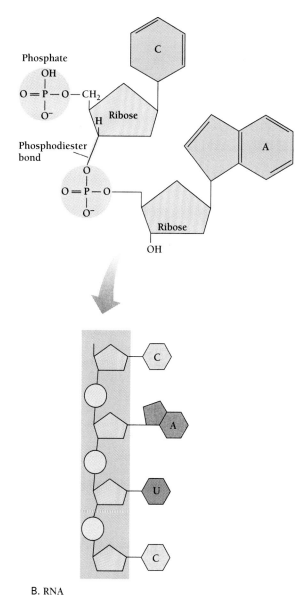

B. RNA

HOW DO AMINO ACIDS LINK TO FORM POLYPEPTIDES?

Proteins perform thousands of jobs. They hold the body together, lend support to bones, transport molecules and ions in and out of cells, and much more. Every organism has hundreds to thousands of different proteins. The human body alone contains at least 80,000 different proteins.

Collagen, the most common protein in the human body, gives strength to connective tissues in tendons, ligaments, bones, muscle, and skin. Dissolved in sugar water, collagen gives Jell-O its jiggle. Yet, in the body, collagen is as strong as steel. Elastin makes skin stretchy, while stringy keratin toughens hair, horns, nails, and claws.

Some proteins taxi materials from one place to another. Hemoglobin, for example, binds to oxygen in the lungs and transports it, by way of the blood, to the cells in the rest of the body. Other proteins act as hormones, antibodies, or poisons (rattlesnake venom, for example). Still others transport molecules across the membranes of cells. Some proteins—often glycoproteins with attached oligosaccharides—act as identification tags on the surfaces of cells. Other proteins help cells to communicate with one another. Protein "signaling molecules" carry messages from one cell to another, and protein "receptors" on cell surfaces pick up those signals.

One of the most important roles for proteins is as **enzymes**—biological molecules that speed up reactions between other molecules. Enzymes are indispensable to life. Without them, many chemical reactions would occur too slowly to sustain life. Enzymes copy and repair DNA, help cells to extract energy from sugars, digest our food, and much more. Enzymes in the mouth, for example, help break down the polysaccharides in bread into simple sugars. That is why bread gets sweeter and sweeter as you chew it. We discuss enzymes in greater detail in Chapter 5.

Figure 3-16 The DNA ladder. The DNA of our genes consists of two long chains of nucleotides linked together by bonds between the nitrogenous bases of each nucleotide. The bonds are often compared to the rungs of a ladder. The whole, long double chain is twisted like a phone cord into a helix.

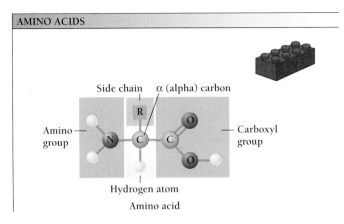

Figure 3-17 What's in an amino acid? An amino acid includes a carboxyl group, an amino group, a hydrogen atom, and a side chain (R). All four are attached to a single, central carbon atom, called the "α carbon."

Cells contain thousands of proteins, each of which has a different role to play.

What Distinguishes the 20 Amino Acids from One Another?

Every protein is made of chains of amino acids called **polypeptides.** A protein may be made of one simple chain or of several different chains. A chain may consist of only a few hundred amino acids or 10,000 or more. Regardless of how long the chains are or how many of them there are, cells build proteins one amino acid at a time.

Each of the 20 amino acids has the same basic structure. Each contains a carboxyl group, which makes it an acid, and an amino group, which makes it a base (Figure 3-17). Both the carboxyl group and the amino group are attached to the same carbon atom, called the **alpha (α) carbon.**

Each amino acid also has a characteristic side chain, called an R group, which is also attached to the α carbon atom. *The α carbon, the carboxyl group, and the amino group are exactly the same in 19 of the 20 amino acids.* (The one exception is proline.) But each amino acid has a different R group. Table 3-2 shows the 20 amino acids found in proteins with their 20 different R groups.

R groups vary in length and in their functional groups. Some are hydrophobic, some hydrophilic. Some have a positive charge, some have a negative charge, and some are neutral. The R group side chains establish the distinctive properties of each amino acid and the properties of the resulting polypeptide.

Just as the 26 letters of our alphabet allow a nearly infinite number of unique sentences, the 20 amino acids allow a nearly infinite number of different proteins. A polypeptide just 100 amino acids long can be made in 10^{130} (or 20^{100}) ways by varying the order and selection of amino acids. Recall that 10^{130} is 1 followed by 130 zeros.

Although amino acids usually end up folded inside of massive proteins, such as hemoglobin, a few perform other roles. For example, some amino acids are neurotransmitters—chemicals secreted by nerve cells that convey information to cells.

All amino acids have a carboxyl group, an amino group, and an R group, all attached to the α carbon. In each of the amino acids, all the parts are identical except for the R group.

How Do Amino Acids Join Together?

Like polysaccharides and nucleic acids, polypeptides are composed of chains of building blocks joined by dehydration reactions. The bond that links the amino acids together into polypeptides is called a **peptide bond.** Two amino acids join when the carboxyl group of one joins to the amino group of the next (Figure 3-18).

Each amino acid in the string is oriented in the same way as the others—amino, carboxyl, amino, carboxyl—forming an unvarying backbone of repeating atoms. The R groups hang off the sides of the polypeptide backbone. Uninvolved in the peptide bond, the R side chains may interact with one another or with other molecules in their environment. The polypep-

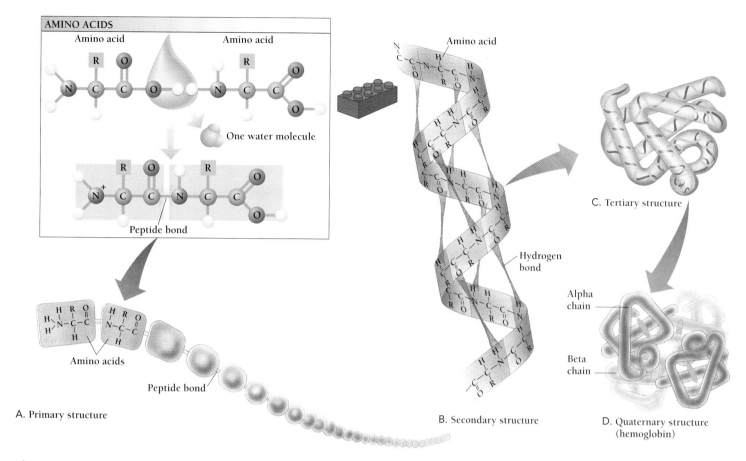

Figure 3-18 Building a protein. The usual dehydration reaction joins two amino acids by a peptide bond. A. A long chain of amino acids forms a polypeptide chain. The primary structure of a protein is simply the sequence of amino acids in its polypeptide(s). B. The secondary structure is the curling of each polypeptide into a three-dimensional form, such as the α helix shown here. C. Tertiary structure is the folding of the curled polypeptide into a more complex three-dimensional shape, such as that of the myoglobin polypeptide shown here. D. Quaternary structure is the fitting together of two or more folded polypeptides, as in the hemoglobin molecule. Hemoglobin is the major oxygen-carrying protein in the blood of many animals.

tide backbone is identical from one end to the other, except for the very beginning and very end.

Peptide bonds link amino acids together into polypeptide chains, in which the R groups hang off the sides.

What Do Proteins Look Like?

Proteins are extraordinarily diverse. For each of the thousands of jobs that proteins perform, there is a unique protein to do it. They can be as short as 50 amino acids or as long as 10,000. They can be as regular in their structure as a stack of Lego bricks or as irregular as a wad of crumpled paper.

Despite their diversity, biochemists have found ways to classify proteins. Most are dense, roundish molecules called **globular proteins.** A classic example of a globular protein is hemoglobin, the protein that carries oxygen in the blood. Like many proteins, hemoglobin is gigantic compared to the small molecules of the cells. Hemoglobin is almost 400 times as big as a molecule of glucose and more than 2000 times as big as each molecule of oxygen that it ferries from the lungs.

In addition to the globular proteins are **fibrous proteins**—long, thin molecules such as the keratin in hair or the collagen in skin and bone. Fibrous proteins provide structural support to the bodies of plants, animals, and other organisms.

Proteins have up to four levels of structure. The **primary structure** of a protein is the sequence of amino acids in each chain (Figure 3-18A). This is like the sequence of letters in a sentence. It is simply the order and kind of amino acids. The **secondary structure** is the coiling, bending, or folding of individual sections of the polypeptide chain. For example, polypeptides can coil into an **α helix,** which resembles the coil of a phone cord (Figure 3-18B).

Other secondary structures include the β pleated sheet and the collagen helix, found in collagen. The **β pleated sheet** is a long chain that folds back and forth against itself—like the coils of a snake or the long lines of people at an airport.

Unlike an α helix, which consists of just one polypeptide chain, a **collagen helix** consists of three polypeptide chains wound around each other (Figure 3-19).

Both globular proteins and fibrous proteins contain α helices and β pleated sheets. But fibrous proteins often have many rigid α helices, whereas globular proteins are made of a mixture of many kinds of secondary structures. In globular proteins, the amino acid sequences of the α helices and β pleated sheets are less regular than those of structural proteins. The polypeptide chain of a globular protein may coil into an α helix for a few hundred amino acids, then switch to a β pleated sheet, then back to an α helix. These different regions of the polypeptides of a protein are called **domains**. For example, globular proteins contain β structures, with two to five parallel sections of a single chain forming a β sheet.

The **tertiary structure**, or **conformation**, of a protein is the three-dimensional folding of an entire polypeptide chain (Figure 3-18C). If you curl a ribbon into a spiral (or helix) by scraping it with a blade, you have produced a secondary structure. If you then take the curled ribbon and tie it around a birthday present, you have given the ribbon a tertiary structure.

Many proteins also have **quaternary structure**—the fitting together of two or more folded chains (Figure 3-18D). This is analogous to piecing together several different ribbons into an elaborate bow. Just as the ribbons can be either all the same color or a mixture of different colors, proteins can be made of several identical polypeptide chains or of several different ones. Hemoglobin, for example, is made of four polypeptide chains, two of one kind of chain and two of another (Figure 3-18D).

A. Collagen molecule

B. Collagen fibers   5 μm

Figure 3-19 What is skin made of? Collagen is the major constituent of skin, tendons, ligaments, cartilage, and bones. A. Collagen is made of three helical polypeptides wound together into a superhelix. B. Collagen fibers. (*B, Professor P.M. Motta and E. Vizza/Science Photo Library/Photo Researchers, Inc.*)

Proteins may be long or short, stringy or globular. One or more polypeptides folded together into a unique shape make a protein.

HOW DO CELLS CONSTRUCT PROTEINS?

The biological role of a protein depends exquisitely on its three-dimensional shape. Because the shape of a protein is so important, the chains that make up proteins must fold in precisely the right way to form biologically active molecules. The way a polypeptide folds depends first on the sequence of building blocks in its chains. By comparison, the exact number and order of sugars in a polysaccharide has little effect on its function.

What determines the three-dimensional shape of a protein is a topic still hotly debated among molecular biologists and biochemists. They call it the "protein-folding problem." Contributions to this worldwide debate come from many sources. But the ongoing work on Stanley Prusiner's scrapie PrP has raised many important questions about protein folding that scientists might not otherwise have asked.

By making different versions of a polypeptide, biochemists have found that tiny changes in amino acid sequence can affect a polypeptide's biochemical behavior. Sometimes changing one amino acid has little effect. For example, a muscle protein from a cow can have a sequence slightly different from that of a human. Yet the difference does not affect the protein's three-dimensional structure or the way it functions in the body. Cow and human proteins differ slightly, but they function the same.

On the other hand, sometimes changing an amino acid can completely alter a protein's structure and abolish its biological activity. The kinds of amino acid changes that make the biggest differences are those that substitute amino acids with very different properties. For example, in the blood protein hemoglobin, the substitution of a hydrophobic amino acid (valine) for a hydrophilic one (glutamate) produces a defective hemoglobin molecule that crystallizes. The result is the disease sickle cell anemia. Sickle cell anemia is caused by a mistake in the nucleotide sequence of the gene for hemoglobin.

A small change in the structure of a polypeptide can enormously alter the way it behaves or have no effect at all.

How Does the Folding of a Protein Depend on Interactions Among Its Amino Acids?

Polypeptides are always linear, never branched like some of the polysaccharides. As we've seen, however, polypeptides can coil or fold into helices, sheets, balls, and other shapes. A protein may consist of a single, folded polypeptide chain or of several

Figure 3-20 What is silk made of? A. The protein fibroin that makes up silk consists of β sheets. B. In a β sheet, polypeptide chains can snake back and forth past each other, held fast by hydrogen bonds between adjacent segments. Biochemists say that the segments that run in one direction are "antiparallel" to those that run the other way. (A, K. Jakes, The Ohio State University)

B. β (beta) structure

chains folded together. In proteins with more than one chain, the individual chains are held together tightly, so that, although made of several separate molecules, the protein behaves like a single, huge molecule.

The α helix, the β structure, and the collagen helix all share an important trait. As regular repeating forms, they are strong (Figure 3-20). Their simple shapes allow them to function as standardized building materials, comparable to the wooden beams and bricks with which a house is built. All fibrous proteins, whose role is to provide mechanical support, consist largely of these simple repeating forms.

In structural proteins, the α helix, β structure, and collagen helix each result from an extremely regular sequence of amino acids in the polypeptide chain. In the collagen helix, for example, every other amino acid is glycine. This smallest of the amino acids fits snugly in the interior of collagen's triple helix.

In globular proteins, the α helix, β structure, and other regular structures are composed of nonrepeating sequences of amino acids. As a result, globular proteins are more flexible in the way they fold and often form intricate shapes that allow them to play more diverse roles than the fibrous proteins. Some globular proteins act as hormones, for example, some act as antibodies in the immune system, some act as transport proteins (e.g., hemoglobin), and some act as enzymes.

The tertiary folding of polypeptide chains results from interactions of the side chains on each amino acid. Each side chain has its particular characteristics. Some are polar, some nonpolar. Some can serve as proton donors in hydrogen bonds. Others serve as proton acceptors. Either a hydrogen bond or the attraction between positively and negatively charged side chains can stabilize a particular shape. Conversely, some interactions, such as the repulsion between two like-charged side chains, prevent certain shapes. The one characteristic of a polypeptide that most determines how it folds is the hydrophobic interactions of the nonpolar side chains.

Singly, any of these interactions contributes only slightly to a given molecular structure. Together, however, the various interactions establish a three-dimensional structure that can be extremely stable in certain chemical environments. Because the internal environment of a cell is relatively constant, each protein usually maintains its characteristic conformation. Subtle changes in the chemical environment of the cell, however, can change the shape of a protein and change the way it behaves.

Structural proteins tend to have regular sequences of amino acids and regular secondary structures, such as α helices and β structures. Globular proteins have nonrepeating amino acid sequences and irregular and unique shapes.

Are the Interactions Among Amino Acids Enough To Determine Protein Structure?

Just seconds after it forms, a polypeptide begins folding into exactly the right shape to fulfill its biological role. A polypeptide chain is as flexible as a snake and could easily take any of thousands of shapes. A short polypeptide just 100 amino acids long has 200 movable angles, not counting the R side chains.

In one study of more than 40 proteins, researchers found that about two-thirds of all hydrogen bonds link the hydrogen of one backbone amino group to the oxygen of a backbone carboxyl group. These hydrogen bonds are very specific. For example, in one protein, the 117th amino acid might link to the 245th amino acid every single time. What ensures that the 117th doesn't link to the 247th amino acid, or even the 10,005th? After all, the elements of the backbone are identical.

Despite all the ways that protein folding can go wrong, these remarkable molecules almost always seem to fold almost inevitably into their "native" functional shape. How do they do it?

The protein-folding problem has plagued biochemists for decades. In the 1950s, biochemists wondered if the sequence of amino acids was enough to exactly determine the precise three-dimensional structure of a protein. Might not some outside molecule somehow help direct the formation of a polypeptide? In that case, the shape of a polypeptide would depend not only on its amino acid sequence but also on the presence of some other "instructional" molecule.

In 1957, the biochemist Christian Anfinsen attempted to answer this question with a now classic experiment. He chose to study an enzyme called "ribonuclease," which breaks the bonds in RNA molecules. Anfinsen first exposed ribonuclease to a chemical solution that interfered with hydrogen bonds and hydrophobic interactions. This treatment changed the enzyme's shape and destroyed its ability to function. It could no longer break the bonds in RNA. When Anfinsen removed the ribonuclease from the harsh ("denaturing") solution, however, the ribonuclease spontaneously reverted back to its "native," functional shape. Once again, it could function properly, breaking bonds in RNA.

The enzyme's recovery required help from no other molecule. The amino acid sequence of ribonuclease and the right chemical environment were enough to determine the enzyme's three-dimensional structure and biological activity. Anfinsen's experiment suggested that proteins spontaneously fold into their biologically active forms, a conclusion that satisfied biochemists for many years.

Christian Anfinsen's experiment suggested that in a normal cell environment amino acid sequence alone determines the structure of a protein.

What Factors Determine How Proteins Fold?

For many years, biologists accepted the idea that in a normal cell environment, the amino acid sequence of a polypeptide implied one and only one three-dimensional structure. Yet biochemists cannot themselves predict the structure of a protein simply by knowing the amino acid sequence. What rules governed the folding has puzzled generations of biochemists.

Gradually, molecular biologists discovered that protein folding is more complicated than Anfinsen's experiment suggested. A pure polypeptide in a dilute solution may ultimately adopt a single, stable conformation. But newly made proteins left to themselves usually do not fold properly—whether in the test tubes of protein chemists or in the interiors of cells.

Several obstacles can prevent correct folding. Inside a cell, the concentration of proteins and other molecules is extremely high. At these high protein concentrations, individual peptide chains can easily form bonds with other peptides, rather than with themselves. Molecular biologists call such bonding between separate molecules *promiscuous interactions*. In addition, polypeptide chains are assembled gradually, one amino acid at a time. So one end of a chain is forming among all these other proteins, while the other end does not even exist yet. This is a recipe for promiscuous interactions.

To prevent the parts of polypeptide chain from linking to the wrong molecule or to the wrong part of itself, a staff of enzymes and special proteins, called **chaperones,** bind to the newly emerging polypeptide and prevent promiscuous interactions. Chaperones not only help other proteins fold into the correct shape, they also help proteins maintain that shape or change shape in order to play a new role.

Amino acid sequence alone was once thought to be sufficient to determine the three-dimensional structure of a protein. But enzymes and chaperone molecules guide the correct folding of a protein in a cell.

In this chapter, we have seen how the structure of a molecule—whether at the level of individual functional groups or at the level of amino acid sequences and tertiary structure—profoundly affects the chemistry of the molecule. The structure of a cell determines its behavior and capabilities just as profoundly, as we will see in the next chapter.

Study Outline with Key Terms

Other than water, nearly all molecules found in living organisms contain one or more carbon atoms. In fact, all **organic** molecules contain carbon. **Organic chemistry** is the study of the structures and reactions of carbon compounds, while **biochemistry** is the study of the chemistry of living organisms.

Most organic molecules are made of **hydrocarbon chains.** But the specific properties of small molecules depend on **functional groups,** such as the **hydroxyl, carboxyl,** and **amino** groups. The properties of **macromolecules** depend on the nature and sequence in which the component building blocks assemble.

Some hydrocarbon chains form rings of carbon. Flat, five- or six-carbon rings are called **aromatic** rings. Most macromolecules are linear, however. The small building blocks all assemble into different kinds of long-chain macromolecules, by the same kind of general reaction. This chemical reaction is called a **dehydration reaction** because the equivalent of one molecule of water is eliminated each time two building blocks join together. Conversely, the breaking down of a macromolecule into its constituent building blocks is called **hydrolysis** because the equivalent of one molecule of water is added as each small molecule is removed from the chain.

The small building blocks of cells fall into four classes—**sugars, amino acids, nucleotides,** and **lipids. Glycosidic bonds** link sugars into macromolecules called **polysaccharides. Peptide bonds** link amino acids together into **polypeptides,** and **phosphodiester bonds** link nucleotides into **nucleic acids.**

Although lipids do not form long chains, some, called **fatty acids,** do contain long carbon chains. Fatty acids may be **saturated, unsaturated,** or **polyunsaturated.** A **glycerol** molecule with two fatty acid chains and a phosphoric acid group is a **phospholipid,** an amphipathic molecule that forms cell membranes. A glycerol molecule with three fatty acid chains and a phosphoric acid group is a **triacylglycerol.** Different kinds of triacylglycerols make up the fats (solid at room temperature) and oils (liquid at room temperature) of animals and plants. A third class of lipids consisting of four interconnected rings is the **steroids,** including both **cholesterol** and steroid hormones such as progesterone and testosterone.

Simple sugars, or **monosaccharides,** are the building blocks of **disaccharides, oligosaccharides,** and **polysaccharides.** Sugars, such as **glucose** and **sucrose,** and **polysaccharides,** such as **starch, glycogen, chitin,** and **cellulose,** are all classed as **carbohydrates.** For labels and signals, cells use **glycoproteins,** which are oligosaccharides attached to proteins.

The building blocks of nucleic acids are nucleotides. Each nucleotide consists of a **nitrogenous base,** either a **purine** or a **pyrimidine,** connected to a sugar, either **ribose** or **deoxyribose,** and (by way of the sugar) a phosphate group.

Many proteins, including **enzymes** and transport molecules such as hemoglobin, are dense, roundish molecules called **globular proteins.** But many structural proteins are **fibrous proteins**—long, thin molecules such as the keratin in hair or the collagen in skin and bone.

All polypeptides are made of strings of amino acids. Each amino acid has a carboxyl group, an amino group, and an R group, all attached to the **α carbon.** In each amino acid, all the parts are identical except for the R group. Proteins can have up to four levels of structure: **primary structure,** the sequence of amino acids; **secondary structure,** the coiling, bending, or folding of individual sections of the polypeptide chain into forms such as the **α helix,** the **β pleated sheet,** and the **collagen helix; tertiary structure,** or **conformation,** the three-dimensional folding of an entire polypeptide chain; and **quaternary structure,** the fitting together of two or more folded chains. Different regions of the polypeptides of a protein are called **domains.**

The folding of a polypeptide into a precise three-dimensional structure generally requires energy. Some of this energy comes from interactions among the atoms of the polypeptide backbone and those of the amino acid side chains. Additional energy may come from **ATP,** which is used by **chaperones** that help proteins fold.

Review and Thought Questions

Review Questions

1. What evidence persuaded biologists that scrapie could not be caused by a bacterium or virus?
2. Where would you find a large concentration of lipids in a cell? Why?
3. What building blocks make up proteins? How many kinds of these building blocks are there?
4. What three components make up a nucleotide?
5. What small molecule is removed during a dehydration reaction? What is the opposite reaction called? What happens in that case?
6. What is the relationship between polysaccharides, sugars, and carbohydrates? What three roles do carbohydrates play in cells?
7. Define peptide bond, polypeptide, protein, and enzyme.
8. Define nucleic acid.
9. Describe the four levels of structure that occur in protein folding.
10. Distinguish between the α helix, the β structure, and the collagen helix, all of which occur in fibrous proteins.

11. What is the major component of each of these materials? Use C (carbohydrate), P (protein), and L (lipid) to answer.
    ____Cotton T shirt   ____Cooking oil
    ____Chicken nugget   ____Wooden toothpick
    ____Apple juice   ____Cheese
    ____Silk scarf   ____Fat on a piece of steak
    ____Tortilla   ____Newspaper
    ____Rice
    ____Lipstick

Thought Questions

12. What is it about the shape of a protein that determines its properties and therefore its functions? Describe the shape of a globular protein and of a fibrous protein. How are these two shapes usually put to different purposes in a cell?
13. The 35 small molecules listed in Table 3-2 are called building blocks. How can the same building blocks be used to construct billions upon billions of different macromolecules?

About the Chapter-Opening Image

This white lab mouse is constructed with building blocks. These represent the building blocks of life.

On-line materials relating to this chapter are on the World Wide Web at **http://www.harcourtcollege.com/lifesci/aal2/**

Quizzing & Testing

Web Bit 3-1 Mad Cows and Promiscuous Fungi

Web Bit 3-2 Molecules with Similar Shapes Can Mimic One Another

Web Bit 3-3 Linus Pauling and the Alpha Helix

4 Why Are All Organisms Made of Cells?

Very Little Animalcules

More than 300 years ago, an uneducated Dutch cloth merchant named Antonie van Leeuwenhoek (1632–1723) discovered something so interesting that he received visits from the Czar of Russia, King James II of England, and King Frederick II of Prussia. Like many young men, Leeuwenhoek had a consuming interest that distracted him from his work (Figure 4-1). He liked to make glass lenses by grinding and polishing bits of glass and then mounting the lenses between gold or copper plates. He used these simple, single-lens microscopes to examine anything he encountered.

To Leeuwenhoek's delight, his lenses—some no larger than a pinhead—revealed hundreds of tiny living beings never before seen by human eyes. Leeuwenhoek astonished his friends and acquaintances with descriptions of red blood cells in blood, sperm in semen, and "very little animalcules" in pond water. With his lenses, Leeuwenhoek also demonstrated that fleas, ants, weevils, and other pests do not arise spontaneously from dust or wheat, as was commonly believed, but developed from larvae that hatch from tiny eggs laid by the adults.

In 1673, when Leeuwenhoek was 41, he began sending reports of his discoveries to the recently created Royal Society in London, then one of the few scientific organizations in the world. Over the next 50 years, until Leeuwenhoek's death in 1723, the Society received and published 375 letters from him, the last when the Dutch lens grinder was 90.

Over the years, Leeuwenhoek described how everywhere he looked, he found tiny organisms whose existence few people had even suspected. In 1683, he became the first person to see bacteria. Leeuwenhoek's great contribution was to demonstrate that life is not limited to organisms visible to the naked eye.

In England, the Royal Society welcomed Leeuwenhoek's fascinating observations. The Society had already hired Robert Hooke to be its "curator of instruments." Hooke's job was to demonstrate the microscope and other new technologies for the entertainment of the gentlemen at the Royal Society's meetings. In 1665, Hooke published his book *Micrographia,* a series of illustrations and descriptions of his observations using new and improved microscopes. Hooke showed, for example, the structure of feathers, the stinger of a bee, and the foot of a fly.

Hooke also coined the term "cellulae" [Latin, *cella* = small room] to describe the boxlike cavities (cells) he saw when he examined a slice of cork under a microscope. "Cells," in those days, referred to the small rooms of a monastery, nunnery, or prison. Hooke later published detailed drawings of cells in other plant tissues, noting that they contained "juices." Leeuwenhoek, Hooke, and the few other microscopists of the day described a whole new world of tiny living organisms, and their research revealed an unsuspected level of structural complexity in larger organisms. These discoveries might well have stimulated the beginning of a new science. Yet, oddly, these discoveries lay dormant for nearly 200 years.

Neither Leeuwenhoek's nor Hooke's studies led directly to any unifying biological principles. Most scientists dismissed Leeuwenhoek's animalcules as little more than amusing natural curiosities. Hooke's "cellulae" were simply bubbly stuff in plants. No one guessed that cells come from other cells by the process of cell division, that cells are the fundamental unit of life, and that all tissues in all organisms are made of cells. No one guessed that some microscopic organisms cause disease. To most people, Leeuwenhoek's microorganisms and Hooke's

Key Concepts

- Every cell has a boundary, a cell body, and a set of genes. Different kinds of cells have characteristic organelles and other structures.
- In eukaryotic cells, internal membranes enclose organelles and help regulate the chemical composition of the spaces inside. Prokaryotic cells have no internal membranes or compartments.
- The cytoskeleton contributes to the organization and internal communication of eukaryotic cells.

cells seemed no more important than the specks of dust in a ray of sunshine.

One reason that the important thinkers took neither Hooke nor Leeuwenhoek seriously was the two men's social class. Leeuwenhoek was an uneducated amateur, and Hooke was an employee of the Royal Society, not a full member. Neither man was, in other words, a "gentleman," so neither commanded respect.

Equally important, 17th-century scientists and philosophers carried a bias, left over from the Middle Ages, that valued theory over experiments and observations. Hooke's and Leeuwenhoek's microscope work was entirely descriptive. Leeuwenhoek described thousands of objects but, probably because of his lack of education, he did not attempt to present any unifying explanation for why things were as they were.

Most scientists saw the microscope as more of a toy than a scientific instrument. By the end of Leeuwenhoek's life, at the beginning of the 18th century, few people were even making lenses anymore. Those simple microscopes already in existence were playthings for ladies, of whose observations there is little record.

> Leeuwenhoek showed that life is not limited to organisms visible with the naked eye.

Another important reason that study of the microscopic details of life languished was the poor quality of 17th- and 18th-century microscopes. Leeuwenhoek's handmade lenses were far better than any others in existence. Some of the 400 lenses he left behind could magnify objects 300 times, to show the largest bacteria. And his mysterious techniques—probably sophisticated lighting—which he never revealed, allowed him to see details that neither Hooke nor any of the other prominent microscopists of the time could match. Still, even Leeuwenhoek's microscopes were crude by today's standards (Box 4.1). His finely polished lenses distorted both the shapes and the colors of the objects under the microscope. It was not until the late 19th century, 100 years after the beginning of the Industrial Revolution, that technology provided microscopes adequate for the study of the insides of cells.

WHY ARE ALL ORGANISMS MADE OF CELLS?

In the 1820s, better microscopes paved the way for rapid advances in cell biology. Within 10 years, biologists discovered two important parts of cells. They found that both animal and plant cells contain a nucleus, a dark mass that we now know contains the genetic material (DNA). Researchers also finally realized that the "juice" observed by Robert Hooke was a living substance, which they named **protoplasm** [Greek, *proto* = first + Latin, *plasma* = a thing molded or formed].

All Organisms Are Made of Cells

In 1838, the German botanist Matthias Schleiden proposed that all plants consist of cells—an idea seconded for animal cells in 1839 by German zoologist Theodor Schwann. Cells, they wrote, are the elementary particles of all living organisms. In addition, Schwann and Schleiden argued that all cells are alive—independent of the organisms to which they belong. In other words, a liver cell is alive in its own right even though it is only a small part of a living organism. Remove the cell from the liver and the cell is still alive.

Recall from Chapter 1 that animals and plants are all multicellular, that the Fungi and Protista include both multicellular organisms and single-celled organisms, and that the Archaea

(Text continued on page 74)

Figure 4-1 Antonie van Leeuwenhoek built primitive microscopes so effective that he saw bacteria 200 years before anyone else. (*Science Photo Library/Photo Researchers, Inc.*)

BOX 4.1

How Do Microscopes Help Biologists Study Cells?

An important way to learn about cells is to look at them. Each major improvement in microscopy has led to a burst of new knowledge and a host of new questions. In the 17th century, the development of the microscope made it possible to see cells for the first time. In the 19th century, successive improvements in microscope design revealed that all organisms are made of cells and that their elaborate adaptations allow specific functions. By the mid-20th century, the development of new types of microscopes had further revolutionized the study of cells—revealing the rich internal structure of cytoplasm, nuclei, and membranes.

How can we see such details? We can see an object only when it is large, clear, and stands out from its background. In order to see an object as small as a cell, first we must use lenses to magnify its image. The ratio between the size of the image and the size of the object itself is called the *magnification*.

Even when we can magnify an object, however, we may not be able to see it clearly enough to learn anything about it. When small objects are too close together, we cannot tell whether we are looking at one or several objects. The *resolution* of a microscope or other optical instrument is the minimum distance between two objects that allows them to form separate images. We can say that resolution is a measure of image clarity. Our eyes have a resolution of about 0.1 mm. A microscope resolves images less than 1/100th that size (0.001 mm or 1 μm). Such resolving power is enough to distinguish individual prokaryotic cells and some internal structure in eukaryotic cells.

Visibility depends not only on magnification and resolution but also on *contrast*—how well an object stands out from its background. Contrast depends on an object's tendency to absorb more or less light than its surroundings. A transparent object is invisible. So is an object that is the same shade as its background.

Microscopists have devised ingenious ways of increasing contrast. Nineteenth-century biologists, for example, found that different dyes would bind to different parts of the cell—one dye to the nucleus, another to the membrane, another to parts of the cytoplasm. Dyes that bind differently to cell components are called stains.

But staining cells usually kills them, so early biologists could study living cells only with difficulty. Beginning in the 1930s, biologists developed techniques for studying living cells. In conventional light microscopy, contrast depends on how different cellular structures absorb light. Most materials in unstained cells absorb little visible light. However, substances that do not absorb or reflect much light can still change the speed at which light passes through. For example, light passes through the cell nucleus more slowly than it does through the watery cytoplasm. The light passing through the nucleus becomes "out of phase" with light waves that pass through the cytoplasm. Special optical arrangements, such as "phase contrast" and "interference contrast" microscopy, can heighten this effect, revealing phase differences among different cell components without the use of stains. These techniques permit the study of living cells.

Figure A(a) shows the workings of an ordinary light microscope. Because a light microscope contains several lenses, it is also called a *compound microscope*. It consists of a tube with lenses at each end, a stage that holds a specimen, and a light source. The light source has its own lens, which allows the user to control the light that passes through the specimen. The lens near the eye is called the *eyepiece*, and the lens at the other end of the tube, near the object being examined, is called the *objective*.

The objective lens magnifies the image of the specimen and projects the image to the eyepiece. The eyepiece magnifies the image again. The final magnification is the product of the magnification of the objective and that of the eyepiece. For example, an objective that magnifies 40 times (40×) and an eyepiece that magnifies 10 times (10×) would produce an image 400 times (400×) as large as the original specimen.

A light microscope often has three objective lenses (for example, with magnifications of 10×, 40×, and 100×) mounted on a rotating turret, so that the viewer can change magnifications. Because eyepieces usually contribute another 10-fold magnification, the final image usually ranges from 100 to 1000 times the size of the specimen.

Even the best lenses, however, have a resolution limited by the nature of light. Light is composed of waves, and lenses cannot resolve structures smaller than the wavelength of light. The wavelength of visible light ranges from about 0.4 μm (for blue light) to about 0.7 μm (for red light). Light microscopes reveal no details of subcellular structures smaller than about 0.5 mm.

One solution is to use wavelengths far smaller than those of visible light. Electron microscopes, which see objects as small as 0.002 μm, depend on the wave properties of moving electrons. In a 100,000-volt electron microscope, the wavelength of an electron is 0.004 nanometers (nm), or 0.000004 μm, less than the diameter of an individual atom.

The lenses of an electron microscope are electromagnets that bend the path of electrons just as a glass lens bends the path of light. In reality, the resolution for looking at cells is only about 2 nm (0.002 μm) but still about 100 times better than the best light microscope (Figure A(b) and (c)).

Biologists commonly use two kinds of electron microscopes—the transmission electron microscope (TEM) and the scan-

Figure A Views of *Paramecium* with different types of microscopes. (a) With a light microscope, we see the internal anatomy of this one-celled organism and a fringe of hair-like cilia, which these creatures use to propel themselves through the water. (b) With a transmission electron microscope, we see a cross section through three rows of cilia at much greater magnification and in great detail. (c) With a scanning electron microscope, set at low power, we see the cilia protruding from the surface in greater detail than with the light microscope but nothing of the *Paramecium*'s internal anatomy. (a, Eric Grave/Phototake; b, David Phillips/Visuals Unlimited; c, Dr. Dennis Kunkel/Phototake)

ning electron microscope (SEM). In each case, the electrons form an image on a phosphorescent screen, like that of a television set. Figure A compares a *Paramecium* seen with a light microscope, TEM, and SEM.

In TEM, a beam of electrons passes through the specimen in the same way that a beam of light would in a light microscope. SEM is different. In SEM, a narrow beam of electrons moves back and forth across the surface of the specimen. Wherever the beam hits the surface of the specimen, it causes the surface to emit more electrons, which produce a detailed, almost three-dimensional, picture of the specimen's surface.

Another powerful technique for examining the structure of cells relies on the immune response. One of the ways that mammals resist infection is by making proteins, called antibodies, that form complexes with foreign molecules, such as those of infectious bacteria. Biologists can induce experimental animals to produce antibodies against specific cell components, such as specific proteins in the cilia of *Paramecium*. These antibodies can form complexes with the specific protein, allowing biologists to visualize its subcellular distribution. Before doing such a study, a researcher may link the antibodies to a fluorescent dye. A special microscope can focus a beam of light from above, and the researcher can look for fluorescence, which shows the distribution of the particular subcellular component.

More About the SEM

and Eubacteria are all single-celled organisms. Noncellular organisms do not exist. With the discoveries of Schleiden and Schwann, biologists recognized the importance of cells. Cells, they saw, were everywhere. But where did cells come from? Where did the new cells that heal a wound come from? Where did the cells that build a child's growing body come from? Where did the burgeoning masses of cells in cancers come from? And where did the one-celled organisms living in a pond come from?

Schleiden and Schwann proposed that cells crystallized spontaneously out of shapeless matter. But ever-improving microscopes allowed more and more scientists to see that individual cells can divide, each giving rise to what biologists call "daughter" cells. In time, biologists accepted the idea that all new cells come from the division of preexisting cells.

In 1858, the widely respected physician and biologist Rudolf Virchow formalized this understanding with the Latin phrase, *omnis cellula e cellula,* meaning "all cells from cells." In a book written for physicians, Virchow summarized his theory about the role that cells play in disease. In doing so, he simultaneously revolutionized both biology and medicine. Virchow argued (1) that cells never arise from noncellular material and (2) that diseases result from changes in specific kinds of cells.

As a result of these arguments, biologists came to accept the principle that cells always come from other cells. Cells healing a wound result when preexisting skin cells begin dividing. A child's growth results from the rapid multiplication of cells. Cancers result from the uncontrolled division of an organism's cells. Single-celled organisms in ponds descend from other single-celled organisms.

Today, we summarize the work of Schleiden, Schwann, and Virchow in the **cell theory,** which states that (1) all organisms are composed of one or more cells; (2) cells, themselves alive, are the basic living unit of organization of all organisms; and (3) all cells come from other cells. As we discussed in Chapter 1, we use the word "theory" to denote a system of statements and ideas that explains a group of related facts or phenomena, *not* to indicate uncertainty.

The "cell theory" says that all organisms are composed of one or more cells, that cells are the basic living unit of organization of all organisms, and that all cells come from other cells.

Every Cell Consists of a Boundary, a Cell Body, and a Set of Genes

The cells of organisms from the three domains of life can differ greatly. Even within a single organism, cells come in wildly different types. Nonetheless, all cells have three common features: (1) a boundary that separates the inside of the cell from the rest of the world, (2) a set of genes, and (3) a cell body.

The **plasma membrane** is a cell's boundary, a highly organized and responsive structure. Like the outer wall of a house, with its windows and doors, the plasma membrane not only defines the limits of a cell, but also helps to regulate the cell's internal environment by selectively admitting and excreting specific molecules.

Each cell also contains a set of genes strung out along one or more molecules of DNA. In eukaryotes (animals, plants, fungi, and protists), the DNA is confined within a membrane-enclosed structure called the **nucleus** (Figure 4-2A). The presence of a true nucleus alone defines the eukaryotes and distinguishes them from the prokaryotes (eubacteria and archaebacteria). In the prokaryotes, the DNA occupies a limited region of the cell, called the **nucleoid,** which has no membrane (Figure 4-2B).

Throughout the 19th century and well into the 20th century, biologists continued to think of the protoplasm as a more

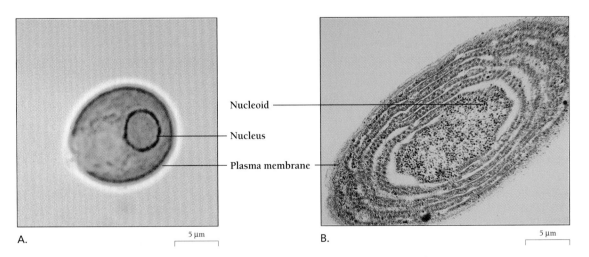

Figure 4-2 Eukaryotes have a nucleus. A. A eukaryotic cell, such as the green alga *Chlamydomonas,* has both a plasma membrane and a nucleus. B. A prokaryotic cell, such as a cyanobacterium, has a plasma membrane, but no nucleus. (*A, Philip Sze/Visuals Unlimited; B, Elizabeth Gentt/Visuals Unlimited*)

or less homogeneous jellylike substance. Improved techniques for looking at cells, however, revealed more and more tiny structures in eukaryotic cells, though not in prokaryotic cells. We now know that each of these tiny structures, called **organelles,** performs a specialized task.

Biologists finally discarded the word protoplasm and renamed the cell body—that part of the cell outside the nucleus but inside the membrane—the **cytoplasm** [Greek, *cyto* = a hollow container (a cell) + Latin, *plasma* = a thing molded or formed]. The part of the cytoplasm not contained within membrane-enclosed organelles was named the **cytosol.** Most of a cell's biochemical work occurs within the cytosol. Running through the cytosol is an intricate network of protein fibers, called the **cytoskeleton,** which gives the cell its shape, holds organelles in place, and participates in cell movement.

Every cell is enclosed by a plasma membrane and keeps its DNA in a nucleus (eukaryotes) or nucleoid (prokaryotes). Eukaryotic cells contain other organelles within the cytoplasm.

How Are Cells Alive?

Cells—from Leeuwenhoek's one-celled "animalcules" to the individual cells of the liver—are the fundamental living units of life. They have all the characteristics of life that we listed in Figure 1–8. Made up of organized parts, cells perform chemical reactions, obtain energy from their surroundings, respond to their environments, change over time, reproduce, and share an evolutionary history. Cell components work together to achieve homeostasis, maintaining an internal cellular environment that is constant enough to carry out the thousands of chemical reactions that allow cells to live and to reproduce.

The ability of cells to obtain energy from their environment depends on a set of chemical reactions, some of which we discuss in Chapters 6 and 7. Almost every cell can capture energy from glucose, usually by oxidizing it to carbon dioxide and water. In eukaryotic cells, many of the energy-producing reactions take place in specialized organelles, while others take place in the cytosol.

Cells change with time, both chemically and mechanically. For example, individual muscle cells shorten as muscles contract. Other cells, too, change their size and shape. To allow for these changes, almost every eukaryotic cell contains a network of fibers, called the cytoskeleton, that supports the cell's components and brings about change in cell shape. The cytoskeleton also serves as train tracks along which cell components can move, pulled by locomotives called *molecular motors.*

In order to reproduce, every cell must be able to copy its genes for its descendants. All cells have elaborate mechanisms for precisely duplicating DNA.

Cell reproduction also depends on the cell's ability to read and to use the coded information in DNA. In a single-celled organism, cells adjust the readout of their genetic information in response to environmental changes. A bacterial cell growing in a glass of milk, for example, makes different proteins from the same type of cell growing in a glass of sugar water. In a multicelled organism, different types of cells make different sets of proteins, according to their specialized characteristics: a red blood cell, for example, makes lots of hemoglobin, whereas a muscle cell makes lots of contractile proteins.

Once a cell produces the appropriate proteins, however, it must deliver them to the right address within the cell. Cells have elaborate mechanisms for directing molecular traffic to different parts of the cell.

Finally, cells must deal with their own garbage—waste products such as carbon dioxide and ammonia, as well as cell remnants damaged by ordinary wear and tear. Special molecules and mechanisms serve as cellular toilets and garbage disposals.

A cell could conceivably employ a wide variety of solutions to its problems. Different cells might use different strategies to derive energy, to distribute ions, and to organize its organelles. But all cells use the same molecules, share the same biochemical pathways, and contain the same types of organelles. This sharing of solutions reflects the common evolutionary heritage of all cells, which almost certainly descended from a common ancestor that lived about 3.5 billion years ago.

What Are the Advantages of Cellular Organization?

Single-celled organisms are nearly all very small, but there are some exceptions. Plants, animals, and other multicellular organisms can be as large as trees and elephants, but a free-living cell never grows that big. In fact, with some remarkable exceptions, cells are nearly all about the same size. Most eukaryotic cells range from 10 to 100 micrometers (μm), while prokaryotic cells are smaller, ranging from 0.4 to 5 μm. (For comparison, a period on this page is about 100 μm in diameter, the size of a large eukaryotic cell, such as those in an onion skin.) No cells are smaller than 0.4 μm, and almost none are larger than 100 μm. Exceptional cells may be enormous. Some plant fiber cells are a meter long (1 million μm), while the nerve cells in the legs of a giraffe may be several meters long.

One reason that cells are nearly all so small is that subdivision into tiny cells offers organisms many advantages. To understand these advantages, we can start by asking a simple question: What problems would you face if you were one very large cell?

A cell's need to regulate its internal environment limits its size

For the biochemical machinery of a cell to function, the cell must maintain a relatively constant internal environment. Otherwise, enzymes and other proteins will not adopt the shapes needed for biological activity. To maintain a constant internal environment, the cell must keep both the concentration of salts and the pH from changing. At the same time, the cell must

If you were one big cell, you would have to maintain just the right environment for each of the different molecular activities in your body. The highly acidic (low) pH that you use to digest food would have to somehow coexist with the neutral pH needed for protein synthesis.

All the materials that come into a cell or leave a cell must go through the thin membrane that envelops the cell. If you were one big cell, your plasma membrane would have to admit or exclude specific molecules selectively. Somehow you'd have to absorb enough pizza to get you through the day.

Protists, such as the *Didinium* pictured in Figure 4-3, perform all these functions and more. In fact, unlike most protists, the *Didinium* even has a sort of mouth, called a cytostome [Greek, *cyto* = cell + *stoma* = mouth], with which it is eating another protist (a *Paramecium*). Protists, however, are all small. Accomplishing all that a protist does wouldn't be so easy for a one- or two-hundred-pound cell like yourself. The membrane's ability to admit or to excrete molecules is limited by its size. A big cell certainly has more external membrane than a small one. As a cell grows, however, its volume increases faster than its surface area.

Let's think, for a minute, about the relationship between the surface area and the volume of any object. Suppose we have a tiny cardboard cube, one centimeter on each side. If we double the linear dimensions of the cube, so that each side is 2 cm by 2 cm, the surface area of the cube increases by a factor of 4. But its *volume* increases by a factor of 8 (Figure 4-4).

The larger cube has proportionately less surface area than the smaller cube. We can see this if we look at the **surface-to-volume ratio**—the amount of surface area for each bit of volume. For the small cube, the surface-to-volume ratio is 6:1. For each cm^3 of volume, the smaller cube has 6 cm^2 of surface. But the surface-to-volume ratio for the larger cube is 24:8 (or 3:1). For each cm^3 of volume, the larger cube has only 3 cm^2 of surface.

Because larger objects have smaller surface-to-volume ratios, larger cells have proportionately less membrane with which to regulate their internal space. As a result, larger cells have a harder time obtaining nutrients, getting rid of wastes, and regulating the internal concentrations of ions and molecules. If you were one big cell, molecules of oxygen and food would take decades to get from your membrane to your interior. You could bury yourself in pizza and you'd still go to bed hungry. You simply would not be able to digest the pizza's macromolecules fast enough to survive.

Knowing how a cell's size affects the regulation of its internal environment, we might wonder how eukaryotic cells, which range from 10 to 100 μm, can be so much larger than prokaryotic cells, which range from 0.4 to 5 μm. The answer is that eukaryotic cells have special adaptations to increase their surface areas. For example, many eukaryotic cells have convoluted surface membranes, and almost all have elaborate internal membrane systems, which we will discuss later in this chapter. Indeed, some eukaryotic cells, such as some white blood cells, have 50 times more membrane inside than outside.

Surface area: 6
Volume: 1
Ratio: 6:1

Surface area: 24
Volume: 8
Ratio: 24:8 or 3:1

Figure 4-4 The relationship between surface area and volume determines many of the properties of cells. In particular, smaller cells have more surface area per volume than do larger cells. Here we can see that a cube 1 cm on a side has a surface area of 1 cm^2 on each of its 6 sides, so that it has a total surface area of 6 cm^2. Its volume is 1 cm^3. When we double the linear dimensions of the cube, we have a cube 2 cm on a side, which gives a surface area of 24 cm^2. Its volume is 8 cm^3. We can see that when the linear dimensions of an object double, surface area increases four times, while volume increases eight times.

The larger a cell, the smaller is its surface-to-volume ratio. A low surface-to-volume ratio limits a cell's ability to absorb nutrients and expel wastes.

Individual cells may specialize for different tasks

In multicellular organisms, individual cells often perform separate tasks. For example, red blood cells carry oxygen, while muscle cells contract and move. Cells in the roots of a tree absorb nutrients from the soil, while cells in the leaves harvest the energy of sunlight. Just as a division of labor makes human societies more productive, the specialization of cells allows an organism to function efficiently.

Figure 4-3 *Didinium* eating a *Paramecium*. *Didinium* has a mouth with which it can consume other protists. (*Stan W. Elems/Visuals Unlimited*)

TABLE 4-1 THE CELLS OF ALL ORGANISMS HAVE BOTH COMMON AND DISTINCTIVE FEATURES

Cell component	Function	Eubacteria Archaebacteria	Protist	Fungus	Plant	Animal
Cell wall	Protects and supports cell	Yes	Yes	Yes	Yes	None
Plasma membrane	Regulates communication with other cells and movement of materials in and out of cell		Yes	Yes	Yes	Yes
Membrane-enclosed nucleus	Isolates DNA from cytoplasma	None	Yes	Yes	Yes	Yes
DNA	Encodes information for proteins	Single molecule	Many molecules			
RNA	Constructs proteins from information in DNA	Yes	Yes	Yes	Yes	Yes
Nucleolus	Synthesizes ribosomes	None	Yes	Yes	Yes	Yes
Ribosome	Synthesizes proteins	Yes	Yes	Yes	Yes	Yes
Endoplasmic reticulum	Modifies proteins, makes lipids	None	Yes	Yes	Yes	Yes
Golgi complex	Tags proteins and lipids, packages them for export	None	Yes	Yes	Yes	Yes
Lysosome	Contain digestive enzymes	None	Yes	Yes	Yes	Yes
Mitochondrion	Synthesize ATP	None	Yes	Yes	Yes	Yes
Chloroplast	Photosynthesize	None	Some	None	Yes	None
Nonchloroplast plastids	Store food, pigments	None	Some	None	Yes	None
Central vacuole	Provides turgor pressure; contains water and wastes	None	None	Yes	Yes	None
Cytoskeleton	Shapes, strengthens, and moves	None	Yes	Yes	Yes	Yes
Flagellum, cilium	Moves cell through fluid or fluid past cell surface	None	Yes	Yes	Yes	Yes
Bacterial flagellum	Moves cell through fluid or fluid past cell surface	Yes	None	None	None	None

man societies more productive, the specialization of cells allows an organism to function efficiently.

If you were one big cell, organizing your body to perform all its different jobs would be difficult. You wouldn't have a heart or arteries to move oxygen and nutrients around your body. You wouldn't have lungs to extract oxygen from the air. You'd just have 125 pounds (or so) of cytoplasm.

Cellular organization allows organisms to make a division of labor among specialized cells.

A multicellular organism can lose and replace individual cells

Subdividing the body into cells has another advantage. Cells often live and die independently of the whole organism. In fact, the death of some cells is a normal part of development. The fingers of the hand, for example, develop from paddlelike appendages as a result of the death of cells between the fingers. The lenses of the eyes, the red blood cells that carry oxygen to all the tissues, and the outer layer of the skin all develop because of the death of specific cells. Equally important, the cells of the skin, blood, and intestines continuously replace themselves. The advantage of such replacement is that the life of a multicellular organism can extend far beyond the life of an individual cell.

Cellular organization allows organisms to outlive the cells that compose them.

WHAT'S IN A CELL?

Eukaryotic cells contain so many different organelles, each with a special set of tasks, that we can imagine each cell as a tiny walled city (Figure 4-5). Inside the wall of the city are power stations, a central library of genetic information, warehouses for packaging proteins, and much more. Table 4-1 summarizes the characteristics of cells from the six kingdoms of organisms.

When scientists first began using the transmission electron microscope in the 1940s and 1950s, one of the great surprises was the huge amount of internal membrane in

78 part I Chemistry and Cell Biology

Figure 4-5 Two eukaryotic cells, a plant cell and an animal cell. A eukaryotic cell, with its plasma membrane and membrane-enclosed organelles, is like a walled medieval city. The plasma membrane regulates the passage of molecules in and out of the cell, just as the city wall regulates the passage of people. Inside, the mitochondria supply power to the cell in the form of ATP. The nucleus is a library, housing all of the information a cell needs to produce proteins. The nuclear membrane determines which molecules have access to the information inside and which bits of information (books) may be checked out for use in the cytoplasm. The ribosomes act as workbenches for the production of proteins. The vacuole serves as a depository for waste materials, much like a toxic waste dump. The Golgi complex is like a mail room. Here, glycoproteins are packaged and labeled for export or intracellular transport. See if you can think of analogies for some of the other structures in these cells. To what would you compare a chloroplast?

Quiz

ANIMAL CELL

- Flagellum
- Ribosome — A workbench for making proteins
- Mitochondrion — A powerhouse
- Lysosome
- Rough endoplasmic reticulum — A transport system for exporting proteins
- Centriole
- Cytoplasm
- Microtubule
- Microfilament
- Plasma membrane — A city wall
- Smooth endoplasmic reticulum

eukaryotic cells (Box 4.1). These membranes divide cells into at least seven kinds of compartments: nucleus, cytosol, endoplasmic reticulum, Golgi complex, lysosomes, peroxisomes, and mitochondria. In addition, a plant cell contains one more set of compartments—plastids.

Eukaryotic cells also contain a variety of membranous sacs with nothing visible inside. The **vesicles** of plant and animal cells are small, apparently empty, sacs—generally about 100 nanometers (nm) in diameter. Eukaryotic cells (especially plant cells) also contain **vacuoles**—large sacs that may occupy as much as 95 percent of the volume of the cell.

What Role Does the Nucleus Play in the Life of a Cell?

The nucleus usually makes up 5 to 10 percent of the volume of the cell. In cells that are not dividing, the nucleus stains as a diffuse mass. This pattern results from the binding of stains

Figure 4-6 **Chromatin in the nucleus of an onion cell.** Chromatin condenses in dividing cells, making the chromosomes visible. *(Runk/Schoenberger from Grant Heilman)*

to **chromatin** [Greek, *chroma* = color], a complex of DNA and protein. In cells that are dividing, the chromatin forms distinct objects, visible in the microscope, called **chromosomes** [Greek, *chroma* = color + *soma* = body] (Figure 4-6).

An electron microscope shows that the nucleus's boundary is a double membrane called the **nuclear envelope** (Figure 4-7). The two membranes are separated by a space about 20 to 40 nm wide. In many places, however, the two membranes are fused to form interruptions called **nuclear pores,** which form channels between the nucleoplasm (the contents of the nucleus) and the cytoplasm. These pores resemble the holes in a tea ball that allow water to enter the tea ball and tea to leak out into the surrounding water. However, nuclear pores play a more active role than mere holes. They control the movement of materials in and out of the nucleus in much the same way that a turnstile controls the movements of people in and out of a subway.

The nucleus serves as the cell's library, and it also provides instructions for the construction of new cells. The nucleus contains the genetic information (coded in DNA) that is passed from one cell to its descendants. When a cell divides, each daughter cell gets a complete copy of the genetic instructions in the nucleus. The DNA in each cell nucleus includes instructions for building every polypeptide the body will ever use. In Part II of this book, we will consider how DNA carries genetic information, how cells use this information, and how they pass it to their daughter cells.

The nucleus of eukaryotic cells contains DNA and protein. The DNA acts as a library of information and a set of instructions for making cell components.

The Cytosol Is the Cytoplasm That Lies Outside the Organelles

The cytosol makes up about half of a cell's volume. Cell biologists usually isolate the cytosol from the organelles by breaking open cells and then spinning the mixture in a centrifuge. A centrifuge is a machine that spins a set of tubes arranged so that when they are spinning their contents come under a force greater than gravity (Figure 4-8). This force—centrifugal force—is the same force that presses clothes in a washing machine to the outside wall of the tub during the spin cycle. In a centrifuge, a tube spins so fast that the contents sink to the bottom, with bigger particles and molecules sinking faster than smaller ones. When cell biologists centrifuge cytoplasm from

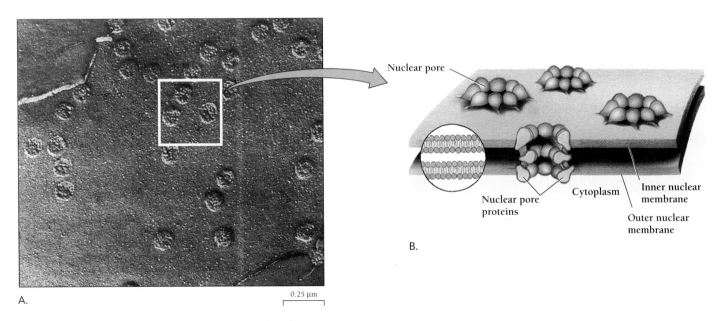

Figure 4-7 **Nuclear pores span the nucleus's double membrane.** A technique called freeze fracture reveals their structure by splitting the membrane. *(A, R. Kessel-G. Shih/Visuals Unlimited)*

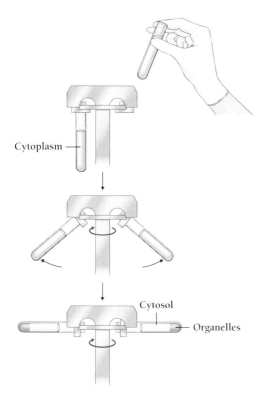

Figure 4-8 A centrifuge spins a tube of macromolecules at up to 80,000 revolutions per minute. Such speed places the contents of the tube under forces more than 100,000 times that of gravity. Large molecules travel to the bottom of the tube fastest.

cells, the organelles, which are heavier than the cytosol, drop soonest to the bottom of each tube.

The cytosol contains thousands of different kinds of enzymes responsible for producing building blocks, degrading small molecules to yield energy, and synthesizing proteins. Although the cytosol is aqueous (mostly water), about 20 percent of its weight is protein, giving it the consistency of Jell-O. Embedded in the cytosol are organelles, vesicles, and vacuoles, as well as smaller structures that are not enclosed by membranes. These include granules of energy-rich glycogen, droplets of stored fat, *ribosomes*—tiny, round organelles, about 15 to 30 nm in diameter—and *proteasomes*—barrel-shaped organelles, slightly smaller than ribosomes.

Ribosomes are tiny assemblies of RNA and protein that are indispensable to protein synthesis and therefore to cell maintenance and cell reproduction. Ribosomes [Latin, *soma* = body + "ribo" because they contain ribonucleic acid (RNA)] act as workbenches on which protein molecules are stitched together with peptide bonds. Ribosomes in the cytosol are called "free ribosomes" (Figure 4-9A). Ribosomes associated with a cell's internal membranes are called "membrane-bound ribosomes."

Proteasomes are protein complexes that degrade old proteins in the cytosol and recycle their amino acids. Proteins, like other machines, eventually wear out or become damaged, and the cell must dispose of them. Proteasomes function as molecular garbage disposals: each contains a hollow central chamber, in which denatured polypeptides are broken down, and feeder assemblies at each end, which help push denatured polypeptides into the central chamber (Figure 4-9B).

The jellylike cytosol, which surrounds the cell's organelles, contains ribosomes, glycogen, fat, thousands of enzymes, and much more.

The Endoplasmic Reticulum Is a Folded Membrane

The **endoplasmic reticulum (ER)** [Greek, *endon* = within + *plasmein* = to mold + Latin, *reticulum* = network] is an elaborate structure needed for cell reproduction, since it is important in the synthesis of proteins and lipids, which the cell exports.

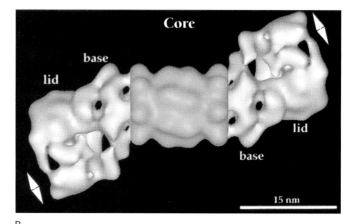

Figure 4-9 Ribosomes and proteasomes. A. A computer-generated model of the three-dimensional structure of a single ribosome. B. A proteasome, the cytosolic garbage disposal. Denatured proteins enter through the "lids" and are digested in the central "core." (A, Harry Noller, University of California, Santa Cruz; B, Wolfgang Baumeister, Max-Planck Institute, Martinsried, Germany)

The ER resembles the marbling in marble cake. When biologists cut a cell into uniform slices, as we would slice a loaf of marble cake, the ER in each slice, or "section," forms an extensive and convoluted network. Careful study of successive sections, like those in Figure 4-10, show that the ER is actually a single sheet of membrane enclosing a network of interconnected cavities and channels called the **lumen** [Latin, *lumen* = light, an opening]. The ER lumen is like a three-dimensional maze. The ER membrane is continuous with the outer membrane of the nucleus. In many eukaryotic cells, the lumen of the ER makes up about 15 percent of the cell's volume, and the membranes of the ER more than half the total membrane.

The ER consists of two parts that are distinct in both appearance and function. The **rough ER** is studded with ribosomes on the cytosolic side of the membrane and is mostly devoted to modifying recently synthesized proteins, especially those destined for export to organelles or out of the cell (Figure 4-10B). The **smooth ER** has no ribosomes and is mostly devoted to the synthesis and metabolism of lipids (Figure 4-10C). Another important function of the smooth ER is detoxification of substances such as alcohol.

Both rough and smooth ER are present in most eukaryotic cells. But since the two kinds of ER serve different purposes, some specialized cells may have especially large amounts of one or the other. Cells specialized for the production of secreted proteins, such as those in the pancreas, have large amounts of rough ER. Cells specialized for the production of lipids, such as the cells that produce steroid hormones, contain large amounts of smooth ER.

The endoplasmic reticulum (ER), with its mazelike internal lumen, consists of rough ER and smooth ER.

The Golgi Complex Directs the Flow of Newly Made Proteins

The **Golgi complex** functions as both a packaging center and a traffic director. During cell reproduction and maintenance, it helps form structures that stay in the cell, such as lysosomes. It also contributes to specialized cell function by helping prepare materials for export outside the cell.

The Golgi complex, usually found near the nucleus, consists of sets of flattened discs, like stacked dinner plates (Figure 4-11). Each stack is about 1 μm in diameter and typically consists of about 6 discs. Surrounding the stacks of flat discs are some smaller, spherical vesicles.

Every eukaryotic cell has a Golgi complex, but the number of stacks varies among different cell types. The cells with the most Golgi stacks are those that secrete large quantities of glycoproteins, which are proteins, such as the albumin in egg white, that have sugars or carbohydrates attached. Mucus is also made of glycoproteins, and the mucus-secreting goblet cells of the intestines have particularly large numbers of Golgi stacks (Figure 4-11). Most of the proteins that are exported to the outside of the cell are glycoproteins. Biologists like to say, "All exported proteins are sugar-coated."

To see what the Golgi complex does, George Palade and Marilyn Farquhar, at Rockefeller University, labeled newly made

Figure 4-10 **The two parts of the endoplasmic reticulum, the rough ER and the smooth ER, have different functions.** A. The rough ER is continuous with the smooth ER. B. The rough ER is studded with ribosomes and is mostly devoted to synthesis and modification of proteins, especially those destined for export to organelles or out of the cell. C. The smooth ER has no ribosomes and is mostly devoted to the synthesis and metabolism of lipids and the detoxification of substances such as alcohol. (*B, Don Fawcett/Science Source/Photo Researchers, Inc.; C, David M. Phillips/Visuals Unlimited*)

glycoproteins with radioactive tracers. Proteins destined for export appeared first in the rough ER and then in the Golgi complex (Figure 4-11). The Golgi complex modifies the glycoproteins, then packages them into membrane-enclosed secretory vesicles. The vesicles eventually fuse with the plasma membrane and discharge their contents to the outside of the cell.

The Golgi complex also manages the flow of proteins to different destinations. The Golgi complex "addresses" glycoproteins by modifying the carbohydrates (which were originally attached to the glycoproteins in the lumen of the ER). Proteins acquire specific "tags" that direct them to specific destinations within the cell—such as lysosomes, or to the cell's export machinery. Other proteins are labeled specifically for export outside of the cell.

The Golgi complex packages and labels glycoproteins for destinations both inside and outside the cell.

The Lysosomes Function as Digestion Vats

Lysosomes, present in all eukaryotic cells, are small vesicles, enclosed by a single membrane. They are the sites of many chemical reactions. In some cells, they are the sites of the first digestion of food.

Lysosomes vary widely in size, but all contain enzymes that break down proteins, nucleic acids, sugars, lipids, and other complex molecules. The large vacuole of a plant cell contains similar enzymes, and we may consider it to be a giant lysosome. Lysosomes are particularly numerous in cells that actively perform **phagocytosis** [Greek, *phagein* = to eat + *kytos* = hollow vessel, or cell], the process by which cells take in and consume large particles of solid food. Free-living amebas and many other single-celled eukaryotes feed primarily by phagocytosis (Figure 4-12). Many of our own white blood cells regularly engulf and digest bacteria, viruses, and assorted cellular debris, for example, and these white cells are packed with lysosomes. Once a cell has engulfed material within a vesicle, the vesicle fuses with a lysosome, whose enzymes then digest the engulfed material.

The membrane of a lysosome keeps its enzymes from digesting the cytosol's own proteins. If the lysosome membrane breaks down, however, the cell begins to digest itself and dies.

Lysosomes are made by other organelles in the cell. Lysosomes first form from membranes derived from the Golgi complex. Their digestive enzymes are made on the ribosomes of the rough ER. They are packaged into lysosomes after passing through the lumen of the ER, where each lysosomal enzyme acquires a special attached sugar. This sugar provides the lysosomal proteins with an address tag, so that the Golgi complex directs them to lysosomes rather than exporting them to the outside of the cell.

Lysosomes contain enzymes that the cell uses to digest food, wastes, and foreign invaders. Lysosome membranes normally isolate these dangerous enzymes from the rest of the cell.

Figure 4-11 Newly made proteins pass through the ER and then the Golgi complex. A. Transmission electron micrograph of a Golgi complex or "stack." B. Vesicles on the surface of the Golgi complex shuttle glycoproteins from the rough ER to the Golgi stack, which "labels" the glycoproteins and packages them into vesicles for secretion. (A, Cabisco/Visuals Unlimited)

Figure 4-12 Phagocytosis. An ameba engulfing algae. *(Michael Abbey/Visuals Unlimited)*

Peroxisomes Produce Peroxide and Metabolize Small Organic Molecules

Peroxisomes are another kind of membrane-enclosed vesicle, widespread in all kinds of eukaryotes (Figure 4-13). **Peroxisomes** contain enzymes that use oxygen to break down various molecules by means of biochemical reactions that produce hydrogen peroxide (H_2O_2), hence their name. Because hydrogen peroxide is highly reactive and can easily damage cells, peroxisomes also contain a protective enzyme, called *catalase*, that breaks down hydrogen peroxide into water and oxygen.

Peroxisomes, whose diameters can range from 0.15 μm to more than 1 μm, can occupy as much as half the volume of a cell. They are especially abundant in cells that synthesize, store, or break down lipids. For example, peroxisomes in the seeds of some plants contain an enzyme that breaks down stored fats and therefore provides energy for the growing plant embryo.

Figure 4-13 Peroxisomes. Peroxisomes contain enzymes that degrade molecules in reactions that produce hydrogen peroxide. *(D. Friend and D. Fawcett/Visuals Unlimited)*

Peroxisomes contain a chemical dangerous to the cell itself—hydrogen peroxide. Cells use enzymes that give off hydrogen peroxide to break down fats and oils.

Mitochondria Capture the Energy from Small Organic Molecules in the Form of ATP

Mitochondria [singular, mitochondrion; Greek, *mitos* = thread + *chondrion* = a grain] are the organelles most responsible for the cell's ability to obtain energy from nutrients. In cells that do not directly harvest sunlight, these subcellular power plants produce nearly all of the ATP that powers the chemical reactions of the cell. This ATP production depends on the organization of the internal and external membranes of mitochondria.

Even under the best light microscope, mitochondria, often less than 1 μm in diameter, can barely be seen. However, the transmission electron microscope (TEM) reveals that mitochondria are some of the most numerous organelles in most eukaryotic cells (Figure 4-14). Different kinds of cells have different numbers of mitochondria. A liver cell, for example, may contain several thousand mitochondria, which together make up about a quarter of the cell volume. Cells that use particularly large amounts of energy, such as those of the heart, may contain still more.

The transmission electron microscope also reveals the mitochondrion's characteristic structure. Two membranes, 6 to 10 nm apart, separate a mitochondrion's interior from the cytosol. The outer membrane of a mitochondrion is smooth, but the inner membrane forms elaborate folds, called *cristae*. Mitochondria come in a variety of sizes and shapes. Time-lapse photography with a phase contrast microscope reveals that mitochondria may be constantly in motion, changing shape, fusing, and dividing.

Mitochondria are unusual among organelles not only in having double membranes: they also have their own DNA and make some of their own proteins. These characteristics have led biologists to think that mitochondria arose by the incorporation of captured bacteria, as we discuss in Chapter 18.

Mitochondria synthesize energy-rich ATP in their mazelike interior.

A Plant Cell's Chloroplast Is Just One Kind of Plastid

Plastids are organelles in plant cells that, like mitochondria and nuclei, are surrounded by a double membrane. Plastids are present in nearly all plant cells. The most important plastid in the **chloroplast,** present only in green plant cells and some protists, which uses the energy of sunlight to build energy-rich sugar molecules in the process of **photosynthesis**

Figure 4-14 Mitochondria manufacture ATP. The outer membrane encloses the whole organelle, and the inner membrane encloses the matrix. The inner membrane is folded into cristae, which extend into the matrix. *(A, Don Fawcett/Visuals Unlimited)*

[Greek, *photo* = light + *syntithenai* = to put together]. Most organisms on Earth therefore ultimately depend on chloroplasts for food. Plants also have other kinds of plastids, which serve as storage depots for various kinds of molecules (Figure 4-15).

During photosynthesis, chloroplasts make sugars. The chloroplasts also serve as temporary storage depots for the newly made sugars (in the form of starch grains). The size, shape, and number of chloroplasts in a cell vary from species to species and from cell type to cell type. A leaf cell in a flowering plant, for example, typically contains several dozen chloroplasts. A stem cell contains only a few, and a flower cell may contain none at all.

A typical chloroplast is a large, round, green organelle. The distinctive shade of green comes from the pigment *chlorophyll*. The chloroplast is 5 to 10 μm in diameter, but only 2 to 4 μm high, a thick disc. Like a mitochondrion, a chloroplast has an intricate internal structure of folded membranes. But, unlike in a mitochondrion, the chloroplast's folded membranes are not continuous with the inner membrane. Instead, the internal membranes lie in flattened, disclike sacs, called **thylakoids** [Greek, *thylakos* = sac + *oides* = like]. The thylakoids are piled like plates (Figure 4-15A). Each stack, of ten or more discs, is called a **granum** [plural, grana; Latin, *granum* = grain] because the stacks resemble the granaries that farmers use to store grain. Each chloroplast contains many grana.

Like mitochondria, chloroplasts have double membranes and DNA, and they perform protein synthesis. Because of these and other clues, biologists have concluded that chloroplasts originated as free-living organisms that were somehow "captured" by early eukaryotic cells.

Two other kinds of plastids are chromoplasts and amyloplasts. **Chromoplasts** contain the pigments that give yellow, orange, and red colors to flowers, fruits, vegetables, and autumn leaves (Figure 4-15B). Chromoplasts arise from chloroplasts through the reshaping of the inner membrane and the breakdown of chlorophyll. **Amyloplasts** do not derive directly from chloroplasts, but from *proplastids*, the common precursors of all plastids. Amyloplasts, which store starches, are abundant in roots such as sweet potatoes and turnips, as well as in wheat, rice, and other seeds (Figure 4-15C).

Chloroplasts turn solar energy into chemical energy in the process of photosynthesis. Other kinds of plastids store pigments or starches.

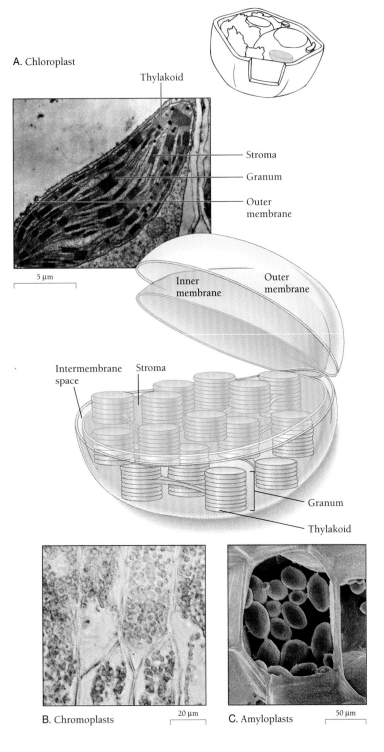

A. Chloroplast

B. Chromoplasts

C. Amyloplasts

Figure 4-15 Plastids. A. A chloroplast uses sunlight to build sugar molecules, which are temporarily stored within the chloroplast. A chloroplast's green color comes from the pigment *chlorophyll*—a central player in photosynthesis. Like mitochondria, chloroplasts have an outer membrane, which encloses the whole organelle, and an inner membrane, which encloses a matrix. In mitochondria, the inner membrane is highly folded, whereas in chloroplasts the inner membrane is smooth. In chloroplasts, a third membrane, derived from the inner membrane, is even more highly folded than the inner membrane of mitochondria. This membrane forms pockets called thylakoids, arranged in stacks called grana. The intermembrane space is called the stroma. B. Chromoplasts in the cells of a yellow petal (from a tulip). C. Amyloplasts in raw potato. (A, Grant Heilman Photography; B, Biophoto Associates/Science Source/Photo Researchers, Inc.; C, Jeremy Burgess/Science Photo Library/Photo Researchers, Inc.)

Quiz

Some of the specialized proteins of the cytoskeleton are closely related to muscle proteins. While muscle proteins are essentially permanent structures, however, the cytoskeleton is constantly dissolving and reforming itself.

Microtubules are hollow cylinders about 25 nm in diameter, which vary in length (Figure 4-16A). Microtubules play a critical role in cell division. When a cell divides, microtubules distribute the DNA and other materials to the two daughter cells.

Although microtubules appear throughout the cytoplasm, and even in the nucleus, their assembly always seems to originate in structures called **microtubule organizing centers (MTOCs)** (Figure 4-17). In the cytoplasm of most eukaryotic cells, the major MTOCs are near the nucleus, in a zone called the **centrosome**—a region that in many cells also contains a distinct organelle called the **centriole.** The MTOCs of the centrosome play an important role in cell division, as we discuss in Chapter 8.

The cylinders of microtubules consist of two globular protein molecules called **tubulins.** Pure tubulin in a test tube can form microtubular structures all by itself. In living cells, however, up to 50 other proteins may associate with microtubules, influencing their distribution and stability. Some of these *microtubule-associated proteins* (or *MAPs*) bundle microtubules together, allowing them to establish the shapes of cells. In plants, for example, the layout of the cellulose fibers that determine the shapes of plant cells depends on the arrangement of microtubules and MAPs.

Other microtubule-associated proteins function as **motors,** which use the energy of ATP to move along microtubules. The motors can attach to other subcellular components and carry them, as freight, along microtubule tracks. Some motor molecules (called *dyneins*) travel away from the MTOC end of microtubules, while others (called *kinesins*) travel toward the MTOCs, as illustrated in Figure 4-17B.

Actin filaments, also called *microfilaments,* are much finer than microtubules—only about 7 nm in diameter (Figure

What Does the Cytoskeleton Do?

Eukaryotic cells have a complex network of protein filaments, called the cytoskeleton, which is visible with the TEM (Figure 4-16). Three types of filaments make up the cytoskeleton: microtubules, actin filaments, and intermediate filaments (Figure 4-16). Like our own bony skeletons, the cytoskeleton provides structure and support. The name "cytoskeleton" is slightly misleading, however, since the cytoskeleton does more than give mechanical support for the cytoplasm and its component organelles. It also provides the force for most of the movements of cells, including the transport of materials and changes in the overall shape of a cell.

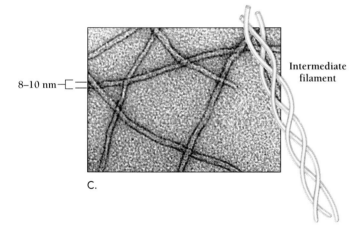

Figure 4-16 The cytoskeleton of a cell. The cytoskeleton is a complex network of three types of protein filaments, each with distinctive appearances and distributions. A. The microtubules emanate from a microtubule organizing center (MTOC) and serve as tracks for freight carried by motor molecules. B. Actin filaments are often closer to the cell surface. C. Intermediate filaments reinforce cells, especially at places of particular stress. (A, Richard Wade; B, Roger Craig; C, Roy Quinlan)

4-16B). Actin filaments are often anchored to the cell surface and provide the force for movement and shape changes. Actin filaments consist of a single kind of protein, called actin, which is also present in muscle fibers.

In muscle, actin filaments are permanent structures that interact with other proteins to produce the force leading to contraction. In nonmuscle cells, actin filaments are constantly being built and destroyed. Even in the test tube, actin molecules can self-assemble to form a two-stranded helix apparently identical to the actin filaments in intact cells. In cells, they form cross-linked cables that provide mechanical support. They also associate with other proteins to furnish musclelike contractile forces. Other proteins associate with actin filaments to act as molecular winches that pull on actin cables.

Intermediate filaments, usually 8 to 10 nm in diameter, are fibrous proteins such as keratin (Figure 4-16C). Keratin strengthens wool, hair, fingernails, the outer layers of the skin, and other epithelial sheets. All intermediate filaments seem to be common in parts of cells that are subject to mechanical stress, suggesting that these filaments strengthen cells and tissues (Box 4.2).

The cytoskeleton is both skeleton and muscle system for the cell. Hollow microtubules, along with the microtubule organizing center near the centrosome or centriole, provide the cell with a shape. Microfilaments supply the dynamic contractile force, and intermediate filaments provide strength.

WHAT DO MEMBRANES DO?

We saw earlier in this chapter that cell size is partly limited by the amount of membrane, which controls the transfer of materials in and out of the cell, just as a system of freeways and rail lines carries materials in and out of a city. Membrane, then, is as important to cells as a transportation system is to a city.

But membranes do more than just move materials in and out of the cell. In the past 30 years, cell biologists and biochemists have found four major roles for cellular membranes:

1. Membranes are essential boundaries that separate the inside from the outside.
2. Membranes regulate the contents of the spaces they enclose.
3. Membranes serve as a "workbench" for a variety of biochemical reactions, especially those involving the metabolism of lipids and the secretion of proteins.
4. Membranes participate intimately in energy conversions (as we discuss in Chapters 6 and 7).

What Kinds of Molecules Do Membranes Contain?

The plasma membranes of all cells look the same, appearing in cross section as two dark lines, 7 to 10 nm apart, with a lighter zone between them. The membrane consists, then, of at least two parallel layers separated by a slight space. Our first question concerns the structure of these layers. What molecules make up the layers, and what holds them in the same characteristic structure in all organisms?

The first insight into the molecular structure of membranes came from studies of red blood cells. Mammalian red

Figure 4-17 Microtubules radiating out from the microtubule organizing center. Most MTOCs are near the nucleus in a zone called the centrosome, which plays a central role in cell division. *(B, K.G. Murti/Visuals Unlimited)*

blood cells are ideally suited for membrane studies: they have no internal membranes, no organelles, and few types of macromolecules other than hemoglobin, the red oxygen-binding protein. Red blood cells do not even have nuclei. Like empty plastic bags, they can be squashed into many shapes. This flexibility allows them to pass more easily through the tiny vessels of the circulatory system.

Biochemists can obtain essentially pure plasma membrane by breaking open the red blood cells to release the hemoglobin and then using a centrifuge to separate the empty plasma membranes, called red blood cell "ghosts." Biochemical analysis has shown that these membranes consist mostly of lipids and proteins, as well as some carbohydrates. In the 1920s, biochemists calculated that a red cell ghost contained just enough lipid to enclose the cell with a continuous layer two molecules thick. This result suggested that the ghosts might indeed consist of a lipid "bilayer."

The most important lipids in the membrane are phospholipids. Recall from Chapter 3 that phospholipids are highly amphipathic; that is, they have a hydrophilic (water-loving) head and a hydrophobic (water-fearing) tail. When pure phospholipids are placed in water, they can spontaneously organize into sheets that are two layers thick. The oily tails of the lipids point inward toward one another, while their hydrophilic heads point outward. Once biologists understood this, they wondered if a biological membrane was in fact just a phospholipid bilayer. If so, where would the membrane proteins lie? Since biologists knew that proteins studded both the inner and outer surfaces of the cell membrane, they hypothesized that the membrane proteins were scattered on the inner and outer surfaces of the lipid bilayer, in contact with the polar groups of the phospholipid. But this is not quite correct.

Researchers have now determined that proteins not only occupy the two layers of the plasma membrane, but also occupy the space *in between* the two layers. Some membrane proteins (called *peripheral proteins*) are located only on the outside or the inside of the membrane; others (called *transmembrane proteins*) span the entire lipid bilayer (Figure 4-18).

Membrane proteins, like membrane lipids, are amphipathic, with both hydrophilic and hydrophobic regions. The hydrophilic regions interact with the aqueous solutions of the cytosol and the cell exterior, as well as with the hydrophilic heads of the lipids. The hydrophobic regions interact with the hydrophobic interior of the membrane.

The two lipid layers of the membrane differ slightly from one another (Figure 4-18B). Certain enzymes, for example, can break down the lipids in one layer, but not the other. Another difference is that carbohydrates appear only on the outer face. Some of these carbohydrates are attached to proteins (glycoproteins), and some are attached directly to membrane lipids (glycolipids). The inner and outer layers also differ in the way they interact with membrane proteins. Proteins that span the entire thickness of the membrane, for example, consistently orient themselves in a certain direction (Figure 4-18). We will see later in this chapter that the orientation of a protein is crucial because some membrane proteins work like turnstiles,

BOX 4.2

The 9 + 2 Structure of Flagella

Many prokaryotes, including *Escherichia coli,* have long fibers on their surfaces. Each of these structures is about 10 to 20 nm thick and up to 10 μm long and is called a **flagellum** [Latin, = whip; plural, flagella]. The flagellum's proportions are similar to those of a half-inch rope 40 feet long. Flagella act as propellers, driving bacteria at speeds of tens of micrometers per second. In terms of the bacteria's own dimensions, this would be equivalent to a 15-foot-long car moving at 100 miles per hour.

The flagella of eukaryotes are different from those of prokaryotes. When these structures are long and few in number, they are called **flagella** (Figure A). When they are short and numerous, they are called **cilia** (Figure B). However, both cilia and flagella have the same characteristic cross section. Each consists of a ring of nine sets of microtubules, surrounding two more sets in the center. The microtubule organizing center of a cilium or flagellum also has a nine-membered ring of microtubular structures and is called a basal body. Protists and some cells of animals (including sperm) use flagella or cilia to propel themselves. Cilia on the surfaces of the respiratory and reproductive tracts push fluids along the surface.

Figure A Flagella. The protist *Euglena* (left) has a long flagellum, which it whips around like a propeller. An electron micrograph (right) reveals the characteristic 9 + 2 organization of the *Euglena* flagellum, which resembles that of the cilia shown in Box 4.1 and the 9-fold symmetry of the microtubule organizing center in Figure 4-17. (Left, David M. Phillips/Science Source/Photo Researchers Inc.; right, Dr. Gopal Murti/Science Photo Library/Photo Researchers, Inc.)

Figure B Cilia. Cilia of a *Paramecium,* as seen with a confocal microscope. (Anne Fleury, Laboratoire de Biologie Cellulaire 4, Université Paris Sud, CNRS)

Figure 4-18 A plasma membrane. A. Carbohydrates stud the outer surface of the membrane (as glycolipids and glycoproteins). Transmembrane proteins pass through the lipid bilayer. B. Edge of cell showing plasma membrane. *(Biophoto Associates/Photo Researchers, Inc.)*

letting molecules pass through the membrane in only one direction.

The plasma membrane consists of an asymmetric phospholipid bilayer. Some membrane proteins pass through the interior of this bilayer, and others associate with the inner or outer surface. Carbohydrates occupy the outer surface only.

How Does the Structure of a Membrane Establish Its Functional Properties?

The plasma membrane has a consistency similar to that of thin oil, such as you might use to lubricate a sewing machine or a bicycle. Water can diffuse rapidly from one side of the membrane to the other. Small, nonpolar molecules can diffuse across a membrane and within it. The membrane selectively pumps other molecules from one side to the other.

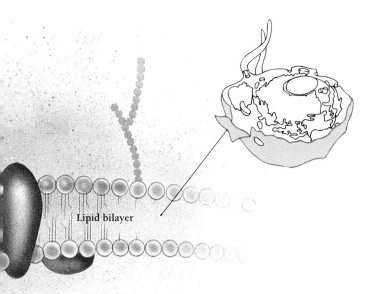
Lipid bilayer

What about the molecules of the membrane itself? Are they also able to move about? To answer this question, biochemists marked certain membrane phospholipids with a chemical group that they could detect with a special method (called electron spin resonance). The biochemists could then study the movements of the marked lipids within the membrane. These experiments showed that membrane lipids move freely within the plane of the membrane but that the lipids in one layer do not usually move to the other layer.

Proteins also move freely within each layer of the membrane. To study the movement of proteins, biochemists marked membrane proteins with a fluorescent (glowing) dye, then used a laser to bleach the dye in a small region of the membrane. They then watched the membrane through a microscope to see if new fluorescent molecules would appear in the bleached area. They found that glowing proteins entered from adjacent areas. These experiments showed that about half of the proteins in a membrane move freely, and about half remain tightly bound in their original position within the membrane.

These experiments showed that both lipid and protein molecules move about with relative ease within each layer of the membrane. The molecules move easily within each layer because the layers are **fluid**. On the basis of this fluidity, two biochemists at the University of California San Diego, Jonathan Singer and Garth Nicholson, proposed a new model of membrane structure, called the **fluid mosaic model.** Singer and Nicholson's model emphasizes that proteins and lipid molecules can usually move freely within the lipid bilayer. Since this model was first advanced in 1972, cell biologists have used it as a basis for understanding all cellular membranes. Today we summarize the fluid mosaic model—somewhat modified since Singer and Nicholson first proposed it—as follows:

1. The basic structure of the membrane is a lipid bilayer, with two sheets of phospholipids arranged tail to tail.
2. Proteins are dispersed through the membrane like the individual pieces of stone or tile in a mosaic. Proteins contribute to membrane structure and function. Some proteins span the membrane, while others are confined to the inner or outer surface.
3. The membrane is fluid; both protein and lipid molecules move freely within the plane of the membrane.
4. The lipid bilayer serves as a hydrophobic barrier, confining hydrophilic molecules to the inside or the outside of a cell (or organelle).
5. Some membrane proteins help transport specific molecules across the membrane.

HOW DO MEMBRANES REGULATE THE SPACES THEY ENCLOSE?

Throw a handful of raisins into a pot of boiling water and the raisins gradually fill with water, swelling until they resemble the grapes from which they were made (Figure 4-19). The raisin skin is a kind of membrane. Water and other small molecules easily pass through the skin, but larger molecules, such as proteins or cellulose, cannot pass through the skin. Large molecules inside the raisin are trapped; those outside are excluded. Such a membrane is said to be **selectively permeable**, meaning that it allows the passage of some molecules (especially of water), but not of others.

Plasma membranes are also selectively permeable. Proteins, for example, cannot directly move through plasma membranes at all. Nor can many ions, such as sodium ions. The regulation of a cell's internal environment—cellular homeostasis—depends on membrane molecules that actively pull specific small molecules and ions through the membrane.

Why Does Water Move Across Membranes?

Before we can discuss how membranes regulate a cell's internal environment, however, we must first talk about the general rules that govern the movement of water and dissolved substances. These rules are the same physical laws that govern the movement of water into and out of raisins.

What is diffusion?

If you pour a teaspoon of salt into a pot of water, most of the salt will at first drop to the bottom of the pot in a little pile. Gradually, the salt will dissolve into sodium ions and chloride ions. At first, the concentration of ions will be very high near the spot where you first poured the salt. The **concentration** is simply the number of molecules in a given volume. However, as time passes, the salt ions will move about randomly in amongst the water molecules until the ions are distributed evenly throughout the pot of water. Even after the ions are distributed evenly, they continue to move about randomly. Since as many ions move out of a given area as into it, however, the net charge is still zero. The random movement of like molecules or ions from an area of high concentration to an area of low concentration is called **simple diffusion** [Latin, *diffundere* = to pour out]. Stirring speeds this process, but diffusion will happen all by itself if you wait long enough.

The presence of an area of high concentration adjacent to an area of low concentration is called a **concentration**

Figure 4-19 Concentration gradient. A. A raisin skin is a selectively permeable membrane. The high concentration of sugar molecules inside the raisin creates a concentration gradient. Water flows inside. B. In a concentrated sugar solution, the water in the swollen raisin flows outward. Because raisins are so sweet, such a sugar solution would have to be very concentrated. *(Photos, courtesy of Amy Dunleavy)*

gradient (Figure 4-19). Concentration gradients are virtually everywhere. Cigarette smoke moving from the smoking section of a restaurant to the nonsmoking section creates a concentration gradient from an area of dirtier air to an area of cleaner air. A teaspoon of sugar poured into a cup of coffee creates a concentration gradient—for a while.

Concentration gradients are always temporary: simple diffusion, ordinary mixing, and turbulence all tend to eliminate concentration differences. How fast diffusion occurs depends on a variety of factors, including temperature, the sizes of the molecules, and the steepness of the gradient. At higher temperatures, molecules move faster and so diffuse faster. Small molecules generally diffuse faster than large ones. Finally, the greater the difference in concentration at the two ends of a gradient, the faster diffusion occurs.

In simple diffusion, solutes move from an area of high concentration to an area of low concentration.

What is osmosis?

Diffusion occurs across membranes as well as through open solutions. For example, raisins immersed in a pot of plain water fill with water because the raisins have a high concentration of sugar and other molecules. The sugar molecules cannot pass easily through the skin of the raisin and out into the water. However, water molecules move in and out of the raisin easily. From the perspective of a water molecule, then, a raisin is an area of low water concentration. The difference between the sugary inside of the raisin and the watery outside constitutes a concentration gradient, along which the water molecules move.

The movement of the water molecules along this concentration gradient into the raisin is called osmosis. **Osmosis** [Greek, *osmos* = push, thrust] is the movement of water across any selectively permeable membrane in response to a concentration gradient.

As water flows in, it creates a force against the inside surface of the membrane. This **osmotic pressure,** comparable to physical pressure, stretches the raisin's skin until it begins to resist. If the skin is weak, it may break. Otherwise, it pushes back, meeting osmotic pressure with mechanical pressure. The physical pressure exerted by the raisin's skin counterbalances the osmotic pressure.

Osmosis is the movement of water across any selectively permeable membrane in response to a concentration gradient. The pressure exerted by osmosis, called osmotic pressure, is comparable to mechanical pressure.

Which way does water flow through a membrane?

We have seen what happens when a closed membrane (the raisin) with a high concentration of solutes (in this case, various sugars) is dropped into a solution with a low concentration of solutes (plain water). What happens if we take the water-filled raisins and put them into a highly concentrated solution of sugar?

The answer depends on how concentrated the sugar is. Look inside a can of fruit cocktail and you'll see what happens when the concentrations of sugar in a grape and the solution are about the same. The grape neither shrinks nor expands, but retains its normal shape. Put the grape into a highly concentrated sugar solution, however, and you will see it actually shrink as the water molecules inside move out through the skin, following the concentration gradient (Figure 4-19).

In the raisin example, the movement of water into the raisin is greater than the movement of water out of the raisin.

The water in which the raisin sits is said to be **hypotonic** [Greek, *hypo* = under + *tonos* = tension] compared to the inside of the raisin. In ordinary canned fruit cocktail, the water movements in and out of grapes are balanced perfectly. In that case, the sugary solution in which the grape is sitting is said to be **isotonic** [Greek, *isos* = equal] with the inside of the grape. If we put grapes into too heavy a syrup, however, the concentration of the syrup solution is so high that water moves out of the grapes. In that case, the solution outside the grape is said to be **hypertonic** [Greek, *hyper* = above] compared to the inside of the grape.

Cells, it turns out, behave in the same way. A red blood cell immersed in an isotonic solution, such as blood plasma, or a suitable salt solution, remains intact (Figure 4-20A). But if we immerse the red blood cell in a highly concentrated (hypertonic) solution, the cell shrivels (Figure 4-20B). The selectively permeable membrane allows water to pass freely out of the cell. When enough water flows out so that the new concentration of molecules within the cell is again the same as outside the cell, no more water flows.

Conversely, a red blood cell immersed in a hypotonic solution such as plain water will tend to fill with water and burst. When we decrease the solute concentration outside the cell, water then flows into the cell, causing it to expand (Figure 4-20C). In principle, water will flow in until the total solute concentration is equal on both sides of the membrane. In practice, however, there may be too few solute molecules on the outside for this equalization to occur, as, for example, when the cells are suspended in pure water. In this case, the water will continue to flow into the cell until the cell bursts and releases its contents. (This is how cell biologists can prepare red cell ghosts.)

When a selectively permeable membrane separates two solutions that differ in their total solute concentrations, water flows in the direction that tends to equalize the solute concentrations on the two sides of the membrane. A solution may be hypotonic, isotonic, or hypertonic, relative to the concentration of solutes on the other side of the membrane.

How does turgor pressure support plants?

When plant cells, such as those in the onion skin shown in Figure 4-21, are exposed to a hypotonic solution, water moves into the cell but does not accumulate in the cytoplasm. Instead, the inflowing water passes from the cytoplasm to a vacuole. The expansion of the vacuole pushes the surrounding cytoplasm against the **cell wall**—the rigid boxlike structure that encloses all plant cells. Conversely, when an onion cell is placed in a hypertonic solution, water flows out of the vacuole and the cell separates from the cell wall, a phenomenon called **plasmolysis** [Greek, *plasma* = form + *lysis* = loosening].

In a hypotonic solution, the flow of water into the cell creates pressure against the cell membrane. The cell wall pushes back, exerting mechanical pressure that prevents the further flow

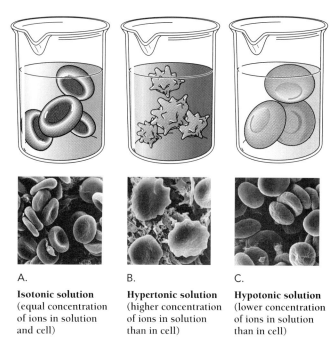

A.
Isotonic solution (equal concentration of ions in solution and cell)

B.
Hypertonic solution (higher concentration of ions in solution than in cell)

C.
Hypotonic solution (lower concentration of ions in solution than in cell)

Figure 4-20 Iso, hyper, hypo. A. A red blood cell in a isotonic salt solution remains intact. B. A red blood cell in a hypertonic salt solution shrivels. C. A red blood cell in a hypotonic salt solution swells with water. In a still more hypotonic solution, such as plain water, the cells in C would fill up and burst open. (*Photos, Dennis Kunkel/Phototake NYC*)

of water into the cell. (Water actually continues to flow equally in both directions through the plasma membrane, but there is no *net* inflow of water.) The mechanical pressure thus balances the osmotic pressure. The pressure of a plant cell's contents against its cell wall is called **turgor pressure** [Latin, *turgere* = to swell].

Turgor pressure supports nonwoody plants in the same way that water supports the shape of a water balloon. This is why plants (and water balloons) begin to droop, or wilt, when they do not have enough water. The same phenomenon accounts for the way lettuce wilts soon after it is covered with salad dressing. Because the dressing is hypertonic, it draws water out of the cells of the lettuce leaves, causing them to wilt.

The flow of water into a plant cell gives mechanical strength to plant tissue.

What Determines the Movement of Molecules Through a Selectively Permeable Membrane?

We can ask two questions about the passage of molecules through a membrane: "Which way?" and "How fast?" In addressing the first question, membrane researchers distinguish between "passive transport" and "active transport." **Passive transport** occurs spontaneously, without the expenditure of energy (Figure 4-22B). The rate of passive transport can vary,

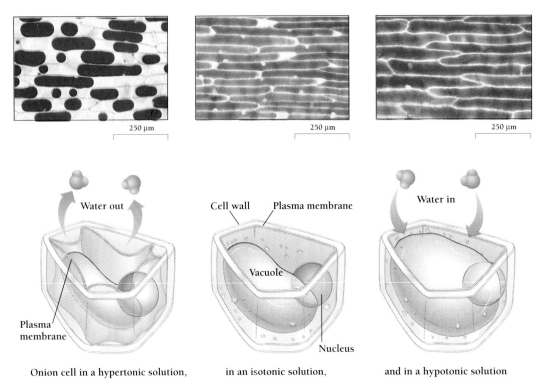

Figure 4-21 Why does water move in and out of a plant cell vacuole? Plant cells like this onion cell behave much like red blood cells (and raisins). The water content of the vacuole increases or decreases according to whether the cell is in a hypertonic, isotonic, or hypotonic solution. A. Hypertonic: the vacuole has emptied, and the cytoplasm has concentrated away from the cell wall. B. Isotonic: there is some space between the cytoplasm and the cell wall. C. Hypotonic: the cytoplasm has pushed against the cell wall, filling all the available space. (*Photos, Michael Abbey, Photo Researchers, Inc.*)

depending on whether the membrane contains specific carriers that participate in a process called "facilitated diffusion," which we discuss shortly. **Active transport** moves molecules against a concentration gradient and requires energy (Figure 4-22C).

The answer to the question "Which way?" is different for passive transport and active transport. Passive transport results in equal concentrations of a molecule on the two sides of a membrane, at least for uncharged molecules. For charged molecules or ions, passive transport results in an unequal distribution of ions. This unequal distribution happens because all cells have an excess of negative charges inside the cell membrane. (This excess results from the unequal pumping of sodium and potassium ions, which we discuss later.) The result is that the cell interior repels negative ions and attracts positive ions. Biologists can calculate how unequal the distribution of ions should be, and any deviation from the predicted distribution indicates that active transport has occurred.

Active transport comes in several forms. In one form, cells move ions across the membrane so that a net charge accumulates on one side. For example, an important active transporter, the **sodium-potassium pump,** simultaneously transports sodium ions (Na^+) out of cells and potassium ions (K^+) into cells, using the energy of ATP. Cells have more negative charges inside than outside. The net negative charge results from the sodium-potassium pump, which moves three sodium ions out of cells for every two potassium ions that it moves in.

Facilitated diffusion accelerates passive transport

Some molecules move across membranes much faster than would be expected on the basis of diffusion alone, even though their final concentrations are consistent with passive transport. Such **facilitated diffusion**—an increased rate of passive transport—depends on the action of specific molecules within the membrane that help, or "facilitate," transport. Glucose, for example, usually moves across cell membranes by means of facilitated diffusion.

A transmembrane protein called the glucose transporter is responsible for moving glucose across the cell membrane, for example, in red blood cells. To show that this protein causes the facilitated diffusion of glucose, researchers studied the rates of entry of radioactive glucose into liposomes—tiny bubbles of artificial phospholipid bilayers that behave like membranes. Pure liposomes allow the entry of scarcely any glucose. But

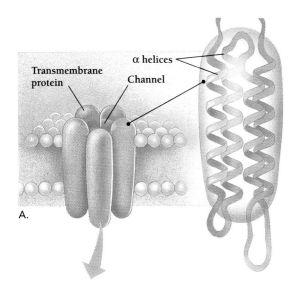

Figure 4-22 **Passive and active transport.** A. Transmembrane proteins help transport molecules and ions across the cell membrane. B. In passive transport, no energy is required. In channel-mediated transport, ions or small molecules pass through the transmembrane channels without regulation or help. In facilitated diffusion, molecules move faster than they ordinarily would, thanks to the action of transporter proteins that carry molecules across the membrane. Transporters bind only to certain molecules, such as glucose. C. Active transport requires energy, as well as a carrier protein that couples transport with energy expenditure. (1) The sodium-potassium pump binds to ATP and Na^+ ions on the *inside* of the plasma membrane and to K^+ ions on the *outside* of the membrane. It then transports Na^+ and K^+ in opposite directions, using the energy of ATP. (2) In cotransport, a cell membrane transports molecules using the energy from a concentration difference across a membrane (instead of using ATP). Cotransporter proteins transport glucose using a concentration difference in Na^+ ions, linking the "uphill" transport of one molecule or ion to the "downhill" transport of another.

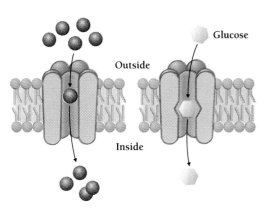

B. Passive transport

C. Active transport

when the glucose transporter is added, the rate of glucose entry increases dramatically, showing that the transporter facilitates the entry of glucose into the liposome.

Transport proteins, such as the glucose transporter, bind to the molecules they transport, then carry the molecules across the membrane (Figure 4-22). Transporters bind only to specific molecules. The glucose transporter, for example, transports only glucose and other similar sugars.

Active transport uses energy to move ions

How can a membrane accomplish active transport? To transport a substance actively across a membrane requires a special "pump" mechanism that couples transport to an energy source (such as ATP). For example, the sodium-potassium pump binds to ATP and Na^+ ions on the inside of the plasma membrane and to K^+ ions on the outside of the membrane (Figure 4-22). When inserted into liposomes, the pure protein transports Na^+ and K^+ in opposite directions, using the energy obtained by hydrolyzing ATP.

Somewhat surprisingly, other examples of active transport do not directly depend on ATP. Instead, the cell harnesses energy from an already existing concentration difference across a membrane, usually of Na^+ or H^+ ions.

The coupled transport of two substances is called **cotransport** and depends on specific transmembrane protein molecules, called cotransporters. Cotransporters couple the "uphill" transport of one molecule or ion to the "downhill" transport of another. This mechanism is comparable to the action of a children's seesaw. As gravity pulls one child down, the child at the other end is raised up against the force of gravity. In a cotransporter used by the cells that line the small intestine, for example, the concentration difference of Na^+ ions supplies the energy for the cells to take up glucose (Figure 4-22).

We can see that membranes are important regulators of a cell's contents. They allow water to pass freely, with the net movement dependent on the relative solute concentrations inside and outside the cell. They also admit slightly larger molecules, but selectively. In some cases, they speed the passive transport of specific small molecules such as glucose with transport proteins. In other cases, they harness the energy of ATP or of existing concentration gradients to accumulate or exclude particular molecules or ions.

While we have spoken almost entirely about plasma membranes, the internal membranes of eukaryotic organelles are similarly important in regulating the contents of the spaces that they enclose. In Chapters 6 and 7, we will discover how crucial the membranes of some organelles are to the production of ATP.

Molecules pass through plasma membranes by means of diffusion, facilitated diffusion, and active transport. Active transport mechanisms use energy from ATP or from a concentration gradient of an ion or molecule different from the one being transported.

HOW DO MEMBRANES INTERACT WITH THE EXTERNAL ENVIRONMENT?

The plasma membrane of a cell functions like the outer surface of an organism. In multicellular organisms, the cell itself creates its own immediate environment. Most nonanimal cells, for example, surround themselves with a rigid carbohydrate structure, called the **cell wall** (Figure 4-23A). The cell wall can continue to function even after the death of the cell. In fact, some plant cells function *only* after death. Many of the vessels that conduct fluids in vascular plants, as well as many structural fibers, are composed of the walls of dead cells. Animal cells can similarly contribute to their own extracellular environment, which in many cases consists of a more diffuse network of carbohydrates and proteins, called the **extracellular matrix** (Figure 4-23B).

Cells, like organisms, must communicate and must consume and excrete materials. Membranes help accomplish all of these activities and ensure that the cell's contents are best suited to the cell's immediate challenges.

Membrane Fusion Allows the Import and Export of Particles and Bits of Extracellular Fluid

The tendency of membranes to avoid water, together with their fluidity, means that biological membranes can change shape rapidly. The ruptured membranes of red cells, for example, can reform as intact "ghosts" of cells that do a remarkably good job of regulating their internal environments by pumping ions and molecules.

The reorganization of membranes is an important part of many cellular functions. In cell division, for example, the plasma membrane of the parent cell divides and reseals itself around each daughter cell. In phagocytosis, cells engulf relatively large particles, such as microorganisms or cell debris. First, part of the plasma membrane extends outward and surrounds the particle (Figure 4-12). The membrane then fuses to form a large vesicle, which travels to the interior of the cell. The vesicle's membrane later fuses with the membrane of a lysosome, thereby exposing the vesicle's contents to the lysosome's digestive enzymes.

In contrast, in **endocytosis,** the cell takes in tiny amounts of materials in vesicles that arise by the inward folding (or invagination) of the plasma membrane. The pouchlike vesicles formed in endocytosis are tiny, about one-tenth the size of those formed in phagocytosis.

Cell biologists now recognize two types of endocytosis. In **pinocytosis** [Greek, *pinein* = to drink], the cell takes up any bits of liquid and dissolved molecules. In **receptor-mediated endocytosis,** the cell takes up only specific substances, which

Figure 4-23 The cell wall. A. A rigid, external cell wall made of carbohydrate surrounds the plasma membranes of plant and fungus cells, most prokaryotes, and many protist cells. B. An extracellular matrix of carbohydrates and proteins encloses many animal cells. (*A, Kevin Collins/Visuals Unlimited; B, Fred Hossler/Visuals Unlimited*)

it recognizes by means of special proteins on the cell's surface, called receptors. Cholesterol, which travels through the blood in complexes with an amphipathic protein called LDL (low-density lipoprotein), enters cells through receptor-mediated endocytosis (Figure 4-24A; Box 3.1).

Exocytosis is the export of molecules from a cell by a process that is approximately the reverse of pinocytosis. Cells accumulate molecules to be exported in vesicles, surrounded by membranes. These vesicles then move to the cell surface, where they fuse with the plasma membrane and release their contents to the outside of the cell (Figure 4-24B). Molecules exported by exocytosis include proteins that function outside of cells, such as the digestive enzymes of the gut, and chemical signals, such as hormones and neurotransmitters.

The regulation of exocytosis is an extremely important part of cellular function. In the nervous system, for example, intercellular signaling depends on the precise timing of neurotransmitter release by exocytosis. Rapid increases in intracellular calcium ions stimulate exocytosis, and dozens of proteins are involved in its coordination. Some of these proteins are part of the vesicle's own membrane. Other proteins are part of the plasma membrane, where the vesicle will fuse, while still other proteins form temporary bridges that allow the two membranes to fuse. Figure 4-24C shows the fusing of a vesicle membrane and a plasma membrane in a nerve cell.

Plasma membranes fuse easily and can spontaneously form vesicles. These abilities allow cells to change shape and to perform phagocytosis, endocytosis, and exocytosis.

How Do Cells Communicate in Multicellular Organisms?

Because the individual cells in all multicellular organisms must cooperate, communication among cells is paramount. In general, cells communicate by means of chemical signals such as hormones. Such signals allow the activities of one cell to influence those of another.

Figure 4-24 Endocytosis and exocytosis. A. Endocytosis plays a central role in the regulation of low-density lipoprotein (LDL) and cholesterol levels in the blood, an important predictor of heart disease. In one form of receptor-mediated endocytosis, LDL particles full of cholesterol bind to LDL receptors on the cell membrane. The cell membrane folds around the LDL particle and engulfs it, then the cell digests it. B. In exocytosis, the reverse appears to happen: vesicles full of proteins, hormones, or neurotransmitters made inside the cell fuse with the cell membrane and diffuse into the outside environment of the cell. Exocytosis depends on the coordinated action of proteins on the membranes that will fuse. C. A new technique, called electron microscopic tomography, has allowed researchers to reconstruct a three-dimensional picture of exocytosis. In this photo, we can see exocytosis of vesicles in a nerve cell. The vesicles contain acetylcholine, which (as we discussed in Chapter 1) stimulates muscle contraction. *(C, courtesy of U. J. McMahan, Stanford University)*

How do cells communicate with their immediate neighbors?

Plant cells are both cemented together and separated from each other by rigid cell walls. Plant cells are nonetheless in intimate contact through fine intercellular channels called **plasmodesmata** (singular, plasmodesma) [Greek, *plassein* = to mold + *desmos* = to bond]. Plasmodesmata, which are between 20 and 60 nm in diameter (Figure 4-25), allow molecules to pass directly from cell to cell. Some kinds of cells, such as those in the tips of a plant's roots, have huge numbers of plasmodesmata. The more plasmodesmata between neighboring cells, the more rapidly materials pass from one cell to the next.

Animal cells are not separated from each other by cell walls, and they communicate with each other in a variety of ways. Some animal cells easily pass ions and molecules, even proteins, through specialized regions of their membranes called **gap junctions.** In gap junctions, the membranes of adjacent cells are only 2 to 4 nm apart, instead of the usual 10 to 20 nm (Figure 4-26A). In addition, gap junctions connect cells with a protein called connexin that allows the cells to exchange chemical signals. Gap junctions, for example, connect heart muscle cells and allow them to coordinate their contractions.

Other animal cells communicate using "adhering junctions." **Adhering junctions** help provide strength to resist mechanical stress without affecting the passage of molecules between them. Adhering junctions usually consist of **desmosomes** [Greek, *desmos* = to bond + *soma* = body], buttonlike structures that hold two membranes together (Figure 4-26B). Desmosomes may be localized in spots, like the rivets on a pair of jeans, or they may be spread in bands or belts.

Several types of specialized junctions connect **epithelial cells** [Greek, *epi* = upon + *thele* = nipple], the cells that line the surfaces throughout the body, for example the lining of the gut. Epithelial cells are linked together tightly to form a sheet, called an **epithelium** [plural, epithelia]. Epithelia enclose many fluid-containing cavities, such as kidney tubules and intestines. One task of these epithelia is to keep the fluids from escaping. Impermeable junctions not only weld cells together but also prevent any molecules from leaking between them. In mammals and other vertebrates, the main kind of impermeable junction is called a **tight junction,** in which the membranes of adjacent cells actually fuse (Figure 4-26C).

Plasmodesmata and gap junctions allow materials to pass directly from cell to cell. Adhering junctions strengthen the connections between cells, and tight junctions keep fluids from passing between them.

How do cells communicate with distant cells?

Much intercellular communication, such as the transmission of information from a root to a leaf or from a nerve cell to a muscle cell, depends on chemical signaling. Molecules made by one cell bind to specific receptors in other cells, triggering a series of specific events.

Chemical signals all act by binding to specific **receptors,** specialized proteins that can lie either on the surface of a cell or within the cell. A particular receptor normally only binds with one signaling molecule.

Some chemical signals enter the cell by passing through the plasma membrane and then binding with receptors within the cytoplasm. The steroid hormone testosterone, for example, acts in this way, actually entering cells and altering the expression of individual genes. At puberty, a young man's hypothalamus (a part of the brain) sends a chemical signal to the pituitary gland, at the base of the brain. The pituitary responds by secreting a hormone that is carried throughout the body by the bloodstream. When the hormone reaches the cells of the testes, the male reproductive organs, it binds to cell surface receptors, changing the cells' properties. The cells of the testes then begin to secrete the hormone testosterone, which enters the bloodstream and binds to cytoplasmic receptors in different cells all over the body. In each cell, the receptorhormone complex moves into the cell nucleus and influences the production of various proteins. The results include a gradual deepening of the voice, a rapid spurt in body growth, and the development of a sex drive.

Other chemical signals, however, stay outside the cell, influencing what is going on inside by means of cell surface receptors. Some surface receptors can open channels in the plasma membrane to allow the passage of new materials into or out of the cell. Other receptors can regulate the levels of an influential chemical—called a *second messenger*—inside the cell. Many extracellular signals, for example, trigger an increase

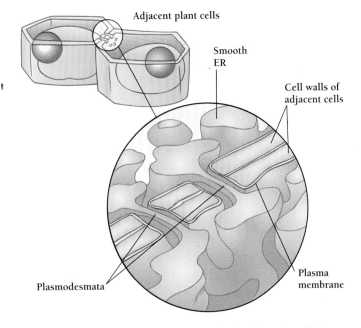

Figure 4-25 Cell to cell. Fine intercellular channels called plasmodesmata connect all the living cells of higher plants.

Figure 4-26 Cell junctions. A. Gap junctions are protein assemblies that form channels between adjacent cells. B. Like rivets, the desmosomes of adhering junctions anchor cells together, but without interfering with the passage of molecules from one cell to the next. C. Tight junctions seal cells together and prevent molecules from leaking between them. (A, Gilulap, Fawcett/Visuals Unlimited; B, Farquhar, Palade, Fawcett/Visuals Unlimited; C, Hull, Staehelin, Fawcett/Visuals Unlimited)

in the concentration of calcium ions, which serve as a common second messenger.

Still other surface receptors act directly on the surface itself. When receptors bind to the hormone insulin, for example, they trigger the incorporation of additional glucose transporters into the cell membrane. The result is an increase in the cell's rate of glucose uptake.

In multicellular organisms, cells communicate both physically (through plasmodesmata and gap junctions) and chemically (through hormones and other chemical signals). These routes of communication allow organisms to coordinate their responses to changes in the external environment.

Cells respond to chemical signals by means of specialized proteins called receptors.

In this chapter, we have seen how biologists first recognized the importance of cells as the fundamental living units of all organisms. We have examined the internal structures of cells and membranes and seen how cells must coordinate the activities of their many internal components. In the next three chapters, we will use some of this knowledge to try to understand the biochemical processes that take place inside of cells.

99

Study Outline with Key Terms

Cells are not only the units of biological structure; they are also the units of function. Individual cells are specialized to perform distinct tasks. Cells are living entities that are highly organized, perform chemical reactions, obtain energy from their surroundings, respond to their environments, change over time, reproduce, and share an evolutionary history.

The **cell theory** states that (1) all organisms are composed of one or more cells; (2) cells, themselves alive, are the basic living units of organization of all organisms; and (3) all cells come from other cells.

The subdivision of organisms into cells and of cells into organelles allows for an efficient division of labor and makes it easier to maintain a constant environment within each cell (because of a higher **surface-to-volume ratio**). Multicellular organisms can survive the deaths of individual cells because of cell replacement.

Cells from all six kingdoms contain three common features: a **plasma membrane** that separates the inside of the cell from the rest of the world, a set of genetic instructions encoded in DNA, and a cell body (or **cytoplasm**). In the cells of all eukaryotes, the genetic material is enclosed in a membrane-enclosed **nucleus.** The nucleus contains the cell's genetic instructions in the form of **chromosomes,** which are made of **chromatin,** a complex of DNA and protein. The two membranes of the **nuclear envelope** are fused in many places to form **nuclear pores.**

The Archaea and Eubacteria, or prokaryotes, lack a membrane-enclosed nucleus. They keep their DNA in a region called the **nucleoid.** The cytoplasm of most eukaryotic cells (formerly referred to as the **protoplasm**) is divided into well-defined compartments, or **organelles,** by internal membranes. The jellylike **cytosol,** which surrounds all the cell's organelles, contains **ribosomes, proteasomes,** glycogen, fat, and thousands of enzymes.

Within the cytoplasm are several well-defined kinds of membrane-enclosed organelles: the endoplasmic reticulum (ER), Golgi complex, **lysosomes, peroxisomes, mitochondria,** and such **plastids** as **chloroplasts, chromoplasts,** and **amyloplasts.** In addition, eukaryotic cells contain **vesicles** and **vacuoles,** which are also enclosed by membranes but whose structures and functions vary widely.

The **endoplasmic reticulum (ER)** is an extensive network of membranes that encloses a space called the ER **lumen.** One part of the endoplasmic reticulum—the **rough endoplasmic reticulum**—is studded with ribosomes and makes proteins mainly destined for export to the outside or the outer surface of the cell. The other part of the ER—the **smooth ER**—contains no ribosomes and appears to function in the synthesis of lipids.

The **Golgi complex** serves as the traffic director in the cell, helping to package some proteins for export and others for inclusion in other membrane-enclosed organelles. Lysosomes and peroxisomes are small membrane-enclosed organelles that are responsible for digestion. They dispose of both material taken in from outside the cell and the cell's own debris.

Mitochondria and chloroplasts are the largest organelles outside the nucleus. Each is surrounded by a double membrane, and each contains complicated internal membranes. Mitochondria are the powerhouses of cells, making most of the cell's ATP by breaking down small, organic molecules in the presence of oxygen. Chloroplasts, which contain stacks of **thylakoids,** called **grana,** are present only in photosynthetic cells. They are the sites of **photosynthesis.**

Together with internal membranes, the **cytoskeleton** provides internal organization for eukaryotic cells. It consists of three sets of filaments: **microtubules** (which are made of two kinds of **tubulin**), **actin filaments,** and **intermediate filaments.** In most cells, the cytoskeleton consists of structures that are assembled and disassembled continually. Their assembly seems always to originate in structures called **microtubule organizing centers (MTOCs). Motors** can carry organelles along microtubule tracks either toward or away from the MTOC. The major MTOCs are near the nucleus in a zone called the **centrosome,** a region that (in many cells) also contains a distinct organelle called the **centriole.** Some cells have permanent structures, such as **cilia** and **flagella,** which are composed of elements of the cytoskeleton.

All cells are surrounded by membranes. Membranes consist mainly of lipids and proteins. The lipids, mostly phospholipids, are arranged in two layers, a "bilayer." Some proteins are located on the external and internal surfaces of the bilayer, in contact with both the hydrophobic interior of the membrane and the hydrophilic cytoplasm or cell exterior. Other proteins pass completely through the membrane. The membrane is **fluid,** both protein and lipid molecules move freely within it. This picture of membrane structure is called the **fluid mosaic model.**

Membranes are **selectively permeable:** they allow some molecules to pass in and out freely and confine others to the inside or the outside. In **simple diffusion,** solutes move from an area of high **concentration** to an area of low concentration. That is, solutes move along a **concentration gradient. Osmosis** is the movement of water across any selectively permeable membrane in response to a concentration gradient. A solution may be **hypotonic, isotonic,** or **hypertonic,** relative to the concentration of solutes on the other side of the membrane. The pressure created by the concentration gradient, called **osmotic pressure,** is comparable to mechanical pressure.

In plants, the central vacuole fills with water, which presses against the **cell wall,** creating **turgor pressure**—the principal mechanical support of nonwoody plants. When a plant cell is immersed in a hypertonic solution, water flows out of its vacuole and the cell separates from the cell wall in **plasmolysis.**

Molecules pass through plasma membranes by means of **passive transport** (which requires no energy), **facilitated diffusion,** and **active transport** (which requires energy from ATP or **cotransport**). The **sodium-potassium pump** is one active transport mechanism. Certain cells engulf large particles by extending their plasma membranes in a process called **phagocytosis.** Most cells are also capable of **endocytosis** and **exocytosis.** Endocytosis may be nonspecific **(pinocytosis),** or it may depend on specific receptors on the membrane **(receptor-mediated endocytosis).**

The outer surfaces of cells are responsible for their interactions with other cells and with other molecules that bind to **receptors** on the cell surface or within the cytoplasm. Cell walls enclose most prokaryotic and many eukaryotic cells (though not animal cells). Animal cells often are surrounded with **extracellular matrix,** a complex of carbohydrates and proteins.

Multicellular organisms depend on coordination of activities among many cells. Plants and animals have a variety of specialized structures by which cells interact. Some of these structures—**plasmodesmata** in plants and **gap junctions** in animals—allow the free passage of some molecules between adjoining cells. Other specialized junctions between cells—**adhering junctions**—rely on **desmosomes** and provide mechanical strength to sheets **(epithelia)** of **epithelial cells. Tight junctions** weld some cells together and prevent the passage of molecules through them.

Review and Thought Questions

Review Questions

1. State the cell theory. Be sure your statement includes the concepts of what a cell is, where cells occur, where they come from, and what they do.
2. Describe three features that all cells have in common.
3. Describe the differences between prokaryotic and eukaryotic cells.
4. Explain how the surface-to-volume ratio of a cell is related to its ability to take in resources and to rid itself of wastes. Why are most cells so small? How can a chicken egg, which is one cell, be so big?
5. Sketch and label a mitochondrion and a chloroplast. What is the principal function of each? Where does each obtain fuel?
6. Describe the fluid mosaic model of membrane structure.
7. A plasma membrane functions to keep the material outside a cell separate from the material inside. What is one property of the membrane that allows this?
8. Explain how membrane-enclosed vesicles move materials into, out of, and within cells.
9. What is the role of the cytoskeleton in eukaryotic cells?
10. Compare the sizes of microtubules, actin filaments, and intermediate filaments. Describe one activity or structure in which each participates.

Thought Questions

11. Between conception and death, most of your cells die while new ones are produced by the division of other cells. What makes you really "you"? What is it that is constant about your structure?
12. Explain how the endoplasmic reticulum might increase the effective surface-to-volume ratio for a cell, even though it is located inside the outer perimeter of the cell. Can it really assist in bringing resources in and getting rid of wastes?
13. Chloroplasts are also called the "powerhouses" of cells. What do they produce to earn this description? What types of cells contain chloroplasts?

About the Chapter-Opening Image

One-celled organisms, such as this diatom, were among the first studied by early microscopists.

On-line materials relating to this chapter are on the World Wide Web at **http://www.harcourtcollege.com/lifesci/aal2/**

Web Bit 4-1 *Caulerpa*, The World's Largest Single-Celled Organism?

Web Links

Quizzing & Testing

5 Directions and Rates of Biochemical Processes

Ludwig Boltzmann: Left Behind, or Ahead of His Time?

In September 1906, the eminent Austrian physicist Ludwig Boltzmann fell into the last of many deep depressions. Hoping to lift his spirits, he took his wife and daughters on a trip to Duino, Italy, a spectacular resort on the Adriatic Sea.

The mercurial Boltzmann had always been emotional (Figure 5-1). He could be moved to tears of joy by a well-played sonata, the color of the sea, or a piece of poetry. At other times he lapsed into long, grief-stricken silences. He wept at the railroad station whenever he left his wife for any extended period.

Although Boltzmann was unstable, his growing despondency also had roots in the scientific milieu of his time. At 60, the physicist was convinced that much of his life's work had been wasted. For years, he had championed the idea that matter is composed of atoms and that the behavior of atoms accounts for all chemical transformations. Other chemists and physicists discussed atoms extensively, but for them, atoms were just a useful way to think about the formation of

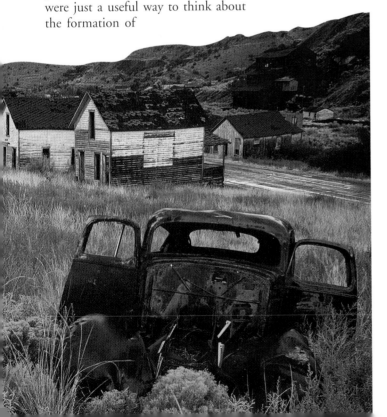

chemical compounds. Almost no one thought atoms were real, and no one before Boltzmann thought of them as the key to understanding energy.

The hypothesis that matter was made of tiny particles called atoms was first popularized by the Greek philosopher Democritus in the fifth century B.C. Not until the 17th century, however, did "atomism" get serious attention. And the same renaissance in thinking that gave birth to atomism also gave birth to another idea that would help drive Boltzmann to despair. According to "positivism," observable facts are the only basis for a true knowledge of the world.

Since nobody could see atoms, 19th-century positivists insisted that there was no reason to believe in atoms. Boltzmann's peers, all eminent scientists and heartfelt positivists, insisted that assuming that atoms were real—actually basing scientific theories on their existence—was no better than mysticism. Indeed, one of Boltzmann's stoutest critics, the physicist Ernst Mach, used to harass Boltzmann as Boltzmann lectured on atoms by calling out from the audience, "Have you ever seen one?" Of course, Boltzmann had not. (Indeed, only with the development of atomic force microscopes in the 1980s could scientists actually "see" atoms.)

But Boltzmann felt certain that atoms were real, and for years he had written scathing critiques of even his closest friends' work. In his personal life, Boltzmann was charming, modest, and earnest, but in debates and in the lecture hall, he was a bulldog, relentlessly attacking arguments he believed were wrong and proclaiming his own ideas.

For almost 40 years, Boltzmann worked to show how atoms accounted for the behavior of all forms of matter. His greatest achievements were to find a way to understand the behavior of matter in terms of large collections of molecules and atoms. He showed that the statistical laws of large numbers could be applied to collections of atoms or molecules.

But Boltzmann's work could be correct only if atoms were real. And, after 40 years of arguing, fewer and fewer people seemed to think that atoms existed. As Boltzmann wrote to a friend, "I am conscious of being only one individual struggling weakly against the stream of time," and later, "I present myself to you therefore as a man left behind."

Key Concepts

- The Second Law of Thermodynamics predicts the direction of any process because energy becomes less available for work as a reaction or process occurs.
- The rate of a chemical reaction depends on the random movements of molecules.
- Enzymes speed up biochemical reactions by lowering their activation energies.
- Cells and organisms regulate their own metabolism by changing the shapes and activities of enzymes.

Boltzmann did not know it, but by 1906, the year he took his family to Italy, the atom was about to take center stage in the world of science. The darkest days for his ideas were over, and bright young scientists were taking a second look at atoms. In 1905, for example, 26-year-old Albert Einstein had published a paper showing that the constant jiggling of particles that scientists noticed when looking down a microscope (called Brownian motion) was most likely due to bombardment by atoms—real, massive objects, not abstract statistical averages. In just a few years, Ernest Rutherford would locate the tiny, central nucleus of the atom. And soon after that, the physicist Niels Bohr would calculate the movements of electrons.

But Boltzmann had given up. On a sunny day in September 1906, while his wife and daughters were swimming, Boltzmann ended his rich and productive life by hanging himself in his hotel room. Had he endured just a few more years of frustration, he would have lived to see the triumph of his own ideas and the humiliation of his intellectual enemies.

Indeed, Boltzmann's intellectual victory was so complete that the material in this chapter—written nearly 100 years after his death—rests heavily on his work. The most important applications of Boltzmann's idea were in the field of **thermodynamics** [Greek, *thermo* = heat + *dynamis* = power, force], the study of the relationships among different forms of energy. For example, as a car converts the chemical energy of gasoline into the energy of motion, the car moves forward. Every event, every reaction, every process is accompanied by a "transformation of energy."

The direction a transformation takes depends on the rules of thermodynamics. A scientist can predict the direction of a reaction without knowing exactly how the reaction takes place. In the same way, we know that water always flows downhill, even if we do not know whether it gets there through a pipe, a stream, or a waterfall.

Figure 5-1 Ludwig Boltzmann (1844–1906). (Science Photo Library/Photo Researchers, Inc.)

Before Boltzmann, thermodynamics was understood as a set of abstract mathematical laws. These two laws could predict which way a reaction would go, but they said nothing about the actual process of the reaction.

Boltzmann showed that the reason for thermodynamics was the behavior of large numbers of individual atoms and molecules. He showed how chemical reactions occur, what heat is, and why some chemical reactions produce heat while others absorb heat. Most important, Boltzmann's work paved the way for the theory of **kinetics** [Greek, *kinetikos* = moving], the study of the rates of reactions. Before Boltzmann, chemists could predict only whether a reaction should take place, not how long it would take. Kinetics, which explains why a reaction takes place, can also predict how fast it will happen.

Understanding a biological or chemical process (or any process, for that matter) requires more than noticing changes in structures. For example, we may observe that most of the gasoline in a car's engine is converted to water and carbon dioxide, with a tremendous release of heat and energy of motion. But we cannot understand what is happening unless we know the answers to two general questions: (1) Which way will the reaction occur? and (2) How fast will it occur?

We must ask, Why does the conversion of gasoline to water and carbon dioxide occur in one direction only? Why don't the carbon dioxide and water present in the air we breathe combine spontaneously to form gasoline? And why does the conversion from gasoline to carbon dioxide and water occur in fractions of a second? What if converting a tank of gasoline to water and carbon dioxide took several years? In that case, gasoline would be a poor fuel. Why does the reaction occur as rapidly as it does? What factors determine how fast any reaction goes?

103

Predicting the speed of a process requires that we know something about the process itself. For example, we know that water plunges quickly downhill in a waterfall and meanders slowly in a gentle stream. The rate at which the water moves depends on the steepness of its descent. Thermodynamics predicts that water will flow downhill. But only kinetics can tell us how fast the water will flow. In this chapter, we will see that thermodynamics tells us which way a reaction will go and that kinetics tells how fast.

WHAT DETERMINES WHICH WAY A REACTION PROCEEDS?

Everyone can tell when a movie runs backward: a waterfall goes up instead of down; a rubber ball bounces up a flight of stairs; a smashed apple spontaneously reassembles (Figure 5-2). We all know the direction of time's arrow. Biological processes also have direction: a child matures into an adult; a bud unfolds into a leaf; dead organisms decay.

The key to predicting direction lies in understanding the conversion of energy. In any process—whether the flow of water, the growth of a tree, or the maturation of a child—energy changes form. The study of thermodynamics has provided rules that govern all energy changes, both biological and nonbiological. These rules establish the direction of time's arrow.

Figure 5-2 **We all know the direction of time's arrow.** If we saw two photos of an apple taken before and after a bullet passed through it, we would know instantly which photo was taken first. The destruction of an apple is a process that increases entropy. The only way to reassemble such an apple would be to return the pieces of apple to an apple tree, which might construct a new apple in a few months with the aid of energy from the sun. *(Photograph by Dr. Harold E. Edgerton, Harold and Esther Edgerton Foundation, ©1999. Courtesy Palm Press, Inc.)*

How May Work Be Converted to Kinetic or Potential Energy?

Most of us think of work as any process that requires energy. There is, however, a scientific definition of work: **work** is the movement of an object against a force. For example, lifting a sack of concrete against the force of gravity is work. So is winding the spring in a child's toy (Figure 5-3).

Work, in this sense, can be stored for later use, the way energy is stored in a battery. The wound-up spring in a toy beetle can propel the toy by operating its legs. When the spring is wound, we say that the work of winding the spring is stored as **potential energy.** When the spring unwinds and the beetle leaps forward, we say that the toy has **kinetic energy**—the energy of moving objects. A boulder at the edge of a cliff has potential energy. The same boulder falling to the bottom of the cliff has kinetic energy. A moving car, a running athlete, and a contracting muscle all have kinetic energy. Fast-moving molecules also have kinetic energy, which we call heat. Like kinetic energy, potential energy exists in many forms, including gravitational, electrical, and chemical.

Rearranging atoms in a chemical reaction is also a form of work. Recall, for example, that the energy-rich molecule ATP consists of a molecule of adenosine linked to three negatively charged phosphate groups (Figure 5-3). Making a molecule of ATP requires the squeezing together of three negatively charged phosphate groups. This squeezing is work against an electrical force. Once the atoms are squeezed together into a molecule, the work is stored as potential energy. Thus, ATP has potential energy. This energy can later be released to do work. If, for example, the third phosphate group is removed, stored energy is released. That energy can drive another chemical reaction, which can make a muscle move, heat the body, or help us understand a sentence.

Work, the movement of an object against a force, can be stored as potential energy. The construction of a molecule of ATP stores potential energy that can be released later as kinetic energy.

How Does Thermodynamics Predict the Direction of a Reaction?

Scientists and engineers first began to understand the relationship among different forms of energy in the late 18th century. The major stimulus for the development of the rules of thermodynamics was the increasing use of the steam engine, which converted the potential energy of coal into kinetic energy capable of powering a locomotive or a factory. As we shall see, the same rules that apply to steam engines also apply to the energy conversions in organisms.

One of the first and most important theories of the new field of thermodynamics was that, in any process, energy is neither created nor destroyed—it only changes form. The **First Law of Thermodynamics** states that the total amount of energy in any process stays constant (Figure 5-4A). In this view,

Chapter 5 Directions and Rates of Biochemical Processes   105

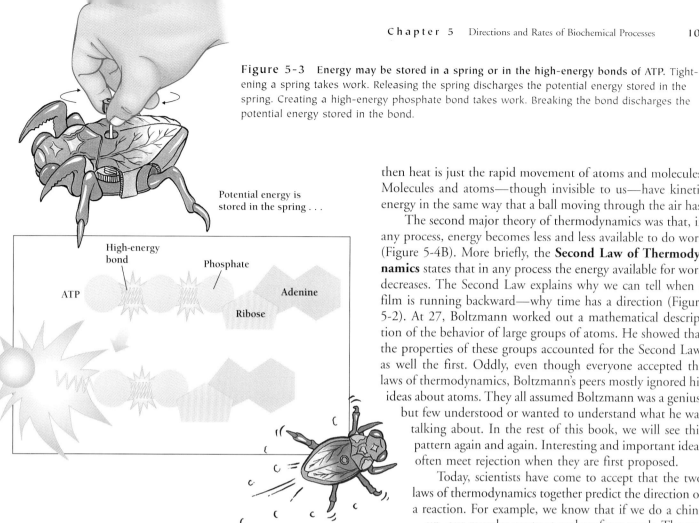

Figure 5-3 **Energy may be stored in a spring or in the high-energy bonds of ATP.** Tightening a spring takes work. Releasing the spring discharges the potential energy stored in the spring. Creating a high-energy phosphate bond takes work. Breaking the bond discharges the potential energy stored in the bond.

energy may switch from potential energy to kinetic energy and back again, but it is neither created nor destroyed. Any heat generated during a process is merely another form of kinetic energy. When Ludwig Boltzmann was only 22, he showed that the movements of atoms nicely explain the First Law of Thermodynamics. If atoms and molecules actually exist, he argued, then heat is just the rapid movement of atoms and molecules. Molecules and atoms—though invisible to us—have kinetic energy in the same way that a ball moving through the air has.

The second major theory of thermodynamics was that, in any process, energy becomes less and less available to do work (Figure 5-4B). More briefly, the **Second Law of Thermodynamics** states that in any process the energy available for work decreases. The Second Law explains why we can tell when a film is running backward—why time has a direction (Figure 5-2). At 27, Boltzmann worked out a mathematical description of the behavior of large groups of atoms. He showed that the properties of these groups accounted for the Second Law, as well the first. Oddly, even though everyone accepted the laws of thermodynamics, Boltzmann's peers mostly ignored his ideas about atoms. They all assumed Boltzmann was a genius, but few understood or wanted to understand what he was talking about. In the rest of this book, we will see this pattern again and again. Interesting and important ideas often meet rejection when they are first proposed.

Today, scientists have come to accept that the two laws of thermodynamics together predict the direction of a reaction. For example, we know that if we do a chin-up, our muscles contract and perform work. The energy for this work comes from the chemical energy in ATP. In the muscle cells, ATP molecules give up part of their stored energy by splitting off one of three phosphate groups. The remaining molecule has two phosphate groups and is then called adenosine *di*phosphate, or ADP [Greek, *di* = two]. ADP still has potential energy, but less than ATP.

But, according to the Second Law, muscle contraction can convert only a part of the energy obtained from ATP into useful work. The rest of the energy from splitting ATP becomes

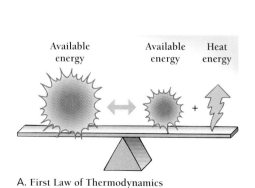
A. First Law of Thermodynamics

B. Second Law of Thermodynamics

Figure 5-4 **Two laws of thermodynamics.** A. First Law: Energy may change from one form to another, but it is never lost. B. Second Law: The energy available to do work decreases as energy is lost as heat.

heat—the random motion of molecules. From an organism's point of view, heat is mostly wasted energy. Even though some energy is lost as heat, however, no energy is actually destroyed. All of the energy released from splitting ATP goes either to contracting the muscle or to heat.

The First Law implies that all forms of energy—including chemical energy, electrical energy, mechanical energy (work), and heat—can be measured in the same units. The unit of energy that we use in this book is the **calorie**—the amount of energy needed to raise the temperature of 1 gram of water by 1°C. The energy contained in food is often reported in Calories (with a capital C). One Calorie equals 1 kcal ("kilocalorie"), which equals 1000 calories. The daily diet of a college student usually contains from 2000 to 3000 kcal (Calories) of energy.

Every time energy changes form—whether from chemical to electrical or from electrical to light energy—a little more of the energy turns into heat (Figure 5-5). Over time, more and more energy turns into heat, and less and less energy becomes available for work (Figure 5-4B).

Heat loss is a natural consequence of all energy transformations. If we ask which way a reaction will run, the answer is that usually it will run in the direction that releases heat. The formation of a chemical bond between two atoms generally releases energy. Conversely, the breaking of a chemical bond usually consumes energy. The only way to reverse such a reaction is to add energy in some form.

The First Law of Thermodynamics states that energy changes form but is neither created nor destroyed. The Second Law of Thermodynamics states that in any process, the energy available to do work decreases as energy is lost in the form of heat. The loss of heat establishes the direction of the process. Without an input of energy, the process cannot reverse itself.

Figure 5-5 A hot bike. In both living organisms and in machines, energy is lost in the form of heat each time energy changes from one form to another. (*Superstock*)

How Do Changes in Free Energy Predict the Direction of a Reaction?

To predict the direction of a chemical reaction, we need a measure of how much energy is available. **Free energy** is the energy in a system available for doing work. Free energy is abbreviated *G,* after the American chemist J. Willard Gibbs, who, in the 1870s, single-handedly worked out the theory of chemical thermodynamics.

Every chemical has a certain amount of free energy. And every chemical reaction involves a "change in free energy." When two molecules (or **reactants**) react to form one or more new molecules (or **products**), the change in free energy is equal to the free energy of the products minus the free energy of the reactants. The change in free energy is abbreviated Δ*G*. (Δ, pronounced "delta," is the mathematical symbol for a change in a quantity.) Chemists abbreviate this statement by writing

$$\Delta G = G_{products} - G_{reactants}$$

This equation might look intimidating, but it is no different from saying that the amount of weight you have gained or lost in a week is equal to the amount you weigh today minus the amount you weighed last week:

$$\Delta lbs = lbs_{today} - lbs_{last\ week}$$

Here, Δlbs ("delta pounds") equals the change in your weight, the amount of weight you lost or gained. In fact, without adding energy to your body's "system" by eating, you will inevitably lose weight, and Δlbs will be negative. That's because your body's activities are energy-releasing processes.

All processes that release energy are said to be **exergonic.** The breaking of ATP into ADP and phosphate, for example, is exergonic. Likewise, a coconut falling from a tree releases energy as it falls. Stand underneath it and you will not doubt that it is an exergonic process.

On the other hand, forcing another phosphate onto ADP to make ATP is an **endergonic,** or energy-consuming, process. Carrying a coconut up a tree is work and an endergonic process. It requires an input of energy, but it increases the capacity of a system to do further work (Figure 5-6).

In all exergonic processes, free energy decreases, so Δ*G* is negative. As a result, exergonic processes, such as ATP changing to ADP, occur spontaneously (given enough time). In endergonic processes, free energy increases and Δ*G* is positive. Endergonic reactions will not occur spontaneously, but require energy from some other source.

Finally, a change in free energy depends only on the beginning and end of a reaction. The change does not depend on anything that happens in between or on how long the process takes. A ball on a staircase might fall down the whole staircase in one big bounce or in several smaller bounces. The ball might even sit unmoving on the top step for years on end until it is jostled by an earthquake, a passing truck, or a curious child. But in all these cases, the change in free energy is the same.

Thermodynamics cannot predict how long it will take for a ball to fall down the stairs or for a muscle cell to break down ATP into ADP and phosphate. All that thermodynamics tells us is that, if we wait long enough, the ball will roll down the stairs—not up—and that ATP will break down into ADP and phosphate. In both examples, the process goes "downhill," from higher free energy to lower free energy.

> *If, during a reaction, the change in free energy is negative, the reaction is said to be exergonic and the reaction can occur spontaneously.*

Figure 5-6 Work and play. Carrying a coconut up a tree is an endergonic (energy-consuming) process that scientists call "work." Dropping a coconut is an exergonic (energy-releasing) process. In the same way, adding a phosphate group to ADP consumes energy and removing the same phosphate group releases energy. It's work to make ATP.

What Is the Source of the Free Energy Released or Consumed During a Reaction?

When two isolated atoms form a chemical bond, energy is released. The tighter the bond, the more energy is released. During a chemical reaction, some of the bonds in the reactants break, while other bonds form to create the products. For example, consider the breakdown of ATP to ADP:

$$\text{ATP} + \text{H}_2\text{O} \longrightarrow \text{ADP} + \text{phosphate}$$
$$(\text{Reactants}) \longrightarrow (\text{Products})$$

In this reaction, the oxygen atom in the water molecule separates from one of its hydrogens and replaces the oxygen atom in ATP's third phosphate group (Figure 5-7). Altogether, two bonds break—an O–H bond (in water) and a P–O bond (in ATP). In addition, one new bond forms—a P–O bond (in phosphate). The new P–O bond (in the phosphate) is more stable than the old one. In general, free energy decreases when unstable bonds break and stable ones form. And, overall, the free energy of the products (ADP and phosphate) is less than the free energy of the reactants (ATP and water).

The ATP → ADP reaction favors the breaking of less stable bonds and the forming of more stable bonds. Biochemists call unstable chemical bonds that give up energy easily **high-energy bonds**. They are sometimes represented as a squiggle (~). The bonds that hold the second and third phosphates of ATP are both high-energy bonds.

Chemists can predict the direction of a chemical reaction by measuring the change in free energy for the making and breaking of chemical bonds. For example, the change in free energy (ΔG) for the breakdown of ATP to ADP and phosphate is -7.3 kcal/mole. (As mentioned in Chapter 2, a mole is a fixed, huge number of molecules—6×10^{23}.) Because ΔG is negative, we can tell that free energy decreases and the reaction is exergonic. It will occur spontaneously.

The bonds between the atoms of a molecule have characteristic energies that change in the course of a reaction. Rearrangements of these chemical bonds may consume or release energy.

How Can One Process Provide the Energy for Another?

We saw earlier that an energy-consuming, or endergonic, reaction needs a source of energy. That energy often comes from an exergonic reaction. The most common source of energy for biological processes is the breakdown of ATP. When ATP breaks down into ADP and phosphate, it releases energy that becomes available to drive other chemical reactions, such as the linking of glucose molecules into a polysaccharide.

When an exergonic reaction drives an endergonic reaction, we say that the two processes are **coupled.** The chimpanzees and coconuts in Figure 5-8 are a coupled system. The falling coconut releases energy that lifts the small chimpanzee. A coupled reaction must be exergonic overall. That is, the exergonic reaction and the endergonic reaction must together release energy. If the picture in Figure 5-8 showed an elephant at the end of the log, we could guess that the coconut could not lift it. In the same way, the energy from splitting a single ATP molecule can drive only small reactions. But most reactions in biology occur in small steps.

How do cells couple endergonic and exergonic reactions? The answer is **"enzymes,"** large molecules, usually proteins, that speed up chemical reactions. Enzymes do not affect the change in free energy of a reaction. That is, enzymes do not change whether a reaction will happen or not, just how fast. You can slide down a hill faster if it's covered with ice than if it's just dirt. But no matter how slippery a hill is, you can't slide up the hill. Enzymes are like the ice on the hill: they speed up a spontaneous reaction but cannot reverse its direction.

Enzymes can also couple endergonic and exergonic reactions. In muscle contraction, for example, an enzyme couples the breakdown of ATP to the contraction of tiny muscle filaments.

Figure 5-7 The most important exergonic reaction in the body. Splitting a high-energy phosphate bond in ATP releases energy that can be used in thousands of other reactions.

der. The Second Law is a formal statement that the amount of disorder in the universe always increases.

Physicists have defined a formal measure of disorder, called **entropy**. Entropy has a high value when objects are disordered or distributed at random, and a low value when they are ordered. Surprisingly, because entropy tends to increase, it can be used to do work (Figure 5-9).

Any process that converts an orderly arrangement to a less orderly one can perform work. For example, when we burn wood (converting well-ordered molecules of cellulose to disordered molecules of carbon dioxide and water), we can use the energy from this exergonic reaction to warm a house, to grill a hamburger, or to power a steam engine.

On the other hand, any process that converts a disorderly arrangement to an orderly one consumes energy. An oak tree uses energy from the sun and carbon dioxide from the air to build cellulose in its massive trunk and branches. Likewise, a cell uses energy to maintain concentrations of molecules inside the cell that are different from those outside the cell. But cells must perform work to keep their internal environments constant. When no energy supply is available to perform this work, the differences between the inside and the outside of the cell disappear, and the cell dies.

The movement of molecules from a more concentrated to a less concentrated state constitutes an increase in entropy, an increase that can be harnessed to do work. The free energy—the energy available to perform work—depends on the concentration of the molecules (Figure 5-9). The greater the difference in concentration between two sides of a cell membrane or other barrier, the more work can be done. In summary, whether a reaction occurs depends not only on the energy in the individual bonds but also on the concentrations of both chemical reactants and products.

Imagine a chemical reaction in which reactant A is converted to a product B. If the reaction is exergonic (energy releasing), then the reaction will proceed, and product B will begin to accumulate. As the concentration of B increases, however, so does B's free energy (the energy available to do work). Gradually, the difference in energy between the reactants and the products decreases. When the free energy of B equals the free energy of A (and $\Delta G = 0$), the reaction is finished. It is

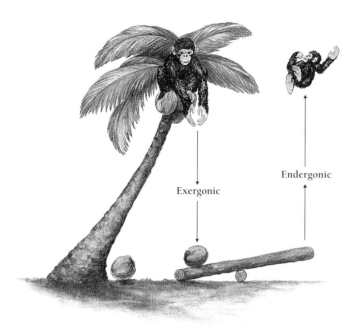

Figure 5-8 **A coupled reaction.** When the adult chimpanzee drops the coconut, the baby chimp is thrown up into the air. The falling coconut is *coupled* to the flying chimp.

Enzymes couple exergonic reactions, such as the splitting of ATP, to endergonic reactions.

How Do Concentration (and Entropy) Affect Equlibrium?

The Second Law of Thermodynamics that the energy available to perform work always decreases—also implies that uncoordinated motion is more probable than coordinated motion. Our common experience with "time's arrow" is that it is more likely that the pieces of a burst balloon will fly apart than it is that they will spontaneously reassemble. It is more likely that the books in your room will become disorganized than that they will remain neatly shelved in alphabetical or-

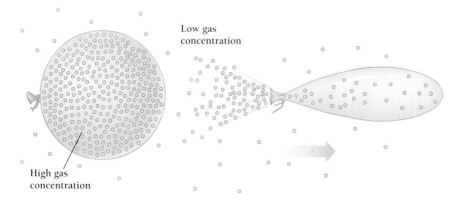

Figure 5-9 **The more you blow it up, the faster it flies.** Any difference in concentration creates potential energy. The greater the concentration of gas molecules inside the balloon, the faster the balloon flies when we release it. More precisely, the greater the difference between the concentration inside the balloon and outside, the faster it flies.

no longer "downhill." The products and reactants have reached **equilibrium.**

Thermodynamics predicts where the equilibrium lies, but gives no information about how long a reaction will take to get to equilibrium. In the next section, we will find out how the rules of kinetics predict the rates of chemical reactions.

The equilibrium of a reaction depends both on the free energy in the bonds between the atoms of each molecule and on the concentration of the reactants and products.

WHAT DETERMINES THE RATE OF A CHEMICAL REACTION?

In the 19th century, archaeologists opened the tomb of an Egyptian pharaoh and found, among other artifacts, a perfectly preserved breakfast laid out for him. If the pharaoh had lived long enough to eat this meal, the food would have been oxidized in a matter of minutes. Yet, even after sitting for several thousand years in an oxygen-filled room, the pharaoh's breakfast was chemically unchanged. What kept the pharaoh's breakfast from being oxidized? In other words, How can cells initiate and speed up chemical reactions that run slowly or not at all outside the body?

We have seen that thermodynamics predicts both whether a reaction will occur and the direction of a reaction, independent of the precise mechanism. But to find out how *rapidly* a reaction occurs, we must understand its mechanism. We need to know what actually happens—what "path" a process takes from start to finish. For chemical reactions, this means going beyond discussing changes in bond energies and concentrations. Predicting the rates of chemical reactions requires an understanding of how molecules behave.

How Does Molecular Motion Help Explain Reaction Rates?

Molecules and atoms are constantly in motion. When a substance absorbs heat, its temperature increases and the atoms and molecules bounce and jostle about more rapidly. This simple fact is the basis for kinetic theory, or kinetics. Kinetics allows us to understand the rates of chemical reactions and how temperature and other environmental factors influence these rates.

For a moment, picture molecules as tiny frogs constantly jumping about at random (Figure 5-10A). Some of these frogs (molecules) are in an area called "Level A," and some are in an adjacent, but lower area, called "Level B." Some frogs will naturally jump down to Level B. But frogs will have a harder time jumping back up to A. Because more frogs can jump from A to B than can jump back up, frogs will tend to accumulate at B. Only the frogs (molecules) with the highest kinetic energy will be able to jump back to Level A. (Of course, real frogs do not hop about at random, but in molecules, the kinetic energy of "hopping" is exactly the same as heat energy.)

Kinetic theory, which is based on the random movements of atoms and molecules, helps predict the rates of chemical reactions.

What Stops a Chemical Reaction?

In the 1870s, Ludwig Boltzmann proposed that molecules are constantly in motion and that their behavior explains many of the properties of matter. Boltzmann's ideas inspired the concept that heat is the kinetic energy of moving molecules. The faster molecules move, the higher their temperature. However, temperature measures only the average energy of the molecules. Molecules may have a range of kinetic energies.

In our frog model, we would say that some frogs jump higher than others. At high temperatures, most frogs can jump the barrier (Figure 5-11). Only a few will be stuck at Level B. After the frogs have jumped for a while, they reach equilibrium. At equilibrium, just as many frogs jump from A to B as from B to A, and the number of frogs at each energy level remains the same.

The equilibrium for a reaction depends on the temperature. At equilibrium at a high temperature, lots of frogs will be at Level A. At equilibrium at a low temperature, more frogs will be at Level B.

We can see that this model leads to the same conclusion that thermodynamics does: molecules move in the direction of less free energy (Level B). In addition, the jumping-frog model allows us to see once more that the concentrations of reactants and products influence chemical equilibrium.

A chemical reaction stops when it reaches equilibrium. Equilibrium depends in part on the temperature and concentration of the reactants and products.

What Starts a Chemical Reaction?

A ball sitting at the top of a flight of stairs will not bounce down the stairs until something starts it moving. Just so, every process—no matter how much it is favored thermodynamically—needs to be started in the right direction. In chemical reactions, we must ask, Where does this push come from? The answer is that the energy for the reactions between molecules comes from the random movements of molecules.

To understand the factors that start chemical reactions, let us return to the jumping-frog model. The difficulty in starting a chemical reaction is comparable to the height of the barrier between the frogs. To get from Level A to Level B, the frogs first must jump over the barrier (Figure 5-10). The molecules or frogs must have a minimum energy to do this. The minimum energy needed for a process to occur is called the **activation energy.**

Figure 5-10 Leaping frogs and bouncing molecules. The random bouncing of molecules resembles random leaping in a group of frogs. A. Even if all the frogs begin at Level A, more will end up at Level B because it is harder for them to jump back up. B. With a decreased activation energy barrier, frogs can more easily jump between the two levels, and it will take them less time to reach equilibrium.

The number of the molecules in a space that are able to react depends first on the temperature and second on the height of the activation energy barrier (Figure 5-10). The higher the temperature, the greater the number of molecules with enough energy to jump the barrier. The lower the activation energy, the greater the number of molecules that can jump the barrier and the more rapidly equilibrium will be achieved.

Laboratory and industrial chemists can speed up a specific chemical reaction by using a **catalyst**—a substance that lowers the activation energy of a reaction but is not consumed or changed in the reaction. A catalyst, in essence, lowers the barrier between the two groups of frogs (Figure 5-10B).

An enzyme is a biological catalyst. Almost all enzymes are proteins. Enzymes allow organisms to lower the activation energy for thousands of specific chemical transformations and so shorten the time required to attain equilibrium. Hastening reactions is essential to life. We all need to be able to digest our breakfast in something less than 5000 years.

Kinetic theory helps us understand why some thermodynamically favored reactions take place while others do not. Spontaneous reactions will actually occur only when most of the molecules have the needed activation energy. The carbohydrates in a piece of bread can be oxidized either by increasing the temperature (who hasn't burned toast?) or by exposing them to enzymes such as those in the human digestive tract.

The presence of an active enzyme allows a cell to use a molecule such as ATP in specific ways. Depending on how particular enzymes couple the splitting of ATP to other reactions, the result may be the contraction of muscle, the

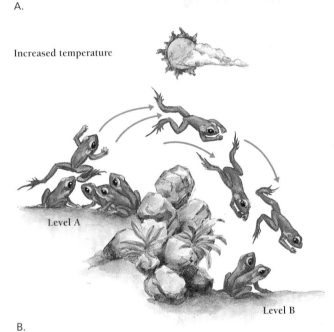

Figure 5-11 Hot frogs, cold frogs. The equilibrium point of a reaction depends on temperature. A. At a low temperature, few, if any, frogs have the activation energy to reach Level B. At equilibrium, all the frogs remain at Level A. B. At a higher temperature, half the frogs have enough energy to get to Level B. C. At an even higher temperature, all the frogs have enough energy to get to Level B, but only some can leap back. The reaction equilibrates with seven frogs at Level B and three at Level A.

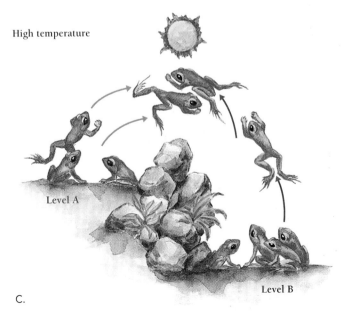

pumping of ions into or out of cells, or the synthesis of other molecules. Cells control their chemistry by regulating the production and activity of individual enzymes.

The initiation of a chemical reaction depends on the temperature of the molecules (their kinetic energy) and on the activation energy of the reaction. In organisms, enzymes can hasten a chemical reaction by lowering the activation energy.

HOW DO ENZYMES WORK?

In the last section, we saw that enzymes speed up biological reactions by lowering activation energies. In this section, we will see how enzymes accomplish this feat.

How Does an Enzyme Bind to a Reactant?

Enzymes speed up chemical reactions by binding to the reacting molecules, which are called **substrates.** Enzymes have enormous *catalytic power.* The right enzyme can make a chemical reaction go a million times faster than it would without the enzyme. For example, at 0°C, the freezing point of water, a single molecule of an enzyme called catalase can break down 5 million molecules of hydrogen peroxide each minute.

Enzymes are also highly *specific.* Each enzyme usually catalyzes only a single chemical reaction. What makes enzymes specific is their shape and also subtle interactions with the substrate. Most enzymes are large, globular proteins with irregular shapes. Every enzyme has an **active site**—a groove or "cleft" on its surface with which it binds to the small substrate molecule (Figure 5-12A). The binding of the enzyme to the substrate lowers the activation energy for a particular chemical reaction. Different kinds of enzymes may be able to bind to the same substrate, but in most cases *each enzyme lowers the activation energy for only a single kind of reaction.*

The enzyme temporarily binds to the substrate by weak noncovalent bonds—including hydrogen bonds, ionic bonds, hydrophobic interactions, and van der Waals attractions. Sometimes, as the two molecules bind, the shape of the enzyme changes (Figure 5-12A). When this occurs, the interaction of enzyme and substrate is called an **induced fit.**

The binding of the enzyme and the substrate is reversible, and the reactant or product molecules are quickly released back into the solution. Some enzymes are capable of binding, altering, and releasing individual substrate molecules more than 100,000 times a second.

Even so, the interactions between the enzyme and the substrate occur one molecule at a time. So although the rate of any biochemical reaction depends on temperature and the concentration of the reactants (as in the case of the jumping frogs or molecules), it also depends on how many molecules of enzyme are present and how fast the enzyme works.

The active site on an enzyme binds to a substrate by means of weak noncovalent bonds. Each enzyme lowers the activation energy for one reaction.

How Does an Enzyme Lower the Activation Energy of a Chemical Reaction?

The formation of noncovalent bonds between enzyme and substrate lowers the activation energy for a reaction. In all cases, the enzyme itself remains unchanged after the completion of the reaction. So, if the enzyme is unchanged, how does it change the activation energy of its substrate? The answer is that the enzyme briefly changes the shape of the substrate.

First, enzymes bring substrates together in a position that favors a reaction (Figure 5-12B). Without the enzyme, substrate molecules in solution bump into each other at random, but only a few of these collisions lead to a reaction. Most molecules do not react. We say the reaction is occurring very slowly.

Second, when an enzyme binds to a substrate molecule, the enzyme may strain and distort the covalent bonds within the substrate. Bonds under such strain can break more easily. The distorted form of a substrate is called a **transition state**—a molecule intermediate in form between the starting reactant and the final product. The transition state exists only for a brief moment—as little as a billionth of a second.

If we use our fingers to flex a spring clip just before clipping together a sheaf of papers, we are performing a role similar to that of an enzyme (Figure 5-12A). Our hand is the enzyme, the spring clip is one of the substrate molecules, and the sheaf of papers—held firmly in our other hand—is a second substrate. By using our hand, we can briefly alter the shape of the substrate (spring clip) so that it can bond with the second substrate (sheaf of papers). We have created a transition state. At the end of the "reaction," our hand is unchanged, just as an enzyme is unchanged.

An enzyme lowers the activation energy barrier. In some cases, enzymes may contribute directly to the chemical reaction—for example, by lending or temporarily accepting an

Figure 5-12 **How enzymes work.** A. Induced fit model of enzyme-substrate interaction. Both the enzyme and the substrate change shape, straining bonds during the reaction. The strained form of a substrate—the transition state—exists for as little as a billionth of a second. B. How does an enzyme catalyze a reaction between two other molecules? The enzyme brings both substrates close together in the proper orientation and also strains the covalent bonds of the substrates, lowering the activation energy of the reaction.

atom or an ion. This change strains the substrate molecule, bringing it to its transition state. Again, the net effect is to lower the activation energy barrier to the reaction.

Enzymes lower the activation energy for a reaction by aligning two or more substrate molecules, by straining the covalent bonds within the substrates, and even by participating directly in the reaction.

How Do Environmental Conditions Affect the Rates of Enzymatic Reactions?

Enzymes work most efficiently at certain temperatures, pH values, and salt concentrations. In mammals, for example, most enzymes work best at body temperature. In our previous discussion of kinetics, we saw that all chemical reactions occur more rapidly at higher temperatures because more molecules possess enough kinetic energy to get over the activation energy barrier.

But increasing the temperature of an enzymatic reaction increases the rate only to a point. Once the temperature reaches a certain point, the bonds that maintain the enzyme's structure begin to loosen and break. When an enzyme loses its shape and activity, the enzyme is **denatured**. When an organism's enzymes denature, it dies. We have all seen denatured proteins in cooked egg white, for example. Because heat denatures enzymes and other protein, most organisms cannot survive temperatures much higher than 40°C. A few unusual organisms live at higher temperatures. Bacteria that live in hot springs, for example, have evolved enzymes that function at high temperatures.

Besides temperature, pH (or acidity) also matters. Most enzymes work best in only a narrow range of pH values. The pH of the environment can, for example, alter the shape of an enzyme and add or remove hydrogen ions from the enzyme, changing the way the enzyme binds to the substrate. Most cellular enzymes work best at a pH of between 6 and 8. (Recall that pure water is neutral, with a pH of 7.) However, pepsin, the protein-digesting enzyme secreted by the stomach, works best at a pH of about 2. The fuller the stomach becomes, the more the stomach acid is diluted and the higher the pH rises, though usually no higher than about 3.5. If the pH of the stomach contents were to rise above 5, the pepsin would no longer digest our food.

Many enzymes work by acting as acids or bases. That is, they donate or accept hydrogen ions to a reaction. For example, when the amino acid histidine is part of the active site of certain enzymes, it can act as a base or an acid. If the pH of the environment is about 6.5 (slightly acid), histidine is an acid. At a pH of about 7.5 (slightly alkaline), however, histidine loses a hydrogen ion and becomes a base. An enzyme with an active histidine side chain can donate a hydrogen ion and help break a particular chemical bond, but only if the pH is right. Such relationships can help a cell regulate the cellular environment and maintain homeostasis.

Most enzymes, however, do not work in highly acid environments. This is why vinegar is such an effective preservative. The acetic acid in vinegar disables any enzymes present either in food itself or in any microorganisms that slip into the food. Without enzymes, reactions take place so slowly that pickled cucumbers, onions, peppers, herring, and cabbage can remain almost unchanged for decades or longer.

The rates of enzymatic reactions increase with temperature, but only up to a point. Most enzymes work in a narrow range of pH values.

HOW DOES A CELL OR ORGANISM REGULATE ITS OWN METABOLISM?

Organisms and cells are enormously flexible in the way they process, or metabolize, food. They function in lean as well as in fat times. They can obtain energy when food is available or break down their own stores when it is not. They can make building blocks not available in their diet, then stop making a particular building block when it becomes available. An organism with a diet rich in protein, for example, need not waste hard-won ATP making its own amino acids.

Organisms do all this by regulating the amounts and activities of enzymes. They both control the total flow of energy and focus that energy on specific tasks.

Why Do Enzymatic Reactions Often Occur in Small Steps?

When organisms and cells transform one molecule into another, they frequently do so by means of a series of small steps or subreactions, each one mediated by a different enzyme. For example, if a chemical reacts to form product B, which reacts to form C, which reacts to form D,

$$A \longrightarrow B \longrightarrow C \longrightarrow D$$

each step is catalyzed by a different enzyme. In humans, separate reactions number in the thousands. For example, ten different reaction steps and ten different enzymes are necessary in the initial extraction of energy from the sugar glucose. Almost all organisms—including all bacteria, all eukaryotes, and most archaebacteria—use the same ten enzymes to extract energy from this sugar (Chapter 6).

The network of biochemical reactions in organisms is complex for at least four reasons. First, some of the chemical transformations that cells perform are quite complex and can only be accomplished in many steps. Second, vast numbers of reactions are endergonic. For these to proceed, each one must be coupled to an exergonic reaction, such as the breakdown of ATP to ADP. Coupled endergonic reactions are especially common when cells synthesize large molecules from small ones, a process called **anabolism.**

Third, many exergonic transformations produce more energy all at once than the cell can spend. If you wanted to buy a soft drink, you couldn't use a $100 bill in most cities (besides Las Vegas). A cell "makes change" by separating a highly exergonic reaction into steps, so that the cell can capture smaller denominations of energy. A series of smaller reactions allow the cell to more easily store the energy for later use and also minimize the buildup of heat. Stepwise transformations of this kind are especially common in the breakdown of food molecules, called **catabolism.**

Fourth, in some stepwise reactions, intermediate products are the starting materials for the synthesis of essential building blocks. For example, some of the intermediate steps in the breakdown of sugars to carbon dioxide and water are precursors for amino acids, the building blocks from which proteins are made.

Many biological reactions occur in series of small steps. Stepwise reactions allow cells to carry out complex transformations, to couple reactions, to capture smaller denominations of energy, and to generate useful intermediate products.

How Do Organisms Regulate Enzyme Function?

The huge variety of molecules in a cell can undergo countless thermodynamically favored chemical reactions. The only reactions that happen rapidly enough to matter, however, are those that are catalyzed by enzymes. These include all the reactions that provide cells with the energy they need for life. Not only do enzymes serve as catalysts for specific reactions, they also regulate the rates of reactions, often in response to environmental changes.

For example, organisms need to regulate the quantities of building blocks they make. For an organism to make unnecessary building blocks would waste energy. Organisms that conserve energy by regulating the activity of enzymes therefore have an advantage over organisms that waste energy by building molecules they do not need. An example of such regulation is the synthesis of the amino acid isoleucine. When isoleucine is not available, cells make it from another amino acid, threonine. Making isoleucine from threonine requires five enzymatic reactions. But this succession of steps is energetically expensive, so the pathway from threonine to isoleucine shuts down when enough isoleucine accumulates. Isoleucine, the end product of the pathway, specifically inhibits the first step of this pathway (Figure 5-13).

This is a classic case of negative feedback, first mentioned in Chapter 1. Other molecules besides end products may also change the activity of enzymes—sometimes inhibiting enzyme action, sometimes increasing enzyme action. Any molecule or ion that changes the activity of an enzyme is called an **effector.** Effectors that inhibit enzymes are called **inhibitors.** But other effectors increase the activity of enzymes.

One way cells regulate enzyme activity is through effectors, which increase or decrease the activity of enzymes.

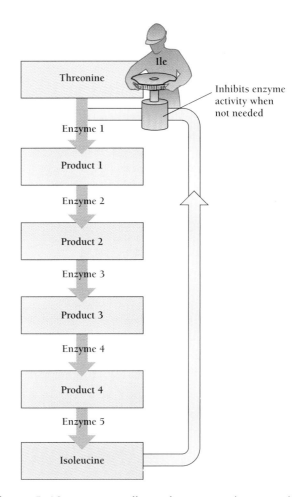

Figure 5-13 How can cells regulate enzyme function? The production of the amino acid isoleucine self-regulates by means of negative feedback. In the absence of isoleucine, a sequence of five enzymatic reactions synthesizes isoleucine from threonine. However, when isoleucine molecules accumulate, they inhibit the enzyme that catalyzes the first of the five reactions, shutting down the isoleucine pathway.

How Do Inhibitors Influence Enzymes?

Because an enzyme recognizes the shape of its substrate, enzymes can be "fooled" into binding to molecules that are similar in size and shape. Molecules that resemble the substrate can block an enzyme's active site and prevent the enzyme from working (Figure 5-14). Such a molecule is called a **steric inhibitor** [Greek, *stereos* = solid, three-dimensional] because it inhibits the work of the enzyme by means of its physical shape.

In carbon monoxide poisoning, for example, carbon monoxide (produced by burning gasoline or other fuel) competes with oxygen for a site on the hemoglobin molecule in our blood. Carbon monoxide binds so tightly to the hemoglobin that the hemoglobin can no longer carry oxygen to the brain and body. Unlike carbon monoxide, however, most steric inhibitors are reversible. They detach from the enzyme, so that a given enzyme sometimes binds to the substrate and sometimes to the inhibitor.

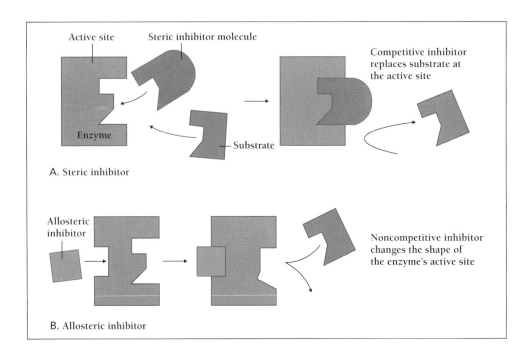

Figure 5-14 **What can prevent an enzyme from catalyzing a reaction?** A. Steric (competitive) inhibition. A steric inhibitor is a molecule that binds to the enzyme's active site, preventing the substrate from binding. A steric inhibitor competes with the substrate for the enzyme's active site. B. Allosteric (noncompetitive) inhibition. An allosteric inhibitor does not compete with the substrate but binds to another site on the enzyme in such a way that the active site no longer functions.

A steric inhibitor can be overcome. If the concentration of substrate is much greater than the concentration of a steric inhibitor, the enzyme will mostly bind to the substrate, not the inhibitor. The substrate and the inhibitor are competing for access to the enzyme. So the more concentrated the substrate, the more effective it will be at binding enzyme. And the more concentrated the steric inhibitor, the more effective it will be at binding enzyme. Because of such competition, a steric inhibitor is also called a "competitive inhibitor."

Many effectors can act as "noncompetitive inhibitors." Such molecules do not compete directly with the substrate. Lead and other heavy metals, for example, work by blocking specific groups outside the active site, sometimes changing the overall shape and chemistry of an enzyme. An effector that binds to an enzyme at a place other than the active site is called an **allosteric effector** [Greek, *allos* = other, different]. For example, an allosteric inhibitor blocks an enzyme by binding to a site on the enzyme other than the active site.

Some noncompetitive inhibitors act only temporarily. Others act irreversibly by forming a permanent bond with the enzyme and altering its shape. The antibiotic penicillin, for example, irreversibly damages an enzyme that bacteria use to build their cell walls.

Many noncompetitive inhibitors are poisons, like lead. Other inhibitors are essential parts of cells. Cells use allosteric inhibitors, for example, to regulate their own chemical processes. Most enzymes have evolved so that their most active shape is also their most stable shape. When an allosteric inhibitor binds to an enzyme, however, it changes the enzyme's shape to a less active form. The allosteric inhibitor acts as a switch, changing the enzyme from its active shape to its less active shape.

Effector molecules increase or decrease the activity of enzymes—either by competitively binding to the active site or by altering the shape of the enzyme.

Study Outline with Key Terms

All processes—biological and nonbiological—are accompanied by changes in the form of energy, which are subject to the laws of **thermodynamics.** The **First Law of Thermodynamics** states that the total amount of energy does not change. The **Second Law of Thermodynamics** states that the total amount of energy in a closed system becomes less and less available for **work;** more and more of it is converted to heat. In short, **entropy** increases.

The laws of thermodynamics allow scientists to predict whether a reaction will occur spontaneously. The making and breaking of chemical bonds between atoms leads to changes in available energy. A reaction will occur spontaneously if the energy available to do work—the **free energy (G)**—decreases. A reaction is **exergonic** if free energy (ΔG) decreases and **endergonic** if free energy increases.

Organisms can accomplish "uphill reactions" by **coupling** endergonic processes to exergonic reactions with protein **catalysts** called **enzymes.** In many cases, the energy that drives endergonic reactions comes from the **high-energy bonds** of ATP.

During chemical reactions and other processes, **kinetic energy** may be converted to **potential energy.** The free-energy change in a chemical reaction depends on the concentrations of **reactants** and **products,** as well as on the number of chemical bonds that are made and broken. Reactions tend to go in the direction of increased disorder, or entropy, in which the concentrations of molecules become more uniform.

To predict how long a process will take to occur, we need to know how it occurs. The study of **kinetics** allows scientists to predict the rates of chemical reactions from the properties of molecules. The basis of kinetics is that molecules are constantly in motion, with the average energy of a group of molecules determined by the temperature, measured in **calories.**

At chemical **equilibrium,** the direction of a reaction is balanced by the opposite reaction. Enzymes and other catalysts speed up specific chemical reactions but do not affect the free-energy changes of chemical reactions. They increase the speed at which equilibrium is approached, but they do not affect the character of the equilibrium.

Enzymes hasten chemical reactions by lowering the **activation energy** of the reaction. Enzymes accomplish this by binding to the reacting molecules by means of noncovalent interactions. An enzyme may join two or more **substrate** molecules. Sometimes, in an **induced fit,** enzymes strain the bonds of a single substrate molecule, forcing it into a **transition state.** Some enzymes actually participate in the chemical reaction by temporarily giving or receiving atoms or ions. The activity of an enzyme is highly sensitive to environmental conditions that can affect its activity. Temperature, pH, and the presence of other molecules may **denature** an enzyme.

Many chemical transformations performed by organisms in **anabolism** and **catabolism** consist of numerous small reactions. Such small reactions more easily allow the cell to capture the energy for later use.

Cells regulate the activity of many enzymes with **effectors,** including **steric inhibitors** and **allosteric inhibitors.** Steric inhibitors block the enzyme's **active site,** while allosteric inhibitors change the enzyme's shape and activity by binding to a nonactive site.

Review and Thought Questions

Review Questions

1. State the First and Second Laws of Thermodynamics and explain their relationships to an example of work.
2. What is free energy? What is its relationship to the Second Law of Thermodynamics? In what direction must it change in all reactions (or processes of work)? Compare this to the meaning of "entropy."
3. Define endergonic and exergonic in terms of the free-energy change associated with reactions. Which reaction involving ATP is endergonic? Exergonic? Which reaction occurs spontaneously?
4. Which part of downhill skiing is endergonic (work), and which part is exergonic?
5. What is the activation energy for a reaction? How might it influence the speed with which a reaction proceeds?
6. Enzymes are biological catalysts. How do they perform the task of speeding up biological reactions?
7. What is meant by the "active site" of an enzyme? How is it related to the substrate of that enzyme? What is meant by "induced fit?" How does this make the substrate more likely to undergo a reaction?
8. Describe four benefits that cells derive from organizing their reactions into pathways.
9. Decide if these reactions are exergonic or endergonic. Mark "X" for exergonic and "N" for endergonic.
   __Growing hair
   __Digesting pizza
   __Baking soda and acetic acid (vinegar) reacting to give off fizzy carbon dioxide that makes a cake rise
   __Liver cells produce fibrinogen, a blood clotting protein
   __The adrenal glands produce cortisol derived from cholesterol
   __ATP splits during muscle contraction

Thought Questions

10. Where does the energy come from when you perform the work of riding your bicycle? Did *all* of the energy in the food you used for this task actually help turn the wheels of your bicycle? Explain using the First and Second Laws of Thermodynamics.
11. Explain why unraked leaves might form a neat pile in a specific spot in your backyard. Does this mean that the Second Law of Thermodynamics is not working? What is the source of energy for this organization of the leaves?
12. Why do you think that some exergonic reactions will *not* go at all in the absence of a catalyst? If the change in free energy is no obstruction, what is?

About the Chapter-Opening Image

A car rusting in a field is a good example of entropy. The Second Law of Thermodynamics says that as entropy (disorder) increases, less and less energy is available to do work.

On-line materials relating to this chapter are on the World Wide Web at http://www.harcourtcollege.com/lifesci/aal2/

Web Bit 5-1 Industrious Enzymes

Web Links

Quizzing & Testing

6 How Do Organisms Supply Themselves with Energy?

Louis Pasteur and Vitalism

In 1835, the French scientist Charles Cagniard de la Tour observed that the yeast in fermenting beer consisted of tiny living cellular organisms. Cagniard de la Tour noted that these tiny organisms multiplied by budding, and he suggested that they were somehow linked to the process of alcoholic fermentation—the conversion of sugar to ethanol and carbon dioxide. In the same year, two other scientific papers described similar observations. One was by Theodor Schwann, the German botanist whose work helped establish the cell theory described in Chapter 4.

Schwann also showed that if grape juice were boiled to kill the yeast, the juice would not ferment to wine unless yeast cells were added back to the juice. Schwann correctly concluded that yeast plays an essential role in fermentation. He then showed that fermentation began at the same time that the yeast cells first appeared, progressed with their multiplication, and stopped as soon as the yeast cells stopped multiplying.

Amazingly, the scientific community rejected these important discoveries out of hand, owing to a handful of powerful personalities and the scientific and philosophical temper of the time. The most influential scientists insisted that yeast was not a living organism at all, but a substance, and probably an unimportant one.

This mistaken idea was held by no less than the father of modern chemistry, Antoine Lavoisier (1743–1794). Lavoisier had already argued that fermentation could be understood as a simple chemical reaction. Fermentation, he and others had written, consisted of a reaction in which the six-carbon sugar glucose broke down into two molecules of ethanol and two molecules of carbon dioxide:

$$C_6H_{12}O_6 \longrightarrow 2\ C_2H_5OH + 2\ CO_2$$
$$\text{glucose} \longrightarrow \text{ethanol} + \text{carbon dioxide}$$

This nicely balanced formulation seemed so simple and so right that chemists, riding high on their tremendous successes (such as Lavoisier's own discovery of oxygen) could not imagine why yeast would be necessary. Even if yeast did turn out to be necessary, where could it possibly go in the equation? They accepted that yeast was present during fermentation. But there was no question of yeast actually taking part in the reaction. At most, they said, dead and dying yeast cells might release some substance that could speed up the chemical reaction. Nowadays, we would call such a substance a "catalyst."

Nineteenth-century chemists firmly believed that all biological processes were chemical in nature. To believe otherwise, to insist on some mysterious role for living organisms that was not purely chemical in nature, was condemned as **vitalism**— the belief that living systems have powers beyond those of nonliving systems.

The idea that a simple chemical reaction needed tiny living organisms to occur seemed absurd. In 1839, two of the most prominent early chemists, Friedrich Wöhler and Justus von Liebig, composed and published a lampoon designed to humiliate Schwann and the other yeast researchers. Wöhler and Liebig pretended to give a scientific description of the anatomy and behavior of yeast cells. A yeast cell, they said, looks exactly like a miniature still (the apparatus used to distill whiskey). Yeast cells suck in sugar and excrete ethanol and carbon dioxide, Wöhler and Liebig joked, writing, "Although teeth and eyes are not to be seen, one can distinguish a stomach, intestine, the anus (a rose-pink spot), and the organs of urine secretion. . . . The urine bladder in the full condition is shaped like a champagne bottle."

The lampoon must have worked, for Schwann retired from the debate, and Liebig and Wöhler's mistaken opinions prevailed unquestioned for 20 years. However, in 1854, young Louis Pasteur, a bright new star in the scientific sky, was appointed professor of chemistry at Lille, France.

Key Concepts

- Cellular respiration converts the chemical energy of food molecules into the chemical energy of ATP.
- Almost all cells can use glycolysis to derive energy, but cellular respiration provides much more energy.
- ATP production from cellular respiration depends on oxidative phosphorylation, which couples electron transport to the production of high-energy phosphate bonds in ATP. This process depends on the intact organization of mitochondria.

Lille was a major center for the manufacture of ethanol, through the fermentation of beet juice, and it was not long before one of Pasteur's students appealed to him to help solve problems at the family distillery. Pasteur immediately became intrigued by the process of fermentation. Just as quickly, he realized that fermentation depended entirely on the presence of yeast.

Pasteur was ambitious, and he must have known that proving the eminent Liebig wrong would be a magnificent coup. In September, 1855, Pasteur began a series of studies in fermentation. As his wife, Marie Pasteur, wrote to her father-in-law, "Louis is up to his neck in beet juice. He spends all his days in the distillery. He has probably told you that he teaches only one lecture a week; this leaves him much free time which, I assure you, he uses and abuses."[1]

By 1860, Pasteur had succeeded in growing the yeast in a medium seeded with tiny numbers of yeast cells. The ethanol produced was proportional to the multiplication of the yeast. Normally, vast numbers of yeast cells are present in fermenting solutions. And Liebig had long insisted that some chemical released by the decomposing bodies of yeast cells initiated fermentation. But Pasteur's yeast cells were multiplying, not dying, and the scientific community soon rallied to Pasteur's side. Liebig stubbornly rejected Pasteur's results to the end of his life. Fermentation, he said, was a chemical process, and vitalism was not science.

Ironically, although fermentation normally depends on cells, Liebig was largely right. In fact, fermentation *is* a chemical process—though far more complicated than that envisioned by early chemists. And fermentation can take place outside of living cells—provided the right enzymes are available.

In 1897, Hans and Eduard Büchner decided to make an extract of yeast, called "zymase," to be used as a patent medicine. However, the extract kept rotting, so the brothers added sugar, for the same reason that fruit canners add sugar to canned peaches and berries. In high concentration, sugar is an excellent preservative. To their amazement, the yeast extract began bubbling furiously, as carbon dioxide gas burst from the surface of the extract. Hans Büchner immediately recognized that the sugar was fermenting in the absence of living yeast cells. The Büchners' zymase contained essential enzymes that catalyzed the alcoholic fermentation reaction—just as Liebig had predicted.

The scientific community rejected these important discoveries out of hand, thanks to a handful of powerful personalities.

Nonetheless, Pasteur also was right. Fermentation normally does not occur outside of living organisms. It is a highly specialized process by which microorganisms extract energy from sugar. And although biologists now all agree with Liebig's idea that life will ultimately be understood in terms of chemistry, organisms are more than mere bags of enzymes. Their organization and complex structures guide and control the chemical reactions that contribute to life. In this chapter, we will see how cells of all kinds extract energy from biological molecules.

HOW DO ORGANISMS SUPPLY THEMSELVES WITH ENERGY?

All organisms need energy, and the ultimate source of energy for most organisms is sunlight. Plants convert light energy from the sun into chemical energy through photosynthesis. Plants and all other organisms that obtain energy and synthesize organic molecules from inorganic material are **autotrophs** [Greek, *auto* = self, same + *trophe* = to nourish]. The vast majority of autotrophs, including most plants and many algae and bacteria, photosynthesize. Nonetheless, not all autotrophs photosynthesize. Autotrophs include a few bacteria (called *chemotrophs*) that obtain energy by oxidizing inorganic substances such as sulfur and ammonia.

Animals, fungi, and other organisms that obtain chemical energy from other organisms are **heterotrophs** [Greek, *hetero* = other + *trophe* = to nourish]. Both heterotrophs and

---
[1] From Dubos, Rene, *Louis Pasteur, Free Lance of Science*, Little, Brown, 1950.

Figure 6-1 When it comes to sustained aerobic activity, the arctic tern, *Sterna paradisaea*, is the champion. This arctic tern nests each summer near the Arctic Circle in North America, then migrates across the Atlantic Ocean to Europe, then south to South Africa, then across the South Atlantic to Antarctica, a distance of 18,000 km (11,000 miles). In the spring, the bird flies all the way back around the world to the Arctic Circle to nest once more. *(Dwight Kuhn)*

autotrophs obtain energy by breaking down organic molecules in the process called cellular respiration. The chemical reactions of cellular respiration are part of a complex network of biochemical conversions that are collectively called **metabolism** [Greek, *meta* = after + *ballein* = to throw]. The difference between heterotrophs and autotrophs is that autotrophs make these organic molecules themselves, while heterotrophs take them from others—either from autotrophs or other heterotrophs. Ultimately, nearly all the energy that powers living organisms comes from sunlight.

What Is the Common Currency of Energy for Organisms?

If you spend all your time today studying biology, you will use the energy from about 10^{25} molecules of ATP in the next 24 hours—some 40 kg (90 lb) of ATP. If you were to run a marathon, you would use the same amount in just 2 hours. Almost every time one of your cells performs an energy-consuming reaction, it derives the energy it needs by splitting ATP, the universal currency of biological energy. Fortunately, however, cells recycle ATP (so you don't really have to carry around an extra 90 lb).

All cells use ATP. Plants produce ATP during photosynthesis, as we will see in Chapter 7. But both plants and animals can produce ATP in another way—by capturing energy from the breakdown of carbohydrates, lipids, and proteins.

The main way that organisms break down organic molecules is **cellular respiration**, the oxygen-dependent process by which cells extract energy from food molecules. Cellular respiration perfectly illustrates how chemical pathways proceed step by step, how some pathways occur in cycles, and how eukaryotes all share the same metabolic pathways. Animals and other heterotrophs obtain almost all of their energy through cellular respiration. Plants also depend on cellular respiration for ATP at night or at other times when they cannot photosynthesize.

Cellular respiration is distinct from ordinary respiration, or breathing. Breathing—inhaling oxygen and exhaling carbon dioxide—is nonetheless closely related to cellular respiration. Breathing supplies our cells with the oxygen needed for cellular respiration and disposes of the carbon dioxide that is its waste product.

Most eukaryotic cells (and many prokaryotic cells) are **aerobic**, that is, they depend on oxygen for life. All aerobic organisms, from earthworms to arctic terns, produce the bulk of their ATP by means of cellular respiration (Figure 6-1).

Many prokaryotes and a few eukaryotes can live **anaerobically**, without oxygen. The microbe *Clostridium botulinum*, which can poison canned meats and other foods and causes botulism (food poisoning), grows anaerobically (Figure 6-2A). Even our own cells, which are mainly aerobic, can produce energy anaerobically for short periods by a process similar to alcoholic fermentation. Some organisms, like the red-eared turtle shown in Figure 6-2B, can sustain themselves on glycolysis for days at a time.

In all organisms, the common currency of energy is ATP.

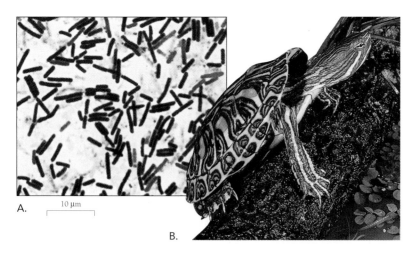

Figure 6-2 Life without air. A. The obligate anaerobe *Clostridium botulinum*, which causes a serious form of bacterial food poisoning, cannot reproduce in the presence of oxygen. It is known, however, for its ability to multiply inside of sealed canned goods. B. Among vertebrates, the red-eared turtle, *Chrysemys scripta elegans*, is unusual in its ability to live without oxygen. It can stay under water for 2 weeks at a time, relying on glycolysis for energy production. *(A, CNRI/Phototake; B, Zig Leszczynski/Animals Animals)*

How Do Heterotrophs Extract Energy from Macromolecules?

Cellular respiration begins with small molecules such as glucose—a simple six-carbon sugar. Cells cannot extract energy directly from complex carbohydrates, proteins, or fats. Large molecules must therefore undergo **digestion,** splitting into smaller units—proteins to amino acids, polysaccharides to glucose and other simple sugars, and fats to fatty acids and glycerol. Digestion occurs through the process of hydrolysis—breaking each link in a polymer through the addition of a molecule of water.

Most digestion takes place outside the cytosol of the cell so that cells are less likely to digest themselves. In animals and fungi, most digestion takes place outside the cell through the action of secreted enzymes, in the intestinal tract of animals, or in the space surrounding a fungus (Figure 6-3). Even in protists and in those animal cells that digest inside the cell, digestion occurs inside lysosomes separated from the cytosol. The lysosomal membrane separates digestive enzymes from the rest of the cell (Figure 4-12A).

Digestion does not provide energy, and it does not occur in a fixed order of reactions. Digestion generates small, energy-rich molecules that move across the cell membrane into the cytosol. One of the most useful of these molecules is glucose. (In vertebrates, for example, the blood carries glucose to all the cells of the body.) Inside the cell, enzymes transfer the energy from glucose into the chemical bonds of ATP.

Digestion also breaks proteins, nucleic acids, and lipids into small molecules that can enter cells. These molecules (or molecules derived from them) ultimately enter the same biochemical pathways that extract energy from glucose, as we will discuss later in this chapter.

Heterotrophs prepare for cellular respiration by capturing macromolecules and by digesting them with enzymes into their component building blocks. Digestion itself does not produce energy.

What Are the Stages of Cellular Respiration?

Cells extract energy from glucose by breaking the bonds between each of its six carbon atoms and moving the energy from those bonds to the high-energy bonds of ATP. During cellular respiration, glucose is completely broken down to carbon dioxide and water. The overall reaction is summarized by the following equation:

$$C_6H_{12}O_6 + 6\,O_2 \longrightarrow 6\,CO_2 + 6\,H_2O$$
$$\text{glucose} + \text{oxygen} \longrightarrow \text{carbon dioxide} + \text{water}$$

The complete breakdown of glucose to carbon dioxide and water and the packing away of the energy into ATP occurs in four stages. In eukaryotes, the first stage, glycolysis, takes place in the cytosol. In this stage, a six-carbon glucose is broken in half, forming two molecules of pyruvate, each with three carbon atoms (Figure 6-4). In the second stage, the three-carbon pyruvate molecules each lose a carbon atom, which is ultimately released as carbon dioxide. The remaining two-carbon molecules go into the third stage, the citric acid cycle. Here, the last remaining carbon-carbon bonds are broken and the single carbon atoms combine with oxygen to form more carbon dioxide. The second and third stages take place in the mitochondrial matrix.

Figure 6-3 **Digestion nearly always occurs outside of the cytosol.** Digestion occurs either outside the cells altogether in some kind of stomachlike cavity, or in lysosomes, separated from the cytosol. Digestion does not produce energy but makes building blocks available to the metabolic pathways.

Figure 6-4 Cellular respiration. The complete breakdown of glucose occurs in four stages. (1) In glycolysis, six-carbon glucose is broken in half, forming two molecules of pyruvate, each with three carbon atoms. Glycolysis produces two ATPs from ADP and two NADHs from NAD^+. (2) In the second stage, each pyruvate molecule loses a carbon atom (as carbon dioxide) to form an acetyl-CoA molecule and one NADH. (3) The citric acid cycle converts two more carbon atoms into carbon dioxide. The cycle generates two high-energy phosphate bonds (in GTP) and uses the eight electrons of each acetyl-CoA molecule to generate six NADHs and two $FADH_2$s. (4) Oxidative phosphorylation captures energy from NADH and $FADH_2$ in the high-energy bonds of ATP.

The fourth stage of cellular respiration is **oxidative phosphorylation.** During this stage, the energy released from the breaking of carbon-carbon and carbon-hydrogen bonds in glucose is transferred to phosphorus-oxygen bonds in ATP. The fourth stage is not actually the last in time, since much of it occurs simultaneously with the other three stages. As we will see later, much of the action of the fourth stage occurs within the inner mitochondrial membrane. Of the four stages of cellular respiration, oxidative phosphorylation (stage 4) makes by far the most ATP.

Cellular respiration occurs in four stages, one of which takes place in the cytosol of eukaryotic cells, and three of which take place in mitochondria.

HOW DO CELLS EXTRACT ENERGY FROM GLUCOSE?

The first of the four stages in the metabolism of glucose is **glycolysis** [Greek, *glykys* = sweet (referring to sugar) + *lyein* = to loosen], a set of ten chemical reactions (or steps). Glycolysis converts glucose, a sugar with six carbon atoms, into two identical smaller molecules, called **pyruvate,** with three carbon atoms each (Figure 6-5). The ten reactions also convert some of the energy of glucose into two high-energy ATP bonds, and some into two high-energy molecules of NADH. All ten reactions take place in the cytosol, outside the mitochondria.

The most remarkable fact about glycolysis is that *all* organisms accomplish it in exactly the same way. This universality suggests that the common ancestors of all present-day organisms performed glycolysis in the same way that we do today. The enzymes that catalyze the ten reactions must have evolved at least 3.8 billion years ago, at the very dawn of life on Earth.

What Is Oxidation?

Most of the energy in food molecules is stored in covalent bonds between carbon and hydrogen atoms. We obtain energy from these bonds by arranging for the electrons shared by carbon and hydrogen atoms to be captured by oxygen atoms. Oxygen atoms are extremely electronegative, so the flow of electrons from carbon atoms in food molecules (such as glucose) to oxygen is highly exergonic, releasing energy that cells use to make more ATP.

In fact, most of the energy-producing processes in cells involve the transfer of electrons from one molecule to another. In cellular respiration, most of the high-energy electrons are first transferred to a molecule called NAD^+, which we will discuss later.

A molecule that accepts an electron is called an electron acceptor, or **oxidizing agent.** A molecule that donates an electron is called an electron donor, or **reducing agent.** Oxidation and reduction reactions always go hand in hand. The oxidation of one atom, molecule, or ion provides the electrons for the reduction of another atom, molecule, or ion. Such reactions are called **reduction-oxidation reactions** or **redox reactions.**

In cellular respiration, electrons from the chemical bonds of glucose molecules combine with oxygen (and hydrogen ions)

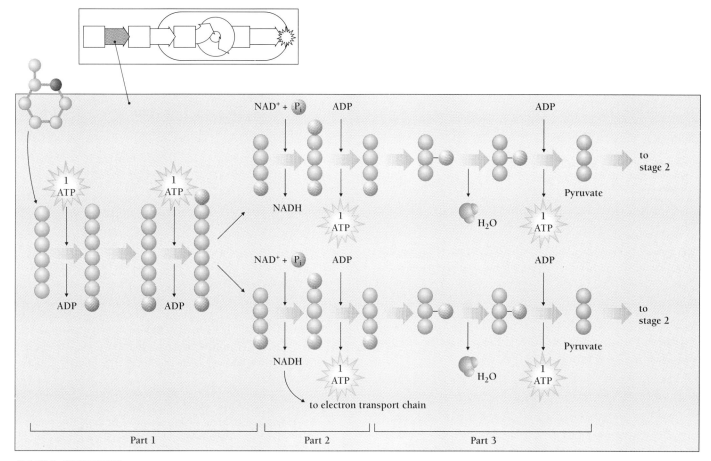

STAGE 1: GLYCOLYSIS

Figure 6-5 **Stage 1: the first stage of metabolism is glycolysis.** Glycolysis is a set of ten chemical reactions that break each six-carbon glucose into two molecules of three-carbon pyruvate. The ten reactions—nine of which are shown here—also convert some of the energy of glucose into high-energy ATP and NADH. All these reactions take place in the cytosol, outside the mitochondria.

to form water. As we have seen, the six carbons in glucose ultimately break down into six molecules of carbon dioxide. Carbon atoms donate electrons, oxygen atoms ultimately accept electrons, and hydrogen ions join the reduced oxygen to form water.

In Chapter 2, we saw that different atoms have different affinities for electrons. Oxygen is the most electron-hungry atom in the environment. Organisms have evolved mechanisms for using oxygen's enormous electronegativity to obtain energy from reactions in which oxygen is the ultimate electron acceptor. As electrons move from glucose and other energy-rich molecules to oxygen, however, they must pass through many intermediate electron acceptors, as we will discuss later.

The two most important electron acceptors in energy metabolism are **NAD$^+$** and **FAD,** which take electrons from glucose and transfer them to oxygen in a series of steps. Each NAD$^+$ can accept two electrons and a hydrogen ion, which converts it to its reduced form, **NADH.** Each FAD can accept two electrons (and two hydrogen ions) to form **FADH$_2$.** Most of the ATP produced by cell respiration comes from energy originally captured in NADH and converted into ATP energy by oxidative phosphorylation.

NAD$^+$ (and NADH) and FAD (and FADH$_2$) are examples of **coenzymes**—organic molecules that are necessary participants in certain enzyme reactions, often as electron donors or acceptors. Examples of coenzymes are most of the B vitamins, which are found in whole grains. NAD$^+$ and NADH participate in many enzyme reactions. NADH donates electrons, and NAD$^+$ accepts electrons.

Glycolysis: How Do Cells Capture Energy in ATP and NADH?

As glycolysis breaks glucose into two molecules of pyruvate, it also converts two molecules of NAD$^+$ to NADH and two molecules of ADP to ATP. The major question about glycolysis is: How do cells capture energy in ATP and NADH? Like other

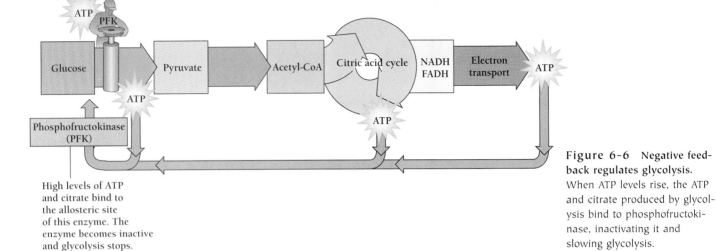

Figure 6-6 Negative feedback regulates glycolysis. When ATP levels rise, the ATP and citrate produced by glycolysis bind to phosphofructokinase, inactivating it and slowing glycolysis.

biochemical pathways, glycolysis proceeds in small steps. But the ten chemical reactions of glycolysis fall into just three parts (Figure 6-5):

Part 1. Splitting of the six-carbon glucose into two three-carbon molecules.

Part 2. Oxidation of the three-carbon molecules and reduction of NAD^+.

Part 3. Further oxidation of the new three-carbon molecules to pyruvate.

Here's how glycolysis is organized: Part 1 of glycolysis converts the six-carbon glucose molecule into two three-carbon molecules. Part 1 requires five separate reactions (or steps), each catalyzed by a different enzyme. This process transfers two phosphate groups from two ATP molecules to glucose and splits glucose in half. The energy realized from the splitting of the high-energy phosphates pushes the reactions in the direction of the three-carbon product, glyceraldehyde-3-phosphate (G3P). At the end of part 1, glycolysis has consumed rather than harvested energy.

In part 2 of glycolysis, NAD^+ oxidizes each molecule of G3P to phosphoglycerate. Each molecule of G3P generates one ATP and one NADH. In addition, a phosphate ion from the cytosol is attached first to G3P, then to ADP to form a new molecule of ATP. For each starting glucose molecule, then, part 2 yields two ATPs and two NADHs, thus recovering the ATP investment of part 1.

The reaction steps of part 2 require both phosphate ions and NAD^+. If either is missing, the reaction simply cannot occur. Phosphate and NAD^+ are essential: cell survival requires a constant supply. Phosphate is usually not a problem, since it is present at fairly high concentrations in most cells and is continuously released as ATP is split. In aerobic organisms, the regeneration of NAD^+ generally requires transfer of NADH's electrons to oxygen. In the absence of oxygen (and in anaerobes), however, the cell must regenerate NAD^+ by other means, as we will discuss later.

Part 3 converts phosphoglycerate to pyruvate and transfers its phosphate to ADP, forming one more ATP molecule for each phosphoglycerate (that is, two ATP molecules for each glucose). Since the reactions in part 2 had already recovered the two ATP molecules invested in part 1, the net gain of ATP during glycolysis is two ATPs per molecule of glucose, as well as two molecules of NADH.

But this is only a small bite out of a very large apple. The pyruvate molecules still contain lots of energy. This energy is extracted only in the later stages of glucose metabolism, which we will discuss later in this chapter. First, however, we will see how cells can use glycolysis to obtain energy even in the absence of oxygen.

Glycolysis, the first stage in the extraction of energy from glucose, breaks glucose into two molecules of pyruvate and produces two molecules each of ATP and NADH.

What regulates the overall rate of glycolysis?

In the cell, the overall process of glycolysis is highly exergonic. This means that glycolysis proceeds spontaneously and irreversibly. Since all the needed enzymes are present in cells, the reactions should also proceed quickly. How, then, does the cell keep all its glucose from being immediately converted to pyruvate?

We can see that glycolysis must be regulated. The question is, Where and how? In general, biochemical pathways are regulated at the first step that is unique to the pathway. This strategy ensures that molecules are not processed needlessly when they could be used for alternate purposes. In glycolysis, regulation takes place in reaction 3, which is catalyzed by the enzyme phosphofructokinase (Figure 6-6).

We would expect glycolysis to proceed rapidly when the cell needs ATP and to proceed slowly or not at all when the cell already has lots of ATP. This is indeed the case. ATP inhibits the action of phosphofructokinase, so when ATP is present, it slows glycolysis at part 3. As a result, glycolysis stops producing ATP when it is not needed.

Because ATP inhibits phosphofructokinase, reaction 3 of glycolysis, cells with plenty of ATP stop extracting energy from glucose. Cells with little ATP can proceed with glycolysis.

BOX 6.1

Sprints, Dives, and Marathons

Respiration yields at least 15 times more ATP than glycolysis. It's no wonder that animals—especially warm-blooded ones—depend on respiration for most of their energy. However, because oxygen is the ultimate electron acceptor in respiration, available energy depends on a steady supply of oxygen. For vertebrates, continuous energy production requires efficient lungs (to take in oxygen-rich air) and hearts (to pump oxygen-rich blood from the lungs to the muscles).

Training can improve the performance of the systems that deliver oxygen, but humans and other animals often need to spend energy more quickly than our oxygen supply will allow. We all need to sprint occasionally, whether we are a pedestrian dashing to avoid an inattentive driver, an outfielder swooping down on a wayward baseball, a cheetah chasing a gazelle, or a gazelle eluding a cheetah. Several energy reservoirs allow such bursts of energy: (1) muscles have stores of ATP and other compounds with similar high-energy phosphate bonds, enough to allow several minutes of work without any new ATP production; (2) muscles can use blood glucose and stored glycogen to produce ATP by anaerobic glycolysis, producing lactate as the end product.

The length of time that an organism can depend on anaerobic glycolysis for ATP production is limited both by training and by genetic capacity. Humans can manage only a few minutes. But some animals, such as whales, seals, and other diving animals, can function by means of glycolysis alone for much longer periods. The freshwater red-eared turtle shown in Figure 6-2B, for example, can stay under water for as long as 2 weeks.

Anaerobic glycolysis provides animals with a way of borrowing energy. But the loan is only short term. After anaerobic exertion, all animals must get rid of the accumulated lactate and restore their supplies of glucose and glycogen. These processes require oxygen. In most animals, the "oxygen debt" must be repaid almost immediately—through heavy breathing and increased circulation. Anyone who has run after a bus or away from an angry sibling has experienced the recovery from an oxygen debt. Animals extract energy from the accumulated lactate by converting it to pyruvate, which can be used in aerobic respiration. In some vertebrates, some of the lactate travels via the blood to the liver, where it can be recycled into glucose. The liver then resupplies the glucose reservoir of the blood.

Anaerobic glycolysis allows untrained schoolchildren to sprint up to 17 miles per hour (27 km/hr) in a 100-yard (or 100-meter) dash. But the necessity for oxygen prevents even champion athletes from averaging more than about 11.5 miles per hour in a 26-mile marathon (Figure A).

Nonmammalian vertebrates (fish, amphibians, and reptiles), which live their lives at a slower metabolic pace than we do, may function without air for long periods. Instead of using oxygen to recycle the accumulated lactate, fish, amphibians, and reptiles may excrete the lactate in their urine. This represents a huge waste of energy, but it allows these animals to feed and function in environments in which they could not otherwise live.

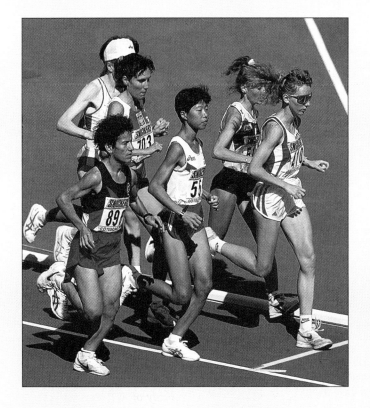

Figure A 1995 World women's marathon. The speed that a marathoner can attain is limited by the need for oxygen. *(David Madison)*

How can glycolysis continue without oxygen?

To sustain glycolysis, a cell must regenerate more NAD$^+$ from NADH. Just how it does this depends on the species of organism and on whether oxygen is present.

In the presence of oxygen, most cells can oxidize NADH to NAD$^+$ by the reactions of oxidative phosphorylation. In the absence of oxygen, however, cells must regenerate NAD$^+$ by using NADH to reduce pyruvate. Their manner of doing so depends on what enzymes they have.

Many microorganisms live by **fermentation** [Latin, *fervere* = to boil]—the anaerobic extraction of energy from organic compounds (Figure 6-7). Yeast cells, for example, ferment the sugars in beer, wine, and bread, converting glucose into carbon dioxide and ethanol. The carbon dioxide makes beer bubble and bread rise. The ethanol gives beer its punch and freshly baked bread its sweet aroma. To perform this feat, yeasts employ all the reactions of glycolysis that we have discussed. In addition, they use the accumulated NADH to reduce pyruvate to form ethanol. In the process, NADH is oxidized back to NAD$^+$. Yeast is a *facultative anaerobe*—an organism that can live either anaerobically, by fermentation, or aerobically, by oxidative phosphorylation. Other microorganisms, such as *Clostridium botulinum*, are *obligate anaerobes*, meaning that they can grow only in the absence of oxygen (Figure 6-2A).

Animals, including humans, often suffer a temporary lack of oxygen in some of their cells. For example, during sprinting or other vigorous exercise, muscle cells need more oxygen (to oxidize NADH) than the blood and lungs can deliver (Figure 6-7B). In this case, and for short periods only, the cells can regenerate NAD$^+$ in the absence of oxygen. Instead of producing carbon dioxide and ethanol (as a yeast cell does), however, muscle cells make lactate (lactic acid, after it has given up a hydrogen ion), a compound with three carbon atoms. During a vigorous sprint, lactate accumulates in the muscles, sometimes causing soreness. Once oxygen becomes available

Figure 6-7 Two kinds of fermentation. In each case, cells regenerate NAD$^+$ for further glycolysis. A. When yeast cells grow in limited oxygen (as in a fermentation vat or a bread dough), pyruvate forms acetaldehyde and ethanol. B. In muscle cells that have exceeded their oxygen supply, pyruvate forms lactic acid (lactate). (*Pizza maker, Jim Erickson/The Stock Market; sprinters, David Madison*)

again, however, cells convert most of the lactate back to pyruvic acid. Heart muscle is especially efficient at extracting energy from lactate by first converting it to pyruvate (pyruvic acid, after it has given up a hydrogen ion).

Glycolysis can continue without oxygen if cells use pyruvate to regenerate NADH from NAD^+.

What Happens to Pyruvate When Oxygen Is Present?

In eukaryotic cells, the steps of energy extraction after glycolysis take place in the mitochondria. The pyruvate produced by glycolysis must enter the mitochondria for cellular respiration to proceed. After pyruvate arrives in the **mitochondria**, it is oxidized through a series of steps to a compound called acetyl-CoA, whose acetyl group contains two carbons (Figure 6-8).

Acetyl-CoA forms from a molecule of acetate (the same as the acetic acid of vinegar) linked by a high-energy bond to a molecule called **coenzyme A (CoA)**. During the formation of acetyl-CoA, the reaction removes one molecule of carbon dioxide from pyruvate, as well as two electrons and two protons. NAD^+ serves as the electron acceptor. This reaction cannot occur if there is no oxygen to drive the fourth stage of respiration, which regenerates NAD^+ from NADH.

Acetyl-CoA has only two principal fates: (1) it can enter the third stage of energy metabolism and generate more ATP, as discussed in the following section, or (2) it can be used to synthesize new lipids, either fatty acids or cholesterol. When a cell's ATP level is high, the cell redirects acetyl-CoA into a set of reactions that make energy-rich fats and oils. This redirection of acetyl-CoA allows the body to store energy for later use. It is the main reason that we accumulate fat when we consume too many energy-rich molecules, whether they are starches in a sandwich or stearic acid in a steak.

When oxygen is present, the pyruvate produced by glycolysis enters a cell's mitochondria, where it is converted to acetyl-CoA. Acetyl-CoA can generate still more ATP.

How Does a Cell Generate ATP from Acetyl-CoA?

Most of the energy in a glucose molecule is still present in the two molecules of acetyl-CoA produced by stages 1 and 2 of cellular respiration. The reactions of stage 3, which we will now discuss, are responsible for capturing most of the high-energy electrons in NADH. Oxidative phosphorylation (stage 4) then harnesses the energy of NADH into the production of ATP.

In nearly all organisms, acetyl-CoA enters directly into a set of reactions that converts the carbon atoms of acetyl-CoA

Figure 6-8 Stage 2: pyruvate to acetyl-CoA. After glycolysis, pyruvate enters the mitochondria for the second stage of energy extraction. In the mitochondria, pyruvate is oxidized to form acetyl-CoA. Acetyl-CoA can either enter the third stage of energy metabolism and generate more ATP or it can be used to synthesize body fat.

into carbon dioxide. These reactions are called the **citric acid cycle**, or the **Krebs cycle**, after Hans Krebs, who began working out the details of the cycle in the 1930s.

In eukaryotes, the citric acid cycle takes place within the mitochondrial matrix. In its barest outlines, the citric acid cycle takes a molecule of the two-carbon compound acetyl-CoA, adds it to a four-carbon molecule (oxaloacetate, or OAA) to form a six-carbon molecule (citric acid), removes two carbons as carbon dioxide, and produces another four-carbon OAA, ready to combine with another molecule of acetyl-CoA.

Altogether, the citric acid cycle removes eight electrons from the two carbons of acetyl-CoA. Three NAD^+ molecules accept six of these electrons, and one molecule of FAD accepts the other two. The citric acid cycle also forms one high-energy phosphate bond. Unlike the other reactions that we have discussed, however, the phosphate acceptor in the citric acid cycle is guanosine diphosphate (GDP), rather than ADP. The resulting guanosine triphosphate (GTP) carries the same amount of energy as ATP.

Oxygen is the ultimate acceptor of electrons and is therefore absolutely essential for the citric acid cycle to proceed. In stage 4 of cellular respiration (the electron transport chain), oxygen accepts electrons from NADH and $FADH_2$, allowing the cell to regenerate more NAD^+ and FAD and permitting both the cycle (stage 3) and the oxidation of pyruvate (stage 2) to continue.

The citric acid cycle not only provides energy but also serves as the major source of building blocks for many amino acids and other small molecules. The flow of molecules through the cycle then responds both to the energy needs of a cell and to the availability of amino acids.

Figure 6-9 summarizes the steps of the citric acid cycle. The central events of the citric acid cycle are its four oxidation-reduction reactions, which employ two kinds of electron acceptors—NAD^+ and FAD. The reactions of the cycle fall into three parts: (1) the formation of the six-carbon citrate (citric acid) from the addition of the two-carbon acetyl-CoA to the four-carbon OAA, followed by the rearrangement of citrate into isocitrate; (2) the conversion of isocitrate into a four-carbon compound called succinate, which is linked to CoA; and (3) the production of another molecule of OAA, which can combine with acetyl-CoA to regenerate citrate and start the cycle all over again.

With each turn of the cycle, citrate loses two carbon atoms as carbon dioxide and regenerates the four-carbon OAA. The carbon atoms of the carbon dioxide from each turn, however, are not those that entered the cycle as acetyl-CoA. Nor is the OAA regenerated by the cycle the same as the one that started the cycle. In each turn of the cycle, citrate loses a total of eight electrons.

The citric acid cycle converts acetyl-CoA to carbon dioxide, transferring, in the process, eight electrons from acetyl-CoA to NADH and $FADH_2$.

How was the citric acid cycle discovered?

Biochemists tried for years to understand how cells use oxygen to obtain energy from organic molecules. But it was the brilliant Hungarian biochemist Albert Szent-Györgyi, a man obsessed by all oxidative processes, who solved the problem. The first observation that Szent-Györgyi made was that extracts of pigeon breast muscle were strikingly efficient at taking up oxygen for cellular respiration. This is not surprising, since the flight muscles of birds require lots of energy and are packed with mitochondria.

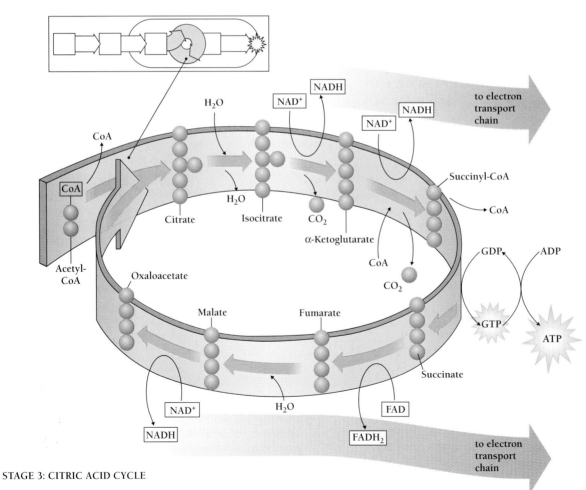

STAGE 3: CITRIC ACID CYCLE

Figure 6-9 Stage 3: the citric acid cycle. During the citric acid cycle, a single molecule of acetyl-CoA generates three NADH, one $FADH_2$, one GTP, and two carbon dioxide (CO_2) molecules.

Szent-Györgyi then added various compounds to his extracts to see how these compounds were oxidized. To his surprise, he found that certain four-carbon compounds stimulated the uptake of oxygen far more than he would have expected if each carbon atom were converted to carbon dioxide and each hydrogen became part of a water molecule. He had calculated the amount of oxygen required for the complete oxidation of each added compound, but the extracts were using more oxygen than complete oxidation would require.

The reason for this discrepancy, we now realize, is that the four-carbon compounds that Szent-Györgyi added were intermediates in the citric acid cycle—they increase the rate at which acetyl-CoA enters the cycle. The higher levels of acetyl-CoA production, in turn, increase the production of NADH and $FADH_2$ and therefore the amount of oxygen needed to regenerate NAD^+ and FAD (Figure 6-9).

By 1937, biochemists knew that pigeon muscle extracts could convert citric acid to these same four-carbon compounds. In 1938, Hans Krebs, a refugee from Nazi Germany working in England, provided the missing link and the final interpretation. Krebs showed that Szent-Györgyi's muscle extracts could make citric acid from the four-carbon compound OAA and acetyl-CoA.

Krebs's crucial contribution was the idea that the conversion of acetate to carbon dioxide involved a cycle. Krebs showed how the particular four-, five-, and six-carbon compounds fit together in a single cycle. He also showed that the same chemical transformations occurred both in cells and in test tubes.

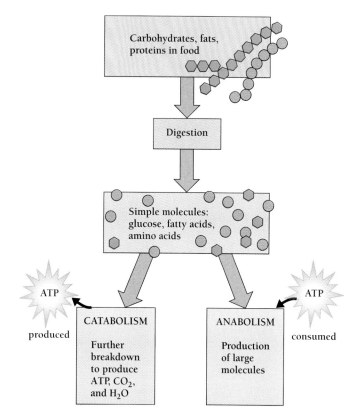

Figure 6-10 **Metabolism consists of two pathways: catabolism and anabolism.** Catabolism produces energy through the oxidation of molecules. Anabolism usually consumes energy in the process of building large molecules.

The first clue to the existence of the citric acid cycle was Szent-Györgyi's demonstration that four-carbon OAA stimulated the uptake of more oxygen than was required to oxidize the OAA.

The citric acid cycle intersects other biochemical pathways

The citric acid cycle lies at the center of the complex network of metabolic reactions. Metabolism is divided into two classes of pathways: **catabolism** [Greek, *cata* = down + *ballein* = to throw]—the breakdown of complex molecules, such as those in food; and **anabolism** [Greek, *ana* = up + *ballein* = to throw]—the synthesis of complex molecules (Figure 6-10). In general, catabolism produces energy and usually involves oxidation—the removal of electrons and the production of NADH. Anabolism, on the other hand, usually consumes energy and involves reduction, so it uses up NADH and other reducing agents.

Figure 6-11 shows that different molecules enter the cellular respiration pathway at different points. Carbohydrates in such foods as bread are broken down into simple sugars that usually take the same path as glucose. Fats and oils are broken down into glycerol, which may enter glycolysis, and into fatty acids, which enter the pathway as acetyl-CoA. Proteins are broken down into component amino acids, which are then converted to pyruvate, acetyl-CoA, or other intermediates of the cellular respiration pathway.

Cells regulate the rates at which pathways or sets of pathways operate. High levels of ATP favor anabolism, and low levels favor catabolism. When ATP levels are high but the products of anabolism are abundant, the energy in food is stored for future use—as carbohydrates, proteins, or fats. Despite the importance of glucose and other carbohydrates in energy metabolism, the preferred form of energy storage in animals is fat.

The end product of the catabolism of fats is acetyl-CoA. Fats are more reduced compounds than sugars; that is, for each carbon atom, fats provide more electrons for oxidative phosphorylation, the most productive stage of cellular respiration. Fats, therefore, contain more free energy—about 9.4 kcal per gram of fat, instead of the 4.1 kcal per gram of carbohydrate and 5.6 kcal per gram of protein. In addition, fats are less polar and less hydrophilic than carbohydrates, which may bind as much as twice their weight in water. As a result of these factors, each gram of stored fat contains about six times as much energy as each gram of stored glycogen. A 50-kg (110-lb) woman, for example, has fuel reserves of about 70,000 kcal in fats, about 18,000 kcal in proteins (mostly in muscle), about 400 kcal in glycogen, and about 30 kcal in glucose. Her fat

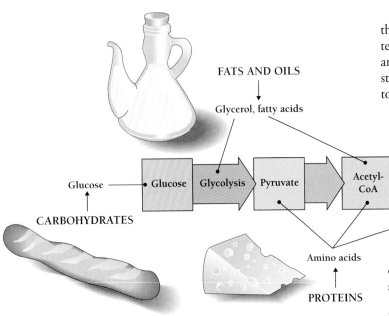

Figure 6-11 How do other food molecules enter metabolism? Fat, carbohydrates, and proteins enter the cellular respiration pathway at different points. The busiest crossroad is at acetyl-CoA, which then enters the citric acid cycle. All fatty acids and many amino acids are converted into acetyl-CoA.

It was the biochemists' rejection of vitalism that misled them. Because biochemists (in Liebig's tradition) were more interested in chemistry than in cells, they tended to take a "grind-and-find" approach to ATP production. Ignoring the complex structure of living organisms, they would simply grind up cells to find out what chemicals are in them. This approach has worked very well. Decades of research have shown that the majority of cellular processes depend on simple reactions that can be reproduced in a test tube. In fact, the grind-and-find approach was so successful that, until the 1970s, most biochemists were blind to the importance of intact cellular structures.

As it turns out, however, the steps of cellular respiration that release the most energy occur in functioning, intact mitochondria. Cellular respiration depends so heavily on mitochondria that we can pick out the cells that are most active in cellular respiration simply by counting the number of mitochondria in them.

Biochemists' "grind-and-find" approach to cell biology prevented them from recognizing the crucial role of cellular organization in cell respiration.

stores, however, weigh only about 8 kg (18 lb). To store the same amount of energy as carbohydrates (with all their bound water), she would need to weigh at least 90 kg (200 lb)!

The citric acid cycle lies at the intersection of the catabolic and anabolic pathways. High levels of ATP favor anabolism, and low levels favor catabolism. When ATP levels are high, the most efficient way to store energy is in the form of fat.

ELECTRON TRANSPORT: HOW DOES THE ENERGY IN GLUCOSE REACH ATP?

A major goal of biologists and biochemists has been to learn how cells convert the energy of sunlight or food into the high-energy bonds of ATP. Until the early 1960s, biochemists believed that they would soon understand how organisms make ATP simply by studying the rearrangements of chemical bonds in reactions catalyzed by enzymes in test tubes. Indeed, decades of ingenious research produced detailed maps of such enzyme-catalyzed transformations, many of which made or used ATP. Yet researchers gradually realized that in living cells most ATP was made in some other way. But how?

Surprisingly, the first clue came not from the properties of enzyme reactions, but from the properties of membranes and subcellular organelles. In producing ATP, cells rely on their internal structure in a way that eluded most biochemists.

How Does the Flow of Electrons from Electron Donors to Oxygen Release Energy to the Phosphate Bonds in ATP?

Altogether, the process of cellular respiration transfers 24 electrons from glucose to oxygen. The first electron acceptor for 20 of the 24 electrons is NAD^+. The other four electrons of glucose are transferred to FAD. Altogether, about 90 percent of the energy stored in the chemical bonds of a glucose molecule is converted to chemical energy in the form of the high-energy compounds, NADH and $FADH_2$.

The two electrons received by each molecule of NADH or $FADH_2$ ultimately go to reduce a single oxygen atom. The reduced atom of oxygen, together with two hydrogen ions from the surrounding water, results in the formation of one molecule of water.

The energy from the oxidation of NADH or $FADH_2$ is used to make high-energy phosphate bonds in ATP by a process that releases the energy in a series of small steps. This stepwise release resembles the breaking of a big waterfall into a gentle cascade.

Electrons move to oxygen as spontaneously as water flows downhill, and for the same reason. The flow of electrons to oxygen is a "downhill"—energy-releasing—process. However, the making of a high-energy phosphate bond to form ATP from ADP is definitely an uphill, energy-consuming process.

One important question, then, is how oxidative phosphorylation couples the downhill flow of electrons to the uphill production of high-energy bonds in ATP. The answer—discovered only in the 1960s—was a complete surprise to the thousands of researchers who were trying to figure it out.

Cellular respiration transfers electrons from glucose to oxygen by way of a series of intermediaries. The first electron acceptor is either NAD^+ or FAD. NADH and $FADH_2$ then release their electrons to the electron transport chain, which transfers electrons to oxygen through a series of small steps. Oxidative phosphorylation couples this transfer to the production of ATP.

Which Molecules Serve as Electron Carriers?

The electrons from NADH and $FADH_2$ make their way to oxygen by way of a series of other molecules, called **electron carriers.** The pathway of the electrons (from one carrier to another) is called the **electron transport chain.** Biochemists like to compare the electron transport chain that carries electrons to oxygen to the "bucket brigade" that old-time firefighters used to carry water to a fire (Figure 6-12D). Each electron carrier passes its electrons to the next carrier in the line. A reduced carrier becomes oxidized as it gives up its electrons, and the next carrier becomes reduced as it receives electrons. In eukaryotes, the transfer occurs in the mitochondrial membrane. In prokaryotes, these reactions occur in association with the cell membrane.

In 1925, the British biochemist David Keilin discovered that cellular respiration depends completely on a set of specific proteins that change color as they accept or donate electrons. These proteins, called **cytochromes** [Greek, *kytos* = hollow vessel, cell + *chroma* = color], have a reddish color that derives from **heme,** the same iron-containing ring structure that gives hemoglobin its color. Heme, like NAD^+ and FAD, may both donate and accept electrons. Keilin and others showed that cytochromes could accept electrons from NADH and also from each other. By comparing the flow of electrons both in isolated cytochromes and in whole cells, biochemists worked out the path of electrons from NADH to oxygen in the electron transport chain.

Iron-containing cytochromes contribute to the transfer of electrons through the electron transport chain.

How Do Cells Harvest the Energy of Electron Transport?

After the discovery of the electron transport chain, the big question still remained: How do cells harvest the energy released by the flow of electrons? Biochemists spent decades looking for intermediate molecules, particularly for small molecules that transferred phosphate groups to ATP in the same way that phosphoglycerate generates ATP in part 3 of glycolysis. (Such a phosphate group transfer to ATP is called "substrate-level phosphorylation.") But no one was able to find such an intermediate. Biochemists (incorrectly) concluded that the intermediate molecules were so unstable that they fell apart before they could be isolated in an experiment.

The situation was this: an electron transport chain in a test tube could pass electrons, but it would generate no ATP. Isolated mitochondria could perform both electron transport and ATP synthesis, but as soon as the mitochondria were broken open, ATP synthesis stopped even though electron transport continued. These experiments clearly showed the importance of intact mitochondrial organization for the production of ATP. But what did the mitochondria do?

The puzzle was solved in the early 1960s by Peter Mitchell, an English biochemist. Mitchell had been studying the way bacteria pump hydrogen ions across their membranes. The inside of a bacterium has a higher pH (a lower hydrogen ion concentration) than its environment. The difference in hydrogen ion concentration across a membrane is called a **proton gradient.**

Mitchell showed that bacteria could use the energy in this proton gradient to transport other substances into the cell. In 1961, he suggested that mitochondria, like bacteria, might be able to make use of the energy stored in a proton gradient. The idea that a proton gradient could be used to make ATP was, at the time, thoroughly unconventional. Mitchell was unorthodox in other respects, as well. For example, he performed his research not in a university laboratory, but in a manor house in the English countryside where he lived.

Mitchell proposed a two-part hypothesis. He suggested first that the flow of electrons through the electron transport chain establishes a proton gradient across the mitochondrial membrane. Second, he argued, some special molecular machinery in the membrane captures the energy stored in the proton gradient by making ATP. Mitchell's hypothesis showed why biochemists had not found the intermediate molecules that supposedly coupled electron flow to ATP production: they simply did not exist. Mitchell argued instead that the formation of ATP was the result of **chemiosmosis** [Greek, *osmos* = to push]—the harnessing of the energy stored in the chemical gradient created by the electron transport chain.

In the 1960s and 1970s, Mitchell, his colleague Jennifer Moyle, and others showed that the coupling of a proton gradient to the synthesis of ATP helps explain not only cellular respiration in mitochondria but also photosynthesis in chloroplasts. As we shall see, Mitchell's radical suggestion, called the **chemiosmotic hypothesis,** has now become the basis for understanding the harvesting of energy both in photosynthesis and cellular respiration.

The English biochemist Peter Mitchell proposed that the energy from the electron transport chain drives ATP production by harnessing the energy in a proton gradient, not by means of intermediate molecules.

Figure 6-12 The electron transport chain is like a bucket brigade. A. Transmission electron microscopy (TEM) of a single mitochondrion. B. The electron transport chain is located in the inner membrane. It consists of four protein complexes and a number of small lipid molecules that carry electrons between the complexes. The intermembrane space, between inner and outer membranes, has a high concentration of hydrogen ions. In contrast, the mitochondrial matrix, inside the inner membrane, has a low concentration of hydrogen ions. C. The electron transport chain drives hydrogen ions out of the matrix, creating an electrochemical gradient. D. The electron transport chain consists of a series of electron carriers that pass electrons from one to the next like the members of an old-fashioned bucket brigade trying to put a fire out. Instead of passing buckets of water, however, electron carriers pass electrons, derived from the breakdown of glucose. The electrons ultimately go to oxygen molecules. (A, D. Friend, D. Fawcett/Visuals Unlimited)

How Do Mitochondria Generate a Proton Gradient?

Recall that a mitochondrion has two membranes. A mitochondrion, therefore, provides four separate locations for the different parts of cellular respiration: the outer membrane; the inner membrane; the space between them, called the intermembrane space; and the **mitochondrial matrix,** the space surrounded by the inner mitochondrial membrane (Figure 6-12A and B).

Most of a cell's NADH and $FADH_2$ is produced within the mitochondrial matrix, which makes up about two-thirds of the volume of a typical mitochondrion. The electron transport chain lies within the inner membrane. The electron carriers are oriented within the inner membrane so that, as they pass electrons from one to another, hydrogen ions move from the matrix out to the intermembrane space by crossing the inner mitochondrial membrane (Figure 6-12B). This movement of hydrogen ions ("protons") out of the matrix establishes a proton gradient across the inner membrane, which lies between the matrix and the intermembrane space. Careful measurements with intact mitochondria confirm that a proton gradient really exists across the inner mitochondrial membrane, with the pH lower outside the membrane than inside. The movement of electrons through the electron transport chain in the inner membrane causes protons to move *out* of the matrix.

Experiments have demonstrated that even fragments of inner membranes can pump protons. In these studies, mitochondria were broken open and their membranes allowed to reassemble spontaneously into empty vesicles. As long as electrons flowed, protons moved. When the experimenter stopped electron flow with a poison, however, proton pumping stopped also. These experiments established the ability of the electron transport chain to pump protons.

The electron transport chain within the inner membrane of the mitochondrion pumps protons out of the matrix and generates a proton gradient.

What Pumps Protons out of the Mitochondrial Matrix?

By taking apart the inner membrane still further, biochemists found four distinct electron transport complexes in the inner mitochondrial membrane. Three of these complexes (complexes I, III, and IV) use the energy derived from transporting electrons part of the way down the electron transport chain to pump protons across the inner membrane (Figure 6-13). (The fourth complex, called complex II, transfers electrons but does not move protons.)

STAGE 4: ELECTRON TRANSPORT CHAIN

Figure 6-13 **Stage 4: the electron transport chain.** Each electron carrier is alternately reduced and oxidized as it gains and loses each electron passing down the chain. As the electrons pass, they lose some of their energy to proton pumps that push hydrogen ions, or protons, out of the matrix. The electrons ultimately combine with oxygen (and two hydrogen ions) to form water.

The complexes have fixed orientations in the membrane, so that they pump protons in a single direction only—from the matrix to the intermembrane space. For example, one of the respiratory complexes, called cytochrome oxidase, passes electrons from the last carrier in the electron transport chain (cytochrome *c*) to oxygen. Purified cytochrome oxidase complex can transfer electrons from reduced cytochrome *c* to oxygen and simultaneously pump protons, even in a vesicle made from artificial membrane.

The outer mitochondrial membrane allows the easy passage of small molecules and ions, so the intermembrane space has the same neutral pH as the bulk of the cytosol, about pH 7. In contrast, the pH of the matrix increases to about pH 8 as the electron transport chain pumps protons out of the matrix. The inner membrane is not permeable to protons (except through special channels), so the pH difference between the matrix and the intermembrane space is stable as long as electron flow continues (Figure 6-13).

Proton pumping establishes a charge gradient as well as a pH gradient. This is because pumping the positive protons out of the matrix leaves behind an excess of negative ions. The proton pump generates an **electrochemical gradient,** a double gradient composed of a *chemical* gradient (the difference in hydrogen ion concentration, or pH) and an *electrical* gradient (the difference in charge). Both components of the electrochemical gradient favor the return of protons to the matrix.

> *Three electron transport complexes in the inner membrane of the mitochondrion pump protons out of the matrix, generating an electrochemical gradient that favors the return of protons to the matrix.*

How Does the Flow of Protons Back into the Matrix Cause the Synthesis of ATP?

Once hydrogen ions leave the matrix, there is only one way back in—through special channels in the otherwise impermeable inner membrane. Protons flow "downhill" back into the matrix—from an area of high concentration to an area of low concentration and from an area of net positive charge to one with a net negative charge. As the protons flow, a protein complex called **ATP synthase** uses the flow of protons to drive the synthesis of ATP (Figure 6-14). ATP synthase works like a windmill, which converts the kinetic energy of wind to electrical energy.

We can actually see ATP synthase protein complexes as components of the inner mitochondrial membrane. Each looks like a miniature lollipop protruding from the inner mitochondrial membrane into the mitochondrial matrix (Figure 6-14). Biochemists have found that each contains two major components in a single complex: (1) the stick, which spans the

A.

B.

Figure 6-14 ATP synthase, the F_0-F_1 complex. A. The F_0-F_1 complex harnesses energy stored in the proton gradient across the inner mitochondrial membrane. F_0 serves as a channel for protons flowing back into the matrix, and F_1 is an enzyme that makes ATP as the protons pass. B. A windmill's blades spin around as it generates electricity. In the same way, the F_1 blades of the F_0-F_1 complex spin around as the complex generates ATP. (B, Alan Reininger/The Stock Market)

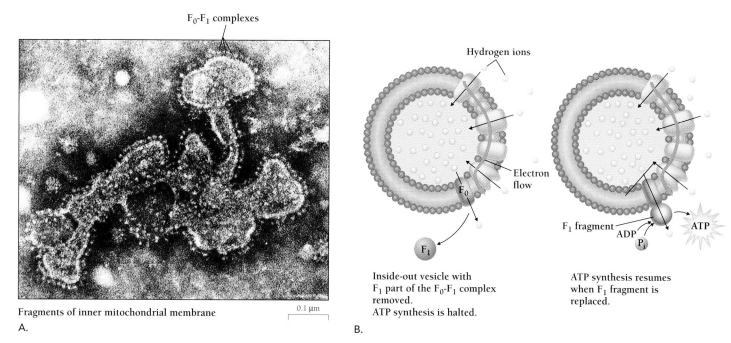

Figure 6-15 ATP synthase, the F_0-F_1 complex. A. This micrograph shows hundreds of F_0-F_1 complexes dotting the surface of an inside-out inner mitochondrial membrane. F_0 is the channel through which protons can move back into the matrix, while F_1 actually produces ATP from ADP and phosphate. The inside of the vesicle corresponds to the intermembrane space, which is continuous with the cytosol. The outside of the vesicle corresponds to the matrix of an intact mitochondrion. B. When the F_1 fragment (the coupling factor) is replaced, ATP synthesis resumes. (A, R. Bhatnagar/Visuals Unlimited)

inner membrane, called F_0, and (2) the top, called F_1. The "F_0F_1 complex" is the same as ATP synthase.

Under certain conditions, fragments of inner mitochondrial membrane form inside-out vesicles from which F_0-F_1 complexes protrude (Figure 6-15). Gentle shaking removes F_1 from the membrane, leaving F_0 behind. After this treatment, the inside-out vesicles still allow protons to flow down an electrochemical gradient, suggesting that F_0 serves as a channel for protons to move back into the matrix. But the removal of F_1 abolishes the ability of the vesicles to make ATP. Replacing the F_1 fragment causes ATP synthesis to resume (Figure 6-15). So F_1 appears to be responsible for making ATP and is called the **coupling factor.** We still do not fully understand, however, exactly how the F_0-F_1 complex harnesses the proton flow to create new high-energy phosphate bonds in ATP. We do know, however, that F_1 spins like a three-bladed windmill as it cranks out ATP.

The F_0-F_1 complex (ATP synthase) channels the flow of protons back into the mitochondrial matrix and uses the free energy from that flow to synthesize ATP.

Are Proton Pumping and ATP Synthesis Really Separate Processes?

Even before the discovery that the F_0-F_1 complex was responsible for the coupling of proton flow to ATP synthesis, biochemists became convinced of the chemiosmotic hypothesis because of a particularly elegant demonstration of the separateness of proton pumping and ATP synthesis. The convincing experiment started with photosynthetic bacteria from the salt flats near San Francisco Bay. In these bacteria, a purple protein called *bacteriorhodopsin* pumps protons across the bacterial membrane in response to light, as part of their process of photosynthesis. Even when researchers put this protein into artificial membrane vesicles, light stimulates the purple protein to pump protons into the vesicles. This proton pumping establishes a pH gradient between the inside and the outside (Figure 6-16).

To these artificial vesicles researchers then added the F_0-F_1 protein complex isolated from the mitochondria of a cow's heart (Figure 6-16). The F_0-F_1 complex allowed protons to flow back out of the vesicles and generate ATP. This experiment demonstrated that the F_0-F_1 complex can harness the

BOX 6.2

The Metabolism of Alcohol

For thousands of years and in many societies, ethanol has provided a popular means of changing mood and behavior. In heavy drinkers, however, ethanol can cause permanent damage to cells in the liver, nerves, brain, and heart. The metabolic pathways described in this chapter help us to understand how this happens.

In humans, liver cells are practically the only cells that can metabolize ethanol. In these cells, NAD^+ oxidizes ethanol, first to acetaldehyde and then to acetic acid, generating abundant amounts of NADH. The NADH then donates its electrons to the electron transport chain. Oxidative phosphorylation then occurs, but without the participation of the citric acid cycle. Sustained use of large amounts of alcohol leads to strange-looking mitochondria, the result of active oxidative phosphorylation and the shut-off of the citric cycle. In addition, carbohydrates that would ordinarily enter the citric acid cycle are instead converted to fat. The fat is secreted by the liver into the blood in little vesicles (produced by the endoplasmic reticulum).

Alcohol accumulates in the blood because it moves from the intestines into the blood more rapidly than the body can eliminate it. The blood carries the alcohol to cells throughout the body. Nerve and brain cells are particularly vulnerable to damage by alcohol. Heavy alcohol use irreparably damages heart muscle after about 10 years in men, sooner in women.

However, the most common organic disease among heavy drinkers is liver failure. After a few years, the liver cells begin to fill with fat and to cease functioning. Eventually, the liver becomes heavily scarred—a condition called *cirrhosis*. Once the damaged liver loses its ability to metabolize alcohol and other toxins, damage to the body accelerates. Cirrhosis of the liver is the seventh leading cause of death in the United States.

Our ability to metabolize alcohol demonstrates how adaptable animals are. Alcohol is just one of hundreds of chemicals from which we can obtain energy. No matter what we eat, biochemical homeostasis keeps the cells supplied with ATP and reducing power. Sources of energy are rarely wasted; they are used or stored. But animals also have evolved within a relatively narrow set of conditions. We cannot tolerate too large a variation of diet for an extended period without damage to the cells of our body.

---

energy of proton flow even in an artificial situation, with separate components from two extraordinarily different sources—a rare microscopic bacterium and a familiar large animal.

The chemiosmotic hypothesis explains how cells transfer the energy from the reduced coenzymes NADH and $FADH_2$ to ATP: (1) the electron transport chain creates an electrochemical gradient across the inner mitochondrial membrane, and (2) the F_0-F_1 complex couples the flow of protons down the gradient to the synthesis of ATP.

The main point is that the creation of the proton gradient and the generation of high-energy bonds in ATP are separate processes. The experiments with *bacteriorhodopsin* and cow F_0-F_1 complex support this idea and further suggest that this strategy for ATP synthesis is nearly universal. In Chapter 7, we will see that proton gradients not only produce ATP in mitochondria but also produce ATP in chloroplasts during photosynthesis.

Proton pumping and ATP synthesis are separate processes.

How Much Usable Energy Can a Cell Harvest from a Molecule of Glucose?

Actual measurements in isolated mitochondria show that the oxidation of each molecule of NADH leads to the movement of 10 hydrogen ions across the inner membrane and to the synthesis of 2.5 molecules of ATP. Similar experiments show that each $FADH_2$ can yield 1.5 molecules of ATP. For each molecule of glucose, then, the NADH produced in the mitochondria can yield 20 molecules of ATP, and the $FADH_2$ yields 3 molecules of ATP (Figure 6-17). The actual yield of ATP is somewhat less than this because the mitochondrion also uses the energy of the proton gradient to drive the transport across the mitochondrial membrane of other molecules and ions, including phosphate.

Glycolysis produces two more molecules of NADH. These are made in the cytoplasm and do not readily cross the inner mitochondrial membrane. The two molecules of NADH made during glycolysis produce either three or five molecules of ATP, depending on the tissue.

Finally, two molecules of ATP are produced in glycolysis and two molecules of GTP in the mitochondria. Since the high-energy phosphates of the GTPs readily produce ATP, we have an additional four molecules of ATP per molecule of glucose.

Cellular respiration thus produces 30 or 32 molecules of ATP for each molecule of glucose. (Until recently, biochemists estimated that each molecule of glucose yielded 36 to 38 molecules of ATP because it was believed that each NADH yielded 3 molecules of ATP, rather than 2.5, and each $FADH_2$ yielded 2 molecules of ATP, instead of 1.5.) All but four of the ATP molecules produced from a glucose molecule are produced by oxidative phosphorylation. In the absence of oxygen, however, glycolysis alone can produce only two molecules of ATP.

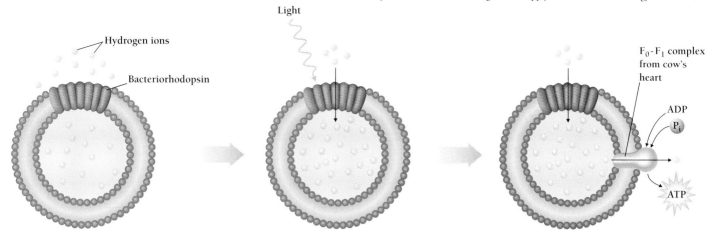

Figure 6-16 Proof that a proton gradient can drive ATP synthesis. Even an isolated F_0-F_1 protein complex can generate ATP, as long as it is embedded in a membrane spanning a proton gradient. In this experiment, researchers took an F_0-F_1 protein complex isolated from the heart muscle of a cow and placed the F_0-F_1 complex into the membrane of a photosynthetic bacterium. A purple protein called *bacteriorhodopsin* pumped protons across the bacterial membrane in response to light. But the F_0-F_1 complex allowed protons to flow back out of the vesicle and generate ATP.

The ATP molecules produced in the mitochondria travel across the mitochondrial membranes to the cytoplasm, where they can be used. Once their high-energy phosphate bond is broken, ADP molecules return to the mitochondria to be reconverted to ATP. Within the mitochondria, the rates of the citric acid cycle and oxidative phosphorylation depend on the relative amounts of ATP and ADP. Because of such regulation, the body catabolizes sugars and fats five to ten times more rapidly during exercise than when at rest.

The process by which energy is extracted from glucose under cellular conditions is highly efficient. A gasoline engine has an efficiency of only 10 to 20 percent. That is, only 10 to 20 percent of the energy in the gasoline is converted to forward motion. In contrast, cells extract energy from glucose with a phenomenal efficiency of 50 percent. It is hard to think of another process with such high efficiency.

Cellular respiration is a highly efficient process for extracting energy—in the form of 30 or 32 molecules of ATP—from glucose.

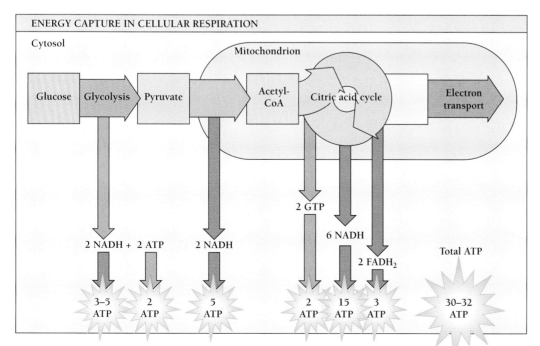

Figure 6-17 Energy capture in cellular respiration. For each glucose molecule, glycolysis produces two ATP and two NADH molecules, and the production of acetyl-CoA yields another two NADHs. The citric acid cycle then yields two GTP, six NADH, and two $FADH_2$ molecules. The electron transport chain (represented by the downward purple arrows at the bottom of the figure) captures the energy of NADH and $FADH_2$ in 26 to 28 ATPs.

Study Outline with Key Terms

Modern biologists reject **vitalism,** the belief that living systems have powers beyond those of nonliving systems. But the production of ATP by the **mitochondria** presents an excellent example of how the chemical transformations of living organisms depend on cellular structures.

Plants and other photosynthesizing organisms that produce their own food from sunlight are called **autotrophs.** Organisms that get their energy from other organisms are **heterotrophs.** Heterotrophs prepare for cellular respiration by capturing macromolecules and **digesting** them into their component building blocks. Digestion itself does not produce energy.

Almost all eukaryotic cells (and many prokaryotic cells as well) are **aerobic,** producing most of their ATP by **cellular respiration.** Most energy-producing reactions in cells are **reduction-oxidation (redox) reactions,** which transfer electrons from the **reducing agent** to the **oxidizing agent.**

Cellular respiration transfers 24 electrons from glucose to oxygen. For 20 of these electrons, the first acceptor is NAD^+, which can accept two electrons and a hydrogen atom to give its reduced form **NADH.** The other four electrons of glucose are transferred to **FAD,** whose reduced form is **FADH$_2$.** These reduced **coenzymes**—the major energy storage molecules in energy metabolism—ultimately transfer their electrons to oxygen in a series of steps. In eukaryotes, the steps occur within the inner mitochondrial membrane. In prokaryotes, these reactions occur in association with the plasma membrane.

Glycolysis is the first stage in the metabolism of glucose. Glycolysis is a set of ten chemical reactions that breaks glucose into two molecules of **pyruvate.** These ten reactions also convert some of the energy of glucose into two high-energy ATP bonds.

Many microorganisms obtain energy from **anaerobic** glycolysis, or **fermentation.** Yeast cells, for example, can avoid the need for oxygen by using NADH to reduce pyruvate to form ethanol.

In eukaryotic cells, energy metabolism (other than glycolysis) takes place in the mitochondria. Pyruvate enters the mitochondria, where cellular respiration proceeds. Pyruvate then undergoes an oxidation reaction, in which NAD^+ serves as the electron acceptor. In acetyl-CoA, acetate is linked, via a high-energy bond, to a molecule called **coenzyme A (CoA).**

The **citric acid cycle,** or **Krebs cycle,** converts the carbon atoms of acetyl-CoA to carbon dioxide. The citric acid cycle is a nearly universal set of reactions. The citric acid cycle adds the four-carbon molecule oxaloacetate (OAA) to two-carbon acetyl-CoA to form a six-carbon molecule, from which two carbons are removed in succession. OAA then reenters the cycle once more.

The citric acid cycle, which not only provides energy but also serves as the major route for the synthesis of many small molecules, lies at the center of the complex network of biochemical conversions that collectively are called **metabolism.** Metabolism is divided into two classes of pathways: **catabolism** and **anabolism.**

Oxidative phosphorylation couples the oxidation of NADH and FADH$_2$ to the production of high-energy phosphate bonds in ATP. The electrons from NADH and FADH$_2$ reach oxygen, their ultimate acceptor, through the **electron transport chain** (or respiratory chain), which consists of a series of **electron carriers.** Five of the electron carriers are **cytochromes,** whose reddish color derives from **heme.**

Oxidative phosphorylation operates by a mechanism (called **chemiosmosis**) summarized in the **chemiosmotic hypothesis:** (1) the flow of electrons through the electron transport chain pumps protons out of the **mitochondrial matrix,** establishing an **electrochemical gradient** across the mitochondrial membrane; and (2) molecular machinery in the membrane captures the energy, as protons flow back into the mitochondrial matrix.

The orientation of the electron carrier complexes establishes the direction of the **proton gradient.** As the protons flow down the electrochemical gradient through channels in the inner membrane, their free energy decreases. A protein complex, the **F$_0$-F$_1$ complex,** or **coupling factor,** uses the free energy from the flow of protons to drive the synthesis of ATP. The F$_1$ protein, also called **ATP synthase,** synthesizes ATP.

Cellular respiration produces 30 (or 32) molecules of ATP for each molecule of glucose. All but four of these ATP molecules are produced by oxidative phosphorylation. In the absence of oxygen, glycolysis alone produces only two molecules of ATP.

Review and Thought Questions

Review Questions

1. Write down a reaction that describes the following situation: substance A carries a net charge of -1, that is, it has one extra electron, while substance B is electrically neutral. During a reaction, the extra electron spontaneously moves from A to B, so that A becomes neutral and B acquires a negative charge. Indicate each of the following: the oxidized substance, the reduced substance, the oxidizing agent, the reducing agent, the more electronegative substance, and the substance with lower affinity for electrons.
2. Name the process in which most of the energy from glucose is converted to ATP. What role does oxygen play in this process? What gas is produced as a waste product? Where in a cell do these reactions occur?
3. Why is the electron transport pathway divided into so many small reactions?
4. Across what eukaryotic membrane is a gradient established during cellular respiration? What is unequally distributed across these membranes? What caused this unequal distribution?
5. Describe and name three stages that aerobic organisms use to extract energy from glucose.
6. How many carbon atoms does a glucose molecule have? Name the end product molecule that is produced during glycolysis. How many carbon atoms does it have?
7. How many ATPs are produced in the citric acid cycle? How does this number compare with that produced in glycolysis?
8. Define metabolism, anabolism, and catabolism, and give examples of catabolic and anabolic pathways. Which type of metabolism is most favored when a cell (a) has abundant ATP? (b) contains little ATP?

Thought Questions

9. Why might you lose weight (a lot of weight) if your mitochondria suddenly lost the ability to couple electron transport to the production of ATP?
10. When you exercise heavily, you consume oxygen faster than you can take it in through your lungs and circulatory system, so your cells have to rely at least partly on anaerobic processes to extract energy from food molecules. Which of the three stages in energy extraction can your cells not perform under these conditions? How much ATP can you form from food molecules under these conditions?

About the Chapter-Opening Image

Louis Pasteur proved that the fermentation of wine and beer depends on the presence of microorganisms (yeasts). Twentieth-century biochemists figured out just what the yeasts were doing.

On-line materials relating to this chapter are on the World Wide Web at **http://www.harcourtcollege.com/lifesci/aal2/**

 Web Bit 6-1 The Little Engine That Could

 Web Links

 Quizzing & Testing

7 Photosynthesis: How Do Organisms Get Energy from the Sun?

The Chemical Evangelist

Early in the 17th century, a Belgian alchemist named Johann Baptista van Helmont (1579–1646) performed a simple and dramatic experiment that scientists have been admiring for nearly 400 years. Although van Helmont dared not publish his results, his studies laid a firm foundation for the modern understanding of **photosynthesis** [Greek, *photos* = light + *syntithenai* = to put together]—the process that plants and other organisms use to transform the energy of light into the chemical bonds of sugars.

Van Helmont was a man who, even in our time, would be considered highly unusual. He was deeply religious, nearly fanatical in his beliefs. Yet he was a born skeptic, always questioning the beliefs of those around him. It was an attitude that would bring upon him the cold fury of the Spanish Inquisition.

As a teenager, van Helmont found his education at the Catholic University of Louvain so disappointing that he refused the title "Master of Arts" because he said he had learned nothing substantial or true. On leaving the university, he then refused a well-paid position as a clergyman because, he said, he opposed "living on the sins of the people." For a while, he tried living the austere life of a Christian Stoic, until the hardship made him ill.

Perhaps illness inspired the young van Helmont, for he then began to study medicine, with particular emphasis on herbal remedies. Again he was disappointed, however. The herbalists, he complained, had discovered nothing new about medicine since ancient Greek times, and argued endlessly over details that had nothing to do with making people well. When van Helmont was 20 years old, he accepted a medical degree from Louvain. Yet he never practiced medicine, believing that he had learned nothing that would actually help anyone who was sick.

His new interest was alchemy, a result, he later wrote, of a call from God. (Some called him the "chemical evangelist.") He had frequently claimed to detest money, but at 28 he married a woman so wealthy that he was freed from ever having to work for a living again. For 15 years, he happily pursued his own researches into alchemy, religious philosophy, medicine, and natural science.

He found, for example, that humans digest food with acid and bile. He recognized gas as a form of matter, defining it as the volatile part of a substance, distinct from the mixture of gases we call air. He recognized that the lung is an organ for gas exchange. By weighing reactants and products, he showed that matter is conserved during chemical reactions.

Van Helmont wrote profusely of his discoveries and investigations, never hesitating to criticize the thinking of others. He especially singled out physicians—for their ignorance and superstition—and the Jesuits, a powerful order of Catholics—for what he viewed as their ignorance and arrogance.

Van Helmont was smart enough not to publish his offensive tirades against physicians and Jesuits. But, eventually, one of his many enemies published one of the worst. The faculty at the Louvain medical school in 1623 denounced the pamphlet as "monstrous." And 2 years later, the Spanish Inquisition declared that 27 statements in van Helmont's manuscript were almost certainly heretical. Between

Key Concepts

- Photosynthesis is the process by which plants use energy from the sun to build sugars from carbon dioxide and water.
- Specialized photosynthetic machinery absorbs light energy, transforms it into the chemical energy of NADPH and ATP, and then stores it in the chemical bonds of sugar.
- Photosynthesis entails two light-dependent reactions—photosystem I and photosystem II.

1624 and 1642, the Spanish Inquisition imprisoned and interrogated van Helmont, then locked him up in a convent, and finally confined him to house arrest for many years. During all of this time, van Helmont published nothing. Yet he continued to work and, equally important, to write.

Much of van Helmont's work was directed at undermining the teachings of the Jesuits. The Jesuits based their teachings on the writings of Aristotle, the most influential of the Greek philosophers. Aristotle had argued that plants derive their sustenance from the soil. A plant's roots, Aristotle said, extract materials from soil in much the same way that an animal's stomach extracts materials from its food. This view seemed logical: farmers had long known that even the richest soils eventually lose the ability to support luxuriant plant growth.

But van Helmont knew Aristotle was wrong. Anyone could see that a potted plant didn't use much soil; it used water. Besides, van Helmont believed that all natural bodies come from water. To knock Aristotle from his pedestal and bolster his own theory, van Helmont performed a now famous experiment (Figure 7-1). He planted a 5-lb willow sapling in 200 lb of soil, which he had carefully dried in a furnace, weighed, and then placed in a large pot. He covered the pot with an iron plate to keep dust from falling into it and watered the growing tree as needed. He did not weigh the leaves that fell from the tree each autumn. After 5 years, he again weighed both the tree and the soil. The willow tree now weighed 169 lb, while the soil weighed only 2 oz less than the starting 200 lb. Van Helmont concluded, "164 pounds of wood, bark, and roots arose out of water alone."

Van Helmont's experiment was excellent, but his interpretation was partly mistaken. His experiment did prove that Aristotle (and the Jesuits) were wrong: the tree's mass could not have come from the soil. However, van Helmont himself was also wrong in assuming that his experiment showed that the tree came entirely from water. It was ironic that van Helmont, the discoverer of gas, never guessed that something in the air might also have contributed to the growth of his willow tree. Only years after he died did other scientists discover the role of gas in plant growth.

◀ **Figure 7-1 Plants are made of water and air.** Van Helmont's willow tree experiment showed that the material that makes up plants does not come from the soil. After 5 years, the willow tree gained 164 lb, but the soil lost only a few ounces.

HOW DO WE KNOW HOW PLANTS OBTAIN CARBON AND OXYGEN?

The first suggestion that the atmosphere contributed to the stuff of plants came not from chemists or alchemists, but from microscopists. With the new microscopes developed in the late 17th century, an Englishman and an Italian almost simultaneously discovered minute pores in the leaves of plants, called **stomata** [Greek, *stoma* = mouth]. Stomata appeared to allow air to pass to the interior of the leaves (Figure 7-2). The activities of plants, they speculated, must somehow depend on interactions with air. By the mid-18th century, scientists had concluded that "vegetation is planted in the air as well as in the earth."

What Do Plants and Air Do for Each Other?

The next big clue to the interaction between plants and air came from a series of experiments in which a mouse or a candle was placed in an airtight container. After some time, the mouse would die from lack of oxygen or the candle would go out for the same reason. This was because the mouse, or the burning candle, consumed the oxygen in the container.

In the 1770s, the English clergyman Joseph Priestley performed an experiment that showed that a sprig of mint could restore the air in the jar and keep a mouse alive. Priestley showed that plants put back what living animals (or burning candles) take out (Figure 7-3). That "something" that plants release is the gas oxygen. Around the same time, the French chemist

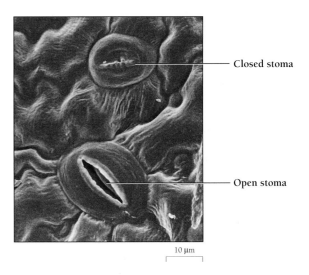

Figure 7-2 Stomata from the underside of a leaf. Tiny pores in the surfaces of leaves called stomata open and close depending on whether plants need to conserve water or collect carbon dioxide. The discovery of stomata in the surfaces of leaves suggested that plants need gases from the air as well as water from the ground. Scientists concluded that "vegetation is planted in the air as well as in the earth." (©Dwight Kuhn)

Antoine Lavoisier independently discovered oxygen, so credit for the discovery usually goes to both men (Figure 7-4).

In the dark, plants act just like animals: they consume oxygen and release carbon dioxide (Figure 7-5A). In light, however, plants put back the oxygen and remove something very different—carbon dioxide. During the day, plants remove carbon dioxide from the air and release oxygen as a waste product (Figure 7-5B).

Today we know that during photosynthesis plants take in water and carbon dioxide with which they synthesize the simple sugar glucose. To a large extent, then, plants are made of water and air. However, as we saw in Chapter 3, proteins and most other biological molecules also include atoms of nitrogen, phosphorus, and other elements. These come, in small quantities, from the soil. Indeed, the 2 oz of soil that van Helmont measured most likely included nutrients essential to the growth of the willow tree.

In this chapter, we will ask how cells accomplish photosynthesis. How does photosynthesis capture light energy? How does it store the captured energy? What molecules and subcellular structures participate?

Figure 7-3 What did Priestley's experiment show? Plants put something into air that mice need. A. The mouse in the jar uses up the oxygen and fills the jar with carbon dioxide, which suffocates the mouse. B. A photosynthesizing plant placed in the jar takes up the carbon dioxide released by the mouse and replaces it with oxygen. The mouse is fine.

During photosynthesis, plants take carbon dioxide from the air and water from the soil and release oxygen back into the air.

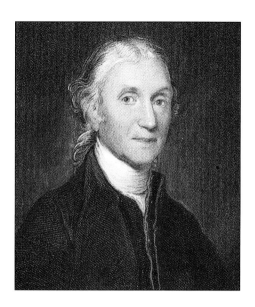

Figure 7-4 Joseph Priestley. Even as modern chemistry took its first breath, its greatest exponents—Joseph Priestley and Antoine Lavoisier—were caught up in the turmoil surrounding the American and French revolutions. In 1791, Priestley's sympathies for the French Revolution inspired an angry mob to destroy his home, library, and laboratory. Within 3 years, he fled to the United States. In the same year, on the other side of the English Channel, the nobleman Lavoisier, who had served King Louis XVI, was guillotined. His executioners proclaimed, "The Republic has no need for scholars," and threw his body into a common grave. (*The Granger Collection, New York*)

Where Do the Atoms Go in Photosynthesis?

Now that we know the raw materials of photosynthesis, we can summarize the overall chemical transformation in the following equation:

$$6\ CO_2 + 6\ H_2O \longrightarrow C_6H_{12}O_6 + 6\ O_2$$
$$\text{carbon dioxide} + \text{water} \longrightarrow \text{glucose} + \text{oxygen}$$

Notice that this is the exact reverse of the overall equation for respiration given in Chapter 6 on page 121. Another way of writing the equation for photosynthesis is:

$$CO_2 + H_2O \longrightarrow C(H_2O) + O_2$$
$$\text{carbon dioxide} + \text{water} \longrightarrow \text{carbohydrate} + \text{oxygen}$$

Here, $C(H_2O)$ is the general formula for a carbohydrate, with each carbon atom associated with the equivalent of one molecule of water.

To early 20th-century chemists, this equation suggested a hypothesis for how photosynthesis might occur: light splits the carbon dioxide into oxygen and carbon. Then oxygen is expelled into the atmosphere, and the carbon combines with water to form carbohydrate. Although this idea is appealing, it is wrong.

The first realization that this view of photosynthesis was wrong came from studies of photosynthetic bacteria. In the early 1930s, a graduate student at Stanford University named Cornelius van Niel was examining photosynthesis in bacteria

A. Nighttime   B. Daytime

Figure 7-5 Photosynthesis A. In the dark, plants respire only. Like animals, they use oxygen to break down sugars. B. In light, plants also photosynthesize, using solar energy to make sugar from carbon dioxide and water.

A. Photosynthesis in plants

B. Photosynthesis in bacteria

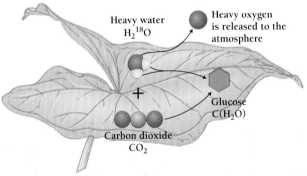

C. Photosynthesis in plants using heavy water

Figure 7-6 Where do the oxygen atoms come from? A. The oxygen released by plants during photosynthesis comes from water, not from carbon dioxide. Two experiments suggested this: B. First, the sulfur released by certain bacteria during photosynthesis comes from hydrogen sulfide, which is chemically analogous to water. C. Second, in plants supplied with "heavy water," the heavy oxygen released during photosynthesis clearly comes from the water.

that used hydrogen sulfide (H_2S) instead of water (H_2O). Instead of producing oxygen as a by-product, these bacteria produce sulfur (Figure 7-6).

$$CO_2 \;+\; 2\,H_2S \;\longrightarrow\; C(H_2O) \;+\; 2\,S \;+\; H_2O$$

carbon dioxide + hydrogen sulfide $\longrightarrow$ carbohydrate + sulfur + water

Van Niel reasoned that these bacteria probably used the same biochemical pathways as other photosynthesizing organisms. Water (H_2O) and hydrogen sulfide (H_2S) probably served the same biochemical role. Yet, the sulfur produced by photosynthetic bacteria clearly comes from splitting hydrogen sulfide. It stood to reason, then, that the oxygen produced by plants might come from splitting water, not carbon dioxide.

However, van Niel had no evidence to support this hypothesis. No one knew which oxygen atoms came from which molecules. More than 10 years passed before scientists found a way to distinguish between the oxygen molecules in water and those in carbon dioxide. The key was the isotope ^{18}O, an oxygen atom that is slightly heavier than usual. Biochemists provided plants with normal carbon dioxide and so-called heavy water, $H_2^{18}O$. The plants split the heavy water and produced heavy oxygen, but the carbohydrates were normal.

$$CO_2 \;+\; 2\,H_2^{18}O \;\longrightarrow\; C(H_2O) \;+\; H_2O \;+\; {}^{18}O_2$$

carbon dioxide + water $\longrightarrow$ carbohydrate + oxygen

This experiment showed conclusively that plants split water, not CO_2 (Figure 7-6C).

The oxygen released by plants comes from water, not from carbon dioxide.

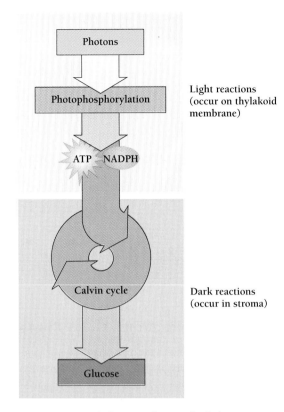

Figure 7-7 The two parts of photosynthesis. The light-dependent reactions, called photophosphorylation, generate energy-transfer molecules such as ATP and occur in the thylakoid membrane of the chloroplast. The light-independent, or "dark," reactions use the energy packaged in the light reactions to make glucose from carbon dioxide. These reactions occur in the stroma, outside the thylakoid membrane.

Biochemists now had a clear understanding of what raw materials plants used to synthesize glucose. But two difficult questions still remained: How do plants channel light energy into this chemical reaction? And how do plants use that chemical energy to synthesize sugar?

Photosynthesis consists of two parts. In the first part, energy from sunlight powers the addition of a phosphate group to ADP to form ATP. Adding a phosphate group is called "phosphorylation." Because light comes in packets of energy called "photons," using light energy to phosphorylate something is called **photophosphorylation.** Sometimes light energy is used to reduce $NADP^+$ to NADPH, another energy-storage molecule. Both kinds of reactions are called photophosphorylation, or **light-dependent reactions.**

The second part of photosynthesis uses the energy from the light reactions (ATP and NADPH) to make glucose. Because these reactions use no photons and can occur in the dark, they are sometimes called the "dark reactions." But the dark reactions do not have to occur in darkness; they just do not require an immediate supply of photons. The dark reactions are more correctly termed **light-independent reactions.** In the rest of this chapter, we will review both the light reactions and the light-independent reactions (Figure 7-7).

HOW DO PLANTS COLLECT ENERGY FROM THE SUN?

Julius Mayer, one of the discoverers of the First Law of Thermodynamics, elegantly summarized the business of photosynthesis: "Nature has put itself the problem of how to catch in flight light streaming to the earth and to store the most elusive of all powers in rigid form."

What Is Light?

In the morning, sunlight streams in your window, shining on the curtains, the furniture, and the floor. All of these objects absorb sunlight and transform it into heat. In photosynthesis, plant cells absorb light and transform it into chemical energy. To understand how plants use the energy in light, we need to understand what light is.

Light is the energy of particles, called **photons,** that travel at the speed of light (186,000 miles per second or 3×10^8 meters per second). Like other moving particles, photons travel in straight lines. They can hit a surface and bounce off at an angle (Figure 7-8). Light can also vary in intensity. The intensity of light shining on the pages of this book corresponds to the number of photons hitting the page each second. If you turn on a bright light, or even six dim ones, you will increase the total number of photons hitting the page.

Photons all travel at the same speed, but they nevertheless have different amounts of energy. How much energy depends on their wavelength, for light behaves like a wave as well as a particle. Light has troughs and crests that interfere with

Figure 7-8 Light is made of particles. Particles of light called photons move in straight lines and bounce off surfaces in the same way as marbles or billiard balls. *(Leonard Lessin/Peter Arnold, Inc.)*

one another, just like the ripples in a rain puddle (Figure 7-9A and B).

One of the most important properties of light is its **wavelength**—the distance between successive crests (Figure 7-9C). Light energy with a long wavelength has long troughs and crests; light with a short wavelength has short troughs and crests. The shortest wavelengths are measured in nanometers (nm), each of which is a billionth of a meter. Wavelengths vary in length from as short as a few nanometers, which we call

Figure 7-9 Light is a wave. A. The circular ripples created by raindrops on a pond make a pattern of troughs and crests. B. Light also forms waves with crests and troughs. C. Wavelength is the distance from the crest of one wave to the next. Light with a long distance from one crest to the next (long wavelength) is red. Light with a short wavelength is blue. *(A, Runk/Schoenberger from Grant Heilman; B, Yoav Levy/Phototake)*

A. Sunlight is a mixture of many wavelengths

B.

Figure 7-10 Sunlight. A. Sunlight is a mixture of many wavelengths of light, from infrared to ultraviolet. B. Irises, tiger lilies, and other living organisms have evolved spectacular pigments that absorb some wavelengths of light and reflect others.

gamma rays, to meters or even kilometers, which we call radio waves.

Most light, including sunlight and the light from most light bulbs, is a mixture of many wavelengths (Figure 7-10). But our eyes and brains allow us to see only those wavelengths between 390 nm and 760 nm—what we call **visible light** (Figure 7-11). Within this narrow range, we can distinguish all the different wavelengths of the rainbow. The shortest wavelengths of light appear to us as blue or purple, and the longest wavelengths appear red. For example, we see 400-nm light as violet, 500-nm light as blue-green, and 600-nm light as orange-red.

A photon is a package of energy. Unlike an electron or a proton, a photon has no mass. It is pure energy. The amount of energy in a photon depends on its wavelength. Photons of shorter wavelengths (more bluish) have more energy than photons of longer wavelengths (more reddish). Blue light, for example, has about twice as much energy per photon as red light. The energy, wavelength, and color of light are all measures of the same thing. (But light intensity, how bright the light seems to us, is different. Light intensity is a measure of numbers of photons, not the energy, or wavelength, of the individual photons.)

Plants photosynthesize by taking advantage of the way light interacts with matter. An atom or a molecule can absorb a photon of light if the energy of the photon matches the energy needed to boost one of its electrons to a higher energy level. Our own vision is the result of visible light exciting electrons in special molecules in our eyes. Visible light can also alter the chemicals in photographic film to create beautiful photographs. A photon can also induce an electrical signal, for example, in a video camera or the photoelectric cell of a solar collector.

In photosynthesis, light excites electrons in special molecules, increasing their free energy. The light and "dark" reactions capture some of this energy and convert it into high-energy chemical bonds.

The wavelength of a photon is a measure of its energy. Short-wavelength photons have high energy, and long-wavelength photons have low energy. Photons with the right amount of energy (the right wavelength) can increase the free energy of molecules by exciting their electrons.

How Can We Tell Which Molecules Help with Photosynthesis?

Much of the light from the sun does not reach the Earth's surface (Figure 7-12). High-energy ultraviolet ("beyond violet") light is absorbed by ozone in the upper atmosphere, and low-energy infrared ("below red") is absorbed by water vapor and carbon dioxide in the atmosphere. Not surprisingly, living organisms have evolved molecules that take full advantage of the energy in visible light—the light that best penetrates our atmosphere. The energy contained in a visible photon stimulates the chemical reactions of both photosynthesis and our own eyes.

Different molecules absorb different wavelengths of light. For example, a red flower absorbs almost all the wavelengths

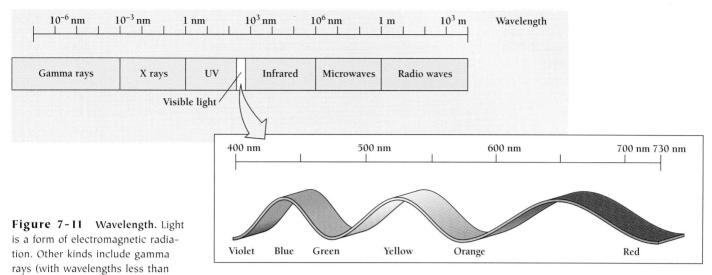

Figure 7-11 Wavelength. Light is a form of electromagnetic radiation. Other kinds include gamma rays (with wavelengths less than 1 nm) to radio waves (with wavelengths measured in meters and even kilometers). Visible light ranges from about 390 nm to about 760 nm. To our eyes, 400 nm light appears violet, 500 nm is blue-green, and 600 nm is orange-red.

in the visible range. The only ones the flower does not absorb are those that we see as red, which the flower's petals reflect into our eyes and elsewhere. Likewise, the green pigment in plants absorbs every color but green.

When any molecule absorbs a photon, the extra energy kicks an electron to a higher-energy orbital. The high-energy electron can return to the lower orbital by losing energy as another photon (fluorescence), or as heat. Alternatively, the high-energy electron may move to another molecule, leaving the first molecule in an oxidized state. We will see that such a transfer of electrons drives photosynthesis.

The particular set of wavelengths that a molecule absorbs is called its **absorption spectrum.** A molecule's absorption spectrum tells us how much energy the molecule needs to move an electron from one orbital to another (Figure 7-13).

Most of the molecules in an ordinary cell absorb ultraviolet light only. This is because the energy required to move an electron to a higher-energy orbital is usually quite high. But some molecules, called **pigments,** can absorb low-energy light from the visible spectrum. Any molecule that absorbs visible light is a pigment.

In plants, the most important molecule of photosynthesis is the green pigment **chlorophyll,** which absorbs every color but green and yellow. Plants use two kinds of chlorophyll—chlorophyll *a* and chlorophyll *b*. Each has a slightly different absorption spectrum. A solution of pure chlorophyll absorbs red and blue photons, while green and yellow photons pass through or are reflected. These unabsorbed photons excite pigment molecules in our eyes, and we perceive a green color.

We can look at plant pigments another way. Every light-dependent process has an **action spectrum**—the relative effectiveness of different wavelengths in promoting the process. In plants, for example, we can measure how well lights of

Figure 7-12 Our protective atmosphere. Because the Earth's atmosphere absorbs most infrared light and reflects most ultraviolet light, most sunlight that reaches the Earth's surface is visible light. Organisms have evolved biochemical pathways that allow us to use this light to build complex molecules.

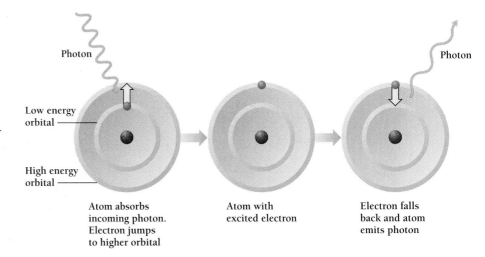

Figure 7-13 Exciting electrons. We say an atom *absorbs* light when a photon excites an electron to a higher energy orbital. This temporary state of excitement ends when the electron falls back to a lower energy orbital, releasing the energy as either light or as heat. A pigment molecule absorbs photons of only a certain range of wavelengths.

different colors promote photosynthesis. Figure 7-14 shows that the rate of photosynthesis increases under blue and orange-red light and decreases under green and yellow light. The action spectrum of photosynthesis roughly follows the absorption spectrum of chlorophyll. We might quickly conclude that the absorption of light by chlorophyll must be responsible for photosynthesis.

But, surprisingly, the action spectrum of photosynthesis does not correspond exactly to the absorption spectrum of chlorophyll *a*. Chlorophyll *a* does not absorb blue-green light (500 nm), yet blue-green light can stimulate photosynthesis because plants have other pigments that can take advantage of other wavelengths. These pigments include both chlorophyll *b* and the yellow and orange **carotenoids,** such as beta (β)-carotene (Figure 7-15). Carotenoids, such as β-carotene, work by transferring the energy of their excited electrons to chlorophyll.

Most chlorophyll molecules transfer energy the same way. Researchers have found that most of the pigment molecules in a leaf do not actually photosynthesize. Instead, hundreds of chlorophyll molecules associate with a few carotenoids to form a weblike **antenna complex,** which traps light energy. The pigments of the antenna complex then transfer energy to a **photochemical reaction center,** where a small number of chlorophyll molecules actually converted the captured light energy to chemical energy.

The action spectrum of photosynthesis corresponds to the absorption spectrum of chlorophyll and other pigments. Chlorophyll and other pigments trap and transfer light energy to the photochemical reaction center.

Why Do Plants Use Two Types of Reaction Centers?

We can now describe a fairly simple model of photosynthesis: (1) chlorophyll and carotenoid molecules absorb light; (2) these pigments transfer the captured energy to the reaction center; and (3) in the reaction center, the energy splits oxygen from water and forms chemical bonds.

If photosynthesis in plants were this simple, each absorbed photon would contribute equally to photosynthesis, which is indeed the case in photosynthesis by bacteria. Photosynthesis in plants, however, appears to operate by a more complicated mechanism.

Some of the wavelengths that are readily absorbed by plant pigments are inefficient at promoting photosynthesis. As the action spectrum in Figure 7-14 shows, when absorbed light has a wavelength longer than 680 nm, efficiency drops steeply.

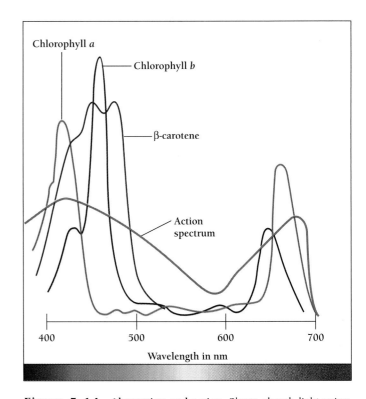

Figure 7-14 Absorption and action. Plants absorb light using two kinds of chlorophyll and carotenoids such as β-carotene. The set of wavelengths that a molecule absorbs is called its absorption spectrum. For example, chlorophylls absorb blue and red light but reflect green light. The range of wavelengths that promote photosynthesis is called the action spectrum. The action spectrum of photosynthesis roughly follows the combined absorption spectra of the chlorophylls and β-carotene.

Yet deep red light (with a wavelength of 700 nm or longer) will promote photosynthesis if shorter-wavelength light (680 nm) also illuminates the plant. And when both 680-nm (red) and 700-nm (deep red) light shine on a plant, photosynthesis is more efficient than with 680-nm light alone. In other words, photosynthesis works better with light of two wavelengths than with either alone.

Plants' need for more than one wavelength of light suggested to researchers that photosynthesis consists of two distinct sets of reactions. Each set of reactions depends on light of different wavelengths. One set of reactions, called **photosystem I**, best absorbs light with wavelengths of about 700 nm (deep red). The other set, called **photosystem II,** requires light with wavelengths of about 680 nm (red light). (While plants use two photosystems, photosynthetic bacteria use only one.)

The existence of the two photosystems in plants explains why indoor plants grow better when lit with both a bluish fluorescent bulb and a yellowish incandescent bulb than with either kind of bulb alone. Later, we will see how the two photosystems each contribute to harnessing light energy.

Where are these photosystems? Recall from Chapter 4 that inside the chloroplasts of each cell are stacks of thylakoids called grana (Figure 7-16). Each thylakoid is a hollow disc of thylakoid membrane. The two photosystems are distinct particles, but both lie in the thylakoid membranes. Each photosystem particle consists of chlorophyll molecules together with carotenoids, specific proteins, and other molecules and ions, and each photosystem performs its own biochemical reactions (Figure 7-17).

But both photosystems are required for plant photosynthesis. The chlorophyll in photosystem I absorbs light most efficiently at 700 nm and is called P_{700} ("P" stands for "pigment"). The chlorophyll in photosystem II absorbs light best at 680 nm and is called P_{680}.

The excited electrons of photosystem I can take either of two paths. The first path, called "cyclic photophosphorylation," produces ATP. The second, called "noncyclic photophosphorylation," produces NADPH but requires electrons supplied by photosystem II. The excited electrons of photosystem II all take one path—to the electron-hungry P_{700} molecules generated by photosystem I.

The two separate photosystems of plants each absorb a different wavelength of light most efficiently.

What path do the excited electrons take in cyclic photophosphorylation?

When P_{700} absorbs a photon from the antenna complex, its free energy increases, and it becomes a powerful electron donor. An excited P_{700} readily gives up a high-energy electron to one of two electron transport chains (Figure 7-18). When P_{700} loses its excited electron, however, it also loses most of the free en-

Figure 7-15 Autumn leaves along the Kancamagus Highway, in New Hampshire. Some of the beautiful colors of autumn leaves result from the carotenoid pigments that help plants photosynthesize sugar. *(Stephen J. Krasemann/Photo Researchers, Inc.)*

ergy it gained from the photon. Because P_{700} has now lost an electron, it is "oxidized."

One of the two electron transport chains associated with photosystem I leads the excited electron back to oxidized P_{700}. This cycle restores P_{700} to its original state and makes it ready to absorb another photon, a process called **cyclic photophosphorylation** (Figure 7-18).

The electron acceptor in cyclic photophosphorylation is a cytochrome complex that passes the electron on to another acceptor and back to oxidized P_{700}, completing the cycle. As in mitochondria, the passage of electrons through the cytochrome complex results in a proton gradient across a membrane. Protons move from a space just outside the thylakoids, called the **stroma,** into the thylakoid space (Figure 7-19). The stroma corresponds to the mitochondrial matrix. In mitochondria, protons move out of the matrix to the intermembrane space. In chloroplasts, protons move out of the stroma into the thylakoid space.

Like the inner mitochondrial membrane, the thylakoid membrane contains spinning, windmill-like ATP synthase molecules, which produce ATP as the protons flow back down their gradient (Figure 7-18). The ATP is made in the stroma, just as ATP is made in the mitochondrial matrix. The net effect of the electron route just described is the production of ATP from the energy of absorbed light. The ATP made in the stroma is later used to synthesize glucose.

In cyclic photophosphorylation, excited electrons from P_{700} pass through an electron transport chain and back to P_{700}. This cyclic passage of electrons creates a proton gradient across a membrane in the chloroplast capable of generating ATP.

Figure 7-16 Anatomy of a chloroplast. A. The cells in an oak leaf contain photosynthetic organelles called chloroplasts. Inside each chloroplast, the inner membrane is folded into pockets called thylakoids, arranged in stacks called grana. The inside of each thylakoid pocket is called the thylakoid space. The space outside the stacks of thylakoids (but inside the chloroplast) is called the stroma. Light-*in*dependent reactions occur in the stroma. Light-*de*pendent reactions occur in the photochemical reaction center in the antenna complex of the thylakoid membrane. B. TEM of a chloroplast. (B, James Dennis/CNRI/Phototake)

What path do the excited electrons take in noncyclic electron flow?

Photosystem I also includes an alternate, noncyclic electron path, which does not produce ATP (Figure 7-19). Instead, the energy of the excited electron of P_{700} is used to make NADPH, a derivative of NADH. NADPH is the electron-carrying coenzyme that chloroplasts use (called $NADP^+$ when lacking its electron). Recall from Chapter 6 that the similar molecules NADH and NAD^+ are used in pathways such as glycolysis and respiration (catabolism). But NADPH and $NADP^+$ are different. They are used in the building of high-energy molecules such as glucose, in chloroplasts, in photosynthetic bacteria, and even in nonphotosynthetic organisms.

The noncyclic route of photosystem I supplies most of the reducing power for glucose synthesis. But after P_{700} has given up its excited electron to form NADPH, P_{700} needs another electron before it can again serve as an electron donor. That is, some other molecule with an available electron must oxidize P_{700} before it can function again.

In the noncyclic route of photosystem I, the excited electron from P_{700} goes to help form NADPH. P_{700} is then missing one of its electrons.

Figure 7-17 The heart of photosynthesis. A. Plants use two kinds of chlorophyll, chlorophyll *a* (shown here) and chlorophyll *b*. B. The photochemical reaction centers are made up of clusters of chlorophyll molecules embedded in the thylakoid membrane. Light excites electrons in the chlorophyll molecules, stimulating electron flow similar to that in the electron transport chain of mitochondria.

How does photosystem II provide electrons to P_{700}?

The chlorophyll molecules of photosystem II absorb light best at 680 nm and are called P_{680}. Just as photon-excited P_{700} is a powerful reducing agent that can donate a high-energy electron, photon-excited P_{680} is an even more powerful electron donor (Figure 7-20). Electrons move from excited P_{680}, through the electron transport chain of photosystem II, to oxidized P_{700} (from photosystem I). By regenerating reduced P_{700}, photosystem II supplies electrons to photosystem I even when there is no cyclic flow of electrons.

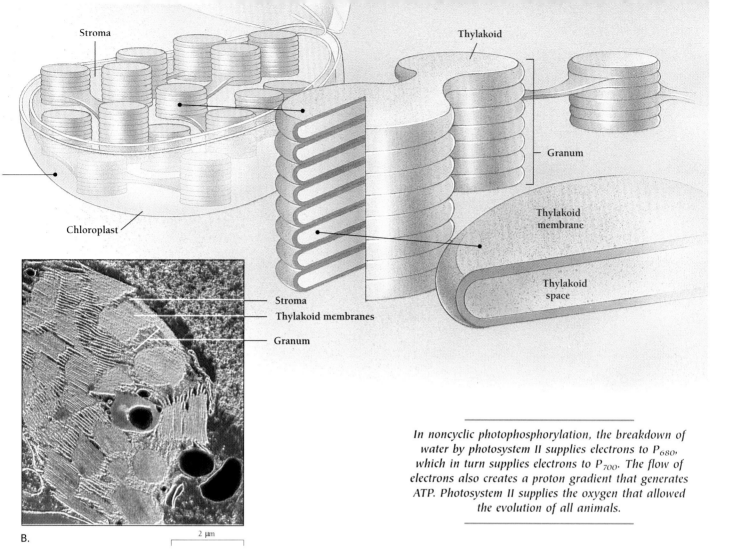

In noncyclic photophosphorylation, the breakdown of water by photosystem II supplies electrons to P_{680}, which in turn supplies electrons to P_{700}. The flow of electrons also creates a proton gradient that generates ATP. Photosystem II supplies the oxygen that allowed the evolution of all animals.

Photosystem II also produces ATP as protons flow down the gradient generated by its electron transport chain. This ATP production is called **noncyclic photophosphorylation**, to distinguish it from the ATP production in the cyclic path of photosystem I.

But we have the same question all over again: If the electron from excited P_{680} is transferred to oxidized P_{700}, how is the reduced form of P_{680} regenerated? How does it get an electron back again? The answer is that the electrons for regenerating P_{680} come from water. Recall from the beginning of this chapter that, using heavy oxygen isotopes, biochemists were able to show that plants split water molecules into oxygen and hydrogen. This fact is central to our existence.

In all plants, photosystem II strips electrons from water molecules, separating the hydrogens from the oxygen molecule. The waste product of this reaction is oxygen gas, which we gladly breathe. Indeed, until photosystem II evolved, at least 3 billion years ago, the Earth's atmosphere contained virtually no oxygen. No animals existed; nor could they have existed. Only when photosystem II evolved did oxygen begin to accumulate in our atmosphere and allow the evolution of animals, fungi, and other heterotrophic organisms.

How Do the Light Reactions Generate NADPH and ATP?

Figure 7-20 shows how photosystems I and II together make up the light-dependent reactions of photosynthesis. The absorption of light by the two photosystems causes (1) the splitting of water to oxygen and hydrogen ions, (2) the production of ATP, and (3) the production of NADPH. Oxygen is a waste product released into the atmosphere through the stomata. ATP and NADPH mostly remain in the chloroplast, where they contribute energy to the synthesis of sugar—the light-independent reactions of photosynthesis.

The mid-20th century biochemist Albert Szent-Györgyi, whose work on cellular respiration we discussed in Chapter 6, characterized this efficient process as "a little electric current driven by the sunshine." Each photon of light excites a single electron. Reducing a molecule of $NADP^+$ to NADPH, however, requires two electrons. To move these two excited electrons along the electron path to $NADP^+$, photosystems I and II must each absorb one photon. The production of a molecule of NADPH requires two photons of light.

As each electron flows from photosystem II to photosystem I, it pumps enough protons to produce a molecule of ATP. So the energy of two photons is transformed into the chemical energy of one molecule of NADPH and two molecules of ATP.

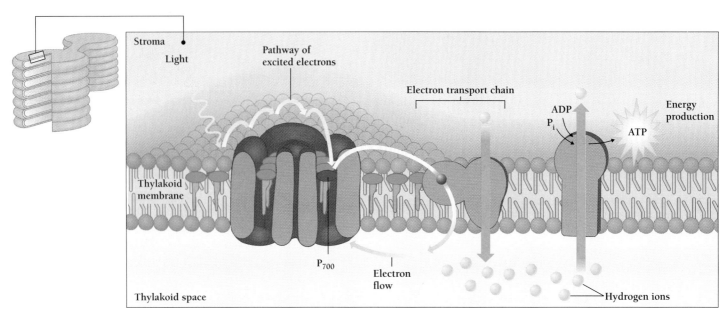

PHOTOSYSTEM I: Cyclic photophosphorylation

Figure 7-18 Cyclic photophosphorylation in photosystem I. Excited electrons from P_{700} pass through an electron transport chain and back to P_{700}. This cycle of electrons creates a proton gradient across the thylakoid membrane that can generate ATP.

PHOTOSYSTEM I: Noncyclic photophosphorylation

Figure 7-19 Noncyclic photophosphorylation in photosystem I. An excited electron from P_{700} goes to help form NADPH. P_{700} then is missing an electron, giving it a positive charge.

Figure 7-20 Photosystems I and II. Photosystem II's P_{680} molecule acquires electrons from water, then supplies these electrons to P_{700} by way of an electron transport chain. The flow of electrons generates a proton gradient that allows ATP synthesis. The breakdown of water into oxygen and hydrogen by photosystem II is the ultimate source of the oxygen we breathe. The evolution of photosystem II underlies the evolution of animals and other heterotrophs.

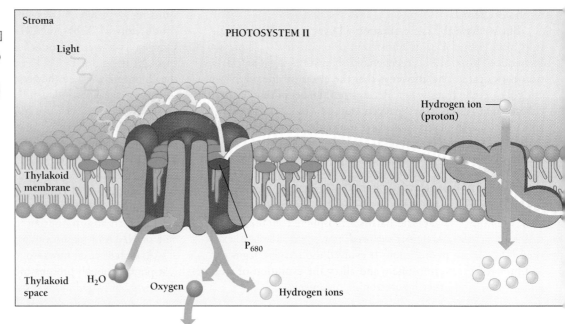

The energy stored in the chemical bonds of NADPH and ATP is used to make glucose, a more stable form in which to store energy. Plants then convert glucose to sucrose, starch, or some other carbohydrate. Starch forms solid deposits in chloroplasts. In contrast, sucrose, or table sugar, remains in solution and can move throughout the plant. Just as animals transport sugar from cell to cell in the form of glucose, plants transport sugar from cell to cell in the form of sucrose.

The process by which plants transform the energy in ATP and NADPH into more stable forms such as glucose is the subject of the next section. The synthesis of sugar is the second part of photosynthesis.

The light-dependent reactions of photosynthesis split water into oxygen and hydrogen ions and generate a flow of electrons that produces ATP and NADPH. These molecules supply the energy to synthesize sugar.

HOW DO PLANTS MAKE GLUCOSE?

Glucose is plants' most important way to store both energy and the building blocks carbon, hydrogen, and oxygen all in one convenient package. We've seen that the hydrogen for glucose comes from water, and the carbon and oxygen come from carbon dioxide in the atmosphere.

The overall reaction of photosynthesis is the reverse of respiration (though the individual steps are not). So we can guess, from what we learned in Chapter 6, that the synthesis of glucose from carbon dioxide and water must consume a lot of energy:

$$6\ CO_2 + 6\ H_2O \longrightarrow C_6H_{12}O_6 + 6\ O_2$$
carbon dioxide + water ⟶ glucose + oxygen

Recall that the individual reactions in the breakdown of glucose divide the huge downhill process into many small steps whose energy can be captured. In the same way, the individual reactions of glucose synthesis allow the energy of ATP and NADPH to be converted to glucose a little bit at a time. The energy "hill" is climbed in a set of small steps.

We know, from the last section, that photosystems I and II split the H_2O in this equation and provide electrons. But we still need to find out what happens to those electrons and what happens to the carbon dioxide. That is, what is the process by which the single carbon atoms in carbon dioxide are combined into six-carbon glucose?

What Is the Calvin Cycle?

The discovery of the radioactive isotope ^{14}C provided a way to answer this question. In 1940, Martin Kamen and Samuel Ruben, then working at the University of California, Berkeley, discovered the isotope ^{14}C and immediately showed that $^{14}CO_2$ (carbon dioxide made from ^{14}C) could be used to study photosynthesis. In 1945, Melvin Calvin, another researcher at Berkeley, began the crucial experiments.

Calvin injected a small amount of $^{14}CO_2$ into a flask of algae that were actively growing in light. Quickly, sometimes within 5 seconds, he killed the algae by adding alcohol. Then he worked out what compounds the algae had formed using the radioactive carbon atoms. To do this, Calvin used a technique called paper chromatography (Figure 7-21). In paper chromatography, the compounds made by the algae were dotted on a piece of

A. Experimental apparatus of Calvin and Benson

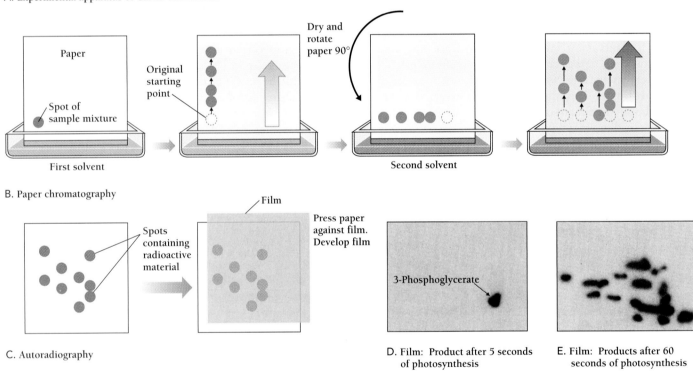

B. Paper chromatography

C. Autoradiography

D. Film: Product after 5 seconds of photosynthesis

E. Film: Products after 60 seconds of photosynthesis

Figure 7-21 How did Melvin Calvin discover the products of photosynthesis? Calvin sorted out the individual products of photosynthesis by using paper chromatography. Calvin then suggested a sequence of reactions that explained the order in which the various chemicals appeared—a sequence that came to be known as the Calvin cycle. A. Calvin added radioactive $^{14}CO_2$ to photosynthesizing algae growing under bright lights. He stopped photosynthesis at different times by killing the algae and then extracted whatever molecules were present. B. Calvin separated the many compounds from one another by two-dimensional paper chromatography. C. The radioactive compounds show up as black spots on photographic film, which is sensitive to radiation. D. After 5 seconds of photosynthesis, most of the radioactivity is in a single spot, corresponding to a three-carbon molecule called 3-phosphoglycerate. E. Later, the pattern becomes more complicated, reflecting the subsequent conversion of 3-phosphoglycerate to other compounds. (A, Melvin Calvin, U. C. Berkeley)

paper and then treated with solvents. As the solvents moved across the paper, the various compounds were carried along, too. But the compounds moved across the paper at different rates, separating from one another like runners in a footrace.

But the compounds were still invisible. To detect them, Calvin pressed the paper against a photographic film. The radioactive compounds made black spots on the film. Wherever ^{14}C was present, the film became black (Figure 7-21). Calvin later wrote that the main data from his experiments were contained "in the number, position, and intensity—that is, radioactivity—of the blackened areas. The [film] ordinarily does not print out the names of these compounds, unfortunately, and our principal chore for the succeeding ten years was to properly label those blackened areas on the film."

The first detectable product made by the algae, Calvin concluded, was a three-carbon compound. When he collected his algae within 5 seconds after the injection of $^{14}CO_2$, all the radioactivity was in a single spot. At successively later times, the pattern of spots became much more complicated, and it took a while for Calvin and his colleagues to discover the order in which the compounds formed (Figure 7-19). Eventually, however, the team worked out a set of reactions that converts the carbon of carbon dioxide into glucose. The reactions form a cycle called the **Calvin cycle,** after its principal discoverer (Figure 7-22).

> *Melvin Calvin used paper chromatography to obtain an ordered list of the chemical products of photosynthesis. Calvin was then able to suggest a reaction sequence that explained the order in which the various chemicals appeared in time. This reaction sequence came to be known as the Calvin cycle.*

What Happens to Carbon Dioxide in Photosynthesis?

Carbon enters the Calvin cycle as carbon dioxide and exits as sugar. The transformation of carbon dioxide into sugar occurs in three parts:

Part 1. Capturing the carbon. In the first part of the Calvin cycle, carbon dioxide combines with a five-carbon compound called ribulose bisphosphate. The resulting six-carbon molecule immediately breaks down into two molecules of phosphoglycerate.

Part 2. Making sugar. In the second part of the Calvin cycle, each of the two molecules of phosphoglycerate is converted into a true three-carbon sugar. These two sugars may either be converted to glucose or other compounds, or they move into part 3.

Part 3. Regenerating ribulose bisphosphate. In the third part of the Calvin cycle, the three-carbon sugars made in part 2 are used to regenerate ribulose bisphosphate, which, recall, captures carbon dioxide.

What is the first thing that happens to carbon dioxide?

The first reaction of the Calvin cycle is the addition of carbon dioxide to a five-carbon compound called ribulose bisphosphate, to form a six-carbon molecule (Figure 7-22). This six-carbon molecule is unstable, however, and immediately breaks into two molecules of three-carbon phosphoglycerate.

The enzyme that catalyzes this reaction is ribulose bisphosphate carboxylase, but it has a friendlier name—**Rubisco.** Rubisco has the world on its shoulders. This small enzyme is almost solely responsible for recycling carbon dioxide from the atmosphere into plants and, ultimately, for supplying carbon to all living things.

But Rubisco is no superenzyme. Most enzymes can process 1000 molecules per second, and some champions even manage 10,000 per second. But Rubisco is so slow that it barely deserves to be called an enzyme, managing only three molecules per second. To compensate for this enzyme's sluggishness, plants typically produce huge amounts of Rubisco. In some plants, Rubisco makes up more than 50 percent of the total protein in the chloroplasts. Rubisco is probably the most abundant protein on the planet.

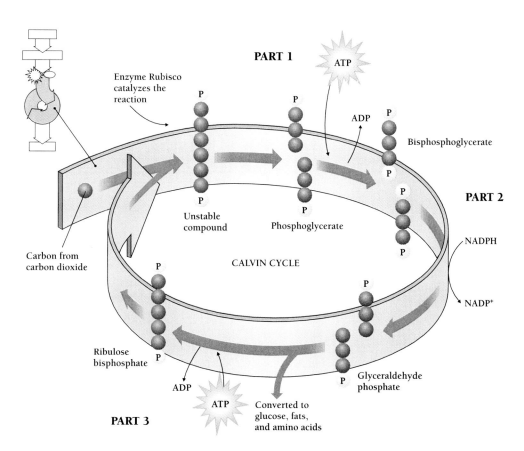

Figure 7-22 The Calvin cycle turns carbon dioxide into sugar. In part 1, CO_2 combines with ribulose bisphosphate to form a three-carbon molecule (phosphoglycerate). In part 2, the three-carbon molecule changes into a true three-carbon sugar (glyceraldehyde phosphate). Part 3 makes more ribulose bisphosphate for part 1.

BOX 7.1

How Do Some Herbicides Kill Plants?

When we think of farmers and gardeners, we imagine them nursing tender seedlings into healthy plants. But growing plants also involves killing other plants, weeds that compete with whatever a farmer or gardener is trying to grow. A weed is simply a plant that is unwelcome for any reason. The easiest way to kill a weed, or any plant, is to spray it with an herbicide, a chemical that kills plants. Farmers began using modern herbicides long before anyone understood how these chemicals hurt plants. In many cases, however, research has revealed precisely the metabolic processes in which herbicides act.

Some herbicides selectively kill different types of plants. For example, 2,4-D—a synthetic compound that mimics a plant hormone—kills C_3 plants more effectively than it kills C_4 plants. Many weeds are C_3 plants, while corn, a major U.S. crop, is a C_4 plant. Thus, farmers can apply 2,4-D to corn fields, killing the weeds, but not the corn.

Most herbicides kill all plants indiscriminately. Such herbicides act by interfering with photosynthesis or other basic processes. Two herbicides, atrazine and diuron, do so by interfering directly with electron transport in photosystem II. Both compounds prevent the transfer of electrons from excited P_{680} to one of the cytochromes. The site of interference is a thylakoid protein, called QB, that helps transfer electrons. Atrazine and diuron bind to QB and prevent electron transfer.

Corn farmers have used atrazine and related compounds heavily, since corn contains enzymes that make it resistant to atrazine. In contrast, diuron is as toxic to corn as it is to weeds. So farmers can only use it when there is no risk of killing their crop. In recent decades, however, many species of weeds have evolved tolerance to atrazine. In other words, these weeds thrive despite being sprayed with massive amounts of atrazine.

Researchers have been able to identify the exact cause of this resistance. Atrazine-resistant weeds have an altered QB protein. In nonresistant weeds, QB becomes disabled when it binds to atrazine. But a single change in one of QB's 300 amino acids can prevent atrazine from binding, rendering the weed safe from atrazine.

Another widely used herbicide that inhibits photosynthesis is paraquat. Paraquat and related compounds accept electrons from photosystem I and transfer them directly to oxygen to produce *free radicals*, highly reactive ions that destroy delicate cellular machinery. These radicals rapidly attack thylakoid membranes, destroying their ability to sustain photophosphorylation and leading to plant death. Because paraquat serves as an electron shuttle—taking electrons from photosystem I and transferring them to oxygen—it continues to do damage for as long as photosystem I is operating. Unfortunately, paraquat is also toxic to humans and must be used with great caution. In humans and other animals, paraquat can potentially damage the digestive tract, the kidneys, and the lungs—thereby causing diarrhea, vomiting, and sometimes death.

Some herbicides act on other basic metabolic pathways that are unique to plants. The widely used herbicide glyphosate (Roundup), for example, interferes with the synthesis of aromatic amino acids (phenylalanine, tyrosine, and tryptophan). Unlike paraquat, however, glyphosate is not particularly toxic to animals. In animals, the synthetic pathways for the aromatic amino acids are completely different from those in plants.

A number of herbicide-producing companies are now using genetic engineering techniques to produce herbicide-resistant crop plants. For example, some crops now carry the atrazine-resistant QB protein. Chemical companies argue that such herbicide-resistant crops allow farmers to destroy weeds selectively by applying a particular herbicide. As a result, farmers do not have to plow their fields to get rid of weeds, and less topsoil is lost to erosion. However, most environmentalists doubt the wisdom of this strategy, since it will ultimately increase herbicide use.

Hundreds of herbicides are used everywhere today: in fields where fruits and vegetables are grown; on crops we feed to animals, themselves bound for our dinner table; along roadsides; and on our lawns and gardens. We find herbicides useful, indeed, hard to live without. Some break down soon after they are applied and are nearly harmless to the environment in the long run. Most do not harm humans. A few are hazardous in high doses, such as those that farm workers might be exposed to. Some herbicides remain in the environment for months or years, continuing to poison plants and, in some cases, animals. All plants regularly exposed to herbicides ultimately evolve resistance, forcing researchers to develop new toxins.

In the first reaction of the Calvin cycle, carbon dioxide combines with ribulose bisphosphate with the aid of the enzyme Rubisco. The resulting six-carbon molecule immediately breaks down into two molecules of three-carbon phosphoglycerate.

What happens to the two molecules of three-carbon phosphoglycerate?

The second part of the Calvin cycle involves two steps that together convert phosphoglycerate into a three-carbon sugar called glyceraldehyde phosphate. Glyceraldehyde phosphate is a true carbohydrate.

These two reactions require one molecule of ATP and one molecule of NADPH for each molecule of phosphoglycerate. Each molecule of carbon dioxide entering the cycle results in two molecules of phosphoglycerate, so each turn of the cycle requires two molecules of ATP and two molecules of NADPH.

Some of the glyceraldehyde phosphate enters the cytoplasm, where it is converted to glucose, fats, or amino acids. And some of it remains in the chloroplast, where it regenerates ribulose bisphosphate, the task of part 3 of the Calvin cycle.

Each of the two molecules of phosphoglycerate is converted into a three-carbon sugar. This sugar is either converted to glucose, fats, or amino acids (in the cytoplasm), or it is used to regenerate the ribulose bisphosphate consumed in part 1 of the Calvin cycle.

How is ribulose bisphosphate regenerated?

The molecule that first binds to carbon dioxide is ribulose bisphosphate. But once it combines with carbon dioxide and splits into two molecules of three-carbon phosphoglycerate, the ribulose bisphosphate is gone. How can a plant cell make more of this important molecule?

The short answer is that a cell uses some of the glyceraldehyde phosphate to make more ribulose bisphosphate. This process is the messy part of the Calvin cycle and involves seven different enzymatic steps. In the last of the seven steps of part 3, ATP contributes a second phosphate group to form more ribulose bisphosphate.

The important molecule ribulose bisphosphate is regenerated in seven steps from glyceraldehyde phosphate.

How Many Photons Does a Chloroplast Need To Make One Glucose?

For each molecule of glucose produced, six molecules of carbon dioxide must enter the Calvin cycle. As the cycle turns six times, it produces 12 molecules of three-carbon glyceraldehyde phosphate. Of these, two form glucose, and the remaining ten regenerate six molecules of five-carbon ribulose bisphosphate.

The cost of making a molecule of glucose is considerable. Each turn of the Calvin cycle requires one molecule of ribulose bisphosphate and two molecules each of ATP and NADPH. The six turns needed to make a single molecule of glucose use 12 molecules of ATP and 12 molecules of NADPH. In addition, the production of the six molecules of ribulose bisphosphate requires an additional six molecules of ATP. The total cost of making a molecule of glucose, then, is 18 molecules of ATP and 12 molecules of NADPH.

All the NADPH and ATP for the Calvin cycle comes from the light-dependent reactions in the thylakoid membrane. To make the 12 molecules of NADPH, a chloroplast must absorb 48 photons of light by means of photosystems I and II. The same 48 photons provide energy for up to 24 ATP molecules—more than enough for the needed glucose synthesis.

To make one molecule of glucose, the chloroplasts need 18 molecules of ATP and 12 molecules of NADPH, or the energy from 48 photons.

HOW DO PLANTS COPE WITH TOO LITTLE WATER OR CARBON DIOXIDE?

The reactions of photosynthesis do not take place in a test tube, but in living, growing plants, subject to storms, drought, and attacks by plant-eating animals. Grasses, small plants, shrubs, and trees all live in complex and demanding ecosystems where many plants die without reproducing. Many factors determine whether a plant survives and even thrives. One of these factors is how well the plant uses sunlight to synthesize sugars.

We have seen that plants use the three-carbon sugar glyceraldehyde phosphate to make glucose, fats, and amino acids. These, in turn, are used to make the carbohydrates and proteins that make up the bulk of the body of the plant. The construction of leaves, stems, trunks, and flowers creates a massive demand for glyceraldehyde phosphate from the Calvin cycle.

If a plant lacks any of the raw materials of photosynthesis—water, carbon dioxide, or photons—meeting this demand becomes impossible. As a result, plants have evolved a variety of adaptations for increasing their productivity, the efficiency with which they collect and use these three raw materials.

The factors that influence plant productivity are not only important to plants but important to humans as well. All the world's major crops—including rice, wheat, and corn—depend on plant breeders' efforts to increase the productivity of those crops. The world food supply and the lives of billions of people depend on these super-productive crops.

What Factors Limit Productivity?

The major factors that affect photosynthesis are the amount and wavelength of incoming photons, the amount of water available, and the amount of carbon dioxide available. All of these can be in short supply at times.

We have already seen how the wavelength of light might influence the efficiency of photosynthesis. Plants cannot absorb light that is the wrong wavelength. Only light in the action spectrum of photosynthesis is of any use at all. Equally important is the *intensity* of light, the number of photons striking a leaf surface per day. Plants cannot produce as much sugar in the shade as they can in full sun.

Many factors influence how much light a plant gets. These include, for example, the length of the day, the length of the growing season, and the amount of cloud cover. In London, for example, pollution and fog reduce by half the total sunlight reaching a plant. Plants also compete with one another for light. In a tropical rain forest, 15 to 20 layers of vegetation may absorb and reflect more than 95 percent of the light before it reaches plants at ground level.

Too little water can also limit photosynthesis. Plants pull water from the soil and transport it through stems and trunks to their leaves. Some of this water is used in photosynthesis, but most passes out of the stomata in the leaves and evaporates into the surrounding air.

When the weather is hot and dry and water is in short supply, plants shut their stomata to avoid losing water. But the stomata need to be open to allow waste oxygen to leave the plant and carbon dioxide to enter. As photosynthesis continues, oxygen levels inside the plant rise, and carbon dioxide levels drop. For desert plants, which must keep their stomata closed as much as possible to avoid drying out, stocking enough carbon dioxide for photosynthesis is a problem.

How Do Plants Prevent Photorespiration?

Plummeting carbon dioxide levels not only slow photosynthesis, they also initiate a wasteful chemical pathway called **photorespiration,** in which plants squander much of their hard-won carbohydrate. The problem occurs because Rubisco requires a lot of carbon dioxide to work well. Normally, in the first step of the Calvin cycle, Rubisco converts ribulose bisphosphate into two molecules of phosphoglycerate. However, when carbon dioxide levels are low, Rubisco converts ribulose bisphosphate into one molecule each of phosphoglycerate and another, almost useless, two-carbon molecule. The result is that during photorespiration plants waste as much as half of the carbohydrate they produce. Photorespiration, unlike cellular respiration, produces no ATP or NADH. Its function, if any, is unknown.

There is only one cure for photorespiration: lots of carbon dioxide. Some plants have evolved a special biochemical pathway that allows them to maintain high levels of carbon dioxide inside their cells, even when their stomata are closed most of the time.

In these plants, the first reaction of carbon dioxide does not produce the usual three-carbon molecule phosphoglycerate. Instead, they make a four-carbon compound called oxaloacetate, or OAA (which is also part of the citric acid cycle of Chapter 6). The cells then convert OAA to another four-carbon compound, which moves into special cells that make carbohydrates. There it breaks down to release carbon dioxide, which immediately enters the Calvin cycle.

Plants that use the three-carbon phosphoglycerate, which includes most plants, are called **C_3 plants.** Plants that use the four-carbon OAA pathway are called **C_4 plants.** C_4 plants, such as corn and sugar cane, have a distinctive anatomy and grow well in hot, dry climates. The C_4 pathway and anatomy are adaptations that allow plants to minimize water loss by keeping the stomata closed, while still providing enough carbon dioxide to supply photosynthesis and prevent photorespiration. C_4 plants waste far less sugar than C_3 plants, but C_4 plants need more ATP (and therefore more light) than C_3 plants.

In hot, dry climates with plenty of light, C_4 plants predominate. In temperate zones with less light and more water, C_3 plants have the advantage. In Kentucky, for example, the humid summers allow plants to leave their stomata open much of the time, and C_3 plants, such as Kentucky bluegrass, thrive. But Kentuckians who move to California are in for a rude surprise. When they plant beautiful Kentucky bluegrass lawns, ugly yellow-green crabgrass gradually takes over. Crabgrass is a C_4 plant that grows better in the bright but dry California summers.

C_4 crops are especially efficient at capturing sunlight, converting as much as 8 percent of the solar energy falling on a field to carbohydrates. As oil and coal reserves decline, some countries have begun to fuel their cars and trucks with alcohol produced from corn and sugar cane.

Desert-adapted plants, such as cacti, use a modified C_4 pathway called crassulacean acid metabolism, or **CAM,** a type of photosynthesis that conserves even more water. CAM plants open their stomata to collect CO_2 at night only, when the air is cool and water loss is minimal. They then store the CO_2 in a four-carbon molecule until the next day, when photosynthesis can proceed. But collecting and storing CO_2 at night requires enormous amounts of ATP. This extra ATP comes from the breakdown of carbohydrates made during the day. The enormous energy cost of CAM plants' nighttime respiration keeps their growth rates extremely low.

C_4 plants minimize water loss and photorespiration and maximize photosynthesis by storing high carbon. C_4 plants are more efficient than C_3 plants, but they need more light than C_3 plants.

In these first few chapters of *Asking About Life,* we have examined some of the basic chemical processes that enable organisms to live. In the next few chapters, we will see how cells multiply and grow. Then we will see how organisms recombine their genetic material in novel arrangements in the process of sexual reproduction. We will learn how the structure of DNA was discovered and find out how genes are expressed. Finally, we will explore the ways that molecular biologists have learned to control the expression of genes and how this dizzying array of new knowledge applies to humans.

Study Outline with Key Terms

In **photosynthesis,** green plants and other organisms use the energy of **photons** to synthesize carbohydrates from carbon dioxide and water. In the **light-dependent reactions (photophosphorylation),** photosynthesis splits water into oxygen and hydrogen atoms, generating oxygen and high-energy electrons. In the **light-independent reactions,** the electrons from the light reactions help link the carbon atoms in carbon dioxide into three-carbon sugars.

Each photon carries a fixed amount of energy, which corresponds to its **wavelength. Visible light** ranges from about 400 nm to about 750 nm. Plants contain **pigments** that absorb light, including the **chlorophylls** and the **carotenoids.** Every molecule has an **absorption spectrum,** and every light-dependent process has an **action spectrum.**

Antenna complexes, consisting of large numbers of chlorophyll and carotenoid molecules, absorb photons and transfer them to one of two kinds of **photochemical reaction centers,** located in the thylakoid membrane of the chloroplast. Each kind of reaction center participates in a separate photosystem.

In **photosystem I,** the main pigment is P_{700}. In **photosystem II,** the main pigment is P_{680}. When P_{700} or P_{680} absorbs a photon, the pigment's energy increases, which excites its electrons. The excited electrons transfer to special electron acceptors.

The excited electrons of photosystem I can take either of two paths. In **cyclic photophosphorylation,** excited electrons from P_{700} pass through an electron transport chain and back to P_{700}. This cyclic passage of electrons creates a proton gradient between the **stroma** and the thylakoid space in the chloroplast that generates ATP. In **noncyclic photophosphorylation,** the excited electrons of photosystem I flow down a different electron transport chain to produce NADPH. But the noncyclic path leaves P_{700} short an electron. Only electrons from photosystem II can make up this deficit.

The breakdown of water by photosystem II supplies electrons to P_{680}, which in turn supplies electrons to P_{700}. The flow of electrons from P_{680} to P_{700} also creates a proton gradient that generates ATP. The breakdown of water by photosystem II supplies the oxygen that allowed the evolution of all animals.

The **Calvin cycle** is the heart of the light-independent reactions of photosynthesis. The Calvin cycle uses the energy of ATP and NADPH (both generated by the light reactions) to make sugar out of carbon dioxide. In the Calvin cycle, the enzyme **Rubisco** synthesizes three-carbon sugars from ordinary carbon dioxide. These small sugars can then be converted to glucose, fats, amino acids, or ribulose bisphosphate. The production of one molecule of glucose from six carbon dioxide molecules requires 48 photons of light.

The efficiency of photosynthesis depends on the amount of light, water, and carbon dioxide. In hot, dry weather, plants close their **stomata** to limit water loss. With the stomata closed, however, waste oxygen builds up, and carbon dioxide levels plummet. Under such conditions, C_3 **plants** may begin **photorespiration,** an unhealthy biochemical pathway that wastes about half of all carbon dioxide. C_4 **plants** and **CAM** plants avoid photorespiration by maintaining high levels of carbon dioxide inside the cells that make glucose.

Review and Thought Questions

Review Questions

1. What is photosynthesis? What material does it synthesize? Where in a cell does this process occur?
2. Write the overall reaction for photosynthesis. How do the reactants get into the leaf and into the cells? How does oxygen get out of the leaf cells and back into the air outside the plant?
3. Which parts of the electromagnetic spectrum do plants use in photosynthesis? Which parts of the spectrum are reflected or absorbed by the Earth's atmosphere?
4. Describe the cyclic photophosphorylation associated with photosystem I. What is the relationship between the absorption of light and the excitation of electrons? How does the transfer of electrons lead to the synthesis of ATP?
5. Describe the reactions that make up the Calvin cycle. What enzyme captures carbon dioxide from the atmosphere? What are the products of this reaction? How is the energy from light introduced in the Calvin cycle? Why must the cycle produce more ribulose bisphosphate?
6. Under what environmental conditions might a plant close its stomata? Compare the rates at which plants can carry out photosynthesis when their stomata are open and when they are closed. Why should the opening of stomata affect the rate?

7. Explain how photorespiration causes C_3 plants to waste some of their photosynthetic efforts under conditions of low carbon dioxide concentration.
8. Describe Priestley's experiment that showed that a plant could restore the air in a bell jar. Make a sketch of a bell jar containing only a mouse, and explain why the mouse dies. Then sketch a bell jar containing both a mouse and a plant, and show with arrows how oxygen is produced and consumed.
9. What advantages do C_4 plants have over C_3 plants? Why don't C_4 plants take over everywhere? What limits them?
10. Put the letter "R" in the blank if the event is part of cell respiration. Use the letter "P" for photosynthesis. Use the letter "B" for both.
    ____Sugar is product
    ____Uses oxygen
    ____Requires light
    ____Uses electron transport chain
    ____Releases carbon dioxide
    ____Produces a proton gradient
    ____Releases oxygen
    ____Uses an ATPase to make ATP

Thought Questions

11. What prevents the roots of a plant from carrying out photosynthesis? By what metabolic process do you suppose roots extract energy from their available resources?
12. Describe in a few sentences the concept of how light energy is used to drive the reaction of glucose synthesis. (Be careful to give the outline of the story, so that a nonbiologist could understand what you mean. Don't just list a lot of details.)
13. Why is the process of photosynthesis important to feeding our entire planet? Can you name any living things that do not depend on plants (directly or indirectly)?

About the Chapter-Opening Image

A large plant in a small pot of soil emphasizes the idea that plants make their bodies primarily from air and water, not from soil.

 On-line materials relating to this chapter are on the World Wide Web at **http://www.harcourtcollege.com/lifesci/aal2/**

 Web Bit 7-1 Photosynthesis Art

 Web Links

 Quizzing & Testing

part II

Genetics: The Continuity of Life

Chapter 8
Cell Reproduction

Chapter 9
From Meiosis to Mendel

Chapter 10
The Structure, Replication, and Repair of DNA

Chapter 11
How Are Genes Expressed?

Chapter 12
Jumping Genes and Other Unconventional Genetic Systems

Chapter 13
Genetic Engineering and Recombinant DNA

Chapter 14
Human Genetics

Photo: Ponderosa pine cones.
(Art Wolfe APR94)

8 Cell Reproduction

The Unfortunate Henrietta Lacks

Early in 1951, an anxious 31-year-old woman checked into the medical clinic at Johns Hopkins University in Baltimore (Figure 8-1). Her anxiety was confirmed when her doctor found a one-inch purple growth on her cervix, the neck-shaped entrance to the womb. He told her he would have to do a biopsy, and he cut out a small piece of the growth and sent it to a pathologist, a medical specialist who can tell, among other things, whether a human cell is cancerous or not. The report came back positive. Henrietta Lacks's growth was dangerously malignant. Its cells were dividing rapidly, and, without treatment, the tumor would kill her.

The news came as a blow. She and her husband had two children to raise. She was too young, she thought, for so serious a disease, one that does not usually afflict women until their 40s or 50s. Within days, the doctors at Johns Hopkins began treating her with high doses of radiation, in what was to be a futile effort to save her life. In the autumn of 1951, the best part of Henrietta Lacks died. Ironically, her tumor cells survived and live today in laboratories all over the world.

The day of her biopsy, Lacks's doctor took a piece of her tumor to a group of researchers who were struggling to grow human cells in plastic dishes containing a mixture of nutrients and blood products. They tried adding different nutrients or changing the pH of the mixtures. No one had yet succeeded. Most human cells died immediately. A few divided several times, then died. Still, the researchers hoped that any day they would find the right cells or the right conditions for growing human cells in the laboratory.

When Henrietta Lacks's cells arrived in the lab, they were different from anything anyone had ever seen before. They grew without restraint, dividing and redividing every 24 hours. The same virulence that would finally kill Henrietta Lacks made these cells unstoppable in the laboratory. Before Lacks's life was over, her cells took up an independent existence, thanks to the work of medical researchers.

Although Henrietta Lacks's cells, named HeLa cells, were the first human cells to be successfully cultured, hundreds of others soon followed. Unlike her cells, most of these were difficult to culture. In general, human cells die quickly outside the body, even in the rich broths that allow many other types of cells to grow. A few stray cells on the side of a glass flask, or in the fluid left in a pipette, expire almost immediately. Beginning a culture of human cells in a plastic culture dish requires care and attention.

Still, in the 1950s and 1960s, growing human cell cultures was exciting work. For the first time, scientists could study human biology without experimenting on humans themselves. They could study the way human cells divide, the way human cells move, the way cancer cells differ from normal cells. The possibilities seemed endless. Every week a lab somewhere developed a new kind of human cell to culture. These new cell lines were rapidly distributed to other labs—some down the hall, others on the other side of the globe.

In the early 1970s, odd things began to happen. The National Institutes of Health (NIH), in Washington, DC, sent five kinds of human cells, which had come from five different research labs in the Soviet Union, to a lab in Berkeley. There, an alert researcher named Walter Nelson-Rees discovered that the five supposedly different cell lines were all HeLa cells.

Soon HeLa cells began to turn up everywhere. Nelson-Rees began checking every cell line he knew of. He published lists of cell cultures that were not what they were supposed to be. By 1981, he had found 90 different cell lines, each of which was HeLa cells. Researchers studying the biology of breast cells,

Key Concepts

- Cell reproduction in eukaryotes involves three distinct but interconnected processes: the duplication of DNA in the nucleus, its equal distribution to each of two daughter cells, and the division of the cytoplasm.

- During mitosis, microtubules and other subcellular components bring about chromosome movements that distribute one copy of each chromosome to each daughter cell.

- Eukaryotic cells, ranging from yeast to human cells, use similar molecular mechanisms to regulate passage through the cell cycle. Failure of these mechanisms can result in cancer.

both normal and cancerous, were horrified to find that their years of research had not been done on breast cells —as they had thought—but on Henrietta Lacks's cervical cells. The story was the same for research on prostate cells, kidney cells, liver cells, and more.

The repercussions were enormous. Years of research by scores of researchers had become meaningless. Much of the biology of cancer had to be rethought. For years researchers had believed that all cancer cells have the same nutritional needs and all cancer cells have abnormal chromosomes, the so-called mark of cancer. Then, as one cell line after another proved to be not prostate cells, not kidney cells, not this and not that, but HeLa cells, biologists came to realize that, in fact, cancer cells are not all alike. The only cancer cells that were all alike were HeLa cells. The abnormal chromosomes were typical of HeLa cells, but not of all cancer cells.

But how had it happened? The answer was that Henrietta Lacks's cancer cells were unbelievably vigorous. If so much as a single one of her cells fell by accident onto a culture dish, her cells could take over, no matter what other cells were living in the dish. One group of researchers found that HeLa cells could float through the air inside the tiny bits of spray created when a rubber stopper was pulled from a bottle containing HeLa cells. The floating cells could then drop down into nearby culture dishes. Within a week, a culture of prostate cells was taken over by cervical cells. It was a scenario that scientists had never imagined.

Today, cell biologists have learned to be more cautious. They work with cell cultures whose ancestry they've checked. And, using DNA-based techniques, they can now make sure that they have exactly the cells that they think they have.

But HeLa cells are still used, and Lacks's cells live on. All over the world her cells continue to divide relentlessly day after day. In the 1950s, her cells were used as hosts for growing poliovirus, enabling researchers to develop the vaccine against polio. Today, genetic engineers insert genes for valuable human proteins into HeLa cells, which then synthesize the proteins. Henrietta Lacks died before she could finish raising her own two children, yet her cells have been instrumental in basic research and have helped save the lives of millions of children all over the world.

Figure 8-1 Young Henrietta Lacks died of cancer in 1951, but biologists have kept her cancer cells alive ever since. In the 1950s, her cells, called HeLa cells, were used to develop a vaccine against polio. Today, genetic engineers insert genes for valuable human proteins into HeLa cells, which then synthesize the proteins. (Science Photo Library/Photo Researchers, Inc.)

CELL DIVISION

Cells are the fundamental living units of all organisms. An organism may consist of a single cell or many cells. Bacteria such as *Lactobacillus*, yeasts such as *Candida albicans*, and protists such as *Paramecium* are all single-celled organisms. In contrast, an adult human contains about 10 trillion (10^{13}) cells. Because all the cells in a multicellular organism originate from a single cell, the continuation of life ultimately depends on cell reproduction.

How Do Cells Divide?

In the five minutes or so that it will take you to read this page, your body will produce a billion (10^9) new cells. Most of these replace battered and dying cells of your skin, intestines, and

BOX 8.1

Cervical Cancer and the Pap Smear

In the 1930s, more women died of cervical cancer in the United States than of any other form of cancer. Today, the rate of cervical cancer is less than one-tenth that of breast cancer and declining rapidly. Even more encouraging, the overall survival rate (more than two-thirds) is one of the highest of all cancers, thanks to a diagnostic test invented by George N. Papanicolaou in 1943.

In the past 50 years, the widespread use of the Pap test, or Pap smear (named after its inventor), has transformed a diagnosis of cervical cancer from a virtual death sentence to a highly survivable disease. The survival rate for cervical cancer would be even higher if more women were tested regularly.

The test consists of a simple office procedure in which a doctor or nurse scrapes a few living cells from the surface of the cervix. The scraping may hurt very slightly, but the life-saving test is well worth the second or two of discomfort. Once the cells have been collected from the cervix, they are sent to a laboratory pathologist for analysis.

In most forms of cervical cancer, cells undergo changes in appearance years before they become cancerous. Such "precancerous" cells have distinctive characteristics that a good pathologist can easily recognize (Figure A). The cervical cells are usually classified into five grades: normal, atypical, low-grade precancerous lesion, high-grade precancerous lesion, and invasive cancer. In the great majority of cases, all the cells collected will be normal. Occasionally, some of the cells will be "atypical," usually because of some temporary infection of the cervix or vagina.

Women with low-grade or high-grade lesions are immediately tested for any sort of growth. Any precancerous growths that are found are removed from the cervix, using cryosurgery (freezing) or lasers. After treatment, a woman is normally cancer free and able to bear children if she wishes.

If the Pap smear reveals cancerous cells, as opposed to precancerous ones, the recommended treatment is a hysterectomy (the removal of the uterus). If the cancer has spread, the surgeon will also remove the upper part of the vagina. If the cancer is caught early, when it is still localized on the cervix, the survival rate is 91 percent. If the cancer is diagnosed late, after it has spread beyond the reproductive organs, the survival rate is less than 9 percent.

Cervical cancer strikes women of all ages and ethnic groups. African-American women, however, are twice as likely as white women to die of the disease. Researchers attribute the difference to lack of good medical care, especially regular Pap tests. Fortunately, increasing numbers of African-American women are receiving the Pap test, and the difference in survival is therefore decreasing.

Because most cervical cancer develops slowly, it is quite easy to diagnose the disease while it is still in one of the precancerous stages. Public health experts recommend that all women have annual Pap smears starting at age 18 or at the onset of sexual activity, whichever occurs first. After three or more negative tests, a woman is safe in having an exam every other year. Women who are at risk for cervical cancer should continue to have the exam every year.

The main risk factor for cervical cancer is infection with human papilloma virus (HPV), also known as genital warts. In 93 percent of all cases of both cervical cancer and precancerous lesions, HPV DNA is present. Indeed, HPV incorporates two genes into the genomes of previously healthy cervical cells. These two genes encode proteins that cause the cells to form tumors. In Chapter 12, we will learn more about how viral DNA infects cells and how tumor viruses cause these cells to form tumors.

Because HPV is a sexually transmitted disease (STD), the best ways to avoid infection are the same as for AIDS and other STDs: abstinence or else barrier methods of birth control (condom or diaphragm) and fewer than five sexual partners over a lifetime.

Figure A Normal cervical cells *(left)* are distinctly different from precancerous *(middle)* and cancerous *(right)* cervical cells. Note the sperm-shaped cells that characterize invasive cancer cells. (©Biology Media/ Photo Researchers, Inc.; Science Photo Library/Photo Researchers, Inc.; M. Rotker/Photo Researchers, Inc.)

blood. Each newly made cell contains both a copy of the genetic information you inherited from your parents and the molecular machinery and materials needed to interpret that information, including membranes, organelles, macromolecules, and small molecules.

Cell reproduction is so common that it is easy to forget how extraordinary it is. Our wonder at this copying process may be heightened by our everyday experience with nonbiological copying—the recording of music, the filming and taping of movies and television programs, and the printing of books. In each case, complicated machinery carries out a complex but ordered process. But cell reproduction is even more remarkable than these other processes, since the cell is both the thing being copied and its own copying machine.

Cell reproduction has different results in single-celled and multicelled organisms. Single-celled organisms divide to create new individuals. Cell division in yeast, for example, leads directly to an increase in the number of individuals. In a multicellular organism such as a frog the same kind of cell reproduction allows one individual to grow from a single cell into a whole organism. Cell reproduction also allows a multicellular organism to replace dying cells. A dandelion root can regenerate its leaves and flowers after they are cut away by a lawnmower. In the same way, children quickly replace the skin cells that they scrape and tear off in their inevitable bangs and falls. Before cells divide, they must first duplicate their contents, including their DNA. Eukaryotic cells must also duplicate their internal membranes and organelles.

An altogether different kind of cell division also occurs in sexually reproducing organisms. Plants, animals, and other sexually reproducing organisms make sperm and eggs (or their equivalents) through an elaborate series of chromosomal divisions and fusions. We discuss this process, called *meiosis*, in Chapter 9. In this chapter, we examine simple cell division, the kind that bacteria use to reproduce. Humans and other multicelled organisms use the same process to grow and to replace the cells of tissues such as leaves, bark, muscle, and hair. In simple cell division, the cell duplicates its DNA, then divides it in a process called mitosis.

Both single-celled organisms and multicellular organisms make more cells by dividing. Just before they divide, cells double their DNA. In eukaryotic cells, they then distribute identical copies of their DNA in a process called mitosis.

What Did Early Biologists Discover About Chromosomes?

In the 19th century, the cell theory established that all organisms are made of cells and that all cells come from other cells. But the cell theory suggested a problem. How does a cell know how to make another cell just like itself? How does the cell pass on the knowledge of what kind of cell it is? How, for example, does a cell from Koko the gorilla know how to make another Koko cell, a cell unlike that found in any other gorilla and unlike that found in other primates such as your classmates?

The answer, we now know, is in the **chromosomes** [Greek, *chroma* = color + *soma* = body]—the threads of DNA and protein that carry the genes. Yet, 19th-century microscopists saw chromosomes and examined them in great detail without recognizing that they were the means by which organisms inherit traits. Early biologists held the key to the problem of inheritance without knowing it.

In the late 19th century, improvements in the lenses of microscopes for the first time allowed microscopists to see the tiny structures within cells. Some of the most important advances, however, came from the use of natural dyes developed for the burgeoning 19th-century textile industry. The two most useful dyes, still in use today, were the purple-and-blue dye hematoxylin (which comes from the logwood tree of Central America) and the synthetic red dye eosin.

Thanks to these dyes, microscopists of the 1870s were able to describe in detail the threadlike chromosomes in a variety of animal and plant cells. In fact, during one 4-year period, researchers published over 200 papers on the complex dance of the chromosomes during cell division. By the 1880s, biologists had discovered that cells reproduce themselves in a process called **cell division,** in which a parent cell gives rise to two **daughter cells.** Cells stained with hematoxylin and eosin showed the cell's nucleus to be full of a grainy material called **chromatin** [Greek, *chroma* = color] (Figure 8-2). Biologists also knew that when a cell divides, its nucleus divides, providing one-half of its material to each daughter nucleus.

The German physician and microscopist Walther Flemming pieced together a detailed description of the process by

Figure 8-2 Onion cells reproduce by dividing in two. Cells stained with a dye that binds to DNA show the DNA in the cell's nucleus. In some cells, the DNA stains as a diffuse grainy material called chromatin. In dividing cells, the chromatin condenses into distinct chromosomes. Chromatin is present in nearly all eukaryotic cells, while chromosomes are visible only in dividing cells. (*Dwight Kuhn*)

which the chromosomes are equally divided and distributed to the two new cells during cell division. Flemming named this elaborate division of the chromosomes **mitosis** [Greek, *mitos* = thread].

One of the first events that Flemming noted was the gathering of the chromatin into visible threads (later called chromosomes). He also observed in salamander cells the duplication of an object that he called an **aster** [Latin, *aster* = star]—the center of a finer set of threads just outside the nucleus. During mitosis, the two asters migrate to opposite ends of the cell. While the asters are migrating, the chromosomes line up in a row in the middle of the cell, and then each one splits into two. Each of the two half-chromosomes then migrates toward one of the asters. Finally, the chromosomes disappear, and the nuclei regain their grainy appearance.

Late 19th-century microscopy revealed all the major details of cell division in eukaryotes without explaining the role of the nucleus or chromosomes in the life of the cell.

How Did Biologists Discover the Function of the Chromosomes?

But what was the point of this careful division? For the moment, no one knew. The nucleus of the cell had been recognized as a standard component of cells as early as the 1830s, but its function was elusive. During cell division, the nuclear membrane breaks down, and the nucleus seems to disappear. Because the nucleus disappears, biologists did not at first consider that it might play an important role in cell division. Many biologists assumed that the cell constructs a new nucleus after each cell division.

The first hint that the nucleus might be important came with studies of *fertilization,* the process by which a sperm and egg fuse to form a *zygote.* In the 1870s, several microscopists noticed that the sperm is little more than a nucleus with a tail to propel it. Further studies showed that a zygote forms from exactly one sperm and one egg and that the sperm nucleus and the egg nucleus fuse to form a single nucleus. In time, the zygote develops into an embryo by dividing into more and more cells.

By 1880, biologists knew that all the nuclei of a developing organism are descended from a single nucleus created by the fusion of the nuclei of the egg and the sperm. This fact did not, however, suggest to biologists that the nucleus contained the hereditary material. First, 19th-century biologists were not necessarily looking for a hereditary material. Many biologists of the time believed that the process of fertilization did not transmit a material substance, but merely provided some form of energy or force that triggered the development of the embryo.

In addition, biologists of the time were unaware that the material of the chromosomes carries information. Biologists assumed that the individual chromosomes in a cell are all alike, no more different from one another than are individual pretzels. Although they might look different from one another, dividing them would be similar to dividing a bag of pretzels between two people. It wouldn't matter who got which pretzel, as long as each person got half.

Flemming and others had demonstrated the presence of individual chromosomes. For example, we now know that every human cell has the same set of 23 pairs of chromosomes. Each member of a pair has a distinctive and unique shape and size. Flemming saw the uniqueness of chromosomes in other organisms and described how each chromosome splits down its length, one-half "predestined for one new cell and [one-half] for the other." If we divided a bag of pretzels in half, however, we would not cut each pretzel down its whole length. We would just give each person half the pretzels in the bag.

It was the great embryologist Wilhelm Roux who first observed that the complex way the chromosomes split during mitosis strongly suggested that the chromosomes were not uniform at all. Why, he asked, would the cell use such a complex process to divide the nucleus, dividing each chromosome neatly in half, if simply cutting the nucleus in two would work? The chromosomes in the nucleus, argued Roux, must contain material of different kinds. The cell must be carefully dividing the material so that each daughter cell receives some of each.

In 1889, August Weismann insisted that the nucleus contained the hereditary material, which he called the *germ plasm.* "Heredity," Weismann wrote, "is brought about by the transference from one generation to another, of a substance with a definite . . . molecular constitution." But Weismann was known for his radical ideas, and most biologists of the time dismissed Weismann's claim.

In the same year, the German cell biologist Theodor Boveri performed an unusual experiment that lent firm support to the idea that the nucleus can determine the characteristics of an organism. Boveri discovered that, by vigorously shaking the eggs of a sea urchin until they broke open, he could knock the nucleus right out of the eggs. He then fertilized these damaged "enucleated" eggs with sperm from a different species of sea urchin. Unlike normal zygotes, these had no genetic material from the mother (Figure 8-3).

Miraculously, these zygotes developed into normal sea urchins. But the resulting adult sea urchins did not look like the species from which the eggs came. Instead they looked like the species from which the sperm came. In a control experiment, Boveri used sperm that belonged to the same species as the eggs. The resulting sea urchins looked like the species from which the eggs and sperm both came. Since the sperm contributes virtually nothing but a nucleus, here was dramatic evidence that the nucleus by itself can determine the characteristics of the organism. In 1891, Boveri wrote, "in all cells derived . . . from the fertilized egg, one half of the chromosomes are of strictly paternal origin, the other half of maternal."

But Boveri's experiment left hardly a ripple in the scientific community. As we will see in the next chapter, evidence that was even more persuasive failed to persuade many biologists that the chromosomes are the hereditary material. And,

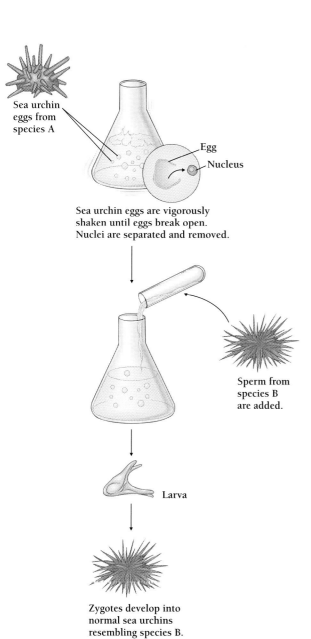

Figure 8-3 The nucleus determines the characteristics of an organism. In an experiment in 1889, Theodor Boveri shook sea urchin eggs with such force that the nuclei fell out. "Enucleated" eggs that were fertilized with sperm from a different species of sea urchin developed into individuals like the species from which the sperm came. Boveri concluded that a nucleus (in this case, the sperm's) determines the characteristics of the organism.

as recently as the 1930s, a few respected biologists still rejected the notion that the chromosome is the vehicle of inheritance.

Today we have much more information about both chromosomes and the nucleus. In later chapters, we will see how biologists learned more about chromosomes, nuclei, and heredity. We will see that each daughter cell carries genetic information inherited from the parent cell and that the genetic material determines how the organism interacts with its environment. In all organisms (and all cells) the genetic material is DNA.

Despite detailed descriptions of the division of chromosomes during mitosis, 19th-century biologists did not realize that the chromosomes carry hereditary information.

HOW DOES A DIVIDING CELL ENSURE THAT EACH DAUGHTER CELL RECEIVES AN EXACT COPY OF THE PARENT CELL'S DNA?

One of the main tasks of cell reproduction is to **replicate,** or copy, the cell's DNA (Chapter 10). After the DNA replicates, the cell's machinery distributes one copy to each of the two daughter cells. When a single molecule of DNA is involved, as in most bacteria, equal distribution is relatively simple. In eukaryotes, where several DNA molecules are involved, the distribution process is more complex. Cell division also splits the rest of the cell's contents between the two daughter cells. Cell reproduction, then, accomplishes three tasks: the replication of DNA, the equal distribution of the DNA to the two daughter cells, and the division of the other cell components.

We can describe the process of cell reproduction in terms of three M's—materials (the small molecules that carry energy or serve as building blocks), machinery (the organelles and macromolecular structures needed to carry on cellular processes), and memory (the information, contained in DNA, for building cellular machinery from the building blocks). Before a cell divides, it usually doubles its materials, machinery, and memory. In most cases, each daughter cell then receives an equal share of each.

Each of the three M's is essential to cell reproduction. DNA, for example, contains the information needed to produce the next generation of cells and organisms. Without the memory encoded in DNA, the cell's materials and machinery are useless. On the other hand, without the cell's materials and machinery, the DNA has no environment in which to express itself.

The information in the DNA comes from both parents. But the materials and machinery come exclusively from the mother. The precise doubling of DNA employs some of the cell's most elaborate molecular mechanisms. We discuss these mechanisms at length in Chapter 10. In this chapter, we see how dividing cells coordinate the duplication and distribution of DNA and cytoplasm.

Each DNA molecule consists of two polynucleotide chains. After the DNA replicates, each daughter DNA molecule contains a new chain and one of the parent chains. The cell's machinery then distributes one copy to each of the two daughter cells.

How Do Prokaryotic Cells Divide?

Cell division in prokaryotes is fairly straightforward. Prokaryotes divide by **binary fission,** in which a cell pinches in two, distributing its materials and molecular machinery more or less evenly to the two daughter cells. Since prokaryotes have no membrane-enclosed nucleus, binary fission is simpler than mitosis. In most prokaryotes, genetic information is stored in a single molecule of DNA, which usually forms a closed loop (Figure 8-4A). This loop is referred to as a "circle" of DNA, even though it looks like a true circle only in a diagram. Before the prokaryotic cell divides, it duplicates its DNA. Each of the two resulting daughter DNA molecules then attaches to the membrane fold at which binary fission begins (Figure 8-4B and C). As the cell divides, the two DNA molecules are pulled apart by their attachments to the plasma membrane, so that each daughter cell receives a single molecule of DNA.

> *Bacteria divide by binary fission in a process that does not involve mitosis.*

How Do Eukaryotic Cells Divide?

The internal membranes and compartments of eukaryotic cells make the process of cell reproduction far more complicated than in prokaryotes. Just the existence of a nucleus poses a problem: DNA cannot be easily distributed to daughter cells as long as it is surrounded by the nuclear membrane. Further, a eukaryotic cell may contain more than 1000 times as much DNA as a prokaryotic cell—far more than any single circle of DNA could include.

We can observe the doubling and distribution of DNA if we stain dividing cells with a dye that binds to DNA. Figure 8-2 shows individual cells from an onion root in various stages of division. We can see that, in some cells, the material of the nucleus has become organized into chromosomes. The cells of each species contain a characteristic number of chromosomes: onions have 16, humans have 46, and the fruit fly *Drosophila melanogaster* has 8. (Don't conclude, however, that 46 chromosomes make you smarter than a fruit fly or an onion. Your baked potato had 48 chromosomes in each cell before it went into the oven.)

Nearly every cell in the human body has copies of the same 46 chromosomes. (One exception is the red blood cell; as a red cell develops, it sheds its nucleus and ends up with no chromosomes at all.) The chromosomes of most sexually reproducing eukaryotes come in pairs, like socks (Figure 8-5). Humans, for example, have 23 pairs of chromosomes, whereas fruit flies have 4 pairs. Each member of the pair is similar but not identical to its mate. Pairs of matching chromosomes are called **homologous chromosomes** [Latin, *homo* = same], and each member of the pair is called a **homolog.**

Everything that happens to a cell from the time that it first forms until it divides is part of the **cell cycle**—the orderly sequence of events that accomplish cell reproduction. The time from one division to the next is one cycle. Biologists distinguish three major stages in the cell cycle:

1. **Mitosis,** the division of the nucleus.
2. **Cytokinesis,** the division of the cytoplasm and formation of two separate plasma membranes.

A. 1 μm

B. 1 μm

Figure 8-4 Prokaryote cells divide by pinching in two. A. In *Escherichia coli* and most other prokaryotes, genetic information is stored in a circular loop of DNA. Here, circular DNA is spilling out in loops from a broken *E. coli* bacterium. B. As the cell divides, the two DNA molecules are pulled apart by their attachments to the membrane. (A, Dr. Gopal Murti/Science Photo Library/Photo Researchers, Inc.; B, K.G. Murti/Visuals Unlimited)

A.

B.

One homolog

Homologous chromosomes

Figure 8-5 The 46 chromosomes in the human karyotype come in pairs, like socks. A. If we take a photograph of a set of chromosomes, then cut out the 46 individual chromosome images and arrange them in pairs of decreasing size, we create a tidy picture called a karyotype. In this micrograph, we can see chromosome bands that provide landmarks for the location of a hemoglobin gene in human chromosomes. B. Each member of a pair is similar, or homologous, but not identical to its mate. In each chromosome, the DNA loops and clusters in characteristic ways to form a pattern of bands that becomes visible under a microscope when the bands are stained. Biologists can detect abnormal chromosomes by comparing the banding patterns of abnormal chromosomes with those of normal chromosomes. *(A, top, Custom Medical Stock; bottom, Paraskevas Photography)*

3. **Interphase,** the time when the DNA replicates and when the chromosomes are not condensed and the cytoplasm is not dividing.

The amount of time that a single cell cycle takes depends on the organism and on its circumstances. Most plant and animal cells are capable of dividing in about 1 day, with mitosis and cytokinesis occupying 1 or 2 hours. But every cell is different. For example, a neuron spends most of its life in interphase, while rapidly dividing cancer cells spend very little time in interphase. The replication of DNA during interphase is invisible, and nothing appears to be happening in the cell. Only when biologists recognized the importance of DNA replication did they realize that what happens during interphase is just as important as what happens during mitosis and cytokinesis.

The amount of DNA in a cell provides a useful way of dividing the cell cycle into different phases (Figure 8-6). The most visually dramatic phase is **M,** for *m*itosis, which includes both mitosis and cytokinesis (the actual division of the cell). After cytokinesis come the three parts of interphase: **G_1,** the *g*ap, or growth phase, between the completion of M and the beginning of DNA synthesis; **S,** the period of DNA *s*ynthesis, when a cell doubles its DNA content; and **G_2,** the *g*ap between the completion of DNA synthesis and the beginning of M (of the next cell cycle). During G_1, a cell accumulates the materials and machinery needed for the DNA synthesis that will occur during S. In G_2, the cell prepares for mitosis and cytokinesis, assembling the molecular machinery needed to sort the chromosomes and divide the cell.

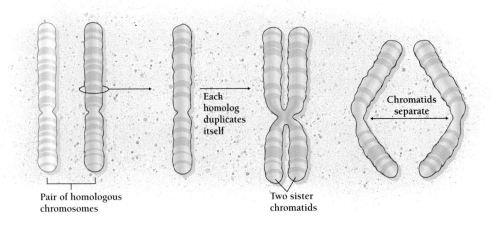

Figure 8-6 **The cell cycle consists of G_1, S, G_2, mitosis, and cytokinesis.** After the first growth phase (G_1), the DNA doubles (S). The cell then increases materials (G_2) in preparation for the separation of identical sister chromatids (mitosis) and the separation of the two daughter cells (cytokinesis).

B. Animal cell (whitefish)

25 μm

Figure 8-7 The stages of mitosis appear clearly in microscopic images of cells whose chromosomes have been stained. Look for the mitotic spindle, sister chromatids, centriole, kinetochores, centromere, kinetochore microtubules, polar microtubules, metaphase plate, and asters. (B, M. Abbey/Photo Researchers, Inc.)

Quiz

More on Mitosis

With each new phase of the cell cycle, the characteristics of the chromosomes change. At the beginning of mitosis, when the chromosomes first become visible, each chromosome consists of two separate but connected bodies, called **sister chromatids** (Figure 8-7). The two chromatids are joined at a single point called the **centromere.** Both the chromatids and the centromere consist of molecules of DNA joined to various proteins. The two sister chromatids that make up a single chromosome are duplicate copies with exactly the same genetic information. In contrast, homologs are similar but different. During mitosis, the sister chromatids within each chromosome separate and go to opposite poles of the cell.

During cytokinesis each chromosome consists of a single chromatid, which gradually unfolds into chromatin and disappears. All the chromosomes remain unchanged throughout G_1, and we cannot see them under a microscope. During S, each chromosome duplicates, a process that takes 8 to 10 hours in most eukaryotic cells. The amount of DNA also doubles, so that each chromosome again consists of two (still invisible) sister chromatids. During G_2, the paired chromatids remain invisible. Only at the beginning of mitosis do the paired sister chromatids again become visible.

Researchers cannot tell whether a cell is in S, G_1, or G_2 by looking. They can sometimes tell what phase a cell is in, however, by measuring the amount of DNA in each cell. One way to measure the DNA content of individual cells is to treat cells with a fluorescent dye that binds to DNA. The amount of fluorescence from each cell is a measure of its DNA content and so of its phase in the cell cycle (Figure 8-7). To see if a cell is in S, researchers supply radioactively labeled precursors called nucleosides. Cells in S take up these radioactive nucleosides (to make new DNA), while cells in G_1 or G_2 do not.

During G_1, the cell and the DNA appear to do nothing. Yet this is the time when the cell does most of the work of reproduction. During G_1, the cell doubles nearly all of its materials and machinery (other than DNA). If the cell runs out of nutrients, it stops the cycle in G_1.

The cell cycle consists of the three parts of interphase (G_1, S, and G_2), mitosis, and cytokinesis. During interphase, a cell doubles all of its materials, including the DNA. During mitosis, the sister chromatids separate. During cytokinesis, the cell divides into two daughter cells.

HOW DOES MITOSIS DISTRIBUTE ONE COPY OF EACH CHROMOSOME TO EACH DAUGHTER CELL?

The major task of mitosis is to distribute one chromatid from each chromosome to each daughter cell. Mitosis normally ensures that each offspring cell inherits two sets of chromosomes. Remarkably, all eukaryotic cells accomplish mitosis with the same chromosomal ballet that Flemming first described more than 100 years ago. Modern optical techniques, however, allow biologists to watch the whole process in living cells far more conveniently than could Walther Flemming.

Mitosis Is a Continuous Process, but Biologists Distinguish Four Phases

Cell biologists divide mitosis into four phases. These are prophase, metaphase, anaphase, and telophase. The boundaries between these phases are arbitrary, since mitosis is a continuous process. Nonetheless, the events described in these four phases always occur in exactly the same order (Figure 8-7).

During prophase, chromosomes condense and the mitotic spindle forms

During the first phase of mitosis, **prophase** [Greek, *pro* = before], the diffuse chromatin condenses into discrete chromosomes, each consisting of two chromatids, joined together at the centromere. As the chromosomes become visible within the nucleus, the nucleoli disappear. This process takes 10 to 15 minutes in mammalian cells growing in culture.

At the same time, a new structure, called a **mitotic spindle,** develops outside the nucleus. The mitotic spindle, shaped like an American football, consists of prominent bands of **microtubules,** hollow tubes constructed of the protein tubulin. Microtubules form the machinery that moves the chromatids apart. The microtubules of the mitotic spindle are identical in appearance to those found in cilia, flagella, and the cellular cytoskeleton, as discussed in Chapter 4.

In animal cells (and in some other eukaryotes as well), each pole of the mitotic spindle contains a pair of small cylindrical **centrioles,** which lie at right angles to one another. Microtubules originate from each pole of the spindle. Some of the microtubules, the "polar microtubules," run toward the equator. Others, called "astral microtubules," extend outward from each centriole to form an aster, as first noticed by Flemming. By the end of prophase, the spindle has started to elongate, and the two poles (with or without asters) begin to move apart. (The cells of most plants usually have neither centrioles nor asters, and mitosis in these cells is said to be "anastral.")

During the transition from prophase to metaphase, so much happens that researchers give this period its own name, "prometaphase." The nuclear membrane disappears, and the chromosomes are free to attach to the mitotic spindle. Throughout mitosis, the homologs in each pair of chromosomes behave independently of one another.

In mammalian cells, the chromosomes take 10 to 20 minutes to attach to the mitotic spindle. Each chromatid develops a **kinetochore,** a specialized disc-shaped structure that attaches the mitotic spindle to the centromere (Figure 8-8). Some of the polar microtubules attach to the kinetochores (and so to chromosomes) and become "kinetochore microtubules." By the end of prometaphase, animal cells have three sets of microtubules, all of which originate at a spindle pole: kinetochore microtubules, polar microtubules, and astral microtubules. Most plant cells contain kinetochore and polar microtubules but no astral microtubules. After the spindle apparatus has formed, the attached chromosomes are moved along the spindle toward the cell's equator.

During metaphase, chromosomes align

By the end of **metaphase,** the second and longest stage of mitosis, the chromosomes have moved halfway between the two poles of the spindle, where they form a disc, called the **metaphase plate.** For about 1 hour, all the chromosomes then lie in a single plane at the equator (at right angles to the spindle fibers). At this stage, the individual chromosomes look different from one another, in size and in the positions of the centromeres.

During anaphase, chromatids separate

The most dramatic stage of mitosis is **anaphase,** the time of chromatid separation. The centromere of each chromosome splits so that each chromatid contains its own centromere con-

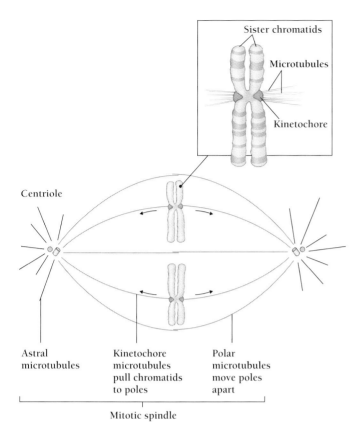

Figure 8-8 Each chromatid develops a disc-shaped kinetochore that attaches the mitotic spindle to the chromosome's centromere. Polar microtubules attach to the kinetochores (and so to chromosomes) and become "kinetochore microtubules." By the end of prometaphase, animal cells have three sets of microtubules originating at a spindle pole: kinetochore microtubules, polar microtubules, and astral microtubules.

nected to the kinetochore microtubules. Then, all at once, the two chromatids that make up each chromosome begin to move away from the metaphase plate. The paired chromatids separate and move in opposite directions. For 5 to 10 minutes, all the chromatids move at the same speed (about 1 μm [millionths of a meter] per minute) toward the two poles of the spindle. Each centromere leads the way, with the rest of the chromatid dragging behind, as if the kinetochore microtubules are pulling the chromatid toward the poles.

Even as the chromosomes move toward the spindle poles, the poles themselves are moving further apart, often doubling the distance between them. By the end of anaphase, the two processes, acting simultaneously, have separated the two sets of chromatids.

During telophase, the mitotic apparatus disappears

During **telophase** [Greek, *telos* = end], the last phase of mitosis, the mitotic apparatus (including kinetochore, polar, and astral microtubules) disperses. The chromosomes once more assume the diffuse appearance of chromatin, losing their distinct identities, and the nucleoli again become visible. The nuclear membranes form and enclose the two new sets of daughter chromosomes. Each daughter nucleus has the same number of chromosomes as the parent nucleus, but now each chromosome consists of a single chromatid. (Each chromosome will replicate during the next S phase.) After the completion of telophase, the cell usually completes cytokinesis, which we will discuss later.

What Propels the Chromosomes During Mitosis?

We have described three sets of movements whose molecular mechanisms we would like to understand: (1) the movement of the chromosomes toward the equator to form the metaphase plate, (2) the movement of the separated chromatids toward the poles, and (3) the movement of the spindle poles away from each other during prophase and again during anaphase. The chromosomes themselves do not play active roles in their own movement. Rather, the major players are the microtubules. These seem to push the chromosomes around by lengthening and shortening, as necessary.

Experiments have demonstrated the importance of microtubules in mitosis. In one experiment, a researcher used a laser to cut the kinetochore microtubules that attach one of the two sister chromatids to one pole at metaphase. The loose chromatid followed its sister to the pole opposite the one to which it normally would have gone. This experiment showed that the kinetochore microtubules normally pull the two chromatids in opposite directions. Cell biologists still do not fully understand the mechanism of mitosis, however, and the identities and roles of the components of the mitotic machinery are the subject of intense research.

During mitosis, microtubules move the chromosomes by attaching to the chromatids and growing or shortening.

HOW DOES A CELL FIT ALL ITS DNA INTO A NUCLEUS?

The DNA in a cell carries information that the cell uses to construct cell machinery and to carry out cell processes. When a cell divides, the cell must duplicate this information and pass it on to its two daughter cells.

Each molecule of DNA contains two polynucleotide strands, wrapped around each other in a double helix. This ropelike structure is so fine that a DNA molecule long enough to circle the Earth at the equator would weigh less than a grain of fine sand. In contrast, the same length of fine cotton thread would weigh about 275 kg (over 600 lb). The DNA in a human cell weighs about 6×10^{-12} g (6 trillionths of a gram) and has a total length of about 2 m. Yet a chromosome is only a millionth of a meter long, and the nucleus is only 5 μm in diameter. A cell must therefore solve a major problem: how to fit 2 m of DNA into the tiny nucleus without the long threads of DNA becoming tangled.

How Does the DNA Fold Up?

Collapsing DNA into so small a structure depends on folding and bundling the DNA with the aid of specialized proteins, the most abundant of which are the **histones** [Greek, *histos* = web], a set of small, positively charged proteins. Most cells have five types of histones (called H1, H2A, H2B, H3, and H4). A cell may contain some 60 million copies of each type of histone.

The positive charges of histones come from their unusually high content of the basic amino acids (arginine and lysine), which attract hydrogen ions and acquire positively charged side chains. Because DNA has a negative charge, the positively charged histones bind to it tightly (Figure 8-9B). In a test tube, purified DNA and four of the five kinds of histones associate with each other to form what look like beads on a string (Figure 8-9A). Each bead, along with its strand of DNA, is called a **nucleosome** and is a DNA-histone complex about 11 nm in diameter. Each nucleosome contains a 146-nucleotide-long stretch of DNA and 8 histone molecules. The formation of nucleosomes is the first stage in the packing of DNA into a nucleus.

Figure 8-9 How does DNA pack so tightly into the nucleus? Histone proteins pack DNA into small, tight nucleosomes; strings of nucleosomes fold and loop to form chromosomes small enough to fit into the nucleus. The positively charged histones attract the negatively charged DNA, and the two molecules form nucleosomes. Each nucleosome contains a 146-nucleotide-long stretch of DNA and 8 histone molecules.

Histones from all eukaryotes resemble one another. For example, histone H4 of cows differs by only 2 amino acids (out of 102) from that of peas. This extraordinary similarity suggests that the structure of histone H4 has remained almost unchanged for more than a billion years. We can conclude that this method of packing DNA evolved long ago.

Proteins called histones pack DNA into small, tight bundles called nucleosomes that resemble beads on a string.

How Do DNA, Histones, and Other Proteins Form Such Compact Structures?

Histone H1 appears to be responsible for holding together groups of nucleosomes to form a fiber that is 30 nm in diameter. As a result of this packing, the DNA of a typical human chromosome could be condensed from an average length of 5 cm to about 1 mm—a terrific achievement, but still not small enough for it to fit into a 5-μm nucleus. We still do not know exactly how DNA is made more compact. In some specialized cells, microscopists can see loops extending from chromosomes (Figure 8-10). Many biologists now think that such loops are common to the chromatin of all cells.

In chromosomes, the DNA is even more condensed than in the diffuse chromatin. Still more loops form, and these loops cluster in characteristic ways in each chromosome. The specific pattern of DNA clustering results in **chromosome banding,** the pattern (visible under a microscope) that results from the selective binding of certain dyes. Biologists can detect abnormal chromosomes by comparing the banding patterns of abnormal chromosomes with those of normal chromosomes. Using DNA technology, it is now possible to determine the location of individual genes on specific chromosomes, using chromosome bands as landmarks (Figure 8-5A).

Strings of nucleosomes fold and loop to form the chromosomes, the highly compact arrangement of chromatin that fits into the nucleus. No one knows how this occurs.

HOW DOES A CELL DIVIDE ITS CYTOPLASM?

So far we have discussed how a cell distributes the two copies of its genetic information. But we still need to describe cytokinesis—the process by which a dividing cell partitions its cytoplasm.

Cytokinesis and mitosis are separate processes, and they depend on different molecular machinery. In some organisms, for example, mitosis sometimes occurs without cytokinesis, leading to two or more nuclei within a single cell, as in the early fruit fly embryo and in the specialized tissues of a plant seed. This shows that cytokinesis and mitosis are truly separate processes. Nonetheless, cytokinesis almost always accompanies mitosis, usually beginning during anaphase and finishing shortly after the end of telophase.

During cytokinesis, the plane of cell division is always perpendicular to the axis of the mitotic spindle (Figure 8-11). In early anaphase, dividing animal cells form a beltlike **contractile ring,** a bundle of actin filaments that surrounds the dividing cell. The contractile ring works like a purse string, pinching the cell into two parts through the spindle's equator. The force for this movement comes from a mechanism similar to that which causes muscle contraction. In contrast to the permanent contractile structures of muscle cells, however, the contractile ring is a temporary structure like the mitotic spindle.

As the beltlike contractile ring tightens, the cell membrane is pulled into a deepening groove, called the **cleavage furrow** (Figure 8-11). As telophase proceeds, the two daughter cells

Figure 8-10 In some cells, the DNA forms loops, obvious in this micrograph. *(Joseph Gall, Carnegie Institution)*

25 µm

become almost, but not quite, completely separated. A thin connection between the daughter nuclei, called the midbody, persists until the end of mitosis. The midbody is packed with microtubules from the spindle apparatus. Finally, at the end of telophase, the microtubules disassemble, and the midbody breaks, leaving the daughter cells completely separated.

As the two daughter cells separate, they must enlarge their cell membranes (since two cells have more surface area than a single cell of the same volume). The additional membrane material comes from a supply of extra membrane made by the parent cell during interphase.

A plant cell, which has a rigid cell wall, cannot simply pinch itself in two the way an animal cell does. Cytokinesis in a plant cell requires the building of a new cell wall between the daughter cells. The new wall begins in telophase as a small, flattened disc, called the early **cell plate,** in the space between the two daughter nuclei (Figure 8-12). The disc grows to become a cell plate, which, except for connecting plasmodesmata, completely seals off the two daughters from one another.

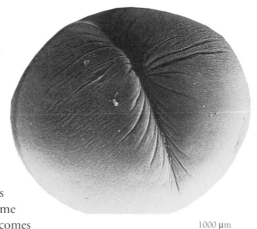

Figure 8-11 During cytokinesis, actin filaments inside the cell form a contractile ring, or cleavage furrow, and pinch the dividing cell into two. *(David M. Phillips/Visuals Unlimited)*

During cytokinesis in animal cells, actin filaments form a contractile ring that pinches the cell in two. In plant cells, telophase includes the building of a new cell wall between the two daughter cells.

HOW DOES A CELL REGULATE PASSAGE THROUGH THE CELL CYCLE?

In this chapter, we have seen how normal cells accomplish the equal distribution of chromosomes by mitosis and the division of cytoplasmic components by cytokinesis. But how do cells know when to begin the cell cycle? How do they coordinate its separate parts? And how do they know when to stop dividing?

If cell division continued without stopping, single-celled organisms would soon run out of space and food. For example, after just 4 days of undisturbed growth, a single yeast cell could have divided 48 times and produced 2^{48} (about 10^{15}, or 10 million billion) descendants, as many cells as are in 100 people.

In multicellular organisms, cells normally subordinate such growth and division to the needs of the body's tissues and organs. Cells divide rapidly during periods of growth and more slowly (or not at all) in mature organisms. The only exception to this rule is the uncontrolled growth and division of cancer cells, such as those of Henrietta Lacks. These multiply at the expense of all the other cells in the body.

How Do Normal Cells Determine When To Stop Dividing?

What normally prevents such runaway cell division? Two general mechanisms appear to operate—cell senescence (aging) and growth control. **Cell senescence** limits the number of

Figure 8-12 In plants, a new cell wall forms between daughter cells. The cell wall begins during telophase as a small, flattened disc, called the early cell plate, then grows into a full-sized cell wall, complete with interconnecting plasmodesmata. *(B, R. Calentine/Visuals Unlimited)*

times a cell can divide: the more times a cell divides (at least under conditions of laboratory culture), the more likely it is to withdraw from the cell cycle. An average cell taken from a newborn baby, for example, will divide about 50 times in a standard culture medium. But cells taken from an 80-year-old stop cycling after about 30 divisions. Despite the general occurrence of cell senescence, however, even a mature mammal contains cells that are capable of unlimited cell divisions in the presence of the right "growth factors."

A second set of regulatory mechanisms—those involved in **growth control**—prevent uncontrolled division by allowing the cell cycle to proceed under some conditions and to stop under others. As an example of growth control, think about what happens when you cut yourself. The skin cells on the edges of the cut begin to divide and fill the space left by the wound. The cells divide until the two edges again touch. When the cells on each side of the cut meet, and the wound is healed, cell division ceases.

In a culture dish, cells regulate their division in much the same way (Figure 8-13). Cells attach to the bottom of a plastic dish and divide until they form a single, continuous layer of cells that occupies the whole surface. Then they stop dividing. If we now make a "wound" in this monolayer by scraping away a swath of cells, the cells on the margins move into the open space and begin to divide. Division stops when the space is filled.

We can summarize the growth control of normal cells in culture with the following rules: (1) cells stop dividing during G1 when they run out of free space on which to spread—that is, when neighboring cells all touch each other; and (2) cells that have proceeded beyond G1 begin to divide when contact with their neighbors ceases. This kind of growth control is called **contact inhibition** of cell division.

Cancer cells provide an excellent example of what happens when contact inhibition fails. Cancer cells do not exhibit growth control in cell culture (Figure 8-13). Instead of forming a continuous monolayer, they pile up on top of one another and grow to much higher densities than normal cells.

Growth control depends on the actions of several types of proteins: (1) growth factors, which stimulate cell division or promote cell survival, depending on the type of cell; (2) proteins that bind to growth factors and contribute to the sequence of events that they trigger; (3) proteins that regulate the expression of genes into RNA and protein; and (4) "tumor-suppressor" proteins, which regulate the passage of cells through the cell cycle. Changes in just one kind of tumor-suppressor protein, called *p53*, occur in more than 50% of human cancers.

Growth control mechanisms such as contact inhibition prevent normal cells from growing out of control the way cancer cells do.

How Do Normal Cells Determine When It Is Time To Divide?

Most cells divide only after they have first doubled their mass. Otherwise, daughter cells would get smaller with every generation. Somehow cells sense when they have reached a critical size and have enough nutrients to supply two daughter cells. We do not know how eukaryotic cells accomplish this sensing, but once they do, they become irreversibly committed to cell division. Once a cell proceeds beyond a certain point in G_1, the cell proceeds through the rest of the cycle, including mitosis and cytokinesis. This "point of no return" is called **Start.** Arrival at Start depends heavily on the environment of the cell—including the availability of nutrients and signals from other cells. Once the cell has gone through Start, however, the rest of the cycle proceeds, independent of the extracellular environment.

One important exception to the general rule that cells double their mass before dividing are egg cells, which grow to enormous sizes without dividing. For example, the ostrich egg is a single cell. Another important exception are the early embryos of many animal species. Embryos derived from large eggs begin life after fertilization with a series of rapid cleavage divisions, which divide the embryo into many cells without increasing its mass. A frog zygote, for example, is an enormous single cell about 1 mm in diameter. Cleavage divisions rapidly divide it into cells that are more nearly the size of cells in the adult, about 10 to 20 μm in diameter. Even a tiny human zygote divides into 50 or 60 cells in the week before it connects to its mother's blood supply. Only after the human zygote connects can it increase the size of its cells.

The cells of mature organisms only divide when they have doubled their mass, unless they are egg cells or the cells of an early embryo.

What Triggers the Main Events of Mitosis?

Cell reproduction requires the coordination of three cycles:

1. The duplication and packaging of DNA, which we discuss in Chapter 10
2. The duplication of the centrosomes and the operation of the mitotic spindle apparatus
3. Cytokinesis, which depends on the contractile ring

Recent research has shown that all eukaryotes, from yeasts to humans, use almost identical mechanisms to coordinate and initiate the events of these three processes. One way researchers study cell cycles is to create **synchronous cell populations,** groups of cells that are all at the same stage of the cell cycle. One source of such synchronous populations are the rapidly dividing cells of early embryos from marine invertebrates; a researcher can start thousands of embryos dividing simultaneously just by mixing sperm and eggs, collected separately. Another way of producing synchronous populations is to block DNA synthesis with a chemical inhibitor that stops the cell cycle just before S. When the inhibitor is removed, all the cells begin S at the same time and continue in lock step through the rest of the cell cycle.

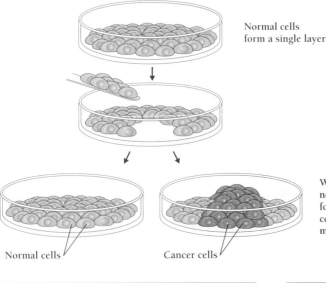

Normal cells form a single layer

When cells are scraped away, normal cells divide until they form a single layer. Cancer cells continue to divide and form a mound of cells

Normal cells   Cancer cells

Figure 8-13 Cells regulate their division by means of contact inhibition. Cancer cells fail to regulate their growth in tissue culture. They pile up on top of one another, unlike normal cells. Normal cells stop dividing when they come into contact with other cells. Cancer cells, in contrast, continue to divide even after they come into contact with other cells. In cell cultures, normal cells divide until they form a single, continuous layer of cells, then stop. If we scrape a "wound" in this monolayer of cells, cells from the edge move into the open space and divide until the wound is filled. *(courtesy of G. Steven Martin)*

Normal cells   100 μm

Cancer cells   100 μm

Studies with all three kinds of synchronized cells have led to the discovery of specific proteins, called **cyclins,** whose concentrations change through the cell cycle. Each type of cyclin accumulates during a particular phase of the cell cycle and stimulates the passage of the cell into the next phase of the cycle.

Each type of cyclin stimulates a specific enzyme, called a cyclin-dependent kinase, which alters proteins of the cell (by adding phosphates from ATP). Among the targets of these enzymes are components of the mitotic spindle. These protein modifications are the "ticks" of the cell cycle clock, propelling the cell to enter the next phase. As the cell moves into the next phase of the cycle, the cell's protein-digesting machines (the proteasomes) digest each type of cyclin, causing the oscillations that give the cyclins their names.

Various events regulate a cell's passage through the cell cycle, including the completion of DNA synthesis and fluctuations in cyclins.

In this chapter we have seen how eukaryotic cells with two sets of chromosomes divide their genetic material to create two daughter cells, each with two sets of chromosomes. Eukaryotes, such as ourselves, employ such mitotic division to create genetically identical copies of cells. In the next chapter, we will see how eukaryotic cells divide a cell with two sets of chromosomes to create gametes (for example sperm and eggs)—cells with just one set of chromosomes. This process, called meiosis, creates cells that are genetically different from the parent cells and also genetically different from one another. Along the way, we will see that sex and meiosis are two sides of the same coin.

Study Outline with Key Terms

Cells reproduce themselves in a process called **cell division,** in which a parent cell gives rise to two **daughter cells,** each with the same genetic information as the parent cell. Dividing cells replicate DNA and distribute it during an orderly sequence of events called the **cell cycle.**

Prokaryotes divide by a simple process called **binary fission,** which distributes one copy of the parent cell's DNA to each daughter cell. Cell reproduction is more complicated in eukaryotic cells, whose DNA is enclosed in a nucleus. Eukaryotes employ a spindle apparatus, but prokaryotes do not.

Chromosomes are highly condensed and visible in dividing eukaryotic cells and more diffuse (**chromatin**) in nondividing cells. Chromosomes come in nonidentical pairs called **homologous chromosomes.** Each member of a pair is a **homolog.** When the DNA in a cell has doubled, each chromosome consists of two identical **sister chromatids.**

Cell biologists divide the cell cycle into five phases: mitosis, cytokinesis, G_1, S, and G_2. The process of evenly distributing chromosomes to daughter cells is called **mitosis.** The process of dividing a cell's cytoplasm is called **cytokinesis.** Mitosis and cytokinesis together make up **M.** During G_1, a cell accumulates the materials and machinery necessary to **replicate** its DNA; during **S,** it synthesizes DNA, making two copies of the parent cell's DNA; during G_2, it prepares for mitosis and cytokinesis. G_1, S, and G_2 together are called **interphase.** The point of no return is **Start.**

Just before a cell begins mitosis, each chromosome consists of two identical chromatids. Mitosis is a continuous process, but biologists distinguish four major phases: prophase, metaphase, anaphase, and telophase. During **prophase,** chromosomes condense and the **mitotic spindle** forms. As the **centrioles** of animal cells move apart, **microtubules** radiating outward from the centrioles form **asters.** Between prophase and metaphase, the nuclear membrane disappears and the chromosomes attach to the spindle fibers. During **metaphase,** the chromosomes align in a disc called a **metaphase plate,** then bind to the **kinetochores** that form on the **centromere** of each chromatid. During **anaphase,** the two chromatids that make up each chromosome separate. During **telophase,** the mitotic apparatus disappears.

During interphase, the cell duplicates its single centriole. During prophase, a mitotic spindle assembles, with microtubules emerging from each centriole to form a mitotic spindle. Microtubules connect to each chromatid in a structure called a kinetochore. The mitotic apparatus is responsible for chromosome movements during mitosis.

The DNA of eukaryotic cells is highly folded. Protein molecules called **histones** bind to DNA and form particles called **nucleosomes** that resemble beads on a string. DNA, histones, and other proteins fold to form still more compact structures. The orderly packing of DNA produces a characteristic pattern of **chromosome banding** in each metaphase chromosome.

During cytokinesis, a **contractile ring** pulls the membrane inward to form a **cleavage furrow,** which deepens and separates the two daughter cells. Cytokinesis in plant cells requires a special mechanism for building a new cell wall. The new wall begins in telophase as a small, flattened disc that grows to become a **cell plate.**

Cells actively regulate passage through the cell cycle. Regulated cell division in multicellular organisms depends on **cell senescence** and **growth control.** Cell senescence limits the number of times a cell can divide. Growth control regulates cell reproduction according to external conditions. For example, many cells exhibit **contact inhibition** of cell division, in which cells divide only when they are not in contact with their neighbors. Growth control may also involve growth factors—extracellular proteins that stimulate cell division. Cancer cells fail to show normal growth control. Studies of **synchronous cell populations** show that changes in specific proteins, called **cyclins,** control a cell's passage through the cell cycle.

Review and Thought Questions

Review Questions

1. Why must a cell double its materials, machinery, and memory before it can divide?
2. Why did the discovery that DNA was the genetic material change the way that cell biologists regarded interphase?
3. Describe a method for measuring the amount of DNA in a cell. How does the amount of DNA in a cell change during the cell cycle?
4. Describe and draw the events of the four stages of mitosis. When does the actual division of the chromosomes occur?
5. Using diagrams, compare the events in prophase to those in telophase. Why do you think they are so nearly opposite?
6. Compare the role of the centriole in mitosis with that of a kinetochore.
7. What factors regulate a cell's ability to proceed to the next round of cell division?
8. Describe the role of histones in the structure of eukaryotic chromosomes. How do histones help DNA fit into a nucleus?
9. How does cytokinesis in a plant cell differ from that in an animal cell?

Thought Questions

10. How would prevention of DNA synthesis affect mitosis?
11. What is the relationship between the materials accumulated in G_1 and G_2 and the processes that follow?
12. How might cancer be associated with improper regulation of the cell cycle?

About the Chapter-Opening Image

Socks come in matched pairs, as do chromosomes.

On-line materials relating to this chapter are on the World Wide Web at **http://www.harcourtcollege.com/lifesci/aal2/**

Web Bit 8-1 Artificial Chromosomes

Web Links

Quizzing & Testing

9 From Meiosis to Mendel

Why Is the Yellow Dog Yellow?

In the spring of 1902, a young graduate student at Columbia University, in New York City, approached his professor with barely suppressed excitement and declared that he had solved the problem of heredity. The student, Walter S. Sutton (1877–1916), was a 6-foot-tall, 215-pound Kansas farm boy who towered over everyone around him (Figure 9-1). Gleefully, he announced, "I know why the yellow dog is yellow." At just 25, Sutton had discovered that tiny bars in cells, called chromosomes, must carry the hereditary material.

Chromosomes, realized Sutton, were the mechanism by which organisms inherit traits. They were thus the key to the then-infant science of genetics. But disappointingly, Sutton's professor only nodded indulgently. Neither he nor anyone else took Sutton seriously.

It was disappointing, especially because Sutton's professor was E.B. Wilson (1856–1939), the most distinguished cell biologist in the United States. But when summer came, Wilson invited Sutton to the seaside at Beaufort, North Carolina—to study marine animals, to sail, to swim, and to talk. For the usually preoccupied Wilson, it was also a slow time when he would really listen to young Sutton. "It was only then," Wilson later wrote, "that I first saw the full sweep and the fundamental significance of his discovery."

Sutton had come to biology partly by chance, and he hadn't yet decided if he would stay. In 1896, he had begun studying engineering at the University of Kansas. When he returned home for the summer in 1897, he took with him a case of typhoid fever that felled his entire family. When it was over, Walter Sutton's beloved 17-year-old brother John was dead. In his grief, Sutton decided to abandon engineering for medicine. That fall, he returned to the University of Kansas at Lawrence and began to study biology.

Before long, Sutton became friends with a young professor of zoology named Clarence McClung. On vacations at home, Sutton collected giant grasshoppers from the wheat fields of his family's farm to send to McClung, who admired the grasshoppers' enormous cells (Figure 9-2). The cells of *Brachystola magna,* the lubber grasshopper, in fact, made perfect subjects for the study of the fine structure of the cell.

McClung was especially interested in the fact that some of the grasshopper sperm cells appeared to have an extra chromosome. At the time, no one

Key Concepts

- Because genes lie on chromosomes, the inheritance of genes parallels the inheritance of chromosomes.
- Sexually reproducing organisms have *pairs* of homologous chromosomes, which meiosis distributes—one chromosome from each pair—to each gamete (egg or sperm).
- At fertilization, two gametes unite to form a zygote, which has paired chromosomes. Each gamete contributes one chromosome to each pair.
- The phenotype of an individual results from the two-way interaction between its genotype and the environment.

Figure 9-1 Walter Sutton. At age 25, Sutton discovered the location of the genes in a cell. Genes, he said, had to be in the nucleus on the chromosomes. (*University of Kansas Archives*)

knew what chromosomes were, but McClung called this chromosome the "accessory chromosome" and hypothesized that it determined sex. As it turned out, he was essentially correct. McClung's idea that a chromosome could give an individual grasshopper a specific trait—its sex—suggested to Sutton that the other chromosomes might do the same thing for other traits.

In April 1900, Sutton published a paper describing in great detail the sex cells of the lubber grasshopper, including 41 drawings and 10 photographs. A year later, Sutton received his master's degree based on his work on cell division in the sperm cells of the grasshopper. In the fall of 1901, encouraged by McClung, he transferred to Columbia University to work with the eminent cell biologist E.B. Wilson.

Within 6 months, Sutton was hanging on Wilson's sleeve trying to persuade him that he had solved the mystery of heredity. That fall, the famous British biologist William Bateson came to Columbia to talk about the new science of genetics. Two years before, Bateson told his audience, three botanists had rediscovered the laws of inheritance first described and published by the Austrian monk Gregor Mendel in 1866. These laws laid a foundation for all future work on inheritance. The years between 1900 and 1910 would be a breathtaking period in biology, and Bateson's inspired talk about inheritance electrified his American audience.

Bateson banished any doubts Sutton may have had about the importance of his idea. Everything Bateson said made perfect sense in the light of Sutton's knowledge of chromosomes. Within one month, Sutton had published a brief paper describing his idea that the chromosomes are the physical basis for heredity. Four months later, he had published a full-length paper describing all of the theoretical implications of his idea.

Research in the 100 years since has confirmed nearly every detail of Sutton's hypothesis. His ideas are so well accepted today that it is hard to imagine why more-experienced biologists

Figure 9-2 Lubber grasshopper. Walter Sutton's interest in biology began early with his research on the chromosomes of the lubber grasshopper, *Brachystola magna*, which he collected on his parents' farm in Kansas. (*Skip Moody/Dembinsky Photo Associates*)

in his time did not notice the role that chromosomes play in inheritance. Yet, as is so often the case in science, older biologists not only did not see what now seems obvious to us, but some continued to resist Sutton's idea for nearly 40 years.

Maybe it was partly the cool response of his elders. Maybe it was Sutton's idea that he could do more immediate good in the world as a doctor. In any event, after doing work so important that would forever influence biology, Sutton dropped out of graduate school, abandoning biology forever. Two years passed before, pressured by his father, he returned to Columbia, where he resumed his long-delayed medical training. He was as outstanding a medical student as he had been a biologist, and in 1907 he received his medical degree. Returning to Kansas in 1909, he practiced as a surgeon for the rest of his short life. Still a bachelor, he died of a burst appendix at 39.

Nevertheless, his research in biology would change the world. At 25, he discovered a fundamental fact that had eluded older scientists for decades. Sutton revealed the basic mechanism by which organisms inherit traits—why, as he put it, "the yellow dog is yellow." His work is the foundation for all of modern genetics and molecular biology. In the next section, we will see why it took a young biologist such as Sutton to recognize what now seems so obvious. In the rest of this chapter, and in succeeding chapters, we will try to find out exactly what the chromosome does.

WHY WAS THE CHROMOSOMAL THEORY OF INHERITANCE SO HARD TO ACCEPT?

Historically, biologists had no trouble posing theories to explain how organisms stay the same between generations. Before the 17th century, a favorite theory was that miniature versions of future organisms were already "preformed" in sperm or eggs, so that the properties of all future generations were already determined. Many philosophers believed that the egg contributed the materials for an individual and that the sperm contributed some life force that determined the form or shape of the individual. Philosophers compared the creative action of the sperm to the divine creation of the universe from formless matter.

Blending Inheritance: A Wrong Turn

In contrast, plant breeders had long known that when two varieties of plants are crossed, both parent plants contribute equally to the form of the next generation. No matter which kind of plant provided the sperm (pollen) and which kind provided the egg, the resulting cross was the same. In the face of such evidence, most biologists reasoned that the different types of offspring somehow resulted from the "blending" of the genes of the parents. We are not surprised, for example, when a black mare and a white stallion produce a gray foal or when a red

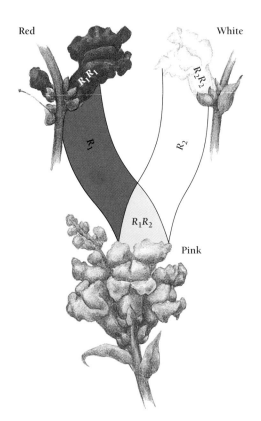

Figure 9-3 Blending inheritance? The offspring of red and white snapdragons are pink. Blending inheritance seemed reasonable to early biologists. But organisms inherit color and other traits by way of discrete units called genes. Here the genes for flower color are represented by R_1 *(red)* and R_2 *(white)*.

snapdragon crossed with a white snapdragon produces a plant with pink flowers (Figure 9-3). The color of the offspring appears to be the result of blending, as if two cans of paint have been poured together and mixed.

The **blending** model of inheritance worked well for traits that varied smoothly, such as height, weight, and even, sometimes, color. But the blending model did not always make sense. Pink flowers can give rise to red or white offspring, and brown-eyed parents often have blue-eyed children. If traits blend, how can unblended characteristics reappear in later generations?

The blending model also created a serious problem in understanding the mechanism of evolution (Chapter 15). If traits blended, variation would eventually disappear, and natural selection would have no inherited variation on which to act. For 19th-century biologists, the problem of blending inheritance was so intriguing that, in time, nearly every important biologist of the second half of the 19th century had a different hypothesis about how inheritance worked.

Most 19th-century biologists mistakenly accepted blending inheritance.

Why Did Biologists Doubt That Chromosomes Could Carry Information?

Recall from the last chapter that a wealth of clues suggested that the chromosomes carry the hereditary material. Walther Flemming had shown that during mitosis, the chromosomes divide down the middle, so that half of each chromosome goes to each cell. Wilhelm Roux had argued that this complex way of dividing the chromosomes at mitosis suggested that each chromosome was unique. Theodor Boveri had shown that a nucleus, which contains the chromosomes, can determine the characteristics of an organism. Boveri also showed that in each and every cell of the body, half the chromosomes come from the father and half come from the mother.

But the idea that the chromosomes were responsible for heredity was still no more than a possibility, and most 19th-century biologists did not take the idea seriously. One exception was the German biologist August Weismann. Weismann hypothesized that organisms inherit traits by means of some information-carrying molecule in the nucleus of the cell. Moreover, he argued, since the hereditary material from the mother and father fuse at fertilization, the egg and sperm must each contain only half the normal amount of hereditary material. Otherwise, said Weismann, the hereditary material would double every generation.

Within 3 years, cell biologists confirmed Weismann's prediction. First, they showed that sperm and eggs contain only half the usual number of chromosomes. Second, they observed the special process of **meiosis,** by which cells produce daughter cells (eggs and sperm) with only half the normal number of chromosomes. By the late 19th century, most cell biologists accepted that the chromosomes were central to the process of both mitosis and meiosis and probably somehow essential for the normal development of an embryo into an adult. But they did not know what the chromosomes did. No one yet knew that the chromosomes were the hereditary material.

In 1900, however, an intellectual earthquake rocked the scientific landscape. In that year, three scientists independently rediscovered the work of an obscure but brilliant scientist named Gregor Mendel. In 1866, Mendel had discovered several important principles of heredity while studying inheritance in pea plants. In particular, Mendel had shown that every organism has two sets of genes.

The rediscovery of Mendel's work gave birth, in just a few months, to an entire new field of science—genetics. Biologists began to suspect that the behavior of chromosomes during the creation of sperm and egg cells (meiosis) might explain Mendel's principles. Chromosomes suddenly claimed the intense attention of a great many biologists.

But it was one thing to suspect that chromosomes *could* carry genes, quite another to show that they *did*. Two important problems stood in the way. First, chromosomes disappear between cell divisions, and it seemed possible, even likely, that cells create new chromosomes after each cell division. In short, biologists had no idea if the chromosomes passed intact from generation to generation. If not, biologists reasoned, chromosomes could not carry information from parent cell to daughter cell.

Second, biologists had no evidence that the individual chromosomes in a cell were different from one another. The chromosomes seemed to come in a variety of shapes, but they all behaved in the same way, and they all seemed to be made of the same material. Like pretzels, they might come in slightly different shapes, but they were all the same stuff. There was nothing to suggest that each one might be carrying unique information.

Weismann's argument that each chromosome always maintains its individuality and integrity was only a guess. Young Walter Sutton, the Columbia University graduate student, was the first person to state explicitly that Mendel's genes were on the chromosomes and to give good reasons why this must be so.

Nineteenth-century biologists doubted that the chromosomes were the genetic material because the chromosomes disappeared during cell division and because biologists had no reason to think that each chromosome carried unique information.

HOW DO ORGANISMS PASS GENETIC INFORMATION TO THEIR OFFSPRING?

Genetics, the study of inheritance, today faces the same contradictory challenges it did 100 years ago. Genetics must explain both how like begets like—why children resemble their parents—and also the origin and maintenance of variation—why children differ from their parents. In this chapter, we will discuss **transmission genetics,** the study of how variation is passed from one generation to the next. We will see that the behavior of chromosomes during meiosis and fertilization does much to explain the inheritance of variation.

Distinct from transmission genetics is **molecular genetics,** the study of how DNA carries genetic instructions and how cells carry out these instructions. Molecular genetics helps explain how a single cell divides and eventually develops into liver cells, skin cells, nerve cells, and more. Molecular genetics also explains how the DNA in these different kinds of cells helps direct day-to-day operations in the cell. We discuss molecular genetics in Chapters 10 through 14. In this chapter, we begin our discussion of transmission genetics by briefly defining how the zygote (the fused egg and sperm) uses the genetic information it receives from its parents to become a unique individual. No two organisms are identical, and the unique collection of traits that define an individual is called the "phenotype."

How Is Phenotype Related to Genotype?

Everything about an organism, from its size and chemistry to its behavior, constitutes its **phenotype** [Greek, *phenein* = to show]. Phenotype encompasses both physical and behavioral characteristics. In humans, eye color, height, and skin color are obvious *phenotypic traits,* the individual aspects of phenotype. In contrast, the **genotype** is the particular collection of genes of a cell or organism. Biologists use the word "genotype" to refer both to all the genes in an organism, its **genome,** or, alternatively, to the subset of genes (or even one gene) that influences a particular trait, such as eye color. A genome, the set of all the genes, is a relatively passive entity within the cell. Like a cookbook, a genome provides lists of ingredients (proteins) and directions for how much of each protein to use.

Phenotype results from the interaction of the genes with each other and with their environment. A **gene** is a region of the DNA that does two things: it specifies a particular protein and it regulates "expression." A gene is *expressed* when the cell makes that protein. How much protein is made, and when, is *regulation.* Each gene responds to signals within a cell that tell the cell when to make more and when to make less of its particular protein.

An organism's phenotype reflects both its genotype (what genes it has) and its environment. In the case of plants and animals, the phenotype reflects the environment in which an individual has developed. "Environment" includes many effects that most people don't associate with the word. For example, an egg cell contains proteins made by the mother that influence the development of an embryo. But which proteins and how much of each depends on the external environment of the mother. In zebra finches, for example, red stripes painted on the father bird's legs induce the mother bird to load up her eggs with an extra dose of the hormone testosterone. The extra testosterone influences the expression of genes and makes the baby birds more aggressive.

Testosterone directly affects the expression of genes throughout the body, but its own production is influenced by the social environment of the individual. In humans and other mammals, hormones and nutrients in the blood supply from the mother similarly affect the development of an individual. After birth, nutrition and many other factors continue to influence the expression of genes. In men, winning a chess game, watching the home team win a basketball game, and picking up a paycheck can all increase the production of testosterone. Losing a game or enduring a chewing-out from the boss can reduce testosterone levels.

Figure 9-4 More than genes. Even identical twins, with exactly the same genes, often look and behave differently. Identical twins occur when a single-celled embryo divides into two separate cells, which then develop into two separate people. Identical twins have identical genetic information. As the two individuals develop in the womb, however, subtle differences in environment—such as a different blood supply—can lead to distinct differences in phenotype. The two 8-year-old twins shown here are genetically identical. Yet the facial bones of one twin are more robust, those of the other twin, more delicate. Prenatal differences in the womb account for the striking differences in facial contours. (*Dr. Nancy L. Segal, Entwined Lives: Twins and What They Tell Us About Human Behavior, 1999, New York, Dutton*)

Studies of genetically identical twins beautifully illustrate that both genes and environment shape individual differences during development. (The relative contribution from genes and environment varies from trait to trait.) Identical twins occur when a zygote divides into two separate cells, which then develop separately. Identical twins thus have identical genetic information. Yet, they can have different phenotypes owing to environmental influence (Figure 9-4). (Fraternal twins, which develop from two eggs fertilized by two sperm, are no more related than ordinary sisters or brothers.) Most of the differences between identical twins result from differences in their environments, in the womb and outside it as they grow to adulthood.

One of the most difficult tasks in genetics is to distinguish between environmental and genetic influences on phenotype. Genes and environment affect the phenotype separately but also interactively. Genes influence the environment of an individual. To take a common case, people treat tall men differently than short men, on average. This different treatment affects their behavior. Similarly, environment influences the way genes are expressed. A dominant male baboon can inhibit the manufacture of testosterone in subordinate males, altering the way the subordinates look and behave.

Although most geneticists study the inheritance of the genes only, the phenotype of an organism is ultimately more interesting to us, first, because we are fellow organisms, and second, because the phenotype determines survival and reproduction. It is the phenotype that is subject to natural selection. The genotype plays an essential but indirect role. The crucial biological questions—How well does a given individual compete with other organisms? and How much will it contribute to the next generation?—depend on the phenotype.

Phenotype includes all of the physical characteristics of an organism. The word "genotype" is used to mean either all the genes an individual has—the genome—or, alternatively, a particular gene or group of genes that produce a trait such as blue eyes.

The Same Laws of Inheritance Apply to All Sexually Reproducing Organisms

All sexually reproducing eukaryotes—from the most primitive algae to the most complex primates—follow the genetic rules that we describe in this chapter. There are some notable exceptions to these rules, including the inheritance of genes carried in the DNA of mitochondria and chloroplasts, which we will discuss in Chapter 12. These rules also do not apply to bacteria and archaebacteria, which do not reproduce sexually. But the rules of genetics are otherwise nearly universal. The reasons for this unity are:

1. All organisms use DNA as the genetic material.
2. The DNA of all eukaryotic organisms is organized into chromosomes.

A.

B.

Figure 9-5 Asexual reproduction. A. The single-celled protozoan *Paramecium* can reproduce by simply doubling its DNA and dividing into two cells, as in mitosis. B. Many plants also reproduce "vegetatively," making genetically identical "clones" through mitosis. Shown here is a head of garlic. Each clove can take root and form a separate individual that is genetically identical to the parent plant. (A, Biophoto Associates/Photo Researchers, Inc.; B, Kevin Schafer/Peter Arnold, Inc.)

3. Almost all chromosomes exist in pairs at some time during a sexual life cycle.
4. These pairs of chromosomes behave in the same ways during meiosis and at fertilization in all eukaryotes.

HOW DO SEXUALLY REPRODUCING ORGANISMS KEEP THE SAME NUMBER OF CHROMOSOMES FROM GENERATION TO GENERATION?

In Chapter 8, we saw how cells distribute two identical sets of chromosomes to their daughter cells. In this chapter, we will see how sexually reproducing organisms provide single sets of chromosomes to their sperm and egg cells. Later, when a sperm and egg fuse, the resulting individual has a full, double set of chromosomes. This is the essence of sexual reproduction.

For single-celled eukaryotes, such as the common freshwater protozoan *Paramecium*, the simplest way to reproduce is to divide in two through mitosis (Figure 9-5A). Even multicelled eukaryotes can reproduce through mitosis, by splitting off a single cell or a group of cells, which then develops into a whole individual. Plants, for example, often reproduce "vegetatively," by sprouting a new individual from a root or branch (Figure 9-5B). Orchardists can create hundreds of genetically identical trees by taking cuttings of branches from a single individual. Indeed, this method has been in use for thousands of years and is called "cloning" [Greek, *klon* = twig].

But all reproduction through mitosis, called **asexual reproduction,** produces offspring with genes from just one parent. All the offspring have the same genes, those of their one parent. We say the offspring that result from asexual reproduction constitute a **clone,** a set of genetically identical individuals.

In contrast, **sexual reproduction** produces offspring that inherit genetic information from two parents. Sexual repro-

Figure 9-6 Like socks, chromosomes come in matched pairs. These are the 23 pairs of chromosomes of a human. Is it a male or a female? How can you tell? *(Leonard Lessin/Peter Arnold, Inc.)*

duction combines two sets of genes in new ways, so that, in every generation, a new set of unique individuals is created. Sexual reproduction creates enormous diversity. But sexual reproduction creates a problem: How can each new organism receive chromosomes from both parents and still maintain the same total number of chromosomes?

The answer depends on the fact that chromosomes, like socks, come in pairs. Figure 9-6 shows the 46 chromosomes of a human being. Notice that for each chromosome with a characteristic size and banding pattern there is another that looks just like it. So the 46 chromosomes in a human cell are actually 23 *pairs* of chromosomes. Cells that contain paired sets of chromosomes are said to be **diploid** [Greek, *di* = double + *ploion* = vessel]. Pairs of matching chromosomes are called **homologous chromosomes,** and each member of the pair is called a **homolog.** Like the individual socks in a pair, homologous chromosomes are basically alike, but they are not perfect copies. One may have a slight defect at one end; the other may have a different color in one spot near the middle.

Organisms that reproduce sexually ensure that their offspring have the right number of chromosomes by giving the offspring only half of the needed chromosomes. Each parent provides one of the chromosomes in a pair. In humans, for example, each parent contributes 23 chromosomes. This is analogous to receiving for your birthday 23 unmatched socks from your mother and 23 unmatched socks from your father. You put them together and discover that, miraculously, you have 23 matching pairs.

An organism that reproduces sexually passes the half-set of chromosomes to its offspring in specialized reproductive cells called **gametes** [Greek, *gamos* = marriage]. Each gamete is **haploid** [Greek, *haploos* = single], meaning that it carries a single set of chromosomes—in humans, 23. (Sometimes the word "haploid" is confusing, since it sounds like "half." But haploid means single, not half.)

Nearly all sexually reproducing organisms produce two kinds of gametes. One kind, called the **egg,** or **ovum** [Latin, = egg; plural, **ova**], is large and usually cannot move itself. The other, called the **sperm,** or **spermatozoon** [Greek, *sperma* = seed + *zoos* = living], is small and able to move under its own power. Males and females of other species may be completely unlike humans and other mammals. But no matter how small or strange an organism seems to us, if it produces sperm, we say it is male. Likewise, anything that produces eggs is female. (A few odd species produce more than two kinds of gametes—up to 13, any two of which can fuse to form a new individual. We say such organisms have 13 sexes, instead of the usual 2.)

Gametes are the genetic link between generations. But where do these haploid cells come from? The answer is that haploid gametes arise from diploid cells by **meiosis,** a process that allots one haploid set of chromosomes to each of four daughter cells. Gametes and the special cells from which they arise are called **germ cells** or the **germ line** (Figure 9-7A). Gametes can come only from germ cells. In adult animals, the germ cells are found in special gamete-producing organs, called **gonads** [Greek, *gonos* = seed]. The gonads are called **ovaries** in females and **testes** in males.

All of the rest of the cells in the body of a multicelled organism are called **somatic cells** [Greek, *soma* = body]. Mature somatic cells may divide by means of mitosis, but not by means of meiosis. In normal sexual reproduction, genetic instructions pass exclusively through the germ line.

The germ cells of sexually reproducing organisms undergo meiosis to produce haploid gametes, which fuse to form a diploid individual.

The First Cell of the New Generation Has Two Sets of Chromosomes

Every new generation begins with **syngamy,** or **fertilization,** the union of the two haploid gametes (one from each parent) to form a single diploid cell called a **zygote** (Figure 9-7B). In humans, syngamy and fertilization are often called "conception." Many biologists prefer the term "syngamy" to "fertilization," because "fertilization" wrongly suggests that the sperm

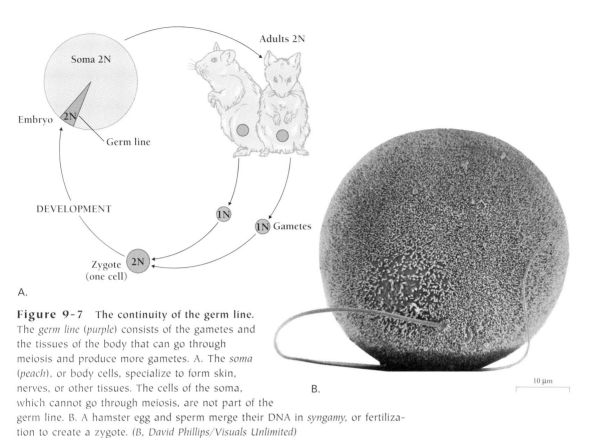

Figure 9-7 The continuity of the germ line. The *germ line* (*purple*) consists of the gametes and the tissues of the body that can go through meiosis and produce more gametes. A. The *soma* (*peach*), or body cells, specialize to form skin, nerves, or other tissues. The cells of the soma, which cannot go through meiosis, are not part of the germ line. B. A hamster egg and sperm merge their DNA in *syngamy*, or fertilization to create a zygote. (B, David Phillips/Visuals Unlimited)

makes the egg fertile. In fact, the egg and sperm are both equally fertile before they fuse to form the zygote. Nonetheless, in this book we will use the older and still commonly used term, "fertilization."

Fertilization depends heavily on adaptations of the gametes. For example, the spermatozoon's small size, whiplike tail, and ATP-packed mitochondria specialize it for rapid movement. The egg cell, in turn, is a virtual supermarket of useful molecules and organelles that are essential for the early development and growth of the embryonic organism. In addition, the egg has special adaptations for moving the sperm nucleus and its own nucleus together, for fusing the two haploid sets of chromosomes into one diploid set, and for dividing rapidly after fertilization.

How Do the Egg and Sperm Find One Another?

Organisms bring the egg and sperm together in many ways. Some aquatic animals produce millions of eggs and sperm, which they "broadcast" into the water. Only a tiny fraction of these eggs and sperm meet and fuse, however. Most of the rest are filtered from the water by filter-feeding animals such as clams and sponges. Many plants broadcast their sperm (but not their eggs) in the form of grains of pollen blown by the wind. Other kinds of plants entice insects, bats, birds, and other animals to carry pollen from plant to plant. Females of many species of fishes and amphibians lay eggs externally while a male sheds his sperm over them.

In many animals, fertilization is "internal," meaning that it takes place inside the female's reproductive system and therefore requires copulation. Mammals, birds, reptiles, insects, and many other animals that practice internal fertilization often rely on elaborate behavioral and structural adaptations to accomplish fertilization (Figure 9-8). Putting the egg and sperm in the same place is only the beginning, of course. Many factors determine whether fertilization occurs, including pH, temperature, and immune responses.

Despite the many different ways that species achieve fertilization, it always serves the same purpose: to combine two haploid sets of chromosomes into one diploid set of chromosomes (Figure 9-9). Meiosis, in turn, always serves to produce the gametes with haploid sets of chromosomes. All sexually reproducing eukaryotes accomplish meiosis in essentially the same way, in an elaborate chromosomal ballet that both resembles and differs from mitosis.

Every species of organism has behavioral and structural adaptations that bring eggs and sperm together. In animals, fertilization may be either external or internal. In all sexually reproducing eukaryotes, meiosis creates haploid gametes, which combine to form diploid individuals at fertilization, or syngamy.

A.

B.

C.

Figure 9-8 Adaptations to internal fertilization. Internal fertilization, or copulation, requires special behavioral and physical adaptations. A. Here a pair of black-tailed rattlesnakes (*Crotalus molossus*) copulate, lying together under a juniper tree in Arizona with their tails crossed (*center of photo*). The male is at the left, and the female, at the right. The most striking physical adaptation to internal fertilization is the penis. Male snakes and lizards have two penises, called "hemipenes," but use only one at a time. In A, the male rattlesnake's left hemipenis is inserted into the female's bulging cloaca. B. A close-up shows the female's cloaca and the base of the male's purple-red left hemipenis. C. The two everted hemipenes of a male rattlesnake under anesthesia. The forking groove along the left organ is a channel through which sperm travels into the female. The circle of downward-pointing spines most likely keeps the two snakes from coming apart during copulation; it's not unusual for the female to begin crawling away, literally dragging the male with her. (*Harry W. Greene/Cornell University*)

▶ **Figure 9-9 Meiosis and fertilization.** In mature organisms, meiosis produces haploid gametes, which have half-sets of chromosomes. At fertilization the two haploid sets of chromosomes, from the egg and the sperm, combine into one diploid set of chromosomes in a zygote. (*Photo, Rob Lang*)

HOW DOES MEIOSIS DISTRIBUTE CHROMOSOMES TO THE GAMETES?

Like mitosis, meiosis is a continuous process. Biologists divide meiosis into separate stages for convenience only. Meiosis consists of two cell divisions, called **meiosis I** and **meiosis II**. These divisions occur only in cells of the germ line during the production of gametes. Biologists divide each of these two meiotic divisions into four phases (Figure 9-10). These are prophase, metaphase, anaphase, and telophase.

In general outline, each phase of meiosis resembles the corresponding stage of mitosis. During each **prophase**, chromosomes condense, and the nuclear membrane breaks down. During each **metaphase**, the chromosomes move to the equator of the spindle apparatus. During each **anaphase**, spindle fibers move the chromosomes toward the poles. And during each **telophase**, the nuclear membrane reforms.

Despite the resemblance to the stages of mitosis, the end result of meiosis is crucially different. In mitosis, each and every division produces two diploid daughter cells. In meiosis, by contrast, DNA replication occurs before the first division (meiosis I), but *not* before the second division (meiosis II). As a result, the two divisions of meiosis produce four haploid daughter cells.

In mitosis, a cell replicates its DNA once and divides once, resulting in two diploid daughter cells. In meiosis, a cell replicates its DNA once but divides twice, resulting in four haploid daughter cells.

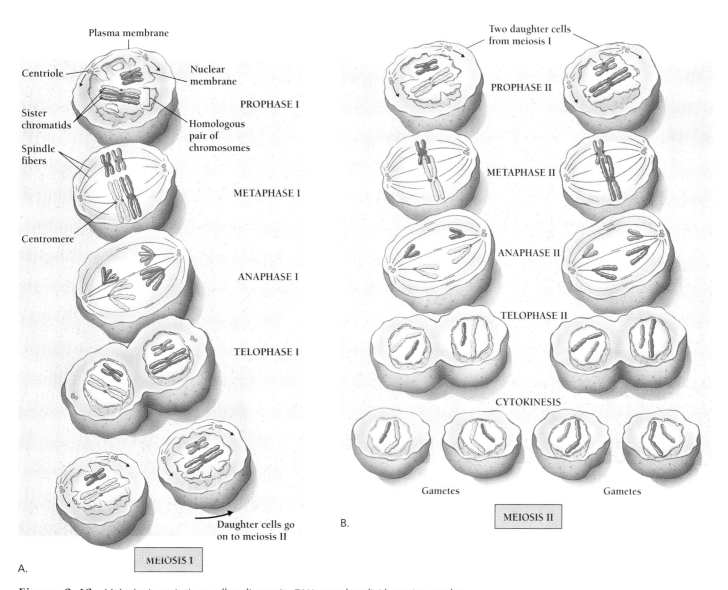

Figure 9-10 Meiosis. In meiosis, a cell replicates its DNA once but divides twice, producing four haploid daughter cells. Each of the four daughter cells (*bottom, right*) receives two of the eight homologous chromosomes in the original cell (*top, left*)—one from each tetrad. In the first division, meiosis I, one chromosome from each homologous pair goes to each daughter cell. Meiosis II, the second division, resembles mitosis, but in a haploid cell.

 Quiz

 Quiz

During Prophase I, Homologous Chromosomes Form Pairs

During interphase, before prophase I, the cell copies, or replicates, the chromosomes. As in mitotic prophase, the (doubled) chromosomes condense, the nucleoli and the nuclear membrane disappear, and the spindle apparatus begins to form. But meiotic prophase differs from mitotic prophase in one essential way: after the chromosomes begin to condense, the homologous pairs come together in the process of **synapsis** [Greek, = union], in which homologous chromosomes align exactly. Each chromosome itself consists of two sister **chromatids,** so that the four chromatids form a **tetrad** (Figure 9-10). Synapsis resembles the process of pairing socks before putting them away. From a disorganized pile of 46 socks, we can put together 23 pairs.

Synapsis is the central event of meiosis. It is the process by which homologous chromosomes assort into two haploid sets. Following synapsis, the homologous chromosomes go their separate ways, while the chromatids of each chromosome stay together. (In both mitosis and in meiosis II, it is the *sister chromatids* that separate, not the homologous chromosomes.)

Another important event that occurs during synapsis is **crossing over,** in which homologous chromosomes break and exchange equivalent pieces. Crossing over results in new

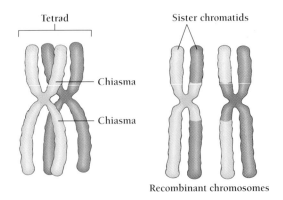

Figure 9-11 Crossing over. When homologous chromatids pair up during synapsis (prophase I), they may break at special places called chiasma and exchange equivalent pieces. The result is new combinations of genes from the two parents. These chromatids are exchanging two pieces.

combinations of genes from the two parents. The visible result of crossing over is a crosslike configuration called a **chiasma** [Greek, = cross; plural, **chiasmata**] (Figure 9-11). At each chiasma, homologous chromosomes may exchange chromatid segments. Crossing over is an important way that sexually reproducing organisms recombine genes and increase variation. In humans, each pair of homologous chromosomes exchanges an average of two or three segments. We will discuss such recombination in detail later.

Prophase I is the longest stage of meiosis, taking up more than 90 percent of the total time. One reason that prophase I takes so long is that during this time, the egg cell builds up materials and machinery to be passed on to the gametes. In the females of some species, meiosis may seem to stop for months or even years during prophase I. In frogs, for example, prophase I may last for several years. In humans, meiosis begins before birth, pauses in prophase I, and does not resume until after puberty. Starting at around age 12 years, one or two eggs complete meiosis each month. Since female humans may continue to ovulate into their 50s, some individual cells may not resume meiosis for more than 50 years.

During prophase I, a cell accumulates materials for its daughter cells and sorts its doubled chromosomes in the process called synapsis. In addition, homologous chromosomes cross over.

During the Rest of Meiosis I, One Chromosome from Each Homologous Pair Goes to Each Daughter Cell

After prophase I comes metaphase I, in which spindle fibers pull each tetrad to the equator (Figure 9-12). Then, during anaphase I, something dramatic happens that is very different from what occurs in mitosis: the sister chromatids go to the *same* pole. (In contrast, in mitosis, the two chromatids go to *opposite* poles.) Telophase I and cytokinesis rapidly follow. Another critical difference is that the daughter cells of meiosis I move right into meiosis II without any interphase and without any further DNA synthesis. The chromosomes may even remain partly condensed.

An important question is how the tetrads sort out during metaphase I. Do all the chromosomes from the mother go to one pole, and all the chromosomes from the father to the other? If so, meiosis would recreate the same sets of chromosome generation after generation. Instead, the tetrads sort themselves randomly. One pair of homologs aligns one way, another pair aligns another way. Biologists say that each pair aligns "independently" of all the other pairs.

Because chromosome pairs sort themselves independently, meiosis is a major source of genetic diversity. Sutton showed that just two pairs of chromosomes can form 4 (2^2) kinds of gametes (Figure 9-13). Similarly, three pairs of chromosomes can form 8 (2^3) kinds of gametes. A single human germ cell, with 23 pairs of chromosomes, can therefore form 2^{23} (about 8 million) kinds of gametes, each one unique. Together, a woman and a man could produce more than 64 trillion genetically unique offspring (8 million kinds of eggs × 8 million kinds of sperm). In addition, crossing over increases the actual diversity of possible gametes and zygotes far beyond 64 trillion. All this diversity comes

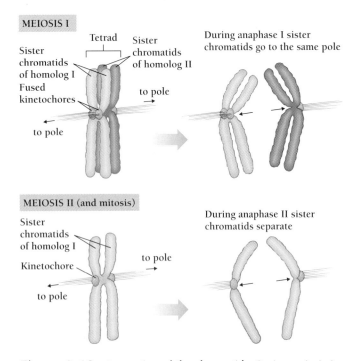

Figure 9-12 Separation of the chromatids. During meiosis I, sister chromatids go to the same pole. During meiosis II, sister chromatids go to opposite poles (as in mitosis).

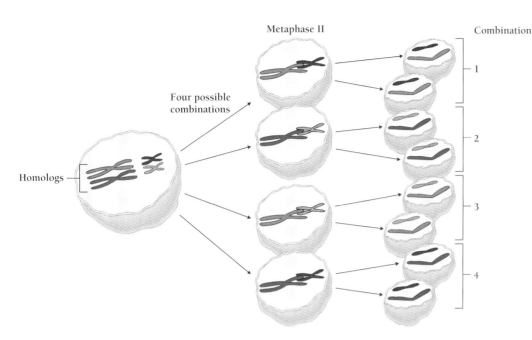

Figure 9-13 Just two pairs of chromosomes can form four kinds of gametes. If the chromosomes arrange themselves as in the first cell in this figure, the resulting gametes have chromosome combinations 1 and 2. If, in other germ cells, the big chromosomes on the left swap places (which occurs at random), then the resulting gametes have combinations 3 and 4.

from (1) crossing over and (2) the random alignment and subsequent separation of tetrads in metaphase I. Given this random recombination, it is no wonder that all sexually reproducing species (including our own) are so wonderfully diverse.

During anaphase I of meiosis, sister chromatids go to the same pole in the cell, but homologous chromosomes go to opposite poles. Telophase I and cytokinesis I resemble their counterparts in mitosis.

Meiosis II Distributes Sister Chromatids to Daughter Cells

Meiosis II looks just like mitosis would look in a haploid cell. At the end of meiosis I, the cell is haploid. After it divides again in meiosis II, it is still haploid. Meiosis II simply separates the sister chromatids, so that each chromosome of the daughter cell contains a single chromatid.

Prophase II (unlike prophase I) is brief, since the chromosomes are still mostly condensed from meiosis I. If a nuclear membrane has reappeared during telophase I, it breaks down during prophase II. The chromosomes attach to newly assembled spindle fibers. Then, during metaphase II, the chromosomes line up across the equator of the spindle apparatus, with each centromere connected to spindle fibers from both poles. During anaphase II, the chromosomes split—one chromatid moves to each pole. In telophase II, the nuclear membrane forms once more. The daughter cells each have a complete haploid set of chromosomes, each consisting of a single chromatid.

Meiosis II resembles mitosis except that each resulting cell is haploid, rather than diploid.

Disjunction and Nondisjunction

The separation of homologous chromosomes in anaphase I and of sister chromatids in anaphase II is called **disjunction.** Occasionally, separation fails to proceed normally, an event called **nondisjunction.** This results in a gamete having too many or too few chromosomes. In humans, for example, a sperm or egg might have 22 or 24 chromosomes instead of 23. After fertilization, the resulting zygote likewise has the wrong number of chromosomes—45 or 47. A cell or individual with the correct number of chromosomes is said to be **euploid** [Greek, *eu* = good, true + *ploion* = vessel], while one with an abnormal number of chromosomes is said to be **aneuploid** [Greek, *an* = not + euploid].

In humans, one of the most common results of nondisjunction (and therefore aneuploidy) is **Down syndrome,** a disorder that leads to mental retardation and the abnormal development of the face, heart, and other parts of the body (Figure 9-14A). Down syndrome is almost always associated with a chromosomal abnormality called **trisomy 21** [Greek, *tri* = three + *soma* = body], the presence of three, rather than two, copies of chromosome 21 (Figure 9-14B).

A. B. 3 copies of chromosome 21

Figure 9-14 Children with Down syndrome have distinctive characteristics. These include mental retardation, heart defects, a flat facial profile, excess skin on the neck, double-jointedness, an unusual crease in the palm of each hand. B. Down syndrome usually results from the presence of three copies of chromosome 21, a condition called "trisomy 21." (A, Hattie Young/Science Photo Library/Photo Researchers, Inc.; B, Kunkel/Phototake NYC)

Trisomy 21 usually results from nondisjunction during meiosis I. If, during anaphase I, the two chromosomes 21 fail to separate, then two of the four possible gametes produced during meiosis will each have two copies instead of one, and the other two will have no copies of chromosome 21 (Figure 9-15). When these gametes unite with normal gametes at fertilization, they produce abnormal zygotes—some with three copies of chromosome 21, and some with only one copy (Box 9-1).

If, during anaphase, homologous chromosomes fail to separate, the gametes will have too many or too few chromosomes.

WHY SEX?

We saw earlier that many organisms can reproduce either sexually or asexually. For example, plants can reproduce sexually, as when pollen finds its way to a flower, or asexually, as when the roots of an aspen tree sprout new trees. Even some animals, such as the many-armed hydra, can reproduce asexually, by budding.

Asexual reproduction is the quickest and easiest way to reproduce. Sexual reproduction is enormously expensive. To reproduce sexually, organisms must make hundreds, thousands, or even millions of gametes to ensure that a few meet for fertilization. A female codfish produces some 9 million eggs per year, of which only a small fraction are fertilized. A typical male human may produce 20 to 30 billion sperm over the course of the weeks or months it takes him to successfully fertilize a single egg. Even a female human is born with 2 million egg cells (oocytes). In addition, humans and other animals invest huge amounts of energy into finding, courting, and mating with other individuals. Many plants construct elaborate and showy

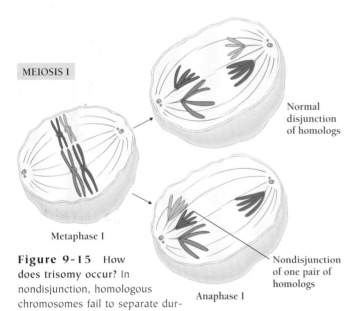

Figure 9-15 How does trisomy occur? In nondisjunction, homologous chromosomes fail to separate during anaphase. The resulting gametes have either two copies of a chromosome or none, but not the usual one. Later, when the gamete combines with another, normal gamete, the resulting zygote has one homolog or three, but not two.

BOX 9.1

What Happens When Meiosis Goes Wrong?

Among human embryos with Down syndrome, or trisomy 21, about 80 percent die before birth in **spontaneous abortions,** natural abortions that occur because the embryo cannot live. Only about 20 percent of trisomy 21 embryos survive until birth, although those that survive may live for many years. For comparison, about 70 percent of all human embryos normally survive to birth. Although the death rate among embryos with trisomy 21 is high, the death rate is even higher—almost 100 percent—among those with only one copy of chromosome 21.

About 1 baby in every 700 has Down syndrome. Although the extra chromosome usually comes from the mother, about one-fifth of the time, the extra chromosome 21 comes from the father. Older women, whose eggs are older, are much more likely than younger women to produce eggs with an extra chromosome 21.

As a rule, physicians test the embryos of women older than 35 years for Down syndrome, but there is nothing magic about age 35. The likelihood of giving birth to a child with Down syndrome increases gradually (Table A).

Cell biologists believe that the chance of nondisjunction (at least for chromosome 21) increases with maternal age because of the pattern of meiosis in humans. The cells that form human eggs begin meiosis while the mother is herself still an embryo in the womb. But meiosis stops in the middle of prophase I, well before birth, and does not resume until after the egg is released from the ovaries, anywhere from 13 to 55 years later. Nondisjunction of chromosome 21 may be the result of interrupting the complicated process of meiosis for so long a time.

TABLE A   RISK OF TRISOMY 21

Age of mother (y)	Risk of giving birth to child with trisomy 21
Under 25	1/1400
Under 30	1/1000
At age 35	1/350
Over 40	1/100

---

flowers stocked with sweet nectar, adaptations for attracting insects and other animals that carry pollen from one plant to another. What makes sexual reproduction worth a great expense?

One answer is genetic variation in one's offspring. We have seen that, because of crossing over and because of the independent assortment of chromosomes during metaphase I, a single pair of humans can produce a seemingly infinite number of unique zygotes. In contrast, asexual reproduction produces offspring that are genetically identical to the parent.

What Good Is Genetic Variation?

In an unchanging environment, a single, perfectly adapted genotype might persist unchanged for millions of years. Most environments, however, change constantly. A polar bear with short hair might survive fine in a period of global warming. But as temperatures drop, its siblings with longer fur do better. Parents that produce a diversity of offspring are more likely to have some of them survive and reproduce than parents that produce offspring that are all identical. Indeed, evolutionary biologists generally find that the less predictable an environment is, the more likely that the organisms living in it will have evolved methods for increasing diversity.

Some species actually change their mode of reproduction depending on how predictable their environment is. For example, a female water flea reproduces asexually when food and space are plentiful, giving birth to huge numbers of genetically identical offspring, but reproduces sexually when food is limited (Figure 9-16). The resulting diversity increases the chance that at least some individuals will survive.

▶ **Figure 9-16** Where's Dad? When food is abundant, the water flea, *Daphnia*, reproduces asexually. Each female is born pregnant with more female water fleas, each one a clone of its mother. A drop in water temperature or a decrease in food supply results in the birth of males and sexual reproduction. *(Oxford Scientific Films/Animals Animals)*

Male   Female

Figure 9-17 Sex chromosomes in humans. In certain organisms, one sex has a pair of unmatched sex chromosomes. For example, a female human has 23 pairs of matched chromosomes, including two X chromosomes, which are called sex chromosomes. A male human has 22 pairs of matched chromosomes, plus two sex chromosomes called X and Y. *(Left, Don Kelly/Grant Heilman Photography; right, Photo Researchers, Inc.)*

In humans and other mammals, the X chromosome in a male is the same as the X chromosome in a female. The Y chromosome, however, is much smaller and bears far fewer genes than the X chromosome. In both females and males, the two sex chromosomes (XX in females and XY in males) pair and segregate during meiosis in the same way as homologous pairs of autosomes.

How can X and Y pair off during synapsis if they differ so much in size and form? The answer is that they share a region that is homologous. Despite their striking differences in size and form, then, they still behave as homologs during meiosis.

Sex chromosomes are chromosomes involved in sex determination. In mammals, the female has two homologous X chromosomes, and the male has one X chromosome and one Y chromosome, which are not homologous.

Many biologists think that the need for genetic variation alone is not a good enough reason for sex. After all, many organisms, including all bacteria, reproduce asexually. Bacteria have survived quite well for billions of years. In Chapter 43, we discuss one alternative hypothesis for why so many eukaryotes reproduce sexually. The reasons for sex remain controversial.

Sexual reproduction is energy expensive, and biologists wonder why organisms bother with it. One reason may be that sex increases genetic diversity, and genetic diversity increases the chances that some individuals will survive, no matter how the environment changes.

What Are Sex Chromosomes?

In mammals, birds, and some other organisms, either the male or the female may have a pair of unmatched chromosomes. In mammals (including humans), for example, the female has two homologous "X" chromosomes, while the male has one "X" chromosome and one "Y" chromosome (Figure 9-17). In birds, the males have two matched sex chromosomes ("WW"), and the females have one "W" chromosome and one "Z." In many insects, including the lubber grasshoppers that Sutton collected, the female has two homologous X chromosomes and the male has only one X chromosome. The **X** and **Y** chromosomes (as well as the W and Z chromosomes) are called **sex chromosomes.** The rest of the chromosomes are called the **autosomes.** A human cell has 22 pairs of autosomes and 1 pair of sex chromosomes.

Protists, fungi, plants, and many animals lack sex chromosomes. In most organisms, males and females both have complete sets of homologous (matching) chromosomes. In these organisms, what sex an individual becomes depends on different factors, including, for example, plant hormones in plants or temperature during development in some animals.

HOW DO THE NUMBER AND MOVEMENTS OF CHROMOSOMES EXPLAIN THE INHERITANCE OF GENES?

If a cell has two copies of each chromosome, then the cell must also have two copies of every gene. This is the key to understanding genetics in higher organisms.

The copy of a gene on one chromosome may differ slightly from its counterpart on the other chromosome. Alternate versions of the same gene are called **alleles.** For example, in humans, both chromosomes in a pair may have a certain gene that influences eye color, but one chromosome might carry an allele for brown eyes, while the other chromosome carries an allele for blue eyes.

Most genes are simply instructions for the production of a particular protein. Alleles are variations in those instructions. Just as your aunt June makes chocolate brownies with extra chocolate and your cousin Ralph makes them with a teaspoon of vanilla, different alleles specify slightly different proteins. Sometimes the differences are subtle and the resulting protein is the same either way; sometimes an allele specifies a protein so different from normal that the protein cannot function at all.

For a given gene, an individual can carry two identical alleles or two different alleles. In snapdragon flowers, for example, one allele of a particular gene specifies white flowers and the other allele specifies red flowers. If the flower has two red-flower alleles, the flowers are red. If the snapdragon has two white-flower alleles, the flowers are white. On the other hand, if the snapdragon has one red-flower allele and one white-flower allele, the flowers are pink (Figure 9-3). If an individual has two copies of the same allele, we say it is **homozygous** for that allele. If an individual has two different alleles of a gene, we say it is **heterozygous.**

Biologists name the alleles of a gene with letters or short abbreviations. In snapdragons, for example, we can call the red-flower allele R_1 and the white-flower allele R_2. We can imagine that the R_1 allele allows the plant to make a red protein pigment, while the R_2 allele allows the plant to make no functioning pigment at all. The genotype of the homozygous red snapdragon would be "R_1R_1," and that of the homozygous white snapdragon would be "R_2R_2." The genotype of the pink heterozygote would be "R_1R_2."

Plant breeders knew about homozygous plants for hundreds of years before the first geneticists. But, knowing nothing about genes, plant breeders called plants that were homozygous for flower color "true breeding" for flower color. In **true-breeding** plants, all the offspring of the red-flowered variety produce red flowers only, generation after generation; and all the offspring of the white-flowered variety produce white flowers only, generation after generation. A line of plants (or other organisms) can be true breeding for other traits besides flower color. In all true-breeding varieties, both chromosomes carry the same allele for a certain trait. (We say they are homozygous.)

Snapdragons normally "self-fertilize." Each snapdragon flower produces both pollen (which produces sperm) and egg cells. The sperm from the pollen fertilizes eggs within the same flower. So, on a red snapdragon, gametes carrying red-flower alleles always pair up with gametes carrying red-flower alleles. The same is true for white-flowered snapdragons.

A plant breeder can cross a true-breeding red-flowered plant with a true-breeding white-flowered plant only by preventing self-fertilization. To cross the two homozygous plants, the breeder must prevent each plant from self-pollinating. This is done by snipping off the pollen-carrying structures and then artificially pollinating the flower with pollen from another plant. This method of breeding two genetically distinct organisms is called **cross breeding** (or **crossing**). The offspring of such a cross are called **hybrids.** In snapdragons, crossing a homozygous red snapdragon with a homozygous white snapdragon results in a heterozygous hybrid that has pink flowers (Figure 9-3).

Geneticists call the original parents in such a cross the **parental,** or **P,** generation, and the offspring the "first filial" [Latin, *filia* = daughter, *filius* = son], or **F1**, generation. The F2 generation are the offspring of the F1 generation. In other words, the F2 generation are the grandchildren of the P generation. In the snapdragon cross, then, the parents are R_1R_1 and R_2R_2 homozygotes, and the F1 generation are R_1R_2 heterozygous hybrids. The genotypes of the heterozygous hybrids are all R_1R_2, and their phenotypes are pink flowers.

Eukaryotes that have two sets of chromosomes (diploid) have two copies of each gene. An individual may have two identical alleles, or two different alleles.

What Are the Genotypes and Phenotypes of the F2 Generation?

Both a plant breeder and a geneticist would want to know what color flowers result from crossing two hybrid snapdragons. We

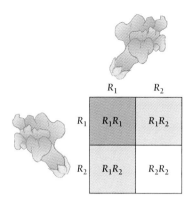

Figure 9-18 Inheritance of flower color in snapdragons. A cross between two R_1R_2 heterozygotes, shown here in a Punnett square, produces three genotypes: snapdragons homozygous for R_1, snapdragons homozygous for R_2, and R_1R_2 heterozygous snapdragons.

can easily figure this out before we even cross the plants. We know that meiosis produces haploid gametes with just one of each chromosome. The F1 hybrids each have one R_1 allele and one R_2 allele, each on a separate chromosome. So each snapdragon gamete receives either R_1 or R_2. Half of all the gametes get the R_1 allele, and half get the R_2 allele. An R_1 can pair up either with another R_1 or with an R_2, and an R_2 can pair up with either another R_2 or with an R_1. So the F2 generation can have R_1R_1 (red flowers), R_2R_2 (white flowers), or R_1R_2 (pink flowers). But how many of each kind will we get?

A convenient way to find out is to use a checkerboard of numbers invented by an early 20th-century British geneticist named Reginald Punnett (Figure 9-18). In a **Punnett square,** we show each kind of gamete made by one parent along the top of the square and each kind of gamete made by the other parent along the left side of the square. Within the small squares, we write the genotypes that would be produced by each combination. Usually, the R_1R_2 genotype is the same as the R_2R_1 genotype. The flowers are pink no matter which allele comes from which parent. But we end up with twice as many pink flowers as white or red.

The Punnett square shows that the three possible genotypes of the F2 generation of our snapdragon experiment should be in the ratio 1:2:1. If we actually cross the flowers, we can directly count the number of red, white, and pink flowers. Because the pink phenotype of the heterozygote is different from the phenotypes of either homozygote (red or white), we can directly count the ratios of the genotypes in the F2 generation. Experiments of this kind always produce a ratio fairly close to the expected 1:2:1.

If we cross two R_1R_2 heterozygotes, one-quarter of the offspring are homozygous for R_1, one-quarter are homozygous for gene R_2, and one-half are heterozygous R_1R_2.

Figure 9-19 Gregor Mendel in his monastery garden. (*The Bettmann Archive*)

HOW DID GREGOR MENDEL DEMONSTRATE THE PRINCIPLES OF GENETICS?

Today, the inheritance of flower color in snapdragons makes sense to us because we know about chromosomes and how they assort during meiosis. But 19th-century biologists knew little about chromosomes and nothing about genes. For them, inheritance was mystifying. Our knowledge of the behavior of chromosomes and genes gives us a great advantage.

The first person to penetrate the mystery of the ratios of different phenotypes in genetic crosses was Gregor Mendel (Figure 9-19). In his monastery's garden, in what is now the Czech Republic, Mendel studied seven different pairs of traits in pea plants (Figure 9-20): round versus wrinkled seeds, yellow versus green seeds, purple versus white flowers, inflated versus constricted pods, green versus yellow pods, axial versus terminal flowers, and tall versus dwarf stems. Each of these pairs of traits showed the same pattern of inheritance. Mendel chose pairs of traits in which each variant was distinct (in other words, tall could not be confused with short).

In each case, Mendel showed that the parental stocks were true breeding. For example, self-pollination of plants with round seeds always produced offspring with round seeds.

Mendel then cross-pollinated flowers with different traits, a purple-flowered pea with a white-flowered pea, for example (Figure 9-21).

But F1 hybrids do not always have an intermediate phenotype like the pink flowers of hybrid snapdragons. Instead, a heterozygous hybrid may have the same phenotype as one of the parents. For example, when Mendel crossed true-breeding, purple-flowered peas with true-breeding, white-flowered peas, all the hybrids had purple flowers (Figure 9-22A). The F1 hybrids are all heterozygotes, but only the allele for purple flowers is expressed. We say that an allele is **dominant** when it alone determines the phenotype of a heterozygote, as in the case of the purple allele in peas. An allele is **recessive** when it contributes nothing to the phenotype of a heterozygote, as is the case of the white allele of peas. (The names of dominant alleles begin with a capital letter, and those of recessive alleles with a small letter.)

Today we know that a dominant allele usually specifies a functional protein, and a recessive allele usually specifies a protein that doesn't function at all. Recessive alleles that are harmless, such as the one that causes type O blood, may be extremely common. Others, such as the allele for the genetic disease Tay-Sachs, which kills children at 3 or 4 years of age, are quite rare. Lethal recessives such as Tay-Sachs are rare because people who are homozygous for this allele never pass the gene on to any offspring.

When two alleles each contribute to the phenotype, as in the case of the R_1 and R_2 alleles of snapdragons (which together gave a color different from either homozygote), the alleles are said to lack dominance. In the snapdragons, for example, both alleles are expressed and we say they each have **partial dominance** or "codominance." But blending inheritance cannot explain the reappearance of red and white snapdragons after crossing the hybrids.

In each of Mendel's seven crosses, all of the F1 hybrids had the same phenotype. Round pea stocks crossed with wrinkled pea stocks gave F1 hybrids with round peas, tall pea stocks crossed with short pea stocks gave tall F1 hybrids, and so on. We say that the purple flower allele is dominant to the white flower allele, the round allele is dominant to the wrinkled allele, and the tall allele is dominant to the short allele.

Mendel began his experiments with true-breeding (homozygous) pea plant stocks. He chose pairs of traits in which the two alternate forms were distinct. Each pair of traits was determined by just two alleles, and each allele in a pair was either dominant or recessive.

What Did Mendel's F1 Crosses Show?

Mendel allowed the F1 hybrids to self-pollinate, then counted the F2 offspring, or *progeny*. In every case, the dominant allele phenotype made up about three-fourths of the progeny, and the recessive allele phenotype made up the remaining one-fourth of the progeny.

MendelWeb

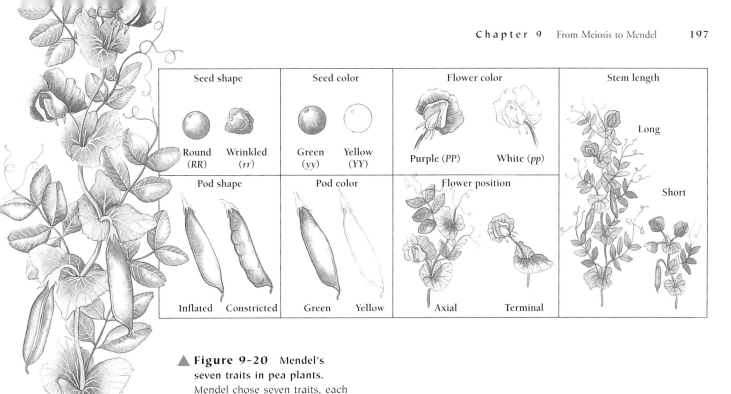

Figure 9-20 Mendel's seven traits in pea plants. Mendel chose seven traits, each with two forms that corresponded to two alternate alleles of the gene.

This is the same result we got with the snapdragons, except that the heterozygotes (which make up half of the offspring) look the same as the dominant homozygotes (which make up a quarter of the offspring). This was because, in each of Mendel's crosses, one allele was dominant and one was recessive.

In the case of the round versus wrinkled cross, for example, $1/4$ were homozygous for the round allele, $1/4$ were homozygous for the wrinkled allele, and $1/2$ were heterozygous. Since the round allele is dominant, the heterozygotes had the same phenotype as the homozygote for the round allele. So $3/4$ ($1/4 + 1/2$) were round, and $1/4$ were wrinkled.

The Principle of Segregation

Mendel had to make sense of his results without the knowledge of genes and chromosomes that we now have. Amazingly, he was able to do it just by studying the ratios of phenotypes in his breeding experiments.

Mendel's first conclusion was the **principle of segregation:** *Each sexually reproducing organism has two genes for each characteristic; these two genes segregate (or separate) during the production of gametes.* (Of course, Mendel didn't use the word "gene," which is a modern term. He called genes "determinants." But his meaning was the same.)

To test the principle of segregation, Mendel performed a different kind of cross, called a **backcross.** Instead of allowing the F1 heterozygotes to self-pollinate, he crossed them with

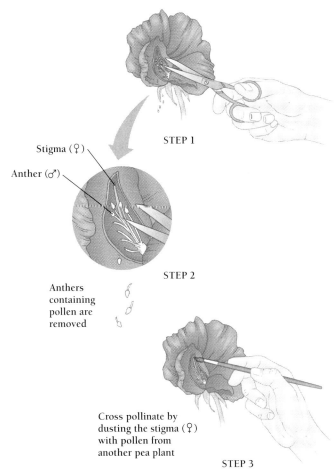

Figure 9-21 Cross-pollinating peas. Pea flowers carry both male and female parts, which allows them to self-pollinate. But Mendel's studies of inheritance required him to prevent the peas from self-pollinating so he could cross different lines. He snipped the male, pollen-producing anthers from each pea flower and then sprinkled the female stigma with pollen from another flower.

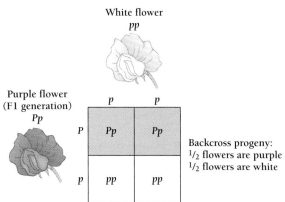

Figure 9-22 A backcross of purple- and white-flowered peas. A. First, a homozygous white flower is crossed with a homozygous purple flower to produce heterozygotes. B. Then, in the backcross, a white homozygote is crossed with a purple heterozygote. Half the offspring are purple heterozygotes and half are white homozygotes. (By a convention begun by Mendel, we begin the names of dominant alleles with a capital letter and those of recessive alleles with a small letter.)

the homozygous recessive parental (P) stock—those which contained only the recessive allele (Figure 9-22B). By crossing the purple F1 plants to a white homozygous recessive, Mendel could find out how many of the F1 plants had recessive alleles and how many didn't. Any cross with a homozygous recessive that is supposed to reveal the genotype of the other parent is called a **testcross.**

In Mendel's testcross, the parents looked just like those in the original parental cross—a purple-flowered plant with a white-flowered plant (Figure 9-22A). But Mendel had discovered that the purple-flowered F1 hybrids could produce two kinds of gametes, one with the purple allele, and the other with the white allele (whereas the original purple parents produced only gametes with purple alleles). The plants with white flowers, on the other hand, would only produce gametes with the white allele. He correctly predicted that half the progeny would have purple flowers and half, white flowers.

Mendel's hybrid crosses and backcrosses produced offspring with ratios of phenotypes that suggested the principle of segregation; each sexually reproducing organism has two genes for each characteristic, which segregate during the production of gametes.

The Principle of Independent Assortment

Mendel next wanted to know if different traits were inherited together. In other words, were traits "linked?" To find out, he took two true-breeding stocks that differed in both the color and the form of the peas they produced. One parent produced round, yellow peas; the other, wrinkled, green peas. From previous crosses, he knew that round was dominant to wrinkled and yellow was dominant to green. As he expected, all the F1 peas were yellow and round. He then determined how these traits were inherited in the F2 generation. The allele specifying round peas is *R*, and the one for wrinkled peas is *r*; the allele specifying yellow peas is *Y*, and the one for green peas is *y* (Figure 9-23). The genotypes of the parents are therefore *RRYY* and *rryy*.

Homozygous parents can produce only a single kind of gamete: *RRYY* plants produce only *RY* gametes, and *rryy* plants produce only *ry* gametes. The genotype of the F1 plants therefore must be *RrYy*.

But what happens in the F2 generation? Mendel allowed the F1 to self-pollinate and then counted 556 peas in the F2. Of these, 315 were yellow and round (representing the action of the dominant alleles), and only 32 were green and wrinkled (Figure 9-23). The rest represented new phenotypes, unlike either of the parental stocks: 101 were yellow and wrinkled, and 108 were green and round. The two crosses had produced new combinations of genes, illustrating again how sexual reproduction provides diversity in a population. But how did Mendel make sense of the numbers?

First, Mendel noted that the traits associated with each gene were in a ratio of about 3:1, just as in his experiments with only one trait. In the F2 generation there were three times as many yellow as green peas (416:140 = 3:1), and three

TWO-FACTOR CROSS

Figure 9-23 Testing two traits at once. In this two-factor cross, Mendel crossed round, yellow peas with wrinkled, green peas. In the first (F1) generation, all the peas were round, yellow heterozygotes. When he crossed the F1 generation, he got an assortment of different peas.

times as many round as wrinkled (423:133 = 3:2:1). Mendel immediately saw that the two traits were behaving independently; whether a pea was yellow or green had no effect on whether it was wrinkled or round.

Mendel could explain the data if he assumed that each F1 plant produces four kinds of gametes: *ry*, *Ry*, *rY*, and *RY*. The Punnett square of Figure 9-23 predicts the distribution of genotypes and phenotypes of the F2 generation of this cross. It shows that 9 out of the 16 combinations of gametes produce plants with round, yellow peas, but only one combination produces plants with wrinkled, green peas. The overall ratio of phenotypes predicted by this analysis is 9:3:3:1.

TWO-FACTOR BACKCROSS

Figure 9-24 Independent assortment. By backcrossing the heterozygous F1 peas with the homozygous recessives (green, wrinkled), Mendel was able to show that each gene was distributed independently of the others.

The correspondence of Mendel's experiments to this prediction is called the **principle of independent assortment**. This principle states that each pair of genes is distributed independently from every other pair during the formation of the gametes. The independent assortment of homologous chromosomes during meiosis causes the independent assortment of genes that are on different chromosomes.

Mendel tested his understanding of the genotypes of the gametes produced by the heterozygous F1 hybrids by doing another testcross with plants that produce green, wrinkled peas (Figure 9-24). Because both *r* (for wrinkled peas) and *y* (for green peas) are recessive, the genotype of the plants used in the backcross must be *rryy*. The gametes produced by these plants all must be *ry*.

If the genotype of the F1 plants is really *RrYy* and the chromosomes bearing the two genes assort independently, then the four kinds of gametes (*RY*, *rY*, *Ry*, and *ry*) should be made in equal numbers (Figure 9-24). When these gametes form zygotes with *ry* gametes, we expect equal numbers of the four phenotypes, in contrast to the 9:3:3:1 ratio expected in the case of self-fertilization of the F1 plants. Again, Mendel obtained the expected ratios, confirming his understanding of the process.

Mendel's principle of independent assortment states that each gene in a pair is distributed independently during the formation of the gametes.

WHAT WAS THE EVIDENCE FOR THE CHROMOSOMAL THEORY OF INHERITANCE?

The rediscovery of Mendel's laws in 1900 hit British biologist William Bateson like a bolt of lightning. Within hours of reading the news, Bateson began devoting himself to spreading the word. As we saw earlier in this chapter, Bateson's talk at Columbia inspired Walter Sutton to quickly publish his own theory of inheritance.

In 1902, Walter Sutton and the German biologist Theodor Boveri independently published papers that showed that the behavior of chromosomes during meiosis perfectly explained Mendel's principles of segregation and independent assortment. Both biologists concluded that the genes had to be on the chromosomes.

What Were Sutton's Arguments?

Sutton's argument consisted of six separate points that together formed an almost fortresslike argument for what came to be known as the **Sutton-Boveri chromosomal theory of inheritance.** Sutton's first three points merely confirm the ideas of others; the second three are his own new ideas.

First, said Sutton, chromosomes come in pairs, one of which comes from the mother and one from the father. Second, synapsis consists of the pairing of homologous maternal and paternal chromosomes. Because the chromosomes of the lubber grasshopper had distinctive sizes and shapes, Sutton was able to see that the chromosomes occurred in pairs and that they separate during meiosis.

Third, the chromosomes retain their distinctive shapes throughout the cell cycle. Sutton argued that the constant differences in size and shape of the chromosomes suggested that the chromosomes are not recreated every cell cycle. The best evidence for this idea came from Boveri, who had studied sea urchin cells with reduced numbers of chromosomes. Only cells with the correct number of chromosomes developed normally. Boveri concluded that since every chromosome had to be present for normal development, each chromosome must be unique and must make a unique contribution to development.

Fourth, said Sutton, the chromosomes contain Mendel's genes. Fifth, meiosis creates new combinations of genes in each generation. During meiosis, when the diploid cell is reduced to two haploid cells, each homolog in a pair orients randomly with respect to the two poles of the dividing cell.

TABLE 9-1 WALTER SUTTON'S SIX RULES OF INHERITANCE

1. Chromosomes come in pairs. One chromosome comes from the mother and one from the father.
2. Synapsis is the pairing of homologous maternal and paternal chromosomes. These pairs separate during meiosis.
3. The chromosomes are not broken down and created anew during each cell cycle.
4. The chromosomes carry Mendel's genes.
5. Meiosis creates new combinations of genes in each generation. During meiosis, either homolog may end up in either new cell, regardless of the way all the other homologous chromosomes divide. This accounts for Mendel's principle of segregation.
6. Each chromosome carries a different set of genes, and all of the genes on one chromosome are inherited together. That different chromosomes are inherited independently accounts for Mendel's principle of independent assortment.

As a result, either homolog may end up in either new cell, regardless of the way all the other chromosomes divide. This, said Sutton, accounted for Mendel's principle of independent assortment.

In addition, because the maternal and paternal homologs of each chromosome migrate to each new cell independently of the other homologs, each parent can produce 2^n kinds of gametes, where n is the number of chromosome pairs. As we saw earlier, and as Sutton himself argued, an organism with two pairs of chromosomes, for example, would have 2^2, or 4,

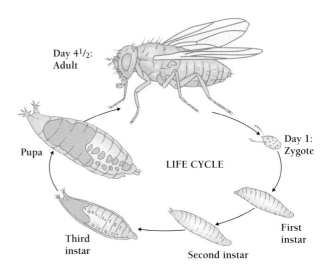

Figure 9-25 Growing up fast. Fruit flies (*Drosophila melanogaster*) mature from egg to adult in just 2 weeks, making them an ideal organism for studies of inheritance.

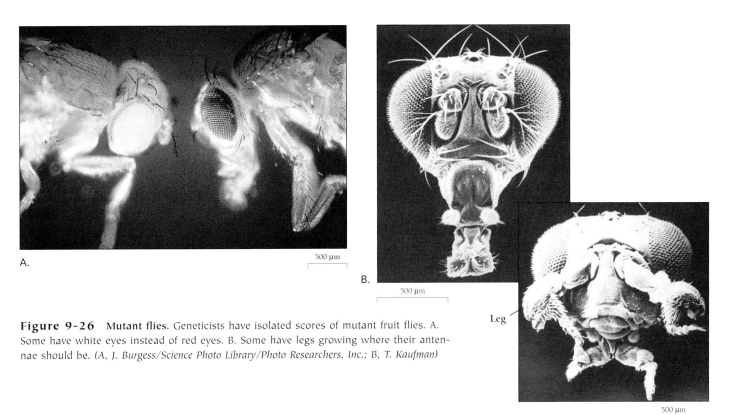

Figure 9-26 Mutant flies. Geneticists have isolated scores of mutant fruit flies. A. Some have white eyes instead of red eyes. B. Some have legs growing where their antennae should be. (A, J. Burgess/Science Photo Library/Photo Researchers, Inc.; B, T. Kaufman)

types of gametes, while a human could produce 2^{23}, or more than 8 million kinds of gametes.

Sixth, argued Sutton, each chromosome carries a different set of genes, and all of the genes on one chromosome are inherited together. That different chromosomes are inherited independently accounts for Mendel's principle of independent assortment. We now see Sutton's insights as the basis for understanding the chromosomal basis of the laws of inheritance. Table 9-1 summarizes the rules of inheritance that we can take both from Sutton's work and from later work. Because of Sutton's synthesis and because of vast amounts of subsequent work in genetics, we are in a better position to understand genetics than were the 19th-century biologists who first discovered the physical basis of heredity.

How Did Thomas Hunt Morgan Bolster the Theory of Chromosomal Inheritance?

With the publication of Sutton and Boveri's papers, the field of genetics took off. One of the most important contributors to the new field was Thomas Hunt Morgan, who began breeding tiny fruit flies called *Drosophila melanogaster* in his laboratory at Columbia University, in New York City, in 1909. Fruit flies have many advantages for genetic research. Among them are: (1) they breed rapidly, growing from eggs to sexually mature adults in less than 2 weeks (Figure 9-25); (2) they are prolific, with each female laying hundreds of eggs; (3) they are small (about 3 mm long) and easy to keep; (4) they are intricately and precisely built organisms, with dozens of easily identifiable traits under genetic control; and (5) they have only four pairs of chromosomes.

Morgan first collected his flies by leaving a piece of ripe pineapple on his windowsill. He kept the flies captive in small milk bottles "borrowed" from his home milk delivery man. Before long he was raising thousands of the tiny flies, and his laboratory, called "the fly room" by other biologists, became the center of genetic research in the United States for more than 25 years. Since then, thousands of geneticists and hundreds of thousands of undergraduate biology students have studied inheritance in fruit flies, demonstrating over and over that traits such as eye color, body color, and wing shape are all inherited according to Mendel's laws.

One of Morgan's first discoveries concerned a gene for eye color, which, he showed, lies on the X chromosome. Most wild fruit flies have red eyes. One day, however, Calvin Bridges, an undergraduate hired to wash the hundreds of dirty milk bottles in Morgan's laboratory, noticed a single fly with white eyes (Figure 9-26). Morgan crossed the white-eyed male with several normal, red-eyed females. All the F1 progeny had red eyes, which showed that the white-eye allele must be recessive and the corresponding red-eye allele, dominant.

Morgan then crossed the F1 males and females and obtained an overall ratio of 3:1 of red-eyed flies to white-eyed

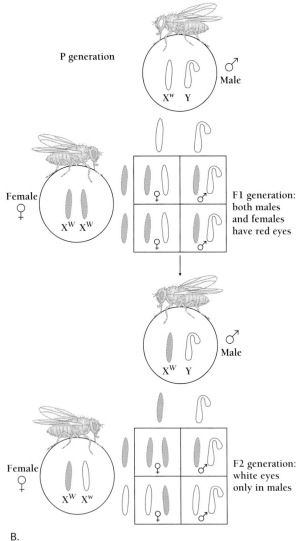

Figure 9-27 Sex-linked genes. A. Fruit flies have four pairs of chromosomes. As in humans, the male has an unmatched set of sex chromosomes. B. The white gene is on the sex chromosome, and we can trace its descent through two crosses by looking at the way the sex chromosomes assort.

flies. But all of the white-eyed flies were male! None of the females had white eyes. Half of the F2 males had red eyes, but all of the females had red eyes. What was going on?

Drosophila have only four pairs of chromosomes (Figure 9-27A). Males have three pairs of autosomes plus one pair of unmatched X and Y sex chromosomes, while females have three pairs of autosomes plus one pair of X chromosomes. As in humans, some genes are on both the X and Y chromosomes. But the Y chromosome is smaller than the X chromosome, so many genes on the X chromosome are not on the Y chromosome. Many genes on the X chromosome are present in two copies in females but in only one copy in males. Biologists say the males are *hemizygous* for genes on the X chromosome.

Morgan hypothesized that a single gene determined eye color in this case and that the gene was on the X chromosome. In Figure 9-27B, Punnett squares explain Morgan's experiment, showing the dominant red allele as X^W (because it is on the X chromosome) and the recessive white allele as X^w (because it too is on the X chromosome). The *W* stands for *white,* the name of the eye-color gene. Geneticists frequently name a gene after a mutant allele that functions badly or not all, rather than after the normal allele. So, ironically, the normal *white* allele (*W*) specifies red eye color!

The males and females in the F1 generation have different genotypes. The females are all heterozygous, with the white-eye allele (*w*) on one X chromosome and the dominant red-eye allele (*W*) on the other X chromosome. So all the females have red eyes. The males are all hemizygous, with the red-eye allele on their single X chromosome and no eye-color allele at all on their Y chromosome. So all the males have red eyes, too, which they got from their mothers.

In the F2 generation, the females receive one X chromosome from each parent. All the females receive the dominant X^W from their fathers and have red eyes. Of the female, half are homozygotes that received dominant X^W from both mother and father, and half are heterozygotes that received a recessive X^w from their fathers. The males receive their Y chromosomes from their fathers and their X chromosomes from their mothers. Because the Y chromosome has no eye-color allele, they receive no allele from their fathers. From their mothers, half of the male F2 flies receive X^W and have red eyes, and half receive X^w and have white eyes.

To double-check this conclusion, Morgan performed a backcross, mating what he thought were heterozygous F1 females with a hemizygous white-eyed male. His results were exactly as he predicted. Half of the males and half of the females had white eyes. This result established that the *white* eye-color gene indeed lay on the X chromosome.

The gene studied in this experiment is one of almost 1000 genes in *Drosophila* that are sex linked. In flies and other organisms with unmatched sex chromosomes, **sex-linked genes** have different patterns of inheritance in males and females because they lie on chromosomes that also determine the sex of the offspring. In mammals and flies, for example, the X chromosome contains many more genes than the Y chromosome,

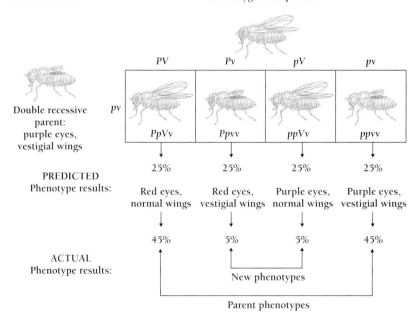

Figure 9-28 Independent assortment? No, genetic linkage. P and V indicate the dominant wild-type alleles, red eyes and normal wings. Small p and v indicate the recessive alleles, purple eyes and vestigial wings. A. In a cross between a red-eyed parent with normal wings and a purple-eyed parent with vestigial wings, all the offspring are heterozygotes with red eyes and normal wings ("wild-type"). B. If the two gene assorted independently, a back-cross between the heterozygous and homozygous flies should produce a ratio of 1:1:1:1. Instead, nearly all the flies had either red eyes and wings or purple eyes and vestigial wings. The two genes do not assort independently because they are on the same chromosome.

and sex linkage therefore nearly always refers to genes on the X chromosome. This experiment and others performed by Morgan and his students helped establish that genes were indeed on chromosomes, just as Sutton had proposed a decade earlier.

Thomas Hunt Morgan's experiments with sex-linked genes confirmed Sutton's theory that the genes lie on the chromosomes.

Do Genes on the Same Chromosome Assort Independently?

Mendel's paper did not attempt to explain what happened when two genes were on the same chromosome, or linked. But Morgan, with his new knowledge of genes and chromosomes, was able to figure it out. The principle of independent assortment depends on the independent assortment of chromosomes, but it cannot hold for genes that are close together on the same chromosome. That is, if Mendel had chosen an eighth trait, it would not have assorted independently. Sutton recognized this problem in 1903, but Morgan and his students were the first geneticists to design experiments around this question. In so doing, they further established the chromosomal basis of inheritance.

Morgan crossed two types of flies. One kind had wings of a more or less standard length and red eyes—characteristics usually found in wild populations. These flies were called *wild type*. The second kind of fly had short, stubby "vestigial" wings and purple eyes (Figure 9-28). Morgan had previously shown that purple-eyed flies with vestigial wings bred true, and he concluded that the purple, vestigial phenotype results from the ho-

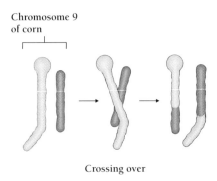

Figure 9-29 Crossing over made easy. In corn, one homolog of chromosome 9 has a knob at one end and a tail at the other. When homologous chromotids cross over, one gets the knob and one gets the tail. In 1931, Harriet Creighton and Barbara McClintock easily showed that what looked like crossing over in chromosome 9 corresponded to the pattern of genetic linkage of genes on chromosome 9.

mozygous condition of two genes. Morgan named the purple allele *pr* and the vestigial allele *vg*. For simplicity we call them *p* and *v* and the corresponding wild-type alleles *P* and *V*.

Purple vestigial flies crossed with true-breeding, wild-type flies produced normal (wild-type) F1 flies (Figure 9-28A). This showed that *p* and *v* were recessive alleles and that *P* and *V* were dominant. Since *p* and *v* were recessive, the genotype of the purple vestigial flies was *ppvv*.

To find out whether the *p* and *v* genes would assort independently, like Mendel's pea genes, Morgan performed a backcross. He mated the heterozygous F1 flies with purple vestigial flies. If the two genes assorted independently, this cross should have yielded four possible phenotypes in equal proportions. Morgan expected: $1/4$ purple eyes, vestigial wings; $1/4$ purple eyes, normal wings; $1/4$ red eyes, vestigial wings; and $1/4$ red eyes, normal wings (Figure 9-28B).

Instead, almost 90 percent of the flies had the same phenotypes as the two original parents—either purple eyes and vestigial wings or red eyes and normal wings. The remaining 10 percent of the flies were divided into flies with purple eyes and normal wings and flies with red eyes and vestigial wings.

Morgan concluded that the two genes did not assort independently. The *v* and *p* alleles almost always stayed together, and the *V* and *P* alleles almost always stayed together. Morgan guessed that these alleles stayed together because they were on the same chromosome and, more generally, that genes assort independently only when they are on separate chromosomes.

The tendency of genes on the same chromosome to stay together is called **genetic linkage.** Morgan and his students found that all of the fly alleles they studied fell into just four distinct **linkage groups,** sets of genes that do not assort independently. The four linkage groups, Morgan realized, corresponded to the four pairs of chromosomes in *Drosophila*. One linkage group consisted of sex-linked traits, indicating that it included all of the genes on the X chromosome. The fact that genes lie on chromosomes explains not only Mendel's laws but also the existence of linkage groups.

When Morgan repeated Mendel's studies of independent assortment, he found that genes on the same chromosome do not assort independently. This discovery further bolstered Sutton's chromosomal theory of inheritance.

Why Did Some Flies Have Normal Wings but Purple Eyes?

Genetic linkage explains why 90 percent of the flies in the cross shown in Figure 9-28B resemble one or the other parental phenotype. But why do some flies in the backcross have purple eyes and normal wings?

The answer is that homologous chromosomes frequently exchange material with each other during meiosis. Earlier in this chapter, we mentioned that crossing over occurs when homologous chromosomes pair up during meiosis I. Single chromatids in homologous chromosomes break, exchange material, and rejoin (Figure 9-11).

After crossing over, a new chromosome contains some genes from each of the two homologs. These new combinations of genes then stay together through the rest of meiosis. In this way, the chromosomes of the resulting gametes can differ dramatically from the chromosomes of the parents.

The physical basis of crossing over was first demonstrated in 1931 by graduate student Harriet Creighton and her mentor, geneticist Barbara McClintock, at Cornell University. McClintock, an expert on the genetics of corn, had noticed a pair of homologous chromosomes that differed in appearance. One homolog had a knob at one end and a long tail at the other, while the other homolog was relatively plain (Figure 9-29). McClintock suggested that Creighton find out whether homologous chromosomes literally traded parts by comparing the inheritance of alleles on the odd pair of chromosomes with their appearance after meiosis. Creighton, not realizing that the work was important, dawdled, and McClintock had to press her to finish the work. In time, young Creighton was able to show that crossing over wasn't just an abstract concept but that the chromosomes actually swapped long sections, so that the new homologs had either a knob but no tail or a tail but no knob (Figure 9-29).

In 1931, Thomas Hunt Morgan came to visit Cornell. When he heard what Creighton had done, he insisted that she write up her results and publish them immediately. Although he didn't mention it, Morgan knew that another famous geneticist was finishing up the same work in *Drosophila* and would soon publish. Morgan sat down on the spot and wrote a letter to the editor of a scientific journal, telling him to expect Creighton and McClintock's paper in 2 weeks. Because *Drosophila* reproduce quickly, each experiment lasts just a couple of weeks, while the same experiment with corn takes a whole summer. As Morgan later admitted, "I thought it was about time to give corn a chance to beat *Drosophila!*"

Crossing over recombines alleles in new ways. Harriet Creighton and Barbara McClintock showed that crossing over is a physical exchange of material between homologous chromosomes.

Can All the Genes on a Chromosome Cross Over and Recombine?

Once Sutton and Boveri had persuaded biologists that genes were on chromosomes, the door was open for a new kind of question: Just *where* on a chromosome did a particular gene lie? At least 1000 genes lie on the *Drosophila* X chromosome, and at least 10,000 or more lie on the human X chromosome. How could they map these genes?

The answer came from Alfred Sturtevant, another undergraduate in Morgan's laboratory. Sturtevant discovered that the farther apart two genes are on a chromosome, the greater the likelihood that they will cross over. Sturtevant expressed the "genetic distance" between two genes as the chance that recombination would occur between them. He defined a **map unit** as the distance between two genes that would produce 1 percent recombinant gametes (and 99 percent parental gametes). He then constructed a **genetic map** (or **linkage map**) that summarized the distances between many different genes. Geneticists refer to the position of a gene on a chromosome as its **locus** [Latin, = place; plural, **loci**]. The distance between the *v* locus and the *p* locus, for example, is about ten map units.

The chance that any two genes on a chromosome will recombine is proportional to the physical distance between them.

In this chapter, we have seen how cells divide to create haploid gametes by means of meiosis and how the twin processes of syngamy and meiosis complement each other. We have also seen that the behavior of the chromosomes during meiosis explains the patterns in which genes are inherited. In the next chapter, we learn what a gene does and what it is made of. We will also see that the structure of DNA explains how it can copy itself during the S phase of the cell cycle.

Study Outline with Key Terms

Genetics is divided into **transmission genetics,** the study of how variation is passed from one generation to the next, and **molecular genetics,** the study of how DNA carries genetic instructions and how cells carry out these instructions. For many years, biologists thought that the characteristics of offspring represented a **blending** of the characteristics of their parents. But most genes are present in two copies, one from each parent, and—most of the time—these genes do not change as they pass from generation to generation. The **Sutton-Boveri chromosomal theory of inheritance** states that the units of heredity are genes carried on chromosomes.

Phenotype encompasses all physical and behavioral characteristics of an organism. **Genotype**—the genetic constitution of a cell or organism—is a collection of genes. The word "genotype" can refer to the entire **genome** or to a subset of genes. A **gene** is a region of DNA that specifies the amino acid sequence of a polypeptide and its pattern of expression.

Asexual reproduction, through mitosis, produces offspring with genes from just one parent, a **clone. Sexual reproduction** produces offspring that inherit genetic information from two parents. This genetic information is contained in two sets of chromosomes, one set from each parent.

The **somatic cells** of an organism all have two copies of each chromosome and are said to be **diploid.** The **sperm,** or **spermatozoon,** and **ovum (egg),** or **gametes,** each carry only one copy of each chromosome and are said to be **haploid.** Gametes arise from the **germ line** or **germ cells** in the **gonads**— **ovaries** in females and **testes** in males.

Male and female gametes come together at fertilization (or syngamy) to form a **zygote. Fertilization** (or **syngamy**) restores the diploid chromosome number. The genetic information in the zygote, however, is a mixture of information from the parents. Adult organisms produce new haploid gametes by means of **meiosis,** a process that allots one haploid set of chromosomes to each of four daughter cells. Because the distribution of chromosomes occurs randomly, meiosis generates combinations of genes that are different from those present in the parents and thus accounts for much of the genetic diversity of individuals within a species.

Meiosis consists of two cell divisions (**meiosis I** and **meiosis II**), each of which consists of **prophase, metaphase, anaphase,** and **telophase.** During prophase I, homologous chromosomes form precisely aligned pairs in a process called **synapsis.** Each chromosome consists of a **tetrad.** During the rest of meiosis I, one chromosome from each pair (with two **chromatids**) goes to each daughter cell. Meiosis II distributes the sister chromatids of each chromosome to daughter cells.

Each member of a matching pair of chromosomes, or **homologous chromosomes,** is called a **homolog.** Homologous chromosomes pair and **cross over** during synapsis, forming **chiasmata. Disjunction,** the separation of homologous chromosomes (resulting in **euploid** gametes), sometimes fails. **Nondisjunction** leads to the production of gametes with an abnormal number of chromosomes, or **aneuploidy.** One example is **Down syndrome,** or **trisomy 21.** Most aneuploid embryos die long before birth in **spontaneous abortions.**

Alternative versions of the same gene are called **alleles.** An organism with two copies of the same allele is **homozygous,** one with two different alleles is **heterozygous,** and one with only a single allele is hemizygous.

An allele is **dominant** when it alone determines the phenotype of a heterozygote and **recessive** when it contributes nothing to the phenotype of a heterozygote. In **partial dominance,** both alleles are expressed.

Mendel made use of **true-breeding** varieties of peas to show his **principles of segregation** and **independent assortment.** Breeding two genetically distinct organisms is called **cross breeding** (or **crossing**), and the progeny of such a cross are called **hybrids.** To test the principle of segregation, Mendel performed a new kind of cross, called a **backcross.** A backcross to a homozygous recessive individual that is intended to reveal the genotype of the other parent is called a **testcross.** The original parents in a cross are called the **parental,** or **P,** generation, and the progeny are the "first filial," or **F1** generation, **F2** generation, and so on. All kinds of crosses can be represented in a **Punnett square,** with each kind of gamete produced by one parent along the top of the square and each kind of gamete produced by the other parent along the left side of the square.

The position of a gene on a chromosome is its **locus.** The tendency of genes on the same chromosome to stay together is called **genetic linkage.** Geneticists can construct a **genetic map** (or **linkage map**), which summarizes the distances between genes in **map units. Linkage groups** are sets of genes that do not assort independently. Compelling evidence that genes are on chromosomes came from the analysis of genetic linkage in *Drosophila*. Genetic maps of *Drosophila* and other organisms identified the location of the genes on the chromosomes and explained deviations from the principle of independent assortment.

In mammals and fruit flies, males have two unmatched **sex chromosomes,** called **X** and **Y,** as well as homologous **autosomes.** Females have the same autosomes plus a pair of matched X chromosomes. The X and Y chromosomes have homologous regions and pair during synapsis. Because males have only one X chromosome, they are susceptible to **sex-linked** genetic defects such as hemophilia and color-blindness that result from being hemizygous.

Harriet Creighton and Barbara McClintock showed that crossing over is a physical exchange of material between homologous chromosomes. Their work was immediately accepted as confirming the chromosomal theory of inheritance.

Review and Thought Questions

Review Questions

1. How are homologous chromosomes similar to each other? How are they different?
2. Describe and draw the steps of meiosis I and explain what they accomplish. Then do the same for meiosis II. Which part of meiosis actually accomplishes the reduction in chromosome number?
3. Explain how crossing over of chromatids in prophase I leads to genetic recombination, so that the gametes produced from one parent are virtually all unique.
4. Describe the main ways in which meiosis differs from mitosis.
5. Define the principles of independent assortment and segregation. What aspects of meiosis explain these principles?
6. Explain the difference between autosomes and sex chromosomes. How many of each do you have? How many of each does a potato have?
7. What causes Down syndrome and other forms of aneuploidy? Why should the frequency of aneuploid eggs increase as a mother ages?

Thought Questions

8. In what way is fertilization the opposite of meiosis? Why is it important for a complete sexual life cycle to contain both processes?
9. If you found a new species in which the sexes look different, what criteria would you use to decide which to call "female" and which to call "male"?
10. Textbooks make a point of contrasting the genotype and the phenotype as mutually exclusive. If the phenotype is every measurable aspect of an organism, including its structure, chemical makeup and behavior, could an organism's genotype be considered a part of the phenotype? Why or why not?

About the Chapter-Opening Image

When young Walter Sutton solved the mystery of heredity in 1902, he cried, "I know why the yellow dog is yellow!"

On-line materials relating to this chapter are on the World Wide Web at **http://www.harcourtcollege.com/lifesci/aal2/**

 Web Bit 9-1 Sex Linkage in humans

 Web Links

 Quizzing & Testing

10 The Structure, Replication, and Repair of DNA

Rosalind Franklin and the Double Helix

On February 28, 1953, James D. Watson and Francis Crick discovered the structure of DNA, the hereditary material of the genes. Their breathtaking discovery laid a foundation for all of the most spectacular advances in biology until the present time. It was "the secret of life," as they announced at the time. The two men were helped in this momentous accomplishment by conversations with many other scientists. Yet, a young scientist named Rosalind Franklin, working in near-total isolation, almost beat them to the punch (Figure 10-1).

In 1951, Jim Watson was a 23-year-old former radio Quiz Kid, cherished by his elders as an up-and-coming genius. He had just received his Ph.D., studying the genetics of bacteria and the viruses that infect them, and his professors had arranged for him to go to Copenhagen to study the chemistry of nucleic acids. But the chemistry of nucleic acids bored him, and he began looking about for something else to work on.

While on a trip to Naples, Italy, Watson found that something—a question so fundamental that its answer seemed guaranteed to bring a Nobel prize. At a lecture by an English physicist turned biochemist named Maurice Wilkins, Watson saw a slide of an x-ray diffraction pattern from what Wilkins said was a crystalline form of DNA. Wilkins argued that his photos, taken with the help of graduate student Raymond Gosling, were the first step in deducing the structure of DNA. The molecular structure of DNA would, of course, shed light on how genes work.

Watson was transfixed. He understood that if DNA could form crystals, it must therefore have a regular and relatively simple shape. Then, within just a few days, he read a sensational paper by famed chemist Linus Pauling on the alpha-helical structures within proteins. Watson longed to do something equally spectacular. With DNA and helices swirling in his head, Watson persuaded his former professors back in the United States to get him out of Copenhagen and into a lab in England where he could study DNA.

In a few months, Watson arrived at the Cavendish Laboratories in Cambridge, in the same building as Max Perutz and John Kendrew, pioneers in the study of macromolecules using x-ray diffraction techniques. Also in the lab was Francis Crick, a 31-year-old graduate student working on his Ph.D. project. Working under Perutz, Crick was studying the x-ray diffraction of polypeptides and proteins, especially the blood protein hemoglobin.

Crick immediately liked Watson, and the two began to do an enormous amount of talking, so much so, in fact, that Perutz and Kendrew finally banished the two younger scientists to a room of their own. Watson knew nothing about x-ray crystallography except that he needed it to solve the structure of DNA. He nagged Crick relentlessly to help him work on DNA.

Crick liked to talk about DNA, but he wasn't about to start working on its structure. With a wife and daughter to support, he had to finish his Ph.D. and get a job. Besides, etiquette forbade Crick from working on the DNA problem. Crick and Wilkins were old friends going back to their joint work for the British Admiralty during World War II. Just as important, the Cavendish and the King's College Laboratory in London, where Wilkins worked, were sister facilities— and the DNA problem formally belonged to King's. At King's Col-

Key Concepts

- DNA consists of two polynucleotide strands, which wind together to form a double helix.
- The two strands of DNA contain complementary versions of the same information. During DNA replication, each strand directs the synthesis of a complementary strand.
- In all organisms, genes are made of DNA, and each allele of a gene usually specifies the structure of a single polypeptide.

lege, Wilkins and Rosalind Franklin were already working on the structure of DNA, and they hadn't asked for help.

Still, it didn't hurt to talk. The first step to solving the three-dimensional structure of DNA was to study its density, size, and other properties using x-ray crystallography. Because DNA is a long, thin molecule, it would be impossible to obtain true DNA crystals until the 1960s, when researchers learned to produce crystals from short pieces of DNA synthesized in the laboratory (Figure 10-2). The DNA Wilkins had acquired was not crystals, but *fibers,* in which hundreds of millions of DNA molecules lay parallel to one another.

In 1951, the best x-ray diffraction photos of DNA fibers were those taken by Rosalind Franklin at the King's Laboratory. In November 1951, Watson took the train down to London to visit Maurice Wilkins and to hear Franklin give a lecture on the structure of DNA.

Franklin had arrived at King's at the beginning of 1951. A talented and recognized authority in industrial physical chemistry, she had just completed 4 years of work on the structure of coals at a lab in Paris. Her work laid the foundation for modern industrial carbon-fiber technology. Today, carbon-fiber technology is used in making strong, lightweight (and expensive) bicycle frames and sailboats, for example. For her successful analysis of coals, Franklin learned and improved on known x-ray diffraction techniques, developed new mathematical techniques for interpreting the resulting photos, and constructed scale models like those that Linus Pauling had used in his work. She returned to her native England eager to apply x-ray diffraction techniques to an important biological molecule.

John Randall, the head of King's Laboratory, invited Franklin to set up and head an x-ray crystallography lab, with all the newest, most powerful equipment available. Once the lab was set up, her assigned task was to work out the structure of DNA. He assigned a graduate student to assist Franklin. It was Raymond Gosling, the same young man who had taken pictures for Wilkins.

What Randall did not explain to Franklin was that Maurice Wilkins, just down the hall, had already begun work on DNA and considered the DNA problem his own. In fact, even

Figure 10-1 Rosalind Franklin on a walking tour of France in about 1950. Franklin produced most of the data needed to discover the structure of DNA. *(Vittorio Luzzati)*

in his letter offering her the job, Randall had never mentioned Wilkins. Worse, at her first meeting with Randall and Gosling to discuss the project, Wilkins was not present. With that peculiar oversight, Randall set the stage for one of the most disastrous personality clashes in 20th-century science.

Wilkins was equally in the dark about Franklin. He viewed her more as a technician than a scientist, and he appears to have believed that she was hired specifically to do his x-ray crystallography work, in which he had no expertise. He handed over his best DNA fibers, hoping she would produce for him the photos that would reveal their structure. But Franklin displayed none of the deference that Wilkins expected. Instead, she treated Wilkins as a colleague, vigorously arguing with his ideas about DNA in the spirited style she had picked up in Paris. It was her way of trying to make friends. Offended by what he perceived as combativeness, he struggled to put her in her place. Franklin found him unaccountably touchy; he was given to turning away in the middle of conversations he didn't like.

Figure 10-2 DNA is a long, thin molecule. A double strand of DNA is only 2 nm thick, but many reach a length of many centimeters. DNA fibers such as those that Rosalind Franklin studied consist of many parallel molecules. A. Atomic force micrograph showing a close-up view of DNA. B. Electron micrograph of a strand of DNA. (A, Science VU/BMRL/Visuals Unlimited; B, K.G. Murti/Visuals Unlimited)

Wilkins spitefully attacked her behind her back, undermining her relationships with other scientists at the lab. This was all too easy, for Franklin was never around when the other scientists gathered to talk at lunchtime or teatime: women were not allowed in the King's dining room.

Over the next 6 months, Franklin set up the x-ray crystallography lab and then, with Gosling's assistance, began working out the structure of DNA. Though she liked her work, the social atmosphere was poisonous, unlike anything she had experienced in other labs. Franklin was used to making warm friendships with her colleagues. But between Wilkins' hostility and her banishment from the dining room, this seemed impossible. She longed for her happy days in Paris. Unable to do the x-ray crystallography himself, Wilkins resentfully moved on to other projects. Randall, who might have reconciled the two researchers, did nothing.

WHAT IS THE STRUCTURE OF DNA?

When Franklin began her work in 1951, the chemical makeup of DNA—but not its three-dimensional structure—had been known for about 30 years. Franklin knew that DNA consisted of long chains of nucleotides linked together into polynucleotides. She knew that each nucleotide consisted of a sugar, a phosphate, and a base, and that the sugar and phosphate units were linked together alternately (sugar-phosphate-sugar-phosphate, etc.) into a sugar-phosphate "backbone" (Figure 10-3A).

Biochemists also knew that from the sugar-phosphate backbone hung four kinds of nucleotide bases. Two of the four bases are **pyrimidines,** which consist of rings made up of four carbon atoms and two nitrogen atoms. The other two bases are **purines,** which contain a double ring of six carbon atoms and four nitrogen atoms (Figure 10-3B). The purine bases are **adenine** and **guanine,** and the pyrimidine bases are **cytosine** and **thymine.**

Each nucleotide has a long, correct name: deoxyadenosine monophosphate, or dAMP, for adenine; deoxyguanosine monophosphate, or dGMP, for guanine; deoxycytidine monophosphate, or dCMP, for cytosine; and thymidine monophosphate, or TMP, for thymine. But biologists rarely use these names. In fact, they generally refer to each nucleotide by a single letter abbreviation: A, G, C, or T. This convention not only saves space, it also emphasizes that the nucleotides are indeed the four letters in the DNA alphabet.

What was *not* known by anyone at the time that Franklin set to work was that a molecule of DNA consists of exactly two polynucleotide chains that are twisted together into a long helix, like a ladder twisted about its long axis (Figure 10-4). The two legs of the ladder are the two sugar-phosphate chains. Each rung of the ladder consists of a pair of bases (A always goes with T, and G always goes with C). Each twist of the ladder contains ten rungs, which are 0.34 nanometer (nm) apart. A single molecule of DNA may be either short or long. In humans, a molecule of DNA may be very long—up to 9 cm (3½ inches) long—and may consist of millions of nucleotide pairs. The width of the ladder, however, is the same from one end to the other—2 nm.

Importantly, each sugar-phosphate chain has a direction, and the two chains—the legs of the ladder—run in opposite directions (Figure 10-5). In each sugar-phosphate backbone,

Figure 10-3 DNA is a chain of nucleotides. A. DNA consists of a backbone of sugars (deoxyriboses) and phosphates. B. Four kinds of bases hang off each sugar group. Just as the letters in this sentence carry information, the nucleotide bases in the DNA carry information. A nucleotide includes a phosphate, a sugar, and a base. (A, Paraskevas Photography)

the third carbon of one sugar is attached by a phosphate group to the fifth carbon of the next sugar. The third carbon is called the 3′ carbon, pronounced "three-prime carbon," while the fifth carbon is called the 5′ carbon, or "five-prime carbon." One end of each polynucleotide has a free 5′ carbon, called the **5′ end,** while the other end has a free 3′ carbon, called the **3′ end.** If we draw an arrow from the 5′ end to the 3′ end of each strand in Figure 10-5, one arrow points up, the other, down.

> *Each nucleotide in DNA consists of a sugar, a phosphate, and a base. The sugar and phosphate units are linked together alternately (sugar-phosphate-sugar-phosphate, etc.) into a sugar-phosphate "backbone." The two parallel backbones (running in opposite directions) resemble a ladder twisted around its long axis, with each rung consisting of a pair of bases (A-T or G-C).*

How Much Did Franklin Discover About the Structure of DNA?

By the time Franklin gave her talk in November 1951, her 5 months of work had already revealed some essential facts about the structure of DNA. She believed that DNA was probably a big helix with two, three, or four parallel chains, with the phosphates wound around the outside of the helix, like the uprights of a ladder. She had accurately measured the density of DNA and the number of water molecules associated with each nucleotide, and discovered that DNA fibers have two slightly different structures, depending on whether they are wet or dry. Finally, she had identified the "symmetry" of DNA. Crystallographers classify crystals into 230 types according to their symmetry. Which of these symmetries DNA had might seem like an obscure point, but it would be crucial to Watson and Crick's solution to the structure of DNA. DNA's symmetry was the clue that the two strands run in opposite directions.

Only the crystallographers working at the Cavendish, particularly Crick, could fully appreciate the meaning of the particular symmetry Franklin had measured. If Crick had heard Franklin speak, he would have understood that Franklin's parallel chains were symmetrical in a particular way. Just as the two ends of a sharpened pencil look different, a chain that ran in one direction only would look different when rotated 180°. In contrast, two pencils running parallel but in opposite directions look the same at both ends, and a pair of symmetrical chains running opposite to each look the same when rotated. The symmetry Franklin had discovered implied that DNA consisted of two chains, one running up, the other running down. The chains could have been four (two up and two down), but Franklin's density measurements ruled that out.

Franklin did not have the extensive training of a full-fledged crystallographer, and unlike Crick, she had never before worked with a large biological molecule, so she couldn't yet deduce the shape of the DNA molecule. Nonetheless, she was on her way.

Watson, who knew almost nothing about crystallography, watched Franklin talk, but he understood little of what she said and took no notes. After the talk, he visited Wilkins, who said that Franklin didn't know what she was doing. Back at Cambridge, Watson could tell Crick almost nothing about Franklin's talk. Crick was annoyed, but it would be 14 months before he learned how much he had missed.

What Watson did bring back was one wrong fact, the amount of water in the molecule, and the mistaken impression, gained from Wilkins, that neither Franklin nor Wilkins was capable of solving the structure of DNA. Clearly, the two at King's were not working together, and Wilkins seemed to be going nowhere by himself. Watson does not appear to have considered that Franklin might solve the structure. It seemed to him ridiculous to allow Franklin to bumble away, not solving this important problem, when he and Crick could be deciphering DNA themselves.

Back in Cambridge, Watson finally persuaded Crick to help him attack the problem. Within days, the two produced a model of DNA consisting of three intertwined helices, sugar-phosphate backbones on the inside, and far too little water. They called Wilkins and Franklin to come up to Cambridge the next day and look over their model. Franklin glanced at the model, knew at once that it was completely wrong, and said so. Crick was humiliated, for he knew instantly that she was right.

Watson had missed everything important that she had said. Worse, Watson and Crick had worked on a problem that belonged to King's, and they had

Figure 10-4 DNA is like a twisted ladder. The two sugar-phosphate backbones run in opposite directions. The bases hanging off the sugars connect in the middle, like the rungs of a ladder.

Figure 10-5 A two-way street. Like the two sides of a street, DNA's two sugar-phosphate backbones run in opposite directions. Biologists call one end the "5′ end" and the other end the "3′ end." Each nucleotide has five numbered carbon atoms, and each nucleotide is oriented with its #3 carbon, called the 3′ carbon, toward the sharpened end of the pencil. The 5′ carbons are all oriented toward the eraser end. By convention, the 3′ end has an arrow because new nucleotides always attach to that end. *(Photos, Paraskevas Photography)*

been wrong. The head of the Cavendish Lab, Sir William Lawrence Bragg, sternly reminded them that the DNA problem was not theirs to work on. Reluctantly, they agreed to stop trying to model its structure, but they continued to talk about DNA. They couldn't help it. They loved to talk (Figure 10-6).

For another year, Franklin studied DNA in near solitude, talking only to Gosling. For a while she thought that maybe the structure couldn't be a helix after all. The dry form of the DNA fiber gave pictures that seemed to suggest DNA was not a helix. There was, in fact, no certain evidence for a DNA helix early on. Pauling had argued that a repeating chain of any kind, whether a molecule or a telephone cord, will tend to twist about its axis into a helix. A helix seemed likely. But the pictures of the dry form did not lend support to a helix.

Franklin played with other possibilities, including repeating figure eights. In the summer she took an excellent picture of wet DNA fibers, showing a pattern that fairly shouted "helix." She labeled it photo number 51. She liked it a lot, and spent several days thinking about it (Figure 10-7). She had been working on the dry form for so long, however, that she soon returned to that. If she had had a collaborator, someone to argue with, she might have given more thought to photo number 51. But the atmosphere at King's was as bad as ever. She was already making plans to go to another lab.

Back at the Cavendish, Watson frittered away his time, playing tennis, going to parties, and fretting vaguely about DNA. He was, in his own words, "totally underemployed." Then, in the fall of 1952, two young researchers arrived at the

Figure 10-6 Francis Crick and James D. Watson share a cup of tea in the early 1950s. *(A, Barrington Brown/Science Source/Photo Researchers, Inc.)*

Cavendish—Linus Pauling's son Peter and one of Pauling's graduate students. In mid-December, Peter announced that his father had discovered the structure of DNA. It was more than Watson could stand. In January, he begged Peter to get his father to send the manuscript describing his structure for DNA.

Franklin, who had also heard the news, wrote to Pauling's lab with the same request. Meanwhile, she began calculating bond lengths and angles for the different building blocks of DNA. Her months of work had yielded the main dimensions

Figure 10-7 An x-ray diffraction photo of DNA by Rosalind Franklin. *(Cold Spring Harbor Laboratory Archives)*

of the molecule. She suspected that DNA was a double helix. She knew for certain that it was 2 nm in diameter and that if it was a helix, each complete turn of the helix would be 3.4 nm long. Now, with these and other hard-won clues at hand, she quietly prepared to build a model of DNA, the same approach that had worked so well for Pauling in deducing the α helix of proteins.

On January 30, 1953, Watson went to King's College and stopped to visit Wilkins. Finding Wilkins busy, Watson dropped in on Franklin to taunt her with his copy of Pauling's manuscript. When she looked at the manuscript she could see as well as Watson that Pauling's three-chain helix was hopelessly wrong.

She also discovered that Pauling and his co-author, Robert Corey, had snubbed her. The previous spring, Corey had come to King's and asked to see her pictures, and she had graciously shown him her work. Wilkins and Gosling had taken earlier pictures of DNA also, but these were vastly inferior to Franklin's. Now Pauling and Corey had not only ignored her letter requesting their manuscript, they had also cited Wilkins' inferior x-ray diffraction pictures of DNA—which they had never seen—and not her own, which they had. Watson later recalled that he then began to goad her again, implying that she was incompetent to interpret her own x-ray diffraction photos. Although few people believe him, Watson claimed that the petite Franklin was so angry that she chased him from her lab.

According to Watson, he then found Wilkins, and the two men took turns complaining about Franklin. Wilkins reminded Watson of Franklin's ideas on DNA, the same ideas she had presented 14 months earlier, then showed Watson her striking picture of the wet form of DNA, number 51. Crick and another scientist had already worked out a mathematical understanding of the way that a helix would interact with x rays. That work, which Crick had published and which Franklin was fully aware of, had shown that a helix should generate a cross-shaped pattern. Now, with that knowledge and without Franklin's scores of other pictures to confuse him, Watson leapt to the conclusion that DNA must be a helix. It was obvious to everyone who had seen it—Franklin, Wilkins, and Watson—that this photo was of a helix. Nonetheless, Wilkins told Watson that Franklin stubbornly insisted that DNA was *not* a helix.

Watson could only conclude that Franklin really was incompetent. He was tired of the British laboratory etiquette that prevented him from working on DNA himself. He would build a model of the wet form of DNA, whose picture looked so simple. Instinctively, he decided to try two backbones. He had a feeling that for DNA to duplicate itself most easily, it should have two chains. Four chains seemed needlessly complex, and three chains seemed illogical. He was still convinced, however, that the sugar-phosphate backbones must be on the inside of the helix, with the bases hanging off the outside, despite Franklin's assertion that the sugar-phosphate backbones were on the outside. Watson rushed back to the Cavendish and at last succeeded in persuading Bragg to let him work on DNA. Feverishly, Watson began working to build a model. Crick, still writing his Ph.D. thesis, watched from across the room with amusement and growing interest.

Watson made little progress until, on February 5, 1953, Crick finally persuaded him to try putting the two sugar-phosphate backbones on the outside, as Franklin had suggested. For the first time, Watson had the beginnings of a model. The following weekend, Wilkins came up to Cambridge and reluctantly gave Watson and Crick permission to work on the DNA problem. Wilkins promised Crick in a letter to "tell you all I can remember & scribble down from Rosie."

No one asked Franklin's permission. Unaware that Wilkins and Bragg had now given away her research problem, she restarted her work on a structure for the easier, wet form of DNA. She was trying to decide if it was one chain or two.

For the first time, Crick felt free to work on DNA. The best source of information about DNA was Franklin, but Crick had no more intention of talking to her than had Watson or Wilkins. Crick soon learned, however, that Franklin and Gosling had summarized their unpublished work in an annual report on research at King's. Max Perutz had a copy of the report, and when Crick asked for it, Perutz handed it over.

Now Crick knew almost everything that Franklin and Gosling had so laboriously worked out. For the first time, he began to take the whole enterprise rather seriously. When he read Franklin's report describing DNA—its density, its water content, and, critically, its symmetry—it gradually dawned on Crick that DNA had to have two chains and that they had to run in opposite directions.

Franklin and Gosling also reported that the distance along the wet form of the molecule through one complete turn was 3.4 nm. Previous work had shown that the distance between the bases was 0.34 nm, so it was easy to see that there were probably ten bases per 360° turn. Wilkins and Gosling had earlier shown that the diameter of DNA was about 2 nm, and Franklin and Gosling had confirmed this. Things were beginning to come together.

Crick tried to persuade Watson to build a model with the chains running in opposite directions. But Watson couldn't figure out how to make it come out right. So, while Watson was off playing tennis, Crick abandoned his Ph.D. dissertation for an afternoon and took over the model. He built a more open double helix than Watson's, with sugar-phosphate chains running in opposite directions as required by the symmetry, and 10 bases per turn of the helix. The building blocks, with their bond angles and lengths, all fit nicely. Each complete turn was 3.4 nm, as Franklin had measured, and the distance from one base to the next was 0.34 nm. The only problem now was to find a way to fit the bases together in the middle.

It wasn't immediately obvious how this could be done. The four bases were of very different sizes and shapes. Some pair combinations would be wider than 2 nm and would make the molecule bulge and strain. Other combinations would have been too narrow and would have left a gap in the middle or pinched the two backbones in. Franklin's x-ray data showed, however, that the helix had a constant diameter. It neither bulged nor pinched in.

Watson began reading all the biochemistry of bases that he could find. If DNA were to carry genetic information, it should be able to accommodate any sequence of nucleotides, just as each line of print on this page must be able to accommodate any sequence of letters. Two weeks after he first put the sugar-phosphate backbones on the outside, Watson guessed that like bases paired (guanine with guanine and so on). But within days, Jerry Donohue, the young chemist from Pauling's lab, completely demolished Watson's hypothesis, explaining that the molecular forms that Watson had used, from standard textbooks, were all wrong. Besides, added Crick, Watson's like-with-like base pairing did not square with Chargaff's rules.

What Were Chargaff's Rules?

In the 1920s and 1930s, chemists believed that DNA contained almost equal amounts of each base. Then, in the 1940s, Erwin Chargaff developed a new method for analyzing the base composition of DNA. Chargaff found that the proportions of the bases could vary widely from species to species. But the amount of A always equaled the amount of T, and the amount of G always equaled the amount of C. Neither Chargaff nor anyone else at the time understood the significance of his findings, summarized in what came to be called **Chargaff's rules.**

In the spring of 1952, 9 months before Watson and Crick discovered the correct structure of DNA, Chargaff had come to Cambridge. Watson and Crick were introduced to Chargaff, and the older scientist took an immediate dislike to the two young men. "It struck me as a typical British intellectual atmosphere, little work and lots of talk," Chargaff later complained to the writer Horace Judson. Crick knew almost nothing about Chargaff's work, but the older man brusquely set him straight.

At the time, Crick saw that Chargaff's rules might imply complementary base pairing. That is, A always pairs with T, and G always pairs with C. Such a relationship could, he reasoned, provide a means for DNA to copy, or replicate, itself. Nine months later, however, Crick had forgotten his own insight.

By Saturday, February 28th, Watson had constructed cardboard cutouts representing the shapes of the four bases. He fiddled with them, fitting them together this way and that. He struggled to come up with pair combinations that would have the same symmetry and diameter, and still allow any order of bases on the backbone. Finally, he tried pairing A with T and G with C. The fit was unbelievably good. An A-T pair was the same width as a G-C pair (Figure 10-8).

Donohue quickly assured Watson that the two sets of bases could in fact fit together that way. The arrangement of atoms in the bases is such that A is perfectly arranged to make two hydrogen bonds with T, while G can form three hydrogen bonds with C. The chemistry worked, and for the first time, Watson and Crick understood the entire structure. It would be another week before they had the materials to build a complete model, but their work was done. Euphoric, they announced to everyone they met that they had "discovered the secret of life."

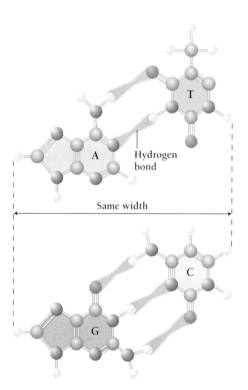

Figure 10-8 Base pairing. In DNA the bases connect in the middle of the twisted ladder by means of hydrogen bonds. The nucleotides A and T can form two stable bonds with each other, while G and C can form three. When matched in this way, the width of a G-C pair is the same as the width of an A-T pair, as predicted by Franklin's x-ray diffraction data.

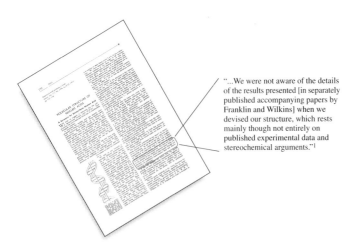

"...We were not aware of the details of the results presented [in separately published accompanying papers by Franklin and Wilkins] when we devised our structure, which rests mainly though not entirely on published experimental data and stereochemical arguments."[1]

On Saturday, March 7, Watson and Crick eagerly took model bases that had been made for them in the Cavendish machine shop and built an accurate model of the DNA double helix. On the same day, Wilkins, unaware of the model, wrote a note to Crick, including the following:

> I think you will be interested to know that our dark lady leaves us next week and much of the . . . data is already in our hands. I . . . have started up a general offensive on Nature's secret strongholds on all fronts: models, a theoretical chemistry and interpretation of data, crystalline and comparative. At last the decks are clear and we can put all hands to the pumps!
>
> It won't be long now.
> Regards to all,
> Yours ever,
> M.

Franklin Comes Within Two Steps of the Correct Model

But the race was over. Wilkins little knew how close Franklin had come to solving the problem on her own, or how late he was in restarting his own work. Nearly 2 weeks earlier, on February 23, Franklin had begun remeasuring photo number 51. By the end of the next day, she had correctly concluded that DNA was a double helix, with a diameter of 2 nm. She was now certain that the sugar-phosphate backbones were on the outside. She guessed that the backbones had some peculiar relationship to each other, but what? She hadn't yet realized that the two backbones ran in opposite directions. She knew about Chargaff's rules, and she had begun playing with the problem of how the bases would fit together in the middle. She was two steps from the answer.

But Franklin's work on DNA was at an end. Randall had told her she could leave King's for another lab if she wanted, but she could not take DNA with her. Accordingly, she left King's and, with Gosling, wrote up the results of their 18 months of work on the structure of DNA. Their work was not complete, but it was substantial. The potent combination of Franklin's experimental work, Crick's theoretical work, and Watson's infectious enthusiasm had provided everything needed to solve the problem. Yet, amazingly, when Watson and Crick wrote up their results for the prestigious British journal *Nature,* Franklin was the one person whose work was not formally cited. This deficiency was apparently no oversight. According to writer Horace Judson, Wilkins asked Watson and Crick to delete a sentence from their manuscript that said, "It is known that there is much unpublished experimental material." Instead, Watson and Crick explicitly asserted that they were unaware of Franklin's work, writing that "We were not aware of the details of the results presented [in separately published accompanying papers by Franklin and Wilkins] when we devised our structure, which rests mainly though not entirely on published experimental data and stereochemical arguments."[1] In fact, according to later accounts, Watson and Crick's structure rested heavily on the unpublished data of Rosalind Franklin and Raymond Gosling, details of which Watson and Crick were well aware.

Not until March 18, when Wilkins first received Watson and Crick's manuscript, did Franklin learn of their achievement. Eagerly, she took the train up to Cambridge to see their model. Watson was astonished at how quickly and graciously she agreed that the model was correct. He had been sure that she would find some grounds on which to criticize it, for he little appreciated how much she knew and understood. She, for her part, had no reason not to be gracious, for she little suspected how much Watson and Crick had depended on her own work. The elegant beauty of the structure delighted her. She was filled with admiration for Crick's genius.

Franklin was glad, in any case, to be hard at work in her new lab. She was now studying the structure of tobacco mosaic virus. Over the next 5 years, Franklin directed a highly productive research team. Later, she collaborated with researchers at Berkeley and Yale in the United States, and at Tübingen in Germany. When she discovered she had cancer, in 1956, she continued to work cheerfully and productively until the very end, so devoted was she to the research she loved. She died in 1958, only 37 years old, without ever learning of her own crucial contribution to the discovery of the structure of DNA.

When, a couple of years later, it appeared that Watson and Crick might get a Nobel prize for their work, Sir Lawrence Bragg, who had headed the Cavendish, made sure that Wilkins received the prize with them. Recounting the story to writer Horace Judson, Bragg said that "when it came to the Nobel Prize, I put every ounce of weight I could behind Wilkins getting it along with them. It was just frightfully bad luck, really." Wilkins was lucky enough, however, to receive in 1963 a third share of a Nobel prize for the discovery of the structure of DNA. Franklin herself was ineligible, for only the living are honored.

In 1968, 10 years after Franklin's death, James Watson wrote a personal account of the discovery of the structure of

---

[1] From Watson, J.D., and Crick, F.H.C., "Molecular Structure of Nucleic Acids: A Structure for Deoxyribose Nucleic Acid," *Nature*, April 25, 1953, vol. 171, pages 737–738.

DNA. In *The Double Helix,* Watson for the first time made clear how essential Franklin's experimental results had been to their discovery. He also falsely depicted her as Wilkins' rebellious and shrewish assistant, who seemed to deserve any bit of ill treatment that came her way. Watson's book was attacked and reviled by nearly everyone who had been present during the discovery. It has never been considered an accurate account. Yet, as a work of literature and as a window into the mind of a successful molecular biologist, *The Double Helix* is a classic.

Modern Science Is a Social Endeavor

The discovery of DNA's structure is the best-documented story in the history of science. Watson, Crick, and others have all written extensively and passionately about both the science and the scientists. It is clear that Franklin's failure to discover the structure of DNA before Watson and Crick was a direct result of her social isolation. Watson himself said in 1993, "[Franklin] would be famous for having found DNA if she'd just talked to Francis [Crick] for an hour."

Watson and Crick had each other to talk to. They had Max Perutz, John Kendrew, Jerry Donohue, Peter Pauling, and even, grudgingly, Erwin Chargaff. All told, at least a half dozen scientists besides Rosalind Franklin herself provided essential clues that guided Watson and Crick toward the correct answer. Franklin, isolated by her feud with Wilkins, as well as her exclusion from the King's dining room, had no one to talk to but Gosling, who had little useful knowledge.

Watson and Crick's discovery of the structure of DNA was a borrowed victory that rested heavily on Franklin's discoveries. Their main contribution to biology was their early recognition that DNA's structure explains how it works. In fact, Watson and Crick's insight into the implications of base-pairing stands as one of the most influential advances in the history of biology. It has provided the sole basis for the twin fields of molecular genetics and genetic engineering, and transformed every other area of biology.

What was so important about their model? We've seen that the structure was like a twisted ladder with the uprights running in opposite directions (Figure 10-9). Each rung of the ladder was a pair of bases held together by hydrogen bonds. Watson and Crick's rules of complementary base pairing meant that the sequence of one polynucleotide strand exactly specified the sequence of the second strand. That is, one strand could give directions for building the other. This explained how a chromosome could replicate itself during cell division. Further, it was clear that the bases could be arranged in a nearly infinite variety of different sequences. That meant that the sequence of bases in DNA, like the sequence of letters on a page, could contain coded information.

In the rest of this chapter, we will examine the structure of DNA in more detail. Along the way, we will find out how biologists found out that the genetic material was DNA, what a gene is, and how DNA replicates and repairs itself. In the next chapter, we will see how cells actually use the information encoded in the DNA.

WHAT IS A GENE?

Defining a gene remains a problem not only to students taking biology examinations but also to scientists searching for biological molecules. To Mendel in 1865, a gene was an abstraction—some form of information that determined the traits of individuals. To Sutton and Morgan in the early years of the 20th century, a gene was a concrete physical structure that was part of a chromosome.

But biologists still did not know what genes actually were or how cells used and copied the information that genes contain. That knowledge came only when geneticists tried to understand the function and the structure of genes in chemical terms. In the rest of this chapter, we will see how researchers established (1) what genes do, (2) what they are made of, and (3) how they replicate. We will learn that a gene spells out the sequence of amino acids in a polypeptide, that a gene is made of DNA, and that the structure of DNA explains its ability to replicate and repair itself.

In this and later chapters we will also see that genes are not only physical entities or places on the chromosomes, but pure information. In this sense, genes are abstractions, as Mendel conceived them. In the same way that information about biology, for example, can be communicated as printed pictures or text, as spoken words in a lecture, or as electrons dispersed on the Internet, genetic information can be relayed in many forms. The physical expression of that information is separate from the information itself.

How Do Genes Affect Biochemical Processes?

The first hint of the biochemical importance of genes came in 1902 from the work of an English physician, Archibald Garrod. Among Garrod's patients at the Hospital for Sick Children in London was a baby whose diapers developed black urine stains. The baby's disease, alkaptonuria, had first been described in 1649 and named for dark "alkapton bodies" that formed after the urine was exposed to air. The precursor of the alkapton bodies was identified as homogentisic acid (HA), which oxidized to form the alkapton bodies.

Garrod, working just 2 years after the rediscovery of Mendel's paper, suspected that alkaptonuria might be an inherited disease because the baby's parents were first cousins (Figure 10-10). After studying the way the disease was inherited, he realized that alkaptonuria behaved as if it were caused by a recessive allele, much like the white flowers of Mendel's peas. Each parent had contributed the same recessive allele to the baby, who therefore had two copies of the same allele. We say the baby was homozygous for that allele.

Like the amino acids phenylalanine and tyrosine, HA contains a benzene ring. Garrod hypothesized that HA was an intermediate in the metabolism of phenylalanine and tyrosine. Most people have an enzyme called *HA oxidase,* which helps oxidize HA so that it can be excreted as carbon dioxide and water (Figure 10-11). Alkaptonuria patients, however, cannot

Figure 10-9 Computer-generated model of DNA. The structure of DNA is now so well known that most of us have seen the familiar double helix somewhere. (*Ken Eward/Biografx/Science Source/Photo Researchers, Inc.*)

One Gene—One Enzyme

Garrod's insight, that genes control the activity of enzymes, at first seemed to apply to only a few rare diseases—alkaptonuria and albinism, for example. In these conditions, the failure to make an enzyme explained an obvious abnormality. But do all genes specify enzymes?

The answer did not come until 1941, when two Stanford University geneticists, George Beadle and Edward Tatum, began looking for other genetic defects in metabolism. Beadle and Tatum chose to study a simple fungus, the pink bread mold, *Neurospora*. Beadle and Tatum induced mutations in *Neurospora* with x rays and then looked for biochemical defects (Figure 10-12). Such biochemical defects were easy to detect because *Neurospora* is a haploid organism—it has only a single copy of each gene. When that single copy is altered, the change immediately shows up in the phenotype. (In diploid organisms such as humans, a normal allele on the homologous chromosome can compensate for a recessive allele.)

Wild-type *Neurospora* can grow on an extremely simple medium (called a minimal medium) that contains glucose, ammonia, and salts. But Beadle and Tatum identified a large number of mutant *Neurospora* molds that could not grow on such a minimal medium. These mutants required the addition of various compounds that the molds were incapable of synthesizing themselves. One mutant strain, for instance, would grow

oxidize HA, and the HA accumulates and turns to alkapton bodies in the urine. Garrod realized that the absence of HA oxidase might cause alkaptonuria. He hypothesized that the recessive allele whose inheritance he had charted could explain the missing enzyme. If an alkaptonuria patient had two copies of a defective allele of the gene for HA oxidase, he or she would not be able to make the enzyme.

Archibald Garrod's study of the genetic disease alkaptonuria suggested that genes affect biochemical processes by specifying enzymes.

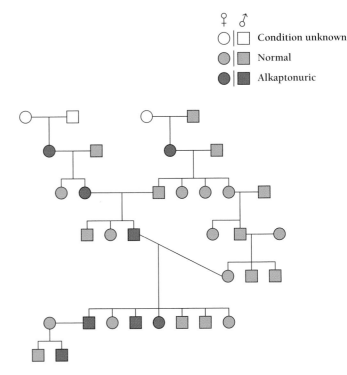

Figure 10-10 How do people inherit alkaptonuria? A pedigree of alkaptonuria from a Lebanese family. The marriage of cousins and other closely related family members increases the incidence of homozygous recessive individuals for both alkaptonuria and other genetic diseases.

Figure 10-11 What causes alkaptonuria? Cells can break down phenylalanine into homogentisic acid (HA), but without a crucial enzyme (#5) that normally breaks down HA to carbon dioxide and water, HA accumulates in the body. After it is excreted in the urine, the HA becomes oxidized and turns to dark alkapton bodies. Alkaptonuria is a relatively harmless genetic anomaly.

only if the minimal medium was enriched with vitamin B_6. Another mutant strain needed the amino acid arginine (Figure 10-12).

Beadle and Tatum then mapped the chromosomal location of each of their mutants by studying genetic linkage, as described in the last chapter. The mutants that needed arginine, for example, fell into three groups. Beadle and Tatum established that the three groups corresponded to mutations in three separate genes, on three separate chromosomes. Each gene coded for a separate enzyme needed to make arginine.

One group of mutants (*arg*-1) would grow when arginine was added and also in the presence of two related compounds, ornithine and citrulline. The second group of mutants (*arg*-2) would grow in the presence of arginine or citrulline, but not ornithine alone. The third group (*arg*-3) required arginine; neither citrulline nor ornithine would do.

Beadle and Tatum realized they could explain their results with the following model of arginine synthesis:

$$\text{precursor} \xrightarrow{\text{enzyme 1}} \text{ornithine} \xrightarrow{\text{enzyme 2}} \text{citrulline} \xrightarrow{\text{enzyme 3}} \text{arginine}$$

Each step in the synthesis of arginine, they reasoned, is catalyzed by a different enzyme. The *arg*-1 mutations, they argued, must lie in the gene for enzyme 1. The absence of enzyme 1, then, could be overcome by ornithine, citrulline, or arginine. Similarly, defects in the gene for enzyme 2 can be overcome by the addition of citrulline or arginine, but not of ornithine, since the organism cannot process it. Finally, defects in the gene for enzyme 3 can only be overcome by the addition of arginine itself (Figure 10-13). Beadle and Tatum's work helped unravel previously unknown biochemical pathways, since the mutations blocked the pathways at different steps and allowed biochemical analysis. Still more importantly, however, their work suggested that every gene somehow affects the function of a single enzyme. Beadle and Tatum captured this concept in a single phrase: **"One gene—one enzyme."**

Beadle and Tatum showed that one gene could specify one enzyme.

One Gene—One Polypeptide

Recall from Chapter 3 that all enzymes are proteins, but cells contain thousands of proteins that are not enzymes. Do genes contain information for making enzymes only? Are enzymes the only link between genotype and phenotype? Or can genes code for other proteins besides enzymes?

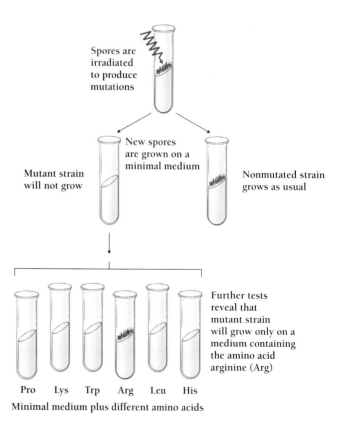

Figure 10-12 Mutant *arg* genes in the bread mold *Neurospora crassa*. Irradiation causes various mutations in bread mold spores. One set of mutant strains of *Neurospora* could not make the amino acid arginine in the normal way and needed either arginine or other amino acid supplements to survive.

Figure 10-13 One gene—one enzyme. In the bread mold *Neurospora crassa*, the three normal *arg* genes (*arg-1*, *arg-2*, and *arg-3*) each specify one enzyme. Together these three enzymes allow the synthesis of arginine. A mutation in any of the *arg* genes that makes the specified enzyme defective disrupts the chain of reactions. *Arg-1* mutants grow in the presence of arginine or two related compounds, ornithine and citrulline. *Arg-2* mutants grow in the presence of arginine or citrulline, but not ornithine alone. *Arg-3* mutants require arginine; neither citrulline nor ornithine will do.

The answers to these questions came with the unraveling of the molecular basis of another genetic disease, sickle cell anemia. The name "sickle cell" comes from the crescent or sickle shape of the red blood cells of people with this disease (Figure 10-14). The distorted blood cells clump and block the small blood vessels of the circulation, causing excruciating pain, anemia, heart trouble, and brain damage. The body's defense mechanisms eliminate the defective cells as rapidly as possible, but this causes severe anemia. People with sickle cell disease almost always die young. Sickle cell disease, like alkaptonuria, results from a recessive allele. Unlike alkaptonuria, however, sickle cell anemia is both serious and common— 1 in 13 African Americans carry the allele, and 1 in 625 have two copies of the allele and have the disease.

In 1949, the biochemist Linus Pauling and his collaborators showed that the red, oxygen-binding blood protein hemoglobin of sickle cell patients was different from that of healthy people. He called normal hemoglobin protein HbA and sickle cell hemoglobin HbS. The difference between HbS and HbA is a matter of a couple of amino acids. But Pauling used a technique called electrophoresis that reveals even tiny differences in the sizes and charges of molecules (Figure 10-15). Pauling and his colleagues showed that the healthy parents of sickle cell patients have a mixture of half HbS and half HbA. Their children, however, lack HbA entirely. A simple explanation of the disease emerged:

1. A recessive allele codes for the abnormal protein, HbS.
2. Heterozygotes such as the parents have genes for both HbA and HbS.
3. Heterozygotes with both HbA and HbS are healthy.
4. Homozygotes for the recessive allele make only HbS and are sick.
5. In the absence of the normal protein, HbA, HbS causes sickle cell disease.

In sickle cell disease, differences in the gene had caused differences in the hemoglobin protein, not in an enzyme. Pauling's work suggested that genes controlled other proteins besides enzymes. Pauling extended "one gene—one enzyme" to "one gene—one protein."

Exactly how does an abnormal gene change a protein? To answer this question, Vernon Ingram, a young biochemist working at Cambridge University in 1956, used new methods of protein chemistry to analyze the structural differences between HbS and HbA.

Normal hemoglobin is made of four polypeptides folded together—two of one kind, α-globin, and two of another, β-globin. Ingram showed that the α-globins in HbS were normal, but the β-globins contained a tiny "point" mutation. In the β chain's entire sequence of 146 amino acids, a single amino acid change produced an altered protein that could cause a deadly disease.

In the β-globin of HbS, amino acid number 6, normally glutamic acid, has been changed to valine. But while glutamic acid has a negative charge, valine is neutral. This single amino

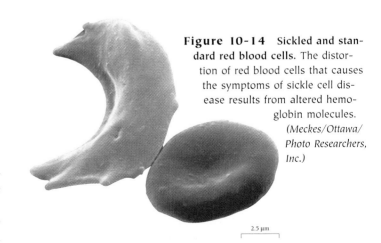

Figure 10-14 Sickled and standard red blood cells. The distortion of red blood cells that causes the symptoms of sickle cell disease results from altered hemoglobin molecules. *(Meckes/Ottawa/Photo Researchers, Inc.)*

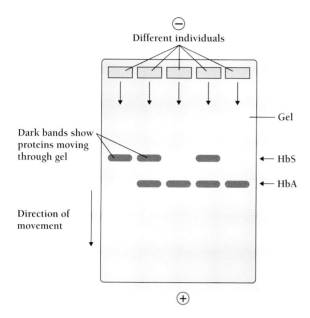

Figure 10-15 Electrophoresis. Biologists can separate macromolecules according to charge, shape, and size. In electrophoresis, an electric current pulls DNA, RNA, or proteins through a gel. Molecules move through the gel at different rates depending on their size and charge. Smaller ones move faster than larger ones, and those with a greater negative charge move faster than those that are less negative. The sickle cell hemoglobin polypeptide, HbS, is less negative than the normal hemoglobin polypeptide, HbA. So HbS moves more slowly through the gel than HbA.

acid substitution alters the way that hemoglobin molecules pack into the crowded red cell. When oxygen concentration drops—due to heavy exercise, for example—the HbS forms a fibrous precipitate that distorts the cell. This same charge difference is what makes HbS move more slowly than HbA during electrophoresis.

Ingram found that the sickle cell gene produced abnormalities only in the β-globin polypeptide, never in α-globin. Biologists concluded that each gene is responsible for the production not of an entire protein, but of a single polypeptide chain. "One gene—one protein" became **"one gene—one polypeptide."** But even this rule has important exceptions. As we'll see later on, some genes specify no polypeptide at all. And in special immune system proteins, several genes determine the sequence of a single polypeptide.

Nonetheless, we can say that, generally, genes influence phenotype by carrying the information necessary for the synthesis of polypeptides. But exactly how does a gene code for a polypeptide? To answer this question, we must look more closely at what a gene is.

Genes carry information necessary for the synthesis of polypeptides.

HOW DID BIOLOGISTS LEARN WHAT GENES ARE MADE OF?

Research on an amazing variety of organisms—including humans, peas, flies, and bread mold—has contributed to our understanding of what genes do. To find out what genes are made of, biologists turned to less complex, smaller organisms: bacteria and the viruses that infect them.

How Could Dead Bacteria Change the Genetic Properties of Other Bacteria?

Before the development of antibiotics during World War II, many people died of bacterial pneumonia. Intense research directed at developing a pneumonia vaccine led, by chance, to the discovery of the nature of genes.

In 1928, Frederick Griffith, an English bacteriologist, was trying to stimulate the immune defenses of mice against the pneumonia bacterium *Streptococcus pneumoniae,* also called pneumococcus. Griffith worked with two strains of pneumococcus. One caused pneumonia when injected into the mice, and one was harmless. When Griffith grew the two strains in dishes of gelatin, the harmless pneumococci formed rough (R) colonies, while the disease-causing strain formed smooth (S), glistening colonies. Today, we know that the virulent S bacteria form smooth colonies because they produce a polysaccharide capsule that protects them from attack by the body's immune cells. The R bacteria are mutants—they lack an enzyme needed to produce the polysaccharide capsule, which is why they cannot cause disease.

Griffith tried to produce immunity to pneumonia in mice by injecting them with different combinations of live and heat-killed pneumococci. As he expected, the mice injected with live S strain bacteria got pneumonia and died, while the mice injected with either live R strain bacteria or with heat-killed S bacteria survived. Griffith was amazed, however, by the results of injecting a mouse with a mixture of heat-killed S strain bacteria and live R strain bacteria (Figure 10-16). The mice died! Equally astonishing, the blood of the sick mice contained live S bacteria. Something in the dead S bacteria had transformed the harmless R bacteria into virulent S bacteria.

Frederick Griffith discovered that dead virulent bacteria could transform harmless bacteria into virulent ones.

What Was the Transforming Factor?

Other scientists soon found that they could convert R strains into S strains in a test tube as well as in a mouse. Extracts of S strains, dead or alive, could give R strains the ability to make the protective polysaccharide coat and become virulent. Miraculously, once the R strains were transformed into S strains, their descendants inherited the properties of S strains. How could the

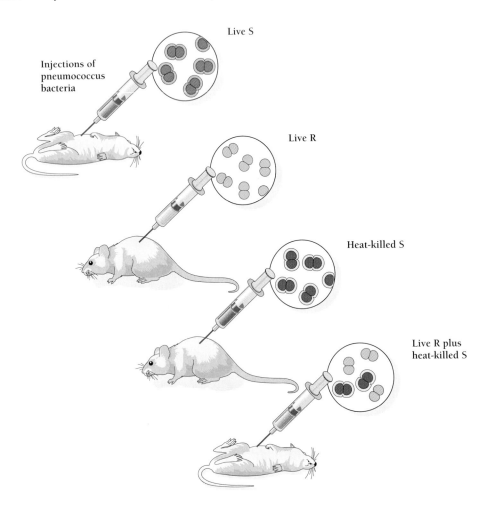

Figure 10-16 Griffith's experiment. Virulent S (smooth) *Streptococcus pneumoniae* bacteria kill mice, while harmless R (rough) bacteria and heat-killed S bacteria do not. Yet, amazingly, a mixture of harmless R bacteria and dead S bacteria results in a deadly infection. Griffith's work showed that DNA from the virulent S bacterium can transform harmless bacteria into deadly ones.

R bacteria "inherit" traits from individuals they were not descended from? What was *transforming* them into S bacteria?

It looked as if whatever was transforming the bacteria was some sort of genetic material. In the early 1930s, Oswald Avery at the Rockefeller Institute in New York set out to determine what that material was. With his coworkers Colin MacLeod and Maclyn McCarty, Avery carefully isolated different kinds of molecules from heat-killed S bacteria and tested to see which ones transformed the R bacteria.

At the time, almost everyone expected that the transforming factor would turn out to be made of protein, since proteins were already known to perform so many other cellular functions. Yet transformation had occurred even when the bacteria had been heat killed. Because heat denatures, or unfolds, proteins, it destroys their ability to function.

Avery and his colleagues worked for almost 10 years, isolating substance after substance from heat-killed S bacteria and then testing each substance to see if it could transform R bacteria. They treated each isolated substance with a variety of enzymes that would break down proteins into amino acids, polysaccharides into sugars, and nucleic acids into nucleotides. Finally, in 1944, they were forced to conclude that the transforming factor was DNA. First, only DNA from the S bacteria transformed the R bacteria. Second, the extract's ability to transform was destroyed only by an enzyme (DNAase) known to break down DNA but not proteins.

Avery's work strongly suggested that genes are also made of DNA. Despite the care with which Avery and his colleagues had carried out their work, most biologists rejected Avery's conclusion. Biologists had two important reasons for being skeptical. First, if DNA were the genetic material, the DNA of bacteria should differ from the DNA of butterflies and bananas. Yet, all DNA seemed to be chemically identical. Second, most biochemists accepted an idea proposed in the 1920s that DNA was a "tetranucleotide," the same sequence of four nucleotides repeated over and over. In that case, DNA would be unable to carry information. So, as late as 1944, most biologists continued to believe that DNA was a simple repeating molecule, no more capable of carrying information than a string of identical beads. DNA was regarded as a structural component of the chromosomes, no more.

By 1944, Oswald Avery and his colleagues had shown that bacteria are transformed by DNA, not proteins. Yet most biologists rejected the idea that DNA could be the genetic material.

What Is a Bacteriophage?

In 1915, long before Griffith's work on pneumococcus, a young French microbiologist named Felix d'Herelle discovered that bacteria themselves could be killed by a tiny infectious agent called a **bacteriophage** [Greek, *bakterion* = little rod + *phagein* = to eat], or simply **phage.**

D'Herelle first became aware of bacteriophages while trying to stop a plague of locusts that was then sweeping the Yucatan, in Mexico. The young biologist noticed that some of the locusts were dying from some disease that gave them diarrhea. He collected bacteria from the diarrhea of dying locusts and cultured it. Then, as hoards of the locusts advanced across the Mexican countryside, d'Herelle dusted the vegetation before them with his bacterium. As the locusts ate, they sickened and died. D'Herelle's plan worked magnificently.

D'Herelle was puzzled, however, by the appearance of clear spots in his cultures, places where the bacteria themselves seemed to have died. What was killing his bacteria? At first, he had no idea. The answer did not come until a moment of inspiration some 5 years later. Back in France, d'Herelle was studying an epidemic of dysentery in a squadron of French cavalry stationed near Paris. The soldiers had Shiga dysentery, and d'Herelle had been culturing the Shiga bacteria when, once more, the bacteriophages made their presence known.

> The next morning . . . I experienced one of those rare moments of intense emotion which reward the research worker for all his pains: at the first glance I saw that the broth culture, which the night before had been very turbid [cloudy], was perfectly clear: all the bacteria had vanished, they had dissolved away like sugar in water . . . in a flash I . . . understood: what caused my clear spots was in fact an invisible microbe, . . . a virus parasitic on bacteria. Another thought came to me also: "If this is true, the same thing has probably occurred during the night in the sick man, who yesterday was in a serious condition. In his intestine, as in my test tube, the dysentery bacilli will have dissolved away under the action of [the virus]. He should now be cured."

And, indeed, d'Herelle dashed to the hospital to find the sick man greatly recovered.

D'Herelle saw that bacteriophages were small enough to pass through the filters he used to remove bacteria. Also, these "filterable agents" could reproduce themselves within bacteria and, therefore, seemed to be alive.

Biologists do not consider bacteriophages alive, however, because they are not made of cells. Bacteriophages are **viruses,** molecular assemblies of protein and either DNA or RNA. Viruses are not cells. They cannot reproduce outside of a host cell, and they themselves cannot obtain energy from their environments. A single virus can, however, infect a cell and subvert the cell's molecular machinery to make more viruses. A virus, then, is a molecular assemblage of nucleic acid and protein that parasitizes cells. Viruses are not confined to infecting bacteria. They may infect plant cells, animal cells, and other kinds of eukaryotic cells.

Bacteriophages are a kind of virus—a parasitic assembly of protein and DNA or RNA capable of forcing host cells to make more viruses.

What Convinced Biologists That Genes Are Made of DNA?

In the 1940s, Max Delbruck and Salvador Luria began to study the genetics of the bacteriophages that infect the common intestinal bacterium (Figure 10-17A). Because bacteriophages reproduce so rapidly, they were, Delbruck and Luria decided, the ideal material for genetic studies. A bacterium such as *E. coli* infected with a single bacteriophage breaks open after about 20 minutes and releases about 200 new phage particles (Figure 10-17B). Each of these new phages can then infect another bacterium. If the bacteria are growing on a solid culture medium, the phages create a clear spot, or "plaque," on the otherwise continuous bacterial "lawn." As the phages multiply, their descendants may eventually consume almost the entire lawn.

Delbruck, Luria, and other members of the "phage group" catalogued many inherited variants of the bacteriophage. The researchers produced many mutant phages with distinct phenotypes. Some phages produced clear plaques, and others produced cloudy plaques. Others infected one strain of *E. coli* but not another.

The different kinds of phages passed their traits to their descendants. Phage particles isolated from cloudy plaques produced viruses that made cloudy plaques. Bacteriophages, then, like organisms, have genes, and alleles of these genes specify alternative phenotypes such as clear or cloudy plaques. The phage researchers knew that bacteriophages consisted of complexes of proteins surrounding a molecule of nucleic acid, usually DNA. Their genes therefore had to be in either the protein or the DNA.

Which one was settled in 1952 by Alfred Hershey and Martha Chase (Figure 10-18A). They performed an experiment specifically designed to discover if DNA was also the hereditary material in bacteriophages. Hershey and Chase knew from microscope studies that during a phage infection only part of the virus enters the host cell. The rest remains outside (Figure 10-18B). They reasoned that the material that actually enters the cell must carry the virus's genetic information. So the question was, What was that material?

To find out, Hershey and Chase used radioisotopes to label the protein and DNA. The work on the atomic bomb during World War II had greatly increased the availability of radioisotopes, so Hershey and Chase grew bacteria ^{32}P-labeled phosphate and ^{35}S-labeled sulfate. Because some of the amino acids that make up proteins contain sulfur but no phosphorus and because nucleic acids have phosphorus but no sulfur, the bacteria therefore had ^{35}S incorporated in their proteins and ^{32}P in their DNA. Hershey and Chase then allowed phages to infect these radioactive bacteria. The newly made phages picked up the same radioactive labels, so their DNA contained ^{32}P and their protein contained ^{35}S.

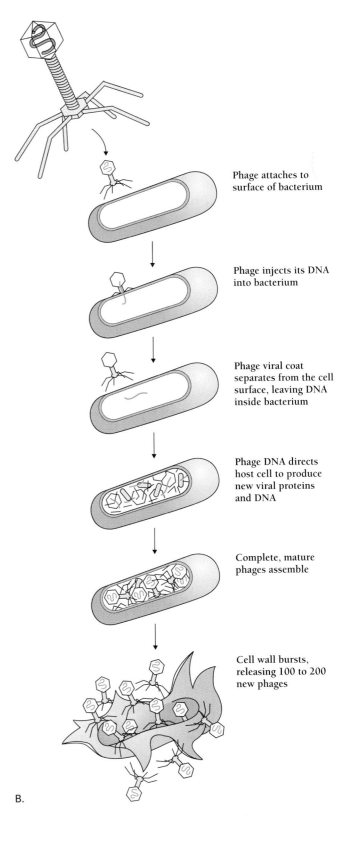

Figure 10-17 The life of a phage. A. Max Delbruck and Salvador Luria examining a petri dish containing bacteria and phages. B. Life cycle of a bacteriophage. (A, Cold Spring Harbor Laboratory Archives)

Hershey and Chase then used the radioactive phage to infect new, nonradioactive bacteria. They waited just long enough for the infection to begin and then interrupted the process by dumping the bacteria into a kitchen blender. The force of the blender's whirling blades knocked the empty phage particles off the outside of the bacteria and separated these particles from the bacteria. Hershey and Chase then examined the infected bacteria to see what part of the phage had entered the bacteria. Was it the DNA or the protein?

It was the DNA. The ^{32}P-labeled phage DNA was inside of the bacteria, and the ^{35}S-labeled phage protein was outside. Hershey and Chase also showed that the phage DNA alone could direct the synthesis of new phages and that these new phages—made and released by the infected cell—contained ^{32}P, but never ^{35}S. Hershey and Chase therefore concluded that the genes of the bacteriophage were made of DNA, not protein.

Why was Hershey and Chase's experiment more persuasive than Avery's? When Avery's paper came out in 1944, most molecular biologists still believed that DNA was a simple repeating structural molecule. But by 1950, Erwin Chargaff had shown that all DNA was not the same—that different species had different proportions of bases. Without equal proportions of each base, DNA could not possibly be a regular repeating tetranucleotide.

In addition, Delbruck and Luria's phage group had convinced many biologists that the genes of bacteriophages must be similar to the genes of organisms. Hershey and Chase's 1952 experiment therefore immediately drew the attention of a large

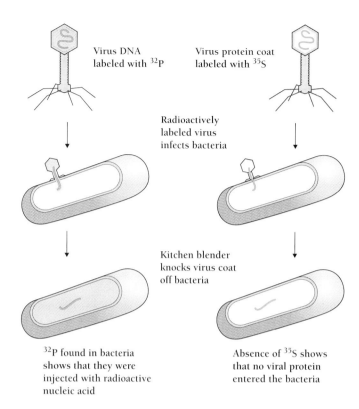

Figure 10-18 What convinced biologists that the genetic material was DNA? A. Martha Chase and Albert Hershey showed that when a phage infects a bacterium, it is the viral DNA, not protein, that enters and infects the bacterium. B. In the Hershey-Chase experiment, radioactive DNA from a phage shows up inside the infected bacterium. Radioactive protein stays outside. This 1952 experiment finally convinced biologists that DNA, not protein, was the genetic material. (Later work showed that a few viruses use RNA as their genetic material.) (A, Cold Spring Harbor Laboratory Archives)

group of influential biologists in a way that Avery's paper, 8 years earlier, had not. Finally, Watson and Crick's 1953 paper on how the structure of DNA might allow it to carry information and replicate silenced any remaining skeptics. After 1953, nearly all biologists accepted the idea that DNA was the genetic material.

It took from 1903 to 1952 to persuade biologists that DNA was the genetic material. The next questions were, What is the structure of DNA? and How does it work? As we have seen, Watson and Crick substantially answered both questions within a month of Hershey and Chase's paper.

Hershey and Chase's experiments with radioactively labeled phages showed that when a phage infects a bacterium, it is the DNA that enters and infects the bacterium, not the protein. This experiment convinced biologists that DNA is the genetic material.

HOW DOES ONE DNA STRAND DIRECT THE SYNTHESIS OF ANOTHER?

Watson and Crick's complementary base-pairing rule established the basis of DNA's ability to replicate itself. Since a base on one strand can match with only one other base, the sequence of bases on each strand implies the sequence of bases in the complementary strand. In their first paper describing their model, the two coyly commented on this property: "It has not escaped our notice that the specific pairing we have postulated immediately suggests a possible copying mechanism for the genetic material." (Years later, they confessed that they wanted to claim the idea without saying so much that they sounded foolish if they turned out to be wrong.) In the same way that a photographer can make a positive print from a negative, or a negative from a positive print, each strand of DNA has the capacity to direct the synthesis of the other.

This complementarity is the basis for understanding both how a chromosome replicates itself during cell division and how genes actually work. The structure of DNA explains both how cells copy their DNA and how genes specify proteins (Chapter 11).

How Do Dividing Cells Copy the DNA?

Each of the four bases in DNA can pair with only one other base—A with T and G with C. The rules of base pairing in DNA mean that the sequence of one polynucleotide strand specifies the sequence of the other strand. The two strands of DNA are complementary, meaning that they fit together to make a unified whole—the double helix.

The basis for DNA replication is that the sequence of bases in one chain specifies the sequence of bases in the other. That

BOX 10.1

How To Sequence DNA

One of the most important questions a biologist can ask about a gene or any stretch of DNA is, What is its sequence of nucleotides? With that information, medical researchers can begin to understand complex genetic diseases such as sickle cell anemia, Tay-Sachs disease, and many others. The most widely used method for DNA sequencing was developed in England in the 1960s by Fred Sanger. Sanger's method depends on DNA polymerase and five "rules" of DNA synthesis:

1. DNA synthesis is semiconservative.
2. A newly made DNA strand grows in a single direction, adding one nucleotide at a time to the 3′ end.
3. The nucleotide sequence of the template strand determines the sequence of nucleotides in the growing strand.
4. DNA polymerase requires a primer polynucleotide.
5. The copying of a single strand into a new complementary strand is catalyzed by a single enzyme, DNA polymerase.

In addition, DNA sequencing requires that a researcher be able to separate DNA fragments of different lengths.

Sanger began by heating the DNA until it separated into two single strands. He then added a specific complementary primer, DNA polymerase, and plenty of A, C, G, and T. The DNA polymerase automatically set to work extending the primer until it produced a complete copy of the template strand.

Sanger then devised a way of stopping the polymerase reaction at different places. He could stop the reaction at A, C, G, or T by replacing each kind of base with one that would prevent further polymerization. For example, Sanger replaced A with a similar molecule that lacked a hydroxyl group at the 3′ end. Once this modified A "base" was added to the growing strand, DNA polymerase could add no more bases. The result was a set of fragments of different lengths, each ending in A. Sanger did the same for C, G, and T.

The next problem was to put all these fragments in order, shortest to longest. Fortunately, electrophoresis does just that. If we run a small electric current through a flat sheet of gel containing the fragments, the smallest fragments speed across the gel, while the longest ones move more slowly as they work their way through the complex structure of the gel. Using electrophoresis, Sanger could put the fragments in order, shortest to longest—even ones that differ in length by no more than one nucleotide (Figure A). Each fragment showed up on the gel as a dark spot. Each column of spots represents the different fragments ending in A, G, C, or T (Figure B). From these four sets of bands, researchers can work out the complete sequence of any newly made DNA.

Figure A In electrophoresis, an electric current causes fragments of DNA to spread across a gel according to the speed with which they move through the gel. The biggest fragments are the slowest. Biologists then "read" the last letter of each fragment, starting with the shortest fragment. Reading from shortest to longest, the nucleotide letters are A-T-C-G-T-T-G-A.

Figure B Photo of an actual DNA sequencing gel. Reading from the 5′ end (bottom of the gel), the sequence is TATCCCGTTGGAAGGTCGTCTGCTCCCTG-GAAGTAG. (James D. Colandene, University of Virginia)

is, the two strands of the DNA double helix contain the same information in two different forms. Picture a molecule of DNA unzipping to form two complementary single strands. If the bases and the proper cellular machinery are available, each strand can act as a template for the manufacture of a new complementary strand. When each strand has created its one complement, the result is two new double-stranded DNA molecules, each a replica of the parent DNA.

According to this model, DNA replication is **semiconservative.** That is, half of each parent molecule is present in each daughter molecule (Figure 10-19). Each new DNA molecule, then, is really half old and half new.

DNA replication is semiconservative: When DNA replicates, each half of the double helix acquires a new mate.

Our discussion has so far focused on how DNA passes information to daughter DNA molecules. But what is the mechanism? How do cells actually produce new DNA? As in the case of other biochemical reactions, DNA replication depends on enzymes. The full process of DNA replication requires at least a dozen enzymes. Even now, more than 40 years after molecular biologists first synthesized DNA in a test tube, they still do not know all the details of DNA replication. The most basic steps of the process, however, are well understood.

How Does a Cell Begin Assembling a New Strand of DNA?

The assembly of DNA involves linking many small molecules (nucleotides) into a long chain, a process called **polymerization** [Greek, *polys* = many + *meros* = part]. The enzyme that strings together the nucleotides is called **DNA polymerase.** In each strand, the nucleotides must polymerize in a fixed order, guided by the order of bases in the complementary strand (Figure 10-20).

The Watson-Crick model for DNA replication suggests how each strand of DNA serves as a **template,** or guide, for the assembly of a complementary sequence. Our understanding of how the template works and how the new chains are assembled depends heavily on the work of Arthur Kornberg and his colleagues. By 1957, Kornberg had demonstrated that extracts of bacterial enzymes could catalyze DNA synthesis. He and his colleagues then isolated DNA polymerase and studied the reactions it catalyzed.

Kornberg found that DNA polymerase can add nucleotides only to an already existing chain of nucleotides called a **primer** (Figure 10-21). DNA polymerase adds nucleotides to the 3' end of the primer, so that the polynucleotide grows in a single direction—from its 5' end toward its 3' end.

The particular nucleotide added at each step depends upon the base sequence of the template strand. Each added nucleotide forms hydrogen bonds with the next base on the template strand. So each growing strand extends in a single direction, starting with the sequence closest to the 5' end of the primer and extending from its 3' end.

Once a free 3' end exists, we can see how the process continues. But if DNA can only add nucleotides to the 3' end of existing chains of nucleotides, how can the replication process ever begin? The resolution of this problem is that DNA replication does not begin with a DNA primer, but with a temporary RNA primer.

The RNA primer arises from the action of **primase,** an RNA polymerase that makes short stretches (fewer than ten nucleotides) of RNA using DNA as a template. Primase does not itself require a primer. With the RNA primer in place, DNA polymerase can then begin synthesizing the new strand of DNA, adding nucleotides step by step to the RNA primer's 3' end.

An RNA polymerase called primase creates an RNA primer that jump-starts DNA polymerase, enabling it to begin making a copy of each strand of DNA.

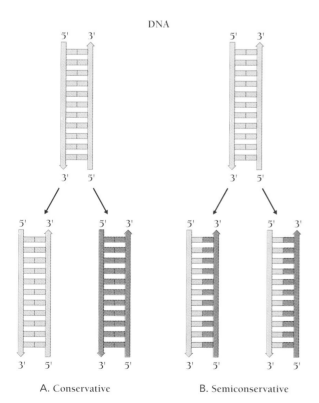

Figure 10-19 Semiconservative replication. A. At one time biologists hypothesized that DNA replication was conservative. Both halves of the old molecule stayed together, they thought incorrectly, while a whole new molecule was created. B. In fact, when DNA replicates, each half of the old double helix acquires a new mate.

How Does a Cell Copy Both Strands of DNA Simultaneously?

Researchers can actually see cellular DNA in the process of replication with the electron microscope. The DNA in a dividing tissue culture cell looks like a long, thin molecule

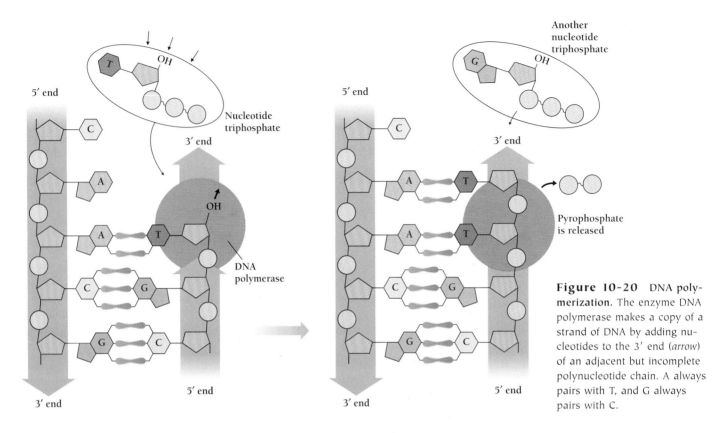

Figure 10-20 DNA polymerization. The enzyme DNA polymerase makes a copy of a strand of DNA by adding nucleotides to the 3′ end (*arrow*) of an adjacent but incomplete polynucleotide chain. A always pairs with T, and G always pairs with C.

punctuated by bubbles. As DNA replicates, these bubbles expand. Beyond each bubble are the two strands of parental DNA. Within each bubble are the four strands of replicated DNA—two parental strands and two newly made strands.

Each bubble consists of two **replication forks,** Y-shaped regions of DNA where the two strands of the helix have come unzipped (Figure 10-22A). At the replication fork, each strand begins to direct the assembly of a new complementary strand. As replication proceeds, the forks move away from each other. In bacteria, the replication forks move away from each other at about 500 nucleotides per second, and, in mammals, at about 50 nucleotides per second.

DNA replication occurs in replication forks, Y-shaped regions of DNA where the two strands of the helix have come apart.

How Does DNA Replicate in the 3′ to 5′ Direction?

As a replication fork moves along both strands of DNA, it moves from the 5′ end to the 3′ end along one strand, but from the 3′ end to the 5′ end on the other strand. We have seen how simply and continuously the first strand is replicated. But the second strand cannot be replicated in this way because nucleotides must be added to its 5′ end, a task that DNA polymerase cannot perform. How is this strand made?

The answer is that the second strand is produced in bits and pieces. After the moving fork has exposed about 1000 nucleotides of template for the lagging strand, primase produces a short piece of complementary RNA. This RNA serves as a primer for DNA polymerase, which attaches nucleotides to its 3′ end. This process produces fragments, called **Okazaki fragments,** after their discoverer Reiji Okazaki. Each Okazaki fragment contains a short stretch of RNA connected to about 1000 nucleotides of a DNA strand (Figure 10-22B). DNA polymerase then removes the RNA primer, and yet another enzyme, called **DNA ligase,** stitches the fragments together.

In one strand of a replication fork, nucleotides add directly to the 3′ end of an RNA primer. The other strand is produced in short fragments that are joined together by DNA ligase.

How Does DNA Synthesis Begin?

DNA always begins replication at a special site called an **origin of replication.** *E. coli,* the organism whose DNA synthesis has been most studied, has a single piece of circular DNA with a single origin of replication. Starting from this single site, two replication forks move in opposite directions away from the origin of replication, and DNA synthesis stops when the fork that is moving clockwise meets the counterclockwise fork (Figure 10-23).

Figure 10-21 How do enzymes replicate DNA? The blue strand of DNA acts as a template for the creation of a new complementary strand of DNA (rust). The enzyme *helicase* unwinds the strands. On the left, DNA polymerase adds nucleotides in a continuous string, working from the 3' end of the template to the 5' end. On the right (*center*), DNA polymerase works in the opposite direction, starting with RNA *primers* (also rust) to create Okazaki fragments, which are later linked together by *DNA ligase*.

WHAT IS THE ULTIMATE SOURCE OF GENETIC DIVERSITY?

Every living organism shares the same chemistry of life. Your DNA is made of the same stuff and follows the same rules as the DNA of the roses nodding in your neighbor's yard. Yet you are different from a rose. If, for example, you want lunch, you must get up and make yourself a sandwich, while the rose simply waits passively for the sun to emerge from behind a cloud. What makes us so different from a rose?

Each species has its own characteristic DNA. But life most likely began on Earth just once. If that is the case, then at one time all organisms were of the one kind and all of their genetic material was the same. How, over billions of years of evolution, did we organisms become so different from one another? The short answer is **mutations,** permanent changes in the sequence of DNA. The different alleles of a gene are created by changes in the sequence of the original gene. The original sequence itself is a mutation of some other gene. All genetic differences ultimately depend on the accumulation of mutations in the genome.

Although the evolution of organisms depends on the accumulation of changes in the DNA, the vast majority of mutations are either harmful or meaningless. If we opened up the back of a television set and switched two wires at random, the chances that this would improve the picture or the general performance of the television are almost zero. The same is true of mutations in highly evolved genomes, such as those of butterflies or mammals. Most gene mutations are a disaster for the phenotype. Still, every now and then, a new mutation is useful under certain circumstances. In that case, natural selection may increase its representation in a population of organisms.

We have already seen in Chapter 9 how the recombination (crossing over) and the mixing of gametes during sexual reproduction create unique combinations of alleles. But, ultimately, all new alleles result from mutations. Our focus in the

Several proteins participate in DNA replication. An enzyme called *helicase* uses energy from ATP to untwist the helix, while another enzyme called DNA gyrase prevents the accumulation of kinks in the stems of the replication forks. Still another protein binds to the unwound DNA and keeps the jaws of the replication fork open.

In eukaryotic cells, DNA replication is more complex. Each cell contains thousands of origins of replication. Groups of 20 to 50 origins, called **replication units,** form replication forks at the same time. Within a replication unit, the forks move in opposite directions from each origin until they encounter a fork from an adjacent origin.

DNA synthesis always begins at special sequences in the DNA called origins of replication. Proteins bind to these origins of replication and open and untwist the double helix.

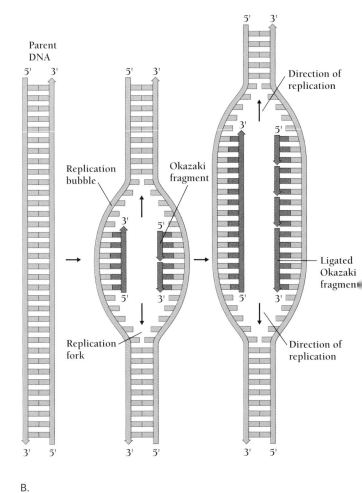

Figure 10-22 Replicating DNA. A. TEM of a replication bubble, with a replication fork at each end. B. A simplified view of a DNA replication bubble. Replication begins in the middle of the bubble and moves away from the middle in two directions at once. (A, D. Hogness and H. Kriegstein, Stanford University, Proceedings of the National Academy of Science, 71, 135–139, 1974)

Figure 10-23 Replication of bacterial DNA. Replication begins at the origin of replication (*arrows*), then works in both directions around the circle.

last part of this chapter will be on the origin and consequences of mutations: How do mutations arise? How do they influence the phenotype of the organism?

Mutations are the ultimate source of all genetic diversity.

What Causes Mutations?

Mutations occur spontaneously as random events in the cell. The structure of any molecule—including DNA—can change as a result of a random collision with another molecule or ion. For example, the bonds between the A and G (purine) bases in DNA and the backbone sugar are slightly unstable. As a result, each cell in the human body loses some 5000 to 10,000 A and G nucleotides every day. (Of course, most of this damage is repaired, as we'll see later.)

Occasionally, mutations occur because DNA polymerase places the wrong nucleotide in a sequence during the synthesis (S) phase of the cell cycle (Chapter 8). In addition, certain agents, called **mutagens,** increase the rate of mutation. Ultraviolet light, high-energy radiation such as x rays, and many chemicals can all act as mutagens. Each mutagen has a characteristic spectrum of effects. Ultraviolet light, for example, may link together adjacent Ts in the same DNA strand. X rays are particularly effective producers of chromosomal mutations, causing DNA to break and rejoin in strange new arrangements. Chemical mutagens, such as alcohol and dioxin, for example, also increase mutation rates.

Mutations occur spontaneously. But they occur more frequently when DNA is exposed to ultraviolet light, x rays, chemicals, or other mutagens.

How Often Do Mutations Occur?

Changes in DNA sequence occur spontaneously all the time. Mutations can initially result from random collisions between molecules, from mistakes in replication (for example, from a failure to form a correct base pair), or from changes in the structure of a nucleotide.

The **mutation rate** is how often mutations occur. More specifically, the mutation rate is the number of changes per nucleotide per generation. The mutation rate depends both on how often a sequence mutates and on how efficiently cells repair these mutations. Some sequences within a gene, called **mutation hot spots,** are more likely to mutate than others. In bacteria, for example, some sequences mutate 25 times more rapidly than others. Even the highest rates of mutation, however, are far less than the error levels to which we are accustomed in everyday life. In the days before computers and word processors, for example, an outstanding typist could type this whole chapter without going back to correct anything and with only one mistake. Yet, the average rate of mutation in bacteria during ordinary DNA replication is only 1 mutation in 10 million base pairs each generation. DNA replication is at least 1000 times more accurate than the best human typist.

Mutations in bacterial DNA occur at the rate of about 1 mutation in 10 million base pairs each generation.

What Kinds of Mutations Are There?

We can categorize mutations on the basis of the sizes of the changes in DNA. Geneticists often distinguish between **point mutations,** those which change one or several nucleotide pairs, and **chromosomal mutations,** those which change relatively large regions of chromosomes. A point mutation may be (1) a **base substitution**—the replacement of one base (nucleotide) by another, (2) an **insertion**—the addition of one or more nucleotides, or (3) a **deletion**—the removal of one or more nucleotides.

Chromosomal mutations may affect large regions of chromosomes or even whole chromosomes. Chromosomal mutations include (1) **deficiencies**—deletions that are larger than a few nucleotides, (2) **translocations**—in which part of one chromosome is moved to another chromosome, (3) **inversions**—in which a segment of a chromosome is flipped over 180° from its normal orientation, and (4) **duplications**—in which part of a chromosome appears twice. **Aneuploidy,** abnormal chromosome number, is regarded as another type of chromosomal mutation. For example, **trisomy** is the presence of three copies of a certain chromosome, and **monosomy** is the presence of only a single copy in a cell that normally has pairs of chromosomes.

A mutation is any change in the sequence or number of nucleotides, ranging from small changes of one or a few nucleotides to chromosomal inversions or aneuploidy.

HOW DO CELLS HANDLE MISTAKES IN THE NUCLEOTIDE SEQUENCE OF DNA?

Because DNA mutates, cells must be able to repair damage to DNA if they are to maintain the continuity of life. This is especially important for the germ line cells that pass information to the next generation by way of eggs and sperm. But mutations in somatic (body) cells are dangerous as well. In skin cancer, for example, a cell with a defective gene for growth control multiplies so fast that immature daughter cells take over the area. More than half of all forms of cancer arise from cells with one or more mistakes in just a few genes. The normal forms of these genes specify proteins called "tumor suppressors."

All cells have sophisticated mechanisms for repairing damaged DNA and correcting mistakes in replication. These mechanisms depend on the redundancy of information in DNA's structure and on special enzymes that recognize common mistakes.

To correct a mistake, an enzyme detects something wrong in one strand of the DNA and removes the error. For example, cytosine may be accidentally converted to uracil, a base that is ordinarily not present in DNA. A special enzyme called uracil-DNA glycosidase can recognize the strange base and remove it. In another example, two T nucleotides may bond to one another instead of to their respective A nucleotides (Figure 10-24).

Once correction enzymes remove the offending nucleotides, DNA polymerase goes to work. It copies the information in the intact second strand and creates a new stretch of DNA that is properly matched. DNA ligase then seals the gap, and the original sequence is restored. Over 50 enzymes "proofread" DNA for errors such as unusual bases, gaps, and bulges.

Errors still can slip through this elaborate system. For example, the genes for error-correction enzymes may themselves be damaged. In that case, large numbers of mutations can accumulate (Figure 10-25). If these uncorrected mutations occur in genes that help control cell division, the cells become cancerous.

Figure 10-24 Healthy cells are quick to repair damaged DNA. Ultraviolet light can damage DNA by linking together adjacent Ts in the same strand. Repair enzymes excise the abnormal nucleotides and restore the original sequence, then DNA polymerase and DNA ligase repair the gap as in normal replication.

Cells have sophisticated mechanisms for repairing damage and for correcting mistakes made during DNA replication.

In this chapter, we have seen that DNA has a special structure that enables it to replicate. We have also seen how we know that genes are made of DNA and that genes are physical regions in the DNA. In the next chapter, we will see how DNA's structure allows the "expression" of genes as polypeptides and how cells control the expression of genes—so that a given polypeptide is synthesized at the right time and in the right amount. We will also further explore how mutations in the structure of DNA change the information in genes.

Figure 10-25 Sunburn damages DNA and leads to skin cancer. The ultraviolet light in sunlight can damage DNA repair mechanisms. In the absence of DNA repair, many genes become damaged, including those that control the cell cycle. In early rounds of sunburn, most cells with damaged DNA self-destruct, and the skin peels. One theory of skin cancer holds that a few cells with damaged DNA remain but do not reproduce because they are surrounded by healthy cells. During subsequent sunburns, however, more cells with damaged DNA self-destruct, clearing the field for cancerous cells to begin growing. (© 1990 Peter Menzel/Stock, Boston)

Study Outline with Key Terms

Biologists first hypothesized that genes specified the structures of enzymes. When an organism cannot make an enzyme, it cannot carry out the biochemical conversions catalyzed by the enzyme. In alkaptonuria, for example, the absence of a specific enzyme causes the accumulation of a compound that turns the urine black. Studies of similar interruptions in biochemical pathways in the bread mold *Neurospora* in the 1940s suggested that every gene specifies the structure of a single enzyme. Researchers Beadle and Tatum coined the phrase, **One gene—one enzyme.** But within just a few years, other research showed that genes could specify the structures of proteins other than enzymes. Genes could specify single polypeptides only, however, not necessarily whole proteins. A more accurate statement of gene function is, **One gene—one polypeptide.**

Different alleles of a gene produce different versions of the same polypeptide. Thus, an alteration in the gene coding for the β-globin polypeptide leads to a **hemoglobin** with a structure different from normal hemoglobin. Sickle cell disease results when an individual inherits two copies of an allele that alters the structure of β-globin at a single amino acid.

The idea that genes are made of DNA came from work on bacteria and bacterial **viruses,** called **bacteriophages,** or simply **phages.** A virulent strain of bacteria can transfer the genetic instructions for virulence to a nonvirulent strain by means of DNA.

DNA consists of nucleotides held together by bonds between the third (**3′**) carbon of the sugar of one nucleotide and the fifth (**5′**) carbon of the sugar of the next nucleotide. Each polynucleotide chain has a direction with a 5′ end and a 3′ end. Hanging off the sugar-phosphate backbone of DNA are four kinds of nitrogen-containing bases. The bases are two **purines** (**adenine** and **guanine,** abbreviated A and G) and two **pyrimidines** (**cytosine** and **thymine,** abbreviated C and T). According to **Chargaff's rules,** the amount of C always equals the amount of G, and the amount of A equals the amount of T in the DNA of a given species.

In cells, DNA consists of two polynucleotide strands twisted around each other in a double helix. In the early 1950s, Rosalind Franklin's research showed that DNA was a helix. She was also the first to describe its density, water content, symmetry, and the number of bases (10) per turn of the helix. Previous work had shown that DNA was 2 nm wide, which Franklin confirmed, further reporting that the molecule was consistently 2 nm along its whole length.

Using Franklin's results, James Watson and Francis Crick built a model of DNA. They found that with the two strands of DNA running in opposite directions, hydrogen bonds could link the base pairs, G with C and A with T. These base pairings, argued Watson and Crick, explain the equal concentrations of G and C, and of A and T. The base pairings also meant that the information contained in the sequence of nucleotides of one strand was equivalent to that in the other strand. Each strand was a template for the other, and each strand could direct the synthesis of a complementary strand during DNA replication.

DNA replication is **semiconservative:** When DNA replicates, each strand of the double helix acquires a new mate. The enzyme **DNA polymerase** assembles nucleotides into a sequence specified by one strand of DNA, the **template** strand, a process called **polymerization.** DNA polymerase can add nucleotides only to a preexisting RNA **primer** made by the RNA polymerase **primase.** DNA polymerase can add nucleotides only to the 3′ end of the primer and continues to add nucleotides to the 3′ end of the growing polynucleotide.

DNA replicates in Y-shaped regions of the DNA called **replication forks.** Nucleotides add directly to the 3′ end of one strand in the fork, called the leading strand. The other strand is made discontinuously in **Okazaki fragments** of about 1000 nucleotides. Each fragment starts with a short RNA primer. **DNA ligase** links the fragments into a continuous strand. Special sequences in DNA serve as **origins of replication,** where DNA synthesis always begins. The sequences bind to proteins that open and untwist the double helix. Groups of 20 to 50 origins are called **replication units.**

DNA molecules suffer inevitable damage in the form of mutations. A **mutation** is any change in the sequence or number of nucleotides. **Point mutations** include multiple **base substitutions, insertions,** and **deletions. Chromosomal mutations** include **deficiencies, translocations, inversions,** and **duplications,** as well as **aneuploidies** such as **trisomy** and **monosomy.** Mutations are the ultimate source of all genetic diversity.

Mutation rates are usually low—occurring in bacterial DNA, for example, at the rate of about 1 mutation in 10 million base pairs each generation. **Mutation hot spots** mutate more often than other parts of the genome. Mutations can occur spontaneously, but they are hastened by ultraviolet light, x rays, certain chemicals, and other **mutagens.** Cells have sophisticated mechanisms to repair damage to DNA and to correct mistakes made during DNA replication. One result of uncorrected mistakes is cancer.

Review and Thought Questions

Review Questions

1. What facts about the structure of DNA did Franklin contribute? What did Watson and Crick contribute?
2. How does Watson and Crick's theory of base pairing explain how DNA replicates itself?
3. Do genes specify enzymes, proteins, or polypeptides? Why did biologists worry about this distinction?
4. What does it mean for a bacterial cell to be *transformed*? What does the transforming?
5. How is semiconservative replication different from conservative replication?
6. What serves as a template for the replication of a single strand of DNA?
7. What enzyme polymerizes nucleotides into a new strand of DNA?
8. Why does the cell need an RNA primer for DNA replication to begin? Where on the DNA does DNA polymerase begin replication?
9. What is the ultimate source of all genetic variation?
10. Why do cells need mechanisms for repairing DNA?

Thought Questions

11. D'Herelle and others thought that bacteriophages might provide a way to kill disease-causing bacteria—an idea that was especially exciting in the days before the discovery of penicillin and other antibiotics. Yet, until recently, no one found a way to use bacteriophages to destroy bacteria that have already infected the body. What problems might prevent medical researchers from making bacteriophages kill bacteria in the body?
12. Why does the dentist place a lead shield over your lap before x-raying your teeth? What might happen if the dentist did not take this precaution?

About the Chapter-Opening Image

Like these two pencils, the two strands of a molecule of DNA are parallel but run in opposite directions.

On-line materials relating to this chapter are on the World Wide Web at **http://www.harcourtcollege.com/lifesci/aal2/**

Web Bit 10-1 The Double Helical Computer

Web Links

11 How Are Genes Expressed?

How Was the Genetic Code Discovered?

As soon as Watson and Crick discovered the structure of DNA, scientists all over the world wanted to become involved in the obvious next phase of the work, solving one of life's oldest secrets—the genetic code. Especially attracted were mathematicians and physicists, who believed they could crack the genetic code through logic alone. Yet, after years of effort, the code was ultimately cracked, not by the theoreticians, but by the rigorous experimental work of two young and unknown molecular biologists.

The first chink in the armor of the coding problem came in 1955, when Marianne Grunberg-Manago, working at New York University, discovered an enzyme that could link nucleotides together into long polymers. Grunberg-Manago took a year to isolate the enzyme. Once she had the enzyme, however, she began making artificial RNA polymers. She could, for example, link together long chains of A (adenine) nucleotides, which she called "poly A," or long chains of C (cytosine) nucleotides, called "poly C." In addition, she could make poly U, poly G, poly AU, and other combinations.

For the moment, however, no one knew what combination of nucleotides could actually be translated by a cell into a polypeptide. It was 6 more years before the next big break came. Then, in 1960, 31-year-old Johann Heinrich Matthei arrived at the National Institutes of Health (NIH) near Washington, DC. Matthei had come from Germany earlier that year looking for interesting scientific work. He had not found anything satisfactory yet, but he was especially interested in working on protein synthesis. At NIH, he got the names of three researchers who were trying to synthesize proteins artificially outside of a cell. Thirty-three-year-old Marshall Nirenberg stood out from the other two (Figure 11-1). His interests and instincts matched Matthei's own, and in November 1960, Matthei and Nirenberg embarked on a productive, if sometimes strained, collaboration.

Although, technically, Nirenberg was Matthei's boss, Matthei had come to NIH for collaboration, not direction. He was fully able to work on his own. Nirenberg's goal was to synthesize the polypeptides that make up proteins. He wanted to make them in a test tube. To do this, he planned to add ATP (for energy) and free amino acids to a mixture of ribosomes, nucleic acids, and enzymes extracted from cells. Other researchers had used this technique and had shown that the amino acids were incorporated into polypeptides. No one, however, knew what substances in the cell extracts were mak-

Key Concepts

- Protein synthesis in test tubes, using cell extracts, revealed how information flows from messenger RNA (mRNA) to polypeptide.
- DNA and RNA carry information in a 4-letter alphabet, while proteins carry information in a 20-letter alphabet.
- Translation of messenger RNA requires correct initiation, elongation, and termination of polypeptide synthesis.
- In both prokaryotes and eukaryotes, many proteins cooperate to accomplish and influence the expression of a single gene.

ing the polypeptides, and no one knew what kinds of polypeptides were being synthesized.

The obvious next step was to try to make a particular polypeptide by putting genetic information into the test tube. Several labs in the United States were hotly pursuing this idea. Before Matthei arrived, Nirenberg had tried to synthesize penicillinase, the enzyme that antibiotic-resistant bacteria use to break down the antibiotic penicillin. Nirenberg added bacterial DNA and RNA to his test-tube system. In experiment after experiment, however, he got no penicillinase. But without knowing how the code worked, no one would ever be able to synthesize any polypeptide, let alone a particularly valuable one such as penicillinase. Nirenberg needed to ask a simpler question.

After Matthei arrived, the two researchers asked a basic question: What kinds of RNA stimulate the synthesis of polypeptides? To answer this question, they took an extremely methodical approach. Before trying to synthesize any protein, they would make sure that their test-tube system could respond to RNA, any RNA. The two spent months perfecting their system. In the end, they could detect even minute amounts of protein synthesis. Their test-tube system was now highly sensitive to added RNA.

Nirenberg and Matthei knew that protein synthesis required ribosomes and that ribosomes contained RNA. Like most other scientists of the period, they believed that RNA must carry a sort of message from the DNA in the nucleus to the ribosomes in the cytoplasm, where protein synthesis normally happened. But although Nirenberg and Matthei's test tubes were packed with ribosomal RNA, they were seeing very little protein synthesis. Instinctively, they felt that, besides ribosomal RNA, there must be another kind of RNA—a kind that carried information that could specify a polypeptide and would make protein in their test tubes. But where was it? What was it?

They drew up a list of some 200 kinds of RNA to test. One of the first possibilities was RNA from tobacco mosaic virus. (In some viruses, RNA carries the genetic information.) The results were spectacular. Nirenberg and Matthei's test-tube system incorporated huge amounts of amino acids into some mystery protein.

Nirenberg instantly recognized a golden opportunity. Another molecular biologist, Heinz Fraenkel-Conrat, at the University of California, Berkeley, had recently worked out the sequence of the tobacco mosaic virus protein. Nirenberg called Fraenkel-Conrat, forged an instant collaboration, and in May 1961 left to spend a month in California to try to synthesize tobacco mosaic virus proteins. If he succeeded, he would gain scientific celebrity.

Matthei stayed behind at NIH, patiently testing the other possibilities on the list of 200 different kinds of RNA. High on the list were the artificial RNAs made using the methods invented by Marianne Grunberg-Manago. A week after Nirenberg left, Matthei set up the test tubes, the different enzymes, the ribosomes, the ATP, and the 16 amino acids he had on hand. Then to the different test tubes he added poly U, poly A, and poly AU. He knew that if he got any synthesis at all, poly U should make a polypeptide consisting entirely of one kind of amino acid. Poly A should too, but it should be a different amino acid.

Figure 11-1 Marshall Nirenberg. Nirenberg and Johann Matthei cracked the genetic code in 1960. (UPI/Corbis-Bettmann)

235

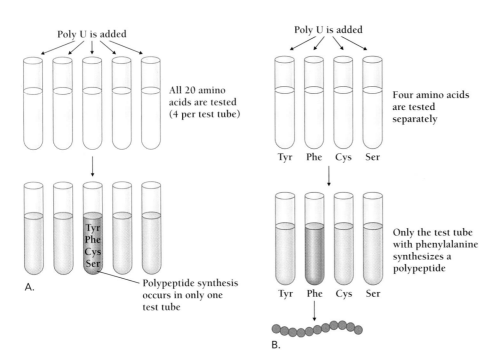

Figure 11-2 A simplified view of Nirenberg and Matthei's experiments. A. Matthei added four different radioactively labeled amino acids to each of five test tubes containing enzymes, ribosomes, and ATP, as well as poly U. One test tube synthesized large amounts of an unknown polypeptide. B. To find out which amino acid had been joined into polypeptides, Matthei added just one amino acid to each of five more test tubes. The test tube that synthesized protein was the one that contained the amino acid phenylalanine. The synthesized polypeptide consisted entirely of phenylalanine.

The results again were dramatic (Figure 11-2). Poly A and poly AU did almost nothing, but poly U stimulated 12 times the protein synthesis Matthei and Nirenberg got with nothing else added. The only question now was, What polypeptide had been made? In other words, which of the amino acids had been polymerized? Matthei attacked the problem aggressively. For 5 days, he worked almost round the clock, testing one amino acid after another.

On Friday night, Matthei stayed up all night again. Early on Saturday morning, bleary eyed but elated, Matthei had his answer. Poly U coded for a polypeptide made exclusively of the amino acid phenylalanine. He now knew that U nucleotides coded for phenylalanine, but how many nucleotides it took—whether one, two, three, or even more nucleotides—was another question. The number of nucleotides required to specify one amino acid was still unknown. Nonetheless, as Matthei stood in the lab, he realized that he alone in all the world knew the first word of the genetic code. He told Nirenberg's boss at NIH, but Matthei did not call Berkeley.

Nirenberg did not hear the news until he returned from California nearly 3 weeks later. His experiments at Berkeley had been inconclusive. The protein that tobacco mosaic virus RNA seemed to be synthesizing could not be identified. He must have heard Matthei's news with a mixture of emotions—delight at Matthei's success, disappointment that he had not been there to share in it.

Everyone at NIH soon knew what Matthei and Nirenberg had done. But the rest of the world either had not heard or did not believe it. Nirenberg was quiet, soft spoken. As one well-placed molecular biologist later said apologetically in reference to Nirenberg, "either a person . . . is someone who's in the club and you know him, or else his results are unlikely to be correct, because he [isn't] in the club." Nirenberg was definitely not in the club. And, as far as the fast-moving world of molecular biology was concerned, Matthei might as well have not existed.

Together Nirenberg and Matthei wrote up the results of all their work so far and sent it to a scientific journal. A week later, NIH sent Nirenberg to Moscow for the Fifth International Congress of Biochemistry. In a small, nearly empty room, Nirenberg presented the results of his and Matthei's months of work. Of the handful of scientists present, only molecular biologist Matthew Meselson really listened.

As Meselson later recalled, "I was bowled over by the results, and I went and chased down Francis [Crick], and told him that he must have a private talk with the man." Crick talked to Nirenberg, then arranged for him to give his talk once more in front of an audience of hundreds of biochemists and molecular biologists. Nirenberg's scientific fortune was made. He might not belong to the club, but he had been heard by nearly everyone who did.

While Nirenberg was in Moscow, Matthei nailed down still another "word" in the genetic code. Poly C, he found, made a polypeptide composed entirely of the amino acid proline.

Nirenberg returned to NIH in triumph. Yet within days of his announcement in Moscow, another research lab began using Nirenberg and Matthei's refined technique to crack the

First letter (5' end)	Second letter				Third letter (3' end)
	U	C	A	G	
U	UUU Phe UUC Phe UUA Leu UUG Leu	UCU Ser UCC Ser UCA Ser UCG Ser	UAU Tyr UAC Tyr UAA Stop UAG Stop	UGU Cys UGC Cys UGA Stop UGG Trp	U C A G
C	CUU Leu CUC Leu CUA Leu CUG Leu	CCU Pro CCC Pro CCA Pro CCG Pro	CAU His CAC His CAA Gln CAG Gln	CGU Arg CGC Arg CGA Arg CGG Arg	U C A G
A	AUU Ile AUC Ile AUA Ile AUG Met	ACU Thr ACC Thr ACA Thr ACG Thr	AAU Asn AAC Asn AAA Lys AAG Lys	AGU Ser AGC Ser AGA Arg AGG Arg	U C A G
G	GUU Val GUC Val GUA Val GUG Val	GCU Ala GCC Ala GCA Ala GCG Ala	GAU Asp GAC Asp GAA Glu GAG Glu	GGU Gly GGC Gly GGA Gly GGG Gly	U C A G

Nucleotide base   Codon   Amino acid

A.

B.

Four different codons code for the same amino acid. Note difference is in last base.

Figure 11-3 The genetic code. A. This table shows which amino acid corresponds to each RNA triplet. All the DNA Ts are replaced by Us, since this is RNA. B. Some amino acids have more than one codon. Usually the first two bases are the same, but the third base is different.

rest of the genetic code. Although initially taken aback by the aggressiveness of this move, Nirenberg soon threw himself into the competition, determined to be first.

In his excitement, Nirenberg began making all the decisions, and Matthei began to feel like a fifth wheel. His friendship and collaboration with Nirenberg soured, and 18 months after they began to work together, Matthei was back in Germany working by himself once more.

Nirenberg never missed a step. He began working with another collaborator, Philip Leder, and the two soon showed that a length of RNA just three nucleotides long could specify an amino acid. They were able to define a **codon** as a group of three nucleotides that specifies a single amino acid within a polypeptide.

Then another researcher stepped into the picture. Organic chemist Gobind Khorana found a way to link nucleotides together in any order he wanted. In one experiment, he put together long chains of poly AC, which contained the sequence ACACACACAC and so on. Khorana found that poly AC RNA stimulated the production of a polypeptide consisting of two alternating amino acids—threonine and histidine.

Khorana showed that ACA specified threonine and CAC specified histidine. Thus ACACACACACAC coded for threonine-histidine-threonine-histidine. From a series of such experiments, Khorana and his colleagues established a dictionary of codons. It took 5 years, but by 1966, Nirenberg, Khorana, and others had written a complete dictionary for the genetic code (Figure 11-3). There were 64 different codons. Sydney Brenner and Francis Crick showed that 3 of the 64 possible codons coded for no amino acids at all. They were "nonsense" codons whose role was to signal the end of a polypeptide chain, just like the period at the end of a sentence.

In 1968, Nirenberg and Khorana won a Nobel prize for their elucidation of the genetic code. In the same year, James Watson's best-selling book, *The Double Helix,* was published. In the world of biology, the first era in the history of molecular biology had come to an end.

The genetic code was deciphered by means of careful laboratory experiments.

What Is the Genetic Code Like?

The nucleotide sequences of DNA and RNA are written in a different language from the amino acid sequences of polypep-

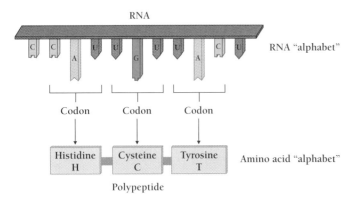

Figure 11-4 Three nucleotides code for one amino acid. The DNA/RNA language is written with an alphabet of nucleotides, while the polypeptide language is written with an alphabet of amino acids.

tides. The DNA/RNA language is written with an alphabet of just 4 nucleotides, while the polypeptide alphabet has 20 amino acid letters (Figure 11-4).

Although the DNA and protein languages are different, the relationship between them is simple. This is because the **genetic code** clearly specifies which nucleotide sequences correspond to which amino acid sequences. Each set of three nucleotides, called a codon, specifies an amino acid.

Since different combinations of any 3 nucleotides can code for 64 different amino acids, and there are only 20 amino acids, what do the extra 44 codons do? The answer is that most of these extra codons also code for the same 20 amino acids. The genetic code has several ways of specifying each amino acid. Because several codons may have the same meaning, we say that the genetic code is "redundant."

Of the 64 triplets, only 61 specify amino acids. The other three codons (UAA, UAG, and UGA) are called **nonsense,** or **stop, codons** because rather than specifying an amino acid, they normally signal the end of a polypeptide chain.

Because the same code is used by all species, the genetic code is said to be **universal.** One demonstration of this universality comes from studies of hemoglobin, the oxygen-carrying molecule in red blood cells. Before red blood cells enter the bloodstream, they synthesize large amounts of hemoglobin from two polypeptides, called globins, made from two kinds of RNA. When this RNA is purified from rabbit red blood cells and then added to wheat germ cells, the wheat cells make rabbit globin (Figure 11-5). This demonstrates that wheat cells and rabbit cells use the same code to translate RNA into protein. Indeed, all eukaryotes not only use the same code but they also use interchangeable molecular machinery.

Nonetheless, the genetic code is not completely universal. Mitochondria, whose DNA is passed on separately from the DNA in the nucleus, use a slightly different code from the one in Figure 11-3. And yeast mitochondria use a different genetic code from human mitochondria.

We have seen how molecular biologists solved the genetic code, which helps explain how genetic information flows from

Figure 11-5 All organisms use the same genetic code. The protein-making machinery of a wheat plant can synthesize blood proteins normally found only in rabbits.

DNA to RNA to proteins. But we must also ask about the mechanism by which this information flows. What happens, step by step?

The genetic code translates the language of DNA and RNA into the language of polypeptides. Three nucleotides—a codon—specify one amino acid. The genetic code is redundant and nearly universal, and 3 of the 64 codons are nonsense, or stop, codons.

GENETIC INFORMATION FLOWS FROM DNA TO RNA TO POLYPEPTIDES

By the mid-1950s, biologists knew that genes were sequences of nucleotides that specify the sequences of amino acids in polypeptides. The next question was, How does a sequence of nucleotides specify a sequence of amino acids? Molecular biologists such as Nirenberg and Khorana confirmed what Watson and others had long suspected: RNA is always an intermediate in the production of proteins.

Chapter 11 How Are Genes Expressed? 239

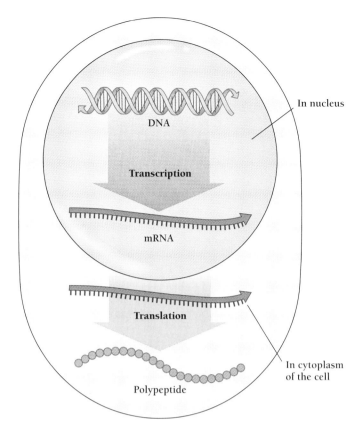

Francis Crick summarized cellular information flow in a statement he called the **central dogma** of molecular biology (Figure 11-6): "DNA specifies RNA, which specifies proteins." According to Crick's central dogma, information can flow *only* from DNA to RNA to proteins. It cannot flow from proteins to RNA or DNA. Crick put this idea another way, writing, "Once 'information' has passed into protein, *it cannot get out again*." Proteins cannot change the information in the genes.

Crick used the word dogma ironically because "dogma" means an opinion, belief, or a tenet of faith. It is not the word to use to describe scientific facts or well-supported theories. At the time Crick formulated it, molecular biologists had no good evidence supporting the central dogma. Crick and others just thought it must be true. Crick's joke was lost on most other researchers, however, and the "central dogma" has now been presented in textbooks *as dogma* for decades. As we will see in the next chapter, Crick had good reason for his twinge of doubt. The central dogma is mostly true, but not completely: some viruses—like the AIDS virus—carry information in RNA, which infected cells can incorporate into their own DNA. In addition, information in RNA and proteins influences the expression of the DNA's information.

Nevertheless, the idea that DNA specifies RNA, which specifies proteins, is a good starting point for understanding how genes are expressed through the two processes of transcription and translation. The process by which DNA transmits its information to RNA is called **transcription** (Figure 11-7A). During transcription, the information in the DNA is

Figure 11-6 The central dogma. DNA is transcribed into RNA, which is translated into protein. In eukaryotes, transcription occurs in the nucleus and translation occurs outside the nucleus, in the cytoplasm. In prokaryotes, which lack a nucleus, both transcription and translation occur in the cytoplasm.

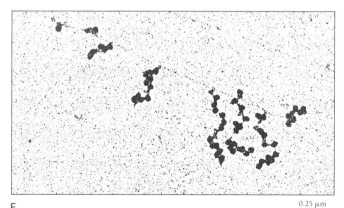

Figure 11-7 Transcription and translation. A–E. In this sequence of micrographs, RNA polymerase transcribes a strand of DNA into mRNA (which is not visible). RNA polymerase, the large, round blob, moves from the middle of the strand (A) to the end (E) in less than 5 minutes. F. In prokaryotes, which lack a nuclear membrane, transcription and translation can occur simultaneously. In this electron micrograph, the DNA runs diagonally from the upper left to the lower right. Transcription began at the upper left end of the DNA, so the longest mRNAs (with the most ribosomes) are at the lower right. Clusters of up to 13 ribosomes read individual strands of mRNA, even as they are being transcribed from the DNA. (A–E, M. Guthold, X. Zhu, and C. Bustamante, University of Oregon; F, Dr. Barbara Hamkalo, University of California, Irvine)

0.25 μm

BOX 11.1

Which Strand of the DNA Is Transcribed?

At the beginning of this chapter, we saw that protein synthesis in a test tube helped solve the genetic code. Such test-tube protein synthesis also allowed researchers to discover how RNA polymerase transcribes the DNA.

Just as DNA polymerase assembles new molecules of DNA one nucleotide at a time, RNA polymerase assembles RNA by adding one nucleotide at a time. However, DNA polymerase copies both strands of the DNA double helix, while RNA polymerase copies only one of the two strands of DNA. The strand that is copied is called the *template strand*. Although the two strands are chemically the same, the information in the template strand is completely different from that in the nontemplate strand of DNA. The template strand, also called the *antisense strand,* is a reverse copy of its mate, the *sense strand*. If this sentence were the sense strand, we might think the template strand would look something like this: :siht ekil gnihtemos kool dluow dnarts etalpmet eht kniht thgim ew ,dnarts esnes eht erew ecnetnes siht fI

Actually, however, the template strand is not only a reverse copy of the sense strand. It is also composed of the Watson-Crick base pairs, as if we had substituted different letters into our reversed sentence. For example, if the sense strand of DNA read CATTAG, then the antisense strand, read forward by RNA polymerase, would be CTAATG.

$$\xrightarrow{\text{sense strand}}$$
$$\text{CATTAG}$$
$$\text{GTAATC}$$
$$\xleftarrow{\text{antisense strand}}$$

The most important result of this reversal and substitution is that the antisense strand cannot code for a protein. The antisense strand is a chaotic sequence of codons that includes so many stop codons that any strange polypeptides made following its instructions could not be more than a few amino acids long. The sense strand, in contrast, has stop codons only at the end of each gene.

The antisense strand is useful, however, because the mRNA created from the antisense strand is not nonsense. The mRNA is an RNA version of the sense strand, the reverse of a reverse. The antisense strand at left, for example, would be transcribed as CAUUAG, since wherever DNA has a T, the mRNA has a U.

RNA polymerase copies the antisense DNA strand by holding in place an RNA nucleoside triphosphate base, which pairs with the next nucleotide in the template. The mRNA strand grows from its 5′ end to its 3′ end, as it copies the template DNA strand from its 3′ end to its 5′ end. The new mRNA molecule has the same sequence of bases as the sense strand of DNA, except that wherever DNA has a T the RNA has a U.

---

rewritten, or transcribed, to RNA. It is that information in RNA that cells use to make polypeptides. Using RNA as a secondary information source is analogous to working from a photocopy of an important document, while keeping the original in a safe place, in this case the nucleus.

The RNA copy of the information from DNA is composed of **messenger RNA (mRNA)**. Molecules of mRNA have the same information as the DNA original. RNA, like DNA, is a polynucleotide. It differs from DNA in just three ways:

1. The backbone sugar is ribose instead of deoxyribose.
2. Uracil (U) replaces thymine (T) as one of the pyrimidine bases.
3. RNA is usually single stranded, whereas DNA is usually double stranded.

Despite these differences, a message in the RNA alphabet (A, G, C, and U) is just the same as in the DNA alphabet (A, G, C, and T), but with U replacing T.

The conversion of the message carried by the RNA into strings of amino acids—actual polypeptides—is called **translation** (Figure 11-7F). Translation takes the mRNA text and translates it into the language of amino acids. This is the essence of the central dogma.

$$\text{DNA} \xrightarrow{\text{Transcription}} \text{RNA} \xrightarrow{\text{Translation}} \text{Polypeptide}$$

In eukaryotes, transcription occurs in the nucleus and translation occurs in the cytoplasm (outside the nucleus). Messenger RNA passes through the nuclear membrane out to the cytoplasm. There, ribosomes, acting like molecular sewing machines, stitch together amino acids into polypeptides.

In the next part of this chapter, we will discuss the details of the molecular machinery responsible for the flow of information from DNA to RNA to polypeptide. The process of building a cell from the information in DNA results from chemical reactions that biologists can now describe in detail. These reactions are subject to the same thermodynamic laws as other chemical processes. The same kinds of forces drive them, and no special force is needed to build the machinery of life.

The message in the DNA is transcribed into mRNA, which is translated into polypeptides. The central dogma says that information flows from DNA to RNA to protein only. Information in protein is not supposed to be able to influence the information in the DNA.

RNA POLYMERASE TRANSCRIBES DNA INTO RNA

When a cell makes a polypeptide specified by a particular gene, we say that the cell is "expressing" that gene. Gene expression begins when the enzyme **RNA polymerase** transcribes the DNA gene into mRNA. Actually, there are several kinds of RNA polymerases, but since they all make RNA in the same way, we will discuss them as if they were a single enzyme.

How Does RNA Polymerase Begin Transcription?

In the last chapter, we saw that DNA polymerase can build DNA in two directions (extending the 3' end of each DNA strand) and that the enzyme begins by adding nucleotides to an existing polynucleotide called a primer. In contrast, RNA polymerase transcribes DNA into RNA in only one direction and does not require a preexisting RNA primer (Figure 11-8). The starting signal for mRNA synthesis is a special sequence of DNA called a **promoter.** RNA polymerase (working with other proteins) recognizes the promoter, unwinds and separates the two strands of DNA near the promoter site, and then attaches to the promoter and begins to "read" the sequence of bases in the gene. Wherever the template strand has a G, RNA polymerase will place a C; wherever the template has an A, it places a U.

In bacteria, the promoter consists of two characteristic sequences, each about 6 nucleotides long, with about 25 other nucleotides between them. RNA polymerase binds to the promoter and begins to make an RNA molecule that is complementary to the template strand.

In eukaryotes, the production of RNA involves an extra step. Unlike prokaryotic RNA polymerases, eukaryotic RNA polymerases do not themselves recognize the promoter. Instead, they depend on the assistance of other proteins, called **activators**, or **transcription factors**, which lead the RNA polymerase and its accompanying proteins to the promoter. One set of activators, for example, guides RNA polymerase to a DNA sequence called a "TATA box," since it usually contains the sequence TATA. Other activators guide RNA polymerase to other specific DNA sequences.

Both eukaryotic and prokaryotic RNA polymerases stop transcription at special sequences, called **terminators**, which stop the enzyme from transcribing any more of the DNA.

With the help of activators, RNA polymerase produces molecules of RNA by copying one strand of DNA starting at a promoter sequence.

How Does the Cell Alter the mRNA Transcribed from the DNA?

In prokaryotes, the mRNA binds to the ribosomes and begins directing protein synthesis as soon as the mRNA is made. Sometimes translation begins even as the DNA is being transcribed (Figure 11-7F). In eukaryotes, however, transcription and translation occur separately. Transcription occurs in the nucleus, and translation occurs—minutes, hours, or even days later—in the cytoplasm.

Transcription and translation in eukaryotes are separated by much more than just the nuclear membrane. After the mRNA is transcribed from the DNA, in the nucleus, it is called pre-mRNA. While still in the nucleus, pre-mRNA undergoes a number of modifications that transform it into a molecule of "mature" mRNA that can be translated into a polypeptide. These modifications include: (1) capping the 5' end of the RNA with a GTP (a relative of ATP), (2) adding a "tail" of 150 to 200 As to the 3' end of most mRNAs, and (3) removing large pieces of the RNA and splicing the remaining pieces together. In addition, using a process called RNA editing (only recently discovered by molecular biologists), cells can add, remove, or change a few selected nucleotides in some regions of the mRNA (Figure 11-9A).

The cutting and splicing of pre-mRNAs was a surprising discovery, since it demonstrated that genes are interrupted by nucleotide sequences that do not code for proteins. These interruptions in pre-mRNA (and the corresponding sequences in DNA) are called **introns,** for intervening sequences. The RNA sequences that directly code for proteins are spliced together to form the mature mRNA molecule. These sequences (and the corresponding sequences in DNA) are called **exons,** for expressed sequences. In a typical eukaryotic gene, most of the sequence consists of introns, with exons making up only a tiny fraction of the total length. A single pre-mRNA molecule may contain as many as 70 introns, each one ranging in size from 80 to 10,000 nucleotides. Yet all the exons in a single mRNA are, together, rarely more than 3000 nucleotides long.

Most of the DNA in a gene, then, consists of introns and does not code for a functioning polypeptide. In fact, more than 98 percent of human DNA does not directly specify polypeptide sequences, and a large part of this apparently useless DNA is introns. No one knows what introns do. Some scientists believe that the DNA in the introns is basically "junk DNA" that has accumulated in the genome in the same way that old keys and odd parts of things accumulate in the back of a kitchen drawer. Others believe that all DNA must be in the genome for a purpose, even if that purpose is not yet apparent.

Pre-mRNA includes long stretches of noncoding DNA sequences, called introns, that must be removed. The remaining exons are spliced together to form mature mRNA.

One Gene—Heaven Only Knows How Many Polypeptides

Most eukaryotic genes consist of exons interspersed with introns. Transcription, we have said, then yields a single kind of

Figure 11-8 How does RNA polymerase begin transcription? Starting at the DNA promoter, RNA polymerase transcribes DNA into mRNA until it reaches the terminator.

pre-mRNA containing all the introns and exons. Molecular machinery within the nucleus cuts and splices the pre-mRNA to produce mature mRNAs without introns.

As it turns out, transcription can be far more complicated than this simple picture suggests. In some cases, cells can actually transcribe many different mRNAs from the same gene.

Figure 11-9 RNA splicing. A. In this example, introns are snipped from the mRNA, the five exons are spliced together, and each end of the spliced mRNA is punctuated. Like a capital letter, the GTP cap indicates the beginning of the gene sentence. Like a period, the poly A tail indicates the end. B. The same pre-mRNA can be snipped and spliced in different ways to produce mRNAs that code for different polypeptides.

For example, the mRNA transcript may start or end at different sites on the same gene. Or a single pre-mRNA may be broken up and spliced together in several different ways, so that one mRNA's exon becomes another mRNA's intron (Figure 11-9B). As a result, a single gene may encode several polypeptides. What is more, the cell may express all of these various polypeptides at once.

Geneticists spent decades arriving at our definition from Chapter 10, "one gene—one polypeptide." But studies of gene expression beginning in the late 1970s have conclusively shown that one gene can actually encode several polypeptide chains, occasionally with very different functions. Most molecular biologists now say that a gene is a DNA sequence that is transcribed as a single unit and encodes either a single polypeptide or a set of related polypeptides.

A single gene can encode several different polypeptides.

HOW DOES A CELL TRANSLATE mRNA INTO A POLYPEPTIDE?

The synthesis of proteins requires a multitude of molecular tools. Among them are three distinct kinds of RNA: (1) messenger RNA carries genetic information from DNA that specifies the amino acid sequence of a polypeptide; (2) transfer RNAs carry each amino acid to a codon in the mRNA; (3) ribosomal RNA, complexed with proteins to form ribosomes, serves as the site for polypeptide synthesis. In addition, an array of enzymes and other proteins assist in the synthesis and folding of new proteins.

What Do Ribosomes Do?

Ribosomes are like subcellular sewing machines that link amino acids into polypeptides. All cells contain ribosomes. Each ribosome consists of a small and a large subunit (Figure 11-10). Each subunit contains both RNA and proteins. The *E. coli* ribosome, for example, consists of 3 molecules of RNA, called **ribosomal RNAs,** or **rRNAs,** and 54 kinds of protein molecules. Eukaryotic ribosomes are even more complicated, with 4 molecules of rRNA and some 82 kinds of protein.

Ribosomal RNA is very different from messenger RNA. The sequence and length of messenger RNA varies, depending on the kind of polypeptide it encodes. But the RNA that makes up the ribosome is always the same, no matter what kind of protein the ribosome is producing.

Ribosomal RNAs from organisms of all kingdoms are highly similar to one another. Such similarity suggests that the exact shape of the ribosome is important to the way it works. We now know that ribosomal RNAs play an important role

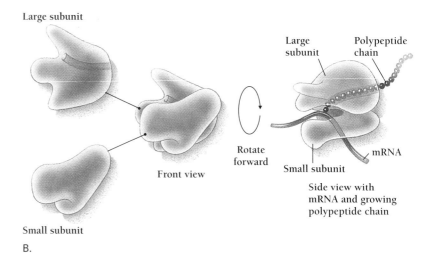

Figure 11-10 How do ribosomes work? A. Electron micrograph of whole ribosome. B. The two parts of a ribosome fit together around the mRNA molecule and the growing polypeptide chain. *(A, courtesy of James Lake)*

 Quiz

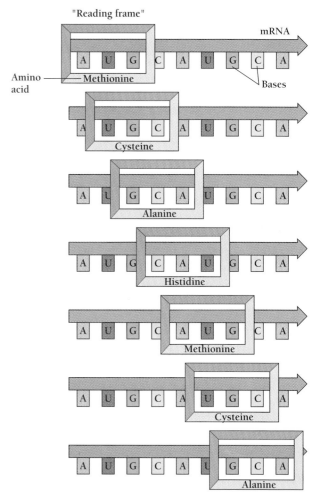

Figure 11-11 The reading frame. The exact base at which the ribosome begins translating the mRNA determines how it interprets all of the three-base codons from there onward.

in holding ribosomal proteins together in just the right orientation. Some researchers argue that ribosomal RNA may also act as a "ribozyme," helping to catalyze the joining of amino acids into polypeptides. We can guess that any individual with even minor changes in the standard ribosome design would be unable to make proteins and would therefore die immediately.

In 1999, after nearly 40 years of work, a number of research groups obtained the first detailed information about the three-dimensional structure of ribosomes. Knowledge of ribosome structure should soon allow biologists to understand the precise roles of individual components of the ribosome, including those of the rRNAs.

Ribosomes help catalyze the formation of peptide bonds between amino acids, in an order determined by mRNA. Ribosomes' main tasks are to assemble amino acids into polypeptides in the order specified by mRNA. In addition, ribosomes must recognize where to start and where to stop reading the mRNA for the production of each polypeptide.

For example, consider the translation of the mRNA sequence AUGCAUGCA. If a ribosome started reading at the first nucleotide, it would assemble the peptide methionine-histidine-arginine. If, however, the ribosome started with the second nucleotide, it would assemble cysteine-methionine. If the ribosomes started with the third nucleotide, it would assemble alanine-cysteine. The grouping of nucleotide triplets for translation is called the **reading frame** (Figure 11-11).

Because starting at the right place is crucial to reading mRNA correctly, mRNAs must contain signals that say "start" and "stop." Start and stop signals are like punctuation that tells where a sentence begins and ends. The smaller ribosomal subunit finds the right place to start, using slightly different mechanisms in prokaryotes and in eukaryotes.

Once a ribosome has found the start site, prokaryotic and eukaryotic ribosomes work in much the same way. Each ribo-

BOX 11.2

How Do the Ribosomes Read Different Kinds of Mutations?

A base substitution in a gene may result in a *missense mutation,* which changes the codon for one amino acid into the codon for another amino acid. The resulting polypeptide may be either nonfunctional or altered in the way it functions. For example, some altered polypeptides result in *conditional mutations,* in which the resulting protein may be able to function under some conditions, but not under others. In one *Drosophila* mutant, for example, the fly functions normally at 20°C but becomes paralyzed at 30°C. This is due to a protein that regulates ion transport across nerve cell membranes, which fails at 30°C. Such temperature sensitivity results when an amino acid substitution reduces a protein's stability at higher temperatures, so that it denatures, or changes shape, more readily than the normal protein.

A base substitution can also result in a *nonsense mutation,* which changes the codon for an amino acid into a nonsense or stop codon, leading to a shortened polypeptide (Figure A). Such a shortened polypeptide almost certainly will not function at all in its former role.

Mutations that have no effect on phenotype are called *silent mutations.* Some are simply single-base substitutions that result in a codon that codes for the same amino acid (because of the redundancy of the genetic code). In that case, the polypeptide is unchanged. Most silent mutations, however, lie in DNA sequences that do not actually code for a polypeptide. Other supposedly silent mutations may merely produce changes in the phenotype so subtle that researchers cannot tell the difference.

Insertions and deletions in structural genes produce *frameshift mutations,* which alter the groupings of nucleotides into codons. The addition or subtraction of a single base can wreak havoc in the ribosomes' reading of mRNA downstream from the mutation. The result is often a string of missense changes, followed by a stop codon.

Sometimes point mutations can reverse themselves in a process called *reversion.* Reversion mutations, sometimes called *suppressor mutations,* consist of a second mutation that reverses the effects of the first mutation. A reversion might be an exact restoration of the original nucleotide sequence. Alternatively, the DNA might have a nucleotide sequence that is not identical to the original but that specifies the same amino acid—a silent mutation.

Figure A Nonsense mutation. A mutation in DNA (A → T) changes an AAG codon (lysine) into UAG (stop) in RNA. This nonsense, or stop, codon shortens the polypeptide.

some contains two grooves. On the small subunit is a groove for reading the mRNA. On the large subunit is a groove for the growing polypeptide chain. Protein synthesis requires the participation of both subunits. The small subunit is responsible for starting synthesis at the right place, while the large subunit contains the machinery for making the peptide bonds that hold together the polypeptide.

Ribosomes, which are made of rRNA and proteins, link amino acids together into polypeptides. The small ribosomal subunit reads the mRNA, and the large ribosomal subunit makes the peptide bonds between each pair of amino acids in the growing chain.

How Do tRNAs Serve as Adapters Between mRNAs and Amino Acids?

Once molecular biologists began to suspect that RNA dictated the sequences of amino acids in polypeptides, they tried to imagine how this could occur. An early idea was that each codon in an mRNA molecule bound directly to a specific amino acid. No one, however, could find any such binding. Besides, it was hard to imagine how a sequence of just three nucleotides could distinguish among amino acids as similar, for example, as leucine, isoleucine, and valine.

The actual explanation, proposed by Francis Crick, is that the amino acids are first linked to special adapter molecules, called **transfer RNAs,** or **tRNAs.** The mRNA codons bind to the tRNA, not to the amino acid itself. Each tRNA consists of 75 to 85 nucleotides and folds to form a secondary structure shaped like a cloverleaf (Figure 11-12). The "stems" of the tRNAs are double strands of RNA held together by base pairing. The "leaves" are loops of RNA with unpaired bases. X-ray diffraction studies of tRNA's tertiary structure show that the whole cloverleaf bends into a shape resembling the letter L.

One loop of each tRNA attaches to the amino acid corresponding to that codon (Figure 11-12). The opposite side of each tRNA molecule binds to a specific codon in the mRNA. In one of the loops is a sequence of three nucleotides, called the **anticodon,** that forms base pairs with the matching codon sequence in the mRNA (Figure 11-12). Each tRNA molecule has only one kind of anticodon and can pick up only one kind of amino acid.

We may wonder how many tRNAs a cell needs—one for each amino acid or one for each codon? A cell must have at least one tRNA for each of the 20 amino acids. Yet we might expect 61—one for every codon except the 3 codons that do not specify any amino acid. In fact, different species have different numbers of tRNAs. However, every species has more than 20 and fewer than 61 kinds of tRNA. Organisms can have fewer than 61 tRNAs because some tRNA anticodons recognize more than 1 codon.

How do we know that mRNA recognizes the tRNA, rather than the amino acid itself? To find out, researchers "tricked" the mRNA. They took the tRNA for cysteine, attached the cysteine, then chemically converted the attached cysteine into alanine so that the tRNA for cysteine actually carried an alanine. Finally, researchers added this incorrectly loaded tRNA to a cell extract that was synthesizing protein. The extract made incorrect polypeptides. Wherever the mRNA specified cysteine, the tRNA inserted alanine instead. This result showed that the mRNA recognizes the tRNA, not the amino acid itself.

Transfer RNA helps assemble amino acids into polypeptides by binding first to an amino acid and then to the correct codon in the mRNA.

How Do tRNAs Recognize the Right Amino Acid?

We have seen how the tRNA anticodon recognizes the correct codon. But how does tRNA recognize and load the correct amino acid? The task is to match the unique shape and charge distribution of each amino acid with the shape and charge distribution of the appropriate tRNA. The linking of tRNAs to their corresponding amino acids to form "charged" tRNAs—tRNAs joined to their amino acids by high-energy bonds—depends upon a set of enzymes. The tRNAs and amino acids recognize each other because the different tRNAs have different three-dimensional structures recognized by the enzymes. Each charging enzyme has two binding sites: one site holds a particular tRNA, and the other site holds the corresponding amino acid (Figure 11-13).

Special charging enzymes help tRNAs hook up with the correct amino acid.

Figure 11-12 Cloverleaf-shaped tRNAs translate codons into amino acids. A. The anticodon is at the tip of the middle "leaflet." B. Like many macromolecules, tRNAs have a complex folded structure. Like so many hairpins, hydrogen bonds hold this complex shape together. C. The anticodon binds to the codon.

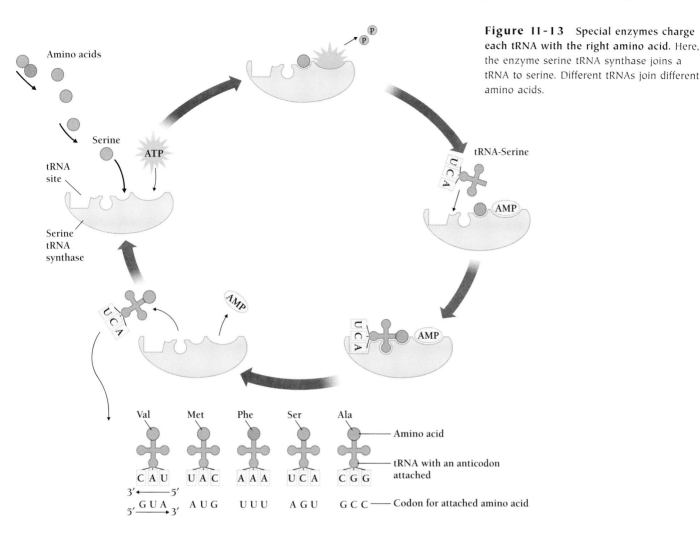

Figure 11-13 Special enzymes charge each tRNA with the right amino acid. Here, the enzyme serine tRNA synthase joins a tRNA to serine. Different tRNAs join different amino acids.

How Do Ribosomes Begin Translation?

We have seen how each molecule of mRNA contains signals that tell the ribosomes where to start and where to stop. We have also seen that starting translation at the right place is crucial, since a mistake of only a single nucleotide would lead to a complete misreading of the information.

The first step of translation is **initiation,** which begins with the attachment of the ribosome's small subunit to the mRNA. The actual initiation site, or **start site,** in mRNA is always the codon AUG, which specifies methionine. This means that *all* recently translated polypeptides begin with methionine. In most cases, however, a special enzyme later clips away the methionine.

But how can the ribosome distinguish between the AUG meaning "start" and the AUGs that specify methionines in the middle of a polypeptide? The answer to this question differs in prokaryotes and eukaryotes. In prokaryotes, a sequence of about six nucleotides, slightly upstream from the initiation AUG, distinguishes between initiation sites and internal methionine. In eukaryotes, each ribosome simply attaches to the 5′ end of the mRNA and moves along until it encounters the first AUG, where translation then begins (Figure 11-14).

The initiation process aligns the mRNA for proper reading and moves the first amino acid into place. After the first tRNA and its attached methionine are in place, elongation of the polypeptide chain can begin.

In the first step of translation—initiation—the ribosome attaches to the start site—the codon AUG. In prokaryotes, a sequence of six nucleotides distinguishes the start AUG from the AUG codon for methionine. In eukaryotes, the start AUG is simply the first one from the 5′ end of the mRNA.

How Do Ribosomes Make Peptide Bonds?

The process of polypeptide-chain **elongation** consists of three steps (Figure 11-14): (1) putting the next amino acid into position, (2) forming a peptide bond between the growing polypeptide and the next amino acid, and (3) moving the ri-

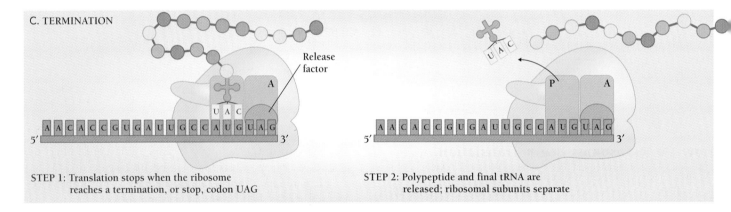

Figure 11-14 Translation: initiation, elongation, and termination. A. During initiation, the small subunit of the ribosome binds to AUG (the start site) of the mRNA and moves the amino acid methionine into place. B. During elongation, a second charged tRNA arrives and the large subunit of the ribosome—holding the two charged tRNAs side by side in the P and A sites—binds the two amino acids together. The ribosome and its attached polypeptide chain then move forward on the mRNA, and another charged tRNA moves into place. The tRNA released from the P site binds temporarily to the E site as it exits from the ribosome. (For clarity, the drawing does not show the E site.) C. When the ribosome reaches a stop codon, a protein called release factor binds to the stop codon, and an enzyme cuts the newly made polypeptide chain from the last tRNA. The ribosomal subunits then separate from each other and become available for another round of initiation and elongation.

bosome to the next codon. Every amino acid is joined to its neighbor in exactly the same way.

Elongation, unlike initiation, requires the participation of the large ribosomal subunit (which binds to the initiator tRNA) as well as the mRNA and the small ribosomal subunit. The large subunit contains three special pockets for tRNAs, called the P site [for *p*eptide], the A site [for *a*mino acid], and the E site [for *e*xit]. The P and A sites bind two tRNAs to adjacent codons in the mRNA; the E site binds each tRNA as it exits from the ribosome. The binding of the two charged tRNAs to the large subunit is the first step of elongation.

After the P and A sites are filled, the ribosome links the attached amino acids to form a peptide bond. This linkage is the second step of elongation. The tRNA in the P site then falls away from the ribosome via the E site. Left behind is the tRNA in the A site and its attached amino acid, which is now

STEP 2: Peptide bond forms between amino acids; ribosome moves one codon to the right

STEP 3: tRNA with leucine is now in the P binding site; new tRNA comes into A binding site

linked to the entire polypeptide chain. In the third step, the ribosome moves forward so that the tRNA in the A site with its attached polypeptide chain ends up in the P site. The repositioning of the tRNA and the attached polypeptide to the P site completes the three steps of the elongation cycle.

The cycle now repeats over and over until all the codons in the mRNA are read. In a bacterium, each elongation cycle (the addition of one amino acid) takes about 1/20th of a second, so the synthesis of an average-sized polypeptide (including 400 to 1200 amino acids) takes 20 to 60 seconds.

> *During elongation, amino acids are joined one after the other in exactly the same way. A tRNA first moves an amino acid into position next to the last amino acid. Once two amino acids are in place, an enzyme in the ribosome joins them together with a peptide bond. The ribosome then moves forward on the mRNA, repositioning the tRNA and its attached polypeptide chain.*

Several Ribosomes Can Read a Single Strand of mRNA Simultaneously

As soon as a ribosome has moved far enough away from the AUG start site, the start site is free for another ribosome to attach and begin translation. As a result, several ribosomes, spaced as little as 80 nucleotides apart, may be attached to a single mRNA at one time, each one overseeing the production of identical polypeptides. This complex is called a **polysome,** short for polyribosome (Figure 11-15). In this way, the cell can synthesize many copies of a polypeptide at once, greatly speed-

Figure 11-15 A large polysome. Individual ribosomes move along a strand of mRNA, translating as they go. The ribosomes on the far left are just beginning translation and the polypeptide strands are not yet visible. The ribosomes on the right and in the middle have produced visible polypeptide strands. (*Oscar L. Miller, University of Virginia*)

ing the mass production of proteins. We will see later in this chapter, for example, that an *E. coli* bacterium can begin producing enzymes to break down milk sugar (lactose) as soon as the bacterium detects the milk sugar.

How Does Polypeptide Synthesis Stop?

The appearance of a stop codon in the mRNA signals the end of a polypeptide chain, a process called **termination.** When the ribosome reaches one of the three stop codons (UAA, UAG, or UGA), there will be no corresponding tRNA. Instead a protein called a **release factor** binds to the stop codon (Figure 11-14). The enzyme that makes peptide bonds during elongation now cuts the polypeptide from the last tRNA, releasing the completed polypeptide into the cytoplasm. The ribosomal subunits then separate from each other and become available for another round of initiation and elongation.

All of the steps of protein synthesis, including transcription and translation, are summarized in Figure 11-16. We can see that once a polypeptide chain is assembled, it must then fold into a functioning protein.

> When the ribosome reaches a stop codon, a protein called release factor binds to the stop codon and an enzyme cuts the newly made polypeptide chain from the last tRNA.

How Do Cells Direct Newly Made Proteins to Different Destinations?

Most of the proteins produced on the ribosomes stay in the cytosol, but many must find their ways to various cytoplasmic organelles, to the plasma membrane, or to the outsides of cells. But how do newly made proteins find their way to the right destinations?

The answer is that many newly made polypeptides have "address tags," additional blocks of amino acids that specify where the protein should go in the cell. In proteins destined for the plasma membrane or for secretion outside of the cell, for example, a **signal peptide** of 4 to 12 amino acids directs the newly made polypeptide into the space (or "lumen") that is surrounded by the endoplasmic reticulum (ER) during protein synthesis. In one experiment, molecular biologists attached a signal peptide to a protein not normally sent to the ER. The protein then entered the ER, demonstrating that the signal peptide indeed acts as an address label. Other polypeptides contain signals that direct them to the mitochondria, the chloroplasts, or the nucleus. In the case of secreted proteins, the signal peptide is removed. In other cases, the address tag remains part of the protein.

> Cells direct proteins to different destinations by addressing the protein with a signal peptide.

HOW DO CELLS REGULATE GENE EXPRESSION?

We have seen now how cells express the information in the DNA. But cells contain thousands of genes. Some of these are expressed as polypeptides all the time, some only occasionally, and many are never expressed. Prokaryotes transcribe almost all of their genes into RNA. But eukaryotes regulate gene expression more closely. Cells, especially eukaryotic cells, do not make all of their polypeptides at once for the same reason that we do not leave all of the water taps in our homes flowing at once. Just as we turn on hot or cold water according to our needs, a eukaryotic cell produces different proteins according to its needs. Further, just as a water tap can be turned on full blast or at a trickle, a cell can turn out thousands of copies of a polypeptide, or only two or three copies.

Eukaryotes, most of which are multicellular, face an additional problem. The dramatic cell specialization that characterizes most multicellular organisms requires the differential expression of genes in cells that have identical genomes. A human, for example, consists of more than 200 distinguishable types of cells. Nearly every one of these cells contains the same set of about 30,000 genes. Yet each kind of cell expresses only a fraction of its 30,000 genes.

A typical eukaryotic cell transcribes only about 20 percent of its DNA into RNA. In a human skin cell, for example, this 20 percent would include information for basic life-support

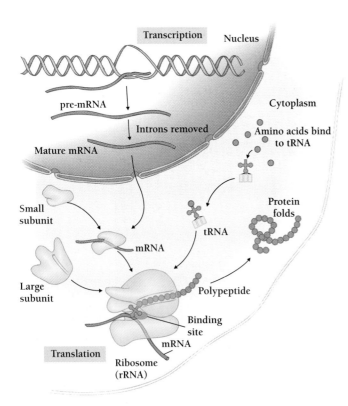

Figure 11-16 Transcription and translation in a eukaryotic cell.

BOX 11.3

Antibiotics and Protein Synthesis

Antibiotics are usually complex organic molecules that allow bacteria and fungi to dispose of (and often to eat) their competitors. The medical uses of antibiotics began at least 2500 years ago, when the Chinese began to use the moldy curd of soybeans to treat boils and similar infections. In the 20th century, the pursuit of antibiotics as a treatment for human disease started with Alexander Fleming. In 1928 Fleming discovered that the growth of bacteria in one of his experiments had been stopped by the presence of a green mold.

Fleming identified the mold as a species of *Penicillium*. Later, other scientists isolated the active substance and named it penicillin. Penicillin and its many derivatives have remained the most frequently used antibiotics. But dozens of other antibiotics have been isolated from molds or bacteria.

Penicillin acts by interfering with cell wall synthesis in many bacteria. Other antibiotics act by interfering with protein synthesis. This characteristic has made them extremely important tools for dissecting the steps of protein synthesis. Each of the steps of elongation, for example, is subject to inhibition by a different set of antibiotics: in prokaryotic cells (bacteria), tetracycline interferes with positioning of the ribosome, chloramphenicol with peptide bond formation, and erythromycin with the movement of the ribosome on the mRNA (Figure A).

Some of the most useful antibiotics interfere with prokaryotic protein synthesis without affecting that in eukaryotic (human) cells. For example, while erythromycin blocks translocation of the growing peptide chain in bacteria, it does not inhibit this step during protein synthesis on eukaryotic ribosomes. This specificity reflects differences in the proteins of prokaryotic and eukaryotic ribosomes.

Antibiotics have dramatically decreased the number of deaths caused by infectious disease. They can not only cure an initial infection but also prevent long-term consequences such as rheumatic fever, which, until this century, often resulted from strep throat. In addition, killing the bacteria that make one person sick prevents the spread of an infection. Finally, the ability to deal with infections has made surgery safer and allowed the development of operations ranging from the removal of cancers to the replacement of vital organs.

Figure A Each of the steps of elongation is inhibited by a different set of antibiotics. In prokaryotic cells, tetracycline interferes with positioning, chloramphenicol with peptide bond formation, and erythromycin with the movement of the ribosome along the mRNA.

proteins that all human cells need—such as membrane proteins, ribosomal proteins, and metabolic enzymes—as well as information for specialized proteins, such as the stretchy collagen that makes skin so tough. In a skin cell, the 80 percent of the DNA not transcribed would include, for example, the genes for α- and β-globin, various muscle proteins, and a host of digestive enzymes.

Cells can regulate the expression of genes at many levels. The most efficient and most usual way is so-called **transcriptional control,** in which cells increase or decrease the amount of mRNA transcribed from the DNA. Every cell in the body has the gene for the globin polypeptides, for example, yet only the developing red blood cells transcribe them. In every other kind of cell in the body, the globin genes sit unused.

Cells can regulate the expression of genes in other ways. A cell may process pre-mRNA in alternative ways—either by altering the patter of exon splicing or by changing the efficiency of export to the cytoplasm (**posttranscriptional control).** Or a cell may vary the rate at which it translates individual types of mRNAs (**translational control**). Finally a cell can modify the structure of a protein and change the rate at which it turns over (**posttranslational control**). The addi-

Figure 11-17 Negative feedback in the *lac* operon. A. Peanuts entice an elephant from the tracks so that a locomotive can pass. In the same way, lactose binds to the repressor so that RNA polymerase can transcribe the DNA. B. The repressor sits on the operator and blocks RNA polymerase. C. When lactose is present, the repressor binds to lactose instead of the operator, and the RNA polymerase transcribes the mRNA for three enzymes that break down lactose.

tion of phosphate groups to particular sites in proteins is an especially important form of posttranslational control, as we will discuss in Chapter 36.

Cells regulate the expression of genes in many ways, but transcriptional control is the most efficient and most common way.

How Do Prokaryotes Regulate Gene Expression?

Bacteria provide the best-understood examples of gene regulation. Most of the time, bacteria regulate gene expression by controlling transcription. Like other organisms, bacteria must respond to their environment. They have evolved sophisticated and sensitive molecular mechanisms that ensure the production of enzymes when they need them. When a bacterium's environment contains high concentrations of amino acids, it devotes most of its energy to rapid division and does not squander energy making more amino acids. If no amino acids are available, however, the bacterium rapidly synthesizes amino acids from simple molecules, usually taking carbon atoms from sugar and nitrogen atoms from ammonia, for example.

To do all this, the bacteria must make the right enzymes at the right time. For example, the synthesis of just one amino acid, histidine, from simple carbon- and nitrogen-containing compounds requires the participation of seven different enzymes. Conveniently, the genes for these seven enzymes lie right next to each other on the bacteria's DNA.

Molecular biologists were surprised to find that the cell transcribes the seven genes into a single huge molecule of mRNA instead of into seven separate lengths of mRNA. Transcribing the seven genes together ensures that all the enzymes needed to make the final product (histidine) are turned on and off at the same time. A stretch of DNA that includes several genes under coordinated control is called an "operon."

Negative regulation in prokaryotes

The best-studied operon in prokaryotes regulates the use of lactose (the main sugar in milk) by *E. coli,* a common bacterium in the intestines of humans and other mammals. This elegant system enables bacteria to produce costly enzymes only

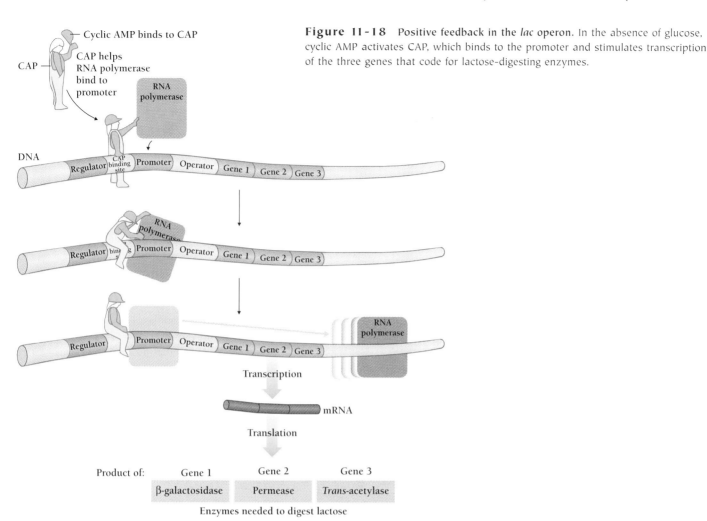

Figure 11-18 Positive feedback in the *lac* operon. In the absence of glucose, cyclic AMP activates CAP, which binds to the promoter and stimulates transcription of the three genes that code for lactose-digesting enzymes.

when they are needed. As long as glucose is available in the intestines, *E. coli* will not digest the milk sugar lactose. When glucose is absent, however, *and* the host organism (that's you) drinks milk, then *E. coli* begins synthesizing three proteins that allow it to digest lactose. One of these is the enzyme β-galactosidase (which splits lactose into the sugars galactose and glucose).

The three genes encoding these three proteins lie next to each other on the DNA of *E. coli* and are part of the lactose operon, or **lac operon.** *E. coli* transcribes the DNA of the *lac* operon into an mRNA that encodes all three proteins.

In the absence of lactose, each bacterium contains only about three molecules of β-galactosidase. However, when glucose is absent and lactose becomes available, the lactose stimulates, or **induces,** the expression of the *lac* operon, and the bacterium synthesizes about 3000 molecules of β-galactosidase. Lactose is called the "inducer" because it induces gene expression.

In 1960, two French molecular biologists—François Jacob and Jacques Monod—discovered how *E. coli* bacteria control these dramatic changes. Jacob and Monod discovered two genes that regulate the expression of the *lac* operon. One gene, which lies a short distance from the gene for β-galactosidase, is the gene that encodes the **lac repressor,** a protein that prevents the expression of the *lac* operon (Figure 11-17A). The other gene, a sequence of 21 nucleotides called the **operator,** lies immediately next to the gene for β-galactosidase and is part of the operon. The operator, we now know, is not actually a separate gene because it does not encode a protein (or RNA). It is a regulatory site to which the *lac* repressor can bind.

The operator overlaps with the operon's *promoter*, which, recall, is the DNA sequence that signals the start of a gene. When the *lac* repressor binds to the operator, it prevents RNA polymerase from binding to the promoter and transcribing the three enzyme genes in the operon (Figure 11-17B). The *lac* repressor was the first example of a protein that regulates gene expression. The operator was the first example of a regulatory site.

How do the repressor and operator allow lactose to control gene expression in the *lac* operon? The key to understanding the induction of the *lac* operon is the realization that the *lac* repressor has two binding sites: one site binds to the operator (a DNA sequence), and the other binds to lactose (a sugar). But the repressor cannot bind to both at the same time. When the repressor binds to lactose, it can no longer bind to the operator and prevent transcription (Figure 11-17C). RNA polymerase can then bind to the *lac* operon's promoter and transcribe the RNA that specifies the three proteins needed to digest lactose. The *lac* repressor is an example of a **negative regulator,** a molecule that inhibits transcription.

When lactose binds to the lac *repressor (a protein), the* lac *operon's promoter becomes unblocked, and RNA polymerase can transcribe the three genes that encode proteins needed to digest lactose.*

Positive regulation in prokaryotes

Bacteria also use **positive regulators,** or activators, molecules that increase the rate of transcription. One such activator is the **catabolite activator protein,** or **CAP** (Figure 11-18). Like the *lac* repressor, CAP has two binding sites. One site binds to a particular sequence in DNA, and the other binds cyclic AMP, a nucleotide derived from ATP. Cyclic AMP is a *signaling molecule,* a molecule that conveys signals from one cell to another. In bacteria, cyclic AMP is a "hunger" signal: high concentrations of cyclic AMP tell the cell that glucose is unavailable and that the cell needs to get energy from another source, such as lactose.

In the presence of lactose, CAP binds to the *lac* promoter and stimulates the transcription of the *lac* operon. When the cell has lots of glucose and little cyclic AMP, however, the *lac* operon is not active even in the presence of lactose.

The combination of negative control (by the *lac* repressor) and positive control (by the CAP activator) allows bacteria to respond quickly to changes in their environment. When lactose is present but glucose is not, *E. coli* quickly begin to make the enzyme β-galactosidase. Similarly, when glucose becomes available or when lactose is exhausted, *E. coli* almost immediately stops production of the enzyme. This system saves the bacteria from expending energy making polypeptides they do not need.

CAP activates transcription not only of the *lac* operon but also of several other genes. All of these genes code for enzymes needed to obtain energy from molecules other than glucose. CAP provides an example of how a single activator can affect the transcription of genes that are physically separated but functionally related.

The activator protein CAP increases the transcription rate for certain mRNAs.

How Do Eukaryotes Regulate Gene Expression?

Until the late 1980s, molecular biologists knew far more about regulation in prokaryotes than in eukaryotes. More recent work, however, has resulted in an explosion of information about gene regulation in eukaryotic cells. We now know that eukaryotes control gene expression at many more levels than prokaryotes. Nonetheless, as in prokaryotes, most regulation occurs at the level of transcription.

The major differences between regulation in prokaryotes and eukaryotes result from the presence, in eukaryotes, of a nuclear membrane and of a much larger amount of DNA. The presence of the nuclear membrane has several important consequences. One is that only proteins in the nucleus can contribute to regulation. Another consequence of the nuclear membrane is that the pre-mRNA can be modified before it moves into the cytoplasm. Posttranscriptional regulation may occur in any of the steps involved in modifying the pre-mRNA.

Eukaryotes also have much more DNA than prokaryotes. Each human cell, for example, contains about 1000 times as much DNA as is in an *E. coli* bacterium. Recall from Chapter 10 that to fit the DNA of a eukaryotic cell into its nucleus, the cell packages the DNA into tightly folded nucleosomes in which proteins called *histones* bind tightly to DNA. The nucleosomes are part of more elaborate chromatin structures, which limit transcription by restricting access to RNA polymerase. In general, the highly packaged DNA of eukaryotes is much less available for transcription than that of prokaryotes. As a result, eukaryotic transcription requires an elaborate set of activators. In eukaryotes, these activators turn on specific genes in specific cell types at specific times. In mammals, for example, genes that promote milk production in the cells of mammary glands turn on during pregnancy and after the mother gives birth, but not at other times of life and not in other parts of the body.

Eukaryotes also have many intracellular compartments and elaborate mechanisms for the distribution of newly made proteins. Sending the right proteins to the nucleus, for example, is especially important for building the chromatin and regulating gene expression. Varying the delivery of these regulatory proteins is another way by which cells regulate gene expression.

The protein tubulin, from which the cell builds its cytoskeleton, provides a nice example of regulation at the level of translation. The microtubules of the cytoskeleton form through the assembly of α- and β-tubulin polypeptides into αβ-tubulin "dimers" [Greek, *dimeres* = have two parts]. When enough dimers accumulate in the cytoplasm, they activate an enzyme that blocks the mRNA on ribosomes that are producing tubulin polypeptides. Once the ribosomes stop producing α- and β-tubulin polypeptides, the concentration of dimers in the cytoplasm drops, and the enzyme stops interfering with the production of α- and β-tubulin.

Eukaryotes can regulate gene expression in more ways than prokaryotes because of the presence of the nuclear membrane, because of the way the DNA is folded and packaged, and because of the many compartments in eukaryotic cells.

How complex is the problem of gene regulation in eukaryotes?

Just as in prokaryotes, the best-studied cases of transcriptional regulation in eukaryotes depend on the interaction of proteins with regulatory sites on the DNA. Unlike prokaryotic DNA, however, most eukaryotic DNA is largely "repressed" because of its tight association with histones. Most eukaryotic regulation therefore depends on specific activators of transcription.

Regulatory proteins may affect transcription in two ways. The first class of regulators change the rate of transcription: some of these resemble the *lac* repressor in decreasing initiation by RNA polymerase; others resemble the CAP protein in increasing initiation. The second class of regulators (sometimes called "chromatin-remodeling machines") produce permanent changes in chromatin that make a particular gene more or less accessible to the transcriptional machinery. Such a change in chromatin structure can persist long after the regulator itself has disappeared.

Many of the proteins that contribute to transcriptional regulation bind directly to specific sequences in DNA. Among these sequences are (1) *promoters,* which specify the starting point and direction of transcription; (2) *response elements,* also called *enhancers* because they usually stimulate the transcription of neighboring DNA without themselves serving as promoters; (3) *silencers,* which help inhibit transcription; and (4) *termination signals.*

Some regulatory proteins bind not to DNA but to other proteins, forming large assemblies of proteins. Such a protein assembly can interact with activator proteins around a set of enhancer elements in a specific gene.

The regulation of the 100,000 or so genes in the human genome is an enormously complex problem. In *E. coli,* researchers have estimated that as many as 10 percent of the approximately 1000 genes contribute to transcriptional regulation. Gene regulation in the mammalian nervous system is certain to be far more complex. Biologists estimate that mammals have at least 5000 transcriptional activators and a corresponding number of response elements (silencers and enhancers). Each transcriptional activator may interact with more than one response element, and each response element can respond to more than one activator. The number of possible interactions at the transcription level alone is huge. Yet the sets of 10,000 to 100,000 kinds of protein molecules in each of the trillion or so nerve cells in the human brain probably require even more complex regulation than that.

> *In eukaryotes, the huge numbers of regulatory proteins and their corresponding binding sites in DNA allow nearly unlimited varieties of interactions. The problem of gene expression is enormously complex.*

A single transcription factor may affect many genes

The hormone testosterone, which works by binding to a transcriptional activator, shows the diverse effects of a single protein. In human and other mammalian embryos, testosterone is one of a group of hormones called androgens that stimulate the formation of the male genitalia (penis, scrotum, and the internal ducts that carry sperm). Later, at puberty, androgens stimulate the development of male secondary sexual characteristics—male distribution of fat, muscle, and body hair; deepening of the voice; and other patterns of growth and behavior.

We know that androgens accomplish all this by binding to an **androgen receptor.** This androgen receptor binds both to testosterone and to response elements in the DNA, thereby influencing the expression of genes related to the secondary sexual characteristics.

In a condition called **testicular feminization,** a chromosomal (XY) male produces testosterone but no androgen receptors. Despite the presence of the Y chromosome and plenty of testosterone, such an individual is insensitive to androgen (testosterone) and develops into a woman, with breasts and female external genitalia. However, she has no uterus, or only a rudimentary one, and no ovaries. She has testes, but these are high in the body cavity, rather than descended. This XY individual is female in nearly every other respect. Testicular feminization shows how the actions of testosterone and its receptor increase the rate of transcription of a host of important genes.

> *The absence of a single transcription factor—the androgen receptor, for example—can greatly alter gene expression and the resulting phenotype.*

In this chapter, we have seen how the coded information in the gene flows from DNA to RNA to polypeptides and how cells regulate that flow. In the next chapter, we will see that genetic information (in the form of individual genes) is not confined to the DNA in the nucleus but can move about within the cell or from cell to cell. The consequences of this genetic mobility are enormous.

Study Outline with Key Terms

DNA and RNA carry information in the sequences of nucleotides. Protein synthesis translates this information into a sequence of amino acids. The **genetic code** is the **univer-** **sal** dictionary that specifies the relationship of the nucleotide sequence to the amino acid sequence. A **codon** is a group of three nucleotides that specifies a single amino acid. There are 64 codons. Of these, 61 specify the 20 amino acids and 3—**nonsense**, or **stop, codons**—signal the end of a polypeptide.

According to Francis Crick's **central dogma:** "DNA specifies RNA, which specifies proteins." Cells **transcribe** information from DNA into RNA and **translate** information from **messenger RNA (mRNA)** into protein. **RNA polymerase** copies information from one strand of DNA into RNA. The starting point for transcription is called a **promoter.** In eukaryotes, **transcription factors** lead the RNA polymerase to the right promoter. Eukaryotic and prokaryotic RNA polymerases stop transcription at **terminators.**

In prokaryotes, RNA polymerase produces mature RNAs directly. In eukaryotes, cells modify the initially transcribed pre-mRNAs into functional mRNAs. Such modifications include cutting out **introns** and splicing together the remaining **exons** to form a patchwork mRNA.

Ribosomes, made of **ribosomal RNAs (rRNAs)** and proteins, are the molecular factories that assemble amino acids into polypeptides. Several ribosomes may associate with a single mRNA molecule to form a **polysome.** Before ribosomes can arrange the sequence of amino acids according to the information encoded in mRNA, however, the amino acids are first linked to adapter molecules called **transfer RNAs (tRNAs).** Each tRNA contains an **anticodon,** a specific nucleotide sequence that forms base pairs with a codon in mRNA. The pairing of specific tRNAs with their appropriate amino acids is the task of enzymes that recognize one amino acid and the corresponding tRNA and link them together.

The translation of information from mRNA into the sequence of amino acids in a polypeptide occurs in three distinct steps, called initiation, elongation, and termination. **Initiation** positions the mRNA in the correct **reading frame.** Initiation always starts with AUG, which is both a signal for methionine and the **start site. Elongation** consists of three steps: (1) putting the next amino acid in place, (2) forming a peptide bond, and (3) moving the ribosome to read the next codon. **Termination** occurs when the ribosome encounters a stop codon to which a protein called a **release factor** has bound.

Eukaryotic cells sort newly made polypeptides to different addresses such as the cytosol, organelles, membranes, or locations outside of the cell. The destination of a polypeptide depends on the presence of **signal peptides.**

Most molecular biologists now say that a gene is a DNA sequence that is transcribed as a single unit and encodes either a single polypeptide or a set of related polypeptides. Cells regulate the expression of genes in many ways, including **posttranscriptional control, translational control, and posttranslational control.** But **transcriptional control** is the most efficient and the most common way.

Two types of transcriptional control in prokaryotes are positive regulation and negative regulation. The *lac* **operon** is an example of **negative regulation.** When lactose binds to the *lac* **repressor** (a protein), the *lac* operon's promoter becomes unblocked, and RNA polymerase can transcribe the three genes that encode enzymes for digesting lactose. We say lactose **induces** the expression of the *lac* operon. When lactose is absent, the *lac* repressor binds to a regulatory site called the **operator** and prevents transcription.

CAP is an example of an **activator,** or **positive regulator.** CAP increases the transcription rate for certain mRNAs, provided that other regulatory mechanisms will allow these mRNAs to be transcribed at all.

Eukaryotes can regulate gene expression in more ways than prokaryotes because of the presence of the nuclear membrane, because of the way the DNA is folded and packaged, and because of the many compartments in eukaryotic cells. In eukaryotes, the huge numbers of transcription factors and regulatory sites allow nearly unlimited varieties of interactions. The problem of gene expression is enormously complex. In **testicular feminization,** the absence of a single transcription factor—the **androgen receptor,** for example—can greatly alter gene expression and the resulting phenotype.

Review and Thought Questions

Review Questions

1. What is the central dogma?
2. What is the difference between the template, or antisense, strand of DNA and its mate, the sense strand? Which one is transcribed into RNA? What would happen if the sense strand were transcribed?
3. What kinds of proteins help RNA polymerase recognize the right promoter? What is a promoter?
4. How do tRNAs mediate between the mRNA and the amino acids? What does "t" stand for? What did Francis Crick call tRNAs before they were discovered?
5. What is the difference between positive regulation and negative regulation of gene expression? Give an example of each.
6. Why do bacteria need to regulate the expression of their genes?
7. What is a regulatory site? Give an example of one.
8. Why do eukaryotes need to regulate the expression of their genes? Why do multicellular organisms need to regulate gene expression?

Thought Questions

9. Why is it hard to clone humans and other mammals? If you could clone someone, whom would you choose to clone? How many copies of the person would you make? Would you expect them to be completely identical in every respect? What would you hope to accomplish by this? Would you send them all to college?
10. Suppose that you spent half your girlhood and all of your teenage years training to be a sprinter. You have qualified for the Olympic team and will compete in the 2000 Olympics. The night before the first heat, your coach tells you that your chromosome test has revealed that you have a Y chromosome and you will not be allowed to compete. You have never thought of yourself as anything except a girl, now a young woman. How do you feel? How will this change your life?
11. If you are a heterosexual man, what would you do if you discovered that your wife of 5 years had a Y chromosome?

About the Chapter-Opening Image

The sewing machine is a metaphor for the ribosome, a tiny molecular machine that stitches amino acids together into polypeptides.

On-line materials relating to this chapter are on the World Wide Web at **http://www.harcourtcollege.com/lifesci/aal2/**

Web Bit 11-1 How Do We Know that a Codon is Three Nucleotides Long?

Web Links

Quizzing & Testing

12 Jumping Genes and Other Unconventional Genetic Systems

Jumping Genes and Indian Corn

In the 1940s, a geneticist named Barbara McClintock made a discovery that would eventually revolutionize molecular biology, a science that did not even exist yet (Figure 12-1). Like Gregor Mendel, McClintock was so much ahead of her time that her research made no impact on the world of biology for 30 years.

McClintock began her career in 1918 at Cornell University, where she was first an undergraduate student, then a graduate student, and finally a researcher. In the 1920s and 1930s, geneticists there noticed that the normal purple color of corn kernels switched to pale yellow or white as a result of a mutation in a pigment gene. Multicolored "Indian" corn is normal corn: the pale corn we buy at the store has been bred to prevent the expression of these purple pigment genes. Oddly, however, the mutation that turns purple kernels yellow or white can reverse itself, or partly reverse itself, from generation to generation.

Each kernel of corn on a cob contains a separate, unique individual—the result of a separate pairing of sperm and egg, a separate pairing of maternal and paternal chromosomes. As a result, some kernels on a cob might have the mutation and some might have it reversed: some kernels are white and some are purple. Still others may be streaked or spotted—the result of partial expression of the pigment gene. When McClintock was at Cornell, no one understood how these "unstable mutations" could flip back and forth so easily.

In 1941, McClintock moved her own research on corn to Cold Spring Harbor Laboratory on Long Island, New York. She was especially interested in broken chromosomes. She had noticed a corn plant in which all of the chromosomes always broke in the same place—an oddity, since chromosomes usually break randomly. McClintock called the place where the chromosome always broke the *Ds* gene. *Ds* was short for "dissociator," the gene that took apart (dissociated) the chromosome. She soon discovered that the *Ds* gene could move around on the chromosome, or "transpose," from generation to generation. Then came an even more startling revelation. McClintock noticed that the same corn plants in which *Ds* had moved also acquired the unstable pigment mutation.

That connection suggested to McClintock that the unstable pigment mutation might be caused by the movement of the *Ds* gene into the middle of the pigment gene. Taking another step, she saw that if the *Ds* gene jumped back out of the pigment gene, the pigment gene would function again, coloring the kernel purple. This would explain how the kernels seem to gain and lose the mutation, switching back and forth between a normal phenotype and an abnormal one, from generation to generation. McClintock's hypothesis not only explained how unstable mutations worked, but introduced a completely new idea to the science of genetics: the idea that genes could move. She coined the word **transposon** to describe these genes that jumped around the genome. Others would come to call them "jumping genes."

In time, McClintock found more complex examples of transposons. She also realized that just as the *Ds* gene could influence the expression of the pigment gene, regulatory genes could finely regulate the expression of structural genes. McClintock was the first person to recognize that one gene could regulate the expression of another. It was McClintock who first discovered promoter and

Key Concepts

- Unconventional genetic systems include all genes that are not a stable part of a cell's nuclear genome.
- A virus is an assembly of protein and nucleic acid that can replicate within a living cell.
- The genomes of both prokaryotes and eukaryotes contain DNA that originally derived from mobile genes.
- Energy organelles—mitochondria and chloroplasts—contain independent genetic systems, which probably evolved from ancient associations of prokaryotes.

suppressor genes. Unfortunately, she discovered them in corn, an organism far more complex than the bacteria in which Jacob and Monod discovered the *lac* operon. The system of regulation that McClintock discovered was far more difficult—both to understand and to explain—than the *lac* operon, and most geneticists did not know what to make of her results.

McClintock was nonetheless a highly respected geneticist. In 1931, recall, she and Harriet Creighton had demonstrated that chromosomes cross over, exchanging genes during meiosis (Chapter 9). The two women had provided the conclusive evidence for the theory of chromosomal inheritance. Yet, in 1951, when McClintock presented her work on transposons to a meeting of geneticists specifically interested in mutations, her talk "met with stony silence," as McClintock's biographer Evelyn Fox Keller wrote in her biography of McClintock, *A Feeling for the Organism*. "With one or two exceptions, no one understood. Afterward, there was mumbling—even some snickering—and outright complaints. . . . It was impossible to understand." McClintock knew her work was complex, but she had not anticipated such rejection. "It was such a surprise that I couldn't communicate," she told Keller. "It was a surprise that I was being ridiculed, or being told I was really mad."

The idea that genes can move from one part of the genome to another is not, in fact, extraordinarily difficult to understand. But it was extraordinarily difficult for geneticists to accept. McClintock's colleagues just assumed that—except for crossing over and other changes that occur during meiosis—the genome is a rigid, stable structure. McClintock might as well have told them that the streets they drove every day moved from place to place. She was viewed as something of a mad genius. Her explanations were extraordinarily complicated; her writing lacked the spare elegance of Jacob and Monod's discussions of the *lac* operon. For 30 years,

Figure 12-1 Barbara McClintock in the early 1920s, at her parents' home in Brooklyn, New York. (*Cold Spring Harbor Laboratory Archives. Permission of Marjorie M. Bhavnani*)

most geneticists did not even try to understand McClintock's work.

Not until the early 1970s, when bacterial geneticists began to discover the same things that McClintock had discovered in corn in the 1940s and 1950s, did biologists begin to recognize the importance of her work. They realized that McClintock's transposons, or jumping genes, are just one example of a whole class of **mobile genes**—genes that move around.

Finally, biologists began to see how broadly McClintock's ideas applied to all sorts of mobile genes, not just jumping genes and not just in corn, but to mobile genes in all living organisms.

By the 1980s, young molecular biologists had begun to use mobile genes as tools for the study of how genomes work. And, as we will see in the next chapter, mobile genes also play a pivotal role in the development of a great range of genetic engineering techniques. In 1983, when McClintock was 81, her great contributions to biology were finally recognized with a Nobel prize.

Since the beginning of the 20th century, geneticists have carefully studied both the information in a gene and the gene's place in the chromosome. While the information in a gene is important, it is a gene's location that determines how the gene will be inherited. We have already seen, for example, that genes that are far apart on the chromosome are more likely to recombine than genes that lie next to one another. Barbara McClintock's work showed that individual genes on the chromosome can actually move from place to place, with profound consequences.

In this chapter we will also examine an odd collection of **unconventional genetic systems.** Unconventional genes either move about within the genome or are not in the nuclear (or nucleoid) genome at all. Mitochondria and chloroplasts have their own DNA. This DNA does not move, but it constitutes an unconventional genetic system because it is passed down from organelle to organelle separately from the DNA in the nucleus and without any kind of sexual recombination.

The other unconventional genes we will discuss are all mobile genes. Jumping genes, or transposons, are the simplest mobile genes. Transposons replicate as part of a host cell's DNA. In contrast, **plasmids** usually exist as simple loops of DNA separate from the cell's own DNA. Plasmids, which exist in prokaryotic cells (and yeast cells), can pass from cell to cell and may sometimes integrate into the cell's DNA. Viruses, familiar to us as the purveyors of colds and flu, are another variety of mobile gene. Viruses move genetic material from cell to cell and can insert genetic information into the genomes of both prokaryotic cells and eukaryotic cells.

As we survey this odd collection of unconventional genes, however, we will see that the replication and expression of all genes, unconventional or conventional, can occur only within a cell. The study of unconventional genes has fundamentally changed cell biology. It has given biologists new insights into diseases ranging from the common cold to cancer and AIDS. It has suggested new views about the origin of eukaryotic cells and of life itself. And it has led directly to the extraordinary growth of the biotechnology industry.

WHAT IS A VIRUS?

Viruses are examples of unconventional genetic systems. A virus is an assemblage of nucleic acid (DNA or RNA) and proteins (and occasionally other components, such as lipids or carbohydrates). Viruses have many properties of life: they are highly organized, they reproduce, they respond to their environments, and they evolve. But viruses cannot obtain energy from their surroundings, and they cannot form the building blocks needed for the synthesis of their components. Because a virus cannot perform metabolic reactions, it can replicate its genetic material and build its proteins only by using the biochemical machinery of a host cell. In short, a virus is not alive.

Most viruses that infect bacteria, plants, and animals do not cause disease. Many are relatively harmless, and some even introduce useful genes into the genomes of other organisms. In fact, in the next chapter, we will see how molecular biologists have used this attribute of viruses to create new strains of bacteria, plants, and animals.

Nonetheless, the study of viruses has, historically, been the study of disease-causing viruses. In the 19th century, the word "virus" [Latin, slimy liquid or poison] came to mean any material associated with death and disease. By the late 19th century, Louis Pasteur and others had shown that bacteria and other small cells cause many infectious diseases and also spoil food. Pasteur's colleague Charles Chamberland developed a way of removing even the smallest cells from liquids by forcing them through a fine filter (Figure 12-2A). But biologists soon discovered that some infectious agents managed to get through the bacterial filters. These mysterious entities were called "filterable viruses." One by one the bacteria specifically responsible for many diseases were identified and named. The only infectious agents left unnamed were the mysterious filterable viruses, which microbiologists simply call "viruses."

In the 1890s, just a few years after Chamberland's filter became available, two men—the Russian pathologist Dmitri Iwanowski and the Dutch microbiologist Martinus Beijerinck—independently demonstrated that a filterable virus was responsible for a disease of tobacco leaves (Figure 12-2B). Because the diseased leaves have a patchy, or mosaic, appearance, the disease was called *tobacco mosaic disease,* and the virus that causes it is now called *tobacco mosaic virus.*

Other microbiologists soon discovered that viruses cause diseases in animals as well, including yellow fever (a deadly human disease carried by mosquitoes) and hoof-and-mouth dis-

Figure 12-2 Viruses are much smaller than bacteria. A. Fine filters remove the bacteria that cause milk to sour. B. Fine filters do not remove the virus that causes tobacco mosaic disease.

A. Porcelain filter removes bacteria from sour milk.   B. Tobacco mosaic virus passes through filter.

Figure 12-3 Three-dimensional model of tobacco mosaic virus. (Drs. Gerald Stubbs and Hong Wang, Vanderbilt University)

ease (a disease of cattle). In 1915, the English bacteriologist Frederick Twort discovered that viruses even infect bacteria. Twort called his bacteria-infecting virus a **bacteriophage** [Greek, *phagein* = to eat].

Discoveries such as these demonstrated that disease-causing agents could be smaller than the 0.5-μm pores in a Chamberland bacterial filter. No one, however, knew what viruses actually were. Most people thought that they were just cells too small to be stopped by a filter and too small to be seen through a microscope.

In 1935, this view fell apart, however, when Wendel Stanley, a biochemist working at the Rockefeller Institute, produced crystals of tobacco mosaic virus (Figure 12-3). This was confusing indeed. A crystal forms only when identical molecules arrange themselves in a regular lattice (as in the case of the familiar salt and sugar crystals on the breakfast table). How could an infectious agent—something that was assumed to be alive—form the precisely ordered arrays of a crystal? One certainly couldn't imagine a crystal of bacteria or of mice.

Stanley's discovery seemed to imply that a virus was just a large molecule and that viral diseases were similar to chemical poisoning. But a poisonous substance would be diluted and broken down if it were passed from individual to individual, while tobacco mosaic virus from a dissolved crystal could be passed indefinitely from plant to plant. Clearly, tobacco mosaic virus was replicating itself—something ordinary molecules are not known to do.

We now understand that viruses are relatively large, regular molecular complexes, some of which can form crystals. By analyzing such crystals by x-ray diffraction, researchers have been able to determine the structure of some viruses in great detail. Most viruses consist of a nucleic acid core (either DNA or RNA) that carries the virus's genes, all surrounded by a protein coat. (A few viruses direct the host cell to make a viral coat out of the host cell's own plasma membrane. This trick helps an infecting virus conceal its identity.) Some of a virus's genes direct the synthesis of proteins and nucleic acids that make up the virus, while other genes specify proteins that regulate the host cell's DNA. Such regulatory genes force the host cell to make new virus particles.

A virus is a large complex of nucleic acid and protein capable of infecting a cell. Infection may be helpful, harmless, or fatal to the cell.

What Is the Form of a Virus?

With the development of the electron microscope in the 1940s, scientists could actually see viruses for the first time. They saw that viruses are large particles, much smaller than cells but larger than single protein molecules. The biggest viruses are the pox viruses such as smallpox virus, which range from 250 to 400 nm in diameter. The smallest ones, including, for example, poliovirus, are 20 to 30 nm in diameter (Figure 12-4).

Viruses have many shapes, but most have one of two forms: a long helix, such as tobacco mosaic virus (Figure 12-4A); or a compact, nearly spherical particle, such as poliovirus (Figure 12-4B). Poliovirus is not actually spherical. It is a 20-sided object, called an **icosahedron** [Greek, *eikosi* = twenty + *hedra* = base], a little like a soccer ball. Unlike the sides of a soccer ball, however, each facet is shaped like a triangle. Other viruses have more complex forms. The bacteriophage T4 is among the most complex of viruses (Figure 12-4C). With an icosahedral head and an intricate tail structure, it looks and functions like a tiny medical syringe, injecting the bacteriophage's DNA into its bacterial host.

The outside surface of a virus is either a coat of protein—a **capsid**—or a **membrane envelope**—a membrane like that which surrounds a cell. Many viruses that infect animals, such as influenza virus and human immunodeficiency virus (HIV, the AIDS virus), direct the host cell to make a viral membrane from the host's own cell membrane, together with proteins specified by the virus's genes (Figure 12-4D). Otherwise, viruses consist almost entirely of proteins and nucleic acid. The nucleic acid may be either DNA or RNA, depending on the virus. Just as in humans and other living organisms, the nucleic acid serves as the virus's genetic material and contains the information for most, if not all, of the virus's proteins.

Viruses have regular shapes, such as an icosahedron or a helix. They consist of a core of genetic material—DNA or RNA—enclosed in a protein capsid or a membrane envelope.

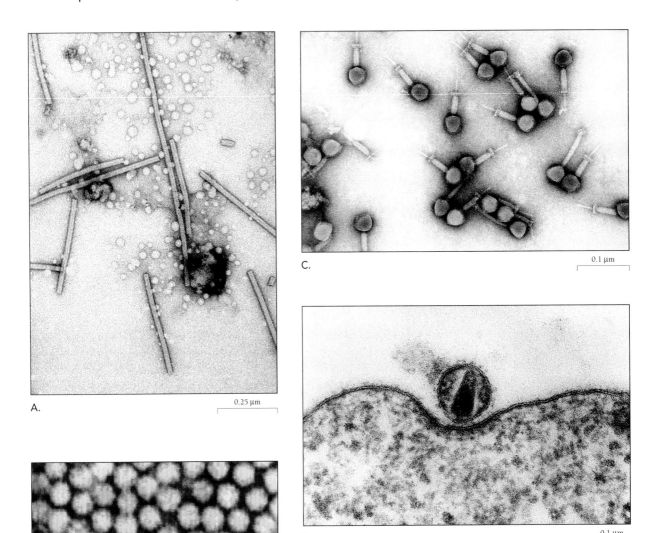

Figure 12-4 Viruses come in many sizes and shapes. A. Tobacco mosaic virus. B. Like the poliovirus, tomato bushy stunt virus is a 20-sided icosahedron. C. T4 bacteriophage. D. HIV, which causes AIDS. (A, Jack D. Griffith, University of North Carolina; B, Andrew O. Jackson, University of California, Berkeley; C, Thomas Broker/Phototake NYC; D, Hans Gelderblom/Visuals Unlimited)

How Do Viruses Make More Viruses?

Virologists' studies of how viruses infect cells in the laboratory have revealed much about how viruses replicate. Although viruses can make more viruses in many different ways, their histories always include the following stages: (1) the virus attaches to the host cell; (2) the viral nucleic acid enters the cell; (3) the cell synthesizes proteins specified by the virus's genes; (4) the cell replicates the virus's DNA or RNA; (5) the new viral protein and DNA or RNA made by the host cell assembles into new viruses, also called virus particles; and (6) the new viruses are released from the cell (Figure 12-5).

Each stage in a virus's multiplication cycle depends on the interaction between the host cell and the virus. A given kind of virus can infect only certain kinds of cells. We say that viruses are adapted to specific hosts. Most cold viruses and the HIV virus, for example, are specific to humans. In this section, we will briefly discuss four of these six stages of viral infection. After that, we will discuss the replication of the viral DNA or RNA in greater detail.

How Does a Virus Attach to a Cell?

The attachment of the virus to a cell depends on an interaction between a viral protein and receptor molecules on the host

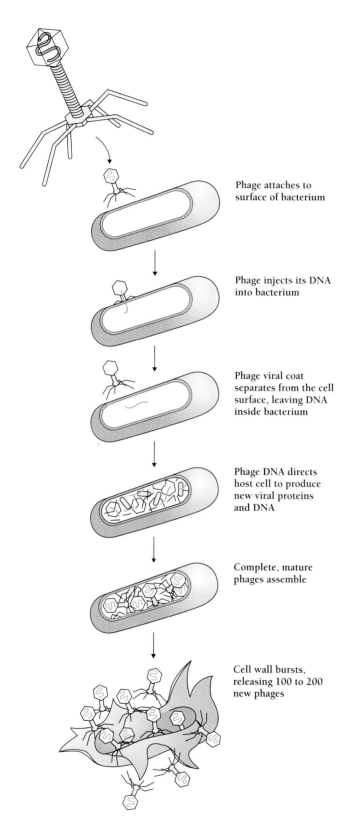

Figure 12-5 The reproductive cycle of a bacteriophage.

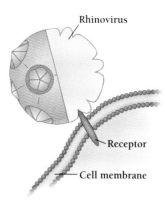

Figure 12-6 Viruses latch onto cell surface receptors.

cell's membrane (Figure 12-6). Specific proteins on the surface of the virus bind to molecules on the surface of the host cell. The attachment site may be on all of the host's cells or just on a special set of cells. Human cold and flu viruses, for example, specifically infect cells of the respiratory tract. Others, such as chicken pox virus and some types of herpes viruses, infect skin and nerve cells. Poliovirus initially attacks cells in the digestive tract, then moves into the bloodstream, where it can gain access to all the cells in the body. In 1 percent of polio cases, poliovirus attaches to and destroys nerve cells, causing paralysis or even death. In comparison, HIV primarily infects cells of the immune system and the brain.

Viruses attach to protein receptors on the surfaces of cells.

How does the viral DNA or RNA enter the host cell?

The mechanism by which the viral nucleic acid enters the host cell depends on the kind of virus and the kind of host cell. Bacteriophage T4, for example, contains an enzyme that digests away part of a bacterial cell wall, allowing the phage to inject its DNA into its host. Other viruses, such as the flu virus, are taken into cells by endocytosis, the process by which the cell's membrane enfolds particles (Figure 12-7). Still other viruses, including HIV, are enclosed in bits of the cell's own membrane, which the virus particles capture as they leave an infected cell. Such viruses easily enter their next host cell, when the virus's stolen membrane envelope fuses with the host cell's membrane (Figure 12-7).

Viruses enter cells either by cutting a hole in the cell membrane, or by inducing the cell to engulf the virus particle, or by fusing with the cell's membrane.

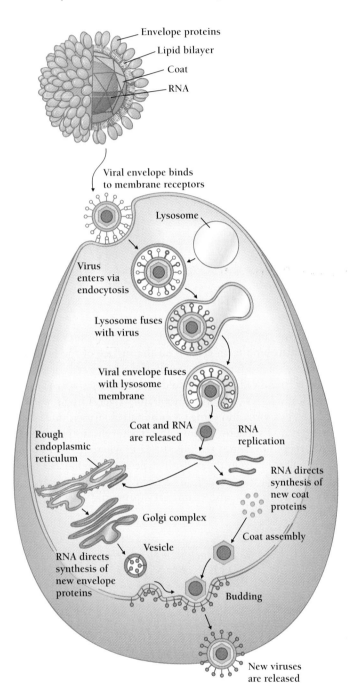

Figure 12-7 Like the Trojan horse, an influenza virus tricks a cell into taking up the virus. The virus attaches to cell surface receptors, which initiate endocytosis of the virus. Once inside, the virus releases RNA, which the cell duplicates and packs in viral protein coats for release to the outside.

How are virus particles manufactured and released?

Once inside the host cell, the virus's genes direct the synthesis of viral proteins in a variety of ways. Some direct the host's protein-synthesizing machinery to make an enzyme that destroys the host's own DNA. Other viruses rely on especially powerful promoters and other regulatory sequences to ensure rapid replication, transcription, and translation. In most cases, viral protein completely dominates protein synthesis.

For a bacteriophage, the whole cycle, from attachment and entry to assembly of new virus particles, may take only 20 minutes and may produce 100 or more new bacteriophages from each original invading bacteriophage inside the cell (Figure 12-5). After these new viruses are produced, a viral enzyme causes the host cell membrane to break open in a process called **lysis** [Greek, lysis = loosening]. The ruptured cell then releases the newly made viruses, which are ready to infect more host cells. Viruses that destroy their host cells in this way are called **lytic viruses**.

Not all viruses cause cell lysis, however. Killing the host—whether the host is a single cell or a large multicellular organism—is not the only strategy by which a virus can successfully replicate. After all, a host that a virus kills may be the last one it encounters for a while, so killing the host could spell the end of the line for a virus. Some viruses have a multiplication cycle similar to that of a lytic virus, but they exit the cell without destroying the host. Other viruses stay inside the cell for long periods, as we will see.

Viruses force the host cell to make viral proteins—sometimes by destroying the host's DNA, sometimes through subtler forms of gene regulation. Many viruses can escape from their hosts by breaking open the cell membrane.

HOW DO VIRUSES AND OTHER MOBILE GENES REPLICATE THEIR GENES?

"A hen is only an egg's way of making another egg," wrote the English novelist and essayist Samuel Butler. Just so, a virus is only a gene's way of making another gene. This statement might sound extreme, but the idea behind it helps us to understand the strategies that viruses and other mobile genes use to replicate.

A virus is the most elaborate of the mobile genes, but there are many other kinds of mobile genes that consist only of DNA or RNA. Such mobile genes dispense with any kind of exterior coat or membrane. Like viruses, however, they include sequences that induce the host cell to make copies of their genes.

Mobile genes, including viruses, plasmids, and transposons, have two reproductive strategies. In **autonomous replication,** the genes are separate from the DNA of the host cell and can replicate even when the host DNA is not replicating. In **integrated replication,** the genes become integrated

BOX 12.1

Can Viruses Make You Well?

Imagine one day going to the doctor with a case of strep throat and getting a spoonful of virus to cure it. As more and more bacterial infections become resistant to antibiotic treatments, that day may not be so far off. In some parts of the world, bacteriophages, the viruses that infect bacteria, are already in use to combat disease.

Many infections that were once easily cured have become resistant to all but a few antibiotics. The list includes some kinds of pneumonia, ear infections, strep throat, urinary tract infections, cholera, tuberculosis, and others. The heavy use of antibiotics during and after surgery has made hospital recovery rooms into breeding grounds for antibiotic-resistant strains of bacteria. In recent years, some hospitals have reported outbreaks of hospital-acquired pneumonia that are resistant to every known antibiotic.

Bacteriophages are natural parasites of bacteria. Once a phage invades a bacterial cell, it reproduces rapidly, and within an hour the cell bursts open to release millions of phage progeny, which then hunt down more bacteria.

Phages have several advantages over antibiotics. Because antibiotics are toxic, they often cause side effects, such as intestinal disorders and hearing problems. Some people are violently allergic to antibiotics. And many people don't take the complete course of antibiotic. In such cases, all of the bacteria may not be eliminated—either by the antibiotic or by the immune system—and the surviving bacteria in such a lingering infection tend to be those that are resistant to the antibiotic. Phages have none of these problems. They function regardless of whether the bacteria are antibiotic resistant or not. They produce no side effects and tend not to be allergenic. And, since they multiply themselves in the process of slaughtering the bacteria, forgetting to take a dose or stopping treatment too early is never a problem.

But phages are no panacea. Each strain is so specific for a particular type of bacteria that doctors have to know exactly what kind of bacterium is causing an infection. That could mean a delay of several days while the species of bacterium is identified. Also, some harmful bacteria have no known phage predators. And, unfortunately, bacteria may evolve resistance to infection by phage, just as they do to antibiotics. Finally, little is known about the efficacy of phage therapy.

In the 1930s and 1970s, researchers made some attempts to treat infections with phages, and many hoped that they would be the ultimate treatment for bacterial infections. In Sinclair Lewis's popular novel *Arrowsmith*, written in 1924, the hero tried to combat a devastating epidemic with a bacteriophage treatment. Nonetheless, the method was all but abandoned with the advent of antibiotics, but only in the West.

In parts of the former Soviet Union, however, phages have been used routinely to treat everything from burn infections to food poisoning. Doctors at a hospital in Tbilisi, Georgia, report using phages successfully to disinfect operating rooms and surgical instruments, as well as to reduce the spread of antibiotic-resistant pneumonia and strep throat. While bacteriophage therapy may be low tech, a few researchers hope that this therapy will be the wave of the future.

---

into the host's DNA and are replicated along with those of the host. Integrated genes can replicate *only* when the host DNA replicates. Some mobile genes can switch back and forth from autonomous to integrated replication depending on circumstances.

In the next few sections, we will discuss plasmids, mobile genes that usually replicate only autonomously; transposons, mobile genes that can replicate only when integrated into the host's DNA or into a plasmid; and **episomes,** viruses and plasmids that can switch back and forth between integrated and autonomous replication.

How Do Plasmids Replicate Autonomously?

A plasmid is a small piece of circular DNA or RNA in a bacterial or yeast cell that includes from 3 to 300 genes. Most plasmids do not integrate themselves into the host cell's DNA but exist as independent entities. Usually, plasmids replicate only when the DNA of their hosts does. However, many plasmids can replicate even when host DNA synthesis has stopped.

Such plasmids are said to replicate autonomously. A plasmid can do this because it contains an origin of replication, the DNA sequence where DNA replication begins. As a result of this autonomous replication, a cell may contain up to 1000 copies of each plasmid. And the cell may carry many different kinds of plasmids.

Plasmids, unlike viruses, do not produce protein coats that would allow them to move easily from one cell to another. However, plasmids have another way to get around. Bacteria sometimes engage in a sexlike process called **conjugation,** during which two bacterial cells temporarily attach to one another and exchange genetic material (Figure 12-8). During conjugation, plasmids can move from one bacterium to another.

Plasmids often contain genes that bring their hosts some advantage. For example, some plasmids—called **resistance factors,** or **R factors**—contain genes for enzymes that make the cell resistant to antibiotics (Figure 12-9). Other plasmids contain genes for resistance to heavy metal poisoning, for the metabolism of unusual compounds (such as those found in oil spills), or for the production of toxins that kill other bacteria.

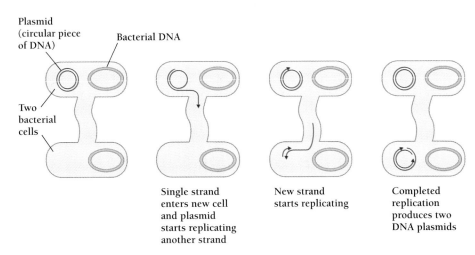

Figure 12-8 Conjugation in bacteria. During a brief encounter, bacteria exchange genes carried on the bacterial DNA or on plasmids. The bacteria may, for example, transfer genes on plasmids that confer antibiotic resistance. By means of conjugation, bacteria can spread genes for antibiotic resistance among different strains and even different species of bacteria. (*Photo, Dennis Kunkel/Phototake NYC*)

In addition, molecular biologists have synthesized thousands of **recombinant plasmids,** plasmids that contain combinations of genes never found in nature. We will discuss the production and uses of such recombinant DNA in Chapter 13.

A plasmid is a circular piece of DNA or RNA capable of autonomous replication that carries genes, such as those for antibiotic resistance, from one bacterial cell to another.

Many Mobile Genes Can Replicate Only When Integrated into a Host Cell's DNA

Unlike a plasmid, a transposon can replicate only when it is part of the host cell's DNA. A transposon, or jumping gene, is a DNA sequence that moves, either by changing its own chromosomal location or by duplicating itself for repeated insertions elsewhere. A transposon always includes a gene for an enzyme called **transposase,** which catalyzes insertion into new sites (Figure 12-10). While plasmids occur mostly in bacteria and in some yeast cells, transposons are common in all cells, including McClintock's corn.

Eukaryotic transposons mostly reproduce only in the integrated form. They are surprisingly mobile, however. In eukaryotic organisms from corn to humans, they can jump from place to place within the genome, duplicating themselves, disrupting other genes, deleting and repeatedly duplicating whole stretches of DNA, breaking chromosomes, and sometimes even attaching part of one chromosome to another. The mapping of eukaryotic genomes has revealed that virtually all genomes are littered with DNA sequences that appear to be relics of ancient transpositions of jumping genes. Corn is especially rich in such transposons.

A transposon may be copied many times and each copy moved about the host's genome separately. For example, at least six different transposons are present in the genome of *E. coli*. Several of these transposons have been copied five to ten times within each *E. coli* genome.

A **simple transposon** specifies transposase and nothing else. As far as biologists know, a simple transposon's only function is to make more simple transposons. Still, simple transposons may help bacteria in an indirect way. Because transposons rearrange the genome and interrupt genes, they increase genetic variability in their hosts. Greater genetic variability could allow species to adapt to new conditions more quickly and more effectively.

When two simple transposons lie near each other on the host genome, they can jump together, carrying with them any

Figure 12-9 A plasmid carrying a gene for antibiotic resistance.

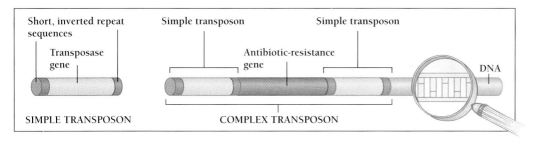

Figure 12-10 Transposons. A simple transposon carries the gene for transposase and little else. A complex transposon is two simple transposons with some other genes carried between them.

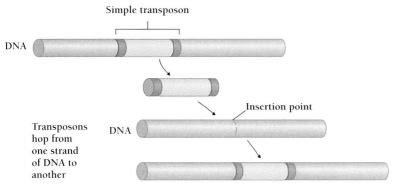

part of the host's genome that lies between them. Such combinations are called **complex transposons.** Some complex transposons carry antibiotic resistance genes, for example (Figure 12-10). Transposons can move these genes into plasmids or into the host cell's chromosomal DNA. In this way, transposons can help a plasmid accumulate genes for resistance to multiple antibiotics (tetracycline and penicillin, for example). The plasmid can then carry this resistance from one bacterium to another, spreading antibiotic resistance even to different species of bacteria.

As we will see in Chapter 20, transposons' ability to move genes ultimately explains how antibiotic resistance spreads so rapidly among bacteria. Some strains of the bacteria that cause tuberculosis (TB) are now resistant to every known antibiotic. The medical treatment for people infected with such bacteria is essentially the same as that used 50 years ago, before the availability of antibiotics. TB patients with multiple-antibiotic-resistant TB are kept isolated to prevent them from passing this dangerous form of TB to others. But there is no specific medical treatment that will help them get well.

A simple transposon codes for nothing but transposase, which is all the transposon needs to insert itself into a new site in the host's DNA. Complex transposons carry other genes as well.

Tumor viruses also insert themselves into chromosomal DNA

Many viruses replicate autonomously, lyse the cell, and escape to infect more cells. But viruses that cause tumors are a special case. Rather than multiplying as rapidly as possible and killing their host cells, tumor viruses multiply only when their eukaryotic hosts do (Figure 12-11A).

Tumor viruses exist independently of the host's DNA only during a brief, independent, mobile phase when the tumor virus finds the eukaryotic cell and infects it. After infection, the genes of most tumor viruses become integrated into the DNA of their hosts and stay there. The virus uses the host's enzymes to express one or more viral genes. The resulting viral proteins ruin the host's ability to control cell division, forcing the cell to divide rapidly, which causes a tumor to form. With each round of cell division, a new generation of viral genes is created.

The viral genes whose expressed proteins interfere with the host cell's ability to control cell division are called **oncogenes** [Greek, *onkos* = mass, swelling], a word that means "genes that cause tumors." Most oncogenes evolved from normal genes in humans or other animals. The original versions of oncogenes, called **proto-oncogenes,** normally control growth and differentiation. Tumor viruses have somehow captured these genes, which have evolved further into **viral oncogenes.** Most oncogenes of tumor viruses provide an overdose of normal cellular proteins that ordinarily stimulate normal cell division.

Tumor viruses integrate their DNA into that of a eukaryotic host, then force the host to replicate both viral and genomic DNA more often than normal. The result is many copies of the viral DNA.

Figure 12-11 Virus life cycles. A. How do DNA tumor viruses cause tumors? DNA tumor viruses integrate their DNA into the DNA of eukaryotic cells, then disrupt the cell cycle, forcing infected cells into excessive growth. B. The lysogenic cycle of the bacteriophage lambda. Lambda (λ) can either integrate into the DNA of its host or initiate a lytic cycle.

Episomes can replicate both autonomously and through integration

Episomes are viruses or plasmids that switch hit between autonomous replication and integrated replication. Episomes can reproduce either autonomously or as part of the host's chromosome, as the occasion requires.

The bacteriophage lambda (λ), for example, is an episome with two alternative reproductive cycles in its *E. coli* hosts. In the **lytic cycle,** λ's DNA is separate from that of its host, and λ lyses its host. Occasionally, however, λ integrates its DNA into *E. coli*'s DNA. The integrated form of a virus is called a **provirus,** or, in the case of a bacteriophage, a **prophage** [Greek, *pro* = before]. The provirus (or prophage) and the lytic cycle together make up the viral **lysogenic cycle** (Figure 12-11B).

When the λ bacteriophage is in the prophage stage, the host cell does not make viral proteins. The reason for this is that the λ DNA directs the synthesis of a powerful repressor, called the λ *repressor,* which prevents the transcription of any other λ genes. Ultraviolet light and other environmental insults that threaten to destroy the host cell trigger an "escape" mechanism in the prophage. In escape mode, the λ DNA directs the synthesis of a protein that represses the λ repressor. Once the λ repressor is no longer preventing the expression of the λ genes, the virus enters a lytic cycle, destroys the host cell, and new λ bacteriophages escape the dying cell.

Because a virus in the provirus stage does not lyse its host cell, it cannot infect other cells. Although a provirus or prophage does not force the cell to make new virus particles, the nucleic acid of the provirus is copied along with the cell's own DNA each time the cell divides or otherwise copies its DNA. As a result, all of the daughter cells of an infected cell are also infected with proviruses, and an infected eukaryotic organism cannot rid itself of the virus.

If the host cell is exposed to ultraviolet light or certain DNA-damaging chemicals, however, the viral DNA then escapes and begins a lytic cycle (Figure 12-11B). This may explain why intense sunlight can activate the herpes virus that may be present in the cells of the mouth and lips. When the virus becomes active, it enters a lytic cycle, lysing cells and causing a painful cold sore.

Other viruses and plasmids have RNA genes. These include the viruses that cause polio, influenza, AIDS, and many animal tumors. How do these genes function? The short answer is that they use special enzymes, encoded in the virus's genome. The particular function of the enzyme varies with the virus, but RNA viruses all depend on standard base-pairing to make a complementary strand.

Figure 12-12 shows several different ways that RNA viruses express their genes. Some viruses carry their genes on a strand of RNA that, like messenger RNA, actually codes for various viral proteins. Such RNA is expressed directly by the host cell's machinery as soon as the virus enters the cell. Such viruses are called "plus-strand" RNA viruses (Figure 12-12B). Other RNA viruses, called minus-strand viruses, use an "antisense" RNA strand. Such viruses contain a viral enzyme that first copies the RNA into a complementary (sense) strand that functions as mRNA and codes for proteins (Figure 12-12C).

> *Plus-strand viral RNA works like mRNA. Minus-strand viral RNA is antisense RNA, which is transcribed into RNA that works like mRNA.*

RNA tumor viruses present special problems

By the mid-1960s, virologists had discovered many of the ways that RNA viruses could reproduce. But the reproductive cycle of RNA tumor viruses, such as Rous sarcoma virus, mystified them. DNA tumor viruses inserted their genes directly into the DNA of their hosts, permanently changing their genetic makeup. But where could an RNA tumor virus leave a permanent copy of its genes? Certainly single-stranded RNA could not insert directly into chromosomal DNA.

The answer was first suggested in the 1960s by Howard Temin, a virologist at the University of Wisconsin. Perhaps, Temin argued, the RNA is recopied into DNA, and the DNA is then inserted into the host cell's genome. Many scientists found this suggestion heretical, since it violated Francis Crick's central dogma: "DNA specifies RNA, which specifies proteins."

In 1970, however, Temin and David Baltimore, a molecular biologist then at MIT, independently discovered that RNA tumor viruses contain an enzyme, called **reverse transcriptase,** which indeed copies RNA into DNA. The enzyme received its name because the direction in which information flows is backward from that found in normal transcription (where information goes from DNA to RNA). As we will see in the next chapter, reverse transcriptase has become an essential tool for molecular biologists. Viruses that use reverse transcriptase in this way are called **retroviruses.** The discovery of reverse transcriptase supported Temin's hypothesis.

Subsequent work showed that Temin was exactly right. In most viruses, only the virus's nucleic acid actually enters the host. In retroviruses, however, the reverse transcriptase enzyme is actually packaged *with* the viral RNA, so the enzyme (not just the gene for the enzyme) enters the cell along with the

Figure 12-12 How do viruses express their genes? A. Single-stranded DNA virus. B. RNA plus-strand virus. C. RNA minus-strand virus.

> *Episomes are viruses or plasmids that switch back and forth between integrated and autonomous replication as the occasion demands.*

How Do RNA Viruses Reproduce and Express Their Genes?

Many mobile genes use genetic material other than double-stranded DNA. These genetic entities must use special tricks to replicate themselves. For example, the virus ΦX-174 uses single-stranded DNA. This virus simply copies the single strand into a complementary strand of DNA to make a double-stranded "replicative form" (Figure 12-12A).

virus's RNA (Figure 12-13). The reverse transcriptase then copies the RNA first into single-stranded and then into double-stranded DNA. The double-stranded DNA then integrates into the host cell's DNA, so that the virus's genes are replicated along with those of the host. Like DNA-containing tumor viruses, many (but not all) retroviruses contain an oncogene, whose expression leads to tumor formation. Retroviruses are known to cause cancer in birds, mice, monkeys, and humans. HIV is a retrovirus that infects cells of the human immune system but does not contain an oncogene.

Retroviruses use reverse transcriptase to transfer information from RNA to DNA, in the opposite direction to that specified by the central dogma.

HOW DO MOBILE GENES EVOLVE?

Viruses and other mobile genes live so closely with cells that we might not realize at first that they are influenced by evolutionary forces independent of those that influence their hosts. Even though mobile genes are not alive, they are still independent entities whose replication is subject to natural selection—the mechanism by which Darwin explained the evolution of organisms.

Although Viruses Are Not Alive, They Are Still Subject to Natural Selection

Because a virus uses energy-expensive materials from its host cell, the cell's ability to grow and reproduce is impaired. As a result, hosts that evolve ways to resist viral infection are more likely to survive and reproduce. This is simply natural selection. Cells have therefore evolved general mechanisms for resisting most kinds of viruses and also resistance mechanisms aimed at specific viruses. Many strains of *E. coli*, for example, guard against viruses by producing **restriction enzymes,** enzymes that cut foreign DNA into little pieces. Most restriction enzymes do not cut the DNA at random, but at special "target sites"—particular sequences of nucleotides that occur commonly in all genomes. Although a restriction enzyme can protect *E. coli* (or another bacterial strain) from infection by a bacteriophage, such an enzyme would be self-destructive if it also digested *E. coli*'s own DNA. To prevent such molecular suicide, *E. coli* modifies its DNA by adding methyl groups that protect the bacterial DNA from digestion.

As hosts evolve mechanisms for resisting a virus, the virus, in turn, evolves mechanisms for getting past the cell's defenses. The kinds of viruses that are able to infect *E. coli*, for example, all have DNA that is missing the sequence targeted by *E. coli* restriction enzymes. Because the viral DNA lacks this target sequence, the cell's restriction enzymes cannot destroy the viral DNA.

Viral multiplication cycles—from infection to release—have evolved to become highly efficient. Cycles are generally

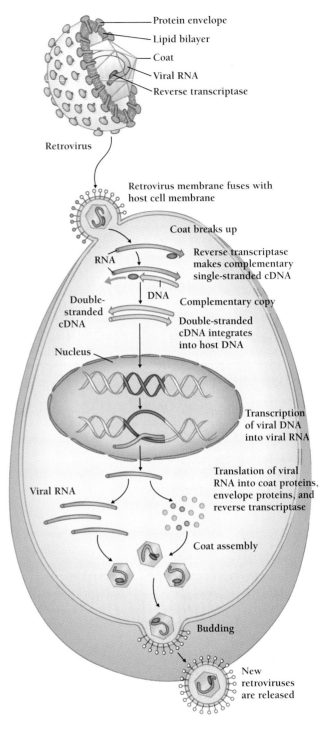

Figure 12-13 How do HIV and other retroviruses reproduce?

BOX 12.2

How Can Researchers Design Drugs That Will Combat AIDS?

The greatest epidemic in recent history is acquired immune deficiency syndrome (AIDS). Most biologists now agree that AIDS is caused by one of two types of human immunodeficiency virus (HIV1 and HIV2). HIV infects specialized white blood cells, called T lymphocytes, that normally help find and destroy invading pathogens. Over several years, the virus reduces the number and effectiveness of these crucial immune cells, making the body susceptible to infections and cancers that a healthy immune system normally fends off. These secondary illnesses, rather than the HIV virus itself, are what kill most people infected with AIDS.

Because HIV grows and replicates in the body for years before producing any symptoms, a person can test positive for HIV antibodies, indicating the presence of the virus, long before the onset of any illness. During this early stage of infection, bodily fluids such as blood and semen harbor active viruses that can transmit the infection to a new host. Most people contract the virus by sharing contaminated hypodermic needles or by having unprotected sex with an infected partner.

Like all retroviruses, HIV carries its genetic material in a single RNA strand accompanied by a molecule of reverse transcriptase. Once inside the cell, the reverse transcriptase generates a DNA strand copied from the RNA template. A second round of synthesis produces a double-stranded DNA copy of the viral RNA, which inserts itself into a chromosome in the host cell's nucleus. Without reverse transcriptase, HIV could still get into the cell, but it would never be able to replicate.

Since human cells do not depend on reverse transcriptase, the virus's enzyme makes a perfect target for anti-AIDS drugs. One type of antiviral therapy deceives the enzyme with synthetic molecules that resemble ordinary nucleotides, the raw materials for DNA synthesis. Reverse transcriptase incorporates these nucleotide impostors into the growing strand of viral DNA. But since the false nucleotides are subtly different from real ones, replication stops. AZT and 3TC, two drugs now commonly used to slow viral replication in patients with AIDS, are examples of false nucleotides. Human DNA polymerase is more picky than reverse transcriptase in selecting nucleotides and rarely incorporates AZT or 3TC. As a result, human cells exposed to the drugs continue to grow and divide normally. Unfortunately, HIV mutates rapidly, and strains that no longer use the false nucleotides quickly emerge. These resistant strains thrive in patients treated with these drugs, rendering the drugs useless.

Another class of HIV drugs works by interfering with a different viral enzyme—one that cuts a viral protein. After viral DNA integrates into the host's chromosome, host proteins transcribe the viral genes. Rather than producing a separate mRNA for each viral protein, however, the messages for several proteins form a single, long mRNA molecule. The resulting mRNA gets translated into a long polypeptide that must be cut into separate proteins before they will function. One of the proteins in the viral polypeptide chain is the enzyme that does the cutting. This enzyme, called a *protease*, begins cutting even while it is still part of the "polyprotein." Without the protease, the polyprotein remains unclipped and inactive, and the virus cannot go on to infect more cells.

In an attempt to stop HIV by knocking out the protease, researchers devised tiny molecules that nestle into the active site of the protease enzyme and prevent it from doing its job. The first of these protease inhibitors, called *saquinavir*, was released in December 1995—with dramatic results. A drug cocktail containing saquinavir, AZT, and 3TC can reduce the amount of virus in a patient's body to undetectable levels. For the first time since the discovery of HIV, this "triple therapy" has transformed AIDS from a death sentence to a potentially survivable disease.

But triple therapy has its problems. The cost of these AIDS drugs, particularly protease inhibitors, is still extremely high, placing an economic barrier between many sufferers of AIDS and good treatment. In the United States, not all insurance covers AIDS therapies, and in Africa, where most of the world's AIDS cases occur, the price is out of reach of almost everyone who has the disease. Also, triple therapy doesn't work well for everyone, and it will be several years before researchers know whether the drug cocktail can permanently clear the virus out of the body. Indeed, researchers worry that HIV may evolve resistance to protease inhibitors, as it has to other drugs. In that case, it could still return in full force, insensitive once more to all drug treatments.

To help stave off that possibility, research is underway on still another class of drugs that may close the door on the virus, preventing it from getting into T cells in the first place. To enter T cells, an HIV virus must bind to a pair of proteins on the cell surface, one of which is the recently identified chemokine receptor CCR5. Chemokines are chemical signals that induce inflammation at a site of injury or infection. Researchers noticed in 1995 that chemokines block HIV from infecting cultured cells. Not long after, more work revealed that CCR5 helps HIV enter cells and that the chemokine was interfering with that process. Drug companies are now racing to develop a drug that binds to CCR5 without setting off the inflammatory pathway. One reason to believe that this approach may work is that studies of people with a defective CCR5 gene have shown that the defect causes no obvious health problems yet provides genetic resistance to many strains of HIV.

The battle against the AIDS epidemic may finally be succeeding. In 1996, the number of patients dying of AIDS in the United States declined for the first time. While this is a significant landmark, the epidemic is far from over. As has been true of many other infectious diseases, the AIDS virus will continue to evolve, and researchers may always be just one step ahead.

Figure 12-14 Circular DNA in energy organelles. A. From a mitochondrion. B. From a chloroplast. Chloroplast DNA is ten times longer than that in animal mitochondria. The smaller circular DNA shown is from a virus. (A, CNRI/Phototake NYC; B, Richard Kolodner, Ludwig Institute for Cancer Research)

rapid, with lots of new viruses produced in each infected cell. Viral regulatory genes are so powerful that the host's transcription and translation machinery recognizes them in preference to the host's own genes.

One way that viruses speed up their cycle times is to keep their genetic material to a minimum. Unlike eukaryotic genomes, which are loaded with "junk DNA," every gene in a virus matters. One kind of cold virus, for example, specifies several different proteins from a single gene merely by arranging to have the mRNA spliced in different ways.

How May Viruses Have Evolved?

Some scientists believe that viruses are descended from free-living cells that became parasites on other cells. According to this view, the viruses progressively (over eons) lost most of their cellular components, letting the host cells do all the work.

Most biologists, however, believe that viruses may have evolved from plasmids, which evolved from transposons. Knowing what we now know about mobile genes, we can imagine that mobile genes first appeared as simple transposons that could move from place to place in the genome. Transposons that also included genes that benefited the host cell—such as genes that allow the host to break down poisons or antibiotics—would have allowed the host cells that had them to reproduce faster than cells without them. As a result, transposons with beneficial genes would have replicated much faster than those without beneficial genes. The differential survival of the rapidly replicating viruses is simply natural selection.

Later, beneficial genes could have acquired the ability to replicate independently as plasmids do. Finally, some autonomously reproducing plasmids could have acquired genes for protein coats that allowed them to move efficiently from cell to cell. Such a protein-wrapped plasmid would look just like a virus.

Over evolutionary time, transposons may have acquired the ability to replicate autonomously, making them plasmids. Some plasmids may then have acquired the genes for making a protein coat, allowing them to evolve into a modern virus.

MITOCHONDRIA AND CHLOROPLASTS CONTAIN THEIR OWN DNA

Chloroplasts and mitochondria contain their own DNA. Since all eukaryotes have mitochondria and many have chloroplasts as well, almost every eukaryotic cell therefore carries at least one genetic system independent of the nucleus. In a mammalian cell, each mitochondrion contains five to ten identical DNA molecules, each about 16,000 nucleotides long. Since a single cell may have hundreds of mitochondria, up to 1 percent of the total DNA in the cell may consist of mitochondrial DNA. Each piece of mitochondrial DNA is circular, meaning that it forms a continuous loop, like the DNA of bacteria and many viruses (Figure 12-14A).

Figure 12-15 An intracellular parasite. *Listeria monocytogenes* is a bacterium that infects human cells, including cells in the nervous system and the genital area and the white blood cells of the immune system. It most frequently causes disease in pregnant women and their babies. *Listeria monocytogenes* is remarkable for its ability to move about inside cells with the aid of one to four actin filaments, which work like flagella. (J.A. Theriot)

Plant mitochondria also contain circular DNA molecules, but usually of several different kinds. The total length and the total information content of mitochondrial DNA in plants can be more than 20 times greater than in animal mitochondria. Chloroplasts contain DNA as well (Figure 12-14B). Each species has a single kind of chloroplast DNA. The chloroplast DNA is circular and usually about 150,000 nucleotides long. In higher plants, chloroplast DNA may constitute up to 15 percent of the total DNA in a cell.

The reproduction of mitochondria and chloroplasts depends on cellular machinery that is specified both by the DNA in the nucleus and by the DNA in the organelle itself. The replication and expression of mitochondrial or chloroplast DNA, for example, requires about 90 proteins—far more than the organelle's DNA can usually encode. These proteins are specified by the DNA in the nucleus and made on cytoplasmic ribosomes. The DNA of mitochondria and chloroplasts nonetheless contains important information. For example, both mitochondria and chloroplasts contain the proton-pumping complex F_0-F_1 (ATP synthase) that harnesses the energy of proton gradients to make ATP. Each complex contains nine different polypeptides, only some of which are specified by the DNA in the nucleus. The other genes are in the mitochondria or chloroplasts.

Mitochondria and chloroplasts contain complete systems for transcription and translation, including not only DNA but also ribosomes, transfer RNAs (tRNAs), and an RNA polymerase similar to that of bacteria. Some components of the machinery of protein synthesis are made in the cytoplasm, and other components are made within the organelles.

Mitochondria and chloroplasts rely on both their own genetic information and that in the nucleus.

Studies of the DNA and RNA of Energy Organelles Support the Theory That They Derive from Ancient Prokaryotes

Like prokaryotic DNA, the DNA of most mitochondria and chloroplasts is circular. The ribosomes of mitochondria and chloroplasts are also more like bacterial ribosomes than like the ribosomes in the cytoplasm of eukaryotic cells. These and many other similarities between the energy organelles and prokaryotes support the **endosymbiotic theory** [Greek, *endon* = within + *syn* = together + *bios* = life]. This theory holds that present-day mitochondria and chloroplasts are descended from prokaryotes that came to live inside ancient eukaryotic cells. In support of this view, biologists point to many modern examples of cells that host endosymbionts (Figure 12-15). Recent studies of DNA sequences have revealed that *all* mitochondrial DNA shares many common features. Researchers now suspect that all mitochondria have evolved from a single type of mitochondrion in an ancient eukaryote.

The genetic code used in mitochondria is slightly different from that used in prokaryotes or in eukaryotic nuclei, with four codon assignments that are different. If mitochondria are descended from free-living prokaryotes, their different genetic code suggests an origin so ancient that it predates the establishment of a universal genetic code. In Chapter 18, we will discuss the origin of mitochondria and chloroplasts in more detail.

The genetic systems of mitochondria and chloroplasts differ from those in both eukaryotic nuclear DNA and in prokaryotes, suggesting that these organelles have an ancient and separate origin.

In this chapter, we have seen that not all genes reside in the chromosomes, nor do all genes pass neatly from generation to generation by means of meiotic divisions. Genes exist in mitochondria and chloroplasts, suggesting that these organelles were once independent organisms. Even more intriguing, however, genes also exist in mobile forms, such as viruses, plasmids, and jumping genes. In the next chapter, we will see that mobile genes are the power behind genetic engineering, a new field that may have profound effects on society in the next century.

Study Outline with Key Terms

Barbara McClintock showed that some genes can move from place to place. She showed that jumping genes, or transposons, cause the curious dappled colors of Indian corn. **Unconventional genetic systems** are DNA or RNA sequences that can replicate and function apart from chromosomal DNA, including the DNA of mitochondria and chloroplasts and mobile genes. **Mobile genes,** which include **viruses, plasmids,** and **transposons,** can move from cell to cell or from place to place within a chromosome. All of these unconventional genes depend on cellular machinery specified by chromosomal DNA.

Viruses, including **bacteriophages,** are assemblages of nucleic acid and proteins that can reproduce within a living cell. Whether shaped like an **icosahedron** or a long helix, viruses consist of an inner core of nucleic acid surrounded by a protein **capsid** or **membrane envelope.** Viruses were first discovered as disease-causing agents that could pass through filters fine enough to remove bacteria. Viruses have many properties of life: they reproduce, they respond to their environments, and they evolve. However, viruses cannot obtain energy from their surroundings and cannot form molecular building blocks.

Although viruses are not alive, they evolve through natural selection. Viruses are able to take over the machinery of their host cells and to produce more viruses. Viral multiplication cycles consist of six steps: (1) attachment to the host cell, (2) entry of viral nucleic acid into the cell, (3) synthesis of viral proteins, (4) replication of viral nucleic acid, (5) assembly of new virus particles, and (6) release of new viruses.

Many viruses kill their hosts in the process of replicating themselves. These viruses, which break open the host membrane in a process called **lysis,** are called **lytic viruses.** Sometimes, viral genes can integrate into the chromosomal DNA of their hosts. The integrated virus, called a **provirus,** does not immediately produce more virus. However, when the host cell replicates, it replicates the viral genes as well, a process called **integrated replication.** Other viruses make new viruses through **autonomous replication.** Actual virus particles are produced only when something induces the provirus to enter a lytic cycle. In a **lysogenic cycle,** a virus or bacteriophage switches back and forth between a **lytic cycle** and an integrated provirus, or **prophage,** stage.

RNA serves as the genetic material in some viruses. In plus-strand viruses, the viral RNA directs polypeptide synthesis. In minus-strand viruses, an enzyme copies the viral RNA into a complementary strand that directs polypeptide synthesis. In AIDS and other **retroviruses, reverse transcriptase** copies viral RNA into DNA, which then integrates into the host cell's chromosomal DNA.

Mobile genes include plasmids, which can replicate only autonomously; transposons, which can replicate only in integrated form; and **episomes,** which can replicate either autonomously or in integrated form. **Simple transposons** carry nothing but the gene for **transposase,** while **complex transposons** carry genes that benefit their bacterial hosts such as those for antibiotic resistance. Plasmids artificially constructed in the lab are called **recombinant plasmids.**

Plasmids called **resistance factors,** or **R factors,** contain genes for enzymes that make the cell resistant to antibiotics. Bacteria engaging in **conjugation** exchange genetic material, including both bacterial DNA and plasmids. In eukaryotic cells, some transposons, or jumping genes, move within the chromosomal DNA, disrupting the genome. Bacteria defend themselves against bacteriophages using **restriction enzymes,** which cut DNA at special "target sites."

Oncogenes specify proteins that help regulate cell division. The **viral oncogenes** of a tumor virus, which stimulate cell division, resemble, and almost certainly evolved from **proto-oncogenes** of eukaryotic host cells.

Energy organelles—chloroplasts and mitochondria—have their own DNA. The circular form of these DNA molecules and the genes they contain suggest that they derived from ancient prokaryotes that lived within ancient eukaryotic cells, a theory called the **endosymbiotic theory.**

Review and Thought Questions

Review Questions

1. What characteristics of a virus makes it nonliving?
2. How can a virus or other nonliving genetic entity evolve through natural selection? Give two examples.
3. How does a virus attach to a cell and transfer its genetic material into the cell?
4. What advantages does integrated replication have over autonomous replication?
5. What is a retrovirus? How does it differ from other RNA viruses?
6. What is the difference between a plasmid and a transposon?
7. What evidence suggests that mitochondria and chloroplasts were once prokaryotes that lived independently of other cells?

Thought Questions

8. If you take antibiotics frequently, you will select for *E. coli* that carry genes for antibiotic resistance. Suppose that you contracted tuberculosis (TB) from your roommate. Could plasmids then transmit antibiotic resistance from the *E. coli* in your intestines to the TB bacteria? Why or why not?
9. Many biologists suspect that transposons move genes among different species of eukaryotes. For example, some evidence suggests that within the past 100 years, a parasitic mite has carried transposons called P elements from one species of fruit fly to another. What evolutionary consequences might result from such transfers?

About the Chapter-Opening Image

Gene Kelly (1912–1996), dancing star of Broadway shows and Hollywood films, jumping over Indian corn, Barbara McClintock's experimental organism. The varied colors of the corn kernels result from the movement of "jumping genes," or transposons.

 Web Bit 12-1 The Evolution of Chloroplasts

 Web Links

 Quizzing & Testing

On-line materials relating to this chapter are on the World Wide Web at **http://www.harcourtcollege.com/lifesci/aal2/**

13 Genetic Engineering and Recombinant DNA

The Maverick

Early in the morning of October 13, 1993, reporters and a photographer showed up at the door of 50-year-old unemployed surfer and molecular biologist Kary Mullis (1944—). The Nobel prize winners for 1993 had been announced in the middle of the night, and all the winners were being interviewed. The reporters gathered around Mullis's beachfront apartment in southern California and caught him on his way out for a morning of surfing (Figure 13-1). They asked him, Did he know he had won? Was he happy? Surprised? How did he feel when the phone call came from Stockholm, Sweden? Mullis knew—he had already gotten his call in the small hours of the morning. Naturally, he was happy. "I realized I did something significant, the kind of thing you get the Nobel prize for," he told reporters, "but never in my wildest dreams did I ever think I'd win."

He wasn't the only one surprised. Explained one colleague to *Esquire* magazine, "He doesn't fit the normal mold of a Nobel-prize winner. . . . He's a wild man. There's a certain amount of politicking involved, and I was not sure his personality and his lifestyle would have allowed him to win." Mullis is indeed a maverick. Eventually, he offends nearly everyone, and he seems to enjoy doing it. At one scientific conference, he began laughing when someone else was speaking. When the speaker chided Mullis for laughing at a serious matter, he responded, "I'm not laughing at that. I'm laughing at you." Mullis is a far cry from the serious, dedicated researcher who still goes to the lab 7 days a week long after retirement. In fact, he hasn't worked in a lab in years.

Mullis, whose personality is equal parts boyish charm and astonishing boorishness, grew up in South Carolina, where, like many a Southern boy, he learned about life by roaming the woods and building rockets from drugstore supplies. At Georgia Tech, he earned a degree and got married for the first time.

In 1966, he arrived at the University of California, Berkeley, with a young wife and baby daughter, to work on a Ph.D. in biochemistry. Berkeley was all he had hoped it would be, and then some. "Six years in the biochemistry department didn't change my mind about DNA," he told UC Berkeley's alumni magazine, "but six years of Berkeley changed my mind about almost everything else." Besides psychedelics such as LSD, Mullis also discovered free love and divorce. "It wasn't a good place to be married."

By 1981, he was on his third marriage, with two young sons. For a while he worked as the night manager at the Buttercup Bakery in Berkeley. When he took a job as a chemist for one of the earliest biotechnology companies, Cetus Corporation, it was as a low-level technician. At Cetus he fell in love with another biochemist and, at 36, ended his third marriage.

One Friday night in the spring of 1983, the two biochemists headed for Mullis's weekend cabin in the coastal mountains of northern California. Late that night, as Mullis drove along a twisting mountain road through vineyards and redwoods, he filled the silence by thinking about DNA. Specifically, he was interested in making his job at Cetus—building short pieces of synthetic DNA—a little easier.

As the winding road unfolded before him, he thought about the way the two strands of DNA untwist and separate so that they can be copied. He designed experiments in his head, wondering how he might single out one particular nucleotide in the DNA and make copies of it. He saw ways his imaginary experiments might go wrong, and twisting his thoughts this way and that, fixed them. His idea seemed foolproof—he couldn't find anything wrong with it. In theory, he should be able to take a test tube filled with thousands of genes, single out a stretch of nucleotides, and make millions

Key Concepts

- Genetic engineering started with the selective breeding of organisms with high levels of natural or induced variation.
- Recombinant DNA technology allows biologists to create genes that do not exist in nature.
- Using recombinant DNA, molecular biologists can engineer both prokaryotic and eukaryotic cells to make specific proteins in large amounts.
- Biologists can also engineer plants and animals to produce new proteins and to have new biological properties.

of copies within a day. For a molecular biologist, it was undreamed-of power.

He realized with a shock that his idea was almost good enough to win him a Nobel prize. There had to be something wrong with his idea, he thought. It was too obvious, too slick, too easy. The idea was this: Mullis could make an oligonucleotide, a short piece of DNA, that binds to a complementary sequence in a far longer strand of DNA (Figure 13-2). Once the oligonucleotide was bound to the DNA, it could serve as a primer for DNA polymerase. The DNA polymerase would copy only the DNA immediately adjacent to the bound oligonucleotide. In the next step, the newly made DNA could itself be used as a template to make still another strand (as we discuss later in this chapter). The result would be a chain reaction: one copy of a specific DNA sequence yielding 2 copies, then 4, then 8, then 16, and so on.

Back at Cetus, Mullis's boss told him to take all the time he needed to develop his idea. Colleagues pitched in and helped him make the idea work. By mid-December, Mullis had worked out all the bugs in a new technique called the **polymerase chain reaction (PCR)**. PCR specifically and repetitively copies any segment of DNA between any two defined nucleotide sequences. Today, specialized machines about the size of a toaster perform this process automatically, so that a single copy of a chosen sequence can be multiplied into millions or billions of copies in less than a day.

Finding a short sequence of nucleotides in a cell's DNA is like searching for a needle in a haystack. PCR finds the right sequence, then copies it repeatedly until, after 30 rounds of replication, the original haystack is dwarfed by a haystack of needles. Not only does PCR help find particular sequences, it also makes them in usable amounts. For molecular biologists to study a particular nucleotide sequence, they need more than one copy of a sequence; they need thousands of identical

Figure 13-1 Kary Mullis, inventor of PCR.
(©1997 Darryl Estrine Photography)

copies. Mullis's idea was a singularly valuable technique that would allow thousands of other molecular biologists to make important discoveries.

The uses for PCR seem limitless. PCR enables crime experts to use DNA fingerprinting on mere specks of blood and other tissues to identify who has been present at the scene of a crime. PCR allows researchers to pick out trace amounts of DNA in blood from HIV and other viruses, as well as pick up early signs of cancer and genetic defects in human eggs and embryos. PCR allows archaeologists and paleontologists to study minute amounts of DNA from ancient animals and plants—from the 130-year-old tissues of Abraham Lincoln to an 18-million-year-old magnolia leaf preserved in a peat bog.

Mullis's discovery might have launched his career, even made him happy. But it did not. Almost as soon as Mullis had perfected the technique, other researchers at Cetus began using the technique to do experiments. Mullis thought his name would be on those papers—as a coauthor—but his colleagues didn't agree. A simple acknowledgment of his invention near the end of the paper seemed like enough to them. Tempers flared, and Mullis was asked to leave if he couldn't get along with his colleagues. In the end, Mullis left Cetus, taking with him a $10,000 bonus for his invention. It was little enough. Cetus made millions of dollars from his invention and ultimately sold the rights to PCR for a mind-boggling $300 million. But even that was a bargain. Today, PCR technology is a $1-billion-a-year industry.

Except for a brief stint as a consultant, Mullis has mostly abandoned conventional science. Friends and colleagues suggest that his talents are being wasted. Modern molecular biology, however, is as much a social endeavor as an intellectual one, and Mullis's intellectual talents are of little value in an environment where getting along with people is paramount. Doing science, especially doing science that others notice, means

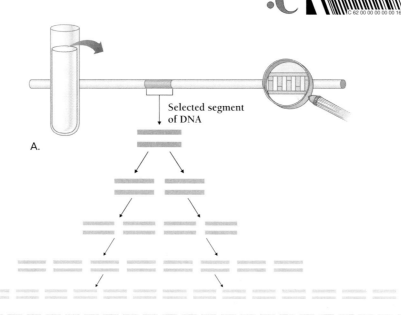

Figure 13-2 PCR is a simple, powerful technique for multiplying specific sequences of DNA. A. When DNA is heated, the two strands uncoil. They are then cooled and replicated. The cycle of heating, cooling, replicating, and then heating again is repeated until millions or billions of copies of the sequence are obtained. B. Short segments of single-stranded DNA called oligonucleotides act as primers and allow researchers to replicate a particular sequence, not just any DNA. The 20 or so bases of the oligonucleotide pair with the correct segment of the DNA and initiate replication.

fitting in, belonging to one or another "club." A few researchers—loners like Barbara McClintock—do not entirely mind if no one appreciates their discoveries. They enjoy their work so much that they continue in a virtual social vacuum.

Mullis will be starved for neither money nor attention. He eventually won $450,000 from the prestigious Japan Prize and a $412,500 Nobel prize for his invention of PCR. And although he may not have a desk at a university or his name on a door at a biotechnology company, his Nobel prize has indirectly provided him with many opportunities for fame and fortune.

WHAT IS GENETIC ENGINEERING?

Thousands of years ago, before cities or farms existed, nomadic hunter-gatherers collected grain and seeds from the plant varieties that yielded the most grain. As they worked with the grain, carrying it or storing it, they spilled some of the grain. In the spring it sprouted and people had only to harvest their crop. In time, they learned to deliberately plant some of their seed in places where it would grow well. They became farmers. Populations that had been nomadic settled in one place to farm.

In time, these first farmers learned to observe closely the individual plants in their fields. Grains from the plants that produced the most seed, the largest seeds, or the best-tasting seeds were saved for the next year's crop. Ten thousand years ago, the first farmers began practicing artificial selection, breeding both plants and animals.

The success of agriculture allowed human populations in the Middle East, northwestern China, and southern Mexico to flourish. As these populations became larger and denser, they built villages and cities. By 5000 years ago, the peoples of the Middle East had mastered enough microbiology to ferment grain into beer and to culture milk into cheese and yogurt.

The early domestication of plants and animals depended primarily on selecting which individuals to breed. Once breeders had chosen desirable varieties of plants or animals, the trick was to preserve and enhance their genetic traits. They did this by allowing individuals with the desired qualities to breed only with each other. Sheep with the thickest wool, for example, were bred to one another. This ancient practice is depicted in the Bible, where Jacob keeps his spotted goats apart from the nonspotted goats of his father-in-law. Brewers and cheese makers likewise learned to select the right strains of microorganisms by starting each batch with the right "starter."

Beer brewing, cheese making, and agriculture itself are early examples of **biotechnology**—the use of living organisms for practical purposes. Modern biotechnology depends as heavily on the selection of organisms for their desirable genetic qualities as the first biotechnology did 10,000 years ago. Genetic engineers can alter the genetic traits of any organism, whether sheep, goats, wheat, corn, yeast, or mold (Figure 13-3). The greatest difference between ancient methods and modern ones is the precision and rapidity with which organisms can be altered. But modern genetic engineering also transgresses the natural boundaries among species. Genetic engineers now happily move genes among all kinds of organisms, including humans, mice, tomatoes, yeasts, and bacteria.

Biotechnology first arose 10,000 years ago when humans began selecting and breeding useful plants, animals, fungi, and microorganisms.

Figure 13-3 Prehistoric corn. A. Cob from a cave in Mexico. Humans have been breeding bigger and bigger corn cobs for thousands of years. Some biologists believe that early corn was bred specifically to be popped. B. A recently popped prehistoric corn kernel compares favorably with the modern version. (A, Science VU/Visuals Unlimited; B, courtesy of C.P. Mangelsdorf)

Genetic Engineering Makes Use of Both Natural and Induced Variation

In 1929, Alexander Fleming discovered penicillin as the result of a chance contamination of one of his bacterial cultures. Fleming isolated the mold responsible for killing his bacteria and grew the mold in pure cultures. In 1938, 9 years after Fleming's discovery, two Oxford University scientists—Ernst Chain, a refugee from Nazi Germany, and his professor, Howard Florey, an Australian pathologist—realized that penicillin might be able to play a role in controlling diseases caused by bacteria. Within 2 years, Florey and Chain announced that penicillin, even in very small amounts, could protect mice from an otherwise fatal injection of bacteria.

But humans need far larger amounts of penicillin than mice. To treat just one adult for a bacterial infection, a doctor needs about 7.5 grams of penicillin (the weight of two and a half animal crackers). But to obtain even a millionth of a gram of penicillin, Florey's team had to process enough mold to fill nearly 30 bathtubs. Penicillin was so rare that, in some cases, the precious drug was extracted from the urine of patients and given to them once again in subsequent doses. It was clear that to make penicillin in the amounts needed to control human disease, scientists had to find a way to make much more penicillin. They did this partly by finding better ways to grow the mold and partly by choosing strains of mold that could produce more penicillin.

To find molds that produced more penicillin, microbiologists searched through, or screened, soil samples from all over the world. The soil sample that produced the most penicillin came from a U.S. Department of Agriculture Laboratory in Peoria, Illinois. Microbiologists in government laboratories, universities, and pharmaceutical companies then increased the genetic variation in the Peoria strain of mold by subjecting it to mutagens—x rays, ultraviolet light, and chemicals. The changes in the DNA were random and most were useless, but because of the great number of mutations, a few turned out to increase the production of penicillin. After each round of mutagenesis, researchers chose the strains with the highest yield of penicillin. By the end of this selection process, researchers had found a mold that produced nearly 100 times as much penicillin as the original Peoria strain.

Biotechnology selects useful traits from a range of variation. This variation can be natural or induced by means of mutagens.

Knowing Biochemical Pathways Helps Molecular Biologists Design Useful Organisms

To find the strain of mold that could produce the most penicillin, Florey and Chain and the other researchers depended on the same general approach that breeders have always used, screening large numbers of individuals for those with the best expression of a particular trait—in this case penicillin production. They improved their chances by treating the molds with mutagens. However, they had no idea why the Peoria strain produced more penicillin than other strains. In the 1940s, researchers could not predict what mutations would increase the yield of penicillin.

Today, however, molecular biologists know so much about the biochemical pathways in some organisms that they can predict what kind of mutation will produce a desired trait. The genetic engineering of tomatoes provides one example. In order for tomato farmers to get tomatoes safely to the supermarket, they must pick the tomatoes when they are green and hard. If the tomatoes were picked when they were ripe and soft, they would be crushed to a pulp by the weight of the other tomatoes in the box.

Normally, green tomatoes ripen into sweet, flavorful red tomatoes before we eat them. Once the tomatoes are ready to be delivered to a store, they are treated with the plant hormone ethylene gas, which triggers their ripening (Chapter 33). The tomatoes also produce ethylene themselves, and, regardless of whether the gas is natural or factory made, the tomatoes ripen just the same. Unfortunately, about 40 percent of tomatoes are picked too soon and can never ripen properly. These are the tomatoes we all know so well that turn red but taste like wet cardboard. Only tomatoes that have developed different "flavor components," chemicals that make them taste good, are actually ready to be picked. But it is impossible for the grower to tell the green tomatoes that are ready to be picked from those that are not. All green tomatoes look the same.

One biotechnology company solved this problem by deliberately damaging a gene that controls the production of ethylene in the tomato. Such tomatoes can be left on the vine until they turn a pinkish orange, which signals that they have developed their flavor components. These tomatoes will never soften (or turn red and sweet) on their own, for they cannot make ethylene. Only after the tomato grower has harvested the tomatoes and the tomato distributor has treated the tomatoes with ethylene will they ripen. Since the tomatoes do not need to be picked until they have both color and flavor components, growers won't pick them too soon.

Biotech companies have also used specific knowledge of biochemical pathways to select hundreds of commercially important strains of bacteria and fungi. Some of the molecules produced by these strains include amino acids such as glutamic acid (used as the flavoring "monosodium glutamate," or MSG), vitamins such as riboflavin, and even industrial chemicals such as ethanol and acetone. Microorganisms also produce commercially important enzymes, such as those used to convert starch into fructose for use as a sweetener and the protein-digesting enzymes used in laundry detergents.

Molecular biologists can design useful organisms by inserting or destroying genes that code for proteins involved in specific biochemical pathways.

WHAT IS RECOMBINANT DNA AND HOW IS IT USEFUL?

The discovery, in the early 1950s, that genes are nothing more than chains of nucleotides led directly to our present ability to both alter individual genes and move them around wherever we want them. By the early 1970s, molecular biologists had learned to cut and paste pieces of DNA from different organisms (as well as from various mobile genes), and by the early 1980s, they had learned how to alter the individual nucleotides in a gene. Earlier work had already revealed how viruses and plasmids shuttle genes around the genomes of bacteria and other host cells. Molecular biologists found it easy to use these tiny molecular parasites to carry along and reproduce pieces of DNA that had been patched together in a test tube.

How Do Molecular Biologists Use Recombinant DNA?

Recombinant DNA is any DNA molecule consisting of two or more DNA segments that are not found together in nature. For example, researchers can join a gene for antibiotic resistance with a plasmid, then infect a cell with the plasmid, giving the cell a gene for antibiotic resistance. Because molecular biologists can now design and make recombinant DNA that fulfills particular requirements—scientific, medical, agricultural, or industrial—the resulting recombinant DNA molecules are sometimes called "designer genes."

Recombinant DNA technology is so powerful that it is difficult to summarize all the ways in which it has proved useful. Recombinant DNA has provided scientists with an irreplaceable tool for studying the structure, regulation, and function of individual genes; for unraveling the molecular bases of genetic diseases; for understanding why individuals differ in their responses to drugs and medicines; and for turning plants, animals, and microorganisms into chemical factories that can churn out vast quantities of proteins and other substances that these organisms would never make on their own (Figure 13-4).

Recombinant DNA promises purer vaccines (which have fewer risks), new drugs, gene therapy for treating genetic

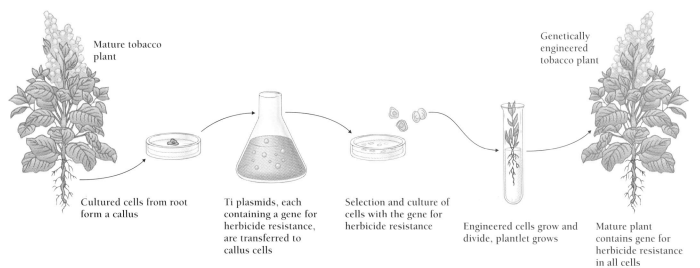

Figure 13-4 Recombinant DNA technology in action. Cells from a tobacco plant are grown in culture, then infected with a Ti plasmid carrying a gene for herbicide resistance and other genes. The growing cells are treated with herbicide, and only the cells expressing the new genes survive. These herbicide-resistant cells can then be grown into mature plants, which will bear seeds carrying the new genes.

diseases such as cystic fibrosis, and genetic tests for hereditary diseases such as Huntington's disease or partly hereditary diseases such as breast cancer. Recombinant DNA means brand new proteins, oils, and carbohydrates—designed by molecular biologists and manufactured by genetically engineered organisms. Finally, recombinant DNA presents a thousand moral and ethical questions that cannot be answered solely by the scientists who raise them. These questions must be understood and answered by all educated citizens.

The development of recombinant DNA technology depended on the union of many different lines of research, including work on (1) restriction enzymes in bacteria, (2) DNA replication and repair, (3) replication of viruses and plasmids, and (4) the chemical synthesis of specific nucleotide sequences. We will see how information and techniques from these different fields all contributed to the ability to make recombinant DNA.

Recombinant DNA techniques join together pieces of DNA in combinations that do not occur in nature.

Figure 13-5 Cutting up DNA. A. DNAases cut DNA at random. Each copy of DNA is cut differently. B. Restriction enzymes cut each copy of DNA at the same places. Matching fragments have the same length and sequence. Restriction enzymes can cut the DNA with blunt ends, as in B, or with sticky ends, as in C.

How Do Restriction Enzymes Cut Up a Genome?

DNAase and RNAase are enzymes that cut the links between the nucleotides in DNA or RNA, respectively. DNAase usually cuts DNA into short fragments at random (Figure 13-5A). In a test tube filled with DNA, each of the many copies of a genome will be cut differently. The resulting fragments are all cut into different sizes and the pieces from one genome overlap those in another. In such a mess, there is little hope of finding a specific sequence. To isolate a specific piece of DNA, a researcher must instead use **restriction enzymes,** special DNAases that cut DNA only at particular sequences (Figure 13-5B).

Happily, such enzymes exist. Bacteria use restriction enzymes to defend themselves against viruses and other mobile genes. When a bit of alien DNA enters a bacterium, the restriction enzymes cut it to bits. Bacteria have many kinds of restriction enzymes, each of which recognizes a different nucleotide sequence. The sequence recognized by a given restriction enzyme is called a **restriction site** and typically consists of four or six nucleotides, though some restriction sites are longer. Some restriction enzymes cut each of the two strands of DNA in the same place, so that the resulting fragment of DNA has "blunt" ends (Figure 13-5B). Other restriction enzymes make staggered cuts, which result in single-stranded, "sticky" ends that easily join with other such fragments that have complementary sequences (Figure 13-5C).

Researchers have prepared restriction enzymes from over 200 different bacterial strains. Each enzyme recognizes a specific sequence, with over 90 such restriction sites now described. One strain of *E. coli*, for example, uses a restriction enzyme, called *EcoRI*, that binds to and cuts DNA wherever it finds the sequence GAATTC (Figure 13-5C). A restriction enzyme such as *EcoRI* will cut multiple copies of DNA everywhere a particular sequence appears. The result is a matching set of **restriction fragments**—pieces of DNA that begin and end with a restriction site.

By comparing the sizes of restriction fragments, researchers can establish a **restriction map,** which shows how the restriction sites are placed within a piece of DNA. More importantly, genetic engineers can join these fragments together in novel ways. A restriction enzyme that makes fragments with sticky ends can be used to make human and mouse fragments that stick together (Figure 13-6A). They can, for example, join a restriction fragment from a human being to a fragment from a mouse or a bacterium.

Restriction enzymes, which bacteria use to defend themselves against viruses, cut DNA molecules at specific sequences. Restriction enzymes allow the preparation of DNA fragments with defined lengths and sequences.

Figure 13-6 Genetic engineering. A. Sticky ends enable researchers to join pieces of unrelated DNA. B. DNA from a mouse and human can be joined, as long as they have been cut at the same sequence by the same restriction enzyme.

How Do Molecular Biologists Join Restriction Fragments Together?

Recall from Chapter 10 that during ordinary DNA replication, the enzyme **DNA ligase** links together the separate pieces of DNA into one continuous strand. Molecular biologists depend on DNA ligase to link restriction fragments together. For example, to make the hormone insulin, researchers initially inserted a sequence coding for human insulin next to the operator and promoter for β-galactosidase from *E. coli* (Figure 13-6B). The manufactured recombinant DNA then contained a piece of a bacterial DNA that regulated the expression of an adjacent mammalian gene. This recombinant DNA, put back into *E. coli*, produces insulin when the bacteria are exposed to the milk sugar lactose. Ordinarily, the presence of lactose would signal the bacteria to prepare to digest lactose. But bacteria carrying the recombinant DNA respond to lactose by producing human insulin.

DNA ligase can link together two pieces of DNA from different sources to produce recombinant DNA.

How Do Molecular Biologists Express Recombinant DNA in Bacteria and Other Hosts?

Once scientists taught themselves to make recombinant DNA using DNA ligase and restriction fragments, they faced two serious challenges. The first challenge was to produce large quantities of particular genes. The second was to induce bacteria or other host cells to express these recombinant genes as usable proteins.

How can bacteria be induced to make great numbers of copies of a gene?

The answer to the first problem (production of large quantities of specific genes) was plasmids, which are capable of forcing a bacterial cell to make large numbers of copies of a single gene. Recall from Chapter 12 that R (resistance) factors are naturally occurring plasmids containing genes for antibiotic resistance. By the early 1970s, researchers had isolated and studied many such R factors. Using restriction enzymes and DNA ligase, they made streamlined plasmids that contained only two parts: (1) an origin of replication—the DNA sequence needed to start DNA synthesis and (2) one or more genes for antibiotic resistance. The antibiotic resistance acted as a marker, allowing researchers to distinguish and select bacteria that carried the plasmid from those without the plasmid.

Using DNA ligase, researchers can hook any gene to such a plasmid. A host bacterium will copy the plasmid DNA as if it were its own, enabling the recombinant plasmid to replicate many times. Researchers then grow the bacteria in an antibiotic. Because the plasmid's antibiotic resistance gene protects the host bacteria against being killed by the antibiotic, only the genetically engineered bacteria survive.

Microbiologists have long used the word **vector** [Latin; carrier, rider] for any organism or virus that carries disease from organism to organism. Molecular biologists therefore use the word "vector" to describe anything that carries genes from organism to organism. Plasmids are not the only vectors. Researchers have also used specially modified bacteriophages, animal and plant viruses, transposable elements, and even fragments of chromosomes.

Vectors are DNA sequences from viruses, plasmids, and other mobile genes that can carry recombinant DNA into cells.

How can bacteria be induced to make eukaryotic genes?

If a researcher wants to express a gene from a virus or a bacterium, pulling out a particular gene for cloning is not too hard. Often enough, the entire sequence is already known. Extracting genes from the chromosomes of plants and animals, however, is not only more difficult, it is often useless. Recall that a eukaryotic gene is interrupted by long sequences of introns. Bacteria, which must express the gene, do not have any machinery for recognizing and removing introns from the resulting pre-messenger RNA. (Prokaryotes, recall, have no introns.) If the bacteria cannot remove the introns, they cannot make a translatable mRNA. So how can researchers acquire a clean, intronless copy of a eukaryotic gene that can be cloned using bacteria?

The solution comes from retroviruses, the viruses that use RNA as their nucleic acid, yet insert themselves into the DNA genomes of eukaryotes. Recall (from Chapter 12) that retroviruses use the enzyme **reverse transcriptase** to copy RNA into DNA. Reverse transcriptase copies almost any RNA into DNA. Researchers who want a particular gene can start with the mature mRNA for that gene, which has already had all the introns removed. (The mature mRNA comes from eukaryotic tissues that express the desired gene.) Reverse transcriptase then copies that mRNA back into DNA. The result is a DNA molecule, called **complementary DNA,** or **cDNA,** that is complementary to the mRNA from which it was copied. Unlike the DNA in the genome, cDNA has no introns, as it is a copy of the mature mRNA (Figure 13-7).

Another enzyme copies the single-stranded cDNA into double-stranded cDNA, the second strand of which is identical in sequence to the original mRNA. The cDNA "gene" is now ready to be joined to a vector and propagated.

Actual genes from eukaryotes cannot be expressed in bacteria. Molecular biologists must make a DNA copy (called cDNA) of a eukaryote's mRNA using reverse transcriptase. The cDNA can then be propagated using standard methods.

Can host cells be induced to express polypeptides in a usable form?

Unfortunately, not all eukaryotic genes can be expressed in bacteria. As we saw in Chapter 11, many genes require extensive modification of the polypeptide itself in order to function. Bacteria cannot perform these modifications. Most membrane proteins, for example, require modifications that can only be made in eukaryotic hosts. Recombinant *E. coli* designed to produce insulin, for example, cannot produce insulin that will function properly in the human body. Nonetheless, hundreds of functional proteins have been expressed from recombinant DNA molecules in bacteria.

The production of other proteins requires that the recombinant DNA be expressed in eukaryotic cells, which are able to modify proteins after translation. One particularly useful cell line comes from a moth that is susceptible to infection by a virus called baculovirus. After infection, these cells devote almost all their protein synthesis to the production of the virus's coat protein. By substituting another protein-coding DNA for the coat protein, molecular biologists can produce large amounts of a desired protein. Other useful cells, which can be grown in large vats under the proper conditions, include yeast, human kidney cells, and hamster ovary cells. Among the commercially important proteins made by eukaryotic cells from recombinant DNAs are insulin (used to treat diabetes), tissue plasminogen activator (TPA, used to treat heart attacks), and erythropoietin (used to stimulate red blood cell production, especially after kidney transplants).

Bacteria and various eukaryotic cells can produce large amounts of specific proteins that would otherwise be available only in small amounts.

How Do Researchers Make Multiple Copies of Recombinant DNA?

Throughout the 1970s and early 1980s, the only practical way to make enough copies of a particular DNA sequence to study it was to introduce a single recombinant DNA molecule (gene plus plasmid vector) into a bacterial host cell. The plasmid can induce the host cell to make up to a thousand copies of the gene it carries. In addition, researchers induce the bacterial host cell to divide rapidly. As the bacteria multiply, so does the recombinant DNA. Each colony of bacteria may grow into 10^{12} or more bacteria. If each one has 1000 (10^3) copies of the plasmid, such a colony may contain 10^{15} identical copies of the recombinant DNA. Researchers call this process "cloning the recombinant DNA."

The word "clone" comes from gardening. Gardeners and fruit growers often grow whole new plants from live twigs cut from mature trees and bushes. All of the plants derived in this way are genetically identical and constitute a **clone** [Greek, *klon* = twig]—a group of genetically identical cells or organisms. In humans, identical twins are a natural clone. Any group of genetically identical individuals constitutes a clone. Similarly, the offspring of a single bacterial cell are called a clone, and the many copies of a plasmid are called a clone. By extension, molecular biologists also refer to the many copies of recombinant DNA as a clone.

Cloning recombinant DNA depends on the DNA replication machinery of bacteria or other cells to copy recombinant DNA. Since Kary Mullis's invention of PCR, however, researchers have been able to make millions, or even billions, of copies of DNA sequences in a test tube, using a DNA polymerase purified from a bacteria. PCR can make huge numbers of individual DNA sequences very rapidly. The sequences must be short, usually no more than 20,000 nucleotide pairs, but the process is much faster than cloning.

PCR depends on the ability to make two specific polynucleotides that correspond to sequences at each end of the sequence to be copied. These two synthetic polynucleotides are typically each about 20 nucleotides long and are examples of **oligonucleotides** [Greek, *oligos* = few]. Each oligonucleotide serves as a primer for DNA polymerase. No enzyme is needed to unwind the DNA. Instead, in PCR, the DNA is simply heated until the two strands come apart. The DNA is cooled, and the two primers then bind to the complementary sequences—one on each strand (Figure 13-2). DNA polymerase then copies each strand until the experimenter (or the machine) stops the reaction by again raising the temperature.

The increased temperature also separates the template strand and the newly made complementary copy. Lowering the temperature then allows DNA polymerase to go back to work, again starting with the oligonucleotide primers. The only region that will be copied—again and again with further cycles—is the segment flanked by the two primers. The result is the rapid and exclusive multiplication of one specific segment of DNA.

Cloning and PCR technologies allow researchers to make thousands or millions of copies of individual genes.

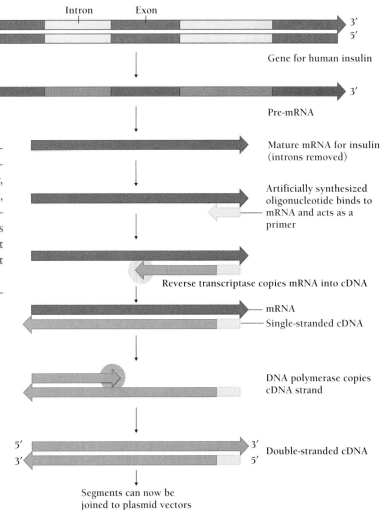

▶ **Figure 13-7** How do researchers derive human genes without introns? In a human cell, the gene for human insulin is transcribed into pre-mRNA. The cell edits out the introns to form mature mRNA. Researchers then isolate the mature insulin mRNA and copy it into DNA, which can be joined to a plasmid vector and expressed by a bacterium.

How Do Biologists Find the Right DNA Sequence in a Recombinant DNA Library?

The genome of any organism—whether bacterium, plant, or animal—can be cut into pieces with restriction enzymes. The collection of thousands or millions of restriction fragments from a single genome is called a **gene library.** These fragments can be combined with plasmid vectors and introduced into populations of bacteria, which can then replicate each fragment in the library indefinitely. If the library contains at least one representative of every sequence in a genome, it is called "a complete genomic library." Alternatively, a recombinant DNA library may be constructed from mRNA alone (using reverse transcriptase), in which case it is called "a cDNA library." In either case, a gene library is useful only if researchers can find the gene they want. Until recently, isolating the right gene was like trying to find a needle in a haystack.

Molecular geneticists have developed techniques for finding and cataloging the tens of thousands of sequences in a gene library. Just as a magnet can help find a needle even in a haystack, molecular biologists can use two kinds of molecular "magnets" to find particular pieces of DNA. One kind of magnet is a **probe,** a short bit of single-stranded DNA whose sequence is complementary to a small part of the sequence in the gene researchers are looking for, much like the DNA primers used in PCR. The other magnets are **antibodies,** proteins from the immune system that recognize and bind to other proteins. As we will see, antibodies do not detect the recombinant DNA itself, but only the proteins specified by the DNA.

A probe, also called a **hybridization probe,** can hybridize by base-pairing with DNA in the library to form **hybrid DNA**—DNA whose two complementary strands come from different sources (Figure 13-8A). Such a probe, dropped into

Figure 13-8 Two techniques for locating a gene. A. A hybridization probe locates a specific DNA sequence. B. Antibodies locate the protein product of the same sequence.

a test tube filled with thousands of restriction fragments, can locate and pair with the one fragment whose sequence complements its own as easily as a bloodhound can track down a lost child. Once the probe binds to the right piece of DNA, it is easy to locate the probe if it has been chemically or radioactively labeled. The technique of hybridizing a labeled DNA probe in order to find the right piece of recombinant DNA in a gene library is called molecular hybridization.

Researchers who know what protein they are looking for can also use antibodies to find the gene for that protein (Figure 13-8B). Antibodies are proteins with surfaces that recognize the shapes of foreign molecules, thereby giving animals an important defense against infection. When an animal is exposed to a foreign protein, it responds by making antibodies that bind specifically to that protein. Researchers start with a complete cDNA library laid out in a half-million colonies of bacteria, each colony carrying and expressing a different cDNA "gene" from the human genome. Over a few days, the 500,000 colonies can be treated with labeled antibody, which binds to and identifies the colony that is synthesizing the protein and therefore carries the gene the researchers are looking for. From the bacterial colony carrying the correct gene, researchers can clone the gene for study.

Figure 13-9 Patterns of gene expression in a benign versus a malignant brain tumor. DNA arrays allow the simultaneous monitoring of the levels of thousands of individual mRNAs. Differences in the expression of specific genes (coded here by green and red spots) give researchers clues about the classification and the treatment of different types of tumors. Red spots indicate individual genes whose expression is higher in a high-grade, malignant brain tumor than in a low-grade, less malignant brain tumor. Green spots indicate genes whose expression is higher in the low-grade tumor. *(Courtesy of Linda Liau, UCLA Department of Neurosurgery)*

Two kinds of molecular probes allow biologists to locate specific genes. Hybridization probes detect genes in recombinant DNA clones, in cell extracts, and in cells. Antibodies detect the synthesis of specific proteins in colonies of bacteria containing recombinant DNA.

HOW CAN BIOLOGISTS STUDY THE EXPRESSION AND STRUCTURE OF MANY INDIVIDUAL GENES?

A DNA probe for a specific sequence can hybridize back to the corresponding gene in cellular DNA and to a recombinant DNA that contains that sequence. The same probe can also hybridize to any mRNA transcribed from that gene. Researchers can therefore use a cDNA or oligonucleotide probe to measure the amount of a particular mRNA in a tissue. Until the early 1990s, researchers performed such hybridizations one gene (one mRNA) at a time, as described in Box 13.1 (northern blots). These experiments showed that different cell types are specialized for the production of certain mRNAs: muscle cells thus have high levels of the mRNA myosin (the contractile protein of the thick filaments), but not of the mRNAs for hemoglobin.

In the last few years, new methods have allowed biologists to measure the levels of thousands of types of mRNA all at the same time. Researchers will soon be able to ask about the expression—at the level of mRNA—of *every* gene in a particular cell type. The simultaneous measurement of so many kinds of RNA depends on the use of thin, dime-sized wafers, containing thousands of tiny DNA spots in a fixed arrangement. Each DNA spot is put in place by a robot or (in some cases) by a lithographic process derived from integrated circuit technology. The whole arrangement of spotted DNA is called a **DNA array** or (in the case of the lithographic wafers) a *DNA chip*. Each spot in the array contains a DNA or an oligonucleotide corresponding to just one gene.

To determine the level of individual mRNAs, a researcher first isolates RNA from a tissue (for example, from a muscle or a tumor) and copies the RNA into cDNA, tagging each of the resulting cDNA molecules with a fluorescent label. Each kind of cDNA molecule now hybridizes to the corresponding DNA spot in the array. The higher the level of a particular mRNA, the greater is the fluorescence from that spot. At the end of the hybridization, the pattern of fluorescent spots tells which genes are expressed in that tissue.

A major problem in this kind of analysis is that there are so many spots. For example, researchers now think that mammals express a total of 80,000 to 100,000 different mRNAs, and they are developing techniques that can determine the levels of each of these at the same time. But no one is yet sure how best to analyze such huge amounts of data. Researchers are developing computer methods that will allow meaningful comparisons of individual spots and of expression patterns in types of tissues.

One of the first applications of this new technique was the study of different kinds of cancers (Figure 13-9). Cancer

BOX 13.1

Southern, Northern, and Western Blotting

Stop by a molecular biologist's lab bench today and chances are you will find a plastic box filled with gel and wired up to run an electric current through the slab of clear gel. Using a technique called **gel electrophoresis,** researchers can separate large molecules on the basis of their size. Gel electrophoresis is one of the most useful techniques available for studying DNA, RNA, and proteins.

In a gel containing sequences of DNA, the current draws the negatively charged DNA through the gel—a porous matrix that slows down the larger molecules as they snake and bump their way toward the positive electrode. A special stain reveals bands of DNA molecules, which group at different positions in the gel according to their size. The smaller fragments have moved farther, and the larger fragments remain closer to the rectangular well in the gel where they were first "loaded."

Sometimes, however, the bands all blur together and it can be hard to tell one fragment from another. For such complex samples, molecular biologist Ed Southern in 1975 developed a technique that lets the experimenter choose which fragments to look at. Usually, only a particular part of the genome is of interest—perhaps a specific gene or a polymorphic region used for DNA fingerprinting. In **Southern blotting** and **hybridization,** the DNA is separated by means of electrophoresis (Figure A). Then, for more convenient handling, the DNA is transferred from the thick, watery gel to special paper. The paper, placed between the gel and a stack of paper towels, wicks the DNA and all the liquid out of the gel. The result is a Southern blot.

Next, the researcher applies a radioactive DNA probe that base pairs, or hybridizes, with the region of interest. (One or more of the phosphorus atoms in the DNA probe is radioactive.) Since the DNA is normally already double stranded, heat is applied to melt the strands apart, allowing the probe to find its target. When the temperature is again lowered, some of the single-stranded probe molecules hybridize with DNA fragments in the blot.

The result of this Southern hybridization is a sheet of paper the size of the original gel with radioactive bands of DNA wherever there were fragments of the gene of interest. The Southern blot is then exposed to a sheet of film, which reveals only the bands of DNA lit up by the probe. The DNA sequences of interest light up, while irrelevant bits of DNA remain invisible.

Several years after Southern developed his blotting technique for DNA, scientists began studying complex samples of RNA with a similar procedure and dubbed the technique **northern blotting.** Other scientists blotted proteins out of protein gels and called it **western blotting.** There is no such thing as an eastern blot, but the other three points of the compass come up in laboratory discussions every day.

Figure A Blotting techniques. Specific nucleotide sequences are detected by hybridization to a blot of fragments separated by electrophoresis. If the fragments are DNA, the technique is a Southern blot. If the fragments are RNA, the technique is a northern blot.

researchers have found that different types of tumors have different patterns of mRNAs. Researchers hope that such differences in mRNA patterns will ultimately help physicians predict how a particular tumor will grow and how best to stop it. DNA arrays also provide a powerful way to do human genetic mapping (as we will discuss in Chapter 14).

DNA arrays allow researchers to study the expression and the structure of thousands of individual genes at the same time.

RECOMBINANT DNA CAN REPROGRAM CELLS TO MAKE NEW PRODUCTS

The ability to manipulate genes has already had important practical consequences. Hundreds of biotechnology, chemical, and pharmaceutical companies have begun to use recombinant DNA to make proteins useful to medicine, agriculture, and the chemical industry. Recent products include newly engineered forms of well-known proteins (such as an improved version of an enzyme used in laundry detergents), as well as previously unknown proteins—such as the protein missing in the lungs of cystic fibrosis patients or the protein growth factors that may prevent nerve death in individuals with neurological diseases. In addition, genetic engineers are breeding plants and animals with new characteristics—plants that are resistant to insects or to particular herbicides, sheep that produce a clotting factor in their milk, and pigs that produce human hemoglobin. Such gene products are then purified and used in medicine.

Figure 13-10 Effects of human growth hormone (hGH). hGH cures certain forms of dwarfism. Formerly hGH had to be isolated from human cadavers, and recipients of the growth hormone too often contacted hepatitis and other viral diseases. Today, hGH is produced by genetically engineering bacteria and is free of viruses that infect humans. (© *Jerry Cooke/Photo Researchers, Inc.*)

Genetically Engineered Bacteria and Eukaryotic Cells Can Make Useful Proteins

Recombinant DNA technology allows the reprogramming of cells to make an extraordinary number of products. These products include insulin, growth hormone, and other protein hormones, as well as ingredients for processed foods and enzymes needed to produce valuable small molecules or to destroy pollutants (Figure 13-10).

Vaccines are another protein product that can now be produced by means of recombinant DNA technology. Recombinant DNA techniques allow the preparation of uncontaminated vaccines because the immunizing molecules are made from viral DNA rather than from active viruses. Instead of growing the virus and then inactivating it, virologists isolate a piece of viral DNA that encodes a single viral protein. A single protein cannot cause a viral disease, but it can provoke an immune response that will guard against infection by the actual virus. Recombinant DNA technology has produced safer vaccines for many kinds of disease-causing viruses. Researchers hope that one day such techniques may also produce vaccines for diseases such as herpes and AIDS, for which no effective vaccines now exist.

Some of the protein products of recombinant DNA are so rare and so hard to isolate that little was known of them before they could be produced from recombinant DNA. *Growth factors,* for example, are rare proteins that keep specific kinds of animal cells alive, stimulate cell division, or trigger particular types of cell specialization. *Cytokines* are other rare proteins that coordinate the immune response. Until recently, researchers knew little about the structure of growth factors or how they work. Recombinant DNA techniques, however, have produced relatively large amounts of pure growth factors and cytokines, allowing researchers to study their structure and mode of action. Many academic and commercial laboratories are now investigating the possibility that growth factors may be able to prevent nerve degeneration in such devastating diseases as Alzheimer's disease, Parkinson's disease, amyotrophic lateral sclerosis (ALS, also called Lou Gehrig's disease), and Huntington's disease.

Genetically engineered bacteria and eukaryotic cells can make useful proteins, including enzymes, vaccines, drugs, and human proteins for treating genetic diseases.

RECOMBINANT DNA CAN GENETICALLY ALTER ANIMALS AND PLANTS

Recombinant DNA can be inserted into the cells of whole plants and animals. Organisms that carry recombinant DNA in their genomes are called **transgenic organisms.** The added DNA is called a **transgene.**

To be expressed in a higher organism, a transgene must contain the appropriate control signals. To identify such regulatory regions, researchers usually join together a segment of DNA that is suspected of being regulatory with a protein-coding gene whose expression can be easily determined.

How Do Researchers Produce a Transgenic Mammal?

For a gene to be expressed in all the appropriate cells of an animal, researchers must put the transgene into the zygote before the beginning of embryonic development. Then all the cells of the organism will contain the engineered DNA.

In the early 1980s, several research groups succeeded in producing transgenic mice. Perhaps the most dramatic of these early experiments produced mice carrying the gene for human growth hormone (hGH). At birth, such transgenic mice were already twice the size of their littermates (Figure 13-11A). In the past 20 years, researchers have produced thousands of kinds of transgenic mice. Because some of the diseases in these mice accurately mimic those in humans (including cystic fibrosis, muscular dystrophy, Alzheimer's disease, and AIDS), these mice provide an opportunity to understand the basis of these diseases and to develop and test new therapies.

To produce a transgenic mammal, researchers must surgically remove eggs from the ovaries in the abdomen of the female. As a result, few eggs are available at one time. These few eggs are put into a dish and fertilized with sperm from a male. The resulting zygote is then injected with the engineered gene. This technique has produced transgenic pigs, goats, and sheep, and it seems likely to yield similar results with other species (Figure 13-11B). So far, however, only one transgenic animal is born for every 100 injected zygotes. Despite tabloid claims to the contrary, no one is known to have made any transgenic human embryos, though the basic steps are no more difficult in humans than in mice or goats.

The engineering of transgenic animals faces serious obstacles. First, few eggs actually incorporate the recombinant DNA. Even genes that initially seem to have been incorporated turn out to be not in the chromosomes, but in some mobile DNA outside the nucleus. Such genes can be lost in subsequent cell division. Perhaps more troubling, current genetic engineering techniques put the recombinant gene into the animal's chromosomes at random, both adding to the host's own genes and, often, disrupting other genes. The disruption of key genes creates mutant animals that die young, which is

A.

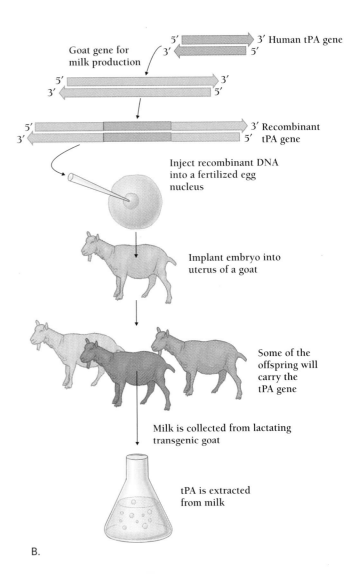

B.

Figure 13-11 Transgenic animals carry foreign genes that they can pass on to their own offspring. A. The first transgenic mammal (*right*) was a mouse carrying the gene for human growth hormone. B. Today, transgenic goats secrete valuable proteins in their milk. The proteins are isolated from the milk for use as pharmaceuticals. This goat carries the gene for tissue plasminogen activator (tPA), a protein that activates the anticlotting system and is used to save the lives of heart attack patients. (*A, R.L. Brinster/Peter Arnold, Inc.*)

one reason why it would be unethical to attempt this technique in human zygotes.

Researchers would like to replace the animal's own genes with the recombinant version instead of merely adding to them. This is relatively easy in prokaryotes and yeast, but more difficult in animal cells. One area of success is the production of genetic "knockouts"—animals in which a particular gene has been inactivated. Knockouts are a useful way of creating research animals with specific genetic defects, such as those found in humans. Usually, knockouts are used as a way of studying what role a gene plays in determining phenotype. The phenotypes that result from knockouts can surprise biologists. Figure 13-12, for example, shows a mouse in which researchers knocked out a gene specifying a growth factor believed to influence early embryonic development. The researchers expected the knockouts to stop developing long before birth. Instead, normal-looking baby mice were born, but with hair that grew so long that the knockouts looked a bit like Persian cats.

Recently, researchers have succeeded in creating "knockin" mice, in which an allele has been *replaced* rather than merely added. Like knockouts, knockins are difficult to make. In a knockin, however, the allele goes exactly where it is supposed to and the old allele is eliminated.

The knockout approach has several drawbacks. Quite often, all of the individual knockouts die as early embryos, and little can be learned about the adults or even the developing embryos. Even more discouraging for researchers, many knockouts show no effects at all from the loss of a functioning gene. Other genes apparently compensate when a particular protein is absent. In one recent example, researchers created knockout mice unable to make oxytocin, a hormone produced during the birth process and commonly given to women to induce labor. Yet, knockout mice unable to produce oxytocin delivered their pups with no problems.

Interpreting the phenotypic results of knockout organisms is also problematic. Knocking out genes to see what they do is like trying to find out what the different objects in a house do by removing them one at a time. If we remove the washing machine, the laundry no longer gets washed. But the same thing occurs if we remove the laundry detergent, or the clothes, or the electrical outlet into which the washing machine is normally plugged. Each of them is essential to getting the clothes done, but no single one of them is solely responsible for clean clothes. Similarly, when molecular biologists claim to have found a "master gene" for a given trait, the gene could equally well be described as the "weak link" in the intricate chain of events and materials that make up the phenotype. Teasing apart that chain of events will continue to be a major challenge.

New techniques now allow researchers to knock out a gene in an adult mouse. Using such "conditional-knockout" mice, researchers can determine what a gene product does in adult life, rather than during development. A related technique allows researchers to knock out a gene just in a particular organ or set of cells. While these experiments are still in their infancy, researchers are using them to unravel the role of individual genes and proteins in a whole living organism.

Figure 13-12 Knockout mouse. Researcher Gail Martin decided to knock out a growth factor gene, called FGF5, believed to play a role in the very early formation of an embryo. Martin and her colleagues expected that none of the homozygous embryos would finish developing. Instead, the knockouts all developed normally, but with hair nearly 50 percent longer than normal. Although FGF5 is normally expressed in muscle cells, brain cells, and motor neurons, all of these tissues were normal. *(Gail Martin, University of California, San Francisco)*

Transgenic mammals present many new opportunities and problems. The opportunities include researchers' new ability to study the function of specific proteins, including some responsible for human diseases. The problems include the insertion of DNA at random in the genome, which means that the genes may not function as researchers hope and may also disrupt other genes. Genetic knockouts and knockins can nonetheless provide clues about how previously mysterious proteins function in the body.

The Genetic Engineering of Plants Is Easier Than That of Animals

Perhaps the most rapidly expanding area of genetic engineering is in agriculture. Plant cells are generally easier to clone than animal cells. Equally important, plants can be grown in vast fields, so that a human protein, such as hemoglobin, produced by a plant, can be grown in hundreds of pounds per acre, instead of in grams per tub of *E. coli*. Tobacco plants, for example, can synthesize mouse antibodies that seem to behave just like mouse-made antibodies. Tomato plants can synthesize human serum albumin, a protein used to treat burn victims. Biotechnology companies have produced crops (sometimes called "nutriceuticals") that can mass-produce otherwise expensive compounds for use as medicines.

Genetic engineering of plants also has the potential to be enormously lucrative. If one of the engineered tomatoes mentioned earlier in this chapter becomes popular, for example, the inventors stand to gain a virtual monopoly in the tomato

BOX 13.2

Can Biologists Clone Dinosaurs?

Scary as *Tyrannosaurus rex* must have been, few people would pass up a chance to see the great flesh-eating dinosaur alive. Is it possible that DNA technology will someday bring dinosaurs back from extinction to populate theme parks everywhere? Can we scrape together bits of prehistoric DNA from fossils, stuff them in a big egg, and watch a baby brontosaurus poke through the shell?

The answer is no. Researchers cloned frog embryos in the early 1950s, and cloning adult animals became reality in 1997 with the birth of Dolly, a lamb whose genetic makeup was taken from the mammary tissue of an adult ewe. But the creation of a living organism from scraps of ancient DNA is an entirely different problem.

The first challenge for would-be dinosaur cloners is to find the genetic material of a race of animals that died out 65 million years ago. Little more remains of the ancient beasts than fossilized bones, in which minerals have replaced the organic material, essentially turning the bone into rock. Fossils therefore contain little or no DNA. Blood from ancient mosquitoes preserved in amber may be a more likely source of Jurassic genes. But tapping into that resource raises other questions: What animal provided the insect's last meal? Did mosquitoes even bite dinosaurs? And if they feasted on several different species, which DNA in the insect's belly came from whom?

Even if scientists found a reliable source of dino DNA, a complete dinosaur genome would be nearly impossible to recover. For one thing, many scientists believe that DNA is not likely to survive intact for so long. One study has shown that DNA in a water solution will break down almost entirely into individual nucleotides after about 50,000 years. Still, under the right conditions, DNA can last longer. In 1990, geneticists successfully extracted DNA from a 17-million-year-old fossilized magnolia leaf dug up in Idaho. Two years later, researchers used PCR to amplify short gene fragments from a termite and a stingless bee, both encased in amber for 30 million years.

Attempts to recover dinosaur genes have been less successful. In 1993, scientists in Montana found blood cells in a *T. rex* bone that had only partly fossilized, thus protecting the central cavity from the elements. Unfortunately, the only sequences isolated from the sample later turned out to be human DNA contaminating the specimen. Other labs have had similar contamination problems in trying to squeeze dino DNA from ancient bones. Nevertheless, some researchers remain confident that PCR will eventually produce authentic dinosaur gene fragments. But even then, those few hundred bases will be a far cry from the millions and possibly billions of bases of sequence that once resided in dinosaur chromosomes.

The lack of an appropriate egg is yet another nail in *T. rex*'s coffin. Even in the unlikely event that scientists discovered a cache of perfectly preserved dinosaur chromosomes, they would need to place them in an egg from some suitable modern animal. Whose egg would they try?

If dinosaurs lie on the evolutionary pathway to modern birds, as many biologists theorize, then the ostrich egg might not be a bad one to start off with. But an egg is more than just a protective coating for DNA. Eggs contain information about how cell division should proceed, or how the embryo will be oriented (which side is up, for example). It seems improbable that an ostrich egg, or any modern egg, would resemble a dinosaur egg closely enough to support the growth of a baby dinosaur, even with the proper genetic instructions.

Cloning living animals is at least technologically feasible. And now that successful clones of sheep have been made, successes in other animals, including humans, may not be far off. But for better or worse, the great dinosaurs are gone for good.

---

market. (Americans eat more tomatoes than any other vegetable, by far.) Already, however, farmers are already planting huge amounts of genetically engineered (and patented) soybeans, corn, and cotton. Some of these seeds, available since 1996, produce a bacterial protein (from an insect pathogen called *Bacillus thuringiensis*) that kills major pests, such as the European corn borer. (In 1998, American farmers planted 2 million acres of this engineered corn.) In 1999, farmers began to plant soybeans engineered to resist a fungus (*Phytophthora*) that causes root rot. Still other seeds, patented by Monsanto Corporation and marketed as "Roundup-Ready," carry a gene that allows crops to resist the herbicide glyphosate (Roundup), also made by Monsanto. These engineered crops allow farmers to spray an entire field with herbicide and kill the weeds but not the crops.

One way that researchers carry genes into plant cells involves a species of soil bacterium called *Agrobacterium tumefaciens*. *A. tumefaciens* helps cause plant tumors, called crown galls. The actual tumor-causing agent is not *A. tumefaciens*, however, but a plasmid called the **Ti plasmid** (for *T*umor *i*nducing), which resides within the bacterium. A tumor results when a piece of plasmid DNA moves into the genome of the host plant. Researchers can engineer the Ti plasmid to carry specific genes into the genome of a plant. As in other forms of genetic engineering, the recombinant DNA usually includes an antibiotic marker that allows researchers to quickly distinguish the cells that have taken up the recombinant DNA from those that have not.

In another technique, sometimes called the **gene gun**, tiny particles of gold or tungsten are coated with fragments of recombinant DNA, then fired into a young plant with a miniature "shotgun" powered by gunpowder or compressed gas. The young plant is then chopped up and its cells grown on a cul-

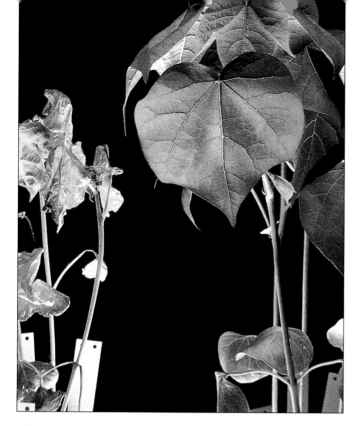

Figure 13-13 Transgenic cotton. Both the normal cotton plant (*left*) and the transgenic cotton plant (*right*) have been sprayed with a potent herbicide. Only the transgenic plant flourishes, however, for it alone possesses a gene for herbicide resistance. (*Calgene*)

ture containing an antibiotic. The antibiotic kills all of the cells except those that have taken up the recombinant DNA introduced with the gene gun.

Genetic engineering and conventional breeding programs have produced new varieties of flowers and remarkable crops (Figure 13-13). Many crops have been engineered or bred for higher yields. Now, however, agronomists and molecular geneticists have begun to collaborate to produce crops engineered to produce well on a specific plot of land, according to the particular farm's microclimate and to the availability of minerals.

Other programs have focused on the production of "value-enhanced" crops, whose value will not come just from higher yields but from specially engineered traits. For example, in 1998, farmers planted more than a million acres of "high-oil" corn, with double the amount of oil in conventional corn. This crop can reduce the cost of feeding livestock and of producing cooking oils. Seed manufacturers are also working on many other crops—soybeans with high levels of lysine and methionine (normally in short supply in grains), canola and sunflowers with high contents of specific fatty acids with desirable properties in cooking or in the manufacture of soap, and colored and wrinkle-resistant cotton, which would reduce the need for chemical dyes and treatments. The production of such specialized crops will require new ways of tracking particular crops from field to market. Such production will also require special testing procedures to ensure that each crop has the properties that it was supposed to have.

Using either the Ti plasmid or the gene gun, molecular biologists can genetically engineer plants that can, for example, synthesize animal or plant proteins, resist herbicides, or resist infection by plant viruses.

What Are the Risks of Recombinant DNA in Agriculture?

The incredible power of recombinant DNA technology lies partly in how precisely and quickly changes can be made and partly in the ability of molecular biologists to transfer genes from one organism to another. The long-term consequences of moving genes from one organism to another are unknown. Some critics argue that the ecological effects, if any, can be global and irreversible. Genetically engineered plants can, for example, exchange their new genes with wild cousins, either through conventional interbreeding or by way of plasmids and other vectors. Will genes inserted into a crop plant insert themselves into unrelated pest plants, creating, for example, herbicide-resistant weeds? Can a wild plant acquire traits that allow it to take over a local ecology or become a serious agricultural pest? Will engineered plants and animals interact with their environments in unforeseen, potentially destructive ways? Biologists and others can only speculate. No one knows if any of these scenarios are real threats.

Many experimental products are also controversial for other reasons. For example, in one laboratory experiment, milkweed deliberately dusted with pollen from insect-resistant corn led to the death of caterpillars of Monarch butterflies, which ordinarily feed on milkweed. Environmentalists worry that genetically engineered crops could inadvertently kill useful (and beautiful) as well as harmful insects.

Furthermore, because genetically engineered seeds are patented, farmers cannot save seed from one year to plant the next (as is common for many crops, such as wheat). Many critics are concerned that the increasing use of patented seeds will make farmers economically dependent on just a few seed producers, who may be more concerned with multinational profits than with the welfare of farmers in any one country.

Criticism of genetically modified crops has been especially loud in the European Union. Part of this opposition derives from ecological concerns, part from economic concerns (especially about U.S.-based corporations), and part from a general suspicion of genetic engineering. In June 1999, the European Union placed a moratorium on importing genetically modified foods.

Critics of genetically modified crops are concerned with the spread of genes that may be ecologically harmful as well as with increasing dependence on a few seed producers.

The Application of Recombinant DNA Technology Poses Moral Questions for Society

At the other end of the spectrum are the moral and ethical issues that come from recombinant DNA's precision. The diagnosis of genetic diseases, for example, is far in advance of treatment. Thus, in the next century, people may know that they are at higher-than-average risk for certain forms of cancer or mental illness, for example, but they will not necessarily be able to do anything about their fate. This is already true for genes that predispose people to certain forms of cancer of the breast, colon, and prostate. People at risk for Huntington's disease can now find out if they carry the disease-causing allele. Those with the Huntington's allele have no way, however, of avoiding the disease, so the test is of very limited value. Yet, potential employers and insurance companies want to know what people's genetic flaws are so that they can avoid hiring or insuring people likely to get sick.

Many people worry about applying genetic engineering to humans. Researchers are already working to treat genetic diseases by inserting genes in human cells. These techniques, called gene therapy, will be discussed in Chapter 14. Some researchers may try to eliminate certain diseases altogether by replacing defective genes even before conception. Such a heritable genetic cure has already been achieved in mice for a neurological disease characterized by abnormal brain development.

However, while few people object to curing deadly diseases, what constitutes a "genetic disease" may become controversial. Societies have often designated people with certain characteristics as "undesirable," and readily embraced arguments that these characteristics are genetic. If biologists develop the capacity to replace genes in humans, will genetic engineers also try to modify genes that affect characteristics other than those responsible for disease? Will future societies attempt to produce more (or less) intelligent citizens? Or less (or more) aggressive citizens? Already, many physicians treat shorter-than-average children with growth hormone, so that the children will "fit in." There is no reason to think that either parents or physicians would hesitate to "treat" such children with gene therapy.

While many aspects of physical appearance and behavior are strongly genetic, it seems unlikely in principle that gene replacement could ever produce a uniform society. Possibly, the wealthiest members of society could afford to have their children engineered for such traits as good teeth and other aspects of health and appearance. Meanwhile, the rest of society would go on as usual. Conventional medicine already suffers from this defect: the wealthiest members of society get better medical care than the poorest. It is clear that all the people in a society need to discuss the issues raised by the future possibilities of genetic engineering. As in the case of any new advance (such as nuclear weapons and nuclear power), the use of technology is a concern for the society as a whole, not just for scientists.

We do not yet know all the limitations of the new technology. Society must carefully evaluate the risks, benefits, and moral implications that arise from the ability to produce new or previously unavailable products, especially engineered plants and animals. In the next chapter, the last chapter of this section on genetics, we will explore some of these issues in more detail.

Study Outline with Key Terms

Until the development of **recombinant DNA** techniques, **biotechnology** consisted mostly of the selection of varieties of organisms with desired traits.

A molecule of recombinant DNA consists of two or more DNA segments that are not found together in nature. For example, by combining DNA pieces from humans and *E. coli*, researchers made a recombinant DNA molecule with the regulatory region of β-galactosidase and the structural gene for insulin. Bacteria that contain this DNA make insulin in response to the presence of lactose instead of β-galactosidase.

Making recombinant DNA depends on the use of enzymes that can cut, link, and modify DNA. **Restriction enzymes** cut DNA at **restriction sites,** where specific sequences occur, allowing molecular biologists to isolate individual DNA fragments, called **restriction fragments. DNA ligase,** an enzyme used in DNA replication, links pieces of DNA from different sources. A **restriction map** shows the positions of the restriction sites in a piece of DNA.

DNA molecules derived from viruses and plasmids can serve as **vectors** for the propagation of recombinant DNAs within host cells. Some of the most widely used vectors derive from R factors, plasmids that carry antibiotic resistance genes. Other vectors include bacteriophages, as well as plant and animal viruses.

Polymerase chain reaction (PCR) is now a basic tool for finding and multiplying, or **cloning,** fragments of DNA. **Oligonucleotides,** short pieces of single-stranded DNA, act as primers for DNA polymerase. Alternating heating and cooling do much of the rest.

The starting materials for recombinant DNA may come from the genomes of viruses or organisms or from DNA copies of mRNA. Most plant and animal genes contain introns, which would interfere with their expression in bacteria. Many recombinant DNAs coding for plant and animal proteins are therefore derived from **complementary DNAs (cDNAs),** copied from mRNA by the **reverse transcriptase** of a retrovirus.

Collections of recombinant DNAs, called **gene libraries,** may contain thousands or millions of different DNA sequences. Finding a desired DNA sequence usually depends on the use of a **probe.** A **hybridization probe**—often a piece of DNA made in the laboratory—bonds with one strand of the DNA to form **hybrid DNA.** By labeling the probe with radioactive nucleotides or a chemical marker, the experimenter can identify recombinant DNA. **Antibodies,** which have surfaces that recognize the shapes of specific molecules, can help identify bacteria that are producing a particular protein.

In **western blots,** antibodies are used to detect specific proteins after **gel electrophoresis.** Hybridization probes are used to identify specific DNA sequences in **Southern blots** and to determine levels of individual mRNAs in **northern blots** or in **DNA arrays.**

In many cases, researchers do not know the function of a particular protein until recombinant DNA techniques make enough of the protein available to study. Recombinant DNAs are likely to be useful in the preparation of safe vaccines for use in disease prevention.

Recombinant DNAs can reprogram organisms to make new products. Products such as insulin and **growth factor** that are made by genetically engineered bacteria are already commercially important as pharmaceuticals. Recombinant DNA can be used to alter animals and plants genetically. **Transgenic organisms** carry deliberately added **transgenes** in their genomes. The current method for making transgenic mammals involves the injection of an engineered gene into a zygote produced by in vitro fertilization. Molecular biologists can also damage selected genes to make genetic knockouts.

Making transgenic plants is easier than making transgenic animals because experimenters can easily add or replace genes in single cells in culture. The engineered cells can then be grown into whole plants. Genes are introduced either by means of a tumor-inducing (or **Ti**) **plasmid,** which normally resides in bacteria that cause plant tumors, or by means of a **gene gun.** The agricultural applications of transgenic plants have been widespread and controversial.

Review and Thought Questions

Review Questions

1. How does modern genetic engineering using recombinant DNA differ from breeding techniques practiced over the last 10,000 years? How are these two kinds of genetic engineering the same?
2. How can knowing a biochemical pathway help molecular engineers design useful organisms?
3. What is recombinant DNA?
4. What two enzymes are basic to assembling recombinant DNA? What do the two enzymes do, and what is their role in nature?
5. Why do molecular biologists need multiple copies of a gene?
6. What two techniques allow molecular biologists to make many copies of a gene? What are the strengths and weaknesses of each technique?
7. Why do molecular biologists want to engineer cells to express large amounts of the gene product, or protein? How do they accomplish this?
8. What prevents bacteria from expressing a gene cut directly from a human genome? How do molecular biologists get around this problem?
9. How do molecular biologists find the right DNA sequence in a recombinant DNA library?

Thought Questions

10. What is your opinion about creating transgenic humans? Suppose that you knew that you carried a gene that would predispose your children to a mental illness. Would you opt to conceive by means of *in vitro* fertilization so that your zygotes could be screened and the gene replaced? Suppose large numbers of other people chose not to use this service. Some of their children would continue to develop severe mental illness, eventually becoming rather expensive wards of the state. Would you feel that everyone should be required to screen his or her zygotes? What if you knew that the families of people carrying this gene tend to be more intelligent and more creative than average? Would that make a difference in your opinion? Why or why not?
11. Would you mind eating a genetically engineered tomato? Why or why not?

About the Chapter-Opening Image

Finding a particular DNA sequence is like finding a needle in a haystack. PCR simplifies the process by making many copies of the needle.

On-line materials relating to this chapter are on the World Wide Web at **http://www.harcourtcollege.com/lifesci/aal2/**

Web Bit 13-1 Human Cloning: A Multiplicity of Sticky Options

Web Links

Quizzing & Testing

14 Human Genetics

To Map the Genome or Hunt the Huntington's Gene?

David Botstein, burly and bombastic, jumped up from the table and paced the small conference room, loudly lecturing the other 12 scientists still seated at the table. It was October 1979, and the 13 scientists had met at the National Institutes of Health near Washington, DC, to discuss the impact of molecular biology on inherited diseases, especially Huntington's disease. But the discussion had come to a critical point of disagreement, and Botstein, a brilliant and temperamental molecular biologist from the Massachusetts Institute of Technology (MIT), could no longer contain himself (Figure 14-1).

Beginning a search for the Huntington's gene now, argued Botstein, was premature and a waste of time. New technology would soon revolutionize human genetics, allowing scientists to map and identify every human gene and find the molecular basis of every genetic disease. Once a general map was established, it would be child's play to find the molecular defects that cause Huntington's disease, cystic fibrosis, muscular dystrophy, and any of the other 3000 inherited diseases.

At the table, still seated, was David Housman, another molecular biologist from MIT. Housman watched Botstein impassively, then quietly disagreed. Yes, eventually, all would be known, he said, but wasn't it better to start first—right now with a single disease? And why not Huntington's disease? Researchers need not wait for a complete genetic map to find the Huntington's gene. There was no reason not to jump ahead, applying knowledge and techniques that were already available. Housman himself had already shown how the abnormal regulation of hemoglobin genes could cause certain inherited anemias called thalassemias.

Botstein scoffed. It was short sighted to jump into problems that might be personally pressing but were not yet scientifically "ripe." Better, said Botstein, for scientists to wait for basic knowledge. Indeed, despite the efforts of dozens of researchers, the Huntington's gene was not sequenced for 14 more years. The early mapping and sequencing of individual disease genes moved excruciatingly slowly. Only when basic research by Botstein and others provided new techniques for faster advances did the mapping of defective alleles begin to proceed quickly.

Yet it was Housman who prevailed at the 1979 conference. The conference had been sponsored by a small foundation called the Hereditary Disease Foundation (HDF), which is based in Los Angeles and dedicated to promoting research on Huntington's disease. The HDF had been started by Milton Wexler, a prominent Beverly Hills psychoanalyst. Wexler's two daughters, Alice and Nancy, were at risk for developing

Key Concepts

- Even though humans are more than 99 percent genetically identical to one another, all humans, except identical twins, are genetically unique.
- Phenotype is a product of a complex interaction between genotype and environment.
- Many aspects of human genetics, including, for example, diagnostic testing and gene therapy, raise serious ethical, legal, and social issues.

Huntington's; in 1968 their mother, Leonore, had been diagnosed with the fatal disease.

Huntington's disease is a devastating neurological disease that causes jerky movements, mood swings, personality changes, and premature death. The most obvious early symptom of the disease is the constant motion of the arms and legs, giving the disease its previous name, "Huntington's chorea" [Greek, *chorea* = dance, as in "choreography"]. Huntington's usually begins in middle age. Over the course of 10 to 20 years, patients gradually lose control of movements, memory, and mood. The bodies of Huntington's patients gradually waste away, and for the last few months of life patients stop moving altogether.

Huntington's disease is a classic genetic disease. An affected parent passes the disease to half of his or her offspring, indicating that the disease is caused by a single dominant allele of some gene. Nearly everyone who inherits the Huntington's disease allele eventually develops the symptoms. This high level of expressivity makes the disease particularly suitable for genetic analysis. In 1979, however, researchers and physicians knew nothing about the gene and almost nothing about the damage it caused at a biochemical and cellular level.

In 1969, Milton Wexler had begun engaging young scientists to brainstorm new approaches to understanding Huntington's disease. He attracted these biologists both by inviting them to parties that included some of the most famous artists and entertainers in the Los Angeles area and by encouraging these young biologists to think freely and creatively. Few young biologists could resist these brushes with Hollywood glamour. But the real purpose of the gatherings was the scientific workshops, in which Wexler brought together small groups of biologists who did not ordinarily meet—researchers in fields ranging from genetics to neurology. These informal workshops were intellectual incubators for new ideas not only about Huntington's disease but about the application of new techniques to the study of all inherited diseases, especially disorders of the brain. Among the young scientists stimulated by these workshops was one of this book's authors, Allan Tobin. In 1979, Wexler appointed Tobin Scientific Director of the Hereditary Disease Foundation. It was Tobin who organized the 1979 workshop on molecular genetics, with help from his former Cambridge neighbor, David Housman.

Figure 14-1 David Botstein. *(Courtesy of Stanford University Medical Center)*

Botstein's fears that attempting to map the Huntington's gene would be difficult turned out to be well founded. To find a particular gene among the 30,000 genes on the 46 chromosomes, geneticists had to look for markers, obvious phenotypic characteristics, such as eye color, that are always inherited with the defective allele and are therefore near the defective gene. Geneticists had constructed linkage maps in *Drosophila* through the painstaking analysis of the selective mating of flies with identifiable phenotypes—such as white eyes or notched wings. But geneticists cannot set up matings of humans; human genetics instead depends on the analysis of matings that, to a geneticist, seem arbitrary and senseless. A few markers in humans, such as those for blood proteins, work well, but none of these were inherited with the Huntington's allele.

The picture changed, however, when molecular biologists began to use recombinant DNA techniques to clone and sequence specific genes. Researchers then discovered numerous small differences among individuals at the DNA level. Even though two people may make exactly the same hemoglobin molecules, for example, their hemoglobin genes are likely to differ from one another. These "silent" differences lie in DNA sequences that are not transcribed and translated into protein—for example in introns or in flanking sequences. By the late 1970s, researchers were convinced that (except for identical twins) every person is genetically distinct from every other. At the DNA level, the homologous genes inherited from mother and father differ at about 1 nucleotide in 500.

The technology that so excited David Botstein was a method for identifying such differences, which he called **restriction fragment length polymorphisms,** or **RFLPs** (pronounced "RIFFlips"), and which we discuss below. Within a few years, differences in DNA sequences would not only allow geneticists to produce a map of the human genome, it would also allow police and military laboratories to match blood and tissue samples with a previously unimaginable accuracy—changing forever the ability to identify criminal suspects, as well as victims of crimes, disasters, and battles. By 1980, Botstein, his collaborators, and his competitors had established the feasibility of making a map of the human genome based on differences in DNA sequence, rather than on differences in phenotypic traits. By 1994, researchers had succeeded in making a relatively complete RFLP map of the human genome, allowing researchers to locate genes within about 1 million base pairs of DNA. More recently, geneticists have begun a much finer map, with a resolution of a few thousand base pairs of DNA, using variations in single bases. These variations are called *single-nucleotide polymorphisms,* or *SNPs* (pronounced "snips").

Until the late 1980s, enthusiasm for making a map of the human genome was not universal, even among molecular biologists. The research required was repetitive and dull. Indeed, the excitement for learning about human genes came from the public's interest in finding the alleles that underlie diseases ranging from Huntington's to cancer. The first success in the application of the new molecular genetics to human disease was, in fact, Huntington's disease.

Huntington's disease provided the key and the momentum for several reasons. First, its inheritance was particularly clear. Second, motivated families, especially those of Leonore Wexler and of the great folksinger Woody Guthrie, who also died of Huntington's disease, encouraged young researchers both psychologically and financially through their support of several foundations concerned with Huntington's disease (Figure 14-2). In particular, Nancy Wexler, the younger daughter of Milton and Leonore, has devoted her professional life to advancing the search for the gene for Huntington's disease. As a researcher herself, she documented the inheritance of Huntington's disease within an extended family that lives in three villages on the shores of Lake Maracaibo in Venezuela.

Within this group of about 10,000 people, several hundred have already developed Huntington's disease, and hundreds more carry the gene. By tracking the simultaneous inheritance of Huntington's disease and of DNA markers within this family, James Gusella (a former student of David Housman), Nancy Wexler, and their colleagues were able—in 1983, years before the construction of a complete human genetic map—to determine the approximate chromosomal location of the Huntington's disease gene. This accomplishment depended directly on the RFLP method advocated by Botstein. The Huntington's allele was the first disease gene ever found with that method.

It took an international consortium, organized by Allan Tobin and Nancy Wexler for the Hereditary Disease Foundation, another 10 years to identify the actual gene for Huntington's disease. By then (1993), recombinant DNA techniques had already led to the precise identification of many other disease-causing alleles, including that for cystic fibrosis, the most common genetic disease among Caucasians and a good target for the development of gene-based therapies. But the Hereditary Disease Foundation's first-ever gene search—undertaken despite Botstein's fierce arguments for patience—dramatically demonstrated the power of molecular genetics to help understand human disease.

Because of the efforts of the HDF, dozens of researchers focused on applying known techniques to the specific problem of Huntington's disease. Choosing a single disease on which to focus their efforts gave biologists a sense of mission and a single, clear-cut goal that was emotionally appealing. They could imagine that their day-to-day research might one day save the Wexler daughters and other sufferers of Hunt-

Figure 14-2 The fight against Huntington's disease. The families of the famous singer Woody Guthrie and biology teacher Leonore Wexler used their considerable influence to initiate and support the search for the Huntington's gene. *(Guthrie, Culver Pictures; Wexlers, courtesy of Alice Wexler)*

ington's disease. In addition, the HDF's intimate workshops promoted many fruitful scientific friendships and collaborations. And finally, Gusella's location of the Huntington's gene in 1983 provided the "proof of principle" that may well have convinced both scientists and scientific administrators that mapping the human genome was feasible. In this chapter we will see, however, that it was the basic research carried out by Botstein and others that invested the science of molecular genetics with the promise it now holds.

The genetics of humans is no different from the genetics of other organisms. For this reason, many biologists object to the very idea of having a chapter on human genetics. The *study* of human genetics, however, does differ in one very significant way from the study of the genetics of fruit flies, pea plants, or corn. Humans cannot be crossed at the whim of geneticists. As a result, the study of genetics in humans (as well as in other undomesticated animals) is much harder than in laboratory organisms in which any two individuals can be crossed at will.

Another difference between the genetics of humans and that of other organisms is that our own traits are, to us at least, infinitely more interesting and subtle. The degree to which a trait is the product of genetic influences or the product of environmental influences rouses people to intense political debate that makes objective discussion nearly impossible. In this chapter we will try to present enough information for readers to make intelligent judgments about what they read and hear. We will learn about genetic diversity in humans, the complex relationship between genotype and phenotype, the Human Genome Project, and, finally, both the promise and limitations of applying knowledge about human genetics.

GENETIC DIVERSITY IN HUMAN BEINGS

On a Saturday evening in the winter of 1983, 5-year-old Alice Brown was abducted just steps from her own front door. She and her mother, Sarah Brown, were about to sit down to dinner with a neighbor in their 200-unit apartment complex in northern California when Alice's 3-year-old brother, Brad, begged for his boots. Alice took Brad home, just two doors down, but when Brad returned moments later he was alone. Alice's mother never saw her daughter alive again.[1]

When Alice's body was found a week later, police discovered that she had been raped and then suffocated. Sarah said that Alice would never have gone with any man except one: a neighbor in the same apartment complex with whom Sarah had gone out several times. Police considered Harry Gordon, then 31, a prime suspect, but they had no concrete evidence to link him to the case. They had to have "probable cause."

All of the evidence was circumstantial. He'd given contradictory stories about his whereabouts on the night of the murder. Police had arrested him in 1981 for exposing himself to a little girl and again, in 1984, on suspicion of molesting a third girl. He was eventually convicted of a misdemeanor for exposing himself while drunk. Gordon's ex-wife said he was a likely suspect, and police pursued dozens of leads. But Gordon denied any involvement in Alice's death, and nothing gave the police the evidence they needed to arrest him.

Sarah and Brad moved away and changed their names. Thirteen years passed. Then, in the spring of 1996, when Alice would have been 18 years old, police finally arrested Gordon for the rape and murder of Alice Brown. It turned out that police had had the concrete evidence they needed all along. Subtle changes in California law and advances in molecular biology had finally made it possible for prosecutors to use that evidence to link Gordon to the crime and to know that they could probably convict him. (The same evidence had cleared a previous suspect.)

Just as every individual has a unique set of fingerprints, so every individual also has a unique set of gene fragments called a **DNA fingerprint.** DNA for a DNA fingerprint can be extracted from any cell that has a nucleus. Such cells can be found in blood stains, in saliva on an envelope, in tears on a cheek, in skin cells, under fingernails, in hair, or in semen. Thanks to the invention of PCR, researchers can now analyze DNA from the barest traces of cells. When police found Alice's body in 1983, they had been able to collect samples of semen from her body and clothes. As evidence, these samples had been carefully stored. But in 1983 they were of little value, for DNA fingerprinting had not yet been invented.

By the early 1990s, prosecutors knew that the DNA in the semen could tell them who had murdered Alice Brown. A legal problem, however, prevented prosecutors from acting. Unlike most states, California had not issued a ruling on the admissibility of DNA fingerprinting as evidence. As a result, lower courts had issued conflicting rulings, and a conviction based on DNA fingerprinting stood a good chance of being overturned on appeal. In 1992, a northern California judge had ruled that DNA fingerprinting was not as reliable as a real fingerprint. If prosecutors tried to convict Gordon and the court rejected the DNA fingerprinting, prosecutors knew they would have lost their only chance to convict Gordon, and at great cost. Just arguing the admissibility of DNA fingerprinting in court could cost $50,000—the price of obtaining testimony from expert witnesses.

How Reliable Is DNA Fingerprinting?

The haploid human genome contains some 3 billion base pairs of DNA. Only 5 percent of this DNA (150 million base pairs) codes for polypeptides. The polypeptide-coding regions are almost identical from one individual to the next because natural selection eliminates most mutations.

The 95 percent of the DNA that does not code for polypeptides is much less subject to natural selection. As far as biologists can tell, such noncoding DNA has no central role in the phenotype, and so it can mutate without impairing an individual's ability to survive and reproduce. Noncoding DNA accumulates large numbers of mutations and contains far more variation than coding DNA. Estimates suggest that each pair of chromosomes differ from each other by about 1 base

---
[1]We have changed the names of those involved.

in 1000, which on a single chromosome could add up to some 250,000 base-pair differences. As a result, except for identical twins, every individual is genetically unique, and DNA sequencing can reliably distinguish every individual in a population.

Researchers create a DNA fingerprint by using restriction enzymes to break up an individual's DNA into restriction fragments (Figure 14-3). Fragments from some regions of the DNA vary in length from individual to individual. The length of a fragment depends on whether the sequence includes a restriction site for the restriction enzyme in use. (Other parts of the DNA are consistently the same length from one person to another.) Fragments that vary in length are said to be polymorphic and are called restriction fragment length polymorphisms, or RFLPs. This long name merely means that for a given RFLP sequence, the presence or absence of a restriction site determines whether the fragment is long or short. Many of the most useful RFLPs include a repeated sequence, with the number of repeats varying greatly among individuals.

In the next step, electrophoresis, the fragments of DNA are exposed to an electric current that moves them through a jellylike medium, called a "gel." The DNA fragments, which are negatively charged, move through the gel toward the positive electrode. Some move quickly, some slowly, depending on their size. The shortest fragments travel through the gel rapidly, while the longer fragments are left behind. After a while, the fragments are spread out across the gel like runners in a race. When hybridized to DNA probes in a Southern blot, the fragments produce a pattern of labeled bands (Box 13.1).

Every genome yields a unique pattern of bands. Since every individual has a different DNA sequence, every individual (other than an identical twin) has a unique pattern of bands. Fragments taken from the whole genome in different individuals would reveal a DNA fingerprint as unique as a real fin-

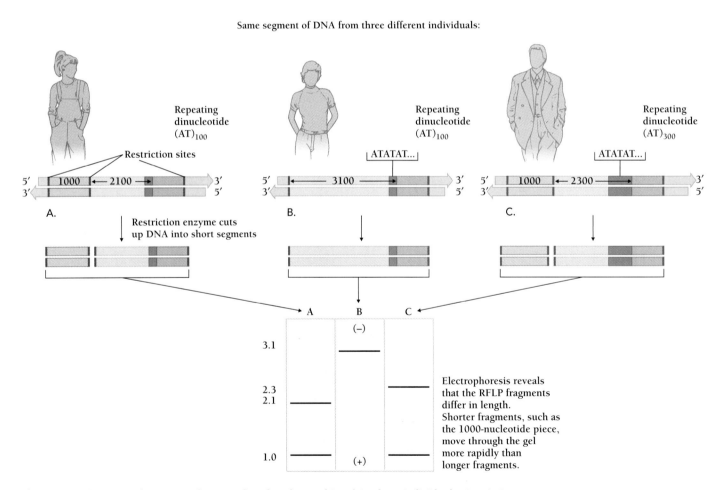

Figure 14-3 RFLPs (restriction fragment length polymorphisms) in three individuals. Restriction enzymes break the DNA of each individual into fragments. For individual A, two fragments result that are 1 kilobase (1000 bases) and 2.1 kilobases in length, respectively. The shorter one moves across the electrophoretic gel faster than the longer one. In B, the middle restriction site is absent and a single long fragment results that is 3.1 kilobases long (1 + 2.1 = 3.1). This long fragment moves across the gel very slowly. In C, the 1-kilobase fragment is the same as the one in A. But a repeating sequence has made the 2.1-kilobase fragment in A 200 bases longer. C's longer fragment moves more slowly than the one in A but faster than the one in B.

gerprint. In practice, however, fingerprinting laboratories examine only a small number of DNA fragments, those known to have many variants. Researchers then calculate the probability that someone else in the general population has that exact set of base changes. Researchers can come up with a probability of uniqueness ranging from 1 in 500 to 1 in several billion.

Two factors affect this number. The first is the number of different markers examined. The greater the number of markers surveyed, the more specific the DNA fingerprint will be, and the more other people in the world can be excluded from consideration. The second factor is how "the general population" is defined. As we will see later, different populations are more or less likely to have certain genotypes. The chances of someone having the same DNA fingerprint as Harry Gordon might be one number for the population of San Francisco Bay Area and another number for the population of all of California. In any case, geneticists do not necessarily know what those probabilities are for different populations. They have estimates for large populations that have been studied, such as those of the United States, Argentina, or Spain, but not for individual cities such as San Francisco or Los Angeles.

Why did police finally arrest Gordon?

In the fall of 1995, prosecutors in the Harry Gordon case began to sense that DNA fingerprinting was coming to be accepted in California. They no longer believed that if they took Gordon (or any other suspect) to court that they would have to prove that a DNA fingerprint is unique. In the fall of 1995, police obtained a court-ordered search warrant and compelled Gordon to give a blood sample. They sent the blood sample and the sample of the semen from Alice Brown's body to a laboratory in Maryland. There, technicians determined that the two samples matched. In six highly variable RFLPs, the pattern of bands was the same.

These six RFLPs vary so much from person to person that it was highly unlikely that a perfect match between the recovered semen DNA and Gordon's DNA could have arisen by chance. The testing laboratory estimated that the DNA recovered from Alice's body could be matched to only 1 in 5.7 billion male Caucasians, some ten times the number of male Caucasians in the world. Prosecutors finally had the tangible evidence for which they had waited so long. In the spring of 1996, they arrested Gordon and charged him with Alice's murder. Many people are hoping that his arrest will bring to an end a string of similar unsolved murders in the area.

What might Gordon's defense attorneys argue?

The use of DNA-based evidence has been much debated in U.S. courts and in the pages of scientific journals. Even if the probability of someone else in the world having Gordon's DNA fingerprint is virtually zero, his defense attorneys can still argue that the evidence is inconclusive. They are likely to bring four kinds of arguments: (1) that the DNA samples (like *any* evidence) could have been mixed up (intentionally or unintentionally), so that the perfect match came from duplicate samples of Gordon's own DNA; (2) that the semen samples are so old that the evidence is unreliable; (3) that another person besides Gordon could have the same RFLP pattern; and (4) that the RFLP analysis was itself faulty. Which of these is arguments is likely to be correct?

Questions about the proper handling of evidence apply to any kind of physical evidence (hairs, threads, or fingerprints, for example). The prosecutor must demonstrate—beyond any reasonable doubt—that the police have carefully collected the biological samples, preserved them appropriately, and not mixed them up.

In some cases, defense attorneys have argued that the tissue samples from the site of a murder have been contaminated by DNA from other sources. Because PCR-based DNA fingerprinting is so sensitive, a few cells from a different person could change the results, they argue. In reality, unless the contaminating cells came from the suspect, they would not match the suspect's DNA. The only way for the suspect's cells to be added or substituted would be if someone deliberately set out to frame the suspect. This is possible, but no more likely with DNA fingerprinting than with other forms of evidence.

Gordon's attorneys are also likely to argue that the semen samples are very old and were improperly stored. But how does improper storage affect the results? The answer is that if the DNA were to break down, for example, as a result of exposure to heat, sunlight, or chemicals, the changes in sequence would be random and extremely unlikely to exactly match Gordon's own DNA. In other words, damaged DNA will not convict an innocent person. Can two people have the same RFLPs by chance? Geneticists are unanimous in the conclusion that everyone (other than identical twins) has a unique genome. So if a laboratory sequenced all the DNA in the blood or semen from two individuals, they would certainly find differences. But, as of the year 2000, no complete human genome has been sequenced, and the cost of total sequencing is likely to remain prohibitive. So DNA fingerprinting depends on a limited sampling of the genome, concentrating on sequences (such as RFLPs) that are highly variable within the population. Prosecutors can calculate the chance that each of these DNA variants (alleles) would appear in a large population (such as that of the United States). In general, however, no one knows the frequency of each variant in a local population. For the RFLPs used for DNA fingerprinting, however, the probability of obtaining six identical DNA patterns by chance is typically 10^{-10}, that is, 1 in 10 billion. And, of course, so far the Earth has only 6 billion people.

Finally, there is the question of faulty laboratory analysis. In fact, some of the analyses presented in evidence in the early 1990s were so poor that expert witnesses for the prosecution ended up arguing for acquittal. Recently, however, new standards have been adopted for the certification of laboratories and technicians, and purely technical errors are much less likely than previously. Still, DNA evidence (like other evidence) must be scrutinized for technical artifacts and misinterpretations.

DNA fingerprinting has changed the way that the police investigate crimes by providing a powerful new kind of physical evidence. DNA analysis has not only led to many convictions but has resulted in the release of suspects and even prisoners unjustly accused. DNA evidence, for example, led to the release of a 32-year-old Baltimore man after 9 years on death row.

DNA fingerprinting also led to the identification and return of kidnapped children. One of the most egregious crimes of the Argentine military dictatorship (1975–1983) was the abduction and killing of some 15,000 people, called "the disappeared." Among these "disappeared" were pregnant women, whose babies were delivered and given to military families. After the fall of the dictatorship, a group of women whose daughters or granddaughters had "disappeared" contacted Professor Mary Claire King, a human geneticist then at the University of California Berkeley. King was using RFLP analysis to track a gene variant responsible for breast cancer. By applying her fingerprinting methods to the DNA of the stolen children and their surviving relatives, she was able to help reunite over 50 children with their families.

Figure 14-4 People from different geographic regions vary in such traits as height, blood type, skin color, and facial features. A. An Indian woman. B. An ethnic minority woman from China. (A, Superstock; B, Paul Grebliunas/Tony Stone Images)

DNA fingerprinting identifies individuals on the basis of variations in DNA sequences.

What Are Human "Races"?

We all know that the human species varies in appearance and in physiology geographically. People of Indian descent, for example, are recognizably different from those of Chinese descent. We can look at the shape of the nose, the eyes, and the ears, for example, and we see differences (Figure 14-4). We can look at the distribution of individual alleles and see differences. Fifty-one percent of Nigerians have type O blood, compared to only 30 percent of Japanese. Twenty percent of Russians have type B blood, while the Amerindians of Lima, Peru, have no detectable levels of type B blood at all.

Nineteenth-century anthropologists struggled to classify human groups into a few major races. Some systems identified only 12 races, while other systems listed 30 or more. One problem was that no matter how anthropologists classified humans, there always seemed to be tribes or nations that would not fit into any known group. The Basques, who live in the Pyrenees mountains between France and Spain, for example, appear European. Yet their language and culture are unlike any other in the world, and some researchers once argued that they are direct descendants of Stone Age Europeans. Similarly, the Bushmen are unique among African groups in both appearance and physiology (Figure 14-5).

A more serious problem with the grouping of humans into races is that most groups do *not* stand out from those around them; they blend. Because groups of humans inevitably mix, through migration, warfare, and trade, human "races" are never pure. Both the Japanese and British, for example, take pride in the purity of their island races. Yet the Japanese are a graded mixture of Korean and Ainu north islanders (a people of possibly European descent). This mix shows up in the distribution of blood types from one end of Japan to the other (Figure 14-6).

The British are even more of a melting pot than the Japanese. The Bronze Age Beaker Folk mixed with the Indo-European Celts in the first thousand years B.C. In the next thousand years, the Angles, the Saxons, the Jutes, and the Picts arrived, followed by the Vikings and their descendants, the

Figure 14-5 Classifying humans into a few discrete races has been unsuccessful. Named races frequently include markedly distinct peoples. A. A Bushman from Namibia. B. A rubber plantation foreman from the Ivory Coast. (A, M.P. Kahl/Photo Researchers, Inc.; B, Charles O. Cecil/Visuals Unlimited)

Figure 14-6 Variation in blood group A in modern Japan. A. Even the Japanese, a relatively homogeneous island race, show variation that demonstrates intermarriage with other groups. Populations closest to mainland Korea have the greatest concentration of A alleles. Populations closest to Hokkaido have the lowest concentrations of A alleles. B. In Hokkaido lives a remnant population of Ainu, sometimes called the "hairy Ainu." The Ainu, who were once a white minority in Japan, are considered primitive by other Japanese. The Ainu had white skin, abundant body hair, and round eyes. Descendants of Causasoid peoples from northern Asia, the Ainu once lived on all four of Japan's major islands. Today they live only in the north. Intermarriage and cultural assimilation by the Japanese have made the traditional Ainu virtually extinct. (A, Superstock; B, from Human Diversity by Lewontin © 1982 by Scientific American Library. Used with permission of W.H. Freeman and Company)

races—in the biological sense—do not exist. This is because, in humans, genetic variation within populations is greater than that between nations or races. Of all human genetic variation, 85 percent is variation among the individuals within a country or a continent. Another 6 percent is variation among populations from the same continent. Only about 9 percent of all DNA reflects genetic differences among peoples ("races") from different continents. The greatest genetic variation among humans is found in Africa. If a disaster killed everyone on Earth except for those living in Africa, the human species would still retain at least 91 percent of its genetic diversity. Less than 9 percent of all genetic diversity would be lost.

In practice, we focus on the small numbers of traits (for example, skin color and eye shape) that vary geographically—a subset of the 9 percent—and we use just those few traits to identify groups of people who differ from us more in their cultural practices than in their genetic constitutions.

Recent work, for example, shows that many differences in skin and hair color reflect variations in a single gene. That gene specifies the structure of a protein, called MC1R. MC1R affects our response to a hormone that regulates the balance of different types of skin pigments (melanins). Caucasians who tan well, for example, have the same form of MC1R that is present in Africans.

Researchers suggest that variations in skin color represent adaptations to different amounts of sunlight. A mere handful of variants are probably responsible for the most obvious traits that have historically been used to distinguish human "races." As geneticists Mary Claire King and Kelly Owens state, "The myth of major genetic differences across 'races' is . . . worth dismissing with genetic evidence."

Geneticists and anthropologists have also begun to combine DNA-based differences among current populations with fossil evidence to study early human history. Current data suggest, for example, that modern humans originated in Africa. By 100,000 years ago, humans were anatomically identical to modern humans, but it was not until 50,000 years ago that substantial numbers migrated out of Africa through the Arabian Peninsula into Asia.

Modern humans appear to have reached Europe only about 40,000 years ago. By analyzing genetic variations among European populations, geneticists have concluded that the modern human colonization of Europe occurred in five distinct waves. The first two of these were from the Middle East, first westward across the Mediterranean Sea to Greece, Italy, and Spain, and then northwest to northern Germany and Britain. Subsequent migrations gave rise to other distinctive populations—first, Finns, Lapps, and other northern European peoples; then, Celtic populations of Britain, Ireland, and Brittany; and, later, the Basques of Spain and France. In each case, the distinctive genetic characteristics of each population reflect either the particular characteristics of the original immigrants or the results of natural selection in distinctive environments.

French Norman invaders. In the last century, Africans, Indians, Pakistanis, and others have lent spice to this already heady mix.

If the Japanese and the British, seemingly isolated by geography, are well mixed, most mainland groups represent a true continuum of traits. In fact, studies of the distributions of different alleles have convinced researchers that human

A surprising conclusion of these genetic-historical studies is that women and men had distinctly different patterns of migration. The traditional view was that women were more likely to stay in one place, while men moved away from their homes in hunts and campaigns. Now, however, it appears that men have been more likely to stay in one place, while women have moved from place to place.

The evidence for this idea comes from studies of the sequence variations in mitochondrial DNA and in the Y chromosome. All of the mitochondria in an egg come from the mother, since the sperm delivers chromosomes only. Because mitochondria have their own DNA, a woman's mitochondria come in a direct line from her mother, her grandmother, and so on back in time.

On the other hand, a man's Y chromosome contains DNA that can come only from the father, who got it from his father, and so on. So every woman has a marker for her maternal line and every man has a marker for his paternal line. (Men also have mitochondria, too, of course. So men also have a marker for their maternal lineage.)

The remarkable results of these studies of mitochondrial and Y chromosome DNA are that Y chromosome variations are more likely to be local, while mitochondrial DNA variations are more likely to be common to several populations.

Human races blur into one another, so that defining races is impossible. Almost all human variation is present in an individual population.

Figure 14-7 **The phenotype of an animal can change with its social environment.** The scalefinned anthias, shown here, lives in the Red Sea. It is one of dozens of fish that can change sex in response to a change in its social environment. If a school of females loses a breeding male, the dominant female changes into a male and begins breeding. Over a period of a few days, a female's ovaries convert into testes, while "her" coloring changes from golden orange to a pale background with magenta head and fins. Four days after the fish's transformation, he begins courting females. His anatomy, coloring, and behavior are indistinguishable from those of any other male's. *(David Hall/Photo Researchers, Inc.)*

THE RELATIONSHIP BETWEEN GENOTYPE AND PHENOTYPE IS COMPLEX

Genes provide the continuity between the generations, but they do not by themselves define individuals. Every individual—as expressed in the phenotype—results from complex interactions between genes and environment. The environment includes both the physical and chemical environment of the developing organism and the ongoing environment of the mature organism. In humans, environmental influences include the quality of the cytoplasm in the egg, hormonal and nutritional influences in the womb, upbringing, and the physical and social atmosphere of the home or workplace.

Studies of other animals illustrate the influence of social environment on gene expression and even brain structure (Figure 14-7). The tiny Japanese goby *Trimma okinawae*, for example, can change sex repeatedly, from female to male and back again. Schools of female fish breed with a single male. If a local breeding male is replaced by a larger male, a juvenile male in the school becomes female and remains to breed with the new male. If the breeding male leaves, however, one of the females becomes male and takes his place. These fish can change the structure of the gonads, genitals, and brain, as well as their behavior, in as little as 4 days. Their anatomy and behavior are the same as that of fish that have always been one sex. In short, the same genotype can impart more than one phenotype. In this section, we will explore the complex relationship between genotype and phenotype.

Genes Are Expressed to Different Degrees

Genes vary enormously in their **expressivity**—the range of phenotypes associated with a given genotype. Two individuals with the same allele, for example, may express it very differently. The human disease neurofibromatosis, for example, is caused by a dominant allele whose expressivity varies so much that in some people it causes blindness or horribly disfiguring growths, whereas in others the only symptom is a coffee-colored spot (Figure 14-8). Still others with the same neurofibromatosis allele have no symptoms at all. Huntington's disease varies, too, in how strongly it is expressed and when. Although most people show no symptoms until middle age or after, a few children may exhibit severe symptoms as early as 2 years of age.

Genes also vary in their **penetrance.** While expressivity is a measure of how strongly a trait is expressed, penetrance describes the likelihood that an individual with a dominant allele will show the phenotype usually associated with that genotype. A dominant allele, such as the allele for purple flowers in Mendel's peas, has "complete penetrance," meaning that it will always be expressed.

A dominant allele with "incomplete penetrance" will, in certain physical or genetic environments, not be expressed. In

Figure 14-8 The symptoms of neurofibromatosis are distinctive. (L. Steinmark/Custom Medical Stock Photo)

frequently sterile, as the *vas deferens,* the tube that carries sperm from the testes, may also be blocked.

Cystic fibrosis results from the failure to produce a polypeptide called cystic fibrosis transmembrane conductance regulator, or CFTR. The disease-causing alleles are recessive, since one normal allele will provide functioning CFTR. In about 70 percent of cases, all of the symptoms of cystic fibrosis are caused by the deletion of just three base pairs in the CFTR gene. CFTR is 1480 amino acids long. The deletion of three base pairs results in the deletion of the 508th amino acid (phenylalanine) from the CFTR protein (Figure 14-10).

The CFTR polypeptide is a plasma membrane protein in *epithelial cells* that pumps chloride ions out of the cell. Epithelial cells are the cells that line the inner and outer surfaces of cavities. They can be found in every kind of duct—in the intestines, the lungs, the nose, and the mouth. Skin cells are also epithelial cells. The altered version of the CFTR polypeptide functions poorly, and epithelial cell fluids do not go where they are supposed to. Thus, one mutation can have wide-ranging effects on the phenotype.

Not all mutated genes produce polypeptides that function as poorly as the altered CFTR polypeptide. Some altered proteins may be able to function reasonably well (and the phenotype will be normal, or nearly so). In between are cases such

flowers, this would simply mean that only a certain proportion (say, 10 percent or 75 percent) of individuals with a given genotype would show the phenotype. But penetrance can be a slippery concept. The allele for Huntington's disease, for example, is said to be completely penetrant. Theoretically, those who have the allele always get the disease. Because the allele may be expressed late in life, however, some people with the allele die of other causes before they experience any symptoms.

Why a gene varies in its expressivity and penetrance is usually not known. Both the environment (temperature and nutrition, for example) and interactions with other genes play crucial roles in determining the expression of individual genes. But the possession of a particular allele usually does not by itself predict what an individual's phenotype will be.

Because alleles vary in their expressivity and penetrance, the genotype does not define the phenotype.

A Single Gene Can Affect Many Traits

When a baby is born with the genetic disease cystic fibrosis (CF), the baby's doctors know that the baby is destined for a life with chronic respiratory infections. Cystic fibrosis sufferers tend to have unusually salty perspiration and thick, sticky mucus that clogs the lungs and leads to repeated bouts of pneumonia, bronchitis, and other respiratory infections (Figure 14-9). In addition, many have blocked pancreatic ducts, which prevents the normal secretion of digestive enzymes. Males are

Figure 14-9 A child who has cystic fibrosis. Children who have cystic fibrosis must be thumped on the chest several times a day to loosen accumulated mucus in the lungs. (©1997 Abraham Menashe)

Figure 14-10 Translation of the gene for CFTR. Most cases of cystic fibrosis result from the deletion of three base pairs, one in the 507th codon and two in the 508th codon. The 507th amino acid is unaffected, but the 508th amino acid is missing and CFTR functions poorly.

as sickle cell anemia, in which the mutation produces a protein that functions much of the time, but not always.

The allele that causes cystic fibrosis affects many aspects of phenotype. The capacity of one gene to have diverse effects is called **pleiotropy** [Greek, *pleios* = more + *trope* = turning]. Every gene that has been studied—in humans and other organisms—has been found to be pleiotropic to some degree.

A small mutation in a single gene, such as the one for the CFTR polypeptide, can affect many aspects of phenotype, a phenomenon called pleiotropy. All genes are pleiotropic.

A Single Trait Can Be Influenced by Many Genes

Phenotype depends on the interactions of many genes, as well as the effects of environment. Most traits are **polygenic,** meaning that they are influenced by more than one gene. Usually, we know little about the identities and effects of each of the genes. In a few cases, however, we know a great deal. In mice, for example, fur color depends on the action (and interaction) of at least five genes: one gene determines the distribution of pigment within individual hairs, a second gene determines the color of the pigment, a third gene allows or prevents the production of the pigment molecules, a fourth gene determines the intensity of the pigmentation, and a fifth gene controls the presence or absence of spots (Figure 14-11). The gene that reg-

ulates the intensity of pigmentation is called a modifier gene because it regulates the action or "expression" of another, separate gene.

Many genes are now thought to be back-up systems for other genes. That is, the genome has several ways of accomplishing the same thing. More and more researchers are finding that knocking out seemingly important genes may have no effect on phenotype.

The long-haired mice mentioned in the last chapter are an example of such redundancy in genetic systems. In 1993, Gail Martin, at the University of California, San Francisco, found that knocking out a growth factor gene (*FGF5*) believed to influence early development in mice actually had no effect on development. The researchers were quite surprised and disappointed to see healthy mice born, but astonished when the mice all grew hair that was nearly 50 percent longer than normal (Figure 13-12). *FGF5* normally is expressed in early embryo tissues, in muscle cells, in the hippocampus of the brain, and in motor neurons. Yet, only the hair follicles were affected by the absence of the gene. Every other aspect of phenotype appeared normal.

Martin and other developmental biologists now believe that nerve, muscle, and brain function survive knocking out the *FGF5* gene, not because *FGF5* has no function in these tissues, but because other genes provide backup systems. If this is true, it is reasonable to ask why other genes do not seem to provide backup for hair length. For the moment, no one really knows.

In humans, as in mice and other organisms, the vast majority of traits are polygenic and probably redundant as well.

Figure 14-11 Several interacting genes determine coat color in mice. Gene A determines where on the hair the pigments go. In wild mice, a band of yellow in the middle of the hair creates an "agouti" or "mousy" coat color. One allele confers a narrow yellow band, which gives a dark coat. Another allele confers a broader yellow band, which results in a lighter coat. In mice with a coat of a solid color, the yellow band is not present at all. The normal allele of gene B encodes a black pigment. Gene C suppresses the expression of pigment in the warmest parts of the body. Only the cooler ears, nose, tail, and paws are dark. Gene D regulates how much pigment each hair receives. Gene S determines the presence or absence of pigment in patches.

Eye color, for example, is controlled by at least two genes. Studying polygenic traits is extremely difficult, since with each added set of alleles, the number of possible genotypes increases geometrically. If these are difficult to study in research animals or plants, they are nearly impossible to study in humans. As a result, almost nothing is known about polygenic traits (most traits) in humans.

The expression of each trait is the result of many genes acting together.

A Single Gene May Have Multiple Alleles

Most of the time we have spoken as if each gene can have only two alleles—a dominant allele and a recessive allele. But a gene may exist in many alternate forms, each of which contains a distinct sequence of nucleotides that specifies a distinct polypeptide. Some of the polypeptides cannot function at all, while others may function differently.

One familiar example of a gene locus with multiple alleles is the one responsible for the ABO blood types. This locus has three alleles, called I^A, I^B, and i, which specify different versions of a surface marker on red blood cells. (Although three alleles exist in most human populations, each individual has only two of the three alleles, one on each homologous chromosome.) Alleles I^A and I^B each specify an enzyme that helps to synthesize a glycoprotein marker on the surface of red blood cells. These two glycoproteins are called, respectively, glycoproteins A and B. Allele i specifies no enzyme and so neither cell surface glycoprotein is synthesized. An individual human may have one of six genotypes (Table 14-1): $I^A I^A$, $I^B I^B$, $I^A I^B$, $I^A i$, $I^B i$, or ii. Because alleles I^A and I^B are dominant to i, individuals with $I^A I^A$ or $I^A i$ genotypes have type A blood, individuals with $I^B I^B$ or $I^B i$ have type B blood, and individuals with ii genes lack these surface glycoproteins and are said to have type O blood. Alleles I^A and I^B are codominant, meaning that in individuals with both alleles, both alleles are fully expressed. $I^A I^B$ individuals have type AB blood.

A single gene may have many functioning alleles. While some amino acid changes result in proteins that do not function at all, other amino acid changes alter the activity of the resulting protein. When both alleles are expressed, they are said to be codominant.

THE HUMAN GENOME PROJECT

If each of the 6 billion human beings on Earth is genetically unique, we can conclude that human beings harbor enormous genetic diversity. At the same time, we must remember that we are still more than 99 percent identical. For this reason, it

TABLE 14-1 HUMAN ABO BLOOD GROUPS

Blood genotype	Red blood type	Plasma cell surface	Antibodies
$I^A I^A$	A	A glycoprotein	anti-B
$I^A i$	A	A glycoprotein	anti-B
$I^B I^B$	B	B glycoprotein	anti-A
$I^B i$	B	B glycoprotein	anti-A
$I^A I^B$	AB	A and B glycoproteins	no antibodies
ii	O	no glycoproteins	anti-A and anti-B

makes sense to most people to study "the human genome" as if it were a definable entity.

What Is the Human Genome Project?

In 1988, the National Institutes of Health (NIH), the federal agency that oversees and funds research in the health sciences, launched the largest and most expensive single project in the history of biology. The Human Genome Project was a $3 billion international project jointly funded by the Department of Energy, NIH's National Human Genome Research Institute and the Wellcome Trust. The project sought to map the human genome in four major ways:

1. The project made linkage maps of all the human genes, restriction enzyme cut sites, or other markers. The resulting map allows geneticists to determine the chromosomal position of any gene. (Recall from Chapter 9 that the closer two genes are on a chromosome, the less likely they are to cross over during meiosis and assort independently.)
2. The project also produced a physical map of each human chromosome by first cutting the chromosomes into restriction fragments, then identifying unique sequences that act as landmarks, and determining the distances to the next landmark.
3. The project sequenced all 3 billion base pairs on one set (23 pairs) of chromosomes for several anonymous volunteers, then loaded this information onto a computer to create an electronic version of the genome.
4. By reducing the cost of DNA sequencing, the project enabled scientists to study variations in the human genome. It will also determine the complete sequence of the DNA of several other species.
5. The project is also developing new methods for studying the functions of newly identified gene products. These studies will depend on studies of the effects of gene expression in cultured cells and in living organisms.

In 1990, work on the project began in laboratories all over the world under the direction of Nobel Laureate James Watson, the codiscoverer, with Francis Crick, of the structure of DNA. In spite of the large number of researchers involved, the task of sequencing the entire human genome seemed insurmountable. At the time, researchers had sequenced fewer than 5 percent of all genes.

Then, just as the project was getting under way, a molecular biologist at NIH named Craig Venter found a way to simplify one part of the job. Since only 3 to 5 percent of the 3 billion bases in the human genome actually code for proteins, he would extract mRNA from human tissues, copy it into cDNA with reverse transcriptase, clone the cDNA, then sequence it. That way, he would sequence only the estimated 150 million nucleotides that actually code for proteins, leaving out all the introns and other junk DNA.

Thousands of researchers all over the world have worked on the Human Genome Project. Technical advances speeded the work, and by the end of 1998, researchers had already completed both the genetic and physical maps of human DNA. By the end of 1999, researchers had complete sequences of several organisms, including *E. coli* and several other prokaryotes, yeast (*Saccharomyces cerevisiae*), the nematode worm *Caenorhabditis elegans*, the fruit fly *Drosophila melanogaster*, and the plant *Arabidopsis thaliana*. They had also determined the complete sequence of one human chromosome (chromosome 22, the smallest). In early 2001, almost three years ahead of schedule, researchers with both the Human Genome Project and with Craig Venter's company Celera Genomics completed the sequence of the whole human genome. Other organisms whose genomes have been targeted for complete sequencing are the mouse *Mus musculus* and the domestic dog *Canis familiaris*.

> *In 2001, the Human Genome Project and Celera Genomics sequenced the entire human genome, and constructed linkage maps and physical maps of each human chromosome.*

How Do Researchers Locate Genes for Specific Diseases?

The drive to sequence the human genome has rapidly provided a gene library of actual gene fragments stored in numbered test tubes, as well as a computer database listing the nucleotide sequence of each human chromosome. Researchers express many of these genes (from mRNA) to find out what sorts of proteins they make. But these libraries have no catalogue to help researchers identify and locate individual genes. The contents of a gene library are therefore still something of a mystery. What genes does the library contain? How do they function in the body? Which chromosomes contain which genes? How can researchers find individual genes among the 80,000 to 100,000 human genes in the genome?

The time-honored way to locate a gene in the genome is by means of experimental crosses. Recall from Chapter 9 that experimental crosses reveal how frequently genes cross over, providing a measure of how far apart they are on a chromosome or whether they are on separate chromosomes. Although researchers cannot arrange experimental crosses in humans, they can obtain helpful information from many families about physical traits or genetic diseases extending back for generations.

Such records, arranged in the form of family pedigrees, show patterns of inheritance. Family pedigrees are useful for analyzing the inheritance of obvious traits that are passed in a very simple manner (Figure 14-12). Traits such as fatal diseases and whether the earlobe is "attached" are easy to identify.

A more sophisticated approach is to look for genetic markers that seem to travel with a disease from generation to generation, a technique called **pedigree analysis.** Researchers first collect blood samples from individuals in families that carry a given disease, then use DNA fingerprinting to look for RFLP alleles that are consistently associated with the disease. In the search for the Huntington's disease allele, for example, James

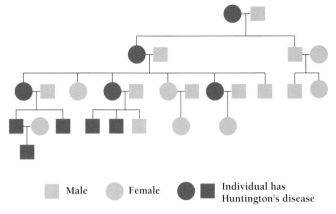

Figure 14-12 Pedigree for a family with Huntington's disease. The marriages and offspring of unaffected individuals have been omitted.

Positional cloning

Once the approximate position of a gene has been located between two RFLP markers some million base pairs apart, researchers can use **positional cloning** to pinpoint the gene. The first step in positional cloning is the assembly of all the clones that cover the region between the two markers on either side of the gene (Figure 14-13). Researchers clone numerous pieces of DNA that are known to overlap, then string them together into a continuous string that extends from one marker to the other, a so-called "contig."

Next, researchers narrow their search down to DNA sequences that are likely to encode a protein. One way of finding such sequences is to hybridize human DNA with labeled DNA from rats, mice, or even flies and worms. DNA from a mouse, for example, can hybridize with human DNA to form a double-stranded molecule, one strand mouse and one strand human. Only sequences that have stayed more or less the same since mice and humans diverged millions of years ago, however, will hybridize. The hybridized sequences mark genes whose exact sequences are so important that they have been conserved in the course of evolution. Most of these highly conserved sequences encode proteins. Mutations in these protein-coding genes are far more likely to cause death or disease than mutations in noncoding DNA.

In the next step, molecular biologists look for mRNA in the tissues where the gene is expressed. A diseased gene that affects the brain, for example, would presumably be expressed in brain tissue. Having identified one or more genes that might cause the disease, researchers then directly compare the sequences from sick individuals with the corresponding sequences from healthy individuals. The purpose of these comparisons is to show that the sequence changes are always associated with disease.

Gusella, at Massachusetts General Hospital, began looking for a particular RFLP allele that traveled with the Huntington's disease allele in some families. Because the human genome is enormous, and because no one knew even on what chromosome the Huntington's gene might lie, the task appeared Herculean. Yet, miraculously, Gusella immediately happened on a polymorphic marker that, in certain families, including the huge family at Lake Maracaibo, in Venezuela, was always inherited with Huntington's disease. This RFLP site came from the short arm of chromosome 4. In 1983, for the first time, the approximate location of the Huntington's gene was known.

Positional cloning is a time-consuming, tedious procedure, taking anywhere from 10 to 100 person-years to locate one gene. Nonetheless, geneticists have identified hundreds of disease-causing genes and determined exactly which sequence changes are responsible for diseases. The diseases caused by such single-gene mutations can affect every system in the body—from the blood (leukemia) to the brain (Alzheimer's disease).

Pedigree analysis allows researchers to find a genetic marker associated with a particular trait or disease that shows approximately where on the genome a gene lies.

Positional cloning is becoming much easier. For example, geneticists no longer have to start from scratch to make "contigs" in a region of interest. Because it will soon be possible to find DNA markers (SNPs) within a few thousand bases of any gene, researchers will be able to determine the possible positions of disease-causing genes much more quickly. Instead of examining the pattern of inheritance of a disease (or any other trait) within single families, researchers will be able to find *all* the DNA sequences shared by any group of people who have a particular disease. Using rapid sequencing, they will then be able to determine which DNA changes go along with a particular disease, even in a group of unrelated individuals.

Figure 14-13 Assembly of a contig. To pinpoint a gene, researchers assemble a single long sequence of DNA between two RFLP markers on either side of the gene.

Geneticists have come to realize that only a small fraction of familial diseases result from alterations in single genes. In addition to finding disease-causing mutations, researchers hope to find the genetic basis of more complex (polygenic) inherited traits. Among the genes that geneticists hope to find are those that affect the susceptibility to diseases, such as schizophrenia, that often run in families but that have no clear pattern of inheritance.

Another set of questions focuses on why people respond differently to particular medicines. Some people, for example, have variant forms of enzymes that break down medicines that would otherwise help them. Researchers, physicians, and pharmaceutical companies hope that this new field of study, called *pharmacogenomics*, will allow physicians to choose more effective treatments more rapidly.

Positional cloning allows researchers to locate a particular gene precisely on a chromosome.

Of What Use Is a Sequenced Gene?

Once a gene has been sequenced, geneticists can test individuals to see if they have a normal allele or a disease-causing allele of the gene. In practice, this may be neither easy nor necessarily useful. For example, in recent years, researchers cloned two genes, called *BRCA1* and *BRCA2,* some alleles of which predispose women to breast cancer. Some of the alleles that cause cancer give a woman an 85 percent chance of developing breast cancer. The disease alleles of these two genes, however, can result from any of hundreds of different mutations, each of which may carry a different risk of causing breast cancer. In addition, other genes, both known and unknown, can alter that risk as well.

In any case, the vast majority of women who get breast cancer do not have an abnormal allele of either *BRCA1* or *BRCA2*. It only makes sense to test women who come from families where breast cancer is strongly inherited and common. Even for that minority of women, however, the test is of limited value. Molecular biologists' knowledge of the sequences of *BRCA1* and *BRCA2* does not enable them to either prevent or cure breast cancer. Finally, once tests reveal that a person has a gene that predisposes him or her to a particular disease, insurance companies will not want to provide health insurance to that person for that disease. Most geneticists argue that no genetic information should be made available to insurance companies. But insurance companies have a powerful economic incentive to overturn or sidestep any rules limiting their access to genetic information. For the average person, then, a genetic test can potentially do more harm than good.

On the other hand, researchers who know the sequence of a disease allele are in a good position to begin to understand how the gene does its damage. Huntington's disease provides a good example of this. In 1993, researchers found that the gene involved in Huntington's disease codes for a polypeptide called huntingtin that is expressed in every cell in the body.

The disease-causing form of huntingtin has a tract of more than 37 glutamines at a particular spot in the protein, whereas the usual (wild-type) form usually has fewer than 34. If researchers can understand how this expanded glutamine tract can lead to cell destruction, then they may be able to devise a treatment for this disease.

Having the huntingtin gene in hand, researchers can transfer the disease-causing version to other cells and organisms. Several laboratories, for example, have produced transgenic mice that express the disease-causing form of huntingtin. The characteristics of these genetically engineered cells and mice have led researchers to unexpected ideas about the causes of cell death not only in Huntington's disease but also in other diseases in which nerve cells progressively die or fail to function properly. For example, the huntingtin-expressing transgenic mice develop normally but begin to show abnormal behavior within the first 2 months of life. When researchers examined the brains of the sick mice, they found abnormal deposits of protein in the nucleus of many brain cells. When other researchers looked in the brains of people who had died from Huntington's disease, they found similar deposits. In each case, the protein deposits contain the protein specified by the disease-causing gene.

Many researchers have hypothesized that these protein deposits are the direct cause of cell death. Other researchers, however, have suggested that the deposits are a cell's way of getting rid of a potentially toxic protein. The best way to decide between these alternatives would be to find a chemical that can prevent the deposits without otherwise damaging the cell.

By using robots that can add individual compounds to hundreds of cell cultures at the same time, researchers hope to screen large sets of chemical compounds. This approach, called *high-throughput screening,* takes advantage of the huge "libraries" of compounds that have been produced by organic chemists, especially within pharmaceutical companies. Compounds that are effective in such cell-based assays could then be used to treat transgenic mice to see if they prevent protein deposits and abnormal behavior. If one or more compounds are successful in mice, researchers could then begin to plan clinical trials on people who have Huntington's disease.

The ability to transfer a disease-causing gene to cells and transgenic mice can allow researchers to develop new therapies.

PREVENTING GENETIC DISEASE

Genetic research has greatly advanced our understanding of how certain genetic diseases are inherited and how they make people ill. Medical researchers have used this information to try to help people avoid having children with serious defects, or, in a few rare cases, to actually treat people with genetic diseases through "gene therapy," or "genetic medicine." In this section we will see how this is done.

Only about 3 percent of all human diseases are caused by defects in a single gene. The most common genetic disease in

BOX 14.1

Can Human Genes Be Patented?

In 1991, Craig Venter suggested that the NIH patent the protein-encoding cDNA, so that any commercial uses for the genes would earn royalties for the federal agency. In June 1991, NIH filed applications on 315 gene fragments. But the scientific community protested. The human genome was viewed as a common heritage, not something that could be patented so that one person or one business could make money from it. Even religious coalitions have made formal objections to the idea of patenting human genes, and whether the patents will be upheld if they are challenged legally is still in doubt.

Ultimately, the applications were withdrawn, but Venter was determined. He started a nonprofit organization called The Institute for Genomic Research (TIGR) that would continue his work at NIH on a massive scale but would be funded by private companies. By 1995, TIGR had sequenced pieces of nearly 90 percent of the genes expressed in human tissues, that is, the cDNA version of all the different kinds of mRNA. A second, for-profit company, Human Genome Sciences, Inc. (HGS), patents and sells whatever interesting genes TIGR clones. Pharmaceutical companies have lined up to give TIGR millions of dollars in exchange for a stake in its discoveries. Drug companies are gambling that human proteins made from cDNA will turn out to be useful and lucrative drugs that will, for example, boost the immune system or dull the appetite. Meanwhile, researchers at many universities and several pharmaceutical companies are freely publishing the full sequences of as many genes as they can, so that at least some human gene sequences will be public property and not patentable. Other universities are themselves patenting DNA sequences.

---

the United States is cystic fibrosis, which affects about 1 person in 10,000. None of these single-gene diseases—several of which we have discussed in this chapter—are the ones that kill most adults. For comparison, cardiovascular disease kills more than one person in three. About half that number die of cancer. People fall prey to one disease or another—lung cancer versus stroke, for example—because of lifelong behaviors and experiences and because a constellation of different genes slightly predisposes them to one weakness or another. Every person carries at least five to ten genes that could make them seriously ill under certain circumstances. As geneticist Michael Kaback puts it, "We are all mutants. Everybody is genetically defective."

Genetic Counseling: How Can Parents Decide?

The first kind of genetic testing is simply to test couples who are considering having a baby to see if they carry alleles for genetic diseases, such as sickle cell anemia, Tay-Sachs disease, cystic fibrosis, or Huntington's disease. Geneticists can now test for more than 100 different genetic defects. Usually, however, geneticists test only for the alleles that are relatively common or for those that the parents seem most likely to carry. A genetic counselor can calculate the likelihood that a child will have a certain disease and help the couple decide whether to risk having children or not.

Tay-Sachs, for example, is a genetic disease that results from a single defective gene for an enzyme normally synthesized by nerve cells. In healthy children, this enzyme breaks down lipids. But in children homozygous for the Tay-Sachs allele, lipids accumulate in the brain, causing mental retardation, blindness, and failure to develop control of the muscles.

All Tay-Sachs children die in early childhood, usually by age 4 years. Tay-Sachs has a simple pattern of inheritance because all children homozygous for the single Tay-Sachs allele become ill. Heterozygous individuals, who have one defective allele and one normal allele, are called "carriers." Carriers can be identified by their low levels of the lipid-digesting enzyme, but they are otherwise healthy.

The Tay-Sachs allele is most common among eastern European Jews and their descendants. If both parents are of eastern European Jewish descent, a genetic counselor will likely recommend that they be tested for the allele. If only one parent is Jewish, however, the counselor will probably not recommend the test, because a baby can only become ill if he or she receives the allele from both parents. This population of people has taken such care to avoid having Tay-Sachs children, however, that Tay-Sachs is now quite rare and, at least in America, the disease may now be more common in the population at large than among those descended from eastern European Jews. This decrease in the prevalence of Tay-Sachs is one of the great successes of genetic testing.

Genetic counselors can tell couples whether they carry certain alleles and the likelihood that their offspring will express certain traits.

Prenatal Testing: Can Parents Tell Before?

Once a woman becomes pregnant, physicians can obtain cells from the fetus in order to search for genetic defects. Besides looking for chromosomal abnormalities by examining the chromosomes, genetics clinics can now detect genetic diseases in the cells of an early embryo. Once the parents know that their baby has a genetic disease, however, they have only two options. They may decide to have the mother carry the fetus to term (birth) and cope as best they can with the baby's condition, or they can decide to abort the fetus.

Amniocentesis

By examining chromosomes from a fetus early in pregnancy, geneticists can tell if the parents have produced a zygote with a chromosomal abnormality. The most common chromosomal abnormality found in humans is trisomy 21, or Down syndrome. Geneticists usually obtain the fetal chromosomes by taking a sample of fetal cells from the **amniotic fluid** [Greek, *amnion* = membrane around a fetus], the watery fluid that surrounds the fetus. This procedure, called **amniocentesis** [Greek, *centes* = puncture], is usually recommended to women over 35 years old and to women with a family history of chromosomal abnormalities or other diseases (Figure 14-14A). Amniocentesis can detect more than 100 genetic abnormalities besides Down syndrome.

At 15 or 16 weeks after conception, a physician inserts a needle through the mother's abdominal wall, just below her navel, and into her uterus. The needle is used to remove from the womb amniotic fluid that contains cells from the fetus. This procedure is virtually painless, but can be emotionally upsetting for some women. Amniocentesis is very safe, but not perfectly so. Approximately 0.5 percent of fetuses exposed to amniocentesis later die because of spontaneous abortion, commonly known as miscarriage. (Some 10 to 15 percent of known pregnancies abort spontaneously before the 20th week, usually because of chromosomal or other abnormalities in the fetus. After 20 weeks, an early delivery is called a "preterm birth.")

To prepare the chromosomes for examination takes 3 to 4 weeks, during which a geneticist grows the fetal cells in culture. During this time, the fetus continues to mature and, indeed, approaches the age when it becomes viable outside of the womb, which is about 20 weeks after conception. (A baby is full term at 30 to 38 weeks after conception. Physicians add 2 weeks to these times, counting from the date of the last menstrual period.) Parents meanwhile think about what they will decide if the fetus turns out to have Down syndrome or some other chromosomal defect.

Amniocentesis allows a geneticist to look for chromosomal or other genetic abnormalities in a 4-month-old fetus.

Chorionic villus sampling

A second procedure, called **chorionic villus sampling (CVS)**, uses fetal cells present in the placenta—the tissue that carries nutrients and oxygen from the mother's blood to the fetus and that carries wastes from the fetus back to the mother (Figure 14-14B). CVS may be done as early as 6 to 12 weeks after conception, much earlier in pregnancy than amniocentesis. Because the results of the test become available so much sooner, a decision to abort becomes more acceptable to some people. Alternatively, news that the fetus is healthy means a shorter period of worry for the parents. The increased risk of spontaneous abortion due to CVS is about the same as in amniocentesis, about 0.5 percent.

Chorionic villus sampling allows geneticists to test for the same genetic defects as amniocentesis, but earlier in pregnancy.

Aborting a fetus

In the next section, we see that researchers are working hard to develop techniques for actually curing genetic diseases, or, at least, mitigating their effects. For now, however, prenatal testing implies only one decision: to abort or not. For many people, abortion is unacceptable under any circumstances. For them, prenatal testing provides time to plan for the care of a child with unusual needs.

For most people in the United States, however, abortion is acceptable under certain circumstances. Each family's circumstances are unique, and what is acceptable varies enormously from individual to individual and from situation to situation. For example, many people find it more acceptable

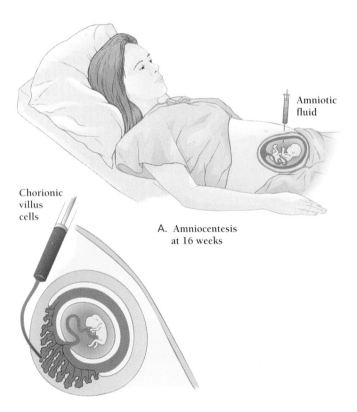

Figure 14-14 Prenatal testing. A. In amniocentesis, a physician extracts amniotic fluid through a needle inserted through the mother's abdominal wall. B. In chorionic villi sampling, a physician extracts cells from the placenta through the mother's vagina and cervix.

to abort a fetus with Tay-Sachs than a fetus with Down syndrome or sickle cell anemia. A child with Tay-Sachs always dies young and cruelly. The effects of Down syndrome, however, vary enormously, and many of its symptoms can be treated. A child with Down syndrome can experience a happy childhood and may live well into adulthood (Figure 14-15).

Likewise, a person with sickle cell anemia may have few or no symptoms much of the time and live well into middle age. New treatments may prevent any significant ill effects. Even now, when a sickle cell crisis does occur, and such a crisis can be both life threatening and agonizingly painful, it usually can be controlled with prompt medical attention. Many symptoms of cystic fibrosis can also be effectively treated with drugs.

Many parents might decide to abort a fetus with Down syndrome, sickle cell disease, or cystic fibrosis, but many others would choose to let the fetus develop into a baby. In neither case is the decision a light one. Parents must weigh their feelings about abortion against their family's ability to marshal both the emotional and the financial resources necessary to provide adequate care for the child (as well as the adult the child becomes) who has special needs. The ethical values of the family, the developmental age of the fetus, the financial status of the family, and the stability of the parents' relationship may all be considerations.

Test-tube babies from selected gametes

Recently, medical researchers have come up with a way for parents to avoid having a baby with a specific genetic defect without confronting the question of abortion. Very simply, a doctor takes unfertilized eggs from a mother who is heterozygous for a particular genetic defect and selects the eggs that lack the allele in question.

The techniques for doing this are rather complex. In one example, the woman must first be treated with hormones that induce her to ovulate several eggs at once. A surgeon then removes these freshly ovulated eggs from the mother's abdomen and takes them to a laboratory. Freshly ovulated eggs have barely completed meiosis, and they are still attached to a sister cell. Because the two daughter cells do not divide evenly, however, the egg cell is very large, while the other cell, called the *polar body*, is very tiny.

The polar body is normally sloughed off, but geneticists have found a use for it. Because homologous chromosomes separate during meiosis, only one of the two cells gets the defective gene. If researchers find that the polar body has the defective gene, then they know that the egg has the normal copy. The normal eggs can then be combined with sperm and implanted in the mother's uterus.

The number of genes that researchers can examine by means of this method will increase ten times in the next decade, since the Human Genome Project is producing potential new genetic tests at the rate of about one a week. Because this method of making babies is rather expensive, only the wealthiest and most motivated couples—such as those carrying alleles for Tay-Sachs—are likely to take advantage of it.

Medical researchers have invented ways to select human eggs that are free of certain genetic defects, but this technology is cumbersome and expensive.

Gene Therapy: Can We Fix It Later?

Gene therapy is a developing technology whose purpose is to deliver genes as a way of treating disease or injury. Researchers have attempted to treat 13 different genetic diseases, as well as AIDS, heart disease, and cancer. They have used dozens of different methods. So far, however, not one variety of gene therapy has resulted in a complete cure of even one person. Neither the genotype nor the phenotype has been "fixed." In the best cases, researchers can effect a partial, temporary cure that must be renewed every few months, often in combination with conventional drug therapy (Figure 14-16).

All human gene therapy has focused on somatic cells rather than germ cells. Human germ-cell therapy would require that researchers apply to humans the techniques now used to make transgenic and knockout mice. Although current technology might make such experiments possible, ethical and legal (as well as scientific) restrictions have prevented human germ-line manipulations. For the foreseeable future, gene therapy research will deal with somatic cells only.

Much gene therapy research has concerned finding vectors to carry DNA into human cells. Almost all of the current approaches depend on modifying vectors that could otherwise cause diseases ranging from colds to cancers to AIDS. Some approaches, however, depend on the ability of tiny lipid globules ("liposomes") to fuse with cells and deliver DNA. The major obstacle to gene therapy is the lack of vectors that carry DNA efficiently into human tissue.

Among the applications of gene therapy to genetic diseases, cystic fibrosis is among the most attractive targets. Cystic fibrosis is the most common serious genetic defect in the United States, afflicting some 30,000 people, or about 0.01

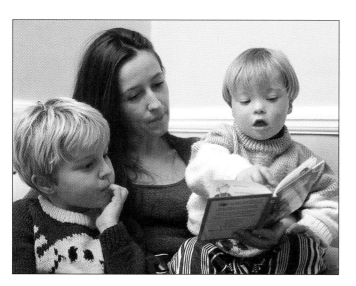

Figure 14-15 A child who has Down syndrome (*at right*). (*Gary Parker/Science Photo Library/Custom Medical Stock Photo*)

Figure 14-16 The most successful example of gene therapy. In 1990, researchers at the National Institutes of Health treated Ashanthi DeSilva for the genetic disease SCID (severe combined immunodeficiency disease), which almost completely disables the immune system. Researchers removed immune cells from the 4-year-old DeSilva, added a functioning allele for the enzyme adenosine deaminase, then injected the genetically engineered cells back into her body. With repeated treatments, DeSilva has been transformed from a chronically sick child with a short expected life span to an evidently healthy young woman. *(Courtesy of Van DeSilva)*

percent of the population. Recall that the many symptoms of cystic fibrosis, from chronic respiratory infections to blocked pancreatic ducts, all seem to result from a defective ion transport molecule in the cell membrane.

In many genetic diseases, the body's cells normally need to express large amounts of the missing protein. In cystic fibrosis, however, even very low levels of expression of the CFTR gene—say 10 to 100 molecules per cell—reduce symptoms dramatically. This fact makes cystic fibrosis more amenable to treatment through gene therapy than other diseases.

One partly successful approach to treating cystic fibrosis through gene therapy has been to spray genetically engineered cold viruses into the noses of patients who have the disease. The engineered *adenovirus* carries a normal version of the CFTR gene into the cells of the inner surface of the nose. The virus is deliberately damaged so that it cannot replicate in the body and make the patient ill. It can only deliver its cargo—the CFTR gene—into the infected cells.

The treated cells express the normal CFTR gene, and the patient's symptoms are alleviated, but only partly. Treating the back of the nose does not, for example, unblock the pancreatic ducts or help alleviate symptoms in any other part of the body. Further, the cells that the virus infects live only a few weeks before they are shed and replaced by new cells. These new cells must be infected with another treatment of virus. Patients must therefore return again and again for more treatments.

Why isn't the gene passed on to daughter cells? A gene can integrate into the genome (into one of the chromosomes) or it can stay unintegrated within the cell's nucleus (along with some regulatory DNA that ensures that it will be transcribed). A gene randomly inserted by a viral vector can affect nearby genes; it could, for example, turn off an adjacent tumor suppressor gene and trigger cancerous growth. So far, however, no such tumors have been observed. But no one has yet succeeded in inserting a gene into dividing progenitor cells in a human.

Many gene therapy researchers are now focusing their attention on how best to get genes into stem cells, dividing progenitor cells that can develop into many alternative cell types. The stem cells in bone marrow, for example, give rise to both red and white blood cells. Other stem cells give rise to muscle and even to brain cells. If researchers could deliver genes to stem cells in the body, those cells could provide a reservoir of health replacements for many types of cells. In addition, because researchers are now able to grow stem cells in the laboratory, they are also able to deliver genes to stem cells in culture. The engineered stem cells can themselves be implanted in the tissues most affected by a genetic disease.

Although many researchers are excited about the prospects of stem cell therapies, the potential use of human stem cells raises serious ethical problems, since most stem cell populations come from embryos. People opposed to abortion are concerned that researchers and physicians will directly or indirectly encourage abortions in order to harvest stem cells. On the other hand, some types of stem cells are available from adult humans, and their use does not raise the same ethical problems. In 1999, after much discussion and debate, the National Institutes of Health proposed guidelines for stem cell research. These guidelines establish strict procedures for obtaining approval for any such research.

Until researchers are able to insert genes into dividing cells, especially into stem cells, gene therapy will be only a temporary cure that can last only as long as the engineered cells live. Such genetic medicines will need to be repeated every few weeks or months. And an individual with a genetic disease will always harbor the defective allele in the germ cells and can continue to pass it on to offspring. For now, gene replacement therapy remains an area of basic research, not of practical results.

Gene therapy is still in its infancy despite expenditures of more than $400 million per year and more than 100 clinical trials in humans. Many of these trials, said the report, were poorly designed, and none has demonstrated even one technique that actually cures genetic disease. Several critical reports have accused researchers of "overselling" gene therapy, of understating its dangers, and of neglecting basic questions about gene regulation, vectors, stem cell function, and other areas. In short, people with genetic diseases should not expect effective gene therapy any time soon.

On the brighter side, many gene therapy centers are turning their attention to conditions where effective therapy should not depend on long-term gene expression or on the continuing proliferation of engineered cells. Most gene therapy trials deliver genes whose products can help mobilize the immune system to combat cancer or AIDS. A particularly promising experimental treatment (called a "biobypass") promotes the growth of new blood vessels in conjunction with bypass heart surgery.

Using new gene therapy techniques, researchers can cause human cells to express a missing or otherwise desired gene. Expression is limited, so far, to the cells that originally take up the engineered DNA. Because the cells usually cannot pass the gene on to new daughter cells in the body, treatments must be repeated. Also, the treated patient cannot pass on a "good" allele acquired in this way to any offspring.

In this chapter on genetics, we have used the knowledge gained from previous chapters to examine genetic diversity in humans, the complex relationship between genotype and phenotype, the Human Genome Project, and the promise and limitations of applied human genetics. Now that we understand inheritance, we are ready to tackle evolution, the most important idea that biology has ever produced.

Study Outline with Key Terms

DNA fingerprinting identifies variations in DNA, called **RFLPs (restriction fragment length polymorphisms),** derived from blood or other tissues. It provides an estimate of how common that particular pattern of variation is in the population. Humans are 99 percent genetically identical. Yet every individual who is not an identical twin has a unique genotype. Human races blur into one another, so that defining consistent races is essentially impossible. The differences among the members of one race are greater than the average differences between races.

A small mutation in a single gene, such as the one that causes most cases of cystic fibrosis, can affect many aspects of phenotype. But the relationship between genotype and phenotype is generally extremely loose. One reason is that alleles vary in their **expressivity** and **penetrance.** In addition, each gene affects many aspects of phenotype—a phenomenon called **pleiotropy.** The expression of each trait is **polygenic,** the result of many genes acting together. Finally, a single gene may have many functioning alleles. Because of all this complexity, the expression of any gene is heavily influenced by both the physical and developmental environment and also by the genomic environment of the gene.

The Human Genome Project set out to sequence an entire human genome, as well as the genomes of a few other organisms, and to construct linkage maps and physical maps of each human chromosome. In 2001, the Human Genome Project and the company Celera Genomics sequenced the entire human genome.

Pedigree analysis allows researchers to find a genetic marker associated with a particular trait or disease that shows approximately where on the genome a gene lies. **Positional cloning** allows researchers to obtain a particular gene on the basis of its precise location.

Genetic counselors can tell couples both whether they carry certain alleles and the likelihood that their offspring will express certain traits. In **amniocentesis,** a geneticist samples a 4 month old fetus's **amniotic fluid** and looks for chromosomal or other genetic abnormalities. **Chorionic villus sampling (CVS)** allows geneticists to test for the same genetic defects as amniocentesis, but earlier in pregnancy.

Using new **gene therapy** techniques, researchers can induce human cells to express a gene that the cells were previously unable to express. Expression is presently limited, however, to relatively few molecules of the desired protein. Because the cells cannot pass the gene on to new daughter cells in the body, treatments must be repeated. Also, the treated patient cannot pass on the "good" allele acquired in this way to any offspring.

Review and Thought Questions

Review Questions

1. What factors ensure that the relationship between genotype and phenotype is a loose one? What is pleiotropy?
2. What are the limitations and strengths of DNA fingerprinting? What is a RFLP?
3. What is the purpose of the Human Genome Project?
4. How do researchers locate genes for specific diseases such as Huntington's disease or cystic fibrosis?
5. How is prenatal testing different from genetic counseling?
6. If a prenatal test reveals a genetic defect in a fetus, what choices do the parents have?

Thought Questions

7. If you and your spouse both carried the gene for cystic fibrosis, would you get prenatal testing? If you decided to have a test, would you prefer amniocentesis or chorionic villus sampling? Why?
8. How would your current knowledge about research on gene therapy for cystic fibrosis influence your decision?
9. If you knew you had an allele that slightly predisposed you to not get heart disease, would that influence your eating and exercise habits? If you had an allele that slightly predisposed you to get liver cancer, would you care if your insurance company and your employer knew?

About the Chapter-Opening Image

Mapping the human genome is symbolized by this woman, wrapped in a map of the world.

 On-line materials relating to this chapter are on the World Wide Web at **http://www.harcourtcollege.com/lifesci/aal2/**

 Web Bit 14-1 Is Genetic Discrimination Unfair?

 Web Links

 Quizzing & Testing

part III

Evolution

Chapter 15
What Is the Evidence for Evolution?

Chapter 16
Microevolution: How Does a Population Evolve?

Chapter 17
Macroevolution: How Do Species Evolve?

Chapter 18
How Did the First Organisms Evolve?

Photo: Ammonite fossils display the multiple-chambered shell of the animal. (Mauritius/Nawrocki Stock Photo, Inc.)

15 What Is the Evidence for Evolution?

The Scopes Trial

It was a hot July morning in Dayton, Tennessee, when high-school teacher John T. Scopes went on trial for teaching evolution in the public schools. In the stuffy courtroom, spectators mopped their brows and swatted at flies. A dozen journalists crowded together stood on top of a table at the back of the room, trying to get a better view. Midway through the interminable testimony the table collapsed, dumping the men on the floor. Cheers and boos from the rowdy audience outside came through the open windows and drowned out the judge's calls for order. Also outside were a score of evangelists. Their trumpeting voices could be heard throughout days of testimony.

The town of Dayton was bedecked as if for a carnival, with banners and posters of monkeys. Newly constructed stands selling hot dogs, lemonade, and sandwiches lined the sidewalks, and a circus man ran a sideshow with two chimpanzees. The animals had been brought to town to "testify for the prosecution," the man said. Two hundred twenty-five newspapermen had arrived from New York, Baltimore, Philadelphia, and points beyond. It was 1925, before television and even radio. These newsmen would put the tiny southern town of Dayton on the map forever.

The law prohibiting the teaching of evolution—passed by the Tennessee legislature in a moment of boredom—was never intended to be enforced. It nonetheless attracted the attention of the American Civil Liberties Union (ACLU), which believed the law violated the First Amendment right to free speech. Eager to test the law in a federal court, the ACLU advertised in the *Chattanooga News* for a teacher willing to be arrested. Dayton boosters jumped at the chance to liven up their sleepy town with some national attention.

John Scopes agreed to be arrested. An affable high-school physics teacher and athletic coach, Scopes had never actually taught evolution, but he'd once substituted for a biology teacher who was ill and assigned the textbook pages that covered evolution. For this crime he was arrested.

The trial—loudly and forcefully argued by two nationally famous lawyers, William Jennings Bryan and Clarence Darrow, and covered by every major newspaper in the country—was, nevertheless, a rather dull affair. The attorneys representing each side wanted the same outcome—Scopes's conviction. The ACLU wanted a clean conviction so they could appeal the case in a higher court. Their goal was to have the antievolution law overturned.

But it was not to be. In an ironic twist, the ACLU won the battle it was trying to lose and so lost the war. Scopes was initially convicted, but then acquitted on a technicality. The judge technically violated the law when he mistakenly levied Scopes's $100 fine himself, instead of letting the jury do it. Because Scopes was acquitted, the ACLU never got to appeal the case, and the Tennessee antievolution law stayed on the books until the U.S. Supreme Court declared such laws unconstitutional in 1968.

Key Concepts

- Evolution is heritable change in populations of organisms, which Darwin called "descent with modification."
- Darwin's theory of evolution described evolution itself, natural selection (a mechanism for how evolution occurs), and the time scales over which evolution occurs.
- The evidence for evolution includes a clear history of evolution in the form of fossils in the ordered layers of the Earth's crust; biogeography; obvious similarities among groups of related organisms; and similarities in structure of all parts of organisms from limbs and organs to individual molecules.
- Natural selection is the differential reproductive success of individuals that leads to the accumulation of genetically inherited traits.

No one was again arrested for teaching evolution. But American textbook publishers—ever shy of controversy—nevertheless took the hint, reducing or even eliminating discussions of evolution in most high-school textbooks. By 1942, fewer than half of all high-school science teachers covered evolution at all. Only with the Soviet Union's 1957 launching of the first satellite, Sputnik, did the United States begin emphasizing better science education and textbooks. But the increasing appearance of evolution in textbooks sparked a new reaction among antievolutionists, who realized that their fight against science was far from won.

In 1973, antievolutionists in Arkansas, Tennessee, and Louisiana passed identical bills calling for "equal time" for teaching evolution and **creationism,** the biblical myth that the universe was created by the Judeo-Christian God in 7 days. But a court ruled that the "equal-time" bill was unconstitutional on the grounds that it violated the separation of church and state. (The United States Constitution outlaws teaching religion in public schools because not everyone has the same, or any, religion; and the United States was founded on the idea of religious freedom.)

Creationists argued that the account given in the Bible's book of Genesis was the basis for "creation science." But creation "science" is not science. In 1982, Federal Judge William R. Overton overturned the Arkansas law, writing, " . . . the evidence is overwhelming that both the purpose and effect of Act 590 is the advancement of religion in the public schools." About creation science, he said, "While anybody is free to approach a scientific inquiry in any fashion they choose, they cannot properly describe the method used as scientific if they start with a conclusion and refuse to change it regardless of the evidence . . . "

On the basis of Overton's decision, another judge struck down the Louisiana law as well. And in June 1987, the United States Supreme Court affirmed this decision in a 7 to 2 vote, effectively ending a dispute that had begun 128 years before.

Creationists of the 1990s have found another way to fight evolution, however. Rather than try to persuade public schools to teach the biblical version of creation, they now simply ask schools not to teach evolution. But it is unconstitutional to outlaw the teaching of ordinary science. So some school boards simply say students cannot be tested on any origin science, including the origin and age of the universe (about 14 billion years), the origin and age of the Earth (4.6 billion years), and the origin of species of organisms.

In December 1999, for example, the Kansas State Board of Education adopted a curriculum stipulating that Kansas schoolchildren not be tested on evolution in any biology or other science class. Students would be expected to learn about natural selection, Darwin's mechanism for evolution, but not that species can evolve one into another or that all organisms on Earth are related.

American scientists expressed bewilderment. And the presidents of Kansas's six public universities wrote a letter stating that the new standards would "set Kansas back a century" and drive away science teachers. "The argument that teaching evolution will destroy a student's faith in God," the university presidents argued, "is no more true today than it was during the Scopes trial in 1925."

Modern debate about evolution began in 1859, when Charles Darwin published his arguments for the evolution of species in *On the Origin of Species by Means of Natural Selection.* In that book, Darwin rejected the idea that each species had been specially created. The book disturbed a great many people. Three of Darwin's arguments offended most often: (1) that the Earth was very old, (2) that one species could change into another, and (3) that humans and apes were related (Figure 15-1).

In this chapter we will discuss the evidence that has persuaded biologists that organisms are all descended from a common ancestor. At the end of the chapter, we will introduce the mechanism by which most evolution probably occurs—natural selection—which Charles Darwin first proposed more than 150 years ago.

> *Eager to test the law in a federal court, the ACLU advertised in the* Chattanooga News *for a teacher willing to be arrested.*

Figure 15-1 Our family tree. Biologists classify humans as primates not only because we look like other primates, but because we are descended from the same stock as they are. We are most closely related to chimpanzees, gorillas, and orangutans. *(Capuchin, Richard Gibson/Natural Selection; baboon, Stephen G. Maka/DRK Photo; lemur, Kevin Schafer; gibbon, Joe McDonald/Bruce Coleman, Inc.; orangutan, Daniel J. Cox/Natural Selection; gorilla, Mimi Forsyth/Bruce Coleman, Inc.; chimpanzee, Frans Lanting/Minden Pictures; Jane Goodall, Jack Dykinga, Bruce Coleman, Inc.)*

WHAT IS EVOLUTION?

Biological evolution is the process that has taken life on Earth from perhaps a few mere flecks of RNA to the breathtaking variety of more than a million species. What is popularly understood to be Darwin's theory of evolution is really a group of several ideas that can be roughly divided into two parts.

The first part is **evolution** itself, the idea that species change over time. Darwin called evolution **"descent with modification."** A concise definition of evolution says it is heritable change in organisms over time. But the theory of evolution also says that all life on Earth is descended from a few types of cells that came into being billions of years ago. Just as the members of a large family that includes hundreds of cousins can trace their family history back to a few great, great grandparents, all species on Earth can trace their lineages back to common origins. Dogs and foxes truly are cousins, of a sort, as are humans and orangutans.

The second part of Darwin's theory is (1) his specific mechanism for evolution, which he called natural selection, and (2) a description of the way he thought evolution occurred, gradualism. **Gradualism** is the idea that species evolve gradually and continuously through the steady accumulation of changes, rather than through sudden changes.

But for Darwin, natural selection was the heart of his theory. It was a mechanism that set his theory of evolution apart from similar theories. Briefly, **natural selection** is the differential reproductive success of individuals due to genetically inherited traits.

As we saw in the genetics section of this book, every individual has unique combinations of genetically influenced traits. Some of these combinations are **adaptive,** which means that they give the individual an advantage in the struggle for survival and reproduction and so increase its chances of perpetuating these useful combinations of genes in future generations.

If, for example, the average beak size of a population of finches increases, the finches have evolved. Whether any trait, such as increased beak size, is adaptive depends on the environment of the organism. When many large, hard seeds are present, a large beak is adaptive. When few large, hard seeds are available but plenty of small soft seeds are available, the large beak may not be adaptive. Birds with smaller beaks may be better at eating the smaller seeds. In that case, smaller beak sizes may slowly return.

Darwin compared natural selection to **artificial selection,** the process by which animal breeders and farmers create new lines of dogs, pigeons, corn, and other organisms. In artificial selection, human beings select the most desirable individuals for breeding, generation after generation. In natural selection, the individuals best adapted to their environment survive and reproduce best, leaving more offspring. At the end of this chapter, we will discuss Darwin's theory of natural selection. In later chapters, we will encounter other mechanisms for evolution. Some evolutionary change can occur through random processes. We will also discuss the rate at which evolution occurs (Darwin's gradualism). But in this chapter, we will focus on the evidence for evolution itself.

Darwin's theory of descent with modification through natural selection rested on four ideas. First, he recognized that the world is ever changing and very old (rather than constant and recently created). Second, he recognized that species change. (As we shall see, neither of these ideas was original with Darwin.) Third, Darwin argued that species are composed of populations of individuals, not identical organisms. Every individual is unique. And fourth, he argued that every species and group of species is descended from a common ancestral species. Ultimately, Darwin said, all organisms—including humans—may derive from a single origin of life on Earth.

These ideas conflicted with the biblical account of creation and with a major idea in Western philosophy, called **essentialism.** According to the Greek philosopher Plato (427–347 B.C.), individuals, whether chairs, daisies, or people, are only distorted shadows of an ideal, or essential, form. Only the ideal form was real. The variations that make every daisy unique, said Plato, are distortions and flaws.

Because of essentialism, most scientists in Darwin's time viewed species in terms of their ideal forms. Like so many Lego bricks, every member of a species was supposed to be identical to every other. Species were supposed to be perfect, unchanging, sharply defined "natural kinds." In contrast, Darwin viewed every individual as an individual, not a distortion of some ideal form. As evolutionary biologist Ernst Mayr has written, "It was Darwin's genius to see that this uniqueness of each individual is not limited to the human species but is equally true for every sexually reproducing species of animal and plant."

Darwin asserted (1) that the world is ever changing and very old; (2) that species are made up of individuals; (3) that species change; and (4) that all organisms are related.

Evolution Before Darwin

Although Darwin was the first to persuade the world that plants and animals had evolved, he was hardly the first to hypothesize that organisms evolve. His own grandfather, Erasmus Darwin (1731–1802), argued in 1794 that all species originated from the same primitive "filament." However, Erasmus Darwin propounded many ideas that seemed outlandish in the 18th century, and no scientist took his ideas on evolution seriously. In fact, in the senior Darwin's time, to be accused of "Darwinizing"—engaging in wild speculation—was an acute embarrassment.

How old is the idea of evolution?

Evolution as an idea actually predates even Plato. In the 6th century B.C., 100 years before Plato, the Greek philosopher Anaximander argued that the world was not created abruptly as most ancient religions said. He believed, instead, that animals evolved, and that humans, like all other vertebrate animals, were descended from fishes. Although we now know that Anaximander was mainly correct, it was 2500 years before his ideas found scientific support and general acceptance.

In large part, no one took evolution seriously because Christian beliefs incorporated Plato's essentialism. And essentialism, along with creationism, established a nearly unbreachable intellectual barrier to the idea of the evolution of species. Another serious barrier to the idea of evolution was the notion—accepted by nearly everyone in 17th- and 18th-century Europe—that the Earth had been created in 4004 B.C. This estimate had been worked out in 1650 by Archbishop James Ussher, who added together the life spans of all the patriarchs named in the Bible, beginning with Adam. Most Christians believed that the Earth was less than 6000 years old.

By the end of the 18th century, however, the biblical account of creation and the notion that species do not change were undermined by three separate lines of evidence. First was the discovery of hundreds of thousands of species not mentioned in the biblical story of creation. Explorers returning to Europe from Africa, the Americas, and the Pacific brought back plants and animals that no European had ever seen or heard of before (Figure 15-2A). Equally disturbing was the great number of fossil plants and animals that were turning up in Europe itself. These fossils were stone remnants of species as alien as any from the far side of the world (Figure 15-2B). Yet, they appeared to have lived in Europe in some previous era. But when? Finally, by the middle of the 19th century, geologists found compelling evidence that the Earth was far older than Ussher's estimated 6000 years.

The enormous difficulty of reconciling these new facts with the story of creation in the Judeo-Christian Bible is apparent in the thinking of five of the greatest scientists of the 18th and early 19th centuries. All of them discovered evidence for evolution, yet because of creationism and essentialism, all of them interpreted the evidence in nonevolutionary terms.

How did Linnaeus's ideas lay the groundwork for evolution?

The first of the five scientists is Carolus Linnaeus (1707–1778), a Swedish physician whose most familiar contribution to biology was the two-part Latin names—genus and species—that biologists assign to all species. The scientific name for a human being, for example, is *Homo sapiens*. *Homo* refers to the **genus** to which we belong, and *sapiens* ("wise") refers to our **species.** Linnaeus named more than 4000 plants and animals, and biologists since Linnaeus have used his system to name some 1.4 million more species. Biologists estimate that 10 to 100 million more remain to be discovered and named. We discuss Linnaeus in more detail at the beginning of Chapter 22.

Even more important than Linnaeus's system of naming organisms was his system of grouping, or classifying, them. For example, the genus *Homo* belongs to the **family** Hominidae. And the family Hominidae belongs to the **order** Primates, one of 16 orders in the **class** Mammalia. Thus, we group species into genera, genera into families, families into orders, and orders into classes (Chapter 19). Each of these groupings is a **taxon** [Greek, *taxis* = arrangement; plural, **taxa**]. The science of classifying organisms is called **taxonomy.**

Linnaeus firmly believed that each species was individually created by God, yet he arranged species into groups that reflected their similarities to one another. He could see, for example, that all the species of mice were more closely related to rats than to horses, and his formal system for recognizing such natural relations was fundamental to Darwin's arguments 100 years later.

Linnaeus avoided asking *why* organisms fall so naturally into families, piously writing, *"Deus creavit, Linnaeus disposuit"*—God creates, Linnaeus arranges. But one of his contemporaries, the great French biologist George-Louis Leclerc, or Count Buffon (1707–1788), couldn't leave the question alone.

Linnaeus introduced a modern system of classifying organisms.

Why was Count Buffon afraid to discuss evolution?

Buffon was an accomplished mathematician as well as biologist. He was also a science popularizer, loved by the public but resented by fellow scientists. In his early manhood, Buffon boldly stated that the Earth must be far older than 6000 years. But when Church elders threatened him with excommunication for this heresy, he quickly took back his arguments.

Figure 15-2 Another world. A. Western grey kangaroo and joey. European explorers and naturalists brought back strange animals and plants no one had ever heard of, let alone seen. These organisms were not mentioned in the Bible. B. Even at home in Europe, paleontologists dug up fossil animals as strange as this fossil pterosaur, a flying reptile of the Jurassic Period discovered in Bavaria. None of these had been named in the Bible nor classified by European scientists. (*A, Tim Flach/Stone; B, Kevin Schafer/Peter Arnold, Inc.*)

In later years, he carefully recanted each heresy as he committed it. In a 36-volume natural history, in which he interspersed descriptions of plants and animals with discussion of natural philosophy, Buffon noted the remarkable similarity of the horse and the ass, or donkey. He wrote, in 1753, "But if we once admit that there are families of plants and animals, so that the ass may be of the family of the horse, and that the one may only differ from the other through degeneration from a common ancestor, we might be driven to admit that the ape is of the family of man." And, as if that were not bad enough, he added, " . . . then there is no further limit to be set to the power of nature, and we should not be wrong in supposing that with sufficient time she could have evolved all other organized forms from one primordial type."

To placate the church, however, he quickly contradicted this insight: "It is certain through Revelation that all animals have shared equally in the grace of creation and that each emerged from the hands of the creator as it appears today."

Buffon's theory of evolution resembled Darwin's, but without natural selection. Fearing excommunication, however, he denied that he believed his own arguments.

What made Hutton think that the Earth was older than 6000 years?

Beginning in the mid-18th century, the Industrial Revolution gave rise to an insatiable demand for coal and metals that turned geology into a practical and hugely popular science. For the first time, scientists began to seriously explore the surface of the Earth. They learned that soil, rock, and mountains were eroded by wind, frost, and running water. In some places, whole mountain ranges seemed to have eroded completely away. But could all this have happened, geologists asked, in just 6000 years?

In 1788, the Scottish geologist James Hutton (1726–1797) suggested that it could not. The Earth, he said, was eternal. The perpetual erosion of mountains was repaired by the uplift of sediments and the expulsion of magma from the Earth's depths. At the time, most geologists doubted continents and mountains could rise from the Earth's surface. But Hutton insisted that the Earth was not a static, passive body but a "beautiful machine," powered by the heat of its interior.

He proposed that endless cycles of erosion, sedimentation, and uplift molded the Earth's surface. The Earth, he said, might be immeasurably old. In his now famous conclusion, he wrote, "The result, therefore, of our present inquiry is, that we find no vestige of a beginning—no prospect of an end."

It was heresy. But if Hutton was a thorn in the side of theologians, Baron Georges Cuvier (1769–1832) was their savior. In 1812, the great French anatomist and paleontologist found an ingenious way to reconcile geology and Genesis—a way to reconcile the great age of the Earth with theological doctrine.

Catastrophism: How did Cuvier try to reconcile Genesis and geology?

Cuvier was a gifted anatomist and the first to study fossils systematically. He invented both **comparative anatomy,** the detailed comparison of the anatomies of different species, and **paleontology** [Greek, *palaios* = ancient + *on* = being + *logos* = discourse], the study of fossils. **Fossils** [Latin, *fossilis* = dug up] are the remains or imprints of past life. The most impressive fossils are the intact skeletons of whole animals, such as 70-million-year-old dinosaurs. Such spectacular finds are rare, however, compared to the hundreds of thousands of small fossilized shells and tiny bone fragments paleontologists have unearthed. Fossils include preserved bones and teeth, petrified trees, bits of fossilized dung, footprints, impressions of the surfaces of ferns, as well as ants and other insects preserved in hardened tree resin (amber).

Fossils had puzzled people for thousands of years. Fossils look like plants and animals or parts of plants and animals. But they are made of stone, and they rarely look like *familiar* plants and animals. Some people thought fossils were God's first experiments with creation or maybe the remains of plants and animals that had literally missed the boat. Failing to board Noah's ark, they had perished in the great flood described in the Bible (Figure 15-3A).

But Cuvier knew that all the different organisms represented by fossils could not have died in the same flood. The different kinds of fossil organisms had lived and died out at different times in the Earth's history. By 1760, geologists had already shown that rocks could be arranged in a series of layers, or strata, according to their age. The oldest layers were usually deepest (Figure 15-3B). The newest layers were on top.

William "Strata" Smith, a geologist and canal engineer, compared these layers to "slices of bread and butter on a breakfast plate." Smith showed that different fossils appeared and disappeared from layer to layer. Older fossils appeared in the deepest layers, then disappeared and were replaced by new fossils—on up to the topmost layers.

Starting with the assumption that species could not evolve, Cuvier concluded that the fossils in the oldest layers had nothing to do with modern life. To account for the different kinds of life that appear in the fossil layers, Cuvier hypothesized that the organisms from the different layers had perished in a series of catastrophes. Noah's flood was merely the most recent and dramatic catastrophe. (The previous catastrophes were not mentioned in Genesis, Cuvier argued, because God didn't think Moses needed to know about them.)

After each catastrophe, argued Cuvier, new species appeared in another creation. **Catastrophism,** as his theory came to be called, allowed for the great age of the Earth and the fossil record without implying that Genesis was wrong. This idea was appealing to a great many pious Christians, and catastrophism remained popular for decades. In fact, the 1981 Arkansas equal-time bill called for a revival of this very theory.

But, although paleontologists have identified several mass extinctions, the evidence does not support catastrophism. First,

A.

B.

Figure 15-3 Catastrophism and uniformitarianism. A. Noah's flood. The deluge as portrayed in the first edition of the Luther Bible, published in 1534. B. The Grand Canyon. James Hutton wrote, "Thus . . . from the top of the mountain to the shore of the sea . . . everything is in a state of change; the rock and solid strata slowly dissolving, breaking and decomposing, for the purpose of becoming soil; . . . without those operations which wear and waste the solid land, the surface of the earth would become sterile." (B, Jack Dermid/Bruce Coleman, Inc.)

although Cuvier had originally proposed only four or five catastrophes before Noah's flood, fossil hunters continued to find so many distinct layers of fossils that more and more catastrophes had to be invented to account for them all. The Swiss-American biologist Louis Agassiz (1807–1873) maintained, for example, that the fossil record revealed 50 to 80 separate catastrophes, each one requiring a separate creation.

And while some organisms appeared once in the fossil record and then disappeared forever, others kept reappearing virtually unchanged in "creation" after "creation" (Figure 15-4). Lastly, archaeologists found human bones and other artifacts that suggested that humans predated not only the biblical flood, but the hypothetical creation of the world in 4004 B.C. Catastrophism didn't seem to fit the facts.

Uniformitarianism: Lyell rejects catastrophism

Catastrophism received its most damaging blow in 1830, when the Scottish geologist Charles Lyell (1797–1875) published the first volume of his three-volume *Principles of Geology.* Lyell rejected catastrophism. Instead, expanding on Hutton's ideas, Lyell argued that the slow processes that now mold the Earth's surface—erosion, sedimentation, and upheaval—are the same ones that have always molded it. Geologic change, he said, is slow, gradual, and steady—not catastrophic. Given enough time, mountains could erode down to mere hillocks, sea beds rise to great heights, and rivers cut deep canyons (Figure 15-3B). Gradual processes alone could create the world we know.

Lyell, who had been trained as a lawyer, provided huge amounts of evidence in his book to back up this hypothesis, later named **uniformitarianism.** Darwin was so influenced by Lyell that he later wrote, "I have always thought that the great merit of the *Principles* was that it altered the whole tone of one's mind, and therefore that, when seeing a thing never seen by Lyell one yet saw it partially through his eyes." But even though Lyell's *Principles* provided fertile soil for Darwin's ideas, Lyell himself believed that species could not evolve. It was Darwin's own book *The Origin of Species,* published almost 30 years after the *Principles,* that changed Lyell's mind.

The Industrial Revolution's demand for fossil fuels created the science of geology. The fossil record discovered by geologists showed the Earth was extremely old.

Figure 15-4 Horseshoe crab. Many groups of organisms have changed hardly at all over millions of years. The modern horseshoe crab remarkably resembles its Jurassic ancestor, which lived nearly 200 million years ago. (Left, Alan Desbonnet/Visuals Unlimited; right, John Cancalosi/Peter Arnold, Inc.)

Why was Lamarck ignored?

Even together, the fossil record and the great age of the Earth were not enough to convince most scientists that species evolve. Among reputable scientists, evolution was at best a dubious and controversial hypothesis. Count Buffon and Erasmus Darwin had proposed that species evolve, but neither man was taken seriously by fellow scientists. In 1809, the great French biologist Jean Baptiste Lamarck (1744–1829) again proposed the idea that species evolve over time.

Lamarck came from an impoverished noble family and served as a tutor to Count Buffon's son. He studied medicine and botany in Paris and published two works on botany before being appointed professor of zoology at the Jardin des Plantes in 1793. A renowned taxonomist, Lamarck named hundreds of species, wrote more than a dozen volumes on fossil "invertebrates" (a word he invented), and argued that plants and animals are all related. To express this union, he invented the word "biology."

In his studies of fossils, Lamarck saw a clear line of descent—from older fossils to more recent fossils to modern species. *The oldest fossils looked the least like modern organisms. Later groups of fossils included more and more familiar organisms.*

Lamarck hypothesized that the accumulated changes were adaptive—that is, the changes made organisms better able to survive under new conditions. However, Lamarck argued that the changes resulted from a drive toward perfection and complexity.

A giraffe, for example, reaching ever higher in the trees for leaves to eat, would constantly stretch its neck and thus acquire a permanently longer neck. The giraffe's offspring would then inherit longer necks. Lamarck hypothesized that adaptive characteristics acquired by the parent could be inherited by the offspring.

In Lamarck's day, most scientists accepted his mechanism for evolution, the theory of the **inheritance of acquired characteristics.** But they vehemently rejected evolution itself, and he was vilified by the entire scientific community. The fiercest opposition came from Cuvier. As both an eminent scientist and a French cabinet minister, Cuvier's opinion carried enormous weight, and Cuvier criticized Lamarck relentlessly. In 1832, 3 years after Lamarck had died—poor and almost forgotten—Cuvier delivered an "elegy" to Lamarck at the prestigious French Academy, in which he once more dismissed not only Lamarck's theory of evolution but also his reputation as a biologist.

So it was not surprising that when Darwin began writing his theory of evolution, he did everything he could to distance himself from Lamarck (and his own grandfather Erasmus) and to anticipate every possible objection to his own ideas.

How Did Charles Darwin Convince Others That Species Evolve?

By the middle of the 19th century, natural science had become enormously popular with the lay public. Every sector of English society was obsessed with the exotic and the unusual in nature. Even as explorers brought back exotic animals and plants from Africa, Australia, and Asia, every middle-class English family acquired an aquarium, a fern case, a butterfly cabinet, or a shell collection. Mothers routinely taught their children the names of ferns and fungi. Aristocrats turned their estates into parks for exotic animals such as elands, beavers, and kangaroos. Books on natural history were so popular that one—*Common Objects of the Country*—sold 100,000 copies in a week, a total that even today would earn the book a place high on *The New York Times* best-sellers list.

Darwin's education

Charles Robert Darwin (1809–1882) came of age in the middle of this national obsession and was himself entirely caught up by it. An upper-class English boy whose father was a doctor, Charles spent his boyhood tramping through the woods and fields of Shrewsbury, hunting, fishing, and collecting insects. As a young man, his interests narrowed to hunting, and his father despaired of Charles ever making anything of himself. Darwin senior once wrote to his son, "You care for nothing but shooting, dogs and rat-catching, and you will be a disgrace to yourself and all your family."

Darwin's utter lack of ambition forced his father to pick a profession for him, and in 1825 Robert Darwin sent young Charles to Edinburgh University to study medicine. In the course of his studies, Darwin attended lectures on geology and zoology and thought they were the most boring thing he'd ever had to endure in his life. His only interest, outside of shooting birds, was the meetings of a natural history club. As for medicine, if he had ever considered becoming a doctor, he changed his mind after watching two horrifying surgical operations done without anesthesia, one on a child. His resistance to studying medicine, or anything else, only hardened when he realized that his wealthy father would probably support him for the rest of his life, whatever he chose to do (or not to do).

Dr. Darwin then decided that Charles should study for the clergy and in 1828 sent him to Cambridge University. Fortunately for modern biology, the clergy was then a haven for naturalists, and 19-year-old Charles Darwin soon shook off his perpetual boredom and rediscovered his boyhood passion for nature.

In 3 years, he managed to complete his B.A., and when he graduated from Cambridge in 1831, one of his professors recommended him for a position as ship naturalist on board the H.M.S. *Beagle*. The *Beagle* was a survey ship about to embark on a trip around the world, its 5-year mission to map the coast of South America. Darwin's duties were to collect specimens of the plants, animals, and rocks from this unexplored new world. In December 1831, just before the *Beagle* sailed, a professor friend sent Darwin the first volume of Lyell's *Principles of Geology*. His course was set.

During the long trip across the Atlantic, Darwin immersed himself in Lyell's arguments about the slow processes that formed the Earth. He thought of nothing but geology for

Figure 15-5 Darwin and Wallace. A. Charles Darwin and Emma Wedgwood Darwin. Even in his most productive years, Charles Darwin was sickly and worked only for short periods. B. Alfred Russel Wallace's ideas about evolution generally matched Darwin's. However, although Wallace believed that humans had evolved from ancestral primates, he did not believe that our capacity for thought and spiritual passion could have evolved in the same way. (A, Bridgeman Art Library; B, The Granger Collection, New York)

months. Lyell formalized two important ideas: first, that the Earth is very old, and second, that geologic processes occur very slowly and gradually. Darwin's theory of evolution, as he would later conceive it, likewise stated that life is very old (millions of years) and change (evolutionary, in this case) has been slow and gradual.

As Darwin explored South America over the next 5 years, he began to suspect that all animals and plants were descended from a common ancestor. He wondered if Count Buffon, Lamarck, and even his own grandfather had been right about evolution. He took notes; he collected specimens. He noticed what animals ate, where they lived, what kind of soil this plant or that seemed to prefer.

Almost as soon as he returned to England in 1836, Darwin began to classify his specimens and to organize his notes. But it was 6 more years before he "allowed himself the satisfaction" of even outlining his theory. In the meantime he read the work of the second great intellectual influence in his life after Lyell—Thomas Malthus.

Malthus: Too many children

In 1798, the English economist and clergyman Thomas Robert Malthus (1766–1834) wrote a highly controversial paper, *An Essay on the Principle of Population*. Malthus argued that human populations tend to increase geometrically, while food supplies and other resources increase only arithmetically, if at all. For example, if a couple had three children, and each of their children had three children, and the nine grandchildren each had three children, the family farm could not possibly support the 27 great-grandchildren. The family may be able to increase the amount of food they can produce on their farm, but they cannot possibly triple production every generation. Poverty, famine, overcrowding, disease, and war are all the inevitable consequences of this excessive population growth, argued Malthus. He predicted that without some check on population growth, human populations would face a continuing, ferocious struggle for existence in the face of limited resources.

When Darwin read Malthus in 1838, he saw for the first time how evolution could work. Malthus's arguments applied to all species, reasoned Darwin, not just humans. Many more individuals of each species are born than can possibly survive, and, consequently, there is a constant struggle for existence.

"It follows," Darwin later wrote, "that any being, if it vary however slightly in any manner profitable to itself, under the complex and sometimes varying conditions of life, will have a better chance of surviving, and thus be *naturally selected*." His theory of evolution through natural selection was beginning to take shape.

Wallace scoops Darwin

But Darwin delayed publishing. Month after month, year after year, he added new facts to bolster his theory. Lyell, who was by then a friend, still didn't accept evolution, but he urged Darwin to publish his idea before someone else thought of the idea and published first. Then, in 1858, Alfred Russel Wallace (1823–1913), a land surveyor and naturalist working in the Malay Archipelago (in what is now Malaysia), independently hit upon natural selection as a mechanism for evolution.

Wallace had written to Darwin the year before asking for information about selective breeding. Darwin wrote back with the information and added tactfully that he had been working on similar ideas for some 20 years (Figure 15-5). Again Wallace wrote, this time outlining his theory of evolution—but without natural selection. Darwin responded politely, but hinted that he knew more. Wallace was obviously hot on his heels, yet Darwin did nothing.

Then, a few months later, in 1858, Wallace wrote once more. In the interim, he had read Malthus and experienced the same flash of insight Darwin had. Eagerly, Wallace wrote to Darwin, outlining his theory of evolution by natural selection.

Darwin, utterly crushed, was ready to give all of the credit to Wallace. In a letter to Lyell, he wrote:

> "I never saw a more striking coincidence; if Wallace had my [manuscript] sketch written out in 1842, he could not have made a better abstract! Even his terms now stand as heads of my chapters. Please return me the [manuscript], which he does not say he wishes me to publish, but I shall, of course, at once write and offer to send it to any journal. So all my originality, whatever it may amount to, will be smashed."

Then worse misfortune struck. Scarlet fever swept Darwin's family, killing his 19-month-old son Charles in a matter of days. Grief stricken, Darwin threw up his hands and turned everything over to Lyell, who gladly rescued Darwin. Lyell presented both Wallace's paper and an extract from Darwin's unpublished outline to the Linnaean Society, giving credit to both men. The following year, 1859, Darwin finally published his theory in a book, *On the Origin of Species by Means of Natural Selection*.

The Origin, as biologists call it affectionately, is the single most influential scientific book ever written. When it appeared in 1859, all 1250 copies of the first printing sold out in a single day. Its arguments persuaded even those, such as Lyell, who had previously rejected the idea that species evolve. What was so convincing?

WHAT IS THE EVIDENCE FOR EVOLUTION?

Evolution is heritable change in populations of organisms, which Darwin called "descent with modification." Evolution is also "speciation," the division of one species into two or more. Such "diversification" is the means by which a few kinds of simple cells living 3.8 billion years ago gave rise to the millions of complex organisms that live on Earth today.

The Origin's great strength was Darwin's gluttony for evidence. Darwin provided evidence for evolution from the fossil record, from the distribution of plants and animals (biogeography), from taxonomy, from comparative anatomy, from comparative embryology, and from domestic breeding. Since Darwin's time, other biologists have further buttressed his arguments with evidence from fields as diverse as comparative molecular biology, classical genetics, population ecology, developmental biology, and animal behavior. In the rest of this chapter we will briefly examine some of this evidence.

The Fossil Record Tells a Story of Evolution

The sequence of organisms in the fossil record tells a story with an obvious meaning. At the bottom, where sediments are oldest, lie the primitive prokaryotes and eukaryotes (Figures 15-3 and 15-6). Above them come simple multicellular organisms, then trilobites and other invertebrates, and then the first vertebrates, the fishes. Many more fish follow. Then, in rapid succession, amphibians and reptiles arrive. Newer, and higher in the sedimentary strata, are the dinosaurs, crocodiles, birds, and mammals.

Three facts stand out. First, fossils are distributed consistently. Rocks of the same age contain approximately the same groups of organisms. Cambrian rocks, which generally contain trilobites, never contain dinosaurs or horses, only trilobites. Recent Pleistocene rocks contain no trilobites. Such a

Era	Period	Epoch	Millions of Years Ago	Major Events
CENOZOIC	Quaternary	Holocene	0.01	
		Pleistocene	1.65	
	Tertiary	Pliocene	5.3	
		Miocene	23.7	
		Oligocene	36.6	
		Eocene	57.8	
		Paleocene	66.4	
MESOZOIC	Cretaceous		144	
	Jurassic		208	
	Triassic		245	
PALEOZOIC	Permian		286	
	Carboniferous	Pennsylvanian	320	
		Mississippian	360	
	Devonian		408	
	Silurian		438	
	Ordovician		505	
	Cambrian		545	
PRECAMBRIAN			2500	Origin of photosynthesis in bacteria
			4600	Approximate origin of Earth

Figure 15-6 A composite canyon showing the geologic eras and periods as discrete layers of rock. The Grand Canyon includes layers from the Precambrian, at the bottom, all the way up to the Permian, near the rim of the canyon. Nearby Zion National Park includes Triassic and Jurassic layers. At both canyons, the most recent layers, from the Cretaceous onward, have all been washed away. But these layers can be found intact in other places. If you wanted to look for the bones of *Apatosaurus*, would you go to the Grand Canyon or to Zion?

pattern is consistent with the idea that once organisms go extinct, they stay extinct. Extinct species do not suddenly reappear millions of years later as a new creation.

Second, the order in which organisms are laid down in the fossil record suggests a sequence of evolution that is, as we shall see, independently confirmed by other fields of biology. For example, the first true mammals do not appear in the fossil record until *after* the appearance of the mammal-like reptiles, whose anatomy is intermediate between early reptiles and mammals. This pattern is consistent with the idea that mammals evolved from these mammal-like reptiles. Similar patterns appear over and over again in the fossil record.

Finally, recent fossils look most like modern organisms, while the oldest fossils generally look least like modern organisms, a pattern that suggests the accumulation of change—or evolution (Figure 15-7).

The order in which organisms appear in the fossil record is consistent with the theory of evolution.

How do fossils form?

An animal or plant becomes a fossil only if it dies under the right conditions. The vast majority of dead plants and animals are consumed by other organisms and leave no trace of their existence. Dead things leave a trace only if their bodies or imprints are rapidly buried in a bog or at the bottom of a sea or lake. There, protected from scavengers, erosion, and decay, organisms may in time become more deeply buried as successive layers of mud and sand settle over them.

Over time, pressure from all these layers of sediment turns the deepest layers of mud and sand into "sedimentary rock." After millions of years, geologic forces raise the rock up into new mountains. Rivers flow down the sides of the new mountains, eroding canyons through the layers of old sediment, revealing the ancient—and to us, amazing—fossils.

Because organisms fossilize most often when buried in sediment at the bottom of a sea or lake, most fossils are marine or freshwater organisms. Terrestrial plants and animals are usually fossilized only if their bodies happen to fall into a lake or sea. Furthermore, only the hard parts of animals—shells, teeth, and bones—and the woody parts of plants are likely to survive long enough to form fossils. The probability is remote that a soft, terrestrial organism such as a slug will be fossilized.

The chanciness of fossilization means that the fossil record is incomplete. For example, most insects are too small and delicate to be preserved, and so the fossil record for insects is poor. However, where insects are preserved, they often appear in great numbers. Large numbers of insects may appear in one layer, disappear—leaving a gap—then reappear millions of years later. Even vertebrates, some of the best preserved of all

Figure 15-7 (part 1) Distinctive fossils from the geologic past. *(Photos from left to right: Dickinsonia, Biological Photo Service; fossil jellyfish, William E. Ferguson; trilobites, James L. Amos/Photo Researchers, Inc.; Devonian fish* Dinichthys terrelli, *James L. Amos/Photo Researchers, Inc.; calamites, Ted Clutter/Photo Researchers, Inc.; fern fossil, J.C. Carton/Bruce Coleman, Inc.)*

BOX 15.1

Radioactive Dating

The fossil record provides a useful chronicle of the sequence of organisms that have lived on Earth. Fossils also provide an accurate way to determine the relative ages of different layers. Still, even if scientists know the relative ages of every single layer, they still want to know the **absolute age**—the age in years—of each of the layers. Scientists had no way of estimating the absolute ages of rocks until the discovery of radioactivity by Marie Sklodowska Curie and Pierre Curie in 1898.

As we discussed in Chapter 2, radioactive atoms, such as uranium, break down into other materials by giving off alpha, beta, and gamma rays. The breakdown, or **decay**, occurs at a steady rate that is unaffected by temperature, pressure, or other environmental variables. Every form of an atom, called an **isotope**, breaks down into a series of other isotopes. Half of the isotope carbon-14, for example, decays to nitrogen-14 every 5600 years. We say that the **half-life** of carbon-14 is 5600 years.

By comparing the relative proportions of different isotopes in a rock sample, scientists can tell the age of the rock. Different radioactive isotopes have different half-lives. Carbon-14, denoted as ^{14}C, is useful for dating objects that are only a few thousand years old. In contrast, rubidium-87 has a half-life of about 5×10^{10} years (50 billion years), so it is useful for dating rocks that originated at the same time as the solar system (about 5 billion years ago) or even earlier.

---

organisms (because of their hard bones), show a discontinuous record. A mammal may disappear from the record, then reappear, slightly altered, ten thousand years later. Despite its limitations, however, the fossil record clearly shows that life has evolved according to the sequence summarized in the next section.

What Does the Fossil Record Show?

In the 18th and 19th centuries, scientists recognized that the fossils in the top layers of sediment were the most recent, while the fossils in the deepest layers were the oldest. For the first time, scientists were able to read the layers of rock as if they were the pages of a history book. They discovered a startling history of the last 570 million years. Twentieth-century paleontologists have now extended this history back to the appearance of the first cells on Earth nearly 4 billion years ago.

Geologists divided this history into chapters. Each chapter, or **period,** is 30 to 75 million years long and contains distinctive forms of life in its fossil record (Figure 15-6). Periods are grouped into four long **eras,** of from 65 million to several billion years each. The most recent chapters are subdivided into relatively short **epochs.** The same distinctive layers appear in rock formations around the world.

Nineteenth-century geologists could not give exact ages to each layer. However, they could tell the order in which the layers had been deposited. And they estimated the ages of the various periods and eras recorded by these fossil layers remarkably well. The modern development of radioactive dating techniques, in the 1940s and 1950s, allowed paleontologists to assign actual dates to evolutionary events and confirm many of the old estimates.

The **Precambrian Era** comprises most of the history of the Earth, beginning with the planet's formation 4.6 billion years ago. Before the discovery of the first Precambrian fossils in the 1950s, paleontologists viewed the Precambrian as one long lifeless era. With the discovery of ancient bacteria and algae, paleontologists now subdivide the Precambrian into six eras. The first fossils, some as old as 3.8 billion years, resemble modern prokaryotes, or bacteria. These ancient fossils vividly illustrate the breathtaking antiquity of life on Earth. Fossils of the first eukaryotes (cells with organelles) appear in rocks about 2 billion years old. And new evidence—in the form of decayed cholesterol molecules in ancient rocks in northwestern Australia—suggests that eukaryotes may have appeared as early as 2.7 billion years ago. In Chapter 18, we discuss the formation of the Earth and how life might have originated.

The Cambrian explosion

The first multicelled organisms appear in the fossil record near the end of the Precambrian, about 800 million years ago. And the first multicelled animals—which resemble jellyfish, worms, and tiny shelled creatures—do not appear until about 640 million years ago. But it is in the Cambrian Period [Latin, *Cambria* = Wales], at the beginning of the **Paleozoic Era** [Greek, *palaios* = ancient + *zoos* = life], that multicelled life first appeared in all its staggering diversity. By 543 million years ago, at the time of the "Cambrian explosion," the Earth was home to trilobites, jellyfishes, corals, segmented worms, fungi, algae, and organisms unlike anything alive today (Figure 15-7). Indeed, by the end of the Cambrian, 11 of the 35 phyla that live today had already evolved, including the first chordates, an-

Figure 15-7 (part 2) Distinctive fossils from the geologic past. *(Photos from left to right:* Archaeopteryx *fossil, James L. Amos/Photo Researchers, Inc.; fossil dinosaur eggs (Protoceratops andrewsi), Ken Lucas/Visuals Unlimited; ichthyosaur fossil, E.R. Degginger/Bruce Coleman, Inc.; fossil flower, William E. Ferguson; plesiosaur fossil, E.R. Degginger/Earth Scenes; orangutan and baby, Tim Davis/Photo Researchers, Inc.)*

cestors of all vertebrates—animals such as fishes and mammals that have backbones (Chapter 23). Geologic studies suggest this burst of evolution occurred within about 30 million years and, possibly, in as little as 5 million years. By the end of the Cambrian, some 500 million years ago, a mass extinction wiped out whole classes and families of organisms. But the first vertebrates—the armor-plated fishes—survived.

Besides the Cambrian, the Paleozoic Era includes five other periods. During the next two periods of the Paleozoic Era, the Ordovician and the Silurian, the first fossil plants, great numbers of starfishes, nautiluslike animals, and fishes with bony skeletons appeared.

During the Devonian Period, the oceans receded, leaving vast new tracts of dry land. On land, fungi, mosses, and other bryophytes evolved, as well as the first "vascular" plants, those with a system of tubes for transporting water and nutrients. Arthropods such as scorpions, centipedes, millipedes, and springtails appeared as well. In the seas and lakes, so many kinds of fishes evolved that the Devonian is called the Age of Fishes.

During the Carboniferous, the seas rose and fell repeatedly, and the Amphibia, the first vertebrates to live permanently on land, evolved. The Carboniferous Period, together with the next period, the Permian, is sometimes called the Age of Amphibia. But it was also the age of the first great forests. The first reptiles appeared, as well as flying insects, such as cockroaches and dragonflies, some with wingspans of 75 cm (2.5 ft). Swamps filled with ferns and horsetails, and cone-bearing trees covered huge areas of the Earth. These fossilized swamps persist today as the coal deposits from which we get coal, oil, and natural gas.

The extraordinary explosion of plants and animals that occurred during the Carboniferous ended with another mass extinction in the Permian. The continents drifted together to form a single supercontinent, called Pangaea; the seas dropped to their lowest level ever; and the climate became extremely dry. About 90 percent of all marine species went extinct within 5 to 8 million years.

The **Mesozoic Era** [Greek, *mesos* = middle + *zoos* = life], often called the Age of Reptiles, began about 248 million years ago and ended 65 million years ago. The Mesozoic comprised three periods—the Triassic, Jurassic, and Cretaceous.

During the Triassic Period, more kinds of reptiles lived than ever before or since. Soon after the reptiles evolved, they split into three groups. The first group were the turtles. The second group were the ancestors of the first mammals. These were small, shrewlike animals that did not diversify much, but, lucky for us, they hung on for 200 million years until the dinosaurs were gone.

The third group were the ancestors of all the other reptiles—including the lizards, snakes, dinosaurs, and birds. In the seas, long-necked plesiosaurs and ichthyosaurs fed on Mesozoic fish. In the air, pterosaurs flapped across the sky, some with wingspans of as much as 10 meters (30 feet), others as small as sparrows. Late in the Triassic, the first dinosaurs appeared, diversified throughout the Jurassic, then disappeared in another mass extinction at the end of the Cretaceous (Figure 15-7).

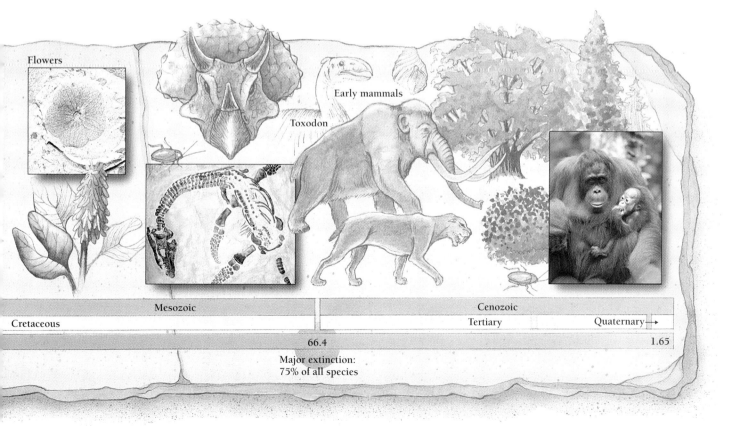

Many organisms survived the mass extinction at the end of the Cretaceous. These included the insects, flowering plants, birds, mammals, and many reptiles. The insects and flowering plants, in particular, diversified enormously during the Cretaceous. But the birds and mammals awaited the arrival of the Cenozoic Era.

The shortest era of geologic time, the **Cenozoic Era** [Greek, *cainos* = recent], extends from about 65 million years ago to the present. The Cenozoic has seen the diversification of the insects and flowering plants, the first grasslands, together with grassland animals, and all of the other modern mammals and birds.

Because so little time has elapsed since the beginning of the Cenozoic—a mere 65 million years—fossils from this era are common and well preserved. Fossils of mammals are especially abundant, and paleontologists have been able to piece together nearly continuous evolutionary histories for many modern species, including those of humans and horses.

Taxonomy

The taxonomist Carolus Linnaeus did not believe that species evolved, yet his classifications of organisms provided Darwin with persuasive evidence for evolution. Linnaeus did not intend for his hierarchical classification to be interpreted literally as a family tree. We now understand, however, that the careful anatomical comparisons that Linnaeus and his successors made usually reflect actual lines of descent (Figure 15-8).

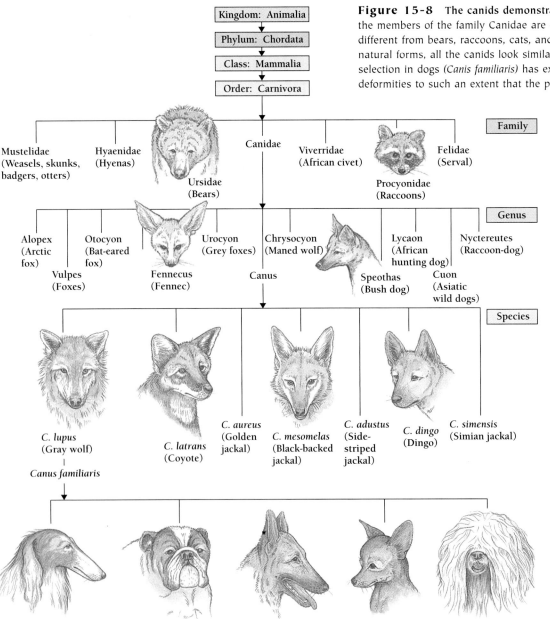

Figure 15-8 The canids demonstrate hidden variation. All the members of the family Canidae are distinctive—recognizably different from bears, raccoons, cats, and other carnivores. In their natural forms, all the canids look similarly doglike. Yet, artificial selection in dogs (*Canis familiaris*) has exaggerated certain traits and deformities to such an extent that the physical variation expressed by just five breeds surpasses that shown by all the other species in the genus *Canis* and even in the whole family Canidae. There are limits to the variation present in a single species, however. No dog could be mistaken for a bear, a cat, a hyena, or any of the other noncanid families in the order Carnivora.

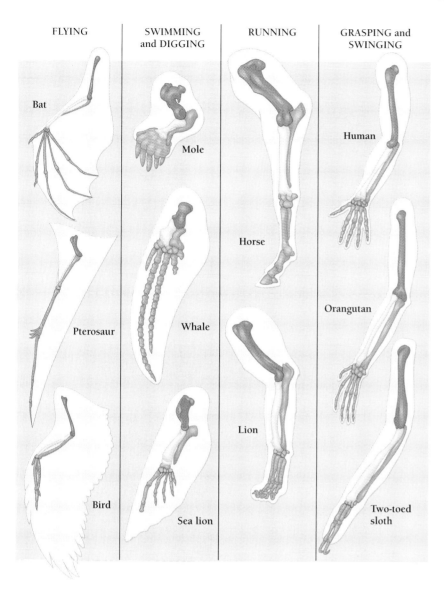

Figure 15-9 Homology in vertebrate limbs. The limbs of bats, moles, horses, and humans contain the same set of bones, modified, respectively, for flying, digging, running, and throwing. This limb homology extends to birds, reptiles, and amphibians.

Comparative Anatomy

Species descended from a common ancestor may evolve in quite different directions and yet retain many of the same characteristics. For example, consider the forelimbs of humans, horses, bats, moles, and whales (Figure 15-9). Each of the different forelimbs possesses the same bones arranged in the same way. But each animal uses its forelimb differently. Humans write, horses stride, bats fly, moles dig, and whales swim. Even though these various bones serve different functions today, they all arose from the same structures in a common ancestor. Similar structures, despite differences in function, imply common ancestry and, therefore, evolution.

Similar structures in two or more species are called **homologous structures.** Homologous structures don't have to perform different functions. For example, the bird, the bat, and the pterosaur evolved different arrangements of the same bones to perform the same function—flying. All these wings evolved separately from homologous structures, and each in its own way allows flight.

And some homologous structures are hidden unless you know what you are looking for. For example, the four tiny bones in the mammalian middle ear are homologous with much larger bones from the jaws of our ancestors (Figure 15-10).

The opposite of "homology" is "convergence." The wings of birds and of insects, for example, evolved from unrelated structures, yet they serve the same function and have a similar shape. All wings must be light and have broad surfaces. The similar wings of a dragonfly and a sandpiper are an example of *convergent evolution* (Figure 15-11).

Evolution is opportunistic. Just as a handyman can use a bit of wire here, a bit of glue there, to fix a bicycle or a light fixture, so evolution may employ a humerus bone here, a radius bone there, to allow one organism to gallop, another to fly.

Paleontologists can trace changes in homologous structures through the fossil record. For example, the first recognizable horse lived about 50 million years ago. The size of a large dog, *Hyracotherium* had four toes on its forefeet and three on its hind feet (Figure 15-12). Its teeth were too weak for it to chew the tough grasses that modern horses live on, so paleontologists believe that these ancestral horses lived on leaves, which are softer and more easily digested.

Subsequent horse fossils show a progressive lengthening of the legs, strengthening of the teeth, and reduction in the number of toes. In the foreleg of a modern horse, the lower bones

We can group species into hierarchies for the same reason that we can group individual humans into families and other hierarchies of relatedness. All organisms are related. We cannot do the same with elements or minerals, which are not related to one another (Chapter 19).

In Figure 15-8, we see that all the dogs, wolves, coyotes, and other species in the genus *Canis* look alike, just as members of the same family should. They are like siblings. The members of related genera—*Canis* and the foxes *Vulpes* and *Fennecus,* for example—are also similar, but not as much. They are like cousins. All the dogs, foxes, and wolves belong to the same family, the Canidae, for the members of the dog family are more closely related to one another than they are to the bears, cats, and other families in the order Carnivora.

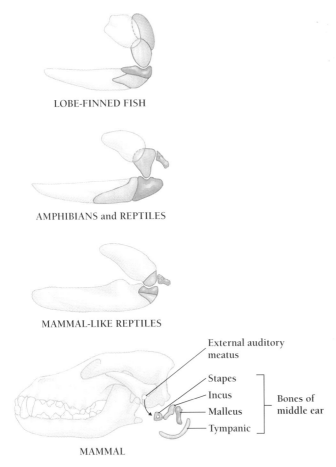

Figure 15-10 Evolution is opportunistic. The four tiny bones of the mammalian middle ear evolved from the jaw bones of early reptiles. The jaw bones evolved, in turn, from the branchial arches (structures that carry gills) of our jawless ancestors.

(radius and ulna) are fused and long. The last segment of a horse's leg is actually a long middle finger, the only one remaining of the five digits found in primitive ancestral mammals. These successive changes in the homologous bones give the modern horse longer legs and extraordinary fleetness. The many fossils that grade from the tiny, horselike animal *eohippus* to modern horse are called **intermediate forms**.

Some homologous structures have no apparent use. The modern horse has a single, useless side toe, called a splint. The splint is an example of a **vestigial structure** [Latin, *vestigium* = footprint], a part of an organism with little or no function but which had a function in an ancestral species. A vestigial structure is a remnant. Our own tailbone, invisible without an x-ray image, is a vestige of another way of life. Similarly, some snakes and whales have vestigial pelvic and leg bones, left over from ancestors that walked (Figure 15-13). Vestigial organs, like other homologous structures, are indicators of biological history and thus offer further evidence for evolution.

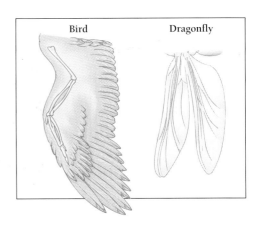

Figure 15-11 Convergent evolution in the wings of bird and dragonfly. Not all similarities in structure are due to homology. Birds and insects have evolved wings separately. The two kinds of wings look similar because flying exerts similar selective pressures on all flying organisms.

Even genes can be vestigial. Molecular biologists call vestigial genes "pseudogenes" because they do not code for functional proteins. Yet these corrupt genes persist, passed from generation to generation. Like a vestigial structure, a pseudogene is a vestige of a common ancestry.

Homologous structures, intermediate forms, and vestigial structures all suggest an evolutionary process whereby ancient structures are redesigned or abandoned.

Comparative Embryology

The early embryos of vertebrates are amazingly alike (Figure 15-14). For example, all vertebrate embryos, including humans, have tails and gill-like *branchial arches*. In fishes and amphibians, the tail develops into a swimming structure. Other groups develop the tail differently. Our own tail is vestigial.

Likewise, the branchial arches of fishes and amphibian embryos develop into gills for breathing under water. Amphibians that change into air-breathing adults lose their gills. In reptiles, birds, and mammals, the branchial arches develop into other structures in the adult ears, mouth, and respiratory tract. These include the bones of the ears and throat, some of the muscles of the face and neck, the parathyroid and thymus glands, as well as a major blood vessel of the lungs (Figure 15-15).

In embryos, the branchial groove or opening between branchial arches may open to form a gill slit (where water exits from the gills). These are obvious in sharks and tadpoles, for example. Amazingly, the opening between the first and second arches forms, in humans, a hole that forms the Eustachian tube leading from the outside of the ear inward to the eardrum (Figure 15-15). The outer hole, called the "external auditory

Figure 15-12 Evolution of the horse family. The fossil record for horses shows many intermediate forms, from the tiny Eocene horse that had four toes to the large modern horse that has one toe on each foot.

Figure 15-13 Vestigial structures. A. The blue whale has vestigial pelvic bones, remnants from the whale's ancestors' life as land animals with legs. B. The South African python's short claw is a vestigial leg. (B, *Anthony Bannister Photo Library*)

meatus," is homologous with a gill slit. The other gill slits normally close up during development, but in rare cases, human babies are born with gill slits in their necks (Figure 15-15). All of these structures can be traced to the embryonic pharyngeal pouches and unambiguously indicate our descent from fishes.

Developing organisms frequently pass through stages that resemble the organisms from which they evolved, a fact consistent with the theory of evolution.

Comparative Molecular Biology

Just as the morphology and embryology of organisms provide clues about their relatedness, so too does their molecular makeup. An organism's genes and the products of those genes, proteins, are a clear record of its lineage.

At the most fundamental level, all cells rely on the same molecular machinery. All cells use DNA to carry the genetic information; RNA, ribosomes, and the same genetic code to translate that information into proteins; the same 20 amino acids to build proteins; and ATP to carry energy. The univer-

Figure 15-14 Early embryos of a mouse (A) and a human (B). These two mammals are at about the same stage of development. The human is only about 4 mm long. Early embryos of vertebrates are remarkably alike, all displaying branchial arches, paddlelike hands and feet, and tails. (A, *courtesy of K. Sulik*; B, *Professor Hideo Nishimura*)

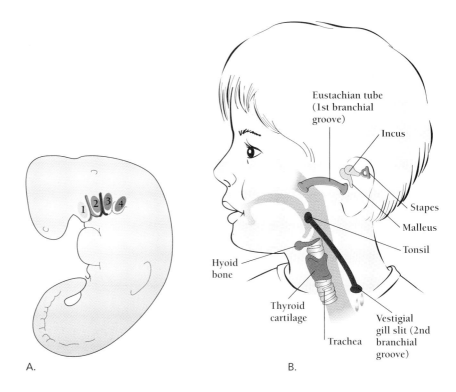

Figure 15-15 Gills in adult humans? Not quite. In some rare individuals, however, the second branchial groove (red) remains open even in adulthood. This opening from the throat to the outside of the neck appears in B as a vestigial gill slit. In all adult humans, the opening into the outer ear and the eustachian tube derive from the first gill slit (blue). In addition, the bones of the inner ear derive from the first and second branchial arches of the gills, as shown in A and B. Bones in the throat derive from the 3rd and 4th branchial arches.

sal homology of this system suggests that it is a common heritage, passed down from some ancient common ancestor.

Other cell proteins are nearly as universal. Organisms as different as yeasts and humans often use similar proteins. By comparing proteins, or the genes that code for the proteins, biologists can construct hierarchies of relatedness much like those derived from comparative anatomy. In general, the more closely related two species are, the greater the similarity between the amino acid sequences of all their proteins.

Today, biologists most often construct phylogenetic trees by comparing genes themselves, rather than proteins. But the results are similar. In fact, biologists can compare not only genes, but also the organization of chromosomes. Such comparisons show that less-related species have more chromosomal rearrangements. A cat genome, for example, differs from a human genome by about 13 rearrangements. These rearrangements occur rarely, and geneticists estimate that in mammals, one rearrangement occurs about every 10 million years.

One of the earliest proteins that biochemists analyzed was cytochrome *c*, a protein in the electron transport chain of the mitochondria (Chapter 6). Cytochrome *c* is a single polypeptide chain of 104 amino acids, whose exact sequence is known in dozens of species. In all of the species examined, 35 of the 104 amino acids are always identical. The remaining 69 amino acids can vary from species to species.

Not surprisingly, similar species have similar cytochrome *c*, while very different species have more changes. For example, humans and chimpanzees have exactly the same sequence of amino acids in cytochrome *c*, while humans and tuna differ by more than a hundred. Evolutionary biologists now routinely use data from such comparisons to construct **phylogenetic trees** (Figure 15-16).

The evolutionary history suggested by amino acid sequences usually agrees rather closely with the history implied by the fossil record. For example, cytochrome *c* sequences suggest that mammals and reptiles diverged from each other long before horses and donkeys did. And the fossil record shows that the first mammal-like reptiles emerged at least 180 million years ago, while the first horselike fossils didn't appear until about 36 million years ago. This is logical: groups of mammals, such as horses, can't diverge from each other if mammals haven't evolved yet. The fossil record, anatomical comparisons, and molecular comparisons are internally logical and independently confirm one another.

Comparative molecular biology has revealed that genes tend to change, or mutate, at a constant rate within certain groups of organisms. Molecular changes seem to accumulate in each type of protein, or in the corresponding piece of DNA, steadily, like the ticks of a *molecular clock*. Molecular clocks help researchers estimate when two species diverged even when no common ancestor exists in the fossil record. By comparing many genes, biologists can construct reasonably reliable phylogenetic trees.

But molecular clocks are imperfect, for individual genes evolve at different rates. And the same gene may evolve faster in one lineage than in another—faster in rodents, say, than in primates. Because genes evolve at different rates in different lineages, molecular clocks have limited value for measuring actual evolutionary time. Nonetheless, molecular clocks can tell us which groups diverged first, second, or third, if not exactly when.

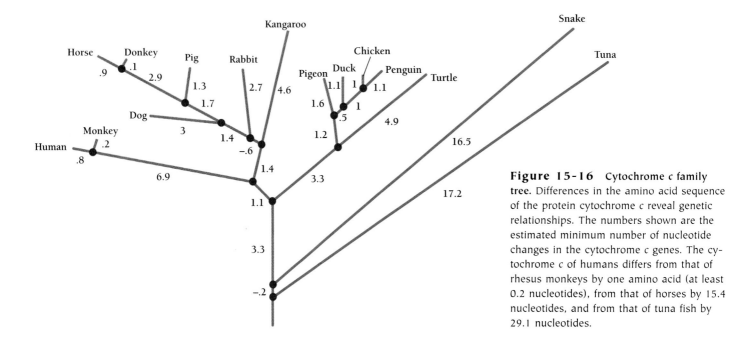

Figure 15-16 Cytochrome c family tree. Differences in the amino acid sequence of the protein cytochrome c reveal genetic relationships. The numbers shown are the estimated minimum number of nucleotide changes in the cytochrome c genes. The cytochrome c of humans differs from that of rhesus monkeys by one amino acid (at least 0.2 nucleotides), from that of horses by 15.4 nucleotides, and from that of tuna fish by 29.1 nucleotides.

The molecules of closely related organisms are more similar than the molecules of distantly related organisms. Comparative molecular biology confirms and elaborates on the story told by comparative anatomy and the fossil record.

Biogeography

When Darwin wrote *The Origin of Species,* most Europeans believed that every species had been individually created—a rhinoceros for this continent, a horse for that continent. If there were no tigers in North America, it was because God had not put any there. But Darwin saw that the distribution of many groups of species made no sense unless they had arisen from common ancestors.

From his studies of the distributions of species, called **biogeography,** Darwin concluded that each species has a definite region of origin, becomes distributed by migrating, and evolves descendant species in the regions to which it migrates. Darwin's view is so well accepted that today, if we ask someone why are there no tigers in North America, we will be told it's because tigers have no easy way to get from India and China across the Pacific Ocean. We all take it for granted that species move from place to place. But 19th-century biologists did not.

One of Darwin's strongest arguments for evolution was the close resemblance of island species to those on nearby continents. A 19th-century biologist would have expected that the closest relative of a finch from the Galápagos Islands, off the coast of South America, would be found in a similar group of islands, matched for climate and geology. But Darwin found that the closest relatives of island species were not on nearly identical islands halfway around the world, but right next door, usually on the nearest continent.

Species all over the world are most closely related to those that live nearby. Darwin's theory of common descent accounts for this pattern.

NATURAL SELECTION: DARWIN'S MECHANISM FOR EVOLUTION

Darwin made a strong case for the theory of evolution. And the evidence that scientists have accumulated since then has strengthened this theory. Indeed, the biologist Theodosius Dobzhansky summed up biologists' feelings best, saying, "Nothing in biology makes sense except in the light of evolution."

We can accept the idea that organisms have changed over time, and we can accept the idea that modern organisms are descended from ancient ones. But what of Darwin's proposed mechanism, natural selection? Darwin believed that natural selection, acting gradually over immense spans of time, created not only new species, but new genera, new families, and all the higher taxa. Why did he think that was possible?

Once Darwin conceived the idea of evolution by natural selection, he developed an obsessive interest in the work of gardeners, farmers, and stock breeders. It was perfectly apparent to him that many of the different breeds of dogs and pigeons, for example, would, if found in nature, be identified as separate species.

Continental Drift

Studies in biogeography reveal that species resemble nearby species more than species that live far away. However, similar organisms can occur in places that are remote from one another. For example, rheas, ostriches, and emus, all large flightless birds, occur in South America, Africa, and Australia, respectively. We may wonder if these birds are related to one another and, if so, how they came to be so widely dispersed.

The fossil record shows many examples of a similar pattern of dispersal, with closely related species in the southern continents. The distinctive fossil plant *Glossopteris*, for example, appears in similar rock formations from the late Paleozoic Era in both Brazil and South Africa. How could organisms that are so closely related live so far from one another?

In 1915, the German geologist Alfred Wegener, observing the jigsaw-puzzle fit of the African and South American coastlines, proposed the theory of "continental drift." Wegener argued that Africa and South America, as well as Australia, Antarctica, and India, were once a single supercontinent, which broke up into separate continents that drifted apart. Geologists rejected Wegener's theory, however, and biologists and paleontologists continued to puzzle over the similarities between the species on the southern continents.

Then, in the early 1960s, geophysicists discovered that the Earth's surface consists of large blocks, or **plates,** which move with respect to one another at the rate of a few centimeters a year. The discovery that both the continents and the sea floor were moving about on the surface of the Earth immediately revived Wegener's theory.

Geologists now understand that this movement comes from the continuous formation of new crust under the ocean. As lava emerges from below the surface, the new rock pushes apart the plates on either side. Powered by this pressure, plates can collide, scrape past one another, or dive beneath one another, pushing up mountains and precipitating earthquakes and volcanic activity. The Himalayas, for example, rise a little higher each year as the Indian-Australian plate crashes—ever so slowly—into the Eurasian plate. Similarly, the Pacific plate, beneath the Pacific Ocean, scrapes northward past the North American plate, creating California's earthquake-prone San Andreas Fault.

Geologists have been able to puzzle out many of the past movements of the continents by studying the magnetic orientation of rocks. This history explains many features of the fossil record and the distribution of modern species as well. About 250 to 200 million years ago, all of the continents were united into a supercontinent called Pangaea. Pangaea was surrounded by one huge ocean, Panthalassa (Figure A). In time, Pangaea began to break apart, allowing independent evolution to occur on the two resulting continents—*Laurasia* (Eurasia, Greenland, and North America) and *Gondwanaland* (India, Africa, South America, Madagascar, Antarctica, and Australia).

By the beginning of the Cretaceous, 135 million years ago, Gondwanaland had also begun to break into the separate continents. By about 65 million years ago, South America had split from Africa and Australia had separated from Antarctica, after which Australian species evolved in isolation. The land connection between North and South America formed only about 3 million years ago.

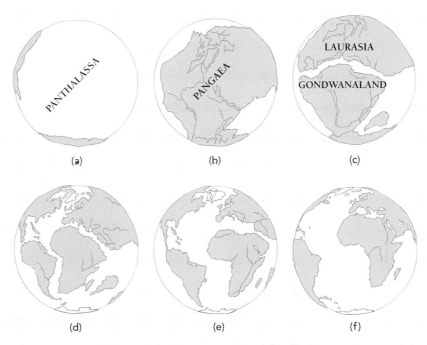

Figure A The history of the continents. (a) and (b) Alfred Wegener's view of the Earth as it existed 200 million years ago. At that time, there was one ocean (Panthalassa) and one continent (Pangaea). (c) By about 20 million years later, the supercontinent had begun to split into northern Laurasia and southern Gondwanaland. (d) Separation continued, and (e) looked this way about 65 million years ago. (f) Further widening of the Atlantic and northward migration of India brings the Earth to its present state.

Artificial Selection

Breeders derive such varied forms from a single species by carefully selecting and breeding only individuals that show certain desired traits most strongly. Under this selective regimen, each generation will show the selected trait more clearly than the last.

Dramatic examples of such intense selection come from all kinds of farm animals and crops. Most amazing are the differences among dogs (Figure 15-8). The English bulldog, for example, is the result of selection for a developmental deformity in which the upper jaw fails to develop at the same rate as the lower jaw. The lesson here is that nearly all species possess tremendous untapped genetic variation.

Darwin's particular interest was the work of Victorian pigeon fanciers. He collected many varieties of pigeons, studied the differences in their skeletons, and bombarded fanciers with questions (Figure 15-17). Ultimately, he concluded that all breeds of pigeons were descended from a single species—the European rock pigeon. Yet, the more than 20 different breeds were so different, Darwin argued, that a zoologist who discovered them in the wild would certainly classify them into a dozen or more species and quite possibly into half a dozen genera, as well.

Clearly, artificial selection by breeders could transform the characteristics of a breed. Success required only (1) that individual organisms vary in their characteristics, (2) that these variations be heritable, and (3) that the breeder consistently select certain traits in each generation.

Artificial selection does not result in the production of new species. At least new breeds are not recognized as species, no matter how distinctive they are. A mammologist from Mars wouldn't hesitate to classify a bulldog as a species separate from a chihuahua or a German shepherd. Morphologically they are distinct. But we know their history and that they can interbreed, so we classify them all as one species. Artificial selection demonstrates how much genetic variation exists in populations of organisms and how much change can occur in just a few years. More importantly, however, artificial selection hints at the much greater degree of change that might occur under the constant pressure of natural selection over millions of years.

Artificial selection provides a model for natural selection, a possible mechanism for evolution.

Darwin's Argument for Natural Selection

How can nature make selections in the same way as human breeders, generation after generation? In Malthus, Darwin found the answer. Malthus, recall, had argued that the production of food and other resources can never keep up with the enormous increases that human populations are capable of, that early death is the fate of a large proportion of individuals born in every generation.

Darwin saw that Malthus's arguments applied to populations of any species. Populations of organisms, although capable of enormous increases, do not continue to grow exponentially, but instead are limited by resources, predation, and disease. Natural selection must occur, Darwin argued, because nature cannot sustain unlimited exponential growth. To illustrate this point, he considered the descendants of a single pair of elephants, among the slowest breeders of all organisms. Even elephants, however, have the potential to breed geometrically. Over the course of 750 years, two elephants could have some 19 million descendants. Yet, historically, the elephant population has been fairly stable. So it is likely that two elephants in the year 1000 had only two descendants alive in the year 1750. The question Darwin posed was, Which two?

According to Darwin's theory, the survivors would be the descendants of whichever elephants were best adapted to the environment over the course of those 750 years. In artificial selection, the decisions of the breeder determine which individuals produce the next generation. In nature, the individuals that manage to survive and reproduce produce the next generation.

We can easily see how this argument might apply to the evolution of horses, for example. Individuals with long legs and teeth capable of chewing the tough grasses so common on the plains apparently survived and reproduced, whereas those with short legs and small teeth died out.

Darwin argued that if large changes can occur in domestic plants and animals over just a few generations, then even larger changes can certainly occur over the millions of years available for evolution.

The rate at which changes accumulate depends both on the intensity of selection and on the extent of inherited vari-

Figure 15-17 Artificial selection in pigeons. Charles Darwin collected different breeds of pigeons to study the variation possible within a single species. Shown here are a pair of baldheads, a carrier cock, two white fantails, two white pouters, and several other varieties. *(The Illustrated London News Picture Library)*

ation. For example, if only a few young horses from each generation survive, because most cannot escape a fast predator or cannot digest grass well enough, selection is strong and advantageous traits would accumulate rapidly. However, if most of the young horses from each generation survive and reproduce, then selection is weak and change occurs slowly or not at all.

The rate of change also depends on the extent of variation within the population. To take an extreme case, if all the individual horses of a generation were identical, then the next generation of horses could not differ from their parents. But, in fact, individuals vary enormously.

Natural Selection Analyzed

Modern evolutionary biologists have broken down natural selection into a series of basic facts and inferences.

1. Fact: **Superfecundity.** All species of organisms are so fertile that their populations would increase exponentially if all offspring survived.
2. Fact: **Not all offspring reproduce.** Populations of organisms do not attain these staggering proportions because many offspring succumb to disease, predation, and other fates before they reproduce. Those that do reach reproductive age find that the world does not contain enough food, territory, and other resources for all of them to reproduce.

    Inference: **Struggle for existence.** Whenever extinction does not occur, some individuals will escape death and disease and avail themselves of the limited resources more successfully than others. In the "struggle for existence," as Darwin called it, some individuals will survive to reproduce, while others will not.
3. Fact: **Individual variation.** Among sexually reproducing species, the individuals of a population differ. Most individuals are unique.
4. Fact: **Heredity.** Many of the differences between individuals are heritable—that is, individual differences can be passed on to offspring. Darwin knew nothing about genetics, but he could see, for example, that the short legs of the dachshund were inherited from its parents.

    Inference: **Adaptive traits may be passed on differentially.** Darwin argued that in the struggle for existence, individuals whose traits best suited them to their environment survive and reproduce more successfully than those less well suited to the environment. Adaptive traits that are heritable are more likely to be passed on to the next generation. For example, in the desert, rain comes infrequently. The faster a seed germinates when rain comes, the more water it can absorb as it forms a shoot and root. Seeds that germinate rapidly will survive and leave more offspring than seeds that germinate more slowly.

    Inference: **Adaptive traits accumulate within a population.** As a result of natural selection, various traits become more or less common in a population, and the overall character of the individuals within that population changes. For example, if the speed with which a certain desert plant can germinate is both heritable and adaptive, then, over time, more and more plants will germinate quickly. In this example, the genetic traits that allow a plant to germinate faster are *adaptive* traits. They are traits that allow it to survive long enough to reproduce.

Biologists accept the reality of both evolution and natural selection. The evidence that natural selection operates in natural populations is now irrefutable, yet some modern biologists question whether natural selection is the primary means by which new species and higher taxa such as genera and families are created. Over the years, biologists have proposed a host of alternatives to natural selection. A few biologists have argued that mechanisms other than natural selection may be responsible for macroevolution, the evolution of species, genera, and higher taxa of organisms. In the next chapter, we will examine natural selection more closely and also consider alternative mechanisms for evolution.

In addition, we will explore the genetic basis of variation. We can see that for natural selection to work, organisms must vary, and that the differences that distinguish individuals must be heritable. But Darwin himself knew nothing about *how* variation was inherited. The details of inheritance would not become clear until after 1900, with the rediscovery of Mendelian genetics and the advent of population genetics. Only in the 20th century did biologists begin to understand how genes, carrying the hereditary information, provide the basis for natural selection.

Study Outline with Key Terms

When Darwin published his book *On the Origin of Species* in 1859, he changed the way we view the world. The theory of **evolution** or **"descent with modification,"** contradicted **creationism,** the Bible's version of the story of creation, the philosophy of **essentialism,** and the Christian doctrine that the Earth was only 6000 years old.

By the middle of the 19th century, Hutton's theory of gradual geologic change and Lyell's **uniformitarianism** had persuaded the scientific community that the Earth was very old, and that change occurs slowly and gradually, not catastrophically. Examination of the fossil record by Cuvier and others suggested a very long history of life on Earth—a history far longer than the 6000 years implied by the Bible. In a theory that came to be called **catastrophism,** Cuvier suggested that fossil species were organisms that had perished in periodic catastrophes, each followed by a creation.

Linnaeus had created a system of naming organisms and a hierarchy that resembled a family tree. Biologists such as Buffon and Lamarck believed in evolution, and Lamarck formally introduced the idea. But Lamarck thought that species evolved in response to "felt needs" and proposed that evolution depended on the **inheritance of acquired characteristics.** Many European scientists accepted the inheritance of acquired characteristics, but rejected evolution.

In 1859, Charles Darwin published the influential *On the Origin of Species,* which defined evolution, and described a mechanism by which evolution could work, **natural selection.** Natural selection is the differential survival and reproduction of individuals with inherited traits that are **adaptive.** Darwin also believed that evolution occurs gradually through the slow accumulation of changes, an idea called **gradualism.**

Darwin argued that natural selection worked like **artificial selection.** In artificial selection, the breeder selects the varieties that will produce the next generation. In natural selection, the individuals who best survive and reproduce are those that will produce the next generation. Natural selection, like artificial selection, favors some forms over others.

Artificial selection suggested a mechanism for evolution, but other lines of evidence suggested evolution itself. One of the most persuasive was the fossil record. **Fossils** are the remains or imprints of past life. Most come from layers of sedimentary rocks, which form from the compression of settling mud, sand, and debris. The order of the layers (or strata) of sedimentary rock establishes the sequences in which ancient life evolved. The simplest organisms lie at the bottom of the record and through **intermediate forms** become both increasingly complex and increasingly like modern organisms, a pattern that suggests evolution. **Paleontologists,** scientists who study fossils, usually divide the history of life into four long eras—the **Precambrian,** the **Paleozoic,** the **Mesozoic,** and the **Cenozoic** eras. Each era in turn consists of several shorter divisions called **periods,** and, more recently, **epochs.**

Another line of evidence for evolution comes from **biogeography.** Species tend to resemble neighboring species in different habitats more than they resemble species that are in similar habitats but far away. This pattern of geographical distribution suggests that species evolved from one another.

Linnaeus's **taxonomy,** a hierarchical grouping of organisms by **species, genus, family, order, class,** and other **taxa,** suggests patterns of relatedness among different species, not separate creations. The relationships implied by taxonomy—so-called **phylogenies**—are derived from studies in **comparative anatomy.** Species that closely resemble one another are considered to be more closely related than species that do not resemble one another. **Homologous structures,** such as limb bones, provide additional evidence for evolution. **Vestigial structures,** such as the horse's splint, indicate evolutionary history.

Embryology provides another highly persuasive line of evidence for evolution. Embryos repeat the development of their ancestors to a limited but nonetheless remarkable degree. The early embryos of fish, amphibians, reptiles, birds, and mammals all have the tails and pharyngeal pouches. Comparative molecular biology confirms the lines of descent suggested by both comparative anatomy and the fossil record.

Natural selection, Darwin's mechanism for evolution, rests on several facts and inferences. All species are **superfecund,** capable of producing far more offspring than the world can support. Not all of these offspring survive long enough to reproduce. In each generation, the offspring of sexually reproducing organisms vary. That is, each individual is unique, and this uniqueness has a distinct heritable component. Those individuals that survive to reproduce determine which traits will predominate in the next generation. Such individuals can be said to be naturally selected.

Review and Thought Questions

Review Questions

1. What three ideas kept most scientists from seriously considering the idea of evolution until the 19th century?
2. Which of these three ideas did Cuvier's theory of catastrophism address? Describe the theory of catastrophism.
3. Describe the theory of uniformitarianism. In what two ways did it influence Darwin?
4. Who wrote the essay that inspired both Darwin and Wallace to invent the theory of natural selection? What did the essay say?
5. Explain why the fossil record is not complete. Describe two possible interpretations for the gaps in the fossil record.
6. During which era did mammals flourish? Were dinosaurs present at the same time?
7. How did Darwin's observations on the distribution of species suggest that species have evolved?

8. If you were a bat, which bones (that you have as a human) would you use for wings?
9. If you were a fish, instead of a human, how would you use the hole that in your human form lets sound into your ear?
10. Describe the molecular clock. What is its use?
11. Darwin's theory of natural selection can be divided into four facts and three inferences. What are they?

Thought Questions

12. The world population is currently about 6 billion. If every couple in the world had 20 surviving children, what would the world population be in two generations?
13. If the individuals of a species did not vary but were all identical genetically, what would be the consequences of natural selection?

About the Chapter-Opening Image

The orangutan represents our close relations with other primates, an idea that disturbs many nonbiologists.

On-line materials relating to this chapter are on the World Wide Web at **http://www.harcourtcollege.com/lifesci/aal2/**

Web Bit 15-1 Mary Anning: the First Woman Paleontologist

Web Links

Quizzing & Testing

16 Microevolution: How Does a Population Evolve?

A Fatal Disagreement

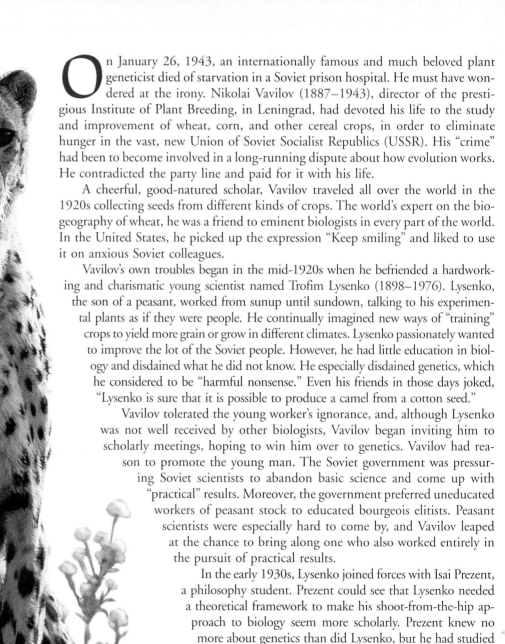

On January 26, 1943, an internationally famous and much beloved plant geneticist died of starvation in a Soviet prison hospital. He must have wondered at the irony. Nikolai Vavilov (1887–1943), director of the prestigious Institute of Plant Breeding, in Leningrad, had devoted his life to the study and improvement of wheat, corn, and other cereal crops, in order to eliminate hunger in the vast, new Union of Soviet Socialist Republics (USSR). His "crime" had been to become involved in a long-running dispute about how evolution works. He contradicted the party line and paid for it with his life.

A cheerful, good-natured scholar, Vavilov traveled all over the world in the 1920s collecting seeds from different kinds of crops. The world's expert on the biogeography of wheat, he was a friend to eminent biologists in every part of the world. In the United States, he picked up the expression "Keep smiling" and liked to use it on anxious Soviet colleagues.

Vavilov's own troubles began in the mid-1920s when he befriended a hardworking and charismatic young scientist named Trofim Lysenko (1898–1976). Lysenko, the son of a peasant, worked from sunup until sundown, talking to his experimental plants as if they were people. He continually imagined new ways of "training" crops to yield more grain or grow in different climates. Lysenko passionately wanted to improve the lot of the Soviet people. However, he had little education in biology and disdained what he did not know. He especially disdained genetics, which he considered to be "harmful nonsense." Even his friends in those days joked, "Lysenko is sure that it is possible to produce a camel from a cotton seed."

Vavilov tolerated the young worker's ignorance, and, although Lysenko was not well received by other biologists, Vavilov began inviting him to scholarly meetings, hoping to win him over to genetics. Vavilov had reason to promote the young man. The Soviet government was pressuring Soviet scientists to abandon basic science and come up with "practical" results. Moreover, the government preferred uneducated workers of peasant stock to educated bourgeois elitists. Peasant scientists were especially hard to come by, and Vavilov leaped at the chance to bring along one who also worked entirely in the pursuit of practical results.

In the early 1930s, Lysenko joined forces with Isai Prezent, a philosophy student. Prezent could see that Lysenko needed a theoretical framework to make his shoot-from-the-hip approach to biology seem more scholarly. Prezent knew no more about genetics than did Lysenko, but he had studied Lamarck's theory of the inheritance of acquired characteristics, finally in its death throes in the West. The theory perfectly supported Lysenko's approach to plant breeding, stating as it did that traits developed during the life of an organism in response to

Key Concepts

- Genetic variation is the fuel that drives evolution.
- Mutations in genes and chromosomes are the ultimate source of all genetic variation.
- Natural selection can bring about evolution by changing gene frequencies in a population.
- Natural selection is only one of several forces that can change gene frequencies in a population. Other forces are mating patterns, random drift, mutation, and breeding between populations.

the environment—long roots grown during a dry season, for example—could be passed on to the offspring. By then, modern genetics had supplanted Lamarckism, and few biologists in the world believed that acquired characteristics could be inherited. Yet, surprisingly, no one had actually disproved Lamarckism.

For this reason, and because Lysenko's peasant background and pragmatic philosophy were above criticism, government officials granted him more and more privileges and opportunities to carry out his research. Lysenko promised spectacular results with his techniques, claiming, for example, that he would double or triple the wheat crop by "training" wheat seedlings to grow in the far North. Grain would be superabundant, Lysenko promised. Soviet bureaucrats responded by meeting his every demand, and newspaper reporters fell all over themselves to interview the young genius.

No one seemed to notice that Lysenko's techniques consistently failed to increase harvest yields. When one idea failed, he came up with another. Each one's promise eclipsed the failures of the last. In the late 1930s, Lysenko was put in charge of all of Soviet agriculture. The whole nation became his "experimental garden," and every peasant farmer became a scientist. Once, Lysenko ordered tens of thousands of collective farms to test a theory about cross-pollinating different varieties of wheat. He requisitioned 800,000 pairs of tweezers for transferring pollen, one pair for each of the peasant farmers who would implement his "experiment."

Joseph Stalin, the totalitarian premier of the Soviet Union, was entranced by the enormous scale of Lysenko's work. Imagine! Eight hundred thousand tweezers! But Vavilov and his fellow biologists were horrified. They knew that cross-pollinating most of the wheat crop in the Soviet Union would eliminate many of the country's best varieties of wheat. And it did.

Despite Lysenko's failures, Lysenko and Prezent became more outspoken than ever in their disdain for genetics. Under their influence, Stalin agreed that genetics would no longer be taught at the universities, medical schools, or high schools. Increasingly, seminars at Vavilov's Institute of Plant Breeding, a world-class research center for genetics, deteriorated into confrontations between geneticists and Lysenko's student supporters. Lysenko came to regard Vavilov's persistent advocacy of genetics as a personal affront. In an era when millions of Soviet citizens were executed for far less than irritating a friend of Stalin's, Vavilov's defense of genetics was courageous but fatal.

In the summer of 1940, Vavilov was arrested for "wrecking" Soviet agriculture and sentenced to be shot. That sentence was later commuted to life in prison. But in the two and a half years he spent in prison, he was never moved to a work camp. Instead, along with thousands of other Soviet intellectuals, he died a slow and humiliating death in prison. In his last year, starving and wasted by dysentery, Vavilov and two other imprisoned professors passed the time by giving a series of lectures on history, biology, and forestry. Vavilov delivered more than a hundred hours of lectures. But his efforts to bring a bit of civilization to the dirty, crowded cells could not stave off the hungry prisoners who stole his bread and the slow death that finally took him.

Vavilov knew that by continually contradicting Lysenko he had signed his own death warrant. Yet he never renounced genetics. Why did the truth not prevail? If Lysenko was wrong in his belief in Lamarckian inheritance, why couldn't Vavilov and his fellow geneticists prove that inheritance was genetic?

Imagine! Eight hundred thousand tweezers!

The short answer to this question is that although geneticists of the 1920s and 1930s knew that genes were somehow passed from generation to generation, they could not prove beyond doubt that genes were real physical objects rather than just theoretical constructs. Nor could they prove Lamarck and Lysenko wrong in their belief that acquired characteristics somehow influenced the genes.

The proof that genes have a physical reality in DNA did not come until the 1950s—too late for Vavilov. Later work demonstrated that information goes from the genes to the body, not the other way around. In general, an individual's behavior—whether reaching for higher leaves, lifting weights, or tanning in the sun—cannot alter the genes it receives from its own parents and passes to its offspring. Nonetheless, actual disproof of Lamarck's theory of acquired characteristics has remained surprisingly difficult to pin down.

Lysenko's influence continued only a few years after Stalin died. In the early 1960s, disastrous crop failures forced the Soviet Union to buy wheat from the United States and Canada, a humiliation that irreparably weakened then Soviet premier Nikita Khrushchev's political power. In 1964, Soviet biologists rejected Lysenko's ideas and resumed the research in genetics that had been suspended for 30 years. Lysenko himself, however, retained his academic titles and his freedom until his death in 1976.

HOW ARE VARIANTS CREATED AND MAINTAINED?

We saw in Chapter 15 that evolution as an idea has a long and venerable history, dating back to the ancient Greeks. Yet even when biologists finally accepted that evolution must have occurred, they couldn't agree on how it could work. Darwin had proposed natural selection. But without an understanding of inheritance, Darwin's mechanism was incomplete. We know now that Darwin's theory of natural selection was, in fact, largely correct. But it took biologists more than 70 years of patient research and bitter dispute to appreciate Darwin's ingenious mechanism for evolution—natural selection.

In this chapter we will see how biologists gradually came to understand how organisms inherit traits. We will study variation in populations, the basis for Darwin's theory of natural selection, the sources of variation, and the means by which that variation is passed from generation to generation. We will examine, as well, some of the theories that have been proposed to account for the evolution of populations.

Why Did Charles Darwin Accept Blending Inheritance?

Darwin wrote the *Origin of Species* primarily to show that evolution had occurred and to discredit the idea that species were specially and individually created by God. His theory of natural selection was a separate idea—advanced to account for how evolution might have happened. Darwin thoroughly understood selection—the differential survival and reproduction of genetic variants. But he did not know where the variation came from.

The gene, as concept or fact, was completely unknown in 1859. Most scientists, including Darwin, believed that inheritance was "blended"—much as yellow and blue watercolors blend to make green (Figure 16-1). Biologists could see that when two races or species were crossed, the resulting offspring generally appeared to possess a smooth blending of the characters of each of

Figure 16-1 Blending inheritance or particulate inheritance? Nineteenth-century biologists believed that the hereditary material from each parent blended together in the offspring much as watercolors blend. Darwin suspected the truth: each parent's genetic contribution remains separate.

the parents. For example, a cross between a horse and a zebra produces an intermediate animal, called a "zebroid," not simply a horse with stripes or a zebra without stripes (Figure 16-2). Many 19th-century scientists assumed that the hereditary material, whatever it was, must blend smoothly also.

Yet Darwin intuitively suspected the true particulate nature of inheritance. As he wrote to Thomas Huxley in 1856, "I have lately been inclined to speculate, very crudely and in-

Figure 16-2 Zebroid. A zebroid is a hybrid of a horse and a zebra with some of the traits of each. How are its stripes not quite like those of a zebra? *(Left, Carl Purcell/Photo Researchers, Inc.; right, John Gerlach/Visuals Unlimited)*

distinctly, that propagation by true fertilisation will turn out to be a sort of mixture, and not true fusion, of two distinct individuals, or rather of innumerable individuals, as each parent has its parents and ancestors." However, Darwin had no evidence for this and did not dispute the theory of **blending inheritance** publicly.

In 1866, just a few years after Darwin's publication of the *Origin of Species,* Gregor Mendel published his groundbreaking research, which showed, among other things, the particulate nature of inheritance (Chapter 9). For evolutionary theory, Mendel's most important conclusion was that the individual units of inheritance, which we now call genes, remained intact and did not blend or fuse with one another. But neither Darwin nor any of his contemporaries ever read Mendel's paper.

Darwin's public acceptance of blending inheritance was unfortunate, for it left him open to a criticism for which he had no answer and led him into a second fundamental misunderstanding about the nature of inheritance.

Why Was the Idea of Blending Inheritance a Problem?

In 1867, Fleeming Jenkin, an enterprising Scottish engineer, announced that natural selection could never work in sexually reproducing organisms. Favorable variants, Jenkin argued, would not be able to pass on their advantages to the next generation. Instead, interbreeding would quickly "dilute" any new traits. His idea was that when a fast cheetah bred with a slow one, for example, the offspring would be only half as much faster as the fast parent. And *their* offspring would be only a quarter as much faster as the fast grandparent. Eventually, the increased speed would disappear altogether. How, he asked, could natural selection operate on variants that faded away as soon as they appeared? Assuming blending inheritance, he had a point.

Thanks to Mendel's work, however, we know that inheritance is particulate, not blending. Sexual reproduction does not necessarily dilute traits, but instead creates an assortment of different traits. In the cheetah example, the offspring would not all be exactly half intermediate in speed between the faster parent and the slower parent. Each young cheetah would be different, ranging from fast to slow, and each one's ability to reproduce would be correspondingly different.

Jenkin's logic was fine, but his grasp of biology was weak, and he was mistaken in nearly everything he said. Darwin could easily have demolished the engineer's argument on a variety of grounds. Yet Darwin was so uncertain about the nature of inheritance that, rather than counter the particulars of Jenkin's arguments, or question the basic assumption behind it—blending inheritance—Darwin made a second mistake. He took refuge in the inheritance of acquired characteristics, an idea that had been universally accepted for hundreds of years.

Darwin knew very well that individuals are unique. He knew that natural populations contain enormous reservoirs of variability on which natural selection can act. If Jenkin was right that variability was constantly blending away, where were new variants coming from? The **inheritance of acquired characteristics,** which Darwin accepted in principle, seemed to provide the only way for new variants to be continuously generated and passed on. According to the theory:

1. Changes in the environment create changes in the needs of organisms.
2. Changes in needs occasion changes in behavior to satisfy these needs.
3. Changes in behavior result in the increased use and development of certain parts, or the use and development of new parts.
4. These changes are passed on to the offspring.

After Jenkin's critique, Darwin began to concede a greater role to the inheritance of acquired characteristics. Darwin died in 1882, one year short of seeing the problem of inheritance solved by one of his most ardent followers.

Fleeming Jenkin proposed that blending inheritance and evolution by natural selection were incompatible. Darwin's acceptance of blending inheritance, as opposed to particulate inheritance, forced him to accept the theory of acquired characteristics.

HOW DID BIOLOGISTS COME TO REJECT THE THEORY OF ACQUIRED CHARACTERISTICS?

Of all the world's biologists, the German biologists were the most enthusiastic about Darwinism. Among them was a man named August Weismann (1834–1914), who became one of the first biologists to unequivocally reject the inheritance of acquired characteristics and the first to suggest a coherent theory of molecular genetics.

As a boy, Weismann happily collected butterflies, beetles, and plants. Like Darwin, he studied medicine, but then returned to natural science, devoting himself to the embryology of insects. Ultimately, an eye disease forced him to give up these studies, which required long hours at the microscope, and he turned increasingly to theoretical work.

Weismann soon realized—as Darwin apparently had not—that the controversy about natural selection could not be resolved until inheritance was fully understood. Like other biologists of the time, Weismann at first accepted the inheritance of acquired characteristics. But in the 1870s he began testing the theory in a series of experiments. By the time he was 49 years old he had completely changed his mind, and, in 1883, he published a paper rejecting Lamarckism entirely.

It was a radical proposal, and biologists of the day nearly universally repudiated it. Only Wallace, Darwin's codiscoverer of evolution through natural selection, publicly supported

Weismann. Wallace and Weismann alone believed that natural selection—without the inheritance of acquired characteristics—was the major mechanism for evolution. Other biologists had no reason to abandon the inheritance of acquired characteristics. Lamarckism was not, in fact, inconsistent with natural selection; it had not been disproved; and it constituted a likely source of variation. On the other hand, as Weismann saw it, no one had any idea how the inheritance of acquired characteristics could work, and Lamarckism was not necessary for natural selection to work.

Furthermore, Weismann had a better idea—another mechanism for inheritance. In 1882, a German cell biologist had discovered the cell's nucleus and noticed the way the chromosomes divide at mitosis. Weismann theorized that the genetic material was all in the nucleus. His theory, he said, was "founded upon the idea that heredity is brought about by the transmission from one generation to another of a substance with a definite chemical, and above all, molecular constitution." He was, of course, exactly right. But no one knew that, and not even he could be sure.

Weismann went on to argue that the hereditary material, once it is formed, is essentially isolated from the rest of the body. Embryologists had already noted that in vertebrate animals, at least, the tissue that eventually forms the ova or testes—the **germ plasm**—separates from the rest of the body early in development. In humans, for example, the cells that will later form the ova or testes appear early in the fourth week of development (Figure 16-3). The ovaries and testes themselves become distinctly formed at about four and a half months.

Weismann concluded that the germ plasm is separate from the rest of the body from the very beginning, and virtually nothing in the development or environment of the body can influence the germ cells or the hereditary material. The genetic material, he said, is passed from generation to generation intact, unblended, and unchanged. This theory came to be known as the "continuity of the germ plasm." Nothing could influence the germ plasm, the location of the hereditary material. And only the germ plasm was passed on to the offspring. Therefore, the inheritance of acquired characteristics must be impossible.

Modern biology has mainly confirmed Weismann's conclusion. A parent who works outside all day and develops a deep tan cannot pass this trait to his or her children. Tanning does not influence the molecular structure of the DNA in the mother's ova, which formed months before she was born, or even in the DNA of the father's sperm. Genetic information flows almost entirely from the DNA to RNA to protein.

August Weismann proposed that the hereditary material was in the nucleus and that it was molecular. He argued that the hereditary material was not susceptible to influence by developments in the rest of the organism and that, therefore, the inheritance of acquired characteristics was impossible.

Figure 16-3 Four-week-old human embryo. At four weeks, the germ plasm, from which the ova or testes will form, is already separate from the other tissues of the developing body. *(Petit Format/Nestle/Science Source/Photo Researchers, Inc.)*

How Did Weismann Explain Variation in Individuals?

Weismann's theory of inheritance ruled out Lamarckism. However, he still needed to explain the origin of the variation on which natural selection acts. Remarkably, he had an answer that was partly correct—sexual reproduction.

By 1883, cell biologists had observed that maternal and paternal chromosomes do not fuse during fertilization, but merely restore the two sets of chromosomes normally found in most cells. From this observation, Weismann leapt to the conclusion that the effect of sexual reproduction was not to blend parental characters (as blending inheritance implied) but to recombine separate "hereditary tendencies."

Weismann rightly concluded that sexual reproduction, far from homogenizing parental traits, as Jenkin had argued, brings together old traits into new combinations. Even though Mendel's work was still buried in obscurity and the word "gene" had yet to be coined, Weismann had, to a remarkable degree, solved the problem of heredity.

Weismann's view remained a minority opinion throughout the last part of the 19th century and into the beginning of the 20th century. But he and Wallace kept alive the idea of natural selection without Lamarckism. Without these two scientists, the idea might have been forgotten. Not until the 1940s did biologists appreciate Weismann's enormous contribution to evolutionary theory.

In the meantime, one major piece of the heredity puzzle remained to be discovered. While sexual reproduction results in an almost unlimited supply of new variants, the mixing of parental genes into new combinations does not account for the origin of new traits. With the rediscovery of Mendelian genetics in 1900, another piece of the puzzle fell into place.

> *Weismann showed how sexual reproduction, by creating new combinations of traits, could act as a source of new variants for natural selection.*

How Did the Rediscovery of Mendel's Work Transform Biology?

In 1900, three biologists working in three different countries independently rediscovered Mendel's laws. In the Netherlands, Hugo Marie de Vries (1848–1935) published a paper showing that if he crossed even numbers of hairy and smooth primroses he got 3:1 ratios of hairy and smooth plants. From this he concluded that the alternative traits—smooth and hairy—were controlled by a single pair of what we now call genes. Each parent, he explained, passes on only one of each of its pairs.

As soon as de Vries's paper came out, he immediately received copies of papers from two younger biologists—Karl Franz Joseph Correns (1864–1935), in Germany, and Erich Tschermak (1871–1962), in Austria—each claiming that they too had discovered the 3:1 ratios, each offering the same genetic explanation as de Vries, and each pointing out to him that Gregor Mendel had anticipated them all by 35 years.

Historians believe that both Correns and Tschermak probably read Mendel's paper before they had their respective "flashes of insight." De Vries himself received a copy of Mendel's paper from a fourth biologist sometime before he published his first paper. Whether there were actually any "independent discoveries" besides Mendel's we may never know. One thing is clear: Mendel's ideas made a splash.

In 1900, the British biologist William Bateson (1861–1926) was traveling to London to deliver a paper to the British Horticultural Society. As the train trundled through the English countryside, Bateson passed the time reading. When he got to de Vries's paper, he was thunderstruck. In London, he cast aside his prepared notes and proclaimed the death of Darwinism (by which he meant gradual evolution) and "the birth of a new science" (later called "genetics").

Bateson, de Vries, and others, who came to be known as the Mendelians, argued that evolution occurred not gradually, through natural selection, but in leaps and bounds, or saltations, by means of genetic mutations. These earliest geneticists firmly rejected the idea that evolution occurred through the slow selection of variants in a population. The Mendelians rejected natural selection and gradualism and any suggestion that evolution occurred among groups of individuals. They believed that evolution occurred suddenly, from one generation to the next, when genetic mutations transformed a single individual into a radically new type. They argued that genetic mutations in an individual were the cause of new species, not to mention new genera, families, and other higher groupings of plants and animals.

The Mendelians contended that every evolutionary change, large or small, was the result of a separate individual mutation, and that natural selection, individual variation, and sexual reproduction played no important role in evolution. Normal continuous individual variation, such as adult height in humans—the very foundation for Darwin's theory of natural selection—was not even heritable, argued the Mendelians.

Naturalists of the era were horrified. They knew, from their close observations of wild populations of plants and animals, that most variation was continuous and that continuous variation was indeed heritable. But they, in turn, overreacted. The naturalists claimed that although the rules of Mendelian inheritance might apply to discontinuously inherited traits such as smooth or hairy plants or blue or brown eyes, such traits were unusual and irrelevant to evolution. Continuous variation, the stuff of evolution, they argued, was not governed by the rules of Mendelian inheritance. Rejecting all of genetics, they clung to a variety of other theories, including Lamarckism. The polarized debate between the Mendelians and the naturalists raged for nearly 40 years.

> *The rediscovery of Mendelian genetics precipitated an intense debate. Early geneticists believed that evolution occurred from one generation to the next as a result of radical mutations in individuals. Biologists who studied natural populations of organisms insisted that Mendelian genetics did not apply to the kind of variation that was the basis of evolution. Both sides were wrong.*

How Did the Modern Synthesis Reconcile Genetics and Natural History?

In the late 1930s, a remarkable marriage of genetics and evolutionary theory occurred, which, for the first time, elucidated the genetic basis of variation and natural selection. This marriage, called the **modern synthesis,** did much to end the international feud between geneticists and naturalists. The modern synthesis fully embraced Weismann's theories and added to them all that was known about genetics. Much of what we will consider in the rest of this chapter comes from a field of genetics that contributed substantially to the modern synthesis—population genetics. Population genetics combined Mendelian genetics with the recognition that evolution can occur at the population level.

Population genetics explains the processes by which variation is generated and passed on within populations of organisms in precise mathematical terms. These processes are sometimes called **microevolution.** Population geneticists define microevolution narrowly as *changes in the frequencies of alleles of genes in a population.* For example, consider a wildflower population that includes some individuals with a gene for red flowers and other individuals with a gene for white flowers. If the relative proportions of red and white flowers change over the course of several generations, the population of flowers is said to evolve. Microevolution is distinct from **macroevolution**—the processes by which species and higher groupings (taxa) of organisms originate, change, and go extinct, a topic we will discuss in the next chapter.

Population genetics explains variation at the population level and provides a basis for natural selection and other evolutionary forces. In the next section, we will discuss what genetic variation is, where it comes from, and how it spreads and persists in populations of organisms.

WHAT IS THE SOURCE OF THE VARIATION THAT FUELS EVOLUTION?

Until the 1930s, few biologists believed that genes were real entities. No one had ever seen a gene, and there was no other evidence for their existence. Geneticists carefully stated that genes were purely theoretical constructs. Gradually, however, geneticists began to notice more and more cases where one genetic trait, such as eye color, was consistently linked to some other genetic trait, such as size. It began to look as if the genes themselves were physically linked together when they were passed from parent to offspring.

As we saw in Chapter 10, genes are very real, consisting of strings of nucleotides in the DNA. Each set of three nucleotides codes for one amino acid. Strings of amino acids, constructed on the ribosomes, fold into proteins.

Recall also that most sexually reproducing organisms have two of each chromosome. The two chromosomes in a pair are said to be **homologous.** One homologous chromosome possesses the same genes in the same positions as its homologue (unless they are unmatched sex chromosomes). Eggs and sperm have only one of each chromosome. When the egg and sperm fuse into a zygote, the two sets of chromosomes pair up. Thus each parent supplies one of the chromosomes in each pair of chromosomes.

Genetic variation originates when a gene mutates. The different forms of a single gene are called **alleles.** If each chromosome in a pair has the same allele at a particular gene position, the organism is said to be **homozygous** for that gene. If each chromosome has a different allele at a particular gene position, the organism is said to be **heterozygous** for that gene (Chapter 9).

What Is Genetic Variation?

Most genetic variability comes from gene mutations. Such mutations change only one or a few nucleotides in a gene and, thus, alter only a single protein. Chromosomal mutations, which change the number of chromosomes or the number or arrangement of genes on a chromosome, can affect a large number of proteins at once. Most chromosomal mutations are lethal or extremely damaging in animals, though less so in plants. But chromosomal rearrangements can, on occasion, benefit the organism. For example, the duplication of a chromosome segment, if harmless, can be passed on. In time, mutations in the new copy of the gene can allow it to take on new functions. As a result, the species has more genes than it did formerly.

In a population at large, a given gene may have just one form, or allele. In that case, every member of the population would have to be homozygous for that gene. On the other hand, a given gene may have two or even hundreds of alleles. If a gene had several alleles, members of the population could be either homozygous or heterozygous in one of several different ways. For example, members of a population possessing a gene with three alleles A, a, and a' could be homozygous in three ways—AA, aa, or $a'a'$—or heterozygous in three ways—Aa, Aa', or aa'. If a population has two or more alleles of a given gene, it is said to be **polymorphic** [Greek, *poly* = many + *morph* = form]. The word "polymorphic" also has a second, related meaning. If the members of a population come in two or more forms, the population is also said to be polymorphic. Humans, for example, are polymorphic for eye color (Figure 16-4).

Figure 16-4 Polymorphism. In many animal species, every individual has eyes of approximately the same color. Human eye color varies from pale blue to nearly black. Most of this variation derives from just two alleles of two genes. The four alleles can combine in nine ways.

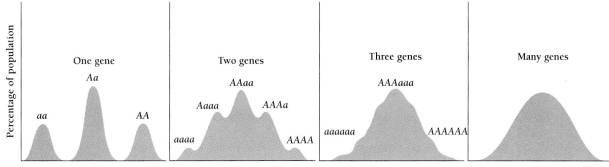

Figure 16-5 Most traits are polygenic. The more genes involved that influence a trait, the more continuously the trait will vary in a population. If only one gene with two alleles is involved, then no more than three basic types will appear. Two genes with two alleles each can result in five phenotypes (as in eye color). However, even those individuals with the same genotype differ from one another because of environmental influences. Most traits are influenced by many genes and display a smooth bell-shaped curve.

The original source of variation is gene mutations or chromosomal mutations. A population is polymorphic for a given gene if that gene has more than one allele.

Are Traits Determined by Single Genes?

The Human Genome Project's attempt to map the entire human genome has focused the public's attention on single genes that, when mutated, cause certain genetic diseases. A single-gene trait with just two alleles creates a pattern of variation like that in the left-hand panel of Figure 16-5. But most genes work in concert with other genes and under the influence of the environment to create complex phenotypes (Chapters 9 and 14). The same figure shows that the more genes are involved, the more a trait can vary.

In a research lab, most genetic research deals with alleles that specify discrete characters such as the presence or absence of a specific enzyme. In the natural world, however, most traits are **polygenic,** influenced by many genes. Everything about individual organisms—height and weight, bone structure, ability to withstand stress or infection, hormone levels, personality, and intelligence—is influenced by multitudes of genes. The constellation of traits we call personality, for example, is probably influenced by thousands of genes, as well as by environment. Height, weight, and skin color vary smoothly and continuously within a population, often forming a bell curve (Figure 16-6).

In Chapter 1, we discussed bell curves in more detail. But two aspects of such curves deserve special attention—the average value, or mean, and the breadth of the curve. The breadth of the curve is a measure of the variability of the trait. We all

Figure 16-6 Height is a polygenic trait. Because adult height in humans is controlled by many genes, the distribution of heights in adults is a smooth curve, as illustrated by this photo of students. The students are arranged according to height, with the shortest on the left and the tallest on the right. Most students are of average height, with very few at the extremes. *(Photo, courtesy of Joiner Associates, Inc.)*

Figure 16-7 Even variation varies. Heights for male African Pygmies, male American Bostonians, and male African Dinkas. Nearly all Pygmies are shorter than Bostonians and Dinkas. Dinkas are, on average, taller than Bostonians, but they have a great range of heights, so that many Dinkas are actually shorter than many Bostonians. Which group varies most? (Pygmies, Klaus Payson/Peter Arnold, Inc.; Bostonians, Frank Siteman/Stock, Boston; Dinkas, F. Jackson/Bruce Coleman, Inc.)

know that the average value of a trait may vary from population to population. African Pygmies are, on average, shorter than people living in Boston (Figure 16-7). But populations may also vary in the variability of a trait. The population of African Dinkas pictured in Figure 16-7, for example, shows a greater range of heights than that of either the Pygmies or the Bostonians. An adult Dinka can be taller than almost any Bostonian, or as short as the tallest Pygmies. Genetic variation, the fuel of evolution, varies from population to population. We should not forget, however, that we cannot tell from the graph if the Dinkas' great variability is all due to genetic variation or also to environmental effects such as diet.

Most traits are influenced by many genes.

What Determines the Genetic Variability of a Population?

Populations that contain many alleles vary more than populations that have few alleles per gene. Highly variable populations can evolve more rapidly than more uniform populations because natural selection has more to select from. Three factors determine genetic variability:

1. How fast mutations accumulate in the DNA
2. How fast such mutations spread through a population
3. How fast selection eliminates mutations from a population, which we discuss later in this chapter

How often do mutations occur?

Mutations cause random changes in protein structure and, thus, most mutations damage the organism. Occasionally, however, random changes may cause no harm or may even give an advantage to the organism that bears them. Such rare advantageous mutations are the fuel of evolution. Without a constant supply, evolution would stop. Instead, mistakes and mutations persist and occasionally provide every species with adaptive variants. For example, on average, each new human baby carries one or two new mutations.

Mutations are not the result of directed molecular surgery but arise constantly as the result of random processes. Because they occur constantly and randomly in all genes, geneticists have been able to estimate the probability that a given nucleotide will mutate in each generation. The allele for eyelessness in the fruit fly *Drosophila melanogaster* occurs in 6 of every 100,000 fly gametes. Because mutation occurs constantly, and because the total number of genes is great (about 100,000 for humans, for example), the total number of variations within a species can be enormous.

How do mutations spread through a population?

The rate at which a population evolves depends on how rapidly new mutations spread through the population. In fact, although evolution could not long continue without new mutations, the spread of those mutations, rather than their original occurrence, is the most important factor in increasing variability in a population. The main way for mutations to spread is through sexual reproduction. Recombination, which occurs during sexual reproduction, mixes mutations that have arisen independently into endless new combinations. Differential reproduction determines which mutated alleles increase in frequency and which decrease.

Genetic variation originates when genes or chromosomes mutate. Genetic variation spreads when sexual reproduction and recombination mix these changes into endless new combinations.

How Much Do Populations Vary Genetically?

Since variability is the raw material for selection, one of the first questions population geneticists ask is How much variation is there in single genes within a population?

To examine the extent of variation of single genes, we can look at the polypeptides and proteins that are the products of genes or at the genes (DNA) themselves. One way of estimating the level of variation within proteins is electrophoresis, a technique for separating molecules according to their charge, size, and shape (Chapter 10). Using electrophoresis, researchers estimate the degree of genetic polymorphism in humans and other species.

Electrophoresis shows that in populations of the fruit fly *Drosophila*, for example, about 53 percent of genes that code for proteins are polymorphic. In a survey of invertebrate animals, nearly half of all proteins examined had more than one form in the population at large.

Another way of expressing the level of genetic variation in a population is to estimate the likelihood that any individual will be heterozygous for a given gene. Any two fruit flies, for example, are likely to differ at about 25 percent of their gene loci. The cheetah, of Africa, has some of the lowest variability found in any animal, with an average heterozygosity of only 0.07 percent. Cheetahs have most likely descended from a small number of individuals, called a "bottleneck." For comparison, invertebrates average 13 percent, and plants are even more diverse (17 percent).

Humans fall somewhere in the middle. In us, about 7 percent of genes that code for proteins are heterozygous. Interestingly, humans appear to possess much less genetic diversity than our closest relatives, the chimpanzees. A 1999 study of a noncoding region of the X chromosome revealed that common chimpanzees possess four times as much variability in this region as do humans. The researchers suggested that humans, like cheetahs, may have passed through a bottleneck at some time in their evolutionary history.

Most natural populations vary enormously.

Is All Genetic Variation Subject to Natural Selection?

The only variants important for natural selection are those that contribute to changes in the whole individual in ways that affect the individual's ability to survive and reproduce. The expressed traits of a whole individual organism are called the **phenotype** [Greek, *phainein* = show]. The phenotype, which includes the appearance, the physiology, and the biochemical makeup, is a result of both the genetic makeup of the individual—its **genotype**—and the effects of the environment on the development of the individual.

Selection acts on the phenotype. For example, an individual homozygous for some lethal, recessive trait, such as cystic fibrosis, expresses the trait and dies. The gene has expressed itself in the phenotype, and selection has occurred. Another individual, heterozygous for the trait, will have only one copy of the deadly allele and will have a phenotype that is perfectly healthy. In the heterozygote, the allele is not expressed in the phenotype, and so selection against the allele cannot operate even though the allele is present in the genotype. Most lethal mutations are recessive and are usually not subject to selection unless they are homozygous.

With the advent of high-speed DNA sequencing machines, evolutionary biologists have begun to accumulate more information on the extent of variation in DNA. On average, about 1 nucleotide in every 500 is different in the two homologous chromosomes. Most of these differences, however, lie in nucleotides that do not code for proteins. They contribute to high variability at the DNA level but do not alter the structure of a protein or change the way the organism functions. Such mutations do not affect phenotype and are therefore not subject to natural selection.

Mutations that do not affect the phenotype are not subject to selection.

Why Is Genetic Variation Essential to Evolution?

If mutations are the fuel of evolution and recombination is its motor, then the driver—the guiding force—of evolution is natural selection. Because some variants survive and reproduce more successfully than others, the mutations that best equip their possessors to survive and reproduce will increase in frequency in each generation.

When biologists study genetic variation within a species, they almost always find lots. In some exceptional cases, however, genetic variation is limited. For example, most thoroughbred horses are direct descendants of a few dozen horses that lived in the late 18th century. Although breeders have allowed some outbreeding—breeding with nonthoroughbreds—the relative lack of genetic variability has led to stagnation. Despite the best efforts of highly skilled breeders, the winning times for the Kentucky Derby have not changed significantly in 50 years. The lack of change in winning Derby times could be considered an example of microevolution that has come to a standstill. Later in this chapter we will discuss some examples of natural populations that lack genetic diversity.

Evolution by natural selection depends on the continued existence of variation. Yet selection tends to reduce variation by eliminating organisms less able to survive and reproduce. We might expect, then, that species would evolve so that all individuals will have the same "best" genotype. If this were to happen, evolution of that species could not continue. (However, environments change constantly, and so do selection pressures. What is "selected" changes over time.)

In the absence of genetic variation, evolution would cease.

In the rest of this chapter, we will see how populations maintain variability, even as their gene distributions change during evolution. Such variability allows populations to adapt to continuing changes in their environment.

How Do Populations Maintain Genetic Equilibrium?

In 1908, G. Hardy, an English mathematician, and G. Weinberg, a German physician, independently reconsidered Fleeming Jenkin's idea that sexual reproduction eliminates variation in light of the fresh knowledge that heredity is nonblending. Recall that Mendel's Principle of Segregation says that alleles do not blend with other alleles; they stay intact. Hardy and Weinberg each reached the same conclusion, now called the **Hardy-Weinberg Principle:** *Sexual reproduction by itself does not change the frequencies of alleles within a population.* Mendelian genetics, argued Hardy and Weinberg, overturned Jenkin's case against natural selection.

To see what this means, we will need to define some terms. We speak of all the genes of all the individuals in a population as the **gene pool.** And we calculate the *frequency* of a given allele within the gene pool by dividing the number of times the allele is present in a population by the number of times it could be present if every organism in the population had the allele on both of the paired homologous chromosomes. In that case, the allele would be present on every single chromosome on which the gene appears. *The allele frequency is the number of times the allele is present in the population divided by the total number of chromosomes on which the gene appears.*

To understand the Hardy-Weinberg Principle, let us consider a gene that has two alleles, A and a. For example, the color of mussels on the north coast of California depends on a single gene with two alleles. One allele, A, gives the mussels a blue color and is *dominant*. The other allele, a, makes the mussels brown and is *recessive*. That is, whether the mussel is AA or Aa, it will be blue. Only when the mussel is aa will it be brown.

The fraction of sperm and ova in the population that carries the A allele we call "p," and the fraction that carries the a allele we call "q." The fraction of the blue allele (A) in a particular mussel population is $p = 0.6$, while the fraction of the brown allele (a) is $q = 0.4$. Since there are only two alleles for this gene, $p + q = 1$. Geneticists call p and q allele "frequencies." Thus the A allele has a frequency of 0.6 and the a allele has a frequency of 0.4.

But the frequencies of the alleles are not the same as the frequencies of the genotypes. Suppose the mussels release their eggs and sperm into the seawater. We know that 60 percent will carry the A allele and 40 percent will carry the a allele. Now the sperm and eggs begin joining up at random and fusing into zygotes with genotypes AA, Aa, and aa. What are the frequencies of those genotypes? We can calculate the frequencies of the three genotypes among the offspring, or F1 generation, using either a Punnett square, as described in Chapter 9, or simple math.

If the frequency with which a sperm or egg carries the A allele is 0.6, then the frequency of an AA homozygote will be the probability that an A sperm will fertilize an A egg. That probability is $p \times p = p^2 = 0.36$. Similarly, the frequency of an aa homozygote is $q^2 = 0.16$. The frequency of a heterozygote is the probability that an A sperm will fertilize an a egg plus the probability that an a sperm will fertilize an A egg, that is:

$$(pq) + (qp) = 2pq = 2(0.4)(0.6) = 0.48$$

The frequencies of AA, aa, and Aa are 0.36, 0.16, and 0.48, respectively. Notice that the sum of the frequencies is $p^2 + 2pq + q^2 = (p + q)^2 = 1^2 = 1$. Likewise, $0.36 + 0.48 + 0.16 = 1$. In other words, we have accounted for 100 percent of the population.

From the genotype frequencies we can derive the allele frequencies in the F2 generation and see if they are the same as in the previous generation, in short, $p = 0.6$ and $q = 0.4$.

First, let's assume that we have 100 individuals, each with 2 alleles, for a total of 200 alleles in the gene pool. What fraction of these 200 are the A allele? Each of the AA homozygotes carries two copies of A, so they contribute $2p^2$ of the A alleles. That's $2(0.36) = 0.72$. So, among the 100 mussel offspring are 72 A alleles from homozygotes. In addition, each of the Aa heterozygotes also carries a copy of an A allele, so they contribute $2pq$, or 0.48, A alleles. That's 48 more A alleles. In Table 16-1, notice that we can derive the frequency for the a allele in much the same way. We see that despite sexual reproduction, the allele frequencies have not changed. In successive generations, the frequencies for the genotypes AA, Aa, and aa—$p^2 + 2pq + q^2$—will also remain the same.

TABLE 16-1 THE ESSENCE OF HARDY–WEINBERG

The frequency of the A allele remains p and the frequency of the a allele remains q. Generally, then, sexual reproduction does not change allele frequencies. The total number of A alleles is $2p^2 + 2pq$. Because there are 2 chromosomes, the *frequency* of the A alleles is this number divided by two, or $(2p^2 + 2pq)/2 = p^2 + pq = p(p + q)$. Remember, however, that $p + q = 1$, so $p(p + q) = p$. So if a population starts with gene frequencies of $p = 0.6$ and $q = 0.4$, the next generation will have the same frequencies.

Genotype	AA	Aa	aa
Frequency	p^2 0.36	$2pq$ 0.48	q^2 0.16
Number of individuals with that genotype in the population	36	48	16
Total alleles	72A	48A 48a	32a

A alleles = 72 + 48 = 120
a alleles = 32 + 48 = 80
Number of individuals is 100.
Total alleles is 200.
So A alleles have a frequency of 120/200 = 0.6
And a alleles have a frequency of 80/200 = 0.4

So far, we have considered only two alleles for each gene, but we know that mutations are possible at every nucleotide of DNA. The number of possible alleles for each gene is large indeed. We could easily accommodate more alleles in our discussion. We might, for example, designate the frequencies of three alleles (A, a, and a') as p, q, and r, so that $p + q + r = 1$. We would find that the Hardy-Weinberg Principle remains the same. The frequencies of the individual alleles in a population do not change as a result of sexual reproduction.

The frequencies of each allele therefore stay constant in successive generations, and sexual reproduction does not lead to the dilution of variation, as Fleeming Jenkin suggested. Equally important, genotype frequencies stay at the same **Hardy-Weinberg equilibrium,** a stable distribution of genotype frequencies maintained by a population from generation to generation.

Such an equilibrium occurs, however, only when the population meets all of the following five conditions:

1. Random mating
2. Large population size
3. No mutations
4. No breeding with other populations
5. No selection

In reality, these conditions are rarely met. Consequently, gene frequencies change and evolution occurs. If a population is not at equilibrium—that is, if the genotype frequencies are changing from generation to generation—then we know that one or more of the five conditions are not being met.

The Hardy-Weinberg Principle shows that sexual reproduction does not lead to the dilution of variation but maintains a stable distribution of genotype frequencies from generation to generation, called the Hardy-Weinberg equilibrium. The Hardy-Weinberg equilibrium is a baseline against which the evolution of populations can be measured and is the foundation for the genetic theory of evolution.

WHAT CAUSES ALLELE FREQUENCIES TO CHANGE?

Of all the causes of changing allele and genotype frequencies—nonrandom mating, small population size, mutations, breeding between populations (gene flow), and natural selection—only natural selection consistently leads to *adaptive* changes in allele frequencies (Figure 16-8). The others tend to cause changes in allele frequencies that are not directly related to the ecology of the population and may or may not be adaptive. These effects are said to be "nonadaptive."

How Can New Mutations Cause Changes in Allele Frequencies?

Slight changes in allele frequencies also occur as the result of mutation. As we saw in the last section, mutation can generate every possible change in the amino acid sequence of each

Figure 16-8 What causes changes in gene frequency? A. Random drift. B. New mutation. C. Nonrandom mating. D. Gene flow. E. Natural selection.

protein during the lifetime of a species. Indeed, mutations provide the raw material of evolution. How do such mutations affect the genetic equilibrium of a population?

The answer is, very little. Random drift—given enough time—can lead to either the disappearance of a new mutation or its spread through the population. However, the number of new mutations in each generation does not much change allele frequencies. Although the rate of mutation is high enough that each new human baby carries one or two new mutations, the change in the frequency of any single allele is low.

Mutation barely alters the overall frequency of any given allele.

How Do Chance Events Change Gene Frequencies?

Flip a coin 1000 times and your chance of getting all heads is virtually zero. You can reasonably expect that the number of heads will be close to 500, the "expected" number. However, if you flip the coin only four times, your chance of getting all heads is rather high (1 in 16), and you're not likely to get heads exactly one-half of the time. In fact, your chance of getting two heads is less than 40 percent. Such chance fluctuations, which affect small samples much more than large ones, are known as "sampling errors." In small populations of organisms, sampling errors can cause allele frequencies to change randomly from generation to generation.

Figure 16-9 A classic bottleneck. In the 19th century, heavy hunting reduced the number of elephant seals to as few as 20 individuals. Legally protected since 1884, the huge seals are now flourishing, but they have become highly inbred. *(Frank S. Balthis Photography)*

For example, in a small population of mussels, the actual proportion of eggs and sperm with the A allele may differ widely from 0.6. The smaller the population, the greater the likelihood that the frequency of various alleles will deviate from what they are "supposed" to be. The effects of chance in small populations can lead to **random drift**—changes in gene frequency not due to selection, mutation, or immigration, but due to random events. Such random events can lead to changes in gene frequency within a small population.

The founder effect

In small populations, random drift can have pronounced effects. In one kind of random drift, called the **founder effect**, a whole population descends from a small number of individuals. For example, in the early 19th century, elephant seals flourished along the California coast from Point Reyes, just north of San Francisco Bay, south to Baja California. By 1890, hunters had virtually eliminated the entire population (Figure 16-9). Best estimates are that the population was reduced to between 20 and 100 individuals. Under federal protection, the elephant seal population now numbers some 100,000 animals. All of these animals have descended from the same few individuals. When, as in this case, a large population is reduced to a few individuals by some disaster, population biologists say the population has passed through a **bottleneck**.

The founder effect produces a population whose gene pool is a tiny sample of the original. In this new population, the frequency of an allele is likely to be quite different from that in the original population. Many alleles will not be represented at all. Since the allele frequencies have changed, the population can be said to have evolved in the absence of natural selection.

Founder effects in the human population—together with the inbreeding and increased homozygosity that occur in small populations—are thought to be responsible for the high incidence of genetic diseases in certain human populations. For example, the founder effect is responsible for the high frequency of otherwise rare alleles among Afrikaaners of South Africa (descended from about 30 Dutch families in the 17th century), the Old Order Amish in the United States (descended from only a few individuals in the 18th century), and Swedes in Lapland (descended from three families in the 18th century).

How can random drift spread new mutations?

Random drift can play an important role in the spread of new mutations. Mathematicians have shown that, even in the absence of selection, a new mutation will either disappear completely or spread through the population. The smaller the population, the less time it takes for either to happen. If a new mutation enters the gene pool, it will at first exist in only a few individuals and will have a significant chance of being eliminated. For example, imagine a population of 25 elephants, 3 of which have a new allele for hairy ears. A volcano that erupts every 10,000 years happens to eliminate the three hairy-eared elephants. Their elimination is not selection, since smooth ears do not protect elephants from volcanoes. It's just chance. Yet, the result is the same as if natural selection had eliminated the hairy-eared elephants—a population of smooth-eared elephants.

Does random drift affect large populations?

The role of random drift in speciation in large populations is not so clear cut. However, even in large populations, random drift will *tend* to cause alleles to disappear over long periods of time. In the absence of natural selection, new mutations, recombination, and migration, random drift would cause the frequencies of all alleles in even large populations to go to either 0 percent or 100 percent. It would happen slowly, but it would happen eventually, leading, ultimately, to a population that was completely homozygous. In such a population, evolution would cease. We can see that new mutations "restock" genetic variation, and the constant mixing of such mutations, within populations and between populations, is key to evolution.

In small, isolated populations, random drift can cause significant changes in gene frequencies. In large populations, drift operates so slowly that its importance may be small compared to the effects of mutation, gene flow, and selection.

How Does Nonrandom Mating Cause Changes in Allele Frequencies?

In deriving the Hardy–Weinberg Principle, we assumed that individuals of one genotype do not mate preferentially, either with their own genotype or with others. Our calculation would be invalid, for example, if blue mussels mated only with other blue mussels and not with brown mussels. In fact, mussels mate by releasing into seawater gametes that then come together essentially at random. The same is true of most plants and many other animals.

But mammals and birds usually choose their mates very carefully. Such nonrandom mating, called **assortative mating,** occurs when individuals choose mates on the basis of the mate's genotype (as reflected in the phenotype). Such choices may be made independent of the individual's own genotype. For example, regardless of their own appearance, humans generally show a preference for mates with average facial structure. Other mammals and birds also prefer mates with a high degree of symmetry.

Individuals may also choose mates on the basis of similarities (or dissimilarities) to the individual's own genotype. Humans show positive assortative mating with respect to skin color and height. That is, on average, they mate preferentially with those of similar height and skin color. Positive assortative mating tends to increase the proportion of homozygotes.

An extreme form of positive assortative mating is **inbreeding,** mating among close relatives, which greatly increases the proportion of homozygotes. For example, Ellis–van Creveld syndrome is a genetic defect frequent among the Amish because of 200 years of inbreeding. Recessive genes may have negative effects that are expressed only in homozygotes. In sexual species, at least, inbred individuals who are homozygous for many genes are, on average, less healthy than individuals with greater heterozygosity.

Neither assortative mating nor inbreeding changes overall allele frequencies. However, any increase in the number of homozygotes facilitates change by natural selection by exposing alleles to selection.

Most species are **outbreeders,** with complex physical or behavioral adaptations that promote crossbreeding with usually not-too-related individuals. That is, individuals choose mates somewhat unlike themselves, a process called negative assortative mating.

How Does Breeding Between Populations Change Allele Frequencies?

Hardy-Weinberg equilibrium requires that a population not interbreed with other populations. But few populations are completely isolated from all others of the same species. Animals immigrate to nearby populations, taking new alleles with them. Plant pollen, carried into a population by wind or insects, introduces new alleles. In either case, the resulting **gene flow** causes changes in allele frequencies. In our own species, we have many obvious examples of gene flow as people move to new countries. Despite a general tendency to find a mate within one's own group, gene flow among different groups is common (Figure 16-10).

Gene flow has two major effects on allele frequencies. Within a given local population, gene flow from outside populations increases genetic variation. The same gene flow makes adjacent populations more like one another.

Gene flow increases variation within a local population but makes adjacent populations more alike.

How Does Selection Decrease the Frequency of Certain Traits?

> A group inept
> Might better opt
> To be adept
> And adopt
> Ways more apt
> To wit, adapt.
> —John M. Burns

A large, isolated population whose members mated randomly and whose mutation rate was low would maintain constant allele frequencies—except for one thing: **natural selection.** Because natural selection reduces the frequency of some alleles and increases the frequency of others, it is a powerful way to change gene frequencies (Figure 16-11).

quency of the *a* allele in the population will decrease by q^2 in each generation. When the lethal *a* allele is present at relatively high frequencies, many individuals are eliminated in each generation. When the allele is present at low frequencies, however, selection eliminates very few individuals. For example, if the frequency (q) of *a* is 0.5, then 25 percent of the population will die young each generation. But if $q = 0.01$, then only 1 infant in 10,000 will die from having the *aa* genotype, and only $2(0.01)(0.99) = 2$ percent of the population will carry the gene. Thus, while selection can dramatically change allele frequencies, even the most lethal recessive alleles can still persist indefinitely as heterozygotes.

Nevertheless, even a small difference in the probability of reproduction can lead, over time, to the effective elimination of a deleterious allele. Suppose, for example, that individuals with the homozygous *aa* genotype are slightly less well adapted than individuals with either *AA* or *Aa* genotypes. Say that each *aa* individual leaves only 999 progeny for every 1000 of the *AA* or *Aa* individuals. Even if we start with a gene frequency for the *A* allele of only 0.00001, we can predict a steady increase in the frequency of the *A* allele, until after 23,400 generations its frequency becomes 0.99. For the fruit fly *Drosophila*, this could occur in less than 500 years. For species with generation times as long as those of humans, it would

Figure 16-10 Gene flow. Humans travel singly or in groups, spreading their customs and genes. In November 1996, hundreds of thousands of Rwandan refugees began returning home after living in Zaire for months. *(Christopher Morris/Black Star)*

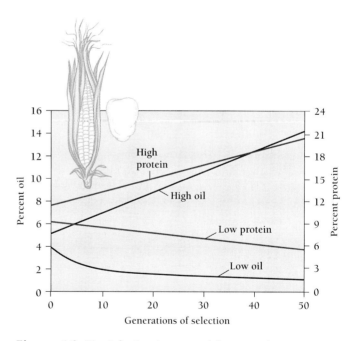

Figure 16-11 Selection is a powerful way to change gene frequencies. In four long-running experiments, researchers select for corn with more protein or less protein and for corn with more oil or less oil. After 50 generations, researchers have continued to produce plants with greater and greater amounts of protein and oil. However, in the experiments selecting for less protein or less oil, the amounts of oil and protein reach a minimum and microevolution stops. Apparently, plants need a minimum amount of oil and protein to survive.

A simple example occurs when an individual carries two copies of a lethal allele, such as that for Tay-Sachs disease. Children with Tay-Sachs die before they are 4 or 5 years old. Having Tay-Sachs therefore eliminates the possibility of reproducing. Although the allele can survive in heterozygotes, homozygotes *never* pass the allele to offspring. As a result, the allele survives only at low frequencies.

Lethal recessive alleles such as Tay-Sachs do not disappear entirely because these alleles "hide" from natural selection in the form of heterozygotes, who are perfectly healthy. (In contrast, natural selection eliminates dominant alleles that kill before the age of reproduction.)

How long does it take to eliminate one allele? To estimate the rate at which selection eliminates recessive lethals, imagine that all *aa* homozygotes die before they reproduce (as in Tay-Sachs). Recall that, according to the Hardy-Weinberg Principle, the frequencies of the *AA*, *Aa*, and *aa* genotypes are p^2, $2pq$, and q^2, respectively. We can say, then, that the fretake about half a million years—certainly a long time, but only a tick in the history of life. Even subtle selection pressures thus have large effects on allele frequencies.

> *Natural selection cannot, by itself, eliminate most alleles, but natural selection is, nevertheless, a powerful way to change allele frequencies.*

How Does Selection Increase the Frequency of Certain Traits?

Just as selection can eliminate or reduce the incidence of one trait, it can increase others. One example is the development of resistance to toxins by natural populations of "pests." Most people are now aware that populations of bacteria, parasites, and pests have developed "resistance" as a result of humans waging chemical warfare against them.

For example, in many parts of the United States, household cockroaches have even evolved an aversion to the glucose (sugar) used to bait roach traps (Figure 16-12). And, after World War II, insects developed resistance to the pesticide DDT. In the 1940s, countries around the world began spraying massive amounts of DDT to kill disease-carrying insects such as flies and mosquitoes. At first, the heavy doses of DDT killed all or most insects. But each time the DDT was applied, a few individual flies would survive. These survivors happened to carry alleles that made them resistant to DDT, and they founded a new population of resistant flies. In time, DDT would no longer kill certain insects in some areas. Because DDT is toxic to many forms of life, its use is now banned in the United States, although not elsewhere.

Of more pressing concern is the evolution of pathogenic bacteria, more and more of which carry genes for resistance to antibiotics. It is now all too common to find patients in big-city hospitals infected with bacteria that have genes for resistance to all known antibiotics. In this case, the antibiotics themselves select for resistant bacteria, killing off all that are not resistant. We discuss this in more detail in Chapter 20.

A classic example of selection in nature concerns the peppered moth *(Biston betularia)*. Because of a long British tradition of collecting moths and butterflies, biologists in Great Britain have been able to trace a change in the moth population over more than 100 years. In the early 19th century, all the peppered moths collected in Britain were pale and mottled. Collectors netted the pale moths as they fluttered through the air. Occasionally, a collector found a dark variant of the peppered moth. These were rare and therefore highly prized (Figure 16-13).

The industrial revolution dramatically changed this scene, however. In the 19th century, thousands of coal furnaces filled the air with dark soot, which blackened the landscape. The

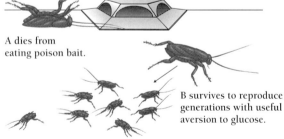

Figure 16-12 Natural selection in action. Because cockroach traps and poisons are commonly baited with the sugar glucose, intense selection pressure favors cockroaches that dislike the flavor of glucose. Researchers have discovered populations of German cockroaches from Florida to California that no longer take such bait because selection has allowed only the glucose-haters to survive.

Figure 16-13 Directional selection. The English peppered moth *Biston betularia* comes in two forms, light and dark. The light ones, whose colors blend well with the pale lichens found on trees, formerly predominated throughout England. The dark form predominates in areas where industrial pollution has blackened the landscape. *(Michael Tweedie/Photo Researchers, Inc.)*

BOX 16.1

Does Selection Act on Preexisting Variants?

Many 19th-century biologists believed, as Lamarck did, that variation arose in response to the needs of the organism. Thus even after mutations and genetics were understood, an important question arose: Do mutations arise spontaneously without reference to the needs of the organism or selective pressure, or do they arise in response to selective pressure? In short, are the variants that selection acts on preexisting ones, or not?

In the early 1950s, the American geneticists Joshua and Esther Lederberg decided to test the idea that variants preexist. They used bacteria, because they grow so fast, and an elegant technique they had devised, called replica plating. Replica plating allowed them to study the descendants of thousands of individual cells in different environments.

After growing bacteria in a nutritious broth, the Lederbergs spread the bacteria onto a petri dish containing the Jell-O–like medium agar. Each single bacterium on the agar divided to produce a visible colony composed of thousands of cells. The Lederbergs then touched the surface of the agar to a piece of velvet. The velvet picked up a sample of each colony. By touching the velvet to another agar plate, they made a replica of the original plate. Every colony of the replica had the same position as the "parent" colony on the original plate. The corresponding colonies all derived from the same cell in the liquid broth and so had the same genotypes.

The Lederbergs used their replica-plating technique to study antibiotic-resistant variants of the human gut bacterium *E. coli*. In plates containing the antibiotic streptomycin, most of the bacteria died, but a few survived and grew into colonies. All of the cells in each of the colonies themselves produced streptomycin-resistant progeny, showing that the resistance to the drug is a genetic property.

But were individuals with genetic resistance present in the original population, or did they arise only as a result of exposure to the drug? Replica plating provided an easy way to answer this question. The Lederbergs simply examined the replicate colonies that had not yet been exposed to streptomycin. Because they knew the identity of each colony from its position on the plate, they could pick out the ones whose brethren, on other plates, had been able to flourish on streptomycin-spiked agar. Here were colonies of the same genotype that had not yet been exposed to streptomycin. Were they resistant to the drug, too? The answer was yes. The Lederbergs proved that the variation preexisted in the original bacterial population, even though the bacteria had no need for such resistance. This result supports the idea that adaptation to changed environment can result from selection among preexisting variants.

---

light peppered moth became rare, and collectors found more and more of the dark variants. While lighter moths continued to prevail in unpolluted rural areas, the frequency of dark moths around factory cities increased to about 98 percent.

Looking at the historical data (from old collections of moths and butterflies), the British evolutionary biologist J.B.S. Haldane calculated that, in industrial areas, the darker variant was twice as likely to survive and reproduce as its lighter counterpart. A reasonable explanation for this was that, against the now dark background of trees and rocks, the dark variant had a selective advantage because it was less likely to be seen and eaten by birds. A fine theory, but the naturalists of the time said that they had never seen a bird eating a peppered moth. Indeed, no one had actually seen a peppered moth resting on a tree or rock. The moths were always caught while flying. The question remained: Was the change in frequency of dark moths the result of selection by birds?

To answer this question, H.B.D. Kettlewell, a biologist at Oxford University, undertook experiments that now provide among the best evidence concerning the operation of natural selection. In the 1950s, Kettlewell captured several hundred dark and light moths and marked them with little spots of paint. He released some of the dark and light marked moths in a smoky industrial area near Birmingham and the rest in an unpolluted woodland near Dorset. He then recaptured as many moths as he could and calculated the percentage of recovered dark and light moths (Figure 16-13). Near blackened Birmingham, the fraction of dark moths recovered was twice that of the light. Near rural Dorset the situation was reversed; the percentage of light survivors was twice that of the dark. This experiment supported the idea that the change in frequency was related to the color of the tree trunks.

But were birds the agents of selection? Kettlewell showed that the moths really were being eaten by birds and that the color of the moth does, indeed, change the chance of being eaten. After placing moths on tree trunks, Kettlewell recorded the menu decisions of the birds in the area with hidden cameras. As expected, the birds preferentially chose the dark moths on the lighter trees of Dorset and the light moths on the darker trees near Birmingham.

Kettlewell's work seemed to demonstrate both the operation and the actual agents of natural selection. The darkening of moths and butterflies to give better protective coloration in sooty environments is now called "industrial melanism." It has occurred in dozens of species not only in Britain, but also near coal-burning industries in the United States and Germany. Beginning in 1956, clean-air laws returned the forests to their normal greens and browns. Gradually, the fraction of dark moths decreased. In Michigan, black peppered moths increased throughout the 1950s, and then, after the United States instituted its first clean-air laws in 1965, the white form came to predominate.

But there is a flaw in this story. At one site where the moths were studied, the pale moths began to return long before the lichens began coming back. And, despite intensive searching, no researcher has ever found a peppered moth resting on a lichen-covered tree (unless the researcher put it there). Some researchers now postulate that the moths roost in the topmost branches of trees and that selection occurs there.

Selection increases the frequencies of certain traits by eliminating or reducing the number of individuals without those traits. Kettlewell showed that in a blackened environment, dark forms of the peppered moth have a selective advantage over lighter forms. Researchers do not yet know, however, whether selection by birds accounts for such color changes in response to pollution.

In What Directions Does Natural Selection Push Populations?

We think of natural selection as pushing populations in some adaptive direction. And, indeed, in many populations that is true. For Galápagos finches, for example, selection enlarges the beaks of birds that have only big seeds to choose from. But selection can have other effects as well. Evolutionary biologists name several kinds. Here we briefly discuss directional selection, stabilizing selection, disruptive selection, and sexual selection.

How does directional selection influence traits?

Directional selection shifts the frequency of one or more traits in a particular direction (Figure 16-14). Industrial melanism is a classic example of directional selection. Selection for resistance, whether to a pesticide or an antibiotic, is also a form of directional selection. Likewise, agricultural breeders have long practiced directional artificial selection to produce such things as chickens that lay more eggs and cows that give more milk.

In nature, directional selection is typical of a changing environment and is a likely explanation for the directional changes that occur during the evolution of new species. The fossil record shows, for example, that the descendants of early humans and horses have become progressively larger. These changes presumably reflect the action of various selective pressures. In the case of horses, paleontologists suggest that the major selective force was the presence of predators. As horses

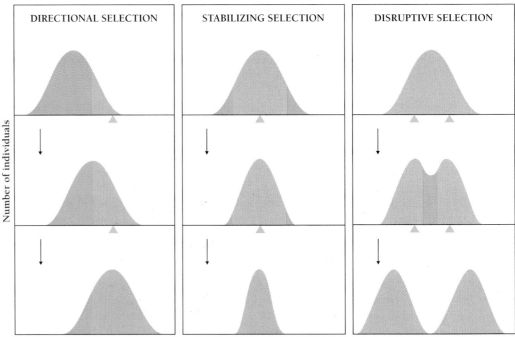

Figure 16-14 Directional, stabilizing, and disruptive selection.

Ginkgo tree

Figure 16-15 They haven't changed a bit. Stabilizing selection probably accounts for organisms that hardly change over millions of years of evolution. Shown here are the ginkgo tree (modern and 250 million years old) and the coelacanth (modern and 350 million years old). *(Fossil ginkgo, Kjell B. Sandved/Photo Researchers, Inc.; fossil coelacanth, John Cancalosi/DRK Photo; modern coelacanth, Tom McHugh/Photo Researchers, Inc.)*

became larger, they could run faster and were better able to avoid being caught and eaten.

It is important to understand, however, that selective pressures can change over time. Traits that are adaptive in one decade or millennium may not be selected for in the next. Thus directional selection can push and pull an organism in many different directions over time.

How does stabilizing selection influence traits?

Stabilizing selection tends to act against extremes in the phenotype, so that the average is favored (Figure 16-14). For example, variability in wasps drops dramatically during the winter, when their mortality is high. An average wasp stands a better chance of surviving a harsh winter than those with unusual phenotypes, which die in great numbers.

Stabilizing selection is characteristic of stable environments because such environments consistently favor the average and rarely or never favor extreme variants. Stabilizing selection is one explanation for the almost unchanging morphological characteristics of such "living fossils" as the horseshoe crab, the coelacanth, and the ginkgo tree (Figure 16-15). If the environment for such organisms were sufficiently stable, the same phenotype could be adaptive for hundreds of millions of years.

How does disruptive selection influence traits?

Disruptive selection is the opposite of stabilizing selection; it increases the frequency of extreme types in a population, at the expense of intermediate forms (Figure 16-14). Like directional selection, it is a likely cause of the evolution of new species. A nice example of disruptive selection is the African swallowtail butterfly *(Papilio dardanus)*, which comes in three distinct variants, each of which lives in a different part of the butterfly's range (Figure 16-16). All three variants appear

Figure 16-16 Disruptive selection. Female African swallowtail butterflies, *Papilio dardanus (top row)*, taste good to birds. The butterflies avoid bird predators by mimicking one of three foul-tasting species of butterflies that also live in the area *(bottom row)*. Intermediate forms of the African swallowtail are snapped up by birds, and only the swallowtails that most closely resemble the bad-tasting species survive and reproduce. *(Hope Entomology Collection, Oxford/Biological Photo Service)*

unappetizing to their predators because of their resemblance to another particularly foul-tasting butterfly. However, each variant resembles a different foul-tasting butterfly that lives in the same area. Predators in each area avoid the three extreme forms, but eat any intermediate variants because they resemble none of the three foul-tasting butterflies.

How does sexual selection influence traits?

Physical or behavioral differences between the sexes are extremely common. Among birds, for example, the male is often showy, with brilliant plumage, a long tail or crest, and a lively song, while the female is drab and inconspicuous. Among mammals, the male tends to be larger and more aggressive, as in humans. Many male mammals have horns, antlers, or other rather theatrical weapons. These are used in displays or fights with other males in competitions for territory or access to females. Such differences between the sexes, called **sexual dimorphism,** are sometimes the result of **sexual selection,** the differential ability of individuals with different genotypes to acquire mates (Chapter 29).

Biologists generally recognize two kinds of sexual selection, both of which revolve around sexual reproduction. These are **female choice** (or, rarely, male choice) and **male competition.** In male competition, a male competes with other males for territory or access to females (Figure 16-17). He is more likely to leave numerous descendants if he can win the contest. Anything that gives him an edge in a contest with other

Figure 16-18 Female choice, in humans. Among the Wodaabe people of Niger, the men dress up in elaborate costumes and makeup for the Geerewal dance, where women observe and judge, selecting the most beautiful men. In America and in many other cultures, it is women who compete in beauty contests and men who do the choosing. (© *Carol Beckwith/Robert Estall Photo Library*)

Figure 16-17 Male competition. Male gemsbok antelopes sparring. Male competition leads to selection for traits that increase the reproductive success of males but are of no value to females. In most animals that have horns, for example, only the males have elaborate horns. (*Fred Bruemmer/DRKPhoto*)

males, whether larger horns or greater aggressiveness, will give him a selective advantage.

The second variety of sexual selection, female choice, involves courtship behavior. The purpose of courtship is to allow the female to assess the male (Figure 16-18). In most cases, females have fewer opportunities to breed than males. Females also invest more time and energy into each of their offspring, on average, than males do. It is important, therefore, for females to be more selective in choosing a mate than males are. In general, females try to mix their own genes with the best set of genes they find. As a result, most species have evolved complex rituals that enable the female to decide if her prospective mate is genetically "good enough" and choose accordingly.

For example, males of outbred and inbred populations of the fruit fly *Drosophila subobscura* differ in their ability to court female flies. To gain a female's favor, a male must perform a series of complex maneuvers, including tapping the female on the head with his front legs and moving around to approach her head on while extending his proboscis, the tonguelike structure with which insects feed and taste. The female quickly sidesteps back and forth, forcing the male to keep adjusting his position. This she does, apparently, to prolong the courtship. The longer the courtship, the more easily she can assess his vigor and skill. Outbred males, which show far greater athletic ability in performing these fly dances, are more successful in their courtship. They have a distinct selective advantage over inbred males.

How can natural selection promote genetic variation?

Given the power of natural selection to change allele frequencies, we may well wonder why significant polymorphism persists at all. The maintenance of genetic heterogeneity in a population is not particularly surprising if different alleles of the same gene are equally useful. For example, within the range of a single interbreeding population, local environments may select for different alleles. The climate of a mountainside may select for short trees that are less likely to be toppled by winter winds, while a milder valley climate may favor larger trees that produce more seeds. Or one allele may be better adapted at one time of year than another. Among the members of one species of ladybug, for example, black ladybird beetles are more frequent in the fall, while red ladybird beetles predominate in the spring. Such a situation creates a balance of different alleles in a population, called a **balanced polymorphism.** The most famous case of a balanced polymorphism in nonhumans is, of course, the current balance between dark and light peppered moths.

Populations may maintain high frequencies of alleles that are deleterious or even lethal when homozygous. In such cases, polymorphism results from the selective superiority of the heterozygote. Sickle cell anemia is an excellent example of a case in which heterozygote superiority creates a balance of different alleles.

As we discussed in Chapter 10, people with two copies of the "β^S-globin" ("Beta S globin") allele suffer from sickle cell anemia. Recall that hemoglobin consists of four polypeptide chains, two βs ("betas") and two αs ("alphas"). In people of African and Middle Eastern descent, the normal allele for the β chain is sometimes replaced by an allele known as β^S. In people with two copies of the β^S allele, low oxygen levels cause the abnormal hemoglobin to precipitate, leading to distortion and stiffening of the red blood cells. The sickled cells are so stiff that they can block capillaries and so fragile that they break open in the strong currents of larger arteries. In the United States, victims of sickle cell disease rarely live beyond age 40, dying of infections, kidney failure, or the blockage of blood vessels in vital areas such as the lungs. Most suffer periodic bouts of pain, increasing weakness, and constant infections.

In Central and West Africa, where health care is worse, homozygotes often die in their teens and 20s and rarely reproduce. Nevertheless, the β^S allele persists in the population, for it has strong selective value in β^A/β^S heterozygotes. Heterozygotes, whose blood contains a mix of normal and abnormal hemoglobins, are not anemic, yet they are resistant to infection by the parasite that causes the deadly disease malaria. The heterozygotes therefore have a distinct selective advantage over people who are homozygous for the "normal" copies of these genes, who are susceptible to malaria.

In some regions of Central and West Africa, the frequency of the β^S allele is about 20 percent (Figure 16-19). Using the Hardy-Weinberg Principle, we can calculate that the frequency of heterozygotes ($2pq$) is then $2(0.2)(0.8) = 0.32$, or 32 percent. The expected frequency of homozygotes would be

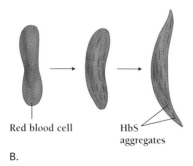

Figure 16-19 A balanced polymorphism. A. The distribution of the sickle cell allele (β^S) closely parallels that of malaria. Blue shows areas of southern Europe, Africa, and the Middle East where malaria is or was once common. Red illustrates areas where sickle cell anemia occurs. Considerable overlap (*purple*) suggests selection for the sickle cell allele (β^S) in areas with malaria. B. The allele for sickle cell alters the shape of a polypeptide in hemoglobin. The altered hemoglobin changes the shape of individual red blood cells.

$(0.2)^2 = 0.04$, or 4 percent. The homozygotes usually die before reproducing, and these deaths represent the "cost" of maintaining the β^S allele in the population. There is a trade-off between the deaths caused by the homozygous condition and the lives saved by the heterozygote condition.

In the United States, where malaria is nearly nonexistent, the β^S allele probably gives people no selective advantage, either in the homozygote or the heterozygote individual. In the United States, the frequency of homozygotes among African Americans is less than one-hundredth the frequency in Central and West Africa, while the frequency of heterozygotes is only about one-third that in Africa.

The reduction in the frequency of the β^S allele is due not only to its selective disadvantage but also to the effects of gene flow. On average, one-third of the genes of contemporary African Americans come from their European ancestry, so the frequency of the β^S allele would be reduced by dilution as well as by selection.

A balanced polymorphism may result from heterozygote superiority or from different selective pressures in time and space within the same population.

In this chapter we have seen where variation comes from, how it is inherited, and some of the ways that allele frequencies can change. Changes in allele frequency in a population, called microevolution, can occur through natural selection or, in small, isolated populations, through random drift. However, explanations for the larger patterns of evolution—the multiplication of species and higher taxa, as well as patterns of extinction, for example—remain controversial. In the next chapter we will consider why it is so hard to define a species, how new species form, and why species multiply and go extinct.

Study Outline with Key Terms

Natural selection was once controversial because heredity was not understood. Darwin believed in **blending inheritance,** which led him to wrongly embrace the theory of **the inheritance of acquired characteristics.**

Weismann knew nothing of genes or mutations, so he could not know the ultimate source of genetic variation. Nevertheless, he was able to correctly guess that inheritance is not blending but particulate, that the genetic material is in the **germ plasm** and is unaffected by the life of the parent, and that sexual reproduction is a means of creating new variants by mixing and recombining old traits.

The **modern synthesis** wedded natural selection and genetics into an essentially modern theory of evolution. Population geneticists attempt to understand evolution in terms of the changes in the frequency of alleles within populations and species. **Microevolution** is shifts in gene frequencies in a population. **Macroevolution,** which we will study in the next chapter, deals with the processes by which species and higher groupings of organisms originate, change, and go extinct.

Genetic variation is the fuel of evolution. Gene and chromosome mutations are the ultimate source of all genetic variation. Genetic recombination, enhanced by sexual reproduction, spreads new **alleles** through a population by recombining different alleles into new genotypes.

Most organisms have two sets of **homologous** chromosomes. When the same allele is on each chromosome in a pair, the organism is said to be **homozygous** for that gene. When a different allele is on each chromosome in a pair, the organism is said to be **heterozygous** for that gene. Traits determined by two or more genes are said to be **polygenic.** Organisms displaying two or more alternative forms are said to be **polymorphic.**

The **Hardy-Weinberg Principle** shows that under certain conditions, a single generation of random mating establishes a distribution of genotype frequencies called the **Hardy-Weinberg equilibrium** ($p^2 + 2pq + q^2$) that will not change in future generations. The Hardy-Weinberg Principle disproves Fleeming Jenkin's assertion that sexual reproduction by itself would change the genetic characteristics of a population. The Hardy-Weinberg Principle applies only if the following conditions are met: random mating, large population size, no mutations, no gene flow between populations, and no selection. **Natural selection** is only one of several forces that can change the frequency of alleles within a population. Others are nonrandom mating (including **assortative mating, inbreeding,** and **outbreeding**), **random drift, mutation,** and **gene flow.** The **founder effect** occurs when a population passes through a **bottleneck,** or a small population founds a larger one. Chance affects gene frequencies in small populations more than in larger ones. A **gene pool** is all the alleles of all the individuals in a population.

Natural selection can bring about evolution by changing gene frequencies in a population. Natural selection can lead to changes in allele frequencies even when individuals with different genotypes have only slight differences in reproductive success. Natural selection acts on the **phenotype** directly, and the **genotype** only indirectly. Although natural selection reduces the frequency of deleterious alleles, deleterious recessive alleles—even when they are lethal in homozygotes—persist at low frequencies for many generations. Natural selection can also act to decrease the genetic variability in a population (**stabilizing selection**), to change the quantitative characteristics of a population in a particular direction (**directional selection**), to increase the frequency of extreme types in a population (**disruptive selection**), or to cause **sexual dimorphism** (**sexual selection**). Biologists believe that sexual selection results most often from **female choice** and **male competition**. Selection for heterozygotes and simultaneous selection for different alleles maintains genetic variation in a population (**balanced polymorphism**).

Review and Thought Questions

Review Questions

1. Darwin understood how natural selection works in the presence of a population with many variant types. What does that variation consist of? How does the variation arise? How does it spread through the population?
2. How much genetic variation do human populations exhibit?
3. Do the eye colors exhibited by your classmates fall neatly into the five colors shown in Figure 16-4? If not, why might they be different?
4. Why doesn't sexual reproduction, by itself, change allele frequencies in a population?
5. Why do the frequencies of the alleles in a gene pool add up to 1?
6. What conditions must be met for a population to be at a Hardy-Weinberg equilibrium?
7. How does chance lead to changes in allele frequencies?
8. Why are certain genetic diseases more common in certain human groups than in others?
9. Give an example of gene flow in a human population and in another animal population.
10. Why would a dominant lethal allele die out in one generation, while a recessive lethal might persist for hundreds of generations?
11. Why is the lethal recessive allele for sickle cell anemia so common in some human populations?

Thought Questions

12. Why doesn't natural selection eliminate variation in a population?
13. Why did Darwin think that sexual reproduction must eliminate variation in a population? What was his mistake?
14. What facts tended to suggest that the theory of the inheritance of acquired characteristics was wrong? What facts were needed to prove that it was wrong?
15. Is it possible that the apparent changes in the proportions of light and dark peppered moths are due to the moths being invisible to collectors, rather than to birds? Discuss your reasoning.

About the Chapter-Opening Image

The cheetah has some of the lowest genetic variability known in a wild organism. Biologists believe that cheetahs passed through a bottleneck within the past few thousand years, in which the entire population was reduced to a few individuals.

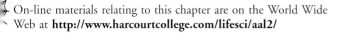
On-line materials relating to this chapter are on the World Wide Web at **http://www.harcourtcollege.com/lifesci/aal2/**

Web Bit 16-1 Battle of the Sex Chromosomes

Web Links

Quizzing & Testing

17 Macroevolution: How Do Species Evolve?

What Darwin Saw at the Galápagos Islands

When Charles Darwin arrived at the Galápagos Islands in the summer of 1835, he found a group of dreary volcanic islands consisting of hundreds of craters, the largest some 3000 or 4000 ft high. Five main islands dominated the little group of 14 islands, 600 miles off the coast of Ecuador. Despite their position on the equator, the Galápagos were not the lush, tropical islands typical of other parts of the Pacific, but barren desert islands populated by a small number of cacti, reptiles, and small, brown birds. It was the dry season.

"Nothing could be less inviting than the first appearance," Darwin wrote in his diary. "A broken field of black basaltic lava, thrown into the most rugged waves, and crossed by great fissures, is everywhere covered by stunted, sunburnt brushwood, which shows little signs of life. The dry and parched surface, being heated by the noonday sun, gave to the air a close and sultry feeling, like that from a stove: we fancied even that the bushes smelt unpleasantly." The most striking animals were the reptiles, large marine iguanas, and immense Galápagos tortoises (*Geochelone elephantopus*), some of them 5.5 ft tall and weighing 500 lb (Figure 17-1).

Darwin spent a month in the islands, exploring and collecting rocks, plants, and animals. He discovered that almost all of the terrestrial animals and at least half of the plants were endemic species, that is, found nowhere else in the world. Yet, all of the Galápagan species resembled species found on the nearby South American mainland—all, that is, but a European rat, probably brought by ship to the islands.

The island-bound plants and animals were related to those on the mainland. But how were they related, and why were they different?

Darwin's observation that so many Galápagan species were unique to the islands did not at first strike him as meaningful. Even when the vice-governor of the islands casually mentioned that tortoises from different islands were so distinctive that he could easily tell which island a particular tortoise had come from, Darwin paid no attention. Then, just days before the HMS *Beagle* was to set sail for Tahiti, the vice-governor's remark struck Darwin with sudden force.

Maybe, Darwin thought, each island supported a different set of organisms. Frantically, he went back through his specimens to see if different species came from different islands. But he had labeled most of them "Galápagos." He had hopelessly mixed up the different kinds of tortoises and the finches. There was no way to tell which islands they had come from.

But one group of birds, the mocking thrushes, "thoroughly aroused" his attention. Of three species, one was unique to Charles Island, one to Albemarle, and one to James and Chatham Islands. Darwin's plant collection, more carefully labeled, was a gold mine. For example, of 71 species found on James Island, 38 were endemic to the Galápagos and 30 of those were endemic to James Island alone. Every island had the same story to tell.

"One is astonished," he wrote in his diary, "at the amount of creative force, if such an expression may be used, displayed on these small, barren, and rocky islands; and still more so, at its diverse yet analogous action on points so near each other."

Key Concepts

- A species is a population of organisms whose members can interbreed to produce viable offspring but that usually do not interbreed with members of other such populations.
- A population may form a separate species when it becomes reproductively isolated by any of a variety of means.
- Adaptive radiation is the generation of diverse new species or other taxa from one or more ancestral species.
- Species probably evolve at different rates at different times.

Two years after his visit to the Galápagos, Darwin began a notebook on the "transmutation of Species." It would be years before he realized the full implications of all that he had seen in the Galápagos. It was no accident, he ultimately realized, that Galápagos species were so closely related to each other and to those on the South American mainland. Each endemic species had evolved on one of the islands, having descended from migrants from other islands or the mainland.

Nowadays, every child knows that different sorts of animals and plants populate each of the different continents. Africa has its lions and camels, South America its jaguars and llamas. Lions and jaguars, although descended from the same ancestral cat, have evolved independently for millions of years. The same is true of camels and llamas.

In Darwin's day, however, the presence of similar animals in very different habitats, or of completely different assemblages of animals in similar habitats, would have been surprising. According to the 19th-century theological concept of special creation, similar habitats should support identical species. Every dark forest should have its bear and every clear stream its trout. And, since each species was thought to have been created by the Judeo-Christian God less than 6000 years ago, species were not supposed to be related to one another. Most biologists, like Linnaeus, could see that animals and plants group naturally into rough hierarchies that resemble family trees. Yet they carefully ignored the suggestion of genealogy suggested by these hierarchies.

But Darwin—scion of a distinguished family—knew the importance of genealogy. He realized that 5 million years ago, the volcanic Galápagos Islands had burst steaming and erupting from the sea, devoid of life. Gradually, plants and animals began to colonize the islands. Seabirds roosted on the barren rocks, leaving seeds that had stuck to their wings or passed through their digestive tracts. Some of these seeds took root and managed to survive. Ocean currents brought more plant seeds and an occasional reptile. Winds also brought seeds, as well as finches and sparrows and other land birds. Some plants and animals survived and reproduced; others did not. Some moved from island to island, spreading and interbreeding with other populations; others remained confined to one island. These slowly evolved away from their cousins on the other islands and the parent population on the mainland. In this way, one species became many.

Frantically, he went back through his specimens to see if different species came from different islands.

In this chapter we will study **macroevolution,** the origin and multiplication of species. We will see what a species is, and explore how it comes into being. We will ask what the patterns in the fossil record tell us about evolution. And we will discover why, after over 150 years, evolutionary biologists are still arguing about the questions raised by Darwin after his visit to the Galápagos Islands. Near the end of the chapter, we will briefly discuss the evolution of humans.

WHAT IS A SPECIES?

In the next section of this book, we will explore the enormous biological diversity of life on Earth. For now, we will merely note that biologists have so far identified about 1.4 million species of living organisms—give or take a hundred thousand. Newly discovered species pour into the museums monthly, so this number is probably less than one-tenth of the total number of species currently living. The actual number could be anywhere from 10 million to 100 million. Of the known species, more than half are insects (Figure 17-2). Nearly half of all insects—290,000 species—are beetles, some 20 percent of all species of organisms combined. The biologist J.B.S. Haldane once ironically remarked that God "must have had an inordinate fondness for beetles."

Figure 17-1 The voyage of the *Beagle*. In 1831, Charles Darwin left England on board the British surveying ship the H.M.S. *Beagle* for a 5-year trip around the world. Darwin and the *Beagle* spent nearly 4 years in South America. After Darwin left the Galápagos Islands, he was struck by the fact that each island had its own kind of tortoise. On each island, the tortoises evolved separately from those on the other islands and so differ with respect to the shape of the shell, the length of the neck, and other traits. The first written account of the differences among Galápagos tortoises came from the captain of the United States frigate *Essex*, which visited the Galápagos during the War of 1812 and, by itself, nearly destroyed the British whaling fleet stationed there. Before the *Essex* was finally captured by the British in 1814, Captain David Porter found time to describe the tortoises and various lava lizards. (*Darwin, The Bridgeman Art Library; H.M.S. Beagle, The Granger Collection, New York; marine iguana, Michio Hoshino/Minden Pictures*)

Why Is It So Hard To Define a Species?

The average person, faced with groups of familiar organisms such as mammals or birds, easily recognizes one species from another. Even when presented with a set of unfamiliar mammals, the average person can generally distinguish most species. In 1927, for example, a young biologist named Ernst Mayr, who later became a leading figure in evolutionary biology, led an expedition into the Arafak Mountains of New Guinea to catalogue new species of animals. After he had identified 138 species of birds, he was surprised to discover that the local Papua natives themselves recognized 137. Two of the species were so similar that the Papuans considered them a single species.

Unfortunately for biologists, such similar-looking species are common. Two species may even look identical. Some

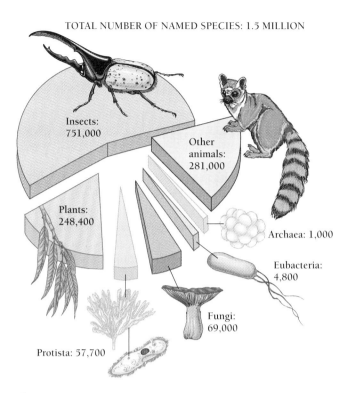

Figure 17-2 **The diversity of life.** At least 1.5 million species now inhabit the Earth.

Darwin concluded that since species gradually evolve one from another, the term "species" is a convenient but arbitrary designation that has no biological meaning. He found this a great "relief," for he no longer felt he had to worry about whether organisms were truly separate species or not.

Modern biology takes a different view from Darwin. We now think that species are, for the most part, discrete and very real entities. Further, biologists have found that in order to discuss evolution, they need to define what a species is. In the 1930s and 1940s, the makers of the modern synthesis produced the **biological species concept,** one modern definition of a species. In 1942, Ernst Mayr defined a **biological species** as follows: *"Species are groups of actually or potentially interbreeding populations, which are reproductively isolated from other such groups."* A biological species, then, is the largest group of similar organisms in which gene flow is possible.

Most species conform nicely both to the biological species concept and to our sense that species should look different from one another. Bullfrogs (*Rana catesbeiana*) and wood frogs (*Rana sylvatica*), for example, look different from each other and do not interbreed (Figure 17-3). They are members of different species.

But the biological species concept is a tricky definition. First, it does not apply to asexual organisms such as prokaryotes, certain fungi, and even some plants.

species of frogs, for example, are distinguishable only by their calls. Adult forms of the malaria-carrying mosquito *Anopheles maculipennis* cannot be distinguished from those of several other species that do not carry malaria. A species that is extremely similar to, but reproductively isolated from, another is called a **sibling species.** We can conclude that appearance alone fails to define a species.

Defining a species is awkward. When biologists believed that organisms were all variants of "ideal types," as 19th-century essentialists believed, the problem was much easier. The definition of a species was a description of the physical features of the ideal type, a sometimes arbitrarily chosen "type specimen." Type specimens are still essential for naming a new species and for identifying other individuals of the same species that are collected later. But we now realize that every individual in a species is unique—ideal types do not exist. Even an individual that happened to be completely average in every respect would not represent its species better than any other individual in the population.

The definition of species must take account of the enormous variation among individuals and recognize that two species can be distinct even if they are very much alike. Darwin found himself exasperated by the problem of classifying species. Writing to his friend J.D. Hooker in 1853, Darwin noted, "After describing a set of forms as distinct species; tearing up my [manuscript], & making them one species; tearing that up & making them separate, & then making them one again (which has happened to me) I have gnashed my teeth, cursed species, & asked what sin I committed to be so punished."

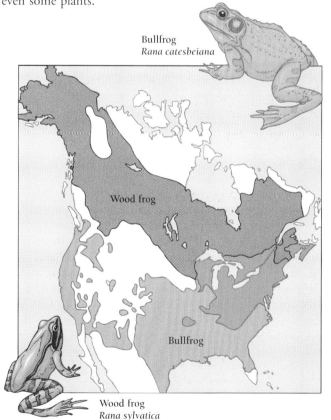

Figure 17-3 **Separate species.** The wood frog *Rana sylvatica* and the bullfrog *Rana catesbeiana* look different and occupy different, though overlapping, habitats. They probably do not interbreed; they are true species. *(After R.C. Stebbins, 1985)*

For example, *E. coli*, the distinctive bacterium that inhabits the human digestive tract, is almost always asexual. Even though it does not satisfy the definition of the biological species, taxonomists have nonetheless assigned it to a genus and species.

Second, the biological species concept does not apply easily to extinct organisms. Paleontologists naming fossils have no way of knowing if two so-called species interbred. Many fossil species gradually give way to increasingly different forms. Paleontologists arbitrarily give an organism a new species name when it looks sufficiently different from its ancestors. But no one knows whether the modern horse *(Equus)*, for example, could interbreed with its predecessor *Hippidium* (Figure 15-12).

Furthermore, paleontologists must give taxonomic species names to fossil organisms on the basis of morphological differences in the hard parts alone. Muscles and other soft parts, as well as clues about physiology and behavior, are almost never fossilized. The two mosquito species mentioned earlier, for example, would never be recognized as two species if they appeared only as fossils. And some paleontologists have begun to suspect that several "species" of dinosaurs may turn out to be different ages and sexes of the same species.

Paleontological species are therefore subjective designations. A group of fossils may be lumped together as a single species or divided into several. But these same organisms might be judged differently if they were alive today. This point will be important to keep in mind when we discuss the evolution of humans at the end of this chapter.

The biological species concept does not even apply well to all living sexually reproducing organisms. It applies poorly to organisms that do not normally live together. Geographically isolated organisms—whether very different or very similar in appearance—may or may not be capable of interbreeding. Lions and tigers interbreed if put together in zoos, which should make them the same species. But in the wild, they never interbreed, so they are effectively separate. Then again, since their habitats do not overlap, there is no way to know whether they would interbreed if they lived closer to one another. In garter snakes of the genus *Thamnophis,* the two species *T. hydrophila* and *T. elegans* live in the same area in northern California and southern Oregon without interbreeding. They seem to be what taxonomists call "good" biological species. Yet, in a small area just south of the Oregon-California border, these two species interbreed. Finally, for the vast majority of living species, no one knows who interbreeds with whom, how often, or where. Detailing the reproductive biology of wild organisms is one of the most challenging areas of biology.

Species are groups of actually or potentially interbreeding populations that are reproductively isolated from other such groups. The best definition of species that we have is the biological species concept, but this definition, good as it is, is often hard to apply.

What About Other Taxonomic Categories?

Many species have **races** or **subspecies,** morphologically distinct subpopulations, which can interbreed. For example, along the western coast of North America, the song sparrow has been divided into as many as 34 subspecies (Figure 17-4). However, it is not always clear that such races are specifically adapted to different environments. For example, six races of carpenter bee inhabit a single island in Indonesia, with no apparent differences in the way they live.

In fact, almost any subpopulation can be distinguished from others on the basis of some cluster of characteristics. For example, the people living in Minneapolis might differ from those living in St. Paul in having, on average, longer noses and a slightly higher frequency of ingrown toenails. But we would not necessarily conclude that the populations of the Twin Cities were separate races. Many evolutionary biologists therefore consider the concept of race or subspecies to be just a normal manifestation of natural variation and without any real biological meaning. The differences among human "races" constitute only about 9 percent of the genetic diversity in the human species. The other 85 percent is individual variation (Chapter 14).

Nevertheless, even though most races or subspecies of organisms interbreed easily, a few do not. Subspecies that do not interbreed may be in the process of separating into species and are sometimes called **incipient species.** Distinguishing ordinary subspecies from incipient species is often a judgment call that only a biologist who has thoroughly studied the population in question can make.

As we will see in Chapter 19, genera and higher taxonomic categories are still harder to define. The hierarchical system of

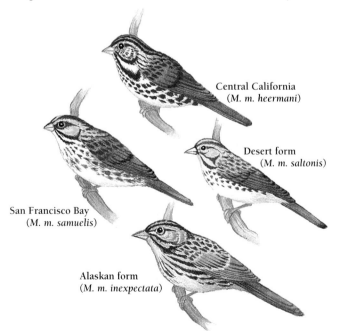

Subspecies of the song sparrow, *Melospiza melodia*

Figure 17-4 Subspecies. Here are just 4 of the 34 known subspecies of Western song sparrow.

classification that taxonomists use to organize biological species into higher groupings generally represents degrees of anatomical difference and so, to some extent, evolutionary distance.

A higher taxon is usually defined by a few characters that consistently distinguish its members from the species in other taxa. Mammals, for example, are distinguished by hair, two sets of teeth, and mammary glands. If we find an animal with these characters, we can be sure it is a mammal.

However, the division of a group of species into separate genera, families, or higher taxa can be very subjective. For example, in 1947, a botanist decided that a recently discovered daisylike plant whose flowers had no petals was so novel that it should be put into its own genus. Only later did botanists discover that the plant was a simple variant of a well-known species that normally had petals. The "new genus" was not even a new species.

Species exist as biological entities independent of human definitions. Higher taxa, as well as subspecies and races, are somewhat subjective groupings.

HOW DO SPECIES FORM?

We have seen that for two populations to become separate species, they must become **reproductively isolated,** unable to interbreed. Often, the process by which one species becomes two begins when one population is geographically isolated from the rest of the species. If that happens, the physically isolated population has a chance of becoming reproductively isolated from the parent species and, with that, a chance of becoming a separate species. Some populations merely happen to become reproductively isolated and then evolve apart.

What Barriers Reproductively Isolate Populations and Species?

Other populations, however, may first evolve separate adaptations and then develop reproductive barriers that allow them to retain their respective adaptations. Many of the most remarkable features of organisms function to maintain reproductive isolation. Such features often include the diverse structures of flowers and the elaborate rituals of animal mating.

The larkspur *(Delphinium)* provides one example of instant reproductive isolation. Most species of larkspur have blue or purple flowers (Figure 17-5). Bees visit blue larkspur (and other blue and purple flowers), carrying pollen from flower to flower. But one northern California larkspur has a mutation that gives it scarlet flowers—a color bees cannot see. Fortunately, hummingbirds see red flowers very well, and they pollinate the red larkspurs. The bees and hummingbirds do not intrude on one another's flowers, and the blue and scarlet larkspurs are reproductively isolated.

For populations to become reproductively isolated, gene flow between them must cease. Barriers to gene flow fall into two broad categories: (1) **prezygotic barriers,** which prevent syngamy, the fusion of the sperm and egg to form a zygote, and (2) **postzygotic barriers,** which make a zygote either inviable (certain to die) or sterile. The red and blue larkspurs are an example of a prezygotic barrier, since the flowers' different colors prevent cross fertilization. Prezygotic barriers also include differences in the ways that species live **(ecological isolation),** in the times at which they reproduce **(temporal isolation),** in their mating behaviors **(behavioral isolation),** in the complementarity of male and female reproductive organs **(mechanical isolation),** and in the compatibility of their gametes **(gametic isolation).**

Postzygotic barriers operate when fertilization has already occurred, but the resulting zygote either dies or fails to reproduce successfully. An individual that results from cross fertil-

Figure 17-5 Reproductive isolation. Some time in the past, the scarlet larkspur, shown here, became isolated from the rest of its species. A mutation produced a generation with red flowers instead of the usual blue. But red flowers tend to attract hummingbirds instead of the bees and other insects that pollinate the blue flowers. Because the hummingbirds do not carry pollen between the red and blue flowers, the scarlet larkspur has become reproductively isolated from its blue relatives.

ization between two different species, or, sometimes, between two distinct populations, is called a hybrid. For example, a cross between a horse and a donkey results in a mule, a hybrid that is nearly always sterile. The mule is an example of **hybrid sterility**. However, hybrids between related species often die during development or early in life because of chromosomal and genetic incompatibilities, a barrier referred to as **hybrid inviability**. In some cases, the hybrids of two species may be viable and fertile, but *their* offspring are weak or sterile **(hybrid breakdown)**.

In plants, changes in chromosome number can simultaneously create new variants and cause their immediate reproductive isolation. For example, if homologous chromosomes fail to separate during meiosis, then each gamete will have the diploid number ($2n$) of chromosomes instead of the haploid number (n). Fertilization can then produce a tetraploid (or $4n$) zygote, with four rather than two sets of homologous chromosomes (Figure 17-6). Such tetraploid plants will produce gametes with $2n$ chromosomes. These gametes cannot form zygotes with gametes of the parental stock, since such zygotes would contain $3n$ chromosomes, which could not pair at meiosis. Tetraploid plants thus become immediately isolated from the main population. Other multiples of the haploid number of chromosomes also occur. Plants that contain more than two complete sets of chromosomes are said to be **polyploid**. Polyploidy has played an important role in the evolution of grasses, ferns, and other vascular plants.

Another common postzygotic barrier in plants results when the zygote combines the chromosome sets of two species. Organisms formed from the gametes of two different species are ordinarily sterile, if not inviable. Such hybrids rarely produce gametes because the maternal and paternal chromosomes are not homologous, and so they cannot pair during meiosis. However, pairing can occur if the hybrid doubles its chromosomes. Such a doubling forms an **allopolyploid**. Each set of chromosomes is present twice and can engage in proper meiotic pairing and gamete formation. But the gametes of the hybrid can no longer form viable zygotes with the gametes of either parental species, so they become reproductively isolated from both parental populations. Since many plant species can self-fertilize, however, a single plant can, potentially, establish a new species.

Populations become reproductively isolated either by prezygotic or by postzygotic barriers.

How Does Geographic Isolation Initiate Speciation?

In most cases, populations must be geographically isolated from the parent population before other barriers to reproduction arise. Oceans, mountain ranges, and rivers, for example, are all capable of sharply reducing or eliminating interbreeding with a parent population. Once a population is isolated, it can follow its own evolutionary course (Figure 17-7). Its gene frequencies can change independently, according to the forces of selection and genetic drift (Chapter 16). In time, the isolated population may acquire new chromosomal arrangements and reproductive strategies that prevent interbreeding with the parent population if the two should again come into contact. These isolating mechanisms preserve the isolated population's differences and allow it to **speciate**—form a new species. Species formation by geographic isolation is called **allopatric speciation** [Greek, *allos* = other + *patra* = country].

Allopatric speciation is speciation that proceeds from geographic isolation. Once two populations are geographically isolated, independent selective pressures and genetic drift can further differentiate the two populations.

How Can Species Form Without Geographic Isolation?

The formation of separate species without geographic isolation is called **sympatric speciation** [Greek, *syn* = together + *patra* = country]. For a single population to spontaneously break into two reproductively isolated subpopulations, something must inhibit gene flow between the two subpopulations. Some barriers to gene flow, such as polyploidy in plants, can isolate subpopulations from their parent populations in one generation. However, most barriers to gene flow would have to evolve more gradually.

Figure 17-6 Reproductive isolation by polyploidy. An error during meiosis produces diploid gametes, which form tetraploid offspring. The offspring can breed only with one another, however, because if the diploid gametes fuse with normal haploid gametes, they form triploid zygotes, which cannot survive.

Figure 17-7 Allopatric speciation. When a single population of mice becomes divided—by a river, for example—the two subpopulations evolve separately and can form separate species.

For example, the ensatina salamander has spread down the two sides of a large valley (Figure 17-8). Increasing geographic distance correlates with both increasing genetic differences and increasing differences in appearance. The populations on either side of the valley are separated from one another and mostly do not interbreed. One exception are the yellow-eyed salamanders on the coast, which can move up a river system to the mountains in the east, where they hybridize with the Sierra Nevada salamanders. Near the bottom, the Monterey salamander and the yellow-blotched salamander, which do not interbreed, are different enough to be considered different species, yet they are linked by a continuous chain of subspecies that can and do interbreed.

Most models of sympatric speciation invoke disruptive selection as a mechanism. The hawthorn fly may present a good example of the first steps in sympatric speciation by disruptive

Figure 17-8 Sympatric speciation. The ensatina salamander *Ensatina eschscholtzii* has diverged into a series of overlapping subspecies along the Coast Range and the Sierra Nevada. Little or no habitat exists in the Great Central Valley, which runs between the two mountain ranges. Within the mountains, adjacent subspecies of the salamander interbreed. The two subspecies at either end of the salamander's horseshoe-shaped range— the Monterey salamander and the yellow-blotched salamander—do not themselves interbreed. But they are linked by a continuum of subspecies. In the south, the yellow-blotched salamander and the large-blotched salamander are phenotypically and genetically similar, and biologists believe that these two species were once linked by a population that has most likely gone extinct. In 1976, one ensatina salamander was seen in this area—the only one since biologists began searching in 1949. (Excerpted from Life on the Edge, copyright 1994 by Bio Systems Analysis, Inc., with permission from Ten Speed Press, P.O. Box 7123, Berkeley, CA 94707)

selection. In 1864, farmers in New York State noticed that the small hawthorn fly *Rhagoletis pomonella,* which normally fed on hawthorn fruits, had begun to feast on apples as well. Because the apples at once became both food source and mating site for some of the hawthorn flies, it has been suggested that the apple-eating flies rapidly became reproductively isolated from the rest of the population. In 1960, some of the flies switched to cherries, possibly creating yet another isolated, but sympatric, race.

The flies move back and forth from one fruit to another, however, and researchers are not certain how different or how well isolated these three populations are. The three subpopulations have divided the same environment into three potentially separate habitats, like three islands of habitat. But so far, biologists still recognize only a single species. Whether these are incipient species remains to be seen.

Sympatric speciation, the formation of species without geographic isolation, is controversial.

Can Species Form with Limited Gene Flow?

A third type of speciation, called **parapatric speciation,** occurs when a species splits into two populations that are geographically separated, but still have some contact. Parapatric speciation is thus intermediate between allopatric and sympatric speciation. Reproductive barriers can isolate parapatric populations if the environments of the two populations impose different selective pressures on each, and if gene flow between the two populations is small (lest it overwhelm the divergent changes).

The growth of grasses around the entrance to a Welsh lead mine provides a good example of parapatric isolation and the probable beginning of parapatric speciation. The soil immediately around the mine has a high concentration of lead—high enough to kill most of the species of grass that flourish in the adjacent pasture. But a variety of one species of grass, *Agrostis tenuis,* tolerates lead and grows well on the mine tailings. This variety has become reproductively isolated from the neighboring parental pasture grass because it happens to flower at a different time.

If we could be sure that the separation were complete—that gene flow had completely ceased—then we would call the two populations separate species. At this point, however, we can only note the different selective pressures of tailings and pasture and the limited gene flow between them. Such a situation could easily lead to true reproductive isolation and speciation.

Parapatric speciation may occur if gene flow between adjacent, nonoverlapping populations is sufficiently limited.

HOW DO SPECIES MULTIPLY?

The separation of one species of organism into two (or more) species is also called **divergent evolution. Convergent evolution,** the independent development of similar features in separate groups of organisms, is not the opposite of divergent evolution, but a separate, unrelated idea. Recall from Chapter 15, for example, the convergent evolution of wings in the unrelated birds, bats, and insects. In some sense, evolution tends to be divergent. From a few simple forms, millions of species have evolved. Species are constantly splitting into two, three, or more new species. These separate species diverge even more to form separate families and orders.

Species often diverge and multiply rapidly when a new place to live becomes available. Biologists call such an event an **adaptive radiation,** the generation of many new species with widely varying adaptations. For example, nearly all the major groups of modern mammals—the carnivores, whales, bats, rodents, and primates, for example—evolved during a 10-to 15-million-year period in the Cenozoic. DNA sequence analysis suggests that these diverse groups of mammals split from one another almost simultaneously from a group of closely related species.

How Does Adaptive Radiation Occur on Chains of Islands?

Adaptive radiation is the key to understanding why there are so many species. We will first discuss examples of adaptive radiation in chains of islands and then try to understand how diversity increases worldwide.

The most dramatic cases of adaptive radiation occur on chains of islands (or archipelagos), such as the Galápagos. We have seen that the formation of new species generally requires either reproductive isolation (so that descendant species do not interbreed with one another) and one or both of two other conditions: either distinct selective pressures (so that different descendant species acquire different adaptations) or else small population size (so that genetic drift can operate).

Archipelagos sometimes provide all of these conditions. Geographic isolation from the mainland ensures reproductive isolation from mainland species. Of course, islands don't have to be surrounded by water to encourage speciation. A series of isolated lakes, mountaintops, or even forests are, for many species, equivalent to islands (Figure 17-9).

Whether the isolation is caused by water or by a surrounding desert, relatively few individuals arrive at each island. Plants may arrive as seeds carried by wind or birds. Animals arrive only if they can fly, swim, or float on a piece of driftwood or other natural raft. In short, founder populations tend to be so small in number that founder effects alone cause gene frequencies in the island populations to differ from those in their mainland parent populations. In addition, the environments of different islands differ enough to exert different selective pressures (Figure 17-10).

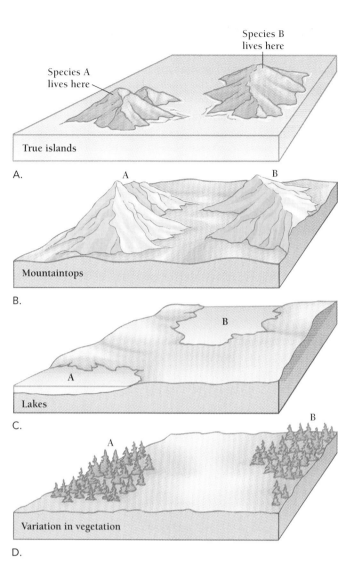

Figure 17-9 Geographic isolation comes in many shapes. Although oceanic archipelagos feature many classic cases of adaptive radiation, other kinds of "islands" also facilitate adaptive radiations. To a species of fish, a series of lakes or ponds is an archipelago. To a mouse, the individual tops of mountains or plateaus can function as islands. A. True islands. B. Mountaintops. C. Lakes, rivers, and oceans. D. Variation in vegetation.

When Darwin visited the Galápagos Islands in the Pacific and the Cape Verde Islands in the Atlantic, he noted three surprising facts (Figure 17-1):

1. The species in each chain of islands resembled those on the closest mainland, rather than those of other islands. Despite the similar climate and terrain of the Cape Verde Islands and the Galápagos Islands, the birds on the Cape Verde Islands were similar to those of Africa and those of the Galápagos to those of South America. This suggested that the islands were colonized by migrants from the mainland.

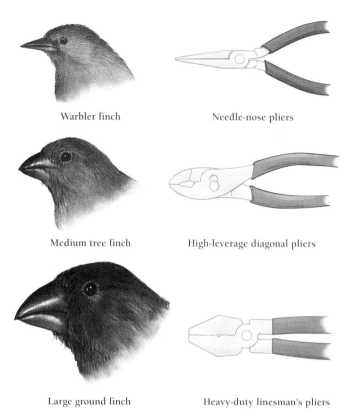

Figure 17-10 Adaptive radiation. Less than 5 million years ago, the Galápagos Islands began erupting from the floor of the Pacific Ocean. Sometime after the islands had been colonized by plants, a single species of finch colonized the islands. This finch then radiated into 13 separate species, which taxonomists have grouped into 4 genera. Here are just three, representing three genera. Each finch is specialized for a different way of life. The finch with the largest beak, for example, breaks open large, hard seeds, impossible for the other two finches to crack.

2. The archipelagos contained a much smaller number of species than the mainland. The islands' biological poverty supported Darwin's idea that only a small number of ancestral organisms had colonized the islands.
3. The individual islands in a chain had distinct species. Tortoises from the drier islands had longer necks and peaked shells that allowed them to reach the cactus and tree branches, while those from wetter islands had shorter necks, better adapted to feeding on ground vegetation.

Darwin recognized that the distribution of species in chains of islands provides a record of the arrival of immigrants from the mainland. For example, of the hundreds of bird species in South America, only eight succeeded in reaching and colonizing the Galápagos. Yet 1 of those 8 gave rise to no fewer than 14 species of finches. The Galápagos finches offer the most famous example of adaptive radiation (Figure 17-11).

The Galápagos finches are all distinct from any on the continent of South America. Yet they were enough alike to

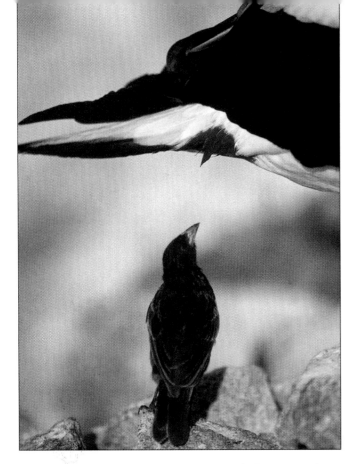

Figure 17-11 Vampire finch. Galápagos finches seem to have evolved to take advantage of any opportunity. This sharp-billed finch (*Geospiza difficilis*), or vampire finch, feeds on the blood of larger birds such as this masked booby. *(Tui de Roy/Minden Pictures)*

suggest to Darwin that they all stemmed from a common ancestor. Contemplating the finches, Darwin wrote his first statement on adaptive radiation: "One might really fancy that . . . one species had been taken and modified for different ends."

The probable mechanism of adaptive radiation on islands such as the Galápagos is relatively easy to understand. Over the course of the 5 million years since the islands first appeared—and possibly as recently as a half-million years ago—a few finches (at least one male and one female) managed to fly the 1000 km (600 miles) from the South American mainland—no easy task, since finches are poor long-distance flyers. The founders were probably ground birds, with short bills useful for crushing seeds. Descendants of these birds managed to colonize the 13 islands in the Galápagos chain, as well as Cocos Island, about 1000 km away. However, because of the finches' limited flying abilities, flights among the islands must have been rare. So the finches on each island mated only with other finches from the same island. In short, the finches on each island became reproductively isolated.

Because each island differs slightly in geology, terrain, vegetation, and insect life, the finch population on each island experienced different selective pressures. Acted upon by both natural selection and genetic drift, the populations came to differ in many ways: size and form of beaks, nesting patterns, songs, plumage, and courtship behavior. Separate species formed through allopatric speciation. Over time, some of the new species managed to colonize the other islands, so that some islands came to sustain several species of finches.

Detailed studies of the finches have shown that pairs of species that share an island are more different from one another—in beak size and other measures—than are species that live on different islands. In short, where several finch species live together, individual species tend to evolve away from each other and become more specialized. Like all organisms on Earth, these birds feel the tug and pull of natural selection and will evolve (or go extinct). Only in these finches and a few other organisms, however, have researchers documented the effects of natural selection so clearly.

Yet, it is the absence of more specialized birds in the Galápagos that has allowed the finches to radiate. For example, because the Galápagos have no true woodpeckers, one species of Galápagos finch has evolved into a sort of imitation woodpecker (Figure 17-12). The Galápagos woodpecker finch does not have the special tongue and beak and other adaptations of a true woodpecker. Yet the woodpecker finch manages to feed on the insects that live in tree bark by using a cactus spine or a twig clamped in its beak to dig the insects from the bark.

Similarly, on the island chain of Hawaii, which also lacks true woodpeckers, a species of honeycreeper digs insects from

North American hairy woodpecker

Galápagos woodpecker finch

Hawaiian honeycreeper (akiapolaau)

Figure 17-12 Opportunity knocks. In the absence of mainland woodpeckers, other kinds of island birds can evolve adaptations for taking grubs from tree bark.

tree bark by chiseling the bark with its lower jaw and picking out insects with its curved upper bill. Neither the Hawaiian honeycreeper nor the Galápagos finch seems as well adapted as a true woodpecker, but each manages to flourish in the "woodpecker" niche in the absence of direct competition from a true mainland woodpecker.

Plants also undergo adaptive radiation. The sunflower family (Asteraceae) provides two especially dramatic examples. In the Hawaiian Islands, 28 species of the most spectacular plants in the islands have all derived from a single small and unremarkable California tarweed. The amazing Hawaiian tarweeds, some of which are shown in Figure 17-13, include the Haleakala silversword, which grows within the cone of a volcano on the island of Maui, as well as the yuccalike iliau, which grows among the scrub vegetation of Kauai. Most of these plants, like their relatives in California, grow in dry areas, but a few have adapted to wet forests and bogs.

An equally dramatic radiation within the sunflower family has occurred on the remote Atlantic island of St. Helena, halfway between Africa and South America. The usually low-growing members of this family have evolved into large "trees." Reproductive isolation and the absence of competitors have evidently allowed the evolution of these distinctive species, which are found nowhere else on Earth.

Adaptive radiation, the evolutionary divergence of a single line of organisms, is the key to biological diversity.

How Do Species Diversify on Continents?

The multiplication of species is not confined to archipelagos. On continents, two ways that species diversity increases are **dispersal,** the spread of a taxon through a large area, and **vicariance,** the fragmentation of an already dispersed species or group of species.

Birds and bats and even insects are, of course, very good at dispersing all over the world. So are plants, whose tough seeds hitch rides on animals and winds. Ants, horses, and other terrestrial organisms may not be able to cross oceans, but they can disperse enormous distances by simply walking a little farther, generation after generation.

In vicariance, a widely dispersed population can become fragmented either through the extinction of populations in between or through the creation of geographic barriers. As we saw in Chapter 15, tectonic plates move around, alternately drifting apart and bumping into one another over millions of years. Such movement not only separates and joins whole continents, it also raises mountain ranges and lifts old sea beds that are then cut by massive canyons. Once populations are isolated by such geologic changes, they can speciate allopatrically.

The fossil record makes sense in terms of the movements of the continents. For example, the giant ground birds of Australia, Africa, and South America are all related and were separated only by the Mesozoic breakup of Gondwanaland. In contrast, the camel family evolved after the continents were already separated. The Camelidae originated in North America during the Eocene. They later dispersed into Eurasia by way of the Bering Land Bridge (long since disappeared) that connected Alaska and Siberia and into South America by way of the Central American isthmus. Today the Camelidae are extinct everywhere except Asia, North Africa, and South America.

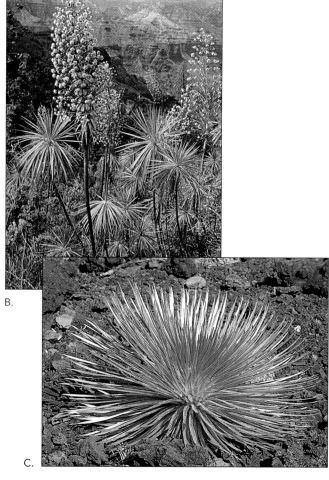

Figure 17-13 Adaptive radiation in Hawaii. In the isolation of the Hawaiian Islands, an unremarkable plant, represented in California by tarweeds in the genus *Madia,* has radiated into 28 species. A. The California tarweed *Madia madiodes.* B. Hawaiian iliau *Wilkesia gymnoxiphium.* C. The Haleakala silversword *Argyroxiphium sandwicense.* (A, N.H. [Dan] Cheatham/Photo Researchers, Inc.; B, Gerald D. Carr; C, C.K. Lorenz/Photo Researchers, Inc.)

Likewise, the modern horse evolved in North America, then migrated to Eurasia and to South America in periods when those continents were connected with North America. Because the current land connection between North and South America was formed only about 3 million years ago, no horse fossils in South America date back before that time.

The Bering and Central American land bridges provided a route for the dispersal of many species between Eurasia and North and South America—including, for example, horses, camels, and humans (Figure 17-14). But there were no such bridges to Australia. Australia became separated from all the other continents, near the end of the Cretaceous Period, some 65 million years ago. After that, Australian species evolved in isolation. At the time Australia became separated, no horses or other grass-eaters had yet evolved anywhere on Earth. Because no horses or similar herbivores reached Australia after its separation, grass-eaters evolved in Australia and populated the plains—the kangaroos.

Kangaroos, like almost all other Australian mammals, are "marsupials"—mammals that bear young as fetuses and carry them in a pouch until the young are able to fend for themselves. In contrast, humans and other "placental mammals" bear more-developed young. Placental mammal mothers nurture their young in the uterus with the aid of a specialized tissue, called the *placenta* [Latin, *placenta* = flat cake], which passes nutrients, oxygen, and wastes between the mother and the developing embryo.

Marsupials and placental mammals diverged from each other in the early Cretaceous, and the marsupials happened to reach Australia first, in the late Cretaceous. With the exception of bats, which evolved elsewhere and later flew there, and various species of rats, which must have rafted from island to island in the East Indies, all the native mammals of Australia are marsupials.

The marsupials of Australia are an extraordinary example of adaptive radiation (Figure 17-15). All appear to have evolved from a primitive, opossumlike marsupial of the Cretaceous. The evolution of the marsupials in Australia parallels that of the placental mammals elsewhere, in what is an equally extraordinary display of large-scale convergent evolution. The present-day marsupials include animals that resemble the placental mammals of other continents, both in appearance and in habit. Although kangaroos do not look much like deer or goats, their teeth are similar, for they live a similar life. Australian marsupials include species that look like rabbits, mice, moles, anteaters, cats, wolves, squirrels, and woodchucks. Distinctly wolflike creatures have separately evolved from both a marsupial and a placental ancestor.

The parallel radiations of marsupials and placental mammals show the opportunism of evolution and help account for some of the enormous diversity we see around us.

Continental diversity increases through the dispersal of organisms to new parts of the world and through vicariance, the fragmentation of dispersed groups into populations that can evolve separately. Australian marsupials illustrate both adaptive radiation and convergent evolution.

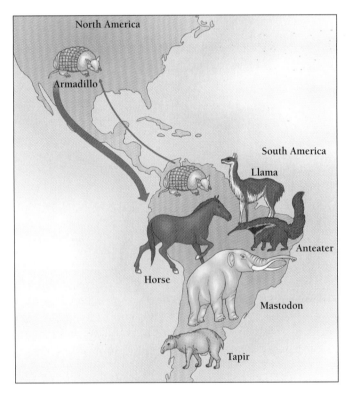

Figure 17-14 Land bridge. The North and South American continents have been isolated and connected at different times in geologic history, depending on the level of the oceans. Land bridges allow many organisms to migrate to new habitats. When sea levels rise and again isolate the continents, separate species arise through allopatric speciation.

THE CAMBRIAN EXPLOSION

Up until the Cambrian Period, almost 600 million years ago, life on Earth seems to have been limited to a few soft, one-celled organisms, such as algae and bacteria. Until quite late in the Precambrian, the only complex organisms that paleontologists find are a few jellyfishlike creatures.

Then, at the beginning of the Cambrian, life bloomed. In a burst of diversification unmatched in the history of the world, all of the modern animal phyla that have fossilizable skeletons appeared, though not necessarily in their current forms. For example, the arthropods, which today include the insects, spiders, crustaceans, and barnacles, were represented by the trilobites, crustaceans, and other marine animals. Also present in the Cambrian were the brachiopods (lamp shells), mollusks (snails, clams, and squid), sponges, echinoderms (starfish and sea urchins), and cnidarians (jellyfish and corals), not to mention the first chordates (today including amphibians and mammals).

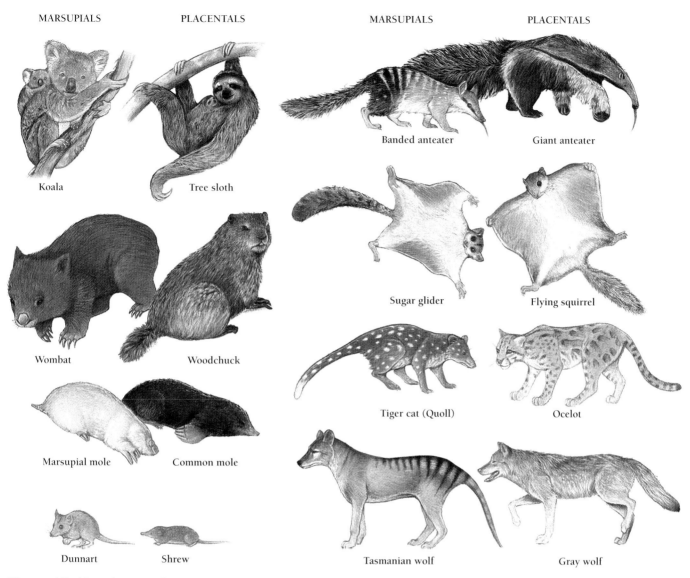

Figure 17-15 Adaptive radiations in parallel. The marsupial mammals of Australia belong to an ancient and very distinct lineage of mammals. Yet, marsupial mammals (*left in each pair*) and placental mammals (*right*) have separately evolved many of the same body types. The sugar glider and the flying squirrel, for example, both have long, bushy tails and webbed limbs—and for the same reason. Both animals quickly cover great distances by sailing from the tops of tall trees down to the ground.

Taxonomists have based the classification of animal phyla on basic body plans—the two-layered radial symmetry of the jellyfish; the three-layered bilateral symmetry of flatworms; the three-layered bilateral symmetry, plus notochord, of chordates, such as ourselves, to name a few. Although enormous evolutionary change has occurred since the early Cambrian and the level of diversity is arguably as great today as it ever was, almost no new body plans have evolved in the half-billion years since these basic forms first appeared.

Evolutionary biologists refer to this one-time evolutionary radiation as the **Cambrian explosion.** Paleontologists remain divided over the exact time span during which this explosion occurred. Some evidence suggests that the major animal phyla may have diverged as many as 900 million years ago, hundreds of millions of years before the Cambrian. Other evidence suggests that all the phyla may have appeared at once at the very end of the Precambrian. Either way, evolutionary biologists are left with a major puzzle. What has prevented major new body plans from evolving since the Cambrian?

One argument is that evolution has somehow slowed since then, that changes simply do not occur as rapidly as they did in the Precambrian and the Cambrian. However, modern populations are capable of evolving quite rapidly. For example, both bacteria and insects evolve resistance to antibiotics and

BOX 17.1

Did It Come from Outer Space?

In 1979, geologist Walter Alvarez discovered a thin layer of mineral that told him that the Age of Dinosaurs ended with a bang, not a whimper. Alvarez was studying rocks in Gubbio, Italy, at what geologists call the K/T boundary. Here, rocks laid down at the end of the Cretaceous (K), 65 million years ago, are overlaid by rocks from the Tertiary (T).

At Gubbio, the boundary is clearly marked by a 2- to 3-cm layer of clay (Figure A). To find out know how many years of Earth's history was represented by that inch-deep clay layer, Alvarez measured its iridium content. The metal iridium is rare on the Earth's surface, but a constant hail of it from space lightly dusts the Earth's surface. Like the sand in an hourglass, the buildup of iridium can be used to determine how much time has passed.

Alvarez expected to find less than 1 part per billion (ppb) of iridium, and, indeed, both the Cretaceous rock below the clay and the Tertiary rock above had the expected amount—1 ppb or less. To his surprise, however, he found a whopping 10 ppb in the clay layer itself. What had caused the spike in iridium in the clay?

Collaborating with his father, Nobel prize-winning physicist Luis Alvarez, and geochemist colleagues Frank Asaro and Helen Michel, Alvarez found similar iridium spikes at two other K/T sites—in Denmark and New Zealand. In 1980, the team published a groundbreaking paper, "Extraterrestrial Cause for the Cretaceous-Tertiary Extinction," in the journal *Science*. In it, the researchers argued that the iridium spike, or *iridium anomaly*, as it is often called, was the result of a gigantic 10-km asteroid that slammed into the Earth, exploded, and drove to extinction all of the dinosaurs and many other forms of terrestrial and marine life.

Could this really happen? Paleontologists initially scoffed dismissively. But the Alvarez team argued back, and the controversy spread, so that soon even nonscientists began taking sides.

The excitement stimulated other scientists to investigate, and to date they have found iridium anomalies at more than 100 different K/T boundary sites, both on land and in ocean sediments. Even more intriguing is evidence that the asteroid may have fallen into the Gulf of Mexico.

If an asteroid 10 km across smashed into the Earth, the pressure generated would liquefy the meteorite and the rock it smashed into, spewing molten meteorite and rock into the air—just as water droplets fly into the air when a boulder is heaved into a pond. As the molten meteorite droplets cooled, they would solidify into smooth round pebbles called *microtektites* and fall back to Earth. Such microtektites, as well as *shocked quartz*—deformed pieces of the mineral quartz—have been found in association with known meteor sites, such as Meteor Crater in northern Arizona.

But the most dramatic evidence would be a huge hole in the ground. Did it exist? Until the late 1980s, no one could find the 150- to 300-km crater that scientists predicted such an asteroid would make. If the crater was deep in the ocean and now filled with sediment, it would not be easy to find. Off the Gulf Coast of Texas, however, researchers found that, just at the K/T boundary, huge boulders had been moved around as if by a tidal wave. Then another researcher reported a subterranean bowl-shaped structure, 180 km across, located in the northwestern corner of the Yucatan Peninsula near the town of Chicxulub (pronounced CHICKS-uh-loob). In 1991, other researchers reported that samples from the Chicxulub bowl contained shocked quartz. The layer of melted rock underneath the huge bowl has now been dated at 65 million years old—the age of the Cretaceous-Tertiary (K/T) boundary.

Few people now doubt that the structure at Chicxulub is the crater from an enormous asteroid that slammed into Earth 65 million years ago. That this impact caused global catastrophe is also indisputable. The impact would have incinerated plants and animals across the North American continent and spewed debris into the atmosphere, blotting out the sun worldwide for months or even years.

The only debate is over the exact consequences of that global catastrophe. Some scientists still insist that the asteroid alone did not cause the mass extinction at the end of the Cretaceous. For one thing, whole lineages of organisms disappeared long before the asteroid hit. This fact suggests that there was a more gradual increase in extinctions during the late Cretaceous, not a sudden catastrophic reduction in species.

But why else might they have gone extinct? The best alternative is massive volcanic activity in India and the Pacific Rim that released enormous amounts of volatile gases and dust into the atmosphere. Such gases could have altered the global climate, caused acid rain, and damaged the ozone layer, and caused a large-scale climatic catastrophe.

For now, paleontologists still disagree about whether the late Cretaceous extinctions were rapid or gradual. If the extinctions were sudden, the asteroid is the most likely culprit. If the extinctions were more gradual, the asteroid may have acted merely as a *coup de grâce*.

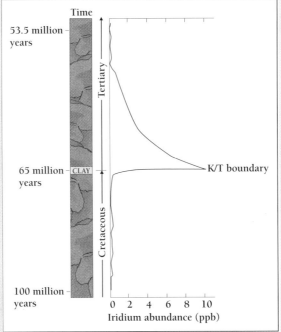

Figure A   An asteroid impact? At the end of the Cretaceous, the amount of iridium present in deposits increases dramatically from 1 part per billion (ppb) to 10 ppb, then gradually decreases again over a period of about 10 million years. Iridium is a metal rare on Earth but common in asteroids and meteors.

pesticides within a few years or even months. Vertebrates, such as fish and birds, are also capable of rapid changes. Research on the finches in the Galápagos has shown that natural selection can change their average beak size, according to what size seeds are available, within a matter of months. All of the Galápagos finches probably evolved from a single species within the last half-million years, a short period in geologic time.

Another argument is that once the major body plans formed, **developmental constraints**—the rules of embryological development that determine the general form of an organism—prevented new body plans from evolving. Many biologists believe that in most organisms, development has become so finely tuned since the Cambrian explosion that any dramatic change results in organisms too defective to flourish. The idea is that modern organisms are so specialized that an organism with an entirely new form could not compete with modern organisms.

But the ancient body plans are not necessarily optimal designs that could never be improved on. Indeed, very likely they are just accidents of history, like the homologous vertebrate limbs illustrated in Chapter 15.

Animal body plans, such as the particular arrangement of bones in the limbs of horses, whales, and humans, may simply be a case of organisms "making do" with what they have. The first airplanes were made of wood, not because this is a better material than light alloy metals, but because that was the best material available at the time. Modern airplanes can be completely redesigned, with new shapes and new materials. But organisms tend to reuse the same parts, reshaping them for new purposes.

Evolution at the species level occurs as rapidly as ever, but the evolution of new phyla seems to have ceased. The Cambrian explosion may have been a one-time phenomenon.

EXTINCTION AND THE RATE OF EVOLUTION

Evolutionary biologists estimate that more than 99 percent of all the species that have ever lived are now extinct and that the vast majority of these went extinct millions of years ago. But why species go extinct is, in most instances, something of a mystery. By looking at species that have gone extinct in historical times, we can say that some species go extinct because of competition from other species, some because their habitat is drastically changed, and some because of a new predator. Human beings, for example, have driven a variety of species to extinction—the American passenger pigeon, for example—through heavy hunting. But in most extinctions, probably no single factor determines whether a species will go extinct. Some species survive such onslaughts, while others do not.

The Mass Extinctions

The fossil record shows that, in the long run, species go extinct at a more or less regular rate. However, superimposed on this ongoing extinction rate are five intriguing mass extinctions, including the death of the dinosaurs (Figure 17-16). Near the end of five geologic periods—the Ordovician, the Devonian, the Permian, the Triassic, and the Cretaceous—the number of extinctions increased by as much as 2.5 times the background extinction rate.

Why these mass extinctions occurred is one of the most controversial questions in paleontology. Some researchers deny that the mass extinctions are qualitatively different from the background extinction rate. They argue that these mass extinctions are just random fluctuations in the usual loss of species over time. Other researchers argue that the five extinctions are the consequence of major worldwide catastrophes. The exact nature of the catastrophes is hotly debated.

The most controversial mass extinction is the one at the end of the Cretaceous. Sixty-five million years ago, all the dinosaurs and half of all species of plants and animals went extinct. Every conceivable catastrophe has been suggested to explain this extinction. But the majority of researchers now seem to believe that this most recent extinction occurred when a giant asteroid hit the Earth with enough force to blast a 180-km crater and plunge the world into almost total darkness for months.

We are now in the midst of a sixth mass extinction. The full extent of this extinction is somewhat controversial. No one knows precisely how many families of organisms have gone extinct, nor how many will go extinct before this extinction event is over. But the cause of this mass extinction is not at all controversial. It is entirely the result of human activity.

After each major extinction in the past, diversity increased dramatically (Figure 17-16). For example, the mammals radiated spectacularly after the Cretaceous extinction of the dinosaurs and other reptiles 65 million years ago. We might guess that the only circumstances under which really new organisms might flourish are where a mass extinction has eliminated much of the competition.

All species go extinct eventually. The rate of extinctions since the Cambrian has been relatively constant except for five mass extinctions. Evolutionary biologists have yet to agree on what caused these mass extinctions.

How Fast Do Species Evolve?

In 1859, Charles Darwin was nearly alone in his belief that evolution was a gradual process. He held that not only individual species but also genera, orders, and other higher taxa all arose through the slow transformation of species. His closest friends tried to persuade him that evolution might very well

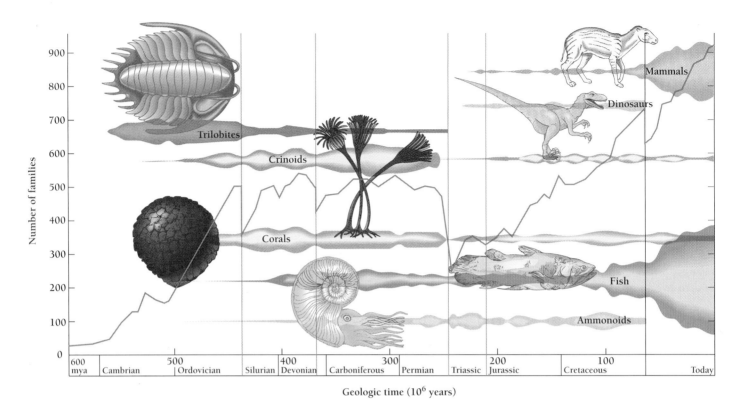

Figure 17-16 Mass extinctions and increasing diversity. This selection of animals shows that numbers of species and families tend to increase over time. Each mass extinction is followed by an enormous adaptive radiation. Recovery from a mass extinction like the one the Earth is experiencing today, however, takes 5 million years or longer. Many groups, such as the dinosaurs and the trilobites, never recover.

occur in jumps, or **saltations.** On the eve of the publication of the *Origin of Species,* in 1859, Darwin's friend Thomas H. Huxley wrote to him, "You have loaded yourself with an unnecessary difficulty in adopting *natura non facit saltum* [nature makes no jumps] so unreservedly."

For Darwin, the conflict was between what he saw everywhere around him in nature and what the fossil record showed. As we noted in Chapter 15, gaps in the fossil record pose a problem in understanding the evolution of species. To demonstrate that a species evolved gradually, we would like to see a fossil representing each small step in an evolutionary line. But, often, the fossil record does not fulfill this wish. Transitional forms between distinct species are disappointingly unusual.

Paleontologists are working with a random sample of organisms through time. They may find a well-preserved fossil of species A in one layer, then no more of that species until 50,000 years later (in the fossil beds). If the next fossil they find is different enough from the first, they may say it is a second species, B, but they won't know how B is related to A. Is B descended directly from A? Are A and B instead descended from a common ancestor C? If so, is C much older than A and B, or just a little older? Did B (or C) gradually turn into A, leaving no unevolved remnants? Or could A, B, and C have all lived at the same time?

The lack of intermediate forms may merely represent the incomplete preservation of ancient organisms. Just as most individual human beings are not named in the pages of recorded history, so most species are not preserved in the layers of rock that make up the fossil record. However, another interpretation for the gaps in the fossil record is that species change suddenly, evolving in short bursts, or saltations. The intermediate forms are not preserved, either because they never existed or because they existed only momentarily in geologic time (less than 10,000 years).

Darwin firmly rejected this latter line of reasoning. He argued that the gaps in the fossil record were the result of the incomplete preservation of ancient organisms.

Darwin had no evidence, and no reason to believe, that the hundreds of species he had studied all his life could change suddenly. First, species were so subtly different that it was hard to decide if they were true species. Three Galápagos Island mockingbird species, each of which came from a different island, were so alike that Darwin assumed they were merely races of a South American species until he took his specimens back to England. There, however, an ornithologist pronounced each a separate species. Later in his life, Darwin spent eight long years struggling to work out the taxonomy of barnacles. Each barnacle species graded so smoothly into the others that he complained bitterly about how hard it was to tell one from another. In short, everywhere Darwin looked, species seemed to grade into one another. Linnaeus also had commented on the difficulty of telling one species from another.

BOX 17.2

Modern Tropical Rain Forests: Mass Extinction or Not?

The fossil record shows five natural mass extinctions during the past 600 million years. Many biologists believe that a sixth, human-induced mass extinction is now under way.

The most species-rich regions of the world are tropical environments, such as the Amazon Basin. For any group of organisms—whether birds, insects, trees, or fungi—the number of species is greatest in the tropics. Although tropical rain forests cover less than 7 percent of the world's land area, one-half to two-thirds of the Earth's 10 million to 100 million species inhabit these rich forests.

Few of these species are known to science—only 1.5 million of the world's species have been named. A disproportionate number of these are European and North American species, where most trained taxonomists have lived and worked. Researchers are struggling to keep ahead of the chainsaws and forest fires that daily destroy the habitats of thousands of tropical forest species, half of which are insects. Tropical rain forests are dense with species. A 1-hectare (2.5-acre) plot of Brazilian forest may contain 200 species of trees, whereas the same size plot in Canada would support fewer than 10. And each species of tree may support five to ten species of insects that specialize only on that one species of tree. Eliminating that tree species causes all the insect species to vanish also. Such cascades of extinctions are not limited to plant-insect interactions. When peccaries (a piglike mammal) were eliminated in one Brazilian forest, so were three species of frogs. All the frogs, it seemed, lived in the mud wallows the peccaries created.

As of today, fewer than one-half of the original tropical forests remain. And logging, agriculture, cattle ranching, and firewood harvesting continue to degrade and destroy tropical forests (Figure A). In some cases, regions of burning forests are so vast that the fires are visible from space. Harvard biologist E.O. Wilson estimates that if tropical forests continue to be destroyed at current rates, fewer than 10 percent of the original forests will remain by the year 2050, and between 2.5 and 25 million species will go extinct.

How does this loss of species compare with the five prehistoric mass extinctions? By examining the fossil record, scientists estimate that the normal, or background, extinction rate may be as high as one extinction per year. In a mass extinction, the number of extinctions increases to at least 2.5 times the background rate. Since 1600, more than 1000 species of plants and animals have gone extinct—yielding a minimum extinction rate of more than 2.5 species per year (Figure B).

However, there were probably at least ten times as many extinctions during that time period. Recall that scientists have, at best, named only about one-tenth of the 10 million to 100 million species thought to exist. Just as the named species are a small fraction of the total species, the tally of documented extinctions includes a similarly small fraction of the total number of extinctions. For the 400 years since 1600, the minimum extinction rate may well approach 25 species per year. In addition, as the remaining tropical forests grow smaller, species are lost at a faster rate. The current loss of tropical forest species is comparable in magnitude to the prehistoric mass extinctions.

Although evolution will continue, the world will not recover its former biological diversity in our lifetime—or in our great-great-great-grandchildren's lifetime. If we can look to the fossil record for an answer, it may be 5 million years before the Earth once more supports the diversity of life that existed when Columbus sailed to the New World.

Figure A Slash-and-burn agriculture. In Brazil, farmers burn small areas of tropical rain forest. They farm for a while and then move on to burn another plot of land as young trees grow up in the last plot. As human population size increases, however, rain forests are being burned faster than they are growing back. (Earth Scenes/© 1996 Dr. Nigel Smith)

Figure B The rate of species loss has increased precipitously in the last 100 years.

Darwin's studies of artificial selection demonstrated that selection was clearly capable of producing enormous changes gradually. The dramatically different races of domestic pigeons, dogs, and sheep, all without question the result of gradual selection, would be considered separate species if found in nature (Figure 15-17).

Nevertheless, the discontinuous nature of the fossil record convinced many 19th-century biologists that evolution must occur by fits and starts, a view known as **saltationism.** After the rediscovery of Mendel's work in 1900, Mendelians argued strenuously that species arose abruptly from one generation to the next because of so-called macromutations—mutations that caused large changes in individual organisms. By the 1940s, however, Darwin's gradualism was finally in vogue and saltationism was out.

By the 1960s, most biologists subscribed to **gradualism**— the view that most species evolved gradually and continuously in direct response to selective pressures from the environment. Paleontologists could object that the fossil record did not support this idea, but geneticists dominated biology at that time, and few of them were interested in accounting for the discrepancies in the fossil record.

Nonetheless, a few renegades held out for saltationism. The most famous of these was the German-American biologist Richard Goldschmidt (1878–1958). He argued that even though macromutations, such as chromosome rearrangements, usually result in wildly deformed organisms, some of these monstrosities have potential if they happen to arise in the right environment. He called these potential species **hopeful monsters.** "A fish undergoing a mutation which made for a distortion of the skull carrying both eyes to one side of the body is a monster," he argued. "The same mutant in a much compressed form of fish living near the bottom of the sea produced a hopeful monster, as it enabled the species to take to the life upon the sandy bottom of the ocean, as exemplified by the flounders" (Figure 17-17). Goldschmidt's term "hopeful monsters" was both memorable and unfortunate. Evolutionary biologist Ernst Mayr's equally memorable term of derision for Goldschmidt's idea—"hopeless monsters"—was quickly picked up by other biologists.

Nevertheless, by 1950, the fossil record, which Darwin had found so imperfect, had improved little. A hundred years of research still revealed few cases with all the intermediate stages in the evolution of a lineage. When new species appeared, they did so suddenly, the intermediate stages leaving no trace. And many groups hardly seemed to change at all over long periods of geologic time. For example, garfishes and sturgeons show only minimal morphological change over 80 million years. Horseshoe crabs have barely changed at all in 230 million years. Such stability over long periods is called **stasis.**

The periods of apparent stasis in the fossil record may not be what they seem, however. Only hard tissues such as bones and shells are preserved. Even extensive changes in soft anatomy, physiology, and behavior of a species would not appear in the fossil record.

Ironically, paleontologists often ignore intermediate forms. Faced with a series of fossils, paleontologists tend to classify individual fossils in a layer as one species or another, not as "something in between." Like Darwin before them, they may have trouble making distinctions between two organisms, but they have to call them something. And classifiers normally say that two fossils are either the same species or two different species.

An intensive study of 15,000 trilobite fossils from Wales provided a good example of what happens when a scientist looks instead for the in-between species. The trilobites, a massively successful group of arthropods that flourished through the Permian, showed so many intermediate forms that classifying individual fossils into species became an exercise in arbitrariness.

Nonetheless, even if we accept that species may not actually evolve quite as abruptly as paleontology suggests, we can still try to explain why speciation should seem sudden at all. Species' abrupt appearances suggest two alternatives. In one scenario, new species appear almost instantaneously, as, for example, in a single generation through saltations. In the other scenario, new species evolve quite rapidly—by geologic standards—over just millions or even mere thousands of years. In either case, the intermediate forms would not appear in the fossil record. The Mendelians and Richard Goldschmidt had hypothesized that new species appeared almost instantaneously.

But the second idea—rapid, but not instantaneous, speciation—is entirely con-

Figure 17-17 Hopeful monster? The flounder lies on the bottom of the sea on its side. To compensate for this position, it has evolved so that its eyes are both on one side of its body (the top side). The flounder begins its life like any other fish, with one eye on each side of its head. As the embryonic fish develops, however, one eye migrates over the top of the head to join the other eye. As an adult, the fish lies on its side in the sand and the mud at the bottom of the ocean and looks up. *(Heather Angel/Biofotos)*

sistent with what population genetics had suggested about speciation in small, isolated populations. The founder effect, allopatric speciation, and even sympatric speciation are all consistent with the idea of rapid speciation in small, isolated populations. Ernst Mayr himself had suggested this idea in 1942. In the case of rapid speciation, intermediate forms would exist in such small numbers and for such a short time geologically that the chances of their being preserved in the fossil record would be almost nil.

Until the 1970s, biologists paid little attention to the connection between rapid speciation and the gaps in the fossil record. Most biologists were interested in microevolution and population studies, not macroevolution. And paleontologists, interested primarily in macroevolution, wondered how biologists could go on believing in gradualism when there were so few examples of it in the fossil record.

In the 1970s, two paleontologists—Niles Eldredge at the University of California, Berkeley, and Stephen J. Gould at Harvard—proposed that the fossil record is exactly what we would expect it to be if the formation of new species was very rapid compared to the accumulation of changes within a species.

This view, called **punctuated equilibrium,** proposes two main ideas. First, *species change very little most of the time* (stasis). Second, *most anatomical or other evolutionary change in individual species occurs during a geologically brief period at the time of speciation.* In this view, speciation events punctuate an otherwise stable equilibrium, and the most important events in evolution are those that lead to reproductive isolation. Discontinuities (even catastrophes) are paramount, and gradual changes within species are mere fine-tuning.

But punctuated equilibrium does not contradict the findings of population geneticists or Darwin's view that evolutionary change occurs gradually. The brief periods during which populations can probably speciate—5000 to 50,000 years—are invisible in the fossil record (and therefore "instantaneous"), but these brief periods in evolutionary history provide ample time for *gradual* change to occur.

How fast can species form? At the extreme end, plants can speciate in one generation through polyploidy. Several species of fish have formed in a lake near Lake Victoria, in Africa, in just 4000 years. And Lake Victoria itself gave rise to 300 species of fish in only 200,000 years. One group of researchers has even suggested that the lake had dried out 12,000 years ago, which would mean that the fish evolved since then, a blink of the eye in geologic time.

Punctuated equilibrium and gradualism are extreme ends of a continuum of possible scenarios (Figure 17-18). The two ideas are not truly alternative theories that can be tested against one another. It's just a question of how much change accumulates how fast. No biologist believes that *all* evolutionary change occurs only gradually over millions of years. And few biologists would argue that all speciation occurs suddenly in just a few generations. Real species can most likely evolve rapidly or slowly, depending on circumstances such as how much genetic variation they have and what their environment is like. Whether *most* species have formed in relatively brief time spans, of 5000 to 10,000 years, or in longer time spans, say 75,000 to 100,000 years, is a question we may never answer. The fossil record is mute on this point.

The gaps in the fossil record between an ancestral species and its descendant species represent periods of rapid evolution that are invisible in the fossil record. However, such rapid evolution—on the order of 5000 to 50,000 years—is consistent with Darwin's ideas about gradual evolution.

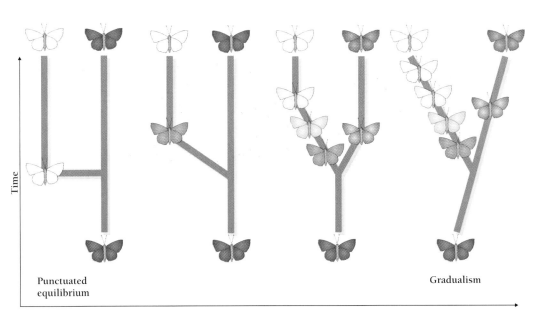

Figure 17-18 Is evolution gradual or abrupt? The answer probably depends on the organism. Punctuated equilibrium and gradualism are the ends of a continuum.

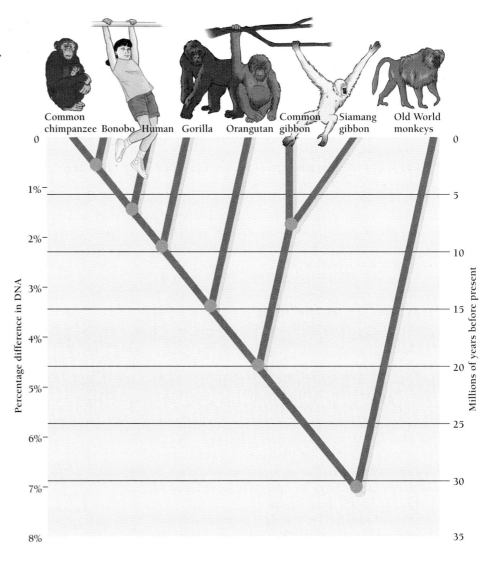

Figure 17-19 Phylogeny of primates. By measuring base-pairing between matching sequences of DNA from different species, biologists can determine relative differences and estimate the time in years since two species diverged. This phylogenetic tree, for example, suggests that humans diverged from bonobos 5 to 10 million years ago.

What Is Phylogeny?

Phylogeny is the history of descent of groups of organisms from their common ancestors. Phylogeny includes the order in which different groups branch off and the time in history when branching occurred. Phylogeny is very like what we humans called "genealogy," except that phylogeny applies to all living organisms instead of just one human family (Figure 17-19).

Biologists classify species according to their phylogeny, how they are related. Groups of related species are put into a genus, groups of related genera are put into families, and so on. (In the same way, we may classify our cousins by which side of the family they're from.)

But species don't always leave a clear record of who's descended from whom, which makes biologists' job difficult. When the fossil record is poor, dozens of large groups of species may appear to emerge all at once. And DNA studies suggest that rapid speciation leaves an equally muddy record in the genes.

Yet, we know that these species did evolve and that there is one true phylogeny. We just don't always know what it is. And sometimes we know what it is, but it's not what we used to think. As we'll see in Chapter 19, biologists now know that crocodiles are more closely related to birds than to lizards. Despite this knowledge, however, birds and reptiles are two separate classes. And the crocodiles are tucked in with the lizards. Even though we still classify crocodiles this way, we know better.

HOW DID HUMANS EVOLVE?

When we visit the zoo, we are often drawn to the animals that most resemble ourselves—the monkeys and apes. Taxonomists say that apes, monkeys, and lemurs all belong to the same order of mammals. This group, called the **primates,** are remarkably unspecialized. Like the earliest mammals, we have five digits on each hand or foot and unspecialized teeth. All primates lack, for example, the sharp cutting teeth and broad grinding surfaces that characterize the teeth of horses, rabbits, and other herbivores. We also lack the long canines and shearing teeth of carnivores such as dogs.

Nonetheless, despite our lack of specialization, we primates have a set of common characteristics that together dis-

Figure 17-20 Platyrrhines and catarrhines. A. Howler monkey. The platyrrhine monkeys (from South and Central America) have nostrils that point sideways. B. Green monkey. The catarrhines—the Old World monkeys and apes—diverged from the platyrrhine monkeys from 17 to 20 million years ago. The catarrhines' nostrils point downward. (*A, Tom & Pat Leeson/Photo Researchers, Inc.; B, Glenn Vanstrum/Animals Animals*)

tinguish us from other mammals. These include large brains and binocular vision, flexible shoulder joints, and hands and feet with five grasping digits, a grasping thumb, and flat fingernails rather than claws.

Most of the 166 species of living primates are tree dwellers. Our flexible shoulder joints and grasping hands and feet enable us to climb trees and swing from one branch to another. As a group, primates are basically herbivores that live on leaves and fruits. Nonetheless, many primates are opportunistic omnivores who will occasionally eat grains, grubs, and even flesh. Most 20th-century humans eat grains and flesh regularly.

How Do We Classify the Apes?

Primates are one of the oldest orders of mammals, dating back some 65 million years. Early primates, which more closely resemble modern lemurs than monkeys or apes, had smaller bodies and longer snouts than modern monkeys and apes (Figure 17-19).

Monkeys and apes consist of two large groups. The more primitive group are the **platyrrhines** [Greek, *platyus* = flat + *rine* = nose], monkeys whose flat noses have widely separated nostrils that point sideways (Figure 17-20A). Many platyrrhines also have prehensile (grasping) tails. All platyrrhines live in South and Central America. Based on fossil and anatomical evidence, zoologists have concluded that they first evolved about 26 million years ago.

The second group, called **catarrhines** [Greek, *kata* = down + *rina* = nose], evolved in Africa and Asia, somewhat later. The catarrhines have downward-pointing nostrils and no prehensile tail (Figure 17-20B). Gibbons and other primitive catarrhines first appeared in East Africa 20 to 17 million years ago (Figure 17-21A). All of the catarrhines except for the humans still live only in Africa or Asia.

Figure 17-21 Two hominoids. The hominoids include the primitive gibbon and the more evolved bonobos and humans. A. White-handed gibbon. The gibbons are the most primitive of the hominoids. B. Bonobo and her baby. (*A, Gerard Lacz/Peter Arnold, Inc.; B, M. Long/Visuals Unlimited*)

Taxonomists group the catarrhines into the catarrhine monkeys, on one hand, and the **hominoids** [Latin, *homo* = human], on the other. Hominoids have large skulls and long arms, and they tend to walk at least partially erect (Figure 17-21B). The hominoids include the gibbons, which are relatively primitive, as well as gorillas, baboons, orangutans, humans, and chimpanzees.

Exactly how all the different apes are related to each other and to humans is a subject of great debate among all kinds of biologists. Molecular biologists compare protein or DNA sequences and draw one conclusion. Zoologists and **physical anthropologists** [Greek, *anthropo* = human], researchers who study humans, compare anatomy and behavior and draw slightly different conclusions. Paleontologists focus on the fossil record and see a subtly different picture. All these researchers agree, however, that humans share a common ancestry with other apes.

Were it not for our tendency to distance ourselves from other animals, humans would probably be lumped into the same family as the gorillas, chimpanzees, and orangutans—together called the **Pongidae**. We are closely related to both gorillas and chimpanzees. Our own DNA differs from that of chimpanzees and their cousins the bonobos ("pygmy chimps"), for example, by less than 1 percent.

Despite the similarities between humans and pongids, we humans have given ourselves a separate family, the **Hominidae**. Fossil and molecular data suggest that the common ancestors of orangutans diverged from those of the other pongids and the hominids about 12 million years ago. Hominids diverged from the chimpanzees and bonobos only 5 to 6 million years ago.

Homo sapiens is not the only hominid. We count as relatives a number of different species of hominids, some of which lived more than 4 million years ago. Two characteristics distinguish all hominids from other apes: (1) hominids are **bipedal**, that is, they consistently walk on two rather than four feet, and (2) hominids have rounded rather than rectangular jaws (Figure 17-22).

All hominid species except ours are now extinct, but anthropologists have discovered the buried bones of hominids throughout Africa, Asia, and Europe. In 1996, researchers digging near the shores of Lake Turkana, in East Africa, discovered the oldest known hominid. *Australopithecus anamensis*, as its discoverers named it, is 4.2 million years old and the earliest example yet of an **australopithecine** [Latin, *australis* = south + Greek, *pithekos* = ape]. *A. anamensis* had a small brain, a squarish jaw, and long apelike arms. But it walked upright.

How are we different from *Australopithecus*? Later hominids such as ourselves have large brain cases. Our backs arch, and our thigh bones attach differently to the pelvis and angle inward, so that we are "knock-kneed" compared to the pongids (Figure 17-23A). We have lost the opposable toe that characterizes other primates. Instead, we have the reduced toes and long feet seen in other running animals, such as horses and dogs (Figure 17-23B).

Figure 17-22 Hominid characteristics. Several characteristics distinguish *hominids* from other *hominoids*. A. Hominids tend to have more rounded jaws than chimps and gorillas. B. In addition, because hominids walk upright, we have arched rather than straight spines.

In addition, *Homo sapiens* combines extremely sparse body hair (compared to other apes) with an extravagant growth of hair on the head and, in the case of males, on the face as well. When illustrating hominids, artists are forced to decide whether to draw our ancestors naked like us, or hairy like pongids. Most artists compromise and draw our ancestors as semihairy, but with short hair on the head. In reality, the degree to which other species of *Homo* shared our strange distribution of hair will probably never be known, as hair is rarely fossilized.

Primates are divided into the lemurs, the platyrrhine monkeys, and the catarrhine monkeys and apes. The catarrhine apes, including every ape from gibbons to humans, are called hominoids. Humans and their relatives are hominids, while chimpanzees, gorillas, and orangutans are pongids.

Why Did *Australopithecus* Stand Up?

The question of how walking hominids evolved from a more primitive ancestor has haunted anthropologists for more than a century. Some researchers have argued that the evolution of

Figure 17-23 Tools for walking. A. Humans and other hominids have evolved the long lower limbs that characterize dogs, cheetahs, horses, and other animals that walk or run. Our legs attach to our hips differently from those of chimps, making us relatively "knock-kneed." B. Our feet are proportioned differently. Our big toe is no longer very good at grasping, and all our toes are much reduced compared to those of other apes. In this respect, we resemble horses, which also have reduced digits.

an upright stance and a hairless body was an adaptation for remaining cool on a hot African savanna millions of years ago. The idea is that a person (or ape) standing up exposes less surface area to the sun. Other researchers have suggested that the first hominids stood up because they could run faster that way—either from predators, or, alternatively, in pursuit of prey. Still others have argued that hominids stood up so that they could carry and wield weapons and other tools. Some now argue that hominids became walking apes when they began scavenging over large territories.

Most of these ideas lack specific evidence in their favor. The only certainty is that an ancestral ape did evolve an upright stance and the long legs of a long-distance walker.

Nearly all of the hypotheses about why hominids evolved an upright stance depend on the idea that hominids left the forest to live in an open grassland, or savanna. One of the most persistent theories is that our upright stance evolved in association with a global cooling 2.5 to 2.0 million years ago. At that time, the world was a colder and drier place. So much water was frozen into glaciers in the far north and south that ocean levels worldwide were much lower than today.

Researchers have theorized for many years that the dry climate caused African forests to turn into savanna. Our herbivorous, tree-dwelling forebears then abandoned both the remaining forest and their vegetarian habits and began hunting other animals out on the savanna, or so goes this widely accepted story. Out on the savanna, hominids evolved an erect posture, stone tools, large brains, and hairless skin. This story has been repeated so many times and with such assurance that most nonscientists are under the impression that it is an accepted scientific fact.

Unfortunately, it is just a story. Recent studies of the remains of plants in 2-million-year-old soils in East Africa suggest no dramatic change from forest to grassland between 2.5 and 2.0 million years ago. Researchers have concluded that hominids of the area most likely never encountered an open grassland. Other research, on the dust in ocean sediments, has shown only a single *dramatic* change in climate in Africa in the last 5 million years. This dry period came about 2.8 million years ago, more than a million years too late to account for the evolution of the first known bipedal hominid, 4.2 million years ago *(A. anamensis)*. This arid period, therefore, could not have been responsible for the evolution of bipedalism.

Later hominids did not necessarily live on the savanna, either. The 3.2-million-year-old hominid *Australopithecus afarensis,* until recently one of the oldest hominids known, was found at two sites in East Africa. One site was a dry, open habitat, as anthropologists had expected, but the other site was a woodland divided by ancient rivers.

Anthropologists trying to save the "savanna hypothesis" argue that bipedalism evolved in a dry, open habitat. It does not matter, they say, where other, later hominids lived or what the climate was like worldwide. They are pinning their hopes for validation on the idea that *A. anamensis,* the oldest known

BOX 17.3

What Is the Nature of Anthropological Evidence?

The kinds of evidence that molecular biologists use when studying living organisms and their relationships to one another are different from the evidence that paleontologists and anthropologists must use. The essence of most scientific evidence is reproducibility, and studies of living organisms tend to yield reproducible results, provided the studies are done carefully. For example, if a molecular biologist finds that chimpanzees share a certain proportion of the same alleles with humans on Tuesday in Los Angeles, a different researcher, using the same material and methods, should be able to get the same result on Friday in Boston.

Anthropologists, however, have limited material with which to work, and much of their evidence is not reproducible. For example, the number of fossils of early apes and humans is minuscule. A species of ape may be represented by a single jaw bone or a few teeth.

One of the most complete skeletons of an early hominid is "Lucy," a 3.2-million-year-old specimen of *Australopithecus afarensis*. Researchers strongly suspect that Lucy was female. Based on that assumption, they have hypothesized what a male of the same species would look like. They have further assumed that the male would be larger by a certain percentage than Lucy, and some textbooks report the height and weight of this theoretical male as if it were a fact. But other researchers have argued that there is no certainty that Lucy is female. Maybe "she" was a male. Further, without hundreds of complete skeletons of the same species, there is no way to know if Lucy was in any way typical of males *or* females. Lucy may have been unusually tall or unusually short.

Nonetheless, the scarcity of evidence in no way detracts from the fact that a long line of species has existed and that those species are morphologically intermediate between modern apes and modern humans. We may never know either the true average dimensions of skulls and limbs of each species or whether males and females were mostly the same size or different sizes. Nor may we ever know with certainty how each ancient hominid species is related to any other. But we do know that we are closely related genetically to modern chimpanzees and other apes. And we do know that the 3.2-million-year-old fossil Lucy walked upright, just as we do. There may be only one Lucy, but Lucy lived.

bipedal hominid, lived in an arid, open habitat. Other researchers argue that huge rivers fed Lake Turkana, where *A. anamensis* lived. These rivers and the lake itself, these researchers assert, must have been rimmed with forests. Anthropologists are now vigorously debating whether the habitat around Lake Turkana was arid or wet 4.2 million years ago.

The hypothesis that the first hominids stood up as a result of moving from the dark forest to a bright, open savanna carries great appeal. So far, however, no one has provided much evidence for it. Other related theories are similarly weak. The idea that standing up and losing the fur were adaptations to the heat of the savanna has little support, as most other animals on the African savanna walk and run on four legs and retain their fur. The ostrich, which is bipedal and runs well, has a broad, horizontal, black back, well suited to absorbing heat. It seems unlikely that the ostrich's bipedal gait is an adaptation for staying cool. Excluding naked mole rats, which live underground, the few African mammals that lack fur are large herbivores—elephants, rhinoceroses, and hippopotamuses, for example.

Did the first hominids stand up in order to carry something? Possibly, but that something was emphatically not stone weapons. The earliest stone tools, which experts say were most likely used to chop and mash vegetables (not to kill zebras), do not appear in the fossil record until about 2.5 million years ago, long after our ancestors stood up. Absent stone tools, it's difficult to imagine our australopithecine ancestors fashioning any other tools or objects important enough to confer a selective advantage on an upright posture. Some possibilities are wooden tools or woven baskets, which usually rot away and so do not appear in the paleontological record.

Another object an ape or australopithecine might have found worth carrying around is a baby. The babies of other monkeys and apes cling to their mother's fur, however. Human babies are born relatively undeveloped compared to other primate infants and are unable to cling. Biologists hypothesize that this is because our heads are so big that if babies waited any longer in the womb, they could not pass through the mother's pelvic bones. But our large heads apparently evolved *after* our upright stance. So researchers would have to hypothesize that hominids lost their fur before they evolved an upright posture, an idea for which there is no evidence.

With luck, researchers may someday agree about what selective pressures caused hominids to evolve an upright stance. For the moment, we can only say with certainty that they did. From the earliest australopithecines to the later ones, the skeleton increasingly became specialized for walking and running while upright.

No one knows exactly why hominids evolved the back, hips, and legs of a bipedal walker and runner. Most anthropologists assume that our ancestors left a life in the trees to live in an open, probably grassy, habitat. Researchers assume that the first hominids had to walk long distances and that this new lifestyle created selective pressures for an upright stance. But we may never know exactly where our ancestors were walking or why.

Who Were the Earliest Known Hominids?

The australopithecines fall into three to six distinct species, which lived at slightly different but overlapping times (Figure 17-24). All of them come from Africa, south of the Sahara desert. In fact, all hominids older than about 2 million years come from Africa, and most anthropologists agree that the first hominids evolved in Africa. The different species of australopithecines appear to have lived between 4.2 and 2.5 million years ago. For example, *A. afarensis* has turned up in Chad and in East Africa, where a nearly complete skeleton was unearthed (Figure 17-25).

This specimen of *A. afarensis,* named "Lucy" by her discoverer, was small, only about 1.1 meters (3 feet 7 inches) tall and weighing less than 30 kg (about 60 pounds). Lucy's adult body was the size of a 9-year-old *Homo sapiens,* but her brain was far smaller. Brain size for most normal adult humans ranges from 1000 to 1600 cubic centimeters (cm^3) (although the full range is 900 to 2000 cm^3). In contrast, the brain of *A. afarensis* was about 400 cm^3, about the size of a chimpanzee's brain (300 to 500 cm^3).

Other australopithecine species, taller and with larger brains, appeared during the next 2 million years. But the first fossils to be recognized as members of the same genus as ourselves—*Homo*—date from 2 to 3 million years ago. These hominids all used crude stone tools, which is one measure that sets them apart from the australopithecines. Members of the

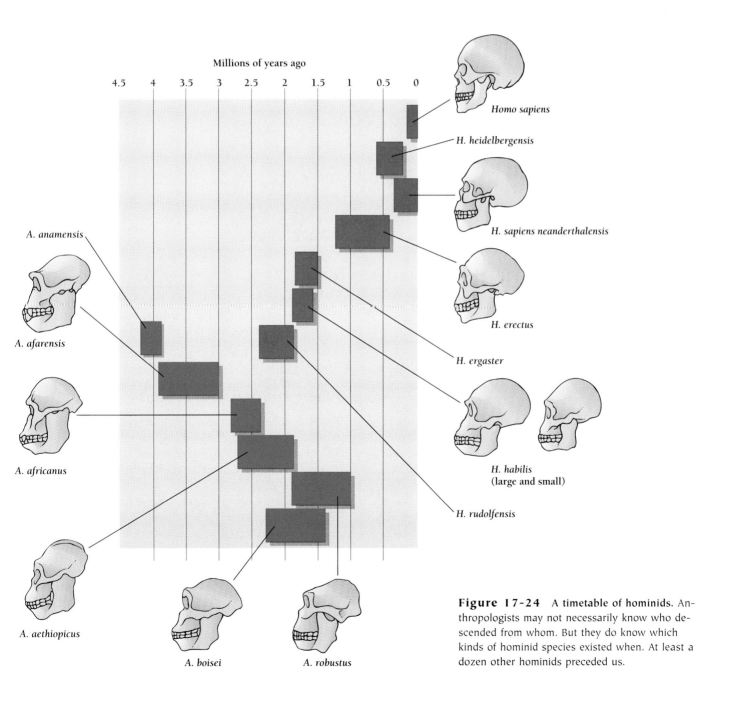

Figure 17-24 A timetable of hominids. Anthropologists may not necessarily know who descended from whom. But they do know which kinds of hominid species existed when. At least a dozen other hominids preceded us.

A.   B.

Figure 17-25 *Australopithecus afarensis.* One of the earliest australopithecines, *A. afarensis* is represented both in the nearly complete skeleton of "Lucy"—discovered in Ethiopia in the 1970s—and in other, more fragmented, fossils from Tanzania. Evidence that this australopithecine was bipedal comes not only from the pelvis, but from the extraordinary footprints found by Mary Leakey and her colleagues. A. Reconstruction of the *A. Afarensis* fossil "Lucy." B. Footprints of *A. afarensis*. A, Dr. Owen Lovejoy and students, Kent State University/Photo © 1985 David L. Brill; B, John Reader/Science Photo Library/Photo Researchers, Inc.)

genus *Homo* include *H. ergaster, H. habilis, H. erectus, H. heidelbergensis,* and *H. sapiens.*

The earliest fossils classified in the genus *Homo* have been discovered in Ethiopia. Estimated dates for these fossils range from 2.7 to 2.2 million years. No one is yet sure exactly how old they are. *Homo ergaster* appears in different parts of Africa from as early as 2.35 million years ago to as late as 1.4 million years ago. Toward the end of their existence, *H. ergaster* peoples used a more advanced method of chipping stones to form tools. Two or three million years ago, they chipped only one side of a stone to form a sharp surface for cutting and chopping—probably plant material. After about a million years, they learned to chip both sides to make a sharper edge.

Homo habilis [Latin, *habilis* = able], the "handy man," used stone tools and also apparently used them to butcher large animals, which *H. habilis* either hunted or scavenged. *H. habilis* stood about 1.7 m (5 ft) tall and had a 590-to 700-cm^3 brain, considerably larger than that of *A. afarensis*. Living side by side with *H. habilis* in at least two sites in East Africa was a hominid that is not included in the genus *Homo*. Researchers now believe that *Paranthropus,* which lived from 2.0 million until 1.5 million years ago, was another branch of the hominid family. Researchers dispute vigorously which of the various australopithecines and hominids are our true ancestors (Figure 17-26).

The early hominids, or australopithecines, were short and small brained compared to modern humans and other members of the genus Homo.

Where Did We Come From?

Nearly 2 million years ago, *Homo* left Africa and migrated to Asia. Asian fossils of *Homo erectus* exist from Java in eastern Asia to northern China. *H. erectus* fossils have also been found in Africa, so it's reasonable to guess that the *H. erectus* in Asia came from Africa.

Homo erectus was by far the longest-lived species of *Homo*. The earliest fossils date from 1.8 million years ago and the last date from just 100,000 years ago, for a total species life span of 1.7 million years. *H. erectus* existed nearly 17 times as long as our own species has.

Homo erectus skeletons are larger than those of *H. habilis,* and the brain case is also proportionately larger, ranging to more than 1000 cubic centimeters (cm^3) in the earliest specimens. In Java, where *H. erectus* lived until just 100,000 years ago, the brain cases are as large as 1300 cm^3, well within the limits of our own brain size (1000 to 1600 cm^3).

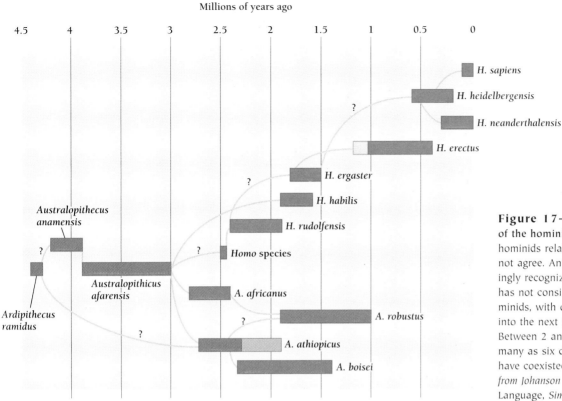

Figure 17-26 Branching evolution of the hominids. How are the various hominids related? Anthropologists do not agree. Anthropologists are increasingly recognizing that human evolution has not consisted of a single line of hominids, with one species neatly evolving into the next to culminate in *H. sapiens*. Between 2 and 3 million years ago, as many as six different hominids may have coexisted. (*Reprinted with permission from Johanson and Edgar: From Lucy to Language, Simon & Schuster Editions*)

Some anthropologists and paleontologists have suggested that *H. erectus* migrated from Africa, starting about 1.5 million years ago, with subsequent waves of migration evolving separately into varieties, or races, that differ from one another. Some of these, they argue, evolved into *H. sapiens*.

Others argue that *H. erectus* living in different parts of the world were replaced first by *Homo heidelbergensis* and later by *H. sapiens*, each of which migrated separately from Africa. In this view, *H. erectus* was a side branch in the evolution of humans, and we are descended from another line of *Homo* that left Africa after *H. erectus*. The most likely candidate seems to be *H. heidelbergensis*, found in Ethiopia and dated at about 130,000 years old. Several skulls that seem transitional between *H. heidelbergensis* and *H. sapiens* have been found in Ethiopia.

Alternatively, *H. sapiens* may have evolved in Africa 200,000 to 100,000 years ago, separately and at the same time as *H. heidelbergensis*, then emigrated to Asia and Europe. The earliest known representatives of *H. sapiens* were found at two sites in Israel and are dated at about 90,000 years old.

The early *H. sapiens* are commonly divided into two kinds, the **Cro-Magnons** and the **Neanderthals.** The Neanderthals, who looked much like us, but with a heavier build, occupied parts of Europe until just 35,000 years ago. Their thick arms and legs and heavy brows provide cartoonists with the classic "caveman" image; and, because of differences in Neanderthal throat bones, some researchers doubt that they spoke as well as we do, if at all. Nonetheless, Neanderthals, who first evolved between 200,000 and 100,000 years ago, probably had a fair amount of mental horsepower. Their brains were as large or larger than those of the Cro-Magnons, and some Neanderthals may have played musical instruments (Figure 17-27).

Figure 17-27 Neanderthal flute? In 1996, a team of researchers discovered what appeared to be a bone flute in a 43,000-year-old Neanderthal cave in Slovenia, in eastern Europe. Apparently fashioned from the thigh bone of a young bear, the hollow bone has four holes spaced as if for a diatonic scale (do, re, mi . . .). Similar flutes have been found in newer, Cro-Magnon caves, and researchers around the world reacted with delight to the idea that Neanderthals might have played music. Some researchers, however, flatly reject the idea. They argue that the holes were made by wolves' canine teeth and that the spacing is purely accidental. (*Blackwell, Queens College*)

Researchers do not know why the Neanderthals became extinct or if they interbred with Cro-Magnons. At one site, in Israel, the Cro-Magnons and Neanderthals appear to have lived within a few miles of each other at the same time. If we knew that they interbred, we could apply the biological species concept and conclude that they were one species. But no one knows, and researchers do not agree whether the Neanderthals were a separate species (*H. neanderthalensis*), a subspecies, or just a now-extinct race of *H. sapiens*. Recent analysis of DNA extracted from a Neanderthal bone suggests that they were genetically distinct from modern humans. But we still do not know if they were equally different from our Cro-Magnon forebears.

Even if we ultimately conclude that the Neanderthals were *H. sapiens*, we probably had other *Homo* cousins in the early days of our existence (Figure 17-24). Both *H. heidelbergensis* and *H. erectus* lived up to 100,000 years ago. What happened to them? No one knows, but anthropologists strongly suspect that *H. sapiens* exterminated them.

Anthropologists do not agree whether Homo sapiens *evolved in Africa. In the last 2 million years, several species of* Homo *appear to have been living on Earth at the same time.*

In the first two chapters of Part III, we reviewed the evidence for evolution and studied the ways that changes in gene frequency allow evolution to occur. In this chapter we discussed how species change and multiply and examined some of the larger patterns of evolution that are evident in the fossil record. In the next chapter, the last in Part III, we will explore a topic that is related, but in many ways quite different—the question of how and where life first originated.

The evolution of life is, like gravity, a simple fact. The origin of life, however, is truly mysterious. We know that life appeared on Earth within a few million years of the cooling of the Earth's crust, but we can only surmise how it got there.

Study Outline with Key Terms

Macroevolution is the origin and multiplication of species. According to the commonly accepted **biological species concept**, a **biological species** is a group of organisms that actually or potentially interbreed to produce viable offspring, but cannot interbreed with members of other such populations. Populations that do not interbreed with other populations are said to be **reproductively isolated. Sibling species** are species that are extremely similar, yet reproductively isolated. **Races** or **subspecies** are subpopulations of a species that are distinct, yet not reproductively isolated. In other words, races can interbreed with other races of the same species. Subspecies that interbreed only rarely are sometimes called **incipient species**.

Populations become reproductively isolated from one another through two kinds of barriers to gene flow: **prezygotic barriers,** which prevent fertilization, and **postzygotic barriers,** which either make the zygote inviable or make the resulting adult sterile. Prezygotic barriers include **ecological, temporal, behavioral, mechanical,** and **gametic isolation.** Once fertilization occurs, postzygotic barriers can come into play. These include chromosomal and genetic incompatibilities that cause the hybrid zygote to die young (**hybrid inviability**); or, if the individual reaches maturity, to produce no gametes (**hybrid sterility**); or, in some cases, if the hybrid produces offspring, to have *its* offspring be weak or sterile (**hybrid breakdown**).

Two other postzygotic barriers, peculiar mostly to plants, reproductively isolate hybrid offspring from the parent population in just one generation. These include the multiplication of chromosomes, or **polyploidy;** and the combining and doubling of nonhomologous chromosomes from different species, or **allopolyploidy.**

Geography frequently isolates populations from one another well before other isolating mechanisms develop. Populations can become isolated through **dispersal** and through **vicariance.** When reproductively isolated populations diverge in morphology or behavior, they are said to **speciate.** Geographic speciation follows several patterns. In **allopatric speciation,** geographic isolation halts nearly all gene flow. Genetic drift or different selective pressures complete the speciation process. In **parapatric speciation,** adjacent populations with limited gene flow manage to speciate. In **sympatric speciation,** populations that occupy the same geographic area become reproductively isolated through barriers to gene flow—temporal, ecological, or other.

Phylogeny is the history of descent of lineages of organisms from a common ancestor. **Divergent evolution** is the separation of a lineage into two or more species or lineages. **Convergent evolution** is a separate idea, referring to the acquisition of similar characters in unrelated lineages. An important kind of divergent evolution is **adaptive radiation,** the generation of diverse new species from a single ancestral species. Because no new body plans evolved after the **Cambrian explosion,** some biologists have suggested that **developmental constraints** limit evolutionary pathways.

Despite the gaps in the fossil record, Darwin believed in **gradualism,** the view that evolution was a gradual process. Early biologists rejected Darwin's gradualism in favor of **saltationism,** the view that evolution occurred in sudden leaps, or **saltations.** The last well-known saltationist suggested that genetic monstrosities, which he called **hopeful monsters,** began new evolutionary lineages.

Although gradualism gained favor from the 1930s through the 1960s, the fossil record seemed to show that many species remained unchanged for millions of years (**stasis**) and then abruptly gave rise to new species. A full record of intermediate stages was usually lacking. Thus, evolution, according to the fossil record, was neither gradual nor continuous. It seemed to happen in fits and starts.

In the 1970s, two paleontologists, Niles Eldredge and Stephen J. Gould, argued that the formation of new species is rapid compared to the rate of accumulation of changes within a species. This view, called **punctuated equilibrium,** rests on two main ideas. The first is that species change very little most of the time. The second is that most anatomical or other evolutionary change in individual species occurs during a geologically brief period at the time of speciation.

Anatomical and molecular evidence suggests that humans are **primates** most closely related to chimpanzees, gorillas, and other pongids. The **Pongidae** are a family of **hominoids,** which are the ape members of the catarrhines. The **catarrhines,** which include Old World monkeys and apes, are a more recently evolved group of primates than the **platyrrhines,** which are tailed monkeys that live in South and Central America.

Humans and their ancestors all belong in the family **Hominidae.**

Hominids are **bipedal** and have rounded jaws. **Physical anthropologists** call the earliest hominids **australopithecines.** Later hominids also have large brains, arched backs, and reduced toes. Two genera of hominids succeed the australopithecines: *Paranthropus* and *Homo.* All species of *Paranthropus* and *Australopithecus* are now extinct, and only one species of *Homo* remains—*H. sapiens.* No one knows exactly why hominids evolved a bipedal gait. Most anthropologists assume that the first hominids left a life in the trees to live in an open, probably grassy, habitat where they had to walk long distances.

Anthropologists do not agree whether *H. sapiens* evolved in Africa, as did all or most other species of *Homo,* or if *H. sapiens* evolved from an intermediate ancestor, which evolved in Africa but then dispersed to Europe and Asia. In the last 2 million years, several species of *Homo* appear to have been living on Earth at the same time. The longest-lived species was *H. erectus,* which lived in parts of Africa and Asia for nearly 2 million years and became extinct only about 100,000 years ago. Early *H. sapiens* are divided into the heavy-boned **Neanderthals** and the more delicate **Cro-Magnons.** Modern humans more closely resemble the Cro-Magnons.

Review and Thought Questions

Review Questions

1. Define the word "species"; then explain why that definition is tricky. What are sibling species? What is a subspecies? Why is a subspecies not a separate species?
2. What conditions are required for speciation?
3. Distinguish between prezygotic and postzygotic barriers to gene flow. Give several examples of each. Is polyploidy a prezygotic barrier or a postzygotic barrier?
4. Distinguish among sympatric, allopatric, and parapatric speciation.
5. What clues from the Galápagos Islands suggested to Darwin the idea of adaptive radiation?
6. Describe the Cambrian explosion. What happened 600 to 900 million years ago that has not happened since? Why hasn't it happened again?
7. Name a postzygotic isolating barrier in plants that is consistent with the saltationist view of evolution.
8. Define gradualism and punctuated equilibrium. What characteristics of the fossil record favor the theory of punctuated equilibrium? Explain why these two ideas are or are not in conflict with each other.
9. What are some weaknesses in the argument for stasis?

Thought Questions

10. Consider an oceanic archipelago consisting of only ten islands. Describe a pattern of colonizations whereby 20 or more species could arise from a colonization by a single mainland species.
11. With Goldschmidt's "hopeful monsters" in mind, make up a science-fiction scenario in which a human baby born with two hearts would have some selective advantage.
12. The current mass extinction will likely result in an adaptive radiation over several million years. What organisms do you think will survive and what niches will they evolve to fill 5 million years from now?

About the Chapter-Opening Image

The Galápagos tortoises are a classic example of the adaptive radiations that lead to speciation, the evolution of new species. The tortoise shown here is a giant tortoise (Geochelone elephantopus vandenburghi).

On-line materials relating to this chapter are on the World Wide Web at **http://www.harcourtcollege.com/lifesci/aal2/**

 Web Bit 17-1 Are Humans Still Evolving?

 Web Links

 Quizzing & Testing

18 How Did the First Organisms Evolve?

Can Organisms Arise Spontaneously from Nonliving Matter?

Until the last few hundred years, most people believed that new life arose spontaneously and daily from nonliving materials. Anyone could see that flies and maggots arose from rotting meat, frogs from swamps, mice from moist grain. Nearly everyone believed that spontaneous generation of life was common among the simple—so-called "lower"—plants and animals. Some people even argued that, under carefully controlled conditions, it might be possible to generate a man from a corpse.

Of course, every farmer knew that wheat and many other plants came from seeds, and that the cows and sheep were born of their mothers. Here and there, a few skeptics also insisted that insects and other creatures come from the eggs or sperm of parent organisms. Not until 1668, however, was this idea put to the test.

The skeptic was Francesco Redi (1626–1697), an Italian physician, poet, and naturalist, and one of the first modern experimental biologists. Heavily influenced by Galileo's belief that the natural world could be understood best through the senses, Redi strove to perform controlled experiments to test his ideas. Among these was a series of experiments challenging the idea of spontaneous generation.

In his own account of these experiments, Redi first states his hypotheses:

> I shall express my belief that the Earth, after having brought forth the first plants and animals at the beginning by order of the . . . Creator, has never since produced any kinds of plants or animals . . . and everything which we know in past or present times that she has produced came solely from the true seeds of the plants and animals themselves, which thus . . . preserve their species. And, although it be a matter of daily observation that infinite numbers of worms are produced in dead bodies and decayed plants, I feel . . . inclined to believe that . . . the putrefied matter in which they are found has no other office than that of serving as a place . . . where animals deposit their eggs at the breeding season, and in which they also find nourishment; otherwise, I assert that nothing is ever generated therein.

Redi then describes how, after preliminary studies of the development of different kinds of maggots into various species of flies, he put several kinds of fresh meat into two sets of jars. One set of jars was open, the other closed. Those that were open all developed maggots and, later, flies. Those that were closed developed no maggots, though, Redi notes, "Outside on the paper cover there was now and then a deposit, or a maggot that eagerly sought some crevice by which to enter and obtain nourishment."

Finally, Redi wanted to prove that the absence of air was not keeping the maggots from developing on the meat. He placed fresh meat into large vases that were either open or covered with very fine mesh. To prevent maggots from crawling through the holes in the netting, he put the covered vases inside a large net-covered frame and carefully removed any maggots that managed to penetrate the outer netting (Figure 18-1). The results were the same. "I never saw any worms in the meat, though many were to be seen moving

Key Concepts

- Life appeared just a few hundred million years after the Earth's crust hardened.
- Scientists can construct a reasonable scenario for the evolution of the first biological molecules and for the generation of abiotic cell-like structures.
- The first cells required a means of coupling metabolism and protein synthesis to genetic instructions. These cells may have used RNA as both genetic instructions and enzymes.
- Eukaryotic cells probably arose through a symbiotic association of prokaryotic cells.

about on the net-covered frame." The meat in the open vases was, of course, riddled with maggots, which, Redi observed, eventually turned into various kinds of flies. This experiment showed that maggots did not appear spontaneously.

Redi's 1668 experiments convinced most scientists that the common conception of spontaneous generation was incorrect. However, within 10 years, van Leeuwenhoek used his new microscope to find little "animalcules" (microorganisms) everywhere. Most people felt that even if maggots and flies could not arise by spontaneous generation, very likely these microorganisms could.

It was not until a hundred years later that anyone showed that they could not. In the 18th century, Lazzaro Spallanzani (1729–1799), an Italian physicist, microscopist, and physiologist, demonstrated that no organisms grew in a rich broth if it were first heated and allowed to cool in a stoppered flask (Figure 18-2). Other experimenters less careful than Spallanzani, however, contaminated their broth and so obtained the opposite result. Moreover, some scientists argued that, by heating the broth and air, Spallanzani may have destroyed some substance needed for spontaneous generation.

For another hundred years, these objections stood. Finally, in 1860, the controversy over spontaneous generation became so intense that the prestigious French Academy of Sciences offered a prize to anyone who could resolve the matter. In 1864, Louis Pasteur, the founder of microbiology, designed a simple but irrefutable experiment that won the prize.

Pasteur repeated Spallanzani's experiment, but with one essential difference. Instead of stoppering the flask to prevent organisms or spores from falling into the broth, Pasteur used a flask with a long neck like that of a swan (Figure 18-3). Air could pass freely in and out of such a flask, but microbes could not get through the long, curved neck. Only when Pasteur broke the neck of the flask was the broth "seeded" by bacteria and fungi falling from the air. Some of Pasteur's open flasks—still sterile after more than 130 years—are on display in Paris.

"All life from life," Pasteur emphatically concluded. "Never," he said, "will the doctrine of spontaneous generation recover from the mortal blow of this simple experiment."

WHERE DOES LIFE COME FROM?

Since Darwin, biologists have known that all modern species have evolved from previous ones. But we must ask, From what did the millions of species on Earth evolve? If Pasteur was correct in saying that all life comes from other life, we must account for the presence of evolved life on a planet that itself came into being a finite number of years ago. Where did life first come from?

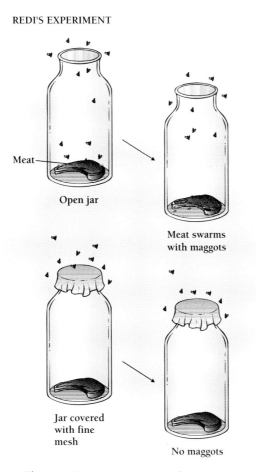

Figure 18-1 Francesco Redi's experiment. Do flies arise spontaneously from rotten meat? Or do they develop from eggs laid on the meat by adult flies?

SPALLANZANI'S EXPERIMENT

Figure 18-2 Lazzaro Spallanzani's experiment. Do microorganisms arise spontaneously in a rich broth?

Every culture on Earth—ancient, primitive, or modern—has its own answer to this question. Many invoke an intelligent, creative force. However appealing many of these myths may be, they do not constitute scientific explanations, and they tell us little more than that "life happened."

Scientists cannot say with certainty how life arose on Earth, either. However, they can speculate, based on a modern understanding of astronomy, planetary science, geophysics, and biology, what might reasonably have happened. Scientists see essentially two alternative answers to the question, Where did life on Earth come from? Either life on Earth arose spontaneously from nonliving organic molecules in the primitive environment or else it arrived from outer space by means of a meteor, comet, or asteroid.

In this chapter, we will see why neither of these ideas is quite as outlandish as it might at first seem. The bulk of our discussion will be devoted to exploring how life might have originated on Earth. As one prominent biologist put it, this discussion resembles a mystery novel with too many suspects.

Evolutionary biologists have inferred part of the story from the shared characteristics of modern organisms. But much of the story must be guesswork. Because the origin of life is not an ongoing, testable process, as evolution is, biologists can only surmise how life might have arisen. The point is to show that life could have arisen by some series of steps, each of which is reasonable in terms of our current understanding of the laws

PASTEUR'S EXPERIMENT

Figure 18-3 Louis Pasteur's experiment. Do microorganisms arise spontaneously in a rich broth provided that air is available?

When and Where Did Life First Appear?

Before 1950, no one had seen a fossil older than about 570 million years (the beginning of the Cambrian Era). Older fossils went unnoticed, it turns out, because they are extremely rare, usually quite small, and found in a kind of rock different from that of later fossils.

Cambrian and later fossils are usually hard bones and shells preserved in shales, sandstones, limestones, and other sedimentary rocks. But most Precambrian organisms had few or no hard parts, and their soft remains usually dissolved or decayed. Their soft, tiny bodies persisted only when caught in an extremely fine-grained rock called chert. Such rocks are now generally far below the Earth's surface, covered by later deposits. Even cherts that have returned to the surface may be hopelessly deformed by heat and pressure. In only a few places on Earth are the cherts that bear Precambrian fossils both accessible and intact.

But for scientists interested in Precambrian life, finding fossil-bearing cherts was only half the battle. In order to see the fossils, paleontologists had to slice the chert into thin sections, polish them until they were transparent, and then examine the sections under a microscope.

The oldest and most primitive fossils known resemble modern prokaryotes—simple rods, spheres, and filaments with no nucleus (Figure 18-4). These ancient fossils, embedded in cherts from southern Africa and western Australia, are about 3.5 billion years old. More recently, researchers have dated fossil organisms that are even older, some 3.8 billion years.

Another important group of ancient fossils are the stromatolites, banded columns of limestone or chert, as old as 3 billion years (Figure 18-5). For many years, scientists did not realize that stromatolites were the remains of living organisms.

Figure 18-4 Filamentous cyanobacteria. Some filamentous fossils closely resemble their modern counterparts. A. Modern filamentous cyanobacterium. B. A 3.5-billion-year-old filamentous cyanobacterium from the Warrawoona rocks of western Australia. *(A, Biological Photo Service; B, Stanley Awramik/Biological Photo Service)*

Figure 18-5 Living stromatolites like these once filled vast shallow seas. These cushions of cyanobacteria at Shark Bay, western Australia, resemble 3.5-billion-year-old fossil stromatolites. *(John Reader/Science Photo Library/Photo Researchers, Inc.)*

of physics, chemistry, and biology. By showing that they can tell such a reasonable story, evolutionary biologists have attempted to refute the idea that there is an uncrossable gulf between living and nonliving matter. We will also see why, despite all this, Pasteur was after all correct: spontaneous generation could not happen today.

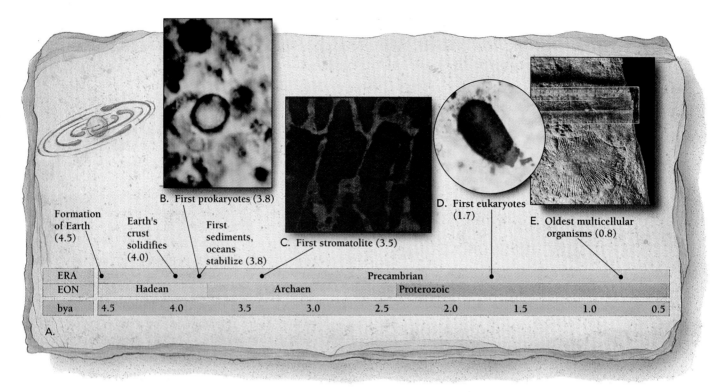

Figure 18-6 A calendar of Earth's early history. Life appeared on Earth almost as soon as conditions would allow it. The first stromatolites (photosynthetic cyanobacteria) appeared 3.5 billion years ago. Eukaryotes appeared nearly 2 billion years after the first prokaryotes. (*B, Science VU-USM/Visuals Unlimited; C, William E. Ferguson; D, Andrew H. Knoll, Knoll and Calder, Palaeontology 26:467, 1983; E, Biological Photo Service*)

This failure was because they consist of an unusually orderly arrangement of sediments, with no trace of the original organisms. But living stromatolites, discovered in a few remote intertidal areas, solved the mystery. Stromatolites are formed when huge, mounded colonies of cyanobacteria and other microorganisms trap fine sediment. The sediment settles in thin layers within each mound. The cyanobacteria in modern stromatolites are photosynthetic. Paleontologists assume that fossil stromatolites were also photosynthetic. In other words, photosynthesis probably evolved at least 3 billion years ago (Figure 18-6). Later in this chapter, we will see how the photosynthetic abilities of early cyanobacteria profoundly changed the world, allowing the evolution of all higher organisms.

As we have noted, there are two general explanations for the relatively sudden appearance of prokaryotic organisms soon after the Earth had cooled: (1) life arose spontaneously from nonliving organic molecules in the primitive environment, or (2) life arrived from elsewhere in the universe. It is almost impossible to imagine how we could prove or disprove either of these explanations. We can, however, examine the reasons for thinking either of them reasonable.

The second hypothesis at first appears to be the simpler explanation for the appearance of life. According to this view, the present diversity of life all derives from a colonization of Earth followed by a huge adaptive radiation. An appealing feature of this hypothesis is that the universe is thought to have originated at least 12 to 16 billion years ago, 8 to 12 billion years before the cooling of the Earth. This far longer history would give life more time to develop. Furthermore, an extraterrestrial origin of life explains some otherwise unexpected chemical properties of organisms. It would explain, for example, the requirement for molybdenum—an extremely rare element on this planet—in key reactions of photosynthesis and nitrogen fixation.

Until recently, no evidence existed for life on the other planets in the solar system. In August of 1996, however, researchers discovered signs of life in a meteorite composed of 3.6-billion-year-old Martian rock, apparently dislodged from the Martian surface by an asteroid collision about 15 million years ago. The rock, which crashed into Antarctica 13,000 years ago, contains what look like tiny fossils and also complex organic molecules that, on Earth, are produced when organisms decay or burn. These molecules are most concentrated in the interior of the rock, suggesting to some researchers that they did not infiltrate the rock from the outside after it landed on Earth.

No one knows if these fossils represent life from Mars. Even if Mars did once harbor living organisms, there is no evidence that life on Earth was seeded by Martian organisms. Earth may have seeded Mars. Or life may have arisen independently on each planet. Another possibility is that life arrived on both planets from some source outside the solar system.

BOX 18.1

Is There Life Elsewhere in the Universe?

In 1959, a young astronomer assigned to the Greenbank Observatory in rural West Virginia began using the big observatory radio telescope to listen for signals from extraterrestrial beings. Frank Drake and his colleagues tried to keep it a secret, but word got out quickly. Newspaper reporters besieged the observatory with calls, but Drake could not honestly report any definite communications from outer space.

Yet, for nearly 40 years, Drake hasn't stopped listening (Figure A). In 1974, he used Cornell University's Arecibo radio telescope, a giant, 300-m telescope in Puerto Rico, to send the first deliberate message from Earth. The message described Earth's position in the Milky Way galaxy, what humans look like, and how many of us there were.

Today, Drake and others are still listening for signs of intelligent life in the universe. This project, called the Search for Extraterrestrial Intelligence (SETI) asks, Is anyone out there? (SETI's work is depicted melodramatically in the recent Hollywood film "Contact.") Biologists tend to be conservative about such things. Many believe that the emergence of life on Earth is so remarkable that it would be astonishing if it occurred elsewhere in the universe. Indeed, the biologist and writer Jared Diamond has argued that the development of the radio on Earth was extremely unlikely. (He adds that, given the way humans treat less advanced civilizations and less intelligent creatures on our own planet, we will be fortunate indeed if we are alone in the universe.) Biologists are also eager to distance themselves from people who claim encounters with intelligent extraterrestrials on Earth.

Astronomers tend to take a different view. They argue that to insist that life could have arisen only on Earth is arrogant and excessively geocentric, or Earth-centered. Of course, they dismiss flying saucers—so-called UFOs (unidentified flying objects). Even the closest stars are too far away for spaceships to get here, let alone home again. And if intelligent life exists, the chances are that it is far away, indeed. Any signals that SETI picks up will be thousands or millions of years old.

Figure A Are we alone? Probably not. Frank Drake, president of the SETI Institute and professor emeritus of astronomy at the University of California, Santa Cruz, first began searching for intelligent extraterrestrials in 1959. (*Bill Lovejoy*)

But we are almost certainly not alone. Drake has suggested that the number of civilizations in the galaxy can be calculated using the following equation, called the Drake Equation:

$$N = R f_p\, n_e\, f_l\, f_i\, f_c\, f_L$$

In English, this equation says, the number of technological (detectable) civilizations in the galaxy (N) is equal to the rate (R) of star formation, times the fraction of stars with planets (f_p), times the number of planets ecologically suitable for life (n_c), times the fraction of those planets where life actually evolves (f_l), times the fraction that develops intelligent life (f_i), times the fraction that communicates (f_c), times the fraction of the planets' life occupied by the communicating civilization (f_L).

Clearly, the answer depends on what values are plugged into this equation. However, many of these values can be estimated with some confidence. For example, astronomers estimate that the Milky Way contains about 400 billion stars. Conservative estimates for most of the rest of the variables give predictions such as 4 billion planets with some form of life and 4000 existing technological civilizations in the Milky Way. And that is in our galaxy only. From astronomers' perspective, the evolution of life was practically inevitable.

But if extraterrestrial life is out there, we haven't found it yet. Soil samples from other planets have revealed no life, and SETI has so far turned up nothing concrete. The only tantalizing hints so far are some complex organic compounds from a meteorite believed to have come from Mars.

Organic molecules seem to abound in the universe. Astronomers have detected the presence of a number of carbon- and nitrogen-containing compounds in interstellar gas clouds. These compounds include those that we think were present in the Earth's primitive atmosphere—hydrogen, hydrogen cyanide, carbon monoxide, and ammonia.

Soil samples brought back from the moon contained several amino acids. Meteors, such as the one that fell in Murchison, Australia, in 1969 (Figure 18-10), contain the amino acids found in organisms as well as other amino acids and more complex molecules. The fragments from the Murchison meteorite contained 1 or 2 percent organic material. If such an endowment is typical of comets, asteroids, and meteors in the rest of the galaxy, it is hard to imagine how life could not have evolved elsewhere.

For life to colonize Earth (and Mars), it would have to have come from at least as far away as the nearest star, some 4 light years away—38 trillion km (23 trillion miles). Living organisms would have had to survive extremely violent events, for nothing else would have propelled them into interstellar space. There they would have had to survive intense radiation for untold millions of years. Recent research has suggested, however, that bacteria might survive the intense radiation if they were encased in ice.

If such a trip were possible, we would still want to understand how (and where) those organisms first arose. The problem of the origin of life would be virtually insurmountable, since we would not know where in the universe such organisms might have come from.

For now, the colonization-from-space hypothesis is only remotely likely. Perhaps more important, the hypothesis is so inaccessible to inquiry that most scientists interested in such things have focused their attention on the possible origin of life on Earth. Most biologists now believe that life must have developed from nonliving molecules on the early Earth's surface. They have accepted the challenge of showing that the development of organisms from nonliving matter—called **prebiotic evolution**—is possible.

Most biologists believe that life originated on Earth. No one knows enough about conditions in the rest of the universe to guess how extraterrestrial life might have evolved or safely arrived on Earth.

PREBIOTIC EVOLUTION: HOW COULD COMPLEX MOLECULES EVOLVE?

By the 1930s biochemists realized that all organisms are made from the same building blocks—the sugars, lipids, amino acids, and nucleotides that make up the "biochemical alphabets" described in Chapter 3. Two influential scientists—Alexander Ivanovich Oparin, a Russian biochemist, and J.B.S. Haldane, the English evolutionary biologist—proposed that these building blocks had been formed in the primitive environment. The original synthesis of these materials, they argued, was **prebiotic** and therefore **abiotic**—before life and therefore without life.

Important differences separate spontaneous generation from theories of prebiotic evolution. First, while spontaneous generation had to occur nearly instantly, the origin of life may have taken as long as 300 million years to unfold. Second, the conditions during which life must have first appeared were dramatically different from those that exist today. One of the most important differences was that the atmosphere contained virtually no free oxygen. Had oxygen been present in any quantity, it would certainly have oxidized the starting materials for life. A second important difference was the absence of life itself. Today, life could not arise on Earth because a plethora of hungry organisms find and consume organic molecules.

Darwin himself grasped this distinction, writing to a friend in 1871:

> It has often been said that all of the conditions for the first production of a living organism are now present which could ever have been present. But if (and oh! what a big if!) we could conceive in some warm little pond, with all sorts of ammonia and phosphoric salts, light, heat, electricity, etc., present, that a protein compound was chemically formed ready to undergo still more complex changes, at the present day such matter would be instantly devoured or absorbed, which would not have been the case before living creatures were formed.

Was Darwin right? In the absence of devouring organisms and destructive oxygen, can the building blocks of life manage to assemble themselves into living organisms? To decide, we must ask whether we can imagine a reasonable scenario for the "spontaneous" production of the building blocks of life and whether laboratory experiments demonstrate that the steps in the presumed process can actually occur. To discuss these questions, however, we must first understand the conditions on Earth shortly after it formed.

The conditions available for the origin of life differ from those for spontaneous generation in three dramatic ways—the long period of time available, the absence of life, and the absence of oxygen.

What Was the Earth Like When It Was Young?

The Earth formed about 4.6 billion years ago, when a spinning cloud of rocks, dust, and gas formed the sun and its planets. Astronomers believe that the spinning cloud gradually spread out and flattened as it rotated, like a lump of spinning pizza dough (Figure 18-7). The bulk of the material, at the center, condensed into an enormous compact mass that became the sun. The rest of the spinning cloud gradually formed the planets, comets, meteors, and other objects that make up the solar system.

The debris orbiting closest to the protosun [Greek, *protos* = primitive] moved fastest, and there the friction of constant collisions generated the most heat. Close to this hot center, pieces of silicon, metals, metal oxides, and other materials that remain solid even at high temperatures fell together, drawn to one another by their own gravities. These hot lumps of solid material gradually formed the inner protoplanets—Mercury, Venus, Earth, and Mars. Water, methane, and other compounds that solidify only at low temperatures formed the cold, outer protoplanets, such as Jupiter, far from the center of the spinning disk. The huge, outer planets attracted and held cold atmospheres of hydrogen, helium, and other gases. But the tiny, inner planets had gravities too weak to hold the hot gases, which escaped into space.

The early protosun was much cooler than today's sun. But Earth was far hotter. Gravity compacted the Earth, and huge

Figure 18–7 Formation of the solar system. Like a spinning lump of pizza dough, a cloud of dust and rocks flattened and then condensed into a central mass that became the sun. Material further out condensed into rotating balls of rock and gas that we now know as the planets. (*Source: Kristin Levier, Science Notes, University of California, Santa Cruz*)

meteors, many of them hundreds of miles in diameter, bombarded the planet. (The impact of one of these collisions was so great that it dislodged a huge chunk of Earth, which became the moon.) The compaction and the bombardment kept the molten Earth from solidifying for another half a billion years. The heaviest elements sank to the center of the Earth, while the lightest ones floated to the top. Iron and nickel sank to the core, the lightest silicates (granite, for example) rose to the surface, while denser silicates of iron and magnesium floated in between.

After 500 to 600 million years, the surface cooled enough to form a thin crust. Dense gases escaped from the interior through cracks and volcanoes in the new crust to form a primitive atmosphere. Among these gases was water vapor, immense quantities of which then condensed into clouds. Torrential rains pummeled the still hot Earth for millions of years, creating the oceans and eroding the Earth's rocky surface. Although the oceans formed relatively early, constant bombardment by meteors, large and small, may have repeatedly vaporized the oceans until about 3.8 billion years ago. About that time, the first sediments and the first life appeared.

The evolution of the atmosphere

The exact composition of the gases in the primitive atmosphere has been a matter of much debate. The modern atmosphere is 78 percent nitrogen, 21 percent oxygen, and about 1 percent argon, with varying amounts of water vapor close to the Earth. Another 0.1 percent of the atmosphere consists of traces of carbon dioxide, carbon monoxide, sulfur oxides, nitric oxides, and other gases.

Scientists long assumed that the early atmosphere contained hydrogen (H_2), ammonia (NH_3), methane (CH_4), and water (H_2O), with traces of carbon dioxide (CO_2) and hydrogen sulfide (H_2S). Recent research has cast doubt on these assumptions, however, suggesting that there was no hydrogen and much more CO_2.

About one important gas there has been little argument. The early atmosphere had almost no oxygen. The strongest evidence for this is the near-total absence of oxidized iron (rust) in the earliest sediments. Had oxygen been present in any quantity, it would certainly have oxidized the starting materials for life. Oxygen, we will see later in this chapter, came much later, probably as a by-product of photosynthesis.

The absence of oxygen had other important consequences. Chief among these was the absence of an ozone layer. The ozone (O_3) layer, about 40 km (25 miles) up, today shields the Earth from intense ultraviolet (UV) light from the sun. Modern organisms need that protection because UV light, extremely good at making and breaking chemical bonds, interferes with normal life functions.

However, 4 billion years ago, any chemical synthesis would have required considerable energy. Recall from Chapter 5 that synthetic reactions are almost all thermodynamically uphill. So UV radiation from the sun would have helped to promote novel chemical reactions. Heat from the hot interior of the Earth provided more energy. But the most dramatic energy source was the lightning accompanying the storms that filled the oceans.

The Earth's crust solidified about 4 billion years ago, but the oceans were not stable until about 3.8 billion years ago. Since the first fossil organisms are about 3.8 billion years old, we can conclude that life on Earth originated relatively quickly—within a 300-million-year window.

Is Prebiotic Synthesis Plausible?

By the early 1950s, scientists had some idea of what the early Earth was like: they believed it had the raw materials for organic synthesis—water, methane, ammonia, and hydrogen—and it had sources of energy—UV radiation, heat, and lightning. But scientists did not know whether the conditions of the Earth's early history would allow the abiotic synthesis of biologically important molecules.

Figure 18-8 Stanley Miller's experiment. Can conditions similar to those on Earth 4 billion years ago give rise to the building blocks of life? A. Miller treated a mixture of water and gases with flashes of artificial "lightning." Among the complex molecules produced after 1 week were five amino acids. B. Miller and his apparatus. (B, © 1988 Roger Ressmeyer/ Corbis)

Then, in 1953, a 23-year-old graduate student working in the laboratory of Harold Urey, at the University of Chicago, undertook to simulate the primitive conditions of the Earth. Stanley Miller simulated the early Earth inside a simple glass apparatus built from the standard equipment found in a sophomore organic chemistry laboratory (Figure 18-8). The simulated Earth consisted of water in a boiler (the ocean) and a chamber fitted with electrodes (the atmosphere, with lightning). In one experiment, Miller mixed together an atmosphere of hydrogen (H_2), methane (CH_4), and ammonia (NH_3). Miller let the gases circulate past the flashing electrodes for a week and then analyzed the contents.

He found that more than 10 percent of the carbon from the methane was in organic molecules. These included at least five amino acids, as well as compounds that are precursors of other amino acids and of the bases of nucleic acids. Among these precursors was hydrogen cyanide (HCN), five molecules of which can directly form adenine (Figure 18-9). The mixture was unexpectedly rich in some (but not all) of the same basic building blocks of life, which we listed in Chapter 3.

Miller and others repeated this basic experiment, with different atmospheres and different sources of energy. As long as free oxygen (O_2) was absent, the same building blocks appeared. Other sources of energy also work—most notably UV light, the principal energy source available on the Earth's surface before the Earth developed its protective ozone layer.

As a result of these experiments, it seemed likely that—within a few million years—the primitive oceans loaded up with organic molecules capable of becoming the building blocks of life. This primordial soup may have contained as much as 1 percent organic molecules in some places, about the same as a watery chicken broth. Remarkably, the amino acids most easily produced in these abiotic experiments are the ones most common in proteins today. And the most easily synthesized base—adenine—is biologically the most important of all the nitrogenous bases.

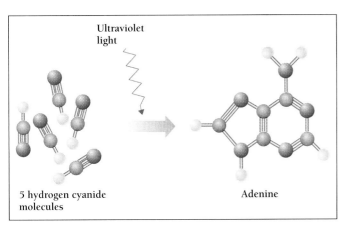

Figure 18-9 **First steps in the formation of life?** Miller's apparatus also produced hydrogen cyanide. In the presence of ultraviolet light, five molecules of hydrogen cyanide form adenine, the nitrogenous base in ATP.

A distinct problem with this scenario, however, is that recent research by geoscientists suggests that the early atmosphere contained neither methane (CH_4) nor ammonia (NH_3). Heavy doses of UV light from the sun would have destroyed such hydrogen-based molecules, and even free hydrogen is so light that it would have escaped into space. Geoscientists say that the major components of the atmosphere were more likely carbon dioxide and nitrogen, with water and traces of carbon monoxide and free hydrogen, the same gases that are ejected from modern volcanoes.

Such an atmosphere would not have allowed the kind of amino acid synthesis suggested by Miller's experiment. Formaldehyde (H_2CO), essential to the formation of sugars, would have formed easily enough. But a likely biochemical pathway for the synthesis of hydrogen cyanide, the building block for adenine and other amino acids, has yet to be found.

Miller, now a professor of chemistry at the University of California, San Diego, defends his early study, saying that smoke and clouds might have shielded the fragile methane and ammonia molecules from UV light. And several Japanese researchers, supporting Miller, argue that solar particles and cosmic rays would have countered the effect of UV light by releasing free hydrogen from water molecules and so promoting the synthesis of methane and ammonia.

Other researchers have suggested another possible source of the first building blocks—the seas surrounding thermal vents in the ocean. These hot springs lie deep in the oceans, at the crests of underwater ridges. Seawater circulates through the vents, where its temperature rises to 350°C and then abruptly plunges to 2°C as it mixes with cold ocean waters. Modern-day vents are the sites of lots of underwater life, sustained by unusual metabolic reactions that do not depend on solar energy.

Some biologists suggest that the abundant hydrogen sulfide and iron sulfide in the superheated water could have provided the perfect environment for the synthesis of primordial building blocks. In this scenario, the building blocks accumulated on the surfaces of underwater rock instead of dissolving in the ocean. Perhaps, these biologists say, there was no primitive soup.

Prebiotic synthesis of biological building blocks may have taken place in the ocean or on underwater rocks.

Was Prebiotic Synthesis Necessary?

Some astronomers have suggested that the debate over the exact composition of the early atmosphere may be moot. They believe that another source of simple organic molecules is possible. Researchers have discovered that asteroids, meteors, and comets are rich in complex organic compounds created during the formation of the solar system (Figure 18-10). Further, based on the calculated number and size of comets and meteors likely to have entered the Earth's atmosphere, astronomers estimate that as much as 10 million kg of complex organic molecules might have showered the Earth each year. At the end of a billion years, they calculate, the Earth could have accumulated as much as a million billion (10^{15}) kg of organic molecules, more than the total mass of all life on Earth today.

The first organic molecules could have come from asteroids, meteors, or comets.

How Can Biological Building Blocks Assemble Outside of Cells?

If we assume that the Earth's early environment either allowed the production of amino acids and other building blocks or received them passively from space, we must still account for the assembly of those building blocks into other amino acids, nucleic acids, ATP, and other macromolecules. Can this process be simulated in the laboratory?

In the Miller atmosphere, adenine easily forms from hydrogen cyanide (Figure 18-9). The other nucleic acid bases require more complex reactions, so adenine could have been the first nitrogenous base to form. If so, it would not be surprising for its activated product, ATP, to have become the universal energy currency of life.

ATP itself may have formed rather easily. Laboratory experiments show that mixtures of adenine, ribose, and phosphate can form ATP when exposed to UV light. Other triphosphates may have arisen in the same way.

Even prebiotic replication of nucleic acids is not at all far fetched. Biochemists have discovered that nucleotides can combine in the test tube to form short RNAs. These RNAs, more-

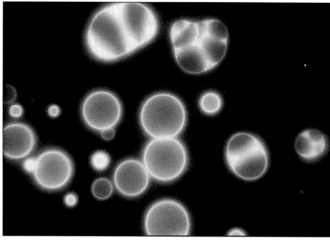

Figure 18-10 Remains of extraterrestrial life? A. This fragment of a meteorite hurtled through the atmosphere and landed in Murchison, Australia, in 1969. Scattered through the rock are particles of organic compounds. B. When extracted, these compounds self-assembled into vesicles. The yellow-green color is the fluorescence of complex organic compounds called polycyclic aromatic hydrocarbons. (*D.W. Deamer, University of California, Santa Cruz*)

over, form double-stranded molecules that follow the Watson-Crick rules for base pairing.

Simulating the polymerization of amino acids into polypeptides is more of a challenge. Many researchers, however, think that ATP in the primordial broth may have somehow contributed to the synthesis of activated amino acids. Solutions of such activated amino acids can spontaneously give polypeptide chains up to 50 amino acids long. Polypeptides also form when researchers heat dry mixtures of amino acids. Sidney Fox, a biochemist at Southern Illinois University, has suggested that such polypeptides could have formed on volcanic cinder cones and washed into the sea.

The earliest polypeptides almost certainly would not have had specific sequences of amino acids. In fact, Fox has dubbed the early polypeptides **proteinoids,** to distinguish them from the true proteins, which are the products of living organisms and whose sequences are defined by the genes in DNA.

The abiotic synthesis of RNA, ATP, and even polypeptides seems plausible.

Some variant of the chemical reactions described earlier may have dissolved in the open seas or may have accumulated in small puddles or on the surfaces of rocks. In any case, most biologists think that the building blocks of life became available for prebiotic evolution, perhaps as long ago as 4 billion years.

But the materials of life are not the same as life. Two important steps would have to have occurred—the evolution of reproduction and the evolution of metabolism. The evolution of reproductive ability depended on establishing a highly specific relationship between the self-replicating nucleic acids and the sequence of amino acids. This meant the development of both the genetic code and a mechanism of translation.

Before this could happen, however, polypeptides and other molecules probably began to come together into organized (and concentrated) associations. As we will see, such primitive cells, or **protobionts,** could concentrate organic molecules and maintain an internal environment different from that of their surroundings. They could, in short, begin to develop the first suggestion of metabolism.

Life required the evolution of reproduction and metabolism.

How Can Prebiotic "Cells" Form?

Polypeptides, nucleic acids, and polysaccharides in solution will spontaneously concentrate themselves into discrete tiny, balloonlike droplets, which the Russian biochemist Oparin called **coacervates.** Coacervates have a discrete inside and outside separated by a membranelike boundary.

If enzymes are present in the solution, these will be bound in the coacervates, which then act like little cells. They absorb molecules from the solution and release the products of the reactions catalyzed by the enzymes. For example, Oparin showed that coacervates containing the enzyme phosphorylase would take up glucose-1-phosphate from the surrounding medium,

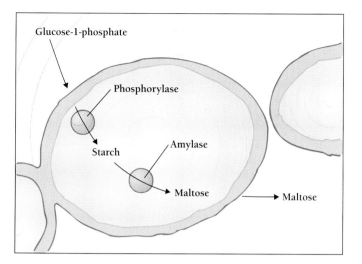

Figure 18-11 Prebiotic metabolism? Macromolecules in solution can form droplets, which Alexander Oparin called coacervates. In one experiment, coacervates that contained the enzymes phosphorylase and amylase could convert glucose-1-phosphate to starch and then to maltose.

Figure 18-12 Proteinoid microspheres. These nonliving polymer bubbles can reproduce themselves by growing and budding. (Steven Brooke and Richard LeDuc)

convert it to starch, and concentrate it inside the droplet (Figure 18-11).

Other mixtures can also give rise to isolated, cell-like structures. For example, Sidney Fox found that the proteinoids formed from dry amino acids would form tiny spheres. These structures, called **proteinoid microspheres,** could absorb more polymers from solution, forming buds or even a new generation of spheres (Figure 18-12). Microspheres dropped in water are differentially permeable. They swell or shrink according to the concentration of salts in the water. Some will even store energy, in the form of a differential electric charge across the membrane. Such microspheres can release a charge in much the same way a nerve cell does. A mixture of phospholipids and proteins can form yet another kind of enclosed structure, called a **liposome.** Some liposome membranes have two layers, much like those in modern cells.

None of these self-organizing structures are cells. Yet they look so much like cells that experienced biologists have occasionally mistaken them for bacteria and even tried to classify them. And protobionts show some of the characteristics of life, including growth, metabolism, and responsiveness to their environment. The existence of protobionts shows that cell-like structures could have formed spontaneously in the primordial soup.

Further, we can envision the evolution of primitive metabolic pathways. As coacervates used up a given compound in the environment, those coacervates able to make the compound from another component of the soup would persist. Once the precursor was exhausted, further growth of protobionts would depend on the ability to produce the needed product from still another precursor. According to this scheme, a number of pathways would evolve—one step at a time (Figure 18-13).

EARLY BIOTIC EVOLUTION: HOW DID THE FIRST GENETIC SYSTEM EVOLVE?

Even with the evolution of primitive metabolic pathways, protobiont "life" would have been severely limited. Despite protobionts' ability to grow and to bud, they could not reproduce. Even if they had accumulated catalysts for specific chemical reactions, protobionts could not make more of these specific catalysts. As unique molecular catalysts were passed on to increasing numbers of "offspring" protobionts, the catalysts became increasingly diluted. To invest the new protobionts with equal concentrations of functional molecules required the invention of a way of "teaching" the next generation to make its own catalysts. Life would need a way of storing, replicating, and translating the information for the building of enzyme catalysts—in short, a genetic system.

Could RNA Have Served as the Original Genetic Material?

A modern cell stores its genetic information as DNA, transcribes the information into RNA, and then translates the message into specific polypeptides. When the cell divides into two cells, it gives a copy of the DNA to each new cell, and so passes on the recipe for each polypeptide.

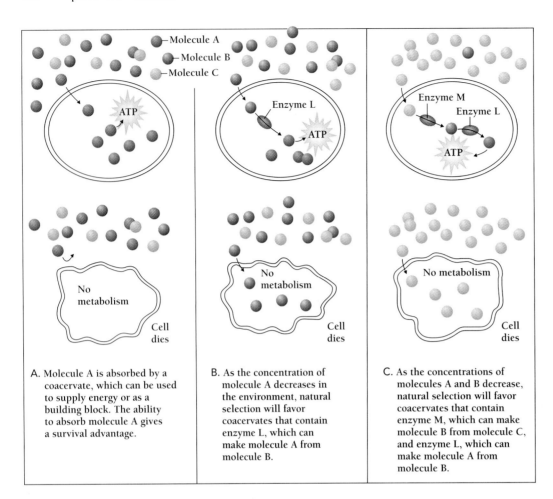

Figure 18-13 How might metabolic pathways have evolved?

The DNA-to-RNA-to-polypeptide scheme is too complicated to have arisen all at once in the first cells. So biologists have tried to guess which of these three kinds of molecules could have first stored information, directed the synthesis of proteins, and also replicated itself. But DNA cannot form without catalytic proteins, and proteins cannot form without DNA.

That leaves RNA. In modern cells, RNA plays a central role in translation, supporting the idea that RNA could have been the primitive genetic material. Still, if RNA acted as the original genetic material, some sort of mechanism had to exist for aligning amino acids along sequenced strands of genetic RNA. Such a mechanism would have to allow RNA to store information, direct the synthesis of proteins, and also replicate itself. Considerable evidence now supports the idea that RNA can do this.

First, scientists have found that some RNA can spontaneously create itself and replicate itself in a test tube. Short strands of nonordered abiotic RNA can self-assemble from individual ribonucleotides. And, in the presence of an RNA template, ordered sequences of five to ten nucleotides will polymerize according to the standard base-pairing rules.

In the early 1980s, molecular biologists Thomas Cech and Sidney Altman showed that certain kinds of RNA could act as enzymes, snipping off a bit at an end or cutting out a bit and splicing the remaining pieces together. It was a dramatic discovery that earned the two researchers a Nobel prize in 1989 and suggested that perhaps RNA could replicate itself without the help of protein enzymes (polymerases).

Indeed, in 1989, Jennifer A. Doudna and Jack Szostak created an artificial RNA enzyme, or **ribozyme,** that could make new RNA molecules. However, although the molecule can splice preassembled strands of RNA along a template, it cannot add individual bases to a growing chain. These discoveries suggested to many researchers that the first cells evolved in an **RNA world,** in which RNA molecules functioned both as enzymes and as carriers of genetic information. The hypothesized RNA world is not without problems, however. RNA is notoriously unstable, and it is unlikely that RNA enzymes could have survived outside of a living cell. Many researchers therefore think that the first enzymes were neither RNA nor protein, but a different type of molecule, perhaps a nucleic acid with a sugar other than ribose.

Cells based on such a system allowed the evolution of RNA-based cells. RNA-based cells in turn allowed the evolution of cells that used proteins as enzymes and DNA as the genetic material. Reverse transcriptase, so important in many tumor viruses and in HIV, may have played an important role in this last process (Chapter 12).

Figure 18-14 Ribosomal RNA. All modern ribosomal RNAs have a similar "stem-and-loop" structure. Notice that the pattern of stems and loops in eukaryote (yeast) RNA is more like the RNA of an archaebacterium than like that of a eubacterium. This supports the theory that we are descended from archaebacteria rather than eubacteria.

How Did Translation Evolve?

We know that RNA is capable of storing information, and it might, perhaps, be able to replicate itself. But the hardest question to answer is how modern translation—of RNA into protein—could have evolved. How a correlation between nucleotide sequences and amino acid sequences could have arisen remains a mystery.

Since the genetic code plays such a pivotal role in translation, we may look to it for clues about the evolution of translation. The first clue is the fact that all organisms use the same triplet genetic code. This suggests that the system originated with life itself. Second, most of the information for each codon is in the first two nucleotides; for 7 of the 20 amino acids, the third "letter" is irrelevant. This does not mean that the code was originally a doublet, however. If the code had started out as a doublet, then all previously evolved genes would have become useless when the code became a triplet. Instead, the present characteristics of the code were almost certainly fixed after its initial success.

As molecular biologist Francis Crick put it, the code is a "frozen accident." Information in mRNA was probably read three bases at a time from the beginning. But initially, the first two letters of each codon contained all the meaning.

Because the code is universal among both prokaryotes and eukaryotes, it must have completed its refinement before the time of the common origin of all modern organisms.

How Did Ribosomes Evolve?

Modern ribosomes contain more than 50 different proteins as well as ribosomal RNAs (rRNAs). Since each protein has a defined amino acid sequence, ribosomal proteins could not have been part of the original translational machinery. It seems more likely that the primitive translational machinery depended primarily on RNA. Indeed, the mRNAs of many prokaryotic and eukaryotic bacteria can form specific base pairs with part of the smaller rRNA.

The anatomy of rRNAs also supports the idea that rRNAs directly participated in protein synthesis. For example, researchers have determined the nucleotide sequences of the smaller rRNAs of many widely divergent organisms, ranging from bacteria to vertebrates, as well as of ribosomes from mitochondria and chloroplasts. For each rRNA, researchers predicted the pattern of stems and loops that could be formed by the formation of duplexes of complementary sequences. While the sequences of the rRNAs vary widely, the patterns of stems and loops are much the same (Figure 18-14). Such conservation suggests an essential function for the pattern of stems and loops, probably as a catalyst for directed protein synthesis.

How Did tRNAs Evolve?

All modern cells attach amino acids to transfer RNAs (tRNAs) before assembling a polypeptide. Transfer RNAs play a critical role in translation, allowing unlike molecules—nucleotides and amino acids—to be connected. All modern tRNAs have similar stem-and-loop structures, and all attach to an amino acid at their 3' ends. In addition, all tRNAs contain an anticodon with common surrounding elements. Thus, like the genetic code, tRNAs probably diverged from a single tRNA early in the history of life.

How Did the First Cells Use Genetic Information?

The first true cell would have had to have the following properties:

1. A boundary, such as that in artificial protobionts
2. Enzymes for extracting energy from chemicals in the environment
3. Energy storage in ATP
4. RNA that specified the structure of specific enzymes
5. RNA replication, using the Watson-Crick rules of base pairing
6. Specification of each amino acid by a triplet codon, with most of the information conveyed by one or two bases
7. Primitive tRNAs to connect amino acids to nucleotides

The first enzymes probably arose by the random assembly of amino acids into short polypeptide segments. Such primitive enzymes were probably neither as active nor as stable as modern enzymes. But if a polypeptide segment could form the active site of a catalyst, then its possessor would have enormous advantages over its competitors.

Similarly, the first genetically useful RNAs probably had only short stretches of information. Once the first genetic system had evolved, however, natural selection would begin to work. Natural selection would favor cells with increased capacity to reproduce, as well as cells better able to extract energy from chemicals in the environment. Reproduction, protein synthesis, and metabolism must therefore have evolved together. The most successful organisms would be those that best coupled their genetic and metabolic systems—those best able to pass on useful catalysts to their descendants.

HOW DID MODERN CELLS EVOLVE?

The first fossils that are clearly eukaryotic are about 1.7 billion years old. More than 2 billion years elapsed between the first appearance of life 3.8 billion years ago and the first eukaryotes. What, we may ask, happened during that period to allow more complex forms to evolve? What biochemical pathways did the earliest organisms develop, and, equally important, how did these developments fundamentally change the Earth's environment?

How Did Early Life Affect the Environment?

Most biologists think that the first cells were **heterotrophs** [Greek, *heteros* = other + *trophe* = nourishment], organisms that obtain energy only by degrading existing organic molecules. Human beings and other modern animals are all heterotrophs, too. We obtain both energy and building blocks, such as amino acids, by eating other organisms. We can extract energy in the process of respiration by oxidizing preexisting organic molecules from other organisms.

The earliest organisms probably obtained both energy and building blocks from preexisting organic molecules in a rich primordial soup. Like all heterotrophs, they consumed what was in the soup but contributed no energy. But unlike animals, they could break down molecules only by means of glycolysis, or fermentation. They were incapable of true respiration, since there was no oxygen. And they were incapable of obtaining energy from the sun by photosynthesis or from inorganic molecules, as plants and other modern **autotrophs** [Greek, *auto* = self + *trophe* = nourishment] do.

Of the three ways to extract energy from molecules—photosynthesis, respiration, and fermentation—only fermentation is common to all bacteria and eukaryotes. It seems likely, then, that fermentation is the oldest metabolic pathway and that it had already evolved in the earliest cells—at least those whose descendants have survived.

The suspected sources of nourishment for early heterotrophs—the organic molecules in the primordial soup—were in limited supply. Early organisms would have gradually exhausted them of their nutritive value, since no autotrophs existed to renew them. Early organisms probably forestalled the exhaustion of the Earth's first resources by developing elaborate chemical pathways for exploiting more and more of the molecules in the soup. Eventually, though, all the original molecules, and all the energy in them, would have been exhausted. One alternative was to begin extracting energy from inorganic chemicals by oxidizing them and to use the energy to build new organic molecules from carbon dioxide. The present-day *Sulfolobus* bacterium, for example, gets its energy by converting sulfur to sulfur dioxide.

The evolution of photosynthesis

The most abundant energy source was not inorganic molecules, however, but sunlight. We may guess at how photosynthesis gradually developed, at least 2.7 billion years ago, by examining the variety of ways in which modern organisms harness solar energy.

The first step in the direct harnessing of solar energy by organisms was probably the development of a light-driven proton pump, such as that in the modern purple bacterium. This bacterium pumps protons across its cell membrane, which allows the bacterium to store energy in ATP, as well as actively transport other substances across the membrane.

More advanced modern bacteria use light to generate reducing power (hydrogen atoms or electrons) to synthesize carbohydrates from carbon dioxide. Some photosynthetic bacteria obtain the needed electrons from such donors as hydrogen gas or hydrogen sulfide.

The largest potential source of electrons, however, is water. Water is difficult to break down, but it is the electron source for nearly all modern photosynthetic organisms. The first ancient organism to synthesize organic molecules from sunlight, water, and carbon dioxide won for its descendants a freedom from want that has lasted more than 2 billion years. Cells that could photosynthesize would no longer have to compete for energy-bearing organic molecules. They could simply make their own from the abundant water, carbon dioxide, and sunlight.

The first global toxic waste problem

However, breaking down water created a serious global toxic waste problem. The very steps needed to separate the hydrogen atoms from water also released huge quantities of molecular oxygen. Oxygen is a powerful oxidant that easily rips apart many molecules. Sensitive organic molecules would have been destroyed by oxygen. Many of the original heterotrophs, which depended on organic molecules, probably starved, their food supply destroyed by the wastes of their own autotrophic cousins.

The development of photosynthesis drastically altered the Earth's early environment. Major contributors to this change were the cyanobacteria making up the stromatolites, mentioned at the beginning of this chapter. The wide distribution of stromatolite fossils suggests that they covered vast areas of the world's shallow waters from about 3 billion years ago until the end of the Precambrian. In the absence of modern grazing sea animals, photosynthetic cyanobacteria flourished nearly unchecked for millions upon millions of years. For perhaps as long as 2.5 billion years, they pumped oxygen into the oceans and the atmosphere. The cyanobacteria began to decline about a billion years ago. However, single-celled algae began to appear throughout the open oceans, their photosynthetic activity further boosting the oxygen content of the seas and the atmosphere.

The availability of oxygen allowed the evolution of a new class of heterotrophic organisms that could use oxygen to break down organic molecules. The complete oxidation of organic compounds to water and carbon supplies more energy than glycolysis. The advantage of exploiting oxygen was so great that today anaerobic organisms persist in only a few isolated environments.

The vast flats of cyanobacteria and oceans of algae must have presented another undeniable opportunity. Some of the new organisms that had developed the ability to breathe oxygen now developed the ability to eat the very organisms that had created this caustic waste. The first autotrophs thus provided the basis of a food chain that still survives today. Most animals that live in the ocean eat smaller animals that eat planktonic animals that eat single-celled algae.

Codon	UGA	AUA	CUA	AGA/AGG
Universal code	STOP	Ile	Leu	Arg
Human mitochondria	Trp	Met	Leu	STOP
Drosophilia mitochondria	Trp	Met	Leu	Ser
Yeast mitochondria	Trp	Met	Thr	Arg
Plant mitochondria	STOP	Ile	Leu	Arg

Figure 18-15 Genetic code in mitochondria. Mitochondrial DNA encodes polypeptides using a slightly different code than that used by nuclear DNA. The top codons translate to stop, isoleucine, leucine, and arginine, respectively, in the universal code. The mitochondria of humans, fruit flies, and yeast each translate one or more of these codons differently. In contrast, plant mitochondria use the universal code. What does that suggest about plant mitochondria?

How Did Mitochondria and Chloroplasts Arise?

Almost all eukaryotes—protists, fungi, plants, and animals—have mitochondria, the organelles that produce most of a cell's energy. Only plants and some protists, however, have chloroplasts. Biologists conclude from this that the first eukaryotes probably had mitochondria but no chloroplasts.

Because mitochondria depend on oxygen to generate useful energy, they could have evolved only as the photosynthetic prokaryotes filled the world with oxygen. Mitochondria are probably extremely ancient, however, as they use a genetic code that differs from the universal code used by all modern eukaryotes and most prokaryotes. In fact, the mitochondria of different organisms use slightly different codes (Figure 18-15).

Mitochondria are probably older than chloroplasts.

Are mitochondria and chloroplasts endosymbionts?

In both size and internal organization, chloroplasts resemble cyanobacteria (Figure 18-16). These similarities have long sug-

Figure 18-16 Endosymbiosis. A. Living inside the photosynthetic protist *Cyanophora* is a photosynthetic cyanobacterium. B. A chloroplast resembles a cyanobacterium. (A, Biophoto Associates; B, Omikron/Photo Researchers, Inc.)

gested that chloroplasts derived from free-living organisms. Indeed, there are many examples today of photosynthetic organisms living within animals or protists. Such a close association of two organisms, one of which lives inside the other, is called **endosymbiosis** [Greek, *endon* = within + *syn* = together + *bios* = life]. Endosymbiosis is mutually beneficial, since it provides an internal source of food for the host and mobility and protection for the inner organism.

Most biologists now favor the view that both chloroplasts and mitochondria are the endosymbiotic descendants of free-living prokaryotes (Figure 18-17). One of the most forceful proponents of this view has been Lynn Margulis, an evolutionary biologist at the University of Massachusetts.

Some biologists still consider the endosymbiosis theory too speculative. Many feel that chloroplasts may well have originated as free-living cyanobacteria but that the case for mitochondria is much weaker. Recent studies of mitochondrial DNA, however, suggest all mitochondria may have evolved from a single kind of ancient bacteria.

Advocates of the endosymbiotic theory are still not agreed about the origin of eukaryotic nuclei, though a strong candidate has recently emerged. Most prokaryotes lack the proteins histone and actin. But the archaebacterium *Thermoplasma acidophilum*, a heat-loving prokaryote, does make histones and actin. The ancestors of this organism, which lacks a cell wall, are a popular best guess for the origin of the nucleus.

The actual origin of eukaryotic cells can never be proven, but Margulis's arguments—together with recent DNA sequence data—have aroused considerable interest in this subject.

Did eukaryotes arise through the proliferation of internal membranes?

Opponents of the endosymbiotic theory argue that eukaryotes arose instead by the proliferation of internal membranes to form new cellular compartments. Lysosomes and the endoplasmic reticulum, for example, form from cellular membranes—not from free-living forms. Perhaps mitochondria and chloroplasts formed when membranes surrounded pieces of mobile DNA (plasmids or transposons) that detached from nuclear genes.

The endosymbiotic theory produces a more complicated evolutionary tree than the traditional view. With this complexity, however, comes a consistency with current views of prebiotic evolution. For the moment, at least, most biologists seem to find endosymbiosis the most convincing general explanation for the origin of eukaryotic cells.

Mitochondria and chloroplasts, which resemble prokaryotes, probably began their association with eukaryotic cells as endosymbionts.

In this chapter, we have explored scientists' hypotheses about how life might have evolved on Earth billions of years ago. Our discussion of such long-ago events has been, necessarily, speculative. In the next section of *Asking About Life*, we return to the present, with a review of the great diversity of modern life. We will examine how biologists classify organisms and then survey the three domains of life.

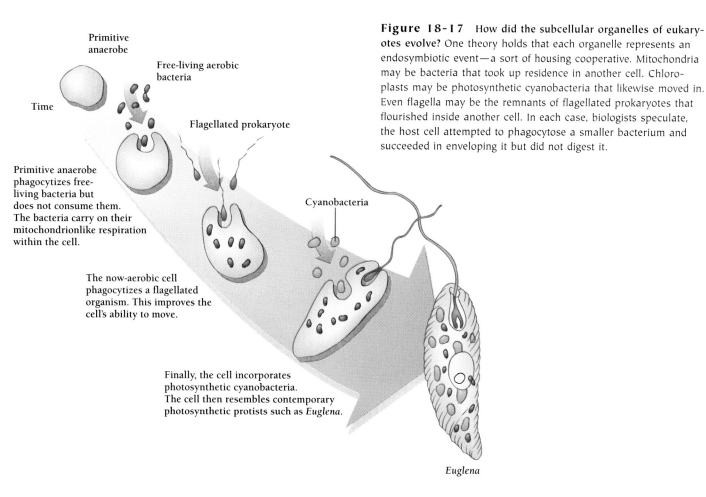

Figure 18-17 How did the subcellular organelles of eukaryotes evolve? One theory holds that each organelle represents an endosymbiotic event—a sort of housing cooperative. Mitochondria may be bacteria that took up residence in another cell. Chloroplasts may be photosynthetic cyanobacteria that likewise moved in. Even flagella may be the remnants of flagellated prokaryotes that flourished inside another cell. In each case, biologists speculate, the host cell attempted to phagocytose a smaller bacterium and succeeded in enveloping it but did not digest it.

Study Outline with Key Terms

By the middle of the 19th century, the work of Redi, Spallanzani, and Pasteur had convinced scientists that life could not arise spontaneously. Yet all life must have some ultimate origin. While some scientists have argued that life may have arrived from elsewhere in the universe, most researchers assume that life evolved from nonliving, or **abiotic,** organic molecules here on Earth.

The theory of the origin of life on Earth is different from the theory of spontaneous generation. Life could not evolve on Earth today because oxygen would oxidize the components and other organisms would eat the components before enough time elapsed for anything to happen. Ideas about the origin of life rest on the demonstration that plausible **prebiotic** chemical reactions could have produced molecules and organized structures like those in modern organisms.

After the Earth formed some 4.6 billion years ago and its crust had hardened half a billion years later, torrential rains began creating the first sediments. Precambrian rocks contain fossils as old as 3.8 billion years, only 300 million years after the formation of the oldest known sedimentary rocks. The Earth's early atmosphere contained no free oxygen. Consequently, no ozone layer protected the atmosphere from ultraviolet (UV) light. UV light and lightning probably contributed the intense energy needed for the formation of the first organic molecules. The early atmosphere may have contained hydrogen, ammonia, and methane, as well as other gases and water. In the laboratory, such a mixture, exposed to electrical discharges, produces many organic molecules, including amino acids and precursors of nucleotides. However, recent research by geoscientists suggests that the early atmosphere was more likely composed of carbon dioxide and nitrogen. Such an atmosphere could have easily given rise to sugars, but to amino acids only with difficulty.

Biologists assume that the primitive oceans became loaded with organic molecules. These organic molecules can form polymers, called **proteinoids.** Mixtures of different kinds of molecules can assemble into organized cell-like structures, called **protobionts.** Polypeptides, nucleic acids, and polysaccharides form **coacervates.** Proteinoids will form tiny spheres called **proteinoid microspheres.** A mixture of phospholipids and proteins form **liposomes,** which have a bilayered membrane similar to that of a true cell.

Although protobionts resemble cells and are capable of cell-like behavior and **prebiotic evolution,** they are not alive because they cannot reproduce themselves. Living cells must be able to store, translate, and replicate genetic instructions. By comparing the characteristics of protein synthesis in contemporary organisms, scientists have inferred that the first organisms used RNA as the genetic material, with an early version of the triplet genetic code. RNA can store information and replicate itself. It can also snip and splice itself like an enzyme. RNA catalysts, called **ribozymes,** can catalyze the synthesis of tRNA, rRNA, and mRNA, the major components of translation. Most evolutionary biologists think that the first cells functioned in an **RNA world,** using RNA both for genes and enzymes.

The first organisms were probably **heterotrophs.** Photosynthetic and other forms of **autotrophic** organisms probably evolved after early life had depleted the primitive oceans of energy-rich organic molecules. The photosynthetic organisms pumped the oceans and the atmosphere full of deadly free oxygen.

The presence of oxygen allowed the evolution of a new class of heterotrophs that could respire. The presence of autotrophs also allowed the development of herbivores and the beginnings of a modern food chain.

Eukaryotes evolved long after prokaryotes, probably after the appearance of oxygen. Most biologists believe that mitochondria and chloroplasts derived from previously free-living prokaryotes that became **endosymbionts.**

Review and Thought Questions

Review Questions

1. Give two reasons why it took paleontologists so long to find Precambrian fossils.
2. Describe the two possible alternative compositions of the Earth's primitive atmosphere, including the disagreements scientists have about them. List two major sources of prebiotic organic molecules.
3. Describe Stanley Miller's landmark experiment to test the idea of prebiotic synthesis. What is the "primordial soup?"
4. What is a coacervate made of, how is it formed, and what can it do? What is a proteinoid microsphere made of, how is it formed, and what can it do? What is a liposome made of, how is it formed, and what can it do?
5. What would a protobiont need to become a living cell?
6. Why do scientists think that RNA was the first genetic material?
7. In what way did the early heterotrophs change the environment? What were some of the consequences of this change?

Thought Questions

8. Can natural selection operate on nonliving structures such as coacervates? Explain.
9. Pasteur's sterile broths contain no organisms that could consume any life that might arise spontaneously. Why have these flasks not given rise to new life in 130 years?

About the Chapter-Opening Image

Until Francesco Redi proved otherwise in the mid-17th century, most people believed that houseflies arose spontaneously.

 On-line materials relating to this chapter are on the World Wide Web at **http://www.harcourtcollege.com/lifesci/aal2/**

 Web Links

 Quizzing & Testing

part IV

Diversity

Chapter 19
Classification: What's in a Name?

Chapter 20
Prokaryotes: Success Stories, Ancient and Modern

Chapter 21
Classifying the Protists and Multicellular Fungi

Chapter 22
How Did Plants Adapt to Dry Land?

Chapter 23
Protostome Animals: Most Animals Form Mouth First

Chapter 24
Deuterostome Animals: Echinoderms and Chordates

Photo: Tail of Meller's chameleon (*Chamaeleo melleri*). (Tim Davis/Stone Images)

19 Classification: What's in a Name?

Is the Red Wolf a Species?

In the 18th century, when European naturalists first began mapping and naming the animals and plants of North America, the destruction of the eastern forests had already driven countless species from the landscape. One of the first to go was the gray wolf, which had, before the time of Columbus, occupied nearly the whole of North America. By the 1970s, the gray wolf had been driven out of virtually all of the lower 48 states, its range reduced to Alaska, Canada, and tiny bits of northern Michigan and Minnesota (Figure 19-1). In the gray wolf's wake were left a rapidly shrinking population of red wolves, which had once lived throughout the American South, from Florida to Texas; and a rapidly expanding population of coyotes, which gradually spread from the Rockies and the Great Plains—north, east, and west—into the once-forested areas vacated by the gray and red wolves.

In 1967, the federal government designated the red wolf an endangered species, and biologists working for the U.S. Fish and Wildlife Service began surveying the South for red wolves (Figure 19-2). The red wolf, they discovered, was all but extinct. By 1972, the last few red wolves clung to the southern edge of North America, where Texas and Louisiana drop into the Gulf of Mexico. These last wolves were actively breeding with coyotes.

Government biologists realized with a shock that the only way to save the red wolf from complete extinction was to capture all the remaining wolves and breed them in captivity. From 1974 to 1980, government biologists trapped more than 400 animals. In 1980, in a marsh edging an industrial area of Galveston, Texas, the last wild red wolves in America walked into traps. Of the 400 captured animals, only 43 actually looked like wolves, however, and only 14 produced descendants that looked like wolves rather than coyotes.

Under the watchful eyes of biologists, this group of 14 animals has produced hundreds of offspring. Many of these captive-bred wolves have been released into places such as the Great Smoky Mountains of Tennessee and the Alligator River National Wildlife Refuge, in North Carolina. The rest of the captive wolves continue to produce offspring to be released in years to come. The Red Wolf Reintroduction Program, as this project is called, is the most successful captive-breeding program in existence. By the early 1990s, the red wolf had been brought back from the very brink of extinction. Today, 70 to 80 red wolves live in the wild in the Great Smoky Mountains and at Alligator River.

Not everyone has applauded the program's success, however. Initially, rural residents worried about the safety of their children and pets. Others questioned the price of saving the red wolf. The red wolf reintroduction program costs about $1 million a year. Yet the program's most vocal opponents are not taxpayers' groups, but sheep and cattle ranchers. For them, the reintroduction of wolves to areas of the country now occupied by cattle and sheep might realize the romantic notions of city-dwelling environmentalists, but at the cost of valuable livestock.

Into this heated conflict stepped David Mech, one of the world's leading wolf experts. In 1989, at an Atlanta meeting of experts on wolf biology, Mech challenged his fellow researchers to tell him how they could justify spending so much money rescuing the red wolf when it might not even be a species.

Since the 18th century, biologists had argued over the status of

Key Concepts

- Naming organisms allows biologists to distinguish one species from another and to group related species.
- Biologists classify organisms into species and also into larger groups such as genera, families, and orders.
- Natural selection shapes organisms in ways that sometimes obscure patterns of relatedness.

the red wolf. In 1791, naturalists designated the red wolf a subspecies of the gray wolf. But later biologists decided that the red wolf was a separate species. In 1979, U.S. Fish and Wildlife Service biologist Ronald Nowak carefully compared the skulls of gray wolves, red wolves, and coyotes and noticed that the size and shape of the red wolf skull fell midway between that of the coyote and the gray wolf. Nowak's interpretation of the fossil record further suggested to him that intermediate skulls like that of the red wolf skull first appeared in North America more than a million years ago, well before the first wolves or coyotes. Nowak concluded that the red wolf was not only a unique species but also the ancient ancestor of both the gray wolf and the coyote.

Additional paleontological evidence suggested that the gray wolves of Europe and Asia originated in North America. Therefore, argued Nowak, the red wolf must be a direct descendant of the first wolf. The rescue of the red wolf from the brink of extinction, then, could be seen as the salvation of a southern species of nearly royal lineage. It was a compelling idea, one that persisted almost unchallenged for 10 years, throughout the early years of the Red Wolf Reintroduction Program.

But David Mech had a different theory about red wolves. In a 1970 book, Mech had proposed that the red wolf was neither species nor subspecies, but a hybrid produced by interbreeding between the gray wolf and the coyote.

Mech's hypothesis accounted for Nowak's data showing that the size and shape of the red wolf skull was midway between that of the gray wolf and the coy-

Historical

Modern

☐ Gray wolf   ☐ Coyote   ■ Red wolf

Figure 19-1 **The disappearance of the gray wolf.** As settlers moved across North America in the 1900s, they used guns, traps, and poison to virtually eliminate the gray wolf from the United States. But the coyote, which has been subjected to similar organized massacres, has flourished. (*Michael Goodman*)

ote. Furthermore, northern gray wolves were known to interbreed with coyotes in some areas, and the offspring had skull measurements similar to those of red wolves. Nowak's skull measurements could be interpreted in two ways, Mech argued: either the red wolf species had split over evolutionary time into three species, the red wolf, the gray wolf, and the coyote; or red wolves were hybrids of gray wolves and coyotes.

In 1989, two University of California biologists, Robert Wayne (of UCLA) and Susan Jenks (of UC Berkeley), approached the U.S. Fish and Wildlife Service and offered to settle the matter once and for all. Like Nowak, Wayne was an expert on the morphology and taxonomy of wolves and other canids (members of the dog family). But unlike Nowak, Wayne had added molecular genetics to his repertoire of research techniques. Wayne and Jenks offered to search the red wolf genome for some unique stretch of DNA that neither the coyote nor the gray wolf had, something that would clearly identify the red wolf as a separate species.

The government agreed to fund the study, and the two biologists began examining DNA from red wolves, gray wolves, and coyotes. Wayne and Jenks began with mitochondrial DNA because, in general, it accumulates mutations more rapidly than DNA from the nucleus. This rapid evolution makes mitochondrial DNA particularly useful for comparing groups of closely related organisms. The base sequence of the mitochondrial DNA of gray wolves (*Canis lupus*) and coyotes (*Canis latrans*), for example, differs by about 4 percent.

The mitochondrial DNA of the gray wolf, Wayne and Jenks found, included

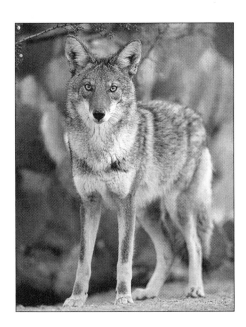

Figure 19-2 The red wolf (*center*) is intermediate between the gray wolf (*left*) and the coyote (*right*) in size, conformation, and behavior. (*Gray wolf, Tom & Pat Leeson/DRK Photo; red wolf, Jane McAlonan/Visuals Unlimited; coyote, C.K. Lorenz/Photo Researchers, Inc.*)

sequences that the coyote did not have. Similarly, the DNA of the coyote included sequences that the gray wolf did not have. The gray wolf and the coyote were clearly separate species. But, to Wayne and Jenks's surprise, the mitochondrial DNA of the red wolf (*Canis rufus*) included *no sequences unique to red wolves*. Red wolf mitochondrial DNA only included sequences identical to those of gray wolves and sequences identical to those of coyotes. The two biologists tentatively and somewhat reluctantly concluded that the red wolf was most likely a hybrid of the gray wolf and the coyote.

Nowak and the other biologists at the U.S. Fish and Wildlife Service could not believe what they were being told. Maybe, argued the government biologists, Wayne and Jenks had simply missed the DNA sequences that distinguished the red wolf. Maybe they had not looked at enough DNA.

To put to rest any lingering doubts, Wayne and other colleagues turned to special repetitive regions of the nuclear DNA called microsatellites. Wayne checked ten microsatellite regions in blood samples taken from hundreds of living animals and skin samples from 16 pre-1930s skins stored in a vast collection of furs at the Smithsonian Institution, in Washington, DC.

The results were the same. Neither the samples of blood from living red wolves nor the samples from the skins of pre-1930s red wolves showed any unique sequences. By 1994, Wayne had found no evidence that the red wolf had ever been reproductively isolated from either gray wolves or coyotes. The red wolf had to be a hybrid of the gray wolf and the coyote.

Wayne's genetic data proved to be an embarrassment to the U.S. Fish and Wildlife Service, which had poured millions of dollars into the reintroduction program in the belief that the red wolf was a unique and endangered species. Yet, the agency had acted in good faith. Until Wayne and his colleagues finished their research, the U.S. Fish and Wildlife Service had no way of knowing that the red wolf was not a species.

Now the government agency was faced with a terrible dilemma. Wayne's results threatened to discredit the reintroduction program, strip the red wolf of its endangered status, and further undermine the increasingly battered public image of the federal Endangered Species Act. The act had been under attack from its very inception by real-estate developers, ranchers, miners, and other industries that find species preservation inconvenient and expensive.

For now, the red wolf hybrid has not lost protection under the Endangered Species Act, but its hybrid status certainly weakened its claim to protection. The American Sheep Industry Association, for example, formally petitioned the U.S. Department of the Interior to take the animal off the endangered list. The Department of the Interior turned down this petition in 1992, arguing that "limited genetic exchange" between groups does not undermine species status. In short, the agency did not acknowledge that the red wolf was probably not a species.

To protect the red wolf, the U.S. Fish and Wildlife Service began pressuring Wayne to avoid the word "hybrid" in his research papers and to substitute the term "intergrade species" and other similar phrases. By 1994, it was clear, however, that the red wolf was not a species engaging in "limited genetic exchange" with other species.

In 1995, the U.S. Department of the Interior issued a legal opinion that said that hybrids would be protected under the Endangered Species Act if *morphological* evidence showed that the hybrids were similar to the endangered "pure" form. In essence, if they looked like red wolves, they would be pro-

BOX 19.1

Foucault's Chinese Encyclopedia

"This book first arose out of a passage in Borges, out of the laughter that shattered, as I read the passage, all the familiar landmarks of my thought—our thought, the thought that bears the stamp of our age and our geography—breaking up all the ordered surfaces and all the planes with which we are accustomed to tame the wild profusion of existing things, and continuing long afterwards to disturb and threaten with collapse our age-old distinction between the Same and the Other.

"This passage quotes a 'certain Chinese encyclopaedia' in which it is written that 'animals are divided into: (a) belonging to the Emperor, (b) embalmed, (c) tame, (d) sucking pigs, (e) sirens, (f) fabulous, (g) stray dogs, (h) included in the present classification, (i) frenzied, (j) innumerable, (k) drawn with a very fine camel-hair brush, (l) et cetera, (m) having just broken the water pitcher, (n) that from a long way off look like flies.'

"In the wonderment of this taxonomy, the thing we apprehend in one great leap, the thing that, by means of the fable, is demonstrated as the exotic charm of another system of thought, is the limitation of our own, the stark impossibility of thinking that."

—From the introduction to *The Order of Things, An Archaeology of the Human Sciences,* by Michel Foucault, Vintage Books, New York, 1970.

Figure A A Chinese hanging scroll depicting hens and roosters in a barnyard. *(Werner Forman Archive, National Gallery, Prague/Art Resource)*

tected. But the genetic data did not support the idea that a "pure" form of the red wolf had ever existed, certainly not in the last 100 years. In issuing this opinion, the agency excluded all the genetic evidence regarding the red wolf's species status. The only question was whether the red wolf looked different from the coyote and the gray wolf. It did, and, therefore, until such time as the government acknowledges the genetic data, the red wolf will be considered a species.

From the point of view of a taxonomist, the agency's decision might seem illogical. But from a political point of view, the decision made sense. The Endangered Species Act is the keystone law for environmental protection. Anything that weakens it, weakens all environmental protection in the United States. Furthermore, the red wolf is enormously popular. One study by researchers at Cornell University found that nearly 80 percent of people surveyed in eight states surrounding the red wolf reintroduction area favored the wolves. Researchers estimated that tourists attracted by the wolves bring $37 million per year to eastern North Carolina and about $132 million per year to the national park area. In addition, the researchers estimated that people in the eight states surveyed are willing to pay about $3 million per year for the red wolf restoration program, and nationally, the public is willing to pay $18 million per year, figures that far exceed the cost of the program.

In any case, biologists themselves continue to disagree about both the status of the red wolf and the political meaning of that status. In a 1998 issue of the journal *Conservation Biology*, Nowak and another biologist wrote, "Studies support the biological hypothesis that the red wolf is systematically valid. . . . Hybridization has been advanced by individuals and groups who oppose reintroduction of the red wolf as a conservation strategy," while in the same issue, Wayne and his colleagues insisted, "Enough scientific evidence exists to suggest that the red wolf is a hybrid of the gray wolf and the coyote. . . . Conservation efforts directed solely at the red wolf may therefore compromise preservation efforts for other wolf species."

WHY IS IT DIFFICULT TO NAME AND CLASSIFY ORGANISMS?

The story of the red wolf illustrates that deciding how to classify organisms into tidy groups is no easy task. Surprisingly, recognizing species is usually the *least* controversial task facing a **taxonomist,** a biologist who classifies organisms. In contrast, determining whether a species should be subdivided into subspecies (and, if so, how many subspecies) nearly always raises an argument. Likewise, grouping species into higher taxa—such as families, orders, or classes—immediately stirs up a host of controversies, which may be scientific, personal, philosophical, or political. In this chapter, we will briefly discuss how biologists classify organisms. Along the way, we will see why classification is nearly always a controversial topic.

Why Do We Name Things?

"What's in a name?" asks Shakespeare's Juliet, "That which we call a rose by any other name would smell as sweet."

Why do we name things? By naming the rose, we distinguish it from other flowers—both in our own minds and in the minds of others with whom we speak. The name allows us to talk with others about the rose's significance—romantic, evolutionary, or otherwise. To the extent that everyone agrees on a name, we can each talk of roses with confidence that we will be understood. Naming things also allows us to communicate what we know about something, whether the something is an individual object or organism or a class of objects or organisms. We can distinguish poisonous plants from palatable ones, and dangerous animals from docile ones.

Further, by naming things, we force ourselves to make distinctions that we might not otherwise have noticed. In thinking about a rose, most people picture a red flower, yet we would not confuse a red rose with a red poppy or red carnation. Nor would we mistake a yellow rose for a buttercup or a dandelion. Naming an object forces us to know it.

Another reason for naming an object is to classify it, or assign it to a group. A couple might name their children David and Nina simply to distinguish them from others, for example. But parents also give their children a last name to identify the children as part of a particular group, in this case a family.

Naming objects allows us to both distinguish one object from another and assign objects to groups of similar objects.

What Did Linnaeus's System of Classification Accomplish?

To classify objects, we usually establish different categories, or levels, of classification and place these categories in a **hierarchy,** an arrangement in which larger groups contain smaller groups, which contain still smaller groups. We can illustrate a hierarchy of classification with a familiar example. Your address probably falls into each of the following categories: apartment number, street number, street, zip code, city, county, state, country, continent, planet, etc. Notice that the categories form a hierarchy, with each category often (but not necessarily) including more than one example of the lower category (Figure 19-3). Each state, for example, includes many counties, and most counties include several cities and towns.

Recall from Chapter 15 that biologists use a hierarchical taxonomy established by the 18th-century Swedish physician and biologist Carolus Linnaeus. Before Linnaeus's work, biologists had already used a category called a **genus** [plural, **genera;** Greek, *genos* = to be born], a group of species that share many morphological characteristics. Linnaeus and his later followers established higher categories, so that today we traditionally recognize the following hierarchy of classification: species are grouped into genera, genera into families, families into orders, orders into classes, classes into phyla, and phyla

Figure 19-3 "Hierarchical classification" means that each level is included within the levels above.

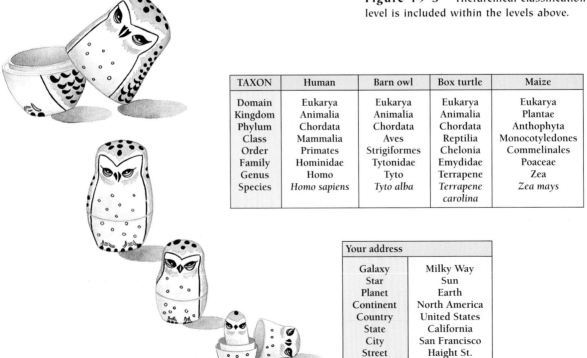

TAXON	Human	Barn owl	Box turtle	Maize
Domain	Eukarya	Eukarya	Eukarya	Eukarya
Kingdom	Animalia	Animalia	Animalia	Plantae
Phylum	Chordata	Chordata	Chordata	Anthophyta
Class	Mammalia	Aves	Reptilia	Monocotyledones
Order	Primates	Strigiformes	Chelonia	Commelinales
Family	Hominidae	Tytonidae	Emydidae	Poaceae
Genus	Homo	Tyto	Terrapene	Zea
Species	*Homo sapiens*	*Tyto alba*	*Terrapene carolina*	*Zea mays*

Your address	
Galaxy	Milky Way
Star	Sun
Planet	Earth
Continent	North America
Country	United States
State	California
City	San Francisco
Street	Haight St.
Number	555

into kingdoms. (Some people remember these categories—in the reverse order—using the mnemonic, "*Kings Play Chess On Fine-Grained Sand.*" But you can invent your own mnemonic.) Recently, biologists have further arranged the kingdoms into three domains: the Archaea, the Eubacteria, and the Eukarya. Figure 19-3 shows the hierarchical classification of several species, including our own. We are members of the **domain** Eukarya, the **kingdom** Animalia, the **phylum** Chordata, the **class** Mammalia, the **order** Primates, the **family** Hominidae, the genus *Homo,* and the species *Homo sapiens.*

Biologists use the word **taxon** [plural, taxa; Greek, *taxis* = arrangement] to mean a group of organisms at any level of the classification hierarchy. So we may speak of all the primates as one taxon and all the bats as another. At the same time, we may speak of all the individuals of a species such as *Homo sapiens* as another taxon.

Linnaeus not only introduced a system of classification but also standardized the naming of organisms. The official name of a species is a **binomial,** a two-part name that contains both the genus (*Homo,* for example) and species (*sapiens,* for example). It is not correct to leave out the genus name or the species name when referring to a species. The fruit fly (or, more properly, the vinegar fly) most often used in genetics experiments is *Drosophila melanogaster.* We may, after writing the name out fully once, abbreviate the genus to its initial letter, as in *D. melanogaster.* But scientists and laypersons alike often refer to *Drosophila melanogaster* as simply *Drosophila.* This is informal and technically incorrect. North America alone has nearly 200 different species that all belong to the genus *Drosophila.* We cannot call all of them *Drosophila* and know what we are talking about.

Binomial names are always Latin or Greek (or at least they are meant to be). Latin was, until this century, regarded as the "universal language" by educated Europeans. The use of Latin names eliminates uncertainty in translation from one language to another (as the Latin name need never be translated). Latin also prevents confusion that results when people in different regions give the same common name to different species. For example, the English robin (*Erithacus rubecula*) bears only passing resemblance and relation to the American robin (*Turdus migratorius*) we know in the United States (Figure 19-4).

Linnaeus standardized both the naming of organisms (with a Latin binomial) and the classification of organisms (into a strict hierarchy).

Who Names a Species?

The first person to describe a new species assigns it its name. Sometimes biologists discover that two people have described and named the same species. The name given first is always considered the correct name and has "priority." This rule is a way of ensuring that species names are stable, that they stay the same from one decade to the next. That way, biologists always know exactly which species is meant.

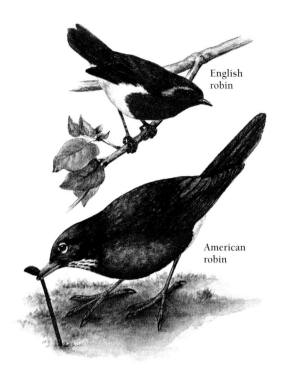

Figure 19-4 Although unrelated, both the English robin and its American namesake are called "robins." Scientific names are intended to minimize such confusion.

Still, sometimes this is too bad. One hundred years ago, the Yale paleontologist Charles Marsh assigned the charming name eohippus ("dawn horse") to the 55-million-year-old ancestor of the modern horse. Marsh did not know that the tiny fossil horse had already been described and named by someone who thought it was a hyrax (an animal that looks like a guinea pig, only bigger). Eohippus, which is both poetic and accurate, lost out to *Hyracotherium,* a mouthful that means hyraxlike mammal. Because of the rules of priority in naming, we are stuck with this unfortunate name—rules are rules.

Taxonomists use their naming privileges for different purposes. Some name species after colleagues they admire. Some try to describe the species: *Rosa alba,* for example, is a white rose (alba, as in albion or albumen, means white), while *Rosa chinensis* is a Chinese rose. Other biologists just want to have fun: the names for some genera of stink bugs (family Pentatomidae), for example, end with the suffix -chisme, pronounced "kiss me."

Species are named by the first person to recognize the species as new.

How Do We Recognize Species?

Long ago, the Greek philosopher Plato described how we make ordinary classifications. Plato argued that the individual objects of the world are all variations of universal "types," forms

BOX 19.2

Are Guinea Pigs Rodents?

One group of animals whose classification has come under close scrutiny recently is the guinea pigs and their relatives. One of the defining characteristics of the order Rodentia is a single pair of razor-sharp incisors with which they can gnaw through the husks of nuts and seeds. In fact, the word "rodent" means "to gnaw." Because guinea pigs possess these distinctive gnawing incisors, as well as other adaptations for gnawing, taxonomists have traditionally classified them in one of three suborders of rodents.

The three suborders—the squirrels, the mice and rats, and the guinea pigs and their relatives—are classified according to the anatomy of their gnawing muscles. The principal jaw muscle is the masseter muscle, which closes the jaws and also pulls the lower jaw forward, a movement critical to gnawing. In mice and rats, the two sections of the masseter—the lateral masseter and the deep masseter—both attach well forward on the jaw, and these animals' ability to gnaw is unparalleled. In squirrels, only the lateral masseter extends forward. In guinea pigs, only the deep masseter extends forward. Consequently, neither squirrels nor guinea pigs gnaw as well as mice and rats.

The suborder to which the guinea pigs belong, called the caviomorphs, differs from other rodents in numerous ways. Caviomorphs are more diverse in their morphology and habits than either the squirrels or rats and mice. They include porcupines, chinchillas, various guinea pigs (called cavys), capybara, pacas, and agoutis. The capybara, for example, is a large, social herbivore. Individuals may reach 150 lb. Caviomorphs all have proportionately larger heads than other rodents and they reproduce differently (Figure A). While a rat may bear litters of 11 every 3 weeks, a guinea pig bears 2 or 3 young every few months.

Like mice and rats, guinea pigs are frequent subjects of scientific experiment. As a result, quite a bit is known about guinea pig physiology and genetics. In 1991, researchers suggested that the protein sequences of guinea pigs are sufficiently different from those of rats and mice to justify giving the guinea pigs and their relatives a separate order.

A series of studies in the early 1990s on both protein and nucleotide sequences gave contradictory answers, however, and the guinea pigs remained classified as rodents. Then, in 1997, a group of Italian and Swedish researchers carefully analyzed differences in the mitochondrial DNA (mtDNA) of 16 animals, whose complete mtDNA sequences were known.

The results were dramatic. Three different methods of classification all put the guinea pigs in a separate taxon from the mice and rats. Indeed, the guinea pigs appeared to be more closely related to rabbits and primates than to rodents. Although the five researchers called for further research—"the phylogeny will be clearer when a larger number of species can be considered in phylogenetic analysis"—they titled their research report in the journal *Nature* decisively "The guinea pig is not a rodent."

Figure A Guinea pigs. Rodents or not? *(Gerard Lacz/Peter Arnold, Inc.)*

that share an ideal underlying plan (Chapter 15). A chair, for example, always has a seat large enough to support a sitter and, usually, has four legs and a back.

Aristotle extended Plato's concept of types to organisms. In this view, called the **typological species concept,** each species consists of individuals that are variants of a fixed underlying plan. The key word here is "fixed." A typological species cannot evolve and become something different. Aristotle would have argued that a robin and a cat each have a fixed underlying plan.

Eighteenth- and nineteenth-century biologists applied the philosophy of types in the following way. When they first recognized a new species, they killed and collected several examples of the organism. One individual was then selected as the "type specimen" for each species and kept in a museum for reference. All other organisms believed to belong to that species were then compared to the type specimen to verify that they were the same. Because of Darwin, however, biologists have recognized that the typological species concept has two important limitations:

1. All species change over time.
2. The individual members of a species vary.

Twentieth-century biologists have mostly defined **biological species** as groups of actually or potentially interbreeding populations that are reproductively isolated from other such groups (Chapter 17). For example, the cardinals in a Virginia suburb can easily breed with one another, as well as with other cardinals in nearby Maryland. But they cannot breed with mockingbirds. In this view, called the **biological species concept,** individuals are members of a species only in relation to other species. In the same way, "brother" is not a category that can be applied to a single individual. Someone is a brother only in relation to someone else.

To decide whether a group of organisms constitutes a separate species, most biologists like to use the biological species concept. For most organisms, however, the biological species concept is impossible to apply. Certainly we can obtain no information about breeding populations for fossil organisms or for organisms that reproduce asexually, including great numbers of plants, fungi, protists, and bacteria.

Even living, well-studied, sexually reproducing organisms pose problems. No one was certain, for example, how frequently gray wolves and coyotes interbred until biologists tested DNA in blood samples drawn from hundreds of wild wolves and coyotes. Such efforts are expensive and unusual. Many sexually reproducing plants and animals readily form hybrids with groups traditionally classified as separate species.

We can see two objections, then, to the biological species concept: (1) the frequent lack of information about whether two groups can interbreed; and (2) the fact that many recognized species can and do interbreed.

Biologists cope with the practical problems presented by the biological species concept by recognizing that a biological species is usually a reproductive unit, a genetic unit, and an ecological unit. A species is a **reproductive unit** in the sense that the species is the group of potential mates. Among organisms that do not reproduce sexually, that definition of a species does not apply. A species—whether sexually reproducing or not—is also a **genetic unit,** because a species can be defined as a gene pool (usually isolated from other "species" by reproductive barriers).

Finally, a species is an **ecological unit:** within a given habitat, a species interacts in a predictable fashion with other species in the environment. Gray wolves, for example, hunt in packs of 7 to 20 if large animals such as deer, moose, and caribou are available. In contrast, coyotes do not hunt in packs and prefer to prey on smaller animals such as mice and hares. Because of their size and behavior, coyotes interact with their environment differently from wolves. Biologists now know, for example, that although under certain circumstances wolves and coyotes interbreed, they also know that the two species generally remain distinct, genetically and ecologically.

Even these rules are not enough to help classify all organisms. As we will see in the next chapter, bacteria frequently exchange genes with other, unrelated species. Thus bacteria are not generally reproductive units or even genetic units, but only ecological ones, and the way that bacteria interact with their environment can change radically with the sudden acquisition of a single gene.

In reality, biologists usually classify species by the way they look, relying on morphology—the form or visual appearance—of organisms for the provisional definition of a species. As we saw at the beginning of this chapter, biologists name even living, sexually reproducing species such as the red wolf on the basis of their morphology, simply because it is easier.

Fortunately, the morphologically defined species usually corresponds exactly to the biologically defined species. The gray wolf and the coyote, for example, look different and behave differently. In large part, they occupy different habitats. In short, they look and act like true species. Genetic data confirm that the two species have unique sequences of DNA, showing that they have been reproductively isolated from each other. That these two species can interbreed extensively does not make them the same species. In general, the morphologically defined species corresponds to the biologically defined species because reproductive isolation tends to lead to the evolution of divergent morphological characters. Despite most biologists' noisy objections to typological species, then, they frequently begin defining a species on a typological basis.

Biologists use both biological and typological concepts to name and classify species.

HOW DO BIOLOGISTS GROUP SPECIES?

Although biologists with different agendas argue whether particular groups deserve the species label, all modern biologists consider species to be real entities, not abstract or arbitrary classifications. Because species are natural entities, recognizing what constitutes a species is obvious most of the time. Biologists disagree surprisingly rarely on the boundaries that separate individual species. The same researchers may bicker constantly, however, about how to subdivide species into subspecies and how to place species into larger groups such as genera, families, orders, phyla, kingdoms, and now domains.

Naming and classification of species is the main business of **systematics,** the scientific study of the kinds and diversity of organisms and of any and all relationships among them. In contrast, **taxonomy** is the theoretical study of classification, including the study of the principles, procedures, and rules of classification. These two words have different meanings, but, for simplicity, we use the word "taxonomy" only.

Biologists would like to classify organisms according to *relatedness,* but they must do so from measurements of *similarity.* In classifying any kinds of objects—organisms, books, or buildings—we naturally group things according to their similarities. We can understand the problems this presents if we imagine attending a big family reunion of 200 people and trying to decide who is related to whom. A group of redheads might appear to be related. We might find, however, that one of them is simply married to someone else in the family and is not related to anyone else at the reunion. Similarity by itself does not always signify relatedness, and which similarities are important ones is often a matter of much dispute.

For example, in Papua New Guinea, the giant tropical island just north of Australia, biologists have compared their own classification of plants and animals with that used by the tribespeople who live there. The biologists and the tribespeople agree nearly exactly on the number and the identity of species. They disagree, however, on the higher classifications. The biologists group organisms according to their biological similarities—lumping all the birds together in the class Aves, for example. The tribespeople classify according to use—lumping together all the species that are edible, all that are poisonous, or the ones that provide materials for clothing. Both systems of classification serve legitimate purposes.

Taxonomy invites dissent. In fact, biologists now use three methods for classifying organisms. The first is classical evolutionary taxonomy, in use for the last 150 years. In addition, two groups of biologists have each proposed a new way to classify organisms. The first of these new approaches—phenetics—focuses on the practical problem of taxonomy and proposes a seemingly objective step-by-step method for classifying organisms. The second new approach—cladistics—ambitiously proposes to classify organisms according to their evolutionary history.

Classical Evolutionary Taxonomy

Classification by means of classical evolutionary taxonomy relies explicitly on the judgment and experience of the taxonomist. In fact, in one definition of a species, biologists like to say, "A species is what a good taxonomist says it is. And a good taxonomist is one whose work has stood the test of time." A good taxonomist examines the similarities and differences in as many traits as possible.

The most commonly used traits in living animals are morphological ones—including external form, internal anatomy, tissue types, and chromosomal morphology. Mammals, for example, are classified into three broad groups according to their skeletons. Adult monotremes such as the duck-billed platypus and the spiny anteaters have no teeth and lack certain bones of the skull that other mammals have. Their limbs stick out to the sides in a fashion that resembles that of lizards and other reptiles. Because of these and other traits, monotremes are considered primitive mammals.

Classical taxonomists also rely on embryological evidence. For example, the early embryos of all mammals, including humans, include a stage with gill-like pharyngeal slits and pouches (Figure 15-15). This fact, with others, suggests that we are descended from aquatic organisms that needed gills to extract oxygen from water. Finally, taxonomists may also classify on the basis of behavioral traits and, increasingly, on the basis of molecular characteristics, either protein sequences or DNA sequences.

Not all similarities between organisms mean that they are related. Birds and insects have both evolved wings, for example, but birds are no more related to insects than are humans. We say that the wings of birds and insects are "homoplasious." **Homoplasy** [Greek, *homos* = same + *plasis* = mold, form] is the possession by two species of a similar trait that is not derived from a common ancestor. ("Homoplasy" has replaced the older word "analogy," which requires that two similar traits perform the same function—a fact sometimes difficult to determine.) In contrast, similarities that are inherited from a common ancestor are said to be **homologous.** The legs of insects are homoplasious with those of humans and other vertebrates. They are jointed and muscled in a similar manner because they perform the same job, not because they were inherited from a common ancestor. In contrast, the leg of a horse and the leg of a human are homologous. They share similar bones and muscles because they evolved from a common ancestor. Supporting evidence comes from embryology: in humans and horses, the bones and muscles of the leg develop from the same embryological tissues.

When classifying organisms, a classical taxonomist must take into account which characteristics are likely to be homologous and which are homoplasious, giving weight to homologous characters and disregarding homoplasious ones. Although the wings of bats and birds seem obviously different to us, homoplasious characters in closely related species are sometimes hard to distinguish from homologous ones.

Homologous characters indicate relatedness, while homoplasious characters indicate similar adaptations but not necessarily relatedness. Taxonomists try hard to distinguish homologous characters from homoplasious ones.

How Do Advocates of Phenetics Classify Organisms?

Phenetics is a method of classification that avoids subjective choices by making none. **Phenetics** classifies species using *all* observable characteristics. The word "phenetics" comes from the same Greek root as "phenotype," *phainein* = to show.

The goal of phenetics is to list as many traits as possible for each organism and then to use computers to sort organisms according to their degree of similarity. Pheneticists, who are also called "numerical taxonomists," argue that the resulting groupings are objective measures of the similarities among organisms. Phenetic classification requires no knowledge of organisms' breeding behavior or ancestry.

A phenetic taxonomist describes the phenotype of an organism with a series of yes-or-no questions. Each of these questions defines a *unit character*—a characteristic of an organism that cannot be split into finer descriptions. For example, some unit characters for an animal are the answers to questions such as: (1) Does it have five digits on its front limb? (2) Does it have four digits on its front limb? (3) Does it have a backbone? The pheneticist records the answers to these questions in a binary (1 or 0) code: 1 for yes, 0 for no.

In phenetics, each unit character is as important as any other. That is, each of hundreds or thousands of characters is considered to be equally important. Whether an animal has spotted fur would be considered as important as whether it has a backbone.

The problem with phenetics, however, is that while it accurately groups organisms according to their similarities and differences, it doesn't necessarily group them according to lines of descent. Unrelated species often evolve similar characteristics (convergent evolution). For example, a marsupial mole looks very like a placental mole, but these two species are not closely related (Figure 17-15).

Because phenetics views all traits as equivalent, it doesn't distinguish between homologous and homoplasious traits. Homologous traits can tell volumes about ancestry, but homoplasious traits (due to convergent evolution) only confuse us. We know birds and bats are related because both groups have backbones, skulls, and other traits that are unique to vertebrates. But their wings and other adaptations for flight might suggest that they are more closely related than they are.

Phenetics is an objective method of classification, but the results may be meaningless biologically. What biologists really want to know is how species are related to one another. At that big family reunion, we want to know which individuals are siblings and which ones are cousins. And sometimes cousins look more alike than siblings, so grouping only by similarity wouldn't necessarily give us the right answers.

Finally, phenetics could never reveal whether the red wolf was a hybrid of a gray wolf and a coyote or a common ancestor of both. Phenetics can only say that red wolves are intermediate in morphology, not *why* they are intermediate.

Pheneticists classify species and higher groupings by assigning equal value to all characters.

How Do Advocates of Cladistics Classify Organisms?

By contrast, **cladistics** [Greek, *clados* = branch] groups organisms according to their evolutionary history, or **phylogeny** [Greek, *phylon* = race or tribe + *geneia* = birth or origin]. Cladistics starts with the assumption that there is one true family tree and that, if we can discover it, we can produce a natural grouping or taxonomy of organisms. In contrast to phenetics, which classifies according to all traits, cladistics constructs family trees using only certain traits. These certain traits are called **shared derived characters.** For example, all mammals share fur and mammary glands, both of which are shared characters. In the context of other vertebrates—amphibians and reptiles—fur is a shared derived character in mammals. But context is everything. In primates, fur is a **primitive character,** one that is shared with all other mammals.

Cladists use shared derived characters to construct a family tree, but it is the tree itself, called a **cladogram,** that is the basis for classification, not the characters. We can construct a cladogram for the primates. The cladogram in Figure 19-5 shows just five primates: a ring-tail lemur, a spider monkey, a rhesus monkey, a chimpanzee, and a human. Constructing the cladogram requires that we note characters in which these species differ. Four of these species (all except the lemur) possess a shared derived character—specialized area in the center of the eye's retina that allows more acute vision. Three species (rhesus monkey, chimp, and human) have nostrils that point downward. Humans and chimps differ from the other three species in that they lack tails.

The cladogram in Figure 19-5 is a classification hierarchy that represents a series of hypotheses about the evolutionary history of the primates. Biologists can test these hypotheses by examining other features of these five organisms and by comparing these organisms with the fossil record. They can also compare the sequences of specific proteins and genes.

According to cladistics, every taxon above the species level—whether it is a genus, a family, or a higher category—should be **monophyletic,** meaning that each taxon includes the ancestral species and all its descendants. Birds, for example, form a monophyletic taxon: all birds are descended from a common ancestor that was itself a bird, and all the descendants of this ancestor are birds.

Cladists reject any taxon that is **polyphyletic**—one that includes descendants of more than one ancestor. As we will see in the following chapters, a number of higher taxa appear to be polyphyletic. For example, protists have undoubtedly evolved from many different separate ancestors. They do not constitute a single lineage. For this and other reasons, cladists reject the traditional grouping of protists into the kingdom Protista.

Another group that are not monophyletic are the fishes, the class Pisces. Traditional taxonomists group the lobe-finned lungfishes with all other fishes. But the descendants of the ancestors of today's lungfishes include not only lungfishes but also amphibians, reptiles, birds, and mammals. Because some of the descendants of lungfish ancestors are not fishes, cladists say the group "fishes," which includes bony fishes and lobe-finned fishes, is not monophyletic but "paraphyletic." A **paraphyletic** group contains some but not all of the descendants of a common ancestor. A strict cladistic interpretation would therefore say that the class of animals we call fish—including lungfishes but not mammals—does not exist. In a cladistic sense, we humans are a kind of fish.

The cladists' argument that phylogeny must be the basis of classification also leads them to argue that the Reptilia is not a meaningful classification unless it includes birds. At first glance, reptiles appear to be a monophyletic group. All reptiles appear to have descended from a common ancestor that was itself a reptile. But, in fact, reptiles are paraphyletic. Some of the descendants of reptiles (dinosaurs) have evolved into birds, organisms not usually classified as reptiles.

From the fossil record, we know that the first organisms with "amniotic eggs" (eggs with layers of surrounding tissues such as are found in reptiles, mammals, and birds) arose about 300 million years ago. These "amniotes" split into several lines: the reptilelike mammals and their descendants, the mammals; the turtles; and the reptiles, including the lizards and snakes and the archosaurs. The archosaurs, in turn, split into various dinosaurs, primitive birds, and crocodiles.

The cladogram in Figure 19-6 shows a nested set of taxa for birds, mammals, and reptiles. According to this scheme, reptiles (understood as turtles, lizards and crocodiles, but not birds) are not a legitimate taxon because it is paraphyletic if we leave out the birds and mammals.

Traditional biologists tend to dislike the idea of including birds and mammals in the reptiles. In part, this is because birds and mammals are "warm-blooded" homeotherms, while the other amniotes are mostly "cold-blooded" heterotherms. But homeothermy is a derived character that has evolved separately in birds and mammals. It does not indicate relatedness. That tradition plays a role in such preferences is supported by the fact that taxonomists have adopted no new paraphyletic groups in recent decades. All paraphyletic groupings are left over from previous centuries.

If birds and crocodiles are more closely related to each other than either is to a lizard, why don't they look more alike? One answer is that birds have evolved faster than either crocodiles or lizards. So birds have effectively left the crocodiles and lizards behind. Cladistics raises such questions and offers insights into the ties among seemingly unrelated organisms. For example, knowing that dinosaurs, birds, and crocodiles are closely related, we can guess that the similar nesting behavior found in all three—including parental care and vocalizations (at least in crocodiles and birds)—is probably another sign of their close family ties. We can conclude that, most likely, the nesting behavior of these dissimilar animals is a homologous trait, not a result of convergent evolution.

Still, the cladistic approach has some limitations. Some groups cannot be classified using cladistics at all. In bacteria, for example, deciding which characters reveal ancestry is nearly impossible. Bacteria exchange genes and traits with such freedom

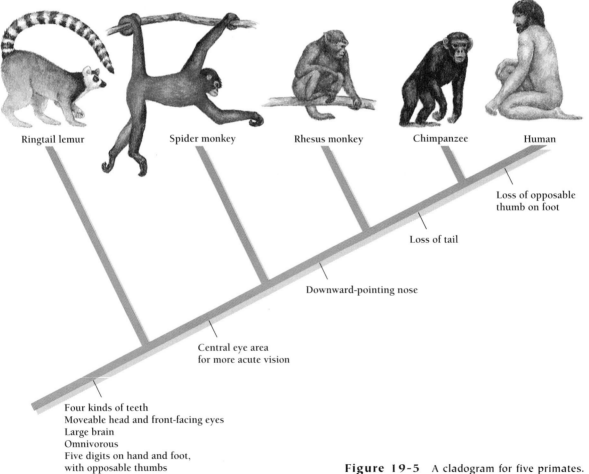

Figure 19-5 A cladogram for five primates.

Cladogram

that the usual lines of descent may not exist. As a result, bacteria are classified phenetically using DNA sequences.

Cladists who study the lineages of plants face similar problems, for separate species of plants often hybridize to form a new species that may or may not be reproductively isolated from the parent species. Even animals hybridize, confusing the search for one true lineage. The red wolf, if it truly is a hybrid, would not have a single, branching lineage but two lines of descent that separate, then come back together.

Cladists group organisms according to their evolutionary relationship to one another.

Do DNA and Protein Sequences Provide "Objective" Criteria for Classification?

As we saw in Chapter 15, the sequence of a protein evolves in much the same way as a morphological feature. For this reason, comparisons of amino acid sequences (or nucleotide sequences) suggest relationships between species. For example, all other things being equal, two species with identical genes for hemoglobin are more likely to be closely related than two species whose genes for hemoglobin differ by several bases.

Because base sequences are linear and unambiguous, it is easy to imagine ranking all genetic differences equally, as pheneticists suggest we do with morphological differences. Indeed, evolutionary biologists have embraced the molecular approach enthusiastically, partly for this reason. The gene sequences that specify proteins, however, evolve in the same way as larger-scale characters, and distinguishing between homologous and homoplasious sequences may be even harder than with morphological traits. Base sequences, then, are not more objective measures of relatedness than are morphological characters.

In addition, just like morphological traits, different sequences evolve at different rates. We can easily see that a structure such as the backbone may persist for 200 million years, while fur color may change dramatically in just a few generations. It makes more sense to group animals according to whether they have backbones than according to their external coloring.

Genes, and the polypeptides they specify, also evolve at different rates. Histone 4, a protein that binds tightly to nuclear DNA, appears to have remained nearly unchanged for

Cladistic Hierarchy

Phylogenetic Tree

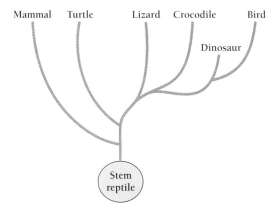

Figure 19-6 Cladistic classification of birds, mammals, and reptiles. In the cladistic view, reptiles are a legitimate taxon only if they include birds and mammals.

over 500 million years, while other protein sequences diverge strongly over relatively short periods. Proteins, like morphological characters, evolve at different rates according to how much selective pressure they experience. The divergence of sequences reflects natural selection. It makes sense when classifying organisms, then, to give priority to the proteins that have evolved most slowly.

Like other traits, nucleotide and amino acid sequences may be homologous or homoplasious, stable or unstable. Molecular approaches to classification have opened up a rich, new vein of evidence, but the sequences themselves are not more "objective" measures of relatedness than morphological and behavioral traits.

Why Is Classification So Difficult?

If all genes and morphological features changed at constant rates and if homoplasy did not exist, then maybe cladists, pheneticists, and traditionalist taxonomists would always agree on how to classify organisms. But evolution does not conform to the desires of biologists. Instead, the characters of organisms change in response to the chance appearance of new mutations, the changing geography of our planet, and the many forces of natural selection.

In the following chapters, we follow a traditional (and flawed) classification. The views taken in this book are that (1) biologists and students must be aware of the difficulties of classification and (2) we must be able to communicate with one another. We therefore will follow the most widely accepted conventions for classification, while occasionally noting the difficulties of conventional wisdom.

Classification is made difficult by the fact that the forces of natural selection shape organisms and their genes in ways that sometimes obscure patterns of relatedness.

HOW MANY GROUPS OF ORGANISMS ARE THERE?

Linnaeus divided all organisms into two kingdoms, the plants and the animals. As biologists learned more about life on Earth, however, they became dissatisfied with this grouping. The fungi might not move about as animals do, for example, but they were clearly not photosynthetic like plants. Furthermore, while Linnaeus classified just 10,000 species, biologists have now named more than a million living species. Classifying them has become a monumental task.

In 1969, the taxonomist Robert H. Whittaker proposed a five-kingdom classification that most biologists accepted. Whittaker put all the prokaryotes into the kingdom Monera, then divided the eukaryotes into four kingdoms: the protists, fungi, plants, and animals. But in recent years, biologists have all but abandoned the five-kingdom classification. Cladists object to it because none of the five kingdoms is monophyletic. Eukaryotes are all descendants of prokaryotes and therefore clearly members of other kingdoms, and the four eukaryotic kingdoms each include lines of organisms with different origins. For example, among the protists, the green algae and the protozoa each arose from different prokaryotic ancestors. We say that the protists are polyphyletic. Among the animals, the descendants of the sponges probably include other animals. Sponges are paraphyletic.

Researchers have suggested several other classifications, including one with eight kingdoms. But most now agree that all the kingdoms, whatever their number, can be assigned to just three domains—the Archaea, the Eubacteria, and the Eukarya.

The classification that we use in this book distinguishes among organisms in two ways: cellular organization and source of energy. First we can distinguish the cells of prokaryotes from those of the eukaryotes. The Eukarya—including the protists, fungi, plants, and animals—all have nuclei and membrane-enclosed organelles. In contrast, prokaryotic cells of the Archaea and the Eubacteria lack a nucleus and other membrane-enclosed organelles.

We distinguish among the three kingdoms of plants, animals, and fungi by the manner in which their members derive energy—plants from photosynthesis, animals from eating other organisms, and fungi from absorbing chemicals from their surroundings. The fourth kingdom of Eukarya, the protists, are mostly single-celled eukaryotes that do not fit into any of the other three kingdoms of eukaryotes.

In the following five chapters, we will survey some of the major groups within each domain and kingdom. In doing so, we will examine the diverse ways by which species survive and flourish.

Using a traditional taxonomy, we group organisms in terms of differences and similarities in cellular organization and lifestyle, rather than strict evolutionary relationships.

Study Outline with Key Terms

A **taxonomist** is a biologist who classifies organisms. Each classification of organisms is a **taxon.** Many biologists now use the word **systematics** for the naming and classification of organisms and **taxonomy** for the study of *how* organisms should be classified. Until this century, biologists usually viewed the members of a species as variants from a fixed underlying plan, in what is now called the **typological species concept.** Today, biologists usually define a species according to the **biological species concept,** which says that a species is a group of actually or potentially interbreeding organisms that do not interbreed with members of other such groups. Besides being **reproductive units, biological species** are also **genetic units** and **ecological units.**

Most of the time, morphologically defined species correspond to the biologically defined species. Classification of species into larger groups, however, may be difficult. Biologists use a **hierarchical** scheme, in which species are grouped into **genera,** genera into **families,** families into **orders,** orders into **classes,** classes into **phyla** (or divisions), and phyla into **kingdoms.** Kingdoms are further arranged into three **domains**—the Archaea, the Bacteria, and the Eukarya. Each species has a two-word designation, called a **binomial,** which names both the **genus** and the species.

The biological species concept is not always useful. Classifiers may know nothing about breeding patterns, and many species interbreed with other species. **Phenetics** gives equal weight to every character, grouping organisms with similar clusters of characteristics. But not all common characters indicate relatedness. Similarities that are inherited from a common ancestor are said to be **homologous** (and are given more weight). Similar traits that evolved separately are said to be **homoplasious.**

Cladistics insists that the only scientifically meaningful classification scheme is one that reflects **phylogeny,** or evolutionary history. The goal of cladistics is to construct evolutionary trees, or **cladograms,** that show the relationship of groups of species. Cladograms are constructed using **shared derived characters** rather than **primitive characters.** Cladists argue that every recognized group should be **monophyletic,** that is, each group should include an ancestral species and all its descendants. Many traditionally accepted groupings are **paraphyletic** or **polyphyletic** and therefore unacceptable to cladists. The classification used in this book reflects differences and similarities in cellular organization and lifestyle, rather than strict evolutionary relationships.

Review and Thought Questions

Review Questions

1. Define a species according to the biological species concept. What two kinds of problems often make this definition impractical?
2. What two purposes are accomplished by Linnaeus' system of classification? Name the seven levels of taxa commonly used by taxonomists.
3. Explain how a numerical taxonomist would decide whether a group of organisms constituted a species. Define phenetics.
4. What is cladistics? Why do cladists require that a lineage be monophyletic to be considered part of a single taxon?
5. Why do cladists insist that the reptiles are not a legitimate taxon?
6. Cladists consider the birds, class Aves, a legitimate taxon because birds are monophyletic. On the other hand, the very first bird was also most likely a dinosaur. Would a cladist argue that the group we call dinosaurs should include the birds? Explain why or why not.
7. Define "paraphyletic."
8. Does DNA sequencing provide a more objective method for classifying organisms? Why or why not?

Thought Questions

9. Do you think that the red wolf should be saved from extinction? Do you think that the decision to save it should be influenced by whether it is a species, a subspecies, or a hybrid?
10. If you found a new species of bird, what would you name the bird? If you were in charge of classifying this bird, what system of classification would you use? Why? Now describe the process, step-by-step, by which you would classify your bird. You may make up characters.

About the Chapter-Opening Image

The red wolf illustrates the confusion that often arises in classifying organisms. How a species is classified can have important consequences. For example, the red wolf's status could profoundly affect the strength of the Endangered Species Act.

 On-line materials relating to this chapter are on the World Wide Web at **http://www.harcourtcollege.com/lifesci/aal2/**

 Web Links

 Quizzing & Testing

20 Prokaryotes: Success Stories, Ancient and Modern

Tracking Down a Killer

In 1983, a young scientist named Scott A. Holmberg showed for the first time that feeding antibiotics to farm animals can sicken and even kill people. Most of us have experienced food poisoning, the severe nausea and diarrhea that result from consuming meat, milk, or eggs contaminated with salmonella bacteria. Farm animals—whether cattle, pigs, or chickens—can carry salmonella, but animals fed antibiotics, Holmberg showed, carry especially deadly strains of the bacteria.

In February 1983, Holmberg was working in Atlanta, at the Centers for Disease Control (CDC), the nation's premier center for the study of all public health problems, from AIDS to smoking among teenagers. His boss, Mitchell Cohen, had received a report that 11 Minnesotans had been hospitalized with salmonella poisoning. Eleven people was hardly an epidemic: after all, 40,000 Americans are hospitalized with salmonella every year, and another 5 million suffer the severe diarrhea and vomiting associated with salmonella poisoning but don't need hospitalization. What was both disturbing and exciting about the Minnesota cases was, first, that most of the patients had been taking antibiotics when they came down with food poisoning and, second, that each patient was extremely sick. One man had died. Cohen strongly suspected that antibiotics were somehow causing or helping to cause the food poisoning. But how?

Cohen called Holmberg on a Saturday, and by the next morning, Holmberg was on a flight to Minnesota to find out. In Minnesota, Holmberg met with other **epidemiologists** [Greek, *epidemia* = how widespread a disease is]—researchers who study the incidence and transmission of diseases in populations. By that afternoon, Holmberg and state epidemiologist Michael Osterholm had two facts in hand. First, all of the victims were infected with the same strain of *Salmonella newport*. This strain of bacteria, resistant to several common antibiotics, including penicillin and tetracycline, was unusual. The fact that all the victims had been made ill by the same rare form of *S. newport* suggested that they had all acquired the infection from a common source. In short, they must have all eaten the same thing. But what?

Key Concepts

- Prokaryotes are successful because of their rapid growth rates and their metabolic versatility.
- The metabolic abilities of prokaryotes provide a basis for their classification, but DNA-based classification reveals still more diversity.
- Plasmids can transfer genes to one another via plasmids and viruses.

Second, 8 of the 11 victims had been taking antibiotics manufactured by the same drug company. At first glance, it looked like the antibiotics themselves were contaminated with salmonella bacteria. However, because 3 of the 11 patients had not been taking antibiotics at all, Holmberg and Osterholm ruled out that possibility.

Nonetheless, another clue suggested that antibiotics played an important role. One woman had been hospitalized with salmonella poisoning the day after she had begun taking amoxicillin for a sore throat. Her husband then developed a sore throat, too, and took some of his wife's amoxicillin. Within 48 hours, he was in the hospital with salmonella poisoning.

Holmberg and Osterholm hypothesized that the antibiotic had actually triggered the couple's salmonella poisoning (Figure 20-1). The couple must have been infected with antibiotic-resistant salmonella that suddenly flourished when the amoxicillin killed off all other competing bacteria. Without competition, the salmonella bacteria multiplied from a few thousand cells into a full-scale infection of millions upon millions of cells in just a few hours. But why were the couple harboring antibiotic-resistant salmonella bacteria in the first place? Where had they come from?

Holmberg spent the following 2 weeks searching for the common source of the food poisoning in the homes of salmonella victims. He looked through refrigerators and cupboards, searching for some sort of food they had all eaten—some common source for the antibiotic-resistant bacteria.

Finally, with still no answer, he returned to Atlanta frustrated and discouraged. He and his boss, Mitchell Cohen, knew that the most likely scenario was that all of the patients had consumed meat, milk, or eggs infected with resistant *S. newport*. In the United States, large numbers of farm animals are fed low doses of antibiotics to increase their growth rate. This practice kills bacteria that are sensitive to antibiotics, encouraging the proliferation of antibiotic-resistant bacteria. By 1983, Cohen strongly suspected that farm animals raised on antibiotics were making people sick, but no one had yet been able to prove it.

Cohen knew of a new technique, however, that might help Holmberg make a direct link between individual farm animals and individual salmonella victims. Molecular biologists had recently developed a way to "fingerprint" bacteria by characterizing their plasmids, simple loops of DNA that sometimes live inside bacteria (Chapter 12). With this new tool, every strain of bacteria could be distinguished from all others according to which plasmids it carried.

Cohen suggested that Holmberg find out where else in the United States *S. newport* existed. Holmberg wrote to epidemiologists all over the United States, and, within a month, he had a clue that would solve the case. The state epidemiologist in South Dakota wrote to say that antibiotic-resistant *S. newport* had recently sickened four people.

From Atlanta, Holmberg interviewed the South Dakota victims by phone. One of the victims, a dairy farmer, told Holmberg that his dairy cows had all had diarrhea the previous November, and that one dead calf had been autopsied and found to have died of salmonella. Researchers at the state agricultural laboratory that had analyzed the dead calf's salmonella told Holmberg that it had the same plasmid profile as the salmonella bacteria in the Minnesota patients. Holmberg soon discovered that three of the four South Dakota victims had eaten beef from cattle raised on a farm adjacent to the dairy farm. The beef farmer said that his cattle had been perfectly healthy, adding that he regularly laced their feed with the antibiotic tetracycline, a handful to every ton of grain. Even if the beef cattle had shown no symptoms of salmonella, however, they could have been carrying salmonella just the same.

"I knew what the story was," Holmberg later told *Science* magazine. "I knew that between this point here in South Dakota

Cohen strongly suspected that antibiotics were somehow causing or helping to cause the food poisoning. But how?

Figure 20-1 How antibiotics can trigger an infection.

DAY 1

Woman eats beef hamburger containing antibiotic-resistant *Salmonella newport* bacteria, which pass into the gut.

DAY 3

The woman takes an antibiotic for a separate infection, which kills off normal gut bacteria.

DAY 4

Antibiotic-resistant *Salmonella newport* multiply and she becomes ill.

and that point there in Minnesota, there was a connection. Nature does not hand you coincidences like this. I was pretty sure that the beef herd was the source [of the *S. newport*].... Now all I wanted to do was connect the two points...."

Connecting the points, it turned out, was not so easy. In mid-January, just days before the salmonella outbreak in Minnesota, the beef farmer had sent all 105 head of cattle to slaughter. Now the beef cattle were gone, and Holmberg had no way to test the cattle for salmonella.

Nonetheless, Holmberg was able to follow the path the tainted meat had traveled to supermarkets in Minnesota and Iowa, where ten of the salmonella victims had shopped. In addition, Holmberg was able to show that the strain of *S. newport* from the dead dairy calf was truly distinctive. He examined 91 samples of different strains of *S. newport* from all over the United States, yet only the bacteria from the dead calf had the same plasmid profile as the salmonella in the sick Minnesotans. Holmberg no longer doubted that the beef cattle in South Dakota were the source of the antibiotic-resistant salmonella in Minnesota.

The plasmid fingerprinting technique had given epidemiologists their first clear evidence that feeding animals antibiotics can make the humans who eat them sick. Antibiotic-resistant bacteria are, it turns out, unusually dangerous. Holmberg showed that people are 21 times more likely to die from an infection by antibiotic-resistant salmonella than from an infection by nonresistant salmonella.

Amazingly, the practice of feeding antibiotics to farm animals continues today. Despite the clear hazards to public health, no laws regulate the practice. As a result of public pressure, only 20 percent of poultry growers now use antibiotics, the poultry industry reports. The beef industry recommends that its members use only antibiotics not used in humans. The pork industry continues to use human antibiotics without restraint.

How bacteria develop resistance to antibiotics, and how they pass that resistance among themselves, is a topic that reveals much about the lives of bacteria. In the rest of this chapter, we will learn a little about how bacteria are different from the rest of us and what kinds of bacteria populate the Earth.

Along the way, we will see why bacteria are so good at becoming resistant to antibiotics.

PROKARYOTES AND PEOPLE

The number of bacteria in your mouth at this moment exceeds the total number of people that have ever lived. You carry many more bacteria on your skin and in your gut than you have cells in your body. From a bacterium's point of view, your body is just another island to be colonized. Still, no human was even aware of their existence until Leeuwenhoek's observations some 300 years ago. And no one thought that they were at all important until about 100 years ago, when Pasteur and others found that they could cause diseases.

Prokaryotes—single-celled organisms with no nuclei—include both eubacteria ("ordinary" bacteria) and archaea, as we mentioned in Chapter 1. Most of the known life (by total mass) of Earth consists of prokaryotes, and some biologists think that there are huge numbers of prokaryotic organisms yet to be discovered in the depths of the oceans and even within the Earth's crust. Despite the incredible diversity of prokaryotic species and environments, most of the prokaryotes we know best are those that interact directly with humans.

But, though we know most about bacteria that live in us, for better or worse, bacteria are not limited to living in or on people or other animals. They live in the bodies of every kind of organism, even inside other cells. They live in soil, ice, and water. A pinch of ordinary soil, for example, contains some 10 billion bacteria. And bacteria can live almost anywhere on Earth, from the ice fields of Antarctica to the steaming cauldron of a Yellowstone hot spring, from the bottom of the ocean to the driest desert soils. In the rest of this chapter, we will see that bacteria are the most numerous, the most diverse, and the most ancient of organisms. We will also see how studies of DNA sequences from prokaryotes have revolutionized our views of the origins of life.

How Do Bacteria Make Us Ill?

Most people think of bacteria as lowly pests—carriers of disease and spoilers of food. In fact, many bacteria are useful to humans in one way or another. The trillions of bacteria in a handful of soil, for example, help recycle carbon and other nutrients. All the organisms on Earth depend on bacteria to pull nitrogen from the air so that we can use it to build proteins. Even the bacteria inside our bodies are useful. Some help us digest various kinds of foods. Others produce vitamins that we would otherwise have to get from special foods. In short, bacteria are a normal and mostly benign part of our lives.

But bacteria can also be **pathogens,** agents that cause disease. To cause a disease, a bacterium usually must accomplish two things:

1. It must invade the host, attaching and multiplying on or in the host's tissues and cells.

Figure 20-2 Koch's postulates. Pathogens such as viruses that can only be grown inside of cells cannot be isolated in a pure culture.

2. It must produce a toxin—a chemical that destroys or interferes with the host's normal processes.

In some cases, disease results just from the excessive growth of pathogenic bacteria. But in most cases, disease results from toxins made by the pathogen. These toxins may be **endotoxins,** toxic substances that are components of the bacterial cells themselves, or **exotoxins,** molecules that are secreted by the bacteria.

In the late 19th century, Robert Koch, a German physician, established rigorous criteria for identifying the pathogen for a given disease. The four rules, now known as **Koch's postulates** (Figure 20-2), are:

1. The same pathogen must be present in every person with the disease.
2. The pathogen must be capable of being grown in a pure culture—one uncontaminated with other organisms.
3. The pure cultured pathogen, introduced into an experimental animal, must cause the disease.

4. The same pathogen must be present in the infected animal after the disease develops.

Although Koch's postulates are now the accepted guidelines for identifying pathogenic organisms, the postulates cannot always be fulfilled. For example, the bacterium *Treponema pallidum,* which causes the sexually transmitted disease syphilis, has yet to be cultured in the lab. In addition, bacteria that live inside other cells (such as the sexually transmitted bacterium *Chlamydia*) cannot grow in pure cultures because these bacteria always require the presence of host cells. Likewise, all viruses must grow inside other cells (Chapter 12), so pathogenic viruses such as HIV (human immunodeficiency virus) and influenza cannot be grown in a "pure culture."

Not all pathogenic bacteria are strangers. Many are normal residents of our bodies that are out of place or have mul-

Figure 20-3 **Microorganisms that normally inhabit the human body.** All of the microorganisms listed here live—usually harmoniously—on the surfaces and in the interiors of human bodies. *Candida albicans,* commonly known as yeast, is a fungus that lives on the skin and in the mouth and vagina. *Candida* is, of course, a eukaryote, not a prokaryote. *(Erich Lessing/Art Resource)*

Figure 20-4 A tuberculosis isolation ward. In the 1920s, sick children were isolated on this ferryboat in New York's harbor. With the advent of antibiotics, hospitals no longer needed isolation wards for patients with tuberculosis. Today, because of strains of tuberculosis that are resistant to all known antibiotics, such wards have returned. (UPI/Corbis-Bettmann)

tiplied when the body's defenses weaken (Figure 20-3). *Escherichia coli* in our intestines, for example, are normal and useful members of the community of organisms in the body. But the very same organism in the bloodstream can kill in a matter of hours.

> Most pathogenic bacteria make us ill by releasing toxic chemicals that kill our own cells. Koch's postulates are guidelines for determining whether a microorganism causes a specific disease.

How Do Bacteria Acquire Resistance to Antibiotics?

Whenever we are sick from a bacterial infection, we and our doctors want to eliminate the pathogenic bacteria as quickly as possible. The easiest way to do that is with an antibiotic. **Antibiotics** are chemicals that kill bacterial cells without harming our own eukaryotic cells. Since the first antibiotics came into use in the 1940s, medical researchers have discovered or synthesized more than 150 different antibiotics.

Amazingly, pathogenic bacteria now exist that are resistant to all antibiotics. In New York City, patients with certain strains of highly resistant tuberculosis, for example, are now treated the same way tuberculosis patients were treated 50 years ago—by isolating them. There is no cure for such people, and health workers try to prevent them from passing their resistant strains of tuberculosis bacteria to other patients or to hospital staff (Figure 20-4).

Where do bacteria acquire resistance to antibiotics? First we must understand how antibiotics work and where that resistance comes from. Most antibiotics are naturally synthesized by soil bacteria or fungi. In fact, soil bacteria produce so many chemicals that are toxic to one another that pathogenic bacteria usually die when exposed to ordinary dirt.

The genes for resistance to these antibiotics come from the very organisms—mostly soil organisms—that make the antibiotics. A soil bacterium needs to be immune to its own toxins, after all, and they have evolved a variety of tools to fend off specific antibiotics.

What are these tools? That depends on the antibiotic. Some antibiotics, such as penicillin, kill bacteria by preventing the bacteria from synthesizing a cell wall. Other antibiotics prevent bacteria from expressing their genes, interfere with DNA folding (which can disrupt gene expression and DNA replication), inhibit the synthesis of folic acid (important in nucleotide synthesis), or block protein synthesis.

Resistance to antibiotics can be classified into two main varieties (Figure 20-5A). Some resistance genes merely enable the bacteria to survive exposure to an antibiotic. Such genes may alter the shapes of bacterial molecules in such a way that the antibiotics can no longer bind to them, pump the antibiotic out of the cell, or destroy the antibiotic as it enters the cell.

Other resistance genes enable the bacteria to release enzymes that actually destroy antibiotics in the environment (Figure 20-5B). This kind of resistance is the most dangerous. If, for example, a college student regularly took the antibiotic tetracycline to treat acne, he could gradually accumulate intestinal and other bacteria able to destroy tetracycline. If he then became infected by nonresistant *Chlamydia trachomatis*, for example, and his doctor prescribed tetracycline, his intestinal bacteria would destroy the tetracycline before it could kill the invading *Chlamydia* bacteria.

Resistance accumulates in populations of bacteria in several ways. In all populations of bacteria, a few individuals carry genes for resistance to antibiotics. Even baboons living in the wild, far from any humans, carry small numbers of intestinal bacteria that are resistant to some antibiotics. Humans have far higher numbers of resistant bacteria. One study of resistance in a group of students showed that one in ten students carried large numbers of *E. coli* that were resistant to a combination of two antibiotics first introduced in the 1970s.

When humans or other animals are exposed to antibiotics, the resistant bacteria flourish in the absence of competing bacteria, multiplying until they are the majority. This is simply natural selection. The longer the antibiotic is present, the more likely the antibiotic-resistant bacteria are to increase in number. Thus, beef cattle regularly fed tetracycline gradually accumulate populations of bacteria that are resistant to tetracycline. Unfortunately, the story does not end there.

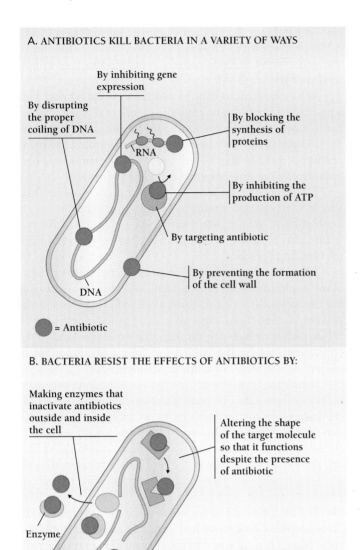

Figure 20-5 How does it work? A. Every antibiotic has a distinctive way of preventing bacteria from reproducing. B. Bacteria have different ways of resisting the effects of antibiotics.

gation, in which one bacterium can pass plasmids and other DNA to another bacterium (Figure 12-8). Plasmids, simple loops of DNA carried separately from the bacterium's DNA, generally carry genes for traits such as antibiotic resistance that are beneficial to the bacteria that they inhabit. Because plasmids themselves pass genes around to one another by means of jumping genes, plasmids accumulate genes for resistance the way a snowball accumulates snow as it rolls downhill. Plasmids can carry resistance to as many as eight classes of antibiotics. Bacteria then pass such plasmids, through conjugation, not only to other bacteria of the same species but also to bacteria that are unrelated.

Unlike most species of organisms, bacteria are not discrete genetic entities, but a vast, integrated microbial world. As a result, multiple antibiotic resistance is the rule, rather than the exception. In 1995, for example, hospital workers in Atlanta found that 25 percent of white children fighting ordinary ear and respiratory infections by *Streptococcus pneumoniae* were infected with strains resistant to two or more antibiotics.

The tendency of plasmids to carry genes for resistance to several antibiotics has important consequences. Cattle treated with tetracycline, for example, can acquire bacteria that are resistant not only to tetracycline but to several other antibiotics as well. This is why the South Dakota beef cattle, which had been treated with tetracycline, carried *Salmonella* resistant to both amoxicillin and tetracycline. The plasmid that protected the bacteria from tetracycline also happened to carry genes for resistance to other antibiotics.

> *Resistance may allow a bacterium simply to survive exposure to an antibiotic or, alternatively, actively to destroy any antibiotic in its environment. Resistance genes accumulate in populations of bacteria, first, through natural selection, and second, by moving from one bacterial strain to another by means of plasmids and other mobile genes.*

If antibiotic-resistant bacteria accumulated only in the presence of antibiotics, through selection, then society's problems with antibiotic resistance would be relatively minor. Strains of bacteria without resistance to a specific antibiotic could never acquire resistance except through chance mutations. Unfortunately, bacteria can pass fully functioning antibiotic resistance genes to completely different kinds of bacteria, so that resistance spreads rapidly. Earlier in this book, we mentioned that bacteria engage in a process called conju-

HOW DO WE KNOW THAT AN ORGANISM IS A PROKARYOTE?

The defining characteristic of prokaryotes that sets them apart from all other organisms is the absence of a true nucleus (Chapter 4). Prokaryotes lack not only a nucleus but also chloroplasts and mitochondria and all other membrane-bounded organelles. Some prokaryotes do have extensively folded plasma membranes, however, and use these large surfaces for respiration, photosynthesis, and DNA replication.

Instead of neatly packing its DNA in a membrane-enclosed nucleus (as a eukaryote does), a prokaryote tucks its DNA into a concentrated mass called a **nucleoid**. The DNA

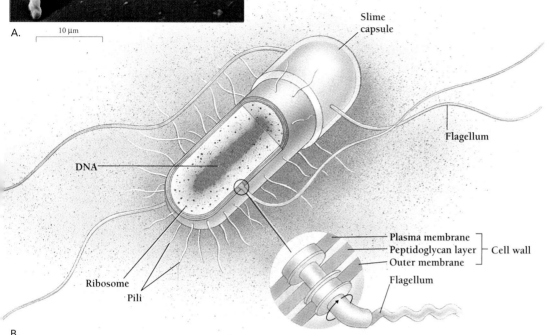

Figure 20-6 The familiar shape of the bacillus typifies both *Salmonella* and *E. coli.* A. *Salmonella sp.* B. *E. coli.* (A, Meckes/Ottawa/Photo Researchers, Inc.)

 Quiz

in the nucleoid consists of a single molecule of double-stranded DNA that forms a closed loop. A prokaryote may also contain one (or more) small loops of DNA called plasmids (Chapter 12). Because a prokaryote's DNA is circular, it can replicate continuously, unlike eukaryotic DNA, which must replicate in pieces.

Also unlike the eukaryotes, prokaryotes undergo neither mitosis nor meiosis. Prokaryotes do not require sex to reproduce. However, as we have seen, they exchange genetic information so freely that reproductive isolation in bacteria is the exception rather than the rule.

Unlike other species of organisms, then, a "species" of bacterium is neither a reproductive unit nor a genetic unit, and classifying bacteria into species has been something of an art. As we will see later in this chapter, recent studies of the complete DNA sequences of prokaryotic genomes have led to new views of the relationships among groups of prokaryotes. Many biologists, for example, now argue that we should not lump eubacteria and archaea together in a single group, even one as broadly defined as the prokaryotes.

Prokaryotic cells come in a variety of shapes (Figure 20-6). The most common are the rod-shaped **bacilli** [singular, bacillus; Latin, = little rod]. In addition are the spiral-shaped **spirilla** [singular, spirillum; Latin, *spira* = coil] and the spherical **cocci** [singular, coccus; Greek, *kokkos* = berry]. Most bacteria are single-celled organisms, though some aggregate to form colonies of unspecialized cells (Figure 20-7).

The defining characteristic of prokaryotes is the absence of a nucleus or other membrane-enclosed organelles. The DNA of prokaryotes forms a single circle of double-stranded DNA.

Figure 20-7 Some bacteria, such as this myxobacterium, are colonial. Cells of this species form a "fruiting body" to reproduce. (Karen Stephens/Biological Photo Service)

from injury. However, while plant cell walls are made of cellulose, bacterial cell walls are made of a mixture of polysaccharides and polypeptides called **peptidoglycan.** The cell wall is a single, giant, bag-shaped macromolecule composed of a network of cross-linked peptidoglycan. The peptidoglycan cell wall is important to biologists for two reasons. It helps distinguish two major groups of bacteria, and it makes some bacteria susceptible to antibiotics. (Archaea, like eukaryotes, do not produce peptidoglycan.)

Biologists divide all of the bacteria into two major groups according to the way their cell walls soak up various dyes. In the 1880s, the Danish bacteriologist Hans Christian Gram found that when he treated bacteria with a particular purple dye, some kinds of bacteria became permanently stained, while others could be washed clean. Those that became permanently stained he called "stain positive." The others were "stain negative." Today bacteriologists call these two categories of bacteria **gram positive** and **gram negative,** after Gram's stain (Figure 20-8).

What sets gram-negative and gram-positive bacteria apart from one another is the structure of the cell wall. Gram-negative bacteria—such as *E. coli* or the gonococcus bacteria that cause the sexually transmitted disease gonorrhea—have a thin, three-layered cell wall, whose outermost layer does not bind the Gram stain (Figure 20-8C). In contrast, gram-positive bacteria—such as the streptococci (which can cause strep throat) and the staphylococci (which can cause boils and other infections)—have a thick, single-layered cell wall containing at least 20 times as much peptidoglycan as the gram-negative bacteria. The peptidoglycan, itself arranged in up to 40 layers, binds the Gram stain so that it cannot be washed away.

Most Bacteria Have Complex Cell Walls

Like plant cells, most bacterial cells have rigid cell walls outside their plasma membranes. Like the wall of a plant cell, the wall of a bacterial cell gives the cell a shape and protects it

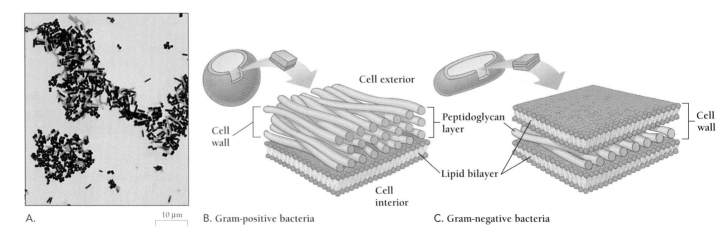

Figure 20-8 It's all in the wall. Bacteria absorb stain or not depending on what kind of cell wall they have. A. A mixture of gram-negative (*light*) and gram-positive (*dark*) bacteria. B. The gram-positive coccus bacterium has a two-layered, peptidoglycan-rich cell wall. C. The gram-negative bacillus has a three-layered, peptidoglycan-poor cell wall. (A, G.W. Willis/Biological Photo Service)

Figure 20-9 Dental plaque bacteria, which coat our teeth in the morning, are one kind of bacteria that have a slime layer.

In gram-negative bacteria, the same outer layer of the cell wall that prevents the Gram stain from penetrating also prevents antibiotics such as penicillin from penetrating. As a result, penicillin and other "narrow-spectrum" antibiotics kill gram-positive bacteria only. "Broad-spectrum" antibiotics such as tetracycline kill both gram-negative bacteria and many gram-positive bacteria.

Outside of the cell wall, some prokaryotes have other materials. The outer surfaces of some prokaryotes possess *pili* [singular, pilus; Latin, = hair], hairlike structures that attach the bacteria to surfaces and to other cells. The F-pilus, for example, provides a bridge for the transfer of DNA during bacterial conjugation.

The bacterium that causes pneumonia is enclosed in a **capsule,** a gelatinous layer of polysaccharides and proteins outside its cell wall. Capsules protect disease-causing bacteria from attack by the body's white blood cells, enabling the bacteria to overwhelm the body's immune system. Such bacteria are more deadly, or virulent.

Other bacteria possess a **slime layer,** a tangled web of polysaccharides that, like a capsule, lies outside the cell wall. A sticky slime layer protects the bacterial cell against dehydration, helps keep food from diffusing away, and also allows cells to attach to and move about on the surfaces on which they grow. Dental plaque is a slime layer made by the bacteria that inhabit the mouth. Plaque helps bacteria grow smoothly over the surfaces of our teeth (Figure 20-9).

The presence of large amounts of peptidoglycan distinguishes the bacterial cell wall from the cell walls of archaea and eukaryotes. Gram-positive bacteria have a thick, peptidoglycan-rich cell wall that leaves them susceptible to penicillin and other narrow-spectrum antibiotics.

How Do Bacteria Move?

Some bacteria move by secreting slime and gliding on it. Others use a propeller—the bacterial **flagellum** (Chapter 4). Bacterial and eukaryotic flagella are entirely different. The eukaryotic flagellum is an extension of the cytoplasm surrounded by membrane that beats back and forth. In contrast, the prokaryotic flagellum is a thin, rotating protein filament attached to the outside of the cell.

Many bacteria use their flagella for directed movement, or **taxis** [Greek, = to put in order]. (A taxicab performs directed movements.) The direction of movement depends on the operation of the flagella. For example, when the flagella of an *E. coli* rotate in a counterclockwise direction, they twist around each other and propel the bacterium forward. When the flagella rotate in a clockwise direction, however, they separate and become uncoordinated. As a result, the bacterium tumbles aimlessly. When the flagella resume a counterclockwise spin, the bacterium moves in a new direction. By controlling the

Figure 20-10 Profiting from prokaryotic adaptations. These tubeworms, *Riftia pachyptila* (A), live near hydrothermal vents, in symbiosis with bacteria, *Staphylothermus marinus* (B), that get energy from inorganic sulfide compounds. (A, David A. Hardy/SPL/Photo Researchers, Inc.; B, Wolfgang Baumeister/SPL/Photo Researchers, Inc.)

timing of clockwise and counterclockwise rotations, bacteria can move toward or away from food, oxygen, heat, light, and other stimuli. Some bacteria even use the Earth's magnetic field to guide them to the lower depths of the ocean. **Chemotaxis** is taxis in response to varying concentrations of specific chemicals.

Bacteria move in directed fashion, by gliding on slime or by rotating propellerlike flagella.

How Do Prokaryotes Grow So Rapidly?

Because bacteria keep their DNA in a nucleoid instead of a nucleus, bacteria can replicate their DNA continuously and divide very rapidly. In fact, the cells in well-nourished colonies of *E. coli* can divide every 20 minutes. Every time the bacteria divide, the colony doubles in size. That is, after one generation, one cell becomes two. After another generation the 2 cells become 4, then 8, 16, and so forth. After 20 such doublings (in less than 7 hours), a single cell would yield 2^{20}, or about a million, new cells; and after 30 doublings (10 hours),

2^{30}, or about a billion, cells. Biologists describe such growth as exponential or logarithmic (Chapter 28).

If a single *E. coli* bacterium could be supplied with enough food to divide every 20 minutes continuously, the combined mass of the bacterium and all its descendants would exceed that of the entire planet after only 2 days. We can see that, in reality, exponential growth can last for only short times. Nonetheless, the rapid proliferation of prokaryotes means that a favorable new mutation such as antibiotic resistance can spread quickly to a huge number of progeny.

A streamlined metabolism and a simple means of replicating DNA together allow bacteria to grow and divide more rapidly than eukaryotes.

Prokaryotes Are Metabolically Diverse

We eukaryotes are monotonous from a biochemical point of view. Nearly all of us use glycolysis and respiration to produce energy. Nearly all of us need oxygen to live. Likewise, all photosynthetic eukaryotes photosynthesize in the same way. In fact, the relentless sameness of metabolic pathways in all eukaryotes strongly supports the idea that we are all descended from a common ancestor.

In contrast, prokaryotes use a great variety of biochemical pathways. Some, for example, perform glycolysis, but with end products other than ethanol or lactate (Chapter 6). Many of these end products contribute to the pleasant (or not-so-pleasant) odors of wines and cheeses. Some prokaryotes do not use glycolysis at all. Instead, they produce ATP by capturing the chemical energy of molecules in their environment. Methane-producing prokaryotes, for example, capture energy released when carbon dioxide (CO_2) reacts with hydrogen gas (H_2) to produce methane (CH_4). Some archaea capture the energy from the reaction of sulfur-containing compounds with hydrogen gas. Hydrogen gas, unstable in the atmosphere, nonetheless accumulates to high levels under the oceans and within the earth's mantle.

Because prokaryotes are so diverse, they flourish in places where eukaryotes could never survive. Over billions of years, prokaryotes have evolved an amazing variety of adaptations that contribute to their continuing success. Some of these adaptations are (1) spore formation, which allows bacteria to survive, multiply, and disperse no matter how hostile the environment; (2) the production of antibiotics that can kill competitors; (3) the use of resistance transfer factors and plasmids to spread resistance to antibiotics from one bacterium to another; and (4) the ability to obtain energy from the waste products of other species (for example, carbon dioxide and methane).

The adaptations of prokaryotes in turn provide new opportunities for eukaryotic life. A 2-meter-long tubeworm, called *Riftia pachyptila,* lives near hydrothermal vents on the ocean floor. *Riftia* gets energy-rich compounds from symbiotic bacteria that use the energy of inorganic sulfur compounds from the vents (Figure 20-10).

The enormous success of prokaryotes depends on their metabolic diversity and their ability to transfer genes to one another.

HOW DO BIOLOGISTS CLASSIFY PROKARYOTES?

Because prokaryotes do not (usually) reproduce sexually, taxonomists cannot use the biological species concept, which defines a species as an interbreeding population, to classify prokaryotes (Chapter 19). Reconstructing the phylogeny of bacteria is equally difficult, as the fossil record for prokaryotes is nearly nonexistent.

Until recently, taxonomists have relied on morphological and metabolic comparisons. This approach is fraught with difficulties, however, since two strains of a single species may look very different. For example, one may possess a distinctive feature such as an external capsule, while the other lacks the capsule. Similarly, metabolic and biochemical differences may not define species, since genes carrying distinctive traits such as antibiotic resistance can spread rapidly—through viruses, plasmids, or transfer factors—among many unrelated bacteria. So far, some 1700 species of prokaryotes have been named, although estimates of the number of species vary widely—from 5000 up into the millions.

How Do Biologists Classify the Archaea?

Although some biologists still put all of the prokaryotes into a single kingdom, the consensus has changed dramatically in the last 5 years. Most biologists now recognize that prokaryotes include two large domains—the Archaea and the Eubacteria—and that each of these domains is enormously diverse. The archaea consist of at least 2 kingdoms and the eubacteria of at least 25 kingdoms (Figure 20-11).

The **Archaea** [Greek, *archaio* = ancient] differ from eubacteria and from eukaryotes in the composition of their ribosomes and in the kinds of lipids in their cell membranes. The Archaea entirely lack peptidoglycan in their cell walls. Different types of archaea derive energy from hydrogen gas in one of two ways—by converting carbon dioxide (CO_2) to methane (CH_4) or by converting sulfur compounds to sulfur. These two groups form the two recognized kingdoms of the archaea.

The first kingdom (called the Euryarcheota) includes the **methanogens,** shown in Figure 20-12, and the **halophiles,** which live in extremely salty environments (Figure 20-13). The second kingdom (called the Crenarchaeota) includes the **thermoacidophiles,** which inhabit hot sulfur springs such as those in Yellowstone National Park (Figure 20-14).

Methanogens, all obligate anaerobes, live principally in two environments: (1) swamps and sewage treatment plants, and (2) the guts of animals, especially of cows, termites, and other animals that digest cellulose. Even the human gut harbors methanogens that produce methane, which, when released, is called flatulence. Methanogens play an essential role in the ecology of the Earth. Where methanogens flourish in swamps made anaerobic by the growth of other bacteria, they release methane, or "marsh gas," into the atmosphere. There, the methane is quickly oxidized to carbon dioxide, which is then available for photosynthesis. If methanogens did not recycle carbon from organic molecules back into the atmosphere, other organisms would gradually run out of carbon. Natural gas, which we use to heat our homes and cook our meals, is about 98 percent methane—all provided by the methanogens.

The Archaea consist of at least two kingdoms. The most widespread and ecologically important archaea are the methanogens.

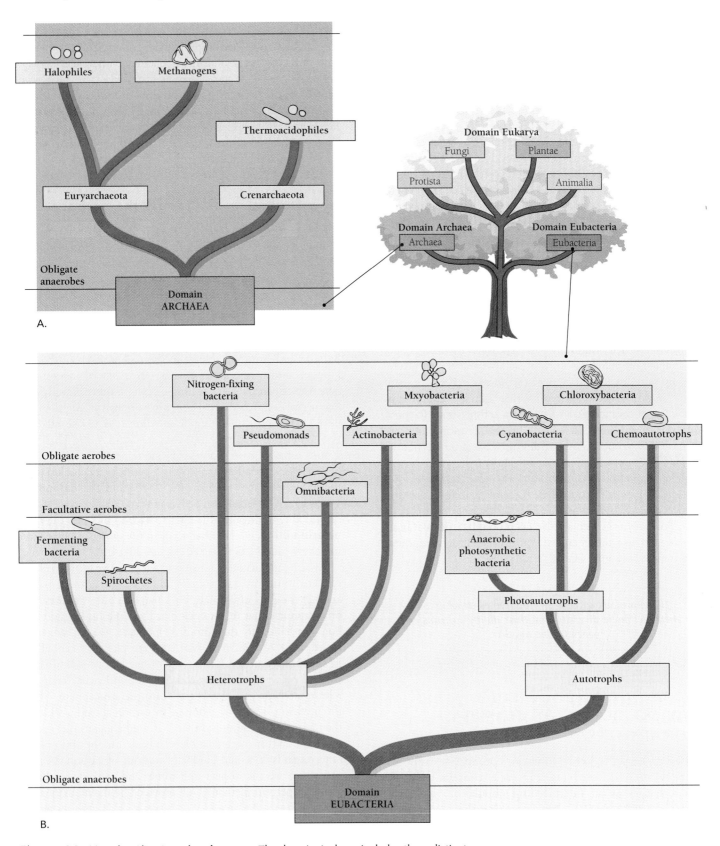

Figure 20-11 Classification of prokaryotes. The domain Archaea includes three distinct groups, which may be kingdoms or phyla. The domain Eubacteria includes at least 25 groups, now given kingdom status, of which a dozen are shown here. The Eubacteria are divided into heterotrophs and autotrophs. The autotrophs are further divided into photoautotrophs and chemoautotrophs.

Figure 20-12 Methanogens. Methanogens, which *generate* the gas *methane*, include bacteria of all three standard shapes—bacilli, spirilla, and cocci. (F. Widdel/Visuals Unlimited)

How Do Biologists Classify the Eubacteria?

Despite the difficulties of classifying prokaryotes, biologists do agree on some broad groupings. In one classification, in 1987, taxonomists divided the **Eubacteria** into 12 kingdoms; a more recent list included more than 25 kingdoms. DNA sequences from each of these kingdoms differ from those of the others more than the DNA from the plant kingdom differs from that of the animal kingdom.

Just as zoologists classify some animals according to whether they eat plants or other animals, microbiologists have

Figure 20-13 Halophiles. Most bacteria cannot grow in high concentrations of salt. Because salt prevents the growth of most bacteria, it acts as a preservative in ham, beef jerky, and other salty foods. Unlike any other bacteria, halophiles such as these *Halococcus* bacteria flourish in briny places the world over—from the Dead Sea to Utah's Great Salt Lake and California's Mono Lake. Halophiles, usually flagellated rods or cocci (like these), form a lavender-pink scum over their salty homes. The pink color comes from a purple pigment in the plasma membrane that captures light energy. (Alfred Pasteka/Science Photo Library/Photo Researchers, Inc.)

traditionally classified Eubacteria into broad categories according to the way they derive food. **Autotrophs** [Greek, *autos* = self + *trophos* = feeder] obtain carbon atoms directly from carbon dioxide, while **heterotrophs** [Greek, *heteros* =

Figure 20-14 Thermoacidophiles. Thermoacidophiles may be chemoautotrophs (reaping energy from sulfur compounds or methane) or heterotrophs (reaping energy and carbon from small organic compounds made by other organisms). The best-known genus, *Sulfolobus*, lives in the hot sulfur springs of Yellowstone National Park. *Sulfolobus* can survive a temperature as high as 88°C (190°F) and an acidity as low as pH 1. Although *Sulfolobus* thrives in heat, it dies of cold at temperatures below 55°C (130°F). (Richard Thom/Visuals Unlimited)

Figure 20-15 Aerobic heterotrophs. A. Nitrogen-fixing bacteria growing in nodules on the roots of a plant. Some nitrogen-fixing bacteria live freely in soil and water, while others associate intimately with specific plants, living, for example, in tiny nodules on the roots of peas, beans, lupines, locusts, and other members of the pea family. The nitrogen-fixing bacteria provide the plant with a usable form of nitrogen, and the plant provides carbohydrates for energy. This relationship allows extraordinarily efficient fixation of nitrogen, as much as 100 times faster than that performed by free-living bacteria. B. *Chlamydia trachomatis*. Besides the enteric bacteria, the omnibacteria also include rickettsia and chlamydia, which live only as parasites inside animal cells and cause a variety of diseases. Shown here is *Chlamydia trachomatis*, which causes the most common sexually transmitted disease in the United States. C. *Streptomyces*. Actinobacteria form branching filaments like those of fungi. They are widely distributed in soils, where they decompose organic matter. *Streptomyces*, for example, produces the valuable antibiotic streptomycin and other substances that kill competing organisms. A few forms of actinobacteria cause human diseases such as tuberculosis and leprosy. (A, C.P. Vance/Visuals Unlimited; B, David M. Phillips/Science Source/Photo Researchers, Inc.; C, Frederick Mertz/Visuals Unlimited)

other + *trophos* = feeder], which include the great majority of prokaryotic species, derive carbon atoms (and also energy) from organic molecules such as glucose. Autotrophs may be divided into **photoautotrophs,** which use light energy to drive the synthesis of organic compounds from CO_2, and **chemoautotrophs,** which derive energy by oxidizing such inorganic substances as hydrogen sulfide (H_2S) or ammonia (NH_3).

Heterotrophs fall into two groups. **Photoheterotrophs** use light energy but also require organic compounds. **Chemoheterotrophs** use no light, relying exclusively on organic molecules for both energy and carbon atoms. Heterotrophic prokaryotes may be **decomposers,** which absorb nutrients from dead organisms, or **symbionts** [Greek, *syn* = together + *bios* = life], which absorb nutrients from living organisms. Symbionts that cause illness, such as *Salmonella newport*, are called pathogens. Others, such as *E. coli*, are usually harmless, or even helpful. Prokaryotic decomposers, like fungal decomposers, recycle the materials that were formerly part of other organisms.

Prokaryotes also vary in the way they use oxygen. Heterotrophs that require oxygen for respiration are called **obligate aerobes.** Those that can grow either with or without oxygen are called **facultative aerobes.** Still others, for which oxygen is a deadly poison, are called **obligate anaerobes.**

The Eubacteria include aerobic heterotrophs, anaerobic heterotrophs, and autotrophs. We will briefly describe examples of each of these three groups and the incredible diversity of environments in which these simple organisms thrive.

The aerobic heterotrophs

The aerobic heterotrophs have tremendous biochemical diversity: some break down all kinds of organic compounds, while others can use nitrate —NO_3^-—instead of oxygen—as the ultimate electron acceptor, and still others make powerful antibiotics that kill competitors. A few aerobic heterotrophs are the nitrogen-fixing bacteria, pseudomonads, omnibacteria, actinobacteria, and myxobacteria (Figure 20-15).

From the perspective of most other organisms on Earth, the most important prokaryotes may be the **nitrogen-fixing aerobic bacteria** (Figure 20-15A). Nitrogen-fixing bacteria provide nitrogen in forms such as ammonia and amino acids from which other organisms can make proteins. Although nitrogen gas (N_2) constitutes some 80 percent of our atmosphere, it is so chemically stable that most organisms are incapable of breaking the bond between the two nitrogen atoms to form complex organic compounds. As a result, all organisms rely on nitrogen-fixing bacteria to provide nitrogen in

BOX 20.1

Carl Woese and the Archaea

In the summer of 1996, a researcher working on an obscure microbe at The Institute for Genomic Research in Rockville, Maryland, announced a finding that immediately hit the television nightly news and the front pages of newspapers all over the United States. In sequencing the microbe's genome, the researcher, Carol Bult, had provided irrefutable evidence that all organisms, historically classified as either eukaryotes or bacteria, actually fall into three basic categories.

The microbe, named *Methanococcus jannaschii*, was only the fourth organism for which researchers had completed such a sequence, but the implications were profound. Bult's work not only supported the theory that the archaea are distinct from both bacteria and eukaryotes, it also provided independent support for two other ideas: (1) that we eukaryotes are more closely related to the archaea than to the bacteria and (2) that both eukaryotes and bacteria are descended from ancient organisms that resembled archaea.

Yet, despite the drama with which these important results were revealed, it was an oddly anticlimactic end to a very long story. Thirty years before, Carl Woese, a young molecular biologist at the University of Illinois, Urbana, began the painstaking task of working out the evolutionary relationships of the bacteria. Earlier researchers had made the attempt and failed. "The ultimate scientific goal of biological classification cannot be achieved in the case of bacteria," wrote the eminent microbiologist Roger Stanier.

Woese admired Stanier's own considerable efforts to classify the bacteria but did not accept that classifying the bacteria was impossible. Indeed, he had an ingenious plan for discovering their relations to one another. He knew that ribosomal RNAs are among the most conserved parts of all organisms. Because the work of translating RNA into proteins is fundamental, organisms with mutations in their ribosomal RNA nearly always die and ribosomes change hardly at all over millions of years. Ribosomes are, in short, some of the best sources of information about long-term evolutionary relationships among organisms.

For 10 years Woese (pronounced woes) catalogued short strings of nucleotides that he cut from the ribosomal RNAs of different microorganisms. It was mind-numbing, labor-intensive work that left him, he later said, "just completely dulled down." Yet it was fruitful work, for he gradually unraveled the relationships among some 60 different kinds of bacteria.

Then, in 1976, Woese's colleague Ralph Wolfe sent him some methanogens on which to work his magic. Methanogens were curious bacteria that came in a variety of shapes but all produced methane as a by-product of their metabolism. No one knew where they fit in with the other bacteria. Woese set to work to find out. To his surprise, however, the methanogens lacked the characteristic sequences of ribosomal RNA that all other bacteria share. As he told Wolfe, the methanogens didn't seem to be bacteria at all.

It was dramatic news, for it meant that in addition to the eukaryotes, and the bacterial prokaryotes, we humans were sharing the planet with a completely different domain of organisms—the archaea—never before recognized. Yet, although the story hit the front page of *The New York Times*, the news faded away immediately. For most people, the news was abstract and meaningless. And most of Woese's own colleagues ignored his discovery. They simply didn't believe him. Although Woese provided the answers to many of the questions that most intrigued the eminent microbiologist Roger Stanier, Stanier, who has since died, never acknowledged Woese. For years, Woese says, he waited in vain for some friendly acknowledgment from Stanier.

R.G.E. Murray, the author of the leading microbiology textbook, didn't even include the archaea until 1986, and then only as a subgroup within the prokaryotes—even though Woese's work had demonstrated that the archaea were distinct from other bacteria. A handful of other researchers stood behind Woese, including Wolfe, arguing with their colleagues on his behalf. Yet, Woese's work was otherwise almost universally snubbed.

Woese merely retreated back to his laboratory, continuing his work, gathering new and better data, and publishing a series of landmark papers on the evolution and classification of bacteria. Today Woese is highly acclaimed. He has been elected to the prestigious National Academy of Sciences and awarded both a MacArthur Foundation "genius" grant and microbiology's top honor—the Leeuwenhoek medal.

Great discoveries in science frequently meet initial rejection. As Wolfe told *Science* magazine in 1997, "We can say, 'Oh, that's just the way science works.' But it was a personal experience for him. He's the one who had to live through it."

Figure A Carl Woese discovered the archaea in 1977. The scientific community did not fully accept his results until 1997. *(Courtesy of University of Illinois at Urbana-Champaign News Bureau)*

a usable form. The bacteria perform this service, called **nitrogen fixation,** at great cost, however, for the task requires a lot of energy. (Conversely, processes that yield nitrogen gas as a product release lots of energy. For this reason, many nitrogen-containing compounds, including commercial fertilizers, are explosives.)

The **pseudomonads,** known as "the weeds of the bacterial world," are a large and aggressive group of bacteria that seem to live everywhere—in soil, ponds, infected wounds, hot tubs, and even medicine bottles. Pseudomonads are typically straight or curved gram-negative rods with flagella. They require oxygen and produce CO_2. The extraordinary success of pseudomonads comes from their metabolic diversity. They can survive on carbohydrates, petroleum products, tough polysaccharides in natural fibers, and even antibiotics and pesticides.

Omnibacteria are the most common organisms on Earth (Figure 20-15B). These bacteria can all perform aerobic respiration and also have the extraordinary ability to use nitrate (NO_3^-) as an electron acceptor instead of oxygen. The most prolific and well known are the common enteric bacteria [Greek, *enteron* = intestine], such as *E. coli,* which inhabit the intestines of animals. Omnibacteria may be rod shaped, as is *E. coli,* or comma shaped, as is the species that causes cholera.

The anaerobic heterotrophs

The anaerobic heterotrophs are mostly obligate anaerobes. They include, for example, the **fermenting bacteria,** which derive energy and carbon atoms from a variety of organic compounds in the absence of oxygen; and the **spirochetes,** which have a distinctive corkscrew shape (Figure 20-16).

The autotrophs

The autotrophs consist of four phyla: the chemoautotrophs, which photosynthesize *without* oxygen, and three phyla of photoautotrophs, which photosynthesize *with* oxygen. The photoautotrophs consist of the **anaerobic photosynthetic bacteria,** the **cyanobacteria,** and the **chloroxybacteria.**

Biologists have traditionally classified prokaryotes according to how they acquire energy and building blocks, whether they photosynthesize, and whether they use or tolerate oxygen. Biologists divide the Eubacteria into three main groups: aerobic heterotrophs, anaerobic heterotrophs, and autotrophs. The most ecologically important Eubacteria are (1) the cyanobacteria, the main photosynthetic prokaryotes, which fix carbon; and (2) the nitrogen-fixing bacteria, which supply the world's organisms with fixed nitrogen for making proteins.

Figure 20-16 Swimming syphilis spirochete. The most famous spirochete is *Treponema pallidum,* which causes the sexually transmitted disease syphilis. Spirochetes contain internal flagella enclosed between layers of the cell wall, by means of which they move differently from any other bacteria. Although some live freely and some are parasites, all spirochetes move by undulating quickly, even through the most viscous liquids. Most spirochetes are finicky about their diets and *T. pallidum* is no exception. Its requirements are so specific, in fact, that although it grows well inside the human body, biologists have been unable to grow virulent strains in the laboratory. *T. pallidum* is transmitted from one person to another by means of direct contact between moist mucous membranes. Until the introduction of penicillin in the 1940s, syphilis was incurable, progressive, and fatal. *(CDC/Science Source/Photo Researchers, Inc.)*

How Has DNA Sequencing Changed the Classification of Prokaryotes?

Two kinds of DNA sequence information have dramatically changed the way that biologists view the prokaryotes. The first type of information initially came from the pioneering work of Carl Woese on the genes for ribosomal RNA (Box 20.1). This work led to the originally revolutionary idea that the archaea represent a separate domain of life, as different from the eubacteria as from eukaryotes. The second type of information started with the determination, in 1997, of the complete genomic sequences of two archaea species, one a methanogen and the other a sulfur-metabolizer.

The complete DNA sequences revealed that, in some important respects, archaea resemble eukaryotes more than eubacteria. The most startling resemblances were in the enzymes responsible for handling of genetic information (DNA replication and RNA synthesis).

Analysis of other archaeal genes, however, suggests that they are more closely related to eubacterial than to eukaryotic genes. Researchers continue to disagree about the family tree of prokaryotes. A major difficulty in interpreting the DNA sequence data is that researchers increasingly think that there has been lots of gene transfer among the archaea.

Biologists are beginning to conclude that the task is far more complicated than originally expected (Figure 20-17). As researchers sequence more prokaryotic genomes, however, we

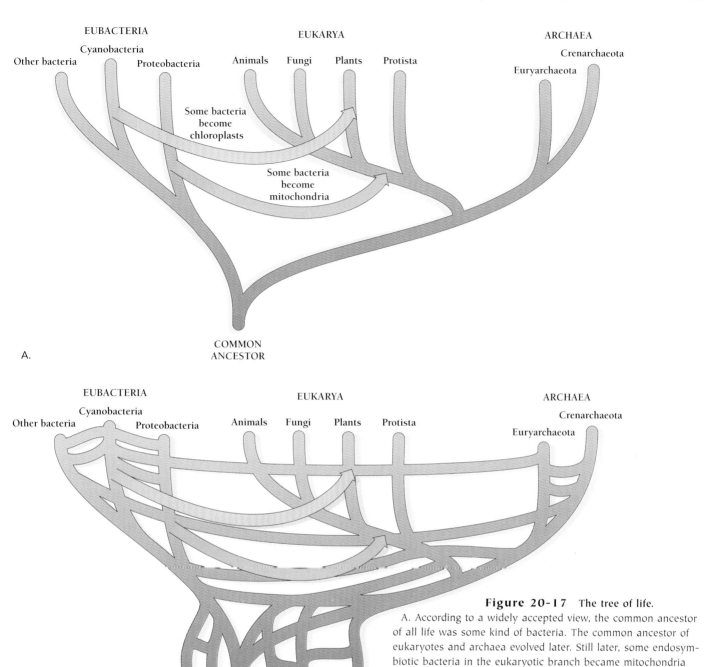

Figure 20-17 The tree of life.
A. According to a widely accepted view, the common ancestor of all life was some kind of bacteria. The common ancestor of eukaryotes and archaea evolved later. Still later, some endosymbiotic bacteria in the eukaryotic branch became mitochondria and chloroplasts. B. According to a more recent view, the history of life must be represented as a complex net, with gene transfer between species a common occurrence. (Drawing based on W.F. Doolittle, Science 284:284, 2124, 1999)

can expect to know more about the evolutionary history of individual genes. Many biologists suspect that the history of prokaryotes will not be one of simple descent from a common ancestor, but will include many instances of gene transfer between species.

In this chapter we have briefly surveyed the two domains of prokaryotes. Just as important, we learned how the biology of bacteria helps them acquire resistance to antibiotics. In the next chapter, we survey the first two kingdoms of eukaryotes, the Protista and the Fungi.

Study Outline with Key Terms

Most **pathogenic** prokaryotes make us ill by releasing toxic chemicals that kill our own cells. These may be **endotoxins** or **exotoxins**. **Koch's postulates** are guidelines for determining whether a microorganism causes a specific disease, a topic of interest to **epidemiologists**. Resistance may allow a bacterium to survive exposure to an **antibiotic** or, alternatively, to actively destroy any antibiotic in its environment. Resistance genes accumulate in populations of bacteria, first, through natural selection, and second, by moving from one strain of bacterium to another by means of plasmids and other mobile genes.

Prokaryotes tuck their DNA into a **nucleoid**, rather than a true nucleus. They also lack other membrane-bounded organelles. Most are single-celled organisms, though some aggregate to form colonies. They come in many shapes, but three forms dominate: the rod-shaped **bacilli**, the spiral-shaped **spirilla**, and the spherical **cocci**. Most have cell walls made of a **peptidoglycan**. Bacteriologists classify bacteria according to whether they are **gram positive** or **gram negative**. Some bacteria have specialized outer surfaces that form a **capsule**, a **slime layer**, or hairlike structures called pili.

Many prokaryotes are capable of directed movement, or **taxis** (including **chemotaxis**), and many have **flagella**, whose structure and operation are different from those of eukaryotic flagella. Prokaryotes grow continuously and rapidly, and usually reproduce asexually.

Prokaryotes are biochemically more diverse than eukaryotes. They may be either **autotrophs** or **heterotrophs**. Some autotrophs (**photoautotrophs**) obtain energy from light, while others obtain energy from inorganic substances, such as hydrogen sulfide or ammonia. Some heterotrophic prokaryotes derive energy from light (**photoheterotrophs**) but also require organic compounds. **Chemoheterotrophs** depend exclusively on organic compounds for both energy and carbon atoms. Heterotrophs may be **symbionts** or **decomposers**. Some heterotrophs require oxygen to live (**obligate aerobes**), others can live either in the presence or absence of oxygen (**facultative aerobes**), and some cannot live when oxygen is present (**obligate anaerobes**).

The metabolic versatility of prokaryotes allows them to live where eukaryotes cannot. Some prokaryotes provide benefits to humans, but others are pathogens, which cause disease. The identification of a pathogenic prokaryote is often an important step in understanding and curing a disease.

The metabolic abilities of prokaryotes provide a basis for their classification. Three groups of prokaryotes, together classified in the domain **Archaea**, differ dramatically from all other organisms. These include the **methanogens**, which make methane from carbon dioxide and hydrogen gas, the **halophiles**, which grow only in high concentrations of salts, and the **thermoacidophiles**, which derive energy by oxidizing sulfur compounds or methane.

The **Eubacteria** include about ten phyla. Aerobic heterotrophs are biochemically diverse but use the same metabolic pathways that are found in eukaryotes. These bacteria include **nitrogen-fixing bacteria, pseudomonads, omnibacteria, actinobacteria**, and **myxobacteria**.

Anaerobic heterotrophs include the **fermenting bacteria** (which often cannot live when oxygen is present) and the **spirochetes** (which contain internal flagella between layers of the cell wall).

Autotrophic prokaryotes include **chemoautotrophs** (which derive energy from a variety of inorganic compounds), **anaerobic photosynthetic bacteria, cyanobacteria**, and **chloroxybacteria**.

DNA sequence analysis of prokaryotic genes, and even of complete genomes, has allowed researchers to learn about the evolution of prokaryotes. The relationships between archaea, eubacteria, and eukaryotes are not simple, since many genes were transferred among species by plasmids and viruses.

Review and Thought Questions

Review Questions

1. What features of eukaryotes do prokaryotes lack?
2. What molecule makes up much of the cell wall of members of the Eubacteria but not of Archaea?
3. What is the difference between gram-positive and gram-negative bacteria? Which kind is susceptible to penicillin?
4. Describe two ways in which antibiotics work against bacteria. Why do these mechanisms not affect eukaryotic cells?
5. Describe the two major ways that antibiotic resistance is spread through populations of bacteria.
6. What features allow prokaryotes to reproduce so much more rapidly than eukaryotes? What advantages does this give bacteria?
7. What criteria do biologists use to classify bacteria?

Thought Questions

8. Are antibiotics helpful against a viral infection such as a cold or flu? Why or why not?
9. Animal breeders began giving animals antibiotics because studies showed that animals raised on antibiotics grew faster. Today, this effect is not as strong as it was 30 or 40 years ago. Why might antibiotics have this effect? Why might antibiotics no longer be so good at increasing growth rates?

About the Chapter-Opening Image

Salmonella bacteria that infect hamburger typify our problems with antibiotic-resistant strains of bacteria, a theme in this chapter about bacteria.

 On-line materials relating to this chapter are on the World Wide Web at **http://www.harcourtcollege.com/lifesci/aal2/**

 Web Links

 Quizzing & Testing

21 Classifying the Protists and Multicellular Fungi

The Fungus Fighters

In 1948, two government researchers set out to find a drug that would ultimately save thousands of human lives. In just two years, Elizabeth Lee Hazen and Rachel Brown succeeded in producing a drug that could safely cure deadly fungal infections in humans—a drug that is still a medical mainstay today. Both women had overcome great obstacles to obtain their scientific training. That they were highly educated, practicing scientists at all was something of a miracle. That they were able to make a major contribution to medicine was a testimony to their lifetime dedication to science and the respect they inspired in those around them.

Elizabeth Lee Hazen, born in August 1885, to a Mississippi cotton farmer and his wife, was an orphan by the time she was two years old. Raised by relatives, Hazen grew up in a large family in Lula, Mississippi. After high school, she enrolled in the Mississippi Industrial Institute and College, the first state-supported college for women in the United States. Without such an institution, she might never have gone to college, might never have studied physiology, anatomy, botany, zoology, and physics. Her grades were unremarkable. Like many people with great contributions to make, her strengths would not become obvious until she was much older.

After graduating from college, Hazen taught high school for six years. When she was 31, she enrolled as a graduate student at Columbia University, in New York, where one interviewer sized her up and then told her that she was so ignorant he doubted she had ever been to college. Nonetheless, she soon proved herself, and the following year Columbia awarded her a Master of Arts degree in biology. During World War I, she worked for the Defense Department diagnosing bacterial infections. After the war, she returned to Columbia, and, in 1927, at the age of 42, she received a Ph.D., becoming one of a small number of women with doctorates in the United States. She stayed at Columbia until 1931, when the New York State Department of Public Health lured her away to supervise its large medical pathology lab. But she maintained a connection with Columbia. In 1944, when the health department needed an expert on fungal infections, she returned to study mycology, at the same time continuing her full-time work in the pathology lab.

For most of us, fungi are no more than a mild curiosity. Mushrooms are a nice addition to a pizza. Mold is an annoyance in the shower. Some of us may have had unpleasant experiences with a fungus, a bout of athlete's foot, a yeast infection, nothing that a simple antifungal medication could not treat. In most people, the immune system's white blood cells quickly engulf and digest any fungal spores that make it into the body, and fungi are not a problem.

Under the right circumstances, however, a fungus can destroy a living human being as thoroughly as it can rot a dead stump in the forest. Diabetics, for example, have tissues that are unusually high in sugar. The warm, sugary environment of a diabetic's lung apparently creates a perfect environment for mucormycosis, the lush growth of a fungus in the order Mucorales. A common, usually harmless food mold, some species of Mucorales can grow through the lungs into the heart. Alternatively, this fungus may infect the sinuses, then run hyphae, long strands of the fungal body, throughout a person's head, destroying nerves and brain function. People with

Key Concepts

- Protists are enormously diverse, but all are eukaryotes. Algae are protists that perform photosynthesis. Protozoa are heterotrophic protists, most of which are motile. Some protists resemble fungi in appearance.
- True fungi lack flagella, have distinctive cell walls, and absorb rather than ingest food.
- Fungi play important ecological roles as decomposers, as partners in lichens, and in association with the roots of higher plants.

---

damaged immune systems—burn victims, transplant patients being treated with immunosuppressive drugs, and those with immune diseases such as AIDS—are all susceptible to fungal infections. For such people, fungi are a potentially life-threatening problem.

In the 1940s, when the first antibiotics came into use, medical experts discovered that patients treated with antibiotics became highly susceptible to fungal infections. It seemed that many of the bacteria killed by the antibiotics were important in keeping fungi at bay. Yet no antifungal drugs existed, and doctors were helpless to stop the spread of the often fatal infections.

Elizabeth Hazen was determined to find a drug that would help. By 1948, she had established an immense collection of pathogenic (disease-causing) fungi with which she and her team of technicians could diagnose fungal infections from people all over the eastern United States. Even more important, Hazen had for years been testing soil samples for bacteria that could kill infectious fungi. Now she had found two different bacteria that produced substances toxic to fungi. But what were the substances? Hazen needed the help of a talented chemist to isolate the active ingredient. One day in 1948, Hazen took a train from New York City to the health department's main office in Albany, New York, 150 miles away. There she had an appointment with senior researcher Gilbert Dalldorf, himself a talented bacteriologist and physician. Dalldorf listened to Hazen's proposal, then took her to see chemist Rachel Brown.

Like Hazen, Brown had succeeded in science against long odds. She was born in 1898 and raised in Springfield, Massachusetts, in a small, middle-class family. In 1912, however, when Rachel Brown was 14 years old, her father abandoned his family. Rachel's mother Annie then supported herself, her two children, and both her parents on a secretary's salary. Despite her poverty, Annie Brown was determined that her two children should go to the best schools she could afford. By Rachel's senior year in high school, however, her heart was set on Mount Holyoke College, a small, private college for women. But Rachel was turned down for a scholarship, and her mother could not possibly afford the tuition.

Miraculously, a wealthy friend of Rachel's grandmother offered to pay all of Rachel's expenses for a full four-year college education. Brown earned a B.A. in chemistry from Mount Holyoke and then put herself through graduate school at the University of Chicago. In 1926, she completed a Ph.D. dissertation in chemistry and shortly after took a job in Albany with the New York State Department of Public Health, an institution that hired an unusual number of women scientists. There, she worked steadily for almost a quarter of a century, isolating proteins and other molecules from infectious bacteria and protists that could be used for diagnostic tests or for vaccines. When Elizabeth Hazen and Gilbert Dalldorf stuck their heads in the door of her lab, it was the beginning of a fruitful collaboration and friendship that would last for more than 25 years.

> The third substance was only mildly toxic and killed virtually any fungus it came into contact with. It looked like a miracle drug, another penicillin.

Hazen explained that she had two strains of bacteria that could kill fungi, but that she needed to isolate the antifungal substances for further testing. In the months that followed, Brown and Hazen sent bulky package after bulky package between Albany and New York City. Hazen would send samples of bacteria, then Brown would extract different compounds for further testing by Hazen. The two researchers quickly isolated three fungicidal molecules from two kinds of bacteria. Two of these molecules were highly toxic to mice and therefore not useful as drugs. The third substance, however, was only mildly toxic and killed virtually any fungus it came into contact with; they called it fungicidin. It looked like a miracle drug, another penicillin.

At a meeting of the National Academy of Sciences in October 1950, Hazen and Brown presented their results. A reporter from *The New York Times* jumped all over the story, and within a day or two, the new life-saving drug was big news. Pharmaceutical companies began calling the state lab, demanding to be let in on the project.

Dalldorf and Brown nearly panicked. Fungicidin had been tested in animals but never in humans. For clinical trials, huge quantities of the drug had to be manufactured cheaply and quickly. Only a pharmaceutical company could do that. But Dalldorf and Brown knew, from previous experience with another drug, that they needed to patent fungicidin quickly before someone else did. Without a patent, to make clear who owned the rights to the drug, no pharmaceutical company would develop it. And Hazen, Brown, and Dalldorf wanted the drug to be used to save lives, the sooner the better.

By February, the researchers had filed a patent application with the United States Patent Office; a month later, they licensed the patent to Squibb Pharmaceuticals. Fungicidin, it turned out, was a name already in use for another chemical, so Hazen and Brown renamed it nystatin (pronounced nye-stat-in), after New York State. Within five years, Squibb had found a way to produce nystatin cheaply and had successfully tested the drug in humans. As the first treatment for fungal infections in humans, nystatin was tremendously useful.

Squibb sold so much nystatin that between 1955 and 1979 alone, royalties from sales of the drug came to $13.4 million. Half of this went to a nonprofit foundation that funded scientific research. The other half went to the nonprofit Brown–Hazen Fund, which supported research and training in the medical and biological sciences and awarded grants to women scientists. Neither Hazen nor Brown ever accepted any of the enormous wealth that resulted from their discovery. Brown apparently helped put several young women through college, as she herself had been helped. Both Hazen and Brown continued to work into their 70s, content in their achievements, honored with medals, honorary degrees, and elections to prestigious societies.

In this chapter we will quickly survey two kingdoms of organisms: the fungi and the protists. Biologists used to define a fungus as a eukaryotic organism that is neither plant nor animal. However, that definition applies to the protists as well. One important difference between the fungi and the protists is their organization. Most fungi are multicellular, while most protists are single celled. A few kinds of protists form colonies, primitive collections of nearly identical cells. A few protists (e.g., seaweeds) have large, complex forms. But the fungi mostly resemble plants and animals in having complex internal structures and specialized cells that serve different purposes.

Another difference between fungi and protists is how easily fungi can be classified. Most of the fungi are remarkably alike and make an easily recognized kingdom of organisms. But the protists are so diverse that many biologists believe that they should be divided into several kingdoms. Indeed, analysis of DNA sequences has led some biologists to suggest that protists form at least 12 kingdoms, each of which is at least as different from one another as plants and animals.

Protists, because they are mostly one celled, are considered more primitive than the fungi. Certainly, the first protists lived in the sea long before the first fungi evolved. However, protists are anything but simple, for they show extraordinary and elaborate specializations at the cellular level.

If we look at a drop of pond water under a microscope, we discover an amazing world of tiny, wriggling creatures

Figure 21-1 How many different protists live in a pond? A variety of protists live in both fresh water and salt water, including algae, dinoflagellates, and diatoms. Some protists are motile; others drift or anchor themselves to some object. Different protists move by waving flagella or cilia or by ameboid movements.

whose existence no one even suspected until Leeuwenhoek's first microscopic observations. Many of these microscopic creatures are members of the kingdom Protista. The protists range from the simple amebas to the most complicated cells on our planet (Figure 21-1). Their variety raises many questions: Why are protists so diverse? How are the various protists related to one another? How are they different from larger, more familiar organisms—such as mushrooms, mosses, or mammals? How are they different from prokaryotes? And how do they derive energy?

WHAT ARE PROTISTS?

Protists [Greek, *protos* = first] include the algae (which photosynthesize like plants), the protozoa (which act more or less like animals), and some funguslike forms. Protists flourish wherever there is moisture. In some marine ecosystems, the brown algae and the red algae are responsible for trapping nearly all of the solar energy that eventually reaches other organisms. The abundance of protists in the fossil record suggests that protists have been successful for a long time.

Protists are eukaryotic. Protist cells have nuclei and other membrane-bounded organelles, and almost all have mitochondria. Most (but not all) also have chromosomes and undergo mitosis.

At a biochemical level, protists are more or less identical to plants, animals, and fungi: all derive energy from glycolysis and respiration, and the photosynthetic protists all use water as a source of electrons. Protists lack the biochemical diversity of the prokaryotes, but they are the champions of cellular diversity. No other kingdom includes so many different kinds of cells.

Why Are Protists So Diverse?

Some protists are autotrophic and have chloroplasts. Others are heterotrophic and possess vesicles and lysosomes. Some are shapeless bags, while others have elaborately sculpted tests, or shells. Some are free living, some parasitic. Some produce spores, others do not. Some are single celled and others are multicellular like plants, animals, and fungi.

Some protists are **plankton** [Greek, *planktos* = wandering], organisms that float in the water, carried by currents (Figure 21-2). Others are **motile**, meaning that they move actively in search of food. Of the motile forms, some propel themselves with one or two long flagella (Figure 21-2). Some use hundreds of cilia, and others use protrusions called **pseudopodia** [Greek, *pseudos* = false + *pous* = foot], temporary extensions of the cytoplasm.

With so much diversity, we may ask, What binds the protists together? The answer is, surprisingly little. What protists have in common is that they do not fit in with plants, animals, or fungi. When biologists regarded all organisms as either plants or animals, they classified all the multicelled protists as plants and all the single-celled protists as either **algae** (which, like plants, photosynthesize) or protozoa (which, like animals, move under their own power).

The common pond-water protists of the genus *Euglena* illustrate why this classification scheme never worked very well (Figure 21-2). In the light, most species of *Euglena* develop chloroplasts and photosynthesize. In the dark, however, *Euglena* lives a heterotrophic life, absorbing nutrients from its surroundings, and may even lose its chloroplasts. *Euglena*, then, is neither plant nor animal, but it is a fine protist.

The green alga *Chlamydomonas*, like *Euglena*, is a single-celled organism capable of both photosynthesis and directed movement. But *Chlamydomonas*'s close relative *Volvox* consists of colonies of hundreds to thousands of cells, each of which looks just like a free-living *Chlamydomonas*. Because the two algae are otherwise identical, it would be unreasonable to put one into the plant kingdom and one into the animal kingdom. As a result, biologists now put both groups into the Protista.

Protists are eukaryotic organisms that are defined not by what they have in common, but by not belonging to any of the other three eukaryotic kingdoms (Figure 21-3). In contrast, we define the other eukaryotic kingdoms, fungi, plants, and animals, partly by the way their cells function and develop. Plants and animals are all multicellular organisms that develop from embryos, organized collections of tissue that, through a highly predictable sequence of steps, develop into more-mature forms. Fungi are organisms, nearly always multicellular, that absorb nutrient molecules directly from their environments. The fungi do not photosynthesize and do not form embryos.

The diversity of the protists is both a joy and a headache. How can we classify them? Biologists do not agree on the number of phyla, or even on whether all the protists should be in the same kingdom: one protist researcher has even suggested that they really comprise some 20 separate kingdoms! Many biologists divide them, according to their cellular organization and mode of nutrition, into plantlike, animal-like, and funguslike organisms. We will follow the classification scheme of Margulis and Schwartz, who divide the protists into 27 phyla, mostly on the basis of their cellular structures and life cycles. We will survey 16 of these phyla. For convenience, we group these into plantlike phyla (the algae), animal-like phyla (the protozoa), and funguslike phyla.

Protists are metabolically more uniform than prokaryotes. However, partly because protists are a grab-bag kingdom (including every eukaryote that is not animal, plant, or fungus), they show enormous diversity at the cellular level.

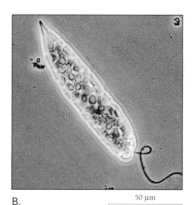

Figure 21-2 Plant or animal? A. Euglena in sunlight are green with chloroplasts. B. Euglena in a dark environment use their flagella to go looking for light. (A, Dwight Kuhn; B, Biological Photo Service)

Figure 21-3 The diverse kingdom Protista.

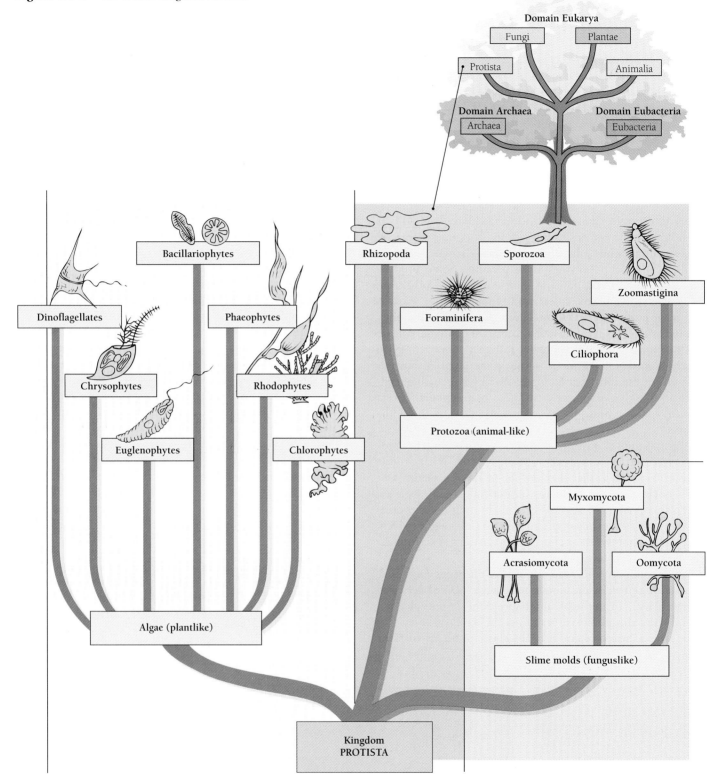

How Do Protists Reproduce?

Every protist species goes through two phases in its life cycle—reproduction and dispersal. Dispersal may take many forms. Some free-living, motile protists swim or crawl. Others scatter seedlike **cysts,** enclosed structures that contain reproductive cells. Still others infect a host that moves, such as an insect, bird, or mammal.

Reproduction may be either asexual or sexual, depending on the species. Some protists are asexual and haploid throughout their life cycles and never undergo meiosis. Other

Figure 21-4 Sexuality in protists. *Chlamydomonas* is a well-studied green alga that grows in ditches, pools, fresh water, and soil. The zygote is the only diploid stage. The mature cell and all other reproductive stages are haploid. Researchers suspect that the first sexually reproducing organism had a life cycle like that of *Chlamydomonas*. (Photo, David M. Phillips/Science Source/Photo Researchers, Inc.)

protists are diploid for most of their life cycle and produce haploid gametes (Figure 21-4). Still other protists are haploid for almost all of their life cycles but produce a short-lived diploid zygote. Finally, some protists undergo **alternation of generations,** passing through both haploid and diploid phases.

ALGAE ARE PHOTOSYNTHETIC (PLANTLIKE) PROTISTS

Algae are relatively simple, plantlike organisms that live in moist environments everywhere: in open water, ice, soil, as well as in puddles or pockets of water in trees or rocks. Most algae are *photoautotrophs,* organisms that supply themselves with energy by means of photosynthesis. Some, such as *Euglena,* can also live as heterotrophs. Just as plants do, algae use chlorophyll and other pigments to capture light for photosynthesis. However, whereas true plants photosynthesize with both chlorophyll *a* and chlorophyll *b,* most protists do not have any chlorophyll *b*. Algal protists also differ from true plants in the structure of their cell walls, and the arrangement of their flagella. In this section we will briefly survey 6 of the 12 phyla of algae: dinoflagellates, chrysophytes, euglenophytes, phaeophytes, rhodophytes, and chlorophytes.

Dinoflagellates

Most **dinoflagellates** [Greek, *dinos* = whirling + *flagellum* = whip] are single-celled organisms that float freely as plankton in warm oceans (Figure 21-5). A few can form colonies, and some live symbiotically with corals, sea anemones, and clams. The ultimate source of food in coral reef and other communities, in fact, is photosynthetic dinoflagellates. Dinoflagellates make up a major part of the phytoplankton, the microorganisms on which most aquatic ecosystems depend.

Of the thousands of species of dinoflagellates, the best known are those that grow rapidly to cause red tides, in which population explosions, or blooms, of dinoflagellates color the ocean red, brown, or other colors, depending on the species (Figure 21-6). Other red tides release deadly toxins. Because many dinoflagellates are bioluminescent, that is, they glow in the dark, they create dramatic nighttime displays.

Almost all dinoflagellates have a rigid wall, or **test,** made of cellulose and coated with silica. Around its circumference, the test contains two characteristic grooves, in which the two flagella are embedded at right angles to each other. Together,

Figure 21-5 Dinoflagellates are renowned for their diversity and beauty. Shown here are the bioluminescent dinoflagellates *Ceratium* and *Peridineum*. (Terry Hazen/Visuals Unlimited)

the two flagella create the spinning that gives these organisms their name.

Dinoflagellates are very old. Because their hard tests are well preserved in the fossil record, we know that they have existed at least since the beginning of the Cambrian Period, some 570 million years ago. They are so ancient, in fact, that their chromosomes are organized differently from those of all other eukaryotes. When a dinoflagellate divides, its chromosomes segregate by attaching to the nuclear membrane rather than to a mitotic spindle. Dinoflagellates have no mitosis, no meiosis, and no sexual reproduction.

Dinoflagellates are among the oldest and most primitive of the eukaryotes.

Chrysophytes (Golden Algae and Diatoms)

Golden-yellow pigments give the golden algae, or **chrysophytes** [Greek, *chrysos* = golden + *phyta* = plant], their name. As dinoflagellates do, many chrysophytes form hard tests containing silica and other minerals that help preserve them in the fossil record as far back as 500 million years ago. Each chrysophyte family contains both single-celled and colonial forms. The colonial forms may be spherical or loosely branched. Most chrysophytes are plankton, widespread in temperate lakes and ponds. Only one group lives in the oceans.

Chrysophytes reproduce asexually in one of two ways. A single cell, called a swarmer cell, may swim away from a colony to found another colony. Alternatively, the hundreds or thousands of cells of a mature colony may split into two groups, which then float away from each other to form two new colonies. Chrysophytes can also form cysts that resist cold temperatures and desiccation.

The *diatoms* are some of the most numerous and beautiful of the chrysophytes. Most of the 10,000 species of diatoms are single cells, although some form simple filaments or colonies. Each species has a unique and elaborate silica test (Figure 21-7). Most diatoms live as free-floating plankton in fresh or salt water. Some, however, can move on slime that they extrude from a slit between the two parts of the test. The tests of diatoms are well preserved in the fossil record, dating from the Cretaceous Period, between 65 and 135 million years ago.

Figure 21-6 Satellite photo of a red tide off the east coast of Italy. Dinoflagellate populations explode into "blooms," or "red tides," when conditions are good. Some species of dinoflagellates make poisons that clams, mussels, and other shellfish accumulate in their bodies, making these ordinarily edible shellfish deadly. Some dinoflagellates secrete toxins that kill whole schools of fish, whose flesh the dinoflagellates then feed on. During a bloom of such dinoflagellates, toxic vapors over the water can make boaters temporarily ill. (Geospace/Science Photo Library/Photo Researchers, Inc.)

Figure 21-7 Diatoms may be even more beautiful than dinoflagellates. *(Robert Brons/Biological Photo Service)*

Diatoms can reproduce sexually, forming gametes by meiosis. Usually, however, they reproduce asexually, dividing their chromosomes through ordinary mitosis. At cell division, each half of the test stays with one of the daughter cells, which then assembles a second half.

Chrysophytes (golden algae) have both single-celled and colonial species. Diatoms have elaborate tests and can reproduce sexually as well as asexually.

Euglenophytes

The **euglenophytes** [Greek, *eu* = true + *glene* = eyeball] derive their name from their light-detecting eye spots, which enable them to live a double life. In the dark, they live heterotrophically, by breaking down dissolved organic matter. When light becomes available, however, euglenophytes use their eyespots, and a single flagellum, to swim into the sunshine. Euglenophytes photosynthesize with both chlorophyll *b* (a pigment also used by higher plants), which gives them their grass-green color, and chlorophyll *a*. Except for the green algae, no other algae or any other protists make chlorophyll *b* (Table 21-1). Unlike plants and green algae, however, euglenophytes have no cell wall and store energy in a polysaccharide called paramylon, instead of in starch. Euglenophytes all reproduce asexually, using a primitive form of mitosis, with no clear mitotic spindle and no well-defined chromosomes. They also do not go through the stages of metaphase or anaphase.

One of the biggest surprises from analysis of protist DNA was the discovery of the close relationship between the euglenophytes and **trypanosomes,** the flagellated protozoa that cause sleeping sickness in Africa and Chagas disease in South

TABLE 21-1 WHICH PROTISTS USE THE SAME CHLOROPHYLLS AS PLANTS?

Chlorophyll:	*a*	*b*	*c*	Storage material
Plants	*a*	*b*		Starch
Chlorophytes	*a*	*b*		Starch
Euglenophytes	*a*	*b*		No Starch
Dinoflagellates	*a*		*c*	Starch or oil
Phaeophytes	*a*		*c*	No starch
Chrysophytes	*a*		*c*	No starch
Rhodophytes	*a*			Starch

America. Trypanosomes, which have complex life histories, are transmitted to people and animals via insect bites (the tsetse fly in Africa and the "kissing bug" in South America).

Euglenophytes can live as either autotrophs or heterotrophs.

Phaeophytes (Brown Algae)

The **phaeophytes** [Greek, *phaios* = dusky brown], or brown algae, together with the rhodophytes (red algae), which we will discuss next, make up most of the organisms that we call seaweeds. Phaeophytes are all multicellular, and most live in the cool waters off temperate coasts. Phaeophytes include the biggest protists in the world—the giant kelps, which in deep water may be up to 100 meters tall (Figure 21-8).

Most phaeophytes reproduce sexually. However, meiosis in phaeophytes produces **spores** rather than gametes. The spores germinate into new multicellular organisms in which all the cells are haploid. This haploid form is called a **gametophyte** because it then produces gametes. The diploid form is called a **sporophyte** because it produces spores. Such alternation of generations also occurs in green algae, red algae, fungi, and, of course, the true plants.

While the size and complexity of the phaeophytes far exceed those of most other protists, the individual cells of phaeophytes resemble those of other protists. In particular, the flagellated sperm of phaeophytes look just like the cells of chrysophytes. Phaeophytes make chlorophylls *a* and *c* and produce other pigments like those of chrysophytes (Table 21-1). But, unlike higher plants, phaeophytes never produce chlorophyll *b* and do not synthesize starch.

Phaeophytes (brown algae), which are related to the chrysophytes, are the largest and most complex of the protists.

Rhodophytes (Red Algae)

Although the brown algae include the most spectacular seaweeds, most of the world's seaweeds are **rhodophytes** [Greek, *rodon* = rose], or red algae. Nearly all of the 4000 species of rhodophytes are marine, and most live in tropical seas, attached to rocks or other algae. Among the most beautiful of the rhodophytes are the coralline algae. These algae deposit calcium carbonate in their cell walls and contribute to the building of coral reefs. Calcium-depositing rhodophytes have been preserved in the fossil record as far back as 500 million years.

The cell walls of red algae occasionally contain cellulose but more often contain another polysaccharide, whose attached sulfate groups give the rhodophytes their characteristic slippery feel. Two such polysaccharides—agar and carrageenan—are useful to humans. Agar helps gel instant desserts, jellies, and baked goods, and is also the most widely used culture medium for growing bacteria in the laboratory. Carrageenan is used to stabilize oil and water emulsions in foods such as mayonnaise, as well as in paints and cosmetics.

All rhodophytes reproduce sexually and have complex life cycles. Meiosis can take place right after fertilization, with no sporophyte stage. Alternatively, the zygote can develop into a diploid organism, which later produces haploid spores that grow into male and female gametophytes. The female gametophyte produces a large ovum, while the male produces a much smaller sperm.

◀ **Figure 21-8 Kelp forests are home to a variety of other organisms.** In shallow coastal waters, the giant kelp *Macrocystus* may grow at enormous rates, producing masses of brown algae that provide both food and shelter for fish, sea urchins, sea otters, and other organisms. Although the giant kelps are simpler than true plants, they have a few specialized structures. Floating photosynthetic blades capture light for photosynthesis. The blades are kept near the surface by gas-filled bladders and anchored to the sea floor by holdfasts. (*Flip Nicklin/Minden Pictures*)

Rhodophytes (red algae) have distinctive pigments and complex sexual cycles.

Chlorophytes (Green Algae)

Chlorophytes [Greek, *chloros* = green], or green algae, are more like plants than any other group of protists. In fact, plants probably arose from a chlorophyte ancestor. Chlorophytes are unlike most of the other photosynthetic protists in having both chlorophylls *a* and *b* and in making starch, which they store within their chloroplasts, as do plants. Among the rest of the protists, only the euglenophytes have both chlorophylls *a* and *b,* and only the dinoflagellates make starch. Although modern chlorophytes include several multicellular groups, biologists do not believe that any of them was the ancestor of the multicellular plants (Figure 21-9).

Chlorophytes come in a great array of colonial and multicellular forms. They range in size and complexity from the single-celled *Chlorella,* about 25 μm in diameter, to the seaweedlike *Codium magnum,* which may be 8 m long. Chlorophytes are mostly free living and aquatic, but some species live on the surface of snow, on tree branches, in the soil, or in close association with other organisms, including fungi, hydras, and other protists (Figure 21-10).

The simplest chlorophyte is *Chlorella,* which lacks flagella and is smaller than *Chlamydomonas. Chlorella* is among the most widespread of the green algae, living in fresh water, salt water, and soil. *Chlorella*-like species live within protozoa as well as within many other invertebrate animals, including sponges, hydras, and flatworms. A close association between two organisms is called **symbiosis** [Greek, *sumbios* = living together]. Organisms living in a symbiotic relationship with one another are called *symbionts. Chlorella* provide their partners with energy-rich compounds, produced by photosynthesis.

Chlorophytes (green algae) are more like plants than other protists and also the most diverse of the plantlike protists.

Figure 21-9 Colonial green alga—blobs of identical cells or multicellular individuals? Members of the colonial genus *Volvox* may contain 500 to 60,000 cells arranged in a single-layered, hollow sphere. The cells in most colonies of green algae are identical. In *Volvox* and some other genera, however, there is some division of labor, suggesting a rudimentary form of multicellularity. For example, only some cells are able to reproduce. In addition, the cells may move smoothly through the water by coordinating the beating of their flagella, much as our heart cells coordinate their contractions. (*Alfred Owczarzak/Biological Photo Service*)

Figure 21-10 The life cycle of a sea lettuce. The marine Ulvophyceae take their name from *Ulva,* or sea lettuce, which grows in shallow seas throughout the temperate zones. Each ulva has a flat, leaflike thallus, which may be a meter or more in length, and a rootlike holdfast, which attaches the thallus to a rock or other object. (*Photo, Brian Parker/Tom Stack & Associates*)

PROTOZOA ARE HETEROTROPHIC (ANIMAL-LIKE) PROTISTS

Even though we no longer classify protists as plants or animals, we still use the old term **protozoa** [Greek, *protos* = first + *zoe* = life] to describe the protists that most resemble little animals. The protozoa include a wide variety of heterotrophs, most of which obtain their food by phagocytosis, ingesting particles or other cells.

Most protozoa are single celled and motile. They are extraordinarily varied and include eight phyla, of which we will survey just five: rhizopoda, foraminifera, sporozoa, ciliophora, and zoomastigina. Three of these—the rhizopoda, the ciliophora, and the zoomastigina—we distinguish by the way they move. Each of the other two has characteristic traits. We identify the sporozoa by their characteristic life cycles and the foraminifera by their distinctive shells.

Rhizopoda (Amebas)

Rhizopoda, or amebas, are common throughout the world, in oceans, in fresh water, in soil, and as parasites of animals. Hundreds of millions of people in the tropics (and about 10 million in the United States) harbor parasitic amebas that infect the lower intestinal tract. Fortunately, only about 20 percent of such infections produce the symptoms of amebic dysentery, which range from mild discomfort to severe diarrhea and death. In the most severe cases, the ameba may cause abscesses that rupture the abdominal wall or penetrate the diaphragm and lungs.

An ameba continuously changes its shape, in an apparently chaotic way (hence the old name of one species—*Chaos chaos*). Amebas do this by means of pseudopodia, temporary extensions of membrane-enclosed cytoplasm. An ameba extends a pseudopod and then moves its cytoplasm into the pseudopod, so that the pseudopod becomes the body. The movement of amebas only appears chaotic, for they actually move in a directed manner toward sources of food. They then envelop the food in their pseudopodia.

All amebas reproduce through simple cell division. Some, such as the ones that cause amebic dysentery, can form cysts immune to desiccation or digestion by their hosts. A few construct distinctive shells (tests) by gluing together tiny grains of sand and other inorganic bits. Similar tests exist in the fossil record, with some dating back to the Precambrian, more than half a billion years ago.

Rhizopoda (amebas) use pseudopodia to move and ingest food.

Foraminifera

The **foraminifera** [Latin, *foramen* = little hole + *ferre* = to bear], or **forams,** are marine organisms whose tests are full of tiny holes (Figure 21-11). The tests of forams are made of organic materials reinforced with grains of sand or minerals and range in size from about 20 μm to several centimeters. Most are tiny and live in the sand or attached to other organisms or stones. Two large groups are free-floating plankton and serve as food for clams, mussels, and other invertebrates that filter food from water.

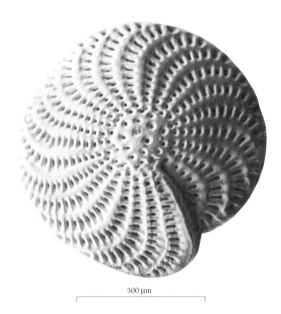

Figure 21-11 The tests of foraminifera are rich in calcium. As these organisms die and drop to the sea floor, their tests form thick, white deposits. (*Biophoto Associates*)

Forams have complex sexual life cycles. Unlike most other protozoans, they have true alternation of generations, with extended haploid and diploid phases—just like plants. Some forams harbor within them photosynthetic protists—dinoflagellates, chrysophytes, and diatoms. Although forams are unicellular, they are subdivided into multichambered shells connected by cytoplasm.

The tests of forams are abundant in the fossil record, especially in the last 230 million years. Forams make up much of the chalky sediment on the bottoms of the seas. The famous white cliffs of Dover, England, are just such foramen sediments, uplifted by geologic processes.

Foraminifera have complex sexual life cycles and tests studded with little pores.

Sporozoans

Sporozoans are all parasites. They reproduce by a combination of asexual and sexual processes, including the alternation of generations. Sporozoans have especially complex life cycles, usually involving two or more different hosts. Sporozoans are a scourge to their many animal hosts but a delight for puzzle-solving parasitologists. The members of one genus, *Plasmod-*

ium, cause malaria, a disease that every year infects some 300 to 500 million people and kills 2 to 3 million, mostly in Africa. *Plasmodium* spreads among humans with the help of its other set of hosts—mosquitoes of the genus *Anopheles* (Figure 21-12). The complex life cycles of sporozoans ensure continued success: at any moment, different phases of each species are present in several forms in several places in two or more hosts. They are virtually impossible to eradicate.

Because of the importance of malaria to worldwide human health, researchers have identified *Plasmodium falciparum* as a priority for complete DNA sequencing. A complete DNA sequence may help researchers to understand *Plasmodium*'s metabolism and life cycle more completely, with the hope of designing vaccines and specifically targeted therapies. The sequence of one of the 14 chromosomes, completed in 1998, revealed that *Plasmodium* diverged from other eukaryotes unexpectedly early in evolution. Another surprise was the presence of organelle DNA that, researchers conclude, could only have come from an ancient plastid. As in the case of the euglenophytes, it is almost impossible to make a clear distinction between "plantlike" and "animal-like" protists.

Sporozoans are nonmotile, spore-forming protozoa, which live as parasites in animals. They are known for their complex life cycles.

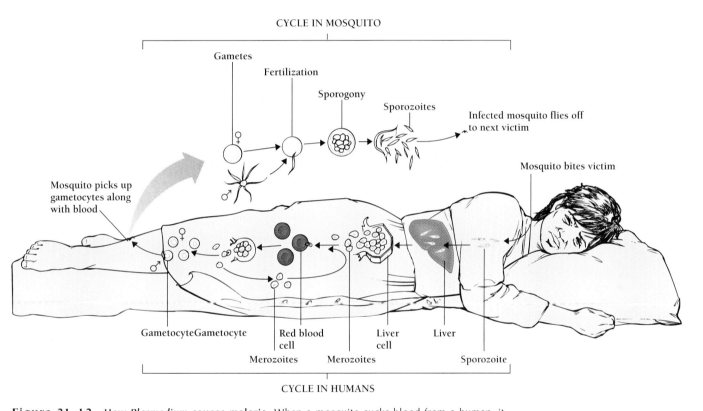

Figure 21-12 How *Plasmodium* causes malaria. When a mosquito sucks blood from a human, it first injects a bit of saliva. The saliva contains an anticoagulant, a chemical that keeps the blood flowing freely and allows the mosquito to drink its fill. A mosquito infected with *Plasmodium* harbors a diploid form of the parasite—called a *sporozoite*—in the wall of its gut and in its salivary glands. When the mosquito injects its saliva, some of these sporozoites enter the human host's blood. In the human bloodstream, the sporozoites travel to the liver, where they divide several times by mitosis and develop into a new diploid phase, called a *merozoite*. The merozoites take up residence inside of red blood cells and divide rapidly. Eventually, the red cells rupture and release merozoites and toxins, which cause the awful fever and chills of malaria. The merozoites infect more red cells or develop into *gametocytes*. The gametocytes, if sucked up by a second hungry mosquito, produce gametes, which form diploid zygotes, which embed themselves in the mosquito's intestinal wall. These develop into sturdy *oocysts*, which undergo mitosis to form thousands of sporozoites, some of which migrate to the salivary gland, where they can begin the cycle again. The simultaneous presence of *Plasmodium* in the gut and salivary glands of mosquitoes and in the blood and liver of humans has made it impossible to eradicate malaria. The draining of swamps and the use of insecticides such as DDT have destroyed breeding grounds for the mosquitoes and reduced the incidence of malaria in some areas. Yet, worldwide, the number of new cases of malaria has more than doubled since the mid-1970s, for the mosquito has evolved resistance to insecticides such as DDT, and *Plasmodium* has evolved resistance to antimalarial drugs.

Ciliophora (Ciliates)

Ciliophora, or the **ciliates,** include about 8000 species of free-living, single-celled heterotrophs that live both in fresh water and in salt water. Their name [Latin, *cilium* = eyelash] reflects the presence, on their surface, of thousands of cilia, shorter and more numerous than eukaryotic flagella. The cilia, which may be arranged in bundles or in sheets, help ciliates move and feed (Figure 21-13). The best-known ciliate is probably *Paramecium,* a common resident of pond water, long used in teaching and research. Most ciliates lack shells, although a few marine ciliates form shells of sand held together by organic cements. The fossil record shows that such shell-forming ciliates have existed for about 100 million years.

Paramecia and other ciliophora (ciliates) are single-celled, free-living heterotrophs coated with cilia.

Zoomastigina (Flagellates)

The **zoomastigina** [Greek, *mastix* = whip], or **flagellates,** are a diverse group of single-celled heterotrophs. All zoomastigotes have at least one long flagellum, and some have thousands. Some are free living in fresh or salt water, while others are parasites in animals. Reproduction of zoomastigotes is mostly asexual, though some species have a sexual cycle as well. Some convert from a flagellated to an ameboid form and back again, depending on the availability of food. DNA sequence studies suggest that the flagellates are far more diverse than previously suspected, as we discussed earlier for trypanosomes, which are more closely related to euglenophytes than to other flagellates.

Zoomastigina (flagellates) are single celled and use flagella to move.

FUNGUSLIKE PROTISTS

We have seen that some protists resemble plants and some resemble animals. The last group of protist phyla most resemble fungi. The names of these phyla reflect the fact that they were initially identified as fungi: all include the Greek root *mukes* [= fungus]. We will describe three of the seven funguslike protist phyla: acrasiomycota, the cellular slime molds; myxomycota, the plasmodial slime molds; and oomycota, which include the water molds, mildews, and potato blight. While these organisms resemble fungi in their general appearance, they more closely resemble other protists in their cellular organization and modes of reproduction.

Acrasiomycota (Cellular Slime Molds)

Acrasiomycota [Greek, *acrasia* = confusing two things + *mukes* = fungus], or cellular slime molds, are peculiar organisms, animal-like at some stages of their lives and plantlike at others. They live in damp soil, in fresh water, on decaying vegetation, and especially on rotting logs. When nutrients are plentiful, they flourish as amebalike cells, consuming bacteria and other food particles by phagocytosis. When food is scarce, the amebas aggregate to form a multicellular stage that looks a bit like a miniature garden slug (Figure 21-14). The slug secretes a slime track and wanders about in search of food and light. When it finds food, it settles down and develops a mushroomlike **fruiting body** that produces spores. Each spore turns

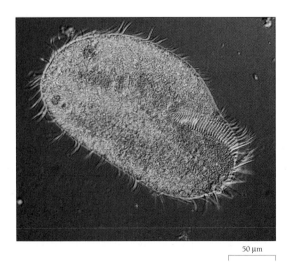

Figure 21-13 The cilia of a ciliate. The cilia surround the oral groove and form oarlike arrangements for movement. (M. Abbey/Photo Researchers, Inc.)

Figure 21-14 The slug of a cellular slime mold. (Cabisco/Visuals Unlimited)

into an ameba. Reproduction is asexual; both the amebas and the slug are haploid.

Acrasiomycota (cellular slime molds) live as amebas and as multicellular slugs.

Myxomycota (Plasmodial Slime Molds)

Myxomycota [Greek, *muxa* = mucus + *mukes* = fungus]—the plasmodial, or acellular, slime molds—closely resemble the cellular slime molds in appearance and in life cycle. The migrating form of a plasmodial slime mold, however, differs greatly from that of a cellular slime mold. Instead of developing from an aggregation of individual hungry amebas, a plasmodial slime mold develops a **plasmodium**—a mass of cytoplasm with many nuclei but no boundaries between cells.

The plasmodial slime molds also differ from the cellular slime molds in having a sexual cycle. The amebas are haploid, and two of them can form a diploid zygote. The nucleus of the zygote then divides repeatedly, but the cell merely grows larger and larger without dividing. The result is a plasmodium with many diploid nuclei. Unlike the cellular slime mold's slug, the plasmodium can both feed and move about. It propels itself by differential growth, growing more in one direction than another. When food or water becomes scarce, the slug forms a fruiting body, undergoes meiosis, and releases haploid spores.

Myxomycota (plasmodial slime molds) live as amebas or as a multinucleated plasmodium.

Oomycota

The **oomycota** most nearly resemble true fungi. They are either parasitic, deriving nourishment from living organisms, or *saprophytic,* deriving nourishment from dead organisms. Like true fungi, the oomycota extend threads into the tissues of their hosts, release digestive enzymes, and absorb the resulting organic molecules. The oomycota differ from the fungi, however, in having flagella on their motile spores. They reproduce sexually, and their life cycles include both haploid and diploid phases. The haploid gametes differ greatly in size; the large ovum is the source of this phylum's name, oomycetes [Greek, *oion* = egg + *mukes* = fungus], meaning "egg fungi."

Among the most common oomycotes are the water molds, which grow in fuzzy masses on the remains of algae or animals. In freshwater ecosystems, water molds decompose and recycle nutrients from dead organisms.

Some oomycotes cause diseases in animals or plants. Historically, the most important of these is *Phytophthora infestans,* which causes blight in potatoes. In 1845 and 1849, this oomycote infected and destroyed the Irish potato crop. A million Irish died of hunger, and another 1.5 million were forced to emigrate to the United States. A similar blight on the German potato crop late in World War I helped force an end to the war.

Oomycota resemble fungi, but their spores have flagella.

What Can We Learn from DNA About the Diversity and Evolution of Protists?

Studies of protist DNA are still in their infancy, and no complete protist genomes have been sequenced. Limited information, from the analysis of a few genes, has nonetheless changed biologists' thinking about protists in two fundamental ways: (1) protists are more diverse than previously thought and (2) endosymbiosis has been a major factor in protist evolution.

Comparisons of the sequences of ribosomal RNAs (and a few other genes) shows that different kinds of protists are much more different from one another than are plants, animals, and fungi. As mentioned earlier, many biologists favor the reclassification of protists into at least 12 kingdoms.

While researchers now think that all mitochondria—in protists and other eukaryotes—had a common origin, protist mitochondria had distinct histories. Most surprisingly, even protists that lack mitochondria have DNA in their nuclei that could only have come from ancient mitochondria. For example, researchers recognize that *Giardia* (a diarrhea-causing protist now common in streams and lakes) has no mitochondria, but it does have genes that, in other eukaryotes, lie on mitochondrial DNA. Researchers have concluded that ancestors of *Giardia* did have mitochondria but that some of the mitochondrial DNA moved to the nucleus before the mitochondria were lost for good.

The other big surprise about organelle DNA was the discovery of genes, for example, in trypanosomes that could only have come from plastids, such as chloroplasts. This result suggests that a trypanosome ancestor was once photosynthetic, and that, during evolution, there was no clear segregation of ("plantlike") algae and ("animal-like") protozoans.

HOW DO TRUE FUNGI DIFFER FROM OTHER ORGANISMS?

We love fungi. We put fungi in our salads and fungi in our sauces. We use fungi to ferment wines and cheeses and to leaven bread. What is more, we need fungi. A walk through a wet wood demonstrates the enormous ecological importance of fungi: they are prodigious decomposers, recycling the nutrients of dead plants and animals. Without fungi, dead trees would fall and accumulate, and the soil, a little bit of which washes away in every storm, would never be replaced.

But fungi also plague us in countless ways. *Cryptococcus neoformans,* for example, is one of many unpleasant fungi that torment humans and other animals. Distributed worldwide, it infects first the lungs, then the membranes surrounding the brain, and sometimes the kidneys, bones, and skin. The first sign of infection is a cough. Usually, the symptoms that bring *C. neoformans*'s victims to a doctor, however, are headache, blurred vision, confusion, depression, or inappropriate speech or dress. Another infectious fungus attacks the sinuses, then the brain, producing intolerable headaches and ultimately consuming its victim alive.

Still other fungi destroy entire harvests of wheat, potatoes, and other crops. Fungi destroy in small ways as well as large. In the kitchen, fungi ruin bread, potatoes, jams, and sauces. In the bathroom, they attack our feet, producing athlete's foot, and grow on the walls of the shower and the porcelain surfaces of the toilet.

Fungi grow wherever there is sufficient warmth, moisture, and energy-rich organic molecules. Sometimes fungi find nourishment in surprising and bothersome places—the lining of a camera lens, clothing, shoes, or the wooden planks of a ship. During the American Revolution, the British lost more ships to dry rot than to the fledgling American Navy. In World War II, fungal infections of the skin took more men out of action in the South Pacific than did battle wounds.

Fungi are a little like plants and a little like animals. As long as biologists classified all organisms as either plants or animals, fungi posed a problem. Like animals, fungi are heterotrophic—they derive their energy from other organisms. But fungi enclose their cells in rigid walls as plants do. Fungi are distinct from both plants and animals, however: they take in food differently from animals, their cell walls are chemically different from those of plants, and they lack the embryonic stages of plants and animals. Everyone now agrees that the fungi deserve their own kingdom.

How Can We Characterize the Fungi?

With the exception of the single-celled yeasts and a few others, most fungi are **molds** composed of masses of threadlike filaments called **hyphae** [Greek, *hyphe* = web; singular, hypha]. The hyphae contain many nuclei (Figure 21-15). In some fungi, the nuclei all lie in a common cytoplasm, rather than in separate membrane-enclosed cells. Such fungi are said to be **coenocytic** [Greek, *koinos* = shared + *kytos* = hollow vessel]. In **dikaryotic** fungi (which have two nuclei per cell), walls called **septa** separate the nuclei within a filament. The septa, however, are perforated by large pores that allow the cytoplasm to flow along the hyphae. Hyphae grow, branch, and intertwine to form a mass, called a **mycelium** [Greek, *myketos* = fungus]. Growth may be so rapid that in the time it takes you to read this page a single mycelium can add meters of hyphae.

Fungi's ability to grow rapidly helps compensate for another odd characteristic of their cells—lack of mobility. All fungal cells, even the gametes, lack flagella and cilia. Fungi can

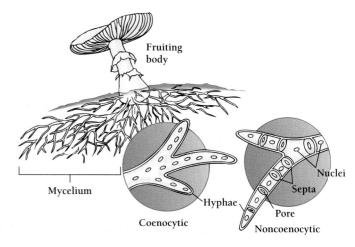

Figure 21-15 Two types of fungal hyphae. The nuclei within hyphae may all share the same cytoplasm (coenocytic), or they may be separated by septa.

move in two ways only. The mycelium can move from place to place by growing, or the mycelium can produce spores that the wind scatters.

Another distinguishing characteristic of the fungi is their cell walls, which are made of a tough polysaccharide called **chitin,** the same substance in the hard shells of insects, spiders, and crustaceans. In plants, the major component of the cell wall is cellulose.

Recall that all fungi are heterotrophs, organisms whose food is organic molecules produced by other organisms. A fungus's food may be either dead or alive. Saprophytic fungi feed on the remains of dead organisms, while parasitic fungi feed on living organisms. Fungi obtain food, whether dead or alive, by infiltrating the bodies of organisms and other food sources with long, thin hyphae, usually just a few micrometers in diameter. The fungi extend their hyphae into the food, secrete digestive enzymes that break down the food, and absorb, rather than ingest, the resulting nutrients.

Fungi may reproduce asexually or sexually. Each cell (or nucleus) within a mycelium is usually capable of generating a new mycelium. Most fungi reproduce vegetatively (asexually). A single mycelium can produce separate individuals by means of simple cell division, by the breaking up of existing hyphae, or by budding. Fungi also reproduce by making and dispersing hardy spores capable of resisting heat, cold, drought, and lack of food. Spores are so small that the wind can scatter them all over the planet.

Spores, which are always haploid, may be made in a variety of intricate structures, called **sporophores.** Ordinary mushrooms are the sporophores of a large mycelium, which may be many meters in diameter (Figure 21-16). In general, asexual spores form when food is plentiful and the environment is otherwise welcoming. Fungi reproduce sexually when food is scarce or the environment has become unfavorable. Recall that sexual reproduction creates a greater diversity of genotypes in the next generation. This diversity increases the chances that a few individuals will survive a hostile environment. Each hap-

Figure 21-16 A fairy ring. An underground mycelium grows outward over hundreds of years. When the mycelium "fruits," it forms mushrooms at its outer edge, which release spores. The grass within the circle of mushrooms is often stunted and pale. For this reason, people once believed that dancing fairies caused these rings. (G. Carleton Ray/Photo Researchers, Inc.)

loid spore can germinate and produce a new haploid mycelium. Fungi differ from both plants and animals in that they do not go through any embryonic stages. Spores form mycelia directly.

Fungi are filamentous, lack flagella, have distinctive cell walls, absorb food, and reproduce by means of spores.

How Do Mycologists Classify Fungi?

Many mycologists still classify the funguslike protists as fungi. In this book, however, we consider the funguslike protists to be distinct from fungi (Figure 21-17). Fungi lack motility, and their cell walls differ from those of protists. Fungi also lack embryos and acquire nutrients differently from plants and animals. We will discuss examples of four phyla of fungi: zygomycetes, ascomycetes, basidiomycetes, and deuteromycetes.

Mycologists group fungi into these four phyla according to whether the hyphae have septa and how the fungi produce sexual spores. Most fungi are capable of reproducing both sexually and asexually. Asexual reproduction may occur vegetatively (as when parts of the mycelium divide to form new individuals) or through the production of spores by mitosis. Sexual reproduction always occurs by means of spores. The manner in which fungi produce their sexual spores is one of the two main characteristics that mycologists use to classify fungi.

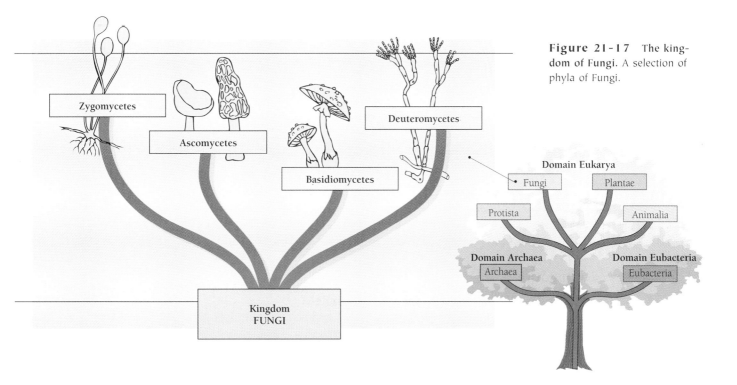

Figure 21-17 The kingdom of Fungi. A selection of phyla of Fungi.

BOX 21.1

Fungal Infections Today

In recent years, fungal infections have become increasingly common. Between 1991 and 1994, for example, fungal infections in one county in southern California increased from 400 per year to 4500 per year. In fact, today nearly 40 percent of all deaths from hospital-acquired infections are due not to bacteria or viruses but to a handful of fungi.

Most hospital infections are due to *Candida albicans,* a virulent relative of baker's yeast. *Candida* infects 70 percent of AIDS patients. Even if science could eliminate *Candida,* however, other fungi would quickly take its place. *Cryptococcus, Histoplasma,* and *Aspergillus* are three of the next most common fungi that infect patients with impaired immune systems. Even ordinary baker's yeast, a normal inhabitant of the human mouth and the stuff we use to make bread rise, can grow out of control in individuals who are already very sick.

Healthy people also fall victim to some seemingly ordinary fungi. In California's San Joaquin Valley, for example, many people become infected by the soil fungus *Coccidioides,* which causes symptoms ranging from mild flulike illnesses to attacks in which the fungus spreads throughout the body (Figure A).

Against this onslaught of rot, health care workers have only a few broadly effective drugs. Nystatin and two other drugs derived from it are still the most effective way to treat fungal infections. One of the most powerful antifungals is amphotericin B, a drug with side effects so nasty that patients sometimes call it "amphoterrible." It can cause fever, chills, low blood pressure, headaches, vomiting, inflammation of blood vessels, and kidney damage.

Safe antifungal drugs with few side effects are much harder to find than safe antibiotics. This is because antibiotics, which kill prokaryotes but not our own eukaryotic cells, work by exploiting any of a variety of differences in cell chemistry between prokaryotes and eukaryotes. But fungi are eukaryotes, like us. Their cells work like our own cells, and chemicals that are bad for fungi are usually bad for us, too.

Nonetheless, the chemistry of fungal cells does differ from that of our own, and any differences can be exploited by researchers who design drugs. For example, the three drugs most commonly used all work by attacking ergosterol, a lipidlike solid that helps form the cell membranes of fungi. Our own cell membranes are made of cholesterol. Fungal cells also differ from ours in having a cell wall. Researchers' latest hope is to find a drug that attacks the cell walls of fungal cells without harming our own cells.

Figure A Fungal spores are everywhere. The tiny haploid particles by which mushrooms and other fungi reproduce float in and out of our nostrils with each breath, land on bread and jams in our kitchens, and grow into dark streaks in our bathrooms. Outside our homes, a rain of fungal spores dusts trees and shrubs, lawns, dead leaves, and soil. Creeks, rivers, and lakes disperse fungal spores from one watershed to the next. And winds carry fungal spores from one end of the Earth to the other. *(Biophoto Associates)*

Zygomycetes (Conjugation Fungi)

The mycelium of a **zygomycete** [Greek, *zygon* = yoke + *mukes* = fungus], such as bread mold, is coenocytic: it consists of hyphae without dividing septa and all the nuclei share a common cytoplasm. The only exceptions are two kinds of spore-forming structures, which form asexual or sexual spores. Most of the time, zygomycetes reproduce asexually by forming spores in spore-containing capsules called **sporangia.**

The common black bread mold, *Rhizopus stolonifer,* illustrates the sexual life cycle of a typical zygomycete (Figure 21-18). When zygomycetes reproduce sexually, they do so by forming a *zygosporangium,* a structure that produces haploid spores from a diploid zygote.

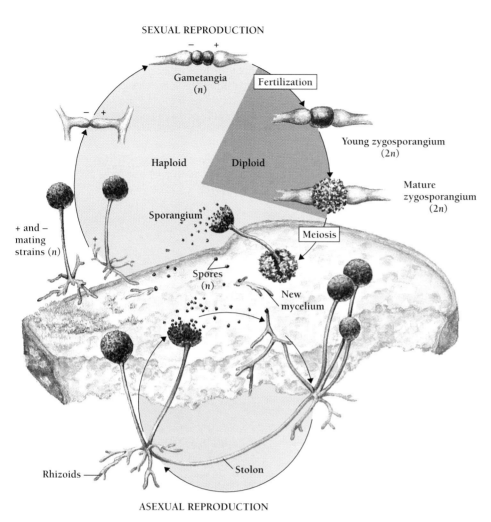

Figure 21-18 Life cycle of black bread mold *Rhizopus stolonifer*.

Whether zygomycetes reproduce sexually or asexually, they always use spores. Both kinds of spores travel on air currents and can, in the right environment, divide to produce a new mycelium. Because spores drift randomly, their arrival on a hospitable host (a piece of bread, for example) is entirely a matter of chance. A few zygomycetes, however, have evolved a means of actively launching themselves into likely new environments (Figure 21-19).

Bread molds and other zygomycetes are coenocytic and usually reproduce asexually by means of spores.

Ascomycetes (Sac Fungi)

Ascomycetes are a large and diverse group, with some 30,000 species. They include most yeasts, the powdery mildews, the molds, the cup fungi, the bread mold *Neurospora* (so useful to geneticists), and the truffles, prized by chefs for their earthy flavor.

Ascomycetes [Greek, *askos* = wineskin + *myketos* = fungus], or sac fungi, are so named because their sexual spores are always produced in little sacs, called **asci** (singular, ascus). They have no sporangia for making asexual spores. Their hyphae usually have septa separating the nuclei. These septa, however, are perforated. As a result, molecules and even organelles can circulate throughout the cytoplasm.

The truffles, yeasts, mildews, and other ascomycetes have septa dividing the nuclei of their hyphae and no sporangia. They produce sexual spores in little sacs called asci.

Basidiomycetes (Club Fungi)

The 25,000 species of **basidiomycetes**, or club fungi, include many of the most familiar fungi: the mushrooms, the bracket fungi, and the puffballs (Figure 21-20 and 21-21). These fungi take their name from the tiny, club-shaped **basidium** [Latin, = little pedestal; plural, basidia] that they use to produce spores. In mushrooms, the basidia develop on the underside of the mushroom's cap, on the sides of the thin gills that radiate from the center like the pages of an open book (Figure 21-22). A

Figure 21-19 Fungi disperse their spores effectively. *Pilobolus*, "the hat thrower," grows in the dung of horses and cattle. To disperse its spores to more dung, the hat thrower relies on a light-sensing mechanism at the base of its sporangia. The fungus bends its sporophores toward bright light, where grass is likely to grow, then fires its spores like cannonballs. The spores travel 2 m or more, with speeds of up to 50 km per hour. Spores that land on blades of grass eaten by horses or cattle pass through the digestive tract with the grass and emerge in the perfect environment—fresh dung. (Photo, courtesy of Michael J. Wynne)

mushroom itself is the equivalent of a fruit and biologists call it a **basidiocarp** [Latin, *carpus* = fruit].

In the common field mushroom, the mycelium spreads underground and may be up to 30 m in diameter (Figure 21-16). The basidiocarps (mushrooms) develop at the periphery of the mycelium in "fairy rings"—often overnight. Folk legends once held that mushroom fairy rings appeared at the bidding of fairies who were tired of dancing. In reality, mushrooms are able to grow so quickly because a huge underground mycelium can quickly mobilize vast amounts of material. As the secondary mycelium exhausts the nutrients in the center of the ring, however, the ring grows outward in search of more nutrients. Because the mycelium typically grows at about 30 cm per year, we can estimate that a fairy ring 30 m in diameter is a century old.

Another group of basidiomycetes, less charming than mushrooms but at least as important, are the rusts and smuts. Rusts and smuts ruin billions of dollars worth of wheat and other grains in the United States and elsewhere every year. As distinct from other basidiomycetes, however, rusts and smuts do not form basidiocarps. Instead, they are parasites with complex life cycles, usually involving hosts of two species.

Mushrooms, rusts, and other basidiomycetes are named for the club-shaped basidia that produce spores. They have septa and distinctive life cycles.

Deuteromycetes (Imperfect Fungi)

Mycologists classify fungi as zygomycetes, ascomycetes, or basidiomycetes, largely according to their mode of sexual spore formation. For some 25,000 named species of fungi, however, mycologists have so far discovered no sexual cycle at all. These are said to be "imperfect." Mycologists disagree, however, about whether the imperfect fungi form a separate phylum (or division).

Figure 21-20 The bird's nest fungus is a basidiomycete. (Michael Fogden/DRK Photo)

BOX 21.2

Ergotism

After *Penicillium* mold, the most famous fungus may be the ascomycete *Claviceps purpurea*, or ergot, which infects the flowers of rye and other grains. This peculiar sac fungus produces tiny spikes, called sclerotia, or ergots, in the grain. The small, black or brown ergots—compact masses of hardened mycelium—replace a few of the seeds that would normally form in the seed head (Figure A).

Each ergot contains a witch's brew of deadly substances that can poison humans and other animals that eat the infected grain. In extreme cases, the symptoms of this poisoning, called ergotism, consist of hallucinations and fatal convulsions. In milder or chronic poisoning, sufferers may have headaches, muscle pains, numbness, and cold fingers and toes—all signs of constricted blood vessels that can result in gangrene, a usually fatal infection when not treated with antibiotics.

During the Middle Ages, ergotism was common in northern Europe and other areas where rye bread was popular. Also called St. Anthony's fire, ergotism could fell thousands of people in a season. In the year 994, for example, 40,000 people died from eating ergot-infected grain. Modern grains are cleaned of the dark masses of mycelium, and today ergotism is rare.

Ergot in small doses is medically useful. The main ingredient is an alkaloid—ergotamine tartrate. Ergotamine stimulates the smooth muscle of the peripheral and cranial blood vessels and also inhibits the neurotransmitter serotonin. As a result, ergotamine causes small arteries and smooth muscle fibers to contract. It is used to induce labor, to control bleeding from the uterus following childbirth, to treat high blood pressure, and—in conjunction with caffeine—to relieve severe migraine headaches. Two other ergot alkaloids, bromocriptine and pergolide, are used in the treatment of Parkinson's disease. These two alkaloids help relieve the symptoms of Parkinson's disease by activating receptors for the neurotransmitter dopamine.

Ergot was also the original source of the psychoactive drug lysergic acid diethylamide (LSD). LSD induces mood changes (euphoria, anxiety, and depression), impaired judgment, and distorted perceptions of many sensations, including time, space, or self-image. Despite LSD's reputation as a powerful hallucinogen, true hallucinations are apparently rare.

Figure A Ergot. At left, the small black ergot (or sclerotium) in the seed head of a rye plant. At right, a normal, uninfected seed head. *(Dennis Drenner)*

Many imperfect fungi, for example, produce conidia (powdery asexual spores) and are probably ascomycetes. Mycologists therefore recognize the **deuteromycetes** as an artificial category. As the mode of reproduction of each species becomes known, mycologists are reclassifying them into one of the other three divisions of fungi.

The deuteromycetes [Greek, *deuteros* = second + *mukes* = fungus] include many species that are important to humans. Some are parasites on the skin of humans and other mammals, causing ringworm, athlete's foot, and other skin diseases. These fungi are able to displace the bacteria that are normally abundant on skin because they are better able to derive nourishment from skin with their powerful digestive enzymes that break down tough skin proteins. These also compete with bacteria by inhibiting the growth of the bacteria with antibiotics. Recall that the first antibiotic used by humans came from a strain of the common deuteromycete mold *Penicillium*.

Some deuteromycetes are predaceous. Using sticky substances on their surface, these fungi trap protozoa and small animals. One species even snares its prey with quickly tightening hoops (Figure 21-23).

Penicillium and other deuteromycetes always reproduce asexually, which makes them hard to classify.

Figure 21-21 Basidiomycetes. Bracket fungi and puffballs have the same sexual life cycle as supermarket mushrooms. *(Left, J. Robert Stottlemyer/Biological Photo Service; right, Michael Fogden/DRK Photo)*

WHAT IMPORTANT ECOLOGICAL ROLES DO FUNGI PLAY?

Fungi have two important ecological roles, both of which depend on the power of the enzymes that fungi secrete. These extraordinarily potent enzymes can reduce substances as different as Wonder Bread and solid rock into small molecules that can be absorbed by fungi or other organisms. Using these enzymes, fungi decompose organic matter and make nutrients available to other organisms. Fungi, together with bacteria, recycle much of the world's organic matter from dead plants and animals, not to mention feces, leaves, and logs. Fungi also participate in close, mutually beneficial symbiotic associations with other organisms. We will discuss two kinds of important fungal associations, lichens and mycorrhizae.

What Role Do Fungi Play in Lichens?

Each of the 25,000 recognized species of lichens looks like a moss or simple plant growing on a rock or a tree trunk.

Figure 21-22 Like the pages of a book. In some basidiomycetes, the spore-producing basidia develop on the paper-thin gills, which radiate from a central stalk.

Figure 21-23 A deuteromycete fungus lassoes a passing nematode worm. *(Ken Barker)*

Lichens may be leafy, shrubby, or crusty. They come in a variety of colors, including black, white, red, orange, yellow, green, and brown. Lichens are not plants at all, however, but associations of fungi with photosynthetic partners (Figure 21-24). In all but about 20 lichen species, the fungal partner is an ascomycete. The photosynthetic partner in a lichen is either a green alga (a protist) or a cyanobacterium (a prokaryote).

The association between a fungus and its photosynthetic partner is not entirely mutual. The various fungi cannot live without the help of photosynthetic partners. The photosynthetic partners, however, are all capable of living without fungi. Most of the diversity of lichens comes from the fungi, for 90 percent of all photosynthetic partners in lichen are the same few organisms (from one of just three genera of green alga or of one genus of cyanobacteria). In contrast, the fungal partner is unique to each lichen. Nonetheless, the symbiotic association between fungi and their photosynthetic partners helps both parties.

The symbiotic association between lichens and their partners makes them especially tough. Fungi extract nutrients from rock and other harsh microenvironments. By doing so, fungi allow the algae or cyanobacteria to grow where they otherwise could not, while the photosynthetic algae provide a continuous source of energy to their fungal partners. Lichens therefore grow where no plant could survive—bare rock, mountaintops, the Arctic tundra, even the seemingly lifeless valleys of Antarctica (Figure 21-25).

Lichens usually reproduce asexually, either by fragmentation or by sporelike starters called soredia, which are clumps of algae enmeshed in a bit of mycelium. The fungal and photosynthetic components can also reproduce independently, either sexually or asexually.

Lichens are mutually beneficial symbiotic associations of fungi and photosynthetic algae or cyanobacteria. In contrast to their photosynthetic partners, lichen fungi cannot live alone.

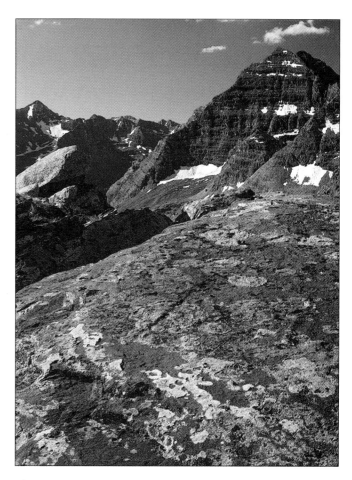

Figure 21-25 Lichens are diverse. They are also able to absorb exquisitely tiny amounts of minerals and nutrients dissolved in rain and dew. While this ability is crucial to their survival in harsh settings, it makes lichens especially vulnerable to pollution. The reduced growth of lichens is a valuable indicator of deteriorating air quality. *(Larry Ulrich/DRK Photo)*

Figure 21-24 Lichen. Like a pie, a lichen is constructed with a protective crust of fungal hyphae above and below and a filling made of mycelium and algal cells.

What Role Do Mycorrhizae Play in the Lives of Plants?

Fungi form close associations not only with photosynthetic bacteria and photosynthetic protists, but also with true plants. **Mycorrhizae** [Greek, *mukes* = fungus + *rhiza* = root] are symbiotic associations between fungi (usually a zygomycete or a basidiomycete) and the roots of plants (Figure 21-26). At least 80 percent of all plant species have mycorrhizae. Some biologists now think that even more plants depend on mycorrhizae. In such associations, the fungus's hyphae penetrate the plant roots and also extend into the surrounding soil.

What are these fungi doing? Although some fungi obtain sugar from the plant's roots, the mycorrhizae are not parasites, for they help the plant flourish by supplying phosphate and metal ions. Indeed, what the mycorrhizae do for plants is more obvious than what the plants do for the mycorrhizae. In one experiment, plants whose mycorrhizae had been removed grew

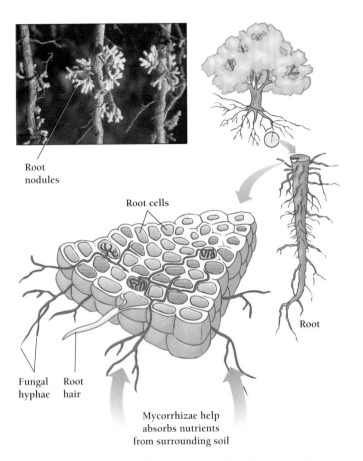

Figure 21-26 Mycorrhizae grow inside of the roots of trees and other plants. *(Photo, John D. Cunningham/Visuals Unlimited)*

more slowly than plants that had mycorrhizae. Plants with no mycorrhizae were then provided with fertilizer, however, and they grew as well as the plants with mycorrhizae. This experiment suggested that the mycorrhizae help plants pull nutrients from the soil.

Mycorrhizae are fungi on the roots of plants that supply minerals to the plant.

Where Did the Fungi Come From?

Because fungi have no hard parts, the fossil record of fungi is not rich. Nonetheless, fossil fungi can be found at least back to the Ordovician Period, 430 to 500 million years ago. The earliest fossils of land plants, which are about 400 million years old, contain petrified mycorrhizae, and some biologists now believe that plants could not have colonized the land without the aid of the fungi. Most mycologists think that the first fungi were zygomycetes, which have the simplest life cycle. Basidiomycetes probably derive from ascomycetes, which are in turn descendants of the more primitive zygomycetes. Most imperfect fungi probably arose from ascomycetes that lost the sexual stages of their life cycles.

Fungi evolved from protists at least 430 million years ago. The fungi may have been responsible for the colonization of land by plants.

In this chapter we have surveyed the simplest eukaryotes, the Protista, as well as the Fungi, the first kingdom that consists entirely of multicellular organisms. In the next chapter, we will survey the plants, which, along with the fungi, are essential for humans and all other animals. On land, plants provide all the energy and simple organic materials that animals and fungi need to live. Plants and fungi are thus inextricably dependent on one another. Animals likewise absolutely depend on plants and fungi.

 Quiz

Study Outline with Key Terms

Protists are eukaryotic organisms that are not fungi, plants, or animals. Protists include organisms that were once called **protozoa** (which move) and **algae** (which perform photosynthesis), as well as some organisms that resemble fungi. Most protists are single-celled organisms, but many are multicellular. Some protists are **plankton,** which drift. Others are self-propelled, or **motile.**

Protist life cycles vary greatly from species to species. Each species goes through two phases: reproduction and dispersal. Some are haploid throughout their life cycles and never undergo meiosis. Other species undergo **alternation of generations,** that is, they pass through both haploid and diploid phases. The diploid phase is the **sporophyte,** which produces **spores** (sometimes encased in **cysts**), and the haploid phase is the **gametophyte,** which produces the gametes.

Algae (plantlike protists) include 12 distinct phyla, of which this chapter surveys 6: **dinoflagellates, chrysophytes** (golden algae and diatoms), **euglenophytes, phaeophytes** (brown algae), **rhodophytes** (red algae), and **chlorophytes** (green algae). Dinoflagellates are among the oldest and most primitive of all eukaryotes and are marked by distinctive **tests.** Chrysophytes have both single-celled and multicellular species, and diatoms can reproduce sexually as well as asexually. Euglenophytes can live as either autotrophs or heterotrophs and are closely related to **trypanosomes,** flagellated protozoa that cause some human diseases. Phaeophytes are the largest and most complex of the protists. Rhodophytes have distinctive pigments, but no flagella. Chlorophytes are the most diverse of the algae.

Protozoa are heterotrophic protists, most of which are motile. They include eight phyla, of which we survey five: **rhizopoda, foraminifera, sporozoa, ciliophora,** and **zoomastigina.** Rhizopods (amebas) move and ingest food with **pseudopodia.** Foraminifera (or **forams**) have tests (shells) studded with little pores. Sporozoans are nonmotile, spore-forming protozoa, which live as parasites in animals. Ciliophora (**ciliates**) are single cells, coated with cilia. Zoomastigina (**flagellates**) are single celled and use flagella to move.

Many protists resemble fungi in appearance but other protists in cellular organization. These protists include seven phyla, of which we described three: **acrasiomycota, myxomycota,** and **oomycota.** Acrasiomycota (cellular slime molds) live as amebas and as multicellular slugs, which can produce **fruiting bodies.** Myxomycota (plasmodial slime molds) live as amebas or as a multinucleated **plasmodium.** Oomycota resemble fungi and are saprophytic. Unlike fungi, however, their cells have flagella.

Despite the tradition of classifying protists into plantlike algae, animal-like protozoans, and fungal-like protists, DNA sequence analysis suggests that evolutionary relationships among the protists are much more complicated. Some DNA-containing organelles persisted and others were lost after some of their genes became part of nuclear DNA.

Fungi are heterotrophic (as are animals) and enclose their cells in rigid walls (as do plants). In contrast to both plants and animals, however, they do not undergo embryonic development. Fungi live mostly on land and, with a few exceptions, consist of a **mycelium,** a mass of intertwined filaments. Their cell walls are made from **chitin,** the same material used to build the external shells of insects. Fungi absorb rather than ingest nutrients after predigesting their food with secreted enzymes.

Most fungi are **molds.** They may reproduce asexually or sexually. They also may reproduce and disperse as spores that resist harsh environmental conditions. Mycologists classify fungi by the way they produce spores (in **sporophores**) and by the appearance of their mycelia. In this chapter we surveyed four phyla of fungi: the zygomycetes, the ascomycetes, the basidiomycetes, and the deuteromycetes. **Zygomycetes** (conjugation fungi) are **coenocytic,** lacking **septa** between the nuclei in their **hyphae. Ascomycetes** (sac fungi) have septa (**dikaryotic**) but no **sporangia.** They produce spores in sacs called **asci. Basidiomycetes** (club fungi) have septa and distinctive life cycles. Each fruiting body of a basidiomycetes mushroom is called a **basidiocarp.** On the thin gills beneath the cap lie the **basidia,** which produce spores. **Deuteromycetes** (imperfect fungi) reproduce asexually.

Fungi play important ecological roles, both as decomposers and as participants in **symbiotic** relationships. **Lichens** are mutually beneficial associations of fungi and photosynthetic algae or cyanobacteria. **Mycorrhizae** are mutually beneficial associations between fungi and the roots of plants. Fungi evolved from protists at least 430 million years ago. Some biologists believe that plants could not have colonized the land without the aid of mycorrhizae.

Review and Thought Questions

Review Questions

1. Why is the kingdom Protista so much more diverse than the plants or animals?
2. How do the algae differ from the protozoa?
3. Which algae contain both chlorophylls *a* and *b*?
4. What features distinguish the slug of the plasmodial slime molds from that of the cellular slime molds?
5. What five features characterize the fungi?
6. To what phylum of fungi do common grocery store mushrooms belong? How is it that mushrooms can grow so quickly after a rain?
7. What are the two partners that compose a lichen? Which partner is more independent?
8. What do mycorrhizae do for plants?

Thought Questions

9. If you wanted to invent a new fungicide that would kill fungal cells but not human cells, what differences between the biology of fungal cells and biology of human cells might you exploit for this purpose?
10. Some paleontologists have speculated that vascular plants could not have evolved to the extent they have without the symbiotic relations with fungi they now depend on. How would you either test such a hypothesis or rephrase it in a way that would make it testable?

About the Chapter-Opening Image

This mushroom represents one of the most familiar and benign fungi.

On-line materials relating to this chapter are on the World Wide Web at **http://www.harcourtcollege.com/lifesci/aal2/**

Web Bit 21-1 Once and Future Kings: The Elm and Chestnut in America

Web Links

Quizzing & Testing

22 How Did Plants Adapt to Dry Land?

The Loves of Plants

How to classify organisms is often a matter of dispute. Prokaryotes are commonly classified according to their metabolic pathways; protists, according to their morphology; and fungi, according to their style of reproduction. When 18th-century biologists first began to classify plants, they could not agree on what traits to use. By 1799, naturalists had suggested at least 52 different systems for classifying plants. One system grouped plants according to the form of the flower, seed, and seed coat. Other systems gave priority to the shape of the fruit.

It was Carolus Linnaeus's system that ultimately came into use, but not without heated arguments from both scientists and clergy. Linnaeus based his classification solely on the reproductive structures of flowers. Indeed, he was fascinated by the "marriages of plants," as he called them (Figure 22-1). The son of a country parson, Linnaeus equated sex with marriage, and his descriptions of "vegetable coitus," as some called it, are simultaneously explicit and romantic.

As the historian of science Londa Scheibinger wrote, "Linnaeus emphasized the 'nuptials' of living plants as much as their sexual relations. Before their 'lawful marriage,' trees and shrubs donned 'wedding gowns.' Flower petals opened as 'bridal beds' for a verdant groom and his cherished bride, while the curtain of the corolla lent privacy to the amorous newlyweds."

Linnaeus divided plants first according to whether their marriages were public or private. Public marriages were those in which the flowers were visible to the naked eye; private marriages were those in which the flowers were too small to be seen. (The division corresponded to the public and private marriages then recognized in most of Europe. In England, the 1753 Marriage Act eliminated clandestine marriages by requiring a public proclamation before a marriage could be considered legal.)

Linnaeus further divided plants according to whether the male and female parts were together on the same flower, "in one bed," as he put it. Otherwise, if male and female parts were in separate flowers, they were in "two beds." Those in one bed, also called "hermaphrodites," were divided according to how many male parts (stamens) were present. These Linnaeus termed "husbands" to the single female pistil, or "bride." Flowers with one stamen were monandrian—having one man. (Linnaeus called his beloved wife Sara "my monandrian lily.")

As most flowers, however, have more than one stamen, indeed several, many naturalists and clergymen were horrified by Linnaeus's terminology. One St. Petersburg professor asked what God Almighty would introduce such "loathsome harlotry" into the plant kingdom. The bishop of Carlisle wrote to the Linnaean Society in 1808 that "a literal translation of the first principles of Linnaean botany is enough to shock female modesty."

Key Concepts

- A plant is a multicellular organism that photosynthesizes and develops from an embryo.

- Evidence for plant evolution comes from fossils and from comparisons of living species.

- The nonvascular plants, or bryophytes (the mosses, liverworts, and hornworts), lack specialized elements for transporting water and nutrients.

- The development of specialized vascular elements enabled plants to transport water and nutrients, to grow large, and to diversify. Life on land offers both advantages and disadvantages to photosynthetic organisms. Seeds allow plants to withstand dry environments. Flowers and fruits of angiosperms have allowed their rapid diversification.

The Englishman William Smellie, who edited the first edition of the *Encyclopedia Britannica* (1771), rejected the "alluring seductions" of plant sexuality. He was especially repulsed by Linnaeus's description of plants that had reproductive structures of both sexes as "hermaphrodites." Smellie compared Linnaeus's work to that of the most "obscene romance-writer."

Most naturalists avoided Linnaeus's more vivid imagery. One exception was Erasmus Darwin, Charles Darwin's maverick grandfather. His *Loves of the Plants* (1789) is a collection of poems depicting flowers as practitioners of free love and even incest.

Linnaeus's taxonomy had other drawbacks. For example, in his system the number of stamens determined the class to which a plant species belonged. But in some groups of plants, the number of stamens can vary from individual to individual. He also decided that the number of stamens (male parts) would determine class, while the number of pistils (female parts) would determine order, the next taxonomic group down. That is, he deemed the stamens more important than the pistils. This system neither corresponded to the phylogeny (relatedness) of plants, nor was the system arbitrary. It was simply Linnaeus's way of importing into science

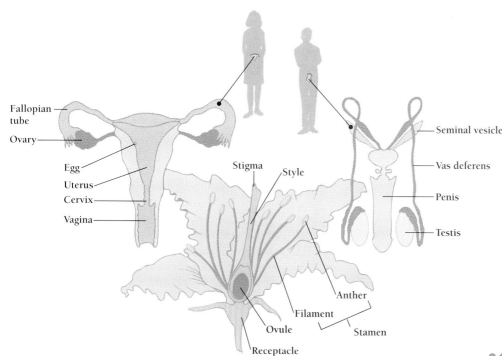

Figure 22-1 Linnaeus equated flower parts with the sexual parts of animals. He argued that the anthers are the testes; the filaments, the vas deferens; and the pollen, the seminal fluid. In the female parts, the stigma is the vulva; the style is the fallopian tube; the ovule is the ovary; and the seeds are the zygotes.

traditional views about the relative positions of males and females.

Although many of Linnaeus's species designations—genus and species—remain in use today, his classification of groups above the genus level have been abandoned in favor of systems that more accurately reflect evolutionary relatedness. As botanists have learned more about the lives of plants, they have acquired a better understanding of how plants live and how they have evolved.

In the rest of this chapter, we will see that plants are surprisingly diverse. Although the animal kingdom includes four times as many species as the plant kingdom, it is plants that first define a habitat—by their shape, size, and behavior. To a large extent, the animal inhabitants of grasslands differ from those of tropical forests because grasses differ structurally and chemically from trees. We can argue, then, that the plants lay the foundation for more habitat diversity than any other kingdom of organisms.

HOW DID PLANTS ADAPT TO LIFE ON LAND?

> The tree is an aerial garden, a botanical migration from the sea, from those earlier plants, the seaweeds; it has a purchase on crumbled rock, on ground. The human, standing, is only a different upsweep and articulation of cells. How treelike we are, how human the tree.
> —Gretel Ehrlich

The beauty of trees and flowers has inspired poets and gardeners alike. Plants delight the eye, tell us the season and even the time of day. They provide us with delicious foods, beautiful clothes, and strong building materials. Botanists like to say that plants provide "food, fiber, feed, fuel, and pharmaceuticals." But even this catchy list doesn't convey the importance of plants.

Through photosynthesis, plants (together with bacteria and protists) provide all of the energy we derive from food, whether that food is itself animal or plant. Plants living millions of years ago captured the energy that we now collect as coal, oil, or gas to heat our homes, to fuel our automobiles, and to drive our power plants. Plants provide the oxygen we breathe. Plants support the entire web of terrestrial life. Without plants, none of the other kinds of organisms could live on land.

A **plant** is a multicellular organism that performs photosynthesis and develops from an embryo. Almost all plants live on the land, and even those that live in water are descended from terrestrial plants. All plants, however, are descended from protists, and plants capture energy from the sun in much the same way as many of the marine protists. Nonetheless, the invasion of the land by plants nearly half a billion years ago depended on the evolution of structures different from anything found among plants' marine ancestors.

What Are the Advantages and Disadvantages of Life on Land?

The land offers three major advantages for photosynthetic organisms: more light (unfiltered by water), more carbon dioxide (which is more concentrated in air than in water), and a more reliable supply of minerals and other nutrients (which are more concentrated in soil than at the surface of a lake or sea).

Plants could not leave the water without solving a host of serious problems. To begin with, the seas provide a continuous supply of water. Water drives photosynthesis and provides a medium in which gametes and spores can disperse or meet for fertilization. Water also provides buoyancy for organisms. Even a 30-meter-tall brown alga can float. So aquatic plants, unlike terrestrial ones, require no skeleton or other structural support system.

To colonize the land, then, plants had to be able to obtain, conserve, and distribute water; to accomplish fertilization with little or no water; and to support their own weight. The evolution of plants depended principally on four adaptations:

1. A waxy **cuticle,** which coats the plant's exposed surfaces and reduces water loss through evaporation
2. The ability to absorb water from dew, rainwater, or groundwater (using roots, for example)
3. Enclosed reproductive organs called **gametangia** in which gametes form
4. Enclosed **sporangia** in which spores form

All plants use some version of a waxy cuticle, an absorption system, and gametangia.

Land offers plants more light, carbon dioxide, and nutrients than the sea. But plants living on land need adaptations for absorbing and retaining water, for fertilization with little or no freestanding water, and, in the case of larger plants, some form of structural support.

What Is Alternation of Generations?

Recall that the cells of humans and most other animals are diploid, containing two of every chromosome, while their gametes (sperm and eggs) are haploid. The life cycle of every sexually reproducing organism includes a diploid phase and a haploid phase. In most animals, the haploid phase, consisting of single-celled sperm or ova, is short lived or barely noticeable. Human sperm, for example, live only a few weeks. Human ova survive much longer but are hidden away as individual cells.

In other kingdoms, however, the haploid stage is multicellular and constitutes a significant part of the life cycle. In plants, the haploid stage is a multicellular **gametophyte** that generates ova and sperm through mitosis, rather than meiosis.

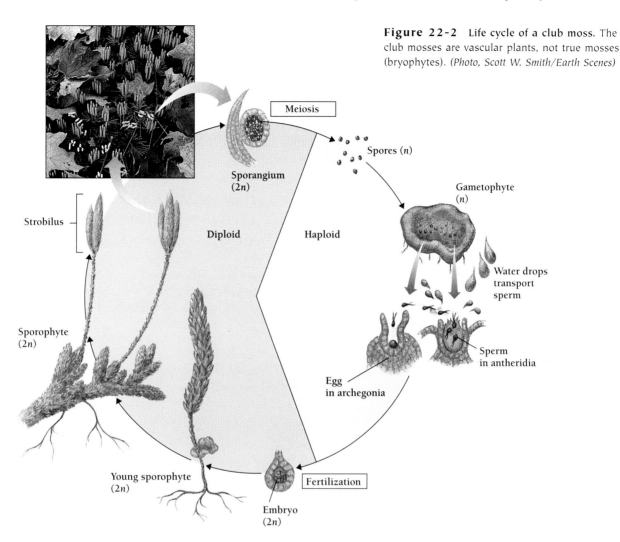

Figure 22-2 Life cycle of a club moss. The club mosses are vascular plants, not true mosses (bryophytes). *(Photo, Scott W. Smith/Earth Scenes)*

Plants, like many protists and fungi, are said to exhibit **alternation of generations,** a sexual life cycle in which the alternating haploid and diploid phases are both multicellular (Figure 22-2).

Plants exhibit alternation of generations, a sexual life cycle in which haploid and diploid phases are both multicellular.

How Are Vascular Plants Different from Nonvascular Plants?

Despite important similarities among all plants, plants may be divided into two groups—the tracheophytes (vascular plants) and the bryophytes (nonvascular plants). Each group has different solutions for the problems of terrestrial life.

The **tracheophytes** [Latin, *trachea* = windpipe] include all the most familiar living plants, from ferns to fir trees and flowering fuchsias. All tracheophytes have pipelike tissues that conduct water and nutrients from one part of the plant to another. Because this system of conducting tissues resembles the vascular system of humans and other vertebrates, the tracheophytes are also referred to as the **vascular** plants. The tracheophytes' vascular system allows these plants to grow far larger than nonvascular plants, sometimes reaching enormous sizes.

The only true plants that lack such a vascular system are the **bryophytes** [Greek, *bryon* = moss], the mosses and their relatives. Because bryophytes lack a vascular system, they are much smaller than the tracheophytes. The bryophytes are also much less diverse, totaling only about 24,000 species, far fewer than the (approximately) 290,000 named species of tracheophytes.

Most botanists refer to the major groups of plants as divisions, equivalent to the phyla of animals that we will survey in the next chapter. However, the term phylum is also becoming acceptable. Each division, like each animal phylum, contains classes, orders, families, genera, and species.

Some botanists divide the plants into only two divisions—Bryophyta and Tracheophyta. That is, they recognize plants

TABLE 22-1 THE DIVISIONS OF PLANTS

Division	Number of species
Bryophyta (mosses, liverworts, and hornworts)	24,000
Psilotophyta	Small
Lycopodophyta	1,100
Equisetophyta (horsetails)	15
Pteridophyta (ferns)	13,000
Coniferophyta (conifers)	600
Cycadophyta (cycads)	100
Ginkgophyta (ginkgo)	1
Gnetophyta	70
Anthophyta (flowering plants)	275,000

Table 22-1 lists ten divisions. Although it gives both a formal Latin name and a common name for many of the divisions, we will use the common name if there is one (for example, "conifers" rather than the Coniferophyta).

Bryophyta is the single division of nonvascular plants. The nine divisions of tracheophytes are all vascular plants.

How Do We Know About the Evolution of Plants?

The fossil record for plants, unlike that for protists, is excellent. Plant fossils are easier to find not only because plants are multicellular and therefore bigger but also because the vascular plants, at least, have many hard parts such as bark, cones, pollen, and veins—all of which nicely survive the process of fossilization.

The oldest plant fossils date from the beginning of the Silurian Period, about 430 million years ago (Figure 22-3). These fossils are not only the oldest plant fossils but also the oldest terrestrial organisms. Until then, organisms apparently lived only in the seas. The first plants appeared on the land, prob-

with specialized vascular systems and those without. Most botanists, however, count 10 or 12 divisions of living plants, of which all but one (the bryophytes) are vascular plants. The divisions of vascular plants vary in their modes of reproduction and dispersal.

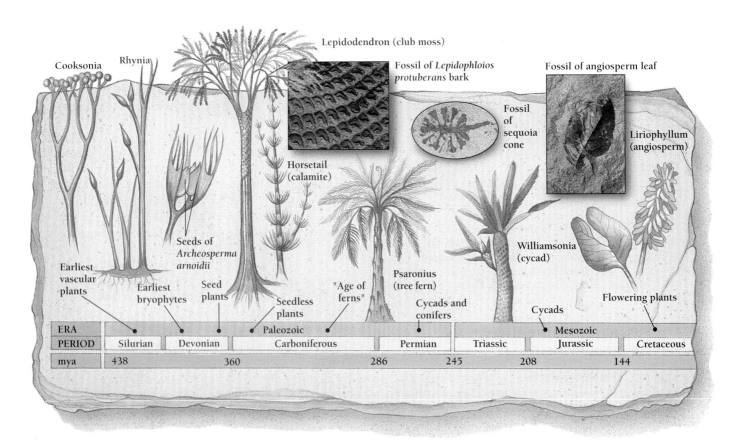

Figure 22-3 The evolution of plants. Notice the first appearance of vascular plants, bryophytes, seed plants, seedless plants, cycads and conifers, and flowering plants. (*Louise K. Broman/Photo Researchers, Inc.; William E. Ferguson; McCutcheon/Visuals Unlimited*)

ably in associations with fungi similar to present-day mycorrhizae (Chapter 21). The successful invasion of the land by plants and fungi allowed the later evolution of terrestrial animals. Among these animals, the most populous and diverse were and continue to be the insects. As we will see, the evolution of the flowering plants (angiosperms) went hand in hand with the evolution of insects and fungi.

We can deduce some of the characteristics of the first plants by noticing what all living plants have in common. All show alternation of haploid and diploid generations, all contain both chlorophylls *a* and *b*, all have cell walls made of cellulose, and all are multicellular. Almost all plants store energy as starch. All develop from embryos that are surrounded by nonreproductive ("sterile") tissue. And, during cell division, all plant cells use a microtubular structure, the phragmoplast, to build a new cell wall between the daughter cells. Because plants share these striking characteristics with some green algae, most botanists believe that both nonvascular and vascular plants descended from the green algae.

True plants, however, have characteristics that set them apart from the green algae. In addition to a waxy cuticle, specialized structures for absorbing water, and specialized enclosed reproductive organs, plants also have **stomata** [Greek, *stoma* = mouth], tiny mouthlike pores that open and close to regulate the flow of carbon dioxide and other gases in and out of the plant.

Evidence for plant evolution comes from fossils and from comparisons of livinf species. The first plants probably evolved from a common ancestor that resembled a green alga.

What Does the Fossil Record Tell Us About the Evolution of Plants?

Although it is tempting to think of nonvascular plants as more primitive than vascular plants, the fossil record of vascular plants is actually older. The earliest bryophyte fossils date from the Devonian Period, about 400 million years ago, while the earliest vascular plants appear at the beginning of the Silurian Period, 430 million years ago (Figure 22-3).

By the time bryophytes first appeared in the early Devonian Period, several types of simple vascular plants were widespread, some of which are strikingly similar to modern species such as ferns and club mosses. Botanists divide the vascular plants into those that produce seeds, such as the flowering plants and the conifers, and those that do not. To the seedless plants, the ferns and their allies, we owe our warm houses and speedy cars. These plants, buried deep beneath layers of sediment, have turned into the famous fossil fuels, mainly oil and coal, on which we are now so dependent. Seedless plants began flourishing in the Carboniferous Period, from 345 to 280 million years ago, when the world's climate was warm and wet. Shallow seas and dense swamps covered most of the land.

The growth of the earliest seedless plants, including ferns, horsetails, and lycophytes, was so luxuriant that a large proportion of these plants did not decay completely before being buried under more dead plants and other sediments. Over millennia, these buried, compressed plants became the major component of all fossil fuels—coal, oil, and natural gas. The large amounts of carbon in these fossils give the Carboniferous Period [Latin, *carbon* = coal + *ferre* = to bear] its name.

During the Carboniferous Period, plants grew taller and taller in an adaptive "light war." The tallest plants had access to the most sunlight and shaded the plants below. The tallest were 40-meter lycophyte trees, as tall as large oaks. Far below this canopy of lycophytes spread vast stands of 18-meter horsetails and 8-meter ferns. Ferns, in fact, became so successful that the late Carboniferous Period is often called the "Age of Ferns."

Although a few Carboniferous plants had seeds, none had flowers. Seed plants, which first appeared in the late Devonian Period, were the *progymnosperms* [Greek, *pro* = first + *gymnos* = naked + *sperma* = seed], the ancestors of the modern gymnosperms. **Gymnosperms** are modern seed plants that do not have fruits or flowers. Gymnosperms include pine trees and other conifers, as well as palmlike cycads.

Besides having seeds, the progymnosperms had more elaborate branching and more complex vascular systems than the earlier vascular plants. The Carboniferous swamps contained conifers and palmlike cycads, as well as lycophytes, horsetails, and ferns. As the climate became drier, the giant lycophytes, horsetails, and ferns began to disappear, and the conifers and the cycads began to dominate the landscape.

At the end of the Paleozoic Era, about 225 million years ago, the Earth's climate changed dramatically, and for 100 million years the dominant plants were the conifers and great palmlike cycads. Then, about 130 million years ago, a new kind of plant appeared—one with flowers.

Flowering plants, which have both seeds and fruits, became the most abundant plants on Earth. Today we live in the Age of Flowers. Of the present 315,000 species of plants, including all bryophytes and all tracheophytes, an astounding 275,000 are flowering plants. All of the flowering plants fall into two groups. These are the **dicots,** or **broad-leafed plants,** such as roses, poppies, oaks, and others; and the **monocots,** such as grasses, lilies, onions, daffodils, and cattails. Of the once dominant gymnosperms, only 529 species remain today.

Vascular plants predate nonvascular plants. To the first seedless plants of the humid Carboniferous Period we owe our current supplies of oil and gas. As the climate dried, first ferns, then conifers and cycads dominated the landscape.

How Did Angiosperms Evolve?

The flowering plants, or **angiosperms** [Greek, *angion* = vessel + *sperma* = seed], make up by far the largest division of

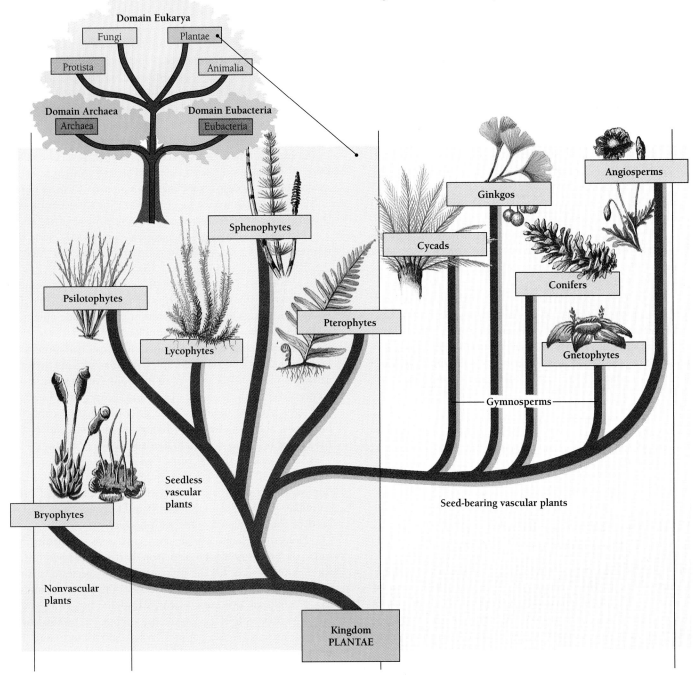

Figure 22-4 The major divisions of the kingdom Plantae.

plants (Figure 22-4). The angiosperms enclose their seeds in a vessel that we know as a fruit. A fresh peach is a fruit, and its pit is an enormous seed encased in a hard shell.

The first fossils of flowering plants date from the Cretaceous Period, about 125 million years ago. These early angiosperm fossils are leaves and pollen grains, rather than flowers.

In the absence of fossil flowers, evolutionary biologists who would like to know how angiosperms evolved from the gymnosperms have had to depend on comparisons among living plants. The magnolia, for example, is considered a very primitive plant with few advanced specializations. The magnolia has simple leaves, woody stems, and insect-pollinated flowers, features that probably characterized the first angiosperms.

In the 125 million years since the angiosperms first appeared, they have diversified both morphologically and biochemically. They have improved their transport system and developed *deciduous* leaves, which fall off seasonally, so that angiosperms are better able to withstand periods of drought than gymnosperms.

In addition, many modern angiosperm species have adaptations that prevent self-fertilization. Such species are more genetically diverse than gymnosperms and therefore better able

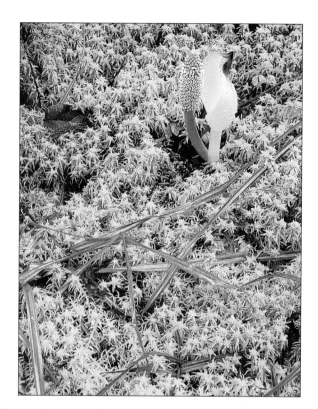

Figure 22-5 Sphagnum moss surrounding a skunk cabbage in the Quinalt Rain Forest, Olympic National Park, Washington. Most mosses grow in habitats that are always wet or damp. *(Pat O'Hara/DRK Photo)*

to survive environmental changes. Angiosperm diversity and their dominance of the land attest to the success of their many evolutionary experiments.

Thanks to improved transport and deciduous leaves, many angiosperms have been able to colonize ever drier habitats. Adaptations preventing self-fertilization have contributed to angiosperms' enormous diversity and success.

In the rest of this chapter, we will review the ten divisions of plants, beginning with the ancient bryophytes and working our way up to the modern angiosperms, the flowering plants.

THE NONVASCULAR PLANTS, OR BRYOPHYTES, INCLUDE MOSSES, LIVERWORTS, AND HORNWORTS

The nonvascular plants, or bryophytes, include the mosses, which give the division its name, and two less common classes—the liverworts and the hornworts (Figure 22-5). Mosses and other bryophytes generally live in wet environments, some tropical and some temperate.

Because all bryophytes lack specialized vascular systems, they depend on at least a film of free-standing water for both photosynthesis and fertilization by free-swimming sperm. Although bryophytes can reproduce only in moist habitats, they nonetheless disperse well and grow all over the world. One species of moss even grows high on an Antarctic mountain just a thousand miles from the South Pole.

The bryophytes play important but little understood roles in the world's ecosystems, recycling carbon and absorbing nutrients and toxic metals. Peat moss alone covers up to 2 percent of all land on Earth, an area the size of the United States. In some areas, bryophytes are the principal harvesters of solar energy, and all other organisms depend on them.

Bryophytes are also remarkable because the most obvious part of the bryophyte is not the diploid phase, but rather the haploid phase. When we look at a moss-covered stump, we are looking at the haploid phase, in which each cell has only one set of chromosomes.

How Do Bryophytes Absorb Water?

Although some bryophytes have elongated cells that conduct water, they lack the specialized transporting tissues of the vascular plants. They anchor themselves to surfaces by specialized structures called *rhizoids* [Greek, *rhiza* = root]. Unlike the roots of vascular plants, rhizoids are not specialized for absorbing water and nutrients. Instead, all parts of a bryophyte are absorbent, and the bryophytes consequently have a spongy feel. Bryophytes can grow on rocks, tree trunks, and other hard surfaces that resist the roots of vascular plants.

Bryophytes are small, usually less than 2 cm high and less than 20 cm long. In a moist environment, however, many plants may crowd together in a large mat. The closeness of the short plants allows them to retain water in the tiny spaces within the tangle. Because direct sunlight dries them out, bryophytes usually grow in the shade.

Although bryophytes lack a specialized transport system, all parts of their bodies are adapted to absorb water, which gives them a spongy feel.

How Do Bryophytes Reproduce?

Like the vascular plants, the fungi, and many protists, the bryophytes alternate generations. In contrast to the life cycles of the flowering plants, however, the most prominent phase of a bryophyte's life is the haploid gametophyte, which produces the gametes. Because the gametophyte is already haploid, it produces gametes—sperm or eggs—through mitosis, not meiosis. The gametes come together in the female

Figure 22-6 Life cycle of a moss.

gametophyte to form the diploid phase, called the **sporophyte** (Figure 22-6). The sporophyte produces haploid **spores** by means of meiosis, just as animals and flowering plants do. But these haploid spores are not gametes. They are unlike any phase in our own life cycle, for once the haploid spore finds itself in a welcoming environment, its cells enter mitosis and the bryophyte grows directly into a leafy, green gametophyte.

The haploid gametophytes may be male or female. Each sex develops *gametangia,* structures in which the male or female gametes develop (Figure 22-6). The male gametangium, called an *antheridium,* produces sperm with two flagella. The female gametangium, called an *archegonium,* produces and houses the ova and is generally shaped like a flask with a long neck. Some bryophytes bear archegonia and antheridia on the same plant.

To reach the ovum in the archegonium, the sperm must either swim through water adhering to the spongy mat or be splashed into the vicinity of the neck of the archegonium (Figure 22-6). Sometimes insects carry moss sperm, suspended in water, from male to female plant. Once sperm are near an archegonium, chemical attractants draw them into the neck, and the sperm swim through a fluid in the archegonium's neck to reach the ovum at the center of the swollen base.

Fertilization occurs within the archegonium. Immediately after fertilization, the diploid zygote begins to divide by mitosis to produce the sporophyte. The sporophyte, which is usually not photosynthetic, consists of a stalk, as tall as 15 to 20 cm, at the top of which develops a fruitlike **sporangium** [Greek, *spora* = seed + *angeion* = vessel]. Cells inside the sporangium undergo meiosis to produce the haploid spores, which can develop into a new mossy gametophyte.

Because bryophyte spores need to germinate in an environment that will sustain them, dispersal is essential to bryophyte reproduction. Some bryophytes propel their spores from exploding capsules. Some depend on splashing rainwater. A few bryophyte species live on dung, and flies carry the spores to fresh dung. Once in a hospitable environment, the haploid spores germinate, divide by mitosis, and develop into mature, leafy, male or female gametophytes.

Bryophytes' life cycles differ from those of vascular plants significantly and from those of animals profoundly: in a bryophyte's life cycle, the haploid gametophyte phase dominates.

Of What Advantage Is a Transport System to a Plant?

The ability of vascular plants to transport water and nutrients allows them to grow to huge sizes and to develop specialized tissues. Vascular plants also have more division of labor than the bryophytes, with distinctive tissues and organs that include roots, leaves, and stems. Their transport systems contain two distinct kinds of conducting tubes: **xylem** [Greek, *xylon* = wood], which carries water and minerals from roots to the photosynthesizing leaves, and **phloem,** which distributes the sugars and other organic molecules made in the leaves to the rest of the plant (Figure 22-7). Phloem consists of elongated, living cells. Xylem consists of the stiff walls of dead cells. Vascular plants also produce a hard, supporting material called *lignin,* which usually lines the walls of the xylem. Lignin, a major component of wood, provides the extra support that allows trees to attain such enormous sizes.

Plants' ability to transport water and nutrients has allowed them to grow larger and develop specialized tissues.

How Do Vascular Plants Reproduce and Disperse?

Recall that in the bryophytes, the haploid gametophyte—the leafy, green part we call moss—is the more prominent part of the plant life cycle. In vascular plants, the diploid sporophytes are more prominent. The seed of a peach tree, for example, develops, in the right environment, into a mature sporophyte—a peach tree. In the spring, the peach tree blooms, producing flowers, which are reproductive structures. Parts of the flowers undergo meiosis to produce short-lived male and female gametophytes, which produce pollen and ova.

Vascular plants that produce seeds have at least one great advantage over the bryophytes. In colonizing new habitats, seed plants do not rely on the chance that a dispersed seed will fall into a benign and nourishing environment. Instead, the seed comes encapsulated in a tough coat that protects the embryo inside.

Within the safety of the plant's female reproductive organ, the plant embryo stops developing, and the mother plant packages the embryo, with a reserve of food, inside a protective seed coat. A seed is able to withstand long periods of cold and drought. When the seed finally encounters water and warmth, it germinates, and the embryo uses the stored food to finish developing into a photosynthetic seedling.

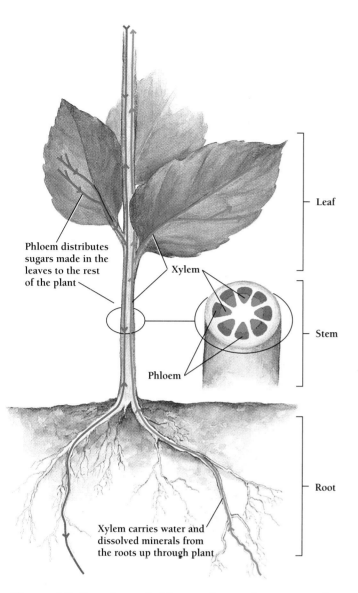

Figure 22-7 Xylem and phloem. Xylem conducts water and dissolved minerals from the roots to the leaves. Phloem conducts sugar from the leaves to the rest of the plant.

In vascular plants, the diploid phase dominates the life of the plant. The protective coat of a seed, and the food stored inside, allow a seed to wait indefinitely for suitable conditions for germination and growth.

THE VASCULAR PLANTS, OR TRACHEOPHYTES, INCLUDE SEED PLANTS AND SEEDLESS PLANTS

The tracheophytes (vascular plants) are both more diverse and more numerous than the bryophytes. Vascular plants include more than 290,000 named species, but some taxonomists say that the number of vascular plant species runs into the millions. We need only look around us to appreciate the success of the vascular plants—wherever we see green, we nearly always encounter a vascular plant.

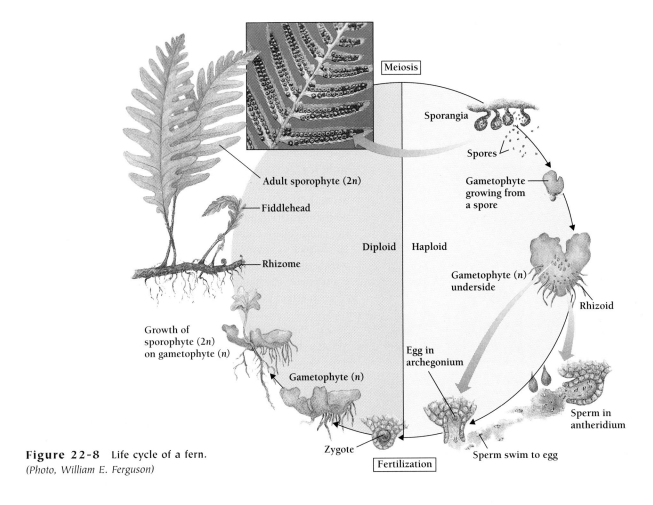

Figure 22-8 Life cycle of a fern.
(Photo, William E. Ferguson)

Four Divisions of Tracheophytes Lack Seeds

Four divisions of living vascular plants lack seeds: the pterophytes (ferns), psilotophytes, lycophytes, and equisetophytes. (Another four divisions of seedless plants exist only in the fossil record.) We will discuss the ferns first because they are the most familiar.

Pterophytes (ferns)

Some two-thirds of the 10,000 species of ferns live in the tropics, and the rest inhabit the temperate zones. Ferns range in size from a few centimeters to giant 30-meter tree ferns, as tall as oak trees. Ferns are common in moist parts of shady forests. Their leaves (called *fronds*) often have a lacy appearance and consist of many *pinnae* [singular, pinna; Latin, = feather], small leaflets that extend from a central stalk (Figure 22-8).

In the large ferns of the tropics, fronds grow from a central stem. In most ferns, however, fronds grow from a *rhizome*, a horizontal stem that spreads on or below the ground. Each frond develops from a characteristic coil called a fiddlehead. Most other seedless plants have **microphylls**—leaves with just one conducting vein. In contrast, fern fronds are **megaphylls**—large leaves with several veins that form complex networks of transport tubes.

Reproduction in ferns and other seedless plants resembles that in the bryophytes (Figure 22-8). Fertilization can occur only if the released sperm can swim to the mature archegonia. Seedless plants therefore require a wet environment to provide at least a film of water in which the sperm can make their way. This need for water, together with the fragility of the gametophyte, limits the environments in which ferns and other seedless tracheophytes can grow.

As in other vascular plants, however, the sporophyte (which we recognize as the fern plant) is much more prominent than the gametophyte. The fern sporophyte has leaves, called **sporophylls,** which are specialized for reproduction. Clustered on the underside of the sporophylls are groups of sporangia called *sori* [singular sorus; Greek, *soros* = heap] (Figure 22-8). The diploid sporangia produce haploid spores by means of meiosis. Some ferns catapult their ripe spores several meters into the air, where breezes carry them aloft. Such airborne spores may disperse over enormous distances.

On landing, each spore germinates to form a tiny nonvascular gametophyte, which is typically flat, thin, heart shaped, and less than 1 cm wide. This gametophyte is green, photo-

Figure 22-9 Horsetails. Equisetophytes contain bits of silica that give them a rough feel. *(William E. Ferguson)*

synthetic, independent, and fragile. It usually bears both male and female reproductive structures.

Ferns have megaphylls, leaves with networks of transport tubes, and independent gametophytes.

The psilotophytes

The psilotophytes contain just two genera, *Psilotum* and *Tmesipteris*. Both have photosynthetic leaflike scales (which lack conducting tissues) instead of true leaves. The common greenhouse weed *Psilotum*, or "whiskfern," is widespread in the tropics and subtropics. Unlike other vascular plants, *Psilotum* lacks true roots as well as leaves. Underground, it has a rhizome, with tiny branches called rhizoids. Above ground, its vertical stem forks several times and bears sporangia and photosynthetic scales.

Psilotophytes, the simplest of the living vascular plants, lack true leaves.

Lycophytes

Lycophytes (also called lycopods) have true roots, stems, and leaves, as well as a simple arrangement of xylem and phloem. Lycophyte leaves are microphylls (with a single conducting vein). As in ferns, some of the leaves are sporophylls, with sporangia on their upper surfaces. The sporophylls may be green and look like ordinary leaves, or they may be distinct and grouped into spore-producing cones, called **strobili** [singular, strobilus; Greek, *strobilos* = cone]. The most common lycophytes are the club mosses, which form evergreen, mosslike mats beneath temperate forests. Their strobili resemble little clubs.

Lycophytes have true roots, stems, and simple leaves.

Equisetophytes

The living equisetophytes are all members of the single genus *Equisetum*, or horsetails (Figure 22-9). These plants grow in moist places along the banks of streams or at the edges of woods. Like the lycophytes, the equisetophytes have true roots, stems, and leaves. Their jointed stems contain many discrete strands of xylem and phloem. The scalelike leaves that grow from the joints are thought to be megaphylls (with branching vessels). The outer cell layer of the plant contains silica (the same material from which glass is made), which gives the plants a rough texture useful to campers and settlers for scouring pots. Indeed, another name for these plants is "scouring rush."

Equisetophytes have true roots, stems, and complex leaves. Their stems are jointed, and their outer cell walls are reinforced with silica.

How Do Seed Plants Reproduce?

Seeds are of great value in a dry, terrestrial environment because they liberate a plant from the need for free-standing water. **Seed plants** manage fertilization without water in the same way we do, through a form of internal fertilization. The gametophytes have no independent lives but develop within the tissues of the sporophytes. The gametophytes of seed plants are tiny—even less prominent than in the seedless vascular plants.

Many seedless plants are *homosporous,* producing just one kind of spore and one kind of gametophyte. In contrast, all the seed plants are *heterosporous*—producing two kinds of gametophytes (male and female), from two kinds of spores on two kinds of sporangia. The sporophytes produce haploid microspores (male) and megaspores (female) in separate sporangia. In some species, the male and female sporangia occur together on the same plant, while in others, the two kinds of sporangia occur on male and female plants. In either case, the microspores and megaspores develop into distinct male and female gametophytes.

The female gametophyte stays within the megasporangium, which lies safe within the main body of the adult plant (the sporophyte), awaiting the arrival of the sperm. The megasporangium, also called a **nucellus** [Latin, *nucella* = a small nut], is in turn covered by one or two additional layers of tissue, called integuments, which will later develop into the protective seed coat. This combined structure—nucellus and integuments—is the **ovule.**

The small, male gametophytes are grains of pollen, released from the microsporangium and carried, usually by wind or insect, to the female gametophyte. The male gametophyte then develops a slender pollen tube, which carries a sperm nucleus to the ovum produced by the female gametophyte. At fertilization, the sperm and ovum fuse to form a diploid zygote. The resulting zygote, still enclosed within the female gametophyte, begins to develop into an embryo.

After fertilization, the ovule and its contents become a **seed.** The seed includes (1) the new sporophyte embryo; (2) a female tissue that nourishes the developing embryo; and (3) the nucellus and integuments (now the seed coat) from the previous sporophyte generation. Inside the seed coat are the embryo and a source of food for its later development. The seed

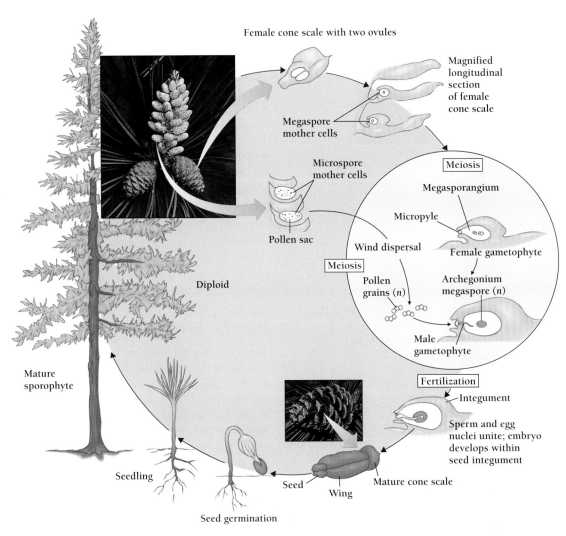

Figure 22-10 Life cycle of a pine. Like other conifers, pines have two kinds of cones: pollen cones and seed cones. In the seed cones, which are the familiar "pine cones," each scale bears two ovules. Each ovule in turn contains a female sporangium, surrounded by a protective integument. A cell within the sporangium undergoes meiosis to produce four haploid cells, one of which undergoes mitosis to produce the female gametophyte. The gametophyte remains within the ovule and develops two or three archegonia, each of which in turn produces a single giant ovum, loaded with carbohydrates and proteins. In the integument surrounding the ovule is an opening for the entrance of pollen. In spring, a sticky fluid in the opening captures pollen grains from the wind. After a few months, the pollen gametophyte develops into a pollen tube, which makes its way to one of the archegonia in the ovule. One of the cells of the pollen tube divides to form two sperm nuclei, which then enter the archegonium. As much as a year after pollination, one of these nuclei finally fuses with the ovum to form a zygote, which develops within the ovule. When the new embryo has a root and several leaves, the seed is ready for dispersal. (*Photos: top, Michael Giannechini/Photo Researchers, Inc.; bottom, R.J. Erwin/Photo Researchers, Inc.*)

coat protects the young embryo from drying out or from being eaten. Once the seed germinates, the embryo resumes its development and growth.

In the following sections, we will discuss the five divisions of living seed plants. The 771 named species of the first four divisions are gymnosperms, which produce seeds that are not enclosed in fruits. In contrast, the 275,000 species of angiosperms produce seeds that are enclosed in a fruit.

Seeds, which enable plants to colonize relatively dry environments, consist of a diploid zygote and a source of food encased in a seed coat. Seed plants are heterosporous.

Gymnosperms: Four Divisions of Seed Plants Without Flowers

The four divisions of gymnosperms—conifers, cycads, ginkgos, and gnetae—are all vascular plants with seeds, but without flowers and fruits. They are not closely related to one another. We first discuss the most familiar division, the conifers.

Conifers

The conifers [Greek, *konos* = cone + *phero* = carry] take their name from their characteristic cones, which are male and female strobili. The conifers include many of the most common trees in the Northern Hemisphere—the pines, firs, spruces, and cedars—as well as the magnificent redwoods of the California and Oregon coasts. Conifers grow well in the short growing seasons of northern latitudes and high altitudes. They are mostly evergreen, keeping their leaves throughout the year. Individual leaves may last through two or three winters. (Bristlecone pines, some of which live to be more than 5000 years old, retain their needles for up to 45 years.) Conifer forests extend almost all the way across North America, Europe, and Asia.

Many conifers, including the familiar pines, have needle-like leaves that are well adapted for dry conditions, including, for example, northern winters, when the air is dry and the groundwater is frozen. A needle's small surface area and heavy cuticle minimize evaporation. Despite its simple shape, however, the needle is a megaphyll, usually with two vascular bundles. The stems of conifers are heavily reinforced with lignin, which gives trees the structural support to reach enormous sizes. The lignin in wood is also what makes it strong enough to support even the largest buildings.

Most conifers bear male and female cones on the same tree. The male, pollen-producing cones are generally small (1 to 2 cm long) and consist of hundreds of sporophylls. Wind carries the pollen grains, often in huge yellow clouds, to the female cones. The female, or ovulate, cone consists of tough scales arranged in a spiral. Figure 22-10 shows the life cycle of a typical conifer, a pine tree.

Conifers produce male and female gametophytes in cone-shaped strobili.

Cycads

The cycads are large-leafed plants that look like palms (Figure 22-11). Some cycads, called "sago palms," are used as ornamental plants in homes and gardens. But true palm trees are flowering plants, while cycads have neither flowers nor fruits: they bear naked seeds on the scales of cones near the top of each plant. Individual plants bear either male or female cones, but not both. Some people eat the seeds and underground

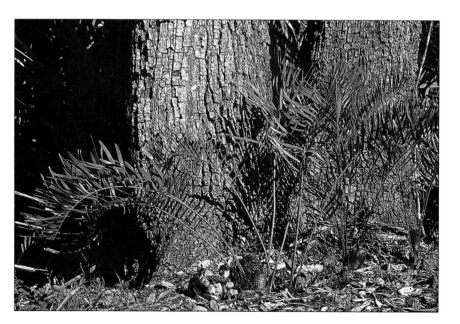

Figure 22-11 *Zamia pumila* is the only species of cycad native to the United States. (*Walter H. Hodge/Peter Arnold, Inc.*)

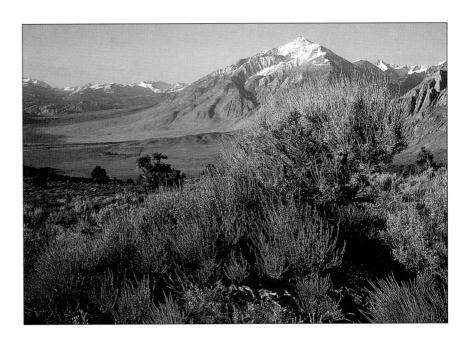

Figure 22-12 The gnetophyte *Ephedra viridas* (mormon tea) resembles an angiosperm. (*Stephen Ingram/Earth Scenes*)

stems of cycads, but cycads contain toxins that can cause gastrointestinal problems, degenerative brain disease, and even paralysis.

The small male cones of cycads consist of microsporophyll scales, each of which contains many pollen-producing microsporangia. Female cones are much larger. The cone of one Australian cycad may measure 1 m in length and weigh 40 kg (90 pounds). As in conifers, pollen travels by wind to the ovule, enters through a micropyle, germinates to produce a pollen tube and sperm, and fertilizes the ovum.

Cycads produce male and female gametophytes in strobili that lie on separate plants.

Ginkgos

The ginkgo is the last living species in the Division Ginkgophyta, the rest of whose members went extinct as long as 280 million years ago. The ginkgo, or maidenhair tree, is a living fossil that has hardly changed in the last 80 million years. Yet this beautiful tree, which decorates many gardens and parks, is itself now extinct in the wild.

Like the cycads, ginkgo trees are either male or female. The male trees have cones (strobili), while the female carries fleshy (and smelly) seeds at the ends of short stalks. The seeds of the female tree have an odor so rancid that most people prefer to plant male trees only.

Ginkgos resemble cycads in their life cycle and conifers in their growth patterns.

Gnetae have characteristics of both gymnosperms and angiosperms

The Gnetophyta, a small but diverse group of cone-bearing desert plants, consist of three distinct genera—*Gnetum, Ephedra,* and *Welwitschia* (Figure 22-12). Of all the gymnosperms, the gnetae most resemble angiosperms. For example, some of their strobili look like flowers, and the vessels of their leaves and stems resemble those of angiosperms. But they have naked seeds and are certainly gymnosperms. Recent DNA studies have led to the rejection of the long-favored hypothesis that angiosperms evolved directly from a gnetophyte. So the origin of angiosperm remains an unsolved, and much debated, problem.

THE ANGIOSPERMS ARE SEED PLANTS WITH FLOWERS AND FRUITS

The diversity of the angiosperms dwarfs that of all other divisions of plants. Angiosperms, also called **Anthophyta** [Greek, *antho* = flower], range in size from the tiny duckweed *Wolffia*, less than 1 mm long, to *Eucalyptus* trees, taller than 100 m. With some 275,000 known species, angiosperms are the most widespread and familiar of plants. We see them daily in our lawns, gardens, parks, fields, and forests. They provide us with furniture, clothes, medicines, and dyes. We stock our kitchens with their seeds, fruits, leaves, stems, and roots. Angiosperms provide us with wine, beer, coffee, and tea. We use the more brilliant angiosperms to brighten our lives and enliven our loves.

Despite their diversity, all angiosperms share two important adaptations: flowers, which promote fertilization, usually by insects or other animals; and fruits, which promote the survival and dispersal of seeds.

Of What Use Are Flowers?

Flowers are reproductive structures that ensure the distribution of pollen, either by animals or by wind. Some flowering plants rely on wind alone to distribute pollen. Grasses, rushes, sedges, plantain, birch, poplar, and oaks, for example, have small inconspicuous flowers and no scent. Because wind pollination is so inefficient, wind-pollinated flowers release mind-boggling amounts of pollen. Individual plants of the weed *Mercurialis annua* may release over a billion grains of pollen.

Other flowers attract insects (especially bees, beetles, butterflies, and moths), birds (especially hummingbirds), and even bats and other small mammals. All these animals' visits are rewarded with a sugary fluid called nectar, a share of the high-protein pollen, or nutritious petals or other flower parts. Animals learn to associate these delicious foods with the seductive perfumes and bright colors of flowers. Not all of a flower's attractions are necessarily attractive to humans: the eastern skunk cabbage (*Symplocarpus foetidus*), for example, attracts pollinating flies by emitting an odor like that of dead fish.

Most flowers are specialized to ensure pollination by just one group of animals. Flowers pollinated by moths, for example, are usually large, white, and overwhelmingly fragrant at night, when moths fly. In contrast, flowers pollinated by butterflies are small and brightly colored, since butterflies fly during the day and rely on sight more than smell. One orchid species produces flowers that look so much like female bees that male bees come to mate with the flowers. As the male bee attempts to copulate with the flower, he is dusted with pollen, which he then carries to another flower in another futile attempt to mate.

Insects are as well adapted to take advantage of flowers as flowers are adapted to take advantage of insects. One South African fly has a tongue nearly 7.5 cm (3 in) long, which allows it to take nectar that is out of reach for other insects. Insects and flowers provide a striking example of **coevolution,** the mutual adaptation of two separate evolutionary lines. Coevolution occurs when two species are significant factors in the lives of one another. In the case of insects and flowers, coevolution ensures a food supply for the insects and pollination for the plants.

Flowers attract animals that distribute pollen.

Of What Use Are Fruits?

A fruit is a mature ovary that encloses and protects seeds. Many fruits promote seed dispersal by animals or wind. The most familiar fruits are those that attract by appearance and taste, for example, peaches, apples, and berries. An animal usually digests the fleshy part of such fruits, but the seeds pass unharmed through the digestive system. (In fact, some seeds cannot germinate until they pass through the digestive tract of an animal.) Seeds excreted by the animal can end up far from the plant that produced them, together with a nourishing supply of fertilizer.

Other fruits use different stratagems to transport their seeds. Thistles and dandelion fruits, for example, soar like kites in the slightest breeze, and maple fruits whirl through the forest like free propellers. Burrs stick to the fur (or socks) of passing animals, while coconuts have spread to countless tropical islands by floating across the ocean. Touch-me-nots (*Impatiens*) propel their seeds from fruits that explode on touch.

A fruit is a ripened ovary that encloses and protects the seeds and usually enhances their dispersal.

How Do Angiosperms Reproduce?

In angiosperms, as in all the other seed plants, the diploid sporophyte generation is more prominent than the haploid gametophyte. In fact, the male and the female gametophytes of angiosperms are even tinier than those of the gymnosperms, and they live their whole lives inside the sporophyte's flowers.

The flowers produce haploid microspores (male) and megaspores (female) in separate sporangia, from which the gametophytes derive. After meiosis, the microspores divide once to give rise to two haploid nuclei—the tube nucleus and the generative nucleus (Figure 22-13). The male gametophyte resumes development only after it arrives on the **carpel,** the flower part that contains the ovule. The pollen tube now grows, and the generative nucleus divides into two sperm nuclei. By the time the pollen tube reaches the ovule, it contains two sperm nuclei.

Meanwhile, within the ovule, the female gametophyte has developed. Meiosis yields four haploid megaspores, only one of which survives. That megaspore divides three times by mitosis to produce the embryo sac, the mature female gametophyte. The embryo sac has eight nuclei and seven cells. One cell (the central cell) contains two haploid nuclei (called the polar nuclei). Another cell becomes the haploid ovum.

Fertilization in angiosperms is different from that in any other organism. As we will discuss in more detail in Chapter 32, there are actually two separate fertilizations, together called a **double fertilization:** (1) one sperm nucleus fuses with the nucleus of the ovum to form the zygote, and (2) the other sperm nucleus fuses with the two nuclei of the central cell to form a triploid nucleus called the primary endosperm nucleus. The zygote then divides by mitosis and develops into an embryo. The primary endosperm nucleus divides to form an **endosperm,** which will provide food for the developing embryo and, in some cases, for the germinating seedling.

Figure 22-13
Life cycle of an angiosperm.

Flowers reproduce by means of double fertilization. Two sperm nuclei from the pollen grain fertilize two ova from the ovary, resulting in a diploid zygote and a triploid cell that forms the nutritious endosperm.

What Are the Parts of a Flower?

Like the gymnosperms, angiosperms are heterosporous. Separate sporangia produce male spores (microspores) and female spores (megaspores). As in the case of the gymnosperms, the sporangia lie on modified leaves, or sporophylls. In angiosperms, however, both male and female sporophylls are flower parts instead of cones. The male sporophylls are **stamens**, and the female sporophylls are carpels.

A flower is a short piece of stem that usually ends with four whorls, or circles, of modified leaves. These are the carpels, the stamens, the petals, and the sepals (Figure 22-13). The innermost parts are the carpels. All the carpels, whether one or more, constitute the "pistil." Each carpel encloses one or more female sporangia (ovules) in a swollen base called the **ovary**. Pollen does not reach the ovules directly but first falls on the stigma, the carpel's receptive area. The style, a slender tube through which the pollen tube enters the ovary, extends from the stigma to the top of the ovary. After fertilization and development, the ovule becomes the seed, and the ovary becomes the fruit.

Just outside the carpels lies the next whorl of modified leaves, namely the stamens, or male sporophylls. Each stamen consists of an **anther**, a thick, pollen-bearing structure, and a **filament**, a thin stalk that connects the anther to the base of the flower. Within the anther are four sporangia, called pollen sacs, where the pollen grains develop (Figure 22-13).

Outside the stamens is a whorl, or **corolla**, of **petals**. In flowers pollinated by insects or other animals, the petals are usually large and brightly colored. In flowers pollinated by wind, however, the petals are reduced or missing altogether. When the petals fuse, the corolla forms a tube.

The fourth and outermost whorl is the **calyx**, a whorl of leaflike parts called **sepals**, which enclose the rest of the flower and protect it before it opens. The sepals are usually small and green, but in some flowers they may be large and colorful.

Flower structures vary widely. Those that lack one or more of the four whorls are said to be incomplete. Those that lack either stamens or carpels are said to be imperfect (as well as incomplete). Flowers may have radial symmetry, so that any cut along the axis of the flower will produce two identical halves. Or they may have bilateral symmetry, so that only one cut through the flower's axis will produce identical (or mirrored) halves.

Fruits are as diverse as the flowers that produce them. Some develop from a single carpel (with one or many seeds) or from several fused carpels with a single ovary. Others, such as raspberries or strawberries, derive from many separate carpels. And still others, like the pineapple, come from many flowers.

The male sporophyll is the stamen—including anther and filament—and the female sporophyll is the carpel—including style and ovary. A corolla of petals and a calyx of sepals generally surround the stamens and carpels.

Why Are There So Many Species of Angiosperms?

Botanists recognize more than 275,000 species of angiosperms, which live in almost every imaginable habitat, from deserts to tropical rain forests, from river valleys to mountaintops. Starting with Linnaeus, botanists have classified flowering plants according to their reproductive organs (flowers), but angiosperms also show bewildering variations in size, woodiness, water transport, seeds, and fruits.

This huge diversity of species depends on mechanisms, discussed in Chapter 17, that maintain barriers to interbreeding. Chromosome incompatibility is usually an important barrier to interbreeding. Chromosomes from different species usually cannot form homologous pairs during meiosis, and so hybrids between two species cannot produce gametes for the next generation.

Many plants, however, have been remarkably adept in jumping this particular barrier. In some cases, plants double the chromosomes of the zygote to form a set of chromosomes that contains a diploid (rather than a haploid) set from each parent. The resulting offspring can then perform meiosis and produce gametes, because each chromosome belongs to a homologous pair.

The coming together of the doubled gametes at fertilization produces a hybrid that is *polyploid* [Greek, *poly* = many + *ploion* = vessel] because it contains several (usually two or four) sets of genes. Polyploid plants, with the combined genetic repertoire of two or more species, are often much better at adapting to new environments than ordinary diploids.

Polyploidy has been a common source of new angiosperm species during evolution. Farmers have also produced many agriculturally important plants by accidentally or deliberately inducing polyploidy. Modern wheats, for example, derive from ancient hybrids of wild ancestors. Studies of angiosperm DNA also suggest that the genomes of many angiosperm species have also expanded during evolution via "jumping genes," (transposons; Chapter 12).

Despite large differences in genome size, diverse plants use genes whose sequences are surprisingly similar, not only among angiosperms but also between angiosperms and gymnosperms. Comparisons of DNA sequences now complement morphological comparisons (especially of flowers) in suggesting the long, obscure origins of modern plants (Figure 22-14).

The terrestrial environment is far more diverse than the marine environment. Extremes of temperature and rainfall, for example, demand a great range of adaptations. As a result, terrestrial organisms are themselves more diverse than their marine cousins. Animals, which we survey in the next two chapters, are especially diverse. But most animal species are insects, and the evolution of insects is closely tied to that of plants. Plants' colonization of the land 430 million years ago may have set the stage for the evolution of insects and other terrestrial animals.

Figure 22-14 *Amborella trichopoda*, a living fossil, grows only on the island of New Caledonia in the South Pacific. Botanists often classify *Amborella* as closely related to avocados and sassafras. Analysis of DNA from hundreds of species suggests that the first flowering plants resembled *Amborella* and not a gnetophyte. *(William E. Ferguson)*

Study Outline with Key Terms

Plants are multicellular organisms that perform photosynthesis. They originated on the land, and most are terrestrial. The land offers three major advantages for photosynthetic organisms: more light, more carbon dioxide, and a more reliable supply of minerals and other nutrients. However, for plants to invade the land, they had to be able to obtain and conserve water.

The first plants probably evolved from green algae at least 430 million years ago. The first seed plants appeared by the late Devonian Period. Fossils of the first flowering plants date from the Cretaceous Period, about 125 million years ago.

The evolution of plants depended principally on three adaptations: (1) a waxy **cuticle,** which coats the plant's exposed surfaces and reduces evaporation; (2) the ability to absorb water and minerals from dew, rainwater, or groundwater; and (3) jacketed reproductive organs. All plants also have **stomata** and life cycles characterized by **alternation of generations** between **gametophytes** and **sporophytes**. The **gametangia** of the gametophytes produce gametes, and the **sporangium** of the sporophytes produces **spores**.

The **bryophytes** have no specialized conduction system for distributing water and nutrients and are therefore limited in the sizes to which they can grow. Bryophytes include mosses and two rarer classes, the liverworts and hornworts. The haploid (gametophyte) phase of bryophytes is more prominent than the diploid (sporophyte) phase.

Vascular plants are far more diverse and more numerous than the bryophytes. Vascular plants contain two distinct kinds of conducting tubes, the **xylem** (which carries water and minerals from roots to leaves) and the **phloem** (which distributes sugars made in the leaves to the rest of the plant). In vascular plants, the diploid phase of the life cycle is more prominent than the haploid phase.

Vascular plants disperse as diploid sporophytes. Most kinds of vascular plants disperse as **seeds,** which contain a diploid embryo, a food reserve, and a protective coat. Many **seed plants** have flowers and fruits, which aid in fertilization and in the dispersal of seeds.

Four divisions of vascular plants, or **tracheophytes,** lack seeds: pterophytes, psilotophytes, lycophytes, and sphenophytes. Pterophytes (ferns) have **megaphylls,** leaves with networks of transport tubes. Other tracheophytes have **microphylls,** leaves with just one vein. Psilotophytes are the simplest of the living vascular plants and do not have true leaves. Lycophytes are more complex than psilotophytes, but less so than ferns. Sphenophytes have jointed stems that are reinforced with silica.

Five divisions of tracheophytes bear seeds, but four of these, collectively called the **gymnosperms,** do not have flowers. A seed contains the embryo, the female gametophyte, the organ that produces the female gametophyte (the **nucellus**), and layers of tissue called the integuments from the previous sporophyte generation. The nucellus and integuments form the **ovule.**

Gymnosperms produce male and female gametophytes in **strobili.** We discussed the four divisions of gymnosperms: conifers, cycads, ginkgos, and gnetae. In conifers the strobili are separate cone-shaped structures on the same tree. Cycads produce male and female gametophytes in strobili that lie on separate plants. Ginkgos resemble cycads in their life cycle and conifers in their growth patterns. Gnetae have characteristics of both gymnosperms and flowering plants.

Anthophyta, or **angiosperms,** are the most diverse group of plants. The gametophytes of angiosperms are even more reduced than those of the gymnosperms. Both male and female gametophytes are part of the flowers of the **sporophyte,** as are the male and female **sporophylls.**

A flower is a short piece of stem that usually ends with four circles of modified leaves: the **carpels, stamens, petals,** and **sepals.** Each carpel—including stigma, style, and ovary—encloses the female sporangia. The stamens are the male reproductive structures. Each stamen consists of an **anther** and a **filament.** A **corolla** of petals, often brightly colored and scented, attracts insects that carry pollen grains (male gametophytes) from flower to flower. A **calyx** of sepals encloses and protects the rest of the flower before it opens.

Fertilization in angiosperms is different from that in any other organism and consists of two separate fertilizations—a **double fertilization.** The zygote then divides and develops into an embryo. The primary **endosperm** nucleus divides to form an endosperm, which will provide food for the developing embryo and, in some cases, for the germinating seedling. Flowering plants are divided into **monocots** and **dicots** (or **broad-leafed plants**), according to the form the embryo takes.

A fruit is a ripened, or mature, **ovary** that encloses and protects the seeds. Flowers and fruits allow angiosperms to interact with animals. Flowering plants have **coevolved** with insects.

Review and Thought Questions

Review Questions

1. Describe six traits that together set plants apart from members of the other kingdoms of organisms.
2. Name several advantages and disadvantages to life on land.
3. What traits enable plants to live on land?
4. Compare alternation of generations with the life cycle of an animal, such as ourselves.
5. How long ago do paleontologists find the first association between plants and mycorrhizae? What does this imply about the coevolution of these two groups of organisms?
6. What group of plants that flourished 300 million years ago provided the fossil fuels oil, coal, and natural gas? What is the name of that geologic period?
7. How do bryophytes absorb water? What important roles do mosses play in the world's ecosystems?
8. What is peculiar about the life cycle of a bryophyte?
9. What are xylem and phloem and what functions do they perform in vascular plants?
10. What is the sporophyte phase of a vascular plant?
11. What is the difference between a megaphyll and a microphyll?
12. What are the advantages of seeds to terrestrial plants?
13. What are the advantages of flowers and fruits to terrestrial plants? Name the parts of a flower.

Thought Question

14. Why did plant taxonomists abandon Linnaeus's system of classification? What characteristics of plants reflect their evolutionary relatedness? Why do these characteristics represent relatedness more than the number of stamens or the number of pistils?

About the Chapter-Opening Image

Artist Georgia O'Keeffe is known for her depictions of flowers as sexual organs. Linnaeus had the same idea almost three hundred years ago.

 On-line materials relating to this chapter are on the World Wide Web at **http://www.harcourtcollege.com/lifesci/aal2/**

 Web Links

 Quizzing & Testing

23 Protostome Animals: Most Animals Form Mouth First

Progress and the Burgess Shale

One day about 530 million years ago, a mud slide the size of a city block suddenly broke loose and buried tens of thousands of small marine animals living in the ooze at the bottom of a shallow sea. Acres of mud sealed off the animals so abruptly and completely that the creatures never had a chance to decay. Over millions of years, the mud slowly turned to rock, perfectly preserving the animals within. Geologic processes slowly lifted the rock thousands of feet high, and today the ancient mud slide, and the creatures entombed within it, sits high in the Canadian Rockies. It is called the Burgess Shale.

When the fossil remains of the tiny animals were finally discovered half a billion years later, they unleashed a bitter debate about the nature of evolution. Within the solid mass of the Burgess Shale, the perfectly preserved creatures revealed a remarkable diversity of animals, a diversity that suggested different things to different people.

The Burgess animals' first discoverer was paleontologist Charles Doolittle Walcott and his family. In 1909, Walcott, together with his wife, Helena Walcott, and several of their children were collecting fossils in the Canadian Rockies when they came across an isolated slab of rock crammed with the small, perfectly preserved organisms. Walcott quickly found the source of this slab, higher up the slope of the mountain, and the following summer, Charles, Helena, and their half-grown children returned and dug more than 80,000 fossils from the mountain.

Walcott was a distinguished leader of the early 20th-century American scientific establishment (Figure 23-1). However, his role as top administrator at the Smithsonian Institution, as well as his positions on the boards of other scientific organizations, left him little time for deep thought or serious research. As discoverer of the Burgess fossils, he was entitled to name and classify the new organisms. In his haste, however, he classified all of the Burgess animals as either trilobites (already well-known ancient organisms) or primitive members of still living groups (such as crustaceans, jellyfish, and polychaete worms). In short, Walcott mistakenly assumed that all his animals were members of already well known phyla.

Walcott well knew how little time he had given the Burgess animals. For the rest of his life, he dreamed of studying them in detail. In 1926, he announced his plans to retire so that he could spend his last years with the Cambrian animals. But he died just 3 months short of his planned retirement, and even though he was one of the leading scientists of his day, he never fully appreciated his own remarkable discovery.

Finally, in the 1970s, three British paleontologists—Harry Whittington, Derek Briggs, and Simon Conway Morris—undertook to reexamine the Burgess fossils. The fossils were not merely flat impressions, but fully formed stone animals whose ancient bodies could be dissected and studied. Using dental drills, Whittington, Briggs, and Conway Morris carefully chipped away layers of

Key Concepts

- An animal is a multicellular organism that eats other organisms and develops from an embryo.

- Zoologists divide animals into two subkingdoms—the Parazoa (sponges) and Eumetazoa (all other animals). The members of the Eumetazoa are grouped according to their symmetry (radial or bilateral) and internal body cavities (coelomate, pseudocoelomate, or acoelomate).

- Arthropods include the greatest number of species and the greatest numbers of individuals of all animal phyla.

stone tissues, revealing the anatomies of some of the most bizarre animals ever found (Figure 23-2). Altogether, argued Whittington, Briggs, and Conway Morris, the Burgess animals represented no fewer than 25 basic body plans, of which only four survive in modern organisms.

In 1989, the Harvard paleontologist and popular writer Stephen Jay Gould wrote a book, *Wonderful Life,* describing the animals of the Burgess Shale and chronicling the work of Walcott, Whittington, Briggs, and Conway Morris. Gould argued that the diversity of life forms in the Cambrian dwarfed anything we know today. The range of anatomical designs among the animals that died in a block of mud 530 million years ago far exceeded that of all the creatures in today's oceans. The Burgess animals, wrote Gould, should have challenged the ablest classifiers of the time. Yet Walcott had classified these bizarre organisms into well-known groups of organisms such as jellyfish and shrimps.

Each of the body plans represented in the Burgess Shale, argued Gould, could have been assigned to a new phylum. Only because many of these types are represented by just a single species had taxonomists hesitated to give each one a whole phylum to itself. Charles Walcott had failed to recognize the true diversity of these creatures, Gould said, for two reasons. Walcott had assumed that the creatures he had found would fit into categories he already knew; and he assumed that evolution, by its nature, inevitably progresses from a few simple forms to a diversity of complex and modern forms (culminating in human beings). This idea—that evolution is progressive—is controversial, however, and depends on how the word "progressive" is defined.

Is evolution progressive? Does diversity increase? Do organisms become more complex? Gould argued that the diversity of the Burgess Shale provided evidence that evolution was *not* a progression from

Figure 23-1 Paleontologist Charles Doolittle Walcott and his wife, Helena Walcott, discovered a rich deposit of ancient organisms in the Canadian Rockies. (*Brown Brothers*)

simple to complex, from few forms to many. He argued that life is less diverse today than during the Cambrian explosion. The Burgess Shale, he said, illustrated that multicellular life began in great variety all at once. Since then, for reasons unknown, most of these forms became extinct. The organisms that survive today, he said, have been merely lucky survivors of various catastrophes in the history of the Earth.

But Gould's arguments were not whole-heartedly accepted by other paleontologists and evolutionary biologists. Soon after the appearance of *Wonderful Life,* further studies of the Burgess fossils by Conway Morris and others suggested that the Burgess animals were not as novel as Conway Morris had initially believed. The animals could be comfortably grouped into smaller numbers of known phyla, although not necessarily the phyla that Walcott had selected.

Were the animals that lived 530 million years ago more diverse than those living today, as Gould had argued? Probably not. While paleontologists have recognized about 11 phyla of animals from the Cambrian, biologists have grouped living animals into 35 phyla. As far as is known, no whole phylum of animals has ever gone extinct. The dinosaurs, for example, were merely a group within the class Reptilia in the phylum Chordata, the phylum to which we humans belong. And the Chordata are represented in the Burgess Shale animals.

Certainly, the history of life suggests that diversity and complexity have increased since the first appearance of life 3.5 billion years ago. All living organisms, in all their current diversity, have descended from the same few simple ancestors. The complex creatures, such as ourselves, that now populate the world have evolved, one generation at a time, from simpler organisms.

In 1995, the philosopher Daniel Dennett attacked Gould, dissecting Gould's arguments about diversity and progress as aggressively as Gould had

Figure 23-2 A few of the animals of the Burgess Shale. Floating in the upper left is a harp-shaped creature named *Marrella*. The large, green creature in the center is *Anomalocaris*, which grew up to a meter long. Just below is *Opabinia*, smaller, but with five eyes and a front-facing "nozzle." The small worm struggling in *Opabinia*'s jaws is a *Burgessochaeta* worm. *Pikaia*, shown lower right, may be the oldest known Chordate—if not our direct ancestor, at least a relative of one. *Odontogriphus* (pink, ribbonlike animal) was a flat, swimming creature. *Hallucigenia* (the green creature walking on the sea floor) had seven pairs of spines and seven tentacles. (Because of an initial misunderstanding by paleontologists, we have depicted *Hallucigenia* upside down.) The three reddish, flame-shaped creatures are *Wiwaxia*, which probably crawled along the sea floor. What look like green plants on stems are actually *Dinomischus*, a sessile animal. *Vauxia*, shown behind *Anomalocaris*, was a sponge. At the lower left is *Aysheaia*, similar to a modern-day onychophoran. Shown in its burrow underneath the sea floor is *Ottoia*, which looked and behaved much like the priapid worms that still live along our coasts today.

scrutinized Walcott's taxonomy of the Burgess animals. Dennett insisted that evolution is, in fact, progressive, in that evolution tends to increase complexity.

As organisms evolve adaptations in response to their environment and to other organisms, complexity increases. For example, the grinding teeth of a horse are an adaptation to tough grasses, which evolved toughness and lack of digestibility in response to grazing by plant-eating animals. In the preCambrian, no land plants existed, so a horse could not have existed.

Increasing complexity may be described as progressive, even though no goal is involved. Such progressive change does not, however, imply that evolution is *designed* to produce increasingly complex organisms. Nor does evolution *necessarily* produce increasing complexity. After all, we are surrounded by relatively simple organisms—including bacteria, amebas, ferns, and flatworms—survivors, like us, of the ravages of major catastrophes, like those that wiped out the dinosaurs, and the constant, local, minor ones that constitute natural selection. All these relatively simple organisms are descendants, as we all are, of a simple ancestor that lived more than 3.5 billion years ago. All organisms, both ancient and modern, are the product of the same evolutionary processes that have continued for billions of years.

WHAT IS AN ANIMAL?

As animals ourselves, we tend to see our own kingdom as the most interesting. Many of us have spent hours at zoos, observing the beautiful and the bizarre. Even the best zoos, however, barely hint at the variety of animals that exist on Earth. A coral reef is a much better place to observe the tremendous diversity of animals. A coral reef swarms with fishes, shelled animals, worms, sponges, corals, sea anemones, and many other unfamiliar organisms. Most land dwellers are awed by the variety of shapes, domes, branches, fans, tubes, and stars, as well as by the range of colors, from dark red to luminous blue.

What Features Characterize the Animals?

An animal is a multicellular, heterotrophic organism that develops from an embryo. All animals are eukaryotes, and most reproduce sexually. A few, such as the corals, can reproduce by asexual processes, such as budding. Unlike plants, fungi, and protists, however, animals never show alternation of generations. With rare exceptions, the only haploid cells are the gametes (sperm and eggs).

Nearly all animals ingest their food—usually other organisms or their remains. Like fungi, a few parasitic animals absorb nutrients directly from their hosts. But, unlike fungi, these parasites develop from embryos, as other animals do. A few animals—including many corals—appear to perform photosynthesis. Careful examination reveals, however, that these animals harbor symbiotic algae that are actually the ones performing photosynthesis.

Like plants and fungi, animals are highly structured. Almost all animals contain many different kinds of specialized cells. Groups of similar cells, along with the materials they secrete, called **matrix,** form **tissues,** units of structure and function. Two or more kinds of tissue in turn can form an **organ,** a structural unit with a distinctive function. For example, one class of cells can form an **epithelium,** a tissue that lines a surface (such as the outside of the body), while another kind of cell, along with its matrix, can form **connective tissue,** a network of loosely connected cells that may help support other body parts. A kidney, which contains both epithelial and connective tissues, is an example of an organ (Figure 23-3). The specialization of cells and the grouping of cells into tissues and organs allow animals to function efficiently.

Unlike plants, most animals can move about in search of food or mates. Some animals, such as the corals, are stationary but can move body parts to capture prey. Many have stages during development when they disperse. Almost all animals coordinate their movements in patterns that we can describe as behavior. For example, animals capture food, avoid predators, and breed. For all but the sponges, the coordination of such behavior depends on the functioning of a network of nerve cells (neurons).

An animal is a multicellular, heterotrophic organism that develops from an embryo. All animals are eukaryotes, and most reproduce sexually.

How Do Zoologists Classify the Animals?

Zoologists (pronounced zo-OL-ogists), biologists who study animals, estimate that living animal species number at least 4 million, and some estimates run as high as 30 million. Zoologists divide these millions of living animals into about 35 phyla (Figure 23-4). The 40,000 species of vertebrates (animals with backbones)—including fish, frogs, snakes, birds, and mammals—are just a subphylum of one of these 35 phyla. Vertebrates include a mere 1 percent of all animal species. Most animals have no backbones. The vast majority are insects (mostly beetles, in fact), snails, jellyfish, and worms.

Zoologists divide all these creatures into two subkingdoms. The subkingdom **Eumetazoa** [Greek, *eu* = true + *meta* = middle + *zoa* = animals] includes almost all of the 4 million named species of living animals. Excluded are the 5000 species of sponges and one other species, which make up the small subkingdom **Parazoa.** The sponges are the simplest animals, for they have neither regular symmetry nor organs.

How do biologists use developmental patterns to classify animals?

The Eumetazoa are divided into major groups of phyla according to how their embryos develop. At fertilization, or syngamy, the egg and sperm fuse to form a **zygote.** An animal zygote divides by mitosis to produce a hollow ball of cells called a **blastula** (Figure 23-5). The blastula folds in on itself to form a **gastrula,** three cell layers surrounding a simple cavity called the **archenteron** [Greek, *arche* = beginning + *enteron* = gut] (Figure 23-6). The three cell layers of the gastrula develop into distinctly different tissues in the adult. The innermost layer, which surrounds the archenteron, is called the **endoderm** [Greek, *endon* = within + *derma* = skin] and gives rise to the intestines and other digestive organs. The outermost layer, called the **ectoderm** [Greek, *ecto* = outside], gives rise to skin, sense organs, and nervous system. The middle layer, the **mesoderm** [Greek, *mesos* = middle], gives rise to muscle, skeleton, and connective tissue. That all animals develop in the same

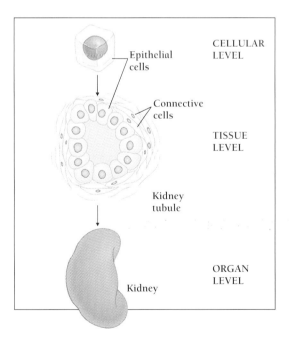

Figure 23-3 The structure of animals. A kidney is an organ, which is made up of two or more tissues. The two tissues in the kidney tubule, for example, are connective tissue and epithelial tissue, which are made of connective cells and epithelial cells, respectively.

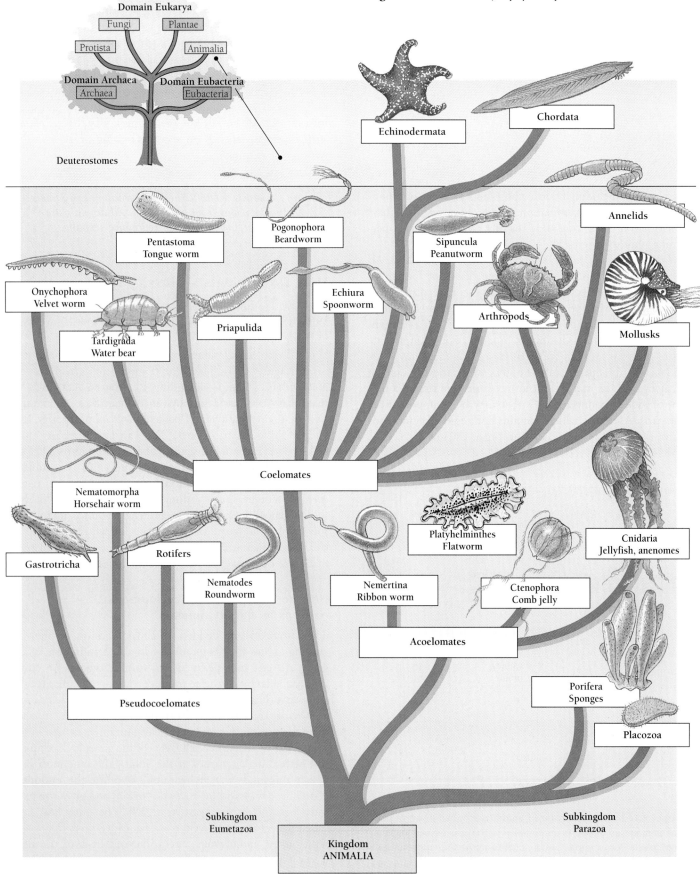

Figure 23-4 The major phyla of protostome animals.

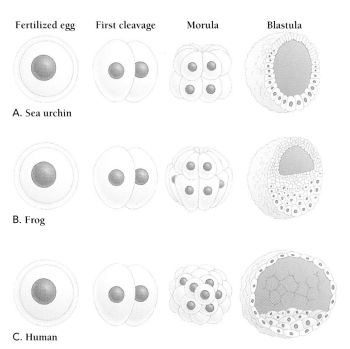

Figure 23-5 All eumetazoan embryos begin development in the same fashion. Whether sea urchin, frog, or human, all true animals begin life by dividing to form a hollow ball of cells called a blastula. Although eggs and their zygotes may differ in size from 1 mm for frogs to 0.1 mm for humans and sea urchins, all share a similar pattern of early development.

the internal organs hang. In the most complex animals (including ourselves), a layer of mesoderm cells lines this space, and the lined cavity is called a **coelom** [Greek, *koilos* = hollow]. Animals with a coelom are called **coelomates.** These include, for example, snails, earthworms, insects, and all of the vertebrates.

The simplest animals, including the jellyfish and flatworms, have no cavity and are called **acoelomates** [Greek, *a* = without]. In between are the **pseudocoelomates** [Greek, *pseudo* = false], which have an organ-containing cavity, but without the mesodermal lining of a true coelom (Figure 23-7). A pseudocoel lies between the endoderm and the mesoderm and derives directly from the cavity in the blastula (Figure 23-5). Almost all the pseudocoelomates are tiny worms called nematodes.

The coelomates are further divided into two major groups according to another peculiarity of the early gastrula (Figure 23-6). At one end of the archenteron is an opening called the **blastopore.** In some animals, the blastopore eventually develops into the adult mouth. In other groups of animals, it develops into the anus, and the mouth develops later. Those in which the mouth develops first, from the blastopore, are called the **protostomes** [Greek, *protos* = first + *stoma* = mouth]. These include, for example, snails, earthworms, and all the arthropods—the spiders, insects, and their evolutionary relatives. In contrast, those in which the anus forms first, from the blastopore, and the mouth forms secondarily are called **deuterostomes** [Greek, *deuteros* = second + *stoma* = mouth]. Most coelomates are protostomes. We vertebrates and our cousins the echinoderms (sea stars and sea urchins, for example) are deuterostomes.

general way provides further evidence that we share common ancestors.

Even in the most complex animals, a gastrula has the same organization, or "body plan," as the adult animal. We may view this body plan as a tube within a tube. The inner tube is the digestive tract, with a mouth at one end and an anus at the other. The outer tube is the body wall. Between the two tubes (of most animals) is a fluid-filled space, within which

---

Zoologists divide the Eumetazoa according to the presence or absence of a coelom or pseudocoel and according to whether the blastopore develops into a mouth (protostomes) or an anus (deuterostomes).

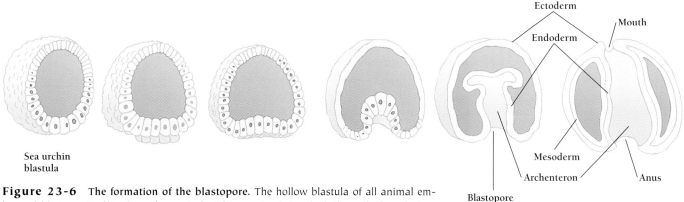

Figure 23-6 The formation of the blastopore. The hollow blastula of all animal embryos folds in on itself to form the gastrula. The opening into which the cells flow during this process is called the blastopore. In some animals (protostomes), the blastopore becomes the mouth; in others (deuterostomes) it becomes the anus. In both kinds of animals, the hollow center (the archenteron) becomes "the tube within a tube," which later forms the gut.

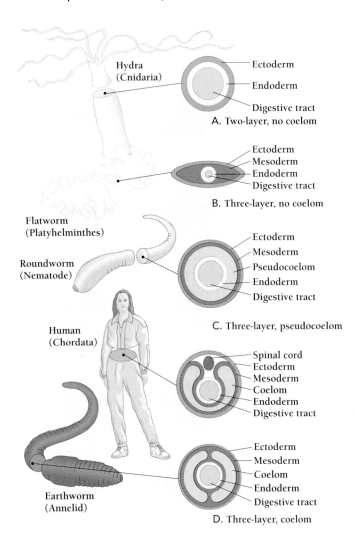

Figure 23-7 Tissue layers. A. A cnidarian has just two tissue layers and no coelom. B. The flatworms have three layers, but still no coelom. C. The roundworms have three layers and a pseudocoel. D. The true coelomates, including both earthworms and humans, have three layers and a true coelom, in which the gut is suspended.

How do biologists use symmetry to classify animals?

Biologists also classify animals according to their symmetry. The Parazoa (sponges) have no symmetry at all. Their lack of symmetry is one of the characteristics that set them apart from the rest of the animal kingdom.

The Eumetazoa have one of two kinds of symmetry. Animals with **radial symmetry**—the simple jellyfish and sea anemones, for example—can be rotated along their central axis without changing their appearance (Figure 23-8A). These animals have a top and bottom, but no front and back and no left and right. Many of these animals live most of their lives either floating passively in the water (jellyfish) or attached to a rock or some other support (sea anemones).

Most animals are **bilaterally symmetrical,** meaning that their left and right halves are (approximately) mirror images of one another (Figure 23-8B). Such animals have a defined

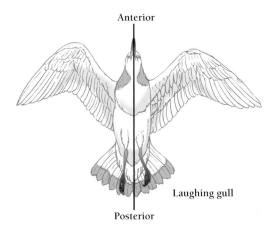

Figure 23-8 Symmetry. A. Radial symmetry in a jellyfish. B. Bilateral symmetry in a laughing gull and a snake. All animals with bilateral symmetry have anterior and posterior ends and dorsal and ventral surfaces. Shown here are the ventral (under) surface of a jellyfish and a gull and the dorsal surface of a snake.

anterior (front) and **posterior** (back end). The *dorsal* surface is the back, or top, which, in most animals, faces the sky. The *ventral* surface is the part facing the Earth.

Bilateral symmetry implies some sort of head and, ordinarily, a preferred direction of movement—usually "head" first. Directed movement, in turn, demands a highly integrated system of muscles, nerves, and sense organs. For these reasons and others, bilaterally symmetrical animals are usually more complex than the radially symmetrical animals.

In general, the sense organs that detect prey and predators, as well as the neural networks that direct and coordinate movement, lie primarily near the front, the anterior end, of the animal (Figure 23-8B). Reproduction, digestion, and excretion tend to lie in the rear, the posterior end.

The Eumetazoa are further classified according to whether they have bilateral or radial symmetry. Bilaterally symmetrical animals, such as ourselves, are usually more complex than radially symmetrical animals.

In this chapter we will begin with the simplest animals, the sponges; move to the acoelomates, first those with radial symmetry, then those with bilateral symmetry; and then move to the pseudocoelomates. We will finish with an examination of the protostome coelomates. In the next chapter, we will survey the deuterostome coelomates.

THE PARAZOA HAVE NO ORGANS

The Parazoa contain two phyla, the Placozoa and the Porifera (sponges). The Parazoa differ greatly from all other animals, in showing no symmetry and minimal organization and cell specialization. The Parazoa have only simple connective tissues and no organs.

The Placozoa consist of a single species, *Trichoplax adhaerens,* which looks like a large, multicellular ameba. Unlike any protist, however, *T. adhaerens* produces gametes, develops as an embryo, and shows cell specialization. So it is a true animal.

The 5000 species of sponges (most of which are marine) come in many sizes, shapes, and colors, and live at all depths of the ocean. Almost all sponges are **sessile,** permanently anchored to rocks, logs, or coral. They range in size from a few millimeters to two meters. Some are shapeless blobs, while others resemble fans, cups, crusts, and tubes (Figure 23-9A). Like the synthetic "sponges" we use at the kitchen sink, the members of Porifera [Latin, *porifera* = hole-bearer] are full of holes.

A sponge's cells lie in three layers surrounding a central cavity, like the gastrula of all other animals. A sponge's outer, epithelial layer of cells is perforated with tiny holes, through which water enters the sponge. Each sponge pumps an enormous volume of water through its simple body. A sponge the size of a fingertip may pump 20 liters of water a day. This water, loaded with nutritious microorganisms and organic matter, enters through the pores, flows into the cavity, and exits through the sponge's large opening (or openings). The driving force for this flow comes from a layer of flagellated cells lining the sponge's central cavity (Figure 23-9B).

Between the outer epidermal layer and the flagellated cells lining the cavity is a layer of **mesenchyme** [Greek, *mesos* = middle + *enchyma* = infusion], loosely attached cells embedded within a jellylike substance. Within this simple tissue, amebalike cells capture food and shuttle it from the inside cells to the outer epithelial cells. These same amebalike cells also give rise to either sperm or eggs. Most sponges are hermaphrodites: a single organism is both male [like the Greek god Hermes] and female [like the goddess Aphrodite].

Although sponges are multicellular, they are extremely primitive. No nerve cells coordinate a sponge's responses, and each cell functions as an independent unit. Nonetheless, single sponge cells, separated from others by a sponge being forced through a piece of cheesecloth, will spontaneously reorganize themselves into a functioning sponge.

Sponges, whose fossils date from the early Cambrian Period, 530 million years ago, almost certainly evolved from flagellated protistan ancestors and diverged from all other animals early on.

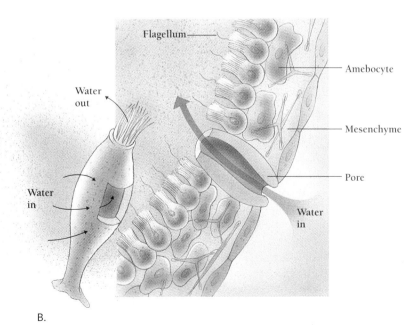

A.

B.

Figure 23-9 Sponges. A. An azure sponge from the Caribbean. B. Flagellated cells inside the sponge pump water through the pores of the sponge. Other specialized cells digest microorganisms and organic matter filtered from the flowing water. *(A, Brian Parker/Tom Stack & Associates)*

The individual cells of parazoans resemble protists. But unlike protists, parazoans develop from embryos and produce sperm and eggs.

RADIALLY SYMMETRICAL ACOELOMATES

The 9000 species of radially symmetrical acoelomates fall into two phyla, Cnidaria (pronounced ni-DAR-ee-uh) and Ctenophora (pronounced ten-OFF-or-uh). The cnidarians, which are simpler and far more numerous, are among the oldest animals in the fossil record, present well before the Cambrian explosion.

Cnidarians

Zoologists recognize four classes of **cnidarians:** Hydrozoa (hydroids), Scyphozoa (jellyfish), Anthozoa (corals and sea anemones), and the jellyfishlike Cubozoans. Almost all are marine (with the exception of *Hydra* and a few other freshwater hydrozoans). By far the largest class is Anthozoa, with about 6200 species. Hydrozoa include about 2700 species, and Scyphozoa only about 200.

Anthozoans [Greek, *anthos* = flower] take their name from their flowerlike appearance. They include sea anemones, sea pansies, sea fans, and sea whips. Each anthozoan has a cylindrical body with a crown of tentacles. Symbiotic algae enhance their plantlike appearance and also contribute to their nutrition, often allowing them to grow in water that is poor in nutrients. Especially impressive are the corals, whose secreted skeletons of calcium carbonate are the foundation of gigantic coral reefs.

The anthozoans have the most complex behaviors of the cnidarians. Some species of sea anemones, for example, feed on crabs, mussels, or even fish. A sea anemone can catch and consume prey with its tentacles. Some sea anemones even attack the tentacles of other sea anemones that intrude too closely.

All cnidarians contain two layers of cells that function as true tissues. But cnidarians have no organs, making them the least complex of the Eumetazoa. Cnidarians come in two basic body plans—**polyps,** which resemble cylinders, and **medusae,** which resemble bells (Figure 23-10). Both forms have a single mouthlike opening to a central cavity, and both use tentacles to capture food. Polyps are usually partly sessile, attached to rocks or other surfaces, with mouths and tentacles pointed upward. Medusae generally float free with the mouth pointed down and tentacles dangling, like the fringe on the edge of an umbrella. Some cnidarians are polyps throughout their life cycles, some are only medusae, and some cycle from one form to the other (Figure 23-11). Most cnidarians start out life as free-swimming larvae and become sessile as adults.

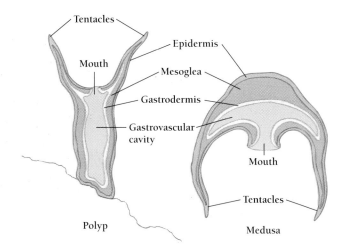

Figure 23-10 The two body plans of the cnidarians. Jellyfish and other cnidarians have just two body plans: the polyp and the medusa. The polyp is the sessile form exemplified by sea anemones and hydras. The medusa, exemplified by the jellyfish, is basically a polyp turned over.

Both polyps and medusae have an internal digestive cavity, which is surrounded by an outer layer of epidermis and an inner layer of gastrodermis. Like higher animals, cnidarians perform most of their digestion extracellularly—that is, within the digestive cavity. Digestion in the cavity, however, is not complete, and the cells lining the cavity engulf fragments of the small animals that compose the cnidarian's diet.

Between the two cellular layers is the **mesoglea** [Greek, *mesos* = middle + *glia* = glue], a jellylike material that contains few if any cells. In some medusae, such as the jellyfish, the mesoglea may be quite thick and gelatinous. In polyps, the mesoglea is usually thin, and in some cases (such as in the sea anemones) consists entirely of cells.

Cnidarians [Greek, *cnide* = nettle] take their name from specialized stinging cells, called **cnidocytes,** that lie on their tentacles (Figure 23-12). Each cnidocyte can use water pressure, at some 140 times the pressure of the atmosphere, to fire a tiny barbed spear called a **nematocyst.** As whalers use barbed harpoons threaded with rope to impale whales and drag the animals back to the whaling boat, cnidarians use nematocysts and their attached threads to capture their prey.

Cnidarians work much more quickly than whalers, however. Only a few milliseconds elapse between the detection of the prey by the tickling of the cnidocyte's flagellumlike cnidocil and the firing of the nematocyst. After it has pierced its prey, the nematocyst can discharge a poisonous protein. The nematocysts of one cnidarian—the Portuguese man-of-war—produces a neurotoxin potent enough to kill a human swimmer.

Once the nematocysts have attached their tiny tethers, the tentacles draw the prey mouthward. Pulling and pushing the prey requires coordination among the tentacles, which is controlled by simple muscle cells and nerve cells. A cnidarian's nerve cells are unlike those in more complex organisms, for they transmit impulses in both directions, rather than just one.

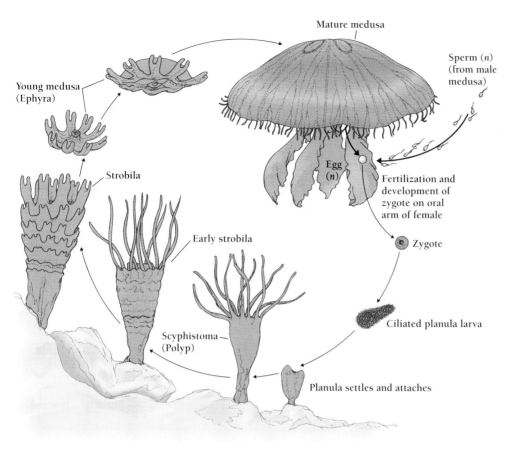

Figure 23-11 The life cycle of a hydrozoan. The class Hydrozoa includes both *Hydra*, a simple polyp often studied in biology classes, and the formidable Portuguese man-of-war (*Physalia*), capable of stinging to death an adult human. Most hydrozoans are jellyfish (also called jellies) that have both polyp and medusa stages. In the polyp stage, they are colonial, with many polyps budding around a branched central cavity. *Hydra* is an exceptional hydrozoan, for it lives as a solitary polyp, with no medusa stage. All hydrozoans shed sperm and eggs from testes and ovaries that project from the body wall.

Cnidarians are radially symmetrical acoelomates with true tissues but no organs. Named for their specialized stinging cells, the cnidarians include the hydras, jellyfish, corals, and sea anemones.

Ctenophores (Comb Jellies)

The **ctenophores** [Greek, *ktenos* = comb + *phora* = motion], or comb jellies, are a small phylum of about 90 living species. Their name refers to the rows of comblike cilia that these beautiful animals carry in bands along their short bodies (Figure 23-13). Ctenophores seem to waltz through the water. By means of the coordinated action of their cilia, they move forward, mouth first, capturing prey with their sticky tentacles, while slowly rotating. They are hermaphrodites (simultaneously male and female) and usually shed both sperm and eggs into the open sea.

Because the combs are arranged symmetrically and because one species of ctenophore has cnidocytes, biologists once classified the ctenophores as cnidarians. But ctenophores' paired tentacles lack the radial symmetry of true cnidarians, and modern zoologists have given the comb jellies their own phylum. The fossil record shows that the ctenophores have existed separate from the cnidarians for at least 400 million years.

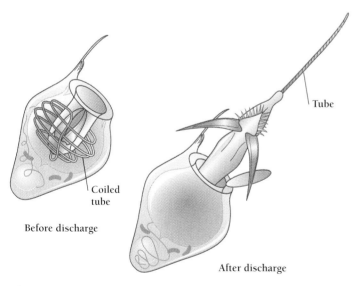

Figure 23-12 Cnidarians are named for stinging cells, or cnidocytes. Coiled inside each cnidocyte is a tiny barbed harpoon, covered by a lid. When the cnidocyte fires, the lid opens and the harpoon ejects, turning inside out as it unfurls. If the barbs strike the flesh of another animal, they catch and inject a paralyzing toxin. When the victim stops moving, the cnidarian pulls its prey to its mouth and begins to digest it.

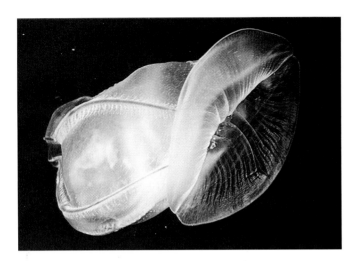

Figure 23-13 A ctenophore, or comb jelly. These graceful animals are noted for their beauty. *(James R. McCullagh/Visuals Unlimited)*

Unlike the cnidarians, the ctenophores (comb jellies) are not radially symmetrical, for they possess paired tentacles. In addition, ctenophores move through the water by means of rows of cilia (the "combs").

BILATERALLY SYMMETRICAL ACOELOMATES

The cnidarians and ctenophores are radially symmetrical, or nearly so, but acoelomates may also have bilateral symmetry, as we do. Two phyla, Platyhelminthes (the flatworms) and Nemertea (the ribbon worms), are especially interesting. Flatworms include a host of parasites, which infect hundreds of millions of people, as well as the planarian *Dugesia,* used in biology laboratories all over the world. The flatworms differ from more advanced animals in having only one opening to their digestive cavities. Mouth and anus are the same.

The ribbon worms, in contrast, are the simplest animals with two openings to their digestive tract. Unlike all other acoelomates, ribbon worms have a separate mouth and anus.

Platyhelminthes (Flatworms)

The 12,000 species of **Platyhelminthes** (flatworms) live in a wide range of environments. Most free-living species are marine, although some (such as *Planaria*) thrive in fresh water, and some are terrestrial. One species is specialized for living in bat guano. Zoologists divide the flatworms into three classes: Turbellaria (free-living flatworms), Trematoda (flukes), and Cestoda (tapeworms). All three classes share the same acoelomate body plan.

All flatworms have three body layers—an endoderm that lines the digestive cavity, an ectoderm on the outside, and a mesoderm in between. They have feeding, digestive, and reproductive organs, a brain, and a simple system for regulating water balance.

One thing flatworms do not have is a circulatory system. Some flatworms nonetheless grow to 60 cm, and tapeworms may be as long as 10 m (over 30 ft). How do such large animals distribute oxygen and nutrients to their cells? Flatworms survive without a circulatory system by keeping diffusion distances small. Their flattened shapes allow gases to diffuse rapidly to and from the outside environment, and the gut cavity of each flatworm is branched and extends throughout the body, so that no cell is far from the gut. The single opening to the flatworm's gut cavity serves as both mouth and anus. Likewise, the gut cavity itself serves both as a reservoir of nutrients and for wastes.

Turbellaria, the free-living flatworms, can reproduce either asexually or sexually. A free-living flatworm's simple nervous system allows it to move efficiently toward food and away from light. Some flatworms detect light with primitive paired eyespots on their heads. They can also detect odor. A good way to collect flatworms is to put a piece of meat in the bottom of a muddy pond. Chemicals diffusing from the meat stimulate receptors on the flatworm's head. When a flatworm smells the meat, it turns its body and moves toward its future meal.

The differences among the three classes of flatworms relate to differences in the way they eat and reproduce. The flukes and tapeworms are parasites. Their outer layer of cells resists the digestive enzymes of their hosts, and they lack eyespots and chemical receptors for detecting food. Flukes take in food through a specialized mouth, while tapeworms lack a digestive system altogether and simply absorb digested food from their hosts.

Parasites are specialized for rapid and prolific reproduction in their hosts, and they often evolve stripped-down bodies, lacking many ancestral characteristics that are not useful for their specialized lives. Flukes and tapeworms illustrate many common adaptations of parasites: (1) rapid reproduction, (2) distinct stages that allow passage and dispersal through more than one host, (3) organs for attachment to their hosts, (4) specialized digestion (in the case of tapeworms, direct absorption), and (5) reduced sense organs.

As a result, the more complex free-living flatworms are probably more similar to the original flatworm ancestors than are the streamlined flukes and tapeworms.

Trematoda (flukes) range in size from less than a millimeter to more than 8 cm. All flukes (more than 8000 species) are parasites. Each attaches to the outside or the inside of its host by a hook or sucker. Like the free-living flatworms, flukes have a digestive system with a single opening.

Many flukes are hermaphrodites and generally reproduce by mutual copulation. They are far more prolific than free-living flatworms, producing 10,000 to 100,000 times as many eggs. Some flukes live their whole lives on a single host. Most, however, have more complicated lives, involving two or more hosts. The human liver fluke *Clonorchis sinensis,* for example, uses three distinct hosts to complete its life cycle—a mammal, a snail, and a fish (Figure 23-14).

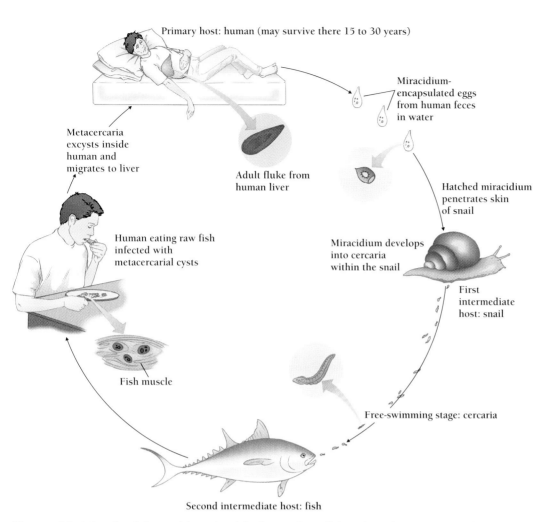

Figure 23-14 The elaborate life cycle of the human liver fluke, *Clonorchis sinensis*. The liver fluke lives its adult life as a 1- to 2-cm worm in the livers of humans, pigs, cats, and dogs. Each worm mates and reproduces, then deposits thousands of fertilized eggs. These develop into larvae, called miracidia, which leave the mammalian host in the feces. Development of the flukes then stops; the cycle continues only if the feces enter water and the larvae are eaten by a snail. Within the snail's digestive tract, each miracidium develops into a swimming larva, called a cercarium, which then reenters the water. The swimming larvae must burrow into the muscles of a fish within two days or they will die. In the fish's flesh, they form themselves into protected cysts, called metacercaria. Development resumes when a human (or other mammal) eats the raw infected fish.

Cestoda (tapeworms) are even more specialized for reproduction than are the flukes. An adult tapeworm has no digestive system at all. It lives in its host's gut, and its cells directly absorb passing nutrients.

A tapeworm maintains its easy life with a specialized attachment organ, called a **scolex,** at its anterior end. Behind the scolex lie a set of repeated segments, called **proglottids.** In the beef tapeworm, the proglottid chain may be as long as 10 meters (Figure 23-15). Each proglottid is a complete reproductive unit, with both male and female organs. If two or more tapeworms live together, they fertilize one another. Otherwise, a tapeworm self-fertilizes. In either case, the proglottids toward the rear end of the tapeworm are filled with embryos, each within a separate shell, which pass, with the feces, to the outside world. If cattle eat vegetation carrying these embryos, they acquire tapeworms. In the United States, about 1 percent of all cattle have tapeworms, an important reason for eating only inspected beef.

Platyhelminthes (flatworms, flukes, and tapeworms) are the simplest animals with heads. They have three body layers, true organs and tissues, a brain, and an organ for regulating water balance. They do not have a one-way digestive system: they take in food and excrete wastes through the same structures. The parasitic members of this phylum possess fewer specializations than the free-living flatworms.

Figure 23-15 Most tapeworms have two or more hosts. A. The broad fish tapeworm, *Diphyllobothrium latum*, infects tiny crustaceans called copepods, copepod-eating fish, and fish-eating mammals, including humans and bears. The broad fish tapeworm occurs worldwide, can reach 20 meters in length, and can shed a million eggs a day. A large tapeworm can live for its host's entire life. B. At the front end of a tapeworm is the scolex, which attaches to the inside of the host by means of hooks (top) and suckers (two holes). (A, Robert Calentine/Visuals Unlimited; B, Oliver Meckes/Photo Researchers, Inc.)

Nemertea (Ribbon Worms)

Ribbon worms, or **Nemerteans,** are a group of about 750 species of free-living acoelomate animals, most of which live inconspicuously in the sea. A few species live in fresh water or on land in the tropics. Their flat, velvety bodies may be less than a millimeter long or as long as 30 meters (Figure 23-16). Each has a characteristic snout, or **proboscis**—a long, sensitive, retractable, and sometimes venomous, tube. A ribbon worm uses its proboscis to explore its environment, defend itself, and capture prey. Ribbon worms are abundant in tidal mudflats, where they spend their days hidden in the mud, emerging only at night to feed.

Figure 23-16 Some ribbon worms grow up to 30 m (100 ft) long. All have a two-ended digestive tract and circulatory system. (Raymond Mendez/Animals Animals)

Ribbon worms are closely related to free-living flatworms, with similar nervous and reproductive systems (though each ribbon worm is usually either male or female, rather than hermaphroditic). Two important advances distinguish ribbon worms from flatworms: (1) a digestive tract with both a mouth and an anus and (2) a circulatory system. The two-ended gut allows continuous eating, uninterrupted by the need to use the same opening for excretion. The second opening also permits more efficient extraction of nutrients from food, as it progresses in a single direction. The gut is like an assembly line, performing different operations at each stage. The anterior end breaks down the food, the posterior processes wastes.

The ribbon worm's circulatory system is relatively simple; in fact, some species have only two vessels. The blood is usually colorless, but may also be yellow, red, orange, or green. There is no heart to pump the blood in a particular direction: instead the vessels and the body wall contract irregularly, sloshing the blood through the circulatory system. These primitive adaptations make ribbon worms the most complex of the acoelomates.

Ribbon worms (Nemertea) have what flatworms have (nervous system and organs, for example). In addition, ribbon worms have a two-ended digestive system, a circulatory system, and a proboscis.

BOX 23.1

The Origin of Feces

Did the evolution of the one-way digestive tract fuel the Cambrian explosion? In a controversial new theory, geochemists have suggested that the development of one-way digestive systems by metazoans provoked drastic changes in the deep ocean, changes that allowed metazoans to colonize the bottom of the sea in an unparalleled adaptive radiation.

Until the Cambrian explosion, some 540 million years ago (mya), life consisted almost exclusively of single-celled bacteria and algae. As photosynthetic algae proliferated, they pumped increasing amounts of oxygen into the atmosphere. To see how this change in the composition of the atmosphere affected ancient life, a group of geochemists led by Graham Logan measured the chemical signature of fossilized organic matter in ancient ocean deposits.

To their surprise, the researchers found that at the beginning of the Cambrian, a dramatic change in the proportion of different carbon isotopes occurred. Before 590 mya, the sea floor deposits were relatively high in ^{13}C, a heavy isotope of carbon. Sometime between 590 and 540 mya, however, the deposits became lower in ^{13}C and higher in ^{12}C. This switch in "isotope signature" coincided closely with the flowering of multicellular life known as the Cambrian explosion. The researchers could not help asking, Were the two events related?

It seemed possible. Organisms tend to have higher ratios of $^{12}C:^{13}C^{12}$ than the atmosphere. During photosynthesis, plants incorporate into sugars and other new compounds a lighter proportion of ^{12}C than ^{13}C. (This is because ^{12}C is lighter than ^{13}C and diffuses more quickly.) The result is that ^{12}C accumulates in photosynthetic organisms and all the organisms that eat them. Organic matter therefore has more ^{12}C than the atmosphere.

But nearly all organisms respire—a process that breaks down organic compounds and releases carbon dioxide. Respiration puts the ^{12}C back into the atmosphere. If photosynthesis and respiration are balanced, the ratio of ^{12}C to ^{13}C is also balanced. In that case, researchers would not see an excess accumulation of ^{12}C in ocean floor sediments.

But Logan and his colleagues had an idea that would account for the change in ratios. Before 590 mya, the researchers hypothesized, bacteria in the upper layers of the ocean decomposed nearly all organic waste, releasing organic ^{12}C back into the atmosphere. Early deep-sea sediments, then, were relatively high in ^{13}C.

Huge numbers of bacteria would have filled the upper layers of the ocean, and their respiration would have depleted ocean waters of oxygen. The surface would have remained oxygenated, but the water near the bottom would have contained very little oxygen. Few organisms could have survived there.

But what if the bacteria were suddenly prevented from decomposing organic matter? Logan and his colleagues suggested that the evolution of a one-way digestive tract made all the difference. Animals with a one-way gut can package wastes in concentrated packets (feces), which fall to the ocean floor before surface-dwelling decomposers can absorb and digest them. With less decomposition occurring in the upper layers of the ocean, the overall oxygen content of the water would have risen. Ultimately, argue the researchers, oxygen would have diffused into the deeper waters of the ocean—allowing colonization by oxygen-requiring metazoans. That, the researchers say, paved the way for the Cambrian explosion.

Did the first fecal pellets fuel the Cambrian explosion? Many researchers remain skeptical of this tentative train of hypotheses. But the idea that the origin of all the major animal phyla depended on the evolution of feces has a certain appeal.

ASCHELMINTHS

Zoologists sometimes jokingly call the **Aschelminths** [Greek, *askos* = bag + *helminthos* = worm] the ash-can phylum, since it includes an assortment of unrelated animals that do not fit anywhere else. Like the kingdom Protista, the Aschelminths are united more by what they do not have than by what they do. Because the members of this group are so different, the Aschelminths are not considered a single phylum, but an assemblage of eight or nine distinct phyla.

All of the bilateral animals we have seen so far have been acoelomates, which have solid bodies with no internal cavity (other than the gut). The Aschelminths also lack a coelom. Some Aschelminths, however, have a cavity called a **pseudocoel** [Greek, *pseudo* = false + *koilos* = hollow]—"false" because it lacks the mesodermal lining of a true coelom. The pseudocoel plays two important roles: (1) it is a distribution route for nutrients, gases, and wastes; and (2) it serves as a hydrostatic skeleton, a support for the body that depends on the rigidity of enclosed fluid under pressure.

Aschelminths share a few traits. Most are tiny, less than a few millimeters long, and covered by a cuticle. The animals all have sticky glands for attaching themselves to some surface; the body cavity, if any, is a pseudocoel; and the sexes are usually separate. Aschelminths also lack true circulatory systems. Finally, Aschelminths have a one-way digestive system, with both a mouth and an anus.

The Aschelminths are a diverse group, and most of the phyla have relatively few species each. However, one phylum, the nematode worms, dominates all the rest. Zoologists have so far named 12,000 nematodes, and many researchers suspect that a million more remain to be identified.

Nematodes (Roundworms)

Nematodes—also called roundworms—are slender, cylindrical, and unsegmented. They range in length from a few millimeters to nearly a meter. Most are free living, however, and quite small. Free-living nematodes in the soil are actually aquatic animals, living in the film of water that surrounds each soil particle. They help to aerate the soil and circulate minerals and organic matter. Nematodes are among the most numerous of animals. A rotting apple in an orchard may contain 100,000 nematodes. An acre of soil contains billions of the little worms.

Many nematodes are parasites. They infect virtually all plants and animals. In fact, it has been said that if all organisms except the nematodes suddenly disappeared, a ghostly image of the world's former occupants would remain in the form of billions upon billions of nematode worms.

A nematode has a large pseudocoel between the body wall muscles and the central tubular gut. Because nematodes lack a circulatory system, fluids move about the body within the pseudocoel. These fluids push against the walls of the pseudocoel and thereby stiffen the whole body in the same way that water stiffens a hose. The nematode's muscles work against this support in the same way that the muscles of a vertebrate pull against a bony skeleton. The hydrostatic skeleton allows the worm's body to return to its original shape after muscle contraction has stopped.

The nervous system of a nematode is more complex than that of an acoelomate. A free-living nematode has sense organs that detect movement and chemicals, nerve cords that run the length of its body, and a primitive brain made up of a ring of nerves just behind its mouth.

Nematodes reproduce sexually, with male and female usually separate. In most species, fertilization is internal and therefore requires copulation. Some species are hermaphroditic, and some can even fertilize themselves. Still others are parthenogenic, meaning that females reproduce without males. The unfertilized eggs develop into normal embryos and adults, just as a diploid zygote would. Nematodes are prolific breeders: a single female can produce more than 25 million eggs in her lifetime.

Nematodes (roundworms) are tiny, cylindrical worms that live everywhere in great numbers. They have a fluid-filled pseudocoel that acts as a hydrostatic skeleton and circulatory system, and they have sense organs and a primitive brain.

The other seven phyla of Aschelminths include the *rotifers* (Figure 23-17), tiny aquatic animals with saclike bodies; the *nematomorpha*, or horsehair worms; and the *gastrotricha*, microscopic, bottle-shaped animals that live in freshwater or marine environments.

Despite the great advantages of the pseudocoel, many more possibilities for animal evolution came with the evolu-

Figure 23-17 Brachionus female with eggs. The 2000 species of rotifers are among the smallest of all animals, ranging in size from 40 micrometers (μm) to 2 mm. These common, freshwater animals are often shaped like tiny trumpets, with a crown of cilia around the open end. The cilia direct water containing bacteria, protists, and small animals into the mouth. The whirling cilia make the crown look like a rotating wheel, giving the phylum its name [Latin, *rota* = wheel + *ferre* = to bear]. Rotifers also have tiny but effective jaws, which grind food after it enters the mouth. The rotifers' success depends on the rapidity of their reproduction: some species can double their population every 2 days. And females commonly reproduce without males, through parthenogenesis. Rotifers are a major source of food for other animals that live in fresh water. *(M.I. Walker/Photo Researchers, Inc.)*

tion of the coelom. In the next section, we will see that the coelom allows animals of much greater size, complexity, and mobility.

PROTOSTOME COELOMATES

All of the more complex animals have internal body cavities outside their intestinal tracts. An additional internal space (whether coelom or pseudocoel) brings several advantages. For example: (1) it prevents muscle movement from interfering with the operation of the digestive and circulatory systems, (2) it can provide a hydrostatic skeleton against which muscles can work, and (3) it provides a protected space for the production of sperm and ova.

An internal cavity allows organs to develop and to work far more freely than in acoelomate animals. Your lungs, for example, could not fill and empty so easily if each breath meant struggling against surrounding muscles instead of just filling an empty space within the chest cavity. Similarly, the coelom isolates the working of the intestinal tract from the movement of the surrounding muscles.

The major difference between a true coelom and a pseudocoel lies in the tissues that surround the digestive tract and the internal cavity. A pseudocoel arises from the space between endoderm and mesoderm, so that the gut is lined with a single layer of tissue—endoderm only. In contrast, a true coelom arises within the mesoderm, so that the gut is lined with endoderm surrounded by mesoderm. This allows the gut to develop more complex digestive organs (such as the vertebrate pancreas or liver). The coelom within the mesoderm also allows the mesoderm itself to develop more efficient organs for circulation and excretion.

How Do Protostomes Differ from Deuterostomes?

Nearly all coelomates have digestive tracts with both a mouth and an anus. Recall that coelomates are divided into two groups according to whether the mouth or the anus develops from the blastopore. Those in which the mouth develops first, from the blastopore of the early embryo, are called protostomes; those in which the blastopore becomes the anus and the mouth forms later are called deuterostomes.

The two groups of coelomates develop in other distinctive ways. In protostomes, for example, the cell divisions immediately after fertilization are usually spiral, while the deuterostomes undergo any of several other patterns of cell divisions. In addition, deuterostome coelomates often form the coelom differently from protostomes. None of the other developmental differences, however, are as clear cut as the developmental fate of the blastopore.

In the rest of this chapter, we briefly survey three large invertebrate phyla of protostomes—the mollusks (which include snails, slugs, clams, oysters, squids, and octopods), the annelids (which include earthworms, segmented marine worms, and leeches), and the arthropods (which include horseshoe crabs, spiders, lobsters, and insects). The arthropods, most of which are insects—some three quarters of a million species—include more named species than any other phylum of organisms.

The annelids, mollusks, and arthropods constitute a group of relatively closely related organisms. The larvae of mollusks, for example, are similar to those of annelid worms. In addition, the segmented parts of earthworms and other annelids recall the repeated body parts found in certain mollusks. The arthropods also resemble annelid worms in certain respects. Like the annelids, arthropods are segmented—think of insects (especially caterpillars), centipedes, and millipedes. Arthropod nervous systems and circulatory systems also resemble those of annelids. Molecular techniques have contributed much to current understanding of how these three phyla are related.

In protostomes, the blastopore develops into a mouth. In deuterostomes, the blastopore develops into an anus.

Mollusks Share a Common Body Plan

If we take the diversity of species as a measure of a phylum's evolutionary success, then, among the animals, mollusks are second only to the arthropods. Over 100,000 species of mollusks are now alive, in virtually every environment where animal life is possible. Mollusks are abundant in the seas, in fresh water, and even on land. The land species, slugs and snails, alone outnumber all species of reptiles, birds, and mammals combined. Mollusks may be as small as a grain of sand (about 1 mm long) or as long as a whale. (The 20-m giant squid is no match for a big whale, however, as it is not nearly as heavy as a whale.) Most mollusks, whether snails or clams, live in protective shells. Indeed, one of mollusks' distinguishing characteristics is their shell. The name **mollusk** [Latin, *molluscus* = soft] refers to the soft body of this invertebrate.

Mollusks are important to humans: oysters, scallops, squid, and other mollusks are among the great delicacies of the culinary world. Pearls and "mother of pearl" (from shells) have long been used to decorate people and their possessions, and the secretion of the marine snail *Murex* long served as the source of the dye royal purple, coveted by kings. Mollusks are also pests: some destroy crops and flowers, others damage docks and boats, and still others are intermediate hosts for such devastating parasites as schistosomes and other flukes.

All mollusks share a common body plan (Figure 23-18A). The intestinal tract, as well as the excretory and reproductive organs, lie within the main body of the mollusk, which is called the **visceral mass** [Latin, *viscera* = internal organs]. Attached to the visceral mass is the **foot,** a muscular extension of the body that the mollusk uses for sensing, grabbing, creeping, digging, and just holding on. The dorsal surface (top) of the visceral mass forms a **mantle,** specialized tissue that, in most mollusks, secretes a shell. Folds of the mantle also enclose a space, called the **mantle cavity,** which contains the mollusk's breathing organs. Both the mouth and the anus usually open into the mantle cavity. Snails and other shelled mollusks can withdraw into their mantle cavities, in some cases pulling behind them a protective flap called the **operculum**, which resembles a trapdoor.

In octopods, squid, sea slugs, and other aquatic mollusks, the mantle cavity contains gills—thin organs that transfer dissolved oxygen to the circulatory system. Air-breathing snails and slugs use the mantle cavity as a lung.

Most mollusks have a **radula,** a rasping tongue covered with teeth made from chitin. Mollusks may use the radula to protect themselves, to capture prey, to scrape vegetation, or to tear food into tiny bits. A snail in an aquarium, for example, uses its radula to scrape algae from the aquarium's glass wall.

All mollusks have a specialized digestive tract, beginning with a mouth and ending with an anus. All have one or two kidneylike **nephridia,** tubular excretory organs that remove nitrogen-containing wastes from the coelomic fluid and regulate water and salt concentration.

All mollusks also have a circulatory system with a heart that receives oxygen-carrying blood from the gills and pumps it to other body tissues. Unlike humans, most mollusks have open circulatory systems, in which the heart pumps blood through spaces between the tissues, rather than exclusively through blood vessels. The only exceptions are the

Figure 23-18 **The body of a mollusk.** A. The basic mollusk has a radula, a foot for locomotion, and gills in a mantle cavity. Cilia on the gills drive water through the mantle cavity, from which the gills can absorb oxygen. The passing water also carries away carbon dioxide and other wastes. Land snails and slugs have no gills; instead, oxygen diffuses directly from the mantle cavity into the surrounding circulatory system. B. Cephalopods such as octopods have specializations not shared by other mollusks, including a siphon, which can be pivoted to force a water stream in different directions, and suckered tentacles. Some cephalopod species are quite large. The largest known fossil cephalopods had straight conical shells up to 5 m (16 ft) long. C. Gastropods such as snails depart from other mollusks in having a twisted body and a distinct head with many sense organs, including touch receptors, chemical receptors, and eyes. D. Bivalves have the foot, mantle, shell, cavity, and coelom of other mollusks, but they also have siphons for directing a current of water through the mantle cavity.

cephalopods, which have a closed, well-developed circulatory system.

Mollusks reproduce sexually. Usually, male and female gonads are in separate individuals, but some (such as land snails) are hermaphroditic (both male and female). Individuals of some species, including sea slugs and oysters, change sexes from male to female and back several times during a mating season.

Cross-fertilization is the rule, even among the hermaphrodites. Self-fertilization, however, does occur—an important adaptation for animals that sometimes move too slowly to find a mate. Among the land snails and slugs, fertilization is internal. In other mollusks, gametes are shed externally.

Mollusks are soft-bodied coelomates, with sophisticated organs for respiration, circulation, digestion, and excretion. All the mollusks share a similar body plan, including a foot, a mantle cavity, and a visceral mass. Most have shells and a rasping tongue called a radula. They reproduce sexually, and their embryos develop into a characteristic larval stage.

Zoologists divide the mollusks into seven or eight classes, of which we will briefly discuss three: the bivalves (which include clams, oysters, and mussels), the gastropods (which include snails, slugs, and abalones), and the cephalopods (which include squid, octopods, and the chambered nautilus).

In addition, the mollusks include *aplacophora* [Greek, *a* = without + *plak* = flat plate], shell-less, wormlike animals that live in the deep seas; *monoplacophora* [Greek, *mono* = one], which have a single dorsal shell; the chitons, or *polyplacophora* [Greek, *poly* = many], which have eight plates on their backs; and the tusk or tooth shells, which have tubular shells open at both ends.

Bivalves

Bivalves [Latin, *bi* = two + *valva* = part of a folding door] include clams, oysters, scallops, mussels, and other mollusks with two shells—a right shell and a left shell. A ligament hinges the two shells (or valves) and a muscular "foot" protrudes from between them (Figure 23-18D). Two large muscles, called adductor muscles, pull the shells together. Although most bivalves lead sedentary lives attached to rocks, some move about a little more. A clam, for example, uses its foot to burrow into the sandy or muddy bottom of the sea or river. The most active bivalves are the scallops, which "swim" by using their adductor muscles to clap their shells together, simulating a human swimmer's "frog kick." A scallop's enormous adductor muscles, tender and delicately flavored, are familiar to seafood lovers (although some restaurants misleadingly give the name "scallops" to circular plugs of shark fin).

Bivalves such as clams, oysters, scallops, and mussels have two shells and usually lead sedentary lives.

Gastropods

The **gastropods** are the most diverse class of mollusks, with some 80,000 named species, including snails, whelks, conches, limpets, abalone, and slugs. They have a distinctive twisted body and, except for the shell-less slugs, asymmetrical spiral shells.

Gastropods differ from all other mollusks in having a mysterious 180-degree counterclockwise twist of the body (viewed from above) that occurs during embryonic development. One result is that both the anus and the mantle cavity end up at the front end (anterior end) of the body, not far from the mouth (Figure 23-18C). Torsion may occur very rapidly, sometimes over the course of just a few minutes. Biologists have many hypotheses about what advantages torsion might confer on gastropods, but they have not yet agreed on an answer.

The shells of gastropods show great variation in form and color. Most gastropods are marine, but many inhabit fresh water or land. Unlike the quiet bivalves, gastropods are active creatures, equipped with a well-developed head and a sensitive nervous system. Gastropods are free living and consume plants, small animals, or decaying organic matter. A few species live as parasites.

Gastropods such as slugs, snails, and abalone are active animals with twisted bodies, single shells (or no shell), and well-developed heads.

Cephalopods

The **cephalopods,** which include nautiluses, squids, cuttlefish, and octopods, are the most active and intelligent mollusks (Figure 23-18B). Although thousands of extinct cephalopods possessed well-developed shells, among the 600 living species only the *Nautilus* possesses an external shell (Figure 23-19). Surrounding the mouth are a set of tentacles that have evolved from the foot. Each species has a characteristic number of tentacles (or arms)—a squid has 10, an octopus 8 (hence its

Figure 23-19 The chambered nautilus. The nautilus shell consists of a series of chambers arranged in a spiral. The nautilus occupies only the outermost chamber, and as the animal grows, it builds a new, larger chamber and moves in. The nautilus uses the remaining chambers to float or sink at will. In the same way that a balloonist makes a balloon rise by adding more hot air to the balloon, the nautilus injects gas into its shell to float higher in the water. To sink, the animal siphons off the gas. Today, only a few species of nautilus exist, all living in the western Pacific Ocean. But relatives of the nautilus, called ammonites, flourished for hundreds of millions of years, from the Devonian Period to the end of the Cretaceous Period. *(Douglas Faulkner/Photo Researchers, Inc.)*

Figure 23-20 A male squid caresses his mate with his many arms. *(Randy Morse/Tom Stack & Associates)*

name), and a nautilus 80 or 90 tentacles. Cephalopods, which are all carnivorous, use their tentacles to capture prey, to pull themselves along the sea bottom, and to steer themselves in open water.

Considering the intellectual limitations of their relatives, the bivalves and the gastropods, cephalopods are amazingly intelligent. In a laboratory aquarium, an octopus (plural, octopods) can learn to distinguish objects of different shapes and even to find its way through a maze. Octopods are very good at escaping, and they think nothing of disassembling an aquarium to do so. Presented with stones or small bricks, an octopus can build a protected enclosure, hide behind it, and bob up and down to look around. An octopus also shows much planning and patience in stalking and luring its prey. Cephalopod mating rituals are charmingly complicated (Figure 23-20).

An octopus's interesting behavior is controlled by the most sophisticated brain of any invertebrate. Information about the outside world comes from touch receptors on the tentacles, as well as from large and complex eyes. When we consider that octopus eyes evolved independently of vertebrate eyes, it is remarkable how similar the eye of an octopus is to our own (Figure 23-21). This similarity is a result of convergent evolution.

Cephalopods respond to their environments not only with directed movements toward prey and away from predators, but also with a complex set of color changes. Most cephalopods have specialized skin cells, usually with three different pigments. The brain controls these cells in immediate response to the environment. An alarmed octopus, for example, may suddenly acquire dark stripes or spots, which presumably confuse potential predators. Yet another escape trick is to squirt a trailing cloud of dark ink. Some cephalopods, especially squids, can move amazingly fast by jet propulsion, squirting water from their mantle cavities.

Cephalopods are mollusks in which the foot has evolved into tentacles. They are predatory, intelligent, and active.

What Do All Annelids Have in Common?

Earthworms and other annelids are common animals, although not nearly as diverse as the mollusks. The 12,000 species of annelids live as both predators and scavengers in salt water, in fresh water, and on land. They range in size from less than 1 mm to over 3 m. They may be red, pink, green, brown, or purple; plain, striped, or spotted.

Annelids have long, segmented bodies consisting of identical (or nearly identical) sections called **metameres.** Segmentation (which also occurs in arthropods and vertebrates) allows animals to achieve larger sizes by repeating an already successful organizational plan. Injuries to individual metameres are less likely to be lethal, since other segments perform the same functions. In addition, the segments can move independently, which gives the animal more flexibility.

Like mollusks, annelids have specialized excretory organs called *nephridia.* Like us, annelids have a closed circulatory system, with blood contained entirely within vessels. Several vessels carry blood from one end of the body to the other, with many fine capillaries within each segment. Some of the larger vessels serve as hearts and pump the blood along its complicated circuits. Annelids have no gills or lungs, however. Gases move directly through the skin to the blood.

Zoologists divide annelids into three classes: (1) polychaetes are free-living bristle worms, most of which are marine; (2) oligochaetes, which include the earthworms and related forms, live in fresh water or salt water or on land; and (3) hirudinea, or leeches, are mainly freshwater parasites that suck blood.

Annelids have segmented bodies, consisting of identical or nearly identical sections, and closed circulatory systems.

Polychaete worms

The exotic and beautiful polychaete worms include many unusual and arresting species, including clamworms, plumed worms, scale worms, peacock worms, and sea mice (Figure 23-22). Most zoologists think that the earliest annelids were marine polychaetes. Each polychaete segment contains a pair of leglike paddles, called **parapodia,** which the worm uses to swim, crawl, or burrow, as well as to respire. On each parapodium is a bundle of bristles called **setae,** which give the polychaetes their name [Greek, *polukhaites* = many hairs, or bristle]. A polychaete worm's head is well developed and contains a variety of sense organs, including two to four pairs of eyes. Unlike the oligochaetes (earthworms), polychaetes have separate male and female sexes.

Polychaetes are primitive marine annelids. Each of a polychaete's segments has a pair of leglike parapodia with setae.

Oligochaetes

Oligochaetes include the common earthworms, the most familiar of the annelids. Each oligochaete segment, in contrast to that of the polychaetes, lacks parapodia altogether and contains just four pairs of bundled setae, leading to the class's name [Greek, *oligo* = few].

Oligochaetes are masterpieces of segmentation. An earthworm, for example, consists of 100 to 175 nearly identical segments, with a few altered segments at the front and the rear. The forward segments contain specialized structures for eating, coordination, circulation, and reproduction, and the most posterior segment contains the anus (Figure 23-23).

Oligochaetes are mostly hermaphrodites and, typically, two worms will mutually cross-fertilize, with their heads point-

Figure 23-21 **The cephalopod eye has long astounded biologists for its resemblance to the vertebrate eye.** Cephalopod eyes can be enormous. In 1933, a squid 21 m (67 feet) long washed up on a New Zealand beach; its eyes measured 40 cm (16 in) across! Scallops have simple eyes that cannot focus. However, they have many of these eyes, which are probably useful for detecting the shadows of starfish and other predators.

ing in opposite directions and their undersides touching. Each transfers sperm to specialized pores in its mate.

Oligochaete segments lack parapodia and have many fewer setae than polychaete segments.

Quiz

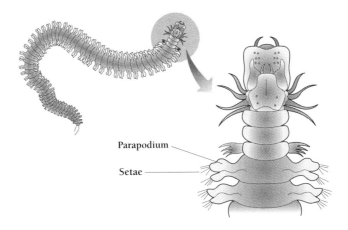

Figure 23-22 Tube worm. Many species of polychaetes—the tube worms—live in hidden places on the shallow sea bottom—under rocks, in burrows, or inside sponges, mollusks, or other animals. Some use their own secretions to build tubes where they live their entire lives. Polychaete feet consist of pairs of leglike paddles called parapodia and tufts of bristles called setae. *(Joyce Burek/Animals Animals)*

Hirudinea

Because most Hirudinea, or leeches, are parasites, they are more specialized than other annelids. Their bodies are flattened rather than cylindrical, and they lack segments and setae. Many species have a sucker at each end of their bodies. Such leeches "walk" like inch worms, attaching first one sucker to a surface and then the other. Like the oligochaetes, from which leeches are thought to have evolved, leeches are hermaphrodites that cross-fertilize.

Leeches range in size from about 1 cm to as much as 30 cm. Most are carnivorous and live in fresh water. Some species—including the medicinal leech, *Hirudo medicinalis*—live on the blood of humans and other mammals. Until the 19th century, leeches were used to treat such conditions as asthma, rheumatism, and drunkenness. Physicians (or barbers, who traditionally performed leech therapy as well as cutting hair) applied leeches, who sucked up to three pints of blood. When the patient fainted from blood loss, the physician re-

▶ **Figure 23-23 A single square meter of meadow soil may contain thousands of earthworms.** As an earthworm moves through soil, it swallows particles of soil. The soil passes through the digestive tract, where any organic matter is ground finely and absorbed. Undigested clay, sand, and other matter pass through the digestive tract and out the anus as small cylindrical "castings." A single worm eats its own weight in soil every day. Because earthworms are so common, they serve to keep soil mixed, which greatly benefits other soil animals and plants.

moved the leeches. Today, physicians use leeches, in a more restrained manner, to drain excess blood from surgical wounds.

A leech's impressive capacity to consume blood depends on powerful muscles of the pharynx (the anterior end of the digestive tract), on sharp teeth, and on a chemical that prevents blood clotting, called an anticoagulant. Physicians use leech anticoagulant to treat heart attack, stroke, and cancer.

Why Are Arthropods So Successful?

Arthropods, which include the insects, spiders, scorpions, shrimps, and many others, are the most successful phylum of animals on the planet. Arthropod species number at least a million, and zoologists estimate that between 2 and 30 million species remain to be described and named. The greatest number of species live in the tropics, but a student could probably collect hundreds (or even thousands) of species on any college campus.

Arthropods are more than just diverse, however. Zoologists estimate that the world today contains some 10^{18} (a billion billion) individual arthropods, hundreds of millions for every human being (a figure that might not startle a mosquito-bitten hiker). In a temperate climate, one square kilometer may contain 20 million arthropods. In the tropics, there are even more.

The enormous success of arthropods depends on both external and internal adaptations. All arthropods share two major external adaptations—an exoskeleton secreted by the cells of the epidermis and jointed limbs.

Jointed limbs, which give the arthropods their name [Greek, *arthron* = joint + *podus* = foot], allow them to move agilely, despite the rigidity of their exoskeletons. Like the annelids, arthropods have segmented bodies. Indeed, both phyla are probably descended from the same segmented ancestor. Some, like the centipedes and millipedes, have legs on nearly every segment. But many of the legs on an arthropod have become adapted for other functions. Over millions of years of evolution, legs have changed into mouthparts, antennae, gills, pincers, claws, and egg depositors.

These specialized appendages often develop from fused segments, called **tagmata** [singular, tagma]. For example, the anterior segments of all arthropods are fused into a highly organized head. Other tagmata include the middle portion of an arthropod, called the **thorax**, and the rear portion, called the **abdomen**.

The **exoskeleton**, a hard external supporting framework, consists of layers of chitin and protein. It helps to protect the animals from predators and parasites and provides mechanical support. The exoskeleton also slows water loss, a feature that has allowed the arthropods to flourish away from water and dominate animal life on land since the end of the Devonian Period, more than 350 million years ago.

An exoskeleton has drawbacks, however. Because a hard exoskeleton prevents an animal from growing, many arthropods (spiders, for example) periodically must shed their exoskeletons and secrete new, larger ones in a process called **molting**. The process of molting is dangerous. After an arthropod sheds its old armor and the new one is still hardening, the animal is highly vulnerable to predators. (Soft-shelled crab is a delicacy in many an East Coast restaurant.)

One solution is for an arthropod to do all its growing as a soft larva (as a caterpillar, for example) that lacks an exoskeleton. Once the animal has reached its adult size, it develops into the adult form in a process called **metamorphosis.** Some insects undergo gradual metamorphosis in stages that are not that different from one another (Figure 23-24). Others, such as flies, butterflies, beetles, and bees, undergo a complete transformation from a grub, caterpillar, or maggot to a flying insect. Metamorphosis and molting divide the life cycle of an insect into many distinct stages, each of which may have different food and lifestyle. In butterflies, for example, the caterpillar generally eats plants, while the adult butterflies drink nectar from flowers. In some arthropods, adults do not feed at all; their only role is to mate and die.

The internal anatomy of an arthropod is similar to that of an annelid. Both arthropods and annelids have a tubular gut that stretches from mouth to anus. The arthropod coelom is less prominent than that of an annelid and consists mostly of a cavity that encloses the reproductive organs.

The circulatory systems of arthropods are open. The heart pumps blood through the spaces of the body, and the blood returns through a series of one-way valves. The respiratory systems of arthropods vary according to taxonomic class and source of oxygen. Aquatic arthropods may have gills, with which they extract oxygen from the water in which they live. Spiders and some other air-breathing arthropods use stacks of modified gills, collectively called a **book lung,** to pull oxygen directly from the air.

Most terrestrial arthropods, including the insects, however, have neither lungs nor gills. Instead, they take in air through regulated openings in the body wall called **spiracles.** Air passes from the spiracles to special ducts called **tracheae.** Each trachea branches into tiny tracheoles, which deliver oxygen to individual cells throughout the body.

Reproduction in arthropods is almost always sexual. Male and female are usually separate animals, though some species are hermaphroditic. For terrestrial arthropods, fertilization must be internal, since there is no external water through which sperm can swim. Millipedes, for example, accomplish fertilization in a multilegged embrace. Elaborate courtship rituals help males and females of the same species recognize one another. Male fruit flies for example, do a little dance, during which the female decides whether she wants to mate with him. In an experiment where fruit flies were kept in the dark for several generations, however, the males abandoned the dance—which the females could not see anyway.

Because arthropods that are predaceous do not always distinguish between mates and meals, copulation among predaceous arthropods can be a dangerous matter. Such animals may have especially protracted courtship rituals, which not only help them to recognize one another but also may serve to prevent one partner from eating the other.

Excretion in most terrestrial arthropods depends on unique organs called **Malpighian tubules** (Chapter 39).

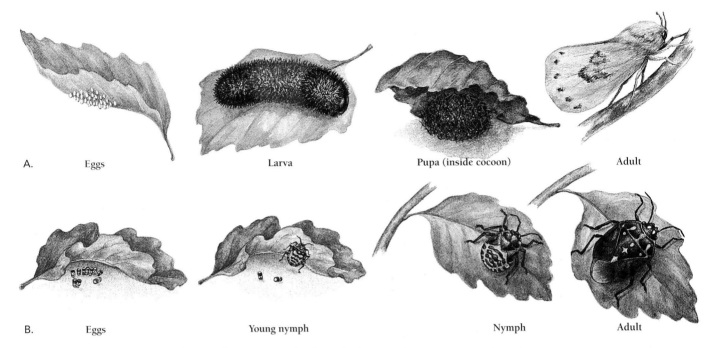

Figure 23-24 Complete and incomplete metamorphosis. A. During complete metamorphosis an insect, such as this isabella moth, hatches from an egg and develops as a legless larva that looks more like a worm than a mature insect. During development, a larva may go through many successive stages, called *instars*, before settling into a usually inactive stage called a *pupa*, or *chrysalis*. The pupa is usually encased in a protective cover, called a *cocoon*. The metamorphosis from larva to pupa and pupa to adult often involves dramatic changes in both external and internal structures. B. In incomplete metamorphosis, these changes occur more gradually. In the harlequin bug (shown here) early stages of development, called nymphs, look much like the adult, and molting functions mostly to increase size rather than to change form. Even in these metamorphoses, however, the juvenile forms may lack wings.

These tubules absorb fluid from the blood and convert the nitrogen-containing waste molecules into insoluble crystals of uric acid or guanine. The Malpighian tubules also reabsorb and recycle salts and water, so that little water is wasted.

The nervous systems of arthropods are often complex. Many arthropods can move, run, and fly quickly, and many have complicated sexual and social behaviors. The nervous system consists of a double chain of ganglia that runs along the lower surface of the body. At the head, the chain curls upward to form a brain—three pairs of fused ganglia. The sense organs—especially the eyes—of arthropods are more sophisticated than those of annelids. While some arthropods (such as the spiders) have simple eyes, most arthropod species have **compound eyes,** made up of numerous simple light-detecting units, called **ommatidia** [singular, ommatidium].

Arthropods include the greatest number of species of all animal phyla. External adaptations of the arthropods include the exoskeleton and jointed limbs. Internal adaptations of arthropods include circulatory, respiratory, reproductive, excretory, and nervous systems.

How Do Biologists Classify Arthropods?

The specialization of appendages, especially in the first few segments, is the basis of classification of living arthropods into three subphyla. (Extinct arthropods such as the trilobites may have their own subphyla.) In one subphylum—the **Chelicerates**—the first pair of appendages are mouthparts called chelicerae. The chelicerae serve as pincers or fangs, often associated with poison glands (Figure 23-25A).

In the other two subphyla, **Uniramia** and **Crustacea,** the first pair (or the first few pairs) of appendages are antennae, and the next pair are jaws, or mandibles (Figure 23-25B and C). These jaws differ from ours, which move up and down. Instead, an arthropod's mandibles crush and grind as they move from side to side. The Crustacea, which include water fleas, shrimps, lobsters, and crabs, are almost entirely aquatic, while the Uniramia, which include the insects, the centipedes, and the millipedes, are mostly terrestrial. The crustaceans all have branched (or biramous) appendages, while the Uniramia all have unbranched (or uniramous) appendages.

Biologists classify living arthropods into three subphyla according to the specialization of the appendages, especially the first two.

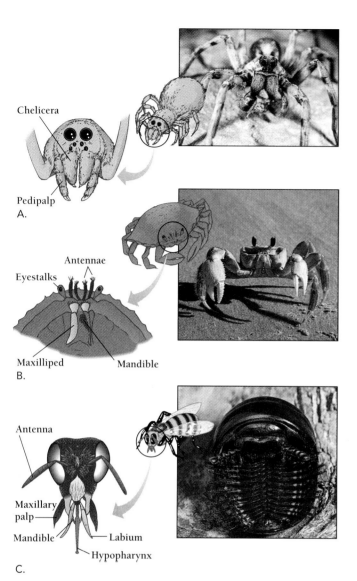

Figure 23-25 The mouthparts of arthropods. A. Chelicerae and pedipalps characterize chelicerates such as spiders and scorpions. B. Crustaceans such as this ghost crab have antennae, mandibles, and biramous appendages. C. The uniramia, including insects, millipedes, and centipedes, have antennae, mandibles, and uniramous appendages. (A, John Cancalosi/Tom Stack & Associates; B, S.E. Georgia/Animals Animals; C, Tom McHugh/Photo Researchers, Inc.)

Chelicerates

The chelicerates are the only arthropods without antennae or jaws. The most anterior pair of appendages are the pincerlike chelicerae, and the second are the pedipalps, which perform different functions in the different classes (Figure 23-25A). The most familiar chelicerates are the horseshoe crabs and the spiders, of which there are some 35,000 named species. The spiders are just 1 of 11 orders of arachnids; arachnids are, in turn, just one of three classes of the chelicerate subphylum.

Spiders, mites, and other arachnids have six pairs of appendages (four pairs of legs, plus chelicerae and pedipalps, but no antennae). Like most other arachnids, spiders are predatory, capturing and eating insects and other small animals. Some spiders hunt and pursue their prey. Others trap their prey in elaborate silk webs. One spider genus, *Mastophora*, "fishes" for its prey with a silk line, tipped with a bit of glue. Spiders also use silk to make balloons (for making long trips), droplines (for making a quick exit), egg cases (for protecting their offspring), shrouds (for storing dead prey), and gifts from males (for luring attractive female spiders).

Spiders and other arachnids have no jaws and cannot chew. They digest only liquefied food. Once a spider has trapped its prey, it injects a paralyzing poison through its chelicerae, tears and grinds a bit of its prey with chelicerae and pedipalps, then regurgitates a potion of digestive enzymes into the body of its prey, which liquefies the prey. The spider then sucks its meal into its gut, where it completes digestion.

The scorpions are the oldest order of arachnids, extending back at least to the Silurian Period, about 425 million years ago. Scorpions differ from most chelicerates in having clearly segmented abdomens. This allows the scorpion to curl its abdomen, holding its stinger aloft. Other arachnids include the distinctive daddy longlegs (or harvestmen) and the mites, of which there are some 30,000 named species and perhaps a million more as yet unnamed. Mites, including the ticks, are usually less than 1 mm long, although the largest are 2 cm long. They live almost everywhere, eating plants, fungi, and animals.

Chelicerates—such as spiders, mites, and horseshoe crabs—have no antennae and no jaws.

Crustaceans

The crustaceans include some 35,000 species of lobsters, crabs, shrimps, barnacles, and crayfish, as well as less conspicuous species such as water fleas, fairy shrimp, pillbugs, and copepods (Figure 23-26). Crustaceans usually have three pairs of chewing appendages (including the mandibles), many pairs of legs, and, unlike other arthropods, two pairs of antennae. Their legs are biramous, or split into two parts at the tip.

The most numerous crustaceans are the copepods. Indeed, next to the nematodes, the copepods are the most abundant animals on Earth. These tiny crustaceans, each a few millimeters long, are a major component of plankton, the microscopic plants and animals that float near the surface of the oceans. Copepods eat marine algae and are a major food in the diet of whales, nurse sharks, and other filter-feeding marine animals.

The largest crustaceans are the decapods, which include lobsters and crabs. Some lobsters may grow to be 60 cm long, and the Japanese spider crab is more than 3 meters across. The decapods [Greek, *deka* = ten + *pous* = foot] have five pairs of legs.

The oddest of the crustaceans are the barnacles. One wit described these creatures as "nothing more than a little shrimp-like animal, standing on its head in a limestone house and

Figure 23-26 Krill. Small crustaceans such as these copepods are the major component of krill, the marine soup on which the great whales feed. *(William E. Ferguson)*

kicking food in its mouth." Unlike other crustaceans, which are active, a barnacle lives its adult life in a single place, submerged in salt water and attached (by its head) to a submerged rock, piling, boat bottom, or even a whale. Barnacles use their appendages to direct passing food particles into their mouths. They secrete calcium-containing plates, which are firmly cemented to the underlying surface.

Crustaceans—such as lobsters, crabs, and shrimps— have jaws, biramous appendages, and, usually, two pair of antennae.

Uniramia

The Uniramia are the most diverse subphylum of arthropods, including 750,000 insects, 10,000 species of plant-eating millipedes, and 2500 species of carnivorous centipedes. Almost all Uniramia take in air through tracheae, use Malpighian tubules to rid themselves of nitrogenous waste, and have unbranched (uniramous) appendages.

The myriapods [Greek, *myrioi* = countless + *pous* = foot], which include the millipedes [Latin, *mille* = thousand], the centipedes [Latin, *centum* = hundred], and two less familiar classes, are the most clearly segmented arthropods.

In insects, segmentation is clearest in the larvae, but adults are segmented as well. The body consists of three tagmata (fused segments): the head, the thorax, and the abdomen. The segments of the head are completely fused. The thorax consists of three segments, each of which carries a pair of legs, and, in many insects, one or two pairs of wings. The abdomen consists of 12 or fewer segments, none of which carries legs or wings.

Entomologists (scientists who study insects) divide the class Insecta into about 30 orders. These orders differ both in adult morphology and in their pattern of development. Some insects (for example, lice, fleas, and silverfish) have no wings, some (flies and mosquitoes) have one pair of wings, some (dragonflies) have two similar pairs of wings, and some (beetles) have two pairs, each with a distinctive structure. The front wings of beetles are hard shells that serve to protect the delicate hind wings, which alone are responsible for flight.

Uniramia have one pair of antennae and one pair of mandibles.

Zoologists divide the animal kingdom into the Parazoa (mostly sponges) and the Eumetazoa (all the rest). The eumetazoans fall into three groups of phyla: the acoelomates (with no coelom), the aschelminths (with a pseudocoel or no coelom), and the coelomates (with a true coelom). We have briefly surveyed all four groups. These animals range from the simple sponges and jellyfish to animals as complex as insects or as alert as octopods. In the next chapter we will see that the deuterostome coelomates are also diverse, ranging in complexity and intelligence from sand dollars to humans.

Study Outline with Key Terms

An animal is a multicellular, heterotrophic, eukaryotic organism that develops from an embryo. Most animals reproduce sexually, producing a single-celled **zygote** from the fusion of an egg and a sperm. The developing embryo goes through a stage called a **blastula,** which is a hollow sphere made up of identical cells. The blastula folds into a three-layered **gastrula** surrounding the **archenteron.** The three cell layers of the gastrula are the **endoderm, ectoderm,** and **mesoderm.** These three layers develop into different types of **tissues** and **organs.** Tissues such as the **epithelial** and **connective** tissues are composed of specialized cells and their **matrix.** Organs are composed of two or more types of tissues.

Zoologists divide animals into two subkingdoms, the **Parazoa** and **Eumetazoa.** The Parazoa, which have no organs, include two phyla, the Placozoa and the Porifera. There is only one known species of Placozoa, but there are about 10,000 species of Porifera, or sponges. Sponges are usually **sessile.** Their cells lie in three layers (epidermal cells, **mesenchyme** cells, and flagellated cells) surrounding a central cavity.

Zoologists divided the eumetazoans according to the presence or absence of a **coelom** or **pseudocoel** and also according to whether they have **radial symmetry** or **bilateral symmetry.** Animals with bilateral symmetry generally have distinctive **anterior** and **posterior** ends. Animals with a coelom are called **coelomates.** Those with no coelom are called **acoelomates.** In between are the **pseudocoelomates.**

Almost all coelomate animals have digestive tracts with both a mouth and anus. The coelomates are divided into the **protostomes** and the **deuterostomes,** according to whether the embryonic **blastopore** develops into the mouth (protostomes) or the anus (deuterostomes). The majority of coelomates are protostomes, including the annelid worms, mollusks, and arthropods. However, two major phyla of coelomates are deuterostomes—the echinoderms and the chordates.

Cnidarians and **ctenophores** are radially symmetrical acoelomates. Cnidarians, which include jellyfish, hydras, corals, and sea anemones, have tissues but no organs. They have two layers of tissues enclosing a layer of **mesoglea.** They may have two body plans: **polyps** (which resemble cylinders) and **medusae** (which resemble bells). Both types have specialized stinging cells, called **cnidocytes,** which fire tiny barbs called **nematocysts.** Ctenophores (comb jellies) resemble the cnidarians but also have distinctive features, including bands of cilia.

Platyhelminthes and **Nemertea** are bilaterally symmetrical acoelomates. Platyhelminthes (flatworms) are the simplest animals that have heads and organs. They consist of three layers of cells. They have no circulatory systems, but their flat shapes allow gases and molecules to diffuse rapidly. There are three classes of flatworms: the **Turbellaria, Trematoda,** and **Cestoda.** The first class consists of the free-living flatworms, while the other two classes include only parasites. Parasites are specialized for rapid and prolific reproduction in their hosts. The tapeworm consists of a **scolex** and many **proglottids.** Nemertea (ribbon worms) have a two-ended digestive system, a circulatory system, and a **proboscis.**

The **Aschelminths** are a group of eight phyla, all bilaterally symmetrical animals with either no coelom or a pseudocoel. **Nematodes** (roundworms) are cylindrical and unsegmented, and have a one-way digestive system with both a mouth and an anus. The prominent pseudocoel of nematodes acts as a distribution route for nutrients, gases, and wastes; it also serves as a hydrostatic skeleton.

Mollusks (which include snails, slugs, clams, oysters, squids, and octopods) are unsegmented animals that typically have shells. Mollusks all have a **visceral mass,** a muscular **foot,** a **mantle,** a **mantle cavity,** a **radula,** and one or two kidney-like **nephridia.** Many shelled mollusks have an **operculum.**

There are seven classes of mollusks, of which we discussed three: **gastropods, bivalves,** and **cephalopods.** Gastropods have single shells, well-developed heads, and twisted bodies. Bivalves have two shells and most lead sedentary lives. Cephalopods have forward-pointing tentacles instead of a foot.

Annelids (which include earthworms, segmented marine worms, and leeches) have segmented bodies that consist of identical or nearly identical **metameres.** Annelids have closed circulatory systems, with blood contained entirely within vessels. Annelids are divided into three classes: polychaetes, oligochaetes, and hirudinea. Each segment of a polychaete (or bristle worm) contains a pair of paddlelike projections, called **parapodia,** each with a bundle of bristles, or **setae.** Oligochaete segments lack parapodia. Hirudinea (leeches) are more specialized than other annelids and lack obvious segmentation.

Arthropods include the greatest number of species and the greatest numbers of individuals of all animal phyla. External adaptations of the arthropods include an **exoskeleton** and jointed limbs. Arthropods have many specialized appendages that develop from **tagmata,** fused segments that form the head, the **thorax,** and the **abdomen.** The specialization of appendages, especially in the first few segments, is the basis of classification of arthropods into three subphyla: **chelicerates, crustaceans,** and **uniramia.**

Because an exoskeleton prevents an animal from growing, many arthropods periodically **molt.** Other arthropods do all their growing in larval form, then change into adults during **metamorphosis.**

Internal adaptations include circulatory, respiratory, reproductive, excretory, and nervous systems. Arthropods breathe by means of a **book lung,** or **spiracles** and **tracheae** and regulate excretion by means of **Malpighian tubules.** Some arthropods have simple eyes, but most have **compound eyes,** made up of numerous **ommatidia.**

Review and Thought Questions

Review Questions

1. How is an animal different from a plant or a protist? Why are sponges grouped with animals? What animal features do sponges and cnidarians lack?
2. What criteria do zoologists use to classify the Eumetazoa?
3. What is a coelom and how is it different from a pseudocoel?
4. What specializations set parasites apart from other members of their phyla?
5. What is the difference between a protostome and a deuterostome?
6. Members of which class of mollusks would make the most interesting pets? Why?
7. Which phylum of protostomes includes the greatest number of species of any phylum of organisms?
8. What specific characteristics of a grasshopper tell you that it belongs in the class Insecta?
9. How do millipedes differ from centipedes? Which class includes more species?

Thought Questions

10. What is progress? In what sense is evolution progressive?
11. Zoologists use the following traits (among others) to classify the animals: number of embryonic cell layers, presence or absence of a true coelom, and symmetry. Do you think that such characteristics truly reflect relatedness between different groups of organisms? If you were going to classify the animals, what criteria would you consider using?

About the Chapter-Opening Image

The succulent katydid of Peru is a classic and beautiful invertebrate.

On-line materials relating to this chapter are on the World Wide Web at http://www.harcourtcollege.com/lifesci/aal2/

Web Links

Quizzing & Testing

24 Deuterostome Animals: Echinoderms and Chordates

Are We Upside Down?

In 1830, two French anatomists staged a public debate before the prestigious French Academy of Sciences on a topic so fundamental that historians of science have returned to it again and again. How many basic body types do animals have?

On one side, the politically powerful French anatomist and paleontologist Baron Georges Cuvier (1769–1832) argued that all animals should be divided into four distinct types. On the other side was Étienne Geoffroy Saint-Hilaire (1772–1844), a renowned comparative anatomist, who argued that animals were of just one type, not four.

Geoffroy was not a man to be intimidated even by a figure as powerful as Cuvier. During the French Revolution, Geoffroy had risked his life to save teachers and colleagues from execution. And when in 1807 Napoleon ordered Geoffroy to obtain collections from Portugal's museums by any means possible, Geoffroy slyly traded specimens from French museums instead of seizing the collections by force.

Geoffroy had long maintained that all vertebrates—mammals, birds, reptiles, and amphibians—had the same basic "body plan," a description of their symmetry and the arrangement of their tissues. In all vertebrates, for example, the backbone is at the back (dorsal) and the heart lies along the chest (ventral). The unity of the vertebrates was an idea that Cuvier and his followers had accepted.

But in 1830, Geoffroy Saint-Hilaire took his argument a huge step further, proposing that chordates (vertebrates and their relatives) have the same body plan as arthropods. In essence, he said, an insect and a mammal are similar. The single main difference, he said, was that we vertebrates are turned upside down: where a grasshopper's nerve cord runs along its belly, a human's nerve cord runs along its back (Figure 24-1). And where a human heart beats against the chest wall, a grasshopper's heart lies up along its back. As one modern biologist has put it, "If you lay down on your back and waved your arms, you would be doing what insects do when they walk."

Geoffroy's hypothesis, which he called *unité de plan* (singleness of plan), was provocative. Not only was it in direct opposition to Cuvier's theory of separate, unrelated groups, but it also implied evolution. Cuvier firmly believed in the fixity of species and had publicly ridiculed Lamarck and other proponents of evolution. According to Geoffroy's hypothesis, however, humans and all other vertebrates were each a simple variation on a basic invertebrate body plan. Geoffroy's idea could be interpreted to mean that vertebrates had evolved from invertebrates.

For Geoffroy, *unité de plan* was a serious scheme, based on his dissections of thousands of animals. But during the debate, the pugnacious Cuvier made *unité de plan* look like a joke, listing, one by one, all of the differences between a duck and a squid.

Key Concepts

- Deuterostomes include both invertebrates and vertebrates.
- Chordates are deuterostomes that, at some time during their development, have a notochord, a dorsal hollow nerve cord, pharyngeal gill slits, and a tail.
- Vertebrates have a segmented spinal column and a distinct head with a skull and a brain.
- Terrestrial vertebrates have special adaptations for life on land, one of which is the amniotic egg.

Historians of science still argue over who won the debate on technical points. But the reality was that Cuvier's scientific and political influence was so profound that for 164 years no biologist resurrected *unité de plan* for serious discussion. In fact, a researcher risked his or her reputation even to introduce the discredited idea.

So the matter might well have rested. In the early 1990s, however, biologists working separately on frogs and fruit flies began to untangle the matrix of genes that underlies the development of the overall body plan. To their astonishment, they discovered that the frog genes and the fly genes were the same genes. Unwittingly, they had resurrected Geoffroy Saint-Hilaire's idea of a single body plan.

In fruit flies, a gene called *dpp* codes for a signaling protein that somehow activates other genes necessary for the formation of **dorsal** structures—all the body parts that lie along the back of an animal. In addition, *dpp* seems to suppress the development of **ventral** structures—all the body parts that lie along the abdominal side of an animal. *Dpp* also suppresses the development of nerve cells. A second fruit fly gene, *sog*, counteracts the effects of *dpp*. *Sog*, which is expressed only in the ventral regions of a developing fruit fly, overrides *dpp* in the ventral parts, allowing nerve cells and other ventral structures to develop.

When researchers looked for similar genes in frogs, they found *bmp–4*. Like *dpp*, *bmp–4* codes for a signaling protein. But, oddly, it operated in reverse. *Bmp–4* activates genes that help form the ventral structures in the frog embryo and suppresses the formation of nerve cells in the dorsal regions (Figure 24-1).

No one made an explicit connection between the three genes, however, until September 1994, when two German researchers pointed out that *dpp* and *bmp–4* do the same thing, only acting in opposite areas of frog and fly. Detlev Arendt and Katharina Nübler-Jung suggested that "the longitudinal nerve cords of insects and vertebrates derive from one and the same centralized nervous system in their common ancestor." After 164 years, they had revived Geoffroy Saint-Hilaire's *unité de plan*.

They were not ridiculed, however. Instead, researchers in California immediately uncovered a fourth gene. Edward De Robertis (at UCLA) and his colleagues found that the gene *chordin* stimulates the organized development of nerve cells in the dorsal region of a frog embryo, overriding *bmp–4*, just as *sog* overrides *dpp* in a fly.

If Geoffroy Saint-Hilaire and his modern supporters were right, then *dpp* and *bmp–4* were essentially the same gene, just expressed in different parts of the embryo. Likewise, *sog* and *chordin* had to be the same gene, separated by half a billion years of evolution.

In a series of dramatic experiments, De Robertis and several other researchers in California and Wisconsin demonstrated that the fly genes (*dpp* and *sog*) work in frogs, and the frog genes (*bmp–4* and *chordin*) work in flies. *Sog* mRNA injected into frog embryos, for example, promoted the development of dorsal structures. Likewise, *chordin* mRNA injected into flies induced ventral development, as *sog* normally does. Further, when researchers sequenced *chordin* and *sog*, they discovered that the base sequence of *chordin* is 47 percent identical to that of *sog*. And biologists have since uncovered several other developmental genes that are the same in both frogs and flies.

Has molecular biology redeemed *unité de plan*? Was Geoffroy Saint-Hilaire right, Cuvier wrong? Many biologists still hesitate to pronounce the matter settled. But nearly all are intrigued. If organisms as different as lobsters and linebackers share genes that help determine basic body plan, these organisms surely share a common ancestor.

De Robertis, for one, is eagerly seeking more genes shared by vertebrates and invertebrates that might shed light on the structure of that common ancestor. Until he and his colleagues succeed, however, even modern molecular biologists will hesitate a bit, reluctant to embrace an idea so long ridiculed.

> "If you lay down on your back and waved your arms, you would be doing what insects do when they walk."

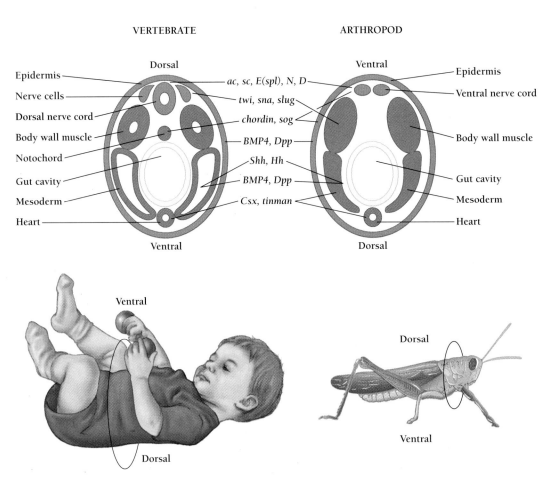

Figure 24-1 Opposites? Cross sections through embryos show that vertebrates and invertebrates share similar body plans. The vertebrate body plan features a dorsal nerve and a ventral heart—an upside down version of the arthropod body plan, which has a ventral nerve cord and a dorsal heart. These similarities go deep. Similar genes operate in both kinds of embryos.

DEUTEROSTOME COELOMATES

As we discussed in the last chapter, the early development of the deuterostome coelomates differs from that of the protostomes in several ways. In all deuterostomes, including ourselves, the tiny hole in the embryo called the **blastopore** develops into an anus instead of a mouth. The mouth develops later (or second), hence the term deuterostome [Greek, *deuteros* = second + *stoma* = mouth].

The deuterostomes consist of four phyla (Figure 24-2): **Echinodermata** ("spiny-skinned" animals such as sea stars), **Chaetognatha** (arrow worms), **Hemichordata** (acorn worms), and **Chordata**. The Chordata include the vertebrates—the fish, the amphibians, the reptiles, the birds, and the mammals—as well as the less well-known sea squirts and lancelets. While sea urchins and humans differ radically as adults, their embryos share many basic features.

Echinoderms

The 6000 species of echinoderms fall into six classes of marine animals. Among these are two of the most familiar inhabitants of tidepools—sea stars and sea urchins. The other four living classes of echinoderms are the sea lilies and feather stars, the brittle stars, the sea daisies, and the sea cucumbers. Another 20 classes of echinoderms are now extinct. The echinoderms are an ancient lineage. Fossil sea stars, sea cucumbers, and sea lilies from the Burgess Shale suggest that the echinoderms had diverged from the protostomes some 530 million years ago.

Echinoderms are plentiful on the bottoms of intertidal zones and deep oceans. The largest is a sea star one meter across. The smallest are just a few millimeters. Most echinoderms have calcium-rich spines that project from their skin, giving the phylum its name [Greek, *echinos* = hedgehog, a small European mammal covered with spines, like a porcupine].

As adults, echinoderms are radially symmetrical, often with five nearly identical parts arranged around a central axis. Like other deuterostomes, however, the larvae of echinoderms have bilateral symmetry. Biologists believe that the echinoderms are descended from bilateral ancestors (Figure 24-3). Neither larvae nor adults show any sign of segmentation.

The sexes of echinoderms are separate, with two sets of gonads (testes or ovaries) in each arm. Males shed sperm and females shed eggs into sea water, where fertilization takes place. In the open ocean, the chances that an individual egg will meet

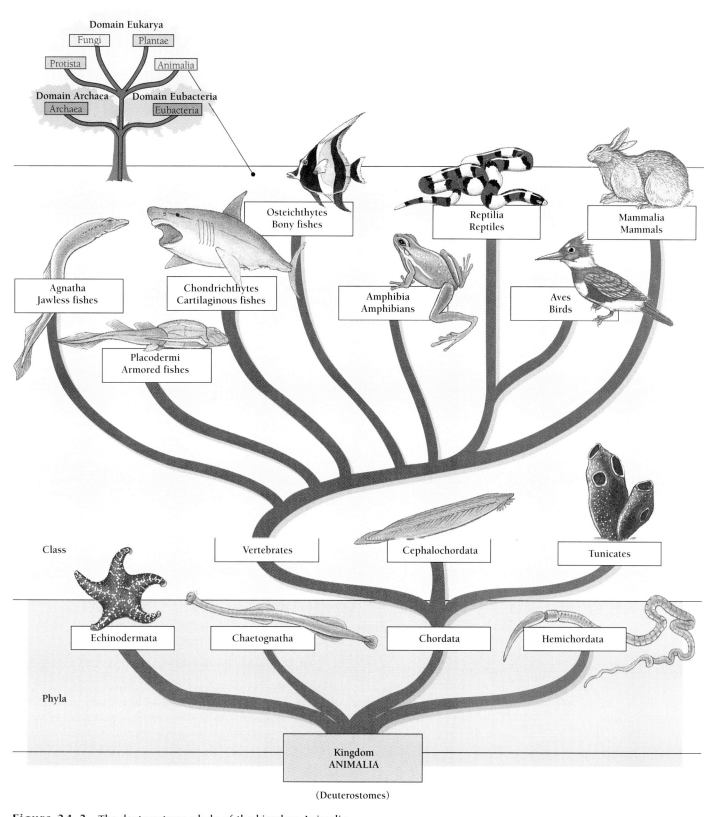

Figure 24-2 The deuterostome phyla of the kingdom Animalia.

A.   B.

Figure 24-3 *Sea urchin adults and larvae.* Although an adult echinoderm appears to have radial symmetry, a trait regarded by zoologists as somewhat backward, the larvae have the same bilateral symmetry as other deuterostomes. *(A, Dave B. Fleetham/Visuals Unlimited; B, Cabisco/Visuals Unlimited)*

a sperm of the right species is small. Therefore, reproductive success depends on sheer numbers of gametes. A single female may produce more than 2 million eggs in a breeding season.

Sea stars (starfish)

Typical echinoderms are sea stars, or starfish. A sea star's body consists of a central region, with arms radiating out from the center. Under the center is the mouth, which opens into a digestive tract (Figure 24-4). Sea stars feed on all sorts of other animals, including sponges, corals, mollusks, crustaceans, worms, and even fish and oysters. However, a sea star prepares and eats an oyster rather differently from the way we would. The sea star first spreads its arms around the oyster and, like a human diner, it pries open the shell. But, in a surprising move, the sea star places its mouth near the gap in the oyster shell and extends its own gut through the opened shell, turning its stomach inside out. A sea star can perform this unusual feat even if the opening in the oyster shell is as small as 100 µm. During the next few hours, the sea star completely digests its prey. Some sea stars spread their stomachs over a piece of coral reef on the sea floor, digesting sponges, algae, and settled organic matter.

A sea star's arms can bend and twist, allowing the sea star to move along the bottom, to grasp prey, and to right itself. Within the coelom is a unique set of canals, called the water vascular system, that helps some echinoderms control the movement of their arms. A primitive skeleton made of calcium-rich plates reinforces the whole body. Over this supporting structure is a delicate skin, with thousands of cells that are exquisitely sensitive to touch. Tiny, suction cup–like tube feet cover the underside of a starfish.

Starfish and some other echinoderms can regenerate arms that are damaged or broken away. Some species of sea stars can

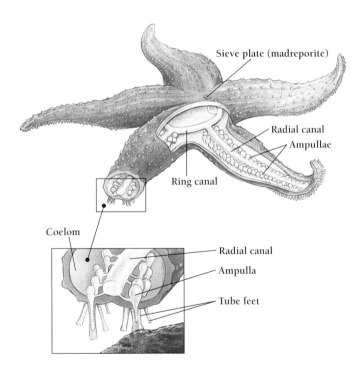

Figure 24-4 *The sea star's water vascular system.* Perhaps the most remarkable feature of sea stars and all other echinoderms is the water vascular system, which serves as both a circulatory system and an elaborate hydrostatic skeleton that allows complex movements of the tube feet. Water enters the system through a perforated sieve plate on the dorsal surface, which filters the water as it enters a tube. The filtered water flows down a tube into the ring canal, which surrounds the mouth (ventral) and from there into the radial canals that extend into each of the arms. In the arms, the radial canals branch to connect to long rows of tube feet. Each tube foot is also connected to a muscular fluid-filled sac, called an ampulla. In sea stars and other echinoderms, each tube foot ends in a sucker. The ampulla regulates fluid pressure, so that each tube foot may extend or retract, hang on or let go. Using the tube feet, a starfish can slowly pull itself along the bottom of the sea.

regenerate a whole new individual from a single arm. In such asexual reproduction, the starfish breaks into two parts, and each regenerates the missing arms. Usually, however, echinoderms reproduce sexually, with males and females releasing gametes into the sea.

Echinoderms are ancient, spiny-skinned animals that are radially symmetrical as adults and bilaterally symmetrical as larvae.

Arrow Worms and Acorn Worms

The 50 to 70 species of arrow worms are common marine organisms that range in length from 5 mm to 15 cm. Arrow worms feed on a variety of small marine animals, which they grasp with moveable hooks. Fossils of arrow worms date from late Cambrian times. Unlike echinoderm larvae or acorn worms, the body of an adult arrow worm extends beyond the anus, forming a tail. The only other deuterostomes with true tails are chordates, the phylum that includes the vertebrates. Because arrow worms have other properties in common with vertebrates, many evolutionary biologists think vertebrates are descended from ancestors that resembled arrow worms.

Still more similar—though not at first glance—to animals of our own phylum are the acorn worms. Most of the 65 species of acorn worms live in U-shaped burrows on the bottom of the sea. They range in length from 2.5 cm to 2.5 m (nearly 8 ft). Zoologists classify these soft-bodied marine animals as hemichordates ("half chordates") because they resemble chordates in two respects: (1) they have **gill slits,** holes that directly connect the throat to the outside; and (2) in addition to a ventral nerve cord, they have the beginnings of a dorsal nerve cord (like our own spinal cord). As we will discuss later, both of these features are fundamental characteristics of chordates.

Arrow worms and acorn worms may resemble the ancestors of the vertebrates because they possess a tail. The acorn worms, in addition, have gill slits and a dorsal nerve cord like that of the chordates.

What Traits Characterize the Chordates?

The chordates include members of three subphyla. Members of two subphyla have no backbones: the tunicates and the cephalochordates. Members of the third subphylum, the vertebrates, have backbones. Members of all three subphyla exhibit four hallmark characteristics at least at some time in their lives: (1) a flexible rod running along the back (dorsal surface), called the **notochord,** which gives the phylum its name; (2) a **dorsal hollow nerve cord,** running between the notochord and the surface of the back; (3) **pharyngeal slits** (sometimes called gill slits), holes in the sides of the body that run from the inside of the gut to the outside surface of the animal; and (4) a segmented body and a "postanal" tail, a tail that extends beyond the anus (Figure 24-5).

The notochord is a long column of fluid-filled cells, with the character of a stiff, but flexible, sausage. In the tunicates and cephalochordates, as well as in some fish and larval amphibians, the notochord provides back support for the adult animal. In many vertebrates, however, the notochord disappears during embryonic development to be replaced by a bony or cartilaginous backbone. Cartilaginous fish have both a notochord and a surrounding vertebral column, while adult reptiles, birds, and mammals have only a vertebral column.

The dorsal nerve cord of embryos and its adult counterparts (the spinal cord and the brain) are partly hollow, and their inner spaces contain fluid. In contrast, the ventral nerve cords of annelids and arthropods are solid. The pharyngeal slits are present in the embryos of all chordates (including humans), but not in air-breathing adults. In early chordates, the slits may

Tunicate larva (urochordate)

Amphioxus (cephalochordate)

Lizard (vertebrate)

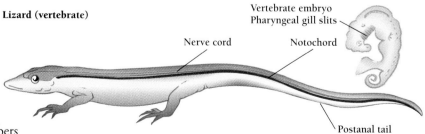

Figure 24-5 The four features of chordates. The three chordate subphyla—the tunicates (sea squirts), the cephalochordates (amphioxus), and the vertebrates—all share four distinguishing features. These are pharyngeal gill slits, a dorsal nerve cord, a notochord, and a postanal tail.

have served to allow filter feeding (as they do in tunicates), but fish, tadpoles, and other aquatic chordates use them primarily for respiration.

All chordates have a notochord, a dorsal hollow nerve cord, pharyngeal slits, and a tail at some time in their lives.

What Have We in Common with Tunicates?

Adult tunicates do not look at all like other chordates (Figure 24-6). They are sessile marine animals that attach to rocks, boats, or the ocean floor. Their name derives from a tough outer coating, or tunic, made, amazingly, of cellulose. (Cellulose is common in the cell walls of plants, but rare in animals.) While adult tunicates have pharyngeal gill slits, they have little else to recommend them as chordates. A tunicate larva, however, has a notochord, a dorsal hollow nerve cord, pharyngeal gill slits, and a postanal tail, most of which are lost when it metamorphoses into an adult. It seems likely that the common ancestor of the chordates resembled a tunicate larva.

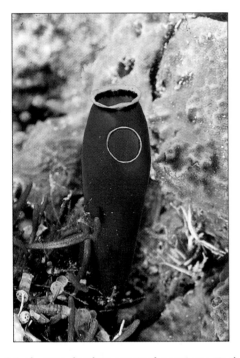

Figure 24-6 Our chordate cousin the tunicate. To the average person, a tunicate might look more like a sponge than a close relative of the vertebrates. Like sponges, adult tunicates feed by filter feeding. But unlike sponges, tunicates have a U-shaped digestive tract. Cilia in the pharynx pull water through an incurrent siphon into the mouth, past the pharyngeal slits, and out through an excurrent siphon. Sticky mucus traps food particles and carries them down into the digestive tract. The anus empties into the excurrent siphon. *(Dave Fleetham/Tom Stack & Associates)*

As larvae, tunicates have a notochord, a dorsal hollow nerve cord, pharyngeal slits, and a tail, all characteristics of chordates.

What Have We in Common with Cephalochordates?

Cephalochordates [Greek, *cephalo* = head], or lancelets, include about 45 species of small, segmented, fishlike animals. Most belong to the single genus *Branchiostoma,* whose common name is amphioxus. In these animals, the notochord, which is present in adults as well as in embryos, extends all the way through the head, hence their name. Sets of chevron-shaped (<<<) muscles pull against this internal skeleton. The segmentation of a lancelet is superficially similar to that of annelid worms and achieves the same advantages. Cephalochordates have all four chordate characteristics.

Cephalochordates have notochords both as larvae and as adults.

What Distinguishes the Vertebrates from Other Chordates?

Vertebrates have many features that distinguish them from other chordates. Two of the most important are (1) a segmented vertebral column made of units called vertebrae and (2) a distinct head, with a cranium (or skull) and a brain. Arches of the vertebrae encircle and protect the dorsal nerve cord, which becomes the adult vertebrate's spinal cord (Figure 24-7).

Vertebrates also have closed circulatory systems and characteristic internal organs—livers, kidneys, and hormone-secreting (endocrine) organs. The heart pumps blood through the circulatory system, through a system of fine capillaries that helps transport gases and nutrients to nearly every cell of the body. The same system carries metabolic wastes away from cells. Liver, kidneys, and other organs closely regulate the chemical composition of the blood, ensuring a nearly constant internal environment for all cells. (Part VII of this book deals with these organs and their operation in some detail.)

Most modern vertebrates have bony skeletons. **Bone** consists of fibers of the protein collagen, embedded primarily with crystals of calcium phosphate. Within this extracellular material are bone-making cells, as well as blood vessels and nerves. Bone is strong enough to support huge vertebrate bodies. Unlike the exoskeletons of arthropods, bone can grow, allowing vertebrates to reach sizes far greater than those of the arthropods.

The embryos of all vertebrates have skeletons made of **cartilage,** also built from collagen, but softer and more elastic than bone, without calcium phosphate. Some parts of the adult skeleton retain cartilage, such as our ears and noses. But, except in the sharks and other cartilaginous fishes, bone mostly replaces cartilage in adults.

clude Agnatha (the jawless fishes), Placodermi (a class of extinct armored fishes), Chondrichthyes (fishes with skeletons of cartilage rather than bone), and Osteichthyes (bony fish). The tetrapods consist of the Amphibia (amphibians), Reptilia (reptiles), Aves (birds), and Mammalia (mammals).

Figure 24-7 Which of these vertebrates has the strongest back? All vertebrates share a vertebral column composed of many vertebrae, as well as a thick skull that protects the brain. In most vertebrates, the vertebral column, skull, and skeleton consist largely of bone. In a few, such as the sharks and rays, these elements are flexible cartilage instead. The hero shrew (*center*), of Africa, is reputed to have the strongest backbone of any vertebrate. According to legend, a man can stand on this tiny insectivore's back without injuring the animal.

Vertebrate bone is so distinctive and so well preserved in the fossil record that biologists know more about the evolution of the vertebrates than about any other group. In this chapter, we survey eight classes of vertebrates. Four of these classes are groups of fishes and four are four-footed animals, or **tetrapods** [Greek, *tetra* = four + *pous* = foot]. The four classes of fishes in-

---

Vertebrates have vertebral columns, skulls, and a well-developed brain. Most vertebrate skeletons are made of bone.

Agnatha Have No Jaws

Most jawless fishes are extinct. These extinct fishes, comprising some 60 families, are called **ostracoderms** [Greek, *ostrakon* = shell + *derma* = skin] because their skin contained a shell-like skeleton of bony plates. Ostracoderm fossils have been found in Cambrian deposits, showing that vertebrates have been around as long as the invertebrates (Figure 24-8). Although, traditionally, the agnathes have been lumped together, ongoing research suggests that they represent several separate lines (or clades).

The living jawless fishes include only about 60 species of **cyclostomes** [Greek, *kyklos* = circle + *stoma* = mouth], soft-bodied fish with round, jawless mouths. The cyclostomes, none of which have bony plates, include the lampreys and the hagfish. Lampreys and hagfish are hardly glamorous animals. Hagfish survive by eating polychaete worms and pieces of dead fish, scavenged from the muddy sea bottom. Lampreys are nearly all parasitic. A lamprey lives by attaching itself to another fish with its circular jawless mouth, then tearing a hole in its living victim with its rasping tongue. The lamprey then sucks whatever fluids flow from the wound.

A lamprey larva, like a lancelet, has a notochord and segmented muscle groups. It differs from a lancelet, however, in possessing

Figure 24-8 The jawless fishes. These fishes have neither working jaws nor paired fins. A. Extinct ostracoderms. B. A modern lamprey. (*B, top: Courtesy of Dr. Kiyoko Uehara; bottom, Berthoule-Scott/Jacana/Photo Researchers, Inc.*)

a brain, a liver, and a kidney. Most zoologists think that the first vertebrate resembled a lamprey larva.

> *Lampreys and hagfish (the living members of the Agnatha) are the only vertebrates that lack jaws.*

Placoderms, Now Extinct, Were Armored Fish with Jaws

At the beginning of the Devonian Period, some 405 million years ago, freshwater fish developed the first vertebrate jaws. These fish rapidly spread into the oceans, diversified, and multiplied. Paleontologists sometimes call the Devonian the "Age of Fishes." Among the prominent jawed fishes in Devonian seas were a class of armored fishes called **placoderms** [Greek, *plak* = plate + *derma* = skin]. The placoderms had jaws, paired fins, and much less armor than the ostracoderms.

Today, all vertebrates except the Agnatha have hinged jaws, with which they can seize, bite, and sometimes chew. Jaws permit a much more diverse diet than that of lampreys or hagfish. Jaws probably enabled placoderms and other jawed fishes to replace the ancient ostracoderms. By the end of the Devonian Period, armored placoderms had disappeared, replaced by the Chondrichthyes and Osteichthyes. These two classes of fishes still dominate the aquatic regions of the Earth's surface.

> *Placoderms were the first vertebrates with jaws.*

Chondrichthyes Are Fish with Cartilaginous Skeletons

In most vertebrates, the soft cartilaginous skeleton of embryos is gradually replaced with bone. In sharks, skates, and rays, however, bone does not replace cartilage during development. Adult Chondrichthyes have lightweight cartilaginous skeletons, giving the class its name [Greek, *chondros* = cartilage + *ichthys* = fish]. Chondrichthyes, with only about 750 species, are much less diverse than the 30,000 species of bony fish (Osteichthyes).

Chondrichthyes also lack a swim bladder, a balloonlike organ that helps the bony fish float. Instead, sharks and their allies increase their buoyancy by storing large amounts of oil in their huge livers. Nonetheless, the Chondrichthyes are heavier than water and sink to the bottom when they are not moving (Figure 24-9).

Although sharks and rays lack the bony scales of the placoderms and ostracoderms, their skin is covered with small toothlike scales (denticles) that give it the feel of sandpaper. In addition, they have teeth (derived from their scales), which they use, along with their jaws and their ability to swim rapidly, to live as successful predators (Figure 24-9).

Sharks have particularly well-developed senses that help them find their prey. Among these senses are keen vision and smell. The lateral line system, a row of tiny sense organs along each side of the body, detects changes in water pressure. Other receptors, originally derived from the lateral line system, detect tiny electrical currents, including those generated by muscular contractions of potential prey.

A.

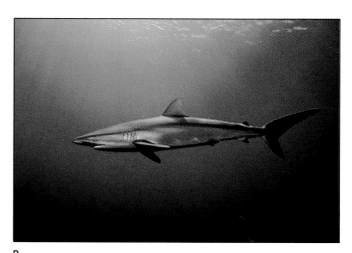

B.

Figure 24-9 The Chondrichthyes. A. Rays and skates have evolved adaptations for a life on the ocean floor, where they feed mostly on mollusks and crustaceans. Their pectoral fins are enormous, giving these fish their characteristic appearance. B. Sharks are mostly fast-moving predators with streamlined bodies, tapered at both ends. Lacking a buoyant swim bladder, a shark keeps from sinking by swimming constantly. (Some sharks take breaks on the sea bottom or in a sea cave.) Paired fins (pectoral and pelvic) on the underside help keep the moving shark afloat. (*A, Jeff Rotman/Tony Stone Images; B, Doug Perrine/DRK Photo*)

The Chondrichthyes have cartilaginous skeletons and no swim bladder.

Osteichthyes Are Fish with Bony Skeletons

The bony fish, Osteichthyes [Greek, *osteon* = bone + *ichthys* = fish], are the most diverse of all the vertebrate classes. More than 30,000 species fill oceans and seas, lakes and streams all over the world. They range in size (as adults) from about 1 cm to more than 6 m (20 feet). The bony fish comprise more than 95 percent of all fish and half of all species of vertebrates.

Like the cartilaginous fishes, the bony fishes first appeared in the Devonian Period, but in fresh water rather than the oceans. Bony fish differ from the Chondrichthyes in having bony skeletons and thin, bony scales. Most bony fish also have a **swim bladder,** an air-filled sac that helps them control their buoyancy (Figure 24-10).

Zoologists divided the living Osteichthyes into two subclasses: the ray-finned fishes, Actinopterygii [Greek, *aktin* = ray + *pterygon* = wing], and the lobe-finned fishes, Sarcopterygii [Greek, *sarkodes* = fleshy]. Nearly all modern fishes are ray-finned fishes: they have paired fins supported by thin, bony rays, originally derived from bony scales. Ray-finned fishes are some of the best swimmers in the world (Figure 24-11A and B). They are extraordinarily diverse and live virtually wherever there is water.

Unlike the fins of ray-finned fishes, the fins of lobe-finned fishes contain muscle as well as bone. Although once abundant, lobe-finned fishes are now unusual, including only four genera of modern fishes (Figure 24-11C). The lungfishes can use their sturdy fins to walk on land, and biologists believe that the first amphibians must have resembled lobe-finned fishes whose fins gradually evolved into limbs. Modern lobe-finned fishes, such as lungfish, live in stagnant freshwater ponds, using their lungs to extract oxygen they get by gulping air from the surface. Lungfishes can radically alter their metabolism to survive long periods of drought by burying themselves in the mud at the bottom of a pond.

Bony fishes (osteichthyes) have bony skeletons, swim bladders, and paired fins.

Figure 24-10 The body plan of a bony fish. Fish share many features with terrestrial vertebrates, including a brain, a liver, and true kidneys. In addition, a bony fish has a gas-filled swim bladder from which evolved the terrestrial lung. By regulating gases in the swim bladder, a fish can adjust its buoyancy, allowing it to float at the level it wants without exertion. Bony fishes also possess a bony operculum over the gills that allows them to pump water across the gills, facilitating the exchange of oxygen and carbon dioxide.

A. B. C.

Figure 24-11 The two subclasses of bony fishes: the ray-finned fishes and the lobe-finned fishes. A. The fast-swimming barracuda typifies the highly evolved ray-finned fishes. B. A pike can achieve accelerations of 12 to 24 g (times gravity), reaching top speeds of 6 meters per second (equal to a 4-minute mile). To find out how fish are able to swim so fast, researchers at MIT have built a robotic pike (*Esox lucius*), named "Robopike," which swims by itself. C. The 350-million-year-old marine coelacanth *Latimeria* is a lobe-finned fish that was thought to be extinct—until museum curator Marjorie Latimer found one in a fisherman's catch in 1938. (A, Rondi/Tami Church/Photo Researchers, Inc.; B, © Sam Ogden; C, Sally Bensusen/Science Photo Library/Photo Researchers, Inc.)

WHAT ADAPTATIONS HAVE VERTEBRATES EVOLVED FOR LIFE ON LAND?

By the end of the Devonian Period, fish had appeared in every conceivable aquatic niche in seas, streams, and ponds. Except for a few lungfish adventurers, however, vertebrates had not found a way to live on the land. Plants and insects had already done so, however, and the land stood all-inviting, with great, lush forests and a diversity of tasty insects and other arthropods. The terrestrial environment offered good meals and few predators.

The movement from water to land was a dramatic event, accompanied by immense changes in vertebrate anatomy. To colonize the land, vertebrates had to solve several problems:

1. They needed to obtain and conserve water in a comparatively dry environment.
2. They needed to extract oxygen from air rather than water.
3. They needed strong skeletons to support their bodies—they could no longer depend on the buoyancy of water for support.
4. They needed to control fluctuations in body temperature, which can be much greater in air than in water.
5. They still needed a watery environment for fertilization and early development.

To say that animals "solved" these problems is a shorthand way of saying that animals adapted by natural selection to the environments in which they found themselves. Such adaptation is not deliberate or goal oriented. It is simply an inevitable result of the differential survival of different kinds. In other words, while individual animals do not solve problems, lineages of animals "solve" certain problems, and they do so by evolving and adapting.

The terrestrial vertebrates—amphibians, reptiles, birds, and mammals—are the survivors that evolved adaptations that solved the problems of life on land.

Amphibians Are Terrestrial Animals That Begin Their Lives in Water

During the Devonian Period, the freshwater lakes and streams dried up and flooded regularly, leaving many aquatic organisms stranded in foul ponds and muddy stream beds. Only fish able to extract oxygen from the air, with a primitive lung, were able to survive. Indeed, the Devonian Period saw the development of the two most important terrestrial adaptations: lungs and limbs.

Nearly all the freshwater fishes that survived the Devonian Period had a kind of lung, a simple sac enhanced with a rich supply of blood. The Devonian fish developed **double circulation,** in which blood circulates between the heart and lungs and between the heart and the rest of the body. All terrestrial vertebrates have double circulation (Chapter 37).

In the Devonian Period, freshwater vertebrates also evolved legs. They used their well-developed fins to actually walk across the bottoms of pools and possibly across land from one deep pool to another. The early amphibians were the first vertebrates to live most of their lives on land. The reptiles, birds, and mammals are all descendants of the earliest amphibians. The first amphibians somewhat resembled modern salamanders and lobe-finned fishes, and dominated the land for about 100 million years, until the rise of the reptiles in the Triassic (Figure 24-12). Some amphibians grew to lengths of three or four meters and were more terrestrial than modern amphibians. Modern salamanders, for example, have evolved several adaptations for moving about in shallow water,

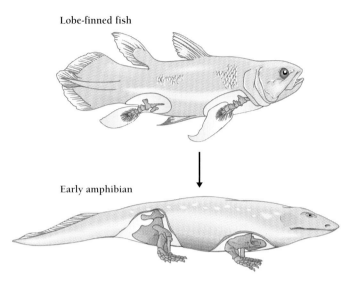

Figure 24-12 The first amphibians. The first amphibians evolved from lobe-finned fishes at the end of the Devonian Period.

including limbs, strong tails, and a flat shape, that earlier amphibians did not possess.

Today, three orders of Amphibia exist (Figure 24-13): Urodela (salamanders), Anura (frogs and toads), and Gymnophiona (caecilians). Anurans lack tails; urodeles have them. The names Anura [Greek, *an* = without + *oura* = tail] and Urodela [Greek, *oura* = tail + *delos* = visible] describe this simple distinction. The anurans are by far the most diverse amphibian order, including about 90 percent of the 3000 amphibian species. The Gymnophiona [Greek, *gymnos* = naked + *ophioneos* = snakelike], which lack feet and resemble snakes or earthworms, are the least familiar amphibians. They are tropical amphibians that live underground and live like moles, pushing through the soil and searching for small invertebrates to eat. About 160 species are known.

Adult toads and frogs have powerful leg muscles for jumping and well-developed ears and vocal cords for communicating through air. They have extendible, sticky tongues with which they can expertly capture flying morsels. Amphibians can gulp air into their lungs, but they also absorb oxygen through their skins, which are moist and thin. In fact, some salamanders breathe entirely through their skins and the linings of their mouths.

An amphibian's life cycle generally consists of two stages, the first of which is aquatic and the second terrestrial—hence their name [Greek, *amphi* = double + *bios* = life]. Amphibians are only partly adapted to life on land. Their need for water dominates their lives. Unlike us, however, amphibians do not drink water. Instead, they must absorb it through their thin, damp skins. Such skin is not good at retaining water, and amphibians must confine themselves to very moist environments or dry out.

Most amphibians start life in water, when the male and female shed sperm and eggs together (Figure 24-14). Because amphibian eggs lack shells and special membranes to prevent water loss, their embryos almost always must develop in a very wet environment.

The fertilized egg (zygote) develops into an aquatic larva, or tadpole, which obtains oxygen through gills and uses its long tail to swim. Later, the tadpole becomes a terrestrial animal by losing its gills and developing lungs and limbs, in a process called **metamorphosis** (Figure 24-14). If it is an anuran, it also loses its tail. The adult amphibian is now capable of surviving on land.

A few species of frogs, salamanders, and caecilians have evolved direct development. The zygote develops directly into the terrestrial form, bypassing the larval form. Other species avoid the need for freestanding water for their larvae in other ways. They lay their eggs in moist places on land, in pockets of water in leaves, or in wet mossy places. Several species carry the developing eggs in special pouches on their backs or in the

A.

B.

C.

Figure 24-13 Three orders of amphibians. A. Jordan's salamander, from North Carolina. B. A river frog from Florida. Frogs differ from toads in the quality of their skins: frog skin is smooth and moist, while toad skin is bumpy and dry. C. Gray caecilian. The caecilians are legless amphibians that live in the tropics. Like earthworms, they move by means of a hydrostatic skeleton—even though they have a backbone. (A, David M. Dennis/Tom Stack & Associates; B, Joe McDonald/DRK Photo; C, Michael Fogden/DRK Photo)

Figure 24-14 How tadpoles turn into frogs. Most amphibians lay their eggs in water. The eggs develop into tadpoles, which eat and grow under water. When the time is right, the tadpoles metamorphose into adult amphibians.

mouth (Figure 24-15). The South American frog (*Rhinoderma*), for example, lays its eggs on moist ground, after which groups of adults guard the developing embryos. When an embryo begins to move, a frog quickly takes it into its mouth. When the embryo finishes developing, the frog yawns, and a tiny froglet jumps out of his mouth.

A few salamander species do not complete metamorphosis. The mudpuppy and the axolotl, for example, retain their gills and continue to live exclusively in water even after they are capable of reproducing. If conditions are right, the axolotl (*Amystoma*) can mature into an adult form with lungs instead of gills. But the mudpuppy (*Necturus*) never loses its gills or develops lungs.

Figure 24-15 Some amphibians care for their young. Here the reticulated poison dart frog carries two tadpoles to water. (*Michael Fogden/DRK Photo*)

Amphibians live a dual life. The larvae are usually aquatic and the adults are usually terrestrial.

Reptiles Can Live Their Entire Lives on Land

Reptiles [Latin, *reptare* = to crawl], unlike amphibians, are entirely adapted to life on land. Although some reptiles, such as the crocodiles and sea turtles, are adapted to aquatic life, they are descended from terrestrial reptiles and must breath air. They do not have gills.

Most of the 6700 species of living reptiles fall into three orders (Figure 24-16A-D): Chelonia (250 species of turtles and tortoises), Crocodilia (22 species of crocodiles and alligators), and Squamata (lizards and snakes). Of these, the Squamata are by far the most successful, with about 3800 species of lizards and 2700 species of snakes. The Squamata also include the Amphisbaenia, which include some 140 species of legless reptiles that resemble the amphibian caecelians. Like moles, the amphisbaenians have lost the use of their eyes and burrow through soil hunting invertebrates. A fourth order of living reptiles includes only a single species, the tuatara, a lizardlike animal that survives only on a few islands off the New Zealand coast (Figure 24-16E).

Some of the most fascinating reptiles are now extinct. Although many of us think of dinosaurs when we think of extinct reptiles, only a fraction of the animals living in the Mesozoic Era, the Age of Reptiles, were actually dinosaurs. Others were swimming ichthyosaurs, and some were flying pterosaurs, for example (Figure 24-17). The long and glorious history of the reptiles began some 300 million years ago in the

BOX 24.1

Why Are Amphibians Disappearing?

Costa Rica's spectacular golden toad (*Bufo periglenes*) has not been seen since 1989, Australia's gastric brooding frog (*Rheobatrachus silus*) was last seen in 1979, populations of the common toad (*Bufo bufo*) have declined on Norwegian coastal islands, and many species of frogs in California have vanished from most of their historic ranges (Figure A). Throughout the world, amphibian species are declining in number and, in some cases, going extinct locally.

Is some unknown phenomenon exterminating toads and frogs?

The question first arose at the First International Herpetological Congress, held in 1989 in Canterbury, England. At that meeting, herpetologists recounted what each thought was a personal sorrow: the disappearance of an amphibian species from a study site. So many scientists reported the same kinds of losses, however, that David Wake, a leading herpetologist in the United States, organized a 1990 meeting in Irvine, California, to discuss whether there was a global decline in amphibian populations. At that meeting and at a follow-up symposium, the herpetological community shared information about declines in amphibian populations in Australia, Canada, the western and southeastern United States, Central America, the Amazon Basin, and the Andes.

For most amphibian populations, scientists do not have long-term data on population cycles. Without information about natural population cycles, it is impossible to determine whether, for any given species, a dip in numbers is natural variation or a decrease induced by human activities. Many frog species have boom-and-bust population cycles—populations explode one year and drop to almost nothing in another year. Such natural variation makes it difficult to interpret declines in some species.

Nonetheless, herpetologists agree that an ongoing, worldwide decrease in amphibian populations is occurring. They further agree that the primary cause is the fragmentation and destruction of habitat, as is unfortunately true for many other organisms.

Herpetologists do not yet agree what other reasons might account for the worldwide disappearance of amphibians. Some researchers say amphibian declines are just part of the worldwide loss of biodiversity caused by destruction of forests, wetlands, and other habitats. The disappearance of amphibian species, they argue, is no different from the extinctions of thousands of other species.

Other researchers insist that amphibians—with their permeable skin, permeable eggs, and complex life cycles—are particularly vulnerable to certain environmental changes. In particular, amphibians seem to be especially sensitive to increases in UV radiation and chemical pollutants. Decreases in numbers of the North American tiger salamander (*Ambystoma*

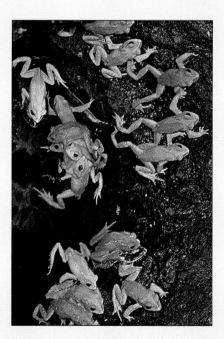

Figure A Golden toads. (*Michael Fogden/DRK Photo*)

tigrinum) have been linked to pollution, especially acid rain. The Tarahumara frog (*Rana tarahumarae*) is extinct in the United States, and researchers have implicated high levels of the toxic metal cadmium. The massive die-offs of fertilized eggs of the Cascades frog (*Rana cascadae*) and the Western toad (*Bufo boreas*) may be due to increases in UV radiation stemming from the destruction of ozone in the stratosphere; the eggs of both of these latter species are low in *photolyase,* an enzyme that helps repair UV-damaged DNA.

A host of nonnative predators have been accused of decimating amphibian populations. In the western United States, bullfrogs (*Rana catesbeiana*), nonnative trout, bass, and sunfish, all predators of amphibians, have been introduced into countless waterways. Losses of the Yavapai leopard frog (*Rana yavapaiensis*), the mountain yellow-legged frog (*Rana muscosa*), and the Chiricahua leopard frog (*Rana chiricahuaensis*) have all been attributed to such introduced predators.

Some researchers argue that an accumulation of environmental stresses, rather than a single cause, may reduce or destroy a population. For example, a population of frogs may become fragmented and reduced by the loss of wetland habitat. If a subsequent drought kills many more individuals, the population may never recover. Or stress from increased UV exposure and pollution may make individuals more susceptible to disease—one explanation for the extinction of 11 distinct populations of boreal toads (*Bufo boreas*) in western Colorado.

Frogs are showing signs of going the way of the dinosaurs. Species in habitats as different as Costa Rican cloud forest, Sonoran desert, and montane pools in the Sierra Nevada are succumbing. Unfortunately, beyond the obvious effects of habitat destruction on all species worldwide—destruction that proceeds as rapidly as ever—there is probably no one-size-fits-all explanation for the loss of amphibian species.

Figure 24-16 The four orders of living reptiles. A. Giant tortoises wallow in a pond on one of the Galápagos Islands. The Chelonia comprise the turtles and tortoises. B. Nile crocodile washing and releasing baby. The Crocodilia comprise the crocodiles, alligators, and caimans. C. Crested dragon lizard, Malaysia. The Squamata comprise the lizards and snakes. D. Two male Panamint rattlesnakes in ritual combat, White Mountains, California. E. The lone species in the order Sphenodonta is the tuatara of New Zealand. (A, Tui de Roy/Minden Pictures; B, Roger de la Harpe/Animals Animals; C, Mike Bacon/Tom Stack & Associates; D, Gordon Wiltsie/Peter Arnold, Inc.; E, John Cancalosi/DRK Photo)

late Carboniferous Period. During the Permian and Triassic periods, the reptiles radiated into about 17 orders.

Reptiles were the first vertebrates to free themselves of the water. As a result, reptiles share with their descendants, the birds and mammals, some common adaptations to life on land. Their dry, watertight skin helps conserve water, serving the same function as the cuticles of insects and plants and our own dry skin. Because reptiles cannot breathe through such tough skins, they must depend entirely on their lungs for oxygen.

Reptiles and their descendants have also evolved more efficient kidneys for conserving water and legs better suited to walking on land. Among reptiles' many adaptations to land life, however, the most dramatic advance was the evolution of the **amniotic egg** (Figure 24-18). The amniotic egg is so well adapted to a dry environment that even crocodiles and turtles, which live primarily in the water, return to the land to lay their eggs. The **amnion** is a membrane that encloses the developing embryo in its own little pond, eliminating the need for a separate aquatic stage. The amnion, in turn, lies within a porous shell that allows the exchange of gases with the surrounding air. Just beneath the shell is another membrane, called the **chorion** [Greek, = skin]. Also within the shell is the **yolk,** a rich food supply surrounded by a membrane called the **yolk sac.** A fourth membrane, the **allantois,** functions in both respiration and excretion.

All reptiles and their descendants have (or had) an amniotic egg. These include all birds and mammals. The amniotic egg is one of the characteristics that distinguishes all of these groups from other animals and shows that we are related.

In both reptiles and birds, a shell forms around the amniotic egg before it emerges from the female, and so fertilization must occur before the shell forms. That is, fertilization must be internal. All male reptiles except the tuatara accomplish fertilization with a penislike copulatory organ. Crocodilians and turtles have a single penis, while snakes and lizards have a pair of "hemipenes." Although "hemi" means "half," snakes and lizards actually have two penises, and in some species the males alternate using one and then the other (Figure 9-8).

In all reptiles, both sexes possess a **cloaca,** a common entrance and exit chamber for the digestive, urinary, and

Chapter 24 Deuterostome Animals: Echinoderms and Chordates 537

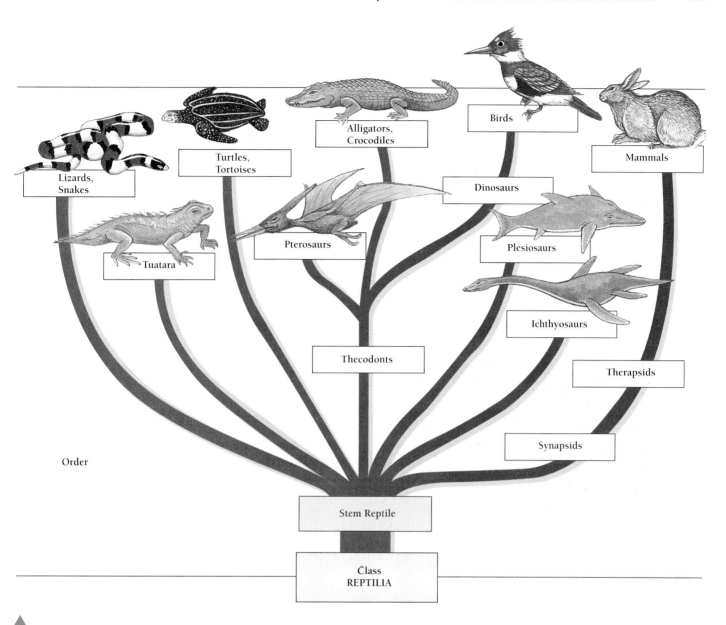

Figure 24-17 Reptiles all. The familiar lizards, snakes, turtles, and crocodiles are not the only animals in the reptile family album.

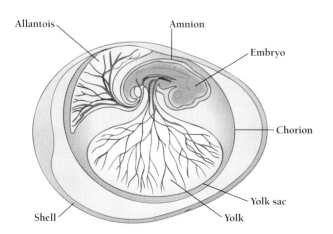

◀ **Figure 24-18 The amniotic egg.** The most obvious feature of an egg is the yolk sac, which serves as a pantry, a storage area for fats, proteins, and other nutrients for the developing animal. But fish eggs and amphibian eggs also have a yolk sac. What distinguishes the amniotic egg are three other membranes: the amnion encloses the embryo in water; the allantois serves as a sort of outhouse, a separate area for wastes; and the chorion, the outermost membrane, encloses embryo, yolk, and allantois. The chorion helps regulate the exchange of oxygen and carbon dioxide between the embryo inside and the world outside.

reproductive systems. In the female, the cloaca serves as an entrance for sperm from the male and an exit for the fertilized eggs. The eggs are usually laid in soil, sand, or leaf mold, although a few reptiles give birth to live young. In either case, the zygote develops into a tiny young reptile that fully resembles its parents.

The evolution of the amniotic egg enabled reptiles to dominate the Earth for millions of years. Reptiles' descendants, the birds and the mammals, also have amniotic eggs.

How Do Terrestrial Vertebrates Regulate Body Temperature?

As we discussed in Part I of this book, life depends on the thousands of biochemical reactions that take place in every cell. Because the rate of every one of these reactions depends on temperature, the temperature of an organism is crucial to the way it lives. In aquatic animals, the temperature of the water around them usually establishes a fairly constant body temperature.

On land, however, air and ground temperatures fluctuate enormously. Animals that have left the water must therefore find a way to regulate temperature. Most reptiles do so by moving into and out of the sun. Because reptiles (as well as amphibians) depend on external sources of heat, they are called **ectotherms** [Greek, *ektos* = outside + *thermos* = heat].

In contrast, mammals and birds warm their bodies by capturing the heat released by metabolism and are called **endotherms** [Greek, *endo* = within]. Endotherms keep nearly constant body temperatures by regulating both the heat produced by biochemical reactions and the heat lost by evaporation. Notice that we have avoided calling reptiles "cold blooded." They are not; they merely get their heat in a different way from endotherms.

The boundaries between ectotherms and endotherms are not always clear. Certain fast-swimming fish keep their muscles warmer than the surrounding water, to increase efficiency and speed. And some scientists speculate that many dinosaurs were endothermic.

Ectotherms allow body temperature to fluctuate with air temperature; they regulate temperature to some extent by basking in the sun or moving into the shade. Endotherms maintain their body temperature within a few degrees of a constant temperature most of the time. They regulate temperature by means of metabolic processes.

What Distinguishes the Birds from the Reptiles?

Birds are wonderfully diverse, with some 8700 species (Figure 24-19). Most species are distinguished by their distinctively colored plumage or their strange and haunting songs and calls. Yet birds are also remarkably uniform. They have three features that clearly distinguish them from the reptiles: feathers, flight, and endothermy. The ability to maintain a constant temperature in cold environments allows birds to live where no reptile could. The Arctic tern, for example, spends half the year north of the Arctic Circle, then flies halfway around the world to spend the rest of the year in Antarctica. Keeping a high body temperature and flying both require enormous amounts of energy. In return for this enormous energy expenditure, birds have access to vegetation, insects, and seafood that rep-

A.

B.

C.

Figure 24-19 Birds are both distinctive and diverse. Shown here are a great blue heron, a blue-throated hummingbird, and an emperor penguin and its young. (A, Jim Zipp/Photo Researchers, Inc.; B, Russel C. Hansen/Peter Arnold, Inc.; C, Barbara Cushman Rowell/DRK Photo) tiles can never reach. Flight also provides terrific protection from most predators.

Most birds find food and shelter on land, though many birds depend on streams, ponds, and seas for their food. Like the reptiles, birds produce amniotic eggs and do not need water for fertilization or early development.

Fertilization is always internal, but unlike the reptiles, only a few species possess a penis. Ducks and flightless birds such as ostriches, herons, flamingos, chickens, and turkeys all have rudimentary penises. In most other birds, both males and females have a cloaca. Birds accomplish fertilization by bringing together their cloacas. As in the tuatara, the sperm from the male's cloaca passes into the female's cloaca. Such intimacy often requires a long courtship, and many bird species have elaborate mating rituals.

Feathers are made of keratin, the same protein that forms a lizard's scales and our own fingernails and hair. Flight feathers are an amazing adaptation to flight, but down, which insulates the bodies of birds, is also important. Many biologists now believe that feathers were originally an adaptation for warmth. Flight came later, as did a host of adaptations that lightened the body and made birds still better flyers. These included the loss of teeth (which lessens the weight of the head), the hollowing of the bones (which lightens the whole skeleton), and the reshaping of the breastbone (which forms a keel to which flight muscles attach) (Figure 24-20).

Feathers, flight, and endothermy distinguish birds from reptiles.

What Distinguishes the Mammals?

Mammals [Latin, *mammae* = breasts] take their name from their **mammary glands,** milk-producing organs in the female that characterize the mammals. No other vertebrates have mammary glands. Mother mammals nurse their young with warm milk, a rich mixture of fats, sugars, proteins, minerals, and vitamins, as well as antibodies that help defend newborns against infections.

Two other traits distinguish the mammals from other animals. Only mammals have hair and two sets of teeth ("baby" teeth and adult teeth). A mammal's hair, like a bird's feathers, helps conserve heat, a trait that is useful for an endotherm. Mammalian teeth, far more specialized than those of other vertebrates, help mammals get the tremendous amounts of energy they need to maintain a constant body temperature.

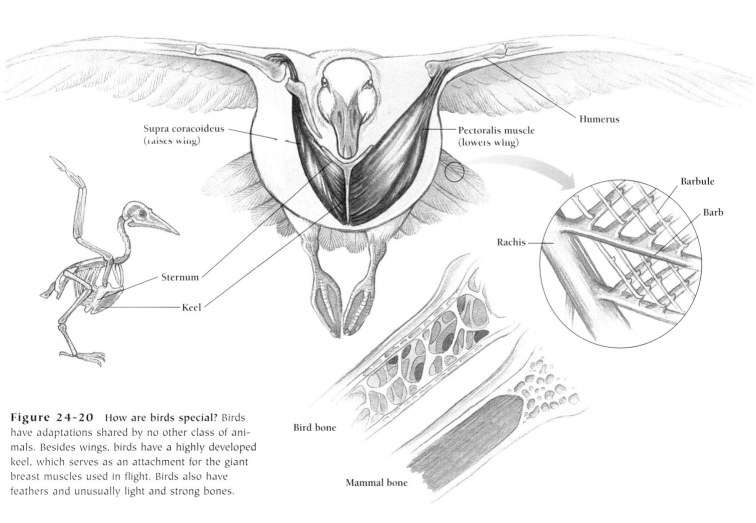

Figure 24-20 How are birds special? Birds have adaptations shared by no other class of animals. Besides wings, birds have a highly developed keel, which serves as an attachment for the giant breast muscles used in flight. Birds also have feathers and unusually light and strong bones.

Mammalian teeth may be specialized, for example, for grinding plants (horses and other herbivores), for shearing flesh (lions and wolves), for breaking bones (hyenas), or for seizing fish (dolphins and whales).

Mammalian teeth are extremely well preserved in the fossil record. Since a single tooth tells volumes about its owner, paleontologists know more about the evolution of mammals than is possible for other classes. In fact, some paleontologists now believe that the efficiency of the mammalian molar is at least partly responsible for the evolutionary success of the mammals since the end of the Cretaceous Period. When the ruling reptiles died off, mammals radiated into about 30 orders, of which about two-thirds still have living representatives (Figure 24-21).

Mammalogists divide the living mammals into three subclasses—prototherians, metatherians, and eutherians. These subclasses have distinctive skulls and teeth that show that they first diverged from each other during the Mesozoic Era, even before the dinosaurs disappeared. These three subclasses differ markedly in the ways they reproduce.

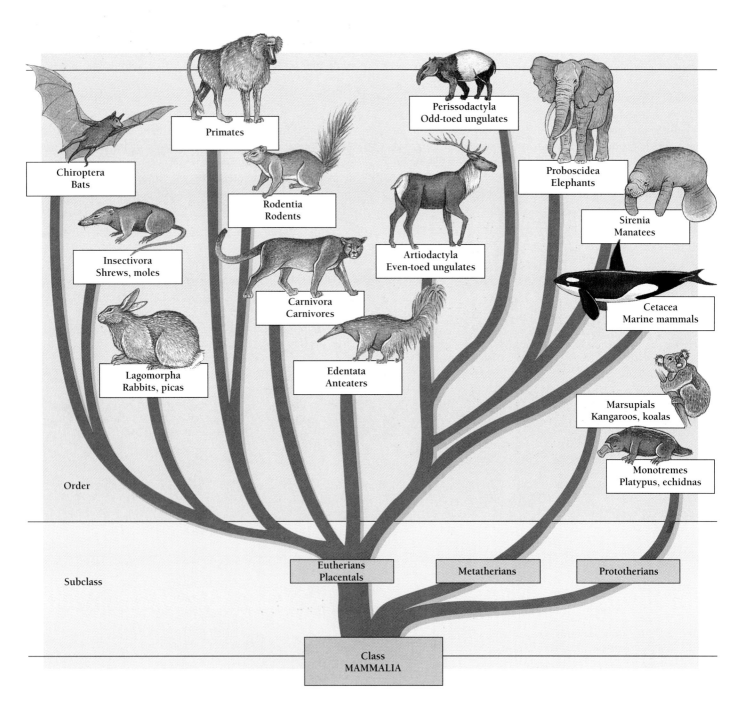

Figure 24-21 Some orders of the class Mammalia.

Most mammals are **eutherians** [Greek, *therion* = wild beast] or **placental** mammals. In humans, elephants, rats, rabbits, and every other kind of eutherian mammal, embryonic development takes place within the uterus of the mother. In fact, the word "placental" refers to the **placenta** [Latin, = flat cake], the flat structure that joins the lining of the uterus with the membranes of the fetus.

Metatherians, or **marsupials** [Latin, *marsupium* = pouch], such as the kangaroo and koala, reproduce differently from eutherians. Newborn marsupials are not babies at all, but undeveloped fetuses. A newborn red kangaroo, for example, is no bigger than a bumblebee (Figure 24-22). Emerging from the uterus only 33 days after fertilization, the tiny creature painstakingly pulls itself along its mother's fur until it finds a nipple. The mother's nipples are in a furry pouch, where the young kangaroo sleeps and nurses for about 6 months. Even after it begins to wander a bit on its own, the young kangaroo regularly returns to the pouch until it is almost 8 months old. Meanwhile, the mother may have another, younger baby in the pouch, as well as an embryo in her uterus.

The **prototherians,** or **monotremes,** are the strangest mammals (Figure 24-23). They are endotherms with hair and

Figure 24-23 Duck-billed platypus. Unlike other mammals, monotremes lay eggs. They are considered the most primitive of all the living mammals. *(Alan Root/Okapia/Photo Researchers, Inc.)*

mammary glands, like other mammals. Like the reptiles from which they and all other mammals are descended, however, monotremes lay eggs that are similar to those of reptiles and birds. In addition, their digestive and reproductive systems empty into a cloaca, hence their name [Greek, *monos* = single + *trema* = hole]. The only living monotremes are the platypus and the two spiny anteaters (or echidnas), which live in Australia and New Guinea. They are believed to be primitive mammals, but exactly how they are related to other mammals is still unknown. Mammals, both modern and ancient, are generally classified according to their tooth structure, and adult monotremes have no teeth.

Figure 24-22 A red kangaroo with its offspring. All marsupials carry their young in a special pouch that contains milk-secreting nipples. A female kangaroo may have a youngster at her side, another in the pouch, and a third in her womb. *(Dave Watts/Tom Stack & Associates)*

Three characteristics set mammals apart from other vertebrates—mammary glands, hair, and two sets of teeth.

In this chapter we have seen how creatures as seemingly different as sea urchins and rabbits can share traits that show they are related. This chapter concludes our survey of the six kingdoms of organisms. In the next section, we see how ecological forces shape these many living organisms.

Study Outline with Key Terms

Research in molecular genetics supports the hypothesis, first advanced in the 19th century, that chordates have an inverted version of the same body plan as most protostome coelomates. Structures that are **dorsal** in frogs and other vertebrates are **ventral** in arthropods. Deuterostomes, however, differ from protostomes in several ways, including, for example, the fate of the embryo's **blastopore.** Deuterostomes include four phyla: **Echinodermata, Chaetognatha, Hemichordata,** and **Chordata.**

Echinoderms, such as starfish, are radially symmetrical as adults and bilaterally symmetrical as larvae. Echinoderms diverged from protostomes more than 500 million years ago. Chaetognatha (arrow worms) and Hemichordata (acorn worms) may resemble the ancestors of the chordates. Hemichordates have pharyngeal slits and a dorsal nerve cord, both of which are also chordate characteristics. They do not have, however, a notochord or a postanal tail, as do the true chordates.

Chordates share four main features at least sometime during their lives: a stiff flexible rod, called a **notochord,** running along the back; a **dorsal hollow nerve cord; pharyngeal slits** (or **gill slits**); and a postanal tail Chordates include the vertebrates and two other subphyla, the tunicates and the cephalochordates. Tunicates resemble other chordates as larvae, but not as adults. Cephalochordates have notochords both as larvae and as adults.

Vertebrates have heads and segmented spinal columns. The embryos of all vertebrates have skeletons made of **cartilage.** Most vertebrates replace the cartilage with **bone** during development. Vertebrates have closed circulatory systems and characteristic internal organs.

Four classes of vertebrates are fishes—Agnatha, Placodermi, Chondrichthyes, and Osteichthyes, and four classes are **tetrapods,** so-called four-footed animals—amphibians, reptiles, birds, and mammals. Agnatha (the extinct **ostracoderms** and the living **cyclostomes**) are the only vertebrates that lack jaws. **Placoderms,** which are now all extinct, were the first vertebrates with jaws. In the Chondrichthyes (sharks and rays), bone never replaces cartilage. The Osteichthyes (bony fish) have bony skeletons and a **swim bladder.** One type of bony fish, with lobe fins and primitive lungs, resembles the probable ancestors of the tetrapods.

Tetrapods have many special adaptations for land life. They have some or all of the following: **double circulation** and lungs, efficient kidneys, water-resistant skin, strong skeletons and jointed limbs, **ectothermy** and **endothermy,** and the **amniotic egg.**

Amphibians are semiterrestrial animals that always begin their lives in water but later develop lungs and other adaptations to live on land, in the process of **metamorphosis.** Reptiles and their descendants, the birds and mammals, can all reproduce away from water as a result of the development of the amniotic egg. The different parts of the amniotic egg—the **amnion,** the **chorion,** the **yolk** and **yolk sac,** as well as the **allantois**—all contribute to the egg's ability to regulate water balance on dry land. Reptiles also developed watertight skins and strong skeletons. Reptiles and birds have a common exit for digestive, urinary, and reproductive systems, called the **cloaca.** Birds added feathers, flight, and endothermy, while mammals developed hair, two sets of teeth, and **mammary glands** for nursing their young. The mammals are divided into the **eutherians,** or **placental** mammals; the **metatherians,** or **marsupials;** and the **prototherians,** or **monotremes.**

Review and Thought Questions

Review Questions

1. Describe three characteristics that distinguish the deuterostomes from the protostomes.
2. What characteristics do humans share with starfish (sea stars)?
3. What three characteristics define the chordates? Can you locate all of these features in your own body? For each feature, explain where it is or where it once was.
4. Rank the following organisms according to how closely related they are to humans: tunicates, lamprey larvae, and arrow worms.
5. Is the vertebral column, or backbone, actually made of bone in all vertebrates? What is bone?
6. Of what value are jaws?
7. Why must amphibians return to water to reproduce?
8. Name the layers of the amniotic egg. What classes of animals have amniotic eggs?
9. What is thermoregulation? Why is it important? In what two ways do vertebrates accomplish thermoregulation?
10. What three mammalian characteristics distinguish a human from a turkey?
11. What is the function of a placenta?

Thought Questions

12. Compare the vertebrate body plan with the arthropod body plan. What advantage(s) might there be to flipping over the arthropod body plan? How would you test your hypothesis?
13. How would the day-to-day lives of humans be different if we were marsupials instead of placental mammals?

About the Chapter-Opening Image

This child illustrates the idea that the vertebrate body plan is the reverse of that of invertebrates such as lobsters and insects. As one biologist put it, "If you lay down on your back and waved your arms, you would be doing what insects do when they walk."

 On-line materials relating to this chapter are on the World Wide Web at **http://www.harcourtcollege.com/lifesci/aal2/**

 Web Links

 Quizzing & Testing

part V

Ecology

Chapter 25
Ecosystems

Chapter 26
Terrestrial and Aquatic Ecosystems

Chapter 27
Communities: How Do Species Interact?

Chapter 28
Populations: Extinctions and Explosions

Chapter 29
The Ecology of Animal Behavior

Photo: Mass of starfish, Rialto Beach, Olympic National Park, Washington. (© *Darrell Gulin/Natural Selection*)

25 Ecosystems

Gaia: Is The Biosphere An Organic Entity?

In the early 1960s, the National Aeronautics and Space Administration (NASA) began planning its first explorations of the solar system. The space agency was especially keen to examine Martian soil for signs of life or conditions that could support life. But analyzing the Martian soil would require new instruments specially designed for the project.

For these, NASA turned to James E. Lovelock, an eccentric English chemist who had invented an ingenious device that detected tiny amounts of chemicals. Lovelock's device was widely used to measure trace amounts of pollutants such as the insecticide DDT.

In the early 1960s, NASA invited Lovelock to the Jet Propulsion Laboratory, in California, to help design instruments for detecting life on Mars. Lovelock's fascination with the search for extraterrestrial life and his insatiable curiosity soon led him beyond this technical chore to reconsider the much greater question of life on Mars. What if Martian or other life didn't look or act anything like terrestrial life? Could it still be detected? If so, how? How could life be described so that scientists could recognize it even in an alien form?

Lovelock approached these questions by considering the concept of "entropy." *Entropy* is a measure of disorder in "systems." A **system** is any collection of interacting parts or objects. If any system—whether a solar system or a dorm room—is left alone, its disorder, or entropy, will increase.

Organisms, by contrast, create order and thus *decrease* local entropy. Plants, for example, convert carbon dioxide and water into highly structured leaves, stems, and flowers. And animals convert amino acids and simple fats into muscle and brain tissue. This decrease in local entropy exacts a cost, however. It requires energy from the sun. And the decrease in entropy is only local. The universe as a whole is running down; its entropy is increasing.

A sure sign of life on Mars, then, Lovelock realized, would be a reduction in local entropy. But what in the world, he wondered, does a "reduction in entropy" look like? Lovelock couldn't stop thinking about the problem. He wondered how an alien being would recognize life on Earth. And it dawned on him that an alien would notice the peculiar chemical composition of Earth's atmosphere.

Earth's atmosphere is loaded with unlikely combinations of chemicals. For example, the atmosphere contains both methane and oxygen, which reacts with methane. Methane is a simple one-carbon hydrocarbon made by bacteria that gives swamp gas its smell. Methane and oxygen readily react with one another to form carbon dioxide and water.

$$CH_4 + 2\ O_2 \longrightarrow CO_2 + 2\ H_2O$$

This reaction constantly destroys methane. If no living organisms were present, methane would soon disappear from Earth's atmosphere. But bacteria in marshes, rice paddies, and cows' stomachs constantly make more methane, so the gas persists.

Lovelock realized that the Earth's chemically unlikely atmosphere could be seen as an extension of our planet's **biosphere**—the system of living things that covers the Earth. Just as a tortoise's shell is part of its body, the Earth's atmosphere is part of the biosphere. Lovelock envisioned the atmosphere, the oceans and soils, as constituting a single, immense living individual. He further hypothesized that this immense organism acts to preserve itself. Lovelock named this entity Gaia after the Greek Earth goddess.

Key Concepts

- An ecosystem is an interacting group of species, together with nonliving components such as sunlight, soil, water, and air. All of the world's ecosystems, together with the atmosphere and the oceans, make up the biosphere.
- Plants and other autotrophs capture energy from sunlight, but only a small fraction of that energy is available to other organisms.
- Although the biosphere constantly loses energy, it recycles materials such as carbon and nitrogen.
- Without the effect of greenhouse gases, Earth would be a lifeless planet. But too much carbon dioxide and other greenhouse gases can damage the biosphere.

Lovelock reasoned that the biosphere regulates the temperature and chemistry of the atmosphere, just as many animals regulate their blood temperature and chemistry. Such self-regulation is called homeostasis (Chapters 1 and 34).

The Gaia hypothesis, first published in the 1960s, vastly entertained the public. TV anchors glibly announced that the Earth was one big organism. Environmentalists seized on Gaia as a cultural metaphor, stressing the need to consider the biosphere and its environment—the atmosphere, the oceans, and the soils—as a whole. NASA's first breathtaking photographs of Earth as a cloud-swathed blue ball floating in space intensified Gaia's already enormous popular appeal.

Yet while biologists found the idea of Gaia amusing, most also found it embarrassing. Despite Lovelock's impressive credentials as an inventor, many scientists viewed him as a crank. Academic ecologists of the 1960s didn't want their careful scientific work confused with the free-wheeling political activity of environmental groups. But Lovelock's credentials as a scientist seemed to blur the line between ecology and the environmental movement. Evolutionary biologists, for their part, dismissed Gaia, arguing that the biosphere cannot be considered an organism because it neither reproduces nor evolves.

Lovelock now denies that he intended Gaia to be understood as an actual organism. He says Gaia is more like a **superorganism,** a group of individuals that cooperate almost like the cells of the body (Figure 25-1). Yet at times Lovelock has described Gaia in almost human terms, as though

Figure 25-1   Do superorganisms really exist? This Portuguese man-of-war consists of a tightly integrated colony of individuals. Each individual in the colony is a kind of polyp (related to the jellyfish) that cooperates with its comrades to form a complex superorganism. A bright blue, gas-filled float doubles as a sail. Wind drives the colony along the surface of the ocean. Beneath hang tangles of polyps bearing stinging tentacles that can reach 20 meters in length. The polyps quickly immobilize and digest fish or other creatures that come in contact with the deadly tentacles. (Kelvin Aitken/Peter Arnold, Inc.)

"she" had intentions and goals. He has referred, for example, to "Gaia's intervention" and "Gaian impatience."

The line between the Gaia metaphor and formal science blurred further when a biologist of national standing stood up for Lovelock. In 1974, Lynn Margulis, a distinguished professor of biology at the Amherst campus of the University of Massachusetts, began enthusiastically promoting a version of the Gaia concept—to the horror of her scientific colleagues.

In Margulis's view, Gaia is not a superorganism but a giant ecosystem, the highest level of organization of life. Margulis, now a member of the prestigious National Academy of Sciences, also proposed a related idea on the cellular level that is now widely accepted (Figure 25-2A). She persuaded her fellow biologists that the mitochondria and chloroplasts of eukaryotic cells were once free-living bacteria that came to live permanently inside other prokaryotic bacteria. In other words, we eukaryotes are a combination of prokaryotes that began cooperating millions of years ago. Such a cooperative arrangement is an example of **symbiosis** [Greek, *syn* = together + *bios* = life], any association between two organisms. Margulis asserts that the different organisms of our biosphere cooperate in similar ways to create a functioning ecosystem. The view that organisms cooperate as well as compete is fundamental to the view that our biosphere is a tightly integrated whole (Figure 25-2B).

Indeed, although most biologists still reject Lovelock's idea that our biosphere is a superorganism, nearly all scientists now accept Margulis's view that the biosphere is a giant, tightly integrated ecosystem. Just don't call it Gaia.

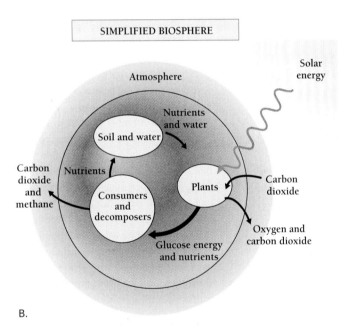

Figure 25-2 Living organisms cooperate. A. Most biologists now agree that the chloroplasts and mitochondria that generate ATP and other energy-rich molecules were once prokaryotic cells that came to live inside our own cells. The idea that eukaryotic cells are symbiotic associations of prokaryotic cells is the brainchild of biologist Lynn Margulis. B. Margulis also asserts that the different organisms of our biosphere cooperate in similar ways to create a functioning ecosystem. Shown here is a simplified biosphere.

WHAT IS ECOLOGY?

The word "ecology" is popularly associated with recycling soft-drink cans, planting saplings in city parks, and campaigning for stricter air and water pollution standards. Yet the science of ecology includes none of these activities. **Ecology** [Greek, *oikos* = home] is the study of the interactions of organisms with one another and with their physical environment, or "home." It is also the study of how those interactions determine the abundance and distribution of organisms.

An individual organism is a member of a number of systems. For example, an individual is a member of a **population**, an interbreeding group of individuals; a **community**, an interacting group of many species that inhabit the same area; and an **ecosystem**, a community of organisms together with the nonliving parts of their environment (Figure 25-3).

In addition, ecologists categorize terrestrial communities into various types of *biomes*, such as desert or tropical rain forest, based on the main kinds of plants that grow there. In Chapter 26 we discuss some of the main aquatic communities and the main biomes (types of terrestrial communities).

What Is an Ecosystem?

An ecosystem consists of parts that are living, or **biotic**, and those that are nonliving, or **abiotic**. The abiotic parts of an ecosystem include forces such as wind and gravity; conditions such as light intensity, temperature, salinity, or humidity; and physical substances such as soil, air, and water. These substances can include both inorganic compounds and organic compounds from living things. Soil, for example, usually contains inorganic nitrogen, minerals, and water. But it also contains organic material such as wood particles from trees and shrubs, chitin from the exoskeletons of dead insects, and disintegrating proteins from all kinds of organisms.

An ecosystem has many of the properties of living organisms that we listed in Chapter 1. An ecosystem obtains energy from its environment, transforms chemicals, changes with time, and responds to environmental changes. And, like individual organisms, ecosystems use energy to maintain a stable state. However, ecosystems differ from organisms in important ways. For example, unlike organisms, ecosystems do not reproduce themselves. Further, an ecosystem does something that no organism does. It recycles. No organism supports its own growth by eating its own waste.

The living and nonliving components of an ecosystem interact with one another as a unit. The atmosphere, the water, and the soil both permit and limit life within an ecosystem. A freshwater lake, for example, provides all of the conditions necessary for certain fish and aquatic plants to flourish. Yet, the same lake would be inhospitable to plants and animals adapted to an estuary or a lake with different physical attributes, such as temperature, pH, or nutrient levels.

Just as the physical environment affects organisms, organisms affect their physical environment. Trees, for example, block sunlight, change the characteristics of soil, and release oxygen into the atmosphere. Animals consume plants, altering the physical structure of the environment. Elephants may uproot whole stands of trees in order to eat their leaves, and rabbits can nibble grasses right down to the ground.

An ecosystem's boundaries may seem well defined: a pond is enclosed by its shores, a grassy meadow by the surrounding forest. But an ecosystem's boundaries are fuzzy: the pond, for example, blends little by little into marsh, and then into a mixture of brush and open meadow. A stream brings nutrients and organisms from other areas into the pond and carries materials to other ecosystems.

Each ecosystem, community, or population is part of a larger one. A pond ecosystem may be part of a meadow ecosystem, the meadow part of a forest, and the forest part of a two-thousand-mile drainage system such as the Mississippi River Valley. We may therefore speak of the frog population of the pond, of the meadow, of the forest, or of the local drainage of streams. How a population is defined depends on the mobility of the organism. Populations of birds may cover vast areas, while populations of frogs may be limited to a few square miles. To some degree, then, biologists choose the boundaries of the populations and ecosystems that they study, always keeping in mind the larger picture.

Most ecosystems interact with other ecosystems. Seeds and spores disperse, animals migrate, and flowing water and air carry organisms—and their products and remains—from one place to another.

All ecosystems taken together make up a much larger ecosystem, the biosphere. The biosphere differs from other ecosystems in having fixed boundaries. Extending over the whole of the Earth's surface, the biosphere begins two miles (3 km) underground and extends into the highest reaches of the atmosphere. It ends where the atmosphere ends and space begins.

In this chapter, we discuss ecosystems as entities, whose parts interact in complex ways. In succeeding chapters of this section on ecology, we will see how ecosystems and communities change over time, how human intervention affects ecosystems, how individual populations behave, and, finally, how individual organisms behave.

Ecology is the study of the interactions of organisms with each other and with their physical environment. An ecosystem is a functional unit consisting of both biotic parts and abiotic parts.

Figure 25-3 Levels of organization. An individual is a member of a population, a community, an ecosystem, and the biosphere.

HOW DOES ENERGY FLOW THROUGH ECOSYSTEMS?

The biotic part of an ecosystem consists of two great groups of organisms, the autotrophs and the heterotrophs. **Autotrophs** [Greek, *auto* = self + *trophe* = to nourish], or **producers,** are organisms that produce their own food. The vast majority of autotrophs are either green plants or photosynthesizing bacteria, both of which obtain energy from the sun.

Heterotrophs [Greek, *hetero* = other + *trophe* = to nourish] are organisms that consume other organisms. Most are animals, fungi, or protists. Biologists divide the heterotrophs into two groups: consumers, such as animals; and decomposers, such as bacteria and fungi. **Consumers** (mostly animals) obtain energy by eating other organisms. Biologists divide consumers according to what they eat. Deer, quail, and caterpillars, for example, eat plants (producers). Such plant eaters are called **herbivores.** Mountain lions, hawks, wasps, and other **carnivores** eat other animals.

Some animals feed from the whole carcasses of dead animals or from large pieces of dead plants. Such **scavengers** include millipedes, sow bugs, earthworms, and termites, which eat dead plant material; as well as flies, lobsters, starfish, clams, and catfish, which often eat animal remains. Many animals, including starfish, lions, and humans, may change from carnivore to scavenger and back, depending on what is available.

But scavengers do not consume all dead material. Most dead organisms are decomposed by **saprophytes** [Greek, *sapro* = putrid + *phyte* = plant]—fungi, bacteria, and even some plants—that live off the energy stored in particles of dead organisms.

Many levels of carnivorous consumers can exist: hawks eat snakes, which eat ground squirrels, which eat lizards, which eat predaceous insects, which eat caterpillars, which eat plants. Such a linear sequence is called a **food chain.** Most food chains have only four or five links (Figure 25-4).

Every ecosystem has numerous food chains, with different numbers of trophic levels. The collection of all the interconnected food chains of an ecosystem is called a **food web** (Figure 25-5). In reality, no food web is static. Instead they are dynamic entities that change with circumstance. In one year, a population explosion of oak moths means that insect predators focus on oak moth caterpillars. In another year, oak moths are rare, and predators eat something else, perhaps mostly one thing, perhaps a diversity of insect prey. Most predators are "generalists," animals that will eat a variety of different foods.

Ecologists assign the organisms in a food web to different "trophic levels," depending on where they get their energy. For example, plants, which get their energy directly from the sun, are in the first trophic level; caterpillars, which get their energy from plants, are in the second; birds that eat caterpillars are in the third. Another way of designating trophic level is to call herbivores "primary consumers," to call carnivores that eat herbivores "secondary consumers," to call animals that eat secondary consumers "tertiary consumers," and so on. **Trophic**

Figure 25-4 A food chain. In a food chain, the productivity of plants and other primary consumers goes to herbivores such as caterpillars, which are eaten in turn by predators such as robins. *(American robin, Gary Meszaros/Dembinsky Photo Associates; orange-tipped oakworm, PSU Entomology/Photo Researchers, Inc.; 500-year-old white oak tree, Jack Rosen/Photo Researchers, Inc.)*

level [Greek, *trophe* = to nourish], then, is an organism's position in the food chain.

The different levels of a food chain are often approximations only. For example, the giant panda normally eats almost nothing but bamboo shoots, which makes it a primary consumer or herbivore. But now and then, this bear will kill and eat a rodent, which makes it a carnivore. Many animals, including humans, bears, pigs, and crows, consistently eat from several levels of a food chain at once. Such animals are called **omnivores** [Latin, *omnis* = all + *vorus* = devouring] because they eat plants, herbivores, and other carnivores. Many ani-

Figure 25-5 A food web. A food web is the sum of many food chains. Organisms at each trophic level may feed from several trophic levels including their own. Decomposers return nutrients back to primary producers.

mals eat one thing as larvae and another as adults. Frogs, for example, are herbivores as tadpoles, but carnivores as adults. Omnivores do not easily fit into any particular trophic level. Unfortunately, for biologists trying to understand food webs, omnivores, far from being the exception, are quite common. One study of a desert food web found that nearly 80 percent of species were omnivorous.

Finally, many species eat each other. The adult fish of species A may eat the young of species B, while the adults of species B eat the young of species A. Which species is at a higher trophic level? Clearly neither. In many food webs, then, trophic level beyond the primary producers is almost meaningless. Still, the *idea* of trophic level is a useful way to think about how energy moves through ecosystems.

Every ecosystem includes one or more food chains that together form a food web. Autotrophs transform energy from the sun into energy-rich molecules. Heterotrophs obtain energy from autotrophs by eating organisms, whether alive or dead.

How Much Energy Flows Through an Ecosystem?

Every ecosystem is unique, whether it is a tiny pond or the vast Serengeti Plain of Africa. Yet, similar ecosystems share fundamental characteristics, including climate, productivity, total mass of living organisms, and numbers of species. We'll discuss many of these characteristics in coming chapters. Here we can look at two important measures of how much energy flows through an ecosystem, *biomass* (the total mass of all living organisms) and *productivity*.

Productivity is a measure of the energy captured in the chemical bonds of new molecules on a square meter of land per unit time, usually per year. Almost all energy for ecosystems comes from sunlight. (The rare exceptions are deep ocean ecosystems fueled by inorganic molecules.) Yet, primary consumers can use only one percent of the energy that comes from the sun. The atmosphere reflects 30 percent back into space and absorbs 20 percent. Most of the rest reaches Earth but is absorbed by rocks, soil, and water.

Different parts of the Earth receive different amounts of sunlight. Each square meter of land near the North and South poles, for example, receives about 700,000 kcal per year. Lands at the equator receive nearly 2.5 times that much, about 2 million kcal per square meter per year. When the sun is low in the sky, as it is at dawn and dusk, more light reflects back into space. At the poles, the sun is never overhead, but always low in the sky. So the poles receive far less sunlight than the equator. You can demonstrate this effect for yourself by holding a piece of paper up to a flashlight. Imagine that the paper is the atmosphere and the flashlight is the sun. If you hold the paper at a right angle to the flashlight, you see more intense light than if you hold the paper at an angle. Equally important, the poles and temperate latitudes receive different amounts of light at different times of year. In contrast, equatorial regions receive the same amount of light all year around. Such differences profoundly affect the ecologies of these regions.

Biologists can easily measure the sunlight that reaches a square meter of forest using a light meter like the ones in automatic cameras. The amount of light is about 1 to 2 million kcal per year, equivalent to the chemical energy in 30 to 60 gallons of gasoline, or in all the food each of us eats in a year.

We can estimate how much of this energy plants and other producers capture by measuring their rate of photosynthesis and their rate of respiration and subtracting one from the other. The result is the "net" energy stored, or productivity. In a pond, for example, we can estimate the total rate of photosynthesis and respiration for all the microorganisms. The productivity of the producers is called the **primary productivity.**

Primary productivity varies enormously from ecosystem to ecosystem. The most productive ecosystems are marshes. A marsh is twice as productive as a temperate forest, 4 times as productive as a wheat field, and 35 times as productive as a desert. In Chapter 26 we discuss some of the reasons that ecosystems differ in their productivity.

Another important characteristic of ecosystems is total **biomass,** the dry weight of all the organisms living in it—including producers, consumers, and decomposers. ("Dry weight" excludes water, so the dry weight of a 120-pound human being, for example, would be about 35 pounds.) Rain forests have more organisms per square meter and therefore more total biomass than other ecosystems, more even than the more-productive marshes.

In terrestrial ecosystems, plant biomass usually exceeds herbivore biomass, which exceeds carnivore biomass. A graph of the total biomass of organisms at each trophic level of a forest ecosystem looks like a pyramid, a **pyramid of biomass** (Figure 25-6).

But a "pyramid" of biomass does not always have a pyramidal shape. In a pond, for example, the biomass of algae may be less than the biomass of fish. How can the biomass of the consumers be greater than that of the producers (Figure 25-7)? The answer lies in the flow of energy from one trophic level to the next.

During the course of a year, the increase in biomass of algae in a pond is far greater than the increase in biomass of fish. The algae always grow faster than the fish. But the fish eat the algae as fast as it grows, and they convert only a fraction of it into fish. The rest is used to power movement and metabolism or is lost as heat.

Once we compare the energy flowing through the first trophic level with the energy flowing through the succeeding levels, it becomes clear that the producers always process more energy than the consumers. In fact, this relationship holds in all ecosystems.

Plants and other producers use the energy of sunlight to build their bodies and to drive life processes. However, every chemical and mechanical process releases energy in the form of

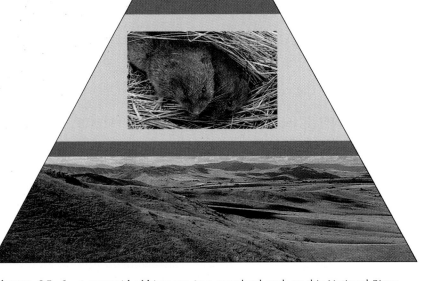

Figure 25-6 A pyramid of biomass. In a grassland such as this National Bison Range, in Montana, the biomass of the primary producers is greater than that of herbivores and carnivores—eastern meadow voles (*Microtus pennsylvanicus*) and bobcat (*Lynx rufus*). (Grassland, Richard J. Green/Photo Researchers, Inc.; voles, Gary Meszaros/Visuals Unlimited; bobcat, R. Lindholm/Visuals Unlimited)

Figure 25-7 Inverted pyramid of biomass in an aquatic community. In aquatic communities, the primary producers may be eaten as fast as they multiply. So, while the accumulated biomass of the consumers is large, that of the producers may be small. Here schools of tuna feed on northern anchovies, which feed on plankton. *(Tuna, David B. Fleetham/Tom Stack & Associates; anchovies, Norbert Wu/Peter Arnold, Inc.; plankton, D.P. Wilson/Science Source/Photo Researchers, Inc.)*

heat. So producers can convert only part of the energy from the sun into chemical energy. Because energy becomes less and less available at successively higher trophic levels, the amount of energy that passes through each trophic level in a certain amount of time (usually a year) looks like a pyramid.

A **pyramid of energy** is always upright. The producers always process more energy than the primary consumers, which always process more energy than secondary, tertiary, and higher-level consumers. Even when the pyramid of biomass is inverted, the pyramid of energy is always upright (Figure 25-8).

Of the 1.5 million kcal that strike a square meter of land in the United States each year, only one percent—no more than 15,000 kcal/yr—is actually available to primary consumers such as deer and cattle. Of this, herbivores convert only 10 percent to tissues. Each trophic level passes on only about 10 percent of the energy of the one below it. The rest is lost as heat in metabolic processes. This generalization is called the **ten percent law.** The ten percent law explains why a pyramid of energy is always upright.

The ten percent law also explains why ecosystems have so few trophic levels and so few individuals at the highest trophic level. If on a square meter of land, primary consumers store 15,000 kcal/yr, secondary consumers (herbivores) will have only about 1500 kcal/yr to live on, and tertiary consumers (herbivore-eating carnivores) will have only 150 kcal/yr to live on, about as many calories as are in a cup of spaghetti.

Worldwide, humans take a huge amount of the Earth's primary productivity. We use 40 percent of terrestrial primary productivity. This includes not only the food we eat and the wood we use to build our houses, but also the areas we use to grow lawns and golf courses and the areas formerly covered by forests and marshes and now covered by asphalt or shopping malls and factories.

Figure 25-8 Pyramid of energy. The pyramid of energy is always upright, no matter whether for a forest or a pond. Here the grassland of the Masai Mara National Reserve, in Kenya, fixes carbon and energy on which zebras feed. These spotted hyenas, from Kruger National Park, South Africa, are the top predators in a similar ecosystem. *(Masai Mara National Reserve, Renee Lynn/PRI; zebras, Jeremy Woodhouse/DRK Photo; hyena, Gil Lopez-Espina/Visuals Unlimited)*

This is why top carnivores must find food over wide areas. Mountain lions, for example, roam over hundreds of square miles. All top carnivores are highly mobile animals, including, for example, sharks, killer whales, and eagles. Their travels force them to expend even more energy, which means that they need even more food. Top carnivores also tend to be "generalists," animals that eat a variety of foods. They can't afford to specialize too much.

Two important measures of energy flow in ecosystems are primary productivity and biomass. A pyramid of biomass may not always be upright. But a pyramid of energy is always upright, because energy dissipates as it moves from one trophic level to the next.

HOW DO ECOSYSTEMS RECYCLE MATERIALS?

We can see that energy constantly escapes from the biosphere. In contrast, the biosphere recycles water, carbon, and other materials (Figure 25-9). As nutrients move from one trophic level to another, they may change form. But, unlike energy, nutrients rarely escape from the biosphere entirely. Sometimes these elements are part of organisms; sometimes they are part of the atmosphere (Table 25-1). A single carbon atom in your fingernail, for example, may have been, at different times, part of an apple, part of a bicarbonate ion in the ocean, and part of a lump of coal. Carbon and other materials pass through many forms—both biotic and abiotic—in a system called a **biogeochemical cycle**.

TABLE 25-1 MAJOR COMPONENTS OF EARTH'S ATMOSPHERE

Molecule	Percent in atmosphere
Nitrogen (N_2)	78.0
Oxygen (O_2)	21.0
Argon (Ar)	0.934
Water (H_2O)	0.10–1.0
Carbon dioxide (CO_2)	0.035
Neon (Ne)	Trace
Helium (He)	Trace
Methane (CH_4)	Trace

Oxygen, nitrogen, phosphorus, and other materials pass through biogeochemical cycles as well. Carbon dioxide (a compound of carbon and oxygen) moves from plants to consumers to water and air and back to plants and other producers. Plants take up carbon dioxide, remove the carbon, and release oxygen. Plants, animals and other organisms then take the same oxygen from the atmosphere, use it to burn carbon compounds, then release carbon dioxide.

The biogeochemical cycles of materials such as carbon and oxygen involve the whole biosphere. A carbon atom may travel all over the world on currents of wind or ocean. In fact, atoms of oxygen and nitrogen become so widely dispersed that each time one of us takes a breath, we breathe in a few atoms from Julius Caesar's dying breath, in 44 B.C. Other materials, such as calcium and phosphorus, tend to stay put and cycle within smaller, more local ecosystems. We will discuss three biogeochemical cycles: those of water, carbon, and nitrogen.

Organisms reuse materials such as water, carbon, oxygen, and nitrogen.

What Drives the Water Cycle?

All life is intimately connected to water. Organisms need a moist environment for the cells of their bodies and plenty of water for the insides of their cells. Vertebrates are about 70 percent water. And plants contain even more water. Plants use some 500 grams of water for every gram of biomass they produce. Some of this water is broken down to help form sugars during photosynthesis. But most of the water passes right through a plant unchanged. Plants pull water from the soil and release it as vapor through tiny pores (stomata) in their leaves, a process called **transpiration** (Chapter 31). On a hot summer day, an average maple tree transpires more than 50 gallons of water—enough to fill a large bathtub—every hour.

Flowing water also refreshes and flushes ecosystems. Running water brings in minerals, organic substances, and even

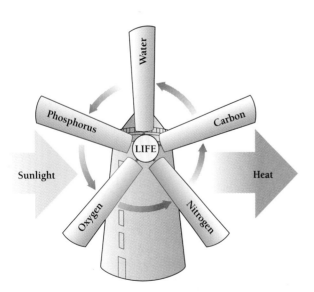

Figure 25-9 Ecosystems recycle material but lose energy forever. Sunlight drives ecosystems the way wind drives windmills. What's left is heat that passes into the atmosphere and, ultimately, out into space. In contrast, carbon, nitrogen, water, phosphorus, and oxygen cycle endlessly through the biosphere, like the vane of a windmill.

Figure 25-10 The water cycle. Far more water evaporates from the oceans than from the land. But rain falls disproportionately on land. Clouds that form over the oceans drift over the land, rise up into the mountain, and drop rain and snow that forms streams, rivers, lakes, and underground water flows.

organisms and carries away wastes. Water cycles through both biotic and abiotic parts of the biosphere (Figure 25-10). More than 97 percent of the water on our planet is in the oceans, which cover 70 percent of the Earth's surface.

Sunlight evaporates large quantities of water from the oceans and the land. Of the water that evaporates from the land, plants transpire nearly 90 percent (Chapter 31). Despite all this "evapotranspiration" (evaporation plus transpiration), our huge atmosphere is only a tenth to one percent water vapor. What water vapor there is, however, easily condenses into clouds that drop rain or snow on to the Earth's surface. Day after day, water evaporates, forms clouds, and rains in a rapid cycle that drops nearly a meter of rain over the entire surface of the Earth each year.

Overall, water moves from the oceans to the continents. Of the water that vaporizes into the atmosphere, just 17 percent comes from the land. Yet 25 percent of all rain falls on the land, so more rain falls on the land than comes from the land.

Why does water move from the oceans to the land? Clouds blowing in off the ocean often encounter tall mountains, which push the clouds up higher into the atmosphere. Up high, temperatures tend to be colder, so the water condenses into rain or snow, which drops onto the mountains, filling lakes, rivers, and underground aquifers. The water then returns to the oceans as runoff in streams and rivers.

Sunlight drives the evaporation of water from the oceans, the largest reservoir within the water cycle.

Where Does Carbon Go?

Plants remove carbon dioxide from the atmosphere and transform it into sugars and other hydrocarbons. These forms of carbon eventually pass to herbivores, carnivores, scavengers, and decomposers, all of which release carbon dioxide back into the air.

Our atmosphere contains only trace amounts of carbon. Still, this adds up to about 635 billion tons, most of it in the form of carbon dioxide. But most carbon is not in the atmosphere or even in living organisms. The oceans contain nearly 500 times as much carbon as the atmosphere, mostly in the form of bicarbonate ions (HCO_3^-). Bicarbonate ions tend to settle to the bottoms of oceans, where they form thick sediments. Over millions of years, these immense layers have turned into sedimentary rocks, which today hold 16 million billion tons of carbon (Table 25-2).

By comparison, we organisms are a mere smear of carbon on the surface of the Earth. Living organisms contain only about 2700 billion tons of carbon, one-tenth of what is dissolved in all of the oceans of the world. Each year, terrestrial

TABLE 25-2 WHERE'S THE CARBON?
Coal, oil, and natural gas that are still in the ground contain more carbon than all other sources combined.

Where	Billions of tons	Percent
Fossil fuels	25,000,000	60.0
Sedimentary deposits	16,000,000	39.0
Oceans	30,000	0.07
Living organisms	3,000	0.007
Atmosphere	635	0.0015
Human carbon released/decade	80	

plants convert about 12 percent of all atmospheric carbon dioxide into sugars and other organic molecules. This carbon then flows through a worldwide food web (Figure 25-11). Plants, herbivores, carnivores, and decomposers break up these carbon compounds by oxidizing them to carbon dioxide. Eventually, the carbon returns to the oceans and atmosphere.

The largest reserves of carbon are in the "fossil fuels" coal, oil, and methane (natural gas). Fossil fuels come from plants and other organisms that lived millions of years ago. Humans have discovered that burning these materials releases huge amounts of energy that can be used to run factories, warm houses, and power vehicles of every description (Figure 25-12).

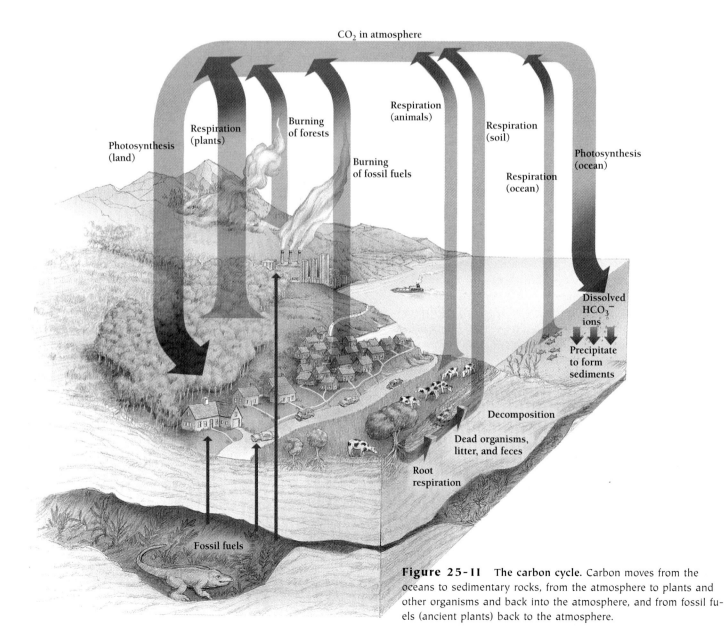

Figure 25-11 **The carbon cycle.** Carbon moves from the oceans to sedimentary rocks, from the atmosphere to plants and other organisms and back into the atmosphere, and from fossil fuels (ancient plants) back to the atmosphere.

Figure 25-12 The 19-foot-long, 7-foot-tall Ford Excursion weighs more than 7000 pounds. Sport utility vehicles (SUVs) now account for 40 percent of new-vehicle sales. Because these gigantic vehicles are technically trucks, they are subject to looser rules than passenger cars. Brakes and other safety features are less strictly regulated than in cars, and SUVs can legally burn more fuel and emit more pollutants per mile than cars can. The Ford Excursion itself, with the same length and heft as a killer whale, is so big that the law requires no mileage tests. *(Ron Kimball Photography, Inc.)*

Carbon atoms cycle through oceans and atmosphere, with large reservoirs in living organisms and in fossil fuels.

How Have Humans Altered the Carbon Cycle?

Less than 200 years ago, humans began burning large amounts of coal, oil, and gas. In the process, we began returning fossil carbon to the carbon cycle. By also burning vast stretches of the world's forests, human societies have increased the Earth's annual carbon dioxide production by 7.8 billion tons per year. As a result, total carbon dioxide in our atmosphere has increased by 30 percent since 1750.

To measure this change, scientists take ice samples from deep in the polar ice caps. The air trapped inside these frozen samples dates back hundreds of years. And the amount of carbon dioxide in this ancient air can be measured. In about 1750, before the Industrial Revolution, the concentration of atmospheric carbon dioxide was about 280 parts per million (ppm). Now, 250 years later, the average concentration of carbon dioxide in the atmosphere is about 365 ppm (Box 25-1). Most of the increase, however, has occurred in the last 50 years.

Because carbon dioxide traps solar energy, increasing concentrations of carbon dioxide can increase the temperature of the atmosphere. Indeed, global temperatures have increased by about 1°C in the last 100 years (Figure 25-13). Is it really warmer than it was in our grandparents' day? Yes. Of the ten hottest years on record, nine were between 1987 and 1997. A

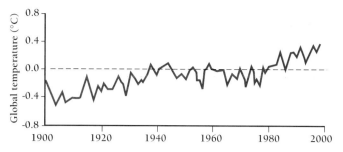

Figure 25-13 Is the temperature rising? The average of global temperatures rose about 1 degree C between 1961 and 1990. *(National Oceanic and Atmospheric Administration)*

few scientists still argue that we can't be sure that the increase in carbon dioxide has caused the increase in global temperatures, but most scientists no longer doubt it.

The tendency of carbon dioxide, methane, and other gases to trap solar radiation is called the "greenhouse effect." The glass walls of a gardener's greenhouse allow light in. Once the light is inside, however, it degrades to heat. The heat does not pass easily back out through the glass. Some of the heat is trapped inside the glass, where it warms the air and the plants inside the greenhouse.

For a gardener this is wonderful, but warming the whole Earth is causing major changes in our biosphere. Scientists have predicted that unless we reduce carbon dioxide emissions, the average temperature of the Earth's surface may increase by 3° or 4°C over the next 75 years (Figure 25-14). Such an increase would melt the polar ice caps and raise sea levels worldwide, inundating coastal cities. Glaciers all over the world are already melting, and the polar ice caps have begun breaking up at the edges. Global sea levels have risen about 2 mm per year for the last 50 years.

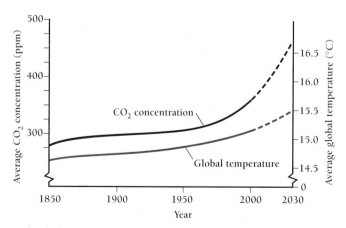

Figure 25-14 Atmospheric carbon dioxide and temperature. Scientists cannot prove that excess carbon dioxide in the atmosphere is causing global warming. Historic increases and long-range forecasts like the ones shown here, however, give most scientists little reason for doubt.

BOX 25.1

Measuring Carbon Dioxide Levels for 40 Years

The first hint that carbon dioxide might be a problem came in 1896, when a scientist noticed that high concentrations of "carbonic acid" in the air increased the temperature of the ground. By the 1930s, scientists had begun to believe that atmospheric carbon dioxide concentrations were increasing. Accurate measuring techniques designed in the 1950s removed all doubt.

In 1958, C.D. Keeling set up a system for measuring carbon dioxide concentrations in the air at Mauna Loa, a 4169-meter active volcano on the island of Hawaii. The measurements were taken from towers erected on a barren lava field 3400 meters above sea level. On the rocky slope, the local influences of vegetation and people were small, and Mauna Loa was and is one of the best places in the world for measuring undisturbed air. (When the volcano itself emits carbon dioxide, the researchers leave out those measurements.)

Keeling was not content to make his measurements for just a few years. He and his colleagues have made continuous measurements for more than 40 years, compiling the longest continuous record of atmospheric CO_2 concentrations anywhere (Figure A). Their precise records at Mauna Loa show a 15 percent increase in the annual concentration of CO_2, from about 315 parts per million in 1960 to about 365 ppm in 2000.

Amazingly, their methods and equipment have remained almost unchanged since 1959. The researchers collect air samples from air intakes at the tops of four 7-meter towers and one 27-meter tower. Four air samples are collected every hour and an infrared gas analyzer registers the concentration of CO_2 in the stream of air.

To ensure that the measurements are accurate, the flow is replaced by a stream of gas of known CO_2 concentration every 20 minutes. The gases used to calibrate the instruments are themselves calibrated against standardized gases whose CO_2 concentrations are measured even more accurately.

Figure A Increasing carbon dioxide concentrations on Mauna Loa, Hawaii. Carbon dioxide levels rise smoothly from 1960 to 1999. The up-and-down squiggles represent seasonal variation in photosynthesis. In summer, plants photosynthesize more than they respire, taking up carbon dioxide and releasing oxygen. In winter, plants respire more, breaking down carbon-rich molecules and releasing carbon dioxide. Overall, carbon dioxide levels decrease during the growing season and increase during the rest of the year. (Source: Dave Keeling and Tim Whorf, Scripps Institute of Oceanography)

The Earth's climate has already changed. In the past 30 years, Europe's growing season has increased by 10 days. More than two dozen species of British birds have begun nesting earlier, and both bird and butterfly species are gradually moving north. In North America, several hot, rainless summers have severely damaged wheat and corn crops, yet, meanwhile, farmers in Alaska say it is so warm, they can now grow rye and even wheat.

Nations around the world have leaped to halt the trend. In December 1997, 174 nations signed a treaty in Kyoto, Japan, agreeing to reduce overall carbon dioxide output. The "Kyoto Protocol" requires industrialized nations to cut their greenhouse gas emissions by 5 percent by the year 2010. Nearly all industrialized nations have promised to do so. Only Canada, Australia, and the United States, the largest emitter of these gases, have so far refused.

The simplest way to reduce the production of greenhouse gases is to reduce the burning of fossil fuels and forests. This could be accomplished through a combination of energy conservation and the use of alternative sources of energy such as solar power, wind power, or nuclear power. But for complex economic and political reasons, these sources of power remain underused.

How can we get rid of carbon dioxide?

Most solutions to the problem of global warming do not include the obvious solution, which is to burn less fuel. Because most nations want to maintain or even increase fuel consumption, researchers are looking for ways to remove carbon dioxide from the atmosphere.

Hide it in the trees

One way is to plant trees that can mop up excess carbon dioxide through photosynthesis. In Canada, for example, plans are underway to plant billions of poplar trees over millions of acres of land. As these trees grow, they will take up carbon from the atmosphere and store it as wood. In Bolivia, a coalition of three American power companies, a U.S. environmental organization, and the Bolivian government are spending $10 million to protect tropical forests there.

Bury it

Meanwhile, a Norwegian oil company has begun pumping carbon dioxide into the ground under the North Sea at the rate of a million tons a year. And on Hawaii's Kona Coast, the Japanese, Norwegian, and U.S. governments together have built a 1.5-mile pipe that can pump liquid carbon dioxide 1000 m below the surface of the ocean.

Fertilize the ocean

Perhaps the most dramatic plan is to stimulate the growth of ocean plankton by fertilizing the ocean with iron. To test this idea, oceanographer John H. Martin experimented with fertilizing the ocean with iron in the early 1990s. The result was dramatic blooms of photosynthetic plankton.

In one study, iron fertilizer caused a 60 percent drop in carbon dioxide released into the atmosphere. Researchers hypothesize that the extra carbon goes into plankton. These either become part of the ocean food chain or die and sink to the bottom of the ocean, carrying their carbon with them.

Expanding on this idea, environmental engineer Michael Markel founded a private company in Virginia that has obtained the rights to 800,000 square miles of ocean around the Marshall Islands, in the eastern Pacific Ocean. Markel's company, Ocean Farming, Inc., plans to seed this area with iron, stimulating the growth of plankton, and drawing carbon dioxide from the atmosphere. Markel estimates that a 100,000-square-mile area could soak up at least one-third of the carbon dioxide released by the United States. The result, he says, would be to stimulate a huge new fishery and cool our fevered planet. As researcher John Martin once wrote in reference to global warming, "Give me a half-tanker of iron, and I'll give you an ice-age."

Will fertilizing the ocean really cool the planet? Markel says yes. He claims that his patented technique can send all excess carbon dioxide generated by burning fossil fuels to the bottom of the ocean for 1000 to 10,000 years. To do this, he wants to continuously fertilize 6 million square miles of ocean with iron. He says his technique is cheap, harmless, and could be implemented immediately.

But many academic scientists worry that this venture is dangerous and pointless. To make a dent in global levels of CO_2, huge amounts of iron would have to be seeded, and no one knows what kind of plankton will grow. Some plankton are extremely toxic, and a 100,000-square-mile bloom of toxic plankton would be an ecological disaster, they say. Oceanographers worry that the blooms may irreversibly damage Marshall Islands coral reefs, which are the foundation for these ocean ecosystems. Once dead, the reefs might not come back.

For now, biologists can only cross their fingers. Markel's plans to seed the oceans near the Marshall Islands are on hold because of political upheavals there. But he hopes to prove his idea by seeding 5000 square miles of ocean somewhere else.

Humans have increased carbon dioxide concentrations in Earth's atmosphere by about 30 percent by burning fossil fuels and cutting down forests. The apparent result is an increase in global temperatures that may dangerously alter the global climate. The Kyoto Accord is an agreement by 174 nations to reduce carbon dioxide in the atmosphere.

Why Is Nitrogen Both Common and in Short Supply?

Molecular nitrogen (N_2) is the most abundant element in the atmosphere, comprising 78 percent of the air we breathe (Figure 25-15). Nitrogen is also an indispensable part of the proteins and nucleic acids that all organisms need. Yet, ironically, usable nitrogen has been in short supply for millions of years. The huge quantities of atmospheric nitrogen are not directly available to animals, plants, and most other organisms.

Atmospheric nitrogen is useless to most organisms because the two atoms of a nitrogen molecule are held together by a powerful, almost unbreakable triple bond. Breaking this bond requires special enzymes that only a few bacteria possess. It should not surprise us, then, that most organisms cannot perform this reaction. Plants, animals, and most other organisms can absorb only **fixed nitrogen** such as **ammonium ions** (NH_4^+) or nitrate ions (NO_3^-). Both plants and the animals that eat them depend on a steady supply of ammonia and nitrates.

Where does fixed nitrogen come from? One important source is lightning. During storms, the high temperatures generated by lightning—greater than the surface temperature of the sun—convert atmospheric nitrogen to NH_4^+ and NO_3^-. Rain washes the fixed nitrogen to Earth, supplying 5 to 10 percent of all usable nitrogen.

Amazingly, however, most nitrogen is fixed by bacteria. These tiny organisms have evolved powerful enzymes that can

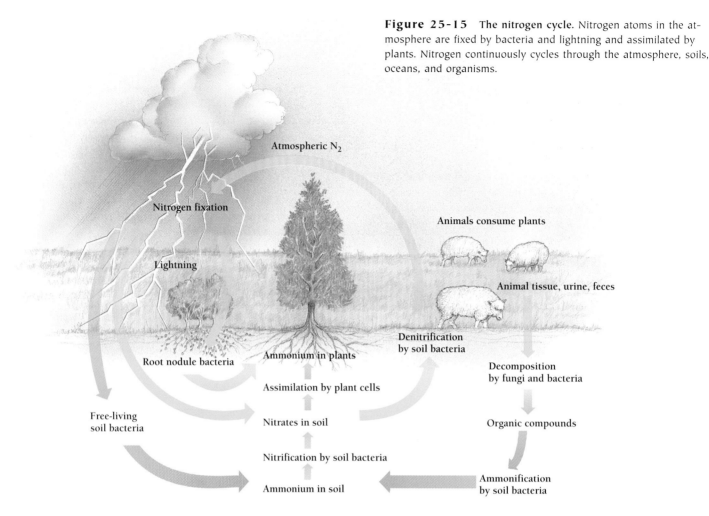

Figure 25-15 The nitrogen cycle. Nitrogen atoms in the atmosphere are fixed by bacteria and lightning and assimilated by plants. Nitrogen continuously cycles through the atmosphere, soils, oceans, and organisms.

break the triple bond in N_2 and convert it to ammonia. Some of the **nitrifying bacteria** are free-living bacteria (such as cyanobacteria) that live in soil or water and convert nitrogen to ammonia. Others live in mutualistic relationships with certain plants. For example, plants in the pea family (peas and beans) play host to *Rhizobium* bacteria, which live inside tiny "root nodules" growing on the plants' roots. The *Rhizobium* bacteria convert atmospheric nitrogen into ammonia that the plants use. The plants, in turn, provide the bacteria with energy-rich sugars, an example of symbiosis.

Plants can use ammonium ion directly in the synthesis of amino acids and proteins. Plants that do not harbor nitrifying bacteria, however, must make do with what is in the soil. Unfortunately, bacteria in the soil break down most of the free ammonium to nitrite (NO_2) and then nitrate (NO_3) in a process called **nitrification.** Other bacteria convert nitrate back to N_2, a process called **denitrification**, and the nitrogen returns to the atmosphere. Plants get whatever nitrate is left.

Nitrate is highly soluble, and heavy rains can wash it out of soils and into lakes and streams. Eventually, it ends up in the oceans. Because nitrogen is often in short supply, plants soak up nitrate very easily and conserve it by removing it from dying leaves, stems, and roots. However, such conservation measures are often thwarted. For example, when squirrels and caterpillars cut living leaves, the nitrogen-filled leaves fall to the forest floor, where their nitrates may wash from the ecosystem.

Animals and other consumers also lose nitrogen. When animals break down proteins during respiration, they convert them to ammonia. Because ammonia is toxic when concentrated, aquatic organisms such as fish dilute the ammonia with large quantities of water and flush it from the body. Terrestrial organisms, with less water available to them, first convert ammonia into something less harmful: **urea** in mammals such as ourselves, **uric acid** in birds, reptiles, and insects. Decomposers obtain energy from urea and uric acid by converting them back into ammonia, a process called **ammonification.** The ammonia (or nitrite or nitrate) can then be used once again by plants, completing the cycle of fixed nitrogen among plants, animals, and decomposers.

In natural ecosystems, most fixed nitrogen moves locally from producers to consumers and decomposers and back to producers. Only small amounts of nitrogen are added to the system through nitrification or lost through denitrification.

Fixed nitrogen cycles locally among all the organisms of an ecosystem.

How Have Humans Altered the Nitrogen Cycle?

The world's organisms have evolved over millions of years in an environment with limited amounts of available nitrogen. Yet, in the last few decades, human activity has doubled the amount of fixed nitrogen entering the biosphere and accelerated the movement of fixed nitrogen through ecosystems. In some areas, such as Antarctica, virtually no human-generated nitrogen enters ecosystems. But in highly developed areas such as northern Europe, the extra nitrogen is ruining ecosystems.

The extra fixed nitrogen comes from several sources. The largest single source is industrially manufactured fertilizers. The United States alone manufactures about 40 billion pounds a year for use on crops. But virtually every country in the world manufactures this cheap fertilizer in order to increase the growth of crops (Figure 25-16).

Another source of extra nitrogen is the crops themselves. On nearly one-third of the Earth's surface, crops have displaced natural vegetation. Many of these crops are soybeans, peas, alfalfa, and other members of the pea family that harbor nitrifying bacteria, which fix atmospheric nitrogen.

Burning fossil fuels in automobiles, factories, and power plants also releases nitrogen in the form of nitrous oxide and other gases. Humans also release large amounts of nitrogen from trees and grasses when we burn forests and grasslands or drain marshes.

The extra fixed nitrogen affects the atmosphere, the carbon cycle, and the functioning of ecosystems. All of the various forms of fixed nitrogen—including nitrous oxide, nitric oxide, and ammonia—are greenhouse gases that are helping to raise global temperatures. In addition, when nitric oxide dissolves in water, it becomes nitric acid, the major component of acid rain. Acid rain kills ponds, forests, and other habitats.

The most profound changes to ecosystems result from excess nitrogen fertilization. In terrestrial ecosystems, nitrogen saturation can leach nutrients from the soil. As ammonium accumulates in the soil, fungal decomposers lose out in competition with bacterial decomposers. The bacteria convert it to nitrate and release hydrogen ions, which acidify the soil. Because the nitrates are water soluble, they leach away into streams and lakes. But the nitrates have negative charges, so when the nitrates leach away, they carry with them a host of positively charged minerals, including calcium, magnesium, and potassium. Thus extra nitrogen can decrease soil fertility.

As calcium leaches away and soil acidity rises, increasing amounts of aluminum dissolve in soil water. The aluminum poisons the roots of trees, and, washed into nearby streams, it also kills fish.

In the long run, soils become less and less fertile, and trees growing in such soils grow very slowly or not at all. In parts of northern Europe, the Front Range in Colorado, and the forests surrounding the Los Angeles Basin, excess nitrogen is so serious that forests are slowly dying off. In many ecosystems, the amount of nitrogen available controls the nature and diversity of plant and animal life and the cycling of carbon and minerals. Excess nitrogen decreases the diversity of species in an ecosystem, as a few species grow wildly out of control and others are crowded out.

Worldwide, 20 percent of all forms of human waste, including sewage and fertilizer, ends up in rivers. In aquatic ecosystems, excess nitrogen, two-thirds of which comes from farms, washes down rivers and accumulates in estuaries, creating "dead zones" in coastal areas such as the Gulf of Mexico and the Adriatic Sea. In a dead zone, massive blooms of algae draw millions of tiny grazing plankton. As the algae die and drop to the bottom, along with the grazers' fecal pellets, staggering numbers of bacteria bloom at the bottom and quickly consume all the oxygen in the water. Fish, shrimp, and crabs swim away from the deoxygenated water, while bottom creatures such as clams, snails, and worms slowly suffocate. In extreme cases, every living organism either swims away or dies.

Dead zones can be enormous. In the summer of 1999, researchers at Louisiana State University mapped a dead zone at the mouth of the Mississippi River that covered 20,000 square kilometers, an area the size of Massachusetts (Figure 25-17).

In this chapter, we have discussed the flow of energy and the cycle of materials within ecosystems. In the next chapter, we will survey a selection of ecosystems. Along the way, we will see how humans are damaging some of these ecosystems.

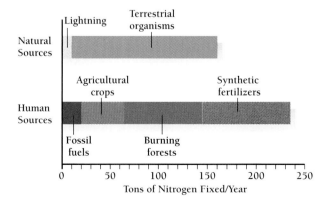

Figure 25-16 Sources of fixed nitrogen. Humans produce half again as much fixed nitrogen as all natural sources combined. As a result, available nitrogen in the biosphere has more than doubled. Excess nitrogen damages soils and kills forests and fish.

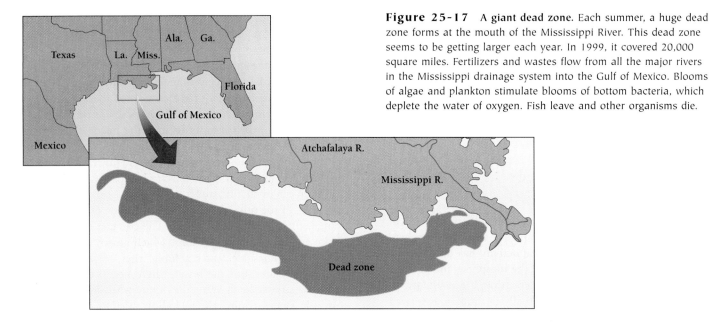

Figure 25-17 A giant dead zone. Each summer, a huge dead zone forms at the mouth of the Mississippi River. This dead zone seems to be getting larger each year. In 1999, it covered 20,000 square miles. Fertilizers and wastes flow from all the major rivers in the Mississippi drainage system into the Gulf of Mexico. Blooms of algae and plankton stimulate blooms of bottom bacteria, which deplete the water of oxygen. Fish leave and other organisms die.

Study Outline with Key Terms

In the 1960s James E. Lovelock proposed that the Earth's atmosphere, oceans, and soils be considered integral with all living organisms, in a single, immense, self-regulating homeostatic superorganism called Gaia. The idea has not been accepted by most scientists, but remnants of his idea persist in the current understanding of what the biosphere is and how it works.

Ecology is the science concerned with understanding the relations among organisms and their environment. Ecologists study **populations, communities,** and **ecosystems.** A **system** is an assemblage of interacting parts or objects. An ecosystem is the living (**biotic**) community together with the nonliving (**abiotic**) components of the community's environment, including soil, water, air, and weather. The **biosphere** is the sum of all ecosystems. An ecosystem shares some properties of an organism, but not others. Like a **superorganism,** an ecosystem is a functional unit with interacting, or **symbiotic,** parts.

The biotic part of an ecosystem consists of **autotrophs,** or **producers,** and **heterotrophs.** Heterotrophs include **consumers** such as **herbivores, carnivores, scavengers,** and **saprophytes.** Autotrophs belong to the first **trophic level** in every **food chain.** Heterotrophs occupy different trophic levels, from the second level up, depending on what they eat. Animals that eat from several levels of a food chain, including humans, bears, pigs, and crows, are called **omnivores.** Decomposers such as saprophytes and scavengers break down whole organisms and detritus.

Food chains are usually interwoven into a complex **food web.** One way of thinking about communities is in terms of a **pyramid of biomass** (dry weight) or a **pyramid of energy.** Only the pyramid of energy is always an upright pyramid.

Almost all the energy that drives ecosystems comes from the sun. Producers capture this sunlight energy, and organisms at higher trophic levels obtain what they can. Each trophic level has access to only about 10 percent of the **primary productivity** available to the one beneath it. This is called the **ten percent law** and is another way of describing the upright energy pyramid.

Ecosystems recycle water, carbon, nitrogen, phosphorus, and many other materials in **biogeochemical cycles.** Water enters the atmosphere from the oceans through evaporation and from the continents and islands through evaporation and **transpiration** from plants. Atmospheric water condenses into clouds, then precipitates as rain, sleet, or snow. Rain and snowmelt form streams and rivers that carry fresh water to the oceans. Along the way, fresh water carries minerals, organic materials, organisms, and wastes into and out of ecosystems.

Plants and other producers absorb carbon dioxide and turn it into carbohydrates through photosynthesis. Producers, consumers, and decomposers all produce carbon dioxide by respiring—oxidizing carbohydrates back to carbon dioxide. In aquatic ecosystems, carbonates tend to precipitate to the bottom to form thick sediments. These sediments form rocks that are eventually lifted up and eroded. Large reserves of carbon also exist in the form of fossil fuels.

The Earth's atmosphere is 78 percent nitrogen. But this nitrogen is not available to most organisms until lightning or **nitrifying bacteria** transform it into **ammonium ions** or nitrate, which plants can use to make proteins. Soil bacteria break down ammonium to nitrite (NO_2) and then nitrate (NO_3) in a process called **nitrification.** Other bacteria convert nitrate back to N_2, a process called **denitrification,** and the nitrogen returns to the atmosphere.

Consumers and decomposers get such **fixed nitrogen** by eating plant proteins, or each other. Most aquatic organisms eliminate waste nitrogen as ammonia. Terrestrial organisms eliminate it in the form of **urea** or **uric acid.** Decomposers obtain energy from urea and uric acid by converting them back into ammonia, a process called **ammonification.** Ecosystems conserve nitrogen, but some is lost when rain washes it from fallen organic matter and carries it to the sea.

Review and Thought Questions

Review Questions

1. What is ecology?
2. What distinguishes an ecosystem from a community?
3. What properties do ecosystems share with organisms? How are ecosystems different from organisms?
4. Distinguish between autotrophs and heterotrophs. Give an example of each.
5. Distinguish between a food chain and a food web.
6. Explain why the pyramid of energy is always narrowest at the top, but the pyramid of biomass is not necessarily narrowest at the top.
7. Explain how the biomass of aquatic producers can be smaller than the biomass of the consumers.
8. Where does the carbon dioxide in the atmosphere come from?
9. In what three forms do organisms eliminate broken-down proteins? What form do aquatic organisms use, and why don't terrestrial organisms use the same form?
10. Humans are omnivores and eat from many levels of the food chain. Make a list of primary producers, herbivores, scavengers, and decomposers that humans commonly eat. Include a plant, an animal, a fungus, a protist, and a bacterium.

Thought Questions

11. Describe the steps of the nitrogen cycle. How is nitrogen lost from ecosystems? How is it replaced? (Mention two ways.) Why does it need to be "fixed"?
12. Suppose a strawberry farmer treated his fields with a fumigant (poison gas) that killed all of the bacteria in the soil. Describe two different things the farmer could do to ensure that the strawberries received an adequate supply of nitrogen.
13. Suppose that all of the primary producers were removed cleanly and instantaneously from the biosphere. List ten immediate consequences (e.g., declines in oxygen levels) and ten indirect consequences.
14. A gallon of gasoline weighs only 6 pounds. Yet each gallon of gasoline burned releases almost 20 pounds of carbon dioxide into the atmosphere. Assuming gasoline is all octane, examine the reaction below and explain why the carbon dioxide released weighs more than the gasoline.

$$2 \, C_8H_{18} + 25 \, O_2 \longrightarrow 16 \, CO_2 + 18 \, H_2O$$

About the Chapter-Opening Image

NASA's first breathtaking photographs of Earth as a cloud-swathed blue ball floating in space reminded Earth's human population that Earth has definite boundaries and that there is only one Earth.

On-line materials relating to this chapter are on the World Wide Web at **http://www.harcourtcollege.com/lifesci/aal2/**

Web Bit 25-1 Measuring Photosynthesis in a Pond

Web Links

Quizzing & Testing

26 Terrestrial and Aquatic Ecosystems

How Many Biomes Can You See in a Day?

One hundred fifty years ago, pioneers in covered wagons struggled to cross a flat, sunbaked desert, the last barrier before reaching fertile California. In the salt-encrusted flats of one 140-mile-long valley, not a single plant grew. Even on the slopes above the valley, plants grew only sparsely and water was rare. The oxen pulling the wagons died one by one, and the pioneers who made it to the other side called the place Death Valley.

In the summertime, temperatures in Death Valley, California, can reach an unbearable 134°F. Average yearly rainfall is a minuscule 1.66 inches. The long valley lies sunken along a geologic fault line, surrounded by rocky mountains nearly as dry as the desert itself. It is the lowest and hottest place in North or South America.

Lowest of all is Badwater, 282 feet below sea level. Badwater is the last place most people would want to go in the summer. Yet this salty pool is the starting line for a footrace that makes the Boston Marathon look like a jog around the block. Some 146 miles to the west and more than 15,000 feet higher lies the top of Mt. Whitney, the highest point in the contiguous United States, and the unofficial finish line for the grueling race. Each summer, a dozen runners attempt the 26-hour race from Badwater to Mt. Whitney (Figure 26-1).

Along the way, the runners pass through one of the most abruptly changing series of biological communities in the world, including most of the communities discussed in this chapter. The runners begin the race at 6 P.M. and run the first 50 miles at night. The flat floor of Death Valley itself supports little more vegetation than the moon. Daytime ground temperatures reach nearly 200°F. Nighttime air temperatures sometimes drop no lower than 95°F. Even in the dead of night the runners must stop for water every mile.

During the night, the runners climb out of Death Valley into the foothills and mountains of the Panamint Range, passing miles of widely dispersed shrubs. Desert salt bush has small, leathery leaves as salty as the saltiest cornchip. In the driest place, creosote bushes grow, each one 10 or 20 feet from the last. As in most **deserts,** only 10 percent of the ground has any vegetation whatever. The runners climb several thousand feet, and temperatures drop slightly.

Key Concepts

- Temperature and rainfall largely determine the character of an ecosystem.
- Latitude and mountain ranges largely determine climate.
- Areas with heavier rainfall support forests. Areas with less rainfall support grasslands or deserts.

The creosote bush gives way to desert sage, a species that tolerates the higher altitude's cold winters. By the time the first runners reach the top of Townes Pass (5000 feet), it is daylight. A blistering sun rises at their backs. Gratefully, the runners descend into the cool but temporary shadow of the Panamints. As the morning wears on, they cross a small desert valley and another range of mountains.

At noon the lead runners enter the broad Owens Valley. A hundred years ago, rich grasslands and orchards surrounded the 15-mile-long Owens Lake. Towns dotted the edge of the lake and a steamboat ferried silver bullion to the south side of the lake. Today, the lake bottom is dust; all but a film of water at the center has been diverted to the millions of people living in the cities of southern California. Owens "Lake" is so

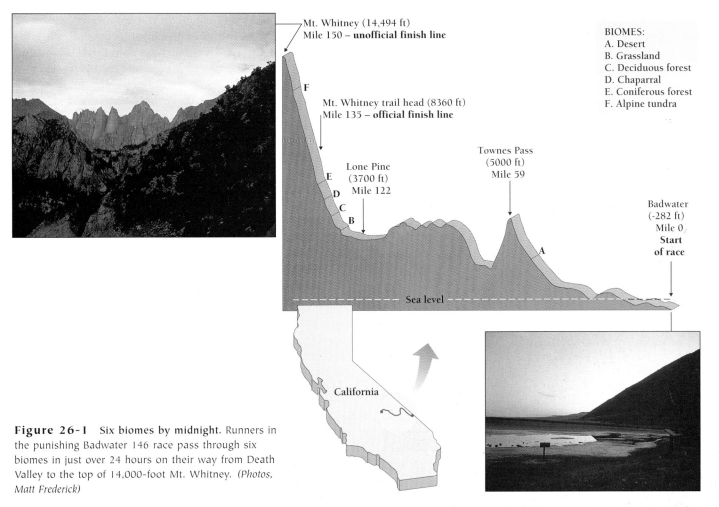

Figure 26-1 Six biomes by midnight. Runners in the punishing Badwater 146 race pass through six biomes in just over 24 hours on their way from Death Valley to the top of 14,000-foot Mt. Whitney. *(Photos, Matt Frederick)*

BIOMES:
A. Desert
B. Grassland
C. Deciduous forest
D. Chaparral
E. Coniferous forest
F. Alpine tundra

563

dry that its dust storms are the single largest source of airborne particulate pollution in the United States.

The little band of runners push past the dead lake and north to the town of Lone Pine, where they turn west. After crossing a narrow strip of fertile **grassland,** they begin to climb into the deep, narrow canyons of the eastern slope of the towering Sierra Nevada. There, the runners catch their first whiff of fresh, tumbling mountain water. Now they climb up through the last of the desert and into the rich bottomland at the base of the Sierra Nevada. Here a mountain stream winds through small meadows and dense groves of deciduous cottonwoods and aspens, its banks overgrown with thickets of willow, its bottom lined with red metamorphic rocks and glistening fool's gold. The runners find relief in the odor of lush vegetation and cool water. But the little **deciduous forest** swarms with mosquitoes, which hum in the runners' ears and cluster on their arms and legs, driving them faster through the close, humid air of the canyon.

The runners have just 12 more miles to go. They have been running almost 24 hours, and it is evening again. Before them, rising above the lush valley, is a 6000-foot wall of red metamorphic rock, cut here and there by narrow, shrubby canyons. As the runners climb higher, they pass through dense thickets of prickly shrubs, or mountain **chaparral.**

Gradually, as they gain altitude, the chaparral gives way to open **coniferous forests** of pine and fir (Figure 26-2). The air is cooler, but also thinner. Breathing becomes difficult. The lead runners climb out of the canyon now and begin following a switchback trail up the side of a granite mountain. A few stop to pant and to look back. In the gathering dusk the Panamint Range looms darkly above Death Valley, and beyond, the purple Funeral Mountains.

As the runners continue to climb, the fir trees thin out, replaced by stunted white bark pines. Finally, above the timber line, the sparse, stunted trees give way to alpine **tundra,** composed of grasses, nonwoody plants (herbs), and knee-high willows and huckleberries. Still higher, even these thin out, and the temperature drops rapidly. It is dark and cold now. An occasional low shrub or wildflower trembles in the wind on the bare slopes of the mountain. But in the last, cold, dark mile, the runners hear only the crunch and chink of running shoes on gravel and rocks.

But even the summit of Mt. Whitney, 14,494 feet high, is not quite bare rock. The summit is too high, too cold, and too exposed to wind and weather to support trees, shrubs, or herbs. But here and there, lichens—symbiotic associations of fungi and photosynthetic algae—grip the cold rocks.

The first runners to make it to the top collapse on the rocks, euphoric with accomplishment and oxygen deprivation. The rest of the runners are scattered across six distinct **biomes**—desert, temperate grassland, temperate deciduous forest, chaparral, coniferous forest, and tundra.

WHAT IS A BIOME?

In every region of the world, ecological succession leads to a characteristic community, or biome. A biome is a geographic region characterized by a distinctive landscape, climate, and community of plants and animals.

Regions with similar climates tend to have similar biomes. A desert biome is the same in the Sahara and in the Mojave. The weather is dry and the plants and animals are adapted to extreme "aridity," or dryness. The deserts of the southwestern United States and northwestern Mexico, central Asia, Africa, and Australia all have plants with spiny or succulent leaves and burrowing animals that emerge only at night. Likewise, the northern parts of Canada, Alaska, Scandinavia, and Russia all share the same kind of forest, called taiga (Figure 26-3).

But biomes differ in the particular species that they include. The grasslands of North America are dominated by deer, elk, and bison, while the grasslands of Africa are dominated by zebras, gazelles, elephants, and giraffes. Differences in the individual species in a biome are partly the result of the unique evolutionary histories of each region and partly the result of differences in geography.

In this chapter, we discuss eight terrestrial ecosystems. We say "biomes" (which really means "communities"), but because we discuss abiotic elements such as temperature, we are really talking about ecosystems (thus the chapter title). In addition, we discuss several kinds of aquatic ecosystems. (Ecologists do not use the word "biome" to refer to aquatic communities, reserving that term for land communities.) Along the way, we will see what effects humans have on each of these ecosystems. We begin with a discussion of what factors determine climate in different parts of the world.

How Do Sunshine and the Earth's Movements Determine Climate?

The most important determinants of biome type are temperature and moisture. These depend, in turn, on the intensity of sunlight and patterns of wind, rain, and ocean currents. All of

Figure 26-2 Coniferous forest. At higher elevations and latitudes, cooler temperatures and moister soils support coniferous forests. *(Superstock)*

A.

B.

Figure 26-3 Taiga here, taiga there. Whether in Canada, Scandinavia, or Russia, taiga is characterized by sparse, stunted coniferous trees. A. Siberian taiga. B. Spruce taiga in central Alaska. (A, © Roland Seitre/Peter Arnold, Inc.; B, Tony Dawson/Words & Pictures/Tony Stone Images)

these depend on the Earth's movements and the distribution of mountains and valleys.

The single most critical factor is sunlight. Of some 10^{21} kilocalories of solar energy that enters the upper atmosphere each year, about half reaches the ground. The rest is reflected back into space or absorbed by the atmosphere. Plants capture less than 1 percent of the light that strikes the Earth's surface. The energy absorbed by the ground and by the atmosphere raises the temperature of the environment and helps determine a region's climate.

The Earth's daily rotation constantly mixes the air, so that all the air along each latitude has a similar temperature. But different latitudes have different climates. This is because the latitudes closest to the poles receive less solar energy and experience more-extreme seasons than those near the equator

(Figure 26-4). As a result, the climate near the equator is both warmer and less variable than the temperate or polar climates.

Differences in air temperature cause air to move in currents that transfer heat away from the equator and toward the poles. Three rules determine the direction of these currents: (1) hot air rises and cold air falls, (2) hot air holds more moisture than cold air, and (3) the rotation of the Earth twists the moving air.

Figure 26-5 shows the pattern of air currents. Right at the equator, not much wind blows, and old-time sailors called this calm area the *equatorial doldrums* [Old English, *dol* = dull]. There, warm, moist air rises, which creates a region of low air pressure. The low air pressure pulls **tradewinds** toward the equator from about 30° latitude (north and south). Because of the Earth's spin, the tradewinds blow westward. The tradewinds were named when sailing ships used them to sail westward.

As the warm air from the equator rises, it cools. Because cool air can hold less moisture than warm air, the moisture condenses into clouds and falls as rain over the equatorial regions, creating the warm, rainy climate of the **tropics.** The equatorial air spreads north and south, raining and drying out as it goes. By the time this air reaches about 30° latitude, the air is quite dry and the lands beneath are nearly all deserts.

As the air loses water, it cools. The cool, dry air drops down toward the Earth's surface, and the Earth's spin gives the air a twist that makes it blow from west to east. These **westerlies** blow at latitudes between 30° and 60° in both the north and south. As this air moves over the Earth's surface, it warms and picks up moisture. At about 60° latitude, the air again rises, cools, and rains (or snows). At this cool but wet latitude are most of the great coniferous forests. Once the air is dry again, it drops down to create a dry, cold region of Arctic tundra. Here, the winds, called the **polar easterlies,** come from the east (Figure 26-5).

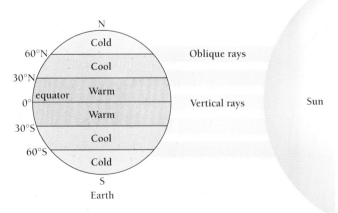

Angle of sun's rays at different latitudes

Figure 26-4 Temperature depends on latitude. Temperature and moisture are the two most important determinants of biome type. The latitudes closest to the equator receive the most sun. The latitudes closest to the poles receive less solar energy than those near the equator because the sun's rays arrive at an oblique angle.

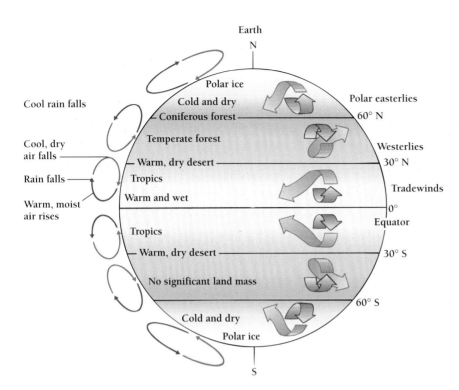

Figure 26-5 Air currents. A pattern of rising and falling air, combined with the Earth's rotation, causes air currents to flow east near the north and south poles, to flow west in the middle latitudes, and to flow east along the equator.

The winds at each latitude depend on the overall movement of warm air from the equator toward the poles. The rotation of the Earth also changes the direction of the air movements, causing clockwise air currents in the Northern Hemisphere and counterclockwise currents in the Southern Hemisphere. The effect of rotation on movement, called the **Coriolis effect,** causes Northern Hemisphere winds (including hurricanes, typhoons, and tornadoes) to twist clockwise and Southern Hemisphere winds to twist counterclockwise. The Coriolis effect also induces ocean currents to turn in great circles—clockwise in the north and counterclockwise in the south.

Ocean currents often move along the edges of continents. The Gulf Stream, for example, carries warm water from the tropics up the eastern coast of the United States and then across the North Atlantic, where it warms Great Britain, and much of northern Europe. The warm Gulf Stream gives northern Europe a far milder climate than would be expected from its latitude.

How Does Topography Influence Climate?

Mountain ranges also influence climate, creating "rain shadows." A **rain shadow** is a land of little rain on the downwind side of a tall mountain range. Mountains force warm surface air to rise to higher, colder altitudes, where moisture condenses into clouds and falls as rain or snow. The prevailing westerlies along the California coast, for example, drive warm, moist ocean air up and over two mountain ranges. When warm, wet air coming in off the Pacific Ocean hits the low Coast Range, the clouds rise, cool, and then dump up to 100 inches of rain a year on the coastal mountains (Figure 26-6A). As the air moves east, down the farther side of the Coast Range, it warms and absorbs moisture, drying the land. The same westerlies then hit the Sierra Nevada, a huge mountain range with some of the tallest peaks in the continental United States. At 3,000 to 14,000 feet, the ocean air cools and drops enough rain or snow to supply water for most of the state's agriculture and most of the state's 35 million residents. By the time the ocean air reaches the far side of the Sierra Nevada, however, it has lost nearly all its water and become very dry. East of the Sierra Nevada, all of the lands in the Owens Valley, Death Valley, most of Nevada, and parts of Oregon, Idaho, Utah, and New Mexico exist in a huge rain shadow, called the Great Basin desert (Figure 26-6B).

Tall mountains have other effects on climate as well. As we climb a mountain such as Mt. Whitney, the climate changes in the same way it does as when we travel from low latitudes to higher ones (Figure 26-7). The Appalachian Mountains, the Sierra Nevada, and the Rocky Mountains all contain extensive coniferous forests, which turn to alpine tundra above the "tree line," the upper limit at which subalpine trees can grow. Even in the tropics, snow covers the highest mountains.

All of these geographic patterns of temperature and rainfall create just a few major types of climate. Within each climate, similar adaptations are useful. For example, the species of a desert biome are recognizably similar, whether the desert is in Asia, North America, Australia, or Africa. A marsupial "wolf" from the forests of Australia has the same long, running legs and powerful jaws as the gray wolf of North America (Figure 17-15).

A.

B.

Figure 26-6 A rain shadow. Rain shadows occur in all parts of the world. One of the most dramatic examples is in the western United States. A. Rain clouds driven inland from the Pacific Ocean hit California's Coast Range and ride up the slope, like skateboarders on a ramp. As the clouds rise, they cool and rain precipitates out, soaking the mountains. The clouds drop down the other side, warm up, and hold their remaining moisture. Over the Sierra Nevada, the clouds drop quantities of rain and snow. B. By the time the clouds reach Nevada's Great Basin, the moving air has lost nearly all its moisture and little rain falls in the desert beneath. (B, Charlie Ott/Photo Researchers, Inc.)

Latitude and the Earth's spin influence intensity and duration of solar radiation, wind, rain, and ocean currents. These factors, as well as terrain, help create distinct climates that strongly determine what biome develops in an area.

HOW DO BIOMES DIFFER?

Biologists divide the world's communities into a great many sets of biomes. But in this book, we have chosen just eight major biomes that we think illustrate both life's diversity and some unifying themes (Figure 26-8). We will begin with the driest biomes, the warm, dry deserts and the cold, dry tundra, and then move through progressively wetter biomes, including chaparral, temperate grassland, and savanna; taiga, coniferous forest, and deciduous forest; and finally, wettest and warmest of all, the tropical rain forest (Table 26-1).

Desert Ecosystems Are Surprisingly Delicate

Deserts have the lowest net primary productivity of all the biomes. These barren regions produce only about 90 grams dry weight of biomass per square meter per year, about the weight of 12 quarters. Deserts may be hot or cold, but all are dry, receiving less than 25 centimeters of rain per year. In the southwestern United States and northeastern Mexico, desert extends east from the Sierra Nevada to the Rocky Mountains and south to the Sierra Madre. Africa, Arabia, central Asia, central Australia, and parts of South America all have deserts.

What plants dominate a desert depends on the temperature. Sagebrush dominates the cold deserts of Nevada and Utah, while cacti and oily creosote bushes dominate the lower, warmer deserts of Arizona and Mexico (Figure 26-9).

Desert plants have special adaptations to conserve water. Sagebrush, for example, has small leaves that minimize evaporation, while succulents (including the cacti) store water in fleshy tissue for future use. Cacti and other desert plants grow prickly spines that protect them from hungry or thirsty browsers.

Figure 26-7 Latitude and altitude have similar effects on biomes.

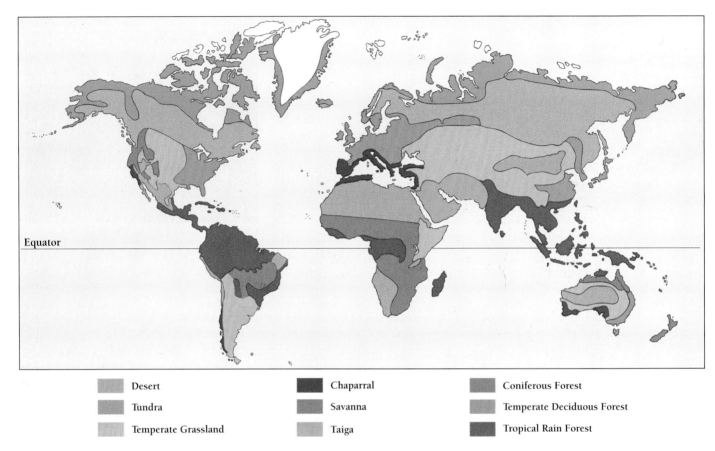

Figure 26-8 World distribution of biomes. The sharp boundaries in this diagram are misleading, as biomes frequently grade together smoothly. In addition, each solid area actually represents an average of many vegetation types. Botanists divide California alone into 24 "floristic provinces."

Like desert plants, desert animals must also be able to survive extreme heat and dryness. In the summer, many are active only during the cool hours of dawn and dusk. Lizards and snakes thrive, as do birds such as cactus wrens and roadrunners, and rodents such as kangaroo rats and mice. Many rodents **estivate,** retiring to underground burrows and entering a sleeplike state called *torpor* during the hottest and driest months of the summer. Round-tailed ground squirrels (*Ammospermophilus tereticaudus*) and several species of pocket mice estivate, as well as the cactus mouse (*Peromyscus eremicus*). Laboratory experiments suggest that lack of food and lack of water can each trigger estivation in the cactus mouse.

TABLE 26-1 TEMPERATURE, PRECIPITATION, AND PRODUCTIVITY OF MAJOR BIOMES

Biomes with the least seasonality and the most rain are also the most productive. Productivity is measured as the dry weight of new plant material that accumulates on a square meter of land in one year.

Biome	Temperature (winter low/summer high)	Precipitation (cm/y)	Productivity (g/m^2/y)
Desert	warm/hot	< 25	90
Tundra	cold/cold	< 25	140
Chaparral	mild/mild	20–40	700
Grassland	cold/warm	25–75	600
Taiga	cold/cool	50	800
Savanna	warm/hot	90–150	900
Temperate deciduous forest	cold/hot	75–125	1200
Tropical rain forest	warm/hot	200–450	2200

any soil, but desert soils are especially vulnerable. Irrigation water from rivers and wells is loaded with dissolved minerals and salts. As it evaporates from the hot desert soil, the minerals and salts are left behind as a deposit. In time, the soil becomes so caked with minerals and salts that farming is impossible. By then, even desert plants cannot grow in the poisoned soil.

Because deserts lack water, they are the least productive of all biomes. Desert plants and animals are adapted to conserve water.

How Can a Dry Biome Be So Wet?

At the North Pole lies an ice desert utterly devoid of plant life. But just a few hundred miles south lies the cold **tundra** [Finnish, *tunturi* = arctic hill], a vast, open land of bogs, ponds, and lakes. Among the lakes are short grasses, shrubs, mosses, and lichens, nearly all less than 20 centimeters tall. The few small shrubs and trees grow on the banks of lakes or streams. Tundra sweeps round the North Pole, covering more than one-fifth of the Earth's land surface, including northern Canada, Alaska, Scandinavia, and Siberia (Figure 26-10). In addition, alpine tundra cloaks the upper slopes of high mountains and plateaus in temperate regions.

The tundra is cold. Freezing temperatures occur in every season, and **permafrost**—permanently frozen ground less than a meter below the surface—persists throughout even the warmest summers. Still, a one- or two-month summer growing season allows topsoil to thaw and plants to grow. As soon as the brief summer warmth and rains arrive, the plants' extensive underground roots allow them to quickly produce leaves, flowers, and fruits.

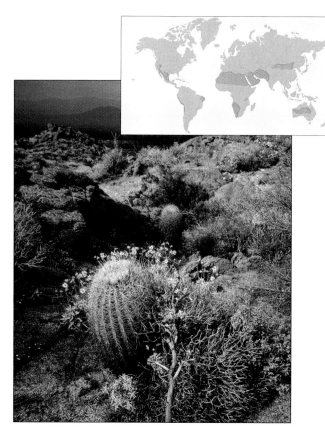

Figure 26-9 Deserts. Barrel cactuses punctuate the many deserts of the southwestern United States. *(John M. Roberts/The Stock Market)*

Although deserts seem tough, they are delicate. The wheels of off-road recreational vehicles destroy fragile desert soils, as well as plants, lizards, snakes, and tortoises. An even greater threat to deserts is the conversion of desert habitat into suburban developments and farms. Excessive irrigation can destroy

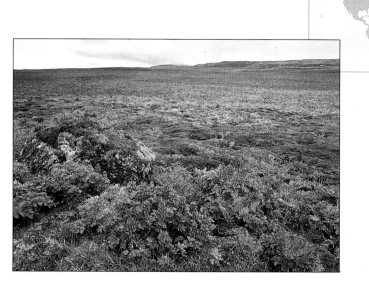

Figure 26-10 Alaskan tundra. Lupine and ferns on the Pribilof Islands, Alaska. *(John Shaw/Tom Stack & Associates)*

Tundra receives no more rain or snow than a desert. Each year, less than 25 centimeters falls, mostly as summer and autumn rains. In winter, the snow pack rarely exceeds 10 to 20 centimeters. Despite this low precipitation, the soil is perpetually wet, for the cold air prevents evaporation and the icy permafrost keeps water from draining away.

Mammals in the tundra include caribou (or reindeer), polar bears, weasels, foxes, and small rodent herbivores called lemmings. In the summer, large numbers of birds arrive from the south to nest and raise young. Swarms of flies and mosquitoes fill the air. In the summer, the tundra supports an enormous amount of life, but the diversity of species is low.

So far, humans have had only limited impact on the tundra. Inuits of Alaska, Greenland, and Canada live on the tundra hunting wild caribou, and the Lapps (Sami) of northern Scandinavia herd domesticated caribou (reindeer). In the past 30 years, however, oil development, oil pipelines, and roads have begun to impinge on life in even this remote biome.

Tundra has a cold, dry climate and a short growing season.

Why Are There No Trees in Grasslands?

Temperate grasslands have well-defined seasons, with hot summers and cold winters (Figure 26-11). Annual rainfall is low, usually 25 to 75 centimeters per year, enough to keep grasslands from turning to deserts but not enough to sustain forests. In addition, grasslands usually have a dry period, during which fires are common. Wildfires and the lack of rain, together with heavy grazing by animals, keep trees from taking hold in this biome.

Grasslands extend over much of the Earth's surface, mostly in the interiors of the continents. They include, for example, the steppes of Russia, the veld of South Africa, the pampas of Argentina, the puszta of Hungary, and the prairies of the central United States and Canada. Before European settlement, more than a million square miles of grassland covered central North America.

The types of grasses vary with the amount of rain and with season. In the eastern American prairie, for example, tall grasses, especially big bluestem, once dominated; in the drier western plains—in the rain shadow of the Rocky Mountains—short grasses dominated.

Grasslands can support large numbers of mammals. For millions of years, African grasslands supported huge herds of wild gazelles and zebras, while the Asian steppes have supported sheep and horses. All of these herbivores in turn supported populations of wolves, lions, humans, and other predators. When Lewis and Clark crossed the American prairie in 1804, some 60 million bison and 40 million pronghorn antelope grazed in vast herds. The explorers pushed through these herds only with difficulty. Yet 19th-century fur traders and sport hunters drove these populations to the brink of extinction in less than 75 years.

Today, in place of the bison and antelope live some 20 million humans, who have virtually eliminated the temperate grassland biome. In place of hundreds of thousands of square miles of American and Canadian prairie, we have created the world's "breadbasket." Corn has replaced tall bluestem, and wheat has replaced the short grass. Mass-produced pigs, cows, and beef cattle are the chief herbivores; humans, the top carnivores. A few thousand bison live in parks and reserves, and a few hundred thousand pronghorn antelope live in the plains adjacent to the Rocky Mountains. Their natural predators,

Figure 26-11 Temperate grassland. This wild prairie at the Living Prairie Museum in Winnipeg, Manitoba, has never been plowed. *(Tom McHugh/Photo Researchers, Inc.)*

however, the Great Plains wolf and the grizzly bear, are gone—shot, poisoned, and trapped to extinction.

> *Temperate grasslands have hot summers and cold winters. In North America, wheat fields have replaced most temperate grasslands.*

Chaparral Always Burns Eventually

Chaparral [Basque, *chabarro* = dwarf evergreen] dominates five widely separated temperate regions—California, central Chile, the shores of the Mediterranean Sea, southwestern Africa, and southwestern Australia. These regions share a "Mediterranean" climate, with mild, rainy winters and hot, dry summers. In the summer, prevailing westerly winds blow cool, moist sea air over the warm land. But the warmed air holds the moisture, so no rain falls.

Chaparral is characterized by dense shrubs and scattered, broad-leafed trees (Figure 26-12). The shrubs are typically spiny, thick, and dense. Each of the five chaparral regions has its own species, but all chaparral species share common adaptations to summer drought and to periodic wildfires. Wildfires clear the way for fresh growth and kill trees that would shade out the chaparral. The seeds of some herbs and shrubs actually require exposure to the intense heat of a brushfire before they can germinate. Perennial shrubs such as chemise and coyote brush are well adapted to fire. When fire burns away their dry branches, these species resprout from still-intact roots.

One of the greatest dangers to chaparral communities is overzealous fire prevention by humans who build houses in them. By putting out small fires, humans ensure the accumulation of dry wood. Inevitably, a fire so big and hot that firefighters cannot put it out sweeps thousands of acres, incinerating the roots of the chaparral vegetation and destroying hundreds of suburban houses (Figure 26-13). Such hot fires also kill seeds and sterilize the soil, so that until the land is recolonized, nothing grows back. In addition, because of the way such superhot fires affect the soil, disastrous erosion may follow (Figure 26-14).

Figure 26-13 Fire. Brush fires in chaparral communities destroy houses in minutes. Here a 20,000-acre fire sweeps through a suburban neighborhood in southern California. *(Reuters/Corbis-Bettmann)*

> *Chaparral biomes are characterized by hot, dry summers and mild, wet winters. Many chaparral species have special adaptations to fire.*

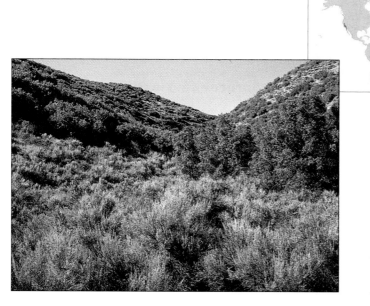

Figure 26-12 Chaparral. Dense shrubs and scattered trees characterize Mediterranean climates. *(Tom McHugh/Photo Researchers, Inc.)*

Figure 26-14 Houses in fire-adapted communities. In chaparral communities, the regular control of brush fires leads to a buildup of dead wood. When a fire finally burns out of control, it burns unusually hot. Super-hot fires not only destroy houses, trees, and shrubs but also damage soil. Chaparral soil contains water-resistant plant compounds that prevent water from penetrating the soil. Heat from a hot fire forces this "hydrophobic layer" deeper into the soil. The uppermost layer no longer sheds water but absorbs it instead. When winter rains drench the upper layer of soil, turning it into mud, the top layer slides away.

Savannas Host Some of the Most Spectacular Grazing Animals

Savannas are grasslands with scattered trees or small clumps of trees (Figure 26-15). The trees are usually deciduous, dropping their leaves in the dry season. Like temperate grasslands, savannas have a regular dry season and periodic wildfires that limit the spread of trees. But savannas generally have more rain (usually 90 to 150 centimeters per year) and hotter climates than temperate grasslands. Savannas are often transitional biomes between temperate deciduous forest and prairies or between evergreen tropical rain forests and deserts. Savannas cover much of south and east central Africa, much of Australia, and parts of the Americas, and Southeast Asia.

Savanna grasses are often tall—2 meters or more in Africa, 1.5 meters in South America—and the trees short—usually less than 10 to 15 meters tall. Deep, dense underground networks of roots allow grasses to obtain water where trees cannot and so survive the dry seasons.

Animal life in savannas varies greatly from continent to continent. The best-known animals are those of the African plains. There, herds of zebras, wildebeest, and gazelles graze while elephants and giraffes browse on the trees. Among these herbivores, lions, cheetahs, hyenas, wild dogs, and other carnivores hunt and scavenge. Savannas also support large, flightless, grazing birds—the ostrich in Africa, the emu in Australia, and the rhea in South America. Termites consume up to one-third of all plant litter and serve as a major food source for birds and mammals. In some areas, termite mounds cover the savanna. These mounds may reach 9 meters in height and number 10 to 150 per hectare.

In most savannas, hunters and habitat destruction are driving many species to the brink of extinction. In East Africa, the seemingly infinite herds of grazing animals are becoming food for rapidly growing human populations. Elephants have become particular targets for poachers, not because of their food value, but because of their valuable ivory tusks. Their slaughter continues despite worldwide efforts to outlaw trading in ivory. Nonetheless, the real threat to African savanna is the continuing conversion of savanna to farms and ranches that support the burgeoning human population. While wildlife sanctuaries protect some of the African and South American savannas, the end of the savanna biome seems inevitable.

Tropical savannas are grasslands with scattered deciduous trees that support vast herds of grazing animals.

Taiga, the Spruce–Moose Biome

South of the tundra, most northern lands are both slightly warmer and wetter. Here sparse conifer forests called **taiga** grow monotonously across thousands of miles of Siberia, Scandinavia, Canada, and Alaska (Figure 26-3). Winters are bitterly cold, and summers are short. The one or two species of trees (often spruce and pine) grow only 5 to 10 meters high. Ferns, mosses, and lichens grow sparsely on the forest floor. Shrubs are rare, except for blueberries and gooseberries that flourish in open areas where trees have temporarily died off.

Moose, mice, squirrels, porcupines, and snowshoe hares are the main vertebrate herbivores. Predators include wolves, grizzly bears, lynxes, and wolverines. Birds migrate to the taiga

Figure 26-15 Savanna. Grasslands and scattered trees in the Serengeti, in Tanzania. *(© Art Wolfe)*

in summer, but leave before the first snows fall. Except for garter snakes, reptiles cannot survive the long, harsh winters. Mosquitoes, black flies, and other insects, however, abound.

The taiga is so inhospitable that human populations are sparse. In some areas, logging has eliminated the original forest, and deer have moved in, replacing the moose. A common parasite of the deer—the nematode brain worm—is fatal to moose and, in many areas, is accelerating the disappearance of the moose.

One of the greatest threats to the taiga comes from the air. Factories, power plants, and automobiles produce particulate pollutants and oxides of sulfur and nitrogen that rise and travel thousands of miles to the taiga. The oxides dissolve in the clouds, dropping acid rain and snow on lakes and rivers. The result has been the wholesale death of fish and forests in many regions of the Northern Hemisphere, including Scandinavia and eastern Canada.

Lichens, in particular, accumulate sulfur dioxide, heavy metals, and other toxins. Ecologists use both the health and chemical composition of taiga lichens to monitor global air pollution.

Taiga has cold winters and short, mild summers.

Why Do the Temperate Deciduous Forests Lack Top Predators?

Temperate deciduous forests typically have warm, rainy summers and cold winters. They receive from 80 to 140 centimeters of rain and snow each year. However, temperature and annual precipitation vary more than in the tropics or the polar regions. As a result, temperate deciduous forests may look rather different from one another. In the northern United States, for example, the dominant trees are maple, oak, hemlock, and birch, while in the southeastern states the dominant trees are pine, oak, and hickory.

Deciduous trees, which shed their leaves each autumn, characterize the temperate forest biome (Figure 26-16). Deciduous plants shed their leaves as a way to minimize water and nutrient loss at times when photosynthesis is minimal. The plants of a temperate forest occupy four vertical layers: trees, shrubs, herbs (nonwoody plants), and ground cover. The trees occupy the highest layer, their tops touching to form a nearly continuous canopy. In the shade beneath the trees are shrubs and bushes whose branches lie closer to the ground. The herb layer may be especially rich and beautiful in the spring. Finally, the ground layer consists of mosses and liverworts, often living among the accumulated leaf litter.

Animal life is abundant in all the plant layers and in the soil beneath the litter (Figure 26-17). Trees and shrubs provide food and nest sites for many species of birds. Woodpeckers and chickadees, for example, nest in cavities in the trees, while warblers and thrushes build nests on their branches. Mammals such as chipmunks and squirrels; reptiles such as snakes, lizards, and wood turtles; and amphibians such as salamanders and tree frogs also find food and homes in the forest.

The eastern third of the United States, as well as western Europe, Japan, Chile, and eastern China, was once dominated by temperate deciduous forests. On every continent, humans have cut down much of the forest for lumber or firewood and to make way for farms, towns, and cities. Before European colonization 400 years ago, forests covered 46 percent of the

Figure 26-16 Deciduous forest. In autumn, the leaves of deciduous trees turn color, as chlorophyll is pulled from the leaves. North County National Scenic Trail, Itasca State Park, Minnesota. (*G. Alan Nelson/Dembinsky Photo Associates*)

Figure 26-17 Forest litter. Biologist E.O. Wilson has written, "It is a failing of our species that we ignore and even despise the creatures whose lives sustain our own." The rich leaf litter of a temperate deciduous forest supports myriad organisms from bacteria, fungi, ferns, and worms to skunks, squirrels, foxes, and mountain lions.

BOX 26.1

How Does Acid Rain Alter Biomes?

Most people now know that severe acid rain can destroy lakes and whole forests (Web Bit 2-2: Life at Low pH). In the worst cases, fish, frogs, plants, and other organisms have disappeared from formerly productive lakes and streams. Yet, increasingly, ecologists are finding that acid rain also has the capacity to transform ecosystems in subtler ways.

Researchers have long known that acid rain increases the nitrogen available to plants. Indeed, some economists have argued that acid rain is a cure for global warming. They say that because nitrates from acid rain can stimulate plants to grow faster, plants in acid environments can take up more carbon dioxide, compensating for the carbon dioxide released when we burn fossil fuels.

Although this is a reassuring thought, ecological research does not support this hypothesis. In one study, researchers at the University of Minnesota applied nitrogen to plots of native Minnesota tallgrass prairie for 12 years to imitate the excess nitrates that result from both acid rain and fertilizer applications by farmers. Because native grasses use nitrogen efficiently, they flourish in places where the nitrogen content of the soil is low. In the nitrogen-treated plots, however, exotic weeds flourished, outcompeting and displacing native Minnesota grasses.

But that was only half of the bad news. As prairie grasses die, their tissues break down very slowly, and over time much of their carbon, nitrogen, and other constituents remain stored in the soil. Such long-term storage is, in fact, the key to the famous, rich soils of the American midwest.

But in the Minnesota study, the weeds behaved differently from grasses. As the weeds died, their huge stores of nitrogen stimulated the growth of nitrogen-hungry soil microbes, which rapidly dismantled the exotic weeds into basic building blocks. The microbes released more nitrates into the groundwater and more carbon dioxide back into the air. The net effect was more carbon dioxide in the air—not less—and more nitrates in the soil, which stimulated the growth of more weeds.

In the spruce forests of Germany's Fichtelgebirge mountains, the excess nitrogen in acid rain supplies the trees with more nitrogen than they can use. Instead of stimulating growth, the acid rain destroys soil. Negatively charged nitrates raining down on forest soils bind with calcium, magnesium, and other positively charged cations in the soil. As the nitrates flow into lakes and streams, they take the calcium and magnesium with them. In some areas, such acid leaching has removed all of the calcium deposited in the soil for the last 500 years.

And, as we saw in the last chapter, as calcium leaches away and soil acidity rises, increasing amounts of aluminum dissolve in soil water. The aluminum poisons the roots of trees and kills fish.

In some areas, all the trees in a forest are dying. In others, ecologists predict, trees deprived of calcium and magnesium will simply habituate and grow quite slowly, like the tiny bonsai trees that gardeners grow in pots. The result may be pygmy forests.

United States, and most of this forest was ancient and deciduous. Although in the eastern United States, forests have returned, they still cover only 32 percent of the land and consist largely of young trees, all the same age and species. More important, these young forests have not brought back the local species of plants, animals, and fungi that went extinct 200 years ago.

Much of the new forest is in small, isolated woodlots, often of 16 hectares (40 acres) or less. These tiny patches tend to hold far fewer species than the extensive virgin forests they replaced (Chapter 28). Such small areas do not provide the large ranges needed to support carnivores at the top of the food chain. A single mountain lion, for example, needs over 75,000 hectares (300 square miles). Even a single family of gray foxes needs some 260 hectares (1 square mile). Gone with these vast areas are nearly all the top carnivores in the eastern United States. Gone, or nearly gone, are the red wolf, the bobcat, and the black bear.

Temperate deciduous forests have warm, rainy summers and cold winters.

Tropical Rain Forests Are the Most Productive and Diverse Biomes in the World

Most **tropical rain forests** are within 10° of the equator—in the Amazon basin of South America; in central and west Africa; and in Malaysia, Indonesia, and New Guinea. The weather in the tropical rain forests is more or less constant throughout the year, with average temperatures of about 27°C (80°F) every month of the year. Rainfall is heavy, ranging from 2 to 4.5 meters per year.

Warmth and moisture are so plentiful in a tropical rain forest that plants compete for light rather than water. Tropical plants are so successful at capturing light that the forest floor is usually rather dim, with little vegetation. Visitors often compare a tropical forest to a cathedral, with its open floor, high vaulted ceilings, and soft filtered light.

Nutrients cycle rapidly through rain forest ecosystems. The soils of tropical rain forests are surprisingly poor in minerals and other nutrients. One reason is the heavy rainfall, which leaches nutrients from the soil. Another is the year-round work of ants, termites, fungi, and bacteria, which rapidly break down leaf litter to simple molecules. In most biomes, a dry season or a long, cold winter slows the breakdown of leaf litter and other organic matter, and soils tend to build up to depths of up to several feet. In the rain forest, only a thin layer of nutrients exists, just beneath the leaf litter at the very surface.

Trees rely on an extensive network of extremely shallow roots to sweep up whatever nutrients are available. The bulk of nutrients and energy in a tropical ecosystem are in the tops of trees and in the other living organisms, rather than in soils or water.

Despite their thin soils, tropical rain forests are the second most productive ecosystem in the world, surpassed only by estuaries and marshes. And tropical rain forests are first in biodiversity. At least half, and perhaps as many as 90 percent, of all the Earth's species live in tropical rain forests. In a hectare (2.5 acres) of temperate forest, we might find two or three dominant species of trees with a scattering of another 10 to 20 species. In contrast, a hectare of tropical forest is home to 40 to 100 different species of trees. More species of butterflies live in one rain forest in Costa Rica, for example, than live in all of the rest of North America.

The diversity of species in a tropical rain forest can be staggering. One ecologist found 163 species of beetles living in just one tree in the Panamanian rain forest. Another ecologist found 445 species of trees on an area of Brazilian rain forest the size of a city block. Fifteen of those trees species were new to science.

More than two-thirds of the plants in a tropical forest are trees. Most are flowering, broad-leaved evergreens (Figure 26-18). Other plants common in tropical forests but rare in temperate forests are **lianas,** vines that are rooted in soil but climb into the tree tops, and **epiphytes,** plants such as orchids that grow entirely on other plants.

As in a temperate forest, the vegetation forms discrete layers. Most of the trees form a continuous layer, called the **canopy,** 30 to 40 meters above the ground (Figure 26-19). Rising above the canopy are occasional, tall, isolated **emergents,** trees with umbrella-shaped crowns extending to a height of 50 meters or more. Below the canopy, a third layer of shorter trees 10 to 25 meters tall catch whatever light they can from gaps in the canopy. Finally, a short shrub layer consists of dwarf palms and giant herbs with especially large leaves, adapted to capture the faint light that filters through upper stories.

All these flowering plants provide a year-round supply of fruits and nectars for an array of birds, insects, snakes, bats, monkeys, and squirrels. Some live permanently in the canopy. Others live in the rivers or on the forest floor. Some eat underground plant parts, others, fruits and nectar. Ants and termites dominate this biome. But anteaters and armadillos, pigs and their relatives, as well as poisonous and constricting snakes, all hold their own in the complex food webs of the tropical rain forest. The top carnivores of the forest floors used to be such large cats as the jaguar in South America and the tiger in Asia. However, these magnificent animals have all but suc-

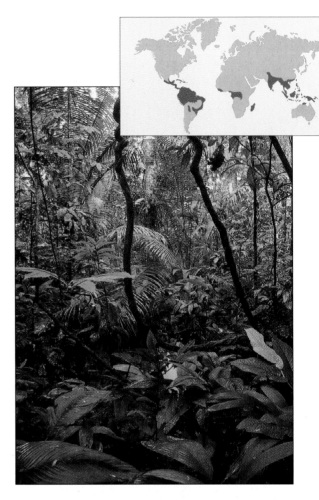

Figure 26-18 Tropical rain forest. Terra-firma Amazonian rain forest with lianas, Peru. *(David M. Schleser/Nature's Images, Inc./Photo Researchers, Inc.)*

cumbed to trophy and fur hunters and to the destruction of their habitat by humans.

How Are Tropical Rain Forests Being Destroyed?

Until recently, Earth contained 25 million square kilometers of tropical rain forests, an area greater than all of North America, Central America, the Caribbean, and Hawaii combined. Today no more than 8 million square kilometers remain. Estimates of the rate of rain forest destruction show that an area of tropical rain forest the size of the state of Washington is logged or burned every year. And even this estimate may be too low.

In 1999, field surveys of wood mills and forest burning in the Amazon showed that the true rate of destruction may be twice as high as previous estimates. As rain forest disappears, watersheds are destroyed, the local climate becomes drier, and remaining forests dry and become susceptible to

Figure 26-19 The layers of a tropical forest. Tropical rain forests are noted for their extensive layering. Shown here are the protruding emergent trees, the canopy, the understory, the shrubs, and the ground-level plants. Because their roots are so shallow, tropical trees often grow huge, supportive buttresses.

fires. In short, rain forest destruction is an accelerating process. How is it happening?

Like other human populations, those in the tropics are doubling every few decades. To raise cash for these expanding populations, governments support the logging and burning of large areas of tropical rain forest. Tropical nations export timber and plywood from rain forest trees, metals mined from rain forest lands, and beef from cattle raised on clear-cut rain forest.

Once cleared, former rain forest soils do not last long. Recall that the nutrients in tropical rain forest are in the vegetation, which is carried away or burned. The average cattle ranch in Central America lasts only 6 to 10 years, after which the soil is too depleted of minerals to support grass for grazing cattle. Each hamburger made from the cattle raised on these ranches costs 5 square meters of tropical rain forest, including about a half ton of trees, flowers, birds, mammals, snakes, lizards, frogs, butterflies, beetles, ants, and termites.

One of the greatest causes of rain forest destruction is small-scale farming. Individual farmers cut and burn a small area of land and raise a few crops. Once the soil is depleted, the farmers move on to another patch of land. This kind of farming is called "slash and burn" farming or "shifting agriculture." Because the individual fires set by farmers are so small, the damage is often invisible in satellite imagery until vast areas have been destroyed by thousands of farmers. In Chapter 28, we discuss how isolating small plots of forest creates "edge effects" that cause many species to go extinct.

Future textbooks may list the tropical rain forest as a biome that has disappeared, for human activity threatens to eliminate all but a few parks within the next 20 to 30 years. Accompanying the destruction of the rain forest will be the disappearance of millions of species of plants, insects, and other animals found nowhere else. One-quarter of the world's 250,000 plant species and one-fifth of the 20,000 named butterflies will likely go. Yet most of the species that disappear will never even have been named.

As if that weren't bad enough, recall from Chapter 25 that the burning of forests releases tons of extra carbon dioxide into the atmosphere, increasing global warming, hastening changes in climate and the melting of the polar ice caps.

Tropical rain forests have the greatest biodiversity of any biome in the world. Unless major changes occur, however, 90 percent of it will be gone within 20 to 30 years.

WHAT ARE AQUATIC ECOSYSTEMS LIKE?

Water covers 72 percent of the Earth's surface—70 percent saltwater oceans and 2 percent freshwater lakes, rivers, and streams. Like terrestrial ecosystems, aquatic ecosystems fall into just a few distinct types. Ocean ecosystems differ from one another according to temperature, depth, and distance from shore.

How Do Oceanographers Divide the Oceans?

Although ocean covers more than two-thirds of the Earth's surface, the oceans contain only about 10 percent of the planet's species—about 160,000 species. Because the oceans are all interconnected, they lack habitat islands that promote geographic isolation and speciation (Chapter 17). Oceans provide fewer opportunities for adaptive radiation than land habitats.

Nonetheless, the oldest, largest, and most stable ecosystems on Earth are those of the oceans. The oceans and their connecting seas extend over nearly all of the Earth's latitudes.

In the last few decades, **oceanography,** the study of the seas and their ecosystems, has revealed the shape of the seas. The ocean bottom, for example, contains vast mountain ranges and rifts, as well as a shallow continental shelf bordering each continent. The average depth of the oceans is about 3 kilometers. But the lowest spot on Earth, the Marianas Trench of the western Pacific Ocean, dips to 11 kilometers (7 miles) below sea level—deeper than Mt. Everest is high.

Biologists distinguish **ocean ecosystems** according to ocean depth and distance from shore. The **benthic division** consists of all organisms that live on the ocean floor. These include those that live in the **intertidal zone,** which lies between high tide and low tide, and those that live at the bottoms of deep canyons in the **abyssal zone** (Figure 26-20). The **pelagic division** consists of all organisms that live in open water, above the bottom.

Organisms living in the intertidal zone are submerged at high tide and exposed to air and pounding waves at low tide. They must be able to withstand drying, hot sun, and crashing waves. Intertidal organisms anchor themselves to rocks or sand, and close shells and other structures tightly to keep from drying out when the tide goes out. Tidepools reveal the rich diversity of the intertidal species (Figure 26-21).

In the pelagic division, microorganisms called **plankton** drift in the water, moving only where currents take them. Free-swimming organisms, such as fish and whales, feed on plankton or on each other. The pelagic division is further divided into the **neritic zone,** which is farther from shore, over continental shelf, and the **oceanic zone,** which is out beyond the continental shelf over the deepest water.

The neritic zone consists of shallow coastal areas near islands and continents. In the neritic zone, algae are the dominant primary producers. While the total area of the neritic zone is much smaller than that of the oceanic zone, it contains many more ecological niches and species.

The oceanic zone comprises most of the world's oceans. Here photosynthetic cyanobacteria, diatoms, and dinoflagellates are the primary producers. Their output is extremely low, however, comparable to that of a terrestrial desert. The oceanic zone lacks mineral nutrients, especially phosphorus and iron. Oceanic ecosystems recycle nutrients poorly, depending on a steady supply from terrestrial biomes. As organisms die, their remains fall from the surface layers to the ocean bottom. Once these nutrients drop to the bottom of the ocean, they remain there.

A few exceptions exist. Off the coast of Peru, the Humboldt Current draws nutrients up from the bottom in a process called **upwelling.** Upwelling creates a rich soup that supports a rate of photosynthesis six times that found in other parts of the open ocean.

Where such currents bring nutrients, grazing protozoa and small animals feed on blooms of photosynthetic microorganisms. Small fish eat the tiny invertebrates, and larger fish eat the smaller fish. A marine food chain typically contains four or five trophic levels. (In contrast, terrestrial food chains seldom have more than three trophic levels.) This enormous productivity supports, in turn, huge populations of fish-eating birds, as well as a large human fishing industry.

Every year, around Christmas, the Humboldt Current reverses itself. Warm water flows down the coast, preventing the nutrient-rich cold waters from rising to the surface. Usually these

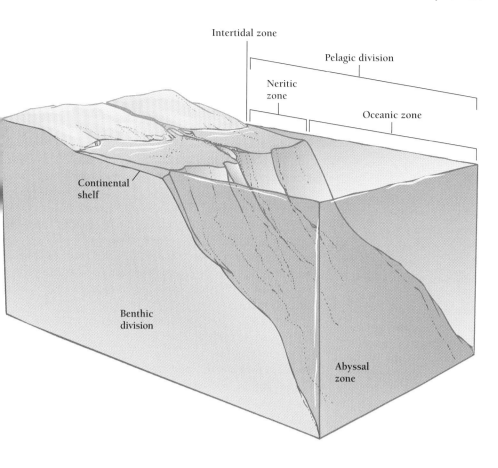

Figure 26-20 Dividing the ocean into zones. Oceanographers divide the ocean into divisions and zones according to the depth of the water and the distance from shore. The benthic division includes intertidal organisms and others that live on the bottom. The pelagic division includes the neritic zone (over the continental shelf) and the oceanic zone (deep water).

changes, which local fishermen call "El Niño" [Spanish, = the Christ child], are temporary and harmless. However, every 2 to 7 years, the current reversal lasts long enough to produce catastrophic effects. In 1982 and 1983, for example, the failure of the Humboldt Current to provide its usual supply of nutrients led to a failure of the sardine industry and the wholesale starvation of fish-eating seabirds in the South Pacific.

Doesn't El Niño Do More Than Kill Fish and Birds?

A big El Niño affects weather around the world. In the winter of 1997–1998, an area in the central and eastern equatorial Pacific Ocean the size of the continental United States increased in temperature by 3.5°C. The warm ocean triggered giant rain storms that battered the entire Pacific Coast, from the Pacific Northwest in the United States and Canada to the Andes of Equador and Peru. Meanwhile, eastward, in Indonesia and Malaysia, a severe drought worsened huge wildfires, which burned desiccated tropical rain forests for months.

El Niño has a sibling weather pattern, called La Niña, that has the opposite effect. During La Niña, the waters of the equatorial Pacific Ocean turn cold, and warm thunderstorms return to Indonesia, carrying heavy rains. Over North America, the cold Pacific jet stream weakens and undulates, sweeping across Alaska one month and down across the central United

Figure 26-21 The intertidal zone. The intertidal zone is characterized by organisms that cling to rocks as the waves crash around them. Shown here is a Pacific Coast tidepool. (*Nancy Sefton/Photo Researchers, Inc.*)

States the next. La Niña brings drier, warmer winter weather to the southeastern United States and wetter, colder weather to the Northwest.

Ocean ecosystems are more homogeneous than terrestrial ecosystems.

Why Are Estuaries and Marshes Valuable?

Estuaries are partly enclosed waters where freshwater streams or rivers meet the ocean. The water is less salty than the oceans, and the salinity is constantly changing. The flow of salt water and fresh water depends on the tides, which gives the estuaries their name [Latin, *aestus* = tide]. In addition, the flow of fresh water from rivers and streams varies with season.

Estuaries are among the most productive of all natural ecosystems, rivaling that of a tropical rain forest. Tides and rivers provide a constant flow of water and nutrients (Figure 26-22). The primary producers in estuaries include plankton, algae, and large plants such as eelgrasses, sea grasses, marsh grasses, and mangrove trees.

Because of this high productivity, estuaries provide breeding and nursery grounds for many species of fish and shellfish. Countries that destroy their coastal wetlands ultimately destroy their fisheries. Estuaries also provide food and breeding grounds for many species of birds, amphibians, reptiles, and mammals.

Estuaries are especially vulnerable to damage by humans. They are the first to receive pollution from streams and rivers, and they are prime targets for oceanfront development. All over the world, farms, condominiums, football stadiums, and industrial parks sit atop filled and drained wetlands.

In the United States, the destruction of wetlands has slowed in the past 20 years. Cities and states have ceased to regard marshes as mere sewers and breeding grounds for mosquitoes. Instead, people are beginning to realize that estuarine ecosystems are not only places of beauty but also irreplaceable contributors to the world food supply.

In the 1940s, Florida drained large sections of the Everglades in an effort to control floods and provide fresh water for a growing population. These efforts destroyed large sections of the Everglades. In places, the Everglades are so dry that wildfires burn out of control. In other places, fish populations are so low that 90 percent of wading birds have vanished, for there is not enough for them to eat. Excess phosphorus from farms has allowed cattails to take over large sections of the Everglades, crowding out species that formerly flourished there.

Fortunately, state and federal governments are now struggling to restore this valuable wetland. Florida farmers have changed the way they irrigate and spray their crops, to minimize the runoff of both pesticides and excess phosphorus. In addition, animal feeds can be formulated with less phosphorus. As one agricultural researcher put it, cows "can't excrete what they don't eat." In addition, giant artificial marshes now remove up to 80 percent of the phosphorus before it reaches the Everglades. And the U.S. Army Corps of Engineers has proposed to Congress an $8 billion plan that would allow water to flow more freely throughout the Everglades.

Estuaries are the most productive ecosystems in the world. They support a diversity of species.

Why Are Lakes, Ponds, Rivers, and Streams Susceptible to Pollution?

Limnology [Greek, *limne* = pool, marshy lake] is the study of freshwater lakes, ponds, rivers, and streams. As rivers and streams flow into lakes, they bring both water and dissolved nutrients. This material, both organic and inorganic, derives from the surrounding land, so the character of lakes and ponds varies from region to region. At the fringes of a lake, marshes and swamps often form transition zones to the adjacent terrestrial ecosystems.

As in ocean ecosystems, primary production in a lake depends on the amount of sunlight penetrating the lake's surface to photosynthesizing microorganisms and water plants. Heterotrophic protists and small animals such as rotifers and crustaceans graze on the photosynthesizing microorganisms, while birds, fish, and invertebrates feed on the larger plants. Fish also eat one another. Despite the efficient aquatic food web in lakes, organic matter falls to the bottom as detritus, where it is consumed by bacteria and in-

Figure 26-22 Estuary. Salt marsh at Assateague Island National Seashore, Maryland. *(Michael P. Gadomski/Dembinsky Photo Associates)*

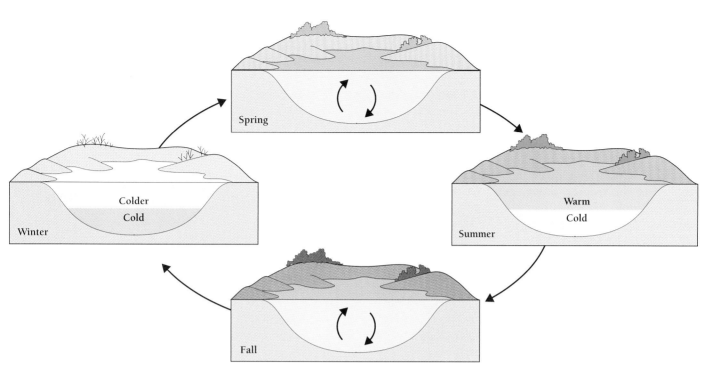

Figure 26-23 Turnover. Deep temperate lakes turn over twice a year. Both turnovers bring nutrients to the surface of the lake, resulting in blooms of algae. Water's highest density occurs at 4°C. In spring, when sun-warmed ice (0°C) melts and reaches 4°C, it sinks below the colder layers, carrying oxygen downward and driving nutrients upward.

vertebrates. Far more energy flows through the detritus food chains than through the chains involving grazing herbivores and carnivorous fish.

In temperate regions, a lake's surface temperature varies greatly from season to season. In the summer, the warm air heats a layer of warm water as much as 20 meters thick (Figure 26-23). As winter approaches, the temperature of the surface layer falls until it is the same as the deeper layer. At that point, the waters of the two layers mix—an event called the *fall turnover*.

A similar turnover occurs in the spring, as the winter ice melts. The two turnovers bring dissolved nutrients from the bottom sediments into the upper waters, where photosynthesis occurs. The result is a seasonal growth of algae. The fall and spring turnovers also move oxygen from the surface down to the bottom-dwelling heterotrophs, which allows them to metabolize settled organic material.

A **eutrophic** lake is one that has abundant minerals and organic matter. The surfaces of eutrophic lakes tend to be clogged with algae and cyanobacteria. As these organisms die and drop to the bottom, they stimulate the lake-bottom decomposers to multiply. During the summer, the decomposers so deplete the deeper layer of oxygen that the lake bottom can no longer support fish and other animals. As in the "dead zone" mentioned in the last chapter, few organisms can live in a eutrophic lake. In contrast, **oligotrophic** lakes have a limited supply of nutrients. These clear lakes have no permanent algal blooms, and they therefore contain more oxygen and support a more diverse community of organisms, including fish.

Eutrophication happens naturally over thousands of years, but humans often accelerate the process by dumping nutrient-rich sewage and other wastes into streams and lakes. Productivity increases at first, but diversity declines, and eventually the lake dies.

Because rain washes both nutrients and poisons into rivers and streams, freshwater ecosystems easily accumulate pollutants.

We have seen how ecosystems move energy and materials and what kinds of ecosystems exist. But ecosystems are made up of individual species. So we should also consider the characteristic roles each species plays in the flow of energy and the use of materials. In the next chapter, we will see how species interact and how ecosystems change in time. We will also examine the role of evolutionary history and current competition in establishing a community's structure.

Study Outline with Key Terms

Our planet contains a limited number of types of ecosystems. These are called **biomes** if they are terrestrial, and simply aquatic ecosystems if they are aquatic. The character of these ecosystems is determined largely by climate, especially temperature and moisture. The average temperature, the total annual precipitation, and the variability in each of these values help determine the character of a biome.

At the equator, hot, moist air rises and flows north and south over the **tropics,** dropping rain. **Tradewinds** blow cooler air from the tropics toward the equator. The **westerlies** are cool, dry winds that cross the Earth's surface between 30° and 60° latitude. These winds warm and pick up moisture as they go. In the Arctic, cold, dry **polar easterlies** blow from the east.

The climate of a region is determined by the total solar energy absorbed and by wind and ocean currents. These depend on latitude and the rotation of the Earth. The Earth's rotation creates the **Coriolis effect,** which causes winds moving north or south to rotate, clockwise in the Northern Hemisphere and counterclockwise in the Southern Hemisphere.

Mountain ranges create **rain shadows** by forcing warm, moist air to drop most of its moisture on one side of the range. Biomes at high elevations resemble those at high latitudes. At high altitudes, **coniferous forest** gives way to **tundra** at the tree line.

Deserts, which receive less than 25 centimeters of precipitation each year, are characterized by widely spaced drought-tolerant plants, such as sage brush (in cold regions) and cacti (in hot regions). Some animals **estivate** during the hot, dry summers. The greatest human threats to deserts are off-road vehicles and agriculture.

The cold, dry tundra, which receives less than 25 centimeters of precipitation each year, is characterized by short, open grassland, punctuated by bogs, ponds, and lakes. **Permafrost** underlies the soil all year long.

Temperate grasslands, which receive 25 to 75 centimeters of precipitation each year, are characterized by tall grasses in wet areas and short grasses in drier areas. Fire, insufficient water, and grazing animals all keep trees from becoming established in this biome. Much of the world's grassland, midwestern North America, for example, has been converted to agriculture, thus eliminating entire food chains.

Chaparral regions, which have mild, wet winters and hot, dry summers, are characterized by dense shrubs and scattered, broad-leafed trees, mostly drought and fire tolerant. Humans build homes in the chaparral, then suppress natural brush fires so long that when fire finally comes, fuel loads create a firestorm. Houses, chaparral, and valuable soil are all damaged or destroyed.

Savannas, which receive 90 to 150 centimeters of rain each year, are characterized by tall grasses and short, widely spaced trees. Savannas are hotter and wetter than temperate grasslands.

The cold **taiga** is characterized by short, homogeneous pine or spruce forests. This biome has suffered less human-caused damage than most other biomes because it has so little to offer economically.

Temperate **deciduous forests,** which receive 80 to 140 centimeters of precipitation each year, are characterized by a heavy cover of deciduous trees and a limited understory. Temperate deciduous forests, found in eastern North America, eastern China, and Europe, are commonly associated with dense populations of humans. These populations have cleared these forests repeatedly and extensively, and hunted to extinction or near extinction all of the major predators, including bears, wolves, and lions or tigers.

Tropical rain forests, which receive 200 to 450 centimeters of precipitation each year, are characterized by a dense **canopy** with a tall understory and little vegetation on the dim forest floor. **Lianas** and **epiphytes** grow on the trees, and **emergents** rise above the canopy. Tropical countries are cutting their rain forests down at a rate that will eliminate this biome within a few decades.

Oceanography has revealed the depth and character of the oceans and the organisms that live in them. **Ocean ecosystems** can be divided into **benthic** and **pelagic divisions.** The benthic division includes the **intertidal** and the **abyssal zones.** The pelagic division includes the **neritic** and **oceanic zones.** **Plankton** float freely and travel passively only. Fish and other large animals propel themselves. Ocean food chains can have more trophic levels than terrestrial food chains, but the total number of species living in the oceans is far less than the number that live on land. Primary productivity is low, except in areas of **upwelling.**

Estuaries are marshes with partly fresh, partly salt waters. Estuaries are extremely productive and provide breeding grounds for many oceanic species. Humans have a long history of draining marshes and estuaries for farms or developments.

The study of freshwater lakes, ponds, rivers, and streams is called **limnology,** one of the oldest branches of ecology. Lakes that are rich in organic matter are **eutrophic.** Those lower in nutrients support a more diverse community and are said to be **oligotrophic.** Humans frequently pollute fresh water with sewage and industrial wastes, as well as the acid rain that results from burning coal and oil.

Review and Thought Questions

Review Questions

1. Explain why the polar regions experience more extreme seasons than the equatorial regions.
2. What kind of biome exists in a rain shadow? Explain why.
3. How might the division of a forest or other habitat into small sections interfere with top predators?
4. What keeps grasslands from changing to forests?
5. Explain the difference between a liana and an epiphyte. Why would tropical plants make good house plants?
6. What percentage of the Earth's surface is covered by water?
7. Why are estuaries ecologically and economically important?
8. What organisms are the primary producers in the oceanic zone? What organisms are the primary producers in the neritic zone?
9. What causes fall turnover, and what benefits does turnover have for freshwater organisms?
10. What causes a lake or pond to become eutrophic?

Thought Questions

11. Explain why the soils of tropical rain forests are poor, even though tropical rain forests are so productive.
12. Why do tropical rain forests have so many species?
13. Why are the oceans species poor?

About the Chapter-Opening Image

Nothing says "desert" more clearly than an organ pipe cactus. Every biome includes characteristic species. Desert biomes, for example, often host species of plants that store water and use spines to defend themselves against thirsty herbivores.

 On-line materials relating to this chapter are on the World Wide Web at **http://www.harcourtcollege.com/lifesci/aal2/**

 Web Links

 Quizzing & Testing

27 Communities: How Do Species Interact?

Volcano: An Ecosystem Is Destroyed

At 8:32 a.m. on May 18, 1980, a medium-sized earthquake shook Mount St. Helens in Washington State. Twenty seconds later, the volcano erupted, and the entire north side of the mountain slid away in a thundering roar. It was the largest landslide in recorded history. Explosions from the volcanic eruption tore through the 2.5 cubic kilometers (0.6 cubic miles) of sliding debris, spewing rocks, ash, gas, and steam across adjacent valleys at velocities approaching the speed of sound. The series of blasts, 500 times as powerful as the bomb that flattened Hiroshima, destroyed an area the size of the city of Chicago—a forest with enough timber to build 300,000 houses. Every tree within 25 kilometers (15 miles) was either blown to bits, burnt to a crisp by 600°C gases, buried under tons of volcanic ash, or blown down. From the north face of the mountain poured a river of melted snow and ice, mud, boulders, and ash. In less than 10 minutes, this colossal river of rocks filled a valley the size of Manhattan to an average depth of 45 meters (deep enough to bury a 15-story building). A black plume of pumice and ash rose 25 kilometers into the air, and 520 million tons of ash began drifting eastward across the United States.

The death toll included an estimated 11 million fish, 27,000 grouse, 11,000 hares, 6000 black-tailed deer, 5200 elk, 1400 coyotes, 300 bobcats, 200 black bears, 57 human beings, and 15 mountain lions. Three more explosions in May, June, and July dropped thick blankets of pumice and ash. In the end, every exposed slope within 10 kilometers (6 miles) of the crater was stripped of trees and other vegetation and buried under 2 meters of ash and rock. In places, the mud, ash, and rock piled up 200 meters (an eighth of a mile) thick. The scene was as gray as a moonscape. For another 16 kilometers (10 miles) from the blast zone, thousands of full-grown fir and hemlock trunks lay flattened in the ash. It was hard to imagine how anything could have survived.

Yet, an enormous amount of life did survive (Figure 27-1). Along snow-covered ridges that faced away from the blast, saplings still bent by the weight of 1 or 2 meters of winter snow, as well as seedlings and shrubs buried under the snow, came through unscathed. Shrubs that were obliterated at the surface, but whose roots survived underground, sprouted and grew anew. From underground, burrowing animals such as pocket gophers and a variety of insects and other invertebrates emerged. Within weeks, plants began growing on the naked slopes of the mountain, wherever rain had washed away the sterile ash to expose the old soil beneath.

Within 10 years, the slopes of Mount St. Helens were covered with dogwood, elder, huckleberry, and other shrubs, as well as seedlings of fir, hemlock, and other conifer trees. Ecologists say that by 2030, the area will begin to look like a young coniferous forest. In 200 years, the area should resemble an old-growth forest. The Mount St. Helens ecosystem—almost completely destroyed—will rebuild itself.

Key Concepts

- Over time, communities undergo predictable changes in species composition, a process called succession.
- Communities of species are both integrated and individual.
- Relationships among species, such as symbiosis, competition, and predation, coevolve over time.
- No two species can occupy the exact same niche and habitat indefinitely.

ARE COMMUNITIES INTEGRATED OR INDIVIDUALISTIC?

The process by which Mount St. Helens renews itself is called **succession,** which is the predictable change in numbers and kinds of organisms in a community over time. How and why succession occurs has been one of the burning questions in ecology for more than a hundred years.

In Chapter 25 we discussed Lynn Margulis's view that our biosphere is a tightly integrated whole that is composed of organisms that cooperate as well as compete. Surprisingly, the philosophical differences that fueled the debate over Gaia are old issues in the science of ecology. Do species cooperate to create a cohesive, functioning environment? Or are species free agents that operate independently of other species? For nearly a hundred years, ecologists have been arguing about whether different species form "integrated communities" or whether each species is "individualistic."

It all began in 1898, when one of America's first ecologists, Henry C. Cowles (1869–1933), published a groundbreaking paper on plant succession. Cowles was a warm, funny man of infectious high spirits who inspired generations of young ecologists. As a graduate student, he studied the succession of plants on sand dunes beside Lake Michigan.

Cowles saw that, over thousands of years, the water level of Lake Michigan has gradually dropped, leaving behind a broad, sandy shore.

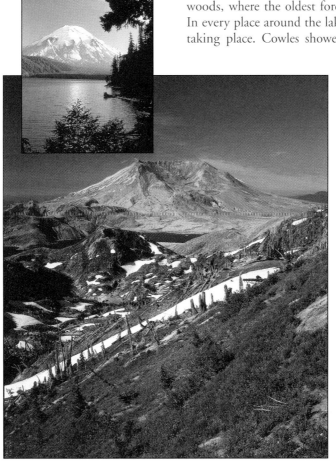

Figure 27-1 Life returns. Mt. St. Helens with Spirit Lake in the foreground before the eruption in 1980 (top photo). Mt. St. Helens in 1994 with Spirit Lake (bottom photo). By 1994, plants had begun to clothe even some of the most devastated areas. Ecologists predict that Mt. St. Helens' ecosystem will return to mature coniferous forest in about 200 years. (*Top, Tom and Pat Leeson/Photo Researchers, Inc.; bottom, John Marshall/Tony Stone Images*)

Fierce winds blow the sand from the former lake bottom into giant dunes. Grasses cover the dunes, growing long roots that stabilize the sand. Soon bearberry, juniper, and other shrubs grow over the dune, displacing the grass. Later on, Cowles found, trees such as jack pine grow among the shrubs. The pines create a moist, shady habitat that kills the shrubs and encourages the growth of black oak trees. Finally, in some places, the black oak forest gives way to beech and maple forests.

This sequence of communities started right by the lake, where the newest sand dunes lay, and ended far up into the woods, where the oldest forests had grown up (Figure 27-2). In every place around the lake, the same series of changes was taking place. Cowles showed that natural communities can change predictably over time. But like the shrubs on the Michigan dunes, Cowles was soon overshadowed by a stronger personality.

Succession's most important booster was Frederick Edward Clements (1874–1945). Clements was Cowles's opposite—arrogant, distant, and ascetic. But he was a prolific researcher and writer all of his life. In one of the earliest automobiles, and before the construction of paved highways, Clements and his wife drove thousands of miles across the roadless western United States, camping and studying. During his life, Clements studied and wrote about every major ecosystem in North America. In 1916, he published the definitive book on plant succession. Clements argued that in every geographic area, as defined by climate and soil, succession always created the same predictable community of plants, the **climax community.**

Clements, sounding rather like James Lovelock of Chapter 25, further argued that

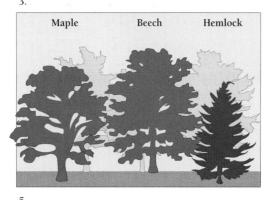

Figure 27-2 Sand dune succession. Henry C. Cowles showed that as the water level in Lake Michigan drops, the sand dunes left behind are gradually colonized by a succession of different plant communities. Beach grasses stabilize sand dunes, allowing the growth of shrubs and the formation of soil. Jack pines grow up through the shrubs, shading the soil. In the cool moist soil, oak trees germinate and grow, eventually shading out the jack pines. Oak forest may, in places, give way to beech and maple forest.

the climax community was an organism. Succession, he said, was a developmental process, like the embryonic development of an organism. Clements wrote, "As an organism, the [climax] formation arises, grows, matures, and dies . . . each climax formation is able to reproduce itself, repeating . . . the stages of its development."

Clements's description is of an **integrated** community, one that consists of characteristic species that always interact with each other in predictable ways. In this view, all maple tree forests, for example, would have approximately the same species of plants and animals, and all these species would interact in similar ways.

Other ecologists found the integrated view compelling. Anyone can see that deserts usually contain cactuses, small mammals, lizards and snakes, while ponds usually contain water plants, frogs, fish, and mosquitoes. Clements's view—that a community is an integrated collection of interdependent species—is one of the most important philosophical themes in ecology. But its simplicity is also its weakness. It invites criticism.

In 1926, a less-famous American ecologist named Henry Allan Gleason published a paper frankly challenging Clements. Gleason argued that communities are not integrated, but composed of separate populations that merely happen to occupy the same habitat. Provocatively, Gleason wrote in part, "In conclusion, it may be said that every species of plant is a law unto itself. . . . [A plant] association is not an organism, scarcely even a vegetational unit, but merely a coincidence." In Gleason's view, communities were **individualistic.**

Gleason's impolite attack on the much-revered Clements so embarrassed other ecologists that most refused to even discuss Gleason's heretical ideas. As Gleason later recalled, "Not one believed my ideas; not one would even argue the matter. . . . For ten years, . . . I was an ecological outlaw."

In time, though, ecologists began to take Gleason's ideas seriously. A healthy and productive scientific debate ensued between Clements's holistic ecologists and Gleason's individualistic ones. Gleason and other individualists argued that every species has an independent distribution, and that, in effect, every community is unique.

The individualistic view garners support from the many careful studies of species distribution patterns that show no obvious clusterings of species. A survey of Wisconsin forests, for example, showed that no two tree species consistently share the same range (Figure 27-3A). Maple trees may be associated with beech trees in one place, elms in another. In the individualistic view, a beech-maple forest is not an organic entity or community, but two species that just happen to share the same moisture requirements. Today, few ecologists accept Clements's idea of a climax community; most research shows that the species composition of forests and other communities varies continuously.

On the other hand, the integrated view of ecological communities rests firmly on thousands of studies that demonstrate interactions among species. To take an obvious example, most woodpeckers and termites could not easily survive in a treeless grassland. Nor could certain plants survive without the fungal mycorrhizae that help them draw nutrients from the soil (Chapter 21). But more subtle interactions also determine who lives where. For example, the presence of sugar maples appears to limit the distribution of basswood trees, perhaps by competition for water or shading (Figure 27-3B). So, even though the unity of a community may not be as well defined as Clements envisioned, ecologists still think of communities as functioning units whose member species all interact.

The integrated and the individualistic views are not mutually exclusive ideas, but the extreme ends of a continuum. Just as the members of a family sometimes behave as individuals and sometimes behave as an integrated whole, the species that make up a community can be both individualistic and integrated. A community is both an integrated set of interacting species and a group of organisms that occur together because their ranges happen to overlap. In the next section, we will examine a few of the ways in which different organisms in a community interact.

The individualistic and integrationist views of ecosystems represent extreme ends of a continuum of thought.

HOW DO SPECIES IN A COMMUNITY INTERACT?

The general nature of a biome depends on abiotic factors such as temperature, moisture, and sunlight, as we saw in Chapter 26. Annual rainfall, for example, largely determines whether an area is a forest or a desert. Abiotic factors alone do not define a community, however. Within each community, the interactions of organisms with one another help to establish the character of a community.

These interactions include those between the consumers and the consumed and between competitors. Interactions between the consumers and the consumed include predator-prey interactions, herbivore-plant relations, and parasitism. The wolf hunts the caribou, and the caribou eats the lichen. Species may also compete—for territory, prey, or other resources—or they may form intimate associations, in which one species benefits from another or where two species help each other. Lichen, for example, turns out to be not one organism, but two, a cooperative association between a fungus and an alga.

Who Eats Whom?

We usually think of consumers as being larger and stronger than the consumed. Hawks, for example, generally prey on mice, squirrels, and other animals smaller than themselves.

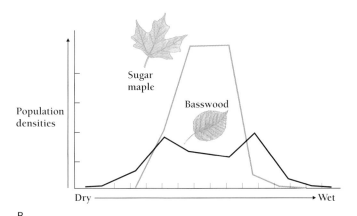

Figure 27-3 Are communities individualistic or integrated? Studies of trees in Wisconsin support both ideas. A. Oaks, maples, and other trees are distributed according to the dampness of the soil, independently of other trees. B. But basswood grows best where sugar maple is absent, even though it is capable of growing in the same soils.

Most predators are indeed larger than their prey, although carnivores that hunt in groups, such as wolves, regularly bring down prey far larger than themselves. But herbivores are often smaller than the individuals they consume. Hordes of tiny oak moth caterpillars can strip all the leaves from a 60-foot oak tree. And parasitic consumers, such as ticks, fleas, and tapeworms, are generally smaller than their hosts.

Consumers, then, come in many forms. The main consumers besides the decomposers and scavengers are herbivores, predators, and parasites. A **parasite**, by definition, consumes only part of the blood or tissues of its host and does not necessarily kill the host. In contrast, a **predator** usually (but not always) kills its prey and consumes most of its prey's body.

In some cases, the differences between predators and parasites may blur. For example, parasites do occasionally kill their host outright, and parasites frequently weaken the host so severely that the host dies of other causes. And some predators consume only part of their prey. For example, some predaceous fish bite chunks of flesh from other fish, without necessarily killing the victim.

Herbivores at first seem to fall outside these definitions, but they actually fit rather nicely. Leaf-eating caterpillars and deer, for example, browse on living plants, just like parasites. And seed-eating birds, which kill and consume an entire plant embryo, may be considered predators. In reality, ecologists rarely stress the parasite-predator distinction in herbivores.

How Do Plants and Animals Defend Themselves?

With notable exceptions, consumers' adaptations for finding and eating their favorite food exhibit a certain sameness: in the case of predators, swiftness, intelligence, acute senses, and sharp teeth; in the case of herbivores, patience and a good digestive system. In contrast, organisms have a diverse arsenal of defenses against potential consumers.

Camouflage

The subtlest defense against being eaten is to remain invisible. The rabbit, for example, simply dives down a hole or freezes, so that it blends into a background of shrubs or grass. But many organisms go to far greater lengths to **camouflage** themselves by mimicking twigs, leaves, stones, bark, and other materials in their environment (Figure 27-4). A walking stick insect, for example, looks just like a twig. Some species even sway slightly, as though an errant breeze were blowing.

Chemical defense

Plants defend themselves against consumers in a variety of ways. Trees protect themselves from termites and other wood-eating creatures with thick layers of bark. Roses and cactuses fend off herbivores with thorns. The single most important defense that plants use, however, is toxic chemicals. The unforgettable wallop of hot chili pepper, the sharp tang of raw

Figure 27-4 Camouflage. Many animals avoid being eaten by blending in with their background. This harlequin crab lives on the underside of a sea cucumber, camouflaged against the background of its host. (*Denise Tackett/Tom Stack & Associates*)

broccoli, and the deadly toxin in poison hemlock all result from certain kinds of **secondary plant compounds**—chemicals that are not essential to a plant's normal metabolism but that often serve a defensive purpose.

Chemical defense is not limited to plants. Bees, ants, and wasps inject a powerful acid into attackers. The bombardier beetle and the skunk both spray would-be predators with noxious chemicals (Figure 27-5). The famous poison-dart frog (*Phyllobates*) of South America and the skin and feathers of the pitohui bird of New Guinea contain the same deadly toxin (Figure 27-6).

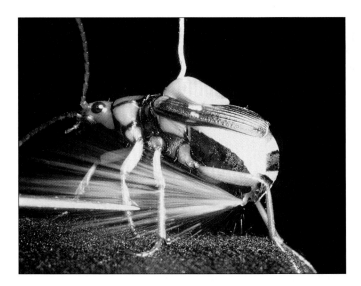

Figure 27-5 Chemical warfare. Everyone knows that skunks fight back with noxious chemicals. So do many plants and insects, such as this bombardier beetle. (*Thomas Eisner and Daniel Aneshansley/Cornell University*)

Figure 27-6 Warning coloration in a pitohui bird? In 1992, researchers discovered three species of birds in the genus *Pitohui*, whose skin and plumage contained the same toxin as that found in poison-dart frogs of South America. Like wasps, skunks, and poison-dart frogs, pitohui birds bear striking colors that may function to warn predators. Research has shown that inconspicuously colored birds taste better than flashy ones. *(W. Peckover/VIREO)*

Warning coloration

If a skunk looked like a rabbit, it would be attacked over and over again, in spite of its potent defense. But because a skunk has a striking and memorable black-and-white pattern, predators quickly learn to avoid the animal. Among all animals, chemical deterrents work best if accompanied by a bright, memorable design called **warning coloration.** Warning coloration, such as the yellow-and-black stripes of the aggressive, stinging yellow jacket, helps predators remember which prey to avoid. Predators will avoid any animal with the colors and patterns they associate with pain, illness, or other unpleasant experiences. Most warning colors are vivid shades of red, yellow, orange, and white, combined with a contrasting shade of blue or black.

Because predators avoid animals with warning coloration, some species copy the warning coloration of toxic or dangerous species. A situation in which one animal *mimics* the color pattern of another is called **mimicry.** Some perfectly harmless and quite edible flies, for example, mimic the black-and-yellow stripes of yellow jackets and bees. A situation in which a harmless animal imitates a dangerous one is called **Batesian mimicry.**

In a second form of mimicry, called **Müllerian mimicry,** two or more dangerous species evolve similar colors. The similar colors signal a similar hazard to any predators the two mimics share in common. For example, the monarch butterfly is toxic to birds. The monarch caterpillar feeds on milkweed, a plant that contains a toxic secondary plant compound. This toxin accumulates in both the monarch caterpillar and butterfly and makes it toxic to birds. A bird that eats a monarch quickly becomes ill (Figure 27-7A). After a few such experiences, the bird avoids monarchs, and the monarch's striking warning colors help the bird remember.

A second butterfly, the viceroy, is unrelated to the monarch but also contains a toxin that makes birds sick. The viceroy looks remarkably like a monarch, and biologists believe these two butterflies have evolved to resemble one another (Figure 27-7B). The advantage to each butterfly is that birds that have learned to avoid the monarch also avoid the viceroy and birds that avoid the viceroy also avoid the monarch.

Müllerian mimicry helps predators learn the same lesson from a variety of prey species. Bees and wasps both use a similar pattern of yellow and black stripes. Anyone who has been stung by a bee also learns to avoid wasps and hornets. Such encounters lead to greater safety for bees, wasps, and hornets. The more species that use these kinds of patterns and colors, the more this universal signal is reinforced.

A.

B.

Figure 27-7 Müllerian mimicry. A. The monarch butterfly is so noxious that birds such as this jay quickly cough them up. B. Viceroy butterflies are unrelated but also toxic. Monarchs (*left*) and viceroys (*right*) have evolved amazingly similar color patterns. Each butterfly's message reinforces the other's: "Don't touch!" *(A, Lincoln Brower; B, Thomas C. Emmel)*

TABLE 27-1 INTERACTIONS AMONG ORGANISMS

Interaction	Effect on A	Effect on B	Example
Competition between A and B	Harmful	Harmful	Otter and mink
Predation by A on B	Beneficial	Harmful	Lion and gazelle
Symbiosis			
Parasitism by A on B	Beneficial	Harmful	Flea and dog
Commensalism of A with B	Beneficial	Harmless	Fish and anemone
Mutualism between A and B	Beneficial	Beneficial	Lichen

Symbiosis: How Do Organisms Live Together?

A species may live in an intimate association with another species, an arrangement called **symbiosis,** meaning literally "living together." Symbiosis may be cooperative or antagonistic. Table 27-1 summarizes some of the possible interactions between any two species. Each species may affect the other positively, negatively, or not at all. In **parasitism,** one species benefits at another's expense. Tiny parasitic wasps, for example, lay their eggs inside of caterpillars. The wasp larvae literally eat the caterpillars alive as both animals mature.

Mutualism describes a symbiotic relationship between two species that mutually benefits each. For example, the bull's horn acacia (*Acacia cornigera*) of Central America has special enlarged thorns that house colonies of ants (*Pseudomyrmex ferruginea*). The acacia's nectar attracts the ants, which use it as their major food source. In return, the ants protect the acacia from being eaten by caterpillars and other herbivorous insects. This arrangement benefits both species. One study showed that acacias grown without ants for 10 months weighed less than one-tenth of those with intact ant colonies.

Another kind of symbiosis is **commensalism,** an intimate relationship between two species that helps one but neither helps nor harms the other (Figure 27-8). For example, the anemone fish lives safely among the poisonous tentacles of sea anemones. The fish, which benefits from the protection of its host, is a commensal. For reasons unknown, the sea anemone, which catches and eats other species of fish, leaves the anemone fish alone.

Species interact as competitors, as predator and prey, and as symbionts.

Do Organisms Coevolve?

The mutualistic relationship between ants and bull's horn acacias demands special adaptations in each species. The acacia has evolved enlarged thorns with a tough, woody exterior and a soft, pithy center, easy for the ants to excavate for nests. The acacia also has "nectaries" at the bases and tips of its leaves to supply the ants with nectar. The ants exhibit special adaptations as well. While most species of ants are active only at night or only during the day, members of the species *Pseudomyrmex ferruginea* remain active and protect the acacia day and night.

These distinctive adaptations in the ant and the acacia appear to result from a long and mutual evolutionary history in which the needs of each species exert selective pressure on the other. The interdependent evolution of two or more species whose adaptations appear to be selected by mutual ecological interactions is called **coevolution.**

Biologists love to suggest likely examples of coevolution. For example, many pollinating insects specialize on flowers that seem, in turn, to be specially constructed to be pollinated by just one kind of insect. Mutualisms such as these present the most likely examples of coevolution. But demonstrating that coevolution has occurred in individual cases is difficult.

Further, all species that have ecological interactions may exert some selective pressure on one another. And virtually every species in a community interacts (however indirectly) with every other species. For example, the bacteria that break down organic matter to nutrients that plants can absorb through their roots are as essential to the herbivores and carnivores as to the plants. Coevolved relationships can be either very specific, as in the case of the acacias and the ants, or very

Figure 27-8 Commensalism. The cleaner shrimp lives safely near the mouth of this predaceous moray eel. *(Mike Severns/Tony Stone Images)*

diffuse. The concept of "diffuse coevolution" is another way of acknowledging that all organisms in a community affect one another.

Interactions between two species influence the evolution of both.

How Does Competition Affect Organisms?

Some species interact as consumers and consumed, and others as commensals or mutualists. Still others compete. In **competition,** one organism uses a resource in a way that limits the availability of that resource to others. For example, if two species of deer both eat the same herbs and live in the same area, and the herbs are in limited supply, the deer are in competition with each other.

Can two similar species share the same habitat?

Measuring competition in natural communities takes careful planning and months or years of research. But laboratory studies can quickly show how competition occurs in simple situations. The Russian ecologist G.F. Gause studied populations of two similar species of *Paramecium*—*P. caudatum* and *P. aurelia*. Each *Paramecium* species, cultured by itself, flourished on the bacterial food that Gause provided. But when Gause mixed the two species, *P. caudatum* disappeared from the culture within 2 weeks. Gause provided only one environment and one kind of food for the two species, and *P. aurelia* was evidently better adapted to Gause's conditions than was *P. caudatum*. One species thus replaced the other.

These and similar experiments have led to a generalization called the **competitive exclusion principle,** which says that when two species compete directly for exactly a resource that is in short supply, one species can eliminate the other. A resource in short supply is a **limiting resource.** For example, territories might be a limiting resource for a pride of lions, but air would not be. Similarly, water is often a limiting resource for terrestrial plants but not for aquatic plants.

The competitive exclusion principle predicts that one species always wins. But the same species does not win under all conditions. For example, in a 40-year series of experiments at the University of Chicago, Thomas Park and his colleagues studied the effects of different environments on competition between two species of grain beetles (*Tribolium castaneum* and *Tribolium confusum*). These beetles were allowed to grow in containers of flour, the favorite food of the larvae. When grown separately, both species did best in warm, moist conditions. When grown together in warm, moist containers, *T. castaneum* drove *T. confusum* to extinction. In a cold, dry container, however, *T. confusum* prevailed.

According to the competitive exclusion principle, no two species can simultaneously exploit the same resources in the same place. The place in which an organism lives is called its **habitat.** In contrast, the way an organism uses its environment is called its **niche.** The habitat of the desert tortoise is the desert. The niche that the tortoise occupies is that of burrowing herbivore.

The competitive exclusion principle states that no two species can indefinitely occupy the exact same niche in the same habitat.

How do competing species coexist?

In the laboratory, species can be made to compete for the same limiting resources. But in nature, two species occupying the same habitat can, in practice, split up the niche in an infinite number of ways. For example, the ecologist Robert H. MacArthur (1930–1972) tested Gause's competitive exclusion principle in field studies of five species of North American warblers. All five bird species hunt for insects and nest in the same kind of spruce tree. The niches of these five birds seem nearly identical, but MacArthur's studies showed that the five warblers had split up the spruce tree niche extremely finely. He showed that each species of bird hunts insects in a different part of the tree and nests at a separate time from the other warblers. As a result, the warblers' niches overlap without being identical. Ecologists call splitting a niche **resource partitioning.**

Another study, on barnacles, shows how two species go about splitting a niche. In tidepools on the coast of Scotland live two species of barnacles. One species, *Balanus balanoides,* lives only on the lowest intertidal rocks, which are wet even during low tide. The other species, *Chthamalus stellatus*, lives higher up, in between the low-tide and high-tide lines.

To find out if the two species were competing, the ecologist Joseph H. Connell removed *Balanus* from some tidepools and *Chthamalus* from others (Figure 27-9). Connell found that where *Balanus* was removed, *Chthamalus* moved down to occupy the newly opened territory and survived there perfectly well. Likewise, where *Chthamalus* was removed, *Balanus* was able to survive on the higher, drier rocks.

Balanus dominated the lower levels because its heavy shell grows under the shell of *Chthamalus* and pries it up off of the rocks, preventing *Chthamalus* from growing on the lowest rocks. Higher up, however, *Chthamalus's* higher tolerance for dry conditions allows it to prevail over the physically stronger *Balanus.*

Connell's two barnacle species illustrate the difference between a species's **fundamental niche,** its potential ability to utilize resources, and its **realized niche,** the resources that it actually uses in a particular community. The fundamental niche of *Chthamalus* included the lower rocks, but its realized niche did not. The fundamental niche of *Balanus* included the higher rocks, but its realized niche did not.

Similar experiments on rat parasites also illustrate competitive exclusion. Although the tapeworm and the spiny-headed worm both prefer the upper intestine of the rat, when the two worms are together, the spiny-headed worm takes the

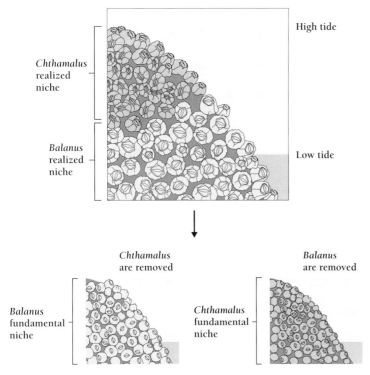

Figure 27-9 Niche splitting in barnacles. In a classic experiment, the ecologist Joseph Connell showed that the two barnacles *Balanus* and *Chthamalus* are capable of growing on the same rocks if the other species is removed. Otherwise they split the tidal niche, *Balanus* below and *Chthamalus* above.

food-rich upper tract and displaces the tapeworm to the lower end of the intestine. All these experiments, and many more like them, suggest that competition in natural populations determines species distributions.

Competing organisms in natural environments may partition resources so that both species persist in the same habitat.

Character displacement: Does competition influence evolution?

We have seen how competition between two species can lead to niche splitting. Species may evolve differences that make niche-splitting strategies permanent. For example, two species of caterpillars that both eat members of the rose family (which includes roses, apples, plums, and other species) might reduce competition by specializing, one on apple trees and one on roses. Any change in morphology, life history, or behavior that results from competition is called **character displacement**.

The finches of the Galápagos Islands provide some of the best-documented examples of character displacement. In the case of the Galápagos finches, character displacement has led to speciation. For example, the small ground finch *Geospiza fuliginosa* and the medium ground finch *Geospiza fortis* occur together on some islands, while on other islands only one of each species lives. On islands where these birds occur separately, both species's beaks are medium sized, and they eat similar-sized seeds. But on islands where the two birds occur together, *G. fortis* individuals have larger beaks and specialize on larger seeds than do *G. fuliginosa* individuals (Figure 27-10). In other words, when the two species occur together,

A. Two species together

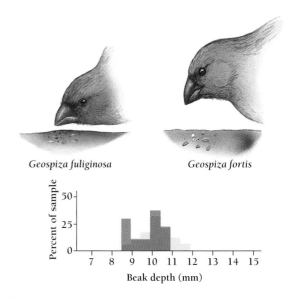

B. Two species separate

Figure 27-10 Character displacement in Galápagos finches. A. On islands where the two species *Geospiza fuliginosa* and *Geospiza fortis* occur together, members of *G. fortis* have much larger beaks and eat larger seeds than do *G. fuliginosa*. B. On islands where the birds occur separately, individuals of each species have beaks of similar size.

they split the niche and specialize on either big seeds or small ones. When they occur separately, they both take the middle ground.

Recent studies have shown that the medium-sized beak is the best option under some circumstances. On islands where the two species occur together, they often hybridize, producing offspring with intermediate-sized beaks. When food is scarce (in times of drought), the hybrids reproduce poorly. When seeds are plentiful, however, the hybrids reproduce *better* than their parent species.

Competition and adversity appear to drive these two species apart, while lack of competition and easy pickings drive them together. Currently, they are separate species. Their distinctive adaptations almost certainly have resulted from previous competition. As Joseph Connell put it, character displacement is "the ghost of competition past."

Competition, over evolutionary time, causes character displacement through natural selection. Strong selection pressure can bring about speciation.

WHAT DETERMINES BIODIVERSITY?

We have seen that for two competing species to coexist, they must exploit different niches. Both resource partitioning and character displacement increase the number of species within an ecosystem.

Each species occupies a unique niche, and each niche has a unique set of properties, such as food supply, sunlight, and other resources. All the unique but overlapping niches fit together to form a kind of multidimensional jigsaw puzzle—the ecosystem. The number of niches is a measure of the richness and complexity of a community.

How Can We Measure Diversity?

The number of species in an ecosystem is called **species richness.** For example, a single river in Amazonia may contain more species of fish than all the rivers of Europe. But species richness is only a crude measure of diversity, since some species may occur in great numbers while others are rare. To take an extreme example, the city of San Diego has an unusually large and well-stocked zoo, and so the city's species richness is high. The vast majority of organisms living in San Diego, however, consist of only a few species of animals—mostly humans, rats, mice, cockroaches, and ants.

A better measure of diversity is **relative abundance,** which takes into account how common each species is. Ecologists use various mathematical summaries of relative abundance. Often, each species is assigned an "importance value" based on the number of individuals in an area, on biomass, or on primary productivity. The relative abundance of species in a tropical rain forest is extremely high. If we find 200 species of trees, most of them will be equally common. In contrast, a temperate forest would be dominated by just a few species. In one West Virginia forest, for example, more than 83 percent of all trees were either yellow poplars or sassafras.

Careful study of a particular ecosystem can reveal the number and kinds of species that compose it. But what determines the particular species makeup of a community? Do latitude, rainfall, and other abiotic components determine the diversity of a biotic community? Or does the biological makeup of each community depend on its unique history—that is, on which plants and animals happen to arrive in what order? Are there any general rules that determine the diversity within a community?

How Do Biotic Factors Influence Diversity?

We have seen how competition and the resulting character displacement can increase the number of species in a community. Predators can also maintain diversity by preventing competitive exclusion. Some predators prevent common species from overwhelming less common ones. When one prey species becomes common, a predator will begin to prey preferentially on that species, switching from prey to prey, according to which is most common. Or the predator may simply keep the population of the dominant competitor at low enough densities that other species can survive. A predator that promotes diversity is called a **keystone species.**

Some ecologists have hypothesized that species diversity increases with productivity and stability. For example, a tropical rain forest has high diversity, enormous productivity, and a stable, predictable climate. On the other hand, some grasslands and deserts are unproductive and unstable, yet they are highly diverse. Overall, there is little correlation between productivity and diversity.

How Do Abiotic Factors Influence Diversity?

The relative importance of biotic and abiotic factors in ecosystem diversity is still a subject of ecological research. That abiotic factors play an important role, however, is certain. Species diversity generally increases from the poles to the equator, and from high elevations to low elevations (Figure 27-11). Physically varied areas (with hills, valleys, and streams) allow more unique niches than flat, featureless landscapes. Finally, the larger the size of an ecosystem, the more species it is likely to contain.

One way that ecologists have studied diversity is by studying the species diversity on islands. They have found that large islands, which have more habitat diversity and more niches, support more species than small islands.

Furthermore, the number of species on an island may remain stable—even as individual species appear and disappear. For example, in 1968, the Channel Islands, off the coast of southern California, had about the same number of bird species as they had had in 1917. But nearly one-third of the species were new.

In the 1960s, Robert MacArthur, and another American ecologist, Edward O. Wilson, argued that the diversity of

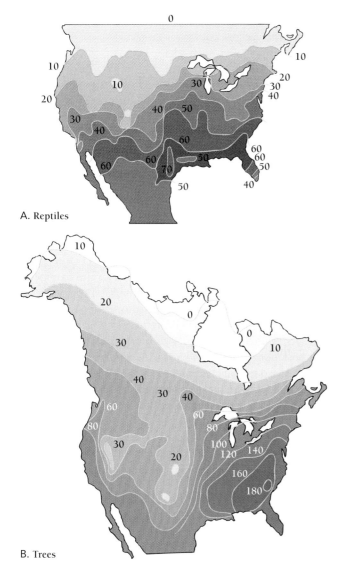

Figure 27-11 How does latitude affect species density? In North and Central America, the number of bird species increases in habitats closer to the equator. (*Source: R. L. Smith*)

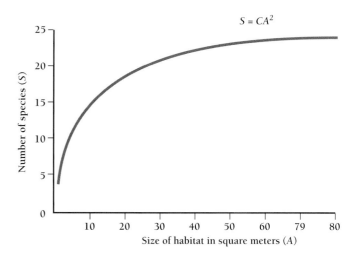

Figure 27-12 **Larger habitats support more species.** A graph of bird species on islands around New Guinea shows that larger ecosystems support more species than smaller ones. We say that species number (S) is proportional to area (A).

island species depends on a balance between the rates of immigration of new species and extinction of old species. According to the MacArthur-Wilson **theory of island biogeography,** a new island, devoid of already established species, should experience a high rate of immigration and a low rate of extinction, since there are fewer species to disappear. As more species arrive, the theory went, newer immigrants would have a harder time establishing themselves because of competition, and the rate of extinction would increase. MacArthur and Wilson also predicted that islands nearer the mainland would be easier for new immigrants to get to and should therefore have a greater rate of immigration and, therefore, more species than more distant islands.

Finally, MacArthur and Wilson formalized what ecologists as far back as Gleason had known—that the number of species in any habitat is proportional to its area. MacArthur and Wilson said the number of species (S) could be predicted by the following equation:

$$S = CA^z$$

where C and z are constants and A is the area of the ecosystem or island (Figure 27-12).

In 1969, Wilson and ecologist Daniel Simberloff devised an experimental test of the MacArthur-Wilson theory. They counted all the arthropod species they could find on several small islands (10 to 20 meters in diameter) in Florida, exterminated all the arthropods (mainly insects) with a chemical pesticide, and then monitored the gradual recolonization of the islands. Within 6 months, the islands had recovered their former arthropod diversity (Table 27-2).

TABLE 27-2 DIVERSITY ON ISLANDS BEFORE AND AFTER DEFAUNATION

Trophic level	Number of species	
	Before	After
Herbivores	55	55
Scavengers	7	5
Detrivores	13	8
Wood borers	8	6
Ants	32	23
Predator	36	31
Parasite	12	9
Unknown	1	3
Total	**164**	**140**

Each island had approximately the same number of species as before it was fumigated. However, the kinds of species that arrived were often different from what had been there before. This study suggested that which species colonize an island may be a historical accident, but the *number* of species depends on the abiotic characteristics of the island. Of course, in this experiment, all of the trees, shrubs, and other plants on the island remained intact. These too probably played a critical role in creating a fixed number of niches for arthropods.

In 1984, Jorge R. Rey repeated this experiment on islands in Oyster Bay, Florida. Rey fumigated islands and then carefully counted the number of insects, spiders, and other arthropod species as they recolonized the islands (Figure 27-13). Rey's work confirmed what Wilson and Simberloff had shown—that an equilibrium number of species exists for each island and that the number of species depends on the area of the island. In addition, Rey showed that, as more species became established on an island, fewer species successfully immigrated to an island and more species on the island went extinct, thus establishing an equilibrium.

How Do Groups of Populations Interact?

Understanding the factors that govern species diversity on islands helps us understand species diversity on the continents. A forest surrounded by croplands or housing subdivisions stands as a kind of island, as does a meadow surrounded by forests or a mountain peak surrounded by valleys. Ecologists call these **habitat islands.** For many species, however, a small island of habitat cannot support a population of sufficient size for plants and animals to find mates or for animals at higher trophic levels to find sufficient prey. Sometimes chance events eliminate all the individuals in such small populations. Eventually, one or more species in these habitat islands goes "locally" extinct.

In the face of such local extinctions, the persistence of a population depends on migration from other habitat islands. A group of local populations that alternately go extinct and recolonize vacant habitat islands are called a **metapopulation** (Figure 27-14). Just as with real islands, the sizes of islands, the number of islands, and their distance from one another all influence the rate of recolonization. In addition, the space between the habitat islands can also influence how easily and quickly species can migrate (Figure 27-15). We discuss extinction and population dynamics in more detail in Chapter 28.

A.

Figure 27-13 Does every island support a certain number of species? A. To find out, Jorge Rey first eliminated all of the invertebrate animals on islands in Apalachee Bay, Florida, by covering the islands with giant tents and filling the tents with pesticides. After all the insects and other invertebrates were dead, he removed the tents and monitored the return of species to each island. B. By about 20 weeks, each island was recolonized by a characteristic number of species. Bigger islands (*circles in top curve*) accumulated more species (from the mainland) than did smaller islands (*triangles in lower curve*). Rey found that the species that colonized an island might not be the same ones that lived on the island before fumigation. The *number* of species, however, was about the same. (*A, Jorge Rey, University of Florida*)

B.

Figure 27-14 Extinction and recolonization in a ghost town. In the desert between Nevada and California is an old gold-mining town called Bodie. Nearby, live small subpopulations of pikas, tiny rabbitlike animals that live in the rocky diggings left over from the old mines. Solid boxes are populated areas, and empty boxes represent extinct subpopulations. Between 1972 and 1991, some subpopulations went extinct (*empty red squares*) and then came back (*green-filled squares*), as pikas from other subpopulations recolonized the empty habitats. Together, all the interacting subpopulations are called a metapopulation. Overall, the number of subpopulations decreased by what percent? (*After Smith and Gilpin; photo, William J. Weber/Visuals Unlimited*).

Species diversity is influenced by both biotic and abiotic factors. Character displacement, keystone predators, and geography all affect diversity. Larger habitat islands support more species than smaller habitat islands. Habitat islands close to a mainland, or other rich source of species, support more species than habitat islands far from such sources.

COMMUNITIES CHANGE OVER TIME

In the last section, we saw that the characteristics of communities depend on the geography and form of the habitat as well as on the history of the community. In this section we will see how communities change over time.

Why Do Communities Change?

After Mount St. Helens blew up in 1980, the near-total destruction of the community of plants and animals that lived on its slopes initiated the gradual development of a new community. Windblown seeds from fireweed, grasses, and other plants first colonized the slopes of the volcano, rooting into the remaining soil wherever the thick volcanic ash had been washed away by erosion. Gradually, these weedy plants covered the barren slopes of Mount St. Helens. In time, shrubs such as dogwood and huckleberry, and small trees such as fir and hemlock, grew in among the first colonizers and replaced them. The trees would replace, or *succeed*, the shrubs, as the shrubs had begun to replace the weeds.

Each set of plants represents a stage in succession, the change in numbers and kinds of organisms in a community over time. Ecologists distinguish between **primary succession,**

Figure 27-15 It's what's in between that counts. A. The islands in MacArthur and Wilson's model of island biogeography were separated by uniform expanses of water. Similarly, the habitat islands of the metapopulation of pikas in Bodie, California, are surrounded by a nearly uniform expanse of uninhabitable desert. Individual pikas must cross this "matrix" to recolonize another habitat island. The greater the distance between habitat islands, the less likely that individuals will be able to cross. B. In nature, most habitat islands are surrounded by habitat that varies and grades from one kind of habitat to another. Depending on what habitat lies between, it may be easier for individuals to cross between islands that are far apart than between those that are closer together. Imagine a frog moving between ponds. Which would be the easier to cross, two miles of stream bed or a quarter mile on a dry, rocky hillside?

the invasion of a completely new environment such as a sandbar or new volcanic island, and **secondary succession,** the sequence of stages in a badly disturbed community. Examples of secondary succession include the reestablishment of a community in a burnt or logged over forest and the return of natural communities to abandoned croplands or vacant lots. In secondary succession, some kind of preexisting community plays a role in the succession.

After the eruption of Mount St. Helens, some areas were entirely sterile. These areas were ready for primary succession to begin. Other areas retained fertile soil with seeds. These seeds sprouted and contributed to the first stages of secondary succession.

The first community in a succession is called a **pioneer community,** while later stages are called **successional communities.** According to the theory of succession, the mix of species in a community changes over time in a way that is cumulative and directional. Climax communities, as they are now understood by ecologists, themselves change over time. Thus, "climax community" has become a relative term.

Pioneer and later communities differ in a variety of other ways, summarized in Table 27-3. For example, pioneer communities usually consist of species that flourish in disturbed areas and breed rapidly. At the other extreme, climax communities often consist of species that breed slowly but gradually take over in undisturbed areas. In general, the short-lived weedy species of pioneer communities (thistles, for example) are gradually replaced, or succeeded, by longer-lived organisms (fir trees, for example).

Most descriptions of succession refer to changes in vegetation because succession is a concept introduced by plant ecologists. However, other kinds of succession exist. For example, a deer's carcass, a rotting tree, or a pile of dung can support a succession of insects, bacteria, and fungi over periods of 2 months or less (Box 27.1). In such cases, there is no climax community, and succession occurs on a small scale in both time and space.

Succession also occurs in patches. If a tree falls in a forest, it leaves a hole in the canopy. Bright shafts of sunlight fall to the forest floor, allowing weedy, light-adapted plant species to crowd out shade-adapted plants. Gradually, the shrubs and trees may take over, and the patch becomes shady forest again. But meanwhile other trees have fallen elsewhere in the forest. The result is a forest with patches of habitat at many different stages of succession and many more species than if the forest were uniform.

TABLE 27-3 HOW PLANT COMMUNITY TRAITS CHANGE DURING SUCCESSION

Trait	Early stages	Later stages
Biomass	Small	Large
Physical structure	Simple	Complex
Role of detritus	Minor	Major
Net primary production	High	Low
Stability (absence of change)	Low	High
Plant species diversity	Low	High
Seed dispersal by	Wind	Animal
Seed longevity	Long	Short

After Barbour, Burke, and Pitts 1980.

BOX 27.1

Succession on a Corpse

"The worms crawl in, the worms crawl out" goes the childhood ditty. But which "worms," and exactly when they crawl in and out, can tell special criminal investigators how long ago, and often where, a person died. By studying the crawling creatures collected from a corpse, forensic entomologists, scientists who apply their knowledge of insects to legal matters, can solve otherwise baffling cases.

The father of American forensic entomology, Bernard Greenberg of the University of Illinois at Chicago, likes to recount the case of a missing 9-year-old girl. After her partly skeletonized body was discovered in an abandoned apartment building, Greenberg collected fly larvae and pupae from the body and took them back to the laboratory. There he raised them at temperatures like those in the apartment building over the days preceding the girl's discovery. He soon learned that the flies took 21 days to mature under those conditions, which meant that the fly eggs had to have been laid 21 days before the girl was found. The last time the little girl had been seen alive was 21 days earlier, near the apartment building in the company of a man in his 20s. As a result of Greenberg's testimony, this man was subsequently convicted of her murder.

Beetles, flies, and other arthropods colonize corpses in a predictable succession. "Flies are the best investigators of corpses," says Carl Olson, an entomologist at the University of Arizona. Metallic blue or green blowflies, drawn by odors that include a compound called cadaverine, can find an uncovered corpse within 10 minutes of death. Blowflies and flesh flies lay eggs in wounds and moist natural body openings, such as eyes, ears, nose, and mouth. Inside the body, bacteria go to work decomposing the body, generating various gases in the process.

As the masses of maggots feed, their activity and that of the bacteria raise the temperature inside the corpse. "It's like when you walk into a room full of people—notice how hot it is?" Olson asks. "The maggots on a corpse create their own environment."

The maggots eventually puncture the skin, initiating what is called the decay stage and releasing what University of Hawaii researcher Lee Goff calls "strong, distinctly unpleasant odors." As the fly larvae complete their feast, they hop or crawl off to pupate in the soil. By that time, the latter part of the decay stage, the fly larvae have created a hospitable environment for the next wave of insects, primarily hide and rove beetles. It is a classic example of ecological succession. Any remaining larvae become food for the visiting beetles, which also finish off the flesh.

By the postdecay stage, only the skin and bones remain. A different set of beetles now arrive. These feed on dried skin and hair. They are the same insects that destroy fur coats and cashmere sweaters. When the body is reduced to bone alone, all of the carrion-feeding arthropods depart.

Any kind of physical disturbance, whether a fallen tree or a collapsed bank along a stream, can create a patch in which secondary succession can begin. Even something as minor as a gopher mound in a meadow creates a disturbed area where a patch of different species can live. A disturbed area can be colonized by seeds already in the soil (a "seed bank") or by organisms from nearby areas.

Succession usually implies a progression through a series of transitional stages to a climax community. Succession can occur in disturbed areas, creating a patchwork of different stages of succession.

Why Does Succession Occur?

Clements argued that succession occurs because organisms degrade their own environment. For example, following a forest fire, the first plants to grow are those that are adapted to live in open sun. But the shade created by these sun-loving plants prevents their own seedlings from growing and favors shade-tolerant plants.

While many of Clements's scientific heirs still contend that each stage in a succession prepares the way for the next, other ecologists argue that this is not always true. Some of the changes in a succession, they say, merely reflect differing rates of growth. Shrubs, for example, appear to come before trees even when both types of seedlings start at the same time because the shrubs take less time to reach maturity.

In reality, **pioneer** species can facilitate succession, tolerate succession, or inhibit succession. In primary successions, however, pioneer organisms always facilitate succession. This is because most organisms cannot colonize bare rock or sand. The formation of soil from bare rock depends on the action of pioneer lichens and plants adapted to thin or nonexistent soils. Once these pioneer species have created a thin layer of soil, other species can grow.

Pioneer or successional species change their environment in such a way as to allow colonization by succeeding species.

What Determines Which Community Is the Climax Community?

The climax community is, by definition, the one that persists the longest without changing. But what factors determine which community will persist? As we saw in Chapter 26, the character of a community depends heavily on such abiotic factors as climate, soil, and terrain. The dry, frozen tundra could never support a tropical rain forest. Early ecologists theorized that an area has only one possible climax community.

Communities assemble themselves flexibly, however, and their particular structure also depends on the specific history of the area. For example, giant sequoia forests in the Sierra Nevada include trees as old as 2000 years. These forests appear to be a classic example of a long-lived climax community. In fact, however, these ancient forests persist only if periodic fires sweep through and kill young white fir and incense cedars growing beneath the mammoth sequoias (Figure 27-16). When humans regularly put out fires, the white fir and incense cedars grow up among the sequoias. Eventually, when a fire too big to put out arises, the smaller trees pass the flames up into the crowns of the giant sequoias, destroying the whole forest. Once that happens, a white fir and incense cedar forest springs up in its place.

Figure 27-16 A grove of giant sequoias. Such rare groves depend on frequent forest fires to prevent succession to white fir and incense cedar. *(Barbara Gerlach/DRK Photo)*

The character of a climax community depends on climate, terrain, and history.

Why Do Communities Stay the Same for Long Periods?

An ecologist who studies a pond today may well find it relatively unchanged a year from today. Individual fish may be replaced, but the number of fish will tend to be the same from one year to the next. We can say that the properties of an ecosystem are more **stable** than the individual organisms that compose the ecosystem.

At one time, ecologists theorized that species diversity made ecosystems stable. They thought that the greater the diversity, the more stable the ecosystem. Support for this idea came from the observation that long-lasting climax communities usually have more complex food webs and more species diversity than pioneer communities. Ecologists concluded that the apparent stability of climax ecosystems depended on their diversity and complexity. To take an extreme example, farmlands dominated by a single crop are so unstable that one year of bad weather or the invasion of a single pest species can destroy the entire crop. In contrast, the many species that make up a native grassland will tolerate considerable damage from weather or pests.

However, research has provided little support for this idea. Indeed, ecologists do not even all agree on what "stability" means. Stability can be defined as simply lack of change. In that case, the climax community would be considered the most stable, since, by definition, it changes the least over time. But, alternatively, stability can also be defined as the speed with which an ecosystem returns to a particular form following a major disturbance, such as a fire. This kind of stability is also called **resilience.** In that case, climax communities would be the most fragile and the least stable, since they can require hundreds of years to return to their former state. As we saw in Chapter 26, tropical rain forests have the greatest biodiversity of any biome and some of the highest primary productivity. Yet once destroyed, they do not quickly return.

Even simple lack-of-change stability is not always associated with maximum diversity. Maximum diversity in some areas is found not in the climax community, but in midsuccessional stages. Once a redwood forest matures, for example, the kinds of species and the numbers of individuals growing on the forest floor are reduced. In general, diversity, by itself, does not ensure stability.

Mathematical models of ecosystems likewise suggest that diversity does not guarantee ecosystem stability—just the opposite, in fact. A more complex system is more likely to break down than a simple one. A computer loaded with conflicting software, for example, is more likely to break down than a simple hand calculator.

Can Ecosystems Bounce Back?

Because ecosystems all over the world are being severely disturbed or eliminated by human activities, ecologists would like to know what factors contribute to the resilience of ecosystems. The severe destruction caused by a volcanic eruption such as Mount St. Helens is nothing in comparison to the huge areas of temperate and tropical rain forests that we humans have destroyed. We need to know what aspects of an ecosystem are most important to its resistance to damage, as well as its recovery.

Many ecologists now think that resilience comes not from diversity of species in one place but from the patchiness of the environment. An environment that varies from place to place supports more kinds of organisms than an environment that is uniform. In an environment with abundant species, a local population that goes extinct is quickly replaced by immigrants from an adjacent community. Even if the new population is of a different species, it can approximately fill the niche vacated by the extinct population and keep the food web intact.

Other factors that contribute to an ecosystem's resilience include resistance to erosion and increased primary productivity. As long as the soil is intact, seeds will quickly sprout, and shrubs, ferns, herbs, and grass will hold the soil in place, preventing erosion. Partly damaged ecosystems often support greater plant growth than mature ones, since younger trees grow more vigorously than trees in a mature climax forest.

Some of the factors that contribute to ecosystem resilience are patchiness of the physical environment, primary productivity, and resistance to erosion.

In this chapter we have seen how communities change over time and what factors, biotic and abiotic, influence the character of a community. In the next chapter, we will examine what factors affect populations and what happens when populations grow too big.

Study Outline with Key Terms

Henry Cowles did the first major American work on **succession**, and Frederic Clements published the first textbook describing succession. Clements also introduced the idea of a **climax community**, a single community of plants that, he said, characterizes a particular area and climate. Clements believed that communities were **integrated**—real organic entities, with species playing the same roles in different communities. In contrast, the ecologist Henry Gleason argued that communities were **individualistic**—mere associations of species that happen to occupy the same area.

The species in communities interact in many ways. They may compete or they may relate as **predator** and prey or as host and **parasite**. Organisms defend themselves from predators and herbivores by using **warning coloration, mimicry** (**Batesian** or **Müllerian**), **camouflage,** and **secondary plant compounds**. Species may share a **symbiotic** relationship, such as **parasitism, mutualism,** or **commensalism.** In evolutionary time, two or more species may **coevolve** mutualisms and other special relations with one another.

Species may also **compete** for a **limiting resource**. The kind of community an organism occupies is its **habitat**. The way the organism uses the resources in its habitat is called its **niche**. The **competitive exclusion principle** says that if two species occupy the exact same niche, one species will eliminate the other through competition. Competition from other species may reduce the **fundamental niche** of an organism to a smaller **realized niche**. In **resource partitioning**, species break up a large niche into several smaller ones. Competition between species can cause them to evolve apart in a process called **character displacement**.

Communities vary in the number of species they contain, **species richness**, and in the **relative abundance** of each kind of species. Ecologists are still trying to discover what determines the diversity of a community. **Keystone species,** productivity, harshness of the environment, latitude, size of ecosystem, and historical accident all seem to play roles.

In the 1960s, Robert MacArthur and Edward O. Wilson proposed a **theory of island biogeography** to account for species diversity on islands of habitat. Tests of this theory suggest that the number of species on an island is a predictable consequence of the characteristics of the island, and not just a historical accident. **Habitat islands** do not have to be conventional oceanic islands. They can be mountaintops or ponds in a forest, for example. A group of local populations that alternately go extinct and recolonize vacant habitat islands are called a **metapopulation.**

Primary succession occurs when organisms colonize a previously unoccupied area, such as a volcanic island or a sand dune. **Secondary succession** occurs when organisms recolonize an area that has suffered some catastrophe, but which still has its soil intact. In secondary succession, the first **pioneer** plant species in a **pioneer community** are annuals, followed by perennials, shrubs, and trees. The final stage in development is called the climax community, but biologists no longer believe that the idea of a final stage makes sense. All stages can change, depending on circumstances. Intermediate stages are called **successional communities.**

Ecosystems are either more **stable** than pioneer or successional ecosystems or less stable, depending on the definition of stability that is used. Patchiness of the physical environment, primary productivity, and resistance to erosion are some factors that seem to contribute to ecosystem **resilience,** the ability of an ecosystem to recover from fires or other cataclysms.

Review and Thought Questions

Review Questions

1. Contrast the integrationist view of communities with the individualistic one. Describe a piece of research that supports one view or the other.
2. Name a commensal and a parasite to humans.
3. Name a mammal, bird, reptile, amphibian, and insect that have warning coloration.
4. Define the competitive exclusion principle.
5. Give an example of a limiting resource for humans.
6. Define the terms "niche" and "habitat." Give an example of each for the same organism.
7. What factors determine the number of species living on an island?
8. Give two definitions of "ecosystem stability." What factors determine the long-term stability of an ecosystem?

Thought Questions

9. Describe a climax community somewhere near where you live. What criteria did you use to determine whether it was a climax community? How would you go about testing your theory that it is a climax community?
10. Describe a series of small habitat islands that exist within 5 or 10 miles of your house and which might be subject to MacArthur and Wilson's theory of island biogeography.
11. You have discovered two butterflies that look nearly identical, but, from other evidence, clearly belong to different families. Assuming that one is mimicking the other, think of an experiment that would show which one was the mimic and which one was the "original."
12. If succession occurs because pioneer or successional species improve conditions for subsequent species, what can we say about climax species?

About the Chapter-Opening Image

Destruction of an ecosystem. When Mount St. Helens erupted in 1980, it destroyed a 230-square-mile ecosystem, about the size of the city of Chicago. What was destroyed?

 On-line materials relating to this chapter are on the World Wide Web at **http://www.harcourtcollege.com/lifesci/aal2/**

 Web Links

 Quizzing & Testing

28 Populations: Extinctions and Explosions

Fighting for Life

In 1998, an eminent biologist gave up his tenured job at the University of California. It was the second time Michael Soulé had left the security of a tenured faculty position—one of the strongest guarantees of lifetime employment in the United States. As before, he did so to follow his heart.

The academic life was, for Soulé, too much thought and too little action. He felt impelled to act, to literally save the world. The problem, as he saw it, was that everything was disappearing. All of the animals and special habitats he had studied and grown to love during his 40 years as a biologist were vanishing. His gardens of discovery were winking out one by one.

"Every field biologist has had this experience," he told writer Robert Coontz in 1990. "You return to a study site you knew as a graduate student, or as an assistant professor, or where as a child you first became interested in biology, and it's been devastated."

He began his career in the normal way of biologists. In his boyhood, he roamed the hills and canyons outside San Diego, catching lizards and snakes, watching birds, rabbits, and other wildlife. In college he studied zoology, then went to graduate school at Stanford. His mentor there was population biologist Paul Ehrlich, whose books on human overpopulation have made Ehrlich far more famous than his decades of research on butterflies.

In 1967, Soulé returned to his boyhood stomping grounds, taking a job as a professor at the University of California at San Diego. He taught biology and studied symmetry in lizards. All animals, including humans, become a little asymmetrical as they develop from embryos to adults. A high degree of symmetry, however, is an indicator of good health. In fact, many animals choose their mates partly on the basis of how symmetrical they are. We could say that symmetry is the biological definition of beauty. Asymmetry, on the other hand, often indicates poor health. Asymmetry, Soulé found, was also an indicator of pollution and other environmental stresses, and he was finding more asymmetry in his animals than he wanted to find.

Before his eyes, the canyons of San Diego

Key Concepts

- Populations may go extinct when habitat is damaged or destroyed. Chance events, such as the disappearance of alleles from the gene pool, can hasten extinction.
- Natural populations rarely grow exponentially for very long. Usually, growing populations reach a plateau, fluctuate, or crash.
- Both density-dependent and density-independent factors affect the population growth of a species.
- The growth of human populations can interfere with the growth and survival of other species and even whole ecosystems.

County were filling with houses. The fragments of land that survived between housing developments were too small to support many individuals of each species. Not only in San Diego, but around the world, species were going extinct at an unprecedented rate.

These shrinking habitats drew Soulé's interest to the problem of inbreeding. He could see that once a population got small enough, the remaining individuals would have only a few others with which to breed. He theorized that inbreeding would bring out genetic defects in the offspring. Fewer would reach maturity, and those that did would be less fit—less able to survive and reproduce themselves. Small populations spiraled downward toward an extinction from which they'd never be able to escape. Soulé called this downward spiral the **extinction vortex** (Figure 28-1).

It was all interesting material for a researcher. Soulé published his experimental work in journals and presented his ideas at scientific meetings. The university promoted him from assistant professor to tenured associate professor and, finally, to full professor.

Yet something was missing. It was one thing to document the slow destruction of study sites and beloved childhood haunts. It was quite another to do something about it. Real biologists were not supposed to act on the grief they felt when yet another study site disappeared beneath a sewage treatment plant or shopping mall.

In the mid-1970s, Soulé's fellow biologists accepted the wall that then divided biology from the environmental movement. For biologists to affiliate themselves with "ecofreaks" who chained themselves to redwood trees was to abandon any pretense of objectivity. At that time, a biologist who expressed sympathy with down-in-the-trenches environmentalists, let alone became one, could lose credibility as a serious biologist.

Nonetheless, Soulé took action. In 1978, he helped organize the First International Conference on Conservation Biology. It was the first time that field biologists acknowledged that they had both a responsibility and a special ability to do more than just document the demise of the habitats they knew so well. Perhaps their research could provide information that would help save some of these habitats.

By then, Soulé was in his early forties and had been on the UC San Diego faculty for 12 years. It's common for middle-aged people to look for change, but most professors would content themselves with a new line of research, a sabbatical, or even a sports car. For Soulé, that wasn't enough. He quit his tenured job at UC San Diego and joined a Zen center in Los Angeles.

Although he stayed at the Zen center for 5 years, he never abandoned biology. Instead, he was reconceiving what it should mean to be a biologist. He helped put together a collection of essays on conservation biology. The two volumes, titled *Conservation Biology: An Evolutionary Ecological Perspective,* galvanized field biologists everywhere. Their 1980 publication marked the informal beginning of **conservation biology** as a field separate from traditional ecology.

For the first time, the majority of biologists began to acknowledge openly that human overpopulation was creating a tidal wave of extinctions unprecedented in the history of humankind. In 1987, a group of research biologists formed the

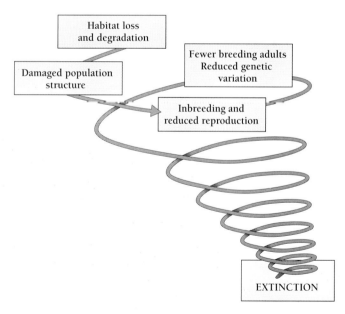

Figure 28-1 The extinction vortex. Biologist Michael Soulé hypothesized that when a population shrinks because of hunting or loss of habitat, the reduced population size—as well as disproportionate numbers of males and females or young and old individuals—causes inbreeding. Inbreeding, in turn, can cause ill health and poor reproduction in each new generation, accelerating the extinction process in a downward spiral. In addition, extinction of one species degrades the environment for other species in the habitat. The feedback loop thus involves more and more species as time passes.

603

Society for Conservation Biology and began publishing a journal devoted to the science of conservation. Soulé returned to the University of California, this time to the Santa Cruz campus, to lead a department of environmental studies.

HOW DO POPULATIONS GO EXTINCT?

Many factors cause species to go extinct. Biologists today lump all these factors into two categories. One category includes **deterministic** events, such as habitat destruction or deliberate extermination through hunting or trapping. The other category includes **stochastic,** or chance, events such as unusual storms.

How Do Deterministic Factors Drive Species to Extinction?

One deterministic factor that can drive species to extinction is competition from other species. In nature, one kind of tree may shade a smaller species, driving the local population to extinction. The larger tree has successfully competed with the smaller one for light. Usually, the result is not extinction of a whole species, however, but only of a particular population. As we saw in the last chapter, individuals from another population can reestablish a population when opportunity arises.

Humans also drive local populations to extinction. But our large numbers and organized approach to taking over resources allow us to drive *every* local population of a species to extinction. How do humans drive other species extinct?

One way is simply hunting and killing as many organisms as possible. In this way, humans have driven to extinction hundreds of animals, including California grizzly bears, Tasmanian wolves, passenger pigeons, dodos, and great auks.

Another way is **habitat destruction.** Sometimes we do this without even realizing we are doing it. If we remove all the trees and shrubs from a forest ecosystem and then plant wheat, we eliminate not only the trees and shrubs, but also habitat for all the other organisms that live in the forest. If we repeat this pattern from Maine to Nebraska, we eliminate not only trees and habitat but also thousands of subpopulations of insects, spiders, worms, fungi, bacteria, and more.

In the 18th century, pioneer families cleared millions of acres of forest in the eastern United States. Most of the farms they built and tended are now long abandoned, and much of that forest has returned. But while many trees and shrubs are back in great numbers, hundreds of other species are gone forever. For many of these species, not a single population is left to recolonize the new forest. For others, the remaining subpopulations, in Canada or the far West, are simply too far away to recolonize the new forest.

Humans drive other species to extinction by destroying their habitats.

How does fragmentation contribute to habitat destruction?

A subtler process that drives other species to extinction is **fragmentation** of habitat. Natural landscapes are a mosaic of different habitats. Forest blends gradually into savanna, chaparral, or grassland. Riparian (river) forests may run through drier habitats, connecting wetter forest ecosystems together. An organism such as a bird or insect can often travel easily from one patch of forest to another. If a population disappears from one habitat patch, individuals from another subpopulation soon arrive to recolonize.

For organisms that cannot fly, however, recolonizing is usually more difficult. Humans create sharply bounded patches of habitat in an uncrossable sea of pavement, houses, and crops. Many species cannot easily move from one patch to another. Even a simple road can permanently divide a habitat for turtles, amphibians, and other slow-moving animals.

Fragmentation also reduces the total area of a habitat. We saw in Chapter 27 that numbers of species depend on the area of the habitat. According to MacArthur and Wilson's species-area curve, a 90 percent loss of habitat (area) results in a 50 percent loss in species. Most species simply cannot survive in a habitat that is too small. For example, numerous studies have shown that scarlet tanagers, pileated woodpeckers, ovenbirds, Acadian flycatchers, and many other birds rarely live in forests that are smaller than 40 acres. Large animals at the top of the food chain, such as hawks, need large areas to supply food.

Some studies suggest that in Brazilian rain forests, 500,000 acres is the minimum size needed to preserve local biodiversity. That is a little more than one-tenth the area of New Jersey. But rain forests in Peru, Venezuela, and other areas would have different sets of species and so would each need a separate 500,000-acre preserve.

Another important consequence of habitat fragmentation is **edge effects.** The smaller the pieces into which a habitat is cut, the more edge the habitat has—relative to interior. At the edges of a habitat, many interesting things happen. At the edge of a forest, for example, more light enters and different kinds of plants grow. Sometimes safe nesting areas are missing. Predators and other species from the adjacent habitat visit to take birds' eggs, kill young animals, or browse on young trees. In areas heavily populated with humans, these visitors may include dogs, cats, and, of course, humans themselves.

Because the edges of a forest or marsh are often dangerous, many species can survive only in the very center of a habitat patch, far from the edge. For some species, many habitat patches are so small that no habitat exists that is far enough away from the edge (Figure 28-2).

One way humans destroy habitat is by breaking it up into small fragments with lots of edge and little central core.

Draining the gene pool

Recall that Michael Soulé's extinction vortex hypothesis suggested that inbreeding accelerates extinction. One source of

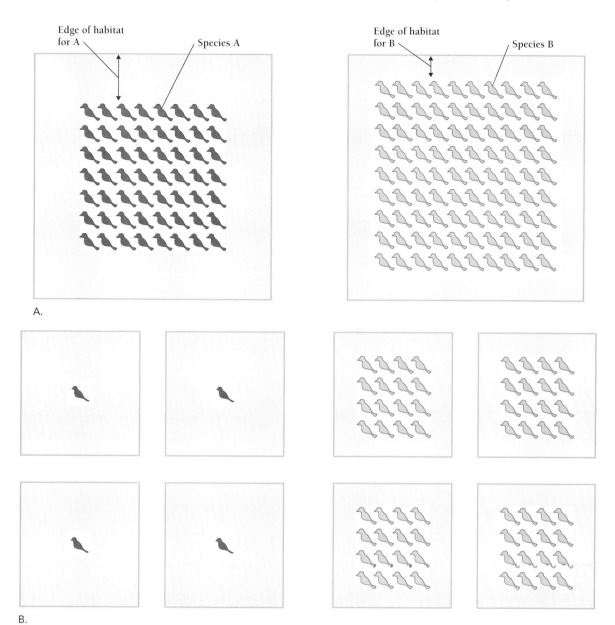

Figure 28-2 Habitat fragmentation and the edge effect. Many organisms cannot survive close to the edge of a habitat. As large habitats are fragmented by roads and other human development, the proportion of edge increases. As a result, a habitat patch that seems large to us is too small to allow many species to survive. Different species have different tolerances for habitat edges. A. An unfragmented habitat patch. B. The same patch broken up into four quadrants is now nearly all uninhabitable edge for species A, but is still habitable for species B.

support for this hypothesis is a 1998 report in the journal *Science* on a long-term study of greater prairie chickens. Beginning in the early 19th century, extensive habitat destruction pushed this native Midwestern bird toward the brink of extinction. In Illinois, the greater prairie chicken is almost gone (Figure 28-3). During a 35-year study of these birds, the population size dropped from 2000 birds in 1962 to 50 birds in 1994. Researchers also recorded a decline in hatching success (hatched eggs per nest) from a high of nearly 100 percent in 1960s to less than 40 percent in 1990. Measures of genetic diversity showed that the Illinois population had less genetic diversity than larger populations in nearby states and that alleles present in the population before the 1970s later disappeared. The data suggest that as the population shrank, so did its genetic diversity.

Biologists hypothesized that genetic defects due to inbreeding were keeping many chicks from hatching (Figure 28-4). When researchers brought in prairie chickens from

Figure 28-3 **The end of the Illinois greater prairie chicken?** One hundred years ago, greater prairie chickens (*Tympanuchus cupido*) lived abundantly throughout Illinois. Today, two tiny populations survive in the southern part of the state. (*Photo, Steve & Dave Maslowski/Photo Researchers, Inc.*)

larger populations elsewhere, both genetic diversity and hatching success increased. Soulé's extinction vortex model predicted that declining population size and declining genetic variability would mutually reinforce each other and increase the likelihood of local extinction. Sadly for the prairie chickens, Soulé's prediction seems to be borne out by this study.

According to the extinction vortex hypothesis, loss and degradation of habitat reduces the size of a population, changes its structure, and leads to a loss of genetic diversity. The loss of genetic diversity leads to inbreeding and reduced reproduction, which leads to smaller population size.

How Do Stochastic Factors Drive Species to Extinction?

Once a population is as ravaged as that of the Illinois greater prairie chicken, almost any chance event can drive it to extinction. In a population of 25 birds, a freak storm could kill several of the healthiest hens in 1 year. The next year, all the chicks could be males, just by chance. Chance, or stochastic, events such as these would not affect a larger population. A storm might kill a few hens, but many others would take their place. And in a population of 2000, all the chicks would not be one sex. Finally, small populations can lose valuable alleles through genetic drift, as appears to have occurred among the greater prairie chickens of Illinois.

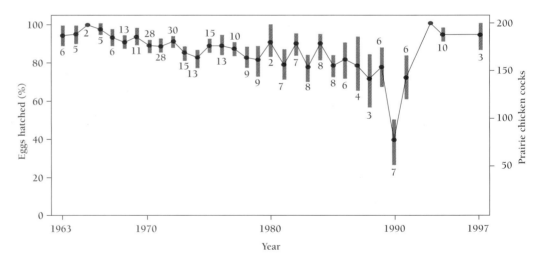

Figure 28-4 **Inbreeding among greater prairie chickens?** As the population of prairie chickens shrank, the percentage of eggs in a nest that successfully hatched dropped from 100 percent in 1963 to less than 40 percent in 1990. Biologists strongly suspect that inbreeding and the expression of deleterious genes may account for this decline. Each data point shown here is the average of hatching success at all the nests in the population. The vertical bars indicate variation in hatching success. Long bars mean more variation. Short bars mean little variation. Increased variation in any trait can be a sign of ill health in a population. Notice that the bars become longer over time. In the late 1990s, biologists introduced prairie chickens from large, genetically diverse populations in other states. The result was a dramatic increase in hatch success in the Illinois prairie chickens. Biologists believe this was likely due to an increase in genetic diversity. Notice that the bar lengths showing variation in hatch success became shorter.

Chance events are more likely to lead to extinction in very small populations.

How Can We Preserve Ecosystems and the Species That Live in Them?

The work of conservation biologists has shown that, unless something dramatic is done soon, approximately half of the world's species will go extinct in the next decade or two. Humankind has now converted one-third of the Earth's land surface to agriculture and covered more with cities, suburbs, roads, and factories. The remaining land includes some of the harshest, least productive areas of the planet, including deserts, jagged mountain ranges, and arctic tundra and taiga. The rate of destruction is increasing. In tropical areas, 29 out of 63 nations have already cut or burned nearly 80 percent of their forests.

As of 1998, 20 tropical nations had committed to saving 10 percent of their land in parks and other preserves by the year 2000. But because far more species live in the tropics than elsewhere, saving only 10 percent of all tropical rain forests means a likely loss of more than half of all the species on Earth.

In the spring of 1998, Michael Soulé expressed his disappointment at this trend in an article in the journal *Science*. He wrote, "Campaigns with targets in this range [10 percent] can create the unintended and false impression that such a paltry tithe to nature is enough to prevent a mass extinction of species."

Not long after the publication of this statement, Soulé left the university once and for all to join the Wildlands Project in Colorado. The ambitious goal of the Wildlands Project is to persuade people and their governments to set aside 50 percent of the North American continent for the preservation of biological diversity. "Our generation is the key," Soulé said in 1990. "Whatever we save in the next few decades is all future humanity will have to use and to appreciate. If we don't do it, it will be too late."

Conservation biologists have discovered several rules that allow preservation projects to save the greatest numbers of species. The best preserves are large areas, whose size minimizes area effects and edge effects in sensitive species. In addition, many conservation biologists recommend **corridors,** narrow strips of protected habitat that allow species to travel from one preserve to another (Figure 28-5). The most likely beneficiaries of corridors are small, slow-moving animals, such as frogs

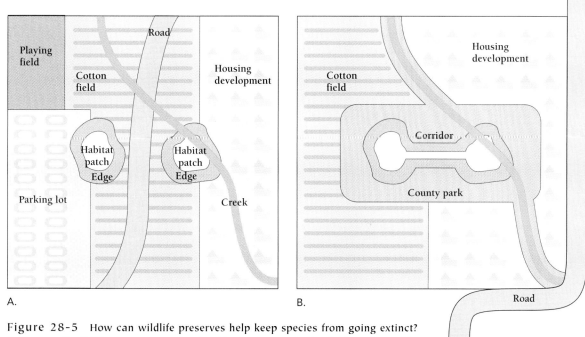

Figure 28-5 How can wildlife preserves help keep species from going extinct?
A. Two small wildlife preserves are bordered by houses, a playing field, and a parking lot. A road and a cotton field separate the preserves. Pesticides from the cotton field and people's backyards flow directly into the stream and then into the right preserve. Dogs, cats, humans, and other animals can enter and destroy plants, animals, nests, and dens. Edge effects are huge. Recolonization would be difficult or impossible for species that cannot easily cross the road and the cotton field. B. In the redesigned preserve, a corridor connects the two habitat patches. The road has been rerouted around the two patches, and the cotton field has been moved to one side. In this arrangement, if a species goes extinct in one habitat patch, individuals can recolonize from the other habitat patch. An inner buffer zone (purple) acts as a wide edge surrounding the corridor and both patches. In this buffer area, human use is restricted to a minimum. An outer buffer zone (green) allows more general human use but still no houses, crops, or industry.

Figure 28-6 Unrestrained growth in elephants. Elephants, like the ones at this water hole, have the lowest r_{max}, or intrinsic rate of natural increase, of almost any species. Even so, one pair of elephants could (hypothetically) produce enough progeny to cover the Earth's surface in only 1000 years. Fortunately, elephants rarely reproduce at the maximum rate. *(Photo, Kennan Ward/The Stock Market)*

and salamanders that need to migrate overland to breeding ponds and streams. If a local population goes extinct, individuals from another preserve can then travel to the empty habitat patch and recolonize it. Other winners would be large predators, such as mountain lions and wolves that need large ranges.

We can preserve many more species if we set aside large preserves, connected by corridors and surrounded by buffer zones.

HOW DO POPULATIONS GROW?

We have discussed how populations go extinct. But all organisms have phenomenal powers of reproduction. Populations of bacteria, cockroaches, and rats sometimes seem to explode out of nowhere. Indeed, if all of the descendants of just 100 California sea stars survived and reproduced for 15 generations, the number of sea stars would exceed the number of electrons in the visible universe.

All organisms are capable of multiplying amazingly rapidly when conditions are ideal. Even elephants, which rarely have more than about 10 offspring in their 60-year life spans, can generate a lot of descendants. For example, if all of the 10 offspring of an elephant survived, and each of them had 10 offspring, and so on, the original elephant could have as many as 100,000 great, great, great grandchildren in just 300 years (Figure 28-6).

In nature, we sometimes see such unrestrained growth when a species invades a new territory, or, for example, when a protozoan discovers an uninhabited human gut. But neither elephants nor sea stars nor protozoans can multiply without end. Even though natural populations are capable of huge growth rates, they rarely grow so fast that they eat themselves out of house and home. Whenever populations begin to grow too fast, they eventually stop. When food or space runs low, or wastes degrade the environment, individuals tend to curb their reproduction. Sometimes, predators or disease bring a population under control.

Natural populations may fluctuate from season to season and from year to year, but most are surprisingly stable over the long run. Two important questions in population ecology are, How do populations grow? and What factors limit population growth?

What Is Exponential Growth?

Under the most favorable conditions, the number of *Escherichia coli* in a nutrient-rich broth can double every 20 minutes. After the first 20 minutes, one bacterium forms two; after another 20 minutes, the two have formed four; after another 20 minutes, there are eight; and so on. We say that the bacteria have a **doubling time** of 20 minutes.

We can calculate the number of bacteria (N) from the number of times the population has doubled since the time at the beginning of the experiment (t) using the following equation:

$$N = 2^t$$

For example, if 60 minutes had elapsed, the population of *E. coli* would have doubled three times. So t would equal 3, and

$$N = 2^3 = 2 \times 2 \times 2 = 8$$

In this equation, t is an "exponent," hence we say the bacteria's growth is **exponential growth,** a growth pattern in which a population doubles in some constant period of time.

Exponential growth means that the larger the population, the faster it increases. For example, we know that 16 *E. coli* cells could double in 20 minutes to 32 cells. The population would increase by 16 cells. But a population of 16 million cells would also double in 20 minutes. That population would increase by 16 *million* cells over exactly the same time period. The large population and the small one would have the same doubling time, but the increase in actual numbers would be much larger in the second case.

When bacteria increase exponentially, a graph of their numbers looks like the letter J. At the bottom of this **J-shaped curve,** the population size (N) increases rather slowly. Then it accelerates. When there are no limits to population increase, N increases faster and faster.

We can describe this ever-increasing growth as follows: During unrestrained exponential growth, the rate of change in the number of organisms is proportional to the number of organisms present in the population. That is, the bigger the population, the faster it grows. Human populations are a good example of a natural population with a J-shaped growth curve (Figure 28-7).

▶ **Figure 28-7 A classic J curve.** The human population has been increasing exponentially for thousands of years. We are rapidly approaching the limit of what our planet can sustain. Already, we consume or use about half of all the solar energy absorbed through photosynthesis on Earth.

The growth rate of a population, called r, is the difference between the average birth rate and the average death rate. **Birth rate** is the number of individuals born each year per 1000 individuals. **Death rate** is the number of individuals who die each year per 1000 individuals. The population will grow only if the birth rate exceeds the death rate. When the birth rate of a population, b, is at its maximum, and the death rate, d, is at its minimum, r is as large as it can be. This is called r_{max}, or the **intrinsic rate of natural increase**. For example, if b is 10 per 1000 and d is 2 per 1000, then r (which equals $b - d$) would be 8 per 1000. In populations that are declining in numbers, d is greater than b, and r is a negative number.

The change in total population size per unit time, sometimes called G, is simply r times N, or rN.

$$G = rN$$

This is just another way of saying that the larger a population is, the faster it grows.

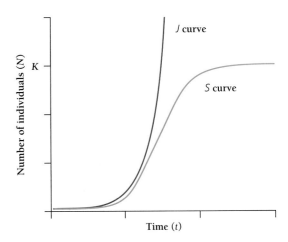

Figure 28-8 How populations grow. All animals are capable of exponential growth for short periods, as represented by the J curve. In natural populations, growth declines as population size increases and approaches the carrying capacity (K) of the environment, as represented by the S curve.

All populations are capable of exponential growth. The larger they are, the faster they grow.

What Limits the Size of a Population?

Even in the laboratory, exponential growth cannot continue for long. We may see a J-shaped population curve when a bacterium has first colonized a glass container of a nutrient-rich medium. But even *E. coli* growing under nearly ideal laboratory conditions soon find limits to their growth. The bacteria run out of food and space, and their own wastes begin to poison them. The concentration of bacteria gradually reaches a plateau. If we look at the whole growth curve, we see that it consists of three parts—an initial "J phase," in which the growth rate is accelerating; then a deceleration phase; and finally a gradual leveling off. The whole curve now looks more like a tilted and stretched out "S," a classic **S-shaped curve** (Figure 28-8).

Ecologists modify the growth equation to describe the decelerating growth in natural populations by adding a multiplier. The new equation must express the slowing growth rate in terms of population size. This is because the larger the population size (N), the faster growth slows down. Ecologists first define the plateau value of population size as the **carrying capacity** of the environment for a particular species. The carrying capacity, also called K, is the maximum population density that the environment can support.

When the environment is not yet filled to its capacity, there remains a fraction of K that is left to be filled. For example, if the carrying capacity for houseflies in Australia were 100 flies per square meter, and the population density N were 70 flies per square meter, then the fraction of K left would be $(100 - 70)/100 = 30$ percent.

Population ecologists take that simple fraction of K and multiply it by rN to get a measure of the change in the population growth over time. The equation they use is called the **logistic growth equation:**

$$G = rN(K - N)/K$$

But it merely describes the **S**-shaped curve. G expresses how much population size (N) changes as over time (t). (G is often expressed, more formally, as dN/dt, but this means something only if you have studied calculus.)

If the equation is plotted as a curve, it starts out like the J-shaped curve, with an ever-increasing rate of population growth. That is, when the population (N) is very small, the unfilled fraction of the carrying capacity—$(K - N)/K$—is about 100 percent, or 1. As the population increases, however, the fraction of unfilled K decreases and the growth rate slows. When population size reaches the carrying capacity, the environment cannot support any more growth. Then $(K - N)/K = (K - K)/K = 0$, the fraction of unfilled K equals 0, and the population stops growing. The population has reached **zero population growth**.

The carrying capacity of the environment limits the size of a population.

What Determines Whether a Population Grows?

The carrying capacity of the environment for a particular species depends on the needs of the species and, therefore, on a multitude of factors—food supply, territories, predators, and competitors, to name just a few. Bees, for example, need nectar, pollen, and a safe place for a hive. Even if a given habitat has plenty of flowers, the carrying capacity for bees may be low if there are only a few good places for a hive within flying distance of the flowers.

Changes in a population size ultimately result from just four processes: birth, death, **emigration** (movement out of the population), and **immigration** (movement into the population). The net change in population size is the sum of births and immigration, minus deaths and emigration.

Anything that influences any of these four factors will cause an increase or decrease in carrying capacity. Competition, disease, number of prey species, number of nesting sites, rainfall, soil fertility, and hundreds of other factors can all affect the value of K—the number of individuals of one species that a given habitat can support.

Density-dependent factors

The rates of birth, death, immigration, and emigration all depend on the population density. As a population's density approaches the carrying capacity, its members must increasingly compete with one another for limited resources. For example, one study of a population of 25 pairs of tawny owls at carrying capacity suggested intense competition. The 50 owls, capable of producing as many as 100 offspring per year, successfully fledged only 18 owlets. Because of limitations in the food supply—in this case, rodents—some owls were unable to feed their offspring, others did not bother to incubate their eggs, and still others did not even breed. Thus competition for food depressed the birth rate. But even 18 owls was too many. Only 11 territories opened up, as a result of deaths among the adults. So only 11 of the 18 fledglings could find territories on which to breed. The other seven young owls had to emigrate.

We can see how a population at carrying capacity can have a low birth rate. Animals run out of food, nesting sites, or territories. Plants competing with one another for water or light cannot get enough. High population density can also encourage emigration and discourage immigration (Figure 28-9). In addition, high population density can increase the death rate from predation, infectious diseases, and parasites.

Predation, parasitism, disease, competition (both intraspecific and interspecific), and emigration all limit growth in proportion to the density of the population. The denser a population, the more slowly it grows. Such factors are said to be **density dependent.**

Density-independent factors

Some populations can also be limited by **density-independent** factors, including fire, drought, storms, tornadoes, volcanic eruptions, and other natural disasters. For example, severe storms can all but eliminate a population of butterflies long before the caterpillar larvae run out of plants to eat. Changes in population density that result from storms and other density-independent limitations do not conform to the logistic growth equation.

Both density-dependent and density-independent factors limit population growth.

Figure 28-9 A plague of locusts. When the population density of certain species of locusts begins to rise, the individuals develop into longer-winged, more gregarious types that fly together. When the proportion of these gregarious types reaches a certain threshold, they rise in great masses and emigrate to a new area, consuming everything in their path—a classic plague of locusts. *(Gianni Tortoli/Photo Researchers, Inc.)*

Do Natural Populations Actually Grow Exponentially?

The growth of bacteria in the laboratory follows the logistic curve closely. Wild populations also follow the logistic curve but usually fluctuate more. Often, population density will temporarily **overshoot** the carrying capacity (K) of the environment. For example, a population of mountain sheep may wildly exceed the carrying capacity, then drop far below the carrying capacity, and then cycle back up to K again.

When a population exceeds the carrying capacity of its environment, it must decrease. The population may then increase again, fluctuating up and down indefinitely. Many populations fluctuate up and down for a while, then gradually level off at the carrying capacity (K) (Figure 28-10). Whether the fluctuations continue depends partly on the extent of the original overshoot. The overshoot depends, in turn, on r_{max}. A species capable of growing rapidly, such as a sea star, is much more likely to overshoot its carrying capacity than a species such as an elephant, whose intrinsic rate of growth is low.

Natural populations do not usually follow the logistic curve exactly.

A.

B.

Figure 28-10 Sea stars approach K. This starfish population grew rapidly, then fluctuated around K. (B, Tom Stack/Tom Stack & Associates)

Why Do Some Populations Cycle Up and Down?

If populations rarely show uncontrolled exponential growth, what limits population growth? The answer seems to vary with the species and the circumstances. Predators sometimes control herbivore populations. For example, the Australian ladybird beetle introduced to California orange groves in 1888 successfully controlled an outbreak of scale insects.

Other factors often play roles, as well. The supply of plant food can limit the growth of herbivores. And plants can limit the growth of the herbivores that eat them. For example, many plants produce toxins in response to damage to their leaves. The toxins may interfere with the digestion, growth, or reproduction of insect herbivores, limiting their population growth.

Biologists have long been fascinated by populations that show regular cycles of population density. In Alaska, for example, populations of rodents called lemmings cycle up and down every 3 to 6 years. These animals can start from a density of less than one lemming per hectare (2.5 acres) and reach levels of more than 25 per hectare. At peak densities, lemmings destroy their food supply and attract throngs of predators. With little food, the lemmings stop reproducing and their population plummets. The cycle then repeats itself. Many other animals, including snowshoe hares, arctic foxes, and ptarmigans, show similarly regular cycles.

Exactly why populations cycle, however, remains a mystery. Biologists have hypothesized that cycling results from fluctuations in the food supply. But that may not always be true. In two groups of meadow voles—relatives of lemmings—researchers found that even though one population had ten times as much food as the other population, it cycled as much as the population with less food.

Population cycles probably result from interactions between species. So far, however, biologists have not discovered exactly how or why cycles occur.

WHAT FACTORS INFLUENCE THE VALUES OF *R* AND *K*?

Species differ greatly in their intrinsic rates of increase, r_{max}. Birds that lay six eggs per clutch, for example, have higher values of r_{max} than those that lay only two. Humans that begin reproduction at age 15 have a greater r_{max} than those who delay until age 30. And a tree that makes seeds for 200 years has greater r_{max} than one that reproduces for only 10 years.

Number of offspring, age of first reproduction, and life span are all traits collectively called **life history traits.** Every species has its own life history. But some species possess all of the traits that tend to maximize r_{max}. Such species are often called *r-selected species.* Examples of classic *r*-selected species are insects and other invertebrates. Such species tend to have short life spans, early first reproduction, and many offspring. Other species have just the opposite traits—long life spans, late first reproduction, and few offspring. Such species are called *K-selected species*. Most mammals and birds are *K* selected compared to most insects.

In the 1960s and 1970s, biologists theorized that natural selection generated species with "*K*-selected" and "*r*-selected traits." Unstable environments, such as those with unpredictable storms or volcanic eruptions, were supposed to produce "*r*-selected species." Stable, predictable environments were supposed to produce "*K*-selected species."

But years of experiments have produced no evidence to support this hypothesis. Species fall along a continuum for these traits, with about half of all species showing no tendency in either direction. Increasingly, biologists suspect that the names "*r*-selected" and "*K*-selected" just describe groups of traits that are associated historically. By analogy, we could divide all animals into those with backbones and those without. Vertebrates all have bony skulls and tails, for example. And, except for a few renegades like turtles and armadillos, none have an external skeleton. But we can't conclude that a certain environment selects for vertebrates and another selects for invertebrates. Both kinds of animals live in just about every habitat on Earth. The same is true of "*r*-selected" species and "*K*-selected" species.

How Do Life Histories Affect Survivorship and Age Distributions?

Elephants invest energy in the production and care of a few, large offspring. As a result, the offspring have a good chance of living long enough to reproduce, usually several times. In contrast, when sea stars reproduce, most of the millions of larvae are filtered from the water by larger animals. Only a tiny fraction of the sea star larvae grow up to be sea stars that have offspring of their own.

One way of representing these two contrasting life histories is a **survivorship curve,** a graph that shows the fraction of a population that is alive at successive ages. The opposite of survivorship is **mortality,** the fraction of a population that dies at a given age. Figure 28-11A shows three kinds of survivorship curves. In a **convex** (or **type I**) survivorship curve, survivorship starts out high and decreases slowly with age. Then, at some point, survivorship begins to decline more quickly. Organisms with convex survivorship curves have a good chance of surviving until they have finished reproducing and caring for their young. Then the chance of dying in a given year increases dramatically. Elephants, dall sheep, adult fruit flies, and humans tend to show convex survivorship curves (Figure 28-11B).

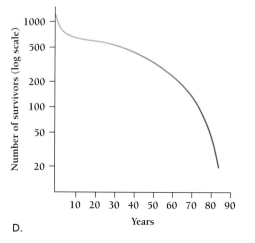

Figure 28-11 Life and death. We all die eventually. But some species are more likely to die young, and some are more likely to die old. Survivorship curves are a way of showing different patterns of mortality and survivorship. A. The members of a type I species survive until they are old and then rapidly die off. The members of a type II species are equally likely to die at every age. Most members of a type III species die soon after they are born, or hatch. Those few that survive live a long time. B. The flower Drummond's phlox shows a classic type I survivorship curve, starting with germination. C. The European lapwing shows a classic straight type II survivorship curve. D. Most natural populations show a combination of shapes, with, for example, high early mortality, a period of steady mortality, and then increased mortality late in life. The survivorship curve shown here is for humans living in the city of Breslau, in Poland in the 17th century. High mortality among infants and children is typical of most human populations in the world today. (*B, Leroy Simon/Visuals Unlimited; C, Roger Wilmshurst/Photo Researchers, Inc.*)

BOX 28.1

Why Are Age Distributions of Males and Females Different?

Age distributions such as those in Figure 28-12 traditionally show males and females separately. We usually think of males and females as being represented in about equal proportions, but in many sexually reproducing organisms, large differences exist in the ratio of males to females.

In certain species of wasps, for example, most offspring are female. All fertilized eggs become females and all unfertilized eggs become males. Consequently, a shortage of males means that many eggs will go unfertilized and more males will hatch. On the other hand, an excess of males means that most of the eggs will be fertilized and will therefore be female. Other organisms use still more complicated negative feedback mechanisms to maintain their sex ratios.

Ants, known for their cooperative behavior, are less cooperative when the subject is sex ratios. A queen ant is equally related to her female and male offspring and bears equal numbers of each. But the males, which develop from an unfertilized egg, are haploid, and the female workers, which develop from fertilized eggs, are diploid. As a result, female workers who share the same father are more related to one another than they are to their brothers.

When Finnish researchers studied the sex ratios of ants from 59 colonies of wood ants, they discovered that in colonies where all the workers share the same father, the males die off disproportionately. Even though the queen lays equal numbers of male and female eggs (the researchers determined the sex of 3000 ant eggs), adult females far outnumber the adult males. What happens to the males? The researchers suspected that female workers either neglect the male larvae or kill them outright. Indeed, in the laboratory, eggs of males added to a nest of closely related females disappeared without a trace.

The situation was very different in colonies in which the queen had mated with two or more males. In such colonies, the females are less related to one another, and the ratio of male and female adult ants remains about equal.

What about humans? Are our sex ratios 50:50? Most people know that women live longer than men, on average. As a result, among older people, women outnumber men. Not everyone knows, however, that in all countries, more baby boys than baby girls are born each day. The difference is not great. In North America, for example, 105 boys are born for every 100 girls.

By the time each cohort reaches reproductive age in the early twenties, however, the sex ratio is nearly exactly 50:50. This adjustment occurs gradually. Baby boys are slightly more likely than baby girls to die of infections and other health problems. Older boys are slightly more likely than girls to die in accidents.

Finally, adolescent boys are dramatically more likely to die in all kinds of accidents than adolescent girls. For example, the leading cause of death for all Americans between 5 and 27 years of age is motor vehicle accidents. But a teenage boy driving a car is more than three times as likely to die in a car accident as a teenage girl driving a car. Among 21- to 24-year-olds, the ratio approaches 4 to 1. It might seem amazing to think that normal 20th-century behavior plays a role in natural selection. But while the kinds of risky behavior that teenage boys engage in may change from century to century, the result—a high death rate—remains the same.

The physiological mechanisms that cause a sex ratio of 105 to 100 are not yet known. We can hypothesize, however, that this skewed ratio has some adaptive value. How would you test such a hypothesis in humans or in some other organism?

Some organisms are as likely to die in midlife as when young or old. When survivorship is plotted logarithmically, birds and many other such organisms show a **diagonal** (or **type II**) survivorship curve—a straight, declining line (Figure 28-11C). Finally, many species are most likely to die early in life. Their survivorship curves are called **concave** (or **type III**).

Few organisms have survivorship patterns that fit any of these curves exactly. The commonest natural pattern is a high mortality rate among the youngest and oldest individuals and a low mortality among those in midlife (Figure 28-11D).

The **age structure** of a population is the fraction of individuals of various ages. We can represent the age structure of a population by a graph such as those in Figure 28-12, which show the number of people in each 5-year age group in Sweden, the United States, and India in 1997. Each such group forms a **cohort,** the set of individuals that enter the population, or are born, at the same time. In this case, we show males and females separately, since their age distributions are slightly different. In all countries, more males are born than females. But in developed countries especially, more females survive in every age category (Box 28.1).

The age distribution in a population is not just a function of survivorship; it also depends critically on the birth rate. A rapidly reproducing population, such as that of India, has a greater fraction of young people than does a population reproducing more slowly, such as that of the United States (Figure 28-12). The age structure also depends on the ages of individuals who move into and out of the population (immi-

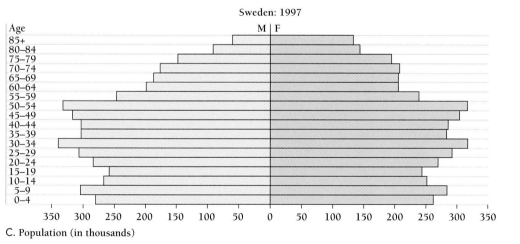

Figure 28-12 Age distributions. A. India, which has a high birth rate, exhibits the classic pyramidal age structure of an expanding population. **B.** In the United States, a slowly declining birth rate and an increasing life span are gradually transforming the age structure from a pyramid into a rectangle. Generation Y is in green. **C.** In Sweden, an almost zero growth rate results in stable population size, with a rectangular age structure. Sweden has as many middle-aged people as young people. If you turn a stable age structure on its side, you see a survivorship curve. What type of survivorship curve does Sweden have? *(Source: U. S. Census Bureau, International Data Base)*

grants and emigrants). As a result, the age structure of the human population differs from country to country, often dramatically.

We can predict the future age structure of a given population if we have the following information: (1) the present age structure, (2) the mortality of each cohort (the chances that individuals of a given age will die each year), (3) the age structure of immigrants and emigrants, and (4) the **fertility** of each cohort (the number of offspring each individual in a cohort is likely to produce). Most such predictions are based on current conditions. When mortality or fertility change, predictions may be wrong.

Differences in survivorship curves lead to differences in age distributions.

Demography: How Do Scientists Measure Populations?

Studying the growth and changes in the structure of human populations is the concern of **demography** [Greek, *demos* = people + *graphos* = measurement], the statistical study of populations. Demography is the population biology of humans. It is important for economists, health planners, and insurance companies, as well as for biologists. Demographers use a variety of statistical measures to describe populations. Mortality is expressed in terms of the death rate, the number of individuals per 1000 who die each year. Fertility is expressed in terms of the birth rate, the number of individuals per 1000 who are born each year. For example, in the United States in 1989, 3.9 million babies were born in a population of nearly 250 million, for a birth rate of 15.6 per 1000. In the same year, 2.1 million Americans died (8.4/1000). Net immigration was 700,000. So the total increase in the population was 2.5 million.

3.9 million − 2.1 million + 0.7 million = 2.5 million

Demographers use several measures to express population growth. One is the **annual rate of increase,** the actual percentage by which a population increases each year (Figure 28-13). For example, the 2.5-million-person increase in the 1989 population of the United States came from a population of 250 million, for an annual rate of increase of 1 percent (10 per 1000). The annual rate of increase for the entire world in 1989 was about 1.66 percent (16.6 per 1000), or 87 million more people.

Another measure of population growth is **completed family size,** the average number of children born to each family that reach reproductive age. For example, the average completed family size in the United States in 1992 was 2.0 children. **Replacement reproduction** is the family size at which each couple is replaced by just two descendants. At the replacement

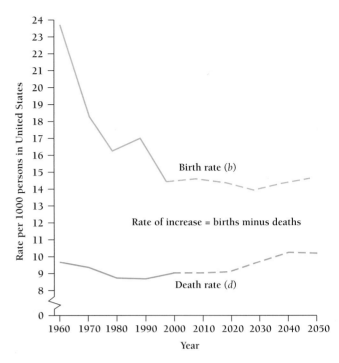

Figure 28-13 What is the natural rate of increase? The birth rate minus the death rate is the natural rate of increase. In the United States, the death rate is increasing as baby boomers age and die, while the birth rate is leveling off. Based on this graph, what was r for the year 1980? What is r today?

level, a population neither grows nor shrinks. Replacement level in the United States is 2.1 children per family. The additional 0.1 makes up for the small number of children who die young or otherwise have no children themselves.

We can see that the population of the United States is not quite replacing itself each generation. But we can also see that the total population of the United States is still growing, at the rate of 1 percent per year. This paradoxical growth is not due only to immigration, as people sometimes argue. Without immigration, the annual rate of increase would still be about 0.7 percent. In the next section, we will see how population growth can continue despite an average family size of 2.0 or less.

Why Is Population Momentum Important?

The population of the United States continues to grow even though completed family size is below replacement level. In fact, even if tomorrow morning human couples the world over began having no more than two children each, the world population would continue to grow for another 50 years, doubling to 10 billion by about 2050. This paradox is called **population momentum** (Figure 28-14).

The reason for population momentum is simple: in a rapidly growing population, a large proportion of individuals will be young. In fact, about 20 percent of the people in the world today are younger than 15 years old. Even if they only replace

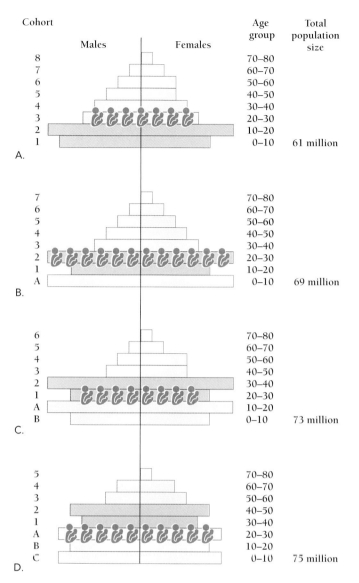

Figure 28-14 **Population momentum.** A population can keep growing even after individuals have no more children than they need to replace themselves. Growth continues when a population has been growing rapidly because the population includes large numbers of children and adolescents. When these cohorts grow up, they replace themselves, each cohort doubling again and again. For simplicity, we have assumed that all reproduction occurs when a cohort is between 20 and 30 years old. A. Rapidly growing population. The members of Cohorts 3 and 4 have more than replaced themselves. Total population size is 61 million. B. Starting here, new parents will have only two children, enough to replace the mother and father. Yet, because Cohorts 1 and 2 are so large, their reproduction causes the total population to increase to 69 million. C. Replacement is still 2.0 children. Yet, as Cohort 1 moves into its reproductive years, the population continues to grow, to 73 million. D. Replacement is still 2.0 children per family. But the many grandchildren of the enormous Cohort 2 are now coming into the world. In addition, more individuals are living into their 70s and 80s than did members of previous cohorts. The population, now 75 million, is still growing.

themselves, they will increase the world population by 20 percent (more than a billion people). And they will meanwhile continue to live alongside their billion-plus children and their billion-plus grandchildren, and they will not begin to die until they are in their 60s, some 50 years from now.

Even if human couples today began having no more than two children each, the world's population would continue to grow for another 50 years.

WHAT IS THE EARTH'S CARRYING CAPACITY FOR HUMANS?

Demographers predict that the world population will increase by 3.3 billion people in the next 50 years. Where will population increase end and how? Just as sea stars cannot go on multiplying exponentially, human beings cannot continue to increase without limit (Figure 28-15). Our growing population raises many questions. Some questions concern our impact on other organisms. Some concern our impact on ourselves. How large, for example, can the human population become before it destroys the biosphere that supports us? If we decrease our birth rate, can we make our population level off or decline to a population size that the Earth can sustain? Or are we headed for a catastrophic population crash caused by famine, disease, or war?

Many people argue about how many people our planet can support. But the Earth's carrying capacity for humans depends partly on what else we want on Earth with us. If all the sunlight that strikes the Earth could be used to feed human beings, we might turn Earth into a giant feedlot for 40 billion or more humans.

Whether we have yet exceeded the Earth's carrying capacity for humans is a matter of debate.

How Well Is Our Energy Supply Keeping Pace with Human Energy Consumption?

The vast majority of humans live in cities and suburbs, where little if any food is grown. Yet each city dweller consumes about a million kilocalories per year. The principal source of this food is farms and ranches that cover one-third of the land. Agriculture provides additional energy to, or **subsidizes,** cities and suburbs.

Agriculture is itself subsidized. The productivity of a cornfield depends not only on sunlight, but also on the fossil fuels needed to run tractors, to pump irrigation water, and to make chemical fertilizers. A farm subsidized with fossil fuel produces about ten times what an unsubsidized, natural ecosystem does (Figure 28-16). All of the extra energy is called an **energy subsidy.**

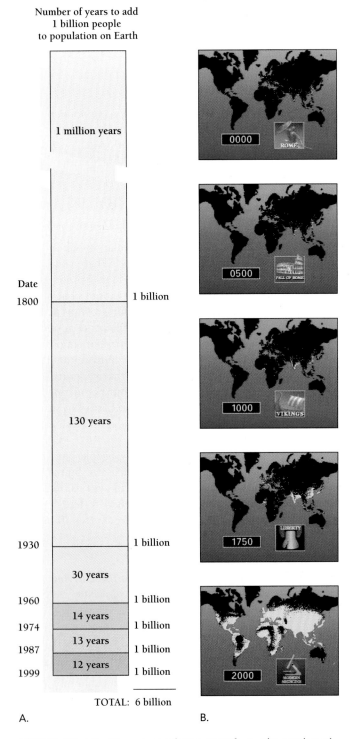

Figure 28-15 **Human population growth.** A. The number of years it takes to add 1 billion new people to our population has decreased from 130 years in the 19th century to a low of 12 years today. B. Population estimates based on historical records show that large population increases began in India and China centuries earlier than in other countries. In just 250 years, the United States has gone from one of the least populous areas on Earth to the third most populous country in the world. (B, Reprinted with permission from Zero Population Growth, Inc.)

Urban humans not only consume a lot of food energy, they also use another 100 to 1000 times more energy—for industry, transportation, heating, lighting, and other activities. This energy goes to buildings, roads, dams, bridges, airports, and all the goods and services we've grown used to. The overall energy flow through an urban area may be 1000 times that of a pond or meadow. Much of this energy ultimately comes from fossil fuels pumped from wells in Alaska, the Middle East, or the ocean floor.

Because humans derive much of their energy and materials from areas far from where they live, the idea of carrying capacity, as applied to natural populations, doesn't quite work. Ecologists talk about the carrying capacity of a particular habitat patch or ecosystem. But technology allows humans to exceed such limits, for humans derive resources from all over the world.

Some ecologists suggest that we think of the human use of resources in terms of **ecological footprint,** the area of land needed to provide all of the resources a single person uses. For example, the average person in the United States uses more than ten times the energy that an average human in the world as a whole does. We use almost ten times the paper, generate five times as much carbon dioxide, and use three times as much fresh water. To fulfill these demands, an average American needs about 5.1 hectares—our ecological footprint. In India, the ecological footprint of the average person is just 0.4 hectares, less than one-twelfth ours (Box 28.2).

When people talk about overpopulation, they often mention the nations with the largest populations—China, India, and the United States—or those with the fastest-growing populations, in Africa and the Middle East. For example, Saudi Arabia's rate of natural increase is 5.5 times that of the United States. But Saudi Arabia has only 20 million people, to our 270 million. So while their population is increasing by 100,000 people per year, ours is increasing by 400,000 people per year.

When we consider total impact on the environment, however, we should look at both population size and ecological footprint together. India's 1.0 billion people each need about 0.4 hectares, or 0.4 billion hectares for the whole country (0.4 × 1.0 billion). But America's 270 million people each need 5.1 hectares, or 1.35 billion hectares for the whole country (5.1 × 0.27 billion).

The area of the United States is only 0.91 billion hectares, however, so we use half again as much land to supply our ever-increasing demand for cars, clothes, and other goods. Even though we are far fewer in number, our impact on the world is more than three times greater than that of India.

Cities depend on energy subsidies and agriculture, which also depend on energy subsidies. Countries with the highest standards of living make the greatest demands on the world's resources.

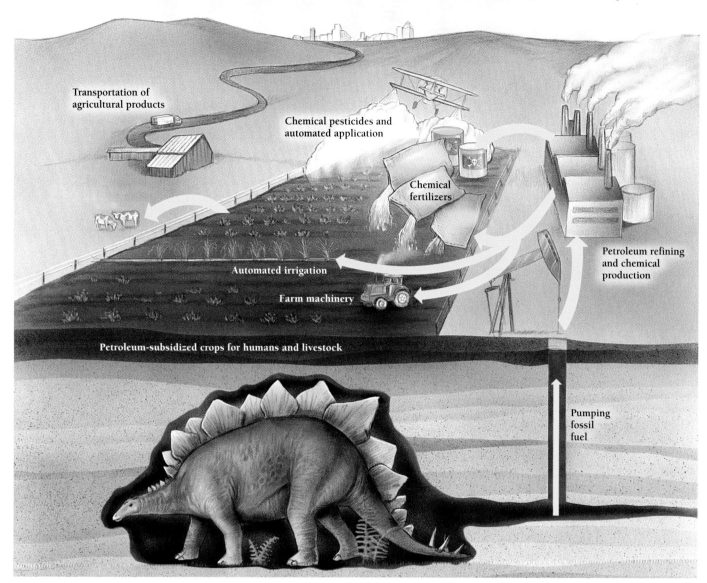

Figure 28-16 **Subsidized agriculture.** Agricultural subsidies and fossil fuel subsidies are the lifeblood of modern cities and suburbs. Modern agriculture, in turn, depends on energy subsidies from ancient stores of fossil fuels—carbon compounds whose energy came from the sun 65 million years ago. Oil-subsidized agriculture can have up to ten times the productivity of a natural ecosystem. However, the supply of fossil fuels is finite.

How Well Has Agriculture Met the Human Demand for Food?

In the late 1960s, agricultural researchers recognized that many countries were outgrowing their food supplies. Every day, hundreds of millions of people were literally starving. To solve the hunger problem, researchers bred rice, wheat, and other crops that could produce up to ten times as much food as older varieties. The effort, called the *Green Revolution,* was a boon to countries such as India, China, and Mexico, where crop increases temporarily outstripped human population growth.

Unfortunately, many of the Green Revolution's "miracle crops" depend on the use of chemical fertilizers, irrigation, and fossil-fuel-powered farm equipment. As American ecologist H.T. Odum wrote, the increased food supply is "partly made of oil." The world's poorest farmers cannot afford to buy the fertilizer, machinery, and fuel they need to grow the newer varieties of crops. But because few farmers now use the old crop varieties—bred over hundreds of years to grow well in particular regions—these old standby varieties are becoming unavailable.

Hunger continues to be a serious problem. Most experts agree that the world's farms currently produce enough to feed the world population. Delivering that food to all who need it, however, has proved to be a staggering problem. Each year, about 10 million children as well as 10 to 15 million adults starve to death—more than the number of people in the whole

BOX 28.2

How Much Do We Have and Need?

"In the 21st century, we are going to pass through a bottleneck. The land required to maintain one American's standard of living, on average, is 12 acres; the amount required to sustain the current standard of living of someone in one of the developing countries, where 80 percent of the world's population lives, is one acre. Now here's the problem: The whole world wants to live like Americans. Thus we have the aftershock following the population explosion: not only too many people, but people attempting legitimately, understandably, to increase the quality of their lives. It's been estimated that to bring the whole world up to America's standard of living would require two more planet Earths. The results, politically, economically, and environmentally, of the rush to achieve that impossibility in the next century will be our greatest problem." —E.O. Wilson, Professor of Biology, Harvard University

Figure A Two families, in India *(left)* and Germany *(right)*, with their possessions. *(Peter Ginter/Material World)*

state of New York. To put it another way, of the 150,000 people who die each day, 70,000 starve to death. At least another half a billion people are profoundly malnourished, with still more living on inadequate diets that leave them susceptible to disease and premature death. In Bangladesh, life expectancy is 48 years, compared to 75 years in the United States and Canada.

Experts doubt the situation will improve. Between 1950 and 1984, global grain production increased enough to stay ahead of population increases. But by 1986, the rate of increase had begun to decline (Figure 28-17). In 1988, severe droughts and crop failures in North America, the former Soviet Union, and China reduced world grain production by 5 percent. Grain stores from previous good years supplied most of the world population during the drought. But the steadily rising demand for food—due to increasing population size—is making it increasingly difficult to stockpile reserves against future crop failures. Meanwhile, global warming has made weather, and therefore crop yields, less predictable than before. Local crop failures are becoming more and more likely.

The Green Revolution increased the world food supply and so Earth's carrying capacity for humans. Today, however, population growth is outstripping the increased food supply.

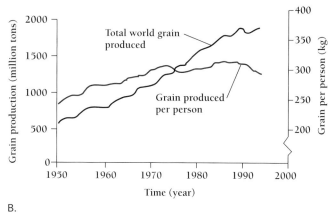

Figure 28-17 Grain production. A. Classic, terraced Asian rice paddy. B. While world grain production is leveling off, the human population continues to grow. As a result, the amount of grain produced per person is declining. (A, R. Ian Lloyd/The Stock Market)

HOW IS HUMAN OVERPOPULATION POLLUTING THE BIOSPHERE?

Humans are not only using up limited resources. We are also damaging air, water, and other renewable resources by polluting them with industrial waste, garbage, and sewage. The more of us there are, the more pollution we generate. In addition, by multiplying, we force more and more people to live in dangerous places. Studies show that as we become more numerous, natural disasters such as earthquakes and floods kill more of us and destroy more buildings and other property.

Pollution Has Degraded the Water

Pollution is most obvious in freshwater communities, although marine ecosystems suffer from pollution as well. Historically, most people have disposed of sewage and other wastes by dumping them into the nearest body of water, to be carried away downstream to the ocean. But such polluted water is dangerous to human health, so government agencies have focused on cleaning up rivers and lakes. The worst damage has not been to people, however, but to other organisms. Several kinds of pollution damage ecosystems, including chemical and thermal pollution and sediments.

Thermal pollution occurs when factories dump hot or warm water into rivers and lakes. For example, when a power plant pumps cool river water through a hot nuclear reactor, the resulting hot water flows back into the river, killing some species and causing others to multiply too fast. Warm water can destroy aquatic ecosystems.

Sediments are particles suspended in water that settle to the bottoms of lakes, streams, rivers, and other bodies of water. They damage aquatic ecosystems by blocking light, burying organisms that live on the bottom, and filling in bodies of water. Lack of light prevents photosynthesis, limiting the productivity of ecosystems. In coastal waters, sediments can inundate coral reefs and shellfish beds.

Sediments result from the natural process of erosion. Indeed, natural erosion working over millions of years cut the Grand Canyon a mile into the Earth. And erosion created the fertile flood plains of the Mississippi and Nile rivers. But virtually all human activities—including the construction of houses and roads, agriculture, logging, and strip mining—enormously accelerate erosion. The result can be too much sediment building up too fast.

Chemical pollutants include a variety of molecules, not all of which are toxic. For example, *biodegradable pollutants* include carbon dioxide, sewage, fertilizers, and other wastes that organisms can consume. Like thermal pollution, these pollutants promote the growth of some organisms and inhibit the growth of others. Nitrogen and phosphorus in fertilizers used in agriculture and home gardening wash into rivers and lakes and promote "eutrophication" (Chapter 26).

Nondegradable pollutants include solid wastes, such as aluminum cans, plastics, glass, and hundreds of other human products that now accumulate in dumps. Even newspaper, technically biodegradable, does not degrade in most landfills. Up to 40 percent of all landfill is newspaper.

At sea, most waste is dumped overboard as a matter of course. Floating objects may stay on the surface for years. Plastic six-pack loops strangle birds and other animals. Lost fishing nets go on catching fish and other animals long after the nets' owners have retired.

The hardest pollutants to deal with are outright poisons. These include heavy metals, such as mercury and lead; reactive gases from smog; radioactive waste; pesticides; and an unknown but growing number of toxic chemicals used in industry and agriculture. Studies have documented the effects of hundreds of these chemicals, but little is known about the effects of thousands of others. Still less is known about the interactions among these chemicals. For example, some substances can be relatively harmless by themselves but toxic in combination with other compounds.

One of the most interesting classes of toxins are *estrogenics,* compounds that happen to imitate the hormone estrogen. Many plastics and other useful modern chemicals contain estrogenics that interfere with the normal development of a variety of animals. Among other effects, estrogenics cause malformations in the reproductive organs of animals, mainly in males. Some estrogenics are softeners found in plastics, such as those in freezer containers. Another group of estrogenics are *surfactants,* molecules that help spread other molecules over a thin surface. Some dishwashers have a dispenser for surfactants, which keep water from beading up on glasses and leaving spots. Surfactants are also added to pesticides to help disperse, or spread, the pesticides when they are sprayed from airplanes onto forests and crops. Recent research has shown that the surfactants wash into streams over many years and interfere with the normal development of salmon. Lab studies show that when salmon get ready to return to the ocean, 20 to 30 percent of those exposed to surfactants die.

Many toxic chemicals become concentrated as they move up the food chain, a phenomenon called **biomagnification** (Figure 28-18). Plants concentrate a compound to many times that in the groundwater. Herbivores that eat the plants further concentrate the toxin in their flesh, and carnivores concentrate it even more. The highest concentrations are found in top carnivores, such as large freshwater fish, fish-eating birds such as pelicans and eagles, and, of course, humans. Trout in the Great Lakes, for example, have concentrations of pesticides and mercury that range from 1000 to 14 million times that in the water in which they swim. The highest concentrations of the pesticide DDT ever found in humans came from men who regularly caught and ate catfish.

Water and soil pollution come in the form of thermal pollution, biodegradable pollutants, nondegradable pollutants, and poisons.

Pollution Has Also Degraded the Air

Industrialization often means a darkening of the air and surrounding landscape, as particles from incompletely burned coal and oil spewed from factories and power plants. Tiny particles of smoke and soot cause lung disease in humans. But while soot may be the most visible and easily removed product of burning fossil fuels, the invisible chemical products of combustion have the most wide-ranging effects.

As we discussed in Chapter 25, burning fossil fuels has doubled the production of carbon dioxide in the atmosphere, contributing to global warming. Burning fossil fuels also generates many other chemicals that alter the chemistry of our atmosphere.

Nitrogen and sulfur oxides may travel thousands of miles with air currents before dissolving in falling rain. The resulting acid rain may destroy forests and aquatic ecosystems hundreds of miles away from the source of the pollution. In addition, the energy of sunlight can convert nitrogen oxides in the atmosphere to the still more reactive compounds of smog.

Another group of dangerous chemicals are the chlorofluorocarbons, chemicals that have eaten a hole in the ozone layer in the upper atmosphere. **Chlorofluorocarbons** are small molecules used as coolants in refrigerators and air conditioners, in the electronics industry as solvents to clean circuit boards, and as propellants in spray cans.

Ozone (O_3) is a highly reactive molecule made of three oxygen atoms, instead of the two in the oxygen (O_2) we breathe. At ground level, ozone destroys living tissue. But 20 to 50 kilometers above the Earth, a layer of ozone called the **ozone layer** absorbs most of the ultraviolet light from the sun, shielding living things from potentially dangerous amounts of ultraviolet light (Figure 28-19). Ultraviolet light that reaches the Earth's surface hurts organisms by increasing mutations in DNA, destroying protein molecules, and increasing the number of chemically reactive molecules in the lower atmosphere. Some biologists suspect that excess ultraviolet light may have driven some species of frogs to extinction. For humans, the most obvious effect of increased ultraviolet light is higher rates of skin cancer.

Almost 20 years ago, atmospheric scientists began to document a thinning of the ozone layer, particularly over Antarctica. They linked the disappearing ozone to the presence of chlorofluorocarbons in the upper atmosphere. By 1988, the evidence for the connection between increased chlorofluorocarbons and decreased ozone had become so clear that chlorofluorocarbon manufacturers agreed to replace these compounds with others.

Scientists initially expected that the hole in the ozone would soon return to normal. Yet despite dramatic reductions in the use of ozone-destroying chemicals, the ozone hole over Antarctica reached a record size of 27 million square kilometers in 1998—larger than all of North America. In the thinnest places the ozone is only 70 percent of normal. Even over the United States, the ozone layer has thinned by 2 or 3 percent. Computer models by atmospheric scientists at NASA now suggest that the ozone layer will continue to thin for another 10 or 15 years.

Air pollution includes visible particles of soot, invisible products of combustion, and chlorofluorocarbons, which damage the ozone layer.

Human population growth has had enormous effects on the biosphere. Our numbers have crowded out whole ecosystems, fragmented ecosystems, driven thousands of species to

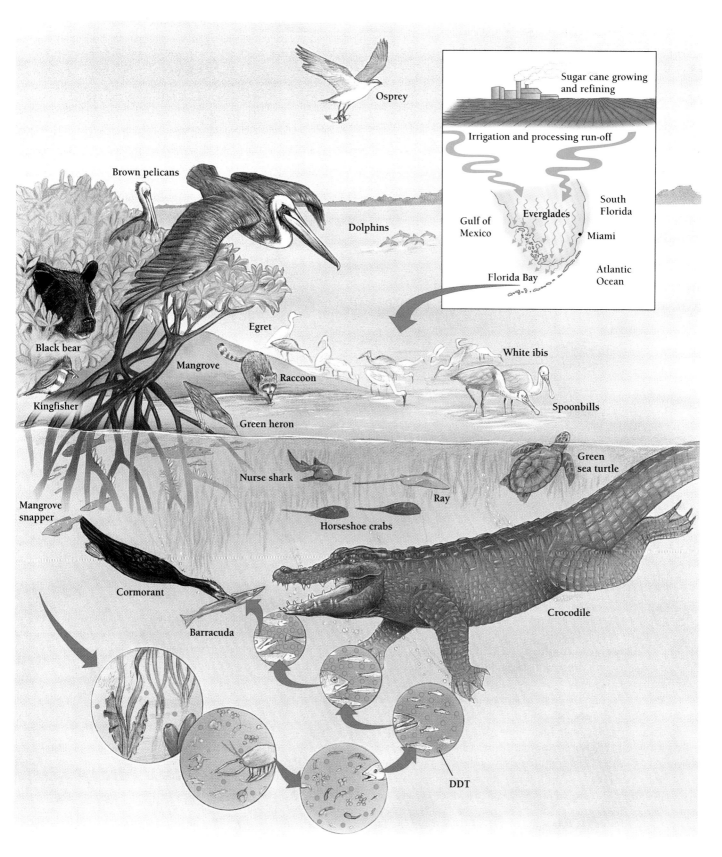

Figure 28-18 Biomagnification. As organisms consume food, they consume whatever toxins are in the food. Toxins that are not broken down or excreted accumulate in the body. At each level of a food chain, the toxin becomes more concentrated—as illustrated here by the increasing concentration of blue dots.

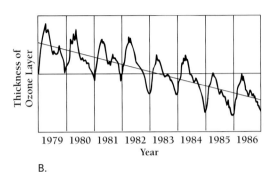

Figure 28-19 How do chlorofluorocarbons break down the ozone layer? Ultraviolet light from the sun breaks down CFC_{l3} into CFC_{l2} and chlorine. The chlorine reacts with ozone to form oxygen and chlorine monoxide, thus depleting ozone in the upper atmosphere. As the ozone layer thins, more ultraviolet light penetrates deeper into the atmosphere.

extinction, polluted the air and water, and altered the global climate.

Many researchers still optimistically believe that we can clean up the mess. We can remove pollutants in water and air, we can reduce the greenhouse effect, and we can close the hole in the ozone layer. We can even stop multiplying so fast. Because of population momentum, however, it may take us as much as a hundred years to halt our growth. One thing we cannot do is bring back the species we have driven to extinction. Once they are gone, they are gone for good. Biologists estimate that the evolution of replacement species will take 5 to 10 million years.

Study Outline with Key Terms

Although a conservation movement has existed in the United States for more than 100 years, 1980 marked the informal beginning of **conservation biology** as a field separate from traditional ecology. Both **deterministic** and **stochastic** factors can drive species to extinction. The way humans drive other species to extinction is through **habitat destruction,** hunting, and the introduction of exotic species. Humans destroy habitat by breaking it up into small fragments with lots of edge and little central core. Roads, suburban developments, and agricultural fields all tend to **fragment** remaining habitat and increase **edge effects.** According to the **extinction vortex** hypothesis, loss and degradation of habitat reduce the size of a population, change its age and sex structure, and lead to a loss of genetic diversity. The loss of genetic diversity leads to inbreeding, reduced reproduction, and smaller population size. In a small population, chance, or stochastic, events are more likely to lead to extinction. We can preserve many more species if we set aside large preserves, connected by **corridors** and surrounded by buffer zones.

All living organisms are capable of **exponential growth.** Bacteria have a **doubling time** of as little as 20 minutes. The human population is currently doubling about every 60 years. Unrestrained exponential growth results in a rapid and limitless increase in total population size over time. A graph of this increase is sometimes called a **J-shaped curve.** The maximum possible rate of increase for a species is r_{max}, also called the **intrinsic rate of natural increase.**

The growth of natural populations is limited by **density-dependent factors,** such as space, resources, and predators, and by **density-independent factors,** such as weather and catastrophes. Density-dependent factors impose a limit to the growth of natural populations called the **carrying capacity** of the habitat, or K. When populations approach the carrying capacity of their habitats, the size of the population levels off, forming an **S-shaped curve.** Such restricted growth can be described by the **logistic growth equation.** When a population stops growing, it is said to have **zero population growth.**

Populations increase or decrease as a function of **birth rate, death rate, immigration,** and **emigration.** When populations **overshoot** their carrying capacity, they may crash, then rise and fall until the carrying capacity is reached or surpassed once more. Some populations **cycle** regularly and somewhat mysteriously. Ecologists used to think that predators drove such cycles. However, the population of prey may cycle as a result of fluctuations in their food supply, and the population of predators may cycle as a result of cycling in the prey population.

Ecologists classify species by **life history traits,** such as life span and number of offspring. One such measure is the **survivorship curve: convex,** or **type I; diagonal,** or **type II;** and **concave,** or **type III.**

If we know the **age structure** of a population, which represents the size of each age cohort in a population, the age structure of immigrants and emigrants, and the fertility and mortality of each **cohort,** we can predict the future age structure of the population. **Demography,** the study of human populations, expresses **mortality** as death rate and **fertility** as birth rate, **replacement reproduction,** or **completed family size.** Demographers express the rate at which a human population increases as the **annual rate of increase.**

The global population of humans has been increasing exponentially for thousands of years. **Population momentum** means that even if humans immediately and permanently reduced their annual rate of increase to zero, the world population would continue to grow for another 50 years. Human agriculture is heavily **subsidized** with energy from fossil fuels, which increase crop yields. Despite the **energy subsidy,** global food production is beginning to fall behind increases in global population size.

Human overpopulation increases water and air pollution, with attendant problems such as **biomagnification,** global warming, and the destruction of the **ozone layer** by **chlorofluorocarbons.** The impact of humans can be expressed in terms of **ecological footprint.**

Review and Thought Questions

Review Questions

1. The world population of humans is about 6 billion, and the doubling time is now around 60 years. Using the equation $N \times 2^t$, calculate the world population after 30 years, 60 years, and 90 years. Graph your results, plotting N against time.
2. Using the information in the table (below), calculate r for Sweden. Show your work.

	Sweden	United States	India
b (births per 1000)	12	14	26
d (deaths per 1000)	11	9	9
r (rate of natural increase [%])		0.6	1.7
Annual rate of growth (%)	0.3	0.9	1.7
Life expectancy (years)	79.2	76.1	62.9
Infant deaths per 1000 live births	4	6	63
Total fertility rate (per woman)	1.8	2.1	3.2
N (population in millions)	9	270	1000

3. What factors make India's r value so much higher than that of the United States?
4. Why does Sweden have a higher death rate than the United States?
5. In your garage live a pair of healthy, mature Norway rats. Assume these animals can produce 10 young every 3 weeks (actually, Norway rats can produce up to 11). Assume also that each litter consists of five females and five males. Females are able to bear young at 12 weeks of age. If none of the adults or offspring die, and if conditions in your garage are so pleasant that none emigrate, how many rats will occupy your garage at the end of 1 year?
6. What is the intrinsic rate of natural increase of this pair of rats, expressed as the rate of increase per year?
7. What factors limit the carrying capacity of a habitat?
8. Give three examples of density-dependent limits to human population growth and three examples of density-independent limits.
9. What is replacement reproduction, and why is replacement reproduction for humans 2.1 instead of just 2.0?
10. If the 1990 annual rate of increase had persisted, what would the global population of humans have been in the year 2000?
11. Why is subsidizing urban populations necessary?
12. Approximately how many people starve to death each year?

Thought Questions

13. If a way could be found for the Earth to support 30 billion people, five times the current density, what other organisms and ecosystems would probably have to go to make room for these people? What could be retained? How would the world be better or worse than it is with a population of 6 billion?
14. Make a list of factors, both density dependent and density independent, that might limit the population of rats in your garage.

About the Chapter-Opening Image

The woolly mammoth—which went extinct 11,000 years ago—led a parade of modern extinctions that now includes millions of species.

 On-line materials relating to this chapter are on the World Wide Web at **http://www.harcourtcollege.com/lifesci/aal2/**

Web Links

Web Bit 28-1 Introduced Plant Species

Web Bit 28-2 The Wildlands Project

Web Bit 28-3 The Human Population is Growing Exponentially

Quizzing & Testing

29 The Ecology of Animal Behavior

Why Is Sociobiology Controversial?

In 1979, a shy and bookish expert on the biology of ants was delivering a scientific talk in Washington, DC, when a group of angry students charged the podium and doused him with water. E.O. Wilson, an eminent Harvard professor, had unwittingly placed himself at the center of a furious controversy by writing an ambitious textbook on animal behavior. The 1975 publication of *Sociobiology: The New Synthesis* almost overnight transformed Wilson into a notorious and, to some people, repugnant political figure.

Wilson became the target of angry public attacks by both scientists and nonscientists that left many people with the impression that sociobiology was promoting racism, sexism, classism, and even Nazism. **Sociobiology,** the study of behavior from an evolutionary perspective, got such a bad name, in fact, that biologists finally renamed it **behavioral ecology.**

The firestorm of debate that Wilson ignited was fueled by several separate but intermingling currents of thought, each with its own long history. These included the ongoing political debate over the role of genetics in shaping human behavior, the related "nature-nurture" debate, and a simmering feud between two groups of scientists.

Until about 1972, two separate disciplines for the study of animal behavior existed—the European science of **ethology,** the systematic study of animal behavior from a biological point of view, and the primarily American science of **experimental psychology.** Both ethologists and psychologists were interested in animal behavior, but their approaches, emphases, and philosophies differed dramatically.

Experimental psychologists focused on how animals learn. For example, the Russian physiologist Ivan Pavlov discovered that he could teach a dog to salivate at the sound of a bell. First, Pavlov would ring a bell each time he fed the dog. When the dog saw the food, it salivated. In time, the bell alone would cause the dog to salivate, a behavior Pavlov termed a **conditioned reflex.**

Experimental psychologists were primarily interested in human behavior, and animals were only substitutes for humans. These researchers thought of behavior as being mostly learned. The adaptive value of behavior was of little interest to them, so all animals were considered more or less equivalent. As the famous experimental psychologist B.F. Skinner wrote in 1959, "Pigeon, rat, monkey, which is which? It doesn't matter."

Ethologists, by contrast, thought of behavior—however much of it was learned—as an adaptation that affected the survival of each individual. Early ethologists preferred to study the normal behavior of animals living in the wild, with special emphasis on behavior that was inherited, or "innate." For example, the tendency of newborn babies to suck milk from a nipple is considered an innate behavior—even though practice improves a baby's ability to suck.

Experimental psychologists were suspicious of ethology for two reasons. First, they believed that because of humans' tremendous ability to learn, genetics and evolution were irrelevant. Second, ethologists' emphasized the idea that behavior can be inherited, which reminded the psychologists of the 19th-century idea that whites were intrinsically (genetically) superior to

Key Concepts

- All behavior develops under the influence of both genetic inheritance and environmental experience.
- Virtually all animals are capable of learning.
- The ecology of an animal determines the adaptive value of its behavior.
- Living in groups has both advantages and disadvantages.

all other races. In the 20th century, this idea was used to justify a variety of evils, including Hitler's human breeding program and the Nazis' mass murder of Jews, Gypsies, and others during World War II.

When E.O. Wilson's book came out in 1975, it brought these conflicts to a head. In his book, Wilson argued that by the year 2000 both ethology and experimental psychology would wither away, to be replaced by sociobiology and neuroscience, the study of the nervous system. Psychologists couldn't help but be offended. For them, acknowledging the value of the evolutionary viewpoint was one thing. Retiring meekly to the sidelines while neuroscientists and ethologists (turned sociobiologists) sorted out the problems of behavior was quite another.

Worse, Wilson speculated freely about the possible genetic basis of various aspects of human social behavior. Many people felt that Wilson's book provided ammunition that could be used to discriminate against the poor, the uneducated, minorities, and women.

A handful of biologists unfairly accused Wilson of being a "genetic determinist," someone who believes that all human behavior is determined by genes. But no biologist takes such an extreme view, and naturally Wilson denied the accusation. Today, the debate has lost much of its heat. Biologists do not argue about whether behavior is controlled by genes or by environment; all agree that behavior is always influenced by both.

HOW IS BEHAVIOR ACQUIRED?

An animal's behavior contributes to its ability to survive and reproduce in the same way that its anatomical, cellular, and molecular characteristics do. Since genes can shape behavior, natural selection can operate on genes that influence behavior, just as it does on genes that influence structure. The adaptiveness of a given behavior depends to a large extent on ecology—the relationship of an individual to its environment. In this chapter, we will examine the ecological significance of behavior, as well as the role of natural selection in shaping behavior.

What Is Innate Behavior?

The human ability to walk is innate. Provided that we are physically able, we all begin walking at about 1 year. Even toddlers whose legs have been in casts during infancy, still begin to walk on schedule (Figure 29-1). **Innate** behavior is behavior that an animal engages in regardless of previous experience. The tendency for a dog to bark at strangers is innate. Most breeds bark at least occasionally. A dog's upbringing and psychological well-being may influence how much it barks and at whom, so that environment plays a role as well, but the dog's tendency to bark is inborn.

By contrast, a dog *learns* who is a stranger and who is familiar. **Learning** is a change in an animal's behavior in response to a specific previous experience. Mammals may be the most sophisticated learners, but essentially all animals—from flatworms to fish—can learn.

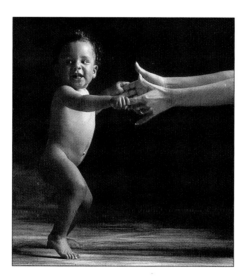

Figure 29-1 Programmed behavior. Humans all begin walking at about the same age. However, the style of walking, whether short steps or long, for example, is learned. (Andy Cox/Tony Stone Images)

Different species of animals are able to learn different things. A grizzly bear is good at learning how to catch salmon in a stream, while a killer whale is good at learning how to catch salmon in the ocean. Every animal has innate tendencies to learn different things. We can see, then, that learned and innate behavior are not completely separable.

One interesting example comes from two races of garter snakes (*Thamnophis elegans*). Garter snakes that live along the coast of California are known to relish slugs, while inland snakes will not touch them. To see if this difference in behavior was innate, researchers hatched baby snakes in the laboratory so the snakes would have no opportunity to learn about

Figure 29-2 A modal action pattern. A greylag goose rolls a wayward egg back to her nest. If the egg is removed in the middle of this procedure, the goose continues the tucking movements until she is back on the nest.

slugs, then offered the young snakes a menu that included slugs. Despite their inexperience, snakes from the coastal population readily ate the slugs, while those from inland populations refused them. This result suggested that genes might determine the snakes' taste for slugs. In a separate experiment, the researchers crossbred the two kinds of snakes. Among the hybrids, more snakes refused slugs than accepted them. The researchers suggested that a gene influencing slug refusal might be dominant.

Any behavior is likely to include both innate and learned aspects.

What stimulates innate behavior?

A highly stereotyped innate behavior is called a **modal action pattern.** Examples of modal action patterns abound. All dogs scratch their ears in the same way, lifting a rear leg above a foreleg. This behavior is so stereotyped that if a human scratches a dog's ear, the dog's hind leg often will rise above the foreleg and move up and down in midair as if scratching. Parrots and many other birds also scratch their heads by bringing a foot over the wing. This mode of head scratching is not necessarily required by their anatomy, however. A parrot is perfectly capable of bringing a foot under the wing, and does so to clean its bill.

A modal action pattern can be partly fixed and partly flexible. For example, when an egg rolls out of the nest of a greylag goose, the goose can roll the wayward egg back into her nest by pushing it with the underside of her bill (Figure 29-2). This bill-tucking movement allows her to pull a loose egg back toward her, but she also must adjust to the sideways wobble of the egg by moving her head from side to side. If the goose is tucking the egg back into its nest and an experimenter suddenly removes the egg, the goose will continue tucking. But the bird's adjustments to the side-to-side movements of the egg will stop. The tucking movement is fixed, but the sideways adjustments are flexible.

Modal action patterns serve a purpose only when performed at the proper time and place. Some signal in the environment must stimulate a modal action pattern. In human infants, a pair of eyes stimulates a smile (Figure 29-3). A loose egg stimulates a greylag goose's chin tucking. A fly provokes the characteristic flick of a frog's tongue. In each case, the animal's senses detect a stimulus and the nervous system issues the order to act. The stimulus for a modal action pattern is called a **sign stimulus.**

Innate behavior often has a learned component. For example, the chick of a herring gull will peck at its mother's beak

Figure 29-3 Releasing a smile. A flat, face-sized mask with one eye spot will not make a young baby smile. The same mask with two eye spots triggers a happy smile. The two eyes constitute a kind of sign stimulus. Older babies learn to prefer human faces.

A.   B.

Figure 29-4 A herring gull chick begins begging for food soon after it hatches, a behavior that is partly innate and partly learned. *(Thomas W. Martin/Photo Researchers, Inc.)*

to beg for food the first time it sees its parent (Figure 29-4). Pecking is therefore regarded as innate. The pioneering ethologist Niko Tinbergen reasoned that since a herring gull chick always pecks at a red spot on its parent's bill, the red spot may be what provokes the pecking behavior. To test this theory, Tinbergen presented chicks with models of gull bills with different colored spots. He found that the red spot elicited the pecking behavior better than any other colored spot, regardless of the shape of the model.

But even this behavior is partly learned. Later research by biologist Jack Hailman showed that a newly hatched gull chick pecks at anything that looks remotely like a beak, including the beaks of its siblings. Over the course of several days, however, the chick *learns* to recognize both its parents and food, so that it pecks in response to a much more limited range of stimuli.

Not all sign stimuli are visual. Touch, sound, and smell can also trigger stereotyped behavior patterns (Figure 29-5). A female silk moth emits a potent chemical, called bombykol, whose odor can attract males from over a mile away. When the male detects as few as 200 molecules of bombykol, he flies upwind until he loses the scent, flies in a zigzag pattern until he detects the bombykol again, and flies upwind again. He continues this pattern until he finds the female. Chemicals like bombykol that influence the behavior of another individual of the same species are called **pheromones.** Pheromones can activate modal action patterns.

Most sign stimuli are quite simple. Even with our own extraordinary brains and sensory systems, we simplify the complexity of the world. As a result, we can easily recognize public figures from simple line drawing caricatures (Figure 29-6). Animals with less developed nervous systems than our own also respond to caricatures. When Tinbergen presented gull chicks with a stick with three red spots, instead of just one, the chicks pecked at it even more than if it had been a real gull's bill. For a herring gull chick, three red spots constitute a **supernormal stimulus,** a stimulus even more stimulating than the normal stimulus.

Figure 29-5 Tick sign stimuli. A female tick needs a blood meal to produce her eggs, but she must wait for a meal to come to her. For days or months, she sits unmoving at the end of a twig. At the moment a warm animal passes beneath her twig, she awakes, lets go of the twig, and drops. If she lands on a suitable animal, she inserts her feeding organ into the skin and sucks herself full of blood. If she lands on the ground, she climbs up onto a nearby plant and waits even longer. A tick responds to three separate sign stimuli. In response to light, a tick moves to the end of the twig. In response to either carbon dioxide (which animals exhale) or butyric acid (found in sweat), a tick lets go of the twig. Finally, in response to any warm surface—even a balloon filled with warm water—a tick will begin feeding. Light, carbon dioxide, butyric acid, and warmth all act as sign stimuli for a tick's behavior.

Figure 29-6 A caricature is a kind of supernormal stimulus. A caricature of Oprah Winfrey exaggerates her features. The drawing doesn't really look like her, but we instantly recognize who it is supposed to be. *(Barbara Ericson)*

> *A modal action pattern is a stereotyped behavior triggered by a sign stimulus. The sign stimulus can be seen, heard, felt, or smelled.*

How Much Can Animals Learn?

In an early episode of the television show *Sesame Street*, Ernie is astonished that Bert has been able to teach Bernice the pigeon to play checkers. "Why, a pigeon that can play checkers!" Ernie exclaims. "That must be the smartest pigeon in the whole world!" "She's really not that smart," demurs Bert. "In the ten games we've played? She's only beaten me twice."

Pigeons may or may not be able to beat Muppets at checkers, but they are surprisingly capable learners. Pigeons and chickens can learn to distinguish between an underwater scene with fish and one without. They can even understand abstract ideas. Pigeons can learn to choose one of two disks according to whether it matches the color of a third disk.

Almost any animal can learn. Pigeons can beat humans at tic tac toe. Rats can learn not only to run a maze accurately and consistently, but to run a mirror image of the maze, switching back and forth on cue. Even planarians, among the simplest of all animals, can learn (Figure 29-7).

Animals pay close attention to one another's behavior. Experiments have shown that blue jays prefer to eat what they have seen other jays eating. Crows, upon seeing another crow sicken or die after eating poisoned bait, will shun the bait. One of the most striking examples of learning in natural populations of animals has occurred in at least 11 species of small English birds, including blue tits, titmice, and others. Starting in around 1920, a few birds learned to open the caps of milk bottles delivered to the front steps of English houses. Beneath the caps was a rich meal of pure cream. The cream was so attractive that the birds attacked the bottles as soon as the delivery person was gone, and, in some areas, flocks of birds followed the milk trucks down the streets. If the bottles were capped with cardboard, the birds pulled the whole cap off or peeled it off in layers of paper. If the bottle was capped with foil, the birds poked a hole in the foil, then peeled the foil off in strips. This behavior first appeared in a few limited areas, then, over a period of years, gradually spread, suggesting that the birds learned from one another.

> *Animals can learn a great deal from their environment and from each other. Even behavior strongly influenced by genes is still open to modification through learning.*

Can old dogs learn new tricks?

Just as humans learn languages most easily when young, many birds learn to sing their species song only when they are young. For many kinds of learning, timing is critical. A chaffinch raised in isolation, so that it cannot hear adult birds singing, never learns to sing an adult song. If a chaffinch hears the song as an adult, he cannot learn it. If, however, the chaffinch hears the correct song during the first few months of life, he can learn to sing a song that resembles the chaffinch song—even if he cannot hear the song when he is actually practicing. The period of time during which an animal can learn a particular behavior pattern is called the **sensitive period.**

Sensitive periods exist for a variety of different kinds of behavior. For example, children born blinded by cataracts can learn to see if the cataracts are surgically removed before about age 10. If the cataracts are removed later in life, however, these people see only a random patchwork of colored shapes, which they cannot interpret. As infants we learn to interpret complex patterns of shading and color as a three-dimensional world. But for newly sighted people, the patterns may be meaningless.

The process by which an animal learns such behavior during the sensitive period is called **imprinting.** The ethologist Konrad Lorenz provided the most famous example. Geese learn to recognize their mother by "imprinting" on the first moving object they see after hatching (Figure 29-8). Usually this

Figure 29-7 Learning in flatworms. The simple planarian, with barely a brain to call its own, learns to find food suspended from a string. Untrained worms search the surface or the bottom of a container for food and miss the meat hanging in the middle. During training, a piece of meat is suspended at the surface and then gradually lowered into the water. The flatworms learn to locate the string at the surface and then to follow it down to the meat. Trained worms quickly find food suspended in the water.

Figure 29-8 Imprinting. Cygnets learn to recognize their mother and follow her soon after they hatch. *(Gerard Lacz/Peter Arnold, Inc.)*

object is the mother goose. But newborn geese will also follow a matchbox on a string, a rubber ball, a flashing light, or even a person.

Just as the development of normal vision depends on visual experience during a critical period, the development of normal social behavior depends on experience at sensitive periods. A young monkey isolated from other monkeys for six months any time during the first year and a half of life does not develop social interactions with other monkeys when later given the chance. Instead, the deprived monkey crouches in the corner of its cage, rocking back and forth, much like a severely disturbed human child. An adult monkey subjected to the same isolation returns to its friends without such serious behavioral problems.

Harry and Margaret Harlow, psychologists at the University of Wisconsin, attempted to find out what aspects of a young monkey's social life normally prevented such behavioral disorders. The researchers isolated young monkeys with and without a cloth-covered wooden dummy. The baby monkeys that had a wooden dummy, or "surrogate mother," showed fewer symptoms of isolation than the one with no dummy.

The Harlows could completely prevent the symptoms of isolation by allowing the young monkeys to spend just 15 minutes a day with a normally socialized monkey that spent the rest of its time with the colony. Such limited social experience usually produced full socialization only if it occurred during the sensitive period in the first year and a half of life. In a few cases, however, monkeys reared in isolation could be partly "cured" by later exposure to persistently friendly monkeys. In monkeys, at least, positive experiences later in life can sometimes partly make up for a horrific early life.

For some complex behavior, then, the sensitive period is not the only time an animal can learn. We can look at humans for another example. Although children learn new languages faster and better than adults do, adults can learn new languages rather well if they are willing to work at it.

Many forms of learning occur only at particular times, called sensitive periods. Other forms of learning take place most easily during sensitive periods, but can occur later as well.

How Do Genes Influence Behavior?

Behavior is the result of both genetic influences and experience (Figure 29-9). The nesting behavior of two small African parrots—the peach-faced lovebird and Fischer's lovebird—offers a perfect example of this. In both species of lovebird the female cuts long strips of bark or leaves for nest building. A peach-faced lovebird can carry up to six strips at a time by tucking them into her rump feathers, while a Fischer's lovebird carries these strips to the nest one at a time in her beak.

A hybrid of the two species tries both methods. When she builds her first nest, she starts out by tucking strips into her feathers, like her peach-faced parent. But she is inept, and the strips fall out before she can get to the nest. With time, she learns to carry just one strip at a time in her beak. After about 3 years, she gives up tucking strips into her feathers. However, just before she flies off with a strip in her beak, she still makes little sideways movements of her head, as if she were about to tuck a strip into her rump feathers.

Still, the exact role of genes in influencing behavior is hard to pin down. In rare cases, a single gene can affect behavior. For example, in humans, a rare mutation in one gene causes Lesch-Nyhan syndrome, in which people compulsively mutilate themselves. The vast majority of physical and behavioral traits, however, are "polygenic," influenced by many genes.

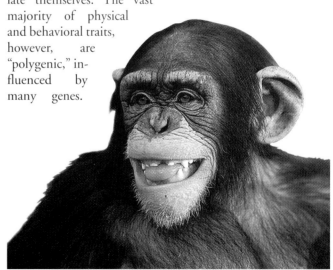

Figure 29-9 Behavior is partly genetic. Specific types of behavior, such as smiling, are often shared by groups of related species. *(Tim Davis/Photo Researchers, Inc.)*

Indeed, biologists have found that only about 2 percent of all traits are under the main influence of just one gene. Two or more—possibly hundreds—of genes influence the expression of almost every trait.

The complex genetics of all phenotypic traits—including behavioral traits—means that biologists cannot easily determine the exact degree to which a given behavior pattern is genetically or environmentally influenced. For example, despite years of research on the heritability of IQ scores—just one aspect of human intelligence— scientists can say only that IQ is somewhere between 30 and 70 percent inherited.

The balance is environmental. Nutrition, health, and environmental stimulation—in the womb, in infancy, and in early childhood—all affect intelligence to some degree. One study found that premature baby boys fed breast milk later had several more IQ points than boys who had been fed formula. Yet, in premature girls, the researchers found no difference.

HOW DOES ECOLOGY INFLUENCE BEHAVIOR?

Economists have debated whether people make rational economic decisions. For example, if food costs twice as much at one store as another, most people go out of their way to shop at the cheaper store. On the other hand, if travel costs are high, they will not spend more money getting to the cheaper store than they would save on the groceries. To a remarkable degree, then, people behave rationally. Animals, with far less intellectual capacity than humans, also behave surprisingly rationally about such decisions.

How Do Animals Make the Best Choices?

Crows open and eat thick-shelled sea snails, called whelks, by picking up the largest one they can find, carrying it to a height of 5 meters, and dropping it on the rocky shore. If the whelk breaks, the crow feasts. If the whelk stays intact, the crow carries it aloft and drops it again, repeating the procedure until the whelk's shell breaks.

Ecologists studying this tedious feeding strategy wondered if this was the best strategy to get the most food for the least energy expended. To see if the crows' strategy was "optimal," researchers themselves dropped variously sized whelks from different heights. The biologists soon found that the largest whelks were the most likely to break and that none of the whelks broke consistently unless dropped from a height of 5 meters or more. The crows' strategy of choosing only the largest whelks and carrying them to about 5 meters was indeed optimal. Further experiments revealed that if a whelk didn't crack open on the first try, persistence with a single whelk was more likely to succeed than starting over with a new whelk. So, a crow's decision to drop the same whelk over and over proved equally sensible. Overall, the crows' feeding behavior is an efficient way of spending energy to get more energy. Such energy-efficient feeding is called **optimal foraging.**

When animals forage, they may efficiently maximize the amount of food they get for the energy expended. Ecologists call this process optimal foraging.

How do competition and predation influence foraging behavior?

Maximizing energy intake is not the only consideration that guides feeding behavior, however. When predators are present, an animal must adjust its behavior to increase its own chances of surviving. One study of chickadee feeding, for example, revealed that the chickadees would eat seeds at a researcher's feeding tray only if the tray seemed safe. When researchers mounted a model of a hawk nearby, the chickadees took the seeds away to a safer place. Taking the seeds away cost the birds time and energy, but given the apparent presence of a hawk, moving the seeds was a good way to both eat and survive.

The presence of competing species can also change feeding behavior. For example, North American mink introduced into Sweden in the 1920s competed directly with European otters for food and caused otter populations to decline. However, the two animals continued to coexist. In America, the mink normally eat almost anything that moves, including crayfish, fish, muskrats, rabbits, birds, and small burrowing mammals. The European otters have a more specialized niche than the mink, feeding on fish, crayfish, crabs, and other aquatic invertebrates only. In Europe, the larger otters defend this narrow niche from the mink aggressively, sometimes killing mink that poach fish and crayfish. As a result, in areas where the otters persist, mink eat less fish and crayfish than those in America.

Foraging behavior depends, in part, on how an individual interacts with members of other species.

How do animals express aggressive feelings?

Reproduction and survival necessitate competition as well as cooperation. Among the most dramatic competitive kinds of behavior are the ritualized fights between pairs of bighorn sheep. The outcome of such combat determines which male will get to mate with a female. The clash of horns is mostly symbolic: usually the fighters do not seriously injure each other. Other species use other symbolic contests—who can roar the loudest, stand the tallest, or stare the longest.

These contests are examples of **agonistic behavior** [Greek, *agonistes* = champion]—all the aspects of competitive behavior within a species, including aggression (outright attacks and fighting), aggressive displays, appeasement, and retreat (Figure 29-10). Ethologists use the term "agonistic behavior"

Figure 29-10 Agonistic behavior in rabbits. Fighting between males is not confined to large animals such as deer or lions. Here, two rabbits fight to establish dominance. Agonistic behavior also includes aggressive behavior and appeasement displays. *(Art Wolfe/Tony Stone Images)*

instead of the word "aggression" to acknowledge that animals never exhibit pure aggression or pure fear. What most animals (humans included) show is a mix of fearful behavior and aggression, sometimes more of one, sometimes more of the other.

Some animals use agonistic interactions to establish a **dominance hierarchy,** a ranking of individuals that fixes which animals have first access to resources or mates. Dominance hierarchies establish rules that allow animals to live in the same territory with limited resources. The classic example of a dominance hierarchy occurs in a flock of hens. Such a flock quickly establishes a "pecking order," which specifies who pecks whom without being pecked in return.

One way that animals compete with others of their own species for mates, resources, and rank is through agonistic encounters.

Territoriality

In many species, an animal engages in competitive behavior to establish and defend a **territory,** an area occupied by an individual (or group of individuals), from which other individuals of that species are excluded. Some species, a few birds and fish, for example, use territories exclusively for mating. More commonly, species use territories for foraging or for raising young as well.

Establishing and defending a territory necessarily involves agonistic behavior. Animals use a wide variety of signals to mark their territories, including chemical scents, brightly colored plumage, or patrolling behavior.

Defending a territory requires energy expenditure. Ecologists have studied the "economics" of territoriality and have shown that some bird species defend smaller territories when food is abundant than when food is scarce. When food is abundant, less territory is necessary to raise the same number of offspring. So, to conserve energy, birds defend a smaller territory—an example of optimal foraging. A few bird species even adjust the size of their territories on a daily basis.

Animals defend territories for raising young, mating, or foraging—as long as defending the territory will increase their reproductive potential.

How does female choice influence the evolution of males?

The long-term result of competition for mates is **sexual selection,** the differential ability of individuals with different genotypes to acquire mates (Chapter 16). Some biologists distinguish sexual selection from other forms of natural selection, arguing that natural selection always results in the accumulation of adaptive structures and behavior, while sexual selection often results in "maladaptive" structures and behavior. But all adaptations to the various demands of survival and reproduction are a constant balancing act for organisms.

Natural selection can act both on inherited variations in structures that are important in mating behavior, such as a buck's antlers or a peacock's tail, and on inherited mating behavior. For example, in the tungara frog (*Physalaemus*), which lives in Panama, female frogs consistently prefer larger males to smaller ones. The females choose the larger males, however, not by comparing male physiques but by listening to the male frogs' mating calls. The song consists of a whine, followed by one or more low-pitched "chucks." Bigger frogs have lower "chucks," so female frogs prefer frogs with deep voices (Figure 29-11A). The lower the "chuck," the more the female frog is attracted, even when the "chuck" comes from a tape recorder.

Unfortunately, the male frog's deep "chucks" also attract the attentions of a frog-eating bat, which, like the female frog, prefers a big frog to a little one (Figure 29-11B). The bat presents the male frog with a serious dilemma. If he chucks very low to attract females, he risks his life by attracting bats. If he does not chuck low enough, he will not be able to reproduce. To maximize his chances of reproducing while minimizing his chances of being eaten, the frog must optimize the number and pitch of his chucks. To whatever extent that variation in pitch is under genetic influence, sexual selection will tend to favor the evolution of low-pitched chucks—but maybe not too low.

Competition for mates over evolutionary time results in sexual selection, the differential ability of individuals with different genotypes to acquire mates.

▲ **Figure 29-11 Cost-benefit analysis by frogs.** The female tungara frog of Panama prefers a male that makes a deep chuck when he calls. Larger males make deeper chucks. Males with deepest voices have the best chance of attracting a female and, unfortunately, the best chance of being eaten by a frog-eating bat. A. A male tungara frog calling while floating in his pond. B. A frog-eating bat swoops away with a frog. (A, M.J. Ryan, University of Texas; B, Merlin D. Tuttle, Bat Conservation International/Photo Researchers, Inc.)

How Do Animals Compete for Resources or Mates?

Animals of the same species may cooperate or compete with one another, according to the demands of their current environment and the constraints of their evolutionary histories. For example, in many species, reproduction requires close cooperation between the male and the female. Such animals exhibit stylized, species-specific courtship behavior, including the well-known mating rituals of birds and fish (Figure 29-12).

Mating rituals allow females to assess the health and vigor of potential mates. Also, because mating season is often a time of heightened aggression, mating rituals allow both sexes to minimize the risk of being injured or killed by their partner. We can see that when such behavior patterns are influenced by the genes, natural selection must operate powerfully—increasing the frequency of those behavioral types that reproduce most successfully.

SOME ANIMALS LIVE IN SOCIAL GROUPS

A tiger lives a solitary life. It meets other tigers rarely and for not much longer than it takes to mate. Yet, the tiger's close relative the lion lives in a tightly knit group called a pride (Figure 29-13A and B). The members of a pride of lions are never far apart. They sleep together, hunt together, and raise their

Figure 29-12 Courtship in the stickleback. Mating rituals allow animals to recognize and evaluate one another.

Figure 29-13 Loners and social animals. A. Tigers hunt alone and kill smaller animals. B. Lions, which hunt in groups, can kill animals much larger than themselves. (*A, Tom Brakefield/DRK Photo; B, Anup & Manoj Shah/DRK Photo*)

young together. Why do some animals live socially, while others live apart?

How Is Social Behavior Advantageous?

Living in groups affords distinct advantages. The greatest advantage to group living is that large groups of animals can more quickly spot predators and other dangers than can solitary animals. Researchers have found, for example, that the more pigeons that are in a flock, the sooner the pigeons will notice a hawk and the less likely that the hawk will catch one of the pigeons.

Groups of animals can also better defend themselves. When threatened, a herd of musk oxen forms a circle with their young in the middle and their horns pointed outward. And the thousands of honeybees in a hive can drive off a predator or an animal bent on stealing their honey where a lone bee would be helpless. Finally, group living can be an advantage in finding food. Predators that hunt cooperatively, such as lions and wolves, can kill prey far larger than themselves (Figure 29-13B).

Social living has disadvantages as well, however. Animals living in groups must compete with each other for food, mates, and breeding areas. Individuals waste time and energy fighting or threatening one another and searching for food in the same places. Groups of animals also transmit diseases to one another and attract predators.

Whether animals live in groups depends on the ecological costs and benefits. Many songbirds stake out territories in the spring and actively repel all but their own mates and offspring. Yet when autumn arrives, and food becomes harder to find, the same birds abandon their territories and roam the countryside in large communal flocks.

Sociality allows animals to spot both prey and predators more easily. But groups of animals are themselves more visible to predators, and individuals in groups must constantly compete for food, mates, and breeding areas.

Honeybees Are an Extraordinary Example of Social Behavior

The coordination of individuals within a colony of social insects rivals that of the cells within an individual. We can view an insect colony as a superorganism, in which thousands or even millions of individuals function as a single unit. Social insects form enormous colonies. An African termite mound, for example, may contain 2 million workers. But the number of individuals alone is not what makes the social insects special. It is social organization that allows social species to take advantage of environmental resources unavailable to more solitary animals.

The familiar honeybee provides an excellent example. Humans long ago domesticated the honeybee to harvest the bees' useful honey and wax. Scientists since Aristotle have studied the workings of beehives and published tens of thousands of papers on bees. The fascination with bees has extended beyond the practical to the almost mystical. Humans admire the fact that the "busy bee" works so hard and loyally and ferociously defends its queen (long thought to be a king). One writer even suggested that honeybees have "voluntarily escaped from the Garden of Eden with poor fallen man for the purpose of sweetening his bitter lot."

A honeybee hive consists of double-layered wax combs that contain thousands of hexagonal cells. In these wax cells, the bees store honey and pollen or house developing larvae. Each bee grows from an egg deposited in a separate cell. The egg develops into a larva, which then develops into a pupa and finally into an adult.

A hive may contain 20,000 to 80,000 workers, each engaged in a succession of tasks. At the beginning of her life, a worker feeds nectar and pollen to the larvae. Later, she makes wax to enlarge the comb, removes the dead and the dying, guards the hive against intruders, and scouts the surrounding territory for food and nesting sites. Finally, toward the end of her short, 6-week life, she undertakes the most hazardous job of all, foraging for nectar and pollen among the flowers.

All the workers in the hive are female, and all are sisters. None reproduce, however. Only a single female—the queen—lays eggs. Most of these eggs develop into sterile workers, but a few may become queens. Workers can create queens by feeding a few, select larvae a special rich diet. Normally, the hive's queen produces a "queen substance," a pheromone that prevents the workers from building special wax cells to raise new queens. In the spring, however, the queen's production of this chemical falls off, and the workers build the royal cells. Workers soon push the old queen out of the hive, and she flies off with a large swarm of workers to start a new hive (Figure 29-14).

About 8 days after the old queen leaves the hive, the new queens begin hatching. Each one can form a swarm and fly off to start a new colony. If a queen elects to stay in the old hive, however, she first seeks and destroys rival queen larvae. She identifies these competitors by making quacking sounds outside the queen cells. If another queen responds to this challenge, a fight to the death ensues. There can be only one queen in each colony.

Once a queen has disposed of her rivals and established her queenship, she mates repeatedly in a series of "nuptial" flights. She leaves the hive, approaches a group of males, and releases small amounts of queen substance. The pheromone attracts males, one of whom will mate with her. The male releases his sperm into the queen's genital chamber and quickly dies. The queen may mate with as many as 17 males over a 4-day period. In this way, she obtains enough sperm for a lifetime. She stores the sperm and uses it to fertilize her eggs, which she deposits one at a time into separate wax cells. A queen may live and lay eggs for more than 5 years.

All of the social insects, including termites, ants, wasps, and honeybees, have three traits in common: (1) they cooperate in raising their young; (2) only a few individuals reproduce, while sterile workers assist and defend the fertile queen; and (3) generations overlap, so that offspring help their mother.

Honeybees display extraordinarily complex social structure.

Is Altruism Adaptive?

In Chapter 16 we learned that natural selection favors traits that increase the likelihood that an organism's offspring will survive. Why, then, we must ask, do the sterile workers in a beehive persist in a behavior that can have no adaptive value for them individually? The workers increase the reproductive output of the queen, but at the expense of their own reproduction. Biologists call such an arrangement **altruism,** behavior that benefits others at the expense of the animal that performs the behavior.

People sometimes jump into frozen lakes to rescue others. We know that this behavior is risky, because many of these would-be rescuers themselves die. Such behavior is clearly altruistic. Similarly, ground squirrels risk their own lives to warn one another of an approaching predator. And the young of many species of birds help their parents rear the next generation. The social insects, however, demonstrate the most extreme example of altruism.

If altruistic behavior has a genetic basis, then biologists must ask how it could possibly evolve. We would expect natural selection to favor sacrifice by parents to help their offspring but to act against the sacrifice of offspring to help their parents. Why do worker bees work so hard to help their mothers reproduce?

Does altruism arise by natural selection?

How could evolution produce workers, however industrious, if they leave no offspring? For a while, Darwin himself thought that this paradox was fatal to his whole theory of evolution by natural selection. To save his theory, Darwin argued that a family (the queen and her offspring, for example) must be the unit

Figure 29-14 Swarming honeybees. A swarm of bees will stay in one place for days at a time while scouts look for a suitable site for a new hive. The scouts return to the swarm and dance the location of a hole in a tree or cave. Each scout, however, is dancing a different location. When all the scouts agree, the swarm sets off to build a new hive. *(Gilbert Grant/Photo Researchers, Inc.)*

of selection, rather than the individual. By their self-sacrifice, the workers perpetuate genes that they share with their sisters, one of whom will become a queen. What is selected is the ability to be altruistic. The altruistic family is successful in passing on its genes.

In the 1960s, the British biologist W.D. Hamilton extended Darwin's idea, suggesting the theory of **kin selection.** Hamilton reasoned that an individual increases its reproductive output by helping relatives, which share its genes, to reproduce. Sisters (and brothers) are just as related to one another as they are to their parents. They share half of their genes. Thus, a worker bee helps ensure the continuation of her genes within her sisters and their descendants (recall that some of the sisters become queens).

Hamilton's work established the concept of "inclusive fitness." Fitness is a measure of selective advantage defined as the contribution to the next generation of one genotype in a population relative to the contribution of other genotypes. **Inclusive fitness** is the sum of an individual's genetic fitness (which includes that of its own direct descendants) plus all its influence on the fitness of its other relatives. Inclusive fitness, then, is the total success in passing alleles to the next generation, either by the efforts of a parent or by the altruistic acts of a relative.

One explanation for the adaptive value of altruism is inclusive fitness, which includes the concept of kin selection.

Why might intricate societies have evolved among certain insects?

Intricate societies evolved at least 11 separate times among the bees, ants, and other members of the order Hymenoptera, yet only once (in the termites) among all other insects. Why the difference? One reason may be that the bees and wasps have an unusual kind of sex determination, called **haplodiploidy.** In haplodiploidy, males develop from unfertilized eggs and are therefore haploid, while females are diploid. Other than providing sperm in a single mating event, males contribute nothing to the economy of the hive. Haplodiploidy dramatically alters the usual genetic relationships among parents and siblings. It means that a worker's inclusive fitness is greater if she devotes her energies to her siblings than if she reproduces herself. Let's see how this works.

First consider the usual case, say in humans. Half your alleles come from your mother, half from your father. Your brothers and sisters also share half of their alleles with your mother.

What, then, is the chance of sharing an allele with your sister? Half of all your alleles came from your mother, and the chance that you and your sister inherited the same allele is 0.5. So the chance that you each inherited a given allele from your mother is $0.5 \times 0.5 = 0.25$. Similarly, the chance that you each inherited the same allele from your father is 0.25. So the chance that you and your sister share any allele is $0.25 + 0.25 = 0.5$, exactly the same as the chance that you share an allele with your mother or father. A similar calculation shows that you share one-eighth of your alleles with your first cousins. The British evolutionary biologist J.B.S. Haldane summarized these relationships by joking, "I would gladly lay down my life for two of my brothers or eight of my first cousins."

As in the case of ordinary sex determination, the queen passes half her alleles to her female offspring. However, since males are haploid, a male passes on all his alleles. That is, his daughters inherit all of his genes. As a result, the (female) workers share 75 percent (rather than 50 percent) of their alleles with one another. Because the queen mates with many different males, not all of her offspring share the same father. However, the average relatedness will be greater than 50 percent.

If the workers were to have their own offspring, they would pass on 50 percent of their alleles. They can therefore pass on more of their genes by putting their energies into making sisters than they can by reproducing and making daughters. To put it another way, the workers more efficiently perpetuate their genes by helping their mother, the queen, produce more offspring (including new queens) than by producing eggs themselves.

Haplodiploidy in some social insects probably facilitates kin selection.

Does inclusive fitness always explain altruistic behavior?

The haplodiploid social insects are not the only animals that help one another. Grazing animals, birds, and other animals that live in groups commonly help protect one another. In as many as 300 species of birds, "helpers" may assist a breeding pair feed and protect the nestlings. Whether such behavior can be explained by kin selection, however, depends on how closely the altruists are related to one another.

A few studies suggest kin selection as an explanation for altruism. For example, when a pair of Florida scrub jays breeds, as many as six other jays may help feed and protect the nestlings. These helpers are usually the breeding jays' offspring from previous clutches, offspring that have yet to mate themselves. Because territories suitable for breeding are in short supply, these helpers may not be able to breed effectively. Like the honeybee workers, the young jays may be better off helping their siblings than trying to breed themselves.

An even more persuasive example comes from Belding ground squirrel populations high in the Sierra Nevada, in California. In the summer, female squirrels defend territories where they raise their offspring. The two biggest dangers to Belding ground squirrels, besides winter snows, are predators and other squirrels. Unrelated or distantly related squirrels regularly kill other ground squirrels, killing up to 8 percent of all pups. At the approach of a coyote or strange ground squirrel, some female squirrels stand up tall and sound a shrill alarm

BOX 29.1

Tit for Tat and Prisoner's Dilemma

The classic game theory analysis of the risks and benefits of cooperation and altruism is presented in the "game" called prisoner's dilemma. Imagine that you and a friend are arrested for committing a crime, and you are held in separate jail cells, unable to communicate. The police immediately commence interrogating the two of you.

Suppose you claim innocence, but your friend confesses and incriminates you. You'll be in prison for a long time, but your friend will get off scot-free. On the other hand, you could get off yourself if you incriminate your friend and your friend pleads innocent. If you each confess (and incriminate each other), the state's job of convicting you will be easy, and your sentences will be light. But if you both plead innocent, the state will have little evidence against you, and your sentences will be even lighter.

What to do? Regardless of what your friend does, you are better off confessing, which is true for your friend, too. On the other hand, if both of you confess, you will be worse off than if both of you pleaded innocent. The safest solution is to confess. Even though pleading innocent might get you off scot-free, it's the riskiest solution to the problem.

The game of prisoner's dilemma seems to show that selfish behavior is safer than cooperative behavior. In 1981, however, a political scientist named Robert Axelrod and a biologist named W.D. Hamilton combined forces to explore what happens when the game is played over and over. They reasoned that in real life, people and animals interact repeatedly. Perhaps cooperation might pay off in the long run.

To make the endeavor more interesting, Axelrod invited game theorists from sociology, political science, economics, mathematics, and biology to design strategies to compete in an ongoing computer tournament. Predictably, players that always cooperated were taken advantage of. But strategies of constant betrayal also did badly.

The winning long-term scheme was a simple strategy called tit for tat. In tit for tat, the player always cooperates (pleads innocent) in the first round. After that, the player does whatever the opponent did in the last round. In short, you reward your friend for cooperating and punish your friend for ratting on you. Over time, tit for tat always won.

In the early 1990s, an even more successful strategy emerged from Axelrod's tournament. "Modified tit for tat" occasionally forgives other players for a betrayal, just often enough to nudge the game in the direction of cooperation and reciprocity. Although the prisoner's dilemma is a simple game, it suggests that reciprocity is a possible explanation for altruism in some situations.

---

call that alerts other nearby squirrels (Figure 29-15). Other individuals who first sight a coyote or a stranger make no alarm and simply dive into the nearest burrow silently, leaving the other squirrels to fend for themselves. The squirrels that sound the alarm increase their own chance of being attacked, so their behavior is altruistic.

Paul Sherman, a biologist now at Cornell University, asked why these squirrels risk their own lives to help others, and why some squirrels behave altruistically, while others do not. Sherman discovered that most of the squirrels that sound the alarm are yearling and older females, who are warning their female relatives—sisters, mothers, and daughters, as well as juvenile sons. However, males and childless females rarely warned other squirrels. The older female incurs a risk that is counterbalanced by the chance that she will save the life of a close relative. Her altruistic behavior may contribute to her inclusive fitness.

In the past 20 years, many biologists, anthropologists, and other enthusiasts have tried to extend the concept of kin selection to virtually all altruistic behavior. But not all examples of altruistic behavior conform to the kin selection model. Dwarf mongooses, for example, help breeding pairs care for their young, but not all of the helpers are relatives. Kin selection does not account for such altruism.

There are other explanations for why animals behave altruistically, but one of the most interesting was first suggested by ethologist Robert Trivers. Trivers argued that animals capable of recognizing one another might engage in altruism on the understanding that the recipient would return the favor sometime in the future. Trivers called this idea **reciprocal altruism**, since the recipient reciprocates. Such a system at first seems too susceptible to cheaters to work. Biologists wondered, How can an animal ensure that the recipient of an altruistic act will return the favor? But research in the area of "game theory," a branch of mathematical logic applied to business, military, sociobiological, and other problems, suggests that altruism can pay off in spite of cheaters (Box 29.1).

Kin selection may explain altruism in other animals besides the social insects. But other explanations, such as reciprocal altruism, are possible, as well.

A. B.

Figure 29-15 How kinship affects behavior. A. At Tioga Pass, California, students watch individual Belding ground squirrels marked with numbers written with hair dye. These ground squirrels live in colonies consisting of related females and their offspring. B. Older females sound a loud alarm when predators approach. Males take no such chances and scurry silently away. *(A and B, George D. Lepp)*

Study Outline with Key Terms

The systematic study of natural (wild) behavior is called **ethology.** In contrast, the laboratory study of behavior common to many animals is called **experimental psychology.** Social and other forms of behavior viewed in the context of evolution and ecology is the concern of the field of **sociobiology,** now called **behavioral ecology.**

Animal behavior is partly influenced by genes, partly by environment. Humans and other animals learn from one another as well as from their environment. Discussions of animal behavior can quickly become politicized.

Biologists are very interested in the relative importance of genes and environment in a variety of animal behavior. If genes help to shape behavior, and natural selection operates on behavioral variants, then biologists must ask, What ecological factors determine the adaptiveness of various kinds of behavior?

Innate behavior includes a baby's first smile and the salivating of a dog over his dinner bowl (a **conditioned reflex**), but all behavior in all animals can be modified through **learning.** A **modal action pattern** is highly stereotyped innate behavior performed in response to a **sign stimulus.** An exaggerated sign stimulus is a **supernormal stimulus.** Stimuli may be sights, sounds, tactile experiences, or smells in the form of **pheromones.** Some forms of learning, called **imprinting,** occur only at special times in development, called **sensitive periods.**

Animals seem to accurately weigh the costs and benefits of different foraging strategies in a process called **optimal foraging.** When animals compete for mates or resources, they engage in **agonistic behavior,** which includes aggression, appeasement, and retreat. Animals such as chickens, dogs, and primates use agonistic behavior to establish **dominance hierarchies.** Competition for mates results in **sexual selection.** Many animals defend a **territory,** which may encompass space and resources for mating, foraging, or raising young.

Social groups allow animals to spot both prey and predators more easily. However, groups of animals are themselves more visible to predators than are solitary animals. Also, the individuals in a group must constantly compete with one another for food, mates, and breeding areas.

The social insects display complex social structure and **altruism.** One explanation for the adaptive value of altruism is **inclusive fitness,** which includes the concept of **kin selection.** All of the social insects except the termites are **haplodiploid,** which means that sisters are more closely related to one another than to their own offspring. Such extreme relatedness facilitates kin selection and altruism. Another explanation for altruism is **reciprocal altruism.**

Review and Thought Questions

Review Questions

1. In the past, what distinguished ethology from experimental psychology?
2. Define a modal action pattern and give an example.
3. What is a pheromone?
4. Give an example of a supernormal stimulus in advertising you have seen.
5. Give an example of a sensitive period.
6. Define and give an example of optimal foraging.
7. Distinguish between aggression and agonistic behavior.
8. Give an example of territorial behavior in dogs.
9. Why, according to the theory of inclusive fitness, are honeybee workers better off not reproducing?
10. Give an example of an innate response modified by learning in an adult human.

Thought Questions

11. Three pizza parlors offer specials on identical giant pizzas. One parlor will deliver the giant pizza for $25. The other two don't deliver. One is just around the corner, and its pizza is $20. The other is 8 miles away and charges $12. It costs you 50¢ a mile to drive your 1980 Buick Skylark, and the car needs a new battery. If you break down in the neighborhood where the $12 pizza is, you might get mugged. What is the optimal solution to this foraging problem?
12. If humans were haplodiploid, how would human societies be different?

About the Chapter-Opening Image

English birds called tits learned from each other how to uncap milk bottles and sip the cream from bottles left on the front steps of English houses.

 On-line materials relating to this chapter are on the World Wide Web at **http://www.harcourtcollege.com/lifesci/aal2/**

 Web Links

 Quizzing & Testing

Appendix A
Periodic Table of the Elements

Appendix B
Units of Measure: The Metric System

Length
Standard Unit = Meter
1 meter (m) = 39.37 in
1 centimeter (cm) = 0.39 in   1 inch = 2.54 cm
1 kilometer (km) = 0.62 mi   1 mile = 1.61 km

Prefixes and Units of Length

Prefix	Meaning	Unit
kilo	thousand	kilometer (km) = 1000 m
centi	one-hundredth	centimeter (cm) = 0.01 m = 10^{-2} m
milli	one-thousandth	millimeter (mm) = 10^{-3} m
micro	one-millionth	micrometer (μm) = 10^{-6} m
nano	one-billionth	nanometer (nm) = 10^{-9} m

These prefixes are also used in units of volume and mass.

Volume
Standard Unit = Liter = 1000 cm³
1 liter (l) = 1.06 qt   1 qt = 0.94 l
1 milliliter (ml) = 0.03 fluid oz 1 fluid oz = 30 ml
1 l = 0.26 gal   1 gal = 3.79

Mass
Standard Unit = Kilogram = 1000 grams
1 kilogram (kg) = 2.21 lb   1 lb = 453.6 grams (g) = 0.45 kg

Temperature
Standard Unit = Centigrade
°C = 5/9 (°F − 32)   °F = 9/5(°C) + 32
Interval equivalents: 1 °Centigrade (°C) = 1.8 °F

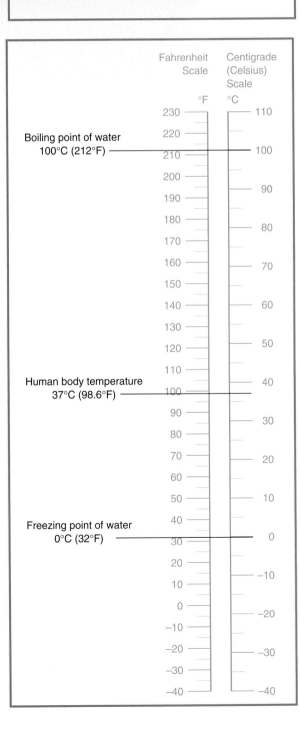

Glossary

3′ end The end of a polynucleotide at which the 3′ carbon of the nucleotide is not attached to another nucleotide but to a phosphate or a hydroxyl group.

5′ end The end of a polynucleotide at which the 5′ carbon of the nucleotide is not attached to another nucleotide but to a phosphate or a hydroxyl group.

9 + 2 pattern The pattern of microtubules in cilia and eukaryotic flagella; the pattern consists of nine doublet microtubules, each consisting of two microtubules fused along their length plus two singlet microtubules in the center.

α (alpha) carbon In an amino acid, the carbon atom to which the carboxyl group and the amino group are both attached.

α (alpha) helix A common secondary structure in proteins in which every carbonyl group of the polypeptide backbone is hydrogen bonded to the amino group four amino acids farther down the polypeptide chain.

A band (for *anisotropic*, meaning that it changes the direction of polarized light) In muscle, the region of a sarcomere that contains myosin thick filaments.

abdomen In a mammal, the part of the body between the chest (thorax) and the pelvis; the rear portion of an arthropod.

abdominal cavity The part of the coelom that encloses most of the intestinal tract, as well as the liver, pancreas, and kidneys.

abiotic Nonliving. Without life.

abscisic acid The plant hormone that causes the abscission of leaves, dormancy in buds and seeds, and the closing of the stomata (by stimulating the flow of potassium ions out of the guard cells).

absolute age The age in years of each of the layers of sedimentary rock.

absorb To take up into cells and into the circulation.

absorption spectrum The relative amounts of light of different wavelengths that a substance or solution absorbs.

absorptive state The state of an animal shortly after a meal, during which cells in many organs take up glucose and other small molecules to make glycogen, fats, and proteins.

abstinence Referring to birth control, refraining from sexual intercourse.

abyssal zone Ocean waters deeper than 1000 meters.

acetyl CoA A compound that consists of the two-carbon acetic acid (acetate) linked to a larger molecule called coenzyme A; acetyl CoA is the end point of glycolysis and a starting compound for the citric acid cycle.

acetylcholine A neurotransmitter used in the neuromuscular junction to stimulate muscle contraction; acetylcholine is also used by parasympathetic nerves to slow the heart.

acetylcholinesterase An extracellular enzyme that destroys acetylcholine soon after it is released into a synapse.

acid A molecule (or part of a molecule) that can give up a hydrogen ion.

acoelomates [Greek, *a* = without] The simplest animals, including the jellyfish and flatworms, that have no coelomic cavity.

acquired immunodeficiency syndrome (AIDS) A disease characterized by a lack of CD4 T_H cells; AIDS patients cannot mount effective immune responses to infections or to certain kinds of cancers.

acrasiomycota [Greek, *acrasia* = confusing two things + *mukes* = fungus] Cellular slime molds—peculiar organisms that resemble animals at some stages of their lives and plants at others.

acridine dye A chemical mutagen; it causes mutations by inserting itself into the backbone of the DNA double helix.

acrocentric Refers to a chromosome in which the centromere is near one end; such a chromosome looks like a bent capital L.

acrosome A specialized lysosome in a spermatozoan; it contains enzymes that allow the sperm to enter the egg.

actin The protein that makes up actin filaments in nonmuscle cells and the thin filaments in muscle cells.

actin filaments The most flexible elements of the cytoskeleton; formed by the association of actin molecules into long filaments about 7 nm in diameter.

action potential (also called a **nerve impulse**) A rapid, transient, and self-propagating change of voltage across the membrane of a neuron or a muscle cell; action potentials allow long-distance signaling in the nervous system.

action spectrum The relative effectiveness of different wavelengths in promoting a specific light-dependent process.

activation energy The minimum energy needed to initiate a chemical reaction or other process.

active site A groove or cleft on an enzyme's surface to which a substrate binds.

active transport Movement of a substance across a membrane in a manner that does not occur spontaneously but requires an expenditure of energy.

adaptation A genetically inherited structure or behavior acquired through the process of natural selection that contributes to the ability of an organism to live in its particular environment.

adaptive Contributing to the fitness of a species.

adaptive radiation The multiplication of a single species into many species, each with a separate ecological niche.

adenosine diphosphate (ADP) [Greek, *di* = two] A nucleotide that consists of adenosine, ribose, and two phosphate groups; ADP is the product of the hydrolysis of a single phosphate from ATP.

adenosine triphosphate (ATP) A nucleotide that consists of adenosine, ribose, and two phosphate groups; ATP is the universal energy currency, providing energy for many biochemical processes in all organisms.

adenylate cyclase The enzyme that catalyzes the conversion of ATP into cyclic AMP.

adhering junction A molecular assembly on the surface of an animal cell that connects the cell surface to actin filaments on the inner surface of the plasma membrane.

adhesiveness The tendency to cling to a surface.

adrenal glands [Latin, *ad* = towards + *renes* = kidneys] Endocrine glands that lie on top of the kidneys.

adrenal virilism The development of male secondary sexual characteristics in women with excessive adrenal gland activity.

adrenergic receptor [from Adrenaline (as in *adrenal* gland), the British trade name for **epinephrine**] One of a set of membrane proteins that binds to epinephrine and related molecules, thereby initiating a biological response.

aequorin A protein, isolated from a luminescent jellyfish, that emits light when it binds Ca^{2+} ions.

aerobic Requiring oxygen, usually for aerobic respiration.

afferent Leading towards, as blood vessels entering an organ or nerve fibers carrying sensory information from peripheral nerves or sense organs to the central nervous system.

age structure In a population, the fraction of individuals of various ages (or cohorts).

aging Progressive developmental changes in an adult organism.

agonistic behavior [Greek, *agonistes* = champion] All the aspects of competitive behavior within a species, including aggression, aggressive displays, appeasement, and retreat.

alchemy An ancient study, one of whose aims was to turn "base" metals, such as lead or copper, into silver or gold.

aldosterone A hormone, made by the adrenal glands, that stimulates sodium and potassium reabsorption in the distal tubules and collecting ducts of the kidneys.

aleurone layer In a seed, a border of cells that surrounds the endosperm.

algae Plantlike protists that perform photosynthesis using chlorophyll *a*.

alkaloid A nitrogen-containing secondary plant compound.

allantois One of the membranes that surrounds an amniotic embryo; the allantois functions in both respiration and excretion.

allele One of several variant versions of the same gene.

allele frequency The number of times the allele is present in the population divided by the total number of chromosomes on which the gene appears.

allergy An inflammatory response to a harmless antigen; allergies involve the production of IgE antibodies and the release of histamine from mast cells.

allopatric speciation [Greek, *allos* = other + *patra* = country] Species formation by geographical isolation.

allopolyploid Description of the chromosome set that occurs when a hybrid doubles its chromosomes.

allosteric effector (because it acts other than sterically) A molecule or ion that changes the activity of an enzyme by binding to a site different from the active site.

altered self In the immune response, nonself antigens presented in the context of an MHC molecule. These occur on the surface of a body cell when the cell is infected by a virus, bacterium, or other microorganism.

alternate (also called **spiral**) The arrangement of leaves on a stem in which the bases of the petioles form a spiral up the stem.

alternation of generations A sexual life cycle in which haploid and diploid phases alternate.

altruism Behavior that benefits others at the expense of the animal that performs the behavior.

alveoli [singular, **alveolus**; Latin, *alveus* = a hollow] The expanded surfaces at the ends of the smallest passageways of the lungs; they provide the huge surface area for the exchange of oxygen and carbon dioxide.

Alzheimer's disease A disease, associated with aging, in which patients suffer massive memory loss and disorientation.

ameboid movement (also called **cell locomotion** or **cell crawling**) The movement of an individual cell over a surface by extension of pseudopodia.

amine A chemical compound that contains an amino group; many chemical signals are amines, derived from amino acids by the removal of a carboxyl group.

amino acid A small molecule that contains both amino and carboxyl groups; the building block of polypeptides and proteins.

amino group A functional group that consists of a nitrogen and two hydrogen atoms (NH_2).

aminoacyl tRNA synthase A "decoding enzyme" that links a transfer RNA to its corresponding amino acid to form a "charged" (aminoacyl) tRNA.

aminoacyl-tRNA A transfer RNA molecule linked at its 3′ end to the appropriate amino acid.

ammonia A small molecule that consists of one nitrogen and three hydrogen atoms; it is secreted as a nitrogen-containing waste by many cells and organisms.

amniocentesis [Greek, *centes* = puncture] Sampling of cells—usually with a syringe—in the amniotic fluid.

amnion One of the membranes that surrounds a reptilian, avian, or mammalian embryo.

amniotic egg An egg in which the embryo is surrounded by an amnion within a porous shell that allows the exchange of gases with the surrounding air.

amniotic fluid [Greek, *amnion* = membrane around a fetus] The fluid that surrounds the fetus.

amphibians Class of four-legged vertebrates that usually must reproduce in water.

amphipathic [Greek, *amphi* = both + *pathos* = feeling] Having both a hydrophilic and hydrophobic region.

amylase An enzyme found in saliva and pancreatic juice that hydrolyzes the polysaccharides of food into shorter fragments of maltose and glucose.

amyloplast In a plant, a plastid that contains large granules of starches.

anabolism [Greek, *ana* = up + *ballein* = to throw] The synthesis of complex molecules from smaller ones.

anaerobic Without oxygen.

anaerobic photosynthetic bacteria Photoautotrophic bacteria that live without oxygen.

anagenesis (also called **phyletic evolution**) Evolution within a single lineage.

analogous Referring to structures that perform the same function though derived from different ancestral structures.

anaphase [Greek, *ana* = up, again] The stage of mitosis during which sister chromatids separate.

anaphase A The part of anaphase during which the chromosomes move.

anaphase B The part of anaphase during which the poles of the spindle move apart.

analogy See homoplasy.

anastral Having neither centrioles nor asters; a characteristic of mitosis in vascular plants.

anchoring junctions In animal cells, molecular assemblies that attach cells to each other or to extracellular matrix; anchoring junctions help provide strength to resist mechanical stress while not affecting the passage of molecules between them.

aneuploid Having an abnormal number of chromosomes.

aneurism Abnormal ballooning of the wall of a blood vessel.

angiosperms [Greek, *angion* = vessel + *sperma* = seed] The flowering plants.

angiotensin A hormone that stimulates the constriction of blood vessels.

Animalia The kingdom of animals—multicellular, heterotrophic organisms that undergo embryonic development.

anion (so called because it will move toward a positively charged electrode, or anode) A negatively charged ion.

annual rate of increase The actual percentage by which a population increases each year.

annuals Plants that complete their life cycles, from seed to mature plant, in a single growing season and then die.

antagonist A molecule that prevents a chemical signal from binding to its receptor and triggering its characteristic response in target cells.

antagonistic pair Muscles that pull in opposite directions.

antenna complex In a chloroplast, an association of chlorophyll and carotenoids that traps light and transfers the energy to the chlorophyll molecules that actually participate in photosynthesis.

anterior Toward the front of an animal.

anterior pituitary gland An endocrine organ that makes and releases a number of peptide hormones.

anther A thick pollen-bearing structure that is part of the stamen.

antheridium In plants and protists, the structure that produces sperm.

anthocyanins The phenolic pigments responsible for most of the red, pink, purple, and blue colors of flowers, fruits, leaves, and other plant parts.

Anthophyta [Greek, *antho* = flower] Angiosperms.

Anthozoa [Greek, *antho* = flower + *zoa* = animal] Sea anemones, corals, and related species.

anthropoid A monkey, ape, or human.

antibiotic A substance that kills (or interferes with the growth of) microorganisms.

antibody A blood protein that forms complexes with molecules (antigens), such as those on the surfaces of microorganisms.

anticodon A sequence of three nucleotides in transfer RNA that forms specific base pairs with the corresponding codon sequence in mRNA.

antidiuretic hormone (ADH) (also called **vasopressin**) A hormone, released by the posterior pituitary gland, that prevents water loss in the kidneys and constricts blood vessels.

antigen processing In the cellular immune response, degradation of an antigen into small peptides that associate with an MHC molecule.

antigen (an *anti*body *gen*erator) A molecule that stimulates the production of an antibody or of another immune response; a molecule that binds to an antibody or other recognition molecule within an animal's immune system.

antigenic determinant (also called an **epitope**) A specific shape and charge distribution to which antibodies or other recognition proteins can bind.

antioxidants Natural plant compounds, such as vitamin E, vitamin C, and β-carotene, that prevent the damaging effects of oxygen.

anus The opening at the far end of the digestive tract through which the undigested and unabsorbed material leaves.

aorta The artery that carries blood from the left ventricle to the rest of the body.

apical dominance In plant development, the suppression of lateral bud development by terminal buds.

apical meristem [Latin, *apex* = top + Greek, *meristos* = divided] In a plant, a self-renewing group of undifferentiated cells, just behind the tip of the shoot or root, that generates differentiated structures.

apoptosis [Greek, *apo* = away from + *ptosis* = fall] Programmed cell death.

aqueous Watery; dissolved in water.

Archaea (also called **Archaebacteria**) [Greek, *archein* = to begin] The kingdom (or "domain") of single-celled organisms that live under extreme environmental conditions and have distinctive biochemical features.

archegonium In ferns and mosses, the female reproductive organ that produces and houses the ova.

archenteron [Greek, *arche* = beginning + *enteron* = gut] The "primitive gut," the innermost tube of an animal embryo; it is lined with endoderm and will become the digestive tract.

aromatic Having a benzene-like planar ring of atoms.

artery One of the vessels that carry blood away from the heart.

arterioles Smaller vessels that branch off from arteries.

artifact A remnant or result of what the researcher does rather than of what exists in nature.

artificial selection The process by which animal breeders and farmers create new lines of plants and animals by selecting the most desirable individuals for breeding.

Aschelminths A group of eight phyla, all bilaterally symmetrical animals with either no coelom or a pseudocoel.

asci (singular, **ascus**) The little sacs in which ascomycetes produce their sexual spores.

ascomycetes (also called **sac fungi**) [Greek, *askos* = wineskin + *myketos* = fungus] So named because their sexual spores are always produced in little sacs.

asexual reproduction A process that produces offspring with genes from a single parent.

associative learning A type of learning in which the subject learns to respond to a stimulus (called the conditioned stimulus) that is not obviously related to the primary stimulus; the response to the conditioned stimulus arises because of its association with a stimulus that leads directly to a physiological response.

assortative mating Nonrandom mating.

aster [Latin, star] A starlike object visible in most dividing eukaryotic cells (other than those of vascular plants); the aster contains the microtubule organizing center.

astral Having an aster and centrioles that participate in mitosis; characteristic of mitosis in animals and in nonvascular plants.

astral microtubules Microtubules that extend from each pole of the mitotic spindle without attaching to any other visible structure.

asymmetric A carbon atom surrounded by four different groupings of atoms.

atherosclerosis Hardening of the arteries.

atom The smallest unit of matter that still has the properties of an element.

ATPase An enzyme that hydrolyzes ATP into ADP and phosphate.

ATP synthase An enzyme responsible for making ATP.

atrial natriuretic factor [Latin, *atrium* = vestibule, referring to the heart chamber, + *natrium* = sodium + Greek, *ouron* = urine] A hormone produced in the smaller chambers of the heart (the atria).

atrioventricular (AV) node A special tissue that conducts electrical signals from one part of the heart to the other.

atrioventricular valves The valves between the atria and ventricles of the heart.

atrium [Latin, courtyard or entry] One of the smaller thin-walled entrance chambers of the heart, through which blood enters the adjacent ventricle.

australopithecine [Latin, *australis* = south + Greek, *pithekos* = ape] An early hominid.

autonomic nervous system Involuntary nervous system; it coordinates the responses of smooth muscles, cardiac muscles, and other effector organs—including those of the endocrine, digestive, excretory, respiratory, and cardiovascular systems.

autonomous replication Propagation of a virus or plasmid in which the replicating DNA remains separate from the DNA of the host cell.

autoradiography A method that detects radioactive compounds by their ability to expose a photographic film or a photographic emulsion.

autosomes The chromosomes that do not differ between males and females.

autotroph [Greek, *autos* = self + *trophos* = feeder] An organism that can make its own organic molecules from simple inorganic compounds (such as carbon dioxide, water, and ammonia).

axillary bud (also called **lateral bud**) [Greek, *axilla* = armpit] A bud that forms just above the point where a leaf joins the stem and which can develop into a branch.

axon A type of neuronal process that usually carries signals away from the cell body and connects to other cells.

β (beta) structure A common secondary structure in proteins in which the amino and carbonyl groups of the polypeptide chain are hydrogen bonded to the other polypeptide chains or to distant regions of the same chain folded back on itself.

B lymphocyte Lymphocytes that when stimulated divide to form antibody-producing effector cells or memory cells.

bacilli [singular, **bacillus**; Latin, little rod] Rod-shaped prokaryotic cells.

backbone A column of hollow bony segments called vertebrae.

bacterial flagellum A long structure, about 10 to 20 nm thick and up to 10 μm long, whose rotation propels a bacterium through a liquid.

bacteriophage (or **phage**) [Greek, *bakterion* = little rod + *phagein* = to eat] A virus that infects bacteria.

balanced polymorphism A balance of different alleles in a population.

basal body The base of an axoneme; a cylinder about 500 nm long that resembles a centriole; the microtubule organizing center of a cilium or a eukaryotic flagellum.

basal metabolic rate The minimum amount of energy required just to stay alive and awake.

base A molecule (or part of a molecule) that can accept a hydrogen ion.

base substitution A type of point mutation in which one base (nucleotide) is replaced by another.

basidiocarp [Latin, *carpus* = fruit] The fruiting body of a mushroom.

basidiomycetes, or **club fungi** Include many of the most familiar fungi: the mushrooms, the bracket fungi, and the puffballs.

basidium [plural, **basidia**; Latin, little pedestal] A tiny, club-shaped structure that produces spores.

Batesian mimicry A deception in which one species mimics another's warning coloration for protection.

behavioral ecology (formerly called **sociobiology**) The study of behavior from an evolutionary perspective.

behavioral isolation Reproductive isolation that results from different mating behaviors.

benthic division All the organisms that live on the ocean bottom.

biceps The major muscle on the front side of the upper arm.

bilateral symmetry Structural symmetry such that one cut through the axis will produce identical (or mirrored) halves.

bile A detergent solution, secreted by the liver; bitter, alkaline, and an ugly green-yellow or brown-yellow color.

binary fission The process of cell division (in prokaryotes) in which a cell pinches in two, distributing its materials and molecular machinery more or less evenly to the two daughter cells.

binding site The region of a protein molecule to which a substrate or a chemical signal binds.

binomial A two-part name that includes both the genus and species.

bioassay A method that estimates the concentration of a substance by measuring its biological activity.

biochemistry (also called **biological chemistry**) The study of the structures and reactions that actually occur in living organisms.

biogeochemical cycle The movement of a substance, such as carbon or nitrogen, through many forms—both biotic and abiotic.

biogeography The study of the past and present distribution of plant and animal species.

biological species The largest unit of a population of similar organisms in which gene flow is possible.

Biological Species Concept A definition of species that says species are groups of actually or potentially interbreeding populations, which are reproductively isolated from other such groups.

biology [Greek, *bios* = life, *logo* = word] The science of life.

biomagnification Toxic chemicals are concentrated at higher levels in the food chain.

biomass Aggregate dry weight of all organisms in a community or ecosystem.

biome A geographical region characterized by a distinctive landscape, climate, and community of plants and animals.

biosphere The system of living things that covers the Earth.

biotic Living.

bipedal Consistently walking on two rather than four feet.

birth control The conscious regulation of reproduction.

birth control pill A formulation of estrogen and progesterone that prevents the anterior pituitary from secreting LH, thereby stopping ovulation and preventing conception.

birth rate The annual number of births per 1000 individuals in a population.

bivalent (also called a **tetrad**) A chromosome pair visible during meiosis I.

bivalves [Greek, *bi* = two + *valva* = part of a folding door] Includes clams, oysters, scallops, mussels, and other mollusks with two shells—a right shell and a left shell.

bladder Where urine is stored until its elimination.

blade The thin, flat part of a leaf.

blastocoel In an animal embryo, the interior of the blastula.

blastocyst [Greek, *blastos* = sprout + *cystos* = cavity] A modified blastula in which the cells do not lie within a single layer, but do enclose an internal cavity.

blastopore An opening at the end of the archenteron.

blastula [Greek, little sprout] A stage of an animal embryo that consists of a sphere with cells on the surface and fluid inside.

blending inheritance The pattern of inheritance in which offspring appear to have characteristics intermediate between those of their parents.

bolus [Greek, *bolos* = lump of earth] In digestion, a ball of macerated food and lubricating mucus.

bone The hard connective tissue of a vertebrate; bone consists of fibers of collagen and crystals of calcium phosphate.

book lung Stacks of modified gills that provide a surface of gas exchange in some animals, including spiders.

bordered pit pair A region of the border between two plant cells in the xylem where the adjacent secondary wall overarches the pit membrane and reinforces the wall of a tracheid.

bottleneck The reduction of a large population to a few individuals.

bottleneck effect The restriction of genetic diversity of a population that results from the reduction of a large population to a few surviving individuals by a random disaster or harsh selection pressure.

Bowman's capsule In a kidney, the bulb that surrounds each glomerulus.

BPG (2,3-bisphosphoglycerate) A relatively abundant 3-carbon molecule in mammalian red blood cells. The binding of BPG to hemoglobin decreases oxygen affinity, allowing it to release more oxygen in the tissues than would hemoglobin without bound BPG.

branchial arches In fishes and amphibians, the tissues between the gill slits.

breathing center A nerve complex on which the coordination of breathing depends.

bronchi In the mammalian respiratory system, the two tubes that branch from the tracheae and carry air to and from the lungs.

bronchioles In the respiratory system, the smallest branches of the bronchi.

Brownian motion (after Robert Brown, a 19th-century British surgeon and botanist) Jerky movements of small particles, reflecting the random thermal motion of molecules.

bryophytes [Greek, *bryon* = moss] The mosses and their relatives.

bud In a plant, the precursor of a leaf or a flower, consisting of several leaf primordia, separated by internodes whose cells have not elongated.

budding yeast (*Saccharomyces cerevisiae;* also called baker's yeast and brewer's yeast) A widely used experimental organism that divides by budding rather than by binary fission.

buffer A molecule that easily converts between acidic and basic forms by donating or accepting one or more hydrogen ions.

bulb of the vestibule One of the corpora cavernosa of the clitoris that corresponds to the penis's corpus spongiosum.

bulbourethral glands The glands that inject a small amount of mucus into the semen.

callus Undifferentiated tissue that forms at the cut surface of a plant.

calorie A measure of energy; the amount of energy needed to raise the temperature of one gram of water 1°C.

calyx The fourth and outermost whorl of leaflike parts called sepals.

Cambrian explosion The burst of diversification—unmatched in the history of the world—in which all of the modern animal phyla that have fossilizable skeletons appeared.

camouflage A defense against being eaten in which organisms mimic materials in their environment in an attempt to be invisible.

canopy In a tropical rain forest, the continuous layer of trees 30 to 40 meters above the ground.

capacitation The process in which the glycoprotein coat that covers the head of each sperm is dissolved by enzymes in the female reproductive tract.

capillary One of the minute blood vessels that brings blood in closest contact with tissues; capillaries connect the finest arterioles with the finest venules.

capsid A coat of protein on the outside surface of a virus.

capsule A gelatinous layer of polysaccharides and proteins outside the cell wall; in disease-causing bacteria, the capsule protects from attack by the host's white blood cells, enabling the bacteria to overwhelm the body's immune system.

carbohydrate A compound that contains the equivalent of one water molecule (one oxygen atom and two hydrogen atoms) for every carbon atom; includes sugars and polysaccharides.

carbonyl A functional group that consists of a carbon atom attached to an oxygen atom by a double bond (C=O).

carboxyl A functional group that contains one carbon atom and two oxygen atoms; the carbon atom forms a double bond with one oxygen atom and a single bond with the other.

carcinogen A chemical that causes cancer.

cardiac muscle The muscles that pump blood through the heart.

cardiovascular system The blood, the heart, and the blood vessels together.

carnivore [Latin, *carn* = flesh] An animal that eats only other animals.

carotenoid A plant pigment, yellow or orange in color.

carpel (also called a **pistil**) In a flower, a female reproductive structure.

carrying capacity The plateau value of the S-shaped curve of population growth. The maximum number of individuals of a species that a habitat can support.

cartilage A type of connective tissue that cushions joints; cartilage consists of collagen, but without calcium phosphate.

Casparian strip In plants, a waxy wall that extends around the walls of endodermal root cells; it restricts the movement of water and dissolved solutes.

catabolism [Greek, *cata* = down + *ballein* = to throw] The reactions of metabolism that break down complex molecules, such as those in food.

catabolite activator protein (CAP) A positive regulator of transcription in *E. coli;* CAP binds to the promoter region of the *lac* operon and increases transcription whenever cyclic AMP is present at sufficiently high levels.

catalyst A substance that accelerates a chemical reaction and is not itself consumed in that reaction.

catarrhines [Greek, *kata* = down + *rina* = nose] Monkeys that have downward pointing nostrils and often no tail at all.

catastrophism The view that the discontinuities in the fossil record result from a series of catastrophes; some catastrophists argued that life arose anew after each catastrophe, rather than deriving from previously existing species.

cation (so-called because it will move toward a negatively charged electrode or cathode) A positively charged ion.

cecum [Latin, *caecus* = blind] A blind sac of the digestive tract, near the junction between the large and small intestine.

cell body The center of a neuron, including its nucleus, as distinct from its axon or dendrites.

cell center (also called the **centrosome**) The microtubule organizing center; a complex of proteins that in animal cells includes a pair of centrioles.

cell cortex A meshwork of actin filaments just below the surface of a eukaryotic cell.

cell crawling (also called **cell locomotion** or **ameboid movement**) The movement of cells over a surface.

cell cycle The orderly sequence of events that accomplish cell reproduction.

cell division The process by which a parent cell gives rise to two daughter cells that carry the same genetic information as the parent cell.

cell locomotion (also called **cell crawling** or **ameboid movement**) The movement of cells over a surface.

cell plate In a dividing plant cell, the precursor of a new cell wall between two daughter cells.

cell theory The summary of the cellular basis of organisms: (1) All organisms are composed of one or more cells; (2) cells, themselves alive, are the basic living unit of organization of all organisms; and (3) all cells come from other cells.

cell wall An external rigid structure surrounding all plant cells and most prokaryotes.

cellular oncogene (also called a **proto-oncogene**) The cellular counterpart of a viral oncogene; cellular oncogenes participate in the normal control of growth and differentiation.

cellulose A polysaccharide made by plants; the major structural material of wood, cotton, and paper.

central cell A cell within the ovule of a flower; it contains two haploid nuclei, derived from the megaspore and develops into endosperm after fertilization.

central dogma The frequently violated principle that "DNA specifies RNA, which specifies proteins."

central nervous system (CNS) The brain and spinal cord.

central vacuole A large membrane-enclosed space within a plant cell.

centrifugation A method for separating macromolecules and subcellular structures according to size (and shape) by subjecting them to high gravitational fields in a spinning tube.

centriole A pair of small cylindrical structures, each about 0.2 μm in diameter and 0.4 μm long, that lie at right angles to one another; centrioles are present at each pole of the mitotic spindle in animal cells and in some other eukaryotes.

centromere The point at which the two chromatids of a single chromosome are joined.

centrosome (also called the **cell center**) A microtubule organizing center that (in many cells) also contains a distinctive organelle called the centriole; the centrosome plays an important role in cell division.

cephalization The concentration of sense organs and ganglia in the anterior end of an animal.

cephalochordates (also called lancelets) [Greek, *cephalo* = head] About 45 species of small, segmented, fishlike animals. Most belong to the single genus *Branchiostoma*, whose common name is amphioxus.

cervical cap A birth control device; a thimble-shaped rubber or plastic cap, about an inch in diameter, that fits tightly over the cervix and prevents fertilization.

cervix [Latin, neck] The narrower, lower part of the uterus.

cestoda Tapeworms.

chaetognatha Arrow worms.

channel A membrane protein that allows the passage of specific molecules or ions.

chaparral [Basque, *chabarro* = dwarf evergreen] The dense growth of shrubs that is characteristic of the American Southwest and other regions, all on the west coasts of continents, that share a "Mediterranean" climate, with mild, wet winters and hot, dry summers.

chaperone One of a number of special proteins that prevent promiscuous interactions by binding to unfolded polypeptides and catalyzing correct folding.

character displacement In evolution, a change in morphology, life history, or behavior that results from competition.

chelicerae The first pair of appendages in spiders and other arachnids; they may serve as pincers or as fangs that are associated with poison glands.

chemical equilibrium [Latin, *aequus* = equal + *libra* = balance] A balance between forward and reverse chemical reactions.

chemical reaction A transformation in which different forms of matter combine or break down.

chemical synapse A distinct boundary between two neurons through which neurotransmitters diffuse.

chemical work The energy-consuming formation or breakage of chemical bonds.

chemiosmosis [Greek, *osmos* = to push] The linking of chemical and transport processes.

chemistry The study of the properties and the transformations of matter.

chemoautotroph An autotroph that derives energy by oxidizing such inorganic substances as hydrogen sulfide (H_2S) or ammonia (NH_3).

chemoheterotroph A heterotroph that uses no light, relying exclusively on organic molecules for both energy and carbon atoms.

chemotaxis The movement of a cell toward a higher (or, in some cases, a lower) concentration of a particular chemical.

chiasma [plural, **chiasmata**; Greek, cross] The sites of exchange of DNA between homologous chromosomes during meiosis; chiasma is visible during prophase of meiosis I.

chitin A tough, nitrogen-containing polysaccharide in the cell walls of insects, arachnids, and crustaceans.

chlorophyll The green pigment of plants and protists; chlorophyll is responsible for absorbing light for photosynthesis.

chlorophytes [Greek, *chloros* = green] Green algae.

chloroplast A large, green, membrane-enclosed organelle that performs photosynthesis.

chloroxybacteria Certain bacteria that live as photoautotrophs.

cholesterol A small molecule that consists of four interconnected rings of carbon atoms; a component of cell membranes and the starting compound in the synthesis of steroids.

chondrichthyes [Greek, *chondros* = cartilage + *ichthys* = fish] Sharks and rays—fish with cartilaginous skeletons.

Chordata All vertebrate animals—fish, amphibians, reptiles, birds, and mammals—and animals that have notochords, including sea squirts and lancelets.

chorion [Greek, skin] One of the membranes that surrounds an embryo; it lies just beneath the shell within an amniotic egg.

chorionic villus sampling A procedure for sampling cells of the early embryo that uses fetal cells present in the placenta.

chromatid One of the two separate but connected bodies that make up a chromosome at the beginning of mitosis, when the chromosomes first become visible; a chromatid contains a single long molecule of DNA.

chromatin [Greek, *chroma* = color] A diffuse material within the nucleus of a nondividing eukaryotic cell; chromatin consists of DNA and proteins.

chromatography A method for separating molecules according to their relative affinities for a stationary support, called the stationary phase, and a moving solution, called the mobile phase.

chromoplast A plastid that contains the pigments of fruits and flowers.

chromosomal mutation A change in a relatively large region of a chromosome.

chromosome [Greek, *chroma* = color + *soma* = body; because it is stained by certain dyes] A discrete complex of DNA and proteins, visible with a light microscope within a dividing cell; originally used only for eukaryotic cells, but now also used to mean a single large molecule of DNA that contains the genes of a bacterium or a virus.

chromosome banding The distinctive pattern, visible in a light microscope, that results from the selective binding of certain dyes to individual chromosomes.

chrysophytes [Greek, *chrysos* = golden + *phyta* = plant] Golden algae and diatoms.

chyme [Greek, *khumos* = juice] A creamy, acidic liquid that passes into the small intestine.

ciliophora [plural, **ciliates**; Latin, *cilium* = eyelash] Some 8000 species of ciliated free-living, single-celled heterotrophs that live both in fresh water and salt water.

cilium [plural, **cilia**; Latin, eyelid, from the hairlike appearance of a cilium] A protein assembly, consisting of microtubules, that can move a cell through a liquid medium (or a liquid medium over a cellular surface); a single cell usually contains many cilia, often arranged in rows; cilia have the same organizational plan as eukaryotic flagella but cilia are much shorter.

circular For DNA, forming a continuous loop; examples of circular DNAs include those of bacteria, energy-producing organelles, and many viruses.

circulation (from "circle") The route of the blood throughout the cardiovascular system.

circumcision The surgical removal of foreskin.

citric acid cycle (also called the **Krebs cycle** or the **tricarboxylic acid cycle**) A set of reactions that converts the carbon atoms of acetyl CoA into carbon dioxide.

cladistics [Greek, *clados* = branch] An approach to taxonomy whose first goal is to describe the groupings of organisms in a way that shows their phylogeny.

cladogenesis Branching evolution.

cladogram A family tree based on shared derived characters. The cladogram itself, which represents the relations among species, is the basis for classifying organisms, not the characters.

class A taxonomic group that consists of one or more orders; a subdivision of a phylum (or, for plants, of a division).

clay A soil that consists only of small particles (less than 2 mm).

cleavage A series of rapid cell divisions following fertilization in early animal embryos; cleavage divides the embryo without increasing its mass.

cleavage furrow A groove formed from the cell membrane in a dividing cell as the contractile ring tightens.

climax community The long-lived community at the end of a succession.

clitoris A highly sensitive erectile tissue that is part of the female genitalia; it is homologous to the penis.

cloaca A common entrance and exit chamber for the digestive, urinary, and reproductive systems in many vertebrates.

clonal selection theory In immunology, a summary of the role of selective proliferation of responding lymphocytes, already programmed, into clones of cells that contain specific gene rearrangements and that recognize specific antigens.

clone [Greek, *klon* = twig] A population of genetically identical individuals or cells descended from a single ancestor.

closed circulatory system A circulatory system in which blood runs only within enclosed vessels.

cnidarians [Greek, *cnide* = nettle] One of two phyla of radially symmetrical acoelomates; cnidarians include three classes: Hydrozoa (hydroids), Scyphozoa (jellyfish), and Anthozoa (corals and sea anemones).

cnidocytes Specialized stinging cells that lie on the tentacles of cnidarians.

coacervate [Latin, *coacervatus* = heaped up] Discrete tiny droplet into which proteins and polysaccharides can spontaneously concentrate.

cocci [singular, **coccus**; Greek, *kokkos* = berry] Spherical-shaped prokaryotic cells.

codominant Two alleles that each contribute to the phenotype of a heterozygote.

codon A group of three nucleotides that specifies a single amino acid.

coelom [Greek, *koilos* = hollow] A cavity lined by a layer of mesoderm cells.

coelomates Animals with a coelom.

coenocytic [Greek, *koinos* = shared + *kytos* = hollow vessel] Fungi in which the nuclei all lie in a common cytoplasm rather than in separate membrane-bound cells.

coenzyme An organic molecule (but not a protein) that is a necessary participant in an enzyme reaction.

coevolution The mutual adaptation of two separate evolutionary lines.

cohesion An attraction between molecules of the same substance.

cohort The set of individuals that enter the population (or are born) at the same time.

coitus interruptus [Latin, *coitus* = sexual intercourse] A method of birth control, also called withdrawal, in which a man attempts to avoid conception by removing his penis from the vagina before ejaculation.

collagen A fibrous protein, secreted by fibroblasts, that provides mechanical strength in cartilage, ligaments, tendons, and bones; collagen is by far the most abundant protein in the extracellular matrix.

collagen helix A regular structure found principally in the structural protein collagen; it consists of three polypeptide chains wound around each other.

collenchyma [Greek, *kolla* = glue] Plant cells that have a thick primary cell wall and often provide mechanical support; collenchyma are usually in the growing regions of stems and leaves.

colon The lower end of the large intestine; it is specialized for the absorption of water and ions.

commensal The species that benefits from the protection of its host in a commensalistic relationship.

commensalism An association between individuals of two species in which one organism benefits without harming the other one.

communicating junctions Membrane-associated structures that allow small molecules to pass freely between two adjacent cells.

community An interacting group of species that inhabit a common area.

compact bone tissue The hard, dense bone that surrounds the spongy interior.

companion cell In plants, a long, nucleated, fully functional parenchyma cell that supplies a sieve tube member with proteins and energy-rich molecules.

comparative anatomy The study of morphological similarities and differences among organisms; it often provides clues to evolutionary relationships.

competition The interaction between organisms or species that depend on the same limited resource; competition occurs when one organism or species uses a resource in a way that limits the availability of that resource to others.

competitive exclusion principle When two species compete directly for exactly the same limiting resources, the more efficient species will eliminate the other.

competitive inhibitor A molecule whose inhibitory effects on an enzyme can be overcome by increased substrate concentration.

complement A set of blood proteins that attack microbial invaders.

complete metamorphosis The pattern of development in which there are distinct larval, pupal, and adult forms; this pattern is characteristic of many insects, including *Drosophila,* fleas, flies, beetles, wasps, moths, and butterflies.

completed family size The average number of children that reach reproductive age born to each family.

complex tissue Tissue consisting of more than one cell type, each with a distinct functional role.

complex transposon A transposable element that contains other genes besides transposase, for example, a gene for antibiotic resistance.

compound A substance produced, by a chemical reaction, from two or more different elements.

compound eye An eye that consists of numerous simple light-detecting units; this type includes the eyes of most arthropod species.

compound microscope A light microscope that contains several lenses.

compression The pushing action of two opposing forces.

concave (or type III) survivorship curve Description of a life history in which individuals have the greatest chances of dying early in life.

concentration The number of molecules in a given volume.

concentration gradient (so named because the concentration changes in a graded manner) A graded difference in the concentration of a substance.

conception Fertilization or syngamy. The union of an egg and a sperm to form a zygote.

condensation (or **dehydration condensation**) **reaction** The linking of two building blocks, accompanied by the removal of a water molecule.

conditional mutation A change in DNA that alters a protein so that it can function under some conditions but not under others.

conditioned stimulus In associative learning, the arbitrary stimulus that is paired with a stimulus (the "unconditioned stimulus") that directly evokes a physiological response.

condom A thin rubber sheath that covers the penis and prevents sperm from entering the vagina; a condom is used both as a birth control device and to prevent the transmission of sexually transmitted diseases.

conformation The three-dimensional arrangement of a polypeptide chain, the equivalent of its tertiary structure.

conjugation A process in which two temporarily attached bacteria exchange genetic material.

connective tissue A relatively sparse population of cells within a bed of extracellular matrix; includes bone and cartilage.

conservative A possible pattern of DNA replication in which each parent molecule would remain intact and both strands of the descendant molecules would be newly assembled; actual DNA replication is not conservative, but semiconservative.

constant region The part of an immunoglobulin light chain or heavy chain that is identical in all antibodies of a particular class.

consumer An organism, usually an animal, that obtains energy by eating producers or other consumers.

contact inhibition of cell division The ability of cells to stop dividing when neighboring cells touch each other.

contractile ring A bundle of actin filaments that surrounds a dividing cell and pinches the cytoplasm in two during cytokinesis.

contrast In microscopy, an object's ability to absorb more or less light than its surroundings.

control In an experiment, the situation in which there has been no experimental manipulation; comparing measurements made in a control versus an experimental situation allows the evaluation of a hypothesis.

convergent evolution The independent evolution of similar features in separate groups of organisms.

convex (or type I) survivorship curve Description of a life history in which survivorship starts out high and decreases slowly with age until a certain point when survivorship begins to decrease more rapidly.

Coriolis effect The twisting effect of rotation on movement, particularly of wind or of water.

corolla In a flower, the whorl of petals.

corpus cavernosa In the male reproductive system of mammals, two spongy cylinders that run along the length of the penis and fill with blood during erection.

corpus luteum [Latin, yellow body] A structure within the ovary that develops from the ruptured follicle; it secretes progesterone and some estrogen.

corpus spongiosum A spongy cylindrical tissue that surrounds the urethra.

corridors Narrow strips of protected habitat designated for the purpose of allowing a species to travel from one preserve to another. Depending on the size of the corridor, the usual beneficiaries are small, slow moving animals such as frogs and salamanders or large predators such as mountain lions and wolves.

cortex The outer region of a cell, a tissue, or an organ.

cortical microtubules Microtubules that encircle a dividing plant cell; they determine the placement of a new cell wall.

corticotropin (also called **adrenocorticotrophic hormone** or **ACTH**) A peptide hormone, made in the anterior pituitary, that stimulates the production of corticosteroids in the adrenal cortex.

cotransport The coupled transport of two substances across a membrane; cotransport depends on specific transmembrane protein molecules.

cotransporter A transmembrane protein that binds to two or more molecules or ions and transports them across a membrane.

cotyledon (also called **seed leaves**) [Greek, *kotyle* = a deep cup] The very first leaves that sprout from a seed; present in the embryos of seed plants, they contain stores of nutrients for use after germination.

countercurrent system A system in which heat, fluids, or gases run past each other in a manner that allows the exchange of heat or matter.

coupling factor (also called the F_0-F_1 **complex**) A protein complex in the mitochondrial membrane that uses the flow of protons to drive the synthesis of ATP.

covalent bond Shared arrangement of electrons that holds atoms together in molecules.

crassulacean acid metabolism (CAM) A variant of the C_4 pathway of the light-independent reactions of photosynthesis; CAM is used by succulent plants such as cacti.

creationism The view of the origin of life that says that the Earth was created very recently and that each species was created individually.

cristae Elaborate folds of the inner membrane of a mitochondrion.

Cro-Magnon An early species of *Homo sapiens.*

crop In the digestive tract of birds, a saclike extension of the esophagus; the crop can store food for later digestion.

cross bridges Extensions of myosin molecules that perform the work of muscle contraction by pulling the thin and thick filaments over one another.

crossbreeding (or **crossing**) The interbreeding of two genetically distinct organisms.

crossing over One type of genetic recombination; it involves the breakage and rejoining of single chromatids of homologous chromosomes.

crustacea A subphylum of arthropods in which the first pair (or the first few pairs) of appendages are antennae, and the next pair are jaws, or mandibles.

crystal A solid that is enclosed by geometrically regular faces; the starting material point for x-ray diffraction analysis.

ctenophores [Greek, *ktenos* = comb + *phora* = motion] Comb jellies, a small phylum of about 90 living species.

cultural energy Energy expended by humans in crop production.

current The movement of electrical charges.

cuticle A waxy covering that keeps the above-ground parts of plants from losing water.

cyanobacteria Photosynthetic bacteria; previously called "blue-green algae"; the most ancient photosynthetic organisms known and the probable ancestors of chloroplasts.

cyclic AMP (cAMP) Adenosine monophosphate in which the same phosphate group is linked to carbon atoms 3 and 5 of ribose; it is used as a "hunger" signal in bacteria and protists and as a second messenger in mammalian cells.

cyclic AMP phosphodiesterase The enzyme that hydrolyzes cyclic AMP to AMP.

cyclic AMP-dependent protein kinase (also called **A-kinase**) An enzyme that, when bound to cyclic AMP, transfers phosphate groups from ATP to sites in particular proteins.

cyclic photophosphorylation The production of ATP from light energy by a series of electron transfers that regenerate the absorbing chlorophyll; in plants, cyclic photophosphorylation depends on the flow of electrons from excited P_{700} in a cycle that regenerates P_{700}, which is then able to absorb another photon of light.

cyst An enclosed structure that contains reproductive cells.

cytochalasin A poison that interferes with the growth of an actin filament.

cytochrome [Greek, *kytos* = hollow vessel + *chroma* = color] One of a set of heme-containing electron carrier proteins that change color as they accept or donate electrons.

cytodifferentiation The development of a specialized cell.

cytokine One of a number of small proteins that regulate proliferation and protein synthesis in cells that participate in inflammation and in the immune response.

cytokinesis [Greek, *kytos* = hollow vessel + *kinesis* = movement] The division of the cytoplasm and formation of two separate plasma membranes.

cytokinin A plant hormone that influences the rate of division and differentiation.

cytoplasm [Greek, *cyto* = a hollow container, i.e., a cell + Latin, *plasma* = a thing molded or formed] In eukaryotes, that part of the cell outside the nucleus but inside the plasma membrane.

cytoplasmic streaming The movements of membrane-bounded organelles along actin filaments.

cytoskeleton A network of protein fibers that runs through the cytosol of eukaryotic cells; it consists of microtubules, actin filaments, and intermediate filaments, along with other associated proteins.

cytosol In eukaryotic cells, the part of the cytoplasm not contained in membrane-bounded organelles.

cytotoxic T lymphocyte (CTL) An effector cell, derived from a T_C cell, that actually kills a target cell.

cytotoxicity In the immune response, the ability of certain T lymphocytes to kill cells with altered surfaces.

daughter cells The cells produced from a single parent cell after cell division.

dead space The volume of air in the air passages that does not come into contact with the surfaces of the alveoli in the lungs.

death rate The number of individuals per 1000 who die each year.

deciduous Shedding leaves annually, as in many trees and shrubs.

decomposer An organism, such as a bacterium or fungus, that lives on the energy in the complex molecules of dead organisms.

degenerate Meaning that several codons are equivalent, that is, that they specify the same amino acid.

dehydration A physiological state in which the body lacks sufficient water.

deletion A type of mutation; the removal of one or more nucleotides.

demography [Greek, *demos* = people + *graphos* = measurement] The statistical study of populations.

denatured A protein that has lost its native, three-dimensional structure and its functional activity, while still having an unaltered primary structure.

dendrite A type of neuronal process that usually carries signals to the cell body.

dendritic cell In the immune response, a type of antigen-presenting cell; a highly extended cell present in many organs and tissues throughout the body.

denitrifying bacteria Bacteria that convert nitrate to atmospheric nitrogen.

density gradient A changing concentration of a dissolved substance—usually sucrose, glycerol, or cesium chloride—with the highest concentration at the bottom and the lowest at the top.

density-dependent Referring to population growth, limited in proportion to population density—the denser the population, the more slowly it grows.

density-independent Referring to population growth, limited by factors other than population density—such as fire, drought, and other natural disasters.

deoxyhemoglobin Hemoglobin that has no bound oxygen.

deoxyribonucleic acid (DNA) The long thin molecules in which organisms store and transmit genetic information.

Depo-Provera (DMPA, depo-medroxy progesterone acetate) A birth-control pill; a progesterone derivative taken as an injection; it suppresses ovulation and inhibits implantation for three months at a time.

depolarizing A voltage change across an excitable membrane that makes the inside of the cell less negative and more apt to produce an action potential.

depurination The loss of a purine base from deoxyribose in DNA.

descent with modification Evolution; change in organisms over time.

desert A dry, relatively barren region, with lower productivity than most other biomes.

desmosome An anchoring junction that consists of transmembrane proteins that attach to a cell's intermediate filaments.

detergent An amphipathic molecule that interacts both with water and with hydrophobic molecules.

determined In development, having a limited potency. A determined cell can no longer develop in accordance with new environmental signals.

detritus Particles of dead organic matter.

deuteromycetes Imperfect fungi.

deuterostomes [Greek, *deuteros* = second + *stoma* = mouth] Those animals in which the anus forms first, from the blastopore, and the mouth forms secondarily.

diabetes mellitus A disease characterized by the overproduction of sweet (sugar-containing) urine.

diagonal (or **type II**) **survivorship curve** Description of a life history in which survivorship decreases in proportion to age, so that the curve is a straight, declining line.

diaphragm (1) A sheet of muscle beneath the lungs that separates the thoracic cavity from the abdomen. (2) A birth control device. A membrane of latex that prevents sperm from entering the cervix.

diarrhea [Greek, *diarrhein* = to flow through] Watery feces.

diastole [Greek, *dia* = between + *systellein* = to contract] The half of the cycle of the heart in which both the atria and the ventricles are relaxed.

dicotyledons (also called **dicots**) Flowering plants that have two cotyledons; there are about 170,000 species, including almost all trees and shrubs, but not the conifers and other gymnosperms, as well as many herbs.

differential gene expression The production of differing amounts of individual proteins (and RNAs) in different types of cells and at different times of development.

differentiation The process by which tissues and cells become specialized and different from one another.

diffraction The scattering of electromagnetic waves by regular structures so that their interference produces a pattern of lines or spots; the diffraction of x rays by a crystal produces a pattern of spots that can be analyzed to reveal the structure of the molecules within the crystal.

diffusion [Latin, *diffundere* = to pour out] Random movements of molecules that lead to a uniform distribution of molecules both within a solution and on the two sides of a membrane.

digestion The process of hydrolyzing large molecules into smaller units such as glucose, amino acids, fatty acids, and glycerol.

digestive tract (also called the **gastrointestinal tract**, the **alimentary canal**, or the **gut**) The tube, extending from mouth to anus, in which animals accomplish digestion, absorption, and elimination.

dikaryotic Having two nuclei per cell, as in certain fungi.

dinoflagellates [Greek, *dinos* = whirling + *flagellum* = whip] Single-celled organisms that float freely as plankton in warm oceans.

diploid [Greek, *di* = double + *ploion* = vessel] Having two sets of chromosomes.

directional selection In evolution, selection that shifts the frequency of one or more traits in a particular direction.

disjunction In meiosis, the moving apart of two homologous chromosomes during anaphase I and of two sister chromatids in anaphase II.

dispersal The spread of a species or group of species through a large area.

disruptive selection In evolution, the opposite of stabilizing selection: it increases the frequency of extreme types in a population, at the expense of intermediate forms.

distal tubule In the kidney, the wide tubule from which the urine flows into the collecting duct and then to the bladder.

disulfide A covalent bond between two sulfhydryl groups; disulfides are a common cross link between two cysteine side groups within a single polypeptide or between two polypeptides.

diuretic A substance, such as caffeine or alcohol, that stimulates water loss.

divergent evolution The separation of one species of organisms into two (or more) species.

DNA polymerase The enzyme that strings together nucleotides into DNA.

dominance hierarchy A ranking of individuals that fixes (at least temporarily) who may dominate whom.

dominant Refers to an allele that alone determines the phenotype of a heterozygote.

dormant [French, *dormire* = to sleep] Referring to a stage in the development of a seed in which growth is suspended until restarted by environmental cues.

dorsal Refers to the top, or back, of an animal.

dorsal hollow nerve cord A flexible rod containing neurons and their processes that runs between the notochord and the surface of the back.

double circulation A pattern of circulation in which blood circulates separately between the heart and lungs and between the heart and the rest of the body.

double covalent bond A covalent bond in which two atoms share two pairs of electrons.

double fertilization In flowering plants, the simultaneous fusion of two sperm nuclei from the pollen tube with two cells of the embryo sac.

doubling time The time it takes a population to double.

Down syndrome (also called **Down's syndrome**) A disorder, resulting from a particular chromosomal abnormality, that leads to mental retardation and the abnormal development of the face, heart, and other parts of the body; it is almost always caused by trisomy of chromosome 21.

downstream In DNA or RNA, toward the 3′ end.

duodenum [Latin, twelve; referring to its length of 12 "fingers"] The first section of the intestine, where most of digestion occurs.

dynamically unstable Meaning that a microtubule or an actin filament loses tubulin or actin monomers and shrinks if it is not actively growing.

early cell plate In a plant cell, a small, flattened disc, formed in mitotic telophase, that is the beginning of a new cell wall.

Echinodermata "Spiny-skinned" animals such as sea stars.

ecological footprint The area of land needed to provide all of the resources a single person uses.

ecological isolation A barrier to reproduction that results from differences in the ways that species live.

ecology [Greek, *oikos* = home] The study of the interactions of organisms with one another and with their physical environments.

ecosystem A community of organisms together with the nonliving parts of the community's environment.

ectoderm [Greek, *ektos* = outside + *derma* = skin] The outermost cell layer of an embryo, including, in vertebrates, the future skin, sense organs, and central nervous system.

ectotherm [Greek, *ektos* = outside + *thermos* = heat] An animal that depends on external sources of heat.

edema [Greek, "swelling"] A physiological state in which water accumulates in the tissues.

edge effects A consequence of habitat fragmentation. Edge effects, such as incursions by light and predators, prevent a species from fully occupying a habitat island.

effector cell In the immune system, a cell that carries out a particular immune function.

efferent Leading away, as blood vessels leaving an organ or nerve fibers carrying information from the central nervous system to the muscles and other organs.

egg (also called **ovum**) A female gamete.

ejaculation The propulsion of sperm out of the penis.

electrical potential (also called **voltage**) A measure of the potential energy that results from a charge separation.

electrical synapse A connection between two neurons through gap junctions.

electrical work Energy change caused by changing the separation of charges.

electrochemical gradient A double gradient composed of a chemical gradient (the difference in hydrogen ion concentration, or pH) and an electrical gradient (the difference in charge).

electrogenic Leading to the accumulation of charge on one side of a membrane.

electromagnetic radiation A form of energy—including light—that is transmitted through space as periodically changing electrical and magnetic forces.

electron A subatomic, negatively charged particle; its charge is exactly equal to that of a proton, but its mass is much smaller.

electron carrier One of the molecules that carry electrons from high-energy, reduced compounds (NADH, NADPH, and $FADH_2$) to oxygen.

electron microscope A powerful microscope that uses electrons instead of light waves to reveal structures; it has much higher resolution than a light microscope.

electron transport chain The pathway of electrons in oxidative phosphorylation or photophosphorylation.

electronegativity A measure of the tendency of an atom to gain electrons.

electrophoresis A method for separating charged molecules (such as proteins and nucleic acids) according to their ability to move in an electric field.

electrostatic interaction The attraction or repulsion of charges.

element A substance that cannot be reduced to simpler substances by chemical means.

elimination The disposal of the remains of digested food.

elongation The adding of additional nucleotides or amino acids to a growing polynucleotide or polypeptide.

elongation zone A region of the growth zone of a plant's root.

embryo [Greek, *en* = in + *bryein* = to be full of] A set of early developmental stages in which a plant or animal differs from its mature form.

emergent property A characteristic that arises only at complex levels of organization.

emergents In tropical forests, trees with umbrella-shaped crowns extending to a height of 50 meters or more.

emigration Movement out of the population.

emphysema A disease of the lungs characterized by irreversible enlargement of the air spaces.

endemic Found nowhere else in the world.

endergonic Refers to a process in which free energy increases.

endocrine [Greek, *endo* = within] Cells or organs that are specialized for secreting specific signaling molecules into the general circulation.

endocrine gland An organ that is specialized for secretion of a hormone into the general circulation.

endocrinology The study of hormones.

endocytosis [Greek, *endon* = within] The process of taking in materials from outside a cell in vesicles that arise by the inward folding ("invagination") of the plasma membrane.

endoderm [Greek, *endo* = inside + *derma* = skin] The innermost cell layer of an embryo, including the future gastrointestinal tract and associated organs such as the pancreas and liver.

endodermis [Greek, *endon* = within + *derma* = skin] In vascular plants, the innermost layer of the cortex in roots and stems.

endometrium [Greek, *endon* = within + *metro* = mother] The inner lining of the uterus.

endoplasmic reticulum (ER) [Greek, *endon* = within + *plasmein* = to mold + Latin, *reticulum* = network] An extensive and convoluted network of membranes within a eukaryotic cell.

endorphin One of three classes of polypeptides that act as natural pain suppressors.

endosperm In flowering plants, a storage tissue that develops during double fertilization.

endosymbiosis [Greek, *endon* = within + *syn* = together + *bios* = life] The close association of two organisms, one of which lives inside the other.

endosymbiotic theory The generally accepted view that present-day energy organelles are descended from prokaryotes that once lived within the early eukaryotic cells.

endothelium In the circulatory system, the epithelial cells that line the capillaries.

endotherm [Greek, *endon* = within] An animal that warms its body by capturing the heat released by metabolism.

endotoxin A toxic substance that is a component of a bacterial cell or other pathogen.

energy The capacity to perform work, to move an object against an opposing force.

energy subsidy The extra energy, usually provided by fossil fuels, that increases the productivity of a farm.

enhancer (because it stimulates the transcription of neighboring DNA, without itself serving as a promoter; also called a **response element** or *cis*-**regulatory element**) A DNA sequence that regulates transcription by binding to a specific regulatory protein called a transcription factor.

entropy A formal measure of disorder; entropy has a high value when objects are disordered or distributed at random and a low value when they are ordered.

enzyme A large molecule, almost always a protein, that accelerates the rate of a specific chemical reaction.

enzyme-substrate complex The association of enzyme and substrate that forms in the course of catalysis.

eon One of several very long periods of time into which geologists and paleontologists have divided the history of life; eons are much longer than eras; a billion years or more.

epicotyl [Greek, *epi* = above] The axis of a developing plant above the attachment point of the cotyledons.

epidemiologist [Greek, *epidemia* = how widespread a disease is] A researcher who studies the incidence and transmission of diseases in populations.

epidermis In animals, the outer layer of the skin; in plants, the outermost layer of the embryo and the plant.

epididymis In the male reproductive tract of mammals, an interconnected network of coiled ducts in which sperm are stored.

epilimnion The warm surface water layer of a lake or pond.

epinephrine (also called **Adrenaline**) A hormone, made in the adrenal medulla, that speeds the heart, dilates the blood vessels, and increases the liver's production of glucose from glycogen.

epiphyte A plant that grows entirely on other plants.

episome A virus or other genetic system that can propagate either autonomously or as an integrated part of the host's chromosome.

epistatic gene A type of modifier gene that limits the expression of another gene.

epithelial tissue [Greek, *epi* = upon + *thele* = nipple] Cells tightly linked together to form a sheet with little extracellular matrix.

epithelium A tissue that lines a surface (such as the outside of the body or the inside of the lungs).

epoch A relatively short period of geologic time; one of the subdivisions of the Tertiary and Quaternary Periods of the Cenozoic Era.

equilibrium In a chemical reaction, the point at which no further net conversion of reactants and products takes place.

era A period of time, ranging in length from 65 million to several billion years, into which paleontologists usually divide the history of life; eras are shorter than eons.

erection The rigid state of the penis or clitoris, caused by the blood filling the tiny spaces within spongy tissues.

erythrocyte [Greek, *erythros* = red + *kytos* = receptacle] A red blood cell.

***Escherichia coli* (*E. coli*)** A common eubacterial resident of the human gut; a favorite experimental organism for thousands of research and industrial biologists.

esophagus The part of the digestive tract that connects the mouth and pharynx with the stomach.

essential amino acids The amino acids that an animal cannot itself produce.

essentialism The view, originally argued by Plato, that individuals, whether chairs, daisies, or people, are only distorted shadows of an ideal, or essential, form.

estivate To pass time in a sleeplike state (torpor) during the hottest and driest months of the summer.

estuary [Latin, *aestus* = tide] A partly enclosed body of water where a freshwater stream or river meets the ocean.

ethology The study of animal behavior.

ethylene A plant hormone; a two-carbon molecule containing a double bond.

etiolated [French, *etioler* = to blanch or whiten] Having a thin, spindly appearance, poor leaf development, and no chlorophyll production.

Eubacteria [Greek, *eu* = true] The commonly occurring prokaryotes that live in water and soil or within larger organisms.

euglenophytes [Greek, *eu* = true + *glene* = eyeball] Protists that have distinctive light-detecting eye spots.

Eukarya The domain that encompasses the Animalia, Plantae, Fungi, and Protista kingdoms.

eukaryotic [Greek, *eu* = true + *karyon* = nucleus] Cells that contain a central nucleus and other membrane-bounded organelles.

eukaryotic flagellum [plural, **flagella**; Latin, whip] A protein assembly, consisting of microtubules, that can move a cell through a liquid medium (or a liquid medium over a cellular surface).

Eumetazoa [Greek, *eu* = true + *meta* = middle + *zoa* = animal] The larger of two subkingdoms of animals, including all of the animals except the sponges and the Placozoa.

euploid Having the correct number of chromosomes.

eutherians (also called **placental mammals**) [Greek, *therion* = wild beast] Animals whose embryonic development takes place within the uterus of the mother.

eutrophic A lake that has excessive minerals and organic matter and insufficient oxygen.

evergreen Keeping leaves throughout the year.

evolution The process by which species arise and change over time. The idea that all organisms have descended from common ancestors.

excitatory In the nervous system, leading to the production of action potentials.

excitement A phase of sexual arousal that is characterized by increased blood flow to the clitoris, the labia minora, and the breasts (in the female) and to the penis and the testes (in the male).

excretion A disposal process by which cells pass materials across a cell membrane.

exergonic Refers to a process in which free energy decreases.

exocrine [Greek, *exo* = outside] A cell or organ that makes products that are carried to specific targets by ducts, such as the duct that carries digestive enzymes into the small intestine.

exocrine organ An organ whose products are carried to specific targets by ducts.

exocytosis The export of molecules from a cell by a process that is approximately the reverse of pinocytosis; molecules to be exported are surrounded by membranes that move to the cell surface

exon A segment of a gene (or of a pre-mRNA) that is also present in mature mRNA; most exon sequences encode polypeptide segments.

exoskeleton [Greek, *exo* = on the outside] External skeleton.

extinction vortex The theory that inbreeding brings out genetic defects that hasten the extinction of a population.

exotoxin A toxic substance secreted by bacteria.

exponential growth A growth pattern in which a population repeatedly doubles in some constant period of time.

expiration Exhalation.

extension The unbending of a limb.

extracellular matrix A network of proteins and polysaccharides found in connective tissue.

F-factor (for fertility factor) A kind of episome in bacteria that can replicate either autonomously or in integrated form; it can move from one bacterium to another during conjugation.

F1 (or **first filial**) **generation** [Latin, son] The initial progeny of a cross.

F2 (or **second filial**) **generation** The progeny of the F1 generation.

facilitated diffusion An increased rate of passive transport; it depends on the action of specific transporter molecules within the membrane.

facultative aerobe A heterotroph that can grow either with or without oxygen.

facultative anaerobe A microorganism that can live either anaerobically (by fermentation) or aerobically (using oxidative phosphorylation).

FAD (flavin adenine dinucleotide) An electron acceptor in oxidative phosphorylation.

FADH$_2$ The reduced form of FAD.

fall turnover In a lake, the annual mixing of the waters of epilimnion and hypolimnion.

fallopian tube Human oviduct.

family A taxonomic group that consists of one or more genera; a family is a subdivision of an order.

fat A triacylglycerol that is solid at room temperature; the component fatty acids are usually saturated.

fate What a cell or a tissue becomes during development.

fatty acid A small molecule consisting of a hydrocarbon chain ending in a carboxyl group; a component of phospholipids and triacylglycerides.

feces [Latin, "dregs"] Waste matter of digestion, discharged through the anus; feces consist of the remnants of food together with bacteria that inhabit the intestinal tract.

feedback inhibition The inhibition of an enzyme reaction by the product of that reaction or of the end product of an entire pathway; a homeostatic mechanism.

female choice A courtship pattern in which the female assesses and chooses the best male.

fermentation [Latin, *fervere* = to boil] The anaerobic extraction of energy from organic compounds.

fermenting bacteria Anaerobic heterotrophs that derive energy and carbon atoms from a variety of organic compounds in the absence of oxygen.

fertility The number of offspring each individual in a cohort is likely to produce.

fertilization The union of two haploid gametes to form a diploid cell or zygote.

fetal masculinization The development of male characteristics in a fetus carried by a woman treated with male hormones during pregnancy.

fetus In mammals, a stage of development in which all organs have formed.

fibril A threadlike structure, made of smaller filaments; the term is used to refer to cross-linked cables of collagen molecules and to assemblies of actin and myosin filaments.

fibrillation In the heart, continuous disorganized contractions.

fibroblast A flat, irregularly shaped cell found in connective tissue.

fibronectin A protein of the extracellular matrix.

fibrous protein A protein with an elongated shape; fibrous proteins include most structural proteins.

filament (1) A small protein fiber, (2) in plants, a thin stalk that connects the anther to the base of the flower.

filtrate In the kidney, the water, urea, and other small molecules in the blood that freely pass into Bowman's capsule from the renal artery.

First Law of Thermodynamics The statement that the total amount of energy stays constant in any process; that is, energy is neither lost nor gained—it only changes form.

fission yeast *Schizosaccharomyces pombe,* a widely used experimental organism that divides by binary fission.

fitness A measure of selective advantage; it is defined as the contribution to the next generation of one genotype in a population relative to the contribution of other genotypes.

fixation A process that preserves cells for microscopic examination.

fixed nitrogen Nitrogen in the form of ammonia (NH_3) or nitrate (NO_3).

flagellum [plural, **flagella;** Latin, whip] In eukaryotes, a protein assembly, consisting of microtubules, that can move a cell through a liquid medium (or a liquid medium over a cellular surface); in prokaryotes, a protein assembly that moves like a propeller.

flexion The bending of a limb.

fluid mosaic model The accepted model of biological membrane structure; the model stresses that proteins and phospholipid molecules can move within each leaf of the lipid bilayer unless they are restricted by special interactions.

fluidity A measure of the ability of substances to move within a membrane.

follicle [Latin, *folliculus* = small ball] In the ovary, the granulosa cells surrounding a primary oocyte.

follicle-stimulating hormone (FSH) A peptide hormone, made in the anterior pituitary that in females promotes the maturation of the follicle during the menstrual cycle and in males stimulates testosterone production.

food chain The sequence in which consumers eat either producers or other consumers.

food web The collection of all the interacting food chains of an ecosystem.

foot In a mollusk, a muscular extension of the body that the mollusk uses for sensing, grabbing, creeping, digging, and holding on.

foraminifera (also called forams) [Latin, *foramen* = little hole + *ferre* = to bear] Marine organisms whose tests are made of organic materials reinforced with grains of sand or minerals and full of tiny holes.

foreskin A loose layer of skin around the outside of the penis, which ends in a flap and folds over the glans.

fossil [Latin, *fossilis* = dug up] An object, usually found in the ground, that represents the remains or imprint of past life.

founder effect The restriction of genetic diversity of a population that results from the founding of a new population by a small subset of a larger population.

frame-shift mutation An insertion or deletion mutation that alters the groupings of nucleotides into codons.

free energy A measure of available energy under the conditions of a biochemical reaction; the term is abbreviated G, after the American thermodynamicist J. Willard Gibbs.

free radical A compound that is highly reactive because it contains an unpaired electron; it is biologically important because it can lead to DNA damage and to cell death; free radicals are formed during normal metabolic processes in both animals and plants.

free ribosomes Ribosomes, present in the cytosol, that are responsible for the synthesis of soluble proteins.

fronds Leaves of ferns that often have a lacy appearance.

fruiting body A mushroomlike growth on cellular slime molds that produces spores.

functional group A standard small grouping of atoms that contributes to the characteristics of an organic molecule.

fundamental niche A species' potential ability to utilize resources.

Fungi The kingdom that includes heterotrophic organisms, both multicellular and single-celled organisms.

G_0 The state of a cell that has withdrawn from the cell cycle.

G_1 The period of the cell cycle that represents the gap between the completion of mitosis and the beginning of DNA replication; it is also called the first growth phase.

G_2 The period of the cell cycle that represents the gap between the completion of DNA synthesis and the beginning of mitosis (of the next cell cycle).

gametangia In plants, the enclosed reproductive organs in which gametes form.

gamete [Greek, *gamos* = marriage] A specialized reproductive cell through which sexually reproducing parents pass chromosomes to their offspring; a sperm or an egg.

gamete formation The production of sperm and eggs.

gametic isolation Reproductive isolation that results from differences in the compatibility of gametes.

gametophyte The haploid form of a life cycle characterized by alternation of generations.

gap junctions Protein assemblies that form channels between adjacent animal cells.

gastrotricha Microscopic, bottle-shaped animals that live in freshwater or marine environments.

gastrula [Greek, little stomach] A stage of an animal embryo in which the three germ layers have just formed.

gastrulation The process of forming a gastrula.

gate The part of a channel protein that opens and closes the channel in response to environmental signals such as voltage or the binding of a neurotransmitter.

gene [Greek, *gen* = to produce] The unit of inheritance. A DNA sequence that is transcribed as a single unit and encodes a single polypeptide, a set of closely related polypeptides, ribosomal RNA, or transfer RNA.

gene flow The movement or flow of genes from one population to another by means of interbreeding.

gene pool All the alleles of all the individuals in a population.

gene rearrangement The cutting and splicing of segments of specific genes; it is known to occur within the immune system.

genetic code The relationship between nucleotide sequence in mRNA and amino acid sequence in polypeptides.

genetic linkage The tendency of two or more genes to segregate together.

genetic map (also called a **linkage map**) A summary of the genetic distances between genes.

genetic recombination Associations of genes that occur in offspring that did not exist in the parents.

genetics The study of inheritance.

genome The collection of all the DNA in an organism.

genotype The genes present in a particular organism or cell.

genus [plural, **genera**; Greek, *genos* = to be born] A group of species that share many morphological characteristics; a taxonomic grouping of one or more species; a subdivision of a family.

germ cells (or **germ line**) Gametes and the cells from which they arise.

germ layer One of the three tubes of the vertebrate embryo—ectoderm, mesoderm, and endoderm.

germ layer formation The movement of embryonic cells to give the three germ layers (ectoderm, mesoderm, and endoderm); gastrulation.

germ line theory The view that antibody diversity results from genetic information that is already present in the zygote.

germination [Latin, *germinare* = to sprout] The resumption of growth by a seed.

gibberellin A plant hormone.

gill slits Openings that directly connect the throat to the outside.

gills Evaginated breathing structures.

girdling Removing a strip of bark around a tree trunk.

gizzard A region of the intestinal tract of birds specialized for the grinding of food.

glans The smooth cap formed by the tissue at the far end of the penis.

glial cell A cell within the nervous system that does not itself transmit electrical and chemical signals, but which provides metabolic and structural support for neurons.

globular protein A relatively compact protein that is roughly spherical in shape.

glomerulus [plural, **glomeruli**; Latin, "little ball"] A tangled network of capillaries in the cortex of a kidney.

glucagon A protein hormone produced in the pancreas; a signal for the postabsorptive state; glucagon inhibits glycogen synthesis and stimulates its breakdown into glucose.

glucocorticoid A steroid hormone that stimulates the production of carbohydrate from protein.

glycerol A polar three-carbon molecule with three hydroxyl groups; a starting compound for triacylglycerides and phospholipids.

glycogen Polysaccharide used in animals for long-term energy storage.

glycogen phosphorylase The enzyme that breaks down glycogen into glucose.

glycolipid A molecule that consists of a lipid attached to carbohydrates.

glycolysis [Greek, *glykys* = sweet (referring to sugar) + *lyein* = to loosen] A set of ten chemical reactions that is the first stage in the metabolism of glucose.

glycolytic fiber A large muscle fiber that derives most of its energy from glycolysis and has few mitochondria.

glycoprotein [Greek, *glykys* = sweet] A protein that contains covalently attached carbohydrates.

glycosidase An enzyme that catalyzes the digestion of glycosidic bonds; it is used to digest starch or glycogen.

glycosidic bond The covalent bond between two sugar molecules in a polysaccharide or oligosaccharide.

glycosylation The process of adding sugars to a newly made protein; it takes place in the ER lumen and in the Golgi apparatus.

glyoxisome A specialized peroxisome present in the seeds of some plants; glyoxisomes provide energy for the growing plant embryo by breaking down stored fats.

goblet cells In the small intestine, specialized cells that produce mucus.

Golgi complex (also called the **Golgi apparatus**) In eukaryotic cells, a set of flattened discs of membrane, usually near the nucleus, involved in the processing and export of proteins.

gonad [Greek, *gonos* = seed] A gamete-producing organ; an ovary or testis.

gonadotrophin One of two hormones (FSH and LH), made in the anterior pituitary, that act on gonadal tissue.

gonadotropin releasing hormone (GnRH) A polypeptide hormone, made by the hypothalamus, that stimulates the secretion of FSH and LH by the anterior pituitary.

gossypol A terpene, made by cotton plants, that is responsible for resistance to insects, bacteria, and fungi; it also works as a contraceptive in human males.

gradualism The idea that species evolve gradually and continuously through the steady accumulation of changes, rather than through sudden changes.

granum [plural, **grana**; Latin, grain] A stack of thylakoids within a chloroplast, bounded by the thylakoid membrane.

gravitropism The response of a plant to gravity.

greenhouse effect The warming of the Earth as the result of the absorption of heat by carbon dioxide and other gases.

ground tissue In plants, the cells that occupy most of the interior of the embryo and the plant.

growth control A regulatory mechanism that prevents cell division by allowing the cell cycle to proceed under some conditions and to stop under others.

growth factor One of a number of protein paracrine signals that stimulate cell division and cell survival.

growth hormone (GH) A peptide hormone, made in the anterior pituitary, that stimulates tissue and skeletal growth, milk production, and other processes in humans, cows, and other mammals.

gut [Anglo-Saxon, "channel"] The gastrointestinal tract, the alimentary canal, or the digestive tract.

guttation [Latin, *gutta* = a drop] The process by which root pressure can push water all the way up the stem and out of tiny holes at the margins of the leaves.

gymnosperms [Greek, *gymnos* = naked + *sperma* = seed] Seed plants that do not have fruits or flowers, including, for example, pine trees and other conifers, as well as palmlike cycads.

habitat The place in which an organism lives, along with the set of environmental conditions that characterize that place.

habitat destruction The process by which humans make a habitat uninhabitable for a species. Habitat destruction can range from bulldozing and paving over the area to the elimination of a single, important (keystone) species.

habitat island A fragment of habitat isolated from other similar fragments.

habituation The decrease in a behavioral response following repeated exposure to a harmless stimulus.

half-life The time required for half the atoms of a radioactive isotope to decay.

halophiles The salt-loving bacteria, one of the three phyla of Archaea.

haplodiploidy An unusual kind of sex determination in which males develop from unfertilized eggs and are therefore haploid, while females are diploid.

haploid [Greek, *haploos* = single] Having a single set of chromosomes.

haplontic Having the haploid phase dominate the life cycle; in a haplontic life cycle, the only diploid cell is the zygote, which itself undergoes meiosis to produce haploid spores.

Hardy-Weinberg equilibrium A stable distribution of genotype frequencies maintained by a population from generation to generation.

heart The muscular organ responsible for pushing the blood through the circulatory system.

heart attack A failure of the blood supply to the heart.

heart rate The number of contractions (beats) per minute.

heat The form of kinetic energy contained in moving molecules.

heavy chain One of the polypeptides present in an immunoglobulin molecule.

heme An iron-containing organic molecule that gives hemoglobin and the cytochromes their red color; heme may donate or accept electrons, as its iron atom changes its charge between Fe^{2+} and Fe^{3+}.

Hemichordata Acorn worms.

hemizygous In male mammals, having one (instead of two) copies of a gene because the Y chromosome lacks the gene present on the X chromosome.

hemoglobin [Greek, *haima* = blood + Latin, *globus* = ball] The oxygen-binding protein that makes red blood cells red.

herbaceous plant (also called an **herb**) A plant that has no woody parts.

herbivore [Latin, *herba* = vegetation + *vorare* = to swallow] An animal that eats only plants.

heterokaryon [Greek, *heteros* = different + *karyon* = kernel, nucleus] A cell that has two nuclei from different sources.

heterosporous Producing two kinds of gametophytes (male and female) from two kinds of spores on two kinds of sporangia.

heterotroph [Greek, *heteros* = other + *trophe* = nourishment] An organism that cannot derive energy from sunlight or from inorganic chemicals but must obtain energy by degrading organic molecules.

heterozygous Having two different alleles for a single gene (in a diploid organism).

hexose [Greek, *hex* = six] A sugar that contains six carbon atoms; glucose is a hexose.

hierarchy An arrangement in which larger groups include smaller groups, which include still smaller groups.

high-energy bond A relatively unstable chemical bond that gives up energy as new, more stable, bonds form.

histamine The amino acid histidine minus the carboxyl group, a major stimulus for the inflammatory response; it is released by cells in damaged tissues; it dilates capillaries and increases their tendency to leak fluid.

histone [Greek, *histos* = web] One of a set of small, positively charged proteins that bind to DNA in eukaryotic cells.

homeostasis [Greek, *homeo* = like, similar + *stasis* = standing] The tendency of organisms to maintain a stable internal environment in the presence of a changing external environment.

homeotic mutation A mutation that causes the cells of an embryo to give rise to an inappropriate structure in the adult, for example, to legs instead of antennae.

homeotic selector gene In a plant, a gene that establishes the fate of one or more whorls of a developing flower.

hominid A member of the human family, Hominidae; the only living hominid is *Homo sapiens*.

hominoid [Latin, *homo* = human] Hominids and apes; all have large skulls and long arms and tend to walk at least partially erect.

homologous Arising from the same structures in a common ancestor.

homologous chromosomes The two matching chromosomes that align during meiosis I.

homologous structures Similar structures in two or more species. Homologus structures may perform different or similar functions.

homoplasy Similar traits that have evolved separately. The possession by two species of a similar trait that is not derived from a common ancestor. "Homoplasy" has replaced the older word "analogy," which requires that two similar traits perform the same function. In contrast, similarities that are inherited from a common ancestor are said to be homologous.

homosporous Producing just one kind of spore and one kind of gametophyte.

homozygous Having two copies of the same allele (in a diploid organism).

hopeful monsters Individuals with macromutations such as chromosome rearrangements whose deformities have adaptive value. A term coined by biologist Richard Goldschmidt to describe a mechanism for a saltationist theory of speciation.

hormone [Greek, *horman,* = to urge on] A substance, made and released by cells in a well-defined organ or structure, that moves throughout the organism and exerts specific effects on specific cells in other organs or structures.

host range For a virus, the set of hosts that a particular virus can infect.

Hox gene Mammalian counterpart of a *Drosophila* homeo domain gene.

human chorionic gonadotropin (HCG) A polypeptide hormone made after conception; the basis of the most common tests for pregnancy.

human immunodeficiency virus (HIV) The retrovirus that causes AIDS; HIV infects and kills T_H lymphocytes by first binding to the CD4 protein on the cell surface.

humus In soils, the residue of decayed dead organisms; a black or brown material that decays slowly.

hybrid The progeny of a cross of two genetically distinct organisms.

hybrid breakdown The weakness or sterility of the progeny of interspecies hybrids that are initially viable and fertile.

hybrid inviability Chromosomal and genetic incompatibilities within hybrids.

hybrid sterility The sterility of hybrids between two species.

hybridoma A hybrid cell line derived from the fusion of a cancer cell (a lymphoma) to another cell, such as an antibody-producing cell.

hydrocarbon chain A chain of connected carbon atoms, with hydrogen atoms sharing other available outer shell electrons.

hydrogen bond A weak attraction between a hydrogen in one molecule that has a slight positive charge and a negatively charged atom in another molecule.

hydrogen ion (H^+) The result of a dissociation of a water molecule; it is also called a proton, though it is actually a hydronium ion (H_3O^+).

hydrolysis [Greek, *hydro* = water + *lysis* = breaking] Breaking the bond between two building blocks by adding a water molecule, reversing the dehydration-condensation reaction.

hydronium ion (H_3O^+) A water molecule that has acquired an extra proton and a charge of +1; the proper name for a hydrogen ion.

hydrophilic [Greek, *hydro* = water + *phili* = love] Water-loving; refers to a molecule (or a part of a molecule) that is soluble in water by virtue of its interactions with water molecules.

hydrophobic [Greek, *hydro* = water + *phobos* = fear] Avoiding associations with water; nonpolar.

hydrophobic interaction The association of nonpolar molecules.

hydrostatic skeleton A rigid fluid-filled space (the coelom) that provides mechanical support in invertebrate coelomates.

hydroxide ion (OH^-) A water molecule from which a hydrogen ion has dissociated; it has a charge of −1.

hydroxyl The OH functional group, which allows molecules that contain it to form hydrogen bonds.

hymen A thin membrane of irregular ragged shape just inside the mouth of the vagina.

hyperpolarizing A voltage change (across the plasma membrane of a neuron) that makes the inside of the cell more negative and less apt to produce an action potential.

hypertension High blood pressure

hypertonic [Greek, *hyper* = above] Having a total concentration of solutes higher than that within a cell.

hypervariable region The polypeptide segments, within the variable region of an immunoglobulin light chain or heavy chain, in which the most sequence variation occurs among antibodies; the hypervariable regions of an immunoglobulin molecule correspond exactly to the complementarity determining regions, where antigens bind.

hyphae [singular, **hypha**; Greek, *hyphe* = web] The threadlike filaments of a fungus.

hypocotyl [Greek, *hypo* = under] The axis of a developing plant below the attachment point of the cotyledons.

hypolimnion In lakes and ponds, the layer of water beneath the epilimnion whose temperature stays about 4°C year-round.

hypothalamus [Greek, *hypo* = under + *thalamos* = inner room] A part of the brain that regulates the expression of many hormone systems.

hypothesis An informed guess—for example, about the way a process works or a structure is organized.

hypotonic [Greek, *hypo* = under] Having a total concentration of solutes lower than that within a cell.

imaginal disc In insect development, a group of larval cells from which adult structures later develop.

imago Adult stage of an insect.

imbibition The taking in of water at the time of seed germination.

immigration Movement into a population.

immune response The defense system by which animals resist microorganisms and other foreign tissues, including cancer.

immunoglobulin An antibody; one of the members of a group of globular blood proteins called globulins.

impermeable junctions (also called **occluding junctions**) Molecular assemblies that connect epithelial cells tightly together; they also prevent molecules from leaking between them.

implantation In mammals, the process in which a zygote burrows into the wall of the uterus.

imprinting The process by which an animal learns behavior during a sensitive period.

in vitro [Latin, in glass] In a test tube.

in vivo [Latin, in life] In a living organism.

inbreeding Mating among close relatives, which greatly increases the number of homozygotes.

incipient species Subspecies that do not interbreed.

inclusive fitness The sum of an individual's genetic fitness (which includes that of its own direct descendants) plus all its influence on the fitness of its other relatives.

individualistic The view that every species has an independent distribution, and that, in effect, every community is unique.

induced abortion The deliberate removal of an embryo from the uterus.

induced fit A change in the conformation of an enzyme brought about by the binding of the substrate.

induction In embryonic development, the process by which one cell population influences the development of neighboring cells.

inflammation A set of responses to local injury; characteristics of injured tissue include redness, heat, swelling, and pain.

inhibiting hormone A peptide hormone, made by the hypothalamus, that inhibits the release of a specific hormone.

inhibitory In the nervous system, tending to prevent the production of action potentials.

inhibitory interneuron A neuron whose stimulation by one neuron prevents the firing of another neuron in the same area.

initiation The start of synthesis.

initiation site The first nucleotide actually transcribed from DNA into RNA.

innate Behavior that an animal engages in regardless of previous experience.

inner cell mass In a mammalian embryo, a small group of cells within a blastocyst that will eventually grow into the embryo itself and subsequently into the adult.

insertion A type of mutation; the addition of one or more nucleotides.

insertion sequence (also called a **simple transposon**) The simplest transposable element; a short length of DNA (up to a few thousand nucleotide pairs long) that can move from place to place in the genome.

inspiration Inhalation.

insulin A protein hormone, produced by the pancreas, that regulates glucose uptake; a signal for the absorptive state; it promotes the synthesis of glycogen and inhibits its breakdown.

integral protein (also called a **transmembrane protein**) A membrane protein that spans the lipid bilayer.

integrated The view that a community consists of characteristic assemblages of species that interact with each other in predictable ways.

integrated replication A method for viral propagation in which the virus's DNA becomes integrated into the host cell's DNA and is replicated along with that of the host.

interferon A cytokine that interferes nonspecifically with the reproduction of viruses.

interleukin-1 (IL-1) A cytokine produced early during inflammation; it stimulates responses both in leukocytes and in organs distant from the site of infection.

intermediate filament In eukaryotic cells, a component of the cytoskeleton that consists of filaments 8 to 10 nm in diameter, thinner than microtubules but thicker than actin filaments.

intermediate form Fossils that grade from one form to another—for example, the fossils intermediate in both shape and in time from the tiny, horselike animal eohippus to the modern horse.

internode In a plant, the region of a stem between nodes.

interphase The part of the cell cycle in which the chromosomes are not condensed and the cytoplasm is not dividing.

interstitial cells In the testes, the matrix amongst the seminiferous tubules that synthesizes the male sex hormone testosterone.

intertidal zone The zone that lies between high tide and low tide.

intrauterine device (IUD) A birth control device; a small piece of plastic or other material that is placed into the uterus, where it interferes with implantation.

intron (also called an **intervening sequence**) A segment of a gene (or of a pre-mRNA) that is transcribed into RNA but excised before the primary transcript matures into functional mRNA.

invagination The local folding of a cell layer to form an enclosed space with an opening to the outside.

inversion A mutation in which a segment of a chromosome is turned 180° from its normal orientation.

involuntary muscle A muscle that cannot be consciously controlled; smooth and cardiac muscles.

ion An atom (or a molecule) with a net electrical charge, the result of a different number of electrons and protons.

ionic bond A bond formed by ions with opposite charges.

iron-sulfur protein One of at least six proteins that participate in the electron transport chain; each contains an iron-sulfur center with two to four iron atoms and an equal number of sulfur atoms.

isomers Molecules that contain the same atoms arranged differently.

isotonic [Greek, *isos* = equal + *tonos* = tension] Having a total concentration of solutes that is the same as a cell's interior.

isotopes Forms of an element that have different numbers of neutrons.

J-shaped curve A graph of the exponential growth of a population; the curve resembles the letter J.

joint The point of attachment between two bones.

K-**selected species** Species with adaptations that increase their ability to maintain populations as close to the carrying capacity (K) as possible; these species tend to live long lives, reproduce late in life, produce few and large offspring, and provide extended parental care to each offspring.

karyotype [Greek, *karyon* = kernel or nucleus + *typos* = stamp] The chromosomal makeup of a cell.

keystone species A predator or other organism whose presence promotes diversity or controls the structure of a community.

kidney In vertebrates, the major organ of excretion and of salt and water balance; regulates the composition of the urine.

kin selection The tendency of individuals in some species to increase their reproductive output by helping relatives, which share their genes, to reproduce.

kinetic energy The energy of moving objects.

kinetics [Greek, *kinetikos* = moving] The study of the rates of reactions.

kinetochore A specialized disc-shaped structure that attaches the mitotic spindle to the centromere.

kinetochore microtubules A subset of polar microtubules that run from a pole of the mitotic spindle to a kinetochore.

Koch's postulates Rigorous criteria for identifying the pathogen for a given disease.

krummholz A forest of stunted trees that grow near timberline on a mountain.

kwashiorkor Protein deficiency disease.

labia majora [Latin, larger lips] In the external genitalia of human females, the skin folds on either side of the labia minora that enclose small amounts of fatty tissue.

labia minora [Latin, smaller lips] In the external genitalia of human females, the two thin skin folds that surround the mouth of the vagina and the end of the urethra.

lac **repressor** A protein that binds to the *lac* operator and prevents the expression of the *lac* operon.

lactose operon (or *lac* **operon**) A region of *E. coli* DNA that encodes three proteins used to derive energy from lactose: the enzyme β-galactosidase (which splits lactose into galactose and glucose) and two other proteins, called permease and acetylase.

lagging strand In replicating DNA, the newly made strand that is extended discontinuously.

larva A feeding form of an animal distinct from the later adult.

larynx The modified upper part of the trachea that contains the vocal cords, folds of membrane that vibrate as air passes over them, producing sound.

lateral bud (also called an **axillary bud**) A bud that forms just above the point where a leaf joins the stem and which can develop into a branch.

lateral meristem A cylinder of actively dividing cells within a root or stem.

leading strand In replicating DNA, the newly made strand that extends continuously, with DNA polymerase adding nucleotides to its 3′ end.

leaf primordium A tiny extension of the apical meristem; it grows into a leaf.

learning Modification of neural activity and behavior as the result of experience.

leukocyte [Greek, *leukos* = clear, white] White blood cell.

liana A vine that is rooted in soil but climbs into the canopy.

ligament A band of connective tissue by which joints may be joined together.

ligand [Latin, *ligare* = to bind] A molecule that binds to a specific binding site in a protein.

light chain One of the polypeptides present in an immunoglobulin molecule.

light source The part of a microscope that illuminates a specimen.

lignin A major constituent of the walls of cells specialized to provide mechanical support or transport water.

limiting resource A resource that is in short supply.

limnology The study of freshwater ecosystems—lakes, ponds, rivers, and streams.

linkage group A set of genes that do not assort independently because they are physically close to one another on the same chromosome.

linkage map A summary of the genetic distances between genes; a genetic map.

lipid A compound that is less soluble in water than in nonpolar solvents.

lipid-soluble signal A chemical signal that can enter a target cell by passing directly through the plasma membrane.

liposome An artificially produced vesicle that is surrounded by a phospholipid bilayer.

loam A soil that consists of a mixture of clay, silt, and sand.

locus [plural, **loci**; Latin, place] The position of a gene on a chromosome.

logistic growth equation An equation that describes the growth of a population.

loop of Henle In a kidney, the hairpin-shaped portion of the renal tubule.

lophophore In three phyla of marine invertebrates, a filtering apparatus that catches food.

low-density lipoprotein (LDL) A carrier protein in the blood that binds to cholesterol.

low-energy bonds Relatively stable chemical bonds.

lumen [Latin, light, an opening] An enclosed space, bounded either by membranes (as in the ER lumen) or by an epithelium (as in the lumen of the gut).

lungs Invaginated breathing structures; localized organs of gas exchange that are always associated with the circulation.

luteinizing hormone (LH) A peptide hormone, made in the anterior pituitary, that in females induces ovulation and stimulates estrogen production and in males increases testosterone production.

lymph Fluid containing dead or foreign cells and waste proteins that passively moves through the lymphatic system.

lymph nodes Regions in the lymphatic veins where filterlike tissue separates cells and other detritus from the lymph.

lymphatic system The network of lymphatic vessels and nodes that provides a secondary route for fluids from the extracellular space to the bloodstream.

lymphocyte One of a class of white blood cells that develop within the lymphoid tissues (including lymph nodes, spleen, thymus, and tonsils); the cells responsible for the immune response.

lymphocyte library The collection of B lymphocytes, each containing a unique random rearrangement of immunoglobulin genes.

lysis [Greek, *lysis* = loosening] The breaking down of a plasma membrane.

lysogenic cycle [Greek, *lysis* = loosen + *genos* = offspring, because it can generate the destructive lytic cycle] The reproduction of a virus along with its host without causing the host cell to lyse.

lysogenic virus A virus that reproduces either in a lytic cycle, in which it destroys its host, or along with the host; the virus can lie in a dormant (lysogenic) state and can be activated to enter a lytic cycle.

lysosome A small membrane-bounded organelle that contains hydrolytic enzymes that break down proteins, nucleic acids, sugars, lipids, and other complex molecules.

lytic cycle Viral reproduction that destroys the host.

lytic virus A virus that destroys its host cell by lysing it.

macroevolution The processes by which species and higher groupings (taxa) of organisms originate, change, and go extinct.

macromolecules [Greek, *macro* = large] Large molecules formed by the polymerization of smaller building blocks.

macrophage [Greek, *makros* = large + *phagein* = to eat] A large phagocytic cell, widely distributed throughout the body.

magnification The ratio of the size of an image to the size of the object itself.

major histocompatibility complex (MHC) A set of 40 to 50 closely linked genes, some of which encode proteins on the surface of every somatic cell; MHC proteins present antigens to the immune system.

male competition Competition among males, usually for territory or access to females.

malnourishment A deficiency in one or more essential nutrients.

Malpighian tubules Excretory organs in arthropods.

mammary glands Milk-producing organs in the female that characterize the mammals.

mantle In mollusks, a specialized tissue that secretes a shell onto the dorsal surface of the visceral mass.

mantle cavity In mollusks, an enclosed space, formed by folds in the mantle, that contains the mollusk's breathing organs.

marsupials (also called **metatherians**) Mammals whose young are born in a fetal stage; marsupials carry the young in a pouch until they can fend for themselves.

mass A measure of the amount of matter.

mass number The total number of protons and neutrons in an atom's nucleus.

mast cell A large round cell; mast cells are distributed throughout the connective tissues and are filled with small histamine-containing vesicles, which are released at the sites of tissue damage.

maternal effect genes Genes that are expressed in the mother during oogenesis.

maternal mRNA An mRNA that is already present in the egg before fertilization.

matrix The intercellular substance within a tissue, or the interior substance of a mitochondrion.

matter Any substance.

mean The average value of a set of numbers, also called the arithmetic mean.

mechanical isolation A reproductive barrier that derives from incompatibilities between male and female reproductive organs.

mechanical work Energy change resulting from the movement of an object against a force.

median eminence A structure that carries hypothalamic hormones to the anterior pituitary.

medulla A part of the brain that regulates the rate of breathing; the inner region of a kidney or adrenal gland.

medusa The bell-like body plan that represents one of the two basic body plans of the cnidarians.

megaphyll A large leaf containing several veins that form complex networks of transport tubes.

megaspore [Greek, *megas* = large] In a flower, the haploid cell within an ovule that develops into the embryo sac, the mature female gametophyte.

meiosis [Greek, *meioun* = to make smaller] The process by which haploid gametes arise from diploid cells; meiosis distributes chromosomes so that each of four daughter cells receives one chromosome from each homologous pair.

meiosis I The first of the two divisions of meiosis, during which homologous chromosomes pair and are distributed into two daughter cells.

meiosis II The second of the two divisions of meiosis, during which sister chromatids are distributed to daughter cells.

membrane envelope A membrane that surrounds a virus particle.

membrane-bound ribosomes Ribosomes associated with a cell's internal membranes; they are responsible for the synthesis of secreted proteins as well as many membrane-associated proteins.

memory cell In the immune system, a cell that can later be stimulated to produce effector cells with particular antigen specificity.

menstruation [Latin, *mens* = month] The monthly shedding of blood, mucus, vaginal secretions, and endometrial tissue out through the vagina.

meristem [Greek, *meristos* = divided] In plants, a region of undifferentiated, actively dividing cells.

mesenchyme [Greek, *mesos* = middle + *enchyma* = infusion] Loosely attached cells embedded within a jellylike substance.

mesoderm [Greek, *mesos* = middle + *derma* = skin] The middle layer of a vertebrate embryo, including the future connective tissues (bones, muscles, and tendons) and the cells of the blood.

mesoglea [Greek, *mesos* = middle + *glia* = glue] In cnidarians, the jellylike material that lies between the two cellular layers.

mesophyll [Greek, *mesos* = middle + *phyllon* = leaf] Green parenchymal cells that are responsible for most of a plant's photosynthesis.

messenger RNA (mRNA) The RNA molecules that carry information from DNA to the ribosomes, where the mRNA is translated into a polypeptide.

metabolism [Greek, *metabole* = change] All the chemical reactions occurring within an organism.

metacentric The term referring to a chromosome in which the centromere is at or near the middle; such a chromosome looks like a V.

metamere One of the identical (or nearly identical) sections that make up an annelid's long, segmented body.

metamorphosis In many animal species, a series of dramatic changes in form leading from a larva to an adult.

metaphase The stage of mitosis or meiosis during which chromosomes move halfway between the two poles of the spindle, where they accumulate in the metaphase plate.

metaphase plate A disc formed during metaphase in which all of a cell's chromosomes lie in a single plane at right angles to the spindle fibers.

metapopulation A group of local populations, which may alternately go extinct and recolonize one another's vacant habitat islands.

metastasize Referring to cancer cells, to spread to other parts of the body.

metatherians (also called **marsupials**) [Latin, *marsupium* = pouch] Animals whose newborns are relatively undeveloped fetuses.

methanogen A methane-producing bacteria, one of the three phyla of Archaea.

MHC restriction In the immune response, the inability of T cells to recognize foreign antigens except in cells with the same MHC type as the responding animal.

micelle A cluster of amphipathic molecules.

microelectrode An electrode that is small enough to enter or maintain contact with a single cell.

microevolution Changes in the frequencies of alleles of genes in a population.

micronutrient A substance required in minute amounts for the life of an organism; examples include molybdenum, copper, zinc, manganese, boron, iron, and chlorine.

microphyll A leaf with just one conducting vein.

microsomes Vesicles derived from fragments of the endoplasmic reticulum after experimental disruption of eukaryotic cells.

microsurgery The physical manipulation of subcellular structures.

microtubule In eukaryotic cells, the largest elements of the cytoskeleton; they consist of tubulin molecules assembled into hollow rods, about 25 nm in diameter and of variable lengths, up to several μm.

microtubule-associated protein (MAP) A protein that binds to microtubules and influences their organization.

microtubule organizing center (MTOC) The region of a eukaryotic cell from which microtubules emanate.

microvillus (plural, **microvilli**) A membrane-covered extension of an epithelial cell.

midbody In a dividing plant cell, the thin connection between daughter cells that persists until the end of cytokinesis; the midbody is packed with microtubules from the spindle apparatus.

midpiece In a spermatozoan, a section that contains a microtubule organizing center and dense concentrations of mitochondria—sources of ATP for the journey to the egg.

mimicry Pretending to be something else.

mineral An inorganic substance, especially one required for the life of an organism.

minipill A birth control pill that contains only progesterone, with no estrogens.

missense mutation A change in DNA that alters the codon for one amino acid into a codon for another amino acid.

mitochondrion [plural, **mitochondria**; Greek, *mitos* = thread + *chondrion* = a grain] In eukaryotes, the major energy-producing organelle; in cells that do not directly harvest sunlight, mitochondria produce nearly all of the ATP the cell needs to power the chemical reactions of the cell.

mitochondrial matrix The compartment surrounded by the inner mitochondrial membrane.

mitosis [Greek, *mitos* = thread] The process of the equal distribution of chromosomes during cell division.

mitotic spindle An elongated structure that develops outside the nucleus during early mitosis; contains the microtubular machinery that moves the chromatids apart.

mittelschmerz [German, *mittel* = middle + *schmerz* = pain] Pain during ovulation.

mobile gene A gene whose chromosomal address changes.

modal action pattern A highly stereotyped innate behavior.

model A simplified view of how the components of a structure operate.

modern synthesis The marriage of genetics and evolutionary theory that elucidated the genetic basis of variation and natural selection.

modifier gene A gene that regulates the expression of another, separate gene.

molarity The concentration of a dissolved substance in moles per liter of solution.

mold A protist or fungus that grows as a downy coating on animal or vegetable matter.

mole The amount of a substance in grams equal to its molecular mass relative to a hydrogen atom.

molecular genetics The branch of genetics that studies how DNA carries genetic instructions and how cells carry out these instructions.

molecular weight The mass of a molecule relative to that of a hydrogen atom.

molecule A specific combination of individual atoms held together by covalent bonds.

mollusk [Latin, *molluscus* = soft] A soft invertebrate inside a shell.

molting The shedding of feathers, fur, or exoskeleton and the secretion of new ones.

Monera The kingdom that, until recently, was considered to include all prokaryotes, including both the eubacteria and the archaea; when used now, the term refers only to the eubacteria.

monoclonal antibody An antibody produced from a single hybridoma clone; each monoclonal antibody has a defined chemical structure and has the ability to recognize a single epitope.

monocotyledon (or **monocot**) One of a large group of flowering plants that have one cotyledon. Monocotyledons consist of about 65,000 species, including all of the grasses, lilies, palms, and orchids. Most are herbaceous.

monoculture The raising of a single crop.

monomer [Greek, *mono* = single] Each of the component parts of a polymer.

monophyletic A taxon that includes an ancestral species and all its descendants.

monosaccharide [Greek, *mono* = one + *saccharine* = sugar] A simple sugar with three to nine carbon atoms; the building block for polysaccharides.

monosomy The presence of only a single copy of a chromosome in a diploid cell.

monotreme (also called **prototherian**) [Greek, *monos* = single + *trema* = hole] An egg-laying mammal; like other mammals, monotremes have hair and mammary glands; their digestive and reproductive systems empty into a cloaca, hence their name.

morning-after pill A birth control method that is taken after unprotected intercourse.

morphogen A substance that specifies the position of a cell within a pattern.

morphogenesis The development of form.

mortality The probability that an individual of a given age will die each year.

morula [Latin, mulberry] A solid ball of cells in an early mammalian embryo.

motor cortex A region of the brain responsible for movements, with each part of the body specified by a distinct area.

motor neuron (also called a **motoneuron**) A neuron that directly connects with a muscle and commands muscle contraction; the final common pathway of instructions that direct movement.

mucigel In plants, the slimy substance that the epidermis and the root hairs secrete that enhances the absorption of water and minerals.

mucus A viscous, slippery substance that coats the food particles and lubricates their movements within the mouth and the digestive system.

Müllerian mimicry The resemblance of two or more equally dangerous species to each other; the similarities in color or form represent similar dangers to common predators, which therefore avoid all the mimicking species.

multicellular Made of many cells.

muscle fiber A giant muscle cell that contains the proteins responsible for contraction.

mustard oil glycosides Compounds that give cabbage, broccoli, radishes, and other plants in the mustard family their characteristic pungent odor and flavor.

mutagen An agent that increases the rate of mutation; a mutagen can be a chemical or a form of radiation.

mutation A change in the nucleotide sequences of DNA.

mutualism An association between organisms of two species from which both organisms benefit.

mycelium [Greek, *mukes* = fungus] In a fungus, the mass into which hyphae grow, branch, and intertwine.

mycorrhizae [Greek, *mukes* = fungus + *rhiza* = root] Symbiotic association between the root of a plant with the mycelium of a fungus.

myelin An insulating structure around a nerve fiber; myelin consists of extensions of the plasma membrane of a Schwann cell or an oligodendrocyte.

myoblast A cell that is a precursor of a muscle fiber.

myocardium [Greek, *myos* = muscle] The muscular walls of the heart.

myofibril [Greek, *myos* = muscle + Latin, *fibrilla* = little fiber] A thread, consisting of actin and myosin and about 1 to 2 μm in diameter, that runs the length of a muscle fiber.

myoglobin An iron-containing muscle protein that pulls oxygen from the blood.

myosin A long, two-headed protein that interacts with actin and generates movement in an ATP-dependent manner.

myxomycota [Greek, *muxa* = mucus + *mukes* = fungus] The plasmodial, or acellular, slime molds; they closely resemble the cellular slime molds in appearance and in life cycle.

NAD^+ (nicotinamide adenine dinucleotide) The major electron acceptor in oxidative phosphorylation.

NADH The reduced form of NAD^+.

Na^+-K^+ ATPase (also called the **sodium-potassium pump**) An important active transporter that simultaneously transports sodium ions (Na^+) out of cells and potassium ions (K^+) into cells, using the energy of ATP.

natural selection The differential survival and reproduction of individuals with certain inherited traits; the major mechanism for evolution.

Neanderthal One of the early species of *Homo sapiens*.

negative assortative mating A process in which individuals choose mates who differ somewhat from themselves.

negative feedback The process of neutralizing external changes.

negative regulator A protein that reduces transcription of a particular gene or operon.

nematocyst In cnidarians, a tiny barbed spear fired with water pressure by cnidocytes.

nematodes Roundworms; slender, cylindrical, and unsegmented.

nematomorpha Horsehair worms.

nemertea Ribbon worms.

nephridia (singular, **nephridium**) Tubular excretory organs that remove nitrogen-containing wastes from the coelomic fluid and regulate water and salt concentration.

nephron [Greek, *nephros* = kidney] The functional unit of the kidney formed by the glomerulus, Bowman's capsule, and the renal tubule together.

neritic zone The zone of the ocean that is out from shore, over the continental shelf.

nerve growth factor (NGF) A protein paracrine signal necessary for the growth and differentiation of specific nerve cells.

neural crest A set of embryonic cells derived from the roof of the neural tube; neural crest cells migrate to different locations in the embryo and develop into a variety of different cell types in the adult.

neural plate A flat plate above the notochord in a vertebrate embryo; the future nervous system.

neural tube A hollow tube that forms from the neural plate, above the notochord in a vertebrate embryo; the precursor of the central nervous system.

neuromuscular junction The chemical synapse between a motor neuron and a voluntary muscle cell.

neuron A nerve cell specialized for the conduction of electrical and chemical signals.

neurotransmitter A signaling molecule that transmits signals from a nerve cell either to another nerve cell or to a muscle or a gland.

neurulation The process of forming the neural tube and neural crest.

neutrons Neutral particles without electric charge found in the nucleus of an atom.

niche The way an organism uses its environment.

nicotine A naturally occurring compound (especially in tobacco) that acts as a stimulant and narcotic; like heroin and cocaine, it is highly addictive.

nicotinic acetylcholine receptor A ligand-gated cation channel that responds to acetylcholine and nicotine; such receptors are found in the neuromuscular junction and elsewhere.

nitrification The process by which energy is obtained by converting ammonia to nitrite (NO_2) and nitrate (NO_3).

nitrifying bacteria Bacteria specialized to convert atmospheric nitrogen (N_2) into ammonia or nitrate.

nitrogen fixation The conversion of atmospheric nitrogen gas into ammonia, which makes nitrogen available to organisms; nitrogen fixation is carried out only by specialized prokaryotes, some of which live within the roots of some plants.

nitrogenous base (so-called because it contains nitrogen and can accept hydrogen ions) A small molecule that contains one or two aromatic rings of carbon and nitrogen, with attached hydrogen atoms; a component of a nucleotide.

nitrogenous wastes Nitrogen-containing products of protein breakdown.

node The region of the stem to which a petiole attaches.

node of Ranvier A gap between myelin wrappings along a nerve axon.

nonassociative learning Learning in which the subject changes sensitivity to a stimulus after repeated exposure.

noncompetitive inhibitor A molecule whose inhibitory effects on an enzyme cannot be overcome by increased substrate concentration; a noncompetitive inhibitor usually binds to an enzyme at a location other than the active site.

noncovalent interaction Attraction (or repulsion) between separate molecules or ions; noncovalent interactions include electrostatic interactions, hydrogen bonds, van der Waals interactions, and hydrophobic interactions.

noncyclic photophosphorylation The production of ATP from light energy by a series of electron transfers that do not directly regenerate the absorbing chlorophyll; in plants, noncyclic photophosphorylation occurs within photosystem II, with electron flow ultimately depending on both photosystems.

nondisjunction The failure of homologous chromosomes or sister chromatids to move apart during meiosis; nondisjunction results in a gamete having too many or too few chromosomes.

noninducible Not made at higher levels in the presence of an inducer that would normally increase production.

nonpolar Having an approximately uniform charge distribution.

nonsense Not specifying any amino acid; a nonsense codon is also called a "stop" codon.

nonsense mutation A change in DNA that changes the codon for an amino acid into a nonsense or termination codon and thereby leads to a truncated polypeptide.

norepinephrine A neurotransmitter derived from tyrosine; norepinephrine is used by the central nervous system and the sympathetic nervous system.

normal distribution A distribution of numbers (data) where equal numbers of data points fall on each side of the mean and all fall within a certain distance of the mean, forming a bell shape. When data have a normal distribution, about 68 percent of the data will fall within one standard deviation of the mean, and about 95 percent will fall within two standard deviations.

notochord A rod that runs from the front to the rear of a vertebrate embryo, beneath its back (dorsal) surface.

nucellus [Latin, *nucella* = a small nut] In a flowering plant, the central portion of the ovule in which the embryo develops; the megasporangium.

nuclear envelope The boundary of a nucleus; it consists of a double membrane separated by about 20 to 40 nm.

nuclear fission The breakup of large nuclei, of uranium or plutonium, for example.

nuclear fusion The fusion of two atomic nuclei to form a larger one—formation of a helium nucleus from two hydrogen nuclei, for example.

nuclear pore An interruption of the nuclear envelope that forms channels between the contents of the nucleus and the cytosol.

nuclear reactions The high-energy transformation of individual atoms; occurs in stars and in nuclear reactors.

nuclear transplantation A technique for moving a nucleus from one cell to another.

nuclear winter A world-wide darkening and cooling that is believed to be the devastating outcome of a large-scale nuclear war; the fine dust from nuclear bombs and from the fires they would ignite could create a dust cloud that would envelop the Earth and reduce sunlight (and photosynthesis) to a few percent of the present level; temperatures would plunge to below freezing in most parts of the world, triggering mass extinctions, similar to those of the Permian and Cretaceous Periods.

nuclease   An enzyme that catalyzes the hydrolysis of a nuclei acid.

nucleic acid   A macromolecule (DNA or RNA) formed by the polymerization of nucleotides.

nucleoid   In the prokaryotes, the restricted part of a cell that contains the cell's DNA; it is not surrounded by a membrane.

nucleolar organizer   A region of one or more chromosomes that contains the genes for ribosomal RNAs.

nucleolus   [plural, **nucleoli**; Latin, a small nucleus] A conspicuous structure within the nucleus, in which ribosomal RNAs are made.

nucleoplasm   The contents of the nucleus.

nucleoside   A nitrogenous base attached to a sugar by a covalent bond between a carbon atom of the sugar and a nitrogen atom of the base.

nucleosome   A DNA-histone complex, about 11 nm in diameter; each nucleosome contains a 146-nucleotide-long stretch of DNA and eight histone molecules (two each of H2A, H2B, H3, and H4).

nucleotide   A small molecule that consists of a nitrogen-containing aromatic ring compound, a sugar, and one or more phosphate groups.

nucleus   In eukaryotic cells, the membrane-enclosed structure that contains most of a cell's genetic information in the form of DNA.

objective   In a microscope, the lens nearest the sample.

obligate aerobes   Heterotrophs that require oxygen to live.

obligate anaerobe   A microorganism that can grow only in the absence of oxygen.

oceanic zone   The zone of the ocean that is beyond the continental shelf over the deepest water.

oceanography   The study of the seas and their ecosystems.

octet rule   The generalization that an atom is particularly stable and chemically unreactive when its outermost shell is full, meaning (usually) that it contains eight electrons.

ocular   The lens of a microscope nearest the eye.

oil   A triacylglycerol that is liquid at room temperature; the component fatty acids are usually unsaturated.

Okazaki fragment (named after its discoverer Reijii Okazaki)   In DNA synthesis, a stretch of DNA that is to be added to the lagging strand.

olfactory receptor   A protein, in the membrane of the olfactory epithelium, that binds to odorant molecules; binding is the first step in informing the brain about smell.

oligodendrocyte   A glial cell in the central nervous system; the oligodendrocyte's plasma membrane wraps around nerve cell extensions in a myelin sheath.

oligotrophic   Referring to a pond or a lake that lacks nutrients; oligotrophic lakes are clear and have no permanent algal blooms; they contain more oxygen and support a more diverse community of organisms than eutrophic lakes and ponds.

ommatidia (singular, **ommatidium**)   The simple light-detecting units that make up compound eyes.

omnibacteria   Bacteria that can perform aerobic respiration and can use nitrate (NO_3^-) as an electron acceptor; the most common organisms on Earth.

omnivores   [Latin, *omnis* = all + *vorus* = devouring] Animals that eat plants, herbivores, and other carnivores.

oncogene   [Greek, *onkos* = bulk, tumor] A gene whose product can change normal cells into cancerlike cells.

oocyte   A precursor of an egg cell.

oogenesis   The production of mature eggs.

oogonium   A precursor cell that is committed to forming female gametes (ova).

open circulatory systems   A circulatory system in which blood mixes freely with extracellular fluids and bathes the organs of the body.

open system   In thermodynamics, a region of space that exchanges both materials and heat with its surroundings.

operator   The DNA sequence in an operon to which a repressor protein binds.

operculum   In fish, a protective bony flap that covers several gill arches on each side of the head; in mollusks, a protective flap that covers the mantle cavity.

operon   In prokaryotes, a set of genes transcribed into a single mRNA.

opiates   Morphine, heroin, and related compounds.

opposite   The arrangement of leaves on a stem so that two petioles extend away from the stem from the same node.

optical isomers   Two molecules that differ only in the arrangement of four different groupings of atoms attached to a single carbon atom.

optimal foraging   Energy-efficient feeding.

orbital (so-called to distinguish from a planet's restricted "orbit")   A limited portion of an atom's space through which an electron moves.

order   A taxonomic group that consists of one or more families; a subdivision of a class.

organ   A structural unit with a distinctive function formed by two or more kinds of tissue.

organ formation   During embryonic development, the movement and specialization of tissues and cells to produce functioning organs such as heart, kidneys, and the nervous system.

organelle   A subcellular structure that performs a specialized task; in eukaryotic cells, many organelles are enclosed by membranes, which isolate the contents of the organelle from the rest of the cytoplasm.

organic   Carbon-containing; the term refers to all carbon-containing compounds, even when they have nothing to do with organisms.

organic chemistry   The study of the structures and reactions of carbon compounds.

organism   A living individual.

orgasm   A complex of changes that often accompany sexual intercourse—smooth muscle contractions in the genital tract, skeletal muscle contractions throughout the body, and feelings of intense pleasure.

origin of replication   A DNA sequence at which replication begins.

oscilloscope   A measuring instrument that displays voltages on a televisionlike screen, showing variations with time.

osmolarity   [Greek, *osmos* = push, thrust] The sum of the concentrations of all the ions and molecules in a solution.

osmosis   [Greek, *osmos* = push, thrust] The flow of water across a selectively permeable membrane as a result of concentration differences.

osmotic pressure   The pressure exerted by osmosis.

osteichthyes   [Greek, *osteon* = bone + *ichthys* = fish] Fish with bony skeletons.

osteoblasts   [Greek, *osteon* = bone + *blastos* = bud] Specialized cells that resemble fibroblasts and manufacture the bone matrix.

osteoclasts   [Greek, *osteon* = bone + *klastos* = broken] Specialized cells that digest collagen and bone matrix.

ostracoderms   [Greek, *ostrakon* = shell + *derma* = skin] Armor-plated jawless fishes, all of which are extinct.

outbreeders   Species with complex physical or behavioral adaptations that promote crossbreeding with individuals who are not closely related.

ova [singular, **ovum**; Latin, egg] Eggs.

ovarian cycle Refers to the events of the menstrual cycle within the ovaries—the production of the oocytes and the growth of the corpus luteum.

ovary [Latin, *ovum* = egg] The organ (in an animal or a flower) that produces female germ cells (or gametophytes); a female gonad.

overlapping genes A single stretch of DNA that contains the information for distinct polypeptides in different reading frames.

overshoot Population growth that exceeds the carrying capacity.

oviduct [Latin, *ovum* = egg + *ductus* = duct] One of two long tubes that lead to the uterus; in humans, it is called a fallopian tube.

ovulation The release of the oocyte from the follicle.

ovule [Latin, *ovulum* = little egg] The structure in a carpel that, after fertilization, forms a seed.

ovum [plural, **ova**; Latin, egg] A female gamete; an egg.

oxaloacetate A four-carbon compound that can combine with the acetyl group from acetyl CoA to produce citric acid, beginning the citric acid cycle.

oxidative fiber A thin muscle fiber that derives most of its energy from respiration.

oxidative phosphorylation The process that couples the oxidation of NADH and $FADH_2$ to the production of high-energy phosphate bonds in ATP.

oxidizing agent The electron acceptor in a redox reaction.

oxygen evolving complex In photosynthesis, a manganese-containing protein that directly participates in the electron transfers that convert coordinates of water to oxygen.

oxyhemoglobin Hemoglobin that is bound to oxygen.

oxytocin A peptide hormone, released by the posterior pituitary, that stimulates contractions of the uterus during childbirth.

ozone layer The layer of ozone gas 20 to 50 kilometers above the Earth; ozone is a highly reactive molecule made of three oxygen atoms, instead of the two in atmospheric oxygen.

pacemaker A group of the cells of the myocardium that are capable of rhythmic spontaneous contractions; they contract slightly more frequently than most myocardial cells; by initiating contraction before other cells in the heart, the pacemaker cells set the pace of contraction for the rest of the heart

paleontology [Greek, *palaios* = ancient + *on* = being + *logos* = discourse] The study of ancient life.

palisade parenchyma In plants, the mesophyll cells that lie just under the leaf's upper epidermis and are elongated and packed with chloroplasts.

paracrine signal A chemical signal that acts only in the immediate region of its production.

parapatric speciation A mechanism of speciation in which a species splits into two populations that are geographically separated, but still have some contact.

paraphyletic Refers to a taxon that contains some but not all the descendants of the ancestral species.

parapodia In polychaetes, a pair of leglike paddles in a single segment; parapodia are used in respiration and to swim, crawl, or burrow.

parasite An organism that consumes parts of a larger organism; it does not necessarily kill the host.

parasitism A symbiotic relationship in which one species benefits at another's expense.

parasympathetic nervous system [Greek, *para* = beside, next to + sympathetic] A division of the autonomic nervous system; it generally acts to husband resources, for example by slowing the heart and increasing intestinal absorption.

Parazoa The smaller of two subkingdoms of animals, which contains two phyla: the Placozoa and the Porifera (sponges); the parazoa differ greatly from all other animals in showing no symmetry and minimal organization and cell specialization—they possess only simple connective tissues and no organs.

parenchymal cells Plant cells with thin cell walls, chloroplasts, large vacuoles, and the machinery to perform photosynthesis.

parental generation (also called **P generation**) The original parents in a genetic cross.

passive transport Movement of a substance across a membrane that occurs spontaneously, without the expenditure of energy.

patch clamping A method that allows the measurement of currents across a tiny piece of membrane rather than across a whole cell membrane.

pathogens Agents that cause disease.

pattern formation The creation of a spatially organized structure during embryonic development.

pedigree A family tree; pedigrees are often used to show the inheritance of a disease within a family.

pelagic division The open waters of the ocean and the organisms within them.

pellet The part of a sample that, during centrifugation, moves to the bottom of the centrifuge tube.

penetrance The fraction of individuals with a particular genotype that show a corresponding phenotype.

pentose [Greek, *pente* = five] A sugar that contains five carbon atoms; ribose and deoxyribose, which are components of nucleotides, are pentoses.

pepsin The stomach enzyme that hydrolyzes proteins into smaller fragments each containing a few amino acids.

peptide A molecule that consists of amino acids linked by peptide bonds; some peptides are water-soluble signals.

peptide bond The covalent bond between two amino acid molecules in a polypeptide; these bonds are formed by the carboxyl group of one amino acid attaching to the amino group of another amino acid.

peptide bond formation During protein synthesis, the transfer of an amino acid from aminoacyl-tRNA to the growing polypeptide chain.

peptidoglycan The cell-wall material of many prokaryotes; peptidoglycan may consist of a single covalently linked molecule formed from polypeptides and polysaccharides.

perennial Refers to a plant that lives and produces seeds for two or more years.

perforation plate In the vascular system of a plant, the end wall of each vessel element; the perforation plate contains one or more holes through which water flows.

pericardium [Greek, *peri* = around + *kardia* = heart] A fibrous sac that encloses the heart itself within a watery lubricating fluid.

pericycle [Greek, *peri* = around + *kykos* = circle] In plant roots, a sheath of parenchyma cells just within the endodermis, but outside the vascular tissue.

periderm In plants, the outer, secondary tissue; periderm is primarily cork.

period In the geological time scale, a time interval of 30 to 75 million years that contains distinctive forms of life in its fossil record; a period is longer than an epoch, but shorter than an era.

peripheral nervous system (PNS) Nerve cells outside the central nervous system (CNS); the PNS carries information from the CNS to muscles and organs and to the CNS from sense organs.

peripheral proteins Membrane proteins located on the outer and inner surfaces of the plasma membrane.

peristaltic waves [Greek, *peristellein* = to wrap around] Coordinated contractions of smooth muscles.

permafrost Permanently frozen ground less than a meter from the surface that underlies the soil during even the warmest summers.

peroxisome In eukaryotic cells, a kind of membrane-bounded vesicle; peroxisomes contain enzymes that use oxygen to break down molecules by pathways that produce hydrogen peroxide (H_2O_2).

petal One of the showy, usually colored, parts of a flower.

petiole The stalk that connects the leaf to the stem.

pH The logarithm (to the base of 10) of the molar hydrogen ion concentration.

pH scale Measures the concentration of H^+ and OH^- ions in a solution.

phaeophytes [Greek, *phaios* = dusky brown] Brown algae.

phagocytosis [Greek, *phagein* = to eat + *kytos* = hollow vessel] A type of cellular ingestion in which the cell's membrane surrounds a relatively large solid particle, such as a microorganism or cell debris.

pharyngeal slits (also called **gill slits**) Holes in the sides of the body that run from the inside of the pharynx to the outside surface of an animal.

pharynx The throat; a common entryway for food into the digestive tract, for air into the lungs, and for water into the gills.

phenetics A method of classification, based on all observable characteristics, that attempts to avoid the subjective choices of classical taxonomy.

phenolic compounds Aromatic substances that play a variety of roles in plants; some phenolics repel herbivores and pathogens, some attract pollinators or fruit dispersers.

phenotype [Greek, *phainein* = to show + *typos* = impression] The collection of all the properties of an individual organism.

phenotypic trait A single aspect of phenotype in which individuals may vary.

phenylketonuria (PKU) A human disease that results from the absence of the enzyme phenylalanine hydroxylase, which converts phenylalanine to tyrosine.

pheromone A substance secreted by one organism that influences the behavior or physiology of another organism of the same species; a pheromone can also activate modal action patterns.

phloem Conducting vessels that distribute the sugars and other organic molecules made in the leaves to the rest of the plant.

phosphate The ion (PO_4^{3-}) formed by the dissociation of hydrogen ions from phosphoric acid (H_3PO_4).

phosphodiester bond The links between nucleotides formed by the phosphate group of one nucleotide attaching to a carbon atom in the sugar component of another nucleotide.

phospholipid An amphipathic derivative of glycerol in which two hydroxyl groups attach to fatty acids and the third to a phosphate ester; phospholipids are principal components of biological membranes.

phosphorylase kinase An enzyme that transfers phosphate groups from ATP to glycogen phosphorylase.

photoautotroph An organism that supplies itself with energy by means of photosynthesis.

photochemical reaction center A complex of the chlorophyll molecules and proteins that convert captured light energy to chemical energy.

photoheterotroph A heterotroph that uses light energy but also requires organic compounds.

photon A package of energy; a light particle.

photoperiodism The response of a plant to the relative lengths of day and night.

photorespiration In plants, an oxygen-dependent process that does not produce ATP or NADPH; photorespiration converts ribulose bisphosphate into CO_2 and serine.

photosynthesis [Greek, *photo* = light + *syntithenai* = to put together] Process in which sugars are synthesized within the chloroplast and temporarily stored there.

photosystem I One of two distinct but interacting sets of electron transfer reactions responsible for storing light energy in high-energy chemical bonds; photosystem I best absorbs and uses light with wavelengths of about 700 nm.

photosystem II One of two distinct but interacting sets of electron transfer reactions responsible for storing light energy in high-energy chemical bonds; photosystem II best absorbs and uses light with wavelengths of about 680 nm.

phototropism The bending of a plant toward light.

phragmoplast [Greek, *phragmos* = fence + *plasma* = mold, form] A set of microtubules that extends between two dividing plant cells at right angles to the cell plate.

phylogeny [Greek, *phylon* = race, tribe + *geneia* = birth, origin] Evolutionary history.

physical anthropologist [Greek, *anthropo* = human] Researchers who study the evolution, anatomy, and behavior of humans.

physiology The study of how living organisms are organized and how their parts work together.

phytochrome [Greek, *phyton* = plant] A plant pigment involved in many processes that depend on the timing of dark and light, including flowering, germination, and leaf formation.

pigment A molecule that absorbs visible light and has color to human eyes.

pilus (plural, **pili**) In bacteria, a long appendage that serves as the means of attachment of conjugating bacteria and a conduit for the transfer of DNA.

pinnae [singular, **pinna**; Latin, feather] In plants, small leaflets that extend from a central stalk.

pinocytosis [Greek, *pinein* = to drink] A type of endocytosis; the nonspecific uptake of bits of liquid and dissolved molecules.

pioneer community The first community in a succession.

pistil (also called a **carpel**) In a flower, a female reproductive structure.

pith In plants, the unspecialized tissue within central nonconducting core of the vascular system.

pituitary gland A pea-sized structure at the base of the brain that releases at least nine hormones.

placenta [Latin, flat cake] Tissue that passes nourishment, oxygen, and wastes between the mother and the developing embryo in placental mammals.

placental mammals Mammals whose embryonic development takes place entirely within the uterus of the mother.

placoderms [Greek, *plak* = plate + *derma* = skin] Extinct armored fish with jaws.

plankton [Greek, *planktos* = wandering] Organisms that float in the water, carried by currents.

Plantae The kingdom that includes plants, autotrophic multicellular organisms that undergo embryonic development.

plasma [Greek, form or mold] The fluid part of blood.

plasma cell A cell specialized for the production of antibodies.

plasma membrane The membrane that surrounds a cell.

plasmid A circular DNA that can replicate autonomously.

plasmodesmata [singular, **plasmodesma**; Greek, *plassein* = to mold + *desmos* = to bond] Fine intercellular channels between plant cells, derived from vesicles trapped in the growing cell plate.

plasmodium A mass of cytoplasm with many nuclei but no boundaries between cells.

plasmolysis [Greek, *plasma* = form + *lysis* = loosening] In a plant cell, the flow of water out of the vacuole in a hypertonic solution and the resulting separation of the plasma membrane from the cell wall.

plastid A plant organelle surrounded by a double membrane; plastids include chloroplasts, chromoplasts, and amyloplasts.

platelet A small, membrane-enclosed element that is a component of mammalian blood; it is formed as a cytoplasmic fragment of a precursor cell in the bone marrow; platelets contribute to clotting.

plate In reference to the Earth's crust, a large block that moves with respect to other blocks at the rate of a few centimeters a year.

platyhelminthes The phyla of the flatworms.

platyrrhines [Greek, *platyus* = flat + *rine* = nose] Monkeys whose flat noses have widely separated nostrils that point sideways.

pleiotropy [Greek, *pleios* = more + *trope* = turning] The capacity of a single gene to affect many aspects of phenotype.

point mutation (also called a **single-base mutation**) A change in a single nucleotide pair of DNA; in some cases, a change in a few nucleotide pairs still qualifies as a point mutation.

polar Having uneven distributions of electrical charge; having positive and negative ends (or poles).

polar body The smaller daughter cell that results from the unequal meiotic division of an oocyte.

polar fronts Regions of low pressure, where the westerlies end and polar easterlies begin.

polar microtubules Microtubules that run from each pole of the mitotic spindle toward the equator.

pollen tube An extension of a pollen grain, which carries the sperm nuclei into the ovule.

polyclonal Referring to antibodies made in the descendants of many B cells, with each B-cell clone producing an antibody that recognizes a different epitope.

polygenic Traits governed by many genes that vary smoothly and continuously within a population.

polymer [Greek, *poly* = many + *meros* = part] A large molecule that consists of smaller identical (or nearly identical) subunits, called monomers.

polymerase chain reaction (PCR) A method that specifically and repetitively copies a segment of DNA between two defined nucleotide sequences.

polymerization [Greek, *polys* = many + *meros* = part] The assembly of many small molecules into a larger one.

polymorphic [Greek, *poly* = many + *morph* = form] Having two forms of a given gene; the term may refer to a gene or to a population.

polynucleotide A chain of nucleotides held together by phosphodiester bonds; DNA and RNA are polynucleotides.

polyp A body plan that resembles a cylinder; one of the two basic body plans of the cnidarians.

polypeptide A chain of amino acids held together by peptide bonds.

polyphyletic Refers to a taxon that includes descendants of more than one ancestor.

polyploid Having two or more complete sets of chromosomes, as in some plants.

polyribosome (also called a **polysome**) Several ribosomes attached to a single mRNA.

polysaccharide A carbohydrate macromolecule formed by the polymerization of simple sugars (monosaccharides).

polyspermy Fertilization by more than one sperm.

polytene chromosomes [Greek, *polys* = many + *tainia* = ribbon] Giant chromosomes, commonly studied in the larvae of *Drosophila* and other flies, that consist of about 1000 chromatids aligned in parallel.

polyunsaturated Referring to a hydrocarbon chain (or fatty acid) with many double bonds.

Pongidae Gorillas, chimpanzees, and orangutans.

population A breeding group of individuals of the same species that inhabit a common area.

population genetics The quantitative study of the processes by which variation is generated and passed on within populations.

population momentum The continued growth of a population even though completed family size is below replacement level.

positional information Chemical cues that establish the position of a cell in a pattern.

positioning During protein synthesis, the binding of the next aminoacyl-tRNA to the A site in the ribosome-mRNA complex; the first step of elongation.

positive assortative mating Choosing mates on the basis of similarities to the individual's own genotype.

positive regulator A protein that increases the rate of transcription.

postabsorptive state The state of an animal in which cells derive energy and building blocks by breaking down stored glycogen, fats, and proteins.

posterior The back (tail-end) of an animal.

postsynaptic Refers to the neuron that is "downstream" from a synapse; a postsynaptic neuron responds to the presynaptic neuron.

postsynaptic potential A voltage change induced by the action of a neurotransmitter.

posttranscriptional processing A series of chemical modifications that convert the primary transcript of a gene to a mature RNA.

posttranslational Occurring after translation is completed.

postzygotic barrier Barrier to gene flow which makes a zygote either inviable (certain to die) or sterile.

potency What a cell or tissue could become during development if it were allowed to develop in another environment.

potential energy A general term for energy that can ultimately be converted into kinetic energy.

pre-mRNA The primary transcript of a protein-coding gene; the term is also used to refer to the partly modified, not yet mature mRNA precursor.

prebiotic evolution The evolution of organisms from nonliving matter.

Precambrian Referring to fossils that date from more than 590 million years ago, that is, before the beginning of the Cambrian Era.

predator An organism that usually (but not always) kills its prey and consumes most of its prey's body.

preformation The theory, now discredited, that all the parts of the adult organism already preexist at the earliest stages of life.

pressure flow hypothesis The accepted explanation for the transport of sap through the phloem.

presumptive Having a specified developmental fate.

presynaptic Refers to the neuron that is "upstream" from a synapse; an action potential in the presynaptic neuron can trigger the release of neurotransmitter into the synapse.

prezygotic barrier A barrier to gene flow that prevents the fusion of the sperm and egg to form a zygote (syngamy).

primary cell wall The wall that lies outside the plasma membrane.

primary compounds In plants, refers to compounds that are essential for day-to-day functioning; primary compounds include glucose, amino acids, ATP, and DNA.

primary growth In a plant, growth from apical meristem tissue; the extension of length.

primary lysosome A nearly spherical lysosome that has not yet fused with an endosome.

primary mesenchyme [Greek, *mesos* = middle + *enchyma* = infusion] In an animal embryo, the first cells to move into the interior of a blastula.

primary oocyte A precursor of an ovum; a diploid cell that is capable of dividing, by meiosis, to form a haploid ovum.

primary organizer The term used by Spemann to describe the dorsal lip of the blastopore, meaning that its action established the organization of the entire early embryo.

primary productivity The productivity of the first trophic level, the energy captured in the chemical bonds of new molecules each year for each square meter.

primary structure The linear sequence of amino acids in each polypeptide chain of a protein.

primary succession The invasion of a completely new environment such as a sandbar or new volcanic island.

primary transcript The RNA first transcribed from a particular gene; the precursor of a mature RNA.

primates The order of mammals that includes humans, apes, monkeys, and lemurs.

primer An already existing polynucleotide to which additional nucleotides are added; all known DNA polymerases add nucleotides to a primer.

primitive characters Traits shared by all the members of a lineage and also shared with more distant relatives. In members of the cat family, for example, long canine teeth are a primitive trait because they are shared with other members of the order Carnivora, such as canids (dogs) and bears.

primordial germ cells The precursors of both sperm and eggs, which arise early in development.

Principle of Independent Assortment (also called **Mendel's Second Law**) The generalization that the alleles for one gene segregate independently of the alleles of another gene.

Principle of Segregation (also called **Mendel's First Law**) The generalization that a sexually reproducing organism has two "determinants" (or genes, in modern terms) for each characteristic, and these two copies segregate (or separate) during the production of gametes.

probability The chance that a given event will happen; it is calculated as the number of times an event has actually occurred divided by the number of opportunities it could have occurred; in the most familiar example, the probability of a tossed coin turning up "heads" is 1 (the number of sides that are heads) divided by 2 (the total number of sides).

proboscis A long, sensitive, retractable, and sometimes venomous, tubelike snout characteristic of ribbon worms; a long, flexible snout or trunk characteristic of elephants. Insects often drink using a proboscis.

producers Organisms, such as plants, that harvest energy directly from sunlight or, rarely, from inorganic molecules.

product rule In the calculation of probability, the statement that the probability of two independent events taking place together is the product of their probabilities.

proglottid One of the repeated segments of a tapeworm; a proglottid contains both male and female reproductive organs.

programmed cell death The death of specific cells as a normal part of development.

prokaryotic [Greek, *pro* = before] Cells that contain neither a nucleus nor other membrane-bounded organelles.

prolactin A peptide hormone, made in the anterior pituitary, that stimulates milk production.

prometaphase [Greek, *meta* = middle] Prometaphase was previously called early metaphase, the stage of mitosis during which the nuclear membrane disappears and the chromosomes attach to the spindle fibers.

promoter A DNA sequence that specifies the starting point and direction of transcription; it is where RNA polymerase first binds as it begins transcription.

prophage [Greek, *pro* = before] The dormant form of a bacteriophage in which the phage DNA has integrated into that of the host cell.

prophase [Greek, *pro* = before] The first phase of mitosis, when the diffusely stained chromatin resolves into discrete chromosomes, each consisting of two chromatids joined together at the centromere.

proplastid Common precursor of all plastids.

prostaglandin One of 16 paracrine signals derived from a 20-carbon fatty acid called arachidonic acid.

prostate gland In the reproductive tract of a male mammal, a gland that secretes a thin milky alkaline fluid into the lumen of the urethra.

protease (also called a **peptidase**) A digestive enzyme that catalyzes the hydrolysis of peptide bonds; used to digest proteins.

protein A macromolecule consisting of one or more polypeptides.

protein kinase An enzyme that transfers the phosphate group from an ATP to a protein.

protein kinase A An enzyme that, when stimulated by cyclic AMP, transfers a phosphate group from ATP to specific proteins.

protein kinase C An enzyme that, when bound to diacylglycerol, transfers phosphate groups from ATP to sites in particular proteins.

proteinoids Polypeptides thought to have been the first to appear in prebiotic evolution; proteinoids are almost certainly without specific sequences of amino acids.

Protista [Greek, *protos* = first] The kingdom that consists mostly of single-celled organisms but that also contains some related multicellular species; protists include algae, water molds, slime molds, and protozoa.

proto-oncogene (also called a **cellular oncogene**) The cellular counterpart of a viral oncogene; proto-oncogenes participate in the normal control of growth and differentiation.

protobiont A primitive cell—thought to represent a stage of prebiotic evolution—that could concentrate organic molecules and begin to evolve the first metabolic pathway.

proton A positively charged particle found in the nucleus of an atom; the word proton is commonly but inaccurately used to refer to a hydrogen ion (H^+, actually a hydronium ion H_3O^+), which comes from the dissociation of a water molecule.

proton channel The route that hydrogen ions follow as they flow down an electrochemical gradient.

proton gradient The difference in H^+ concentration across a membrane.

protoplasm [Greek, *proto* = first + Latin, *plasma* = a thing molded or formed] A term used by nonbiologists to refer to the living material of organisms.

protostomes [Greek, *protos* = first + *stoma* = mouth] Those animals in which the mouth develops first, from the blastopore; protostomes include mollusks, annelids, and arthropods.

prototherian (also called **monotreme**) Egg-laying mammal; like other mammals, prototherians have hair and mammary glands, but their digestive and reproductive systems empty into a cloaca.

protozoa [Greek, *protos* = first + *zoe* = life] The protists that most resemble little animals.

provirus [Greek, *pro* = before] The dormant form of a virus in which the viral DNA has integrated into that of the host cell.

proximal tubule The tubule that lies just next to Bowman's capsule, in the kidney's outer cortex.

pseudocoel [Greek, *pseudo* = false + *koilos* = hollow] A cavity in certain invertebrates that lacks the mesodermal lining of a true coelom.

pseudocoelomates [Greek, *pseudo* = false] Animals that have an organ-containing cavity that lacks the mesodermal lining of a true coelom.

Pseudomonads A group of bacteria that are typically straight or curved gram-negative rods with flagella; they seem to live everywhere—in soil, ponds, infected wounds, hot tubs, and even medicine bottles.

pseudopodia [Greek, *pseudos* = false + *pous* = foot] Temporary extensions of the cytoplasm.

pulmonary [Latin, *pulmo* = lung] Relating to the lungs.

pulmonary artery The artery that carries deoxygenated blood from the heart to the lungs.

pulmonary circulation Circulation from heart to lungs to heart.

pulmonary veins The veins that carry oxygenated blood from the lungs to the heart.

punctuated equilibrium The idea that species change very little most of the time (stasis), and that most anatomical or other evolutionary change in individual species occurs during a geologically brief period at the time of speciation.

pupa An inactive phase, following a larval stage, of insect development; a pupa may be enclosed in a cocoon.

purine A nitrogenous base that contains a particular nine-membered double ring, with five carbon and four nitrogen atoms; adenine and guanine are purines.

pyramid of biomass A graphic presentation of the total mass of organisms at each trophic level of an ecosystem.

pyramid of energy A graphic presentation of the amount of energy used at each trophic level in a specified time (usually a year).

pyramid of numbers A graphic presentation of the total numbers of organisms at each trophic level.

pyramid of productivity An energy pyramid, which shows how much sunlight energy enters the system how much is stored in the biomass of each trophic level, and how much flows between trophic levels.

pyrimidine A nitrogenous base that contains a particular six-membered ring, with four carbon and two nitrogen atoms; thymine, uracil, and cytosine are pyrimidines.

pyrophosphate A molecule that consists of two phosphate groups linked together in a high-energy bond.

pyruvate A three-carbon compound that is the end point of the first stage of glycolysis.

quantitative character A phenotypic trait that may be described numerically within a range of values rather than as clear-cut alternatives.

quaternary structure The relationship among separate polypeptide chains in a protein.

R group (also called a **side chain**) Each amino acid's characteristic group of atoms.

r-selected species A species that has adaptations that support high growth rate (r_{max}).

races (or **subspecies**) Morphologically distinct subpopulations that can interbreed.

radial symmetry Structural symmetry in which rotation along the central axis doesn't change the appearance.

radioactive isotope A radioactive (unstable) isotope of an element; examples include ^{14}C and ^{3}H.

radula In mollusks, a rasping tongue covered with teeth made from chitin.

rain shadow The area adjacent to a mountain range, away from the prevailing winds, where little rain falls.

random drift Changes in gene frequency not caused by selection, mutation, or immigration, but by random events.

random sample A sample chosen at random with respect to any differences among the members of a population.

reabsorption In the kidney, the transport of specific substances such as salt, water, and glucose from the filtrate back into the blood.

reading frame The grouping of nucleotides into codons that specify an amino acid sequence.

realized niche The resources that a species uses in a particular community.

receptor A protein to which a chemical signal first binds.

receptor-mediated endocytosis A cell's uptake of specific substances that are recognized by receptor proteins on the plasma membrane.

recessive The term referring to an allele that does not contribute to the phenotype of a heterozygote.

reciprocal altruism An explanation for apparently selfless behavior; the hypothesis that animals engage in altruism with the expectation that the recipient would return the favor some time in the future.

recombinant Containing a combination of genes not found in nature.

rectilinear Movement in a straight line, especially in snakes.

rectum [Latin, straight] The straight portion at the end of the descending colon in which feces are stored until their elimination through the anus.

reducing agent The electron donor in a redox reaction.

reductionism The effort to understand the whole in terms of the parts.

reflex The most automatic behavior pattern, with motor activity directly responding to a sensory stimulus.

refractory period In an axon, the period during which the further excitation does not result in channel opening and impulse propagation.

regulatory cascade A set of reactions that amplify a signal from a relatively small number of molecules into a response that affects many more molecules.

regulatory mutation A change in DNA that affects the amount (rather than the structure) of a polypeptide.

relative abundance A measure of species diversity that takes into account how common each species is.

releasing factor (also called **releasing hormone**) A peptide hormone, made by the hypothalamus, that regulates the release of specific hormones by the anterior pituitary.

renal artery [Latin, *renes* = kidney] The artery that carries blood to each kidney.

renal tubule In the kidney, a long narrow tube leading away from Bowman's capsule.

renal vein The means by which blood leaves the kidneys.

renin An enzyme secreted by the kidneys.

replacement reproduction The family size at which each couple is replaced by just two descendants.

replication [Latin, *replicare* = to fold back] Copying of a single DNA molecule (or a single set of DNA molecules) into two copies.

replication fork A Y-shaped region of DNA where the two strands of the helix have come apart during DNA replication.

replication unit In eukaryotic DNA replication cells, a group of 20 to 50 origins of replication that form replication forks at the same time.

reporter A protein (or gene) whose activity serves as a measure of the ability of a test sequence to regulate expression in a particular type of cell.

reproductive unit A species as the group of potential mates.

reproductively isolated The term referring to populations that are unable to interbreed.

reptile [Latin, *reptare* = to crawl] An air-breathing, egg-laying vertebrate with an external covering of scales or horny plates; reptiles include snakes, lizards, crocodiles, turtles, and dinosaurs.

resilience One definition of ecosystem stability; the speed with which an ecosystem returns to a particular form following a major disturbance such as a fire.

resistance factor (also called an **R factor**) A plasmid that contains a gene for an enzyme that inactivates an antibiotic.

resolution (1) The minimum distance between two objects that allows them to form distinct images; (2) the part of the sexual response cycle in which blood flow and muscle tension return to normal after orgasm.

resource partitioning Splitting an ecological niche.

respiration The oxygen-dependent extraction of energy from food molecules.

respiratory chain The pathway of electrons in oxidative phosphorylation.

respiratory system All the structures responsible for the exchange of gases between the blood and the external environment.

response element (also called a *cis*-**regulatory element** or **enhancer**) A DNA sequence that regulates transcription by binding to a specific protein, a transcription factor.

resting potential The voltage across the cell membrane when a cell is not producing action potentials.

restriction enzyme An enzyme that cuts DNA at a particular sequence.

retrovirus An RNA virus that uses reverse transcriptase for viral replication.

reuptake The pumping of transmitter from the synaptic cleft either into the presynaptic neuron or into surrounding glial cells.

reverse transcriptase An enzyme that copies RNA into DNA; RNA-dependent DNA polymerase; reverse transcriptase is present in RNA tumor viruses and in HIV.

reversion A mutation that restores the previous version of a gene sequence.

rhizoid A thin, rootlike filament that enables mosses, liverworts, and fern gametophytes to anchor and to absorb nutrients.

rhizome A horizontal stem that spreads on or below the ground from which fronds grow.

rhizopoda Amebas.

rhodophytes [Greek, *rodon* = rose] Red algae.

rhodopsin The protein that detects light in the retina.

rhythm method A birth control method; a form of modified abstinence, in which a couple refrains from sexual intercourse at times in the monthly cycle when conception is likely to occur.

ribbon model A way of depicting the tertiary structure of a protein that stresses the secondary structures of a globular protein; in the ribbon model, an α helix is depicted as a coil, while β sheets are sets of arrows.

ribosomes [Latin, *soma* = body + "*ribo*" because they contain *ribo*nucleic acid (RNA)] Complex assemblies of RNAs and proteins, about 15 to 30 nm in diameter, that are responsible for carrying out protein synthesis.

ribozyme An artificial RNA enzyme.

RNA polymerase The enzyme responsible for transcribing DNA into RNA.

root The part of a plant below the ground.

root cap A region of the growth zone of a plant's root; the root cap produces a polysaccharide slime that lubricates the path of the growing root through the soil and serves as a mechanical shield for the apical meristem, which lies just behind.

root hairs Tiny projections, each of which is the extension of a single epidermal cell, through which most of the water that enters a root comes.

root pressure Osmotic pressure generated by the difference in solute concentrations between root tissues and soil solution.

root system An underground system that anchors the plant and absorbs water and minerals from the soil.

rotifers Tiny aquatic aschelminths with saclike bodies.

rough ER Endoplasmic reticulum that is associated with ribosomes.

RU 486 A drug used to induce abortions within the first nine weeks after conception.

rudiment [Latin, *rudimentum* = beginning] Initial embryonic stage of an organ, from which the final form will develop.

ruminants [Latin, *rumen* = throat] Hoofed, horned herbivores that can regurgitate partly digested food, called cud, for further chewing.

S (for DNA *s*ynthesis) The period of the cell cycle during which DNA replicates.

salivary glands Exocrine organs in the mouth that secrete both enzymes and mucus.

saltationism The view, held by many 19th-century biologists, that evolution occurred by fits and starts.

saltatory conduction The jumping movement of an action potential down a myelinated axon, from one node of Ranvier to the next; saltatory conduction occurs at rates up to 100 times faster than conduction down an unmyelinated axon.

sampling errors Chance fluctuations that affect small samples but not large ones.

sands Soils of large particles (from 20 mm to 200 mm).

saprophyte [Greek, *sapro* = putrid + *phyte* = plant] A plant that decomposes dead material.

saprophytic Deriving nourishment from dead organisms.

sarcomere [Greek, *sarx* = flesh + *meros* = part] In muscle, a small repeating cylinder within a myofibril; each sarcomere is about 2.5 μm long; the contractile unit of a striated muscle.

sarcoplasmic reticulum In a muscle fiber, a system of membrane-lined channels and sacs where Ca^{2+} ions are stored; a specialization of the endoplasmic reticulum.

saturated A hydrocarbon chain (or fatty acid) in which all the bonds between carbon atoms are single bonds; the chain is called saturated because it contains the maximum number of hydrogen atoms.

savanna A tropical or subtropical grassland punctuated by solitary trees or small clumps of trees.

scanning electron microscope (SEM) A type of electron microscope in which electrons are reflected from the surface of the observed object.

scavenger An animal that feeds from whole carcasses.

Schwann cell A glial cell in the peripheral nervous system; its plasma membrane wraps around nerve cell extensions in a myelin sheet.

scientific method A formal manner of formulating, testing, and eliminating hypotheses.

sclerenchyma [Greek, *skleros* = hard] A plant tissue that consists of cells, with thick cell walls; may be fibers or sclereids.

scolex A specialized attachment organ at the anterior end of a tapeworm.

scrotum [Latin, bag] A pouch that lies outside the body that holds the testes.

Scyphozoa The class of animals containing the jellyfish.

Second Law of Thermodynamics The statement that, while the total energy in the universe does not change, less and less energy remains available to do work.

second messenger An intracellular molecule whose synthesis and degradation depends on the action of an extracellular signal and which stimulates changes inside a target cell.

secondary cell wall The wall that lies between the cell membrane and the primary cell wall.

secondary growth Growth at the lateral meristems that increases the thickness of a shoot or a root.

secondary immune response A robust immune response triggered by the reappearance of a previously encountered antigen or microbe.

secondary lysosome An irregular lysosome that results from the fusion of a primary lysosome and an endosome; a secondary lysosome contains material that is in the process of being digested.

secondary mesenchyme In an animal embryo, cells that become mesoderm as gastrulation proceeds.

secondary oocyte A precursor of the ovum; a haploid cell produced from the primary oocyte by meiosis.

secondary plant compounds Chemicals that are not essential to a plant's normal metabolism, but which often serve a defensive purpose.

secondary sexual characteristics Hallmarks of sexual differentiation that appear at sexual maturity; in humans these include an increase in body size, a deepening of the voice, hair growth, and the development of sexual drive.

secondary structure Regular local structures resulting from regular hydrogen bonding within adjacent stretches of a polypeptide backbone; secondary structures include α helices and β structures.

secondary succession The sequence of stages in a community that has suffered serious damage.

secretion In the kidney, transport into the filtrate; secreted substances include potassium and hydrogen ions, ammonia, and organic acids and bases.

secretory vesicles In eukaryotic cells, small vesicles formed on *trans* side of the Golgi apparatus; they contain glycoproteins that are to be secreted.

section A thin slice of tissue prepared for microscopy.

sedimentary rock Rock formed by pressure from succeeding layers turning layers of mud, sand, and other sediment into rock.

segmentation gene A gene that regulates the development of segments within the early embryo.

segregation The separation of two homologous chromosomes during meiosis.

selective theory The view that the immune response depends on an animal's ability to select the set of antibodies to make in response to the presence of a particular infection.

selectively permeable Allowing the passage of some ions and molecules (especially of water) much more rapidly than others.

self-splicing Removal of intervening sequences from an RNA without the participation of other molecules.

selfish gene A sequence of DNA whose only role is to reproduce itself.

semen [Latin, seed] The sperm cells, together with the fluid from the seminal vesicles, the prostate gland, and the bulbourethral glands.

semiconservative The pattern of DNA replication in which half of each parent molecule (one strand) is present in each daughter molecule.

seminal vesicles In the male reproductive tract, organs that secrete sugars and other nutrients into the semen.

seminiferous tubules [Latin, *semen* = seed + *ferre* = to bear] The tightly coiled ducts in the testis in which the sperm matures.

senescence Aging; deterioration; in plant cells, the breakdown of cellular components leading to cell death.

sensitive period The period of time during which an animal can learn a particular behavior pattern.

sensitization The increased behavioral response to a noxious stimulus.

sepal One of the parts at the base of a flower; sepals are usually small and green, but in some flowers they may be large and colorful.

septa In a fungus, the walls that separate the nuclei within a filament.

Sertoli cells In the testes, specialized cells that surround the developing sperm; they nourish the developing sperm, regulate the passage of nutrients from the blood, and secrete a fluid that fills the lumen.

sessile Permanently anchored to rocks, logs, or coral.

seta (plural, **setae**) In annelids, a stiff, bristlelike projection.

sex chromosomes The chromosomes that differ between males and females; in humans, the 23rd pair of chromosomes.

sex steroid A steroid hormone that stimulates, maintains, and regulates reproductive organs and secondary sexual characteristics.

sex-determining region In male mammals, a gene on the Y chromosome that encodes a regulatory protein that promotes the sexual differentiation of the embryonic gonads into testes.

sex-linked gene A gene that has a different pattern of inheritance in males and females because the gene lies on a sex chromosome (usually the X chromosome in mammals).

sexual dimorphism A phenotypic difference between the sexes.

sexual imprinting A special form of imprinting that helps animals recognize potential mates.

sexual reproduction A process that produces offspring that have inherited genetic information from two parents rather than one; because the genes from each parent are likely to differ, sexual reproduction provides new combinations of genes.

sexual selection The differential ability of individuals with different genotypes to acquire mates.

shared derived characters Traits shared by a lineage of organisms but not shared with more distant relatives. Among mammals, for example, fur is a shared derived character because other vertebrates do not have fur. In primates, fur is a primitive character, one that is shared with all other mammals. Cladists construct maps of relatedness, called cladograms, using shared derived characters.

shear The twisting action created by forces that are not opposite one another.

shell In an atom, a group of orbitals whose electrons have nearly equal energy.

shoot The part of a plant above the ground.

shoot system Stems, photosynthetic leaves, and flowers and other organs of reproduction.

sibling species Two species that are extremely similar but reproductively isolated from one another.

sickle cell disease (also called **sickle cell anemia**) A blood disorder that gets its name from the curled appearance of red blood cells in sickle cell patients; a genetic disease caused by a mutation in the gene encoding the β-polypeptide of hemoglobin.

sidewinding Movement in which a snake moves across loose or sandy soil by throwing successive coils sideways.

sieve plate In sieve tubes of a plant's phloem, the top and bottom end walls of each sieve tube member.

sieve tube members In the phloem of a plant, the cells that actually conduct fluid.

sieve tubes In the phloem of a plant, pipelike channels that carry organic matter from the leaves.

sign stimulus The key aspect of an object that triggers a modal action pattern.

signal hypothesis The statement that a hydrophobic signal peptide directs a growing polypeptide chain through the membrane of the endoplasmic reticulum into the ER lumen.

signal peptide A polypeptide segment of 16 to 30 amino acids at the amino terminal end of the newly made protein; a signal peptide directs the newly made protein into the lumen of the endoplasmic reticulum.

silent mutation A change in DNA that has no effect on phenotype.

silt A soil of medium-sized particles (from 2 mm to 20 mm).

simple diffusion [Latin, *diffundere* = to pour out] The random movement of like molecules or ions from an area of high concentration to an area of low concentration.

simple pit pair A region between plant cells where the secondary wall is absent, allowing small molecules to pass.

simple transposon (also called an **insertion sequence**) The simplest transposable element, a short length of DNA (up to a few thousand nucleotide pairs long) that can move from place to place in the genome.

sink The site of storage or consumption of a substance.

sinoatrial (SA) node The pacemaker, which lies near the top of the right atrium in the mammalian heart.

sinus An open cavity; in insects and other arthropods and in most mollusks, it is a space through which blood travels.

sister chromatids The two chromatids that make up a single chromosome; the sister chromatids are duplicate copies of the same genetic information.

site-directed mutagenesis A technique that specifically alters a chosen nucleotide sequence.

skeletal muscles The muscles that are attached to bones.

skull The bony (or cartilaginous) case that encloses and protects the brain; it lies at the forward (or, for humans, the top) end of a vertebrate's skeleton.

sliding filament model The accepted view of muscle movement, which holds that movement results from changing the relative positions of thick and thin filaments rather than from filament contraction.

slime layer A tangled web of polysaccharides that, like a capsule, lies outside the cell wall.

smooth ER Endoplasmic reticulum that is not associated with ribosomes.

smooth muscle The involuntary muscles that line the walls of hollow internal organs such as intestines and blood vessels.

sociobiology (renamed **behavioral ecology**) The study of behavior from an evolutionary perspective.

solute A substance that dissolves within a solvent.

solvent Any fluid in which other substances dissolve.

soma The somatic cells of an organism, in contrast to germ cells or germ line.

somatic cells (also called the **soma**) [Greek, *soma* = body] In a multicelled organism, the cells that do not give rise to germ cells or gametes.

somatic theory The view that antibody diversity arises during development, rather than being inherited.

sori [singular, **sorus;** Greek, *soros* = heap] In a fern, groups of sporangia clustered on the underside of the sporophylls.

source The site of production (for example, of sugars in a plant).

speciation The formation of a new species.

species A biological species is a group of actually or potentially interbreeding populations that are reproductively isolated from other such groups. Species may also be defined differently, for example, according to phenetics or cladistics.

species diversity A measure of diversity that takes into account how common individuals of each species are.

species richness The number of species in an ecosystem.

sperm (also called a **spermatozoan**) A male gamete.

spermatid The four haploid cells produced during the two meiotic divisions (meiosis I and II).

spermatogenesis The development of spermatogonia into mature sperm.

spermatogonium [plural, **spermatogonia;** Greek, *sperma* = sperm + *gonos* = offspring] A diploid cell that is committed to forming sperm cells, but can still undergo mitosis.

spermatozoan (also called a **sperm**) [Greek, *sperma* = seed + *zoos* = living] A fully differentiated male germ cell.

spermicide [Latin, *cidere* = to kill] A cream or jelly that kills sperm; an essential part of the effectiveness of both cervical caps and diaphragms.

sphincters Rings of muscles that surround a tube.

spiracle In the insect respiratory system, an opening through which air enters the body; a spiracle leads directly to the tracheae.

spirilla [singular, **spirillum;** Latin, *spira* = coil] Spiral-shaped prokaryotic cells.

spirochetes Anaerobic heterotrophs that have a distinctive corkscrew shape.

spliceosome A complex of pre-mRNA, proteins, and small nuclear RNAs that catalyzes the removal of introns and the splicing of exons.

spongy bone tissue Tissue that generally lies at the ends and inside the bones.

spongy parenchyma In a plant leaf, the mesophylls that lie just under the palisade parenchyma that are irregular but rounded in shape and separated by numerous air spaces.

spontaneous A reaction that occurs without any external input of energy.

spontaneous abortion (also called **miscarriage**) In mammalian development, embryonic or fetal death, usually resulting from congenital defects.

sporangium [Greek, *spora* = seed + *angeion* = vessel] In plants and fungi, a structure within which cells undergo meiosis to produce the haploid spores.

spore   A cell that divides by mitosis to produce new individuals.

sporophyll   A leaf or leaflike structure that is specialized for the production of spores; in flowering plants, the term refers to carpels and stamens; in ferns sporophyll refers to a leaf that bears sporangia.

sporophyte   The diploid form of a life cycle characterized by alternation of generations; a sporophyte produces haploid spores that give rise to haploid gametophytes by meiosis.

sporozoans   Parasitic protists, including those that cause malaria; sporozoans undergo alternation of generations.

stabilizing selection   Selection that tends to act against extremes in the phenotype, so that the average is favored.

stage   On a microscope, a moveable platform that holds the observed specimen.

stain   A dye that binds differently to cell components and thereby increases contrast for microscopy.

stamen   In a flower, a male reproductive organ; the male sporophyll.

standard deviation   An estimate of the variation in a set of data calculated from how much the data deviate from the mean.

starch   Polysaccharides used in plants for long-term energy storage.

Start (also called the **restriction (R) point**)   The "point of no return" in the G_1 phase of the cell cycle; once a cell proceeds beyond Start, it proceeds through the rest of the cycle, including mitosis and cytokinesis.

stasis   Evolutionary stability over long periods of time.

statistics   The science of collecting and analyzing numerical data.

statolith   [Greek, *statos* = standing + *lithos* = stone] Gravity detector in a plant or animal.

stele   In roots and stems, a central cylinder through which the vascular bundles frequently run.

stem cell   A cell that can either produce more of itself by cell division or undergo differentiation to one or more specialized cell types.

stereoisomers   Molecules with the same atoms and functional groups; two stereoisomers differ only in the spatial arrangements of their atoms.

steric inhibition   Inhibition of an enzyme by a molecule whose shape resembles that of the substrate.

sternum   [Greek, *sternon* = chest] The breastbone.

steroid   A lipid-soluble molecule derived from cholesterol; many steroids are used as hormones.

stoma [plural, **stomata**; Greek, *stoma* = mouth] In a plant leaf, a tiny mouthlike pore that opens and closes to regulate the flow of carbon dioxide and other gases to the interior of the leaf.

stomach   [Greek, *stoma* = mouth] The most dilated and most muscular section of the digestive tract.

stop codon   A codon in mRNA that signals the end of a polypeptide chain.

striated   Striped.

striations   Cross stripes, 2 to 3 μm apart, that lie perpendicular to the long axis of a skeletal or cardiac muscle.

strobilus [plural, **strobili**; Greek, *strobilos* = cone] A conelike structure (such as a pine cone) that consists of overlapping sporophylls grouped around a central axis.

stroke   Failure of the blood supply to the brain.

stroke volume   In the action of the heart, the volume of blood delivered by a ventricle.

stroma   [Latin, mattress] The region within a chloroplast bounded by the inner chloroplast membrane.

structural isomers   Molecules that contain the same atoms, grouped in different ways to produce different functional groups.

structural mutation   A change in DNA that leads to an alteration in the amino acid sequence of a polypeptide.

suberin   The waxy substance, which is impermeable to water, that surrounds each cell of the endodermis.

subsidize   Referring to the growth of crop plants, to provide additional energy.

substrate   A reacting molecule in an enzyme reaction; a substrate is usually (but not always) much smaller than an enzyme.

succession   Progressive, predictable change over time in kinds of species and numbers of species in a community.

successional communities   Intermediate stages of succession.

sucrose   Table sugar; a disaccharide consisting of two monosaccharides, glucose and fructose, linked together.

sugar   A simple carbohydrate; a molecule that has the equivalent of one molecule of water (that is two hydrogen atoms and one oxygen atom) for every atom of carbon.

superfecundity   A state of a population in which organisms are so fertile that their populations would increase exponentially if all offspring survived.

supernatant   The part of a sample that, during centrifugation, remains in suspension.

supernormal stimulus   A stimulus even more stimulating than anything normally encountered in nature.

superorganism   A group of individuals that cooperate more closely than the individuals of a colony and almost as closely as the cells of an individual.

suppressor mutation   A mutation in one gene that prevents the phenotypic changes caused by a mutation in another gene.

surface tension   The tendency of a substance to form a smooth round surface.

surface-to-volume ratio   The amount of surface area for each bit of volume.

survivorship curve   A graph that shows the fraction of a population that is alive at successive ages.

suture   An immovable joint that fuses separate bones.

swim bladder   An air-filled sac that helps fish control their buoyancy.

symbionts   [Greek, *syn* = together + *bios* = life] Organisms living in a symbiotic relationship with one another.

symbiosis   [Greek, *sumbios* = living together]   A close association between two organisms.

sympathetic nervous system   A division of the autonomic nervous system; the sympathetic nervous system initiates the "fight or flight" reaction.

sympatric speciation   [Greek, *syn* = together + *patra* = country] The splitting of one species into two without geographical isolation.

synapse   The distinct boundary between two communicating neurons.

synapsis   [Greek, union] The pairing of homologous chromosomes in prophase I.

synaptic cleft   A space of about 20 nm that separates the presynaptic and postsynaptic cells in a chemical synapse.

synchronous cell populations   Cells that are all at the same stage of the cell cycle.

syngamy   The coming together of an egg and a sperm at fertilization.

system An assemblage of interacting parts or objects.

systematics The scientific study of the kinds and diversity of organisms and of any and all relationships among them.

systemic circulation The route of the blood through the body, minus pulmonary circulation.

systole [Greek, *systellein* = to contract] The part of the cycle of the heart in which the atria contract, followed by the ventricles.

T cytotoxic cell (T_C cell) A T lymphocyte that is responsible for the killing of cells recognized as nonself.

T helper cell (T_H cell) A T lymphocyte that activates both the humoral and cellular immune responses.

T lymphocyte (because it matures in the *t*hymus gland) One of the cells responsible solely for cellular immunity or of the cells that participate in both humoral and cellular immunity.

T-cell receptor A protein on the surface of a T lymphocyte that can bind to a complex of an antigen and MHC molecule.

tagmata (singular, **tagma**) Fused segments.

taiga The broad band of coniferous forest that extends across Canada, Alaska, Scandinavia, and Siberia.

tannin A secondary compound commonly found in woody plants.

taproot A single vertical root found in many dicots.

tar An oily, viscous material, consisting of hydrocarbons, found in cigarette smoke and the insides of chimneys.

target organ A structure upon which a hormone acts.

TATA box (after the first four nucleotides in the sequence) A consensus sequence in eukaryotic promoters.

taxis [Greek, to put in order] Directed movement.

taxol A compound extracted from the bark of yew trees; taxol increases the stability of microtubules.

taxon [plural, **taxa**; Greek, *taxis* = arrangement] A general term for any group of organisms at any level of the classification hierarchy.

taxonomist A biologist who classifies organisms.

taxonomy [Greek, *taxis* = arrangement] The naming and grouping of organisms.

telocentric Refers to a chromosome in which the centromere is at one end.

telophase [Greek, *telos* = end] The last phase of mitosis, during which the mitotic apparatus (including kinetochore, polar, and astral microtubules) disperses and the chromosomes lose their distinct identities.

temperate deciduous forests Temperate forests that receive 80 to 140 centimeters of precipitation each year, composed of trees that lose their leaves in the winter.

temperate grasslands Regions with well-defined seasons, with hot summers and cold winters; low annual rainfall keeps grasslands from turning to deserts but is not enough to sustain the growth of trees.

temperature-sensitive mutation A change in DNA that alters a protein so that it functions normally at one temperature (called the permissive temperature, usually relatively low), but not at a second temperature (the restrictive temperature, generally higher than the permissive temperature).

template A guide for the assembly of a complementary shape; in the context of DNA or RNA synthesis, one strand acts as a template for the assembly of a complementary sequence.

temporal isolation Reproductive isolation that results from differences in the times at which two populations reproduce.

ten percent law The generalization that the organisms of any trophic level provide the next higher trophic level with only 10 percent of the energy that they have assimilated from the lower trophic level.

tendon A type of connective tissue that attaches the muscles to the bones.

tension The pulling action of two opposing forces.

teratogen A compound that causes birth defects.

terminal bud A bud that lies at the tip of a shoot.

termination The ending of chain growth.

termination signal A DNA sequence that determines the end of transcription.

terpene One of a class of unsaturated hydrocarbons that form the largest class of secondary compounds in plants.

terrestrial Living on land.

territory An area occupied by an individual (or group of individuals), from which other individuals of that species are excluded.

tertiary structure (or **conformation**) The complete arrangement of all the atoms of a polypeptide.

testable Refers to a hypothesis for which an experiment could be devised that would disprove the hypothesis if it were incorrect.

testcross A cross between an individual that is homozygous for a recessive allele of a particular gene and an individual whose genotype (homozygote or heterozygote) for that gene is unknown; the phenotype of the progeny reveals the presence of the recessive allele in the parent.

testicles Human testes.

testis (plural, **testes**) A male gonad, which produces sperm.

tetrad, (also called a **bivalent**) A united chromosome pair visible during meiosis I; a tetrad consists of four chromatids.

tetrapod [Greek, *tetra* = four + *pous* = foot] A four-footed animal.

tetrodotoxin A poison, isolated from a species of puffer fish, that blocks the passage of Na^+ ions through voltage-gated channels.

theory A system of statements and ideas that explains a group of facts or phenomena.

theory of island biogeography MacArthur and Wilson's theory that species diversity on an island is a function of the size of the island and its distance from a source of colonizing species, whether the mainland or other islands.

thermacidophiles One of the three phyla of Archaea; the bacteria that inhabit hot sulfur springs such as those in Yellowstone National Park.

thermodynamics [Greek, *thermo* = heat + *dynamis* = power, force] The study of the transformations and relationships among different forms of energy.

thick filament In striated muscle, a myosin filament, which is 14 nm in diameter.

thin filament In striated muscle, an actin filament, which is 7 nm in diameter.

thoracic cavity Chest cavity; that part of the coelom that encloses the heart and lungs.

thorax The middle portion of an arthropod.

thylakoid [Greek, *thylakos* = sac + *oides* = like] A flattened disc surrounded by the innermost membranes of a chloroplast.

thylakoid membrane A membrane, within the stroma of a chloroplast, that delineates stacked vesicles, called grana.

thymine dimers Two adjacent T nucleotides linked together in the same DNA strand; they are formed as a result of exposure to ultraviolet light.

thyroid stimulating hormone (TSH) A peptide hormone, made in the anterior pituitary, that stimulates the production of thyroxin in the thyroid gland.

tidal volume The amount of air drawn in and then expelled in a single breath.

tight junction The main kind of impermeable junction in vertebrates; they are formed by the fusion of the membranes of adjacent cells.

tissue A group of similar cells and associated intercellular material that performs one or more specific functions; tissues are often integrated with other tissues to form an organ.

tolerance In the immune system, a state of induced unresponsiveness to proteins and other molecules; tolerance prevents immune attacks on the body's own components.

torus A thickened region of the cell wall between two plant cells; a buttonlike thickening in the middle of a pit.

totipotent Able to develop into a whole organism.

trachea [plural, **tracheae;** from Greek, *arteria trakheia* = rough artery] A tube, which arises by invagination, that carries air within an animal's body; in vertebrates, the windpipe.

tracheids Long, thin, spindle-shaped cells in the xylem of plants.

tracheophytes [Latin, *trachea* = windpipe] Vascular plants; tracheophytes include all the most familiar living plants.

tradewinds Winds that blow from about 30° latitude, steadily toward the equator.

***trans*-acting factor** (also called a **transcription factor**) [Latin, *trans* = across, because they regulate the expression of other molecules] A protein that binds to DNA and regulates transcription.

transcription The production of RNA from DNA.

transcription factor (also called a ***trans*-acting factor**) A protein that binds to DNA and regulates transcription.

transfer RNA (tRNA) A small RNA molecule that serves as an adaptor in protein synthesis; each tRNA contains an anticodon that allows it to bind to a codon in mRNA and each becomes linked to a specific amino acid.

transformation The transfer of one or more genes from one organism to another.

transgenic Containing a particular recombinant DNA incorporated within the genetic material.

transition state A distorted form of a substrate that is intermediate between the starting reactant and the final product.

translation Protein synthesis; the conversion of information from an mRNA molecule into a polypeptide.

translational control Alterations in the rate at which specific mRNAs are translated.

translocation In chromosomes, a mutation in which part of one chromosome is moved to another chromosome; in protein synthesis, the movement of the growing polypeptide chain with its attached tRNA from the A site to the P site; the third step of the elongation cycle.

transmission electron microscope (TEM) A type of electron microscope in which the electron beam passes through a specimen.

transmission genetics The branch of genetics that deals with patterns of inheritance.

transpiration The process by which plants pull water from the soil and release it as vapor through stomata in their leaves.

transpiration-cohesion theory The accepted explanation for the movement of water from roots to leaves; transpiration in the leaves pulls water up the stem in continuous columns.

transpiration-photosynthesis compromise The trade-off between saving water and maintaining photosynthetic productivity.

transposable element A mobile gene that can replicate only in integrated form, as part of a host's chromosome.

transposase An enzyme that catalyzes insertion of a transposable element into new sites.

tree line The upper limit where subalpine trees grow.

trematoda Flukes.

triacylglycerol (also called a **triglyceride**) A nonpolar derivative of glycerol in which all three hydroxyl groups are attached to fatty acids.

triceps The major muscle on the back side of the upper arm.

triple covalent bond A covalent bond in which two atoms share three pairs of electrons.

trisomy [Greek, *tri* = three + *soma* = body] The presence of three, rather than two, copies of a chromosome.

trophic level Each level of a food chain.

trophoblast [Greek, *trephein* = to nourish + *blastos* = germ] In a mammalian embryo, a prominent outer cell layer in a blastocyst.

tropical rain forest A biome with relatively constant weather, with an average temperature of about 27°C (80°F) every month of the year and heavy rainfall, ranging from 2 to 4.5 meters per year.

tropism [Greek, *trope* = turning] The bending or curving toward or away from a stimulus.

tropomyosin In muscle, a rigid, rodlike protein, about 40 nm long, that stiffens the actin filament.

troponin In muscle, a large protein that binds both to tropomyosin and to free calcium ions.

true-breeding A breeding line in which offspring have the same phenotype, generation after generation.

tubal ligation An irreversible method of birth control; cutting or blocking the oviducts, or fallopian tubes, preventing eggs from reaching the uterus after ovulation.

tubulin One of two globular proteins, called α and β tubulin, from which microtubules are assembled.

tumor suppressor gene (also called an **antioncogene**) A gene that antagonizes the action of an oncogene; this gene is intimately involved in the normal regulation of cell growth.

tumor virus A virus that causes its host cell to lose its normal ability to regulate cell division.

tundra [Finnish, *tunturi* = arctic hill] A vast, open land, with a cold, dry climate; it is dominated by grasses and low shrubs and covers more than a fifth of the Earth's surface.

turbellaria The free-living flatworms; they reproduce either asexually or sexually.

turbinate Elaborately folded bones inside the noses of many vertebrates.

turgor pressure [Latin, *turgor* = swelling] In plant cells, osmotic pressure against the cell walls resulting from the higher concentration of dissolved molecules and ions in cytoplasm than in the surrounding fluid.

typological species concept The view that each species consists of individuals that are variants of a fixed underlying plan.

ultracentrifuge A high-speed centrifuge that can spin tubes at speeds up to 80,000 rpm, subjecting their contents to forces up to 500,000 $\times$ gravity.

ultrastructure Subcellular structures visible only with an electron microscope.

unconditioned stimulus In associative learning, the stimulus that directly leads to a physiological response.

unconventional gene A gene without a stable cellular address.

uniformitarianism The view, originated by Lyell, that the processes that now mold the Earth's surface—erosion, sedimentation, and upheaval—are the ones that have always molded it; geologic change is slow, gradual, and steady, not catastrophic.

Uniramia A subphyla of arthropods in which the first pair (or the first few pairs) of appendages are antennae and the next pair are jaws, or mandibles.

universal Used by all species.

unsaturated A hydrocarbon chain (or fatty acid) that contains at least one double bond between two carbon atoms; the chain is called unsaturated because it could accept two more hydrogen atoms per double bond.

upstream In a nucleic acid, toward the 5′ end.

upwelling The upward movement of water that draws nutrients from the bottom of the ocean.

urea The compound that mammals convert to ammonia to make it less harmful.

ureter [Greek, *ouron* = urine] One of a pair of tubes through which urine flows from the kidneys to the bladder.

urethra The means by which the bladder drains urine to the outside of the body; the single exit tube of the human urethra.

uric acid A nearly insoluble organic compound produced by insects, land snails, most reptiles, and birds as nitrogenous waste.

uterus [Latin, womb] The womb; in mammals, the portion of the female reproductive tract that is specialized to receive, nurture, and deliver a developing embryo.

vacuole Within a eukaryotic cell, a membrane-bounded sac, without any obvious internal structure; a vacuole is larger than a vesicle.

van der Waals interactions Weak attractive forces between molecules that are very close together.

variable region In an immunoglobulin (antibody) molecule, the amino-terminal polypeptide segment, of about 110 amino acids, that contains the sequence variation responsible for antibody diversity.

vas deferens [Latin, *vas* = vessel + *deferre* = to carry down] In the male reproductive tract, two large, thick-walled ducts that carry sperm from the testes.

vascular Having a conducting system.

vascular system [Latin, *vas* = vessel] In plants, the structures and cells that serve as a transport system; the vascular system forms the central core of the plant.

vasectomy An irreversible method of birth control; the cutting and tying of the vas deferens, which prevents sperm from entering the urethra.

vasoconstriction Contraction of the smooth muscles surrounding the arterioles.

vasodilation A process causing arterioles to increase in size with consequent blood flow increase and reddening on the skin.

vasopressin (also called **antidiuretic hormone** or **ADH**) A peptide hormone, released by the posterior pituitary, that stimulates water reabsorption by the kidneys.

vegetative reproduction The process of producing a new plant without fertilization or seed production.

veins In vertebrates, one of the vessels that carry blood back to the heart. In plants, the vascular tissue in the leaves.

velocity In an enzymatic reaction, the rate of appearance of product.

vena cava (**superior** and **inferior**) The two largest veins, which run up through the center of the body and carry deoxygenated blood from the body to the right atrium.

ventilation The flow of air into and out of the alveoli.

ventral The part of an animal facing the Earth.

ventricle In the heart, a thick-walled pumping chamber.

ventricular fibrillation Cardiac arrest.

venules The smallest veins.

vertebrae [singular, **vertebra**; Latin, *vertebratus* = jointed] The hollow, bony segments that make up the backbone.

vesicle A membrane-bounded sac; some vesicles are present inside living eukaryotic cells and others form after a cell is broken open.

vessel elements In flowering plants, structural elements of the phloem that are open at each end and connect end to end to form long open channels.

vestigial structure [Latin, *vestigium* = footprint] A part of an organism, with little or no function, that reflects evolutionary history.

vicariance The fragmentation of an already dispersed species or group of species.

villus [plural, **villi**; Latin, shaggy hair] The highly folded mucosa of the small intestine that each extend about 1 mm into the lumen.

viral oncogene A viral gene that affects growth control in a host cell.

virus An assembly of nucleic acid (DNA or RNA) and proteins (and occasionally other components, such as lipids or carbohydrates) that can reproduce only within a living cell; viruses depend on cells to obtain energy and perform chemical reactions.

visceral mass [Latin, *viscera* = internal organs] The main body of a mollusk; the visceral mass contains the intestinal tract, as well as the excretory and reproductive organs.

visible light Electromagnetic radiation, with wavelengths from about 400 to about 750 nm, that can be perceived by human eyes and brains.

vitamin An organic compound that an animal cannot itself synthesize but that is required in minute amounts for normal growth and metabolism; an essential component of the diet; many vitamins are cofactors in enzymatic reactions.

voltage clamping A method that allows the control of the voltage across a cell membrane.

voltage (also called **electrical potential**) A measure of the potential energy that results from a charge separation.

voltage-gated Open or closed according to the voltage across the membrane.

voluntary muscles Skeletal muscles; voluntary muscles enable voluntary control over body movements.

vulva [Latin, *volvere* = to wrap] In animals, the collective name for the external genitalia of the female.

warning coloration A bright, memorable design that helps the predator remember which prey to avoid.

water potential The potential energy of water per gram (or kilogram); the difference between osmotic pressure and physical pressure on either side of a selectively permeable membrane.

water-soluble signal A chemical signal that cannot pass through the plasma membrane and must act on its surface.

wavelength The distance between the crests of two successive waves.

westerlies Winds that come from the west caused by the rotation of the Earth, characteristic of latitudes between 30° and 60° in both hemispheres.

western blotting (after a similar blotting method, called "Southern blotting" because it was invented by Edward Southern) A method for detecting a particular protein after electrophoresis by blotting the contents of the electrophoresis gel onto a paperlike support and incubating this "blot" with a specific antibody.

whorl One of four concentric circles of flower parts at the end of a specialized stem.

whorled The arrangement of leaves on a stem so that there are three or more petioles at each node.

wild type An allele that is most highly represented in wild populations.

wobble Nonstandard base pairing between the first base in the anticodon of transfer RNA and the third base of the codon in mRNA; wobble allows some tRNAs to recognize more than one codon for the same amino acid.

woody plants Trees, shrubs, and other perennial plants that have woody parts that provide support independent of turgor pressure.

work The movement of an object against a force, or the conversion of energy into electrical energy, chemical energy, or concentration energy.

x-ray crystallography The analysis of the scattering, or "diffraction," of x rays by crystals.

xylem [Greek, *xylon* = wood] Conducting vessels that carry water and minerals from roots to the photosynthesizing leaves.

yield The amount of grain per acre.

yolk In an amniotic egg, a mixture of proteins, lipids, and carbohydrates that nourishes an embryo until it can feed itself.

yolk sac The membrane that surrounds the yolk in an amniotic egg.

Z band In muscle, the boundary of a sarcomere.

zero population growth The state of a population when its size reaches the carrying capacity and the environment cannot support any more growth; the state when a population reaches a stable size.

zoomastigina (also called **flagellates**) [Greek, *mastix* = whip] A diverse group of single-celled heterotrophs.

zygomycetes [Greek, *zygote* + *mukes* = fungus] Conjugation fungi.

zygosporangium A structure that produces haploid spores from a diploid zygote.

zygote [Greek, *zygon* = yoke] The first cell of an embryo, formed by the union of egg and sperm.

Index

Note: Page numbers in *italics* indicate a figure; page numbers followed by "t" indicate a table.

A

Abdomen, arthropod, 517, 521
Abdominal cavity, 713, 732
Abiotic factors, 546, 560
Abiotic synthesis, 404, 415
Abortion, induced, 903, 905
    spontaneous, 192–193, 205, 897, 905, 918, 928
    in U.S., history of, 904
Abscisic acid, 667, 674, 702, 707, 708
Abscission, 706, *707*, 708
Absolute age, 329
Absorption, 739, 750, 752
    by digestive tract, 743, 752
    in small intestine, 750
Absorption spectrum, 147, *148*, 159, 699
Absorptive state, 758, *760*, 768
Abstinence, sexual, 900–901, 905
Abyssal zone, 578, *579*, 582
Acacia, bull's horn, 590
Acacia cornigera, 590
Accommodation, 874, 880
Acetaminophen, 768
Acetone, 281
Acetyl-CoA, 127, *127*, 129
    generation of ATP from, 127–130
Acetylcholine, 7, 785, 787, 854, 855, *855*, 856, 861, 869
    functions of, 856
Acetylcholinesterase, 857, 861
Acid, definition of, 42, 45
Acid rain, in biomes, 575
Acoelomates, 501
    bilaterally symmetrical, 506–508, 520
        Nemertea, 508, *508*
        Platyhelminthes, 506–507, *508*
    radially symmetrical, 504–506, 520
        cnidarians, *504*, 504–505, *505*
        ctenophores, 505–506, *506*
Acorn worms, 527, 542
Acquired behavior, 627–632
Acquired characteristics, inheritance of, 347–348, 366
    theory of inheritance of, 325
Acquired immunodeficiency syndrome, 841
    design of drugs to combat, 271
    symptoms of, 841
Acrasiomycota, 464–465, 475
Acrosome, 887, 892, 904
Actin, 87, 729, 733
Actin filaments, 86–87
Action potential(s), 849, 850–851, 861
    generation by neurons, 849–850
    movement of, 852–853
    production of, 852
    transmission of, nerves and, 853, *853*
Action spectrum, 147–148, *148*, 159, 699, 700
Activation energy, 110–111, *111*, 117
Activators, 241–242, 764
Active site, 112, *113*, 117, 763
Active transport, 94, *95*, 101

Adaptations, 17–18, 20
Adaptive radiation, 376, *377*, 396
    plants and, 379
Adaptive traits, 321, 342
    accumulation of, 341
    in inheritance, 341
Adenine, 210, 233
Adenosine, 854
Adenosine diphosphate, as energy source, 105
Adenosine triphosphate, 61, *62*, 69
    as energy source, 105, 106, 116, 120
    generation of acetyl-CoA from, 127–130
    in glucose synthesis, 153
    light reactions and, 151
    phosphate bonds in, electrons and, 130–131
    photosystem II and, 151
    potential energy in, 104, *105*
    synthesis of, 135
        proton gradient and, 135–136, *137*
Adhering junctions, 98, *99*, 101
Adhesion, of water, 664, 674
Adrenal cortex, 760, 768
Adrenal glands, 760–761, 768, 821, 825
Adrenal medulla, 760–761, 768
Adrenaline, 760, 785
Adrenocorticotropic hormone, 762
ADT. *See* Adenosine diphosphate
Adventitious roots, 691
Aerobes, facultative, 446, 450
    obligate, 446, 450
Aerobic cells, 120, *120*
Aerobic heterotrophs, 446, *446*, 448
Afferent nerves, 868, 880
Agamous mutation, 688
Agar, 460
Agassiz, Louis, 324
Age, absolute, 329
Age distributions, life history and, *613*, 613–616, *615*
    male and female, 614
Age of Reptiles, 331
Age structure, 614, *615*, 625
Agent Orange, 698
Aging, genes and environment influencing, 926
    in vertebrates, 909, 928
Agnatha, 529–530, 542
Agonistic behavior, 632–633, *633*, 639
Agriculture, demand for food and, 619–620, *621*
    recombinant DNA in, 293
    slash-and-burn, 385, *385*
    subsidized, *619*
Agrobacterium tumefaciens, 292
Agrostis tenuis, 376
AIDS. *See* Acquired immunodeficiency syndrome
Air, entering lungs, path of, 796
    plants and, 142, *142*
Air currents, 565, *566*
Air pollution, 622
Air sacs, posterior, 800, *800*
Airways, epithelial cells in, 799, *799*
Alchemist, by David Teniers the Younger, 25

Alcohol, as diuretic, 823
    metabolism of, 136
Aldosterone, 760, 821–823, 825
Aleurone layer, 683, 692
Alexander, R. McNeill, 710–711
Algae, 455, 457, 474
    brown, 460, *460*
    golden, 458
    green, 461, *461*
        in plant evolution, 481
    red, 460–461, 474
Alimentary canal, 741
Alkaloids, 657–658, 659
Alkaptonuria, 217, *218*, *219*, 220
Allantois, 536, 542
Allele(s), 194, 205, 350, 366
    dominant, 195–196, 205
    frequencies of, breeding between populations and, 358, *359*
        changing of, 355–366, *356*
        nonrandom mating and, 358
    hemizygous, 200
    multiple, genes with, 307, 307t
    recessive, 195–196, 205
Allergy, 836
Alligators, 534
    in DDT-laden lake, 754
Allopatric speciation, 374, *375*, 396
Allopolyploidy, 374, 396
Allosteric effector, 116
Allosteric inhibitors, *116*, 117
Alper, Tikvah, 46–47, 49
Alpha (α) carbon, 64, 69
Alternation of generations, in protists, 457, 474
Alternate leaves, 654, *655*, 658
Altitude, high, oxygen delivery at, 804
Altruism, 636
    inclusive fitness and, 637–638, 639, *639*
    natural selection and, 636–637
    reciprocal, 638, 639
Alvarez, Luis, 382
Alvarez, Walter, 382
Alveoli, 796
    gas exchange in, *801*
    surface tension and, 800
Amacrine cells, 876, 880
Amborella trichopoda, 490
Ambystoma tigrinum, 535
Amebas (Rhizopoda), 462, 475
Ames, Bruce, 642, 643
Ames test, 642
Amines, 763, 764
Amino acid balance, 737
Amino acids, 52, 54–55t, 68. *See also* Base(s); Nucleotides; Polypeptides
    definition of, 64
    essential, 737, 737t, 752
    linkage to form polypeptides, 63–66
    peptide bond joining, 64–65, *65*
    protein structure and. *See* Protein(s), structure of

Amino acids, *(continued)*
R groups of, 54–55t, 64
recognition of, by transfer RNA, 245, *246*
structure of, 64, *64*
transfer RNA and, 244, 244–245, 255
Amino group, 51, 68
Aminoethane, 51
Ammonia, 813, *814*, 824
Ammonification, 558, 560
Ammonium ions, 557, 560
Ammospermophilus tereticaudus, 568
Amniocentesis, 312, *312*, 315
Amnion, 532, 538, 911
Amniotic egg(s), 428, 536, *537*, 542, 897
Amniotic fluid, 312, 315, 911
Amphibians, development of, primary organizer in, 921–923
disappearing, 535
first, 532, *532*
gastrulation in, *912*, 912–913
life cycle of, 532, 533, *534*
metamorphosis in, 758
orders of, 533
Amphipathic molecules, 40, *41*, 44–45
Amphotericin B, 460
Amygdala, 868, 880
Amylase, 744, 752
α-Amylase, 683, 692
Amyloplasts, 85, *86*, 100, 698, *698*, 708
Amystoma (Axolotl), 534
Anabolism, 129, *129*, 138
definition of, 114, 117
Anaerobes, 120, *120*
facultative, 126
obligate, 126, 446, 450
Anaerobic heterotrophs, 448
Anaphase, chromatids during, 172–173
in meiosis, 188
Anaximander, on evolution, 321
Androgen receptor, 255
Androgens, 763
Anemia, sickle cell. *See* Sickle cell anemia
Aneuploid cell, 191, 205
Aneuploidy, 231
Aneurism, 782, 787
Anfinsen, Christian, 68
Angiosperms, 490–493, 494
definition of, 481
evolution of, 481–482
fertilization in, 491
life cycle of, 491, *492*
reproduction by, 491
species of, 493
Angiotensin, 821, 825
Animal(s), agonistic behavior of, 632–633, *633*, 639
aquatic, dissolved oxygen for, 795
behavior of, ecology of, 626–640
body composition of, 713
body plan of, 713
body temperature regulation in, 731
breathing by, 788–806
carbon dioxide exchange for oxygen in, 791–785
cells and organs of, coordination of actions of, 754–769
characteristics of, 498–503
choices made by, 632
classification of, 499–503, *500*
developmental patterns and, 499–501, *501*
symmetry and, 502, *502*

cloning of, 920–921, *921*
defenses of, 588–589
of deserts, 568
water conservation by, 812
disposal of nitrogenous wastes by, 813, *814*, 824
eating by, 735–739
form of, 711–718
determinants of, 712
and function in, 710–733, *712*
grazing on savannas, 572
hearing by, 871–873
internal environment of, regulation of, 755, *756*
jaws, teeth, and mouth parts of, 744, *744*
land, water and salt regulation in, 812–813
learning by, 627, 630–631, 639
sensitive period for, 630
locomotion of, style of, as influence on shape, 718–721
movement of, skeleton and muscle in, 721–730
phenotype of, social environment and, 304, *304*
salt and water adjustment by, 821–823
sensing environment, 870–871
size of, gravity and, 714
as influence on shape of, 713–714
metabolic rate and, 714, *714, 715*
surface area of, and shape of, 713–714
in social groups, 634–638
sociality among, 635
structural and physiological adaptations of, 709–928
structure of, 499, *499*
surface areas of, use of, 714–715, *715*
of temperate deciduous forests, 573, *574*
vision of, 873–877
walking and running by, 720
water, salt, and waste management in, 808–825
water and salt regulation by, 810–813
Animal pole, 910, 928
Animalcules, 70
Animalia, 16, 20
Annelids, 514–517
Anomalocaris, *498*
Anopheles, 463
Anopheles maculipennis, 371
Antacids, 2
Antagonist muscles, 722, *722*
Anteaters, spiny, 541
Antenna complex, 148, 159
Antennapedia, 924
Anterior pituitary gland, 892, 904
Anther, 492, 493
Antheridium, 481, *482*
Anthocyanins, 657, 659
Anthophyta, 490, 493
Anthozoa, 504
Anthropologists, physical, 390, 397
Antibiotics, and bacteriophages, 265
broad-spectrum, 440
definition of, 436
functions of, 437
fungal infections and, 453
inhibition of elongation by, *250*
narrow-spectrum, 440
natural synthesis of, 436–437
and protein synthesis, 251, *251*

resistance to, 360, 436–438, *438*, 447
classification of, 437, *438*
and salmonella poisoning, 432–434, *434*
in ulcers, 4
Antibodies, 286, 295, 829
cells and molecules and, 835–836, *835, 836*, 842
immune system and, 837–839
production of, 837, *838*
Anticlotting agents, 768
Anticoagulants, 832, 843
Anticodon, 245, *246*, 256
Antidiuretic hormone, 761, 823, 825
Antigens, 834–835, 842
on blood cell surfaces, 841–842
endogenous, 841, 843
exogenous, 841, 843
Antihistamines, 832
Anti-inflammatory agents, 768
Antioxidants, 739
food sources of, 643t
Antisense strand, 240
Ants, and bull's horn acacia, 590
sex ratios in, 614
Anura (frogs and toads), 533, *533*
Anus, 741, 752
Aorta, *772*, 776, 786
Apes, classification of, 389
Aphids, 671, *671*
Apical meristems, 646
Aplacophora, 513
Aplysia, gill withdrawal in, 859, *860*
learning by, 860, 861
Apoplast, 651
Apoptosis, 827, 915–916, *916*, 928
Appendix, 751
Aquatic animals, dissolved oxygen for, 795
Aquatic ecosystems, 578–581
Aqueous environment, interactions in, 35, *35*, 44
Aqueous humor, 874
Arabidopsis thaliana, 687–688, 692, 698
genetic analysis of, 689
mutations of, 689, *689*
Arachidonic acid, 768
Archaea, 16, 20
classification of, 443, *444*, 450
Archegonium, 484, *484*
Archenteron, 499, 520, 911, 927
Archipelagos, 376
Arctic tern, *120*, 538, 719
Arendt, Detlev, 523
Arginine, 219, *220*
Aristotle, 141, 906
on classifications, 424
Aromatic compounds, 49, 50, 68
Arrow worms, 527, 542
Arterioles, 779, 787
Artery(ies), 775, 779, 787
pulmonary, 776, 786
renal, 814, 825
Arthropods, 522, 541
chelicerates, 518, 519
classification of, 518–520
crustacea, 518, 519–520, *520*
mouth parts of, 518, *519*
success of, 517–518
uniramia, 518, 520, 521
Artificial kidney, 818, *818*
Artificial selection, 340, 342
in pigeons, 340, *340*

Asaro, Frank, 382
Aschelminths, 509–510, 521
    nematodes, 510, 521
Asci, 469
Ascomycetes (sac fungi), 469, *470*, 475
Ascorbic acid, 739
Asexual reproduction, 185, *185*, 192, 204
Aspergillus, 468
Aspirin, 768
Association cortex, 868, 880
Associative learning, 859
Assortative mating, 358, 366
Assortment, independent, principle of, 198–200, *199*, 206
Aster, 166
Asteroids, 382, *382*
Atherosclerosis, 59
Atmosphere, components of, 552, 552t
    evolution of, 405
    oxygen in, 791–792
    protective, 146, *147*
Atom(s), 23, *25*, 44
    composition of, 26
    electrons in, 30
    internal structure of, 27, *27*
    in ionic bonds, 33–34
    Ludwig Boltzmann and, 102–103
    masses of, 26
    nucleus of, 27, 44
    properties of, 26–30
    weight of, 26
Atomic force microscope, 26
ATP. *See* Adenosine triphosphate
ATP synthase, *134*, 134–135, *135*, 138
Atrazine, 156, 754
Atrial natriuretic factor, 823, 825
Atrioventricular node, 784, 787
Atrioventricular valves, 782–783, 787
Atrium, 776, 786
Atropa belladonna, 857
Atropine, 857
Audition, 870–872, 880
Auditory canal, 871
Auditory cortex, primary, 872
Auditory system, 869, 880
Australian marsupials, 380, *381*
Australopithecines, 390, 393, *393*, 397
Australopithecus, standing by, 390–392
Australopithecus afarensis, 391, 392, 393, *394*
Australopithecus anamensis, 390, 391–392
Autonomic functions, 865
Autonomic nervous system, 726, 733, 869, *869*
Autonomous replication, 264, 274
Autosomes, 193
Autotrophs, 119–120, 138, 412, 416, *444*, 448, 450, 548, 560
Auxin(s), 691, 694–696, 696t
    and cytokinin, 705
    discovery of, 694
    in gravitropism, 698
    as herbicides, 696, 704
    mechanisms of action of, 694–695, 696t
    naturally occurring, 702
    in phototropism, 694–696
    in plant growth and development, *702*, 702–704, 708
    synthetic, 695–696
    transport of, *703*, 703–704
Averted vision, 874
Avery, Oswald, 222, 224

Avogadro, Amadeo, 27
Avogadro's number, 27
Axelrod, Robert, 638
Axillary buds, 687, 692
Axolotl (*Amystoma*), 534, 926
Axons, 847, 861
AXR1 wine grapes, 661–662
Aysheaia, 498

B

B-cell receptors, 838, 843
B cells, *834*, 835, 843
B lymphocytes, *834*, 835, 843
Bacilli, 439, 450
Bacillus thuringiensis, 292
Backbone. *See* Spinal column
Backcross, 195, 197, *197*, 205
Bacteria, antibiotic resistance in, 437–438, *438*, 447
    cell walls of, 440–441, *440*
    colonial, 439, *440*
    as common ancestor of life, 448, *449*
    conjugation in, 265, *266*, 274
    fermenting, 448, 450
    functions of, 435
    gene cloning in, 284
    gene expression in, eukaryotic, 284
    genes in, 221
    genetically engineered, proteins from, 289
    gram negative, 439, *440*, 450
    gram positive, 439, *440*, 450
    kingdoms of, 16
    movement of, 441
    nitrifying, 558, 560
    nitrogen-fixing aerobic, *446*, 446, 448, 450
    normally in human body, 435–436, *436*
    as pathogens, 435, 450
    recombinant DNA and, 283–285
    two major groups of, 439
    in ulcers, 2
    viruses and, 260
Bacteriophage(s), 223, 232, 261, 274
    antibiotics and, 265
    to combat disease, 265
    life cycle of, 223, *224*, 268, *268*
    reproductive cycle of, *264*, 264
Bacteriorhodopsin, 135, 136
Badwater race, 562–564, *563*
Bakanae, 704
Bakker, Robert, 710
Balance, 870
    inner ear in, 872
Balanced polymorphism, *365*, 365–366
Balanus balanoides, 591, *592*
Baltimore, David, 269
Bark, 690, 692
Barnacles, 519–520, 591, 592
Barometer, mercury, 792, *793*
Barometric pressure, 792, *793*
Barracuda, 532
Barrier methods, 901, 905
Bartholin's glands, 891, 904
Basal ganglia, 868, 880
Basal metabolic rate, 736, 752
Base(s), definition of, 42, 45
    substitution of, 231, 233
Base pairing, 215, *215*
Basidiocarp, 470, 475
Basidiomycetes (club fungi), 469–470, *472*, 475

Basidium, 469, 475
Batesian mimicry, 589, 600
Bateson, William, 181, 200, 349
Beadle, George, 218, 219, 676
Beaumont, William, 734, *735*, 745
Becquerel, Henri, 29
Beef cattle, salmonella poisoning associated with, 432–434, 438
Beetles, grain, 591
    ladybird, 365
Behavior(s), acquisition of, 627–632
    agonistic, 632–633, *633*, 639
    ecology and, 632–634
    genes and, 627, *631*, 631–632
    homeostasis and, 865
    innate, *627*, 627–630, 639
    learned, 627, 628–629
    neural networks and, 857
    programmed, *627*
    social, 634, *635*
Behavioral ecology, 626, 639
Behavioral isolation, 373, 396
Beijerinck, Martinus, 260
Békésy, Georg von, 871
Belding ground squirrel, 637–638, *639*
Belladonna, 857
Benthic division, 578, 582
Benzene, *50*
Benzodiazepines, 857
Beriberi, 738–739
Bernays, Edward, 790
Bicoid gene, 924
Bile, 750
Binary fission, 167–168, *168*
Binding site, 763, 768
Binomial, 423, 431
Bioassay, 703, *703*, 708
Biobypass, 315
Biochemical processes, directions and rates of, 102–117
Biochemical reactions, 23
    water in, 40
Biodegradable pollutants, 621
Biodiversity, determination of, 593–596
Biogeochemical cycles, 552, 560
Biogeography, 338, 342
Biological molecules, 46–69
Biological species, 371
Biological species concept, 371, 396, 425, 431
Biologists, questions asked by, 4–13
Biology, conservation, 603, 624
    definition of, 13, 20
    molecular, comparative, 336–337
Biomagnification, 622, *623*, 625
Biomass, 549, 550
    inverted pyramid of, 550, *551*
    pyramid of, 550, *550*, 560
Biome(s), 546, 562, *563*, 564, 582
    definition of, 564
    determinants of type of, 564–565, *565*
    differences in, *567*, 567–578
    latitude and altitude effecting, 566, *567*
    major, temperature, precipitation, and productivity of, 567, 568t
    world distribution of, 567, *568*
Biosphere, 544, *545*, 547, *547*, 560
Biotechnology, 278, *280*, 294
Biotic factors, 546, 560
Bipedalism, 390, *390*, 397
Bipolar cells, 875, 880

Birds, *538*, 538–539
    adaptations of, 539, *539*
    breathing by, 800, *800*
    cladistic classification of, 427
    convergent evolution in, 333, *334*
    distances flown by, 719
    evolution of, 428
    mammals, and reptiles, cladogram for, 428, *429*
    marine, kidneys of, 821
    migration of, chemical signaling and, 756
    naming of, 423, *423*
    skeletal adaptations of, 720, *720*
    wings of, 719
Birth control. *See also* Contraception
    approaches to, 899–903
    worldwide, *899*
Birth control pill, 901–902, 905
Birth rate(s), 609, 624
    factors decreasing, 899
    factors increasing, 899
2,3-Bisphosphoglycerate, 804, 805
Biston betularia, *360*, 360–361
Bithorax, 924–925, *925*
Bivalves, *512*, 513, 521
Bladder, 814, 825
Blade, 654, 658
Blastocoel, 910, 927
Blastocyst, 910, 927
Blastopore, 501, *501*, 524, 913, 927
Blastula, 499, 520, 909, 927
Blending inheritance, 182, *182*
Blood, 770–776
    carbon dioxide in, 804, 805
    circulation of, 773–774, 775, *775*
    composition of, 773–774
    functions of, 771–772
    in oxygen delivery at high altitudes, 805
    oxygen delivery by, 782
    pH of, 42–43
    renal filtration of, 814–817, 825
Blood banking, 770
Blood-brain barrier, 865
Blood cells, antigens on surfaces of, 841–842
Blood clot(s), 831, 842
    prevention of, *830*, 831–832
Blood flow, regulation of, 785–786
    renal, filtration rate and, 821
Blood glucose, levels of, 758–759
Blood groups, human ABO, 307t
Blood loss, body control of, 830–831
Blood-nerve barrier, 865, 866
Blood pressure, 785, *785*
    high. *See* Hypertension
    hormones and, 821–823, *822*
    salt consumption and, 808–810
Blood pump, small animal, 776
Blood transfusions, 843
Blood types, variation of, among races, 302
    in Japan, 303, *303*
Blood vessels. *See also* Artery(ies); Veins
    major, *772*
    pressure in, 785, *785*
    structure of, *781*, 781–782
Blotting, 288, *288*
Body, human, three areas of, 730, *731*
Body defense, in broken skin, 830–833
Body plan, tube, *908*, 908–909
Bohr, Niels, 103
Boltzmann, Ludwig, 102–103, *103*, 105, 110

Bolus, 744, 752
Bombykol, 629
Bond(s), high-energy, 108, 116
Bone(s), 524, 538
    forces on, 724, 725
    responses to use, 725–726
    skeletal muscle and, 726
    structure of, 723, *724*
Bone marrow, red, 723, *724*, 733
Bone tissue, compact, 723, *724*, 733
    spongy, 723, *724*, 733
Bonner, James, 699–700
Bonnet, Charles, 906
Bonobos, 390
Book lung(s), 517, 521, *794*, 795
Bordered pit pair, 647–648, 658
Botstein, David, 296, *297*, 297–298
Bottleneck effect, 357, *357*, 366
Boveri, Theodor, 166, *167*, 182, 199
Bowman's capsule, 816, *816*, 825
BPG, 804, 805
Brachial arches, 334, *336*
Brachiostoma, 524
Bragg, Sir William Lawrence, 213, 214, 216
Brain, fluid intake and, 821
    in invertebrates, 864–865
    nerves, and sense organs, 865
    neurons of, 865
    regions in sensation, 863–864
    regions involved in various processes, 863, *863*
    sensory information and, 871–872
    of vertebrates, 866–868
Brain stem, 866
Brassica oleracea, 19, *19*
Brazilian rain forests, 385
Bread mold, black, 468, *469*
Breathing, 120
    by animals, 788–806
    by birds, 800, *800*
    by insects, 801, *801*
    mammalian, 796–798
Breathing center, 805, 806
Brenner, Sydney, 237
Bridges, Calvin, 200
Briggs, Derek, 496–497
Briggs, Robert, 919
Briggs, Winslow, 702–703
Broca, Pierre Paul, 863
Bromocriptine, 471
Bronchi, 796, 806
Bronchioles, 796, 806
Brown, Rachel, 452, 453
Brush borders, 817–819, *818*
Bryan, William Jennings, 318
Bryophytes, 479, 481, 483–484, 645, 649. *See also* Plant(s), nonvascular
    absorption of water by, 483, 494
    reproduction by, 483–484
Büchner, Eduard, 119
Büchner, Hans, 119
Bud(s), axillary, 687, 692
    flowering plant, 685
    terminal, 687, 692
Buffers, 43, 45, 771
Buffon, Count, 322–323, 325, 342
Bufo boreas, 535
Bufo bufo, 535
Bufo periglenes, 535
Building blocks, links of, 52–53
    molecular, 52, *53*, 54t

Bulb of the vestibule, 889, 904
Bulbourethral glands, 888, 904, 892
Bulk, 750
Bulk flow, in animals, 716, 732
Bullfrogs, 371, *371*, 531
Bult, Carol, 446
Burgess shale, 496, *498*
Burnet, Frank MacFarlane, 827
Butler, Samuel, 264
Butterfly, swallowtail, *363*, 363–364

C

C_3 plants, 158, 159
C_4 plants, 158, 159
    anatomy of, *652*, 652–653, *653*
Cabbage, sea, 19, *19*
Cacti, 158, 668, *668*
Cadaverine, 598
Caecilians (Gymnophiona), 533, *533*
Caenorhabditis elegans, 915–916
Cagniard de la Tour, Charles, 118
Calcitonin, 766
Calcium ions, in chemical synapse, 856
Callus, 690–691
Calorie(s), 106, 117
Calvin, Melvin, 153–155
Calvin cycle, 153–155, *155*, 159
Calyx, 489, 491
Cambium, cork, 689–690, *690*, 692
    vascular, 689, 690, *690*, 692
Cambrian explosion, 330–332, 380–383, 396
    feces in, 509
Cambrian Period, 330–331, 380, 381, 499
Camouflage, 588, *588*
Campbell, Keith, 920–921
Canavanine, 642
Cancer(s), cervical, and Pap smear, 164
    damage to DNA and, 231, *232*
    of lungs, 789, 789–791, *791*
    patterns of gene expression in, *287*, 287–289
Candida albicans, 468
Canidae, hidden variation in, *332*
Canines, 744
Canis familiaris, 332
Canopy, 576, *577*, 582
Capacitation, 892, 904
Capillaries, 779, *779*, 787
    water movement in, 663–664, *664*
Capillary action, 39, 40, 663–664
Capsid, 261, 274
Capsule, 440, 447
Capybara, 424
Carbohydrates, 60, 69
Carbon, 553–554, 554t
    alpha (α), 64, 69
    capturing of, in Calvin cycle, 155, *155*
    isotopes of, 28
Carbon-14, 329
    half-life of, *28*, 28–29
Carbon atoms, structures formed by, 49–50
Carbon cycle, 554, *554*
    humans and, 555–557
Carbon dioxide, 553–554, 555
    atmospheric, temperature and, 555, *555*
    in blood, 804, 805
    in Calvin cycle, 155, *155*
    changing delivery of, breathing regulation in, 805
    exchange for oxygen, 791–795
    measuring levels of, for 40 years, 556, *556*
    movement of, 782, *782*

in photosynthesis, 155–157
to prevent photorespiration, 158
removal from atmosphere, 557
Carbon-fiber technology, 209
Carbon monoxide poisoning, 115
Carbonated beverages, 794
Carboniferous Period, 331, 481, 534, 536
Carboxyl group, 51, 68
Carcinogens, 800
Cardiac arrest, 784
Cardiac muscle, 713, 726, *728*, 733
Cardiac output, 786
Cardiovascular system, 776, 786
Carnivores, 548, 560, 736, 752
digestive tracts of, 712, *712*
β-Carotene, absorption spectra of, *148*
Carotenoids, 148, *148, 149*, 159
Carpels, 491, 494, 679, 692
Carrageenan, 460
Carrying capacity (K), 609
of Earth, 617–620, *618*, 624
factors influencing, 612–613
Cartilage, 528, 542
Cascades frog, 535
Caspar, Timothy, 698
Casparian strip, 651
Catabolism, 129, *129*, 138
definition of, 114–115, 117
Catabolite activator protein (CAP), *253*, 253–254, 256
Catalase, 84
Catalyst, 111, 116
Cataracts, 874, 880
Catarrhines, 389, *389*, 397
Catastrophism, 323–324, *324*, 341
Caviomorphs, 424
Cecum, 751, 752
Cell(s), 15
aerobic, 120, *120*
boundary, body, and genes of, 74–75
changes in, chemical and mechanical, 75
communication of, in multicellular organisms, 96, 99, *98*
components of, 77–79, *78–79*
cytoplasm division in, 175
death of, programmed, 914–916, *916*, 928
depolarized, 851
determined, 919, 929
differentiated, totipotency of, 919–920, *920*, *921*
differentiation of, pattern formation and, *924*, 924–925
embryonic, differentiation of, 919–923
epithelial, of small intestine, 747–749, *749*
eukaryotic, 15, 16, 20
transmission electron microscope to assess, 77–79
fate of, 919, 929
features of, 77t
first, of new generation, chromosomes in, 186
functions of, 75
good gone bad, defense against, 839–843
growth control in, 177, *178*
loss and replacement of, by multicellular organism, 77
modern, evolution of, 412–414
of nervous system, 844–861
normal, division of, 176–177
determination of, 177
nucleus of, 74, *74*, 79–80, *80*, 100

organisms composed of, 70–101
organization of, 75–77
passage through cell cycle, 176–177
plant, 84–85
potency of, 919, 929
prokaryotic cells, 15, 16, 20
properties of, surface-to-volume ratio and, 76, *76*, 100
regulation of enzyme function by, 115, *115*
regulation of internal environment by, 75–76
reproduction of, 75
responses in, chemical signals and, 755–757
single, digestion in, 739
specialization by, 76
in small intestine, 747, *747, 748*
stem. See Stem cells
study of, using microscope, 72–73
use of genetic information by, 412
Cell biology, chemistry and, 21–160
Cell body, of neuron, 847, *848*
Cell cycle, passage of cells through, 176–177
stages of, chromosomes and, 168–172, *170*
Cell division, 163–167
cellular regulation of, 177, *178*
chromosomes in, discovery of, 165
daughter cells in, 167
determination in normal cells, 177
DNA and, 173
in eukaryotes, 168–172
mechanisms of, 163–165
of normal cells, 176–177
in plants, 686
in prokaryotes, 167–168, *168*
Cell expansion, in plants, 686, 703
Cell membranes, 772
Cell plate, early, 176, *176*
Cell reproduction, 162–179
cell division in, 163–167
chromosomes in, 165
mechanisms of, 163–165
coordination of cycles in, 177
process of, 167
Cell sap, 645
Cell senescence, 176–177
Cell theory, 74, 100
Cell wall(s), 93, 101, 645
primary, 645, *646*, 658
secondary, 645, *646*, 658
Cellulae, 70
Cellular respiration, 120
definition of, 120
energy capture in, 136, *137*
stages of, 121–122, *122*
Cellular slime molds, *464*, 464–465
Cellulose, 60, 69, 528, 645, 658
Cenozoic Era, 332
Central canal, 866
Central dogma, of molecular biology, 238, 239, *239*, 255
Central nervous system, mammalian, 866, *867*
in vertebrates, organization of, *865*, 865–868, 880
Centrifuge, 80–81, *81*
Centriole(s), 86, 100, 172
Centromere, 171
Centrosome, 86, 100
Cephalization, 865
Cephalochordates, 528
Cephalopods, *512–513*, 513–514, 521
eye of, 514, *515*

Cerebellum, 862, 866, 880
Cerebral cortex, 862–863, 867, 868, 880
Cerebral hemispheres, 867, 880
Cerebrospinal fluid, 865
Cervical cancer, and Pap smear, 164
Cervical cap, 901, 905
Cervical cells, normal, *164*
precancerous and cancerous, 164, *164*
Cervix, 886, *890*, 905
Cestoda, 506, 507, *508*, 521
Chaetognatha, 524, 541, 542
Chain, Ernst, 280
Chamberland, Charles, 260
Channel(s), ligand-gated, 856
sodium, voltage gated, 851, 852, 861
Channel proteins, of neurons, 848, 861
Chaos chaos, 460
Chaparral, 564, 571, *571*, 582
fires in, 571, *571*, 572
Chaperones, 68, 69
Character(s), primitive, 427, 431
shared derived, 427, 431
Character displacement, 592, *592*, 600
Chargaff, Erwin, 215, 217, 224
Chargaff's rules, 215–216, 233
Chase, Martha, 223–225, *225*
Cheetah, 721
genetic variability in, 353
Chelicerates, 518, 519, 521
Chelonia (turtles and tortoises), 534, *536*
Chemical bonds. See Bond(s)
Chemical defense(s), 588
plant, 656–657
Chemical equilibrium, 41, 45, 107–108, 115
Chemical reaction(s), 23
activation energy of, enzymes and, 113–114
coupled, 108, *109*
direction of, 104
prediction of, 104–106, 107, 116
rate of, determination of, 110–112
molecular motion and, 110, *111*
starting of, 110–112
stopping of, 110, *111, 112*
Chemical signaling, cellular responses to, 755–757
development and, 923–925
migration of birds and, 756
signals in, detection of, 763
six steps in, 763
types of, 756, 756–757
Chemical synapses, 854, 861
action of psychoactive drugs on, 857
mechanism of action of, 855–856
structure of, 855, *855*
Chemical warfare, 588, *588*
Chemiosmosis, 131, 138
Chemiosmotic hypothesis, 131, 138
Chemistry, and cell biology, 21–160
definition of, 23, 44
of life, 24–26
Chemoautotrophs, 446, 450
Chemokines, 271
receptors for, AIDS and, 271
Chemotaxis, 442, 450
Chemotrophs, 119
Cherts, 401
Chewing, 744
Chiasma(ata), 189, *189*, 205
Chicxulub, 382
Chihuahua whiptail lizard, 883

Chimpanzees, 390, 392
   pygmy, 390
Chitin, 60, 69, 466, 475
Chlamydia, 435
Chlamydia trachomatis, 437, *446*
Chlamydomonas, 455, *457*, 461
Chloracne, 696
Chlordane, 754
Chlorella, 461
Chlorofluorocarbons and, 622, *624*
Chlorophyll(s), 85, 147, 159, 699, 701
   in chlorophytes, 461
   P_{680}, 149, 150, 151, 159
   P_{700}, 149, 150, 151, 159
Chlorophyll *a*, absorption spectrum of, 148
   photosynthesis and, *150*
Chlorophytes, 461, 474, 644
Chloroplast(s), 100
   anatomy of, 149, 150–151
   DNA in, *272*, 273
   as endosymbionts, 413–414
   plant cell, 84–85, *86*
      functions of, 85
      structure of, 85
   requirements of photons for glucose production, 157
Cholecystokinin, 759
Cholera, cyclic AMP in, 764
Cholera toxin, 764
Cholesterol, biochemistry of, *58*, 58–59
Chondrichthyes, *530*, 530–531, 542
Chordata, 524, 531
   tissue layers of, *502*
Chordates, 522–542
   characteristics of, 527–528
Chordin, 923
Chordin gene, 523
Chorion, 536, 542, 911
Chorionic gonadotropin, human, 896, 897, 906
   pregnancy tests and, 898
Chorionic somatomammotropin, human, 898, 906
Chorionic villi, 911
Chorionic villus sampling, 312, *312*, 315
Choroid, 873
Chromatids, 188, 204–205
   during anaphase, 172–173
   separation of, 190, *190*
   sister, 171, 188, 190
Chromatin, 165, *165*, 166
   in cell, 79–80, *80*, 100
Chromatin-remodeling machines, 254
Chromatography, paper, 153–154, *154*
Chromoplasts, 85, *86*, 100
Chromosome(s), 100, 489, 490
   assortment of genes on, 201–202, *203*
   carrying of information by, 182–183
   definition of, 165
   discovery of, 165–166
   distribution by meiosis, to gametes, 188–192, *191*
   in first cell of new generation, 186
   function of, discovery of, 166–167
   homologous, 168, *169*, 185, 350, 366
      in meiosis, 188–190, 205
   inheritance from. *See* Inheritance, chromosomal theory of
   during metaphase, 172
   movement of, during mitosis, 173
   mutations in, 231
      deficiencies, 231
      duplications, 231, 233
      inversions, 231
      monosomy, 231, 233
      translocations, 231
      trisomy, 231, 233
   number of, from generation to generation, 185–188
   pairs of, 185, *186*
   recombination of two genes on, 204
   sex, in humans, 192–193, *194*
   of sexually reproducing eukaryotes, 168, *169*
   stages of cell cycle and, 168–172, *170*
   X, 192–193, 205
   Y, 192–193, 205
Chromosome banding, 175
Chrysemys scripta elegans, *120*, 125
Chrysophytes, 458–459, *459*, 474
Chthamalus stellatus, 591, *592*
Chyme, 746, 752
Cigarettes, addiction from, 791
   sales of, death rate and, *789*
   history of, 789
Cilia, 89, *89*, 100
   function of, 798–799, 806
Ciliary body, 874
Ciliates (ciliophora), 464, *464*, 475
Ciliophora (ciliates), 464, *464*, 475
Circuits, convergent, 858
   divergent, 858
Circulation, 776
   closed, 778–779, 787
   double, 532, 542
   open, 778, *779*, 787
   pulmonary, 776, 786
   systemic, 776
Circulatory system, 775, *775*
   pathogens sealed from, 830–831
Circumcision, 889, 904
Cirrhosis, 136
Citric acid cycle, 127–128, *128*, 138
   discovery of, 128–129
   other biochemical pathways and, 129–130
Citrulline, 219
Cladistics, 427–429, 431
Cladogram, 427, 431
   for birds, mammals, and reptiles, 428, *429*
   of primates, 427, *428*
Class, in hierarchy of classification, 423, 431
Classification, 418–431
   Aristotle on, 424
   difficulty in, 430
   kingdoms in, 430
   Linnaeus's system of, 422–423
   naming in, 422
   Plato on, 423–424
   red wolf controversy in, 418–421
   species grouping in, 421–430
      cladists and, 427–429
      classical evolutionary taxonomy and, 426
      disagreement over, 430
      DNA and protein sequences in, 429–430
      Foucault's Chinese encyclopedia and, 421
      guinea pigs and, 424
      pheneticists and, 426–427
      of reptiles, 428
Claviceps purpurea, 471
Clays, 669, 674
Cleavage, in vertebrates, 909, 910, 927
   in zygotes, *910*, 910–911, *911*
Cleavage furrow, 175, *176*

Clements, Frederick Edward, 585–587
Climate, of deserts, 564
   Earth, changing, 556
   sunlight, and movements of Earth, 564–566
   topography and, 566–567
Climax community(ies), 585, 597, 599, 600
Clitoris, 890, 905
Cloaca, 536–538, 542, 814, 825, 888, 904
Clonal selection theory, 837, 843
Clone, 185, 204, 285
Cloning, of animals, 920–921, *921*
   positional, *309*, 309–310, 315
Clonorchis sinensis, 506, *507*
Clostridium botulinum, 120, *120*, 126
Club fungi (Basidiomycetes), 469–470, *472*, 475
Club moss, life cycle of, *479*
Cnidarian(s), *504*, 504–505, *505*, 521
   tissue layers of, *502*
Cnidocytes, 504, *505*, 521
Coacervates, 408, *409*, 415
Cocaine, 857
Cocci, 439, 450
Coccidioides, 468
Cochlea, 873, *873*, 882
Cochlear implants, 871
Cockroaches, resistance in, 360, *360*
Codium magnum, 461
Codominance, 196, 205
Codon(s), definition of, 237, 255
   nonsense, 238, 255
   stop, 238, 255
Coelom, 501, 520, 713, 732
Coelomates, 501, 510–520, 524–531
   tissue layers of, *502*
Coenocytic fungi, 466
Coenzyme A, 127, 138
Coenzymes, 123, 138
Coevolution, 491, 590–591
Cohen, Fred, 48
Cohen, Mitchell, 432, 433
Cohen, Stanley, 846
Cohesion, water, in tall plant, 665
   of water, 663, 674
Cohesion-tension theory, 664, *665*, 674
Cohort, 614, 625
Coitus interruptus, 901, 905
Cold, tolerance by animals, 732
Collagen, 63, 725, 733
   helix in, 66, 67
   structure of, 65–66, *66*
Collagen fibers, 725, *725*
Collagen fibrils, 725, *725*
Collenchyma, 646, *647*, 658
Colon, 750, 752
Color blindness, 875
Coloration, warning, 589, *589*, 600
Colostrum, 898, 905
Comb jellies (ctenophores), 505–506, *506*, 521
Commensalism, 590, *590*, 600
Communication, between neurons, 853–857, *854*
Community(ies), 546, *547*, 560, 584–601
   changes in, 596–600
   climax, 585, 597, 599, 600
   individualistic, 586–587, *587*, 600
   integrated, 586, 600
   pioneer, 597, 597t, 600
   resilience of, 599, 600
   species in, interaction of, 587–593
   stability of, 599, 600
   successional, 597, 600

Compact bone tissue, 723, *724*, 733
Companion cells, 649, 672, *673*, 674
Comparative embryology, 334–336, *336*
Competition, 591–593, 600
　　evolution and, 592–593
Competitive exclusion principle, 591, 600
Competitive inhibitor, 116, *116*
Complement, 832, 843
Complementary DNA (cDNA), 285, *285*, 286, 295
Complementary proteins, 737, *738*
Complex molecules, building of, 52
Compound eye(s), 518, 521, 873
Compound microscope, 72, *73*
Compound(s), 24, *25*, 44
Compression, 724, 732
Concentration, 91, 100
Concentration gradient, 848–849, 861
　　in diffusion, 91–92, *92*, 100–101
Conception, 883
　　avoidance of, 900
Conditional mutation, 245
Conditioned reflex, 626, 639
Conditioned stimulus, 859
Condoms, 901, 905
Conduction, saltatory, 853, 861
Cones, 874, *875*, 880
Coniferous forests, 564, *564*
Conifers, 489
　　tracheid pit connections in, *647*
Conjugation, in bacteria, 265, *266*, 274
Conjugation fungi (Zygomycetes), 468–469, *470*
Connective tissue, 499, *499*, 520, 713
　　forces on, 724–725, *725*
Connell, Joseph H., 591, 592, 593
Conservation biology, 603, 624
Constipation, 750, 752
Constraints, developmental, 383, 396
Consumers, 548, 560, 588
Contact inhibition, 177, *178*
Contig, assembly of, 309, *309*
Continental drift, 339
Continent(s), history of, 339, *339*
　　species diversification on, 379
Contraception, 899–902, 905
　　abstinence as, 900–901, 905
　　barrier methods as, 901, 905
　　hormones as, 901–902, 905
　　intrauterine devices as, 902, 905
　　methods of, 900t
　　sterilization as, 902–903, 905
　　withdrawal as, 901, 905
Contraceptive sponge, 901
Contractile proteins, 728, *729*
Contractile ring, 175, *176*
Contrast, 72
Conventional genes, 259
Convergent circuits, 858
Convergent evolution, 376, 396
Coontz, Robert, 602
Copepods, 519
Copulation, adaptations to, *187*
Corals, 504
Corey, Robert, 214
Coriolis effect, 566, 582
Cork cambium, 689–690, *690*, 692
Corn, 687
　　crossing over in, 203
　　high-oil, 293
　　jumping genes in, 258–260
　　origin of, 676–677, *677*
　　pod, 676
　　prehistoric, *280*
Cornea, 873, 880
Corolla, 492, 494
Corpora allata, 758
Corpse, succession on, 598
Corpus callosum, 868
Corpus cavernosum, 889, 904
Corpus lutem, 895, 904
Corpus spongiosum, 889, 904
Correns, Karl Franz Joseph, 349
Corridors, in wildlife preserves, *607*, 607–608, 624
Cortex, 651, 658
　　renal, 814–817, 825
Corticotropin, 762
Cortisol, 760, 768, 898, 905
Cotransport, 95, *95*
Cotton, transgenic, *293*
Cotyledons, 649, 658, 680, 692
Cough, "smoker's," 799
Countercurrent heat exchange, 716, *717*
Countercurrent heat exchanger, 716, 731
Countercurrent oxygen exchange, *717*, 717–718
Countercurrent systems, 716–718, 732
Coupling factor, 135, 138
Courtship, of stickleback, *634*
Covalent bond(s), 30, *32*, 44
　　atoms in, 30
　　double, 30, 44
　　triple, 30
Cowles, Henry C., 585, 586, 600
Crab, horseshoe, *324*
Cranium, 865
Crassulacean acid metabolism, 158, 159, *667*, 667–669, 674
Creationism, 319, 341
Creighton, Harriet, 203–204, 259
Creosote bush(es), 567, *569*, 668, *668*
Cretaceous Period, 331–332, 339, 380, 382, 479, 536
Creutzfeldt-Jakob disease, 46
Crick, Francis, 208, 211, 212–213, *213*, 214, 216, 217, 225, 233, 234, 236–237, 239, 245, 269, 308, 411
　　The Double Helix by, 237
Cristae, 84
Cro-Magnons, 395–396, 397
Crocodiles, 534
　　Nile, *536*
Crocodilia (crocodiles and alligators), 534
Crop, 745, 752
Cross breeding, 194
Cross bridges, 729, 733
Crossing over, 188–189, *189*, 204
　　in corn, 203
Crustacea, 519–520, *520*, 521
Cryptochromes, 699
Cryptococcus, 468
Crystallins, 916
Ctenophores (comb jellies), 505–506, *506*, 521
Cubozoans, 504
Cud, 746
Curie, Marie, 29, *29*, 329
Curie, Pierre, 29, 329
Current(s), ocean, 578–579
Cuticle, 478, 494
　　of flowering plant, 646, 658
Cutin, 654
Cuvier, Baron Georges, 323, 325, 522, 523
Cyanobacteria, filamentous, 401, *401*, 402

Cycads, 489–490, *489*
Cyclic AMP, 764, 768
　　in cholera, 764
　　synthesis of, 765, *765*
Cyclins, 178
Cyclostomes, 529, 542
Cystic fibrosis, *305*, 305–306, 310–311, 314
　　transmembrane conductance regulator, 305
Cysts, in protists, 456, 474
Cytochrome *c*, phylogenetic tree of, 337, *338*
Cytochrome oxidase complex, 134
Cytochromes, 131, 138
Cytokines, 289, 833, 839, 843
Cytokinesis, 168, *170*, *171*, 175–176, *176*
　　in plant cells, 176, *176*
Cytokinins, 696t, 702, 705, 708
Cytoplasm, 75, 100
Cytoplasm division, 175
Cytosine, 210
Cytoskeleton, 75, 86–87, *87*, 100
　　function of, 86
Cytosol, 75, 80–81, 100
Cytotoxicity, 840

D

D & E, 903, 905
Daddy longlegs, 519
Daktulosphaira vitifoliae, 660
Dalldorf, Gilbert, 453
Dalton, John, 26
Dalton (unit), 26
Danger model, 826–829, *828*
Darrow, Clarence, 318
Darwin, Charles, 18–19, 319, *326*
　　altruism and, 636–637
　　artificial selection and, 321
　　Beagle voyage of, 325–326, *370*
　　blending of inheritance and, 346–347
　　on conditions for production of living organism, 404
　　on defining of species, 371
　　distributions of species and, 338
　　education of, 325
　　evolution of species and, 325–327
　　on Galápagos Islands, 368–369, 377–378
　　gradualism and, 320
　　natural selection and, 321, 338–341, 342
　　phototropism and, 694, *695*
　　speed of evolution of species and, 383–387
　　theory of evolution of. *See* Evolution, Darwin's theory of
Darwin, Erasmus, 321, 325, 477
Darwin, Francis, 694
Darwin, Robert, 325
Daughter cells, 74, 165
　　in cell division, 167
　　in meiosis, 190
Daylight, plant responses to, 699–702, *700*, *701*
DDT, estrogenic effects of, 754
　　poisoning from, 622
　　resistance to, 360
de Vries, Hugo Marie, 349
Dead space, 796, 806
Dead zones, 559, *560*
Death rate, 609, 624
　　smoking and, 789, *789*
Death Valley, 562
Decapods, 519
Decay, 329
Deciduous forests, 564, 582

Declarative memory, 868
Decomposers, 446, 450
Defenses, of animals, 588–589
    chemical, 588
    of plants, 588–589
Defibrillation, 784
Dehydration, 813, 824
Dehydration reaction, 52–53, *53*, 68
Delbrook, Max, 223, 224, *224*
Delphinium, 373, *373*
Democritus, 26, 102
Demography, 616
Denaturing, of enzyme, 114, 117
Dendrites, 847, 861
Denitrification, 558, 560
Dennett, Daniel, 497–498
Density-dependent population growth, 611
Density-independent population growth, 611
Deoxyribonucleic acid. *See* DNA
Deoxyribose, 55t, 59, 69
Depo-Provera, 902, 905
Depolarized cell, 851
Dermal tissue system, 646, 658
Dermis, 830, 842
DeRobertis, Edward, 523
Desert(s), animals of, 568
    water conservation by, 812
    climate of, 564
    ecosystems of, 567–569
    plant water conservation in, 668
    plants of, 568
    soils of, 569
DeSilva, Ashanthi, *314*
Desmosomes, 98, *99*, 101, 784
Detergents, 40
Deterministic events, 604, 624
Deuteromycetes (imperfect fungi), 470, 471, *472*, 475
Deuterostomes, 501, 522–542, *525*, 541
    coelomates, 524–531
Development, chemical signaling and, 923–925
    vertebrate, stages of, 909
Developmental constraints, 383, 396
Devonian Period, 331, 479, 526, 527, 528
d'Herelle, Felix, 223
Diabetes mellitus, 759, 768
    insulin-dependent, 759
    juvenile, 759
    non-insulin-dependent, 759
    type 1, 759
    type 2, 759
Diacylglycerol, 765
Diaphragm, 798, 806, 901, 905
Diarrhea, 462, 750, 752
Diastole, 782–783, 787
Diatoms, 458–459, *459*
Dickens, Charles, 862
Dicots, 481, 649–651, *650*, 658, 680, 692
    germination in, 682–683, *683*
Dicotyledons, 649–651, *650*, 658
Didinium, 76, *76*
Diencephalon, 867, 880
Diet, low-salt, 808, 809–810
Differentiation, 914, 928
Diffusion, 91–92, 794
    facilitated, passive transport and, 94, 101
    simple, 91, 100
    in tiny animals, 716, *716*
Digestion, 121, *121*, 138, 739, 752
    behavioral adaptations and, 742
    in cavities with one opening, 741, *741*
    conditions of, 740

    early studies in, 734–735
    by individual cells, 740, *741*
    in two-ended digestive tract, 741–742
Digestive tract, 752
    coordination of processes in, 751
    as "disassembly line," 741–742, *742–743*, 743
    human, 741, 743–751
    two-ended, 741–742
    vertebrate, organization of, 741–742, *742–743*, 743
Digits, programmed cell death and, 917, *917*
Dikaryotic fungi, 466
Dinkas, African, 352, *352*
Dinoflagellates, 457–458, *458*, 474
    test of, 457–458, 474
Dinomischus, 498
Dinosaurs, cloning of, 292
    speed of, 710–711, *711*
Dioxin, 696
Diphyllobothrium latum, 508
Diploid cells, 185, 204
Directional selection, *360*, 360–361, *362*, 362–363, 367
Disaccharides, 59, 69
Diseases, specific, location of genes for, 308–310
Dispersal, 379, 396
Disruptive selection, 363, 367
Distributed processing, 863
Distribution, normal, 12, *12*, 20
Diuretic, 823
Diuron, 156
Divergent circuits, 858
Divergent evolution, 376, 396
Diversity, 418–542
    abiotic factors influencing, 593–595
    biotic factors influencing, 593
    immune system recognition of, 834
    increasing, mass extinctions and, 383, *384*
    measurement of, 593
Dixon, Henry, 664
DNA, 16, 20
    alphabet, 239
    cell division and, 173
    as chain of nucleotides, 210, *211*
    in chloroplasts, *272*, 273
    chromosomal, tumor virus insertion into, 267, *268*
    circle of, 168, *168*
    cloning of, 922
    computer-generated model of, *218*
    damage to, consequences of, 231, *232*
        repair of, 231, *232*
    direction of DNA synthesis in, 225–229
    of energy organelles, 273
    in eukaryotes, 253
    fibers, 209, *210*
    folding of, 174
    in genes, 222, 223–225
    hybrid, 286, *286*, 295
    information in, 167
    loops of, 175, *175*
    in mitochondria, *272*, 272–273
    mitochondrial, of wolves and coyotes, compared, 419–420
    multiplication of, polymerase chain reaction and, 277, *278*, 286
    nucleotides of, 211, *213*
    in nucleus, 173–175, *174*
    phases of cell cycle and, 170, *170*
    and protein sequences, criteria for classification of, 429–430

    protists and, 465
    recombinant. *See* Recombinant DNA
    replication of, 167, 225–227
        by dividing cells, 225–227
        origin of replication and, 228, *230*, 233
        polymerization in, 227, *228*, 233
        replication forks in, 227–228, *230*
        in 3′ and 5′ direction, 228
        Watson-Crick model for, 227
    sequencing of, 226, *226*
    structure of, 210–217
        discovery of, by Franklin, 211–216, *214*
        replication and repair of, 208–233
        3′ and 5′ ends in, 211, *213*
    synthesis of, rules of, 226
    site of, 228, *230*
    transcription of, 240
    as transforming factor, 222
    as twisted ladder, 210–211, *212*
    viral, entry into host cell, 263, *264*
DNA array, 287, 295
DNA chip, 287, 295
DNA fingerprinting, 299, 315
    electrophoresis in, 300, *300*
    as evidence in U.S. courts, 299, 300–301
    in identification of kidnapped children, 302
    noncoding DNA and, 299
    reliability of, 299–302
    restriction enzymes in, 300, *300*
DNA ladder, *64*
DNA ligase, 228, 233, 283, 294
DNA polymerase, 227, *227*, 233
    in DNA assembly, 227, *229*
    in DNA sequencing, 226, *226*
DNA/RNA language, 237
DNA sequencing, and classification of prokaryotes, 448
DNA sequencing gel, 226
Dobzhansky, Theodosios, 338
Dog, cool nose of, *39*
Dolk, Herman, 694
Dolly, 292, 921, *921*
    cloning of, *920*, 920–921
Domains, 16, *17*, 20, 66
    in hierarchy of classification, 423, 431
Dominance, partial, 196, 205
Dominance hierarchy, 633, 639
Dominant alleles, 195–196, 205
Donohue, Jerry, 217
Dopamine, 854, *855*
Dormancy, plant embryo, 681, 692
Dorsal hollow nerve cord, 527, 542
Dorsal horns, 866
Dorsal root ganglia, 866
Dorsal structures, 523
Doubling time, 608, 624
Douche, 901, 905
Doudna, Jennifer A., 410
Down syndrome, 191, *191*, 193, 205, 312, 313
Drake, Frank, 403, *403*
Drew, Charles, 770–771, *771*
Driesch, Hans, 907, 918
Drosophila, imaginal discs in, 925, *925*
    mutant, 245
    pattern formation in, 922, 923–924, *924*, 928
Drosophila melanogaster, eye color gene in, 200, *201*, 202, *202*
    genetic research on, 352, 353
    life cycle of, 200, *200*
    mutant, *201*
    normal, *925*

sex-linked genes in, 201
wild type of, 201, *203*
Drosophila subobscura, courtship of, 364–365
Drugs, actions of, physical basis of, 858
Duck-billed platypus, 541, *541*
Duodenum, 747, 752
Dyneins, 86
Dynorphins, 858
Dysjunction, 190–191, 205, 190–191, 205

E

Ear, inner, 871, *872*, 872–873, 880
    and balance, 872–873
    mammalian, 872–874, *873*
    middle, 872, *873*, 882
    outer, 872, *873*, 882
    microorganisms inhabiting, 436
Eardrum, 871
Earth, carrying capacity of, 617–620, *618*, 624
    life of, origin of, 399–404, *402*
    theories of, 401–402
    movements of, sunlight, and climate, 564–566
Earthworms, 514, *516*
    salt and water balance in, 823–824
Easterlies, polar, 565, *566*, 582
Eccles, Sir John, 879
Ecdysone, 758
Echinodermata, 524, 541
Echinoderms, 522–542, 524–527, 541–542
Ecological footprint, 618, 620, *620*, 625
Ecological isolation, 373, 396
Ecological unit, species as, 425, 431
Ecology, behavior and, 632–634
    definition of, 546, 560
Ecosystem(s), 544–561, 546–547, *547*, 560
    aquatic, 578–581
    boundaries of, 546–547
    components of, 546
    of deserts, 567–569
    energy flow through, 548–552
    interactions of, 547
    nitrogen fertilization and, 559
    organisms cooperating in, 545, 546, *546*
    preservation of, 607–608
    properties of, 546–547
    recycling by, *552*, 552–559
    resilience in, 600
    terrestrial and aquatic, 562–583
Ectoderm, 499, 520, 908, *908*, 927
Ectotherms, 538, 542
Edema, 813, 824
Edge effects, 604, *605*, 624
Eels, 795
    electric, sodium-channel protein in, 851
Effector, allosteric, 116
    definition of, 114, 117
Efferent nerves, 868, 880
Egg(s), 186, 204
    amniotic, 428, 536, *537*, 542
    fertilization of, 897
    union with sperm, 186–188, *187*
Egg cells, growth of, 177
Ehrlich, Paul, 602
Eijkman, Christian, 738–739
Einstein, Albert, 4, 101
Ejaculation, 888, 891, 904
El Niño, 579
Elastin, 63
Eldredge, Niles, 387
Electrical potential, 849, 861
Electrical pulse, generation by neurons, 848–850

Electrical synapse, 854, 861, 854, 861
Electrochemical gradient, 133, 138
Electrochemical impulses, nerves and, 847
Electrodes, 850
Electron(s), 27, *27*, 44
    in atoms, 30
    in noncyclic electron flow, 150
    in photophosphorylation, 149
Electron carriers, 131, 138
Electron donors, 130
Electron microscope, 72–73, *73*
Electron orbitals, 30
Electron transport, 130–137
    energy of, harvesting of, 131
Electron transport chain, 131, *132*, *133*, 138
Electronegativity, 34, 44
Electrophoresis, 220, *221*, 226
    in DNA fingerprinting, 300, *300*
    gel, 288, 295
Elements, 23–24, 44
    of human body, *24*
    Periodic Table of, 30, 31, Appendix A
Elephants, unrestrained growth in, *608*
Elimination, 813, 824–825
Eliot, George, 862
Ellis-van Creveld syndrome, 358
Elongation, 247, *248–249*
    inhibition of, *250*
Elongation zone, 686, 692
Embryo(s), cells of, and cells of adult, 908–916
    differentiation of, 919–923
    development of, fertilization and, 908, 927
    flowering plants, body plan of, 678–684
    development of, 680–683, *682*, *683*
    nourishment of, 683, *684*
    formation of, historical views on, 906
    human, *336*
    mouse, *336*
    of vertebrates and invertebrates, compared, *524*
Embryo sac, 679, 692
Embryology, comparative, 334–336, *336*
Embryonic stem cells, 926
Emergent property, art as, 9, *9*
Emergents, 576, *577*, 582
Emigration, 610, 624
Emphysema, 800
Emulsion, 750
Encephalopathies, transmissible spongiform, 46
Endangered Species Act, 420–421
Endergonic reactions, 107, *107*, 116
    and exergonic reactions, coupling of, 108, 109, 116
Endocrine organs, 766, *766*, *767*
Endocrine signals, 766
Endocrine system, 757
Endocrinology, 757
Endocytosis, 96, *97*, 101, 263, 741, *741*
    receptor-mediated, 96–97, 101
Endoderm, 499, 520, 908, 908–909, 929
Endodermis, 651–652, 658, 659
Endogenous antigens, 840, 844
Endometrium, 886, 904, 911
Endoplasmic reticulum, 81–82, 100
    lumen of, 82, 100
    rough, 82, *82*, 100
    smooth, 82, *82*, 100
Endorphin(s), 762, 858
Endoskeleton, 721, 732
    vertebrate, parts of, 722–723, *723*
Endosome, 741
Endosperm, 491, 494, 680, 692

Endosymbionts, 413–414
Endosymbiosis, 414, *414*
Endosymbiotic theory, 273, 274
Endothelial tissue, 713
Endothelium, 22, 781, *781*, 787
    oxygen movement through, 801, *801*
Endotherms, 534, 538
Endotoxins, 435, 447
Energy, 30, 44
    activation, 110–111, *111*, 117
    extraction from glucose, by cells, 122–130
    extraction from macromolecules, by heterotrophs, 121
    flow through ecosystems, 548–552
    free, changes in, and direction of reaction, 107, 116
    concentration of molecules and, 109, *109*
    definition of, 107
    source of, during reaction, 108, *108*
    harvested from glucose, by cell, 136–137, *137*
    kinetic, from work, 104, 117
    loss of, during ecosystem recycling, 552, *552*
    as heat, 106, *106*
    potential, in work, 104, *105*, 117
    provision of, for biological processes, 108
    pyramid of, 551, *551*, 560
    subsidies of, 617–618, *619*, 625
    sugars for storage of, 60
    supply to organisms, 118–139
    transformation of, thermodynamics and, 103
Energy organelles, DNA and RNA of, 273
Energy requirements, of animals, 736
Energy supply, and energy consumption, 617–618
Enhancers, 255
Enkephalins, 858
Ensatina eschscholtzii, 375, *375*
Entropy, 109, *109*, 116, 544
Environment, adaptation to, 17–18
    animals sensing, 870–871
    conditions in, enzymatic reactions and, 114
    early life and, 412–413
    effects on sexual development, 754–755
    external, interaction of membranes with, 96 99
    as influence on aging, 926
    internal, of animals, regulation of, 755, *756*
    nervous system response to, 865–871
    nonmammalian vertebrate kidneys and, 820–821
    in phenotype, 183–185
    responses of plants to, 697–702
Enzyme(s), 108
    actions of, 112–114
    activation energy of chemical reaction and, 113–114
    active site of, 112, *113*, 117
    binding to reactant, 112–113, *113*
    as catalyst, 111, 112, 116
    denatured, 114, 117
    functions of, regulation of, 115, *115*
    genes and, 218–219, *220*
    influence of inhibitors on, 115–116, *116*
    to link nucleotides, discovery of, 234
    prevention from catalyzation of reaction, 116
    proteins as, 63, 69
    reactions of, environmental conditions and, 114
    subreactions in, 114–115
    restriction, 270, 274, 283, 294
    in DNA fingerprinting, 300, *300*
    recombinant DNA and, *282*, 282, *283*, 294

Enzyme(s), *(continued)*
  specificity of, 112
  temporary binding of, 112, *113*
Eohippus, 334
Ephedra, 490
Ephedra viridas, *490*
Epiblast, 913, 927
Epicotyl, 680
Epidemiologists, 432, 447
Epidermis, *829*, 829–830, 842
  of flowering plant, 646, 658
Epididymis, *887*, 888, 904
Epiglottis, 744, 796
Epinephrine, 760, 761, 763, 764, 765, 768, 785, 787, 869
Epiphytes, 576, 582
Episomes, 265
  definition of, 268
  replication of, 268, 274
Epithelial cells, 97, 100
  in airways, 799, *799*
  of respiratory tract, 798, *799*
  of small intestine, 747–749, *749*
Epithelial tissue, 713
Epithelium, 98, 499, *499*, 817, 825
  in gut lumen, 743, 752
Epitopes, 835, *835*, 843
Epochs, 330, 342
Equatorial doldrums, 565
Equilibrium, chemical, 41, 45, 109–110, 117
  concentration and entropy affecting, 109–110
  factors influencing, 110, *111*
  of gas in solution, 793, 806
  punctuated, 387, *387*, 397
Equilibrium potential, 849–850
Equisetophytes, 487, *487*
Equisetum, 487
Eras, 330
Erection, 890–891, 904
Ergot, 471, *471*
Ergotamine, 471
Ergotism, 471
Erosion, soil, from chaparral fires, *572*
Erythrocytes, 772–773, *773*, 787
  "ghosts," 88
  membranes of, 87–88
  oxygen transport by, 802, *802*
  sickled, *220*
  standard, *220*
Escherichia coli, 244, 252, 254, 255, 291, 361, 371–372, 435, 437, 439, *439*, 440, 441, 442, 446
  doubling time of, 608
  limitation of growth of, 609
  recombinant, 284
  restriction enzymes in, 270
  in study of DNA synthesis, 223, 228
Esophagus, 744, 752, 745–746, 752, 796
Esox lucius, 532
Essential amino acids, 737, 737t, 752
Essentialism, 321, 341
Estivate, 568, 582
Estrogen(s), 766, 894, 895, 896
  environmental, 754–755
  natural, 755
  during pregnancy, 897
  regulation of, 762
Estrogenics, 622, 754
Estuaries, 580, *580*, 582
Ethane, 50–51, *51*
Ethanol, 37, 51, *51*

in fermentation, 119
in genetic engineering of foods, 281
metabolism of, 136
properties of, 36t
water, and oleic acid, compared, 38t, 39
Ethology, 626, 639
Ethyl alcohol, 37
Ethylene, 30
  as plant growth hormone, 702, 706–707, 708
  in ripening, 706
  triple response to, 706, *706*
Ethylene dibromide, 642
Etiolation, 684, *684*
Eubacteria, 16, 20
  classification of, *444*, 445–448, 450
Eucalyptus trees, 490
Euglena, 89, 455, *455*, 457
Euglenophytes, 459–460, 474
Eukarya, 16, 20
Eukaryotes, 414, 430
  cell division in, 168–172
  DNA in, 253
  gene regulation in, 253–254
  nucleus of, 74
  protein production and, 250
  in regulation of gene expression, 254
  sexually reproducing, of chromosomes, 168, *169*
  subcellular organelles of, *415*
  transcription and translation in, 241–242
Eukaryotic cells, 15, 16, 20
  transmission electron microscope to assess, 77–79
Eumetazoa, 499, *501*, 502, 520
Euphorbiaceae, 642
Euploid cell, 191, 205
Eustachian tube(s), 334, *336*, 871
Eutherians, 541, 542
Eutrophic lakes, 581, 582
Everglades, 580
Evolution, 16, 18–19, 317–416
  biotic, early, 409–412
  competition and, 592–593
  convergent, 376, 396
    in birds, 333, *334*
  before Darwin, 321–325
  Darwin's theory of, 18–19, *19*, 20, 319–321, 326–327
  divergent, 376, 396
  evidence for, 318–367, 327–338
  fossil record of, 327–332, *328*
  genetic variation and, 353–354
  of horse, 333–334, *335*, 372, 380
  human, 388–396
  of males, female choice and, 633
  as opportunistic, *334*
  of organisms, 398–416
  preboitic, 404–409, 415
  process of, 320–327
  rate of, extinctions and, 383–388
Evolutionary taxonomy, classical, 426
Excitatory postsynaptic potential, 856, 861
Excitatory synapse, 854
Excitement, 890, 904
Excretion, 813, 824–825
Exergonic reaction(s), 108, *108*, 116
  and endergonic reactions, coupling of, 108, *109*, 116
Exhalation, 798, 806
Existence, struggle for, 341

Exocrine, 757
Exocytosis, 97, *97*, 101
Exogenous antigens, 840, 844
Exons, 242
Exoskeleton, 721–722, 732
  of arthropod, 517, 521
Exotoxins, 435, 447
Expansin, 703
Experiment(s), 6–7, 19
  "control," 7, 20
  design of, 7
  number of individuals in, 10
  statistics to plan and evaluate, 10
Experimental psychology, 626–627, 639
Expiration, 798, 806
Exponential growth, 608–609, 624
Expressivity, gene, 304, 315
Extinction(s), mass, 383, *384*
  and rate of evolution, 383–388
  and recolonization, *596*
Extinction vortex, 603, *603*, 604–606, 624
Extracellular fluid, 771–772, *772*, 811
Extracellular matrix, 96, 724–725, 732–733
Eye(s), 874–878, *875*
  brain information processing and, 876–878, *878*
  of cephalopods, 510, *511*
  compound, 518, 521, 873
  formation of, 916, *916*, 928
  of humans, *515*, 873
  microorganisms inhabiting, 436
  sight defects of, *875*, 875–876
Eye-cup, 873
Eyeglasses, 874, *874*
Eyepiece, 72

F

F_0-F_1 complex, *134*, 134–135, *135*, 138
F1 generation, 194, 205
F2 generation, 194, 205
  genotypes and phenotypes of, 194–195
Facilitated diffusion, passive transport and, 94, 101
Facilitating interneuron, 859–860, 861
Factor X, 831–832
Facultative aerobes, 446, 450
Facultative anaerobes, 126
FAD, 123, 138
$FADH_2$, 123, 138
Fairy rings, *467*, 470
Fall turnover, 581
Fallopian tubes, 886, 904
Family, in hierarchy of classification, 423, 431
Family size, completed, 616, 625
Farlow, James, 711
Farquhar, Marilyn, 82–83
Farsightedness, 874, 880
Fate maps, 919
Fats, acetyl-CoA and, 129
Fatty acids, functions of, *57*
  linking of, 56–59
  polyunsaturated, 56
  saturated and unsaturated, 53–56, *56*
Feces, 741, 752
  origin of, 509
Feedback, negative, 13, 730–731, 733, 762, 768
  positive, 898, 905
Female, choice of, evolution and, 633
Female choice, 364, *364*, 367
Fermentation, alcoholic, 118–119
  kinds of, 126, *126*, 138

Fern(s), 486–487
　　life cycle of, 486
　　reproduction in, 486
Fertility, 614, 625
Fertilization, 166, 186, 204, 908, 909, 927
　　double, 491, 494, 680, 680–681, 692
　　embryonic development and, 908
　　internal, adaptations to, 187
　　in mammals, 888–889
　　meiosis and, 188, 188
　　in plants, flowering, 679–680, 680–681
Fertilizers, manufactured, 559
Fetus, 883, 917, 928
　　full term, 886, 904
　　oxygen for, 805
　　stages of development of, 917–919, 918, 929
Fiber(s), 750
　　collagen, 725, 725
Fibrillation, 783, 787
　　ventricular, 784, 787
Fibrils, collagen, 725, 725
Fibrin, 831, 842
Fibrinogen, 831, 842
Fibroblasts, 724, 725, 732–733
Fibronectin, 913
Fibrous proteins, 65, 69
"Fight-or-flight" reactions, 760, 761, 869
Filament, 492, 494
Filamentous cyanobacteria, 401, 401, 402
Filtrate, 817, 825
Filtration rate, renal blood flow and, 821
Filtration slits, 816
Finch(es), adaptive radiation and, 377, 377–378, 378
　　of Galápagos Islands, 592, 592–593
　　vampire, 378
Fish, bony, body plan of, 531, 531
　　cladists and, 427
　　countercurrent system in, 717, 717–718
　　freshwater, kidneys of, 820
　　　　salt and water balance in, 811, 811
　　jawless, 529, 529
　　lobe-finned, 531, 532
　　ray-finned, 531, 532
　　saltwater, kidneys of, 820
　　　　salt and water balance in, 811, 811
　　　　water retention by, 812
Fission, binary, 167–168, 168
Fitness, inclusive, 637
　　altruism and, 637–638, 639, 639
Flagellates (zoomastigina), 464, 475
Flagellum(a), 100, 441, 450, 887, 904
　　9+2 structure of, 89, 89
Flame cells, 823
Flatworm(s), 506–507, 521
　　excretory organs of, 823
　　free-living, 506
　　learning in, 630
　　nervous system of, 865, 865
　　tissue layers of, 502
Flavor components, 280–281
Fleming, Alexander, 250, 280
Flemming, Walther, 165–166, 172, 182
Florey, Howard, 280
Flounder, 386
Flowering, seasonal control of, 699
Flowering plants. See Plant(s), flowering
Flower(s), complete versus incomplete, 687
　　development of, buds and, 687
　　　　genes in, 687–689

　　functions of, 487–488, 678
　　parts of, 492–493, 678, 678, 679, 680–681
　　perfect versus imperfect, 687
　　pollination of, 491
　　whorls of, 687, 688, 692
Fluid intake, brain control of, 821
Fluid mosaic model, 91, 100
Flukes, 506, 507, 521
Fluorens, Pierre, 862–863
Fluoxetine, 857
Flyers, 719–720, 720
Fly(ies), on corpse, 598
　　fruit. See Fruit fly(ies)
Follicle-stimulating hormone, 762, 892, 894, 895, 904
Follicles, 884–885, 885, 904
Food(s), animal nourishment from, 734–753
　　demand for, agriculture and, 619–620, 621
Food chain, 548, 548, 560
Food intake, animal, amount of, 736
　　too much or too little, consequences of, 736–737
Food web, 548, 549, 560
Foot (feet), of mollusk, 511
　　webbed, 715
Foraging, competition and predation influencing, 632
　　optimal, 632, 633, 639
Foraminifera, 462, 462, 475
Forams, 462, 462, 475
Forebrain, 867–868
Foregut, 743, 745–746
Foreskin, 889, 904
Forest(s), coniferous, 564, 564
　　deciduous, 564, 582
　　deciduous, temperate, 573–575, 574
Fossil(s), 14, 18, 18, 20, 323, 330–331, 342
　　of Cambrian Period, 401
　　formation of, 328, 328–329
　　of Precambrian Period, 401
Fossil fuels, 554, 555
Fossil record, 327
　　of evolution, 327–332, 328
　　factors revealed by, 327–332
　　for plants, evolution and, 481
Foucault's Chinese encyclopedia, 421
Founder effect, 357, 366
Fovea, 874–875, 880
Fox, Sidney, 409
Fraenkel-Conrat, Heinz, 235
Fragmentation, habitat destruction and, 604, 624
Frameshift mutation, 245
Franklin, Rosalind, 208, 209, 209–210, 211–215, 233
　　discovery of structure of DNA by, 211–216, 214
Free energy (G). See Energy, free
Free radicals, 29, 156
Frog(s) (Anura), 533, 533
　　cascades, 535
　　Chiricahua leopard, 535
　　early development in, 912
　　gastric brooding, 535
　　genes of, 523
　　poison dart, 588
　　South American, 534
　　tarahumara, 535
　　Yavapai leopard, 535
Fronds, 486
Fructose, 60

Fruit(s), development of, 493
　　functions of, 491
Fruit fly(ies), 352, 353, 364–365. See also Drosophilia melanogaster.
　　advantages for genetic research, 200, 200
　　genes of, 523
Fruiting body, of cellular slime mold, 464–465
Fuchs, Ephraim, 827, 828
Functional groups, 68
　　definition of, 50, 51
　　properties and structures of, 51, 52t
Fungal infections, 468
　　antibiotics and, 453
　　human diseases from, 452–453
Fungicidin, 453–454
Fungus(i), 16, 20, 466
　　characterization of, 466–467, 467, 475
　　classification of, 454, 467, 467, 475
　　ecological importance of, 465, 472–474
　　infections from, 466
　　multicellular, and protists, classification of, 452–475
　　origin of, 474
　　reproduction of, 467
Furchgott, Robert, 22–23
Fusion, membrane, 96–97, 97

G

G protein, 764
GABA, 854, 855
GABA transminase, 857
$GABA_A$ receptor, 856, 861
Gaia, 544–545, 585
Galápagos Islands, 368–369, 376–379, 377
　　finches of, 592, 592–593
Galápagos tortoises, 368, 368
Galen, 773–774
Galileo, 398
Gall, Franz Joseph, 862
Gallbladder, 750, 752
Gametangia, 478, 484, 484, 494
Gametes, 185–186
　　distribution of chromosomes to, by meiosis, 188–192, 191
　　formation of, 909, 927
　　mammalian, 883–888, 904
　　production of, control of, 892–896
Gametic isolation, 373, 396
Gametocytes, 463
Gametophyte(s), 474, 478, 483–484, 487–488, 678, 692
　　in phaeophytes, 460
Gamma-aminobutyric acid (GABA), 855, 856, 861
Ganglion(a), 859, 864–865
Ganglion cells, 876, 880
Gap junctions, 98, 99, 101, 784
Garrod, Archibald, 217, 218
Garter snakes, 372, 627–628
Gas(es), in atmosphere, 405
　　dissolution of, temperature and, 793–794
　　dissolved in solution, 793–794
　　greenhouse, reduction of, 556
　　in lungs, 794
　　pressure exerted by, 792, 793
Gas exchange, in alveoli, 801
　　in lungs, 796, 801, 801
　　surface area and, 794–795
Gastrin, 759
Gastrointestinal tract, replenishment of, 751
Gastropods, 512–513, 513, 521

Gastrotricha, 510
Gastrula, 499, 520, 909, 927
Gastrulation, 909, 927
    in amphibians, *912*, 912–913
    in mammals, 913–914, *914*
    process of, 912, *912*
    in sea urchin embryo, 910, *911*
Gates, of neurons, 848, 861
Gause, G.F., 591
Gel electrophoresis, 288, 295
Gene(s), in alkaptonuria, 217, *218*, *219*, 220
    in bacteria, 221
    behavior and, 627, *631*, 631–632
    bicoid, 924
    cloning of, in bacteria and viruses, 284
    composition of, 221–225
    continuity and, 16
    conventional, 259
    definition of, 183, 217
    "designer." *See* Recombinant DNA
    Ds, 258
    effects pf, on biochemical processes, 217–219
    and enzymes, 218–219, *220*
    eukaryotic, 242
    expression of, 234–256
        and structure of, study of, 287–289
        by viruses, 269, *269*
    expressivity of, 304, 315
    in flower development, 687–689
    functions of, 183
    homeotic selector, 688, 689, 692, 924, 928
    Hox, 924, 928
    human, patenting of, 311
    as influence on aging, 926
    inheritance of, 16
        chromosomes and, 194
    jumping, 258–275
        in corn, 258–260
    mobile, 259–260, 274
        evolution of, 270–272
        replication by, 264–265, 267–269
    with multiple alleles, 307, 307t
    one gene-one enzyme, 218–219, 232
    one gene-one polypeptide, 219–221, 232
    penetrance of, 304–305, 315
    pleiotropy of, 306, 315
    polypeptide, 219–221
    on same chromosome, assortment of, 201–202, *203*
    sequenced, use of, 308–310
    sex-linked, in *Drosophila*, 201, *202*
    for specific diseases, location of, 308–310
    structure of, 16, 20
    traits expressed by, 305
    traits influenced by, *306*, 306–307
    transforming factor in, 221–222
Gene expression, cellular regulation of, 250–254
    eukaryotes in regulation of, 254
    prokaryotes in regulation of, 251–253
Gene flow, 358, *359*, 366, 376
Gene frequencies, 355–366
    chance events and, 356–358
    selection and, 358–360, *359*
Gene gun, 292, 295
Gene library, 286, 295
Gene pool, 354, 366
    draining of, extinction in, 604–606
Gene regulation, eukaryotes in, 253–254
Gene sequences, 39, *838*
Gene therapy, 313–314, *314*, 315

Generations, alternation of. *See* Alternation of generations
Genetic code, *237*
    description of, 237–238, 255
    discovery of, 234–237
    in mitochondria, 273
    of mitochondria, 413, *413*
    as redundant, 237
    as universal, 238, *238*, 255
Genetic counseling, 311
Genetic disease, prevention of, 310–315
Genetic diversity, in humans, 299–304
Genetic engineering, of animals, 291–293
    definition of, 278–281
    of grain, 278–280, *280*
    natural and induced variation in, 280
    of plants, 291–293, *293*
    and recombinant DNA, 276–297
    of tomatoes, 280–281, 293
Genetic equilibrium, in populations, 354–355
Genetic linkage, of *Drosophila*, 200–201, 202, *203*, 205
    Morgan's work on, 200–202, *202*
Genetic map, 204, 205
Genetic systems, evolution of, 409–412
    unconventional, 258–275
Genetic unit, species as, 425, 431
Genetic variation, 192, *193*, 350–351
    and evolution, 353–354
    from mutations, 229–231
    natural selection and, 353, 365–366
    in populations, 353
Genetics, 161
    definition of, 183
    human, 296–316
    Mendelian, rediscovery of, 349
    Mendel's work on, 183
    molecular, 183, 204
    rules of, 185
    transmission, 183, 204
Geniculate nucleus, lateral, 876, *876*
Genital warts, 164
Genitalia, external, female, 889–890, *890*
    male, 889, *889*
    male and female, common origin of, 890, *891*
    in sexual intercourse, 889–890
Genome, 14. *See also* Human Genome Project
    definition of, 183
    restriction enzymes cutting, 282
Genotype, 353, 366
    definition of, 183, 204
    and phenotype, relationship between, 304–307
    phenotype and, 183–185
Genus (genera), 332
    in hierarchy of classification, 422, 431
Geochelone elephantopus, 368, *370*
Geographic isolation, 376–378, *377*
    speciation and, 374
Geospiza difficilis, 378
Geospiza fortis, 592, *592*
Geospiza fuliginosa, 592, *592*
Germ cells, 186, *187*, 204
    primordial, 883–884, 904
Germ layer, formation of, 909, 911–912, 927
Germ line, 186, *187*, 204
    continuity of, *187*
Germ plasm, 348, *348*, 366
Germination, 682–684, *683*, *684*, 692
    environmental cues triggering, 684

Giardia, 465
Gibberella fujukoroi, 704
Gibberellins, 696t, 702, *704*, 704–705, 708
Gibbons, 389, *389*
Gibbs, J. Willard, 107
Gill(s), *337*, *794*, 795, 806
    fast-swimming fish and, 795
Gill slits, 527, 542
Gill withdrawal reflex, 859–860, *860*
Ginkgos, 490
Girdling, 670, *671*, 674
Gizzard, 739, 746, 752
Glands, identification of, 757
Glans, 889, 905
Glantz, Stanton, 788–789
Glaucoma, 874
Gleason, Henry Allan, 586–587
Glial cells, 713, 732, 847, 861
Globular proteins, 65, 67, 69
Glomerulus, 814, *816*, 825
Glossopteris, 339
Glossypol, 657, 659
Glottis, 745, 796
Glucagon, 758–759, 764, 768
    production of, 759
Glucocorticoids, 763
Glucose, *50*, 59, *60*, 69
    blood, levels of, 758–759
    in Calvin cycle, 154–155, *155*
    extraction of energy from, by cells, 122–130
    plant synthesis of, 153–157, 644, *644*
    during pregnancy, 898
    production from NADPH and ATP, 153
    regulation of, failure of, 759
    synthesis of, photons required for, 157
    uptake of, 864
Glucose transport, liposomes and, 94–95
Glutamate, 854, *855*
Glutamic acid, 281
Glyceraldehyde phosphate, 156
Glycerol, 56, 57, 69
Glycine, 854, *855*
Glycogen, 60, 69
Glycogen phosphorylase, 764
Glycolipids, 88
Glycolysis, 122, *123*, 123–124, 125, 138
    anaerobic, 138
    regulation of rate of, 124, *124*
    without oxygen, 126–127
Glycolytic muscle fibers, 728, 733
Glycoproteins, 61, 69, 88
Glycosidases, 739
Glycoside bonds, 53
Glycosides, mustard oil, 658
Glycosidic bond, 59
Glyphosate, 156, 292
Gnetae, 490
Gnetophyte, 490
Gnetum, 490
Goheen, Austin, 661
Golden toads, 535, *535*
Goldschmidt, Richard, 386
Golgi, Camillo, 846
Golgi complex, functions of, 82, 100
    structure of, 82, *83*
Gonadotropin(s), 762, 892
    human chorionic, 895, 896, 897, 904
    pregnancy tests and, 896
Gonadotropin-releasing hormone, 762, 893, 894, 904

Gonads, 186, 204, 766, 883
Gondwanaland, 339, *339*
Goose (Geese), greylag, 628, *628*
  imprinting by, 630–631, *631*
Gosling, Raymond, 208, 209, 213, 214, 216
Gould, Stephen J., 387, 497
Gradualism, 386, 387, *387*, 396
  Darwin's theory of, 320, 342
Grafting, 691, *691*, 692
Grain, production of, *621*
Gram, Hans Christian, 439
Grana, 149, *150*
Granulosa cells, 884
Granum, 85, 100
Grapevines, phylloxera and, 660–662
Grass, 376
Grasshopper, lubber, 180, *181*
Grasslands, animals of, 570–571
  temperate, 570, *570*, 582
Gravitropism, 697–698, *698*, 708
Gravity, plant responses to, 697–698, *698*
  size of animals and, 714
Green algae, in plant evolution, 481
Green ball, 790
Green glands, 824
Green Revolution, 619
Greenberg, Bernard, 598
Greenhouse effect, 555
Greenhouse gases, reduction of, Martin, John H., 557
Grey matter, 866
Griffith, Frederick, 221
Griffith, J.S., 47
Griffith's experiment, 221, *222*
Ground squirrel, round-tailed, 568
Ground tissue system, 646, 658
Growth, and development, programmed cell death in, 916–917, *917*, 929
  exponential, 608–609, 624
  primary, 646, 658
  secondary, 647, 658
  in vertebrates, 909, 927
Growth factor(s), 289, 295, 752, 757, 766, 769, 846
Growth factor gene, 290, 307
Growth hormone, 289, 762
Growth rings, 691, 692
Grunberg-Manago, Marianne, 234, 235
Guanine, 210, 233
Guanosine diphosphate, in citric acid cycle, 127
Guanosine triphosphate, in citric acid cycle, 127
Guinea pigs, 424, *424*
Gurdon, John, 921
Gusella, James, 298, 299, 308–309
Gustation, 870, 880
Gut, 741, 752
  actions of, hormones and nerve cells in, 751
  primitive, 911
Guthrie, Woody, 298, *298*
Guttation, 663, *663*, 674
Guzman, Rafael, 677
Gymnophiona (caecilians), 533, *533*
Gymnosperms, 481, 489–490, 494, 649
Gyri, 868, 880

H
HA oxidase, 217
Habitat, 591, 600
  size of, and number of species, 594, *594*
  two species sharing, 591

Habitat destruction, 604, *605*, 624
  fragmentation and, 604, 624
Habitat islands, 595, 600
Habituation, 859
Hagfish, 525
Hailman, Jack, 629
Hair cells, 871, *872*, 880
Haldane, J.B.S., 361, 369, 404, 637
Hales, Stephen, 670
Half-life, 329
  of ^{14}C, *28*, 28–29
Hallucigenia, *498*
Halophiles, 443, *445*, 450
Hamburger, Viktor, 845, 846
Hamilton, W.D., 637, 638
Hamner, Karl, 699–700
Haplodiploidy, 637, 639
Haploid, 185–186, 204
Hardy, G., 354
Hardy-Weinberg equilibrium, 366
Hardy-Weinberg Principle, 354–355, 358, 359, 365–366
Harlow, Harry and Margaret, 631
Harlow, Jean, 790
Harvey, William, 774, 774–775
Hawaiian Islands, adaptive radiation in, 379, *379*
Hawaiian tarweeds, 379, *379*
Hawthorn fly, 376
Hazen, Elizabeth Lee, 452, 453
Headgut, 743, 744–745
Hearing, 870, 880
  mammalian, 871–873
  and perception of sound, 862
Heart, 771, 786
  anatomy of, 776, *777*
  chambers of, 774, 776, *777*, 778, *778*
  early studies of, 774–775
  evolution of, *778*
  functions of, 776
  muscles of, 784
Heart attack, 782, 787
Heart beat, 782–785
  coordination of, 783–784
  systole and diastole in, 782–783
  three parts of, 783, *783*
Heart murmur, 783
Heart rate, 785, 787
  parasympathetic nervous system and, 785, 787
  sympathetic nervous system and, 785, 787
Heat, capacity of water, 37, *39*
  loss of, during energy transformation, 106, *106*
  tolerance by animals, 732
Heat exchange, countercurrent, 716, *717*
Heat exchanger, countercurrent, 716, 731
Heinrich, Bernd, 731
Hela cells, 162–163
Helicase, 229, *229*
Helicobacter pylori, 2–3
Heme, 131, 802, 806
Hemichordata, 524, 541, 542
Hemizygous alleles, 200
Hemoglobin, 63, *65*, 66, 772–773
  affinity for oxygen, 804–805
  normal, 220
  as oxygen carrier, *802*, 803
  in oxygenation, 801–802, *802*
  of sickle cell patients, 220, 232
Hemoglobin F, 804
Hemostasis, *830*, 830–831, 842
Hendricks, Sterling, 701

Heparin, 832, 843
Herbaceous plants, 645
Herbal teas, 642
Herbicide(s), actions of, on plants, 156
  auxins as, 696, 704
Herbivores, 548, 560, 588, 736, 752
  digestive tracts of, 712, *712*
  plant defenses against, 656–658
Herbs, 645
Hereditary Disease Foundation, 296–297, 298–299
Heredity, 180, 341
Heron, blue, *538*
Herring gull chick, behavior of, 628–629, *629*
Hershey, Alfred, 223–225, *225*
Heterosporous plants, 487
Heterotrophs, 119–120, 138, 412, 416, *444*, 445–446, 450, 548, 560
  aerobic, 446, *446*, 448
  anaerobic, 448
  in energy extraction from macromolecules, 121
Heterozygous, 350, 366
Heterozygous alleles, 194, 205
Hierarchial classification, *422*, 423, 431
Hierarchy, 422
  dominance, 633, 639
High altitude, oxygen delivery at, 805
High-throughput screening, 310
Hilts, Philip, 793
Hindbrain, 866, 880
Hindgut, 743, 750–751
Hip, ball-and-socket joint of, 723, *723*
Hippocampus, 868, 880
Hirudinea, 512–513
Hirudo medicinalis, 512
Histamine, 786, 787, 832, 836, 842
Histidine, 236
Histone 4, 429–430
Histones, 174, 175, 254
Histoplasma, 468
HIV. *See* Human immunodeficiency virus
H.M.S. *Beagle*, 370
Hodgkin, Alan, 851
Holmberg, Scott A., 432–434
Homeostasis, 13, 20, 751, 752, 755–756, 768, 771
  behaviors and, 865
  definition of, 730, 733
  oxygen, 805
  two stages of, 730–731, 733
Homeotherms, 776
Homeotic selector genes, 688, 689, 692, 924, 928
Hominidae, 322, 390, 397
Hominids, branching evolution of, 394, *395*
  earliest known, *393*, 393–394
Hominoids, *389*, 390, *390*, 397
  stance of, evolution of, 391–392
Homo, 322
Homo erectus, 394–395, 397
Homo ergaster, 394
Homo habilis, 394
Homo heidelbergensis, 395
Homo sapiens, 332, 390, 395, 397
Homolog, 168, 185, 205
Homologous characters, 426, 431
Homologous chromosomes, 168, *169*, 185, 350, 366
  in meiosis, 188–190, 205
Homologous structures, 333, 342
Homoplasty, 426, 431
Homosporous plants, 487
Homozygous, 350, 366

Homozygous alleles, 194, 205
Homunculus, *879*, 906, 927
Honeybees, hive of, 635–636
  social organization of, 635, 636
  swarming, 636, *636*
Hood, Leroy, 48
Hoof-and-mouth disease, 260–261
Hooke, Robert, 70–71
Hooker, J.D., 371
Horizontal cells, 875, 880
Hormones, 736
  blood flow regulation by, 785–786
  and blood pressure, 821–823, *822*
  as chemical signals, 756
  as contraception, 901–903, 906
  control of gamete production by, 892–896
  in coordination of distant organs, 758–762
  definition of, 702
  in digestion, 751, 752
  identification of, 757
  ovarian cycle and, 894–896, *895*
  plant responses to, 702–707
  of plants, 702–707, 708
  production of, by testes and ovaries, *892*, 893–894
  in regulation of menstrual cycle, 896
  targeting of, 762–766
Hornworts, 480
Horse(s), evolution of, 333–334, *335*, 372, 380
  thoroughbred, 353–354
Horsehair worms, 510
Horseshoe crab(s), *324*, 519
Horsetails, 487, *487*
Housman, David, 296, *297*, 298
Hox genes, 924, 928
Human(s), classification as primates, 319, *320*
  development of, *913*
    after implantation, 916–918
  digestive tract of, 741, *741*, 743–751
  eye of, *515*
  genetic diversity in, 299–304
  genetic variability in, 353
  learning by, 869
  modern, origin and migration of, 303
  respiratory system of, 796, *797*, 806
  stem cells in, 926
  tolerance in, 731, *731*
  water and salt balance in, 812, *813*
Human chorionic gonadotropin, 895, 896, 897, 904–905
  pregnancy tests and, 896
Human chorionic somatomammotropin, 897, 905
Human evolution, 388–396
Human genetics, 296–316
Human Genome Project, 307–310, 315
  location of disease-specific genes in, 308
  mapping goals of, 308
  positional cloning in, *309*, 309–310
  sequenced gene applications in, 310
Human growth hormone, effects on dwarfism, *289* in mice, 290, *291*
Human immunodeficiency virus, 261, *262*
  drugs to combat, 271
  replication of, 271
Human papilloma virus, 164
Humboldt Current, 578–579
Hummingbird, blue-throated, *538*
Humus, 669, 674
Huntingtin polypeptide, 310
Huntington's disease, 294, 296, 297–298

gene expressivity and penetrance in, 304, 305
gene mapping of, 297
huntingtin gene and, 310
Milton Wexler and, 297
pedigree of family with, 308–309, *309*
Hutton, James, 323
Huxley, Andrew, 851
Huxley, Thomas, 346–347, 384
Hybrid breakdown, 374, 396
Hybrid DNA, 286, *286*, 295
Hybrid inviability, 374, 396
Hybrid sterility, 374, 396
Hybridization, 288
Hybridization probe, *286*, 286–287, 295
Hybrids, 194, 205
Hydra, 504, *505*
  nerve nets in, 865, *865*
Hydrocarbon chain(s), 49, *50*, 68
Hydrocarbons, polycyclic aromatic, *408*
Hydrochloric acid, in digestion, 734–735
Hydrogen bonding, in water, 39
Hydrogen bond(s), 35–36, *36*, 44
Hydrogen cyanide, 406, *407*
Hydrogen ion(s), 40, 45
  concentration of, 41–42
Hydroids, 504
Hydrolysis, 53, *53*
Hydronium ion, 40–41, 45
Hydrophobic interaction(s), 35, *35*, 44
  van der Waals attractions and, 33
Hydrostatic skeleton, 721, *722*, 732
Hydroxide ion, 40, 45
Hydroxyl group, 51, 68
Hydrozoa, 504, *505*
Hymen, 890, 904
Hyperopia, 875, *875*
Hypertension, prevalence of, in U.S., 809, 809t
  sodium consumption and, 808
Hyperthyroidism, 766
Hypertonic solution, 93, *93*, 101, 811
Hyphae, fungal, 466, *466*
Hypoblast, 913, 927
Hypocotyl, 680
Hypothalamic hormones, 762
Hypothalamus, 731, 736, 761, *761*, 762, 768, 833, 867, 880, 892
Hypothesis(es), definition of, 5, 6, 20
  disproving of, 6
  model of, 5, 5–6
  related, 13
  testable, 6, 20
Hypothyroidism, 766
Hypotonic solution, 93, *93*, 101, 811
Hyracotherium, 333, *335*

I

Ibuprofen, 768
Ice, floating, 36, *36*
  winter, pond life under, 36, *37*
Icosahedron, 261, 274
IgA, 836, *836*
IgD, 836, *836*
IgE, 836, *836*
IgG, 835–836, *836*, 843
IgM, 836, *836*
Ignarro, Louis, 22–23
Iltis, Hugh, 677
Imaginal discs, in *Drosophila*, 925, *925*
Imago, 925, *925*
Immigration, 610, 624

Immune response, cells in, *834*, 835
  mediation of, 835
  secondary, 834
  structure of cells and, 73
Immune system, 829
  antibodies and, 837–839
  attack on body, 842
  memory of, 833–834
  molecules and, 833–836
  of newborn mice, 826
  recognition of diversity by, 834
  specificity of, 833, 834–835
  toleration by, 834
Immunity, attributes of, 833
  inflammation and, 826–844
  nonspecific, 833
  specific, 833
Immunoglobulins, 835–836, 843
Imperfect fungi (Deuteromycetes), 470, 471, *472*, 475
Implantation, 894
  human development after, 916–918
Imprinting, 630–631, *631*, 639
In vitro fertilization, 903–904
Inbreeding, 358, 366
Incisors, 744
Inclusive fitness, 637
  altruism and, 637–638, 639, *639*
Incus, 871
Independent assortment, principle of, 198–200, *199*, 206
Individual variation, 348
Individualistic communities, 586–587, *587*, 600
Induced abortion, 903, 905
Induced fit, 112, *113*, 117
Induction, 253, 256
Industrial melanism, 362
Infant, nursing of, 898
Infections, causing diarrhea, 750
  fungal. *See* Fungal infections
  resistance to, inflammation and, 832–833
Infertility, female, 903–904
  male, 903
  treatment of, 903–904
Inflammation, 829, 832, 842
  immunity and, 826–844
  resistance to infection and, 832–833
Influenza virus, 263, *264*
Ingram, Vernon, 220
Inhalation, 798, 806
Inheritance, of acquired characteristics, 347–348, 366
  adaptive traits in, 341
  blending model of, 182, *182*
  blending of, 346, *346*, 347, 366
  chromosomal theory of, 180–185
    acceptance of, 182–183
    discovery of, 180–181
    evidence for, 199–204
    Sutton-Boveri, 199, 204
  laws of, 185
  particulate, 346, *346*
  Sutton's six rules of, 200t
Inhibin, 893
Inhibiting hormones, 762, 768
Inhibition, 682, 692
Inhibitor(s), allosteric, *116*, 117
  competitive, 116, *116*
  definition of, 115
  noncompetitive, 116, *116*
  steric, 115–116, *116*, 117
Inhibitory postsynaptic potential, 856, 861

Initiation, 246, 256, *248–249*
Innate behavior, *627*, 627–630, 639
Inner cell mass, 910–911, 927
Inositol triphosphate, 765
Insect(s), 520
    breathing by, 801, *801*
    circulatory system of, *779*
    excretory system of, 824, *824*
    genetically engineered crops and, 294
    intricate societies of, 637
    metamorphosis in, 758, *758*, 925, *925*
    oxygen distribution by, 801
Insect repellent, mustard oil glycosides as, 658, *658*, 659
Inspiration, 798, 806
Institute for Genomic Research (TIGR), 311
Insulin, 99, 758, 768
    production of, 759
Insulin-dependent diabetes mellitus, 759
Integrated community, 586, 600
Integrated replication, 264–265, 274
Integration, 870, 880
Integrin, 913
Integuments, 679, 692
Intercalated discs, 784, *784*
Interferons, 833, 842
Interleukin-1, 833, 843
Interleukin-6, 833, 843
Intermediate filaments, 87, *87*
Intermediate forms, 334, 342
Interneuron(s), 858, 876
    facilitating, 859–860, 861
    inhibitory, 858, 861
Internode, 653, 658, 685
Interphase, 170
Interstitial cells, 887, 904
Intertidal zone, 578, *579*, 582
Intestine(s), microorganisms inhabiting, 436
    small, 747–749, 752
        absorption in, 750
        cell specialization in, 747, *747*, *748*
        epithelial cells of, 747–749, *749*
Intrauterine devices, 902, 905
Intrinsic rate of natural increase, 609, 624
Introns, 242, 254
Invagination, 912
Invertebrates, brain in, 864–865
    digestion in, 739
    marine, salt and water balance in, 819
    nervous system in, 866
    water, salts, and wastes in, 823–824
Inviability, hybrid, 374, 396
Involuntary muscle, 726, 733
Ion, 30
Ionic bond(s), 30, 44
    atoms in, 33–34
Ionizing radiation, 29
Ionotropic receptors, 857
Iridium anomaly, 382
Iris, 873, 880
Island biogeography, theory of, 594, 600
Islands, adaptive radiation on, 376
    diversity on, 595, *595*
        defaunation and, 594t
Islets of Langerhans, 759
Isolation, behavioral, 373, 396
    ecological, 373, 396
    gametic, 373, 396
    geographic, 376–378, *377*
        speciation and, 374
    mechanical, 373, 396
    reproductive, *373*, 373–374
        by polyploidy, 374, 396
    temporal, 373, 396
Isoproterenol, 764
Isotonic solution, 93, *93*, 101, 811
Isotope(s), 28, 44, 329
    decay of, 28, 29
Iwanowski, Dmitri, 260

J

J-shaped curve, *608–609*, 609, 610, *610*
Jack rabbit, blacktail, *715*
Jackson, John Hughlings, 863
Jacob, François, 251, 259
Japanese puffer fish, 851, *852*
Jasmonic acid, 707
Jawless fishes, 529, *529*
Jaws, mammalian, 744, *744*
Jellyfish, 504
    nerve nets in, 865, *865*
Jenkin, Fleeming, 347, 366
Jenks, Susan, 419
Joints, 722, 723, 732
Joliot, Frèdèric, 29
Joliot-Curie, Irène, 29
Judson, Horace, 215, 216
Junctions, adhering, 98, *99*, 101
    gap, 98, *99*, 101
    tight, 98, *99*, 101
Jurassic Park, 710, 711
Juvenile diabetes mellitus, 759
Juvenile hormone, 758

K

Kaback, Michael, 311
Kamen, Martin, 153
Kandel, Eric, 860
Kangaroo, red, 541, *541*
    western grey, *322*
Kangaroo rat, water balance in, 812–813
Keeling, C.D., 556
Keilin, David, 131
Kelps, 460, *460*
Kendrew, John, 208, 217
Keratin, 63, 87
Kettlewell, H.B.D., 361–362
Keystone species, 593, 600
Khorana, Gobind, 237, 238
Khrushchev, Nikita, 345
Kidney(s), adaptable, in turtles and tortoises, *810*
    anatomy of, 814, *814*, *815*, 824
    artificial, 818, *818*
    blood filtration by, 814–817, 825
    of freshwater fish, 820
    functions of, 810, 814–817, 825
    of nonmammalian vertebrates, environment and, 820–821
    of saltwater fish, 820
    sodium consumption and, 809
    urine concentration by, 819–820
Kidney dialysis, 818, *818*
Kin selection, 637, 639
Kinesins, 86
Kinetic energy, from work, 104, 117
Kinetics, application of, 103–104
    definition of, 103, 117
Kinetochore, 172, *173*

King, Mary Claire, 302, 303
King, Thomas, 919
Kingdom(s), 16
    in hierarchy of classification, 423, 431
Knee jerk reflex, 858, *859*
Knee joint, 723, *723*
Knockouts, 291, *291*
Koala, 541
Koch, Robert, 435
Koch's postulates, 435, *435*, 447
Kornberg, Arthur, 227
Krebs, Hans, 127, 129
Krebs cycle. *See* Citric acid cycle
Krill, *520*
K/T boundary, 382
Kuru, 46
Kwashiorkor, 737, 752
Kyoto Protocol, 556

L

La Niña, 579–580
Labeled line, 871
Labia majora, 889–890, 904
Labia minora, 889, 904
lac operon, 252–253, *252*, *253*, 256
lac repressor, 253, 256
Lacks, Henrietta, 162, 163, *163*
Lactate, in oxygen absence, 126–127
Lactation, 897–900
Lake Michigan, sand dune succession and, 585, *586*
Lake(s), 580–581
    temperature turnover in, 581, *581*
Lamarck, Jean Baptiste, 325, 342
Lamarckism, 348
Lampreys, 529
Land bridge, 380, *380*
Larkspur (Delphinium), 373, *373*
Larva(ae), 758, 924–925
Larynx, 796, 806
Lateral geniculate nucleus, 876, *876*
Lateral line system, 873
Lateral meristems, 646, 658
Latimeria, 532
Latitude, temperature and, 565, *565*
Lavoisier, Antoine, 118, 142
Leaf(ves), arrangements of, 654–655, *655*
    autumn, *149*
    diversity of, 656, *656*
    functions of, 654–656
    water evaporation from, 664
    water path in, 654–655, *655*
    water transport in, 662
Leaf primordium, 687, 692
Leakey, Mary, 394
Learning, by animals, 627, 630–631, 639
    sensitive period for, 630
    associative, 859
    by humans, 869
    neural networks and, 859
    nonassociative, 859
    by sea hare, 860, 861
    sensitive period for, 630, 639
Leclerc, George-Louis, 322
Leder, Philip, 237
Lederberg, Joshua and Esther, 361
Leeches, 516–517
Legumes, scarification in, 684
Lemmings, population cycles of, 612
Length, measurement of, Metric System, Appendix B

Lens, 873, 880
Leptin, 736
Lesch-Nyhan syndrome, 631
"Let-down" reflex, 899
Leukocytes, 772, 773, *773*, 787, 829, 843
    types of, 829t
Levi-Montalcini, Rita, 844–846, *845*
Lewis, Sinclair, *Arrowsmith* by, 265
Lianas, 576, 582
Lichens, 573, 587
    description of, 473, *473*
    role of fungi in, 472–474, *475*
Liebig, Justus von, 118
Life, chemical foundations of, 20–43
    continuity of, 161
    early, effects on environment, 412–413
    origin of, 399–404, *402*
    theories of, 401–402
    unity and diversity of, 2–20
Life history, survivorship and age distributions and, *613*, 613–616, *615*
Life history traits, 612, 625
Ligaments, 723, *724*, 732
Ligand-gated channel, 856
Ligand-gated ion channel, 857, 861
Ligase, DNA, 282–284, 295
Light, absorption of, 146–147, *147*
    definition of, 145
    photons in, 145, *145*
    plant productivity and, 157
    visible, 146, *147*, 159
    wavelength of, *145*, 145–146, *147*, 159
    waves of, 145, *145*
Light-independent reactions, *144*, 145, 159
Light microscope, 72, *73*
Lignin, 482, 645, 658
Limbic system, 879, 880
Limiting resource, 591, 600
Limnology, 580, 582
Linkage groups, 202
Linkage map, 204, 205
Linnaeus, Carolus, 332, 342, 422, 423, 476
    classification of plants by, 476–477, *477*, 493
    classification system of, 422–423
    evolution and, 322
Lions, 634, *635*
Lipases, 739
Lipid-soluble signals, 762, *763*, 768
Lipids, 52, 54–55t, 68
    processing of, 750
Liposomes, 409, 415
    glucose transport and, 94–95
Listeria monocytogenes, 273, *273*
Litter, forest, 573, *574*
Liver, cirrhosis of, 136
Liverworts, 483
Lizard(s) (Squamata), 534, *536*
    chihuahua whiptail, 883
    crested dragon, *536*
Loams, 669, 674
Lobe-finned fishes, 531, *532*
Lobster, digestion in, 742
Locomotion, animal, style of, as influence on shape, 718–721
    in snakes, 719, *719*, *720*
Locus, 204
Locusts, plague of, *611*
Loewi, Otto, 7
Logan, Graham, 509
Logistic growth equation, 609, 624

Long-term memories, 869, 882
Loop of Henle, 817, 825
Lorenz, Konrad, 630
Loudness, 871, 880
Lovelock, James, 544, 545, 560, 585–586
Lubber grasshopper, 180, *181*
Lumen, 781, 787
    of endoplasmic reticulum, 82, 100
    of gut, 743, 752
    ion content of, tubules and, 819
Lunar month, 894
Lung(s), *794*, 795, 806
    air entering, path of, 796
    cancer of, *789*, 789–791, *791*
    damage due to smoking, 799–800
    gas exchange in, 796, 801, *801*
    gas in, 794
    of land animals, 795
    of mammals and birds, 715
Luria, Salvador, 223, 224, *224*
Luteinizing hormone, 762, 892, 893, 894, *894*, *895*, 904
Lycophytes, 487
Lycopods, 487
Lyell, Charles, 324, 325–326, 327
Lymph, 781, 787
Lymph nodes, 781, 787
Lymphatic system, 780, *780*, 787
Lymphocyte library, 837–838, 843
Lymphocytes, 780, 787, 829, 835
Lysenko, Trofim, 344–345
Lysergic acid diethylamide (LSD), 471
Lysis, 264, 274
Lysogenic cycle, 268, 274
Lysosomes, formation of, 83
    functions of, 83, *84*
Lysozyme, 830, 843
Lytic cycle, 268, *268*, 274
Lytic viruses, *263*, 264, 274

M

MacArthur, Robert H., 591, 593–594
MacArthur-Wilson theory of island biogeography, 594
Mach, Ernst, 100
MacLeod, Colin, 222
Macrocystus, 460
Macroevolution, 350, 366, 368–397
    definition of, 368–397
Macromolecules, 49, 68
    extraction of energy from, by heterotrophs, 121
Macrophage(s), 799, 827, 833, *833*, *834*, 835
Mad cow disease, 47
Maggots, 398–399
Magnification, 72
Magnolia, 482
Maidenhair tree, 490
Major histocompatibility complex, Class I proteins of, 841, 843
    Class II proteins of, 841, 843
    T cells and, 840
Malaria, 463, *463*
    sickle cell disease and, *365*, 365–366
Male competition, 364, *364*, 367, 633
Malleus, 871
Malnourishment, 736–737, 752
    causes of, 737–739
Malpighi, Marcello, 670
Malpighian tubes, 517, 521, 824, *824*

Malthus, Thomas, 326
Mammalia, 322, *536*
Mammals, 539–541
    birds, and reptiles, cladogram for, 428, *429*
    breathing by, 796–798
    central nervous system of, 866, *867*
    gamete formation by, 883–889, 905
    gastrulation in, 913–914, *914*
    marine, water for, 812
    marsupial, 380
    placental, 380, 537, 538
    teeth of, 539–540
    transgenic, production of, *290*, 290–291
Mammary glands, 539, 540
Manglesdorf, Paul, 676
Mangold, Hilde, 921, 923
Mantle, 511, 521
Mantle cavity, 511, 521
Map unit, 204
Marathon, 1995 World women's, 125, *125*
Margulis, Lynn, 414, 545, 585
Marine birds, kidneys of, 821
Marine invertebrates, salt and water balance in, 819
Marine mammals, water for, 812
Markel, Michael, 557
Marrella, 498
Mars, life on, 544
Marsh, Charles, 423
Marshall, Barry, 2, *3*, 4
Marshes, 580
Marsupials, 537, 538
    Australian, 380, *381*
Martin, Gail, 307
    and growth factor gene, *311*, 307
Masiarz, Frank, 47
Mass, 26
    measurement of, Metric System, Appendix B
Mast cells, 832, *832*, 842
Mastophora, 519
Maternal mRNA, 908, 929
Mating, assortive, 358, 366
    nonrandom, and allele frequencies, 358
Mating rituals, 634, *634*
Matrix, 499, 713
Matter, definition of, 23
Matthei, Johann Heinrich, 234–236
    and Nirenberg, experiments by, 235–236, *236*
Matthew, Andrew, 808, 809
Matzinger, Polly C.E., 826, 827, 828
Mayer, Julius, 145
Mayr, Ernst, 370, 371, 386, 387
McCarty, Maclyn, 222
McClintock, Barbara, 203–204, 258–259, *259*, 274, 278
McClung, Clarence, 180
Mean, 11, 20
    and standard deviation, 12
Measure, units of, Appendix B
Mech, David, 418–419
Mechanical defenses, plant, 656, *657*
Mechanical isolation, 373, 396
Medawar, Peter, 827, 828
Medulla, renal, 814–817, 825
Medulla oblongata, 866, 880
Medusa(ae), 504, *504*, 521
Megaphylls, 486, 494
Megaspores, 679, 692
Meiosis, 165, 183, 186
    distribution of chromosomes to gametes by, 188–192, *191*

fertilization and, 188, *188*
homologous chromosomes in, 188–190, 205
Meiosis I, phases of, 188, *189*, 204
Meiosis II, phases of, 188, *189*, 204
Melanism, industrial, 362
Melanocyte-stimulating hormone, 762
Membrane envelope, 261, 274
Membrane fusion, 96–97, *97*
Membranes, cellular, functions of, 87
  interaction with external environment, 95–99
  plasma. *See* Plasma membrane(s)
  regulation of, space enclosed by, 91–96, *92*
  selectively permeable, 91, 100
  sodium ion permeability in, 851
  water moving across, 91–93
    direction of, 92–93
Memory, 868, 880
  declarative, 868
  of immune system, 833–834
  long-term, 868, 880
  reflexive, 868
  selection theory and, *838*, 838–839, *839*
  short-term, 868, 880
Memory cells, 838, 844
Mendel, Gregor, 19, 180, 183, 195, *195*, 217, 258, 348, 349
  genetics discoveries of, 195–199
    cross-pollinating and, 195, *196*, 197, *197*, 206
    principle of independent assortment and, 198–200, *199*, 206
    principle of segregation and, 197–198, *198*, 206
    traits of pea plants, 195–197, *196*
Mendelian genetics, rediscovery of, 349
Meninges, 865
Meningitis, 865
Menstrual cycle, 894
  hormonal regulation of, 896
  irregular, 894
Menstrual extractions, 903, 905
Menstruation, 894, 896, 904–905
Mercurialis annua, 488–489
Mercury barometer, 792, *793*
Meristem(s), 646
  apical, 646, 684, *685*, 692
  elongation and differentiation of, 686
  lateral, 646, 658
    secondary growth and, 689–691
  plant production by, 684, *685*, 692
Merozoite, 463
Meselson, Matthew, 236
Mesenchyme, in sponges, 503, 520
Mesenteries, 747
Mesoderm, 496, 520, 908, *908*, 927
Mesoglea, 504
Mesophyll, 656, 659, 672, *673*
  leaf, 664
Mesozoic Era, 331, 342, 534, 540
Messenger RNA. *See* RNA, messenger (mRNA)
Metabolic pathways, evolution of, 409, *410*
Metabolic rate, size of animals and, 714, *714*, *715*
Metabolism, 120, 138
  of alcohol, 136
  of ethanol, 136
  food molecules entering, 129, *130*
  pathways of, 129, *129*
  regulation of, by cell or organism, 112–114
Metabotropic receptors, 857
Metallurgy, 24–25

Metameres, 514, 521
Metamorphosis, 517, *518*, 521, 533, *534*, 542, 909, *924*, 924–925, *925*, 928
  complete, 925, *925*, 928
  gradual, 925, 928
  in insects and amphibians, 758, *758*
Metaphase, chromosomes during, 172
  in meiosis, 188, 190
Metaphase plate, 172
Metapopulation, 595, *596*, 600
Metatherians, 541, 542
Meteorite, 403, *408*
Methane, 544
Methanococcus jannaschii, 447
Methanogens, 443, *445*, 450
Metric System, Appendix B
Mice, allopatric speciation of, *375*
  coat color in, genes and, *306*, 306–307
  human growth hormone in, 290, *291*
  knockin, 291
  knockout, 291, *291*
  newborn, immune system of, 826
  transgenic, 290, *291*
Michel, Helen, 382
Michod, Richard E., 882–883
Microelectrodes, 850
Microevolution, 344–367
  definition of, 349, 366
Microfilaments, 86–87
Microphylls, 486, 494
Micropyle, 679
Microsatellites, 420
Microscope(s), atomic force, 26
  compound, 72, *73*
  early history of, 70, 71
  electron, 72–73, *73*
  light, 72, *73*
  study of cells using, 72–73
Microspheres, proteinoid, 409, *409*, 415
Microspores, 678, 692
Microtektites, 382
Microtubule organizing center, 86, *88*, 100
Microtubules, 86, *87*, *88*, 100, 172
Microvilli, 747, *748*, 752, 817–819, *818*
Midbrain, 866, 880
Midgley, Julian, 808, 809
Midgut, 743, 747–749
Midpiece, 887, 904
Mifepristone, 903, 905
Milk, human, 898
Milk ejection, 898, 905
Miller, Stanley, experiment by, 406, *406*, 407
Mimicry, 589, 600
Mineral deficiencies, 737–738
Mineralocorticoids, 762
Minerals, from animals, for plants, 670
  in plant growth, from soil, 669
Mini-pills, 903
Miscarriage, 898, 918, 928
Missense mutation, 245
Mitchell, Peter, 131
Mites, 519
Mitochondria, 84, 100, 127, 138
  DNA in, *272*, 272–273
  as endosymbiont, 413–414
  generation of proton gradient by, 133
  genetic code and, 238
  genetic code in, 273
  genetic code of, 413, *413*
  structure of, *85*, 85

Mitochondrial matrix, 133, 138
  pumping protons out of, *133*, 133–134
Mitochondria, transmission electron microscope to assess, 84, *85*
Mitosis, 165, 168, 170, 175, 188
  movement of chromosomes during, 173
  stages of, 171, *171*, 172–173
  triggering in, 177–178
Mitotic spindle, 172
Mittelschmerz, 900–901, 905
Mixture(s), 24, *25*, 44
Mobile genes. *See* Gene(s), mobile
Mockingbirds, 384
Modal action pattern, 628, *628*, 639
Model, 5, 20
Modern synthesis, 349, 366
Molars, 744
Molds, 468, 475
  bread, 468, *469*
  fungi as, 466
Mole, 42
Molecular biology, central dogma of, 238, 239, *239*, 254
  comparative, 336–337
Molecular clocks, 337
Molecular genetics, 183, 204
Molecular motors, 75, 100
Molecular weight, 26, 44
Molecule(s), amphipathic, 40, *41*, 44–45
  biological, 45–69
    organisms building, 49–53
    size of, 49
  bonding in, 30–35
  definition of, 24, *25*
  forces holding together, 35
  hydrophilic, 35
  hydrophobic, 35
  immune system and, 833–837
  in living organisms, 49
  motion of, rate of chemical reactions and, 110, *111*
  movement through plasma membrane, 93–95
  nonpolar, 34, *35*
  organic, 25–26
  polar, 34, *34*
  shape of, 32, *32*–33
  signaling, 254
Mollusks, 511–514, *512*–*513*, 521
  circulatory system of, *779*
Molting, 517, 521
Monarch butterfly, 589, *589*
Moncada, Salvador, 22–23
Monkey(s), 389, *389*
  socialization of, 631
Monocots, 479, 649–651, *650*, 658, 680, 692
  germination in, 682–683, *683*
Monocotyledons, 649–651, *650*, 658
Monod, Jacques, 251, 259
Monophyletic taxon, 427, 431
Monoplacophora, 513
Monosaccharides, 59, 69
Monosomy, 231, 233
Monotremes, 541, *541*, 542
Monsters, hopeful, 386, *386*, 396
Morgan, Thomas Hunt, 200–202, *202*, 203, 217
Morning-after pill, 902, 905
Morphogenesis, 914, 928
Morphogens, 923, 928
Morris, Simon Conway, 496–497
Morula, 909, 927

Mosquito, malaria-carrying, 371
Moss(es), 483
  club, life cycle of, *479*
  sphagnum, *483*
Moth, peppered, *360*, 360–361
Motile, 455, 474
Motoneuron(s), 858
  L7, 859–860, 861
  of spinal cord, 866, 867, 880
Motor axis, 726, 733
Motor instructions, 865
Motor neurons. *See* Motoneuron(s)
Mount St. Helens eruption, 584, *585*
  succession following, 585, *585*, 596, 597
Mouse, cactus, 568
Mouth, 744, 752
  microorganisms inhabiting, 436
Mouth parts, of arthropods, 518, *519*
Movement, in digestive tract, 743
Moyle, Jennifer, 131
Mt. Whitney, 562, 563, 564, 566
Mucigel, 651, 686, 692
Mucosa, of small intestine, 747
Mucus, 744, 752
  in respiratory system, 799, 806
Mudpuppy (*Necturus*), 534
Müllerian ducts, 890
Müllerian mimicry, 589, *589*, 600
Mullis, Kary, 276–278, *277*
Multicellular organism, loss and replacement of cells by, 77
Multicellular organisms, communication of cells in, 97–99, *98*
Münch, Ernst, 672
Murad, Ferid, 22
Murex, 507
Murray, R.G.E., 447
Muscarine, 856
Muscarinic receptors, 857
Muscle(s), antagonist, 722, *722*
  antagonistic pairs of, 726, 733
  cardiac, 713, 726, *728*, 733, 784
  in movement of animals, 721–730
  reductionist study of, *8*, 8–9, *9*
  skeletal. *See* Skeletal muscle
  smooth, 713, 726, *728*, 733
  striated, contraction of, *729*, 729–730
Muscle fibers, glycolytic, 728, 733
  oxidative, 728, 733
Muscle tissue, 713
Mushrooms, 469–470
Mustard oil glycosides, 658
Mutagens, 230–231, 233
Mutation(s), allele frequency and, 355–356
  causes of, 230–231
  chromosomal, 231
  conditional, 245
  definition of, 233
  frameshift, 245
  frequency of, 231
  genetic variation from, 229–231
  missense, 245
  new, random drift and, 357
  nonsense, 245, *245*
  point, 231, 233
  rate of, 231, 233, 352
  reading of, by ribosomes, 245
  silent, 245
  spread through population, 352–353
  suppressor, 245

Mutation hot spots, 231, 233
Mutualism, 590, 600
Mycelium, 466, *467*, 475
Mycorrhizae, 473–474, *474*, 475
Myelin, 853, 861
Myelin sheath, 847, 853, 861
Myocardium, 776, 786
Myofibrils, 728, 733
Myoglobin, 728, 733
Myopia, 875, *875*
Myosin, 729, 733
Myxobacterium, *440*, 450
Myxomycota, 465, 475

N

NAD^+, 123, 138
NADH, 123, 138
NADPH, in glucose synthesis, 153
  light reactions and, 151
Naming, of organisms, 422
  of species, 423
National Aeronautics and Space Administration (NASA), 544
National Institutes of Health (NIH), 308
Natural killer cells, 835, 843
Natural selection, 360, *360*, 366, 633
  altruism and, 636–637
  analysis of, 341
  Darwin's theory of, 19, 20
  directional, *360*, 360–361, 366
  frequency of traits and, 358–362, *359*, *360*
  genetic variation and, 353, 365–366
  preexisting variants and, 361
  pushing of populations by, 362–366
  stabilizing, *362*, 363, 366
Nautilus, 513, *514*
Neanderthals, 395–396, 397
  flute of, *395*
Nearsightedness, 874, 880
Necturus (mudpuppy), 534
Negative feedback, 730–731, 733, 762, 768
Negative regulators, 253, 256
Neljubov, Dimitry, 706
Nelson-Rees, Walter, 162
Nematocysts, 504, 521
Nematodes (roundworms), 510, 521
  apoptosis in, 917
Nematomorpha, 510
Nemertea (ribbon worms), 508, *508*, 521
Nephridia, 511, 514, 521, 824, 825
Nephron(s), 816, 825
  filtration, reabsorption, and secretion by, 817, *817*, 825
  solutes in, 819, *820*
Neritic zone, 578, 582
Nerve cells, *846*, 846–853, *847*
  in digestion, 751
  information transmission by, 847–848
  Levi-Montalcini's work on, 844–846
  organelles in, 846, *847*
  stimulation of, cellular response related to, 847, *849*
  threshold response of, 847
  types of, 846–847, *848*
  voltage of, sodium channels and, *849*, 849–850
Nerve cord, dorsal hollow, 527, 542
Nerve growth factor, discovery of, 846
  functions of, 846
Nerve nets, 864, *864*, 880

Nerves, afferent, 868, 880
  blood flow regulation by, 785–786
  efferent, 868, 880
  parasympathetic, 726, 869, 880
  sense organs, and brain, 865
  sympathetic, 726, 869, 880
Nervous system, autonomic, 726, 733, 869, *869*
  cells of, 844–861
  central, mammalian, 867, *868*
  in vertebrates, organization of, *865*, 865–868, 880
  formation of, 914–916
  information moving through, 869–870
  in invertebrates, 866
  parasympathetic, 726
    heart rate and, 785, 787
  peripheral, 868, 880
    organization of, in vertebrates, 870
  response to environment, 865–871
  and sense organs, 862–883
  signaling by, 757
  somatic, 869
  sympathetic, 726
    heart rate and, 785, 787
Nervous tissue, 713, 732
Neural crest cells, 914, 928
Neural networks, behavior and, 857
  experience and, 859
  learning and, 859
Neural plate, 914, 928
Neural tube, 866
  formation of, 914, *915*, 927–928
Neurofibromatosis, 304, *305*
Neurons, 713, 732, 847, 861
  of brain, 864
  communication between, 853–857, *854*
  electrical pulse generation by, 848–850
  generation of action potential by, 850–851
  maintenance of resting potential by, 849–850
  motor. *See* Motoneuron(s)
  postsynaptic, 853, 861
  presynaptic, 853, 861
  reflex mediation by, 858–859
  relay, 869, 870, 880
  sensory, 858, 859, 861
  trigger, 858
Neuroscience, 627
Neurospora, 218, *219*, 220
Neurotransmitters, 757, 766, 768, 785, 787, 854, 855
  receptors for, 856–857
  synapse clearing of, 857
Neurulation, 914, 928
Neutrons, 28, 44
Neutrophils, 832–833, 842
Newts, cellular potential experiments on, 919
Niche, 591
  fundamental, 591, 600
  realized, 591, 600
Nicholson, Garth, 91
Nicotine, 856
  addiction to, 793
  in tobacco smoke, 791, 792, 799
Nicotinic acetylcholine receptor, 857
Nicotinic receptors, 857
Nightshade, 857
Nirenberg, Marshall, 235, 235–236, 238
  and Matthei, experiments by, 235–236, *236*
Nitric oxide, 22, 23, 786
Nitrification, 558, 560

Nitrifying bacteria, 558, 560
Nitrogen, fixed, 557–558
    sources of, 559, *559*
    supply of, 557–559
Nitrogen cycle, 557, *558*
    alteration by humans, 559
Nitrogen fixation, 448
Nitrogen-fixing aerobic bacteria, 446, *446*, 448, 450
Nitrogenous bases, 61, *62*, 69
Nitrogenous wastes, 813, *814*, 824
Nitroglycerin, 22
Nobel, Alfred, 22
Node, 653, 685
Nodes of Ranvier, 853, 861
Non-insulin-dependent diabetes mellitus, 759
Nonassociative learning, 859
Noncompetitive inhibitor, 116, *116*
Nonpolar compounds, small, 53–59
Nonsense mutation, 245, *245*
Norepinephrine, 785, 786, 787, 869
Norplant, 903
Northern blotting, 288, 295
Nose, microorganisms inhabiting, 436
Notochord, 527, 542, 914, 928
Nourishment, animal, from food, 734–753
Nowack, Ronald, 419, 421
Nübler-Jung, Katharina, 523
Nucellus, 487, 494, 679, 692
Nuclear envelope, 80, *80*, 100
Nuclear pores, 80, 100
Nuclear reactions, 24, 44
Nucleases, 739
Nucleic acid(s), linkage of nucleotides of, 61–62, *63*, *64*
    linking of nucleotides in, 61–62
    proteins and, 52, 53
Nucleoid, 74, *74*, 100, 438, 447
Nucleosome, 174, *174*
Nucleotides, 52, 54–55t, 68
    DNA, 211, *213*
    linkage of, to form nucleic acids, 61–62
    of nucleic acid, linkage of, 61–62, *63*, *64*
Nucleus, of atoms, 26, 44
    of cells, 74, *74*, 100, 79–80, *80*
    and characteristics of organisms, 166, *167*
    differentiated, development and, 920
    of eukaryotes, *74*
Numa, Shosaku, 851–852
Nursing, 898
    ovulation and, 900
Nutriceuticals, 291
Nutrients, from soil, 669t
Nymphs, 925, 928
Nystatin, 454, 468

O

Objective, 72
Obligate aerobes, 446, 450
Obligate anaerobes, 126, 446, 450
Observation, 4–5
    models of, 5, *5*
Ocean, ecosystems of, 578, 582
    fertilization of, 557
    zones of, 578, *579*, 582
Oceanic zone, 578, *579*, 582
Oceanography, 578, 582
Octane, 50
Octet rule, 30
Odontogriphus, 498

Odum, H.T., 619
Oils, 35
Okazaki, Reiji, 228
Okazaki fragments, 228, 233
Oleic acid, properties of, 38t
    water, and ethanol, compared, 38t, 39
Olfaction, 870, 878–879, 880
Olfactory bulb, 878, *879*, 880
Olfactory receptors, 880, *881*, 882
Oligochaetes, 515, *516*
Oligodendrocytes, 847
Oligonucleotides, 285, 294
Oligosaccharides, 59, 69
Oligotrophic lakes, 581, 582
Olson, Carl, 598
Ommatidia, 518
Omnibacteria, 446, 448, 450
Omnivores, 548–549, 560, 736, 752
Oncogenes, 267, 274
    proto-, 267, 274
    viral, 267, 274
Oocysts, 463
Oocytes, passage of, 885–886
    primary, 883–884, *885*, 904
    production of, 894, *894*
    secondary, 885, 888, 904
Oogenesis, 883, *884*, 904
Oogonia, 883–884
Oomycota, 465, 475
Opabinia, 498
Oparin, Alexander Ivanovich, 404, 408
Operator, 253, 256
Operculum, 511, 521
Operon, *lac*, 251, *252*, 255
Opiates, 858
    natural, 858
Opposite leaves, 654, *655*
Opsins, 875, 880
Optic chiasm, 876, *876*
Optic nerve, 876, *876*, 880
Optimal foraging, 632, 633, 639
Orbitals, 30, 44
Order, in hierarchy of classification, 423, 431
Ordovician Period, 472, 644
Orexin, 736
Organ of Corti, 871, *872*, 880
Organelles, 15, 74–75, 100
    energy, DNA and RNA of, 273
    in nerve cells, 846, *847*
Organic chemistry, 49, 68
Organic molecules, definition of, 49
Organism(s), air-breathing, oxygen for, 796–801
    cellular structure of, 15, *15*
    characteristics of, 13, *14*
    classification of, phenetics in, 426–427
    complex, 906–930
    composed of cells, 70–71
    differences in, 16–19
    domains of, 16, *17*
    energy supply to, 118–139
    evolution of, 398–416
    groups of, numbers of, 430
    study of, 7–8
    individual, as unique, 7–10, *10*
    naming of, 422
        and classification of, 421–425
    self-regulation by, 13–15, 730–731
    transformation in, 16
    useful, design of, 281
Organs, 499, 713

    formation of, in vertebrates, 909, 927
    systems of, 713, 732
Orgasm, 891, 904
Ornithine, 219
Oscilloscope, 850, *850*
Osmolarity, 811
Osmoreceptors, 821
Osmosis, 92, 101, 663, 664, 674, 811
Osmotic pressure, 92
Osteichthyes, 530, 531, 542
Osteoblasts, 725–726, 733
Osteoclasts, 726, 733
Osteoporosis, 726, 733
Osterholm, Michael, 432–433
Ostracoderms, 529, 542
Ostrich, 392
Ottoia, 498
Outbreeders, 358, 366
Oval window, 871
Ovarian cycle, hormonal regulation of, 894, 894–896, *895*
Ovary(ies), 186, 204, 489, 491, 766
    of flowering plants, 679
    in mammals, 883
Overnourishment, 736
Overpopulation, human, 621–624
Overton, William R., 319
Oviducts, 883, 904
Ovulation, 885, 904
    nursing and, 900
Ovules, 488, 494, 679, 692
Ovum(a), 186, 204, 883, 884
    production of, mammalian, 884–885
    sperm reaching, 892
Owens, Kelly, 303
Oxaloacetate, 652
Oxidation, 122–123
Oxidative muscle fibers, 728, 733
Oxidative phosphorylation, 122, 138
Oxidizing agent, 122, 138
Oxygen, 24
    for air-breathing organisms, 796–801
    in animals, source of, 791–792
    atmospheric, 791–792
    carbon dioxide exchange for, 791–795
    carried by hemoglobin, *802*, 803
    carried to tissues, 801–802
    delivery by blood, 782, 802, *802*
    delivery of, 786
        changing, breathing regulation in, 805
        at high altitudes, 805
    dissolved, for aquatic animals, 795
    distribution of, by insects, 801
    hemoglobin affinity for, 804–805
    movement of, 715, 782, *782*
Oxygen debt, 125
Oxygen dissociation curve, *803*, 804, *804*
Oxygen exchange, countercurrent, *717*, 717–718
Oxygen homeostasis, 805
Oxygen saturation, 803
Oxygen ship, *802*
Oxyhemoglobin, 802, *802*, 806
Oxytocin, 761, 898, 905
Ozone layer, 622, *624*
    chlorofluorocarbons and, 622, *624*, 625

P

P_{680}, 149, 150, 151, 159
P_{700}, 149, 150, 151, 159

Pacemaker, 784
　　artificial, 784
Palade, George, 82–83
Paleontology, 323
Paleozoic Era, 330–331, 342, 339, 481
Palisade parenchyma, 655, 656, 659
Pancreas, 749–750, 752, 758, *759*
Pancreatic islets, 759
Pangaea, 339, *339*
Pap smear, cervical cancer and, 164
Papanicolaou, George N., 164
Paper chromatography, 153–154, *154*
Papilio dardanus, *363*, 363–364
Paracrine signals, 757, 768, 766, 840
Paramecium, 73, *73*, 76, *76*, 89, 462
　　reproduction by, 185
Paranthropus, 394, 397
Parapatric speciation, 376
Paraphylatic taxon, 427, 431
Parapodia, 515, 521
Paraquat, 156
Parasaurolophus, *18*
Parasites, 588, 600
Parasitism, 590, 600
Parasympathetic nerves, 726, 869, 880
Parasympathetic nervous system, 726
　　heart rate and, 785, 787
Parathyroid hormone, 766
Parazoa, 499, 503, 522
Parenchyma, 646, *647*, 658
　　palisade, 655, 656, 659
　　spongy, 655, 656, 659
Parental (P) generation, 194, 205
Park, Thomas, 591
Parkinson's disease, 866
Parrots, African, 631
Parthenogenesis, 883
Partial pressure, 792, *793*, 806
Partitioning, resource, 591, 600
Parturition, 898, 905
Passive transport, 93–94, *95*
Pasteur, Louis, 118–119, 260, 399, 415, 434
　　experiment of, 399, *400*
Pasteur, Marie, 119
Pathogens, sealed from circulatory system, 830–831
　　skin as barrier to, 829–830
Pauling, Linus, 34, 208, 209, 220
Pauling, Peter, 213, 217
Pavlov, Ivan, 859
Pea plants, backcross of, 195, 197, *197*
　　cross pollinating of, *196*
　　Mendel on, 195–196, *196*
　　testcross of, 197, *197*
　　two-factor cross of, 198, *198*
Pectoral girdle, 713
Pedigree analysis, 308–309, 315
Pelagic division, 578, *579*, 582
Pelvic girdle, 713
Penetrance, gene, 304–305, 315
Penfield, Wilder, 869
Penguin, emperor, *534*
Penicillin, 250, 280, 281, 437, 440
Penicillinase, 234
Penicillium, 471
Pepsin, 734–735, 745–746, 752
Peptic ulcers, 2
Peptidases, 749
Peptide bonds, 53, 64, 69
　　production of peptide bonds by, 247

Peptide hormones, 764
Peptidoglycan, 438–439, 447
Peptids, 763
Perception, 871, 881
　　sensory pathways and, 878–879
　　sound as, 862
Perforation plate, 648, 659
Pergolide, 471
Pericardium, 776, 786
Pericycle, 652
Periderm, of flowering plant, 646, 658
Period, geologic, 330
Periodic Table of Elements, 30, *31*, Appendix A
Peripheral nervous system, 868–869, 880
　　organization of, in vertebrates, 868–869
Peristaltic waves, 745, *745*, 752
Permafrost, 569, 582
Permian Period, 331, 536
Peromyscus eremicus, 568
Peroxisomes, 84, *84*, 100
Pesticides, 857
Petals, 492, 494, 687, 692
Petiole, 654, 658–659
pH, of blood, 42–43
　　changes in, resistance to, 43–44
　　of solution, 42–43
pH scale, 41–42, *43*, 45
Phaeophytes, 460, *460*, 474–475
Phage. See Bacteriophage
Phagocytosis, 83, *84*, 101, 741, *741*
Pharyngeal slits, 527, 542
Pharynx, 744, 752, 796, 806
Phenetics, 426–427, 431
Phenolic compounds, 657
Phenomones, 756, 768
Phenotype(s), 353, 366
　　definition of, 183, 204
　　environment in, 183–184
　　environmental and genetic influences on, 184
　　and genotype, 183–185
　　　　relationship between, 304–307
Pheromes, 629
Pheromones, 636, 639
Phloem, 485, *485*, 494, 645, 658
　　conducting elements of, 649, *649*
　　contents of, 671–672
　　osmotic flow in, 672, *672*
　　sugar and water in, 672–674, *673*
　　translocation of, 672, *672*
Phosphates, organic, 857
Phosphatidylinositol biphosphate, 765
Phosphodiester bonds, 61
Phosphoglycerate, 652
　　three-carbon, 156–157, 158
Phospholipase C, 765
Phospholipid(s), *57*, *58*, 69
Phosphorylation, oxidative, 122, 138
　　substrate-level, 131
Photoautotrophs, 446, 450
　　algae as, 457
Photochemical reaction center, 148, 159
Photoheterotrophs, 446, 450
Photolyase, 535
Photon(s), 145, *145*, 146, 159
Photoperiodism, 699, 700, 708
Photophosphorylation, *144*, 145, 159
　　cyclic, 149, *152*, 159
　　noncyclic, 149, 151, 159
Photoreceptor(s), 873, 876, 880
Photorespiration, 158, 159

Photosynthesis, 84–85, 100, 140–160, *143*, *144*, 159, 667, 669
　　atoms and, 143–145, *144*
　　carbon dioxide in, 155–157
　　chlorophyll α and, *150*
　　evolution of, 412–413
　　factors affecting, 157–158
　　herbicide interference with, 156
　　light reactions of, 150, *152*–*153*
　　molecules in, 146–148, *147*
　　by plant, *142*
　　products of, plant transport of, 670–674
　　　　sorting of, 153–154, *154*
Photosystem I, 149, 159
　　electron transport in, *153*
　　noncyclic photophosphorylation in, 150, *152*
Photosystem II, 149, 159
　　electron transport in, 150–151, *152*
Phototropin, 699
Phototropism, 694, *695*, *699*, 708
　　auxin in, 694–696, 696t, 708
　　pigments in, 699
Phrenology, 862
Phyllobates, 588
Phylloxera, 660–661
Phylogenetic tree, 337, 342
Phylogeny, 427, 431
　　of primates, 388
Phylum, in hierarchy of classification, 423, 431
Physalaemus, 633, *634*
Physalia, 505
Physiology, 730, 733
Phytochrome, 700–702, 708
Phytoestrogenics, 754
Phytophthora infestans, 465
Phytoplankton, 457
Pigeons, artificial selection in, 340, *340*
　　learning by, 630
Pigments, 147, 159
　　in phototropism, 699
Pika, 596
Pikaia, 498
Pike, *532*
Pilobolus, 470
Pine, life cycle of, *488*
Pinnae, 486
Pinocytosis, 96, 101, 741
Pioneer community, 597, 597t, 600
Pioneer species, 598, 600
Pistil, 679, 692
Pit pair, bordered, 647–648, 658
　　simple, 647, *648*, 658
Pitch, 871, 880
Pith, 649, 658
Pitohui bird, 588, *589*
Pits, tracheid, 647, *647*
Pituitary gland, 761, *761*, 768
　　anterior, 761, 762, 768, 892, 904
　　posterior, 761, 768
Placenta, 380, 541, 804, 896, 898, 905, 911, 917, 927
Placental mammals, 380
Placoderms, 530, 542
Placozoa, 503
Planaria, 506
Plankton, 455, *455*, 474, 578, 582
Plant(s), actions of herbicides on, 156
　　adaptation to dry land, 478–481
　　adaptation to life on land, advantages and disadvantages of, 478

alternation of generations in, 478, *479*
evolution in, 478, 480–483, *482*
fossil record of, 481
adaptive radiation and, 379
and air, 142, *142*
bending toward light. *See* Phototropism
broad-leafed, 479
C₃, 158, 159
C₄, 158, 159
anatomy of, *652*, 652–653, *653*
carbon and oxygen obtained by, 142–145, *143*
cells in, cytokinesis in, 176, *176*
classification of, 476–477
day-neutral, 699
defense against herbivores, 656–658
defenses of, 588–589
definition of, 478, 494
desert, water conservation by, 668
of deserts, 567, *569*
dioecious, 687
divisions of, 479, 480t, 481, *482*
embryos of. *See* Embryo(s), flowering plants
environmental responses of, 697–702
to gravity, 697–698, *698*
hormones and, 702–707
to season and daylight, 699–702, *700*, *701*
first, characteristics of, 480
flowering, alternation of generations in, 678, *678*, 692
anatomy of, *685*
egg cell production by, 679
fertilization in, 679–680, *680*–*681*
groups of, 481
growth and development of, 676–708
pollen production by, 678, *679*
roots of, growth of, 686, *686*
seed formation in, 680–681, *682*
sperm production by, 678
structure of, 646–651
genetic engineering of, *293*, 293–294
growth and development in, regulation of, 694–708
growth in, secondary, 689–691, *690*, 692
herbaceous, 645
herbicide-resistant, 156
hormones of, 702–707, 708
lacking materials for photosynthesis, 157
long-day, 699, 700, *700*, *701*
minerals from animals for, 670
monoecious, 687
mycorrhizae and, 473–474, *474*
nonvascular, 479, 483–484
photosynthesis by, *142*
photosystems of, 148–149
physical constraints on, 644–645
prevention of photorespiration by, 158
production of, by meristems, 684, *685*, 692
productivity of, factors limiting, 157
light and, 157
water and, 157–158
seed, reproduction by, 487, 494
short-day, 699, 700, *700*, *701*
structural and chemical adaptations of, 642–659
during succession, 597, 597t
sun as energy source for, 145–153
terrestrial, evolution of, 644–645
water and minerals for, 651–656
transgenic, *293*, 293–294

transport of products of photosynthesis by, 670–674
transport system of, 485
true-breeding, 194
turgor pressure and, 93, 101
vascular, 479, 485–490
lacking seeds, 486–487
reproduction and dispersion of, 485
vascular and nonvascular, compared, 479–480
vascular evolution of, 644–645
water conservation by, 667
Plant cell, chloroplasts of, 84–85, *86*
Plant compounds, primary, 657
secondary, 657
Plantae, 16, 20
major divisions of, *482*
Plaque, dental, bacteria causing, *441*
Plasma, 770–771, 772, 787, 811, 831, 842
Plasma cells, 838, 843
Plasma membrane(s), 74, 90, *90*, 100
fluid layers of, 91
molecules in, 87–90
movement of molecules through, 93–96
proteins in, 88, 91
as selectively permeable, 91, 100
structure of, and functional properties of, 90–91
Plasma proteins, in nonspecific defense, 832
Plasmid(s), 260, 274
recombinant, 266, 274
replication by, 265
Ti, 292, 295
Plasmin, 832, 842
Plasminogen, 832, 842
Plasmodesmata, 98, *98*, 101, 647, 672, *673*
Plasmodial slime molds, 465
Plasmodium, 462–463, *463*
Plasmodium falciparum, 463
Plasmodium, definition of, 465, 475
Plasmolysis, 93, 101
Plastids, 84–85, 84, *86*, 100
Plateau, 891, 904
Platelets, 772, *773*, 831, 842
Plates, 339
Plato, 321
on classifications, 423–424
Platyhelminthes (flatworms), 506–507, *502*, 521
Platypus, 541, *541*
Platyrrhines, 389, *389*, 397
Pleiotropy, 306
Plumule, 680
Pluteus, 925
Pneumococcus, 221
Poison ivy, and cytotoxic T cells, 836, *836*
Poisons, as pollutants, 621–622
Polar body, 885, 904
Polar easterlies, 565, *566*, 582
Polarity, 34, *34*
Poliovirus, 261, 263
Pollen grains, 678, *679*
Pollen tube, 679–680, *680*, 692
Pollutants, biodegradable, 621
nondegradable, 621
Pollution, air, 622
overpopulation and, 621–624
thermal, 621
water, 621–622
Polychaete worms, 515, *516*
Polygenic traits, *306*, 306–307, 315

Polymerase chain reaction (PCR), 277, 295
DNA multiplication and, 277, *278*, 286
Polymerization, in DNA replication, 227, *228*, 233
Polymorphism(s), 350, *350*
balanced, *365*, 365–366
restriction fragment length, 297–298, 300, *300*, 315
single-nucleotide, 298
Polypeptide(s), address tags of, 247–249
cellular translation of, into messenger RNA, 243–249
domains of, 66
linkage to aminoacids to form, 63–66
recombinant DNA and, 284–285
synthesis of, stimulation of, 234
termination of, 249–250
Polypeptide alphabet, 237, *237*
Polypeptide chains, 835–836, 837, 843
Polypeptide genes, 219–221
Polyphyletic taxon, 427, 431
Polyplacophora, 513
Polyploid, 493
Polyploidy, 493
reproductive isolation by, 374, 396
Polyp(s), 500, *500*, 517, 752
Polysaccharides, 52, *53*, 59, 60, *61*
digestion of, 750
formation from sugars, 59–61, 69
use by organisms, 60–61
Polysome, 247, *249*, 256
Polyspermy, 892
Pond life, under winter ice, 36, *37*
Pongidae, 390, 397
Pongids, 390
Pons, 866, 880
Popcorn, 676
Population(s), 20, 546, *547*, 560
age structure of, 614, *615*
annual rate of increase of, 616, *616*, 625
breeding between, allele frequencies and, 358, *359*
cycling up and down of, 612
density-dependent factors and, 610, *611*
density-independent factors and, 610–611
differences in, checking for, 11
evolution of, 344–367
extinctions of, 604–608
deterministic factors and, 604–606
and explosions of, 602–625
genetic equilibrium in, 354–355
genetic variability in, 353
growth of, 608–612
interactions of, 595–596
large, random drift and, 358
measurement of, 616
natural growth of, 612, *612*
and overpopulation, 621–624
pushing of, by natural selection, 362–366
sampling of, 10–11, *11*
size of, limitation of, 610
Population momentum, 616–617, *617*, 625
Porifera (sponges), 502, 503, *503*
Portuguese man-of-war, 504, *505*, 545
Positional cloning, *309*, 309–310, 315
Positive feedback, 898, 905
Positive regulators, 253–254, 256
Positron emission tomography, 863
Postabsorptive state, 758, *760*, 768
Posterior air sacs, 800, *800*

Postsynaptic neuron, 853, 861
Posttranscriptional control, 251, 256
Posttranslational control, 251, 256
Postzygotic barriers, 373, 396
Potassium channel, *852*
Potential energy, in work, 104, *105*, 117
Prairie chicken, extinction of, 605, 606, *606*
Prebiotic evolution, 404–409, 415
Prebiotic synthesis, 405–407, 416
Precambrian Era, 330, 342, 380, 381
Precipitation, of major biomes, 567, 568t
Predators, 588, 600
Preformation, 907, *907*, 927
Pregnancy, hormone levels during, 897, *897*
  and lactation, 896–898
  length of, 896
  and parturition, 898, 905
  Rh factor and, 843
  trimesters of, 897, 905, 918, 928
Pregnancy tests, 897
Premolars, 744
Prenatal testing, 311–313
  aborting of fetus and, 312–313
  amniocentesis in, 312, *312*, 315
  chorionic villus sampling in, 312, *312*, 315
  test-tube babies and, 313
Pressure, exerted by gases, 792, *793*
  partial, 792, *793*, 806
Pressure flow hypothesis, 672, *672*, 674
Presynaptic neuron, 853, 861
Prezent, Isai, 344
Prezygotic barriers, 373, 396
Priestley, Joseph, 142, *143*
Primary auditory cortex, 872
Primary cell wall, 645, *646*, 658
Primary growth, 646, 658
Primary oocytes, 883–884, *885*, 904
Primary organizer, 921, *922*, 928
Primary plant compounds, 657
Primary productivity, 550
Primary visual cortex, *876*, 876–877
Primase, 227, 233
Primates, 322, 388, 397
  cladogram of, 427, *428*
  phylogeny of, 388
Primer, 227, 233
Primitive characters, 427, 431
Primitive gut, 911
Primitive streak, 913, *914*, 927
Primordial germ cells, 883–884, 904
Principle of independent assortment, 198–200, *199*, 206
Principle of segregation, 197–198, *198*, 206
Prions, 46–49
  scrapie and, 48, *48*
Prisoner's dilemma, 638
Probe, *286*, 286–287, 295
Proboscis, 508, 521
Producers, 548, 560
Productivity, 549
  of major biomes, 567, 568t
  primary, 550
Products, free energy of, 105, 115
Progeny, 197
Progesterone, 766
  functions of, 897
  in menstrual cycle, 895–896
  during pregnancy, 898
  in pregnancy, 896
Proglottids, 507, *508*, 521

Progymnosperms, 481
Prokaryotes, 430, 432–451
  adaptations by, 442–443
  biochemical pathways of, 443
  cell division in, 167–168, *168*
  characteristics of, 438
  classification of, 443–449, *444*
    DNA sequencing and, 448
  growth of, 442
  metabolic diversity of, 442
  negative regulation of, 252–253
  nucleoid in, 438, 450
  outer surfaces of, 441
  and people, 435–438
  positive regulation of, *253*, 253–254
  in regulation of gene expression, 251–253
  replication of, 439
  shapes of, 439, *439*
Prokaryotic cells, 15, 16, 20
Prolactin, 762, 898, 905
Prometaphase, 172
Promiscuous interactions, 68
Promoters, 241, 253, 254, 255, 256
Prophage, 268, 274
Prophase, in meiosis, 188–190
  in mitosis, 172
Proplastids, 85
Propranolol, 763, 764
Prostaglandins, 757, 766, 768, 769, 899, 906
Prostate gland, 888, 904
Protease(s), 271, 739
Protease inhibitors, 271
Proteasomes, 81, *81*, 100
Protein(s), cellular direction of, 250
  complementary, 737, *738*
  conformation of, *65*, 66, 69
  construction of, 66–68
  contractile, 728, *729*
  of cytoskeleton, 86
  in endoplasmic reticulum and Golgi complex, 82–83, *83*
  as enzymes, 63, 69
  fibrous, 65, 69
  folding of, 48, 49, 66–67, 68
  globular, 65, 67, 69
  growth control in cells and, 177
  microtubule-associated, 86
  and nucleic acids, 52, *53*
  peripheral, 88
  in plasma membranes, 88, 91
  structure of, 65, 65–66, 67–68
    α helix, 65, 67, 69
    β pleated sheet, 65, 67, 69
    primary, 65, *65*, 69
    quaternary, 66, 69
    secondary, 65, *65*, 69
    tertiary, 65, 66, 69
  synthesis of, antibiotics and, 251, *251*
    steps in, 247, *249*
  transmembrane, 88, *90*
Protein deficiency, 737
Protein-folding problem, 66, 68
Protein kinases, 764, 768
Protein sequences, DNA and, criteria for classification of, 429–430
Proteinoid microspheres, 409, *409*, 415
Proteinoids, 408, 415
Proterenol, 763
Prothoracic glands, 758
Prothoracicotropin, 758

Prothrombin, 831, 842
Protista, 16, 20, 456
Protists, 427, 454
  algae, 455, 457
  characteristics of, 454–455, 474
  chlorophylls in, 459, 459t
  chlorophytes, 461, *461*
  dinoflagellates, 457–458, *458*
  diversity of, *454*, 455, *455*, 456
  funguslike, 464–465
  and multicellular fungi, classification of, 452–475
  protozoa. *See* Protozoa
  reproduction in, 456
  rhodophytes, 460
  size of, 76
Proto-oncogenes, 267, 274
Protobionts, 408, 415
Proton gradient, 131, 138
  ATP synthesis and, 135–136, *137*
  generation by mitochondria, 133
Proton pumping, 135
Protons, 28, 40, *42*, 44
Protoplasm, 71, 74–75, 100
Protostomes, 496–521
  coelomates, 510–520
Protosun, 404–405
Prototherians, 537, *537*, 538
Protozoa, 462–464, 474, 475
  characteristics of, 462
  ciliophora, 464, *464*, 475
  foraminifera, 462, *462*, 475
  as protists, 462
  rhizopoda, 462, 475
  sporozoans, 462–463, 475
  zoomastigina, 464, 475
Proust, Marcel, 870–871
Provirus, 268, 274
Prusiner, Stanley, 46, *47*, 47–48, 66
Pseudocoelomates, 501
Pseudocoel, in aschelminths, 509, 520
Pseudomonads, 448, 450
Pseudopodia, 455, 475
Psilotophytes, 487
Psilotum, 487
Psychoactive drugs, action on chemical synapses, 857
Psychology, experimental, 626–627, 639
Pterophytes (ferns), 486–487
  life cycle of, *486*
  reproduction in, 486
Pterosaur, *322*
Puffer fish, 851, *852*
Pulmonary artery, 776, 786
Pulmonary circulation, 776, 786
Pulmonary veins, 776, 786
Pump, of neurons, 848, 861
Punctuated equilibrium, 387, *387*, 397
Punnett, Reginald, 195
Punnett square, 195, *195*, 205
Pupa(ae), 758, 925, 928
Pupil, 873, 880
Pure substances, 23
Purines, 61, 69, 210, *211*, 233
Pyramid, of biomass, 550, *550*, 560
  inverted, of biomass, 550, *551*
Pyramid of energy, 551, *551*, 560
Pyrimidines, 61, 69, 210, *211*
Pyruvate, 122, 138
  in presence of oxygen, 127, *127*

Q

QB, 156
QB protein, 156
Quartz, shocked, 382

R

R factors, 265, *266*, 274
r_{max}, 609, 624
Rabbit, digestion in, 751
Races, 302–304, 372, 396
  genetic variations within, *302*, 302–303
Radiation, adaptive, 376, *377*, 396
  plants and, 379
  benefits and dangers of, 29–30
  ionizing, 29
  from radioisotopes, 29
Radicle, 680, 682–683, 692
Radioactive dating, 329
Radioactive isotopes, 29, 44
Radioactivity, living organisms and, 29–30
Radioisotopes, 29, 44
Radula, 511, 521
Rain forests, tropical, 385, 575–578, *576*, *577*, 582
Rain shadow, 566, *567*, 582
Ramón y Cajal, Santiago, 845, 846, 853
Rana cascadae, 535
Rana catesbeiana, 371, *371*, 535
Rana chiricahuaensis, 535
Rana muscosa, 535
Rana sylvatica, 371, *371*
Rana tarahumarae, 535
Rana yavapaiensis, 535
Randall, John, 209, 210
Random drift, 356, *356*, 357, 366
  founder and bottleneck effects in, 357, *357*
  large populations and, 358
  new mutations and, 357
Random sample, 10, 20
Rattlesnake, sidewinder, 719, *720*
Ray-finned fishes, 531, *532*
Rays, *530*, 530, 542
Reabsorption, renal tubular, 817, 825
Reactants, enzymes binding to, 112–113
  free energy of, 107, 117
Reactions, chemical. See Chemical reaction(s)
Reading frame, 244, *244*, 256
Receptor(s), 762, 768
  androgen, 255
  cell surface, 98, 101
  $GABA_A$, 856, 861
  ionotropic, 857
  metabotropic, 857
  muscarinic, 857
  for neurotransmitters, 856–857
  nicotinic, 857
  nicotinic acetylcholine, 857
  olfactory, 880, *881*, 882
  sensory, 870, 880
  stretch, 858, *859*
Receptor potentials, 869, 880
Recessive alleles, 195–196, 205
Reciprocal altruism, 638, 639
Recombinant DNA, 281, 285
  and bacteria, 283–285
  cloning of, 285, 294
  definition of, 281
  development of, 281, *282*
    research involved in, 283–284
  DNA ligase and, 283
  genetic alteration of animals and plants by, 290–295
  genetic engineering and, 276–297
  herbicide resistance and, 293
  moral questions associated with, 294
  protein products of, 290
  reprogramming of cells by, 289, *289*
  restriction enzymes and, 281, *282*
  risks of, in agriculture, 293
  uses of, 281–282
Recombinant plasmids, 266, 274
Recombination, of two genes on chromosome, 204
Rectilinear movement, 719, 732
Recycling, by ecosystems, *552*, 552–559
Red blood cells. See Erythrocytes
Red bone marrow, 723, *724*, 733
Red tide, 457, *458*
Red Wolf Reintroduction Program, 418, 420
Redi, Francesco, 398–399, 415
  experiment of, 398–399, *399*
Redox reactions, 122, 138
Reducing agent, 122, 138
Reduction-oxidation reactions, 122, 138
Reductionism, in study of muscles, *8*, 8–9, *9*
Redwoods, 662
Reflex(es), conditioned, 626, 639
  definition of, 858
  "let-down," 899
  monosynaptic, 858, *859*
  neuronal mediation of, 858–859, 861
Reflexive memory, 868
Refractory period, 851, 861
Regulators, 254
  negative, 253, 256
  positive, 253–254, 256
Relative abundance, 593, 600
Relay neurons, 869, 870, 880
Release factor, 248–249, *249*
Releaser, 858
Releasing hormones, 762, 768
Renal artery, 814, 825
Renal cortex, 814–817, 825
Renal medulla, 814–817, 825
Renal tubules. See Tubule(s), renal
Renal vein, 814, 825
Renin, 821, 825
Replacement reproduction, 616
Replication, autonomous, 264, 274
  of HIV virus, 271
  integrated, 264–265, 274
  of mobile genes, 267–269
  by plasmids, 265
  of transposons, 267, 272
  in viruses, 262, *263*, 264–265, 274
Replication fork, 228, *230*, 233
Replication units, 229, 233
Repressor, *lac*, 251, 255
Reproduction, asexual, 185, *185*, 192, 204
  human, control of, 899–903
  sexual, 185, 192, 204
    number of chromosomes and, 185
Reproductive isolation, *373*, 373–374
  by polyploidy, 374, 396
Reproductive tract, female, in humans, 884
  male, in humans, 887
Reproductive unit, species as, 425, 431
Reptiles, 530–534, *532*, *533*
  birds, and mammals, cladogram for, 428, *429*
Reptilia, cladists and, 428

Residual air, 796, 806
Resilience, 599, 600
Resistance, to antibiotics, 360
  in cockroaches, 360, *360*
  to DDT, 360
Resistance factors, 265, *266*, 274
Resolution, 72, 891, 904
Resource, limiting, 591, 600
Resource partitioning, 591, 600
Respiration, 125, 138
  cellular. See Cellular respiration
Respiratory system, epithelial cells of, 798, *799*
  human, 796, *797*, 806
  mucus in, 799, 806
Response elements, 255
Resting potential(s), 849, 861
  maintenance by neurons, 849–850
Restriction enzymes, 270, 274, 283, 294
  in DNA fingerprinting, 300, *300*
  recombinant DNA and, *282*, 283, *283*, 294
Restriction fragment length polymorphisms, 297–298, 300, *300*, 315
Restriction fragments, 295
  joining of, 282–284, *284*
Restriction map, 282, 295
Restriction site(s), 283, 294
Reticular formation, 866
Retina, 873, 880
  image detection by, 874–875, *875*
  organization of, 875, *875*
Retinal, 875, 880
Retroviruses, 269, 274
  reproduction of, 269–270, *270*
Reuptake, 857, 861
Reverse transcriptase, 269, 274, 284–285, 295
Reversion, 245
Rey, Jorge R., 595
Rh factor, 841–842
Rhagoletis pomonella, 376
Rheobatrachus silus, 535
Rhinoderma, 534
Rhizobium bacteria, 558
Rhizoids, 483
Rhizome, 486
Rhizopoda (amebas), 462, 475
Rhizopus stolonifer, 468, *469*
Rhodophytes, 460–461, 474
Rhodopsins, 875, 880
Rhythm method, 900–901, 905
Ribbon worms (Nemertea), 508, *508*, 521
Riboflavin, 281
Ribonucleic acid. See RNA
Ribose, 55t, 59, 69
Ribosomal RNA, 244, 411, *411*
Ribosome(s), 81, *81*, 100
  beginning of translation by, 246–247
  definition of, 243
  evolution of, 411, *411*
  functions of, 243, 243–244, 256
  production of peptides bonds by, 247
  reading of mutations by, 245
  reading of single strand of mRNA by, 249
  structure of, 243
Ribozyme(s), 243, 410, 416
Ribulose bisphosphate, 155, *155*, 157
Riftia pachyptila, 442, 443
Riley, C.V., 660
RNA, alphabet, 239
  of energy organelles, 273
  messenger (mRNA), 240

RNA, alphabet, *(continued)*
    cellular translation of, into polypeptide, 243–244
    maternal, 908, 927
    reading of ribosomes by strand of, 249
    transcribed from DNA, cellular alteration of, 241
    as original genetic material, 409–410
    ribosomal (rRNA), 244, 411, *411*
    in synthesis of polypeptides, 234–235
    transcription of RNA polymerase into, by DNA, 240–243
    transfer (tRNA), as adapter between mRNAs and amino acids, *244*, 244–246, 256
        evolution of, 411–412
        recognition of amino acid by, 246, *247*
    viral, entry into host cell, 263, *264*
RNA editing, 242
RNA polymerase(s), 254
    transcription by, beginning of, 240–241, *241*
    of DNA into RNA, *241*, 241–243
RNA splicing, *242*, 242–243
RNA tumor viruses, 269–270
RNA world, 410, 416
Robins, *423*
Robinson, John, 711
Rodents, guinea pigs as, 424
Rods, 874, *875*, 880
Root cap, 686, 692, 698
Root hairs, 651
Root pressure, 662–663, *663*, 674
Root system, 645, 658
Roots, adventitious, 691
    of flowering plants, growth of, 686, *686*
    functions of, 645, 651–652
    oxygen requirements from, 670
    water flow through, 651, *654*
Rootstock, 691
Rotifers, 510, *510*
Roughage, 750
Roundup (glyphosate), 156, 292
Roundworms (nematodes), 510, 521
    tissue layers of, *502*
Roux, Wilhelm, 166, 182, 907
RU 486, 903, 905
Ruben, Samuel, 153
Rubisco, 155, 158, 159
Rudiments, 917, 928
Ruminants, 746, 752
    stomachs of, *747*
Runners, adaptations of, 720–721
    stride length of, 721, *721*
    stride rate of, 721
Running, by animals, 720
Rusts, 470
Rutherford, Ernest, 27, 30, 103
Rutherford's experiment, *26*

S

S-shaped curve, 609, *610*, 624
Sac fungi (Ascomycetes), 469, *470*, 475
Sagebrush, 567
Saint-Hilaire, Etienne Geoffroy, 522, 523
Salamander(s), 375, *375*, 532–533, *533*, 534, 926
    North American tiger, 535
Salicylic acid, 707
Salivary glands, 744, 752
Salmonella, *439*, 830
Salmonella newport, 432, *434*

Salmonella poisoning, 432
    antibiotics and, 432–434, *434*
Salt, 33, *33*
    consumption of, blood pressure and, 808–810
        in U.S., 808
    and water, regulation of, in animals. See Water and salt regulation
Saltationism, 386, 396
Saltations, 383–384, 396
Saltatory conduction, 853, 861
Sampling, of population, 10–11, *11*
Sands, 669, 674
Sanger, Fred, 226
Saprophytes, 548, 560
Saprophytic oomycota, 465
Saquinavir, 271
Sarcomeres, 728, 729, 733
Savanna(s), 572–573, *573*, 582
    animals in, 572
Savanna hypothesis, 390–392
Scallop(s), eye of, *511*
Scanning electron microscope, 73
Scarification, 684
Scavengers, 548, 560
Scheibinger, Londa, 476
Schleiden, Matthias, 71, 74
Schultes, Richard, 677
Schwann, Theodor, 71, 74, 118
Schwann cells, 847
Scientific method, 4, 20
Scion, 691
Sclera, 873
Sclerenchyma, 646, *647*, 658
Scolex, 507, 521
Scopes, John T., 318
Scopes trial, 318
Scorpions, 519
Scrapie, 46–49
    prions and, 48, *48*
Scrotum, 886, *886*, 904
Scurvy, 739
Scyphozoa, 504
Sea anemones, 504
Sea hare, gill withdrawal in, 859, *860*
    learning by, 860, 861
Sea lettuce, *461*
Sea stars, 524, *526*, 526–527
Sea urchin, larva of, *924*, 925
Sea urchin embryo, cleavage in, 910, *911*
    gastrulation in, 910, *911*
Seabirds, excretion of salt by, 812
Search for Extraterrestrial Intelligence (SETI), 403
Second messengers, 98–99, 764, 765, 768
Secondary cell wall(s), 645, *646*, 658
Secondary growth, 647, 658
Secondary oocytes, 885, 889, 905
Secondary plant compounds, 588, 600, 657
Secretin, 759
Secretion, by digestive tract, 743
    by tubule, 819
Sediments, 621
Seed(s), formation of, in flowering plants, 680–681, *682*
    germination of, 682–683, 684
    in plant fertilization, 487, 488
    size of, *684*
    vascular plants and, 485
Seed plants, reproduction by, 487, 494
    without flowers, 487

Segmentation, 747
Segregation, principle of, 197–198, *198*, 206
Seizures, in epilepsy, 864
Selection, kin, 637, 639
    natural, 633. See Natural selection
        altruism and, 636–637
    sexual, 633, *634*, 639
Selection theory, memory and, *838*, 838–839
    self-nonself recognition and, 838–839, 844
Selective serotonin reuptake inhibitors, 857
Self-nonself model, 826, 828
Self-nonself recognition, 834, 837–838
    selection theory and, 837–838, 843
Semen, 888, 904
Semicircular canals, 873, 880
Semilunar valves, 782–783, 787
Seminal vesicles, 888, 904
Seminiferous tubules, 887, *887*, 904
Senescence, 705, 708
Sensation(s), 870, 880
    brain regions in, 863–864
    classes of, 871
    somatic, 879, *879*
Sense organs, 870, 880
    nerves, and brain, 865
    nervous system and, 862–883
Sense strand, 240
Sensitive period, for learning, 630, 639
Sensitization, 859
Sensory information, 865
    brain and, 870–871
Sensory neurons, 858, 859, 861
Sensory pathways, perception and, 878–879
    smells and tastes, *878*, 878–879
    somatic, 877, *877*
Sensory receptors, 870, 880
Sepals, 492, 687, 692
Septa, fungal, 466
Sequoia forests, *40*, 599, *599*
Sequoia sempervirens, 662
Serosa, 747
Serotonin, 854, *855*, 860, 861
Serotonin reuptake inhibitors, selective, 857
Sertoli cells, 887, 904
Sertraline, 857
Sessile, 503, 520
Set point, 731, 733
Setae, 511, 517
Sex chromosomes, in humans, 192–193, *194*
Sex-linked genes, in *Drosophilia*, 201, *202*
Sex steroids, 763, 766, 768
Sexual development, environmental effects of, 754–755
Sexual dimorphism, 364, 367
Sexual intercourse, external genitalia in, 889–890
Sexual reproduction, 185, 192, 204, 882–907
    evolution of, 882
    human control of, 900–905
    number of chromosomes and, 185
    variation and, 348
Sexual responses, stages of, 890–891
Sexual selection, 364–365, 367, 633, *634*, 639
Shared derived characters, 427, 431
Sharks, 526, 538
Shear, 724
Sheep, cloning of, *921*, 921–923, *922*
Shells, electron, 30, 44
Sherman, Paul, 638
Shock, 770
Shocked quartz, 382

Shoot system, 645, 658
Shoots, functions of, 645
    growth of, 687
Short-term memories, 868, 880
Sibling species, 371, 396
Sickle cell anemia, 66, 220, 305–306, 313
    malaria and, *365*, 365–366
Sidewinding, 719, *720*, 732
Sierra Nevada, 566, *566*
Sieve plate, 649, 659
Sieve tube members, 649, *649*
Sieve tubes, 649, 659, 670, 671–672
Sight, 871, 874–878, 882
    defects of, *875*, 875–876
Sign stimulus, 628, 639
Signal peptide, 249, 255
Signal transduction pathway, 764
Signaling molecule, 254
Signals, chemical. *See* Chemical signaling
Silencers, 255
Silent mutations, 245
Silk, structure of, *67*
Silk moth, behavior of, 629
Silts, 669, 674
Silurian Period, 519
Simberloff, Daniel, 594–595
Simple pit pairs, 647, *648*, 658
Singer, Jonathan, 91
Single-nucleotide polymorphisms, 298
Sinoatrial node, 784, 787
Sinuses, 779, 787
Sister chromatids, 171, 188, 190
Skates, 530, *530*
Skeletal muscle, 713, 726, *727, 728*, 733
    and bones, 726
    neural control of, 726
Skeleton, of birds, 720, *720*
    hydrostatic, 721, *722*, 732
    in movement of animals, 721–730
    vertebrate, parts of, 722–723, *723*
Skin, as barrier to pathogens, 829–830
    broken, body defense in, 830–833
    microorganisms inhabiting, 436
    transplantation between twins, 827
Skinner, B.F., 626
Skoog, Folke, 696
Skull, 722, 732
Skunk cabbage, 491
Sliding filament model, 729, 733
Slime layer, 441, *441*, 450
Slime molds, cellular, 464, 464–465
    plasmodial, 465
Slug, 464, *464*
Small intestine. *See* Intestine(s), small
Smell, 870, 878–879, 880
Smellie, William, 477
Smiling, by infants, 628, *628*
Smith, William "Strata," 323
"Smoker's cough," 799
Smoking, causes of death related to, 792
    damage to lungs due to, 799–800
    history of, 789
    lung cancer and, *789*, 789–791, *790*
    reasons for, 793
Smooth muscle, 713, 726, *728*, 733
Smuts, 470
Snakes, 534, *536*
    garter, 627–628
    internal fertilization by, *187*
    locomotion in, 719, *719, 720*

Snapdragons, inheritance of flower color in, 194–195, *195*
Social behavior, 634, *635*
Social groups, animals in, 634–638
Sociobiology, 626, 639, 627
Socrates, 658
Sodium channels, voltage gated. *See* Voltage gated sodium channels
Sodium chloride. *See* Salt
Sodium ions, in membranes, 851
Sodium-potassium pump, 94, 101, 850
Sog gene, 523, 923
Soil(s), bacteria and fungi in, 435, 437
    desert, 569
    nutrients from, 669t
    of tropical rain forests, 576
Solar system, formation of, 404, *405*
Solutes, 39, 811
    in nephron, 819, *820*
Solution, pH of, 42–43
    water molecules in, 40–42
Solvent, water as, 39–40
Soma, of neuron, 847, *848*
Somatic cells, 186, 204
Somatic sensations, 870, 880, 877, *877*
Somatomammotropin, human chorionic, 897, 905
Soule, Michael, 602, 603, 604, 607
Sound, as perception, 862
Southern blotting, 288, 295
Spallanzani, Lazzaro, 399, 415
    experiment of, 399, *400*
Sparrow, Western song, 372
Speciation, allopatric, 374, *375*, 396
    geographic isolation and, 374
    parapatric, 376
    sympatric, 374–376, *375*
Species, 322
    biological, 371
        definition of, 425
    biological concept of, 425, 431
    in community, interaction of, 587–593
    competing, coexisting, 591–592
    definition of, 369–373
    density of, latitude affecting, 593, *594*
    diversification of, on continents, 379
    diversity of, 369–371, *371*
    as ecological unit, 425, 431
    evolution of, speed of, 383–388
    formation of, 373–376
        gene flow and, 376
        geographic isolation and, 374–376
    as genetic unit, 425, 431
    grouping of, 425–430
    incipient, 372, 396
    interactions among, 590t
    K-selected, 612–613
    migration of, 595, *597*
    morphologically defined, 425
    multiplication of, 376–383
    naming of, 423
    r-selected, 612
    recognition of, 423–425
    reproductive isolation of, *373*, 373–374, 396
    as reproductive unit, 425, 431
    sibling, 371, 396
    two, sharing habitat, 591
    typological, limitations of, 425
    typological concept of, 424, 431
Species richness, 593, 600
Specificity, of immune system, 834

Spectrum, absorption, 147, *148*, 159
    action, 147–148, *148*, 159
Spemann, Hans, 907–908, 918, 921
Sperm, 186, 204
    development and maturation of, *887, 888*
    human, 887, *888*
    passage of, mammalian, 888
    production of, 754
        hormones and, *892*, 893–894
        infertility and, 903
        by mammals, 887–888
    reaching ova, 891–892
    union of egg with, 186–188, *187*
    union with oocyte, 888
Spermatids, 887, 904
Spermatogenesis, *884*
Spermatogonia, 887, 905
Spermatozoon. *See* Sperm
Spermicides, 901, 905
Sphagnum moss, *483*
Sphincters, 744
Spiders, 519
    book lungs of, 795
    digestion in, 742
Spinal column, 722, 732
    anterior, 713
    posterior, 713
Spinal cord, *865*, 865–866, *866*, 880
Spiracles, 517, 521, 801, 806
Spirilla, 439, 450
Spirochetes, 448, *448*, 450
Sponges (Porifera), 502, 503, *503*
Spongy bone tissue, 723, *724*, 733
Spongy parenchyma, *655*, 656, 659
Spontaneous abortion, 192–193, 205, 897, 905, 918, 928
Sporangium(a), 468, 475, 484
Spores, 481, 490
    fungal, 467, *468*
    in phaeophytes, 460, 474
Sporophores, 466, 475
Sporophylls, 486, 494
Sporophyte(s), 474, 484, *484*, 494, 678, 692
    diploid, 484
    in phaeophytes, 460
Sporozoans, 462–463, 475
Sporozoite, 463
Sport utility vehicles, 555
Squamata (lizards and snakes), 534, *536*
Squid, 514, *514*
    neurological experiments on, 851
St. Anthony's fire, 471
St. Hilaire, Geoffrey, 908
St. Martin, Alexis, 734, *735*
Stability, of community, 599, 600
Stabilizing selection, *362*, 363, 367
Stalin, Joseph, 344
Stamens, 492, 494, 678, 692
Standard deviation(s), 11, 12, *13*, 20
    calculation of, 12
    mean and, 12
Stanier, Roger, 447
Stanley, Wendel, 261
Stapes, 871
Starch, 69
    in plants, 60, *61*
Starfish, *526*, 526–527, 612, *612*
Start, 177
Start site, 246, 256
Stasis, 386, 397

Statistical analysis, 20
Statistics, in estimation of value, 11
Statoliths, 698, 708
Stele, 649, 652, *654*, 659
Stem cells, 723, 751–752, 921, 926, 928
    embryonic, 926
    human, 926
Stems, water flow through, 665
    water transport in, 653–654
Stereocilia, 871, *872*
Steric inhibitor, 115–116, *116*, 117
Sterility, hybrid, 374, 396
Sterilization, 900, 902–903, 905, 903–904, 906
Sterna paradisaea, *120*
Sternum, 776, 786
Steroids, 58, 59, 763, 764, 768
Stickleback, courtship of, *634*
Stigma, 679, 692
Stimulus(i), characteristics of, 870
    conditioned, 859
    sign, 628, 639
    supernormal, 629, *629*, 639
    unconditioned, 859
Stochastic events, 604, 624
    extinction of species and, 606
Stomach, 745–746, 752
    of ruminant, *747*
    tissues and cells of, *746*
Stomata, 142, *142*, 158, 159, 481, 494, 644, 655, 659
    opening and closing of, *666*, 666–667
    transpiration regulation by, 666–670
Storage, by digestive tract, 743
Streptococcus pneumoniae, 221, *222*, 221, 438
Streptomyces, 445
Streptomycin, 361
Stress, response to, 760–761, 762
Stressors, 760
Stretch receptors, 858, *859*
Striated muscles, contraction of, *729*, 729–730
Striations, 726, 733
Stride length, 721, *721*
Stride rate, 721
Strobili, 487, 494
Stroke, 782, 787
Stroke volume, 785, 787
Stroma, 149, 159
Stromatolites, *401*, 401–402
Sturtevant, Alfred, 204
Style, 679, 692
Stylet, 671, *671*
Subcutaneous tissue, 830, 842
Suberin, 651
Subspecies, 372, 396
Substantia nigra, 866
Substrates, 112, 113, *113*, 117
Succession, 585, 600
    on corpse, 598
    plant traits during, 597, 597t
    primary, 596–597, 600
    reason for, 598
    secondary, 597, 600
Successional communities, 597, 600
Sucrose, 59, 69
Sugar(s), 52, 54–55t, 68
    formation of polysaccharides by, 59–61
    making of, in Calvin cycle, 155, *155*
    movement of, 660–675
        in plants, 670, 674
    for storage of energy, 60

Sulci, 868, 880
Sulfolobus bacterium, 412
Sun, as energy source for plants, 145–153
Sunlight, *146*
    evaporation of water by, 553
    as influence on ecology, 549–550
    movements of Earth, and climate, 564–566
Superfecundity, 341, 342
Superorganism, 545, 560
Suppressor mutations, 245
Surface area, gas exchange and, 794–795
Surface tension, 38t, 39
    alveoli and, 800
Surface-to-volume ratio, 794–795
Surfactants, 622, 800, 806, 918
Surrogate motherhood, 905
Survivorship, life history and, *613*, 613–616, *615*
Survivorship curves, *613*, 613–614, *615*
Sutton, Walter S., 180–181, *181*, 183, 199, 217
    six rules of inheritance of, 200t
Sutton-Boveri chromosomal theory of inheritance, 199, 204
Sutures, 722, *723*, 732
Swim bladder, 531, 542
Swimmers, adaptations of, 718
    shape of, 718, *718*
Symbionts, 446, 450
Symbiosis, 461, 545, 560, 590
Symmetry, in animal classification, 502, *502*
    bilateral, 502, *502*
    radial, 502, *502*
Sympathetic nerves, 726, 869, 880
Sympathetic nervous system, 726
    heart rate and, 785, 787
Sympatric speciation, 374–376, *375*
Symplast, 651
Symplocarpus foetidus, 491
Synapse(s), 847
    chemical. *See* Chemical synapses
    electrical, 854, 861
    excitatory, 854
Synapsis, 188, 204
Synaptic cleft, 854
Synchronous cell populations, 177
Syngamy. *See* Conception; Fertilization
Synthesis, abiotic, 404, 415
    prebiotic, 405–407, 416
Syphillis spirochete, swimming, *448*
Systematics, grouping of, 425, 431
Systemic circulation, 776
Systemin, 707
Systole, 782–783, 787
Szent-Györgyi, Albert, 128–129
Szostack, Jack, 410

T

T-cell receptors, 840, 843
T cells, 827–828, 829, *834*, 835, 843
    altered, self-recognition of, 841
    cytotoxic, 840, 843
        poison ivy and, 836, *836*
    elimination of, 841
    helper, 840, *840*, 843
T lymphocytes. *See* T cells
Tagmata, 517, 521
Taiga, 564, *565*, 572–573, 582
Tannins, 657, 659
Tapeworms, 506, 507, *508*, 521
Tarahumara frog, 535

Tardigrade, *716*
Target orders, 751, 752
Target tissues, 702, 708
Targets, 756, 768
Tarweeds, 379, *379*
Taste, 870, 878, 880
    taste buds and, 878, *878*
Taste buds, 878, *878*
Tatum, Edward, 218, 219
Taxis, 441, 450
Taxon(s), 322, 423, 431
    monophyletic, 427, 431
    paraphyletic, 427, 431
    polyphyletic, 427, 431
Taxonomic categories, 372–373
Taxonomist, 421, 431
Taxonomy, 322, 332–333, 342, 425–426, 431
    classical evolutionary, 426
Tay-Sachs disease, 196, 311, 312, 359
Tears, salty, 812, *812*
Teeth, human, 744
    mammalian, 539–540
Telophase, in meiosis, 188, 190
    mitotic apparatus during, 173
Temin, Howard, 269
Temperate deciduous forests, 573–575, *574*
Temperature, atmospheric carbon dioxide and, 555, *555*
    body, regulation of, 731
        tolerance of, 731, *731*
    changes in water, 36–37
    dissolution of gases and, 793–794
    equilibrium point of reaction and, 110, *112*
    global, 555, *555*
    latitude and, 565, *565*
    of major biomes, 567, 568t
    measurement of, Metric System, Appendix B
    regulation of, in vertebrates, 538
Temperature turnover, in lakes, 581, *581*
Template, 227, 233
Template strand, 240
Temporal isolation, 373, 396
Ten percent law, 551, 560
Tendons, 723, *724*, 732
Teosinte, 676–677
Teratogens, 917, 928
Terminal buds, 687, 692
Termination, *248–249*, 249
Termination signals, 255
Terminators, 241, 254
Tern, arctic, *120*, 534, 719
Terpenes, 657, 659
Territoriality, 633, 639
Test, 457
Test-tube babies, 313
Testcross, 197, *197*, 205
Testes, 186, 204, 766, 883, 887, 904
Testicles, 883, 904
Testicular feminization, 255
Testosterone, 98, 184, 755, 766, 893, 904
    aging and, *893*, 893–894
    hierarchy of, 893, *893*
    release of, 893
    role of, 893
Tetracycline, 441
    resistance to, 437
Tetrad, 188, 204
Tetrapods, 525, 538
Tetrodotoxin, 851, *852*
Thalamus, 867, 880

Thalidomide, 917
Thamnopis elegans, 627–628
Theory, 13, 20
Theory of island biogeography, 594, 600
Thermal pollution, 621
Thermoacidophiles, 443, *445*, 450
Thermodynamics, 103, 104, 107
    First Law of, 104–105, *105*, 106, 116
    prediction of direction of reaction by, 104–106
    Second Law of, 105, *105*, 109, 116
Thermoplasma acidophilum, 414
Thiamine, 739
Thick filaments, 729, 733
Thimann, Kenneth, 694–697, *696*
Thin filaments, 729, 733
Thoracic cavity, 713, 732, 798, 806
Thorax, 517, 521
Threonine, 236
Threshold response, of nerve cells, 847
Throat, 744
Thrombin, 831, 842
Thylakoid space, 149, *152*
Thylakoids, 85, 100, 149, *150*
Thymine, 210
Thyroid gland, 766
Thyroid hormones, 764
Thyroid-stimulating hormone, 762
Tick, sign stimuli of, 629
Tidal volume, 796, *798*, 806
Tiger(s), 634, *635*
Tight junction, 98, *99*, 101
Time, direction of, 104, *104*
Tinbergen, Niko, 629
Tissue(s), 499, *499*, 520
    connective. *See* Connective tissue
    endothelial, 713
    epithelial, 713
    muscle, 713
    oxygen carried to, 801–802
Tissue culture, 691–692
Tissue plasminogen activator, 832, 843
Tmesipteris, 487
Toad(s), 535
    boreal, 535
    clawed, nuclear transplantation in, 919, *919*
    golden, 535, *535*
    Western, 535
Tobacco. *See* Smoking
Tobacco industry, 788–791
Tobacco mosaic disease, 260
Tobacco mosaic virus, 260, *262*
    crystals of, 261, *261*
    filtration of, 260, *260*
Tobin, Allan, 297, 298
Tolerance, 730, 731–732, 733
    in humans, 731, *731*
Toleration, by immune system, 834
Tomato bushy stunt virus, *262*
Tomatoes, genetic engineering of, 280–281, 293
Tonegawa, Susumu, 837
Tongue, taste buds on, 880, *880*
Topography, climate and, 566–567
Torpor, 568
Torricelli, Evangelista, 792
Tortoise(s), 534, *536*
    desert, *810*
    excretion of salt by, 812, *812*
    Galápagos, 368, *368*
    giant, *536*
Torus, 648

Totipotency, differentiated animal cells and, 918–919, *919*
    in embryonic cells, 906–907, *907*, 927
Toxic waste, global problem of, first, 413
Trachea(e), 513, 517, 796, *797*, 801, *801*, 806
Tracheids, 647, 659
Tracheophytes, 479, 485–490. *See also* Plant(s), vascular
Tradewinds, 565, 582
Trait(s), adaptive, 321, 342
    accumulation of, 341
    in inheritance, 341
    expressed by genes, 305
    frequency of, selection and, 358–360, *359*
    influenced by genes, *306*, 306–307
    passing of, 19
    polygenic, *306*, 306–307, 315, 351, *351*
    sexual selection and, *364*, 364–365
Transcriptase, reverse, 269, 274, 284–285, 295
Transcription, 239–240, *239*, *241*, 241–242, 256
Transcription factors, 241–242, 256, 764
Transcriptional control, 250, 256
Transduction, 870, 871, 880
Transfer RNA. *See* RNA, transfer (tRNA)
Transgene(s), 290, 296
Transgenic organisms, 290, 295
Transition state, 113, *113*, 117
Translation, 238–239, *239*, *250*, 255–256
    evolution of, 411
    initiation in, 246, *248*
Translational control, 251, 256
Transmissible spongiform encephalopathies, 46
Transmission, 870
Transmission electron microscope, 72–73, *73*
    in assessment of mitochondria, 84, *85*
    in investigation of eukaryotic cells, 77–79
Transmission genetics, 183, 204
Transpiration, 552, 560, 655, 662, 664–666, 674
    regulation of, by stomata, 666–670
Transpiration-photosynthesis compromise, 666, 667, 674
Transport, active, 94, *95*, 101
    passive, 93–94, *95*
Transposase, 266, 274
Transposon(s), 258, *266*, 267, 274
    complex, 267, 274
    replication of, 267, 272
    simple, 266–267, 274
Trees, carbon dioxide in, 557
    secondary growth in, 689–691, *690*, 692
    of tropical rain forests, 576
Trematoda, 506, *507*, 521
Treponema pallidum, 435, *448*
Triacylglycerol, 56, *57*, 69
Triassic Period, 331, 532, 536
Tribolium castaneum, 591
Tribolium confusion, 591
Trichoplax adhaerens, 503
Trigger neuron, 858
Tripsacum, 677
Trisomy, 231, 233
Trisomy 21, *191*, 191–192, *192*, 193, 205
    risk of, 193t
Trivers, Robert, 638
Trophic level, 548, 560
Trophoblast, 910, 927
Tropical rain forests, destruction of, 576–578
    productivity and diversity of, 575–576, *576*, *577*, 582
Tropics, 565, 582

Tropism, 697
True breeding, 194
Trypanosomes, 459–460, 474
Tschermak, Erich, 349
Tuatara, *536*
Tubal ligation, *902*, 902–903, 905
Tube body plan, *908*, 908–909, 912
Tube worm, 516
Tuberculosis bacteria, resistant strains of, 437
Tuberculosis isolation wards, 437, *437*
Tubule(s), Malpighian, 824, *824*
    renal, 816, 825
        distal, 817, 825
        lumen ion content and, 819
        proximal, 817, *818*, 825
        reabsorption by, 817, 825
        recycling by, 817–819, 825
        secretion by, 819
        structure of, 817–818, *818*, 825
Tubulins, 86, 254
Tumor necrosis factor-α, 833, 843
Tundra, 564, *569*, 569–570
Tungara frog, 633, *634*
Tunicates, 528, *528*
Turbellaria, 506, 521
Turbinate bones, 716, 732
Turgor pressure, 93, 101, 663, 674
Turtle(s) (Chelonia), 534, *536*
    eastern chicken, *810*
    green sea, *810*
    red-eared, *120*, 125
    water and salt balance in, *810*
Twins, identical, 184, *184*
    transplantation of skin between, 827
Twort, Frederick, 261
Tympanic membrane, 871
Tympanuchus cupido, 605, 606, *606*
Typological species concept, 424, 431
Tyrannosaurus rex, 710

U

Ulcer(s), 2
    peptic, 2
Ulva, 461
Umbilical cord, 898
Unconditioned stimulus, 859
Uniformitarianism, 324, *324*, 341
Uniramia, 518, 520, 521
Unite de plan, 522, 523
Universality, 238
Updwelling, 578, 582
Uracil-DNA glycoside, 231
Urea, 558, 560, 813, *814*, 824
    flow of, 819, *820*
Uremia, 818
Ureters, 814, 825
Urethra, 814, 825, 889, 904
    microorganisms inhabiting, 436
Urey, Harold, 406
Uric acid, 558, 560, 813, *814*, 824
Urine concentration, by kidneys, 819–820
Urodela (salamanders), 532–533, *533*, 534
Ussher, Archbishop James, 321
Uterus, 883, 886, 904

V

Vacuoles, 79
Vagina, 886–887
    microorganisms inhabiting, 436
Value, estimation of, statistics in, 11

van der Waals attractions, 44
    and hydrophobic interactions, 35
van Helmont, Johann Baptista, 140–141
    plant experiment of, 141, *141*
van Leeuwenhoek, Antoine, 70–71, *71*, 399, 435, 906
van Niel, Cornelius, 143–144
Variant(s), creation and maintenance of, 346–347
    preexisting, natural selection and, 361
    in variation, 351–352, *352*
Variation, genetic. *See* Genetic variation
    individual, 341, 348
    natural and induced, in genetic engineering, 280
    sexual reproduction and, 348
    source of, evolution and, 350–355
    variants in, 351–352, *352*
Vas deferens, 888, 904
Vascular cambium, 689, 690, *690*, 692
Vascular system, 645, 658
    in plants, organization of, 647–649
    of sea stars, *526*, 526–527
Vascular tissue system, 646, 658
Vasectomy, 902, *902*, 905
Vasoconstriction, 786, 787, *830*, 831, 842
Vasodilation, 786, 787
Vasopressin, 761, 823, 825
Vauxia, 498
Vavilov, Nikolai, 344
Vector(s), 284, 295
Vegetal pole, 910, 928
Vegetative reproduction, 691, 692
Veins, 780, 787
    pulmonary, 776, 786
    renal, 814, 825
    valves of, *774*, 775, 779
Venae cavae, *772*, 776, 786
Venter, Craig, 308, 311
Ventilation, 798, 806
Ventral horns, 866
Ventral structures, 519
Ventricle(s), 776, 786, 865
Ventricular fibrillation, 784, 787
Venules, 779–780, 787
Venus flytrap, 670, *670*
Vertebrae, 722, 732
Vertebrates, adaptations evolved by, 532–541
    apoptosis in, 917
    back strength in, *529*
    brain of, 866–868
    central nervous system in, organization of, *865*, 865–868, 880
    development of, stages of, 909
    distinguished from other chordates, 528–529
    endoskeleton of, parts of, 722–723, *723*
    limbs of, homology in, 333, *333*
    nonmammalian, kidneys of, 820–821
    peripheral nervous system organization in, 878–869
    temperature regulation in, 538
Vesicles, 79
Vessel elements, 648, *648*, 659
Vestibular sensation, 870, 880
Vestigial structures, 334, *336*, 342
Vicariance, 379, 396
Villus (Villi), 747, 752
Virchow, Rudolf, 74
Virus(es), attachment of, to cell, 262–263, *263*
    and bacteria, 260

bacteriophages as, 223, 232
definition of, 260–264, 274
evolution of, 272
expression of genes by, 269, *269*
forms of, 261, *262*
gene cloning in, 284
life cycles of, *268*
lytic, *263*, 264, 274
multiplication cycles of, 271–272
natural selection of, 270–272
particles of, manufacture and release of, 264
replication of, 262, *263*, 264–265, 274
RNA tumor, 269–270
tumor, insertion into DNA, 267, *268*
Visceral mass, 511, 521
Vision, 870, 873–877, 880
    of animals, 873–877
    averted, 874
    defects in, *875*, 875–876
Visual cortex, primary, *876*, 876–877
Visual illusions, 878, *878*
Visual system, 869, 880
Vitalism, 118, 138
Vitamin(s), as antioxidants, 643, *643*
    deficiency of, 738
    essential, in human diet, 739, 740t
    fat-soluble, 739
    water-soluble, 739
Vitamin A, 643
Vitamin B_1, 739
Vitamin C, 643, 739
Vitamin E, 643
Vitreous humor, 874
Viviparity, 897, 906
Vocal cords, 796
Volcano, 584
Voltage, 849, *850*
Voltage gated sodium channels, 851
    opening and closing of, 851–852, 861
    pumping of, 852
Voltmeter, 850
Volts, 850
Volume, measurement of, Metric System, Appendix B
Voluntary muscle. *See* Skeletal muscle
Volvox, 461
Vomiting, 746
von Baer, Karl Ernst, 906
Vulva, 889, 904

W

Wake, David, 535
Walcott, Charles Doolittle, 496, 497, *497*
Walcott, Helena, 492
Walking, 390, *391*
    by animals, 720
Wall cess, 689
Wallace, Alfred, 326, *326*, 348
Warfarin, 832
Warren, J. Robin, 2
Warts, genital, 164
Wasps, sex ratios in, 614
Wastes, nitrogenous, 813, *814*, 824
Water, in biochemical reactions, 40
    as coolant, *39*
    as essential, 36–44
    ethanol, and oleic acid, compared, 38t, 39
    heat capacity of, 37, *39*

hydrogen bonding in, 39
movement of, 660–675
    roots in, 662–663
    from soil to air, 662
    upward, 662–663
in plant growth, from soil, 669
plant productivity and, 157–158
polarity of, 34, *34*
properties of, adhesion, 38t, 39
    cohesion, 38t, 39
    density, 38t
    polar, 40
retention of, by saltwater fish, 812
and salt, regulation of, in animals. *See* Water and salt regulation
in solution, 40–42
as solvent, 39–40
temperature changes and, 36–37
Water and salt regulation, in aquatic animals, 811, *811*
    by kidneys, 814, 817, 819, 825
    in land animals, 812–813
    water flow in, 810–811
Water cycle, 552–553, *553*
Water intoxication, 813
Water pollution, 621–624
Water potential, 664, 674
Water-soluble signaling molecules, 764, 765
Water-soluble signals, 762, *763*, 768
Water supply, limited, adaptations to, 668
Watson, James, 234, 237, 308
Watson, James D., 208, 211, 212–214, *213*, 216, 217, 225, 233
Watson-Crick base pairs, 240
Wavelength, *145*, 145–146, *147*, 159
Wayne, Robert, 419
Webbed feet, 715
Weed, 156
Wegener, Alfred, 339
Weinberg, G., 354
Weismann, August, 166, 182–183, 347–348
Weissman, Charles, 46
Welwitschia, 490
Went, Frits, 694, 695–696, 702
Wernicke, Carl, 863
Westerlies, 565, 582
Western blotting, 288, 295
Wexler, Leonore, 298, *298*
Wexler, Milton, 296–297, 298, *298*
Wexler, Nancy, 298, *298*
Whales, countercurrent system in, 716, *717*
White blood cells. *See* Leukocytes
White matter, 866
Whittaker, Robert H., 430
Whittington, Harry, 496–497
Whorled leaves, 654, *655*
Whorls, 687, 688, 692
Wildlife preserves, 607–608
Wilkins, Maurice, 208–210, 212, 214, 216
Wilmut, Ian, 920–921
Wilson, E.B., 180
Wilson, E.O., 385, 574, 593–594, 626, 627
    Sociobiology: The New Synthesis by, 626–627
Wine blight, 660, *661*
Wings, 715
    of birds, 719
Withdrawal, 902
Wiwaxia, 498
Woese, Carl, 447, *447*

Wöhler, Friedrich, 120
Wolf, gray, disappearance of, 418, *419*
    red, as hybrid, 418, 420, *420*
        mitochondrial DNA of, 420
        origin of, 419, *419*
        species status of, 418–421
Wolfe, Ralph, 447
Wolffia, 490
Wolffian ducts, 890
Wolpert, Jim, 662
Woodpeckers, imitation, *378*, 378–379
Work, definition of, 104
    endergonic reactions and, 107, *107*
    kinetic energy from, 104
    potential energy in, 104, *105*, 116
Worm(s), acorn, 527, 542
    arrow, 527, 542
Wound healing, 830, 843

X

X chromosomes, 192–193, 205
Xylem, 485, *485*, 494, 645, 658
    function of, 647–648
    water in, 666, 669

Y

Y chromosome, 192–193, 205
Yeasts, in fermentation, 118, 126
Yellow fever, 260
Yolk, 536, 542
Yolk sac, 536, 542

Z

Z band, 728, 733
Zamia pumila, 487
Zea diploperennis, 677
Zea mays, 687
Zebra fish, discoidal cleavage in, 911
Zebroid, 346, *346*
Zero population growth, 609, 624
Zoomastigina (flagellates), 464, 475
Zygomycetes, 468–469, *470*, 475
Zygosporangium, 468
Zygote(s), 166, 186, *187*, 204, 290, 499, 520, 884, 889, 905
    cleavage of, 909–910, *910*
        mechanisms of, 910–912, *911*
    development of, 908
    division of, 909–910, *910*
Zymase, 119